P9-ARH-024

HELICOPTER THEORY

WAYNE JOHNSON

DOVER PUBLICATIONS, INC.
New York

Copyright

Bibliographical Note

This Dover edition, first published in 1994, is an unabridged and slightly corrected republication of the work first published by the Princeton University Press, Princeton, New Jersey, in 1980.

Library of Congress Cataloging-in-Publication Data

Johnson, Wayne, 1946–
Helicopter theory/Wayne Johnson.
p. cm.
"This Dover edition...is an unabridged and slightly corrected republication of the work first published by the Princeton University Press, Princeton, New Jersey, in 1980"—T.p. verso.
Includes bibliographical references and index.
ISBN-13: 978-0-486-68230-3
ISBN-10: 0-486-68230-7
1. Helicopters. I. Title.
TL716.J63 1994
629.133'352—dc20 94-26727
CIP

Manufactured in the United States by Courier Corporation
68230706
www.doverpublications.com

CONTENTS

Acknowledgements xiii

Notation xv

1. Introduction 3

1–1 The Helicopter 3

1-1.1 The Helicopter Rotor 6

1-1.2 Helicopter Configuration 9

1-1.3 Helicopter Operation 10

1–2 History 11

1-2.1 Helicopter Development 11

1-2.2 Literature 20

1–3 Notation 20

1-3.1 Dimensions 20

1-3.2 Physical Description of the Blade 21

1-3.3 Blade Aerodynamics 22

1-3.4 Blade Motion 23

1-3.5 Rotor Angle of Attack and Velocity 24

1-3.6 Rotor Forces and Power 25

1-3.7 Rotor Disk Planes 26

1-3.8 NACA Notation 26

2. Vertical Flight I 28

2–1 Momentum Theory 28

2-1.1 Actuator Disk 29

2-1.2 Momentum Theory in Hover 30

2-1.3 Momentum Theory in Climb 32

2-1.4 Hover Power Losses 34

2–2 Figure of Merit 34

2–3 Extended Momentum Theory 36

2-3.1 Rotor in Hover or Climb 37

2-3.2 Swirl in the Wake 40

2-3.3 Swirl Due to Profile Torque 45

2–4 Blade Element Theory 45

2-4.1 History of the Development of Blade Element Theory 46

2-4.2 Blade Element Theory for Vertical Flight 49

2-4.2.1 Rotor Thrust 51

2-4.2.2 Induced Velocity 52

2-4.2.3 Power or Torque 53

2–5 Combined Blade Element and Momentum Theory 56

2–6 Hover Performance 57

2-6.1 Tip Losses 58

2-6.2 Induced Power Due to Nonuniform Inflow and Tip Losses 61

2-6.3 Root Cutout 62
2-6.4 Blade Mean Lift Coefficient 62
2-6.5 Equivalent Solidity 63
2-6.6 The Ideal Rotor 64
2-6.7 The Optimum Hovering Rotor 65
2-6.8 Effect of Twist and Taper 68
2-6.9 Examples of Hover Polars 69
2-6.10 Disk Loading, Span Loading, and Circulation 72
2–7 Vortex Theory 72
2-7.1 Vortex Representation of the Rotor and Its Wake 74
2-7.2 Actuator Disk Vortex Theory 76
2-7.3 Finite Number of Blades 81
2-7.3.1 Wake Structure for Optimum Rotor 82
2-7.3.2 Prandtl's Tip Loading Solution 83
2-7.3.3 Goldstein's Propeller Analysis 87
2-7.3.4 Applications to Low Inflow Rotors 87
2-7.4 Nonuniform Inflow (Numerical Vortex Theory) 88
2-7.5 Literature 91
2–8 Literature 91

3. Vertical Flight II 93
3–1 Induced Power in Vertical Flight 93
3-1.1 Momentum Theory for Vertical Flight 94
3-1.2 Flow States of the Rotor in Axial Flight 98
3-1.2.1 Normal Working State 98
3-1.2.2 Vortex Ring State 99
3-1.2.3 Turbulent Wake State 101
3-1.2.4 Windmill Brake State 101
3-1.3 Induced Velocity Curve 102
3-1.3.1 Hover Performance 105
3-1.3.2 Autorotation 105
3-1.3.3 Vortex Ring State 106
3-1.4 Literature 107
3–2 Autorotation in Vertical Descent 107
3–3 Climb in Vertical Flight 114
3–4 Vertical Drag 116
3–5 Twin Rotor Interference in Hover 118
3–6 Ground Effect 122

4. Forward Flight I 125
4–1 Momentum Theory in Forward Flight 126
4-1.1 Rotor Induced Power 126
4-1.2 Climb, Descent, and Autorotation in Forward Flight 132
4-1.3 Tip Loss Factor 133
4–2 Vortex Theory in Forward Flight 134
4-2.1 Classical Vortex Theory Results 136
4-2.2 Induced Velocity Variation in Forward Flight 139
4-2.3 Literature 141
4–3 Twin Rotor Interference in Forward Flight 142
4–4 Ground Effect in Forward Flight 146

5. Forward Flight II 149

5–1 The Helicopter Rotor in Forward Flight 149
5–2 Aerodynamics of Forward Flight 167
5–3 Rotor Aerodynamic Forces 171
5–4 Power in Forward Flight 179
5–5 Rotor Flapping Motion 184
5–6 Examples of Performance and Flapping in Forward Flight 194
5–7 Review of Assumptions 205
5–8 Tip Loss and Root Cutout 206
5–9 Blade Weight Moment 206
5–10 Linear Inflow Variation 207
5–11 Higher Harmonic Flapping Motion 210
5–12 Profile Power and Radial Flow 213
5–13 Flap Motion with a Hinge Spring 222
5–14 Flap Hinge Offset 227
5–15 Hingeless Rotor 234
5–16 Gimballed or Teetering Rotor 235
5–17 Pitch-Flap Coupling 238
5–18 Helicopter Force, Moment, and Power Equilibrium 243
5–19 Lag Motion 250
5–20 Reverse Flow 255
5–21 Compressibility 262
5–22 Tail Rotor 264
5–23 Numerical Solutions 265
5–24 Literature 266

6. Performance 278

6–1 Hover Performance 280
6-1.1 Power Required in Hover and Vertical Flight 280
6-1.2 Climb and Descent 282
6-1.3 Power Available 282
6–2 Forward Flight Performance 284
6-2.1 Power Required in Forward Flight 284
6-2.2 Climb and Descent in Forward Flight 286
6-2.3 D/L Formulation 286
6-2.4 Rotor Lift and Drag 288
6-2.5 P/T Formulation 289
6–3 Helicopter Performance Factors 290
6-3.1 Hover Performance 290
6-3.2 Minimum Power Loading in Hover 291
6-3.3 Power Required in Level Flight 293
6-3.4 Climb and Descent 295
6-3.5 Maximum Speed 296
6-3.6 Maximum Altitude 298
6-3.7 Range and Endurance 299
6–4 Other Performance Problems 301
6-4.1 Power Specified (Autogyro) 301
6-4.2 Shaft Angle Specified (Tail Rotor) 302
6–5 Improved Performance Calculations 303
6–6 Literature 304

7. **Design** 313

7–1 Rotor Types 313
7–2 Helicopter Types 315
7–3 Preliminary Design 318
7–4 Helicopter Speed Limitations 321
7–5 Autorotational Landings after Power Failure 325
7–6 Helicopter Drag 331
7–7 Rotor Blade Airfoil Selection 332
7–8 Rotor Blade Profile Drag 337
7–9 Literature 340

8. **Mathematics of Rotating Systems** 344

8–1 Fourier Series 344
8–2 Sum of Harmonics 347
8–3 Harmonic Analysis 348
8–4 Fourier Coordinate Transformation 349
8-4.1 Transformation of the Degrees of Freedom 350
8-4.2 Conversion of the Equations of Motion 355
8–5 Eigenvalues and Eigenvectors of the Rotor Motion 361
8–6 Analysis of Linear, Periodic Systems 365
8-6.1 Linear, Constant Coefficient Equations 366
8-6.2 Linear, Periodic Coefficient Equations 369

9. **Rotary Wing Dynamics I** 378

9–1 Sturm-Liouville Theory 378
9–2 Out-of-Plane Motion 381
9-2.1 Rigid Flapping 381
9-2.2 Out-of-Plane Bending 384
9-2.3 Nonrotating Frame 390
9-2.4 Bending Moments 392
9–3 In-plane Motion 393
9-3.1 Rigid Flap and Lag 393
9-3.2 In-Plane Bending 397
9-3.3 In-Plane and Out-of-Plane Bending 399
9–4 Torsional Motion 403
9-4.1 Rigid Pitch and Flap 403
9-4.2 Structural Pitch-Flap and Pitch-Lag Coupling 408
9-4.3 Torsion and Out-of-Plane Bending 412
9-4.4 Nonrotating Frame 421
9–5 Hub Reactions 422
9-5.1 Rotating Loads 423
9-5.2 Nonrotating Loads 429
9–6 Shaft Motion 435
9–7 Coupled Flap-Lag-Torsion Motion 443
9–8 Rotor Blade Bending Modes 443
9-8.1 Engineering Beam Theory for a Twisted Blade 443
9-8.2 Modal Equations 454
9-8.3 Bending Natural Frequencies 456
9-8.4 Literature 459
9–9 Derivation of the Equations of Motion 460

9-9.1 Integral Newtonian Method 461
9-9.2 Differential Newtonian Method 461
9-9.3 Normal Mode Method 462
9-9.4 Galerkin Method 464
9-9.5 Lagrangian Method 466
9-9.6 Rayleigh-Ritz Method 467
9-9.7 Lumped Parameter Methods 468

10. Rotary Wing Aerodynamics I 469

10–1 Lifting-Line Theory 469
10–2 Two-Dimensional Unsteady Airfoil Theory 471
10–3 Near Shed Wake 484
10–4 Unsteady Airfoil Theory with a Time-Varying Free Stream 492
10–5 Two-Dimensional Model for Rotary Wing Unsteady Aerodynamics 498
10–6 Approximate Solutions for Rotary Wing Unsteady Aerodynamics 513
10-6.1 Lifting-Line Approximation 513
10-6.2 Two-Dimensional, Continuous Wake Approximation 514
10-6.3 Rotary Wing Actuator Disk Model 515
10-6.4 Perturbation Inflow Model for Rotor Unsteady Aerodynamics 520
10–7 Unsteady Airfoil Theory for a Rotary Wing 526
10–8 Vortex-Induced Velocity 535
10-8.1 Straight, Infinite Line Vortex 536
10-8.2 Finite-Length Vortex Line Element 540
10-8.3 Rectangular Vortex Sheet 544

11. Rotary Wing Aerodynamics II 548

11–1 Section Aerodynamics 549
11–2 Flap Motion 556
11–3 Flap and Lag Motion 560
11–4 Nonrotating Frame 564
11–5 Hub Reactions 574
11-5.1 Rotating Frame 574
11-5.2 Nonrotating Frame 579
11–6 Shaft Motion 583
11–7 Summary 590
11–8 Pitch and Flap Motion 596

12. Rotary Wing Dynamics II 601

12–1 Flapping Dynamics 601
12-1.1 Rotating Frame 602
12-1.1.1 Hover Roots 603
12-1.1.2 Forward Flight Roots 605
12-1.1.3 Hover Transfer Function 612
12-1.2 Nonrotating Frame 613
12-1.2.1 Hover Roots and Modes 615
12-1.2.2 Hover Transfer Functions 617
12-1.3 Low Frequency Response 622
12-1.4 Hub Reactions 628
12-1.5 Two-Bladed Rotor 632

12.1.6 Literature 636
12–2 Flutter 637
12-2.1 Pitch-Flap Equations 638
12-2.2 Divergence Instability 640
12.2.3 Flutter Instability 642
12.2.4 Other Factors Influencing Pitch-Flap Stability 646
12-2.4.1 Shed Wake Influence 646
12-2.4.2 Wake-Excited Flutter 647
12-2.4.3 Influence of Forward Flight 648
12-2.4.4 Coupled Blades 649
12-2.4.5 Additional Degrees of Freedom 650
12-2.5 Literature 650
12–3 Flap-Lag Dynamics 653
12-3.1 Flap-Lag Equations 653
12-3.2 Articulated Rotors 657
12-3.3 Hingeless Rotors 658
12-3.4 Improved Analytical Models 663
12-3.5 Literature 664
12–4 Ground Resonance 668
12-4.1 Ground Resonance Equations 669
12-4.2 No-Damping Case 673
12-4.3 Damping Required for Ground Resonance Stability 681
12-4.4 Two-Bladed Rotor 685
12-4.5 Literature 693
12–5 Vibration and Loads 694
12-5.1 Vibration 694
12-5.2 Loads 699
12-5.3 Calculation of Vibration and Loads 706
12-5.4 Blade Frequencies 706
12-5.5 Literature 707

13. Rotary Wing Aerodynamics III 710
13–1 Rotor Vortex Wake 710
13–2 Nonuniform Inflow 713
13–3 Wake Geometry 735
13–4 Vortex-Induced Loads 749
13–5 Vortices and Wakes 753
13–6 Lifting-Surface Theory 754
13–7 Boundary Layers 755

14. Helicopter Aeroelasticity 756
14–1 Aeroelastic Analyses 756
14–2 Integration of the Equations of Motion 760
14–3 Literature 767

15. Stability and Control 768
15–1 Control 768
15–2 Stability 774
15–3 Flying Qualities in Hover 775
15-3.1 Equations of Motion 775
15-3.2 Vertical Dynamics 782
15-3.3 Yaw Dynamics 784

15-3.4 Longitudinal Dynamics 787
15-3.4.1 Equations of Motion 787
15-3.4.2 Poles and Zeros 788
15-3.4.3 Loop Closures 794
15-3.4.4 Hingeless Rotors 800
15-3.4.5 Response to Control 803
15-3.4.6 Examples 804
15-3.4.7 Flying Qualities Characteristics 807
15-3.5 Lateral Dynamics 808
15-3.6 Coupled Longitudinal and Lateral Dynamics 810
15-3.7 Tandem Helicopters 813
15–4 Flying Qualities in Forward Flight 822
15-4.1 Equations of Motion 822
15-4.2 Longitudinal Dynamics 827
15-4.2.1 Equations of Motion 827
15-4.2.2 Poles 829
15-4.2.3 Short Period Approximation 831
15-4.2.4 Static Stability 838
15-4.2.5 Example 840
15-4.2.6 Flying Qualities Characteristics 841
15-4.3 Lateral Dynamics 843
15-4.4 Tandem Helicopters 848
15-4.5 Hingeless Rotor Helicopters 851
15–5 Low Frequency Rotor Response 852
15–6 Stability Augmentation 854
15–7 Flying Qualities Specifications 862
15–8 Literature 869

16. Stall 873

16–1 Rotary Wing Stall Characteristics 874
16–2 NACA Stall Research 883
16–3 Dynamic Stall 888
16–4 Literature 899

17. Noise 903

17–1 Helicopter Rotor Noise 903
17–2 Vortex Noise 909
17–3 Rotational Noise 915
17-3.1 Rotor Pressure Distribution 917
17-3.2 Hovering Rotor with Steady Loading 920
17-3.3 Vertical Flight and Steady Loading 927
17-3.4 Stationary Rotor with Unsteady Loading 929
17-3.5 Forward Flight and Steady Loading 931
17-3.6 Forward Flight and Unsteady Loading 934
17-3.7 Thickness Noise 939
17-3.8 Rotating Frame Analysis 943
17-3.9 Doppler Shift 952
17–4 Blade Slap 952
17–5 Rotor Noise Reduction 956
17–6 Literature 957

Cited Literature 961
Index 1085

ACKNOWLEDGMENTS

Figure 2-15 reprinted by permission of David R. Clark and the American Helicopter Society.

Figure 5-39 reprinted by permission of Franklin D. Harris and the American Helicopter Society.

Figure 12-4 reprinted by permission of James C. Biggers and the American Helicopter Society.

Figure 12-10 reprinted by permission of Robert A. Ormiston and Dewey H. Hodges, and the American Helicopter Society.

Figure 16-1 reprinted by permission of Frank J. Tarzanin, Jr., and the American Helicopter Society.

Figure 16-4 from Alfred Gessow and Garry C. Myers, Jr., *Aerodynamics of the Helicopter,* copyright 1952 by Alfred Gessow and the estate of Garry C. Myers; published by Frederick Ungar Publishing Co., Inc. Reprinted by permission.

Figure 16-6 reprinted by permission of Norman D. Ham and Melvin S. Garelick, and the American Helicopter Society.

Figure 17-2 reprinted by permission of Sheila E. Widnall and the American Institute of Aeronautics and Astronautics.

Results of rotor airloads calculations presented in Chapter 13, sections 13–2 and 13–3 used by permission of Michael P. Scully.

NOTATION

Listed below alphabetically are the principal symbols used in this text. Not included are symbols appearing only within one chapter. Very often dimensionless quantities are used in this text; these are based on the air density, the rotor rotational speed, and the rotor radius (ρ, Ω, and R). See also section 1–3.

a	blade section two-dimensional lift-curve slope
A	rotor disk area, πR^2
A_b	rotor blade area, $NcR = \sigma A$
B	tip loss factor
c	blade chord
C	Theodorsen's lift deficiency function
C'	Loewy's lift deficiency function
c_d	section drag coefficient, $D/\frac{1}{2}\rho U^2 c$
C_H	H-force coefficient, $H/\rho A(\Omega R)^2$
c_ℓ	section lift coefficient, $L/\frac{1}{2}\rho U^2 c$
c_m	section pitch moment coefficient, $M_a/\frac{1}{2}\rho U^2 c^2$
C_{M_x}	roll moment coefficient, $M_x/\rho AR(\Omega R)^2$
C_{M_y}	pitch moment coefficient, $M_y/\rho AR(\Omega R)^2$
C_P	power coefficient, $P/\rho A(\Omega R)^3$
C_{P_c}	climb power loss
C_{P_i}	induced power loss
C_{P_o}	profile power loss
C_{P_p}	parasite power loss
C_Q	torque coefficient, $Q/\rho AR(\Omega R)^2$
c_s	speed of sound
C_T	thrust coefficient, $T/\rho A(\Omega R)^2$
C_T/σ	ratio of thrust coefficient to solidity
C_Y	Y-force coefficient, $Y/\rho A(\Omega R)^2$
D	section aerodynamic drag force; helicopter drag
e	flap or lag hinge offset
EI, EI_{zz}	flapwise bending stiffness

EI_{xx}	chordwise bending stiffness
f	equivalent drag area of helicopter fuselage and hub, $D/\frac{1}{2}\rho V^2$
F_r	section radial aerodynamic force
F_x	section aerodynamic force component parallel to disk plane
F_z	section aerodynamic force normal to disk plane
g	acceleration due to gravity
GJ	torsion stiffness
h	rotor mast height, distance of hub above helicopter center of gravity
H	rotor drag force, positive rearward; blade aerodynamic in-plane shear force coefficient (with subscript)
I_b	characteristic inertia of the rotor blade, normally $\int_0^R mr^2\,dr$ or the flapping moment of inertia
I_f	blade pitch inertia, $\int_0^R I_\theta\,dr$
I_{p_k}	generalized mass of kth torsion mode, $\int_0^R \xi_k^2 I_\theta\,dr$
I_{q_k}, I_{β_k}	generalized mass of kth out-of-plane bending mode, $\int_0^R \eta_{z_k}^2 m\,dr$
I_x	helicopter roll moment of inertia; inertial flap-pitch coupling, $\int_0^R x_I r m\,dr$
I_y	helicopter pitch moment of inertia
I_z	helicopter yaw moment of inertia
I_β	generalized mass of fundamental flap mode, $\int_0^R \eta_\beta^2 m\,dr$
$I_{\beta\alpha}$	inertial coupling of flap and hub motion, $\int_0^R r\eta_\beta m\,dr$
$I_{\beta\zeta}$	Coriolis flap-lag coupling, $\int_0^R \eta_\beta\eta_\zeta m\,dr/(1-e)$
I_ζ	generalized mass of fundamental lag mode, $\int_0^R \eta_\zeta^2 m\,dr$

I_{ζ_k} generalized mass of kth inplane bending mode, $\int_0^R \eta_{x_k}^{\ 2} m\,dr$

$I_{\zeta a}$ inertial coupling of lag and hub motion, $\int_0^R r\eta_\zeta m\,dr$

I_θ section moment of inertia about feathering axis

I_0 blade rotational inertia, $\int_0^R r^2 m\,dr$

k reduced frequency, $\omega b/U$ (ω is the frequency, b the airfoil semichord, and U the free stream velocity)

k_x helicopter roll radius of gyration, $I_x = Mk_x^{\ 2}$

k_y helicopter pitch radius of gyration, $I_y = Mk_y^{\ 2}$

k_z helicopter yaw radius of gyration, $I_z = Mk_z^{\ 2}$

K_P, K_{P_β} pitch-flap coupling, $\Delta\theta = -K_P\beta$ ($K_P = \tan\delta_3$), positive for flap up, pitch down

K_{P_ζ} pitch-lag coupling, $\Delta\theta = -K_{P_\zeta}\zeta$, positive for lag back, pitch down

K_β flap hinge spring constant

K_ζ lag hinge spring constant

K_θ control system spring constant

L section aerodynamic lift force; helicopter roll moment stability derivative (with subscript)

ℓ_{tr} tail rotor distance behind main rotor shaft

m blade index, $m = 1 \ldots N$; aerodynamic pitch moment coefficient (with subscript); blade mass per unit length

M figure of merit, $C_T^{\ 3/2}/\sqrt{2}C_P$; blade section Mach number; helicopter mass, including rotor; helicopter pitch moment stability derivative (with subscript); blade aerodynamic flap moment coefficient (with subscript)

$\dot{m}$ mass flux through the rotor disk (momentum theory)

M_a section aerodynamic pitch moment

M_b blade mass, $\int_0^R m\,dr$

M_f aerodynamic pitch moment

M_F aerodynamic flap moment

M_L aerodynamic lag moment

M_{tip} blade tip Mach number, $\Omega R/c_s$

M_x rotor hub roll moment, positive toward retreating side

M_y rotor hub pitch moment, positive rearward

$M_{1,90}$ blade advancing-tip Mach number

N number of blades; helicopter yaw force stability derivative (with subscript)

N_* longitudinal-lateral coupling parameter of flap dynamics, $N_* = (\nu_e^2 - 1)/(-\gamma M_{\dot\beta}) = (8/\gamma)(\nu^2 - 1) + K_P$

N_F blade root flapwise moment

N_L blade root lagwise moment

p sound pressure

P rotor shaft power

p_k generalized coordinate of kth torsion mode (p_0 is the rigid pitch degree of freedom)

Q rotor shaft torque, positive when external torque is required to turn rotor; blade aerodynamic torque or lag moment coefficient (with subscript)

q_k, q_{z_k} generalized coordinate of kth out-of-plane bending mode

q_{x_k} generalized coordinate of kth in-plane bending mode

r blade or rotor disk radial coordinate

R rotor radius; blade aerodynamic radial shear force coefficient (with subscript)

s eigenvalue or Laplace variable

S_b blade first moment of inertia, $\int_0^R rm\,dr$

S_r blade root radial shear force

S_x blade root in-plane shear force

S_z blade root vertical shear force

S_β first moment of flap mode, $\int_0^R \eta_\beta m\,dr$

S_ζ first moment of lag mode, $\int_0^R \eta_\zeta m\,dr$

t time

T rotor thrust, positive upward; blade aerodynamic thrust force coefficient (with subscript)

T/A	rotor disk loading
T/A_b	rotor blade loading
U	section resultant velocity, $(u_T^2 + u_P^2)^{1/2}$
u_G	longitudinal gust velocity component
u_P	air velocity of blade section, perpendicular to the disk plane
u_R	radial air velocity of blade section
u_T	air velocity of blade section, tangent to the disk plane
v	rotor induced velocity (positive down through the disk)
V	rotor or helicopter velocity with respect to the air
v_G	lateral gust velocity component
v_h	ideal hover induced velocity, $\sqrt{T/2\rho A}$
w	rotor induced velocity in the far wake
W	helicopter gross weight
w_G	vertical gust velocity component
x	rotor nonrotating coordinate axis, positive aft; blade in-plane deflection; blade chordwise coordinate
X	helicopter longitudinal force derivative (with subscript)
x_A	chordwise offset of blade aerodynamic center behind pitch axis
x_B	helicopter rigid body longitudinal degree of freedom
x_h	hub longitudinal displacement
x_I	chordwise offset of blade center of gravity behind pitch axis
y	rotor nonrotating coordinate axis, positive to right (advancing side)
Y	rotor side force, positive toward advancing side; helicopter side force stability derivative (with subscript)
y_B	helicopter rigid body lateral degree of freedom
y_h	hub lateral displacement
z	rotor nonrotating coordinate axis, positive upward; blade out-of-plane deflection
Z	helicopter vertical force stability derivative (with subscript)
z_B	helicopter rigid body vertical degree of freedom
z_h	hub vertical displacement
α	blade section angle of attack; rotor disk plane angle of attack, positive for forward tilt
α_x	hub roll perturbation
α_y	hub pitch perturbation

α_z	hub yaw perturbation
$\alpha_{1,270}$	blade retreating tip angle of attack
$\alpha_{\mu+.4,270}$	blade angle of attack at $r/R = \mu + .4$ and $\psi = 270°$
β	blade flap angle (positive upward)
β_p	precone angle
β_0	coning angle
β_{1c}	longitudinal tip-path-plane tilt angle, positive forward
β_{1s}	lateral tip-path-plane tilt angle, positive toward retreating side
γ	blade Lock number, $\rho acR^4/I_b$
Γ	blade bound circulation
$\delta_0, \delta_1, \delta_2$	coefficients in expansion for section drag: $c_d = \delta_0 + \delta_1\alpha + \delta_2\alpha^2$
δ_3	pitch-flap coupling ($K_P = \tan\delta_3$)
ζ	blade lag angle, positive opposite the direction of rotation of the rotor
ζ_p	prelag angle
η, η_β	mode shape of fundamental flap mode
η, η_ζ	mode shape of fundamental lag mode
η_k, η_{z_k}	mode shape of kth out-of-plane bending mode
η_{x_k}	mode shape of kth in-plane bending mode
θ	blade pitch or feathering angle, positive nose upward
θ_B	helicopter rigid body pitch degree of freedom
θ_{con}	pitch control input (collective and cyclic)
θ_e	elastic torsion deflection
θ_{FP}	flight path angle, climb velocity = $V \sin\theta_{FP}$
θ_{tw}	linear twist rate
θ_0	collective pitch angle
θ_{1c}	lateral cyclic pitch angle
θ_{1s}	longitudinal cyclic pitch angle
$\theta_{.75}$	collective pitch angle at 75% radius
λ	rotor inflow ratio, $(V\sin\alpha + \nu)/\Omega R$, positive down through disk
λ_c	climb inflow ratio
λ_i	induced inflow ratio, $\nu/\Omega R$
λ_x	coefficient of longitudinal variation of induced velocity

λ_y	coefficient of lateral variation of induced velocity
λ_0	rotor mean induced velocity
μ	rotor advance ratio, $V \cos\alpha/\Omega R$
ν, ν_β	rotating natural frequency of blade fundamental flap mode
ν_e, ν_{β_e}	effective flap frequency including pitch-flap coupling, $\nu_e^2 = \nu^2 + (\gamma/8)K_P$
ν_k, ν_{z_k}	natural frequency of kth out-of-plane bending mode
ν_{x_k}	natural frequency of kth in-plane bending mode
ν_ζ	rotating natural frequency of blade fundamental lag mode
ξ_k	mode shape of kth elastic torsion mode
ρ	air density; blade radial coordinate in spanwise integration
σ	rotor solidity, $Nc/\pi R$
ϕ	section inflow angle, $\tan^{-1} u_P/u_T$
ϕ_B	helicopter rigid body roll degree of freedom
ψ	azimuth angle of the blade or rotor disk; dimensionless time, Ωt
ψ_B	helicopter rigid body yaw degree of freedom
ψ_m	azimuth position of mth blade ($m = 1...N$)
$\omega, \omega_0, \omega_\theta$	natural frequency of rigid pitch motion (control system stiffness)
ω_k	natural frequency of kth elastic torsion mode
Ω	rotor rotational speed (rad/sec)

SUBSCRIPTS AND SUPERSCRIPTS

$0, 1c, 1s, \ldots, nc, ns, \ldots\infty$	harmonics of a sine/cosine Fourier series representation of a periodic function
$0, 1c, 1s, \ldots, nc, ns, N/2$	degree of freedom of the Fourier coordinate transform (total number N)

c	climb
CP	control plane
h	hover
HP	hub plane
i	induced

m	blade index, $m = 1$ to N
mr	main rotor
NFP	no-feathering plane
o	profile
p	parasite
p	helicopter stability derivative due to roll rate
q	helicopter stability derivative due to pitch rate
r	helicopter stability derivative due to yaw rate
TPP	tip-path plane
tr	tail rotor
u	helicopter stability derivative due to longitudinal velocity
v	helicopter stability derivative due to lateral velocity
w	helicopter stability derivative due to vertical velocity
β	rotor aerodynamic force due to blade flap displacement
$\dot{\beta}$	rotor aerodynamic force due to blade flapping velocity or hub angular motion
ζ	rotor aerodynamic force due to blade lag displacement
$\dot{\zeta}$	rotor aerodynamic force due to blade lagging velocity or hub yawing motion
θ	rotor aerodynamic force due to blade pitch motion
$\dot{\theta}$	rotor aerodynamic force due to blade pitch rate
λ	rotor aerodynamic force due to hub vertical velocity or induced velocity perturbation
μ	rotor aerodynamic force due to hub in-plane velocity
$\dot{(\ )}$	$d(\ )/dt$ or $d(\ )/d\psi$
$(\ )'$	$d(\ )/dr$
$(\ )^*$	normalized: rotor blade inertias divided by I_b, and helicopter inertias divided by $\frac{1}{2}NI_b$

HELICOPTER THEORY

Chapter 1

INTRODUCTION

1–1 The Helicopter

The helicopter is an aircraft that uses rotating wings to provide lift, propulsion, and control. Figures 1-1 to 1-3 illustrate the principal helicopter configurations. The rotor blades rotate about a vertical axis, describing a disk in a horizontal or nearly horizontal plane. Aerodynamic forces are generated by the relative motion of a wing surface with respect to the air. The helicopter with its rotary wings can generate these forces even when the velocity of the vehicle itself is zero, in contrast to fixed-wing aircraft, which require a translational velocity to sustain flight. The helicopter therefore has the capability of vertical flight, including vertical take-off and landing. The efficient accomplishment of vertical flight is the fundamental characteristic of the helicopter rotor.

The rotor must efficiently supply a thrust force to support the helicopter weight. Efficient vertical flight means a low power loading (ratio of rotor power required to rotor thrust), because the installed power and fuel consumption of the aircraft are proportional to the power required. For a rotary wing, low disk loading (the ratio of rotor thrust to rotor disk area) is the key to a low power loading. Conservation of momentum requires that the rotor lift be obtained by accelerating air downward, because corresponding to the lift is an equal and opposite reaction of the rotating wings against the air. Thus the air left in the wake of the rotor possesses kinetic energy which must be supplied by a power source in the aircraft if level flight is to be sustained. This is the induced power loss, a property of both fixed and rotating wings that constitutes the absolute minimum of power required for equilibrium flight. For the rotary wing in hover, the induced power loading is found to be proportional to the square root of the rotor disk loading. Hence the efficiency of rotor thrust generation increases as the disk loading decreases.

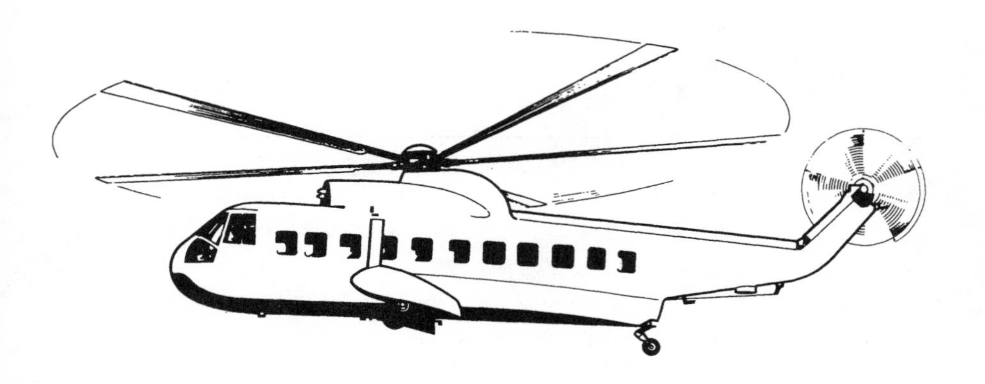

Figure 1-1 A single main rotor and tail rotor helicopter

Figure 1-2 A two-bladed single main rotor helicopter

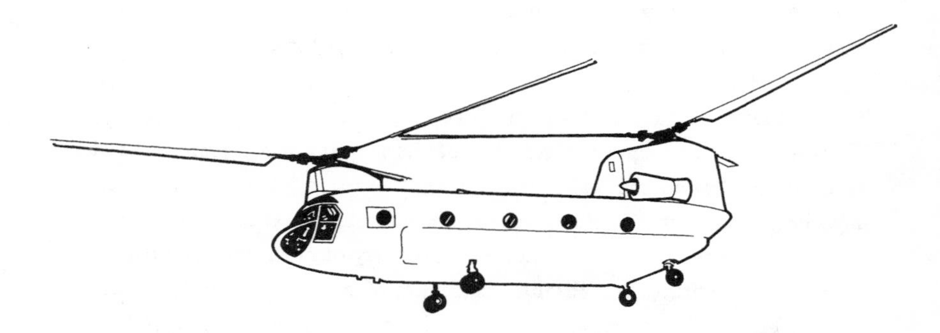

Figure 1-3 A tandem main rotor helicopter

For a given gross weight the induced power is inversely proportional to the rotor radius, and therefore the helicopter is characterized by the large disk area of large diameter rotors. The disk loading characteristic of helicopters is in the range of 100 to 500 N/m^2 (2 to 10 lb/ft^2). The small diameter rotating wings found in aeronautics, including propellers and turbofan engines, are used mainly for aircraft propulsion. For such applications a high disk loading is appropriate, since the rotor is operating at high axial velocity and at a thrust equal to only a fraction of the gross weight. However, the use of high disk loading rotors for direct lift severely compromises the vertical flight capability in terms of both greater installed power and much reduced hover endurance. The helicopter uses the lowest disk loading of all VTOL (vertical take-off and landing) aircraft designs and hence has the most efficient vertical flight capability. It follows that the helicopter may be defined as an aircraft utilizing large diameter, low disk loading rotary wings to provide the lift for flight.

Since the helicopter must also be capable of translational flight, a means is required to produce a propulsive force to oppose the aircraft and rotor drag in forward flight. For low speeds at least, this propulsive force is obtained from the rotor, by tilting the thrust vector forward. The rotor is also the source of the forces and moments on the aircraft that control its position, attitude, and velocity. In a fixed wing aircraft, the lift, propulsion, and control forces are provided by largely separate aerodynamic surfaces. In the helicopter, all three are provided by the rotor.

Vertical flight capability is not achieved without a cost, which must be weighed against the value of VTOL capability in the desired applications of the aircraft. The task of the engineer is to design an aircraft that will accomplish the required operations with minimum penalty for vertical flight. The price of vertical flight includes a higher power requirement than for fixed wing aircraft, a factor that influences the first cost and operating cost. A large transmission is required to deliver the power to the rotor at low speed and high torque. The fact that the rotor is a mechanically complex system increases first cost and maintenance costs. The rotor is a source of vibration, hence increased maintenance costs, passenger discomfort, and pilot fatigue. There are high alternating loads on the rotor, reducing the structural component life and in general resulting in increased maintenance cost. The stability and control characteristics are often marginal, especially in hover, unless a reliable automatic control system is used. In particular, good instrument flight characteristics are lacking without stability augmentation. Aircraft noise is an increasingly important factor in air transportation, as it is the primary form of interaction of the system with a large part of society. The helicopter is among the quietest of aircraft (or at least it can be), but utilization of its VTOL capability often involves operation close to urban areas, leading to stricter noise requirements in order to achieve its potential. All these factors can be overcome to design a highly succsesful aircraft. The engineering analysis required for that task is the subject of this book.

1-1.1 The Helicopter Rotor

The conventional helicopter rotor consists of two or more identical, equally spaced blades attached to a central hub. The blades are maintained in uniform rotational motion, usually by a shaft torque from the engine. The lift and drag forces on these rotating wings produce the torque, thrust, and other forces and moments of the rotor. The large diameter rotor required for efficient vertical flight and the high aspect ratio blades required for good aerodynamic efficiency of the rotating wing result in blades that are considerably more flexible than high disk loading rotors such as propellers. Consequently, there is a substantial motion of the rotor blades in response to the aerodynamic forces in the rotary wing environment. This motion can produce high stresses in the blades or large moments at the root, which are transmitted through the hub to the helicopter. Attention must therefore be given in the design of the helicopter rotor blades and hub to keeping these loads low. The centrifugal stiffening of the rotating blade results in the motion being predominantly about the blade root. Hence the design task focuses on the configuration of the rotor hub.

A frequent design solution that was adopted early in the development of the helicopter and only recently altered is to use hinges at the blade root that allow free motion of the blade normal to and in the plane of the disk. A schematic of the root hinge arrangement is given in Fig 1-4. Because the bending moment is zero at the blade hinge, it must be low throughout the root area, and no hub moment is transmitted through the blade root to the helicopter. This configuration makes use of the blade motion to relieve the

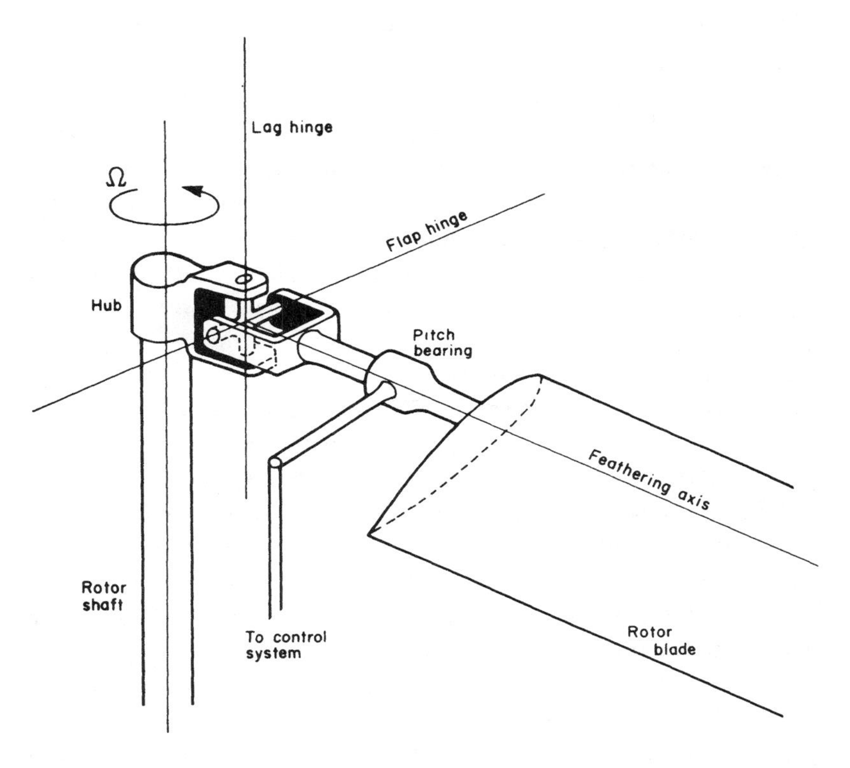

Figure 1-4 Schematic of an articulated rotor hub and root, showing only one of the two or more blades of the rotor.

bending moments that would otherwise arise at the root of the blade. The motion of the blade allowed by these hinges has an important role in the behavior of the rotor and in the analysis of that behavior. Some current rotor designs eliminate the hinges at the root, so that the blade motion involves structural bending. The hub and blade loads are necessarily higher than for a hinged design. The design solution is basically the same, however, because the blade must be provided with enough flexibility to allow substantial motion, or the loads would be intolerable even with advanced materials and design technology. Hence blade motion remains a dominant factor in rotor behavior, although the root load and hub moment capability of a hingeless blade has a significant influence on helicopter design and operating characteristics.

The motion of a hinged blade consists basically of rigid body rotation about each hinge, with restoring moments due to the centrifugal forces acting on the rotating blade. Motion about the hinge lying in the rotor disk plane (and perpendicular to the blade radial direction) produces out-of-plane deflection of the blade and is called flap motion. Motion about the vertical hinge produces deflection of the blade in the plane of the disk and is called lag motion (or lead-lag). For a blade without hinges the fundamental modes of out-of-plane and in-plane bending define the flap and lag motion. Because of the high centrifugal stiffening of the blade these modes are similar to the rigid body rotations of hinged blades, except in the vicinity of the root, where most of the bending takes place. In addition to the flap and lag motion, the ability to change the pitch of the blade is required in order to control the rotor. Pitch motion allows control of the angle of attack of the blade, and hence control of the aerodynamic forces on the rotor. This blade pitch change, called feathering motion, is usually accomplished by movement about a hinge or bearing. The pitch bearing on a hinged blade is usually outboard of the flap and lag hinges; on a hingeless blade the pitch bearing may be either inboard or outboard of the major flap and lag bending at the root. There are also rotor designs that eliminate the pitch bearings as well as the flap and lag hinges; the pitch motion then occurs about a region of torsional flexibility at the blade root.

The mechanical arrangement of the rotor hub to accommodate the flap and lag motion of the blade provides a fundamental classification of rotor types as follows:

a) Articulated rotor. The blades are attached to the hub with flap and lag hinges.

b) Teetering rotor. Two blades forming a continuous structure are attached to the rotor shaft with a single flap hinge in a teetering or seesaw arrangement. The rotor has no lag hinges. Similarly, a gimballed rotor has three or more blades attached to the hub without hinges, and the hub is attached to the rotor shaft by a gimbal or universal joint arrangement.

c) Hingeless rotor. The blades are attached to the hub without flap or lag hinges, although often with a feathering bearing or hinge. The blade is attached to the hub with cantilever root restraint, so that blade motion occurs through bending at the root. This rotor is also called a rigid rotor. However, the limit of a truly rigid blade, which is so stiff that there is no significant motion, is applicable only to high disk loading rotors.

1-1.2 Helicopter Configuration

The arrangement of the rotor or rotors on a helicopter is perhaps its most distinctive external feature and is an important factor in its behavior, notably its stability and control characteristics. Usually the power is delivered to the rotor through the shaft, accompanied by a torque. The aircraft in steady flight can have no net force or moment acting on it, and therefore the torque reaction of the rotor on the helicopter must be balanced in some manner. The method chosen to accomplish this torque balance is the primary determinant of the helicopter configuration. Two methods are in general use; a configuration with a single main rotor and a tail rotor, and configurations with twin contrarotating rotors.

The single main rotor and tail rotor configuration uses a small auxiliary rotor to provide the torque balance (and yaw control). This rotor is on the tail boom, typically slightly beyond the edge of the main rotor disk. The tail rotor is normally vertical, with its shaft horizontal and parallel to the helicopter lateral axis. The torque balance is produced by the tail rotor thrust acting on an arm about the main rotor shaft. The main rotor provides lift, propulsive force, and roll, pitch, and vertical control for this configuration.

A twin main rotor configuration uses two contrarotating rotors, of equal size and loading, so that the torques of the rotors are equal and opposing. There is then no net yaw moment on the helicopter due to the main rotors. This configuration automatically balances the main rotor torque without requiring a power-absorbing auxiliary rotor. The rotor-rotor aerodynamic interference losses absorb about the same amount of power, however. The most frequent twin rotor arrangement is the tandem helicopter configuration—fore and aft placement of the main rotors on the fuselage usually with significant overlap of the rotor disks and with the rear rotor raised vertically above the front rotor. A side-by-side twin rotor arrangement has also found some application.

1-1.3 Helicopter Operation

Operation in vertical flight, with no translational velocity, is the particular role for which the helicopter is designed. Operation with no velocity at all relative to the air, either vertical or translational, is called hover. Lift and control in hovering flight are maintained by rotation of the wings to provide aerodynamic forces on the rotor blades. General vertical flight involves climb or descent with the rotor horizontal, and hence with purely axial flow through the rotor disk. A useful aircraft must be capable of translational flight as well. The helicopter accomplishes forward flight by keeping the rotor nearly horizontal, so that the rotor disk sees a relative velocity in its own plane in addition to the rotational velocity of the blades. The rotor continues to provide lift and control for the aircraft. It also provides the propulsive force to sustain forward flight, by means of a small forward tilt of the rotor thrust.

Safe operation after loss of power is required of any successful aircraft. The fixed wing aircraft can maintain lift and control in power-off flight, descending in a glide at a shallow angle. Rotary wing aircraft also have the capability of sustaining lift and control after a loss of power. Power-off descent of the helicopter is called autorotation. The rotor continues to turn and provide lift and control. The power required by the rotor is taken from the air flow provided by the aircraft descent. The procedure upon recognition of loss of power is to set the controls as required for autorotative descent, and establish equilibrium flight at the minimum descent rate. Then near the ground the helicopter is flared, using the rotor-stored kinetic energy of rotation to eliminate the vertical and translational velocity just before touchdown. The helicopter rotor in vertical power-off descent has been found to be nearly as effective as a parachute of the same diameter as the rotor disk; about half that descent rate is achievable in forward flight.

A rotary wing aircraft called the autogiro uses autorotation as the normal working state of the rotor. In the helicopter, power is supplied directly to the rotor, and the rotor provides propulsive force as well as lift. In the autogiro, no power or shaft torque is supplied to the rotor. The power and propulsive force required to sustain level forward flight are supplied by a propeller or other propulsion device. Hence the autogiro is like a fixed-wing aircraft, since the rotor takes the role of the wing in providing only lift for the vehicle, not propulsion. Sometimes the aircraft control forces and moments are supplied by fixed aerodynamic surfaces as in the airplane, but it is better to obtain the control from the rotor. The rotor performs much like a wing, and has a fairly good lift-to-drag ratio. Although rotor performance is not as good as that of a fixed wing, the rotor is capable of providing lift and control at much lower speeds. Hence the autogiro is capable of flight speeds much slower than fixed-wing aircraft. Without power to the rotor itself, however, it is not capable of actual hover or vertical flight. Because autogiro performance is not that much better than the performance of an airplane with a low wing loading, it has usually been found that the requirement of actual VTOL capability is necessary to justify the use of a rotor on an aircraft.

1–2 History

The initial development of rotary-wing aircraft faced three major problems that had to be overcome to achieve a successful vehicle. The first problem was to find a light and reliable engine. The reciprocating internal combustion engine was the first to fulfill the requirements, and much later the adoption of the turboshaft engine for the helicopter was a significant advance. The second problem was to develop a light and strong structure for the rotor, hub, and blades while maintaining good aerodynamic efficiency. The final problem was to understand and develop means of controlling the helicopter, including balancing the rotor torque. These problems were essentially the same as those that faced the development of the airplane and were solved eventually by the Wright brothers. The development of the helicopter in many ways paralleled that of the airplane. That helicopter development took longer may be attributed to the cost of vertical flight, which required a higher development of aeronautical technology before the problems could be satisfactorily overcome.

1-2.1 Helicopter Development

A history of helicopter development is usually begun with mention of the Chinese top and Leonardo da Vinci. The Chinese flying top (c. 400 B.C.) was a stick with a propeller on top, which was spun by the hands and released.

Among da Vinci's work (late 15th century) were sketches of a machine for vertical flight utilizing a screw-type propeller. In the 18th century there was some work with models. Mikhail V. Lomonosov (Russia, 1754) demonstrated a spring-powered model to the Russian academy of sciences. Launoy and Bienvenu (France, 1784) demonstrated a spring-powered model to the French academy of sciences. It had two contrarotating rotors of four blades each (constructed of feathers), powered by a flexed bow. Sir George Cayley (England, 1790's) constructed models powered by elastic elements and made sketches of helicopters. These models had little impact on helicopter development.

In the last half of the 19th century many inventors were concerned with the helicopter. There was some practical progress, but no successful vehicle. The problem was the lack of a cheap, reliable, light engine. A number of attempts to use a steam engine are known. W.H. Phillips (England, 1842) constructed a 10 kg steam-powered model. Viscomte Gustave de Ponton d'Amecourt (France, 1863) built a small steam-driven model; he also invented the word "helicopter." Alphonse Penaud (France, 1870's) experimented with models. Enrico Forlanini (Italy, 1878) built a 3.5 kg flying steam-driven model. Thomas Edison (United States, 1880's) experimented with models. He recognized that the problem was the lack of an adequate (meaning light) engine. Edison concluded that no helicopter would be able to fly until engines were available with a weight-to-power ratio below 1 to 2 kg/hp. These were still only models, but they were beginning to address the problem of an adequate power source for sustained flight. The steam engine was not successful for aircraft, especially the helicopter, because of the low power-to-weight ratio of the system.

Around 1900 the internal combustion reciprocating gasoline engine became available. It made possible airplane flight, and eventually helicopter flight as well. Renard (France, 1904) built a helicopter with two side-by-side rotors, using a two-cylinder engine; he introduced the flapping hinge for the helicopter rotor. The Breguet-Richet (France, 1907) Gyroplane No. 1 had four rotors with four biplane blades each (rotors 8 m in diameter, gross weight 580 kg, 45 hp Antoinette engine). It made a tethered flight with a passenger at an altitude of about 1 m for about 1 min. Paul Cornu (France 1907) constructed a machine that made the first flight with a pilot (Cornu). It had two contrarotating rotors in tandem configuration with two fabric-covered blades each (rotors 6 m in diameter, gross weight 260 kg, 24 hp

Antoinette engine connected to the rotors by belts). Control was by vanes in the rotor slipstream and was not very effective. This helicopter achieved an altitude of about 0.3 m for about 20 sec; it had problems with mechanical design and with lack of stability. Emile and Henry Berliner (United States, 1909) built a two-engine coaxial helicopter that lifted a pilot untethered. Igor Sikorsky (Russia, 1910) built a helicopter with two coaxial three-bladed rotors (rotors 5.8 m in diameter, 25 hp Anzani engine) that could lift 180 kg but not its own weight plus the pilot. Sikorsky would return to the development of the helicopter (with considerably more success) after building airplanes in Russia and in the United States. Boris N. Yuriev (Russia, 1912) built a machine with a two-bladed main rotor and a small antitorque tail rotor (main rotor 8 m in diameter, gross weight 200 kg, 25 hp Anzani engine). This helicopter made no successful flight, but Yuriev went on to supervise helicopter development in the Soviet Union. Petroczy and von Kármán (Austria, 1916) built a tethered observation helicopter that achieved an altitude of 50 m with payload.

The development of better engines during and after World War I solved the problem of an adequate power source, at least enough to allow experimenters to face the task of finding a satisfactory solution for helicopter control. George de Bothezat (United States, 1922) built a helicopter with four six-bladed rotors at the ends of intersecting beams (gross weight 1600 kg, 180 hp engine at the center). It had good control, utilizing differential collective of the four rotors, and made many flights with passengers up to an altitude of 4 to 6 m. (Collective pitch is a change made in the mean blade pitch angle to control the rotor thrust magnitude.) This was the first rotorcraft ordered by the U.S. Army, but after the expenditure of $200,000 the project was finally abandoned as being too complex mechanically. Etienne Oemichen (France, 1924) built a machine with four two-bladed rotors (two 7.6 m in diameter and two 6.4 m in diameter) to provide lift, five horizontal propellers for attitude control, two propellers for propulsion, and one propeller in front for yaw control—all powered by a single 120 hp Le Rhone engine. It set the first helicopter distance record, 360 m. Marquis Raul Pateras Pescara (Spain, 1924) constructed a helicopter with two coaxial rotors of four biplane blades each (180 hp; a 1920 craft of similar design that used rotors 6.4 m in diameter and a 45 hp Hispano engine had inadequate lift). For control, he warped the biplane blades to change their pitch angle. Pescara was the first to demonstrate effective cyclic for control of the main rotors. (Cyclic pitch is a sinusoidal, once-per-revolution change made in the blade pitch to tilt the rotor disk.) Pescara's helicopter set a distance record (736 m), but had stability problems. Emile and Henry Berliner (United States, 1920-1925) built a helicopter using two rotors positioned on the tips of a biplane wing in a side-by-side configuration. They used rigid wooden propellers for the rotors and obtained control by tilting the entire rotor. Louis Brennan (England, 1920's) built a helicopter with a rotor turned by propellers on the blades, to eliminate the torque problem; the machine was mechanically too complex. Cyclic control was obtained by warping the blades with aerodynamic control tabs. A.G. von Baumhauer (Holland, 1924-1929) developed a helicopter with a single main rotor and a vertical tail rotor for torque balance (two-bladed main rotor 15 m in diameter, gross weight 1300 kg, 200 hp rotary engine). A separate engine was used for the tail rotor (80 hp Thulin rotary engine mounted directly to the tail rotor). The main rotor blades were free to flap, but were connected by cables to form a teetering rotor. Control was by cyclic pitch of the main rotor, produced using a swashplate. Flights were made, but never at more than 1 m altitude. There were difficulties with directional control because of the separate engines for the main rotor and tail rotor, and the project was abandoned after a bad crash. Corradino d'Ascanio (Italy, 1930) constructed a helicopter with two coaxial rotors (rotors 13 m in diameter, 95 hp engine). The two-bladed rotors had flap hinges and free-feathering hinges. Control by servo tabs on the blade was used to obtain cyclic and collective pitch changes. For several years this machine held records for altitude (18 m), endurance (8 min 45 sec), and distance (1078 m). The stability and control characteristics were marginal, however. M.B. Blecker (United States, 1930) built a helicopter with four wing-like blades. Power was delivered to a propeller on each blade from an engine in the fuselage. Control was by aerodynamic surfaces on the blades and by a tail on the aircraft. The Central Aero-Hydrodynamic Institute of the Soviet Union developed a series of single rotor helicopters under the direction of Yuriev. The TsAGI I-EA (1931) had a four-bladed main rotor (rotor 11 m in diameter, gross weight 1100 kg, 120 hp engine) with cyclic and collective control, and two small contrarotating antitorque rotors.

The development of the helicopter was fairly well advanced at this point, but the stability and control characteristics were still marginal, as were the forward flight and power-off (autorotation) capabilities of the designs. It was in this period, the 1920's and 1930's, that the autogiro was developed. The autogiro was the first practical use of the direct-lift rotary wing. It was developed largely by Juan de la Cierva (Spanish, 1920's-1930's; he also coined the word "autogiro"). In this aircraft, a windmilling rotor replaces the wing of the airplane. Essentially, the fixed wing aircraft configuration is used, with a propeller supplying the propulsive force; the initial designs even used conventional airplane-type aerodynamic surfaces for control (ailerons, rudder, and elevator). With no power directly to the rotor, hover and vertical flight is not possible, but the autogiro is capable of very slow flight and in cruise it behaves much like an airplane.

Juan de la Cierva designed an airplane that crashed in 1919 due to stall near the ground. He then became interested in designing an aircraft with a low take-off and landing speed that would not stall if the pilot dropped the speed excessively. He determined from wind-tunnel tests of model rotors that with no power to the shaft but with a rearward tilt of the rotor, good lift-to-drag ratio could be obtained even at low speed. The best results were at low, positive collective pitch of the rotor. In 1922, Cierva built the C-3 autogiro with a five-bladed rigid rotor and "a tendency to fall over sideways." He had a model with blades of flexible palm wood that flew properly. It was discovered that the flexible rotor blades accounted for the successful flight of the model; suggesting the use of articulated rotor blades on the autogiro. Cierva consequently incorporated flapping blades in his design. The flap hinge eliminated the rolling moment on the aircraft in forward flight due to the asymmetry of the flow over the rotor. Cierva was the first to use the flap hinge in a successful rotary-wing aircraft. In 1923, the C-4 autogiro was built and achieved successful flight. It had a four-bladed rotor with flap hinges on the blades (rotor 9.8 m in diameter, 110 hp Le Rhone engine). Control was by conventional airplane aerodynamic surfaces. In 1924, the C-6 autogiro with flapping rotor blades was built (four-bladed rotor 11 m in diameter, 100 hp Le Rhone rotary engine). An Avro 504K aircraft fuselage and ailerons on outrigger spars were used. The demonstration of this autogiro in 1925 at the Royal Aircraft Establishment was the stimulus for the early analysis of the rotary wing in England by Glauert and Lock. The C-6 is generally regarded as Cierva's first successful autogiro (1926).

In 1925, Cierva founded the Cierva Autogiro Company in England, which was his base thereafter. In the next decade about 500 of his autogiros were produced, many by licensees of the Cierva Company, including A.V. Roe, de Havilland, Weir, Westland, Parnell, and Comper in Britain; Pitcairn, Kellett, and Buhl in the United States; Focke-Wulf in Germany; Loire and Olivier in France; and the TsAGI in Russia. A crash in 1927 led to an appreciation of the high in-plane blade loads due to flapping, and a lag hinge was added to the rotor blades. This completed the development of the fully articulated rotor hub for the autogiro. In 1932, Cierva added rotor control to replace the airplane control surfaces, which were not very effective at low speeds. He used direct tilt of the rotor hub for longitudinal and lateral control. Raoul Hafner (England, 1935) developed an autogiro incorporating cyclic pitch control by means of a "spider" control mechanism to replace the direct tilt of the rotor hub. E. Burke Wilford (United States, 1930's) developed a hingeless rotor autogiro that also used cyclic control of the rotors.

By 1935, the autogiro was well developed in both Europe and America. Its success preceded that of the helicopter because of the lower power required without actual vertical flight capability and because the unpowered rotor is mechanically simpler. In addition, it was possible to start with most of the airplane technology, for example in the propulsion system, and initially even the control system. Lacking true vertical flight capability, however, the autogiro was never able to compete effectively with fixed wing aircraft. Autogiro developments, including experimental and practical experience, had some influence on helicopter development and design. The autogiro had a substantial impact on the development of rotary wing analysis; much of the work of the 1920's and 1930's, which forms the foundation of helicopter analysis, was originally developed for the autogiro.

Meanwhile, the development of the helicopter continued. Louis Breguet and Rene Dorand (France, 1935) built a helicopter with coaxial two-bladed rotors (rotors 16.5 m in diameter, gross weight 2000 kg, 450 hp engine). The rotors had an articulated hub (flap and lag hinges); control was by cyclic for pitch and roll, and differential torque for directional control. The aircraft had satisfactory control characteristics and held records for speed (44.7 kph), altitude (158 m), duration (1 hr 2 min), and closed circuit distance (44 km). E.H. Henrich Focke (Germany, 1936), constructed a helicopter with two three-bladed rotors mounted on trusses in a side-by-side configuration (rotors 7 m in diameter, gross weight 950 kg, 160 hp Bramo engine). The rotor had an articulated hub and tapered blades. Directional and longitudinal control were by cyclic, roll control by differential collective. Vertical and horizontal tail surfaces were used for stability and trim in forward flight, and the rotor shafts were inclined inward for stability. This helicopter set records for speed (122.5 kph), altitude (2440 m), and endurance (1 hr 21 min). It was a well-developed machine, with good control, performance, and reliability. Anton Flettner (Germany, 1938-1940) developed a synchropter design, with two rotors in a side-by-side configuration but highly intermeshed (hub separation 0.6 m). The FL-282 had two-bladed articulated rotors (rotors 12 m in diameter, gross weight 1000 kg, 140 hp Siemens-Halske engine). C.G. Pullin (Britain, 1938) built helicopters with a side-by-side configuration, in 1938 the W-5 (two-bladed rotors 4.6 m in diameter, gross weight 380 kg, 50 hp Weir engine), and in 1939 the W-6 (three-bladed rotors 7.6 m in diameter, gross weight 1070 kg, 205 hp de Havilland engine) for G. & J. Weir Ltd. Ivan P. Bratukhin (TsAGI in the USSR, 1939-1940) constructed the Omega I helicopter with two three-bladed rotors in a side-by-side configuration (rotors 7 m in diameter, gross weight 2300 kg, two 350 hp engines). There was considerable effort in rotary wing development in Germany during World War II, including the Focke-Achgelis Fa-223 in 1941. This helicopter, which had two three-bladed rotors in the side-by-side configuration (rotors 12 m in diameter, gross weight 4300 kg, 1000 hp Bramo engine), had an absolute ceiling of 5000 m, a range of 300 km, a cruise speed of 120 kph with six passengers, and a useful load of 900 kg.

Igor Sikorsky (Sikorsky Aircraft Co. in the United States, 1939-1941) returned to helicopter development in 1938 after designing and building airplanes in Russia and the United States. In 1941, Sikorsky built the VS-300, a helicopter with a single three-bladed main rotor and a small antitorque tail rotor (rotor 9 m in diameter, gross weight 520 kg, 100 hp Franklin engine). Lateral and longitudinal control was by main rotor cyclic, and directional control was by means of the tail rotor. The tail rotor was driven by a shaft from the main rotor. The pilot's controls were like the present standard (cyclic stick, pedals, and a collective stick with a twist grip throttle). Considerable experimentation was required to develop a configuration with suitable control characteristics. The first configuration had three auxiliary rotors (one vertical and two horizontal) on the tail for control and stability. In 1941, the number of auxiliary rotors was reduced to two, a vertical tail rotor for yaw and a horizontal tail rotor for pitch control. Finally the horizontal propeller was removed, main rotor cyclic replacing it for longitudinal control. This version was Sikorsky's eighteenth, the single main rotor and tail rotor configuration that has become the most common helicopter type. Sikorsky also tried a two-bladed main rotor. It had comparable performance and was simpler, but was not pursued because the vibration was considered excessive. In 1942 the R-4 (VS-316), a derivative of the VS-300, was constructed (three-bladed rotor 11.6 m in diameter, gross weight 1100 kg, 185 hp Warner engine). This helicopter model went into production and several hundred were built during World War II. Sikorsky's aircraft is generally considered the first practical, truly operational helicopter, although a possible exception is the work of Focke in Germany during World War II; the latter effort reached a dead end in the early 1940's because of the time and place of its development, however. Igor Sikorsky's R-4 was successful because it was mechanically simple (relative to other helicopter designs of the time at least) and controllable—and because it went into production.

Sikorsky's success gave great impetus to the development of the helicopter in the United States. Many other designs began development and production in the next few years. Since World War II there has been considerable progress in the mechanical and technical development of the helicopter, with production to support further development. Lawrence Bell (Bell Helicopter Company in the United States, 1943) built a helicopter with a two-bladed teetering main rotor and a tail rotor, using the gyro stabilizer bar developed by Arthur Young in the United States during the 1930's. In 1946, the Bell Model 47 (rotor 10.7 m in diameter, gross weight 950 kg, 178 hp Franklin engine) received the first American certificate of airworthiness for helicopters. Frank N. Piasecki (Piasecki Helicopter Corporation in the United States, 1945) developed the PV-3, a tandem rotor helicopter (three-bladed rotors 12.5 m in diameter, gross weight 3100 kg, 600 hp Pratt and Whitney engine). Piasecki's company eventually became the Boeing Vertol Company, with the tandem configuration remaining its basic production type. Louis Breguet (France, 1946) built the G-II E, a helicopter with two coaxial contrarotating rotors (three-bladed rotor 8.5 m in diameter, gross weight 1300 kg, 240 hp Potex engine). The rotors had fully articulated hubs with flap and lag dampers. Stanley Hiller (United States, 1946-1948) experimented with several types of helicopters, eventually settling on the single main rotor and tail rotor configuration. Hiller developed the control rotor, a gyro stabilizer bar with aerodynamic surfaces that the pilot controlled in order to adjust the rotor orientation. He built the Model 360 helicopter in 1947 (two-bladed rotor 10.7 m in diameter, gross weight 950 kg, 178 hp Franklin engine). Charles Kaman (Kaman Aircraft in the United States, 1946-1948) developed the servotab control method of rotor pitch control, in which the rotor blade is twisted rather than rotated about a pitch bearing at the root. Kaman also developed a helicopter of the synchropter configuration. Mikhail Mil' (USSR, 1949) developed a series of helicopters of the single main rotor and tail rotor configuration, including in 1949 the Mi-1 (three-bladed 14 m diameter, gross weight 2250 kg, 570 hp engine). Nikolai I Kamov (USSR, 1952) developed helicopters with the coaxial configuration, including the Ka-15 helicopter (three-bladed rotors 10 m in diameter, gross weight 1370 kg, 225 hp engine). Alexander Yakolev (USSR, 1952) developed the Yak-24, a helicopter with two four-bladed rotors in tandem configuration. A number of dynamic problems were encountered in the development of this helicopter, but it eventually (1955) went into production.

An important development was the application of the turboshaft engine to helicopters, replacing the reciprocating engine. A substantial performance improvement was realized because of the lower specific weight (kg/hp) of the turboshaft engine. Kaman Aircraft Company (United States, 1951) constructed the first helicopter with turbine power, installing a single turboshaft engine (175 shp Boeing engine) in its K-225 helicopter. In 1954, Kaman also developed the first twin-engine turbine powered helicopter, an HTK-1 synchropter with two Boeing engines (total 350 shp) replacing a single 240 hp piston engine of the same weight in the same position. Since that time the turboshaft engine has become the standard powerplant for all but the smallest helicopters.

The invention of the helicopter may be considered complete by the early 1950's, and so we conclude this history at that point. In the years that followed, several helicopter designs achieved extremely successful production records, and some very large helicopters were constructed. The operational use of the helicopter has grown to a major factor in the air transportation system. Helicopter engineering is thus now involved more with research and with development than with invention.

1-2.2 Literature

On the history of the development of the helicopter: Warner (1920), NACA (1921a, 1921b), Balaban (1923), Moreno-Caracciolo (1923), Oemichen (1923), Klemin (1925), Wimperis (1926), Breguet (1937), Kussner (1937), Focke (1938), Sikorsky (1943, 1967, 1971), Gessow and Myers (1952), Hafner (1954), McClements and Armitage (1956), Stewart (1962a, 1962b), Focke (1965), Izakson (1966), Anoshchenko (1968), Legrand (1968), Gablehouse (1969), Free (1970), Lambermont and Pirie (1970), Kelley (1972).

On the history of helicopter analysis and research: Glauert (1935), von Kármán (1954), Gustafson (1970), and the original literature.

1–3 Notation

This section summarizes the principal nomenclature to be used in the text. The intention is to provide a reference for the later chapters and also to familiarize the reader with the basic elements of the rotor and its analysis. Only the most fundamental parameters are included here; the definitions of the other quantities required are presented as the analysis is developed. A number of the fundamental dimensionless parameters of helicopter analysis are also introduced. An alphabetical listing of symbols is provided at the end of the text.

1-3.1 Dimensions

Generally the analyses in this text work with dimensionless quantities. The natural reference length scale for the rotor is the blade radius R, and the natural reference time scale is the rotor rotational speed Ω (rad/sec).

For a reference mass the air density ρ is chosen. For typographical simplicity, no distinction is made between the symbols for the dimensional and dimensionless forms of a quantity. New symbols are introduced for those dimensionless parameters normalized using quantities other than ρ, Ω, and R.

1-3.2 Physical Description of the Blade

R = the rotor radius; the length of the blade, measured from hub to tip.

Ω = the rotor rotational speed or angular velocity (rad/sec).

ρ = air density.

ψ = azimuth angle of the blade (Fig. 1-5), defined as zero in the downstream direction. This is the angle measured from downstream to the blade span axis, in the direction of rotation of the blade. Hence for constant rotational speed, $\psi = \Omega t$.

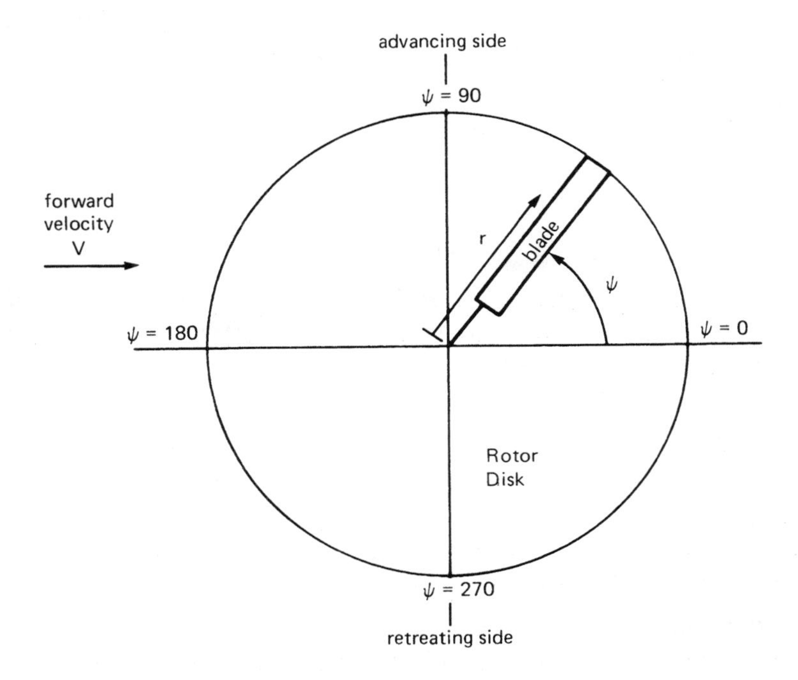

Figure 1-5 Rotor disk, showing definition of ψ and r.

r = radial location on the blade (Fig. 1-5), measured from the center of rotation ($r = 0$) to the blade tip ($r = R$, or when dimensionless $r = 1$).

It is conventional to assume that the rotor rotation direction is counterclockwise (viewed from above). The right side of the rotor disk is called the advancing side, and the left side is called the retreating side. The variables r and ψ will usually refer to the radial and azimuthal position of the blade, but they may also be used as polar coordinates for the rotor disk.

c = blade chord, which for tapered blades is a function of r.

N = number of blades.

m = blade mass per unit length as a function of r.

$I_b = \int_0^R mr^2\,dr$ = moment of inertia of the blade about the center of rotation.

The rotor blade normally is twisted along its length. The analysis will often consider linear twist, for which the built-in variation of the blade pitch with respect to the root is $\Delta\theta = \theta_{tw}r$. The linear twist rate θ_{tw} (equal to the tip pitch minus the root pitch) is normally negative for the helicopter rotor. The following derived quantities are important:

$A = \pi R^2$ = rotor disk area.

$\sigma = Nc/\pi R$ = rotor solidity.

$\gamma = \rho ac R^4/I_b$ = blade Lock number.

The solidity σ is the ratio of the total blade area (NcR for constant chord) to the total disk area (πR^2). The Lock number γ represents the ratio of the aerodynamic and inertial forces on the blade.

1-3.3 Blade Aerodynamics

a = blade section two-dimensional lift curve slope.

α = blade section angle of attack.

M = blade section Mach number.

The subscript (r, ψ) on α or M is used to indicate the point on the rotor disk being considered; for example, the retreating-tip angle of attack $\alpha_{1,270}$ or the advancing-tip Mach number $M_{1,90}$.

1-3.4 Blade Motion

The basic motion of the blade is essentially rigid body rotation about the root, which is attached to the hub (Fig. 1-6).

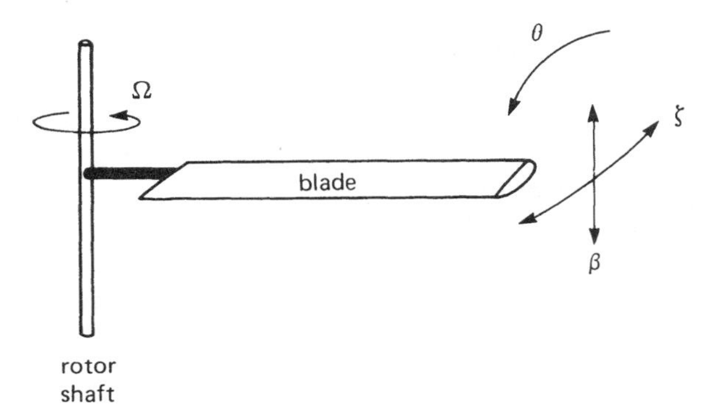

Figure 1-6 Fundamental blade motion.

β = blade flap angle. This degree of freedom produces blade motion of the disk plane (about either an actual flap hinge or a region of structural flexibility at the root). Flapping is defined to be positive for upward motion of the blade (as produced by the thrust force on the blade).

ζ = blade lag angle. This degree of freedom produces blade motion in the disk plane. Lagging is defined to be positive when opposite the direction of rotation of the rotor (as produced by the blade drag forces).

θ = blade pitch angle, or feathering motion produced by rotation of the blade about a hinge or bearing at the root with its axis parallel to the blade spar. Pitching is defined to be positive for nose-up rotation of the blade.

The degrees of freedom β, ζ, and θ may also be viewed as rotations of the blade about hinges at the root, with axes of rotation as follows: β is the angle of rotation about an axis in the disk plane, perpendicular to the blade spar; ζ is the angle of rotation about an axis normal to the disk plane, parallel to the rotor shaft; and θ is the angle of rotation about an axis in the disk plane, parallel to the blade spar. The description of more complex blade motion, for example motion that includes blade bending flexibility, will be introduced as required in later chapters.

In steady-state operation of the rotor, blade motion is periodic around the azimuth and hence may be expanded as Fourier series in ψ:

$$\beta = \beta_0 + \beta_{1c} \cos\psi + \beta_{1s} \sin\psi + \beta_{2c} \cos 2\psi + \beta_{2s} \sin 2\psi + \ldots$$

$$\zeta = \zeta_0 + \zeta_{1c} \cos\psi + \zeta_{1s} \sin\psi + \zeta_{2c} \cos 2\psi + \zeta_{2s} \sin 2\psi + \ldots$$

$$\theta = \theta_0 + \theta_{1c} \cos\psi + \theta_{1s} \sin\psi + \theta_{2c} \cos 2\psi + \theta_{2s} \sin 2\psi + \ldots$$

The mean and first harmonics of the blade motion (the *0*, *1c*, and *1s* Fourier coefficients) are the harmonics most important to rotor performance and control. The rotor coning angle is β_0; β_{1c} and β_{1s} are respectively the pitch and roll angles of the tip-path plane. The rotor collective pitch is θ_0, and θ_{1c} and θ_{1s} are the cyclic pitch angles.

1-3.5 Rotor Angle of Attack and Velocity

α = rotor disk plane angle of attack, positive for forward tilt (as required if a component of the rotor thrust is to provide the propulsive force for the helicopter).

V = rotor or helicopter velocity with respect to the air.

ν = rotor induced velocity, normal to the disk plane, and positive when downward through the disk (as is produced by a positive rotor thrust).

The resultant velocity seen by the rotor, resolved into components parallel and normal to the disk plane and made dimensionless with the rotor tip speed ΩR, gives the following velocity ratios (Fig. 1-7):

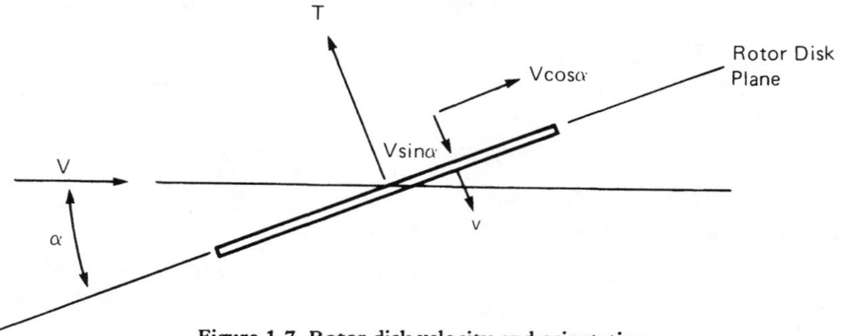

Figure 1-7 Rotor disk velocity and orientation.

$\mu = V\cos\alpha/\Omega R$ = rotor advance ratio.

$\lambda = (V\sin\alpha + \nu)/\Omega R$ = rotor inflow ratio (defined to be positive for flow downward through the disk).

$\lambda_i = \nu/\Omega R$ = induced inflow ratio.

The advance ratio μ is the ratio of the forward velocity to the rotor tip speed. The inflow ratio λ is the ratio of the total inflow velocity to the rotor tip speed.

1-3.6 Rotor Forces and Power

T = rotor thrust, defined to be normal to the disk plane and positive when directed upward.

H = rotor drag force in the disk plane; defined to be positive when directed rearward, opposing the forward velocity of the helicopter.

Y = rotor side force in the disk plane; defined to be positive when directed to the right, toward the advancing side of the rotor.

Q = rotor shaft torque, defined to be positive when an external torque is required to turn the rotor (helicopter operation).

P = rotor shaft power, positive when power is supplied to the rotor.

In coefficient form based on air density, rotor disk area, and tip speed these quantities are:

C_T = thrust coefficient = $T/\rho A(\Omega R)^2$

C_H = H force coefficient = $H/\rho A(\Omega R)^2$

C_Y = Y force coefficient = $Y/\rho A(\Omega R)^2$

C_Q = torque coefficient = $Q/\rho A(\Omega R)^2 R$

C_P = power coefficient = $P/\rho A(\Omega R)^3$

Notice that since the rotor shaft power and torque are related by $P = \Omega Q$, it follows that the coefficients are equal, $C_P = C_Q$. The rotor disk loading is the ratio of the thrust to the rotor area, T/A, and the power loading is the ratio of the power to the thrust. The rotor blade loading is the ratio of the thrust to the blade area, $T/A_b = T/(\sigma A)$, or in coefficient form the ratio of the thrust coefficient to solidity, C_T/σ.

1-3.7 Rotor Disk Planes

The rotor disk planes (defined in Chapter 5) are denoted by:

TPP	tip-path plane
NFP	no-feathering plane
HP	hub plane
CP	control plane

1-3.8 NACA Notation

No true standard nomenclature is used throughout the helicopter literature, so one must always take care to determine the definitions of the quantities used in any particular work, including the present text. One system of notation which is common enough in the literature to deserve some attention is that proposed by the National Advisory Committee for Aeronautics (NACA). The primary deviations from the practice in this text are:

b = number of blades.

x = r/R = dimensionless span variable.

θ_1 = linear twist rate (from the expansion $\theta = \theta_0 + \theta_1 r$).

I_1 = rotor blade flapping inertia.

λ = rotor inflow ratio, defined to be positive when upward through the disk = $(V \sin\alpha - v)/\Omega R$.

α = rotor disk angle of attack, defined to be positive for rearward tilt of the rotor disk and thrust vector.

In addition, λ and α are assumed to refer to the no-feathering plane if there are no subscripts or other indication that another reference plane is being used.

The blade motion is represented by Fourier series with the following definitions for the harmonics:

$$\beta = a_0 - a_1 \cos\psi - b_1 \sin\psi - a_2 \cos 2\psi - b_2 \sin 2\psi - \ldots$$

$$\theta = A_0 - A_1 \cos\psi - B_1 \sin\psi - A_2 \cos 2\psi - B_2 \sin 2\psi - \ldots$$

$$\zeta = E_0 + E_1 \cos\psi + F_1 \sin\psi + E_2 \cos 2\psi + F_2 \sin 2\psi + \ldots$$

A subscript s is used for quantities measured with respect to the shaft or hub plane, for example A_{1_s} and B_{1_s}.

The differences in sign from the present notation arise because the NACA notation was designed for autogiro analysis, and quantities were defined so that the parameters would usually have a positive value. The complete NACA notation system for helicopter analysis is given by Gessow (1948b) and by Gessow and Myers (1952).

Chapter 2

VERTICAL FLIGHT I

Hover is the operating state in which the lifting rotor has no velocity relative to the air, either vertical or horizontal. General vertical flight involves axial flow with respect to the rotor. Vertical flight implies axial symmetry of the rotor, and hence that the velocities and loads on the rotor blades are independent of the azimuth position. Axial symmetry greatly simplifies the dynamics and aerodynamics of the helicopter rotor, as will be evident when forward flight is considered later. The basic analyses of a rotor in axial flow originated in the 19th century with the design of marine propellers and were later applied to airplane propellers. The principal objectives of the analysis of the hovering rotor are to predict the forces generated and power required by the rotating blades, and to design the most efficient rotor.

2–1 Momentum Theory

Momentum theory applies the basic conservation laws of fluid mechanics (Conservation of mass, momentum, and energy) to the rotor and flow as a whole to estimate the rotor performance. It is a global analysis, relating the overall flow velocities and the total rotor thrust and power. Momentum theory was developed for marine propellers by W. J. M. Rankine in 1865 and R. E. Froude in 1885, and extended in 1920 by A. Betz to include the rotation of the slipstream.

The rotor disk supports a thrust created by the action of the air on the blades. By Newton's law there must be an equal and opposite reaction of the rotor on the air. As a result, the air in the rotor wake acquires a velocity increment directed opposite to the thrust direction. It follows that there is kinetic energy in the wake flow field which must be supplied by the rotor.

This energy constitutes the induced power loss of a rotary wing and corresponds to the induced drag of a fixed wing.

Momentum conservation relates the rotor thrust per unit mass flow through the disk, $T/\dot{m}$, to the induced velocity in the far wake, w. Energy conservation relates $T/\dot{m}$, w, and the induced velocity at the rotor disk, v. Finally, mass conservation gives $\dot{m}$ in terms of the induced velocity v. Eliminating w then gives a relation between the induced power loss and the rotor thrust, which is the principal result of momentum theory. Momentum theory is not concerned with the details of the rotor loads or flow, and hence alone it is not sufficient for designing the blades. What momentum theory provides is an estimate of the induced power requirement of the rotor, and the ideal performance limit.

2-1.1 Actuator Disk

In the momentum theory analysis the rotor is modeled as an actuator disk, which is a circular surface of zero thickness that can support a pressure difference and thus accelerate the air through the disk. The loading is assumed to be steady, but in general may vary over the surface of the disk. The actuator disk may also support a torque, which imparts angular momentum to the fluid as it passes through the disk. The task of the analysis is to determine the influence of the actuator disk on the flow, and in particular to find the induced velocity and power for a given thrust. Momentum theory solves this problem using the basic conservation laws of fluid motion; vortex theory uses the Biot-Savart law for the velocity induced by the wake vorticity; and potential theory solves the fluid dynamic equations for the velocity potential or stream function. For the same model, all three methods must give identical results.

The actuator disk model is only an approximation to the actual rotor. Distributing the rotor blade loading over a disk is equivalent to considering an infinite number of blades. The detailed flow of the actuator disk is thus very different from that of a real rotor with a small number of blades. The flow field is actually unsteady, with a wake of discrete vorticity corresponding to the discrete loading. The actual induced power loss will therefore be larger than the momentum theory result because of the nonuniform and unsteady induced velocity. The approximate nature of the actuator disk model imposes a fundamental limit on the applicability of extended momentum or vortex theories. The principal use of the actuator disk model is to obtain a first estimate of the wake-induced flow, and hence the total induced power loss.

2-1.2 Momentum Theory in Hover

Consider an actuator disk of area A and total thrust T (Fig. 2-1). It is assumed that the loading is distributed uniformly over the disk. Let v be the induced velocity at the rotor disk and w be the wake-induced velocity infinitely far downstream. A well-defined, smooth slipstream is assumed, with v and w uniform over the slipstream cross-section. The rotational energy in the wake due to the rotor torque is neglected. The fluid is incompressible and inviscid. The mass flux through the disk is $\dot{m} = \rho A v$; by conservation of mass, the mass flux is constant all along the wake. Momentum conservation equates the rotor force to the rate of change of momentum, the momentum flowing out at station 3 less the momentum flowing in at station 0 (Fig. 2-1). Thus since the flow far upstream is at rest for the hovering rotor, $T = \dot{m}w$. Energy conservation equates the work done by the rotor to the rate of change of energy in the fluid, the kinetic energy flowing out at station 3 less the kinetic energy flowing in at station 0; hence $Tv = \frac{1}{2}\dot{m}w^2$. Eliminating

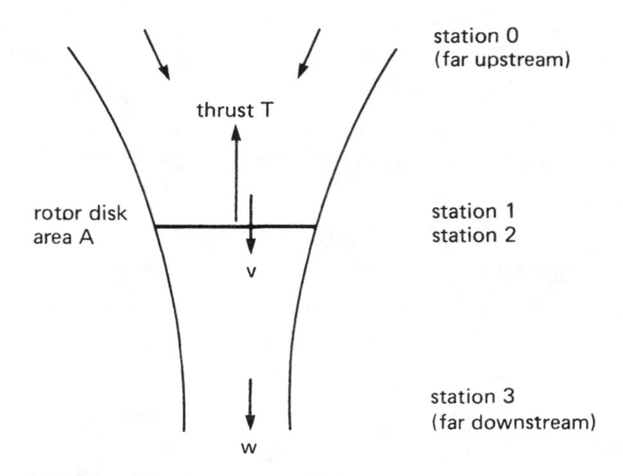

Figure 2-1 Momentum theory flow model for hover.

$T/\dot{m}$ from the momentum and energy conservation relations gives $w = 2v$; the induced velocity in the far wake is twice that at the rotor disk. Note that this is the same result as for an elliptically loaded fixed wing. Since the mass flux and density are constant, it follows that the area of the slipstream in the far wake (station 3) is $\frac{1}{2}A$.

Alternatively, this result can be obtained using Bernoulli's equation, which is an integrated form of the energy equation for the fluid. It is assumed that the pressure in the far wake (station 3) is at the ambient level p_0; this is equivalent to neglecting the swirl in the wake as before. Applying Bernoulli's equation between stations 0 and 1 gives $p_0 = p_1 + \frac{1}{2}\rho v^2$; between stations 2 and 3 it gives $p_2 + \frac{1}{2}\rho v^2 = p_0 + \frac{1}{2}\rho w^2$. Combining these equations, we obtain

$$T/A = p_2 - p_1 = \tfrac{1}{2}\rho w^2 .$$

With $\dot{m} = \rho A v$, this becomes

$$Tv = \tfrac{1}{2}\dot{m}w^2$$

as before. Note that the total pressure in the fully developed wake is $p_0 + \frac{1}{2}\rho w^2 = p_0 + T/A$. The increase in total head due to the actuator disk is equal to the disk loading T/A, which for helicopters is very small compared to p_0. Therefore, the over-pressure in the helicopter wake is small, although the wake velocities may still be fairly high. The pressure in the slipstream falls from p_0 to $p_1 = p_0 - \frac{1}{2}\rho v^2 = p_0 - \frac{1}{4}T/A$ just above the disk, and from $p_2 = p_0 + (3/2)\rho v^2 = p_0 + (3/4)T/A$ just below the disk to p_0 in the far wake. Thus, there is always a falling pressure except across the rotor disk, where the pressure increase accelerates the flow.

Momentum theory thus relates the rotor thrust and the induced velocity at the rotor disk by $T = \dot{m}w = 2\rho A v^2$. The induced velocity in hover v_h is therefore

$$v_h = \sqrt{T/2\rho A}.$$

The induced power loss for hover is

$$P = Tv = T\sqrt{T/2\rho A}.$$

In coefficient form, based on the rotor tip speed ΩR these results become $\lambda_h = \sqrt{C_T/2}$ and $C_P = C_T\lambda = C_T^{3/2}/\sqrt{2}$.

Momentum theory gives the induced power per unit thrust for a hovering rotor:

$$P/T = \nu = \sqrt{T/2\rho A}.$$

This relation determines the basic characteristics of the helicopter. It is based on the fundamental physics of fluid flow, which imply that for a low inflow velocity and hence low induced power loss the air must be accelerated through the disk by a small pressure differential. To hover efficiently requires a small value of P/T (for low fuel and engine weight), which demands that the disk loading T/A be low. With $T/A = 100$ to $500\ N/m^2$, the helicopter has the lowest disk loading and therefore the best hover performance of all VTOL aircraft. Note that the parameter determining the induced power is really $T/\rho A$, so the effective disk loading increases with altitude and temperature, that is, as the air density decreases.

As for fixed wings, uniform induced velocity gives the minimum induced power loss for a given thrust. This may be proved using the calculus of variation, as follows. The problem is to minimize the kinetic energy of the wake $KE \sim \int \nu^2 dA$ for a given thrust or wake momentum $\int \nu dA$. Write the induced velocity $\nu = \bar{\nu} + \delta\nu$ as a mean or uniform value $\bar{\nu}$ plus a perturbation $\delta\nu$ for which $\int \delta\nu\, dA = 0$. Then $\int \nu^2 dA = \bar{\nu}^2 A + \int (\delta\nu)^2 dA$, and minimum kinetic energy requires that $\delta\nu = 0$ over the entire disk, hence that there be uniform inflow. Basically, with nonuniform inflow the areas on the disk with high local loading cost more power than is gained from the areas with low loading.

2-1.3 Momentum Theory in Climb

Now consider momentum theory for a rotor in a vertical climb at velocity V (Fig. 2-2). The basic assumptions are the same as for the hover analysis: actuator disk model, uniform loading, a well-defined and smooth slipstream, uniform induced velocity, slipstream rotational velocities neglected, and ideal fluid. The mass flux is now $\dot{m} = \rho A(V + \nu)$. Momentum and energy conservation give $T = \dot{m}(V + w) - \dot{m}V = \dot{m}w$ and $T(V + \nu) = \frac{1}{2}\dot{m}(V + w)^2 - \frac{1}{2}\dot{m}V^2 = \frac{1}{2}\dot{m}w(w + 2V)$ respectively; note that the momentum conservation equation is independent of V. Eliminating $T/\dot{m}$ gives again $w = 2\nu$, just as for hover; the induced velocity in the far wake is twice that at the

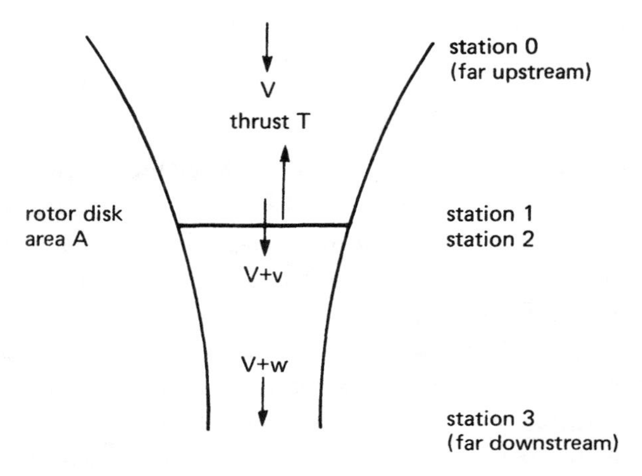

Figure 2-2 Momentum theory flow model for climb.

rotor disk. The total pressure in the wake is now $p_0 + \frac{1}{2}\rho(V + w)^2 = p_0 + \frac{1}{2}\rho V^2 + T/A$.

For the climbing rotor, the relation between the thrust and induced velocity becomes $T = \dot{m}w = 2\rho A(V + v)v$. Now define the parameter v_h as

$$v_h = \sqrt{T/2\rho A}$$

so that $v(V + v) = v_h^2$, which has the solution

$$v = -\frac{V}{2} + \sqrt{\left(\frac{V}{2}\right)^2 + v_h^2}\,.$$

Thus the climb velocity reduces the induced velocity v. The induced and climb power loss is then

$$P = T(V + v) = T\left(\frac{V}{2} + \sqrt{\left(\frac{V}{2}\right)^2 + v_h^2}\right).$$

Finally, the velocity in the far wake is $V + w = V + 2v = \sqrt{V^2 + 4v_h^2}$. For very large climb rates the induced velocity v is approximately v_h^2/V, and the power approaches just the climb work TV. For small rates of climb, however ($V \ll v_h$, which is generally true for helicopter rotors), the induced power loss P is approximately equal to $T(V/2 + v_h) = P_h + \frac{1}{2}TV$. The power required increases with V, but the climb power increment is reduced by the induced power decrease.

2-1.4 Hover Power Losses

Momentum theory gives the induced power loss of an ideal rotor in hover, $C_{P_i} = C_T^{3/2}/\sqrt{2}$. A real rotor has other power losses as well, in particular the profile power loss due to the drag of the blades in a viscous fluid. There is also an induced power loss due to the nonuniform inflow of a real, non-optimum rotor design. The swirl in the wake due to the rotor shaft torque is another loss, although it is usually small for helicopter rotors. Finally, the hovering rotor has tip losses as a result of the discreteness and periodicity in the wake, when the number of blades is finite. The distribution of the power losses of the rotor in hover is approximately as follows:

a) Induced power		60%
b) Profile power		30%
c) Nonuniform inflow	5% to	7%
d) Swirl in the wake	less than	1%
e) Tip losses	2% to	4%

The main rotor absorbs most of the helicopter power, but there are other losses as well. The engine and transmission absorb 4% to 5% of the total power with turbine engines, or 6% to 9% with reciprocating engines. The turbine engine has larger transmission losses since its high rotational speed requires more reduction, while the piston engine has significant losses for cooling. The tail rotor absorbs about 7% to 9% of the total helicopter power, and there is an additional loss of about 2% due to aerodynamic interference (rotor-fuselage and rotor-rotor). A tandem rotor helicopter has about the the same total loss of 9% to 11%, which is primarily due to aerodynamic interference but also includes some additional drive train losses. The tail rotor and aerodynamic interference power losses are much smaller for the helicopter in forward flight.

2–2 Figure of Merit

The figure of merit is a measure of rotor hovering efficiency, defined as the ratio of the minimum possible power required to hover to the actual power required to hover. Thus the figure of merit compares the actual rotor performance with the performance of an ideal rotor, which has only the inescapable induced power loss:

$$M = \frac{P_{\text{ideal}}}{P}.$$

Momentum theory gives the optimum induced power loss as $P_{\text{ideal}} = Tv = T\sqrt{T/2\rho A}$. Hence the figure of merit is given by

$$M = \frac{T\sqrt{T/2\rho A}}{P} = \frac{C_T^{3/2}/\sqrt{2}}{C_P}.$$

The figure of merit $M = Tv/P$ is similar to the propulsive efficiency for a propeller, η = useful power/input power = TV/P. The latter is appropriate for a propulsive device but not for a hovering rotor, where the useful power is that required to produce the thrust. Note that by generalizing the efficiency factor to $\eta = T(V + v)/P$ it can be used over the entire range of axial flow.

In terms of the induced and profile power contributions to the rotor power, the figure of merit can be written as $M = C_{P_{\text{ideal}}}/(C_{P_i} + C_{P_o})$. Usually the profile power C_{P_o} is at least 25% of the total power, and the induced power C_{P_i} is 10% to 20% higher than the ideal loss. Thus the figure of merit may be considered a measure of the ratio of the profile power to the induced power. The figure of merit can be misleading, since it is not directly concerned with the total hover power. By increasing the disk loading T/A the induced power is increased relative to the profile power, resulting in a higher figure of merit. However, the total power required also increases, and this is unlikely to be considered an improvement in the rotor efficiency. The use of the figure of merit for comparing rotor efficiencies should thus be restricted to constant disk loading only. Within this limitation the figure of merit is a valuable measure of the rotor aerodynamic efficiency. It is particularly useful for comparing rotors with different airfoil sections and for examining the influence of other design parameters such as twist or planform variations.

The ideal figure of merit is $M = 1$; it is lower for a real rotor because of profile and non-optimum induced power losses. The figure of merit for a given rotor is typically presented as a function of the ratio of the rotor thrust coefficient to the solidity, C_T/σ, which is a measure of the mean angle of attack of the blade. For current well-designed rotors the maximum figure of merit is typically $M = 0.75$ to 0.80. An inefficient rotor would have a maximum figure of merit around $M = 0.50$. The figure of merit decreases for low C_T/σ because of the reduced disk loading, and at high C_T/σ because of stall (which increases the profile losses). At the design loading of the rotor, a figure of merit of $M = 0.55$ to 0.60 is typical. For sea level density, the definition of the figure of merit gives $T/P = 1170M/\sqrt{T/A}$ when the power loading T/P is in N/hp and the disk loading T/A is in N/m^2. Thus a helicopter disk loading of 250 to 500 N/m^2 implies a power loading of 30 to 40 N/hp.*

2–3 Extended Momentum Theory

The most important and most useful results of momentum theory can be obtained by quite simple analyses, as in section 2-1. A more detailed analysis is not easily justified because of the basic limitations of the actuator disk model, yet there are some useful things to be learned from an extended momentum theory for axial flow. Momentum theory was quite extensively developed in the early part of the 20th century for airplane propellers, for which the actuator disk model is more reasonable because of the high axial speed; see for example Glauert (1935). Here we shall examine momentum theory more rigorously for the rotor in climb or hover (the same problem that was considered in sections 2-1.2 and 2-1.3), include the effects of swirl in the rotor wake, and include the rotor profile power in the energy balance.

The integral conservation laws of fluid dynamics are as follows:

$$\rho\int \vec{q}\cdot\vec{n}\,dS = 0$$

$$\rho\int \vec{q}\,\vec{q}\cdot\vec{n}\,dS + \int p\vec{n}\,dS = \vec{F}_{\text{body}}$$

$$\rho\int \vec{r}\times\vec{q}\,\vec{q}\cdot\vec{n}\,dS + \int p\vec{r}\times\vec{n}\,dS = \vec{M}_{\text{body}}$$

$$\int (p + \tfrac{1}{2}\rho q^2)\vec{q}\cdot\vec{n}\,dS = dE/dt$$

for mass, momentum, angular momentum, and energy conservation respectively. It is assumed that the flow is steady (in a frame moving with the rotor) and incompressible, and that there are no viscous losses on the surface of a body in the fluid. Here dS is the differential area at $\vec{r}$ of a surface

* $T/P = 38.0\,M/\sqrt{T/A}$ when T/P is in lb/hp and T/A is in lb/ft^2; so a disk loading of 5 to 10 lb/ft^2 implies a power loading of 7 to 10 lb/hp.

enclosing the fluid, $\vec{n}$ is the outward normal to the surface, and $\vec{q}$ is the velocity of the fluid. The force and moment on the body in the fluid (the rotor in this case) are $\vec{F}_{\text{body}}$ and $\vec{M}_{\text{body}}$, and dE/dt is the power being added to the flow. Note that application of the mass and energy conservation laws to a stream tube gives Bernoulli's equation: $p + \frac{1}{2}\rho q^2 = \text{constant}$, if no energy is added. The momentum theory extensions to be developed here are based on the application of these conservation relations to a rotor in axial flow.

Momentum theory is a calculus of variations problem. It is necessary to find a function of the rotor radius r, such as the induced velocity $v(r)$, that minimizes the power loss for a given thrust. Consider expressions for the power and thrust as integrals over the rotor disk: $P = \int F(r,v)\,dA$ and $T = \int G(r,v)\,dA$, where $dA = 2\pi r\,dr$. Using a Lagrange multiplier λ, let $I = P - \lambda T$. The solution $v(r)$ for minimum P subject to the constraint T is given by the stationary values of the first variation of I, namely

$$oI = \delta\int(F - \lambda G)dA = \int(\partial F/\partial v - \lambda\partial G/\partial v)\delta v\,dA = 0.$$

Thus the optimum $v(r)$ is given by the solution of the Euler equation,

$$\frac{\partial}{\partial v}\left(F - \lambda G\right) = 0.$$

If the integrands F and G are independent of r, the Euler equation is of the form

$$\text{function}(v) = \text{constant}$$

which has the solution $v = \text{constant}$.

2-3.1 Rotor in Hover or Climb

Consider a rotor with thrust T operating in hover, or in vertical ascent at speed V (Fig. 2-3). The rotor is represented by an actuator disk, which can support a pressure jump but not an axial velocity discontinuity. A well-defined, smooth slipstream is assumed, and for now the energy losses due to the angular momentum in the wake are neglected. Consider a control volume bounded by the slipstream and the disks of area S_0 and S_1 far upstream and downstream. The pressure at station 0 is p_0, and the velocity is V. At station 1, far downstream, the pressure is again p_0 since the wake swirl is neglected.

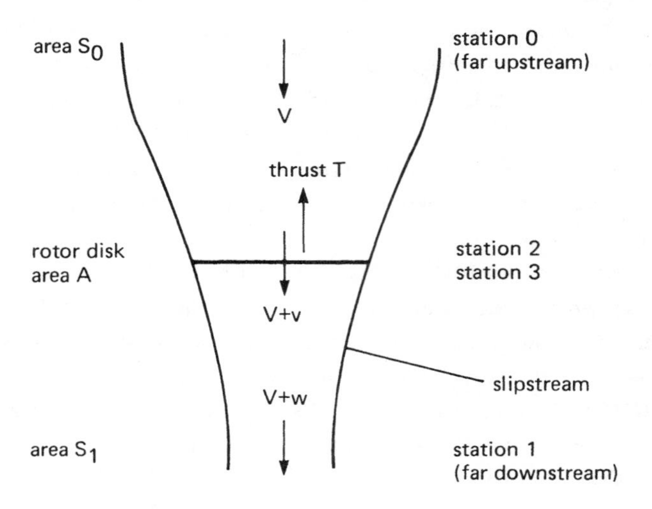

Figure 2-3 Flow model for rotor in hover or climb.

Energy conservation then shows that outside the slipstream at station 1 the velocity is everywhere equal to V. Mass and momentum conservation give

$$VS_0 = \int (V+v)dA = \int (V+w)dS_1$$
$$T = \int \Delta p dA = \int \rho (V+w) w dS_1$$

where Δp is the pressure difference across the rotor disk. There is a net pressure reaction on the ends of the control volume equal to $p_0(S_0 - S_1)$, which is exactly cancelled by the pressure on the slipstream, as may be established by considering momentum conservation for the fluid outside the slipstream. Energy conservation gives

$$P = \int \Delta p (V+v) dA = \int \tfrac{1}{2}\rho (V+w)^3 dS_1 - \tfrac{1}{2}\rho V^3 S_0$$

or, using mass conservation,

$$P = \int \Delta p (V+v) dA = \int \tfrac{1}{2}\rho (V+w)(2Vw+w^2) dS_1 .$$

The first expression is the work done in moving the air through the disk, and the second is the kinetic energy added to the slipstream. Note that subtracting TV from P gives $\int \Delta p\, v\, dA = \int \tfrac{1}{2}\rho (V+w) w^2 dS_1$, which may be interpreted as $\int v\,dT = \int \frac{1}{2} w\,dT$. Thus, the rotor thrust and power have been expressed in terms of the induced velocity in the far wake, $w(r)$, which in general may vary over the wake section. Consider now the following optimization problem: find the function $w(r)$ that minimizes the power $P = \int \frac{1}{2}\rho\,(V+w)(2Vw+w^2)\,dS_1$ for a given thrust $T = \int \rho(V+w)w\,dS_1$. This is a calculus of variations problem with a constraint. Since the integrands of P and T are independent of r, it follows as discussed above that the solution of the Euler equation is simply $w =$ constant.

Bernoulli's equation applied along streamlines above and below the rotor gives

$$p_0 + \frac{1}{2}\rho V^2 = p_2 + \frac{1}{2}\rho(V+v)^2$$

$$p_2 + \Delta p + \frac{1}{2}\rho(V+v)^2 = p_0 + \frac{1}{2}\rho(V+w)^2 .$$

Combining these equations, we obtain $\Delta p = \frac{1}{2}\rho(2Vw + w^2)$. Since the wake-induced velocity w is uniform, it follows that the disk loading Δp is also uniform. Note however that $p_2 + \frac{1}{2}\rho\,(V+v)^2 =$ constant, so that the pressure and especially the induced velocity at the rotor disk are not necessarily uniform. Thus, although the assumptions of uniform Δp and w made in section 2-1 are validated, momentum theory does not say anything about the distribution of the induced velocity at the rotor disk. This result is analogous to the Trefftz plane analysis of a fixed wing, which shows that the minimum induced drag is obtained with uniform downwash in the far wake and elliptical loading, but tells nothing about the induced angle of attack at the wing. Lifting-line theory or lifting-surface theory is required to find the induced angle of attack, which is needed to design the wing to achieve the optimum loading. From lifting-line theory the downwash at the wing is found to be one-half the downwash at the far wake, and therefore the optimal solution may be used directly in the wing design. Similarly with rotors it is frequently assumed that the inflow at the disk is also uniform, and that $v = \frac{1}{2}w$. While not strictly valid, the assumption of uniform inflow for the rotor in axial flight is generally consistent with the accuracy of the actuator disk model.

For uniform loading Δp and far wake inflow w the conservation relations become

$$(V+\bar{v})A = (V+w)S_1$$

$$T = \Delta p A = \rho(V+w)wS_1$$

$$P = (V+\bar{v})T = (V+\tfrac{1}{2}w)T$$

where $\bar{v} = \int v\, dA/A$ is the mean induced velocity at the rotor disk. The energy balance then gives $\bar{v} = \frac{1}{2}w$, so that although nothing is determined about the distribution of v, the mean $\bar{v}$ has the same value as that obtained earlier assuming uniform inflow. When the far wake parameters S_1 and w are eliminated, the rotor thrust and power are given by

$$T = 2\rho A(V+\bar{v})\bar{v}$$

$$P = T(V+\bar{v}).$$

This is the same result as in section 2-1.3, here in terms of the mean induced velocity.

It is customary to replace the integral conservation relations by their differential forms:

$$(V+v)dA = (V+w)dS_1$$

$$dT = \Delta p dA = \rho(V+w)wdS_1$$

$$dP = \Delta p(V+v)dA = \tfrac{1}{2}\rho(V+w)(2Vw+w^2)dS_1$$

The energy equation then gives $v = \frac{1}{2}w$, and eliminating dS_1 and w gives

$$dT = 2\rho dA(V+v)v$$

$$dP = dT(V+v).$$

There is no strict justification for this differential form of momentum theory. Basically it assumes that there is no mutual interference of the disk elements. The key to the result is the assumption that $v = \frac{1}{2}w$ is valid for individual streamlines, which allows the thrust and power to be expressed entirely in terms of v. The usefulness of the differential form of momentum theory is that it may be applied to rotors with nonuniform loading and inflow.

2-3.2 Swirl in the Wake

Consider next the effect of the swirl velocities in the rotor wake, which are due to the rotor induced torque. For shaft-driven rotors, the power and

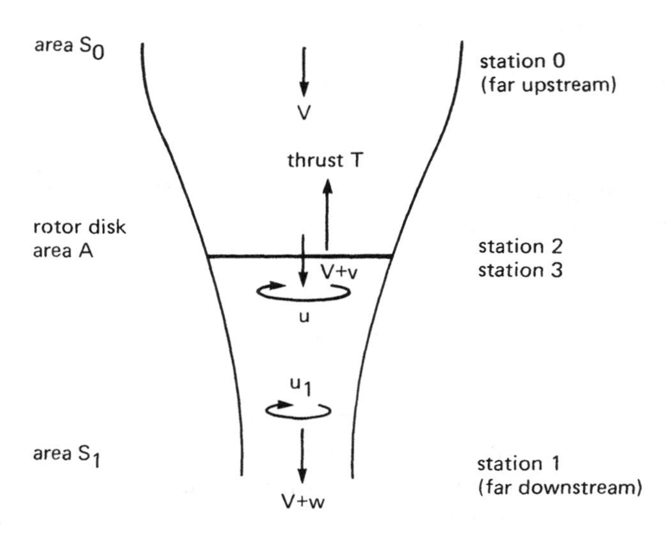

Figure 2-4 Flow model including swirl velocities in the wake.

torque are related by $P = \Omega Q$, where Ω is the rotor rotational speed. The rotor must therefore add rotational kinetic energy to the wake corresponding to the torque. For helicopter rotors the swirl energy is small compared to the axial downwash energy, so only a small correction to the induced power is sought here; an approximate solution will therefore be satisfactory. Fig. 2-4 shows the flow model considered. There is a rotational velocity $u(r)$ just below the rotor disk, and another, $u_1(r_1)$, in the far wake. Angular momentum conservation inside the slipstream shows that there can be no swirl velocities above the disk; that is, the flow remains irrotational until it passes through the rotor. When there is a rotational velocity in the far wake, the pressure no longer has the static value p_0; instead, $dp_1/dr_1 = \rho u_1{}^2/r_1$, and $p_1 = p_0$ at the boundary of the slipstream (r_1 is the radial coordinate at station 1). This pressure gradient provides the centripetal force required to support the rotational velocity of the fluid inside the wake.

Requirements for the conservation of mass, axial momentum, angular momentum, and energy give the following relations:

$$VS_0 = \int (V+v)dA = \int (V+w)dS_1$$

$$T = \int \Delta p dA = \int \rho (V+w)w dS_1 + \int (p_1 - p_0) dS_1$$

$$Q = \int \rho(V+v)urdA = \int \rho(V+w)u_1 r_1 dS_1$$

$$P = \int \Delta p(V+v)dA + \int \tfrac{1}{2}\rho u^2(V+v)dA$$

$$= \int \tfrac{1}{2}\rho(2Vw + w^2 + u_1^2)(V+w)dS_1 + \int (p_1 - p_0)(V+w)dS_1$$

The thrust, torque, and power can be expressed as functionals of the far wake velocities w and u_1 alone, by using

$$p_1 - p_0 = -\int_{r_1}^{R_1} u_1^2/r_1 \, dr_1$$

The resulting optimization problem—finding w and u_1 to minimize P subject to the constraints of the given T and of $Q = P/\Omega$—is more complex than is needed to estimate the power losses due to swirl, however. To formulate a simpler optimization problem, note that $P = \Omega Q$ gives $P = \int \rho(V+v)u\Omega r\, dA$. Then equating the expressions for power gives

$$\int \Delta p(V+v)dA = \int \rho(V+v)(\Omega r - \tfrac{1}{2}u)udA.$$

This relation may be interpreted as equating alternative expressions for the work performed, $\int (V+v)dT = \int (\Omega - \tfrac{1}{2}u/r)dQ$. Now, based on the results of section 2-3.1, the approximation $dT = \Delta p\, dA \cong \rho(V+v)2vdA$ will be used; the thrust may then be written $T = \int 2\rho(V+v)vdA$ and the differential form of $P = \Omega Q$ becomes simply $2(V+v)v = (\Omega r - \tfrac{1}{2}u)u$.

Thus the momentum theory for the rotor with swirl in the wake is formulated as follows: minimize the power $P = \int \rho(V+v)u\Omega r\, dA$ for a given thrust $T = \int 2\rho(V+v)vdA$, subject to the constraint $P = \Omega Q$, or $2(V+v)v = (\Omega r - \tfrac{1}{2}u)u$. The Euler equation for this calculus of variations problem,

$$\frac{(V+v)\Omega r}{\Omega r - u} + \frac{u\Omega r}{2V+4v} = \text{constant},$$

together with $2(V+v)v = (\Omega r - \tfrac{1}{2}u)u$ determines the inflow and swirl velocities at the rotor disk. As an approximate solution, consider $(V+v) = (V+v_0)(1 - \tfrac{1}{2}u/\Omega r)$, where v_0 is a constant; the $P = \Omega Q$ relation may then be solved for u and v as follows:

$$\frac{u}{\Omega r} = \frac{2(V+v_0)v_0}{(\Omega r)^2 + (V+v_0)^2}$$

$$\frac{v}{v_0} = \frac{(\Omega r)^2}{(\Omega r)^2 + (V+v_0)^2}.$$

This solution is found to satisfy the Euler equation to first order in $v_0/\Omega r$ for a low inflow helicopter rotor. After substituting for u and v, the thrust and power may be evaluated:

$$T = 2\rho(V+v_0)v_0 \int \frac{(\Omega r)^2 [(\Omega r)^2 + (V+v_0)V]}{[(\Omega r)^2 + (V+v_0)^2]^2} dA$$

$$\cong 2\rho A(V+v_0)v_0 \left[1 + \frac{(V+v_0)(V+2v_0)}{(\Omega R)^2} \ln\left(\frac{V+v_0}{\Omega R}\right)^2 + \frac{(V+v_0)v_0}{(\Omega R)^2}\right]$$

and $P = T(V+v_0)$. Thus the induced power is obtained as a function of the rotor thrust in terms of the parameter v_0. For hover, after writing $v_h^2 = T/2\rho A$ as usual, the constant v_0 may be evaluated as

$$v_0^2 = v_h^2/(1 + C_T \ln C_T/2 + C_T/2)$$

$$\cong v_h^2/(1 + C_T \ln C_T/2)$$

and the induced power is

$$P = T\sqrt{T/2\rho A} \Big/ \sqrt{1 + C_T \ln C_T/2}.$$

There is about a 2% increase in the induced velocity and induced power because of swirl in the wake, or about a 1% increase in the total rotor power. Since the Euler equation has not been satisfied exactly, this is not the true optimum solution, but it is sufficient to estimate the small losses due to swirl.

The rotational velocity at the rotor disk in hover has been obtained as

$$u = v_h \frac{2v_h \Omega r}{(\Omega r)^2 + v_h^2}$$

(for the velocities it is consistent to use $v_0^2 \cong v_h^2 = T/2\rho A$). The swirl velocity distribution has a peak value of $v = v_h$ very near the root, at $\Omega r = v_h$; on the outboard portion of the blade the velocity is much smaller. Vortex theory (see section 2-7) shows that for a uniformly loaded rotor the wake vorticity is distributed on the slipstream boundary, and in a line vortex along the axis with circulation $\gamma = 2\pi T/\rho A\Omega$. Indeed, the present solution outboard of about 20% radius is $u \cong 2v_h^2/\Omega r = T/\rho A\Omega r$. For hovering, the axial induced velocity at the disk is

$$v = v_h \frac{(\Omega r)^2}{(\Omega r)^2 + v_h^2}$$

which is significantly different from v_h only inboard of about 20% radius. Fig. 2-5 is a sketch of the typical radial distributions of the inflow and swirl velocities in hover according to this momentum theory analysis. The rotor thrust

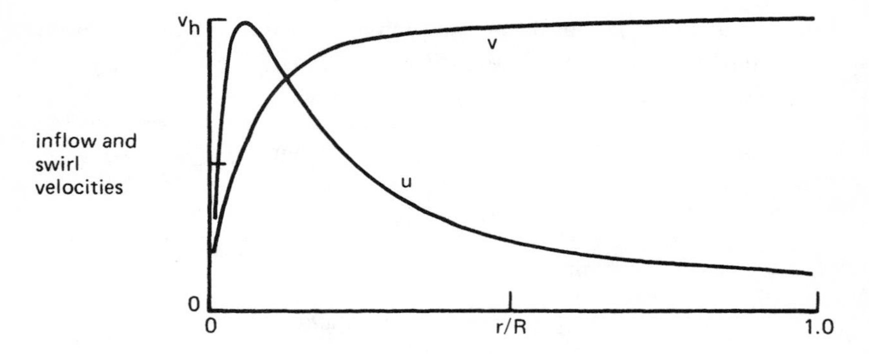

Figure 2-5 Radial distribution of the inflow (v) and swirl (u) velocities of the hovering rotor.

$$\Delta p = \frac{dT}{dA} = 2\rho v^2 = 2\rho v_h^2 \frac{(\Omega r)^4}{((\Omega r)^2 + v_h^2)^2}$$

shows a significant departure from uniform loading only at the blade root. In summary, except for very near the rotor axis, the momentum theory solution consists of uniform inflow v, uniform loading Δp, and swirl u due to a line vortex on the axis. The influence of the swirl velocities in the wake is not significant outboard of about 20% radius, and so may generally be neglected for helicopter rotors.

2-3.3 Swirl Due to Profile Torque

The previous section derived the wake swirl velocity due to the rotor induced torque. The rotor also has a profile power loss, which is due to the viscous drag of the blades, and hence a profile torque that will add more rotational kinetic energy to the wake. In terms of the blade section drag-to-lift ratio c_d/c_l, the profile power may be written

$$P_o = \int \Omega r dD = \int \Omega r \frac{c_d}{c_l} dT = \int \Omega r \frac{c_d}{c_l} 2\rho(V+v)v dA$$

Then the differential relation $P = \Omega Q$ becomes

$$2\frac{c_d}{c_l} v\Omega r + 2(V+v)v = (\Omega r - \tfrac{1}{2}u)u$$

Considering hover, let $v = v_0(1 - \tfrac{1}{2}u/\Omega r)$ again. The profile torque has negligible effect on the inflow velocity v, and the solution for the swirl u becomes

$$u = v_0\left[\frac{2v_0\Omega r}{(\Omega r)^2 + v_0^2} + 2\frac{c_d}{c_l}\right]$$

Thus the profile torque significantly increases the swirl velocities on the outboard portion of the blade, but the profile torque contribution is small compared to the inflow velocity v over the entire blade.

2–4 Blade Element Theory

Blade element theory calculates the forces on the blade due to its motion through the air, and hence the forces and performance of the entire rotor. Basically, blade element theory is lifting-line theory applied to the rotating wing. It is assumed that each blade section acts as a two-dimensional airfoil to produce aerodynamic forces, with the influence of the wake and the rest of the rotor contained entirely in an induced angle of attack at the section.

The solution thus requires an estimate of the wake-induced velocity at the rotor disk, which is provided by momentum theory, vortex theory, or nonuniform inflow calculations. Lifting-line theory is based on the assumption that the wing has a high aspect ratio. For a rotor, the aspect ratio of a single blade is related to the solidity and number of blades by $A\!\!\!R = R/c = (N/\pi)\sigma$. For low disk loading helicopter rotors the assumption of high aspect ratio is usually valid. However, while the geometric aspect ratio may be large, in areas where the loading or induced velocity has high gradients the effective aerodynamic aspect ratio can still be small. Examples of such high gradients for the rotating wing include blade sections near the tip or near an encounter with a vortex from a preceding blade.

Blade element theory is the foundation of almost all analyses of helicopter aerodynamics because it deals with the detailed flow and loading of the blade and hence relates the rotor performance and other characteristics to the detailed design parameters. In contrast, momentum theory (or any actuator disk analysis) is a global analysis, which provides useful results but cannot alone be used to design the rotor.

2-4.1 History of the Development of Blade Element Theory

The early development of rotary wing theory followed two separate lines, momentum theory and blade element theory, which were finally brought together in the 1920's. (The names "momentum theory" and "blade element theory" in fact had somewhat different meanings than the current usage, referring in the early work to separate and seemingly independent approaches to airscrew analysis.) The key factor was the concept of induced drag, which fluid dynamicists were still working to understand for both fixed and rotating wings in the early decades of the 20th century. The concept of an induced power loss, the power required to produce lift on a three-dimensional wing, and its association with the velocity induced at the wing by the wake vorticity had to be fully developed before an accurate calculation of the rotor loading was possible.

The origins of blade element theory can be traced to the work of William Froude in 1878, but the first major treatment was developed by Stefan Drzewiecki between 1892 and 1920. Drzewiecki considered the blade sections to act independently, but he was uncertain of the aerodynamic characteristics that should be used for the airfoils. Thus he proposed to obtain the required airfoil characteristics from measurements on a series of propellers. This was typical of the early approaches to blade element theory. They used only the velocities Ωr and V at the blade section, which are due to the rotation and axial velocity of the rotor respectively, and then considered what airfoil characteristics to use. Momentum theory describes the velocity at the rotor disk as $V + v$, which is greater than the free stream velocity V because of the rotor lift (and also a rotational velocity at the disk due to the rotor torque). However, Drzewiecki maintained that there was no logical connection between the momentum theory axial velocity and the velocity actually experienced by the blade section. The former is a mean velocity, while the latter is the local value, and we have seen that a rigorous momentum theory analysis does not in fact give information about the induced velocity at the rotor disk (momentum theory is really concerned with the velocities in the far wake). Lacking a sound theoretical treatment of the velocities at the rotor disk, Drzewiecki considered only the terms Ωr and V. When two-dimensional airfoil characteristics were used in such an analysis, the calculated performance exhibited a significant error that was therefore attributed to the airfoil characteristics. It was clear for fixed wings at least that the effective aerodynamic characteristics varied with aspect ratio, so Drzewiecki proposed that three-dimensional wing characteristics (for the appropriate aspect ratio) be used in the rotor blade element theory, with any remaining discrepancies to be established from tests on a series of propellers. The results of this theory had the right general behavior, but were quantitatively inaccurate.

There were several attempts from 1915 to 1919 to use the increased axial velocity from momentum theory in a blade element analysis; none developed to the point of using the two-dimensional airfoil characteristics, however, so all resorted at some stage to experiments to establish what characteristics to use. A. Betz in 1915 used the $V + v$ result of momentum theory and remarked that the appropriate aspect ratio to use was higher than that of the actual blade. While recognizing that the aspect ratio was tending toward infinity, he still considered the correct value to depend on the blade planform. G. de Bothezat in 1918 also used the $V + v$ result of momentum theory (and the corresponding rotational velocity at the disk), but he adopted Drzewiecki's plan of a series of special propeller tests to determine the airfoil characteristics. A. Fage and H. E. Collins in 1917 used an empirical fraction of $V + v$; they retained the airfoil characteristics of a wing with aspect ratio 6, and hence the empirical correction to the induced velocity was required to handle aspect ratio variations. Thus blade element theory remained on an empirical basis with regard to both the magnitude of the interference flow and the appropriate airfoil characteristics.

A correct accounting for the influence of the propeller wake on the aerodynamic environment at the blade section followed the development of Prandtl's wing theory, which gave a clear explanation of the role of the wake-induced velocity at the wing. Prandtl, Lanchester, and others developed the concept that the lift on an airfoil is due to a bound circulation, resulting in trailed vorticity in the wake that induces a velocity at the wing. The lifting-line theory for fixed wings involved a calculation of the induced velocity from the vortex wake properties. Thus rotary wing theory also turned to consideration of the vortex wake to define the velocities seen by a blade section. The resulting analysis is called vortex theory, and it was through this approach rather than momentum theory that the induced velocity was finally incorporated correctly into blade element theory. For a rotor or propeller, the vortices in the wake are trailed in helical paths rather than straight back as for fixed wings. This transcendental geometry makes the mathematical task of calculating the induced velocity much more difficult than for fixed wings. Consequently vortex theory, like momentum theory, frequently used the actuator disk model of the rotor, for which analytical solutions were possible.

A general airscrew theory was developed in the early 1920's on the basis of vortex theory and Prandtl's wing theory. By applying the concept of induced velocity, the aerodynamic environment at the rotor disk was established from the vortex theory results. The appropriate airfoil characteristics for this analysis are those of the two-dimensional wing. Later work established that for the same model the momentum theory and vortex theory results are indeed identical, so blade element theory is now usually derived from the momentum theory results for the induced velocity. In the early development of rotary-wing analysis, however, the vortex concepts of Prandtl had so great an impact that vortex theory completely superseded momentum theory. Momentum theory lacked the basis for understanding the induced velocity at the rotor disk, which was required to complete the development of blade element theory. As a result, vortex theory became regarded as the more reliable and logical foundation for both fixed and rotary wing analyses.

2-4.2 Blade Element Theory for Vertical Flight

Blade element theory is based on the lifting-line assumption; for the present derivation we shall also assume low disk loading and neglect stall and compressibility effects in order to obtain an analytical solution. Fig. 2-6 defines the geometry, velocities and forces of the blade section. The blade

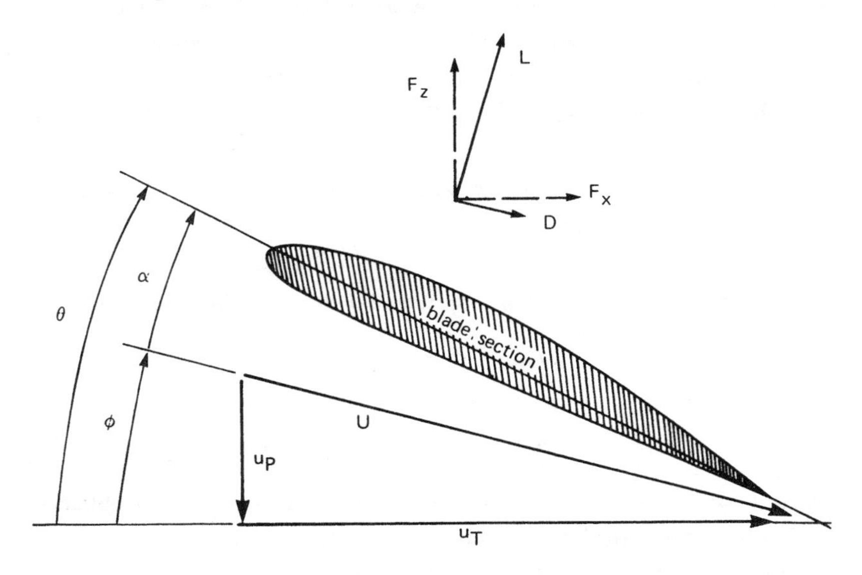

Figure 2-6 Blade section aerodynamics.

section has a pitch angle θ measured from the plane of rotation to the zero-lift line. The air velocity seen by the blade has components u_T and u_P which are tangent to and perpendicular to the disk plane, respectively. The resultant velocity magnitude and inflow angle are then given by

$$U = \sqrt{u_T^2 + u_P^2}$$

$$\phi = \tan^{-1} u_P/u_T$$

The aerodynamic angle of attack of the blade is $\alpha = \theta - \phi$. The air flow at the blade section produces lift and drag forces, L and D, which are normal to and parallel to the resultant velocity, respectively. The components of the total aerodynamic force normal to and parallel to the disk plane are F_z and F_x. Writing the section forces in terms of the lift and drag coefficients gives

$$L = \tfrac{1}{2}\rho U^2 c c_l$$

$$D = \tfrac{1}{2}\rho U^2 c c_d$$

where ρ is the air density and c is the blade chord. In general, the section coefficients c_l and c_d are complicated functions of the angle of attack, Mach number, and other parameters, but quite simple forms will be used here. Resolving the aerodynamic forces normal and parallel to the disk plane gives

$$F_z = L\cos\phi - D\sin\phi$$

$$F_x = L\sin\phi + D\cos\phi$$

Finally, the elemental thrust, torque, and power on the rotor blade are

$$dT = NF_z dr$$

$$dQ = NF_x r dr$$

$$dP = \Omega dQ = NF_x \Omega r dr$$

where N is the number of blades. The total forces on the rotor are obtained by integrating over the blade span from root to tip.

For the rotor in hover or vertical flight, the normal velocity u_P consists of the climb velocity V (zero for hover) and the induced velocity v; the in-plane velocity u_T is due only to the rotation of the blades at rate Ω. Therefore $u_P = V + v$ and $u_T = \Omega r$. Now from the assumption of low disk loading for the helicopter rotor it follows that the inflow ratio $\lambda = (V+v)/\Omega R$ is small (the momentum theory result for hover typically gives $\lambda_h = 0.05$ to 0.07). Then $u_P/u_T = (V+v)/\Omega r = \lambda(R/r)$ is also small, except near the blade root, where the dynamic pressure is low and thus the loads are negligible anyway. Therefore the small angle assumption is appropriate for helicopter rotors, namely $\phi, \theta, \alpha \ll 1$, from which it follows that $\phi \cong u_P/u_T$, $\cos\phi \cong 1$,

$\sin\phi \cong \phi$, and $U \cong u_T$. The next assumption is that stall and compressibility effects are negligible, so that the lift coefficient is linearly related to the angle of attack: $c_l = a\alpha$. Here a is the slope of the blade two-dimensional lift curve (typically $a = 5.7$, including real flow effects). Then the blade section forces reduce to

$$L \cong \tfrac{1}{2}\rho u_T^2\, c a(\theta - u_P/u_T)$$

$$D \cong \tfrac{1}{2}\rho u_T^2\, c c_d$$

and

$$dT \cong NL dr$$

$$dQ \cong N(L\phi + D) r dr$$

Next, all quantities are made dimensionless, normalized with respect to the air density, rotor speed, and rotor radius (ρ, Ω, and R). In coefficient form, the results for the contribution of a blade section to the rotor thrust and power are

$$dC_T = \frac{\sigma a}{2}\left(\theta u_T^2 - u_T u_P\right)dr = \frac{\sigma a}{2}\left(\theta r^2 - \lambda r\right)dr$$

$$dC_P = dC_Q = \left[\frac{\sigma a}{2}\left(\theta u_T u_P - u_P^2\right) + \frac{\sigma c_d}{2}\, u_T^2\right] r dr$$

$$= \left[\frac{\sigma a}{2}\left(\theta r\lambda - \lambda^2\right) + \frac{\sigma c_d}{2}\, r^2\right] r dr$$

where $\lambda = (V + v)/\Omega R$ is the inflow ratio and $\sigma = Nc/\pi R$ is the solidity ratio, which in general is a function of radius, except for constant chord blades. For a general rotor, these expressions may be integrated numerically over the blade span. With certain additional assumptions the integration may be performed analytically—for example, with uniform inflow, constant chord, and constant drag coefficient.

2-4.2.1 *Rotor Thrust*

Blade element theory gives the rotor thrust coefficient as

$$C_T = \int_0^1 \frac{\sigma a}{2}\left(\theta r^2 - \lambda r\right)dr$$

For a blade with constant chord and linear twist ($\theta = \theta_0 + r\theta_{tw} = \theta_{.75} + (r - 0.75)\theta_{tw}$), and assuming uniform inflow (λ = constant), we obtain

$$C_T = \frac{\sigma a}{2}\left(\frac{\theta_{.75}}{3} - \frac{\lambda}{2}\right)$$

where $\theta_{.75}$ is the pitch of the blade at 75% radius.

For uniform inflow, constant chord, and a twist distribution given by $\theta = \theta_t/r$, the thrust coefficient is

$$C_T = \frac{\sigma a}{4}\left(\theta_t - \lambda\right)$$

or with $\phi = \lambda/r = \phi_t/r$

$$C_T = \frac{\sigma a}{4}\left(\theta_t - \phi_t\right) = \frac{\sigma a}{4}\alpha_t$$

where the subscript "t" refers to the value at the blade tip. This twist distribution, while not physically realizable at the root, is of interest because it will be found to give uniform inflow with the constant chord blades. It is called the ideal twist distribution, since as momentum theory shows the minimum induced power loss is obtained with uniform inflow.

2-4.2.2 Induced Velocity

Blade element theory gives the rotor thrust as a function of the pitch angle and inflow ratio. The induced velocity is required if C_T is to be expressed as a function of θ alone. Momentum theory gives the following induced velocity for the rotor in hover or vertical climb:

$$\lambda = \frac{\lambda_c}{2} + \sqrt{\left(\frac{\lambda_c}{2}\right)^2 + \frac{C_T}{2}}$$

where $\lambda_c = V/\Omega R$. In hover $\lambda = \sqrt{C_T/2}$, so that for a constant chord, linearly twisted blade the induced velocity is

$$\lambda = \sqrt{\frac{C_T}{2}} = \frac{\sigma a}{16}\left[\sqrt{1 + \frac{64}{3\sigma a}\theta_{.75}} - 1\right]$$

or

$$\theta_{.75} = \frac{6C_T}{\sigma a} + \frac{3}{2}\sqrt{\frac{C_T}{2}}$$

The first term in the expression for $\theta_{.75}$ corresponds to the mean angle of attack of the rotor blade, while the second term is the additional pitch required because of the induced inflow angle ϕ. These relations allow λ and C_T to be obtained for a given collective $\theta_{.75}$, or alternatively λ and $\theta_{.75}$ for a given thrust.

For a constant chord, ideally twisted blade the momentum theory value for the inflow ratio gives

$$\lambda = \frac{\sigma a}{16}\left[\sqrt{1 + \frac{32}{\sigma a}\theta_t} - 1\right]$$

or

$$\theta_t = \frac{4C_T}{\sigma a} + \sqrt{\frac{C_T}{2}}$$

2-4.2.3 Power or Torque

The differential power coefficient can be written as

$$dC_P = \left[\lambda\frac{\sigma a}{2}\left(\theta r^2 - \lambda r\right) + \frac{\sigma c_d}{2} r^3\right] dr$$

$$= \lambda dC_T + \frac{\sigma c_d}{2} r^3\, dr$$

hence

$$C_P = \int \lambda dC_T + \int_0^1 \frac{\sigma c_d}{2} r^3\, dr$$

The first term in C_P is the induced power loss, $C_{P_i} = \int \lambda d C_T$, which arises from the in-plane component of the lift due to the induced angle of attack $[dP_i = (V + v) dT]$. The second term is the profile power loss which is due to the viscous drag forces on the rotor blade.

For uniform inflow the induced power is simply $C_{P_i} = \lambda C_T$, which agrees with the momentum theory result. (Note that for vertical flight λ includes the inflow due to the climb velocity, $\lambda_c = V/\Omega R$ so that here C_{P_i} includes the climb power $P_c = VT$.) In hover, the momentum theory result for λ

gives $C_{P_i} = C_T^{3/2}/\sqrt{2}$. This induced velocity value is for an ideal rotor. A real rotor with a practical twist and planform and a finite number of blades will have a higher induced power loss than the minimum given by momentum theory. One way to calculate the true induced power is to integrate $\int \lambda \, dC_T$ using the actual induced velocity distribution, which will in general be highly nonuniform as well as larger than the ideal value. An alternative approach is to use the momentum theory expression for the induced power, but with an empirical factor to account for the additional losses of a real rotor:

$$C_{P_i} = \kappa \lambda_h C_T = \kappa C_T^{3/2}/\sqrt{2}$$

Typically the factor κ has a value around 1.15 (see section 3-1.3).

For a constant chord blade, and assuming a constant drag coefficient, the profile power coefficient can be evaluated as

$$C_{P_o} = \frac{\sigma c_{d_o}}{8}$$

For an accurate calculation of the profile power loss, the variation of the drag coefficient with angle of attack and Mach number should be included (which will probably require a numerical integration). Consider a profile drag polar of the form

$$c_d = \delta_0 + \delta_1 \alpha + \delta_2 \alpha^2$$

By properly choosing the constants δ_0, δ_1, and δ_2 the variation of drag with lift for a given airfoil can be well represented for angles of attack below stall. (This representation for c_d was used by Bailey (1941), and his numerical example $c_d = 0.0087 - 0.0216\alpha + 0.400\alpha^2$ is frequently found in helicopter calculations; see section 7-8 for a further discussion.) Then the profile power coefficient is

$$C_{P_o} = \int_0^1 \frac{\sigma}{2}\left[\delta_0 + \delta_1\left(\theta - \frac{\lambda}{r}\right) + \delta_2\left(\theta - \frac{\lambda}{r}\right)^2\right] r^3 \, dr$$

For a constant chord, ideally twisted rotor with uniform inflow this integrates to

$$C_{P_o} = \frac{\sigma\delta_0}{8} + \frac{\sigma\delta_1}{6}\left(\theta_t - \lambda\right) + \frac{\sigma\delta_2}{4}\left(\theta_t - \lambda\right)^2$$

$$= \frac{\sigma\delta_0}{8} + \frac{2\delta_1}{3a} C_T + \frac{4\delta_2}{\sigma a^2} C_T^2$$

using $\theta_t - \lambda = 4C_T/\sigma a$. Similarly, for a constant chord, linearly twisted blade with uniform inflow, the profile loss is

$$C_{P_o} = \frac{\sigma\delta_0}{8} + \frac{\sigma\delta_1}{8}\left(\theta_{.75} + \frac{1}{20}\theta_{tw} - \frac{4}{3}\lambda\right)$$

$$+ \frac{\sigma\delta_2}{8}\left(\theta_{.75}^2 + \frac{1}{10}\theta_{.75}\theta_{tw} + \frac{7}{240}\theta_{tw}^2 + 2\lambda^2 - \frac{8}{3}\theta_{.75}\lambda\right)$$

The simplest relation for the total hover power of a real rotor is

$$C_P = \frac{\kappa C_T^{3/2}}{\sqrt{2}} + \frac{\sigma c_{d_o}}{8}$$

This result gives the basic features of the hover performance and is reasonably accurate when the appropriate empirical factor κ is used for the induced power and an appropriate mean drag coefficient c_{d_o} is used for the profile power. A plot of the power coefficient as a function of thrust coefficient (or C_P/σ as a function of C_T/σ) is called the rotor polar. For an ideal rotor—no profile power loss and minimum induced loss, hence a figure of merit of $M = 1$—the polar is given by $C_P = C_T^{3/2}/\sqrt{2}$. The polar for a real rotor has an offset compared to the ideal polar because of the profile power loss, and the power increases faster with C_T because of the larger induced power. Numerical examples of the hover polar are given in section 2-6.9. The figure of merit corresponding to the above expression for the rotor power is

$$M = \frac{C_{P_{\text{ideal}}}}{C_{P_i} + C_{P_o}} = \frac{C_T^{3/2}/\sqrt{2}}{\kappa C_T^{3/2}/\sqrt{2} + \sigma c_{d_o}/8}$$

Even with such a simple result, some conclusions about the rotor blade design may be reached. Recall that the proper use of the figure of merit is for comparisons of rotors at constant disk loading. For a given C_T, then, high M requires a low value of σc_{d_0}. If the rotor solidity is too low, however, high angles of attack are needed to achieve the required lift, and as a result the profile drag increases. Therefore the rotor should have as small a solidity as possible (small chord) with an adequate stall margin. The blade loading (hence twist and chord) distribution influences both the induced and profile power losses, but a more detailed calculation is required to examine such effects.

2–5 Combined Blade Element and Momentum Theory

The rotor performance calculations in the preceding section used the momentum theory result for the induced velocity, which was assumed to be uniform over the rotor disk. A nonuniform inflow distribution may be obtained by considering the differential form of momentum theory for hover or vertical flight. The resulting analysis is called combined blade element and momentum theory. Blade element theory describes the differential thrust on an annulus of the disk (on all N blades) of width dr at radial station r as

$$dC_T = \frac{\sigma a}{2}\left(\theta - \lambda/r\right)r^2 dr$$

From section 2-3.1, the differential form of momentum theory is $dT = 2\rho dA(V+v)v$, or

$$dC_T = 4\lambda\lambda_i r\,dr$$

where $\lambda_i = v/\Omega R$ is the induced inflow ratio, $\lambda_c = V/\Omega R$ is the climb inflow ratio, and $\lambda = \lambda_i + \lambda_c$. Basically, by using the differential form of momentum theory it is assumed that the induced velocity at radial station r is due only to the thrust dT at that station. Equating the blade element and momentum theory expressions for dC_T then gives

$$\lambda^2 + \left(\frac{\sigma a}{8} - \lambda_c\right)\lambda - \frac{\sigma a}{8}\theta r = 0$$

which has the solution

$$\lambda = \sqrt{\left(\frac{\sigma a}{16} - \frac{\lambda_c}{2}\right)^2 + \frac{\sigma a}{8}\theta r} - \left(\frac{\sigma a}{16} - \frac{\lambda_c}{2}\right)$$

For hover, $\lambda_c = 0$, the solution for the induced velocity is

$$\lambda = \frac{\sigma a}{16}\left[\sqrt{1 + \frac{32}{\sigma a}\,\theta r} - 1\right]$$

This is the sought-for nonuniform inflow distrubution (compare the uniform inflow results in section 2-4.2.2). For a given blade pitch, twist, and chord the inflow may be calculated as a function of r, and then the rotor thrust and power may be evaluated. Although the resulting rotor performance will be more accurate than that obtained with uniform inflow, differential momentum theory is still only an approximate model of the rotor. A further refinement of the inflow calculation requires a consideration of the details of the rotor vortex wake.

Observe that for a constant chord blade, uniform inflow is obtained if θr = constant; that is, if the blade has the ideal twist distribution $\theta = \theta_t/r$. From the uniform inflow it follows that the ideally twisted rotor also has uniform disk loading, and the minimum possible induced power loss.

2–6 Hover Performance

To summarize the equations involved in the calculation of rotor hover performance, blade element theory gives the thrust and power as

$$C_T = \int_0^1 \frac{\sigma}{2} r^2 c_l\, dr$$

$$C_P = \int \lambda dC_T + \int_0^1 \frac{\sigma}{2} r^3 c_d\, dr$$

where the section lift and drag coefficients are functions of the angle of attack $\alpha = \theta - \lambda/r$ and Mach number $M = rM_{\text{tip}}$. In general, the chord and pitch may be functions of the radial station r. The most frequent case encountered is a constant chord, linearly twisted blade: σ = constant, $\theta = \theta_0 + \theta_{tw} r$. If actual airfoil characteristics are not available, the simple relations $c_l = a\alpha$ and c_d = constant can be used. From combined blade element and momentum theory, the inflow distribution is given by

$$\lambda = \frac{\sigma a}{16}\left[\sqrt{1 + \frac{32}{\sigma a}\,\theta r} - 1\right]$$

Alternatively, uniform inflow with an empirical factor may be used, $\lambda = \kappa\sqrt{C_T/2}$. In general it is necessary to integrate the rotor loads numerically over the span; in this formulation, stall and compressibility are easily included by use of the appropriate section airfoil data. The limitations in the performance calculated according to these expressions arise principally from the neglect of three-dimensional flow effects at the tip and the use of the differential momentum theory for the induced velocity.

For future reference, note that by using combined blade element and momentum theory the rotor thrust and induced power may equivalently be written in terms of the induced velocity as $dC_T = 4\lambda^2 r\,dr$ and $dC_{P_i} = 4\lambda^3 r\,dr$.

2-6.1 Tip Losses

Lifting-line theory is not strictly valid near wing tips. When the chord at the tip is finite, blade element theory gives a nonzero lift all the way out to the end of the blade. In fact, however, the blade loading drops to zero at the tip over a finite distance because of three-dimensional flow effects (Fig. 2-7). Since the dynamic pressure is proportional to r^2, the

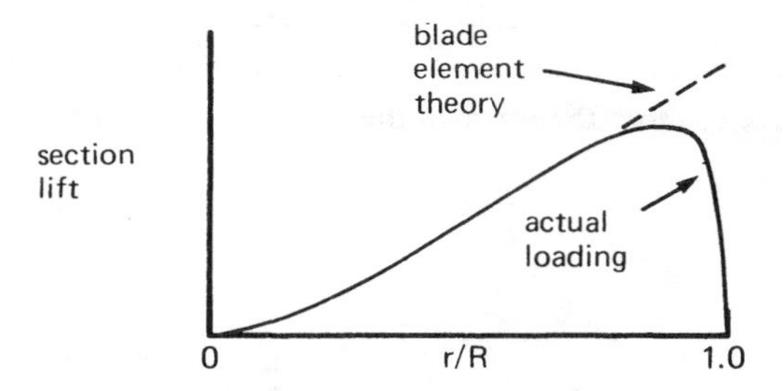

Figure 2-7 Sketch of rotor blade loading, showing the loss of lift at the tip.

loading for a rotary wing is concentrated at the tip and drops off even faster than that for fixed wings. The loss of lift at the tip is an important factor in calculating the rotor performance; if this loss is neglected, the rotor thrust for a given power or collective will be significantly overestimated. A rigorous treatment of the tip loading would require a lifting surface analysis, however, so here we consider an approximate representation of the tip loss effects.

The tip loss may alternatively be considered in terms of the influence of the rotor vortex wake. With an actuator disk model, a nonzero loading extending to the edge of the disk is perfectly acceptable (see section 2-7). Thus the tip loss may be viewed as the influence of the finite number of blades. It is the loading's being concentrated on a finite number of blades, rather than distributed around the disk, that introduces the three-dimensional flow effects. Figure 2-8 sketches the influence of the discrete wake on the flow through the rotor. With a finite number of blades, the discrete vortices in the wake constrain the flow to a volume smaller than the nominal wake boundary. The tip loss in this sense is like having a smaller effective area in the wake, or equivalently a higher effective disk loading, which implies a higher induced power loss.

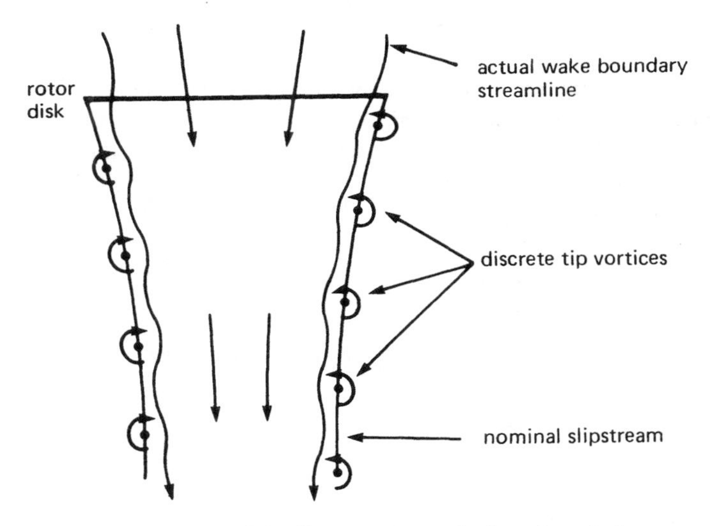

Figure 2-8 Influence of the discrete wake on the flow through the rotor.

An approximate method to account for tip losses is to assume that the blade elements outboard of the radial station $r = BR$ have profile drag but produce no lift. The parameter B is called the tip loss factor. A number of methods are available for calculating the appropriate value of B. Prandtl gives an expression based on a two-dimensional model of the rotor wake; for low inflow rotor it is

$$B = 1 - \frac{\sqrt{2C_T}}{N}$$

where N is the number of blades. (The tip loss depends on the spacing of vorticity sheets in the wake, which is proportional to λ/N; this result is derived in section 2-7.3.2 below.) Prandtl's result is recommended by Sissingh (1941); it typically gives $B = 0.96$ to 0.98. Wheatley (1934d) suggests

$$B = 1 - \frac{1}{2}\frac{\text{tip chord}}{R} = 1 - \frac{c(r=1)}{2R};$$

the outer half-chord length of the blade therefore developes no lift. Similarly, Sissingh (1939) also suggests

$$B = 1 - \frac{2}{3}\frac{c(r=0.7)}{R}$$

Frequently, the tip loss factor is simply set to $B = 0.97$, which generally gives good correlation with experimental data.

When the tip loss factor is included in the rotor thrust, the result from blade element theory becomes

$$C_T = \int_0^B \frac{\sigma}{2} r^2 c_l \, dr = \int_0^B \frac{\sigma a}{2}\left(\theta r^2 - \lambda r\right) dr$$

Then for the constant chord, linearly twisted blade with uniform inflow

$$C_T = \frac{\sigma a}{2}\left(\theta_0 \frac{B^3}{3} + \theta_{tw}\frac{B^4}{4} - \lambda \frac{B^2}{2}\right)$$

and for the ideally twisted blade

$$C_T = \frac{\sigma a}{4} B^2 \left(\theta_t - \lambda\right)$$

There is about a 6% to 9% reduction in the rotor thrust for a given collective pitch due to tip losses. The tip loss affects the required rotor power by increasing the induced velocity. The effective disk area of the rotor is reduced by a factor of B^2, and since the induced velocity is proportional to the square root of the disk loading, it follows that the induced velocity is higher than the momentum theory result by a factor of B^{-1}. Thus the rotor induced power becomes

$$C_{P_i} = \frac{1}{B}\lambda_h C_T = \frac{1}{B} C_T^{3/2}/\sqrt{2}$$

There is then about a 3% induced power increase due to the tip loss ($\kappa = B^{-1} \cong 1.03$). Other factors, particularly nonuniform inflow, also increase the induced power.

There are more rigorous approaches to the calculation of rotor performance including tip losses, such as vortex theory for a finite number of blades (see section 2-7.3), or lifting-surface theory for the rotary wing. Such analyses are very complex, however, and in some cases are not consistently more accurate than simple estimates. The tip loss factor is certainly a crude representation of the three-dimensional flow effects, but because of its simplicity and reasonable accuracy it has found widespread use.

2-6.2 Induced Power Due to Nonuniform Inflow and Tip Losses

In section 2-4.2.3 the hover induced power was expressed as $C_{P_i} = \kappa C_T^{3/2}/\sqrt{2}$, where κ is an empirical factor accounting for the additional losses of a real rotor. The losses due to nonuniform inflow and tip loss may be estimated by using the momentum theory results:

$$C_{P_i} = \int_0^B 4\lambda^3 r\,dr$$

$$C_T = \int_0^B 4\lambda^2 r\,dr$$

For uniform inflow, these relations give $C_{P_i} = 2\lambda^3 B^2$ and $C_T = 2\lambda^2 B^2$, or $C_{P_i} = B^{-1} C_T^{3/2}/\sqrt{2}$. Hence tip loss alone gives $\kappa = B^{-1} \cong 1.03$ as discussed in the preceding section.

Considering a linear inflow distribution, $\lambda = \lambda_t r$, we obtain $C_{P_i} = (4/5)B^5\lambda_t^3$ and $C_T = B^4\lambda_t^2$, or $C_{P_i} = (4/5B)C_T^{3/2}$. Hence

$$\kappa = \frac{4\sqrt{2}}{5B} = \frac{1.13}{B} \cong 1.17$$

Other simple nonuniform inflow distributions give similar results. Thus the hovering rotor is expected to have around an 8% to 12% increase in induced power due to nonuniform inflow, and 2% to 4% due to tip loss. The parameter

κ is actually obtained by correlation with measured rotor performance (see section 3-1.3).

2-6.3 Root Cutout

Performance losses also arise from the root cutout. The lifting portion of the blade starts at radial station $r = r_R$, which is typically 10% to 30% of the blade radius. The area inboard of this station, called the root cutout, is taken up with the rotor hub, flap and lag hinges, pitch bearing, and blade shank. Since the root cutout is aerodynamically an area of high drag coefficient and low lift, the blade element theory evaluation of the thrust should use integration from $r = r_R$ to $r = B$:

$$C_T = \int_{r_R}^{B} dC_T$$

The dynamic pressure is very low in the root cutout area, however, so generally the correction to the performance calculation is minor.

In hover, an effect of the root cutout is to reduce the effective rotor disk area and hence to increase the disk loading and induced velocity. With both root cutout and tip loss factor, the effective disk area gives the induced power parameter κ as

$$\kappa = \sqrt{\frac{A}{A_{eff}}} = \left(B^2 - r_R^2\right)^{-½}$$

For usual values of r_R, the root cutout effect on the induced power is small compared to the tip loss effect.

2-6.4 Blade Mean Lift Coefficient

A useful measure of the aerodynamic operating state of the rotor is a mean lift coefficient for the blades. The mean lift coefficient $\bar{c}_l$ is defined to give a thrust coefficient with the value $C_T = \int ½\sigma r^2 c_l\, dr$ when the entire blade is assumed to be working at $\bar{c}_l$. So

$$C_T = \int_0^1 ½\sigma r^2 \bar{c}_l\, dr = ½\bar{c}_l \int_0^1 \sigma r^2\, dr = \frac{1}{6}\sigma\bar{c}_l$$

and then

$$\bar{c}_l = 6\frac{C_T}{\sigma}$$

Thus, C_T/σ, the ratio of the rotor thrust coefficient to the solidity, is a measure of the blade lift coefficient. Correspondingly, $6C_T/\sigma a$ may be interpreted as the mean angle of attack of the blade. Note that

$$C_T/\sigma = \frac{T/[\rho A_{\text{rotor}}(\Omega R)^2]}{A_{\text{blade}}/A_{\text{rotor}}} = T/[\rho A_{\text{blade}}(\Omega R)^2]$$

is the dimensionless blade loading, whereas C_T is the dimensionless disk loading. The parameter C_T/σ has an important role in rotor aerodynamics, since many of the characteristics of the rotor and helicopter depend on the blade lift coefficient. The rotor figure of merit, using the simple power expression of section 2-4.2.3, may be written

$$M = \frac{\lambda_h C_T}{\kappa\lambda_h C_T + \sigma c_{d_o}/8} = \frac{1}{\kappa + \frac{3}{4}[(c_{d_o}/\bar{c}_l)/\lambda_h]}$$

showing that a high section lift-to-drag ratio is required for a good hover figure of merit.

2-6.5 Equivalent Solidity

In the expressions for the rotor hover performance it has been convenient to account for the rotor chord and number of blades by using a local solidity, $\sigma = Nc/\pi R$, which varies along the blade span if the chord is not constant. The rotor solidity then is

$$\sigma_{\text{rotor}} = \frac{\text{blade area}}{\text{rotor area}} = \int_0^1 \sigma\, dr$$

For constant chord blades, the local solidity and rotor solidity are identical. When comparing the performance of two rotors with different blade planforms, it is desirable to use an equivalent solidity that accounts for the major effects of the varying chord.

It is conventional to compare rotors with tapered blades to a rotor with rectangular blades and an equivalent solidity ratio, operating at the same thrust coefficient. The equivalent solidity σ_e is defined by $C_T = \int \frac{1}{2}\sigma r^2 c_l dr = \frac{1}{2}\sigma_e \int r^2 c_l dr$, or (assuming constant lift coefficient)

$$\sigma_e = 3\int_0^1 \sigma r^2 dr$$

Similarly, for rotors compared on the basis of the same power or torque, the equivalent solidity is

$$\sigma_e = 4\int_0^1 \sigma r^3 dr$$

For linearly tapered blades ($\sigma = \sigma_0 + \sigma_1 r$), the equivalent solidity is

$$\sigma_e = \begin{cases} \sigma(r = 0.75) & \text{thrust basis} \\ \sigma(r = 0.80) & \text{power basis} \end{cases}$$

The thrust-based (r^2-weighted) equivalent solidity is generally used for comparisons of rotor power at a given thrust.

2-6.6 The Ideal Rotor

Consider a rotor with constant chord and the ideal twist distribution $\theta = \theta_t/r$. In section 2-5 it was shown that this twist results in uniform induced velocity over the rotor disk and hence corresponds to the minimum induced power loss. With the ideal twist, the blade loading is triangular:

$$dC_T = \frac{\sigma a}{2}\alpha r^2 dr = \frac{\sigma a}{2}\left(\theta_t - \lambda\right) r\, dr$$

The corresponding bound circulation and disk loading are

$$\frac{N}{\pi}\Gamma = \frac{1}{r}\frac{dC_T}{dr} = \frac{\sigma a}{2}\left(\theta_t - \lambda\right)$$

$$\frac{dT}{dA} = \frac{\pi dC_T}{2\pi r\, dr} = \frac{\sigma a}{4}\left(\theta_t - \lambda\right)$$

Thus the ideal twist gives constant bound circulation and uniform disk loading, which is indeed the loading required by momentum theory to produce uniform induced velocity.

From section 2-4.2, the performance of the ideal rotor (constant chord, ideal twist, uniform inflow) is given by

$$C_T = \frac{\sigma a B^2}{4}\left(\theta_t - \lambda\right) = \frac{\sigma a B^2}{4}\alpha_t$$

$$C_P = \lambda C_T + \frac{\sigma}{8}\left(\delta_0 + \frac{4}{3}\delta_1 \alpha_t + 2\delta_2 \alpha_t^2\right)$$

Momentum theory gives the induced velocity $\lambda = \sqrt{C_T/2B^2}$ for hover; the pitch is $\theta_t = \alpha_t + \lambda$. The local angle of attack and lift coefficient of the blade section are thus

$$\alpha = \frac{\alpha_t}{r} = \frac{4C_T}{B^2 \sigma a}\frac{1}{r}$$

$$c_l = a\alpha = \frac{4C_T}{B^2 \sigma}\frac{1}{r}$$

Actually, the section lift coefficient will be limited by stall at the blade root. Moreover, the ideal twist distribution is not realizable at the root, but the inboard sections of the blade have only a minor role in the rotor performance anyway. The real practical difficulty is that a different twist distribution is required for every operating condition of the rotor; from $\alpha = (\theta_t - \lambda)/r$ it follows that

$$\theta_t = \alpha_t + \lambda = \frac{4C_T}{\sigma a B^2} + \frac{1}{B}\sqrt{\frac{C_T}{2}}$$

(see section 2-4.2.1). The ideal rotor is useful as a limiting case, if not as a practical design, for it indicates the form the twist distribution must approach if the best rotor hover performance is to be achieved.

2-6.7 The Optimum Hovering Rotor

The ideal rotor is designed to have minimum induced power. The angle of attack is $\alpha = \alpha_t/r$, however, so only one blade section can be operating at the best lift-to-drag ratio and the ideal rotor will not have the least profile power possible. Consider now a rotor optimized for both induced and profile power losses. Minimum induced power requires uniform inflow. Minimum profile power requires that each blade section operate at its optimum condition, $\alpha = \alpha_{opt}$, where the best c_l/c_d is achieved. These two criteria determine the twist and taper for the optimum rotor, which has the best hover performance.

Combined blade element and momentum theory gives

$$dC_T = \frac{\sigma a}{2} \alpha_{opt} r^2 \, dr = 4\lambda^2 r dr$$

or

$$\lambda^2 = \frac{\sigma a}{8} r \alpha_{opt}$$

Assuming that α_{opt} is the same for all blade sections, uniform inflow requires σr = constant, hence a blade taper distribution given by $\sigma = \sigma_t/r$ $(c = c_t/r)$. Then the blade twist required is

$$\theta = \alpha_{opt} + \lambda/r = \alpha_{opt} + \sqrt{\frac{\sigma_t a \alpha_{opt}}{8}} \frac{1}{r}$$

The rotor thrust is

$$C_T = \int_0^B \tfrac{1}{2} \sigma r^2 a \alpha_{opt} dr = \frac{\sigma_t a B^2}{4} \alpha_{opt}$$

and the profile power is

$$C_{P_o} = \int_0^1 \tfrac{1}{2} \sigma r^3 c_d dr = \frac{\sigma_t c_{do}}{2} \int_0^1 r^2 \, dr = \frac{\sigma_t c_{do}}{6}$$

since the blade drag coefficient is constant over the span for the optimum rotor. The total rotor power is then

$$C_P = \frac{C_T^{3/2}}{B\sqrt{2}} + \frac{\sigma_t c_{do}}{6}$$

To summarize the optimum rotor design, for a given airfoil section (which defines α_{opt}), the taper and twist required are

$$\sigma = \frac{4C_T}{B^2 a \alpha_{opt}} \frac{1}{r}$$

$$\theta = \alpha_{opt} + \sqrt{\frac{C_T}{2B^2}} \frac{1}{r}$$

As for the ideal rotor, the design depends on the operating state, besides having both the chord and twist singular at the blade root. The optimum rotor solution is useful because it shows the maximum benefits that are attainable with twist and taper of the blade and indicates the design trends required for real rotors. In general, a washout of the blade pitch at the tip, i.e. negative twist, is required, and the blade should be tapered, although the performance gains with taper do not often justify the added manufacturing cost. Usually blades have been designed with linear twist and constant chord, and occasionally with linear taper as well. With modern materials and manufacturing techniques, nonlinear twist and nonconstant-chord designs are being produced. The optimum rotor solution implies that the design of a real rotor must always be a compromise, since a fixed chord and twist distribution can not be optimal for all operating conditions.

On a thrust basis, the equivalent solidity for the optimum rotor is $\sigma_e = 3\int \sigma r^2 \, dr = (3/2)\sigma_t$, so the profile power is

$$C_{P_o} = \frac{\sigma_e c_{d_o}}{9}$$

When this value is compared to the profile power loss with rectangular blades, $C_{P_o} = \sigma c_{d_o}/8$, it is evident that there is at least an 11% reduction in the profile power for the optimum rotor. (The difference will be even greater because of the higher value of the mean drag coefficient for the constant chord rotor.) The figure of merit becomes

$$M = \frac{C_T^{3/2}/\sqrt{2}}{C_T^{3/2}/B\sqrt{2} + \sigma_e c_{d_o}/9} = \frac{1}{\frac{1}{B} + \frac{2}{3\lambda_h} \frac{c_{do}}{c_l}}$$

2-6.8 Effect of Twist and Taper

For a given rotor thrust, radius, and tip speed, both the induced and profile power losses can be minimized by a proper choice of the blade twist and taper. A linear variation of chord or pitch can closely approximate the optimum distributions over the outer portions of the blade, where the loading is most important. In fact, -8° to -12° of linear twist produces most of the induced power gains of blades with ideal twist compared to untwisted blades. Such a twist distribution is easily built into the blades, with little production cost increase for a significant performance benefit. Taper improves the rotor performance also, but is usually justified only for very large rotors because of production costs. The table below, based on data from Gessow (1948a), gives the percentage reduction in power required for various combinations of twist and taper, compared to values for an untwisted blade with constant chord.

twist (deg)	taper ratio	power reduction at $C_T/\sigma = 0.07$
0	1	— —
− 8	1	2.5%
−12	1	4.0%
0	3	2.0%
− 8	3	5.5%
−12	3	5.5%
ideal	1	5.5%

See Gessow (1948a) and Gessow and Myers (1952) for further discussions of the effects of twist and taper.

Negative twist is also found to improve rotor performance in forward flight, since unloading the tips delays stall on the retreating blade and compressibility effects on the advancing blade. However, twist also increases vibration in forward flight and has some effect on autorotation performance. Thus selection of the rotor twist and taper is a complex task requiring consideration of all the operating conditions of the helicopter.

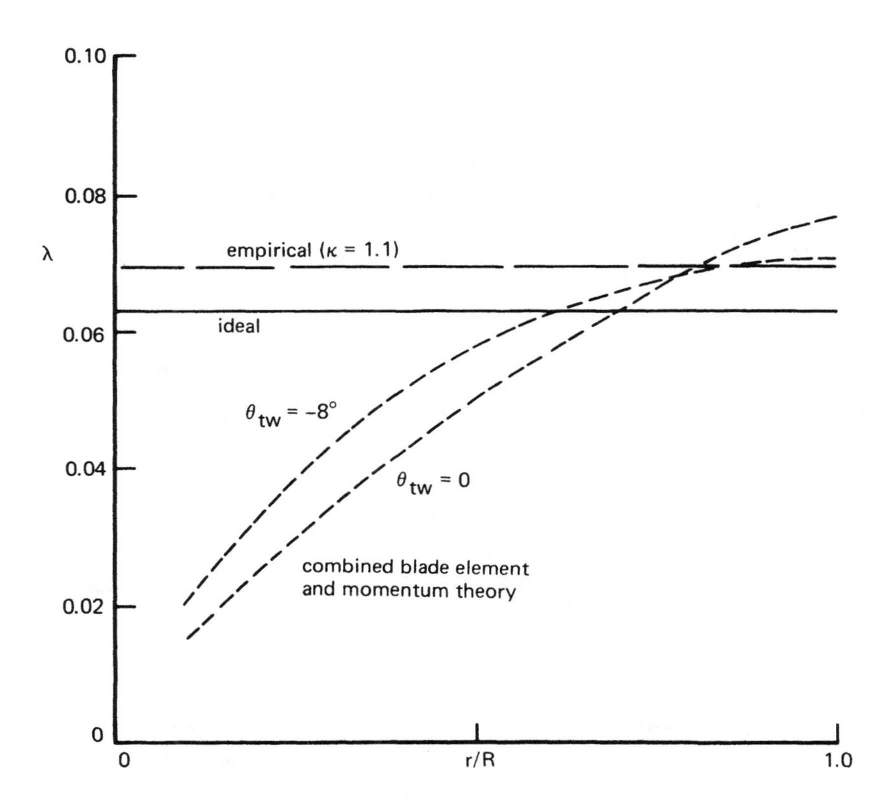

Figure 2-9 Calculated inflow distribution for a hovering rotor at $C_T/\sigma = 0.08$.

2-6.9 Examples of Hover Polars

A number of expressions have been obtained in the preceding sections for the hover performance of both real and ideal rotors. Here we shall present a numerical example and compare the calculated performance for the various cases. Three cases of limiting rotor behavior are considered: the rotor with a figure of merit of unity, which has no profile power losses and minimum induced power losses, so $C_P = C_T^{3/2}/\sqrt{2}$; the optimum rotor, which has twist for uniform inflow and taper for constant section angle of attack, and therefore minimum profile power and induced power; and the ideal rotor, which has a constant chord and, because of its twist for uniform inflow, minimum induced power. For the performance of a real rotor, the simple expression

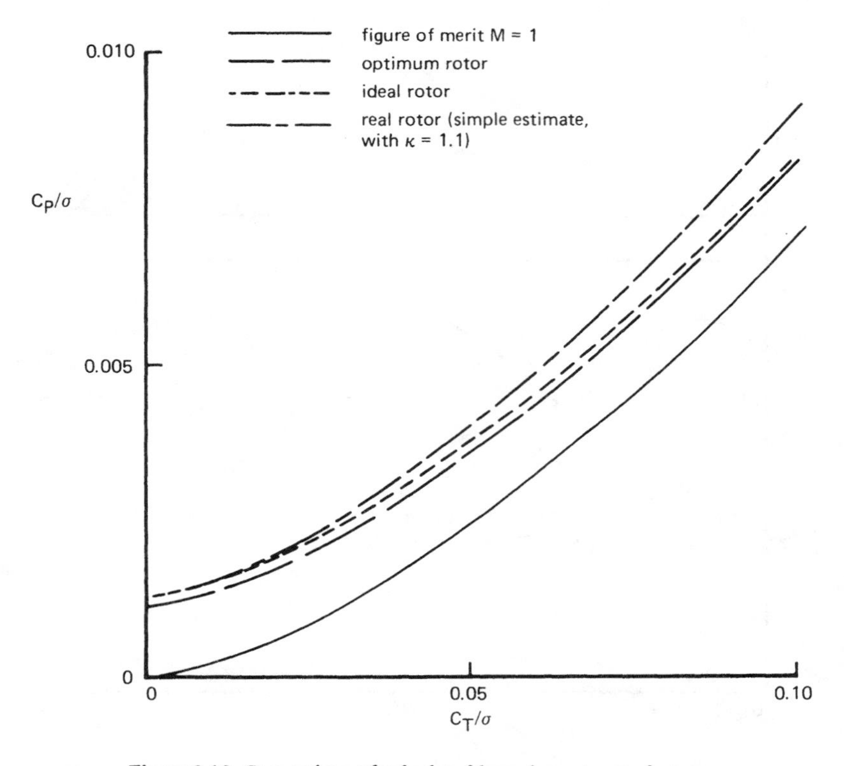

Figure 2-10 Comparison of calculated hovering rotor performance.

$$C_P = \kappa C_T^{3/2}/\sqrt{2} + \sigma c_{d_o}/8$$

is considered. This is essentially an empirical relation that requires a proper choice of κ and c_{d_o} to obtain a good performance estimate. Blade element theory is considered for constant chord and linearly twisted blades with uniform inflow. Finally, combined blade element and momentum theory is used for a constant chord, linearly twisted blade with nonuniform inflow.

The rotor polar, C_P/σ as a function of C_T/σ, was calculated for a rotor with solidity $\sigma = 0.1$, twist $\theta_{tw} = -8°$, section lift curve slope $a = 5.7$, and tip loss factor $B = 0.97$. For the simple performance estimate, $\kappa = 1.1$ and $c_{d_o} = 0.01$ were used; for blade element theory, $c_d = 0.0087 - 0.0216\alpha + 0.400\alpha^2$ was used. Figure 2-9 shows the spanwise distribution of the induced

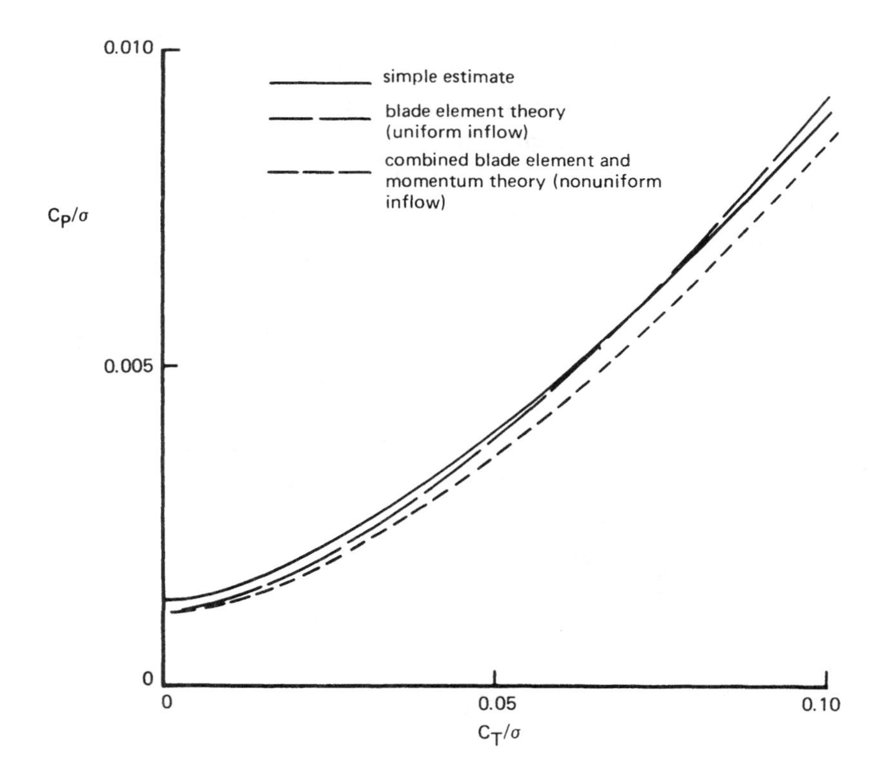

Figure 2-11 Comparison of calculated hovering rotor performance

velocity obtained for $C_T/\sigma = 0.08$ by the various methods. The ideal value $\lambda_h = \sqrt{C_T/2}$ is used for the optimum and ideal rotors. The empirical value $\lambda = \kappa\,\lambda_h$ is used for the basic and blade element theory performance calculations. The nonuniform inflow distribution is obtained by combined blade element and momentum theory.

Figure 2-10 compares the rotor hover performance calculated for the limiting cases and for a real rotor using the simple expression above with $\kappa = 1.1$. The $M = 1$ case has only the minimum induced power loss; the optimum rotor adds the minimum profile power loss; the ideal rotor increases the profile loss slightly because of the constant chord; and for the real rotor the induced power has been further increased by the factor κ.

Figure 2-11 compares the performance calculated by the simplest expression, blade element theory, and the combined blade element and momentum theory. The differences between the results of the simple expression and blade element theory are due to the profile power calculation. The differences between the blade element theory and the combined blade element and momentum theory calculations are due to the nonuniform inflow distribution of the latter. For additional comparisons of calculated rotor hover performance, see Gessow (1948a).

2-6.10 Disk Loading, Span Loading, and Circulation

The actuator disk analysis, particularly with vortex theory (section 2–7), requires a relation between the rotor disk loading, blade span loading, and blade bound circulation. The section loading dL/dr and circulation Γ are related by $dL/dr = \rho \Omega r \Gamma$. Hence

$$\frac{dT}{dA} = \frac{N dL}{2\pi r\, dr} = \frac{\rho \Omega N}{2\pi} \Gamma$$

It follows that uniform disk loading corresponds to triangular blade loading and constant circulation. In dimensionless form then, for uniform disk loading, $\Gamma/\Omega R^2 = (2\pi/N)C_T = 2(c/R)C_T/\sigma$.

2–7 Vortex Theory

Since the lift on a wing is associated with bound circulation, vorticity is trailed into the wake from a three-dimensional wing. The change in the blade loading occurs mostly at the tip on the rotary wing; and the rotor wake vorticity is therefore concentrated in tip vortices that lie in helices below the rotor disk. Unlike the fixed wing, the rotary wing has close encounters with its own wake and the wake from preceding blades. These encounters have a significant impact on the induced velocity and blade loads. Vortex theory is a rotor analysis that calculates the flow field of the rotor wake, in particular the induced velocity at the rotor disk, by using the fluid dynamic laws governing the action and influence of vorticity (the Biot-Savart law, Kelvin's theorem, and Helmholtz's laws). The simplest version of vortex theory uses an actuator disk model. The actuator disk neglects the discreteness in the rotor and wake associated with a finite number of blades, and distributes the vorticity throughout the wake volume. The actuator disk model produces a tractable mathematical problem, at least for vertical flight. When considering the same model as momentum theory, vortex theory must of course give identical results. Vortex theory is better suited than momentum theory to extensions of the model (such as to a nonuniform disk loading), since it is based on a consideration of the local flow characteristics rather than global properties.

If the discreteness in the wake is retained for the vortex theory model, the wake consists of lines and sheets of vorticity trailed behind each blade. Because of the fundamentally transcendental geometry of the rotor wake, integration to evaluate the induced velocity for such a model must be performed numerically. The result is a very large numerical problem, which became practical to solve only with the availability of high speed digital computers to helicopter engineering. With the current availability of computers, use of a discrete vortex model to represent the rotor and wake has become nearly universal when detailed information about the flow field and loading is required. The name vortex theory is generally restricted now to the classical work, which primarily used the actuator disk model. The use of a vortex wake model in numerical calculations of the induced velocity is discussed in Chapter 13 under the heading of nonuniform inflow.

N. E. Joukowski laid the foundations for vortex theory from 1912 to 1929. He investigated the induced velocity due to the helical wake system of a propeller, but had to use the infinite blade model because of the mathematical complexities. The results of momentum theory were duplicated using this vortex theory and actuator disk analysis. In 1918, Joukowski proposed the use of airfoil characteristics for a cascade of two-dimensional airfoils with the induced velocity taken from vortex theory. This approach essentially gave the elements of modern blade element theory since the cascade effect is negligible for helicopter rotors.

In 1919, A. Betz analyzed the vortex system of the propeller wake in detail, determining the minimum power and best thrust distribution by vortex theory. In an appendix to Betz's paper, L. Prandtl gave an approximate correction for the tip effect on the thrust distribution of a rotor with a finite number of blades. Around 1920, investigations furthering vortex theory were made by R. Wood and H. Glauert, and by E. Pistolesi. In 1929 S.

Goldstein considered more accuarately the vortex wake of a propeller with a finite number of blades.

The velocity $\vec{u}(\vec{x})$ induced by a line vortex of strength κ is given by the Biot-Savart law:

$$\vec{u}(\vec{x}) = -\frac{\kappa}{4\pi}\int \frac{(\vec{x}-\vec{y}) \times d\vec{\ell}(\vec{y})}{|\vec{x}-\vec{y}|^3}$$

where the integration is along the entire length of the vortex and $d\vec{\ell}$ is the tangent to the vortex at $\vec{y}$. This result can also be written as

$$\vec{u}(\vec{x}) = -\frac{\kappa}{4\pi}\vec{\nabla}\Sigma$$

where Σ is the solid angle subtended by the line vortex at $\vec{x}$. For a straight line vortex, the induced velocity is entirely circumferential, with magnitude $|\vec{u}| = \kappa/2\pi h$, where h is the perpendicular distance to the vortex. In a real fluid, viscosity eliminates the infinite velocity at the vortex line by diffusing the vorticity into a tube of small but finite cross-section radius called the vortex core. Stokes' theorem equates the flux of vorticity through a surface S with the circulation about the boundary of that surface. Kelvin's circulation theorem states that for an inviscid, incompressible fluid of uniform density the circulation $\Gamma = \oint \vec{u} \cdot d\vec{\ell}$ is constant moving with the fluid. Helmholtz's laws of vorticity then follow: a fluid initially irrotational remains so; a vortex tube (in particular, a line vortex) moves with constant strength with the fluid; and vortex lines must either be closed or end at solid surfaces. By means of these laws, vortex theory determines the flow of the helicopter rotor.

2-7.1 Vortex Representation of the Rotor and Its Wake

Associated with the lift L at a wing section is a circulation Γ about the section such that $L = \rho U \Gamma$ (where U is the free stream velocity and ρ the air density). Thus the rotor blade may be modeled by bound vorticity with strength determined by the rotor lift distribution. Since vortex lines cannot end, this bound vorticity must be trailed into the rotor wake from the blade tips and trailing edges.

With constant blade circulation (which corresponds to uniform loading), vorticity is trailed into the wake only from the blade root and tip. The tip vortex is trailed in a helix because of the combination of the rotational motion of the blade and the axial velocity of the flow through the rotor disk (see Fig. 2-12). In hover, this axial velocity is entirely due to the wake-induced inflow. There is a tip vortex from each blade, trailed in interlocking

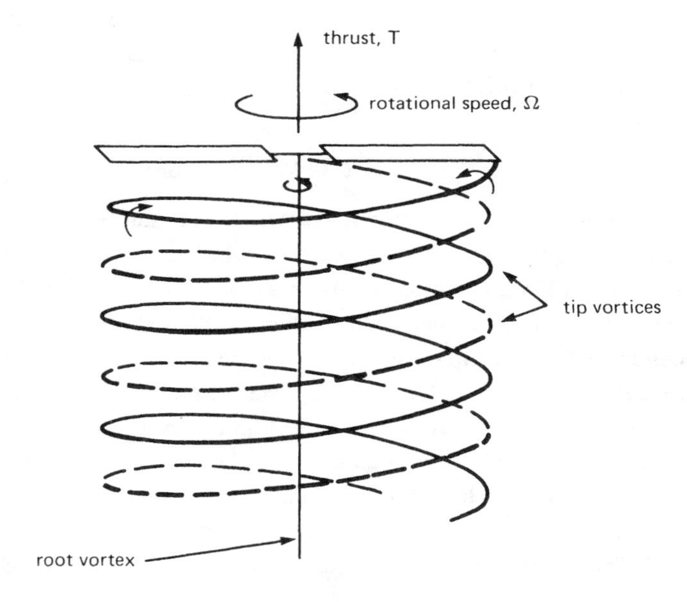

Figure 2-12 Rotor vortex wake in vertical flight.

helices. The root vortices are trailed along the axis of the rotor in a straight line (ignoring any root cutout). With positive thrust on the rotor, the signs of the vorticity are such that the root vortex and the axial components of the tip spirals induce a swirl in the wake in the same direction as the rotor rotation; and the circumferential components of the tip vortex spirals (ring vortices) induce an axial velocity inside the wake in the opposite direction to the thrust. Thus the wake vortex system produces the velocities we have found are required by conservation of axial and angular momentum.

More generally, the blade bound circulation will vary along the span, requiring that vorticity be trailed from the entire trailing edge. The wake then consists of helical vortex sheets behind each blade. For the real rotor, the edges of the vortex sheet quickly roll up into concentrated tip vortices, which are well represented by line vortices. There will also be considerable self-induced distortion of the wake geometry from the nominal helical form. Classical vortex theory usually ignores the roll-up of the vortex sheets, which is a successful approach for propellers, where the high axial velocity sweeps the wake downstream, but for low inflow helicopter rotors a more detailed model for the wake is preferable. In forward flight the blade loading varies with azimuth as well as radially, so radial vorticity in addition to the axial and circumferential vorticity is shed into the wake. Radial vorticity can be present in vertical flight if the blade motion is unsteady.

2-7.2 Actuator Disk Vortex Theory

Consider now vortex theory for the actuator disk model of the rotor in hover. The bound vorticity of the blades is distributed in a sheet over the rotor disk in this infinite-blade approximation. It follows that the wake vorticity is distributed throughout the volume of the wake rather than being concentrated in helical sheets or lines. With this model the difficulties in calculating the velocity induced by the wake are greatly reduced. We have considered this model already in the momentum theory analysis of the rotor. While the results are not new, vortex theory gives more information about their source, which is valuable background for more sophisticated analyses.

Consider first a uniformly loaded actuator disk, for which dT/dA = constant. The blades then have triangular loading and constant bound circulation,

$$\Gamma = \frac{1}{\rho\Omega r}\frac{dT}{dr} = \frac{2\pi}{\rho\Omega}\frac{dT}{dA} = \frac{2\pi}{\rho\Omega}\frac{T}{A}$$

(here Γ is the bound circulation of all the blades). Therefore, the wake consists of a vortex sheet at the boundary of the slipstream and a line vortex on the axis (Fig. 2-13). The line vortex on the axis is the root vortex. Since it is the sum of all the bound vorticity, it has strength Γ. The rotor disk is a sheet of radial vorticity. Since the bound circulation of the rotor is spread over the entire disk. the radial vorticity has strength $\gamma_b = \Gamma/2\pi r = T/\rho A\Omega r$. With uniform bound circulation the wake consists of only tip and root vortices, and in the actuator disk limit of infinite blades the interlocking tip vortex spirals become a vortex sheet on the boundary of the wake, with axial and

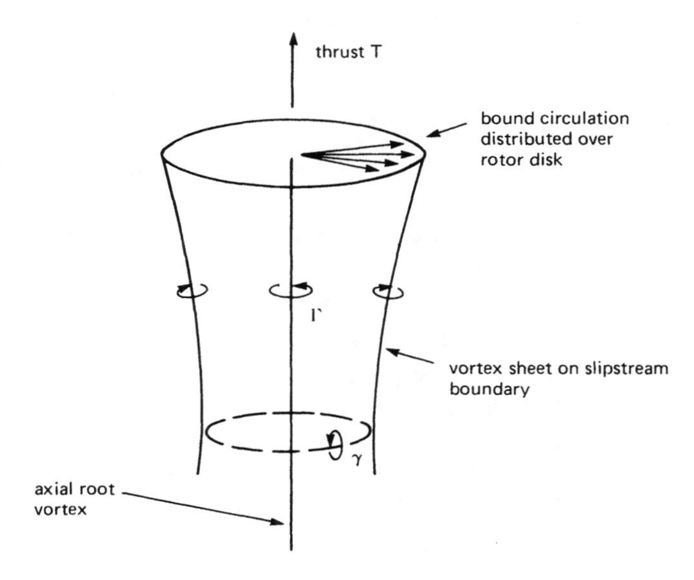

Figure 2-13 Vortex theory for the actuator disk model.

circumferential components. The axial component of the tip vortex sheet has strength $\gamma = \Gamma/2\pi R_1$, where R_1 is the radius of the wake. The vortex lines form a continuous path (as required by Helmholtz's law) consisting of the root vortex, the radial bound circulation of the disk, and the axial vorticity components of the tip vortex sheet. Because of the helical geometry of the tip vortices, in the infinite blade limit the wake also contains a circumferential component of vorticity, which may be viewed as consisting of ring vortices. The ring vortex strength is $\gamma = \Gamma/h$, where h is the distance the wake moves during one rotor revolution. Relating h to the axial velocity at the wake boundary gives $h = 2\pi v/\Omega$, and so $\gamma = T/\rho A v$.

The ring vorticity in the wake produces the axial velocity inside the slipstream. The axial velocity at the rotor disk and in the far wake is due to the wake vorticity alone, with no contributions from the bound circulation. If the wake contraction and swirl velocity are ignored, the induced velocity at the disk is due to a semi-infinite vortex cylinder, and the velocity in the far wake is due to an infinite cylinder. It follows that the induced velocity at the disk is one-half the velocity in the far wake, $v = \frac{1}{2}w$. Since the fluid is irrotational far upstream of the rotor, the flow must always be irrotational unless it passes through the rotor disk. Thus there can be no circulation about any path lying entirely outside the wake, and in particular a rotational velocity can only exist inside the rotor wake. Just above the rotor disk there is no rotational velocity then, while just below the disk there is a rotational velocity u due to the rotor torque. The root vortex induces a rotational velocity component $u_1 = \Gamma/4\pi r$ both above and below the rotor disk; there is no contribution to the swirl inside the wake from the vorticity on the slipstream boundary (Stokes' theorem). The bound vorticity induces a rotational velocity u_b just below the disk and $-u_b$ just above the disk. Satisfying the requirement of no swirl outside the wake then requires $u_b = u_1$, and the total swirl just below the disk is thus $u = 2u_1$. Indeed, since the jump in velocity across the vortex sheet of the rotor disk equals the vortex strength, we have $2u_b = \gamma_b = \Gamma/2\pi r$ again. Note that the velocity seen by the blade due to its own rotation and the wake-induced swirl is then $(\Omega r - ½u)$, which explains the appearance of this factor in the expression for the rotor torque in section 2-3.2.

To examine the axial induced velocity further, consider the relation

$$\vec{u}(\vec{x}) = -\frac{\kappa}{4\pi}\vec{\nabla}\Sigma$$

where $\vec{u}$ is the velocity induced by a line vortex of strength κ that subtends a solid angle Σ at $\vec{x}$. [See also Knight and Hefner (1937).] The rotor axial velocity is due to a semi-infinite cylinder of ring vortices with strength $\kappa = \gamma dz_1$. The axial component of the induced velocity is thus

$$v(\vec{x}) = -\int_0^\infty \frac{\gamma}{4\pi}\frac{\partial}{\partial z}\Sigma\, dz_1$$

where Σ is the angle of the ring vortex at z_1 as seen at z; the rotor disk is at $z = 0$. Now if the wake contraction rate is slow, the change in Σ as the observer moves will be primarily due to the distance change $(z - z_1)$ and only secondarily to the change in the ring size. Motion of the observer and ring are then equivalent, $\partial\Sigma/\partial z = -\partial\Sigma/\partial z_1$, or

$$v = \int_{z=0}^{z=\infty} \frac{\gamma}{4\pi} d\Sigma$$

Next neglect any change in the spacing of the wake spirals, so that the ring vortex strength is constant. With these assumptions, the induced velocity is given by

$$v = \frac{\gamma}{4\pi} \Delta\Sigma$$

where $\Delta\Sigma$ is the total solid angle covered by the wake surface, as seen at the location of v. We shall use this result to evaluate the induced velocity at several points in the flow. For any point on the rotor disk, $\Delta\Sigma = 2\pi$, so $v = \gamma/2$. Recalling that the ring strength is $\gamma = T/\rho A v$, we obtain again for the induced velocity at the rotor disk

$$v = \sqrt{T/2\rho A}$$

Moreover, the induced velocity is constant over the disk for this uniformly loaded rotor. Consider the points still in the disk plane but now outside the rotor disk; then $\Delta\Sigma = 0$ and $v = 0$, so there is no axial induced velocity except at the disk itself. For a point inside the far wake, $\Delta\Sigma = 4\pi$, so $w = \gamma$; the induced velocity is uniform in the far wake, and $w = 2v$, as in momentum theory. Finally, for an arbitrary point on the axis of the wake and at a distance z below the rotor, the induced velocity is

$$v = \frac{\gamma}{4\pi}\left[4\pi - \Sigma_0\right]$$

where Σ_0 is the angle subtended by the rotor disk,

$$\Sigma_0 = 2\pi\left[1 - \frac{z/R}{\sqrt{1 + (z/R)^2}}\right]$$

On the wake axis, the axial velocity is therefore

$$v = v(0)\left[1 + \frac{z/R}{\sqrt{1 + (z/R)^2}}\right]$$

which has the proper limits far above and far below the rotor (at $z = -\infty$ and $z = \infty$, respectively).

Now consider an actuator disk with nonuniform loading. With varying bound circulation on the blade, the trailed vorticity is distributed throughout the wake cylinder rather than concentrated on the boundary. The wake may be viewed as constructed from shells consisting of the cylindrical sheet at radius r plus the corresponding inboard bound vorticity and root vortex required for conservation of vortex lines. The bound vorticity then is built up from the contributions from all shells outboard of r, and the change in bound vorticity at r is due to the trailed wake there. From the previous paragraph it follows that only the shells outboard of r contribute to the induced velocity $v(r)$, since only for these is the point inside the disk. Hence the axial induced velocity is

$$v = \int_r^R \tfrac{1}{2}\gamma\, dr$$

where γ is the strength of the trailed vorticity, which is related to the change in the bound circulation Γ by

$$\gamma = -\frac{d\Gamma}{dr}\frac{\Omega}{2\pi(V+v)}$$

Then

$$v = -\int_r^R \frac{\Omega}{4\pi(V+v)}\frac{d\Gamma}{dr}\, dr$$

$$= \frac{\Omega}{4\pi(V+v)}\Gamma + \int_r^R \frac{\Omega}{4\pi}\Gamma\frac{d}{dr}\left(\frac{1}{V+v}\right)dr$$

In terms of the loading distribution $dT/dA = \rho\Omega\Gamma/2\pi$, this vortex theory result for the induced velocity becomes

$$2\rho(V+v)v = \frac{dT}{dA} + (V+v)\int_r^R \frac{dT}{dA}\frac{d}{dr}\left(\frac{1}{V+v}\right)dr$$

Compare this result with the differential form of momentum theory, $dT = 2\rho(V+v)v\,dA$, which was obtained (without proof) by application of the conservation laws to the annulus of the rotor disk at r. The induced velocity obtained from differential momentum theory (such as in the combined blade element and momentum theory) while not exact, appears to be reasonably accurate as long as the inflow is at least moderately uniform. Similarly, it is recalled that the relation $w = 2v$ between the induced velocities at the disk and in the far wake is not an exact result of momentum theory. The assumptions required in vortex theory to duplicate these momentum theory results give a better idea of the approximations involved in their application.

2-7.3 Finite Number of Blades

Vortex theory for vertical flight is elementary with the actuator disk model, especially for uniform loading. With a finite number of blades, vortex theory models the wake by vortex lines and sheets trailed in helices behind each blade. This problem is mathematically much more difficult than the case of distributed wake vorticity, but in axial flow some analytical solutions are still possible. Finite-blade vortex theory is analogous to the Trefftz plane analysis of fixed wings. The wake is studied far downstream, where the wing has negligible influence on the flow. The solution for the wake vorticity determines the loading on the wing also. By solving the simpler flow problem in the far wake (where there is no axial dependence), an exact loading distribution that includes tip effects is obtained. The accuracy of the solution depends on the wake model used. In the classical analyses approximate models are used, with vortex sheets rather than concentrated tip vortices and no self-induced wake distortion. Moreover, a far wake analysis does not tell how to design the blades to obtain the desired loading; for that the induced velocity at the rotor disk is needed.

For helicopter rotors the primary application of classical finite-blade vortex theory is in calculating the loading near the blade tip. The solutions of Prandtl and Goldstein were developed for high inflow propellers and thus were based on wake models that are not entirely appropriate for low inflow rotors. The key step of these analyses is the choice of the wake structure, which completely determines the solution for the flow field. Specifically, it is assumed that the vortex sheets in the rotor wake move like rigid surfaces. Then the boundary condition of no flow through the sheets determines the flow field, from which the vorticity strength in the wake and hence the loading on the blade is obtained. Prandtl considered a simple two-dimensional model for the rotor wake, while Goldstein used helical vortex sheets for the wake geometry, requiring a much more complicated analysis. For low inflow the two solutions are very close, but both neglect the blade-wake interactions that are important for low inflow rotors.

2-7.3.1 Wake Structure for Optimum Rotor

Let us examine the wake model implied for an optimum rotor. [This discussion follows Betz, as given by Glauert (1935).] For a lightly loaded rotor, the wake contraction near the disk can be neglected. Then the pitch angle of the helical wake is

$$\phi = \tan^{-1}(V+v)/(\Omega r - \tfrac{1}{2}u)$$

where v is the axial velocity and u the swirl velocity induced by the wake at the rotor disk. Momentum theory (section 2-3.2) gives the optimum solution $V+v = (V+v_0)(1 - \tfrac{1}{2}u/\Omega r)$; hence

$$\phi = \tan^{-1}\left(\frac{V+v_0}{\Omega r}\right)$$

The wake of the optimum rotor is thus a helix of constant pitch, undistorted by the induced velocities v and u. For a constant pitch (screwlike) helix, wake elements shed from the trailing edge at a given instant always remain in the same horizontal radial line. The pitch angle ϕ is the angle of the wake surface with respect to the horizontal. This is the wake geometry for the rotor with minimum induced power at a given thrust.

Consider a rotor with a wake consisting of helical vortex sheets that move as rigid surfaces. The rotor has an upward axial velocity V, the wake a downward velocity v_0, and the pitch of the helices is $\phi = \tan^{-1}(V+v_0)/\Omega r$. The wake movement at v_0 imparts a velocity to the fluid at the wake surface. Since there is no flow through the sheet, the velocity of the sheet and the fluid in a direction normal to the wake surface must be equal, with value $v_0 \cos\phi$. With a finite number of blades, this normal velocity will decrease between the sheets and there will be radial flow as well, which decreases the lift at the blade tip. In the limit of infinite number of blades, the sheets are very close and as a consequence all the fluid is carried with the wake. There are then no losses due to flow around the edges. From the velocity $v_0 \cos\phi$ normal to the wake surface, it follows that the induced axial and rotational velocities are

$$v = (v_0 \cos\phi)\cos\phi = v_0 \frac{(\Omega r)^2}{(\Omega r)^2 + (V+v_0)^2}$$

$$\tfrac{1}{2}u = (v_0 \cos\phi)\sin\phi = v_0 \frac{(V+v_0)\Omega r}{(\Omega r)^2 + (V+v_0)^2}$$

which agree with the momentum theory results (section 2-3.2).

Primarily, this discussion justifies the rigid wake model used in classical vortex theory. Since the induced losses will not be too far from the optimum values, the use of the simple rigid geometry even for non-optimum loading is a reasonable approximation. To summarize, the wake of a rotor or propeller with minimum induced power loss consists of spirals of wake vorticity moving axially as rigid sheets, with uniform velocity and no distortion. The transport velocity of the wake is determined by the rotor disk loading, and the pitch of the helical surfaces is determined by the axial and rotational velocities of the blades.

2-7.3.2 Prandtl's Tip Loading Solution

The rotor induced velocity and loading can be obtained by considering the wake far downstream from the rotor disk, with the solution depending on the model used for the wake. Distributed wake vorticity implies distributed loading on the rotor disk, hence an actuator disk model. In fact, the rotor consists of discrete lifting surfaces. The simplest wake model including the effects of a finite number of blades consists of helical vortex sheets trailed from each blade. The major effect of the finite number of blades is a reduction of the loading at the blade tip. In terms of the wake flow, there is a flow around the edges of the vortex sheets from the lower surface to the upper surface that reduces the total downward momentum. An exact solution for the helical wake of a rotor was obtained by Goldstein (see section 2-7.3.3). Prandtl [see Glauert (1935)] obtained an approximate solution, deriving a tip loss correction for a finite number of blades by using a two-dimensional model for the vorticity in the far wake.

Let us replace the system of helical vortex sheets by a series of semi-infinite, parallel vortex lines (Fig. 2-14), thereby replacing the axisymmetric wake by a two-dimensional flow model that can be solved using complex potential methods. Since this model is equivalent to considering the flow only near the edges of the wake helices, for low inflow (small spacing of the sheets) it should be fairly accurate. A coordinate system moving downward with the wake at vorticity v_0 is used, such that the wake sheets are stationary and the external flow is moving upward at v_0. The fluid velocities are

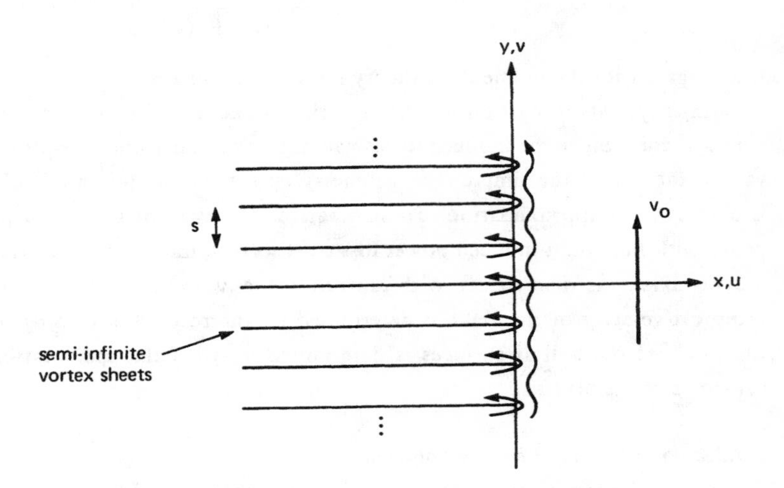

Figure 2-14 Two-dimensional model for the rotor wake.

u and v (see Fig 2-14), and dimensionless quantities are used (based on ρ, Ω, and R). For a lightly loaded rotor, the wake spacing is

$$s = \frac{2\pi}{N}\frac{\lambda}{\sqrt{1+\lambda^2}} \cong \frac{2\pi\lambda}{N}$$

where λ is the inflow ratio and N is the number of blades.

The complex potential satisfying the condition of no flow through the vortex sheets and the requirement that v and u approach v_0 and zero respectively as x approaches infinity is

$$w = -v_0 \frac{s}{\pi} \cos^{-1} e^{\pi z/s}$$

where $z = x + iy$. Then the velocity is

$$u - iv = \frac{dw}{dz} = v_0 \frac{e^{\pi z/s}}{\sqrt{1 - e^{2\pi z/s}}}$$

For example, at $y = 0$ (one of the sheets)

$$u - iv = v_0 \frac{e^{\pi x/s}}{\sqrt{1 - e^{2\pi x/s}}}$$

In the fixed frame, the sheets are moving downward at v_0 and the fluid far away from the wake is at rest. There is air flowing up around the edges of the sheets, however, which reduces the mean downward velocity of the air between the sheets. Momentum conservation implies that there must be a reduction c• the lift near the tips. In the fixed frame, the average vertical velocity between the sheets is

$$\frac{1}{s}\int_0^s (v_0 - v)dy = v_0 \frac{2}{\pi} \cos^{-1} e^{\pi x/s}$$

or $\bar{v}(x) = v_0 F$, where for the helicopter rotor with $\pi x/s = (r - 1)N/2\lambda$,

$$F = \frac{2}{\pi} \cos^{-1} e^{(r - 1)N/2\lambda}$$

The function F is the principal result of this analysis. The vorticity γ in the wake sheet (which is related to the rotor bound circulation distribution) is

$$\gamma = v_{y=0} - v_{y=s} = 2v_0 \frac{e^{\pi x/s}}{\sqrt{1 - e^{2\pi x/s}}}$$

Then the blade bound circulation is

$$\Gamma(x) = \int_x^0 \gamma dx = v_0 s \frac{2}{\pi} \cos^{-1} e^{\pi x/s} = v_0 s F$$

Substituting for $s = 2\pi\lambda/N = 2\pi(\lambda_c + \lambda_i)/N$ and using $v_0 = 2\lambda_i$ gives $\Gamma = (4\pi/N)(\lambda_c + \lambda_i)\lambda_i F$, or

$$\frac{dC_T}{dr} = 4(\lambda_c + \lambda_i)\lambda_i r F$$

which is simply the momentum theory result, corrected for the effect of the blade tip by the factor $F(r)$. The function F is significantly less than unity only over the outer 5% to 10% of the blade. To account for the blade root, the factor $r^2/(r^2 + \lambda^2)$ should also be influded in F, based on the momentum theory with swirl velocities.

In combined blade element and momentum theory, the hover induced velocity now becomes

$$\lambda = \frac{\sigma a}{16F}\left(\sqrt{1 + F\frac{32}{\sigma a}\theta r} - 1\right)$$

The effect of the tip, expressed through the factor F, is to increase the induced velocity and thus reduce the loading at the blade tip and increase the induced power. The factor F also affects the chord distribution required for the optimum rotor, necessitating the introduction of rounded tips.

Rather than using the factor F to correct the span loading near the tips, we may instead use this model to obtain an equivalent tip loss factor B for the rotor loading and performance calculations. An equivalent infinite-blade model (with a smaller effective disk area) will be found that produces the same thrust for a given power as the finite-blade rotor. If the vortex sheets were infinitely close, the fluid between the sheets would all be carried downward at velocity v_0 and the fluid outside would be at rest. With a finite distance between the sheets some fluid flows up around the edges, reducing the downward momentum. By equating the momentum reduction $v_0(1 - B)$ of an infinite-blade model with reduced wake area with the momentum reduction due to the finite number of blades, the tip loss factor B may be evaluated:

$$1 - B = \frac{1}{v_0}\int_0^\infty (v - v_0)dx = \int_0^\infty \left[\frac{e^{\pi x/s}}{\sqrt{e^{2\pi x/s} - 1}} - 1\right] dx$$

$$= \frac{s}{\pi}\ln 2 = \frac{\lambda}{N} 2\ln 2$$

$$= 1.39\frac{\lambda}{N}$$

Here the quantity λ is the inflow velocity which determines the wake spiral spacing. For hover with linear inflow, $\lambda = r\lambda_t = r\sqrt{C_T}$ (section 2-6.2),

$$B = 1 - 1.39\frac{\sqrt{C_T}}{N} \cong 1 - \frac{\sqrt{2C_T}}{N}$$

This is the result normally used; see Sissingh (1941) and Gessow and Myers (1952). This tip loss factor gives fairly good correlation with measured rotor performance.

2-7.3.3 Goldstein's Propeller Analysis

Goldstein (1929) developed a vortex theory for propellers with a finite number of blades in axial flow. The wake was modeled as helical trailed vortex sheets moving axially at a constant velocity like rigid surfaces. The boundary condition of no flow through the sheets completely defines the vortex strength in the wake, which may be related to the bound circulation distribution on the blade. Goldstein solved the potential flow problem of N intermeshed helical surfaces of infinite extent axially (i.e. in the far wake) but finite radius moving with axial velocity v_0. The solution takes the form of a tip loading factor F, which is a function of the inflow ratio, the number of blades, and the blade radial station. Goldstein gives tables and graphs of F as a function of r for propellers with two and four blades (Goldstein uses the notation K rather than F). This factor F is used in the same manner as in Prandtl's solution described in the previous section. Generally it is found that Prandtl's function is a good approximation to Goldstein's more complete result for low inflow, specifically when $\lambda/N < 0.1$ or so. Thus Prandtl's solution is good for helicopter rotors, but for propellers it is necessary to use Goldstein's solution.

Lock (1930) summarizes Goldstein's vortex theory results and their application to propeller design. He compares the solution with vortex theory results for an actuator disk model, showing that the limit of Goldstein's solution for $N \to \infty$ must be $F \to r^2/(r^2 + \lambda^2)$. Lock points out that Goldstein's wake model is really the optimum solution, so the use of this theory requires the assumption that the wake model is good for practical loadings as well. Lock and Yeatman (1934) give tables of the results from Goldstein's vortex theory and discuss the theory and its use (including Prandtl's approximation). Kaman (1943) discusses Goldstein's vortex theory, particularly its application to the helicopter rotor in hover or vertical climb.

2-7.3.4 Applications to Low Inflow Rotors

A wake model consisting of undistorted vortex sheets may be adequate for high inflow propellers, where the wake is quickly convected away from the rotor disk. For low inflow helicopter rotors, where the interference between the rotor blade and the wake vorticity, and the self-induced distortion in the wake, become important, a simple wake model such as used in Goldstein's analysis is not entirely satisfactory. The trailed vorticity quickly rolls up into concentrated tip vortices that remain close to the disk with low inflow rotors and strongly influence the loading near the tips of both the generating and following blades. Such effects must be included in the vortex theory if an accurate calculation of the blade loading is to be achieved.

Thus the classical vortex theories for a finite number of blades are most accurate with high inflow propellers, for which they were originally developed. For low inflow helicopter rotors a more sophisticated analysis is required. Such an analysis has to be numerical because of the complexities of the vortex wake structure and geometry of a real rotor. Prandtl's approximate solution (either the loading factor $F(r)$ or the result for the tip loss factor) is simple enough to justify its use when a more detailed calculation is not possible or not required.

2-7.4 Nonuniform Inflow (Numerical Vortex Theory)

The modern variant of vortex theory is a numerical solution for the rotor induced velocity, loads, and performance that uses a detailed model of the vortex wake, including a representation of the discrete tip vortices and frequently even the distorted wake goemetry. Such an analysis is only practical using high-speed digital computers. While potentially more accurate than the classical solutions, numerical vortex theory faces a difficult task in seeking to improve the prediction of hovering rotor performance. Often it is only a small, but important, improvement in correlation that is sought by using a more detailed flow model, which demands a very accurate analysis. Many of the complex aerodynamic phenomena of the helicopter rotor are still poorly understood, however, and others are not easily analyzed. In addition, it is necessary to be consistent when improving the model, which means that the analysis must simultaneously advance the aerodynamic, dynamic, and structural models. Progress has been made in the calculation of rotor hover aerodynamics, but there are still deficiencies in the current capability. A detailed description of the nonuniform inflow calculation is given in Chapter 13.

Jenny, Olson, and Landgrebe (1968) compared a number of methods for calculating hover performance, including the simple uniform inflow and constant drag coefficient relation, combined blade element and momentum theory, the Goldstein-Lock vortex theory, and a numerical vortex theory calculation of the nonuniform inflow, both with an uncontracted wake and with a contracted one (in the latter case using a prescribed wake geometry based on experimental measurements). They found that the classical methods and the numerical analysis with an uncontracted wake tend to underestimate the hover power required, the error increasing with blade loading C_T/σ (and with the tip Mach number, blade solidity, and reduced twist). The errors were attributed to not accounting for the contraction of the slipstream; in other words, to not accounting for the actual tip vortex geometry. The blade loading is greatly influenced by the tip vortex from the preceding blade and hence is very sensitive to the radial and vertical position of that vortex relative to the blade. The effect of the vortex is to increase the angle of attack outboard' of its position, and decrease it inboard. At moderate rotor loading ($C_T/\sigma = 0.06$ to 0.08) and above, the vortex tends to induce stall on the tip of the blade, limiting the tip loading achieved and increasing the drag, thus producing a loss of efficiency. Since the tip is the region of highest loading, rotor performance is quite sensitive to this effect, and therefore to small changes in the vortex position (and also airfoil or planform changes). Compressibility effects play an important role also, since the Mach number is highest at the tip. In the absence of compressibility and stall effects, the effects of tip vortices on the loading distribution would still be significant, but the effects on the performance would tend to cancel out. If the wake contraction is neglected, then all blade stations are inboard of the vortex and nowhere does the vortex increase the angle of attack; with a distributed vorticity wake model, or even simpler theories, the tip vortex effects can not be observed at all. Thus an accurate wake geometry is crucial to refinements in the calculation of hover performance. The radial and vertical position of the tip vortex at its encounter with the following blade is most important. Clark and Leiper (1970), among others, considered a calculation of the distorted tip vortex geometry for use in hover performance analysis.

Clark (1974) compared the loading of a hovering rotor blade calculated by a theory using nonuniform inflow and a free-wake model with the loading calculated by combined blade element and momentum theory (Fig. 2-15).

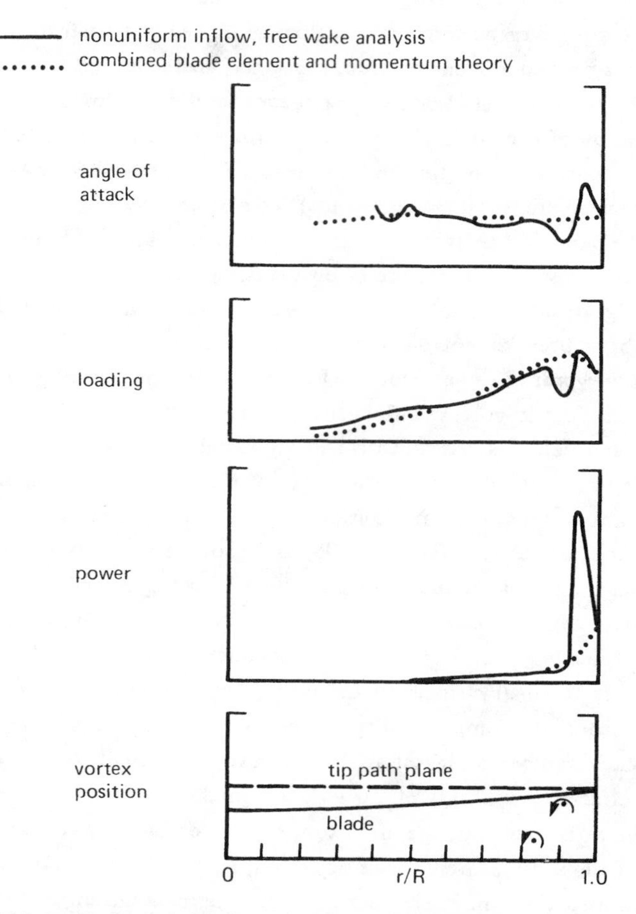

Figure 2-15 **Angle of attack, loading, and power distribution on a hovering rotor, comparing a nonuniform-inflow, free-wake analysis with combined blade element and momentum theory; from Clark (1974).**

The tip vortex, being located at about $0.92R$ when it encountered the following blade, induced an angle-of-attack increase outboard and decrease inboard that the combined blade element and momentum theory misses. Because there is a high angle of attack and Mach number at the tip, the vortex induces shock stall and drag divergence outboard of its station. The lift is reduced inboard because of the vortex, but since stall limits the lift outboard it is about the same as predicted by combined blade element and momentum theory. The drag is significantly increased outboard as a result of the induced compressible drag rise. The net result is reduced lift and increased power for this rotor. These considerations suggest that increased negative twist at the tip would be beneficial, which was confirmed by analysis, wind tunnel tests, and flight tests. Airfoil and planform changes at the tip will also significantly affect this phenomenon.

2-7.5 Literature

Work on vortex theory for the hovering rotor also includes: Helmbold (1931), Knight and Hefner (1937), Reissner (1937, 1940), McCormick (1955), Hough and Ordway (1965), Theodorsen (1969), Erickson and Hough (1970), Gibson (1974), and Baskin, Vil'dgrube, Vozhdayev, and Maykapar (1976). See also the vortex theory references in Chapter 4.

2–8 Literature

On propeller theory (representative work only): de Bothezat (1919), Betz (1920, 1922, 1928a, 1928b), Glauert (1922, 1935), Margoulis (1922), Munk (1922, 1923a, 1923b), Lock and Bateman (1923), Weick (1926), Reissner (1937, 1940), Kramer (1939), Losch (1939), Theodorsen (1944, 1969), Ribner (1945a, 1945b), McCormick (1955), Hough and Ordway (1965), Erickson and Hough (1970), Borst (1973).

On the helicopter rotor in hover or vertical flight: Klemin (1925), Munk (1925), Glauert (1927b), Wheatley and Bioletti (1936a), Knight and Hefner (1937), Prewett (1938), Sissingh (1939, 1941), Bennett (1940), Dingeldein and Schaefer (1945), Gustafson (1945b), Gustafson and Gessow (1945, 1946, 1948), Migotsky (1945), Lipson (1946), Fail and Squire (1947), Carpenter (1948, 1958), Castles and Ducoffe (1948), Squire, Fail, and Eyre (1949), Carpenter and Paulnock (1950), Harrington (1951), Carpenter (1952), Laitone and Talbot (1953), Dingeldein (1954, 1961), Powell (1954, 1957, 1959), Powell and Carpenter (1956), Rabbott (1956), Shivers and Carpenter (1956, 1958), Jewel and Harrington (1958), Jewel (1960), Shivers (1960, 1961, 1967), Sweet (1960b), Shivers and Monahan (1962), Rorke and Wells (1969), Yatsunovich (1969, 1970), Cassarino (1970a), Johansson (1971, 1973, 1978), Bellinger (1972a), Fradenburgh (1972), Gilmore and

Gartshore (1972), Goorjian (1972), Landgrebe and Cheney (1972), Wu, Sigman, and Goorjian (1972), Zimmer (1972), Dietz (1973), Riley and Brotherhood (1974), Young (1974), Wu and Sigman (1975), Bramwell (1977), Landgrebe, Moffitt, and Clark (1977), Moffitt and Sheehy (1977), Rao and Schatzle (1978).

Chapter 3

VERTICAL FLIGHT II

Vertical flight of the helicopter rotor at speed V includes the operating states of hover ($V = 0$), climb ($V > 0$), and descent ($V < 0$), and the special case of autorotation (power-off descent). Between the hover and autorotation states, the helicopter is descending at reduced power. Beyond autorotation, the rotor is actually producing power for the helicopter. The principal concern of this chapter is the induced power of the rotor in vertical flight, including descent. An interpretation of the induced power losses requires a discussion of the flow states of the rotor in axial flight.

3–1 Induced Power in Vertical Flight

In Chapter 2, momentum theory was used to estimate the rotor induced power P_i for hover and vertical climb. Momentum theory gives a good power estimate if an empirical factor is included to account for additional induced losses, particularly tip losses and losses due to nonuniform inflow. In the present chapter these results will be extended to include vertical descent. It will be found that momentum theory is not applicable in a certain range of descent rates because the assumed wake model is not correct. Indeed, the rotor wake in that range is so complex that no simple model is adequate. In autorotation, the operating state for power-off descent, the rotor is producing thrust with no net power absorption. The energy to produce the thrust (the induced power P_i) and turn the rotor (the profile power P_o) comes from the change in gravitational potential energy as the helicopter descends. The range of descent rates where momentum theory is not applicable includes autorotation.

Momentum theory gives the rotor power as $P = T(V + v)$(not including the profile power loss). Here TV is the power input to the rotor for climb at vertical speed V or for descent at speed $|V|$, in which case the airflow supplies the power $T|V|$ to the rotor. The induced power is $P_i = Tv$, where v is the induced velocity at the rotor disk. The induced loss is always positive, $v > 0$. Since the induced velocity is seldom uniform, especially in vertical descent, it is preferable to view v as being equivalent to the induced power by the definition $v = P_i/T$. This view is consistent with the way v is obtained from measured rotor performance. The induced velocity or power is a function of the speed, thrust, rotor disk area, and air density:

$$v = f(V, T, A, \rho)$$

For forward flight, an additional parameter is the disk angle of attack α (see Chapter 4); and there are other parameters influencing the induced velocity that are not considered here, such as the distribution of the loading over the rotor disk. From dimensional analysis it follows that the functional form for v must be

$$\frac{v}{v_h} = f\left(\frac{V}{v_h}, \alpha\right)$$

where $v_h^2 = T/2\rho A$ (the momentum theory result for the hover induced velocity). Note that the induced power and the momentum theory hover power are $P_i = Tv$ and $P_h = Tv_h$, so $v/v_h = P_i/P_h$. The function $f(V/v_h, \alpha)$ may be obtained by analysis (such as momentum theory) or by experiment. A measurement or calculation of P_i and T for a given V is plotted in the form of v/v_h as a function of V/v_h. Any discrepancies in the empirical correlation of measured performance by this function are due to other factors influencing the induced power, such as the twist distribution, number of blades, planform and airfoil shape, and tip Mach number. For the purposes of obtaining a first estimate of the induced power in vertical flight, $v/v_h = f(V/v_h)$ covers the primary functional dependence.

3-1.1 Momentum Theory for Vertical Flight

As in section 2–1, consider momentum theory for an actuator disk model of a uniformly loaded rotor. The rotor is climbing at velocity V, and therefore the flow is downward through the rotor disk (Fig 3-1). It is assumed that the induced velocities v and w at the rotor disk and in the far wake

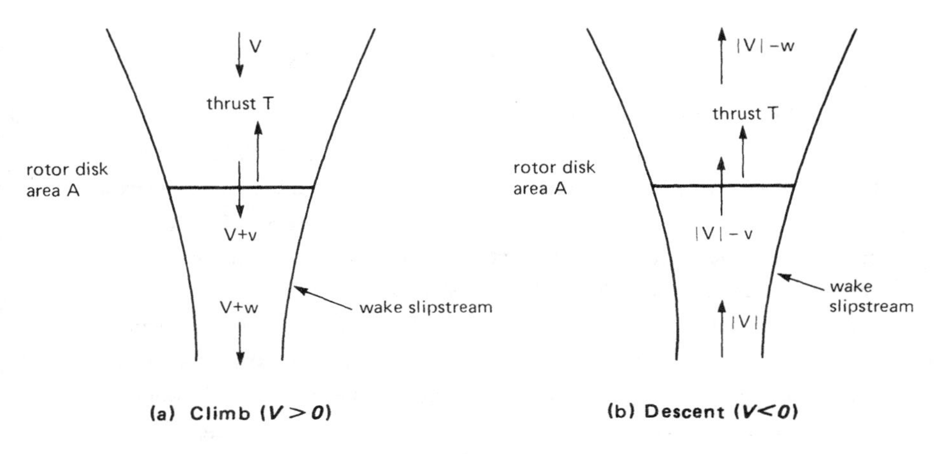

Figure 3-1 Flow model for momentum theory in climb or descent.

respectively are uniform. The sign convention (important when the descent case is considered) is that the thrust is positive upward and the velocities positive downward. The mass flux is $\dot{m} = \rho A(V + v)$. Momentum conservation gives $T = \dot{m}(V + w) - \dot{m}V = \dot{m}w$, and energy conservation $P = T(V + v) = \frac{1}{2}\dot{m}(V + w)^2 - \frac{1}{2}\dot{m}V^2 = \frac{1}{2}m(2Vw + w^2)$. Eliminating $T/\dot{m}$ gives $w = 2v$, and hence $T = 2\rho A(V + v)v$. On writing $v_h^2 = T/2\rho A$, the momentum theory result for the rotor in climb becomes

$$\frac{v}{v_h}\left(\frac{V}{v_h} + \frac{v}{v_h}\right) = 1$$

with solution

$$v = -\frac{V}{2} + \sqrt{\left(\frac{V}{2}\right)^2 + v_h^2}$$

since v must be positive. The net velocities at the disk and far downstream are then

$$V + v = \frac{V}{2} + \sqrt{\left(\frac{V}{2}\right)^2 + v_h^2}$$

and

$$V+w = V+2v = \sqrt{V^2+4v_h^2}$$

The key to the momentum theory analysis is to use the correct model for the flow. The climb model cannot be used with $V<0$, for in descent the free stream velocity is directed upward and therefore the far downstream wake is above the rotor disk. The flow model for descent is also shown in Fig. 3-1. The mass flux is still $\dot{m} = \rho A(V+v)$. Now momentum and energy conservation give $T = \dot{m}V - \dot{m}(V+w) = -\dot{m}w$ and $P = T(V+v) = \frac{1}{2}\dot{m}V^2 - \frac{1}{2}\dot{m}(V+w)^2 = -\frac{1}{2}\dot{m}(2Vw+w^2)$. V is negative now, while T, v, and w are still positive. Since $V+v$ is negative (upward flow through the disk), $P = T(V+v)$ is negative and the rotor is extracting power from the airstream in excess of the induced loss. This flow condition is called the windmill brake state. Eliminating $T/\dot{m}$ gives $w = 2v$ again. The momentum theory result for the induced velocity in descent is $T = -2\rho A(V+v)v$, or

$$\frac{v}{v_h}\left(\frac{V}{v_h}+\frac{v}{v_h}\right) = -1$$

with solution

$$v = -\frac{V}{2} - \sqrt{\left(\frac{V}{2}\right)^2 - v_h^2}$$

The net velocities at the disk and in the far wake are

$$V+v = \frac{V}{2} - \sqrt{\left(\frac{V}{2}\right)^2 - v_h^2}$$

and

$$V+w = V+2v = -\sqrt{V^2-4v_h^2}$$

(the other solution of the quadratic for v gives $v>0$ and $V+v<0$ as required, but has $V+w>0$. Thus the flow in the far wake would be downward, contrary to the assumed flow model.)

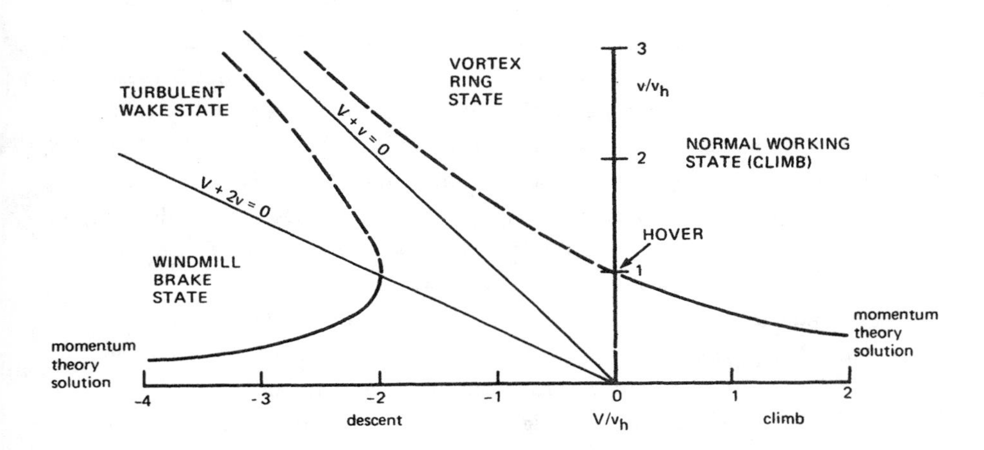

Figure 3-2 Momentum theory results for the induced velocity in vertical flight

Figure 3-2 shows the momentum theory solution for the rotor in vertical climb or descent. The dashed portions of the curves are branches of the solution that do not correspond to the assumed flow state. The line $V + v = 0$ is where the flow through the rotor disk and the total power $P = T(V + v)$ change sign. At the line $V + 2v = 0$ the flow in the far wake changes sign. The lines $V = 0$, $V + v = 0$, and $V + 2v = 0$ divide the plane into four regions, where the rotor operating condition is named the normal working state (climb and hover), vortex ring state, turbulent wake state, and windmill brake state (see Fig. 3-2). For climb, it was assumed that the air is moving downward throughout the flow field (V, $V + v$, and $V + 2v$ all positive). For the branch of the solution given by $V < 0$, however, the flow through the disk and in the wake are downward while the flow outside the slipstream is upward; this is not a physically realizable condition. The climb solution may be expected to be valid for small rates of descent, however, where at least near the rotor the flow is all downward. Thus the region of validity for the momentum theory solution does include hover. For the rotor in descent, it was assumed that the air is moving upward throughout the flow field (V, $V + v$, and $V + 2v$ all negative). For the upper branch of the descent solution, however, $V + 2v > 0$, so the flow is downward in the far wake while it is upward everywhere else, including outside the wake slipstream. Again this is not a physically realizable condition. Thus, in the vortex ring and turbulent wake states the flow outside the slipstream is upward while the flow inside the far wake is nominally downward. Because such a flow state is not possible, there is no valid momentum theory solution for the moderate rates of descent between $V = 0$ and $V = -2v_h$. The line $V + v = 0$ corresponds to ideal autorotation, $P = 0$, and is in the center of the range where momentum theory is not valid. The momentum theory results become infinite at $V + v = 0$ because the theory implies that thrust is produced without mass flow through the rotor disk ($\dot{m} = 0$).

In summary, momentum theory is based on a wake model consisting of a definite slipstream and a well defined wake downstream, with the air moving in the same direction throughout the flow field. Since this a good model for the rotor in climb or a high rate of descent, in the normal working state and windmill brake state momentum theory gives a good estimate of the induced power loss. The momentum solution for climb is actually valid for small rates of descent as well and hence is valid in a range including hover; the flow model is really incorrect, but near the rotor there is no drastic change in the flow until perhaps $V/v_h < -\frac{1}{2}$. For moderate rates of descent, $-2v_h < V < 0$, there is no valid wake model for momentum theory. The flow would like to be upward everywhere except in the far wake, where it wants to be downward. The result is an unsteady, turbulent flow with no definite slipstream. Thus the induced velocity law for the vortex ring and turbulent wake states must be determined empirically from a correlation of measured rotor performance.

3-1.2 Flow States of the Rotor in Axial Flight

3-1.2.1 Normal Working State

Now let us examine in more detail the flow states of the rotor in vertical flight. The normal working state includes climb and hover (see Fig. 3-3). For climb, the velocity throughout the flow field is downward with both V and v positive. From mass conservation it follows that the wake contracts downstream of the rotor. A wake model with a definite slipstream is valid for this flow state (although the wake really consists of discrete vorticity), and momentum theory gives a good estimate of the performance. There will also be

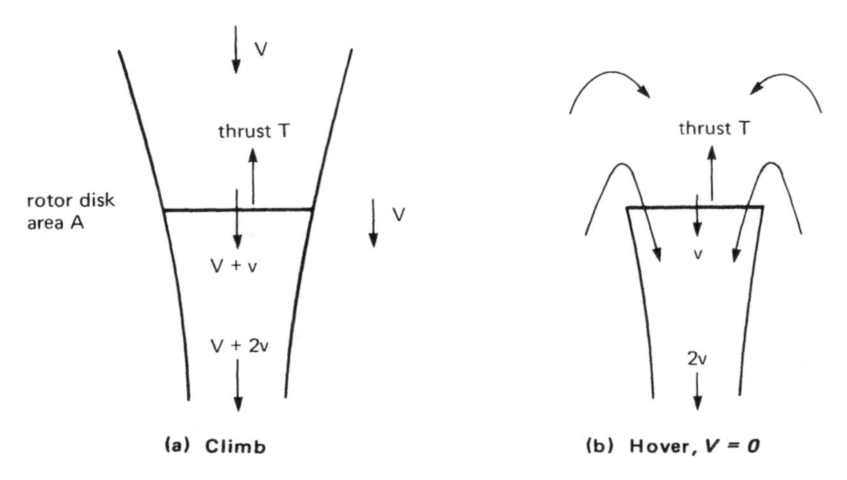

Figure 3-3 Rotor flow in the normal working state.

entrainment of air into the slipstream below the rotor and some recirculation near the disk, particularly for hover. Although such phenomena are not included in the momentum theory model, their effect on the induced power is secondary.

Hover ($V = 0$) is the limit of the normal working state. By mass conservation, the area of the slipstream becomes infinite upstream of the rotor. Still, momentum theory models the flow well in the vicinity of the rotor disk and hence gives a good performance estimate even though hover is nominally a limiting case.

3-1.2.2 Vortex Ring State

When the rotor starts to descend, a defnite slipstream ceases to exist because the flows inside and outside the slipstream in the far wake want to be in opposite directions. Therefore, from hover to the windmill brake state the flow has large recirculation and high turbulence. Sometimes this entire region is called the vortex ring state. The convention here, however, is that the vortex ring state is defined by $P = T(V + v) > 0$, so that the power extracted from the airstream is less than the induced power. The region with $P = T(V + v) < 0$ is called the turbulent wake state. Partial power descents occur in the vortex ring state. Equilibrium autorotation will usually occur in the turbulent wake state.

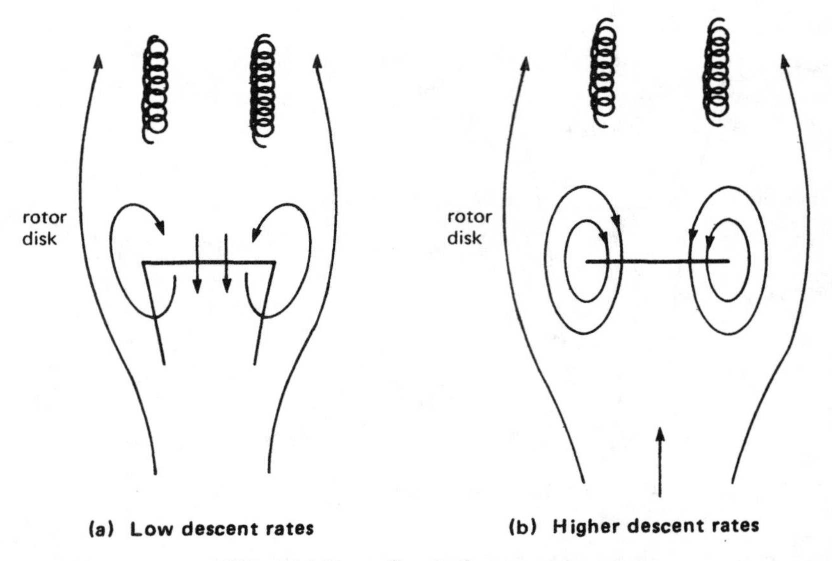

Figure 3-4 Rotor flow in the vortex ring state.

Fig. 3-4 sketches the flow about the rotor in the vortex ring state. At small rates of descent, recirculation near the disk and unsteady, turbulent flow above it begin to develop. The flow in the vicinity of the disk is still reasonably well represented by the momentum theory model, however. Because the change in flow state for small rates of climb or descent is gradual, the momentum theory solution remains valid for some way into the vortex ring state. Eventually, at descent rates beyond about $V = -\frac{1}{2}v_h$, the flow even near the rotor disk becomes highly unsteady and turbulent. The rotor in this state experiences a very high vibration level and loss of control. As will be seen below, in the vortex ring state the power required is not very sensitive to vertical velocity, and hence it is difficult to control the descent rate in this region.

The flow pattern in the vortex ring state is like that of a vortex ring in the plane of the rotor disk or just below it (hence the name given the state; the flow is highly turbulent as well, however). The upward free-stream velocity in descent keeps the blade tip vortex spirals piled up under the disk, forming the ring. With each revolution of the rotor the ring vortex builds up strength until it breaks away from the disk plane in a sudden breakdown of the flow. The flow field is thus unsteady, the vortex ring periodically being allowed to escape and rise into the flow above the rotor. This behavior is a source of very disturbing low-frequency vibration. In the turbulent wake state, $V + v < 0$, so the flow is nominally upward through the rotor disk. The tip vortices are then carried upward, away from the disk again.

3-1.2.3 Turbulent Wake State

Figure 3-5 shows the flow state for ideal autorotation, $V + v = 0$. If the rotor had no profile power losses, power-off descent would be in this condition, since $P = T(V + v) = 0$ for it. While nominally there is no flow through the disk, actually there is considerable recirculation and turbulence. The flow state is similar to that of a circular plate of the same area (no flow through the disk, a turbulent wake above it).

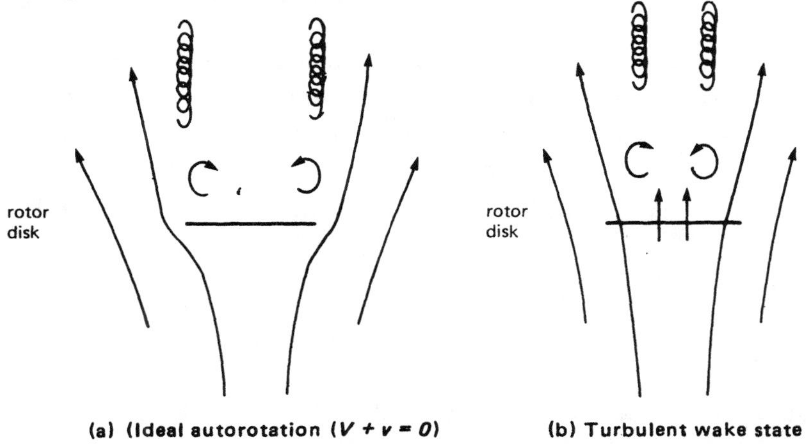

Figure 3-5 Rotor flow in the turbulent wake state.

Figure 3-5 also sketches the flow for the turbulent wake state. The flow still has a high level of turbulence, but since the velocity at the disk is upward there is much less recirculation through the rotor. The flow pattern above the rotor disk in the turbulent wake state is very similar to the turbulent wake of a bluff body (hence the name given the state). The rotor in this state experiences some roughness due to the turbulence, but nothing like the high vibration in the vortex ring state.

3-1.2.4 Windmill Brake State

At large rates of descent ($V < -2v_h$) the flow is again smooth, with a

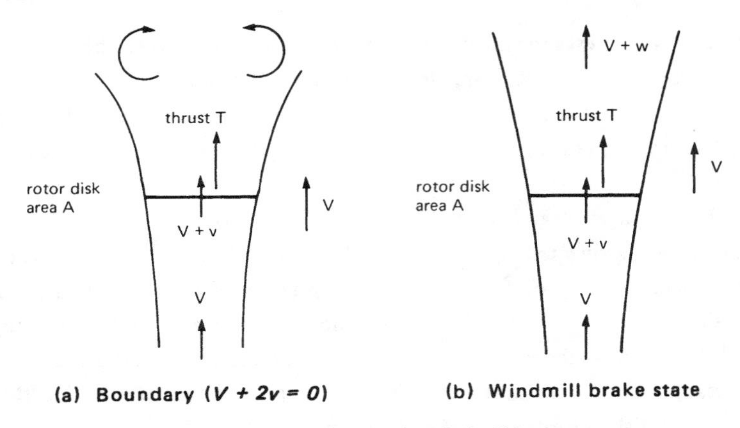

Figure 3-6 Rotor flow in the windmill brake state.

definite slipstream. Figure 3-6 shows this flow condition in the windmill brake state. The velocity is upward throughout the flow field, the slipstream expanding in the wake above the rotor. In the windmill brake state the rotor is producing a net power $P = T(V + v) < 0$ for the helicopter by the action of the airstream on it. The simple wake model of momentum theory is again applicable, and a good performance estimate is obtained.

At the windmill brake state boundary ($V + 2v = 0$ at $V = -2v_h$) the velocity in the far wake above the rotor is nominally zero. Thus the slipstream area approaches infinity above the disk as the flow tries to stagnate. The flow outside the slipstream is still upward, however, so in contrast to the hover case this limit is an unstable condition. At the boundary between the windmill brake and turbulent wake states the flow changes rather abruptly from a state with a smooth slipstream to one with recirculation and turbulence as the nominal velocity in the far wake changes direction. Thus the validity of the momentum theory solution ceases abruptly at the windmill brake boundary.

3-1.3 Induced Velocity Curve

Figure 3-7 presents the universal law for the induced power in vertical flight in terms of v/v_h as a function of V/v_h (a form originated by Hafner). The induced velocity v is not measured directly, but rather the law is a correlation of measured rotor power and thrust at various axial speeds.

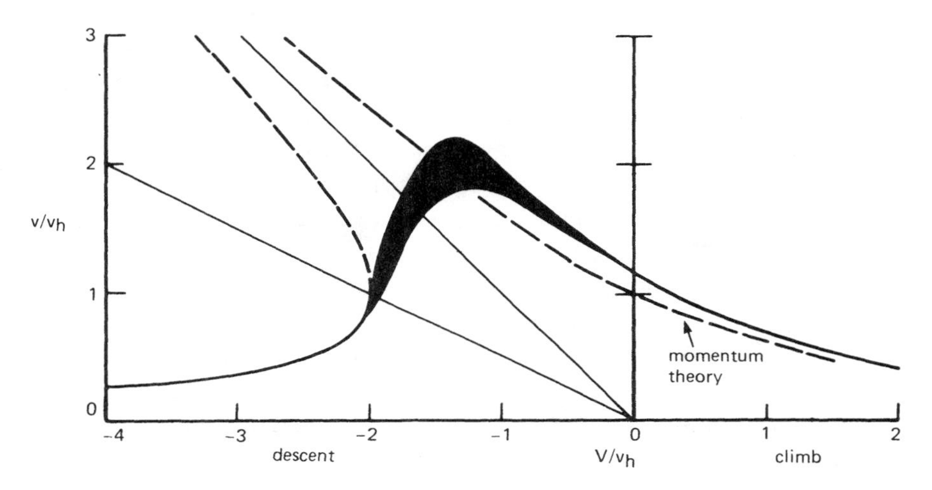

Figure 3-7 Rotor induced power in vertical flight.

The ordinate is therefore best interpreted as P_i/P_h. The measured rotor power also includes profile losses ($P = T(V + v) + P_o$) which must be accounted for to obtain the induced loss:

$$\frac{V+v}{v_h} = \frac{P-P_o}{T\sqrt{T/2\rho A}} = \frac{C_P - C_{P_o}}{C_T^{3/2}/\sqrt{2}}$$

Obtaining the induced velodity thus requires an estimate of the profile power coefficient. The simple result $C_{P_o} = \sigma c_{d_o}/8$ might be used, but a more detailed calculation of C_{P_o} is desirable since any errors in C_{P_o} will result in corresponding scatter in the induced power correlation. By this means the universal induced velocity curve may be constructed, as in Fig. 3-7. The curve presented is based on the available experimental data, particularly from Lock (1947), Brotherhood (1949), Castles and Gray (1951), Gessow and Myers (1952), and Washizu, Azuma, Koo, and Oka (1966a, 1966b). Momentum theory indeed gives a good performance estimate in the normal working and windmill brake states. In hover and climb, the measured induced power is higher than the momentum theory result by a small, relatively constant factor. This power increase is due to the additional induced losses of the real rotor, particularly nonuniform inflow and tip losses. The induced velocity correlation always shows some scatter, due to errors in the profile power calculation, variations in the nonoptimum losses, and the influence of other design parameters, such as tip Mach number and blade twist. At hover, for example, the result could be 5% or 10% different from that shown in Fig. 3-7. It is in the vortex ring state that the scatter must really be taken into account. Because of the highly turbulent and unsteady flow condition, the induced velocity law is not well represented by a single line in this range of descent rates. Moreover, since the vortex ring state is basically an unstable flow condition, it is very sensitive to factors such as ground proximity and wind or ground speed, making good performance measurements in this region difficult to obtain.

An alternative presentation of the induced velocity law, developed by Lock (1947), is in terms of $(V + v)/v_h$ as a function of V/v_h (Fig. 3-8). In this case the total power $P/P_h = (V + v)/v_h$ is given, rather than just the

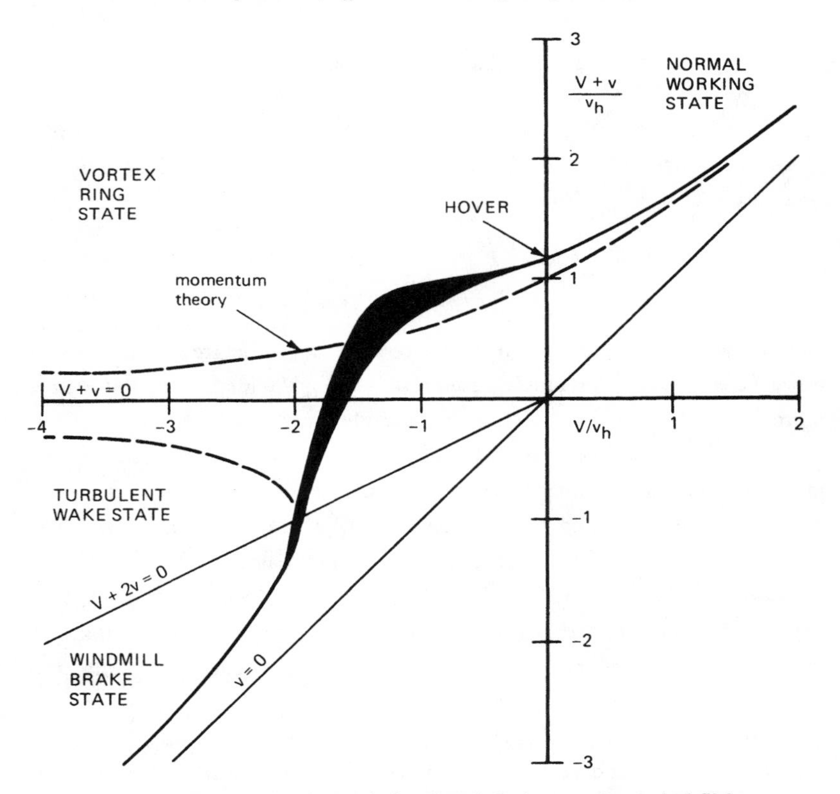

Figure 3-8 Combined rotor induced and climb power in vertical flight.

induced loss. Such a presentation is more consistent with the way the curve is obtained and the way it is used, since it is the total power which is of interest in the rotor performance calculation. Figure 3-8 also shows the lines $V + v = 0$ (the abscissa) and $V + 2v = 0$, which define the four flow states of the rotor in axial flow. The line $v = 0$ goes through the origin at 45°; the induced velocity v is given by the vertical distance between the inflow curve and the $v = 0$ line. The abscissa, $V + v = 0$, is the ideal autorotation case here; for points above ideal autorotation the rotor is absorbing power and for points below it is producing power for the helicopter.

To interpret the scale of these inflow curves, note that for sea level density $v_h = \sqrt{T/2\rho A} = 0.64\sqrt{T/A}\ m/sec$ when the disk loading is in N/m^2. For the disk loading range typical of helicopter rotors, $T/A = 100$ to $500\ N/m^2$, the velocity $v_h = 6$ to $15\ m/sec$.* In the early British literature, the induced velocity curve is often plotted in terms of $1/F = (V + v)^2/v_h^2$ as a function of $1/f = (V/v_h)^2$.

3-1.3.1 Hover Performance

The measured rotor hover performance indicates that the induced power is consistantly higher than the momentum theory result by about 10% to 20%. The momentum theory power estimate is the best possible performance. The additional induced power is due to nonuniform inflow, tip losses, swirl, and other factors. Thus, in hovering performance calculations (such as in section 2-4.2.3) the induced power may be obtained using the momentum theory result with an empirical correction factor,

$$C_{P_i} = \kappa C_T^{3/2}/\sqrt{2}$$

A number of values for the factor κ are recommended in the literature, but $\kappa = 1.15$ is typical.

3-1.3.2 Autorotation

The universal inflow curve crosses the ideal autorotation line $V + v = 0$ at about $V/v_h = -1.71$ (the scatter extends over roughly $V/v_h = -1.6$ to -1.8; see Fig. 3-8). Real autorotation occurs at a higher rate of descent, in the turbulent wake state. In the turbulent wake state the induced velocity

---

*With the disk loading in lb/ft², $v_h = 14.5\sqrt{T/A}\ ft/sec = 870\sqrt{T/A}\ ft/min$; hence when $T/A = 2$ to $10\ lb/ft^2$, $v_h = 20$ to $45\ ft/sec = 1200$ to $2750\ ft/min$.

curve can be approximated fairly well by a straight line on the $(V+v)/v_h$ vs V/v_h plane. Joining the ideal autorotation intercept ($V+v=0$ at $V/v_h = -x$) and the windmill brake state boundary ($(V+v)/v_h = -1$ at $V/v_h = -2$) gives

$$\frac{V+v}{v_h} = \frac{x}{2-x} + \frac{1}{2-x}\frac{V}{v_h}$$

So for $V/v_h = -1.71$ at ideal autorotation, we obtain

$$\frac{V+v}{v_h} = 6 + 3.5\frac{V}{v_h}$$

in the turbulent state. This relation is useful in estimating the descent rate in real autorotation (see section 3–2).

3-1.3.3 *Vortex Ring State*

While no theoretical inflow curve is available in the vortex ring and turbulent wake states, a fairly accurate approximation is given by the cubic relation

$$\frac{V+v}{v_h} = a\left(\frac{V}{v_h}\right)^3 + b\left(\frac{V}{v_h}\right)^2 + c\frac{V}{v_h} + d$$

Matching to the momentum theory results at the windmill brake state boundary ($(V+v)/v_h = -1$ at $V/v_h = -2$) and in the vortex ring state ($(V+v)/v_h = (\sqrt{5}-1)/2$ at $V/v_h = -1$) gives the constants; a good fit is obtained with $b = d = 0$. If the empirical factor κ is then included,

$$\frac{v}{v_h} = \kappa\frac{V}{v_h}\left[0.373\left(\frac{V}{v_h}\right)^2 - 1.991\right]$$

which describes the inflow curve quite well in the range $-2 < V/v_h < -1$. For climb, hover, and low rates of descent ($V/v_h > -1$) and for high rates of descent in the windmill brake region ($V/v_h < -2$) the momentum theory results with an appropriate empirical correction are valid.

In the range $V/v_h = -0.4$ to -1.4, the flow is characterized by a high level of roughness. There are large periodic variations in the velocity at the disk and hence in the rotor loads as the vortex ring alternately builds up and then escapes the rotor disk. The low frequency thrust variations produce a very disturbing vibration of the helicopter that is the dominant feature of the vortex ring state. Since the slope of $(V + v)$ as a function of V is small in this region, there are large changes in descent rate for only small power changes. The result is reduced vertical damping and increased control sensitivity that makes the helicopter descent rate difficult to control in the vortex ring state. In the turbulent wake state, however, a power change produces a small variation in the descent rate, so autorotative descent has much better control characteristics.

3-1.4 Literature

On the flow states in axial flight: de Bothezat (1919), Lock, Bateman, and Townend (1925), Lock (1928), Stewart (1948), Brotherhood (1949), Drees and Hendal (1951), Castles and Gray (1951), Yeates (1958).

On the induced power or induced velocity in vertical flight, especially descent: Lock, Bateman, and Townend (1925), Glauert (1926a), Bennett (1932), Castles (1945), Lock (1947), Stewart (1948), Brotherhood (1949), Drees (1949), Nikolsky and Seckel (1949a), Castles and Gray (1951), Slaymaker, Lynn, and Gray (1952), Katzenberger and Rich (1956), Payne (1956), Castles (1958), Washizu, Azuma, Koo, and Oba (1966a), 1966b), Azuma and Obata (1968), Bramwell (1971), Shaydakov (1971b, 1971c), Wolkovitch and Hoffman (1971), Shupe (1972), Wolkovitch (1972), Heyson (1975), Bramwell (1977).

On the wake flow field in hover and vertical flight: Ross (1946), Brotherhood (1947), Carpenter and Paulnock (1950), Taylor (1950), Gessow (1954), Falabella and Meyer (1955), Castles (1957), Bolanovich and Marks (1959), Heyson (1959, 1960c), Jewel (1960), O'Bryan (1961), Timm (1965), Azuma and Obata (1968), Miller, Tang, and Perlmutter (1968), Boatwright (1972, 1974), Landgrebe and Bennett (1977).

3–2 Autorotation in Vertical Descent

Autorotation is the state of rotor operation with no net power requirement. The power to produce the thrust and turn the rotor is supplied either by auxiliary propulsion (the autogyro) or by descent of the helicopter. In an autogyro the rotor is functioning as a wing. A component of the aircraft forward velocity directed upward through the rotor disk supplies the power to the rotor, so the autogyro requires a forward speed to maintain level flight. In the autorotative descent of the helicopter, the source of power is the decrease of the gravitational potential energy. More directly, the descent velocity upward through the disk supplies the power to the rotor. Although the lowest descent rate is achieved in forward flight, the helicopter rotor is also capable of power-off autorotative descent in vertical flight.

The net rotor power is zero for vertical descent in autorotation: $P = T(V + v) + P_o = 0$. The decrease in potential energy (TV) balances the induced (Tv) and profile (P_o) losses of the rotor. Neglecting the profile losses gives ideal autorotation, $P = T(V + v) = 0$. When the profile losses are included, autorotation occurs at $(V + v) = -P_o/T$. Thus the descent rate may be obtained from the universal inflow curve in the form $(V + v)/v_h$ vs. V/v_h by finding the intercept of the curve with $-P_o/P_h$. In coefficient form,

$$\frac{V+v}{v_h} = -\frac{C_{Po}}{C_T^{3/2}/\sqrt{2}}$$

This intercept typically is at $(V + v)/v_h \cong -0.3$, which is in the turbulent wake state at a descent rate slightly higher than ideal autorotation. Because the slope of the inflow curve is large in this region, the increase in descent rate required to supply the profile power is small. Tail rotor and aerodynamic interference losses should also be included in finding the power $(V + v)/v_h$ of a real helicopter; such losses are only 15% to 20% of the profile power, and therefore make only a slight correction to the descent rate. The limit of the descent rate in vertical autorotation may be obtained from the boundary of the turbulent wake state, at roughly $V/v_h = -1.71$ to -2. Then for sea level density and a disk loading in N/m^2, the descent rate is $V = 1.1\sqrt{T/A}$ to $1.3\sqrt{T/A}$ m/sec.

For a more quantitative estimate of the autorotative performance of real rotors, recall the definition of the figure of merit for hover:

$$M = \frac{C_T^{3/2}/\sqrt{2}}{\kappa C_T^{3/2}/\sqrt{2} + C_{P_o}}$$

so that

$$\frac{C_{Po}}{C_T^{3/2}/\sqrt{2}} = \frac{1}{M} - \kappa$$

Assuming now that C_{P_o} and C_T do not change from hover to autorotation (hence that the blade drag coefficient and tip speed are the same), this is exactly the quantity required to define the autorotation point on the inflow curve. So

$$\frac{V+v}{v_h} = \frac{1}{M} - \kappa$$

which typically gives $(V + v)/v_h = -0.3$ to -0.4. Note that low profile power gives both good hover performance (high figure of merit) and good autorotation performance (low descent rate). Now we shall use the expression obtained in section 3-1.3.2 for the inflow curve in the turbulent wake state: $(V + v)/v_h = 6 + 3.5\,V/v_h$. Combining the two relations for $(V + v)/v_h$ gives the descent rate

$$\frac{V}{v_h} = -\left[1.71 + 0.29\left(\frac{1}{M} - \kappa\right)\right]$$

Typically, then, vertical autorotation takes place at $V/v_h = -1.81$, or $V = 1.16\sqrt{T/A}$ m/sec when the disk loading is in N/m², which gives $V = 15$ to 25 m/sec for the range of disk loadings typical of helicopters.*

The autorotation performance may be considered in terms of a drag coefficient based on the rotor disk area and the descent velocity:

$$C_D = \frac{T}{\frac{1}{2}\rho V^2 A} = \frac{T/2\rho A}{V^2/4} = \left(\frac{2}{V/v_h}\right)^2$$

Hence, a low rate of descent corresponds to a high drag coefficient. This parameter is a useful description of the performance since it is independent of the helicopter disk loading. At the descent rates typical of real helicopters, the drag coefficient has a value in the range $C_D = 1.1$ to 1.3. For comparison, a circular flat plate of area A has a drag coefficient of about $C_D = 1.28$, and a parachute of frontal area A has $C_D \cong 1.40$. The helicopter rotor in power-off vertical descent is thus quite efficient in producing the thrust to support the helicopter. The rotor is nearly as good as a parachute

*With the disk loading in lb/ft², $V = 26.2\sqrt{T/A}$ ft/sec $= 1570\sqrt{T/A}$ ft/min; hence $T/A = 5$ lb/ft² gives $V = 3500$ ft/min.

of the same diameter. The descent rate in vertical autorotation is high because it is a rather small parachute for such a weight. A much lower descent rate is possible in forward flight, however. The rotor flow state in autorotation is similar to that of a bluff body of the same size, so it is not surprising that comparable drag forces are produced. Because the rotor efficiency is about as high as possible, a low descent rate can be achieved only with very low disk loadings. Usually, the disk loading is selected primarily on the basis of the rotor performance; the design of the helicopter for good autorotation characteristics is usually concerned with the ability to flare at the ground (see section 7–5).

Consider now power-off descent in terms of the blade aerodynamic loading. The inflow ratio $\lambda = (V + v)/\Omega R$ is directed upward through the disk, so there is a forward tilt of the lift vector (Fig. 3-9). For power equilibrium at the blade section, the inflow angle must be such that there is no

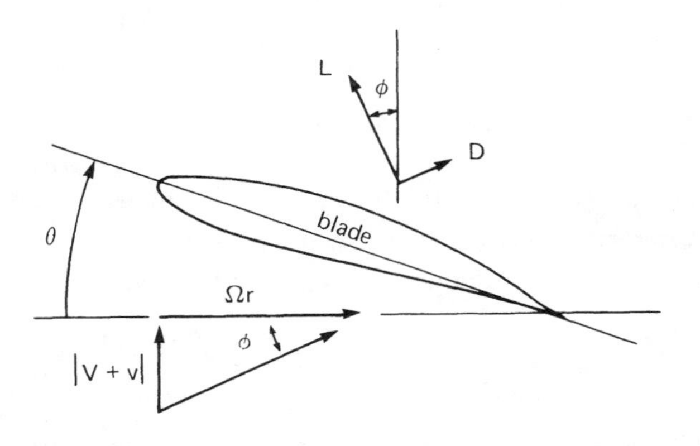

Figure 3-9 Rotor blade section aerodynamics in autorotation

net inplane force and hence no contribution to the rotor torque; $dQ = r\,dr(D - \phi L) = 0$. Because autorotation involves induced and profile torques of the entire rotor, generally only one section will be in equilibrium itself, while the others are either producing or absorbing power. Since $\phi = \tan^{-1}|V + v|/\Omega r$, it follows that the inflow angle is large inboard and decreases toward the tip. Then $dQ < 0$ on the inboard sections, which produce an accelerating torque on the rotor and absorb power from the air; and $dQ > 0$ on the outboard sections, which produce a decelerating torque and

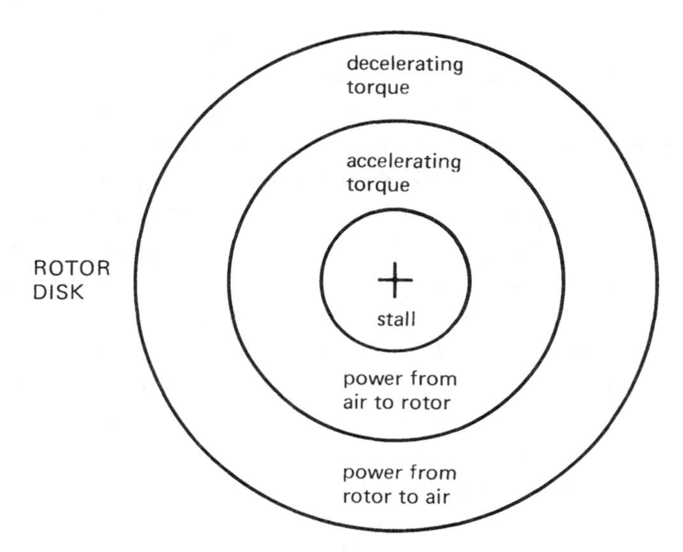

Figure 3-10 Aerodynamic environment of the rotor blade in autorotation

deliver power to the airstream. Since there is no net power to the rotor, the accelerating and decelerating torques must balance. For a given descent rate, the rotor tip speed ΩR will adjust itself until this equilibrium is achieved. Figure 3-10 illustrates the section aerodynamic environment on the rotor in autorotation. If the equilibrium rotor speed is decreased slightly, the inflow angle ϕ will increase. Then the accelerating region moves outboard, increasing in size, and there is a net accelerating torque on the rotor that acts to increase the rotor speed back to the equilibrium value. Thus the rotor speed in autorotation is stable. The angle of attack $\alpha = \theta + \phi$ increases inboard because of the inflow angle increase. At the blade root, then, the sections will be stalled. The negative twist that rotors generally have to improve hover and forward flight performance further increases the angle of attack of the inboard sections. Although negative twist is undesirable for autorotation, most of the work of the rotor is done at the blade tips, where the velocity is high, so the stall at the root does not usually have a particularly adverse effect on autorotation performance.

In hover the inflow is downward through the disk, while in autorotation $(V + v)$ is upward. Hence between hover and autorotation, there is a net increase in angle of attack due to the inflow change if the collective pitch is not changed after the loss of power in hover. The excess decelerating torque attributable to this angle-of-attack change will decrease the rotor speed. In addition, the stall region increases in extent, limiting the blade lift, which is required for the accelerating torque, and increasing the drag, which produces the decelerating torque. With a stalled rotor autorotation may therefore not be possible. To avoid excessive blade stall and rotor speed decrease it is necessary to reduce the blade pitch as soon as possible after power failure. The best collective pitch for autorotation is usually found to be a slightly positive angle; the rotor speed can then be held near the normal value. The rate of descent does not actually vary much with collective or rotor speed as long as large regions of stall are avoided, because the profile power does not change much and the inflow curve is steep in the turbulent wake state.

For section equilibrium, recall that $D - \phi L = 0$, or

$$\frac{D}{L} = \frac{c_d}{c_\ell} = \phi$$

Consider a plot of the blade airfoil characteristics, drawn to show c_d/c_ℓ as a function of α (Fig. 3-11). Section equilibrium requires that $c_d/c_\ell = \phi = \alpha - \theta$, which for a given θ is a line on the c_d/c_ℓ vs. α plane. The intersection

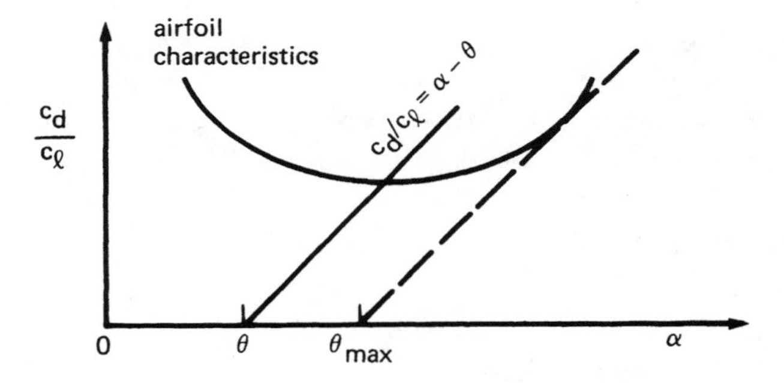

Figure 3-11 Autorotation diagram.

of this line with the curve of the airfoil characteristics determines the angle of attack for which equilibrium is achieved at this section. A plot like Fig. 3-11 is called an autorotation diagram. While only one blade section will be in true equilibrium, the inboard sections working at higher angle of attack and the outboard sections at lower, the autorotation diagram does give a good indication of the characteristics of the entire rotor. Minimum descent rate means minimum ϕ, which therefore requires that the blade operate at the angle of attack with lowest c_d/c_ℓ, resulting in minimum profile power. The collective pitch for this optimum operation is easily determined from the autorotation diagram. At both higher and lower collective, the blade drag-to-lift ratio is higher and therefore the decent rate is higher. At low angles of attack, c_d/c_ℓ increases because c_ℓ is low, and at high angles it increases because of stall. However, for many airfoils the drag-to-lift ratio tends to be fairly flat around the minimum, so the descent rate is not too sensitive to θ near the optimum value. It also follows that while it is not possible for the entire blade to be working at the optimum angle of attack for autorotation, most of the blade will still be at a low value of c_d/c_ℓ. The rotor tip speed is more sensitive than the descent rate to collective pitch changes. The relation $c_d/c_\ell = \phi = |V + v|/\Omega r$ indicates that the maximum rotor speed is obtained at the minimum c_d/c_ℓ, and that the rotor slows down at higher or lower collective pitch values. The autorotation diagram further shows that there is a maximum collective pitch value θ_{max}, above which equilibrium is not possible (see Fig. 3-11). When the angle of attack is high because of the high collective, the rotor stalls and not enough lift becomes available to balance the decelerating torque created by the high drag. The importance of reducing the collective pitch soon after power loss derives from the necessity of avoiding this collective limit, where the rotor speed decreases and the descent rate increases with no possibility of achieving equilibrium.

Blade element theory gives for autorotative descent

$$C_P = \lambda C_T + \frac{\sigma c_{do}}{8} = 0$$

and

$$C_T = \frac{\sigma a}{2}\left(\frac{\theta_{.75}}{3} - \frac{\lambda}{2}\right)$$

Solving for the inflow ratio then gives

$$\lambda = \frac{\theta_{.75}}{3} - \sqrt{\left(\frac{\theta_{.75}}{3}\right)^2 + \frac{c_d}{2a}}$$

For a given collective pitch then, λ and C_T may be calculated. The disk loading gives the rotor speed from C_T, and the inflow curve gives the descent rate from λ. Thus the autorotation descent rate as a function of collective may be plotted, and the optimum collective pitch value determined. A more detailed numerical analysis is desirable, however, because of the importance of blade stall in the autorotation behavior of the rotor. Blade element theory can at least be used to estimate the collective pitch reduction required between hover and autorotation. From $2C_T/\sigma a = (\theta_{.75}/3 - \lambda/2)$, and assuming the tip speed ΩR is not changed, it follows that

$$\Delta\theta_0 = \frac{3}{2}\left(\lambda - \lambda_h\right) = \frac{3}{2}\left(\frac{V+v}{v_h} - 1\right)\lambda_h$$

$$= -\frac{3}{2}\left(\frac{C_{P_o}}{C_T^{3/2}/\sqrt{2}} + 1\right)\lambda_h$$

$$= -\frac{3}{2}\left(\frac{1}{M} + 1 - \kappa\right)\sqrt{C_T/2}$$

Literature on vertical autorotation: Toussaint (1920), Wimperis (1926), Bennett (1932), Gessow (1948d), Nikolsky and Seckel (1949a, 1949b), Slaymaker, Lynn and Gray (1952), Slaymaker and Gray (1953), Katzenberger and Rich (1956). See also section 7–5.

3–3 Climb in Vertical Flight

The momentum theory result for the power required in vertical climb is

$$V+v = \frac{V}{2} + \sqrt{\left(\frac{V}{2}\right)^2 + v_h^2} \cong \frac{V}{2} + v_h$$

where the last approximation is valid for small climb rates (roughly $V/v_h < 1$; see Fig 3-8). Then the induced velocity $v \cong v_h - V/2$ is reduced by the climb velocity because of the increased mass flow through the rotor disk. The power required in climb is $P_c = T(V + v) + P_o$. Assuming that the profile power is unchanged by the climb velocity, the power increment between climb and hover is

$$\Delta P = P_c - P_h = T(V + v - v_h)$$

Using the small climb rate result for $V + v$, the excess power for climb is given by

$$\frac{\Delta P}{T} \cong \frac{V}{2}$$

and the climb rate for a given power increase is

$$V \cong 2\frac{\Delta P}{T}$$

Flight data are in good agreement with this expression, since the approximation involved is good to about $V = v_h$. That would be a very high climb rate for helicopters, which do not usually have much excess power available in vertical flight. Note that the power required just to increase the potential energy of the helicopter is $\Delta P/T = V$. Hence the reduction in the induced power required doubles the climb rate possible with a given power increase.

For an exact formulation, the excess power $\Delta P = T(V + v - v_h)$ gives

$$V = \frac{\Delta P}{T} + v_h - v$$

Now the momentum theory result for climb, $(V + v)v = v_h^2$, can be written

$$v = \frac{v_h^2}{V + v} = \frac{v_h^2}{\Delta P/T + v_h}$$

Eliminating v then gives

$$V = \frac{\Delta P}{T}\frac{2v_h + \Delta P/T}{v_h + \Delta P/T}$$

from which the climb rate may be obtained if the excess power and rotor thrust are specified. For small V this reduces to $V = 2\Delta P/T$ again.

Blade element theory can be used to estimate the collective pitch increase required in climb. From $2C_T/\sigma a = (\theta_{.75}/3 - \lambda/2)$ it follows that

$$\Delta\theta = \frac{3}{2}\left(\lambda - \lambda_h\right) \cong \frac{3}{4}\lambda_c$$

for small climb rates where $\lambda \cong \lambda_h + \frac{1}{2}\lambda_c$ ($\lambda_c = V/\Omega R$). Alternatively, without the small climb assumption,

$$\Delta\theta = \frac{3}{2}\frac{V+v-v_h}{\Omega R} = \frac{3}{2}\frac{\Delta P/T}{\Omega R} = \frac{3}{2}\frac{\Delta C_P}{C_T}$$

3–4 Vertical Drag

The rotor downwash acting on the fuselage produces a vertical drag force on the helicopter in hover and vertical flight. This drag force requires an increase in the rotor thrust for a given gross weight and hence degrades the helicopter performance. To estimate the vertical drag force, consider the downwash velocity in the fully developed rotor wake. In hover, $w_h = 2v_h$, and in vertical flight

$$V + w = \sqrt{V^2 + 4v_h^2} \cong 2v_h$$

So $V + w \cong w_h$ independent of the climb velocity, when $V^2/v_h^2 \ll 1$. The vertical drag characteristics of the fuselage may be described by either an equivalent drag area f or by a drag coefficient C_D based on some relevant area S (so $f = SC_D$). Then the vertical drag produces a rotor thrust increase

$$\Delta T = \tfrac{1}{2}\rho w_h^2 f = \frac{T}{A} f$$

or

$$\frac{\Delta T}{T} = \frac{f}{A} = \frac{S}{A} C_D$$

The fuselage is very near the rotor, and hence may not really be in the far wake; moveover, the downwash field will be highly nonuniform and unsteady. Such effects may be included in an empirical factor. Thus assume that the downwash velocity at the fuselage is nv_h, where the parameter n theoretically varies from 1 at the disk to 2 in the far wake. Then

$$\frac{\Delta T}{T} = \frac{n^2}{4}\frac{f}{A} = \frac{S}{A}\left(\frac{n^2 C_D}{4}\right)$$

The parameter $(n^2 C_D/4)$ may then be obtained from measurements of the force on bodies in the rotor wake. Typically $(n^2 C_D/4) \cong 0.7$, but the value depends highly on the position of the body in the wake, its size relative to the rotor, and its shape. Similarly, for climb

$$\frac{\Delta T}{T} = \frac{S}{A} C_D \left(\frac{V + nv}{2 v_h} \right)^2$$

It is probably consistent with the overall accuracy of such estimates to simply use the hover value for the vertical drag force in climb as well.

Glauert (1935) suggests using the following expression for the vertical drag:

$$\frac{\Delta T}{T} = \frac{S}{A} C_D \left(1.22 + 0.254/C_D \right)$$

The last factor is the effect of the pressure gradient in the wake on the forces acting on the body. Makofski and Menkick (1956) suggest

$$\frac{\Delta T}{T} = 0.66 \frac{S}{A} \frac{b}{2R}$$

bsaed on measurements with rectangular panels $0.2\,R$ to $0.64\,R$ below the disk. Here b is the panel span, so the factor $b/2R$ accounts for the radial variation of the downwash. Another approach is to estimate n and C_D separately for the components of the fuselage in the wake. From vortex theory, at a distance z below the disk

$$n = 1 + \frac{z/R}{\sqrt{1 + (z/R)^2}}$$

and the appropriate drag coefficient can be found from the standard literature.

These approaches are rather crude, but fairly large errors can be tolerated since $\Delta T/T$ is small. A good analysis of the problem is difficult, since an accurate model for the helicopter wake is required, including the interference between the body and wake; and there is not a great deal of experimental data. It has been determined that there is a significant radial variation of the downwash in the wake, which must be accounted for. There are also large periodic variations in the drag, a possible source of helicopter vibration. In fact, the drag is largest when closest to the rotor disk, diminishing rapidly as the body moves from the disk plane. This behavior is due to the periodic variation of the wake downwash. While the mean downwash does increase from the rotor disk to the far wake about as expected from the vortex theory results, the mean dynamic pressure is significantly increased near the disk because of the periodic components in the flow. If the object in the wake is large enough, wake blockage must also be considered. A reduction of the effective disk area, particularly near the tips, will decrease the rotor efficiency. Forward speed of the helicopter sweeps the wake rearward, so there is little vertical drag above transition speeds.

Literature on vertical drag: Castles (1945), Fail and Eyre (1949a), Makofski and Menkick (2956), McKee and Naeseth (1958), McKee (1959), Bramwell (1966), Cassarino (1970b).

3–5 Twin Rotor Interference in Hover

The operation of two or more rotors in close proximity will modify the flow field at each, and hence the performance of the rotor system will not be the same as for the isolated rotors. Examples of such configurations are the coaxial helicopter, the tandem rotor helicopter (typically with 30% to 50% overlap), and the side-by-side configuration. We shall compare the performance of two rotors of the same diameter with that of the isolated rotors operating at the same thrust. A limiting case is the coaxial rotor system, which has just one-half the disk area of the isolated rotors and hence twice the disk loading. It follows that by operating the rotors coaxially the induced power required is increased by a factor of $\sqrt{2}$, a 41% induced power increase. This result is based on the actuator disk model, which is applicable as long as the vertical separation is less than about 10% of the rotor radius. Consider the case of two rotors operating close together but with no overlap. According to vortex theory, in the disk plane but outside the rotor disk circle itself there is no normal induced velocity component and hence no interference power loss. With some vertical separation there may be an interference, favorable or unfavorable, even with no overlap of the rotors. The experimental data on this matter are conflicting. Dingeldein (1954) found about a 15% reduction in the induced power for a separation of $2.06R$, while Sweet (1960b) found no significant difference from isolated rotor performance.

For rotors with some overlap and small vertical separation (less than around $0.1R$), there is a common rate of flow through the overlapped portions of the disk. For the same total thrust, then, the overlap area has a higher disk loading than the isolated rotors, which increases the local induced power loss. As the separation decreases, the increase in power approaches the 41% of coaxial rotors. As the vertical separation of coaxial rotors is increased, the wake of the top rotor contracts and thus effects less area of the lower rotor, reducing the interference power loss.

Consider coaxial rotors with large vertical separation, so that the bottom rotor is operating in the far wake of the top rotor. The top rotor is not influenced by the bottom rotor; its induced velocity is therefore $v_U^2 = T/2\rho A = v_h^2$. The wake from the top rotor has velocity $2v_U$ and area $A/2$ at the position of the bottom rotor. Thus there is a velocity v_L over one-half the area of the lower rotor, and velocity $v_L + 2v_U$ over the other half. Momentum and energy conservation then give $T_L = \rho A(v_U + v_L)w_L - 2\rho A v_U^2$ and $P_L = T_L(v_U + v_L) = \rho A(v_U + v_L)\frac{1}{2}w_L^2 - 2\rho A v_U^3$, assuming a uniform velocity w_L in the far wake of the lower rotor. Eliminating T_L gives $w_L = 2v_U + v_L$, and then $v_L = v_h(\sqrt{17} - 3)/2 = 0.56v_h$. The power of the two rotors is $(P/T)_{\text{upper}} = v_U$ and $(P/T)_{\text{lower}} = v_U + v_L$; hence for both rotors $P/T = 2.56v_h$, compared to $P/T = 2v_h$ for the isolated rotors. There is a 28% increase in the induced power due to interference, which increases to 41% as the vertical separation is reduced.

For a momentum theory analysis of overlapped rotors, consider two rotors of the same radius but perhaps with different thrusts. Let mA be the overlap area; T_1 and T_2 the thrusts on the two rotors, with $T_1 + T_2 = T$ fixed; P_1 and P_2 the induced power losses outside the overlap area and P_m the induced power loss of the overlap area; and v_1, v_2, and v_m the corresponding induced velocities. With uniform loading, $T_1(1 - m)$ and $T_2(1 - m)$ are the thrusts of the areas outside the overlap, and $m(T_1 + T_2)$ the thrust of the overlap area. Negligible vertical separation is assumed, so that in the overlap area both rotors have the common induced velocity v_m. Based on the differential momentum theory results $dT = 2\rho v^2 dA$ and $dP = v\,dT$, it follows that

$$v_1 = \sqrt{T_1/2\rho A}, \quad v_2 = \sqrt{T_2/2\rho A}, \quad v_m = \sqrt{(T_1 + T_2)/2\rho A}$$

$$P_1 = T_1(1 - m)v_1, \quad P_2 = T_2(1 - m)v_2, \quad P_m = m(T_1 + T_2)v_m$$

Then the total power is $P = P_1 + P_2 + P_m$. For the isolated rotors the total power is

$$P|_{m=0} = (P_1 + P_2)|_{m=0} = (T_1^{3/2} + T_2^{3/2})/\sqrt{2\rho A}$$

The interference power is therefore

$$\begin{aligned} P &= (P_1 + P_2 + P_m) - (P_1 + P_2)|_{m=0} \\ &= m[(T_1 + T_2)^{3/2} - (T_1^{3/2} + T_2^{3/2})]/\sqrt{2\rho A} \end{aligned}$$

or, as a fraction of the power for the isolated rotors,

$$\frac{\Delta P}{P} = m\left[\frac{1}{\tau_1^{3/2} + \tau_2^{3/2}} - 1\right]$$

where $\tau_1 = T_1/T$ and $\tau_2 = T_2/T$ (so that $\tau_1 + \tau_2 = 1$) give the distribution of thrust between the two rotors. When the thrust on the two rotors is the same, $\tau_1 = \tau_2 = \frac{1}{2}$, the interference loss becomes $\Delta P/P = 0.41m$, which indeed gives 41% for coaxial rotors (100% overlap, hence $m = 1$). In general, the interference power is directly proportional to the fraction of area overlapped.

Alternatively, for hovering twin rotors with overlap area mA, the performance estimate can be based on the effective disk loading of the system as a whole: $P = T\sqrt{T/2\rho A_{sys}}$, where the total rotor area is $A_{sys} = A(2 - m)$. The ratio of the total power to that of the isolated rotors then is

$$\frac{P}{P_{isolated}} = \left(\frac{T}{T_{isolated}}\right)^{3/2}\left(\frac{2}{2-m}\right)^{1/2}$$

and for the same thrust the interference power is

$$\frac{\Delta P}{P} = \left(\frac{2}{2-m}\right)^{1/2} - 1$$

In the coaxial limit this gives $\Delta P/P = 0.41$ as required. For small overlap, however, $\Delta P/P \cong 0.25m$, and initially the power does not increase with overlap as quickly as in the previous result. The difference is that the present model has a lower disk loading in the overlap region than the previous model, and hence a greater efficiency for small overlap. The larger estimate of the interference loss is probably more representative of tandem rotor behavior. Finally, note that with a shaft separation l the overlap fraction is

$$m = \frac{2}{\pi}\left[\cos^{-1}\frac{l}{2R} - \frac{l}{2R}\sqrt{1-\left(\frac{l}{2R}\right)^2}\right]$$

For small overlap ($l = 2R - \Delta l$, with $\Delta l/R \ll 1$), $m \cong 1.20(\Delta l/2R)^{3/2}$.

Stepniewski (1950, 1952, 1955) developed a combined blade element and momentum theory for twin rotors in hover. The theory assumes that the vertical separation of the rotors is small, so that the overlapped area has a common rate of flow through both rotors. Outside the overlap region, the induced velocities v_1 and v_2 are given by the usual combined blade element and momentum theory expression (see section 2-5). Inside the overlap region, consider an area dA located at radial stations r_1 and r_2 on the two rotors; let $\lambda_m = v_m/\Omega R$ be the inflow ratio in the overlap region. Momentum theory gives $dT = 2\rho v_m^2 dA$ or $dC_T = (2/\pi)\lambda_m^2 dA$. Blade element theory gives $dC_{T_1} = (\sigma_1 a/4\pi)(\theta_1 r_1 - \lambda_m)dA$ and $dC_{T_2} = (\sigma_2 a/4\pi)(\theta_2 r_2 - \lambda_m)dA$, where θ_1 and θ_2 are here the pitch of the two blades at r_1 and r_2, respectively. Equating dC_T and $dC_{T_1} + dC_{T_2}$ gives a quadratic equation for λ_m with the solution

$$\lambda_m = -\left(\frac{\sigma_1 a}{16} + \frac{\sigma_2 a}{16}\right) + \sqrt{\left(\frac{\sigma_1 a}{16} + \frac{\sigma_2 a}{16}\right)^2 + \frac{\sigma_1 a}{8}\theta_1 r_1 + \frac{\sigma_2 a}{8}\theta_2 r_2}$$

Using v_1, v_2, and v_m, the thrust and power can be evaluated from

$$T = \int 2\rho v_1^2 dA + \int 2\rho v_2^2 dA + \int 2\rho v_m^2 dA$$

$$P = \int 2\rho v_1^3 dA + \int 2\rho v_2^3 dA + \int 2\rho v_m^3 dA$$

where the three integrals cover respectively the first and second rotors outside the overlap area, and the overlap area. Alternatively, the blade element theory expressions for the two rotors can be used if integration is performed azimuthally as well as radially along the blade. Stepniewski obtained good results from this analysis, based on a comparison of the downwash and power prediction with test data. He found no aerodynamic interference of practical importance in hover with no overlap; and for overlap in the range $\Delta l/2R = 0$ to $0.4R$, the thrust and power were $T/T_{\text{isolated}} \cong 1.0$ to 0.94 and $P/P_{\text{isolated}} \cong 1.1$ to 1.2. Here P is just the induced power, and P_{isolated} is the induced power for the isolated rotors with uniform inflow; hence the interference power also includes the non-uniform inflow losses of the isolated rotors.

The literature on twin rotor performance in hover includes: Fail and Squire (1947), Harrington (1951), Dingeldein (1954), Sweet (1960b), Baskin, Vil'dgrube, Vozhdayev, and Maykaper (1946.

3–6 Ground Effect

The proximity of the ground to the hovering rotor disk constrains the rotor wake and reduces the induced velocity at the rotor, which means a reduction in the power required for a given thrust; this behavior is called ground effect. Equivalently, ground proximity increases the rotor thrust at a given power. Because of this phenomenon, a helicopter can hover in ground effect (IGE) at a higher gross weight or altitude than is possible out of ground effect (OGE). The thrust increase near the ground also helps flare the helicopter when landing. Ground effect must also be considered in testing helicopter rotors in hover, since the rotor must either be far enough above the ground for its influence to be neglected or the data must be corrected for the influence of the ground. Ground effect has been examined analytically using the method of images, where a mirror-image rotor is placed below the ground plane so that the boundary condition of no flow through the ground is automatically satisfied. Most of the useful information about the phenomenon has come from rotor performance measurements, however.

The influence of the ground can be viewed as a reduction of the rotor induced velocity by a factor κ_G, so that at constant thrust the ratio of the induced power required to that out of ground effect is $C_P/C_{P_\infty} = \kappa_G$. Alternatively, ground effect can be expressed in terms of the increase in thrust at constant power, C_T/C_{T_∞}, as sketched in Fig. 3-12. Constant power implies that $\lambda C_T = \lambda_\infty C_{T_\infty}$ or $T/T_\infty = v_\infty/v = 1/\kappa_G$. Therefore, the thrust results can also be interpreted as a change in induced velocity. The basic parameter is the height above the ground z, expressed as a fraction of the rotor radius or diameter. Ground effect is generally negligible when the rotor is more than one diameter above the ground, $z/R > 2$. It is also found that there is a secondary dependence of ground effect on the rotor blade loading, C_T/σ. Ground effect decreases rapidly with forward speed, since the wake is swept backward rather than being directed at the ground. It follows that

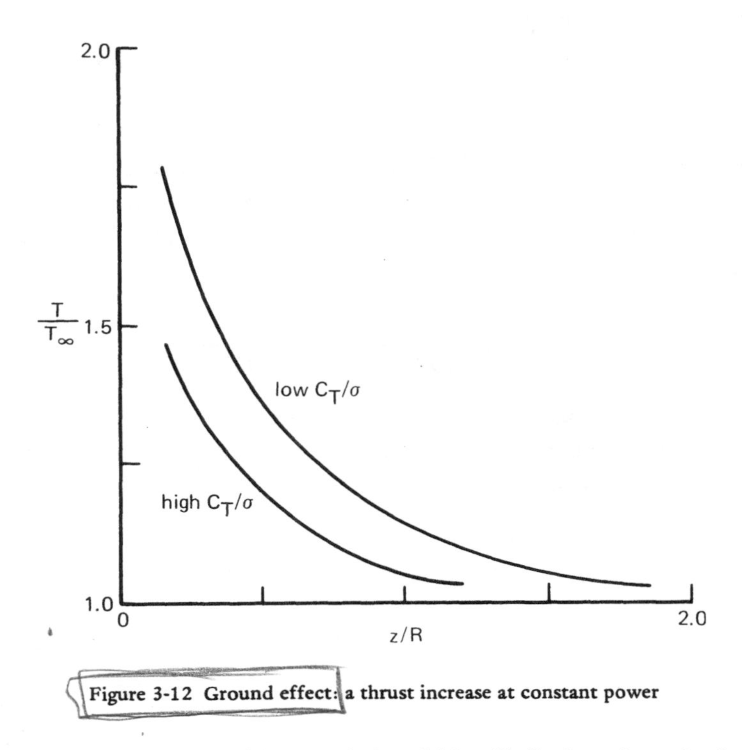

Figure 3-12 Ground effect: a thrust increase at constant power

ground effect is also sensitive to winds, which will displace the wake from under the rotor.

Zbrozek (1947) employed model and flight test data to express the influence of the ground in terms of the thrust increase T/T_∞ at constant power as a function of rotor height and C_T/σ. Betz (1937) analyzed the performance of a rotor in ground effect. For small distances above the ground ($z/R \ll 1$), he obtained the power at constant thrust: $P/P_\infty = 2z/R$. Knight and Hefner (1941) conducted an experimental and theoretical investigation of ground effect. They modified vortex theory to account for the ground by including image vortices below the ground plane. For a uniformly loaded actuator disk, then, the wake consisted of a cylindrical vortex sheet from the rotor to the ground and the corresponding image vortex cylinder below the ground. They obtained good correlation with measurements of the effect of the ground on the rotor performance. Cheeseman and Bennett (1955) made a simple analysis based on the method of images, representing the rotor by a source. They obtained for hover

$$\frac{T}{T_\infty} = \frac{1}{1 - (R/4z)^2}$$

which correlates reasonably well with experimental data. Hayden (1976) correlated flight test data to obtain the influence of the ground on hovering performance. He expressed the correlation in the form $C_P = C_{P_o} + \kappa_G (C_{P_i})_\infty$ where

$$\kappa_G = \frac{1}{0.9926 + 0.03794(2R/z)^2}$$

Other work considering ground effect in hover: Gustafson and Gessow (1945), Fradenburgh (1960), Mil' (1966), Koo and Oka (1971), Newman (1971), Law (1972).

Chapter 4

FORWARD FLIGHT I

The analysis of the helicopter rotor in forward flight begins with this chapter. During translational motion of the helicopter, when the rotor is nearly horizontal, the rotor blades see a component of the vorward velocity as well as the velocity due to their own rotation (Fig. 4-1). In forward flight the rotor

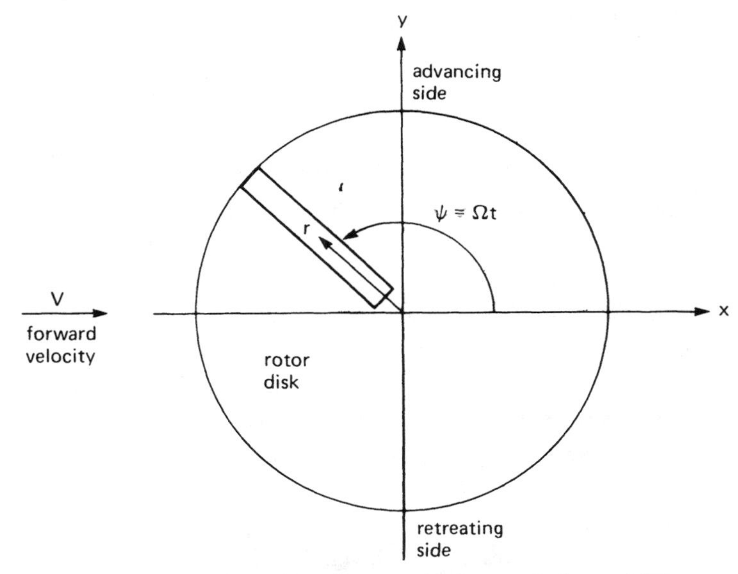

Figure 4-1 Aerodynamic environment of the rotor in forward flight.

does not have axisymmetry as in hover and vertical flight; rather, the aerodynamic environment varies periodically as the blade rotates with respect to the direction of flight. The advancing blade has a velocity relative to the air higher than the rotational velocity, while the retreating blade has a lower velocity relative to the air. This lateral asymmetry has a major influence on the rotor and its analysis in forward flight. It follows directly that the rotor blade loading and motion are periodic with a fundamental frequency equal to the rotor speed Ω. The analysis is much more complicated than for hover because of the dependence of the loads and motion on the azimuth angle ψ.

As a consequence of the axisymmetry, the analysis of the hovering rotor primarily involves a consideration of the aerodynamics. In forward flight, however, the lateral asymmetry in the basic aerodynamic environment produces a periodic motion of the blade, which in turn influences the aerodynamic forces. The analysis in forward flight must therefore consider the blade dynamics as well as the aerodynamics. The subject of rotor blade motion and its behavior in forward flight is taken up in Chapter 5. The present chapter considers a number of aerodynamic topics that are already familiar from the analysis of the rotor in vertical flight. In particular, we are concerned with the momentum theory treatment of the induced velocity and power in forward flight.

4–1 Momentum Theory in Forward Flight

4-1.1 Rotor Induced Power

The rotor induced power in forward flight may be obtained by a momentum theory analysis. As in hover, the power loss is represented by an induced velocity $v = P_i/T$. When used in blade element theory, it is assumed that the induced velocity is uniform over the disk; while that is not as good as an assumption in forward flight as in hover, at high forward speed the induced velocity is small compared to the other velocity components at the rotor blade. At low forward speeds the variation of the inflow over the disk is important, particularly for vibration and blade loads. A uniformly loaded actuator disk is again used to represent the rotor. In forward flight such an actuator disk may be viewed as a circular wing.

Fixed wing theory gives the minimum induced drag for a thin, planar wing of span b, operating at velocity V and lift T:

$$D_i = T^2/2\rho A V^2$$

where $A = \pi(b/2)^2$ is the area of a circle with diameter b (a more familiar form perhaps is $C_{D_i} = C_L{}^2/\pi AR$). In terms of the induced velocity, then,

$$v = \frac{P_i}{T} = \frac{VD_i}{T} = \frac{T}{2\rho A V}$$

This minimum drag is achieved with elliptical loading of the wing. The uniformly loaded rotor has a circular span loading, which is a special case of elliptical loading; at high forward speeds the rotor wake vorticity is swept back in the plane of the disk, like the fixed wing wake. Moreover, the induced drag solution is based on a Trefftz plane analysis in the far wake, so it is valid for wings of arbitrary aspect ratio. Therefore, $v = T/2\rho AV$ is an appropriate solution for the induced velocity of the helicopter rotor in high-speed forward flight. For the rotor, the wing span is the rotor diameter, so A is simply the rotor disk area. Lifting-line theory interprets v as the actual induced velocity at the wing, uniform over the span for high aspect ratios. For the circular wing, which has the aspect ratio $AR = 4/\pi = 1.27$, considerable variation of the induced velocity over the disk may be expected, however.

Expressions for the rotor induced power are now available for the rotor in vertical flight and in high-speed forward flight. A connection between the two regions is required if the inflow for all operating conditions of the rotor is to be specified. Note that the forward flight result may be written as $T = \dot{m}2v$, where $\dot{m} = \rho AV$ is the mass flux through an area equal to the rotor disk area. This is exactly the form of the momentum theory results for vertical flight; in hover and climb, for example, we found that $T = \dot{m}2v$ and $\dot{m} = \rho A(V + v)$. Thus a uniformly valid expression for induced velocity may be obtained by considering the mass flux through the area A for all operating conditions. This observation was first made by Glauert (1926b).

Consider a rotor operating at velocity V, with angle of attack α between the free stream velocity and the rotor disk (Fig. 4-2). The induced velocity at the disk is v; in the far wake, $w = 2v$ and is assumed to be parallel to the rotor thrust vector. Momentum conservation gives the rotor thrust $T = \dot{m}2v$, where the mass flux is $\dot{m} = \rho AU$. Following Glauert (1926b), the resultant velocity U is given by

$$U^2 = (V\cos\alpha)^2 + (V\sin\alpha + v)^2 = V^2 + 2Vv\sin\alpha + v^2$$

Hence $T = 2\rho Av\sqrt{V^2 + 2Vv\sin\alpha + v^2}$. Energy conservation gives the rotor power $P = \dot{m}\left\{\tfrac{1}{2}[(V\sin\alpha + 2v)^2 + (V\cos\alpha)^2] - \tfrac{1}{2}V^2\right\} = T(V\sin\alpha + v)$. For high forward speeds ($V \gg v$) we have $T = \rho AV2v$, and in hover ($V = 0$) $T = 2\rho Av^2$, so this expression does have the proper limits. While there is no strict theoretical justification for this approach at intermediate forward speeds, good agreement has been found with measured rotor performance and with

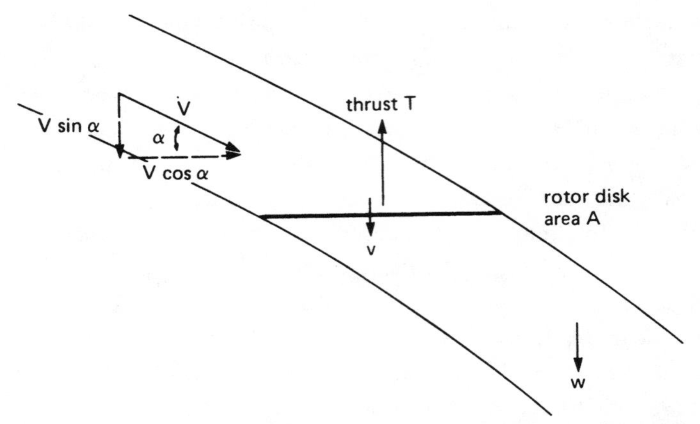

Figure 4-2 Flow model for momentum theory analysis of rotor in forward flight.

vortex theory results; thus the result may be accepted over the entire range of rotor speeds. In the expression for the rotor power, $P = T(V \sin\alpha + v)$, the term Tv is the induced loss and the term $TV \sin\alpha$ is the power required to climb and to propel the helicopter forward (the parasite power loss). As for vertical flight, we may write $(V \sin\alpha + v)/v_h = P/Tv_h = P/P_h$.

The solution for the induced velocity is

$$v = \frac{v_h^2}{\sqrt{(V\cos\alpha)^2 + (V\sin\alpha + v)^2}}$$

where $v_h^2 = T/2\rho A$ as usual. Define now the dimensionless components of the velocity parallel to and normal to the rotor disk, the advance ratio μ and the inflow ratio λ, respectively:

$$\mu = \frac{V\cos\alpha}{\Omega R}$$

$$\lambda = \frac{V\sin\alpha + v}{\Omega R} = \mu\tan\alpha + \lambda_i$$

Then in coefficient form the induced inflow ratio λ_i is

$$\lambda_i = \frac{C_T}{2\sqrt{\mu^2 + \lambda^2}}$$

In general, then, it is necessary to solve a quartic equation for ν or λ_i. An iterative procedure for calculating λ may be obtained by considering the Newton-Raphson solution of $f(\lambda) = \lambda - \mu \tan\alpha - C_T/2\sqrt{\mu^2 + \lambda^2} = 0$, namely $\lambda_{n+1} = \lambda_n - (f/f')_n$, or

$$\lambda_{n+1} = \left(\frac{\mu \tan\alpha + \dfrac{C_T}{2} \dfrac{(\mu^2 + 2\lambda^2)}{(\mu^2 + \lambda^2)^{3/2}}}{1 + \dfrac{C_T}{2} \dfrac{\lambda}{(\mu^2 + \lambda^2)^{3/2}}} \right)_n$$

Three or four iternaions are usually sufficient, starting from the value $\lambda = \mu \tan\alpha + C_T/(2\sqrt{\mu^2 + C_T/2})$.

For high forward speeds, $\mu \gg \lambda$, the momentum theory solution becomes $\lambda_i \cong C_T/2\mu$ (or $\nu = T/2\rho A V \cos\alpha$, which is just the circular wing result). The usefulness of this approximation lies in the fact it is not necessary to iterate to obtain λ_i. Figure 4-3 shows the induced velocity in forward flight for the case $\alpha = 0$ (for which an exact analytical solution is possible). The forward speed reduces the induced power loss as a result of the increased mass flux. Figure 4-3 also shows the approximation $\lambda_i \cong C_T/2\mu$, which is quite good when $\mu/\lambda_h > 1.5$ or so. The singularity in $\lambda_i \cong C_T/2\mu$ at low advance ratio

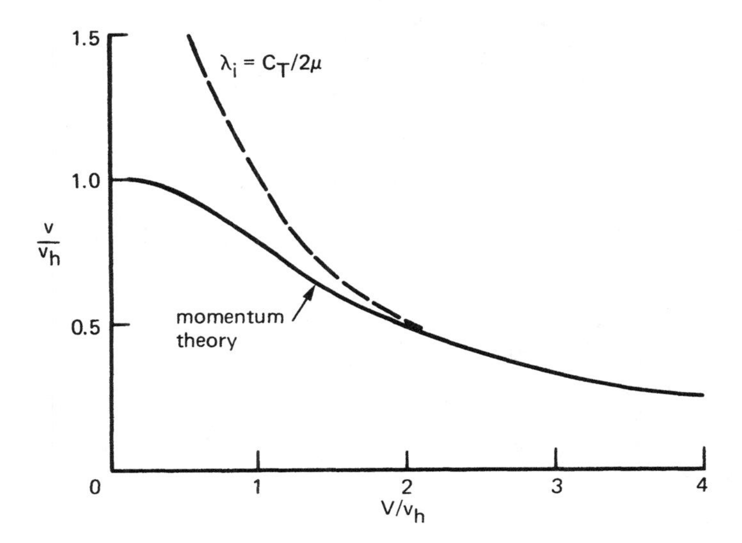

Figure 4-3 Rotor induced velocity in forward flight with $\alpha = 0$.

can be removed by using $\lambda_i = C_T/(2\sqrt{\mu^2 + C_T/2})$ instead, which gives an induced velocity comewhat lower than the correct solution; it would be better to use the iterative procedure, however, to obtain the exact value.

In Fig 3-8 the momentum theory solution was plotted in the form of $P/P_h = (V + v)/v_h$ as a function of the vertical velocity V/v_h. To generalize this presentation, consider $P/P_h = (V \sin\alpha + v)/v_h$ as a function of the normal component of the velocity, $V \sin\alpha/v_h$, with the in-plane velocity component $V \cos\alpha/v_h$ as a parameter (or λ/λ_h vs. $\mu \tan\alpha/\lambda_h$ for a given μ/λ_h; since the rotor disk will not be exactly horizontal, $V \sin\alpha$ and $V \cos\alpha$ are not quite the vertical and horizontal velocity components). The result is shown in Fig 4-4, which is constructed by writing the induced velocity expression as

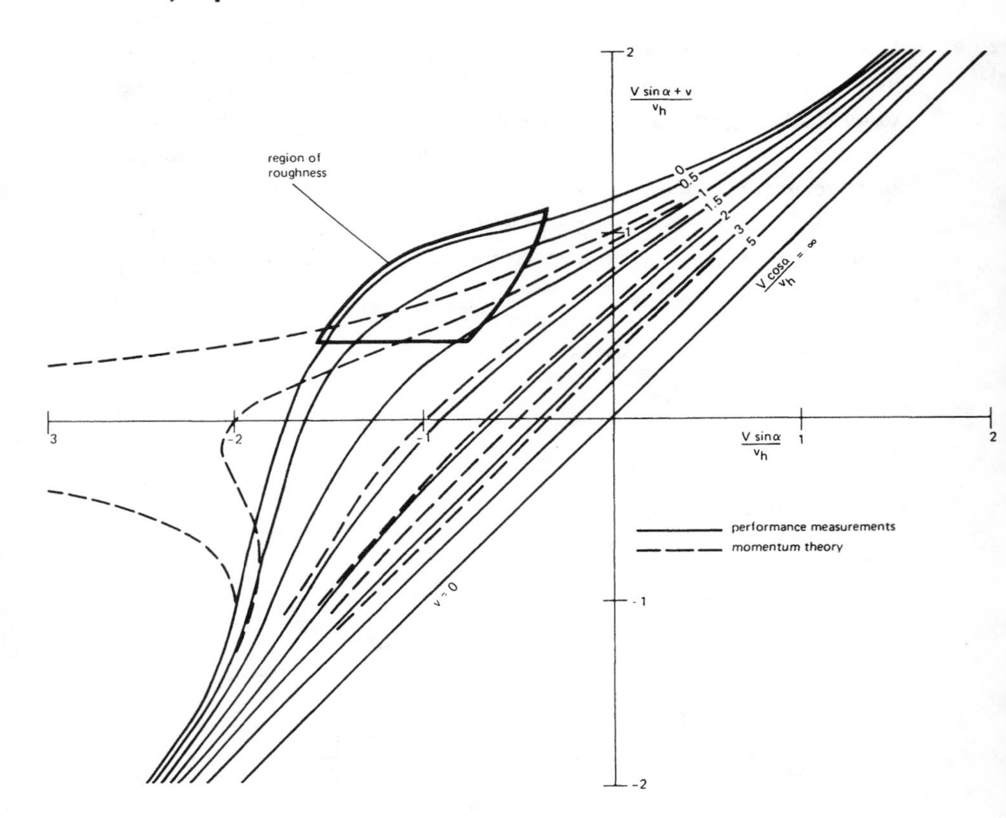

Figure 4-4 Rotor power in forward flight.

$$V \sin\alpha = (V \sin\alpha + v) - \frac{v_h^2}{\sqrt{(V\cos\alpha)^2 + (V\sin\alpha + v)^2}}$$

The effect of forward flight (the addition of the in-plane component $V \cos\alpha$) is always to reduce the induced power. The results in Fig. 4-4 have been corrected in two respects on the basis of empirical data; the corresponding momentum theory results are shown also. First, the measured performance data show that the actual induced power loss is 5% to 20% higher than the momentum theory estimate. Thus an empirical correction factor κ should be included in the induced power calculation, $P_i = \kappa T v$. Secondly, in the vortex ring state of vertical flight the measured performance is the only means of defining the induced velocity curve. It is seen, however, that above about $\mu/\lambda_h = 1$ the momentum theory result does not exhibit the vortex ring singularity. With sufficient forward velocity a moderate descent rate of the helicopter presents no problems, since the wake of the rotor is swept back instead of being allowed to build up under the disk. Consequently, in forward flight the momentum theory result is satisfactory if the correction factor κ is included. Fig. 4-4 shows in addition the general boundaries of the region of roughness in the vortex ring state, which also disappears with sufficient forward speed. Finally, note that the forward speed scale is defined by $v_h = 1.24\sqrt{T/A}$ knots when the disk loading is in N/m^2; typically, $v_h = 15$ to 25 knots.

The high-speed approximation $\lambda_i \cong C_T/2\mu$ may be written $v \cong v_h^2/V \cos\alpha$, which in Fig. 4-4 is a straight line parallel to the $v = 0$ line. It can be seen that such an approximation is quite good for $V \cos\alpha/v_h > 1.5$ or so, which corresponds to forward speeds above $V = 25$ to 35 knots for the disk loadings typical of helicopters. In terms of the rotor advance ratio, $\mu/\lambda_h > 1.5$ typically gives $\mu > 0.1$. Thus the rotor wake system is like a circular wing except at very low speeds. The speed range in which the rotor wake is no longer directly under the rotor but still has a significant vertical extent, roughly $0 < \mu < 0.1$, is called the transition region of the helicopter. The transition region has a number of special characteristics besides the requirement for the general induced velocity expression, notably a high level of blade loads and vibration due to the rotor vortex wake.

4-1.2 Climb, Descent, and Autorotation in Forward Flight

The power required in forward flight, including now the profile losses P_o, is

$$P = P_o + TV\sin\alpha + \kappa T v$$

The term $TV \sin\alpha$ combines the rotor parasite power and climb power, which require the thrust component $T \sin\alpha$ in the direction of v. To determine the rotor disk inclination angle α, consider the equilibrium of forces on the helicopter, as shown in Fig. 4-5. The forces acting on the helicopter are the rotor thrust T, the helicopter weight W, and the helicopter drag D.

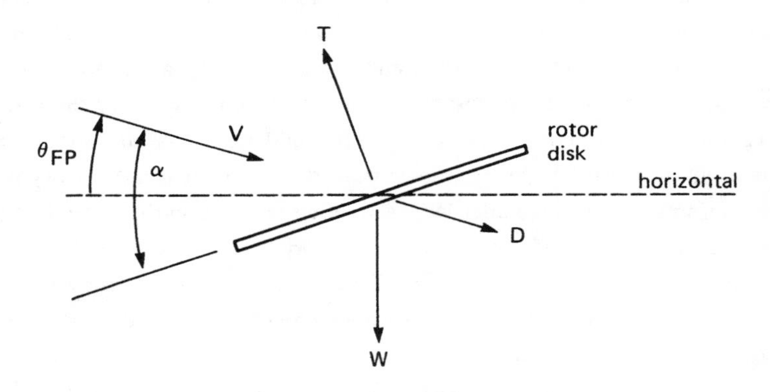

Figure 4-5 Forces acting on the helicopter in forward flight.

Here θ_{FP} is the flight path angle, so the climb speed is $V_c = V\theta_{FP}$. For small angles, vertical and horizontal force equilibrium gives $\alpha = \theta_{FP} + D/T$ and $T = W$. Hence

$$TV\sin\alpha = TV_c + DV$$

where the first term is the climb power and the second is the parasite power. (A more detailed derivation of helicopter force equilibrium and performance is given in Chapter 5.) Now for high enough forward velocity, the rotor induced velocity is $v \cong T/2\rho AV \cos\alpha \cong T/2\rho AV$. Thus the power equation may be solved for the climb velocity:

$$V_c = \frac{P - (P_o + VD + \kappa T^2/2\rho AV)}{T}$$

Because the induced power in forward flight is independent of the climb or descent rate, a simple and direct expression for V_c has been obtained. Assuming that the rotor profile power and the helicopter drag are not influenced by the climb or descent velocity, we have

$$V_c = \frac{P - P_{\text{level}}}{T} = \frac{\Delta P}{T}$$

where P_{level} is the power required for level flight at the same forward speed. The helicopter climb or descent rate is determined simply by the excess power ΔP. The climb and autorotation characteristics of the helicopter in forward flight may then be obtained from the power available and the power required for level flight. In particular, the maximum climb rate is achieved with maximum available power at the speed for minimum power in level flight, and the minimum power-off descent rate is achieved at the same forward speed. The helicopter performance characteristics are considered in more detail in Chapter 6.

4-1.3 Tip Loss Factor

As in hover, the finite number of blades leads to a rotor performance loss not accounted for by the actuator disk analysis. The lift at the blade tips decreases to zero over a finite radial distance, rather than extending all the way out to the edge of the disk. Thus there will be a reduction in the thrust, or increase in the induced power of the rotor. The reduced loading at the tip may be accounted for by using a tip loss factor B such that for $r > BR$ the blade has drag but no lift. A number of expressions for B are given in Section 2-6.1; a typical value is $B = 0.97$.

In momentum theory for forward flight, the tip loss may be viewed as giving an effective disk area $A_e = B^2 A$. Since the induced velocity is proportional to the disk loading in forward flight, it follows that the empirical factor in the induced power calculation ($P = \kappa T v$) has a value $\kappa = B^{-2}$ due to the tip losses alone, giving $\kappa \cong 1.05$ at least (compared with $\kappa = B^{-1}$ for hover, where the induced velocity is proportional to $\sqrt{T/A}$). For the general momentum theory expression, the tip loss factor may be included by using $v_h{}^2 = T/2\rho A_e = T/2\rho AB^2$.

Root cutout does not directly influence the induced power in forward flight, since wing theory shows that the induced velocity depends on the wing span squared and not on the wing area. Root cutout will influence the effective span loading of the rotor and hence increase the induced losses above the optimum for elliptical loading. The root cutout is not the dominant factor distorting the span loading in forward flight, however. The lift limitation on the retreating blade, which must work at lower velocities than elsewhere on the disk, results in a concentration of the loading on the front and rear of the rotor disk, effectively lowering the span of the lifting system.

4–2 Vortex Theory in Forward Flight

In forward flight the helical vortices trailed from the blade tips are carried rearward by the free stream velocity component parallel to the disk (μ) as well as downward by the component normal to the disk (λ). Thus the wake geometry consists of concentrated vortices from each blade trailed in skewed, interlocking helices (Fig. 4-6). The wake skew angle $\chi = \tan^{-1}\mu/\lambda$ may be estimated fairly well using momentum theory. The helicopter transition operating region $0 < \mu/\lambda_h < 1.5$ corresponds approximately to wake angles from $\chi = 0°$ to $\chi = 60°$. The relative positions of the rotor blade and the individual wake vortices vary periodically as the blade rotates, producing a strong variation in the wake-induced velocity encountered by the blade and hence in the blade loading. The induced velocity is thus in fact highly nonuniform in forward flight. The interaction between the blades and the wake is particularly strong on the advancing and retreating sides of the disk, where the tip vortex from the preceding blade sweeps radially along the blade. Under certain flight conditions where the wake is close to the rotor disk, the vortex-induced loads are very high.

The vortex wake of the rotor in forward flight rolls up in a two-stage process. The individual tip vortices quickly roll up into concentrated lines as they are trailed from the blades. Then the interlocking, overlapping spirals in the far wake interact, and roll up to form two vortices like those behind a circular wing. Such behavior has been observed experimentally; the two tip vortices from the edges of the disk are seen forming several rotor radii

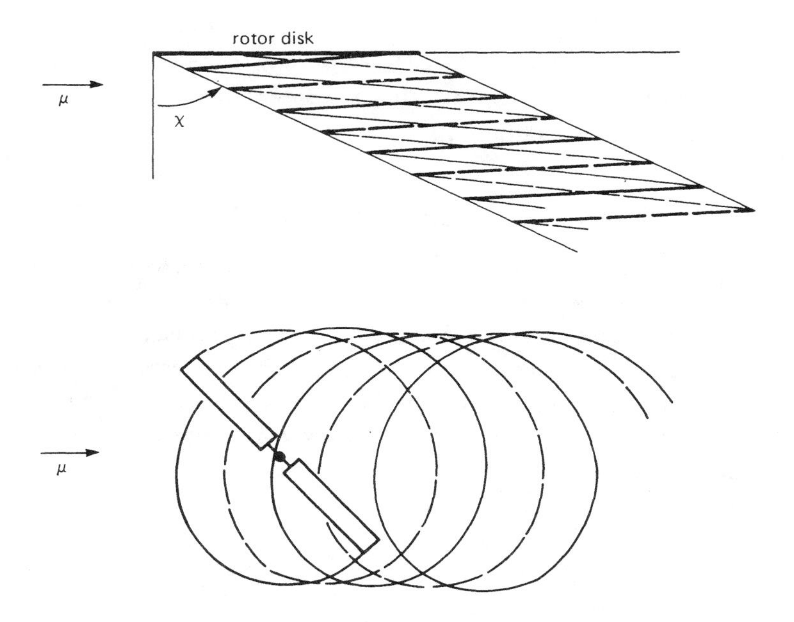

Figure 4-6 Tip vortex geometry of the rotor wake in forward flight, neglecting the self-induced distortion.

downstream from the disk. This behavior is of little consequence as far as the downwash and loading at the disk are concerned, but it can be significant for interference effects of the rotor far wake. It also demonstrates the validity of viewing the rotor as a circular wing in high speed flight.

Classical vortex theory for forward flight is based on the actuator disk model, so the vorticity is distributed throughout the wake rather than being concentrated in discrete lines. Often uniform loading is also assumed, so that the vorticity is only on the surface of the wake cylinder and in a root vortex. These assumptions yield the simplest wake model, but in contrast to hover the mathematical problem is still not trivial, because of the skewed cylindrical geometry. Usually numerical calculations are required to obtain the induced velocity at or near the rotor disk (with the exception of a few special points). With uniform loading the results are the same as from momentum theory; in particular, the high speed results must approach the wing theory solution. Because of the limitations of the wake model, vortex theory results based on the actuator disk are presently useful primarily to indicate the general features of the induced velocity, and the rotor flow field in general. Detailed calculations of the induced velocity are best obtained from a nonuniform inflow analysis, including a representation of the discrete vorticity in the wake (see Chapter 13).

4-2.1 Classical Vortex Theory Results

Coleman, Feingold, and Stempin (1945) conducted a vortex theory analysis of the induced velocity along the fore-aft diameter of the rotor disk. They considered an actuator disk with uniform loading, and decomposed the vorticity into rings and axial lines (and they neglected the latter) to calculate the induced velocity. Along the fore-aft diameter of the disk the normal component of the induced velocity can be obtained in closed form, but it involves elliptic integrals even there. A good approximation to the numerical results was $\nu = \nu_0(1 + \kappa_x r \cos\psi)$, where ν_0 is the usual momentum theory result and

$$\kappa_x = \tan \chi/2$$

based on the slope of the downwash at the center of the disk. Using $\tan\chi = \mu/\lambda$, this result becomes

$$\kappa_x = \sqrt{1 + (\lambda/\mu)^2} - |\lambda/\mu|$$

(Coleman, Feingold, and Stempin give a somewhat different result, based on the velocities in the far wake.) Note that for high speeds ($\mu \gg \lambda$) $\kappa_x = 1$.

Drees (1949) calculated the rotor induced velocity using vortex theory. He considered an actuator disk with radially constant bound circulation, but allowed an azimuthal variation of the form $\Gamma = \Gamma_0 - \Gamma_1 \sin\psi$. The trailed vorticity is still only on the surface of the wake cylinder, but now the cylinder is filled with shed vorticity as well. The dimensionless velocity seen by the blade is $(r + \mu \sin\psi)$, so the total blade lift is

$$L = \int_0^1 \rho U \Gamma dr = \rho \Omega R \int_0^1 (r + \mu \sin\psi)(\Gamma_0 - \Gamma_1 \sin\psi) dr$$

$$= \rho \Omega R^2 \tfrac{1}{2}\Gamma_0 \left[1 + (2\mu - \Gamma_1/\Gamma_0)\sin\psi - 2\mu\Gamma_1/\Gamma_0 \sin^2\psi)\right]$$

and the flap moment is

$$M = \int_0^1 \rho U \Gamma r dr$$

$$= \rho\Omega R^3 \, 1/3 \, \Gamma_0 \left[1 + (3/2\mu - \Gamma_1/\Gamma_0) \sin\psi - 3/2\mu\Gamma_1/\Gamma_0 \sin^2\psi \right]$$

Requiring that the mean blade lift equal the rotor thrust per blade ($L = T/N$) and that the first harmonic of the flap moment be zero (for moment equilibrium on the articulated blade; see Chapter 5) gives the distribution of the blade bound circulation:

$$\rho\Omega R^2 N\Gamma = \frac{2T}{1 - 3/2\mu^2}\left(1 - 3/2\mu \sin\psi \right)$$

Drees found the induced velocity due to the bound, trailed, and shed vorticity associated with this circulation distribution. The induced velocities at $r = 0$ and $r = 0.75$ were

$$\lambda(0) = \frac{C_T}{2\mu(1 - 3/2\mu^2)} \sin\chi$$

$$\lambda(0.75) = \frac{C_T}{2\mu(1 - 3/2\mu^2)}\left[\sin\chi + \left(1 - \cos\chi - 1.8\mu^2\right)\cos\psi - 3/2\mu \sin\chi \sin\psi \right]$$

where $\chi = \tan^{-1}\mu/\lambda$ is the wake skew angle. Note that since $\sin\chi = \mu/\sqrt{\mu^2 + \lambda^2}$, the mean induced velocity is $\lambda_i = C_T/[2\sqrt{\mu^2 + \lambda^2}(1 - 3/2\mu^2)]$. The factor $(1 - 3/2\mu^2)$ will be dropped, since it arises only because some of the wake vorticity was neglected. Assuming a linear variation of the velocity over the rotor disk, this result may be generalized to

$$\lambda_i = \frac{C_T}{2\sqrt{\mu^2 + \lambda^2}}\left(1 + \frac{4}{3}\,\frac{1 - \cos\chi - 1.8\mu^2}{\sin\chi}\, r\cos\psi - 2\mu r \sin\psi\right)$$

Drees also suggests an empirical correction for the momentum theory results in order to remove the singularity at ideal autorotation in vertical flight:

$$\lambda = \mu\tan\alpha + 1.2\frac{1}{\sqrt{\mu^2 + \lambda^2}}\left(\lambda_h^2 - \frac{(\mu\tan\alpha)^2 C_{W_0}}{4(1 + 8\lambda^2/\lambda_h^2)(1 + 8\mu^2/\lambda_h^2)}\right)$$

where $\lambda_h^2 = C_T/2$, and C_{W_0} is the drag coefficient of the rotor in ideal autorotation. Drees suggests $C_{W_0} = 1.38$, which gives $V/v_h = -1.70$ for ideal autorotation. Davis, Bennett, and Blankenship (1974) give a modified form of this expression.

Mangler (1948) calculated the induced velocity for a lightly loaded actuator disk by considering span loadings of the form $\Gamma \sim \sqrt{1-r^2}$ and $\Gamma \sim r\sqrt{1-r^2}$. He gives results for the downwash at the disk and in the far wake for wake skew angles from 0° to 90°. Mangler and Squire (1952) extended this analysis to obtain the Fourier series harmonics for the induced velocity at the disk. The first harmonic gives

$$v = v_0\left[1 + \left(\frac{\pi/2}{\sqrt{1-r^2}} \tan\chi/2\right)\cos\psi\right]$$

which has the same variation with χ as found by Coleman, Feingold, and Stempin.

Castles and De Leeuw (1954) present tables and graphs for the normal component of the induced velocity in the longitudinal plane of symmetry of the flow (the vertical plane through the center of the disk and the wake axis) and on the lateral axis in the disk plane. The velocities were calculated numerically using vortex theory for a uniformly loaded actuator disk. In general they concluded that the downwash reaches its maximum far-wake value about one rotor radius downstream of the center of the disk for high speed flight, that is, about at the trailing edge of the rotor disk. For hover and low speed flight the far wake value is achieved about $2R$ downstream of the center of the disk. Castles and Durham (1956b) extended these calculations to the lateral plane of the rotor disk in forward flight.

Willmer (1959) developed a vortex theory for forward flight and hover in which the effects of the finite number of blades were accounted for. The discrete helical vortex sheets of the wake were modeled as rectangular sheets of appropriate orientation and position under the blade, since for rectangular sheets the induced velocity can be obtained in closed form.

Vortex theory analyses have frequently obtained a result for the mean induced velocity at the disk that is the same as the momentum theory expression except for an additional factor $(1 - 3/2\mu^2)^{-1}$ attributed to the azimuthal variation of the blade loading. Heyson (1960b) showed that a consistent vortex theory analysis in high speed forward flight in fact gives the momentum theory result without an additional factor, regardless of the azimuthal loading variation.

Other works on vortex theory in forward flight: Heyson and Katzoff (1957), Castles, Durham, and Kevorkian (1959), Jewel and Heyson (1959), Heyson (1960d, 1961b), Baskin, Vil'dgrube, Vozhdayev, and Maykapar (1976).

4-2.2 Induced Velocity Variation in Forward Flight

For a first (and very rough) approximation to the nonuniform inflow distribution at the rotor in forward flight, consider a linear variation over the disk:

$$\begin{aligned} \nu &= \nu_0(1 + \kappa_x x + \kappa_y y) \\ &= \nu_0(1 + \kappa_x r \cos\psi + \kappa_y r \sin\psi) \end{aligned}$$

(see Fig. 4-1 for a definition of the x and y axes). Here ν_0 is the mean value of the induced velocity, which may be obtained from momentum theory. The form $\nu = \nu_0(1 + \kappa_x r \cos\psi)$ was first suggested by Glauert (1926b). Typically κ_x is positive and κ_y is negative, so that the induced velocity is larger at the rear of the disk and on the retreating side. At high speeds κ_x is roughly 1, which gives a velocity near zero at the leading edge of the disk and about twice the mean value at the trailing edge; κ_y is generally smaller in magnitude. Both κ_x and κ_y must be zero in hover. An induced velocity variation of this form is easily incorporated in the analysis of the rotor behavior in forward flight (see Chapter 5). At best, though, it can only be expected to improve the estimate of the mean and first harmonic quantities, assuming that good values for κ_x and κ_y are available. The actual nonuniform induced velocity distribution in forward flight is much more complicated, and the higher harmonics of the inflow can be very important.

A net aerodynamic moment on the rotor disk will produce an inflow variation. To estimate this inflow, consider the differential form of the momentum theory result, $dT = 2\rho V \nu dA$ or

$$\nu = \frac{dT/dA}{2\rho V}$$

where dT/dA is the local disk loading. Assuming a linear variation of the loading due to the pitch and roll moments on the rotor gives

$$\frac{dT}{dA} = \frac{T}{A} - 4\frac{M_y}{RA} r\cos\psi + 4\frac{M_x}{RA} r\sin\psi.$$

Then

$$\lambda_i = \frac{C_T}{2\mu} - \frac{2C_{M_y}}{\mu} r\cos\psi + \frac{2C_{M_x}}{\mu} r\sin\psi$$

where C_{M_y} and C_{M_x} are the pitch and roll moment coefficients. It follows that $\kappa_x = -4C_{M_y}/C_T$ and $\kappa_y = 4C_{M_x}/C_T$. Thus the inflow variation is proportional to the offset of the rotor thrust vector from the center of rotation, which may be significant for hingeless rotors.

Consider now a momentum theory estimate for the longitudinal variation of the inflow in forward flight. Recall that the induced velocity v at a wing may be associated with the mass flux through a cylinder circumscribing the span of the wing (for a circular wing the cylinder has the same area as the wing itself). Extending this concept to differential form, the mass flux at each spanwise station y is $\dot{m} = \rho V 2\sqrt{1-y^2}\,dy$, and then the induced velocity is $dv = dT/\dot{m}$. Assuming a uniform loading over the disk so that $dT = (T/A)dxdy$, then

$$\frac{dv}{dx} = \frac{T}{2\rho A V\sqrt{1-y^2}}$$

Here dv is the induced velocity increment at x due to the thrust element dT at x; the total induced velocity at x is then due to all the elements upstream (see Fig. 4-1 for the definition of the disk plane coordinates). Integrating from the leading edge of the disk then gives

$$v = \frac{T}{2\rho A V}\left(\frac{x}{\sqrt{1-y^2}} + 1\right) = v_0\left(1 + \frac{r\cos\psi}{\sqrt{1-y^2}}\right)$$

where $x = -\sqrt{1-y^2}$ at the leading edge and $x = \sqrt{1-y^2}$ at the trailing edge. The induced velocity is thus linear in x and equal to zero all along the disk leading edge and $2v_0$ all along the trailing edge. Based on the inflow variation through the center of the disk, $\kappa_x = 1$. This result is often used for high speeds, but it is also necessary to establish the variation with μ since κ_x must be zero at hover.

The classical vortex theory analyses give a number of estimates of the parameters κ_x and κ_y defining the induced velocity variation in forward flight. Coleman, Feingold, and Stempin (1945) suggest

$$\kappa_x = \tan\chi/2 = \sqrt{1 + (\lambda/\mu)^2} - |\lambda/\mu|$$

where χ is the skew angle of the wake at the disk. Here κ_x indeed approaches unity for high speed. This expression can also be obtained from the results of Mangler and Squire (1952). Drees (1949) gives

$$\kappa_x = \frac{4}{3}\,\frac{1 - \cos\chi - 1.8\mu^2}{\sin\chi} = \frac{4}{3}\left[\left(1 - 1.8\mu^2\right)\sqrt{1 + (\lambda/\mu)^2} - \lambda/\mu\right]$$

$$\kappa_y = -2\mu$$

Hence κ_x is zero at $\mu = 0$, has a maximum of about 1.1 at $\mu = 0.16$, and is approximately 1 at $\mu \cong 0.3$. Castles and De Leeuw (1954) give numerical results that Payne (1959a) suggests can be approximated by a linear expression,

$$\kappa_x = \frac{(4/3)\mu/\lambda}{1.2 + \mu/\lambda};$$

from which $\kappa_x = 4/3$ at high speed.

4-2.3 Literature

On the inflow curve in forward flight: Carpenter (1948), Carpenter and Paulnock (1950), Gessow (1954), Payne (1956), Heyson (1961c, 1975), Washizu, Azuma, Koo, and Oka (1966a, 1966b), Curtiss and Shupe (1971), Shaydakov (1971a, 1971c), Wolkovitch and Hoffman (1971), Wolkovitch (1972), Peters (1974).

On the induced velocity and wake flow field in forward flight: Wheatley and Hood (1935), Castles (1945), Coleman, Feingold, and Stempin (1945), Ross (1946), Mangler (1948), Brotherhood and Stewart (1949), Drees (1949), Fail and Eyre (1949b), Drees and Hendal (1951), Mangler and Squire (1952), Castles and De Leeuw (1954), Gessow (1954), Heyson (1954, 1958, 1960b, 1960d, 1961b), Falabella and Meyer (1955), Castles and Durham (1956b, 1962), Heyson and Katzoff (1957), Castles, Durham and Kevorkian (1959), Ham and Zvara (1959), Jewel and Heyson (1959), Willmer (1959), Gray (1960), Connor and O'Bryan (1962), Miller, Tang, and Perlmutter (1968), Levinsky, Thommen, Yager, and Holland (1969),

Joglekar and Loewy (1970), Levinsky and Strand (1970), Bramwell (1971), Johansson (1972, 1973), Biggers and Orloff (1974), Boirun, Jefferis, and Holasek (1974), Landgrebe and Johnson (1974), Maresca, Favier, and Rebont (1974), Jenkins and Marks (1975), Biggers, Lee, Orloff, and Lemmer (1977a, 1977b).

4–3 Twin Rotor Interference in Forward Flight

In general, the mutual interference of a multi-rotor system may be accounted for by writing the induced velocity at the mth rotor as

$$v_m = \kappa_m v_{im} + \sum_{n \neq m} \chi_{mn} v_{in}$$

Here v_{in} is the ideal induced velocity for the isolated nth rotor; κ_m is the correction for the additional induced losses of a real rotor; and χ_{mn} is the interference downwash at the mth rotor due to the thrust of the nth rotor. For a power loss the interference factor $\chi_{mn}$is positive, while χ_{mn} is negative for favorable interference. This expression is applicable to all speeds, including hover, although the interference factors χ_{mn} will depend on speed. In high speed forward flight, momentum theory or wing theory gives the induced velocity $v_{in} = T_n/2\rho AV$. The total induced power loss in forward flight is therefore

$$P = \sum_m T_m v_m = \Big(\sum_m \kappa_m T_m^2 + \sum_m \sum_{n \neq m} \chi_{mn} T_m T_n\Big) \Big/ 2\rho AV$$

(it has been assumed that all the rotors have the same area A, which is usually the case.) Since the isolated rotor power is just $P_{\text{isolated}} = \Sigma \kappa_m T_m^2/2\rho AV$,

$$\frac{P}{P_{\text{isolated}}} = 1 + \frac{\sum_m \sum_{n \neq m} \chi_{mn} T_m T_n}{\Sigma \kappa_m T_m^2}$$

The second term is the interference power, which is usually a significant positive increment. For some configurations a modest favorable interference is possible.

Consider now the case of twin main rotors of equal area. Let $\kappa_m = 1$, since we are primarily concerned with the interference losses here. Then the induced power of the individual rotors is

$$P_1 = (T_1^2 + \chi_{12} T_1 T_2)/2\rho AV$$

$$P_2 = (T_2^2 + \chi_{21} T_1 T_2)/2\rho AV$$

and the total induced power is

$$\frac{P}{P_{\text{isolated}}} = 1 + \frac{(\chi_{12} + \chi_{21})T_1 T_2}{T_1^2 + T_2^2}$$

For equal thrusts ($T_1 = T_2$), then

$$\frac{P}{P_{\text{isolated}}} = 1 + \frac{\chi_{12} + \chi_{21}}{2} = 1 + \chi$$

where χ is the interference factor for the entire rotor system.

Wing theory for a single lifting surface shows that the induced power is proportional to the thrust squared divided by the span squared, $P \sim (T/\text{span})^2$. Consequently, the total induced power of a multi-rotor system depends on the span of the effective lifting surface. For twin isolated rotors of thrust T and span $2R$, $P = 2(T^2/2\rho A V)$. The same two rotors in coaxial configuration act like a single rotor with twice the span loading; hence the induced power is doubled, or $\chi \cong 1$.

Wing theory also shows that the total induced drag loss is independent of the longitudinal separation of lifting elements. As a result, tandem rotors with no vertical separation have about the same loss as coaxial rotors, $\chi \cong 1$. The distribution of the loss between the two rotors is the property that changes with longitudinal separation. For the coaxial configuration, the two rotors are identical, so $\chi_{12} = \chi_{21} = 1$. For large longitudinal separation, however, the front rotor is not influenced by the rear rotor, while the rear rotor sees the fully developed wake of the front one. Hence for the tandem configuration, $\chi_{FR} = 0$ and $\chi_{RF} = 2$ is expected as a limit. As the vertical separation of the tandem or coaxial rotors increases, they approach isolated rotors in forward flight; hence $\chi < 1$, decreasing to $\chi = 0$ for a vertical spacing of about one rotor radius. Note that it is the vertical spacing of the rotor wakes, not the disks themselves, that determines the interference.

Consider the side-by-side configuration. For zero lateral spacing (coaxial rotors), $\chi \cong 1$ again. When the shaft spacing is $2R$ (rotor disks just touching) the system is like a single rotor with the same span loading as the two isolated rotors. Hence the total induced power should be reduced by a factor of two, or $\chi \cong -\frac{1}{2}$. This favorable interference is due to each rotor operating in the upwash flow field of the other. The span loading of the side-by-side configuration will be far from elliptical, however, even with uniformly loaded rotors. Thus the actual interference, while still favorable is not as great as indicated by the span loading. As the lateral separation increases further, χ approaches zero again.

Consider a momentum theory analysis of the tandem rotor helicopter in forward flight. Assuming that rear rotor wake has no influence on the front rotor, and that the rear rotor is operating in the fully developed wake of the front rotor, the total induced velocities at the front and rear rotors are $v_{\text{front}} = v_F$ and $v_{\text{rear}} = v_R + 2v_F$ respectively, where $v_F = T_F/2\rho A V$ and $v_R = T_R/2\rho A V$. The total induced power is then

$$P = T_F v_F + T_R(v_R + 2v_F) = (T_F^2 + T_R^2 + 2T_F T_R)/2\rho A V$$

$$= (T_F + T_R)^2/2\rho A V$$

The interference factor is

$$\chi = \frac{\Delta P}{P_{\text{isolated}}} = \frac{2T_F T_R}{T_F^2 + T_R^2}$$

which has a maximum of $\chi = 1$ for equal thrust on the two rotors. (Although unequal thrust reduces the interference power, the minimum total power is obtained with equal thrust.) Thus the tandem helicopter in forward flight is less efficient than the isolated rotors, requiring about twice the induced power when there is no vertical separation. Usually, however, there is substantial vertical separation, since the rear rotor is raised to minimize the effect of the front rotor wake. Moreover, in forward dlight the induced power losses are only a small fraction of the total power required. From empirical data $\chi \cong 0.9 (P/P_{\text{isolated}} \cong 1.9)$ for usual tandem rotor configurations; hence the induced velocities at the front and rear rotors are v_F and $v_R + 1.9v_F$ respectively. There are also some results indicating that $P/P_{\text{isolated}} \cong 2.2$ to 2.3 when tip losses and nonuniform inflow losses as well as the interference losses are accounted for. Vortex theory results for a single rotor can also be used to estimate the interference effects.

Stepniewski (1950, 1955) developed a momentum theory analysis for tandem rotors in forward flight that includes the effect of vertical separation. Figure 4-7 shows the configuration considered. Because of the elevation of

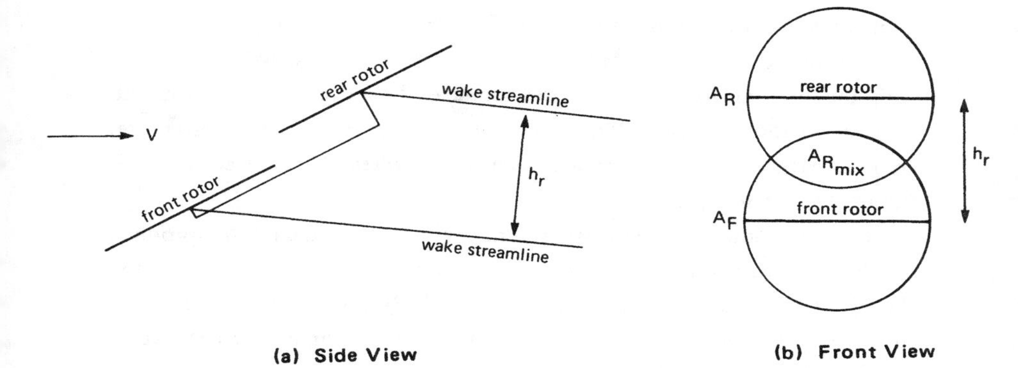

Figure 4-7 Momentum theory analysis of tandem rotors in forward flight.

the rear rotor on a pylon and the forward tilt of the helicopter, the rear rotor wake is a distance h_r above the front rotor wake; typically $h_r \cong 0.3R$ to $0.5R$. The rear rotor sees an interference velocity less than $2v_F$ because of the separation of the wakes, and as a consequence the efficiency of the rotor system is improved. To estimate the interference at the rear rotor, recall the interpretation that the induced power in forward flight is due to the mass flux through a cylinder circumscribing the wing span. For $h_r = 0$, the cylinders of the two rotors coincide, corresponding to a full $2v_F$ interference velocity at the rear rotor. For $h_r > 0$, the cylinders overlap only partially. Let us use the overlap fraction as a measure of how much of $2v_F$ is seen at the rear rotor. The overlap area is $A_{R_{mix}} = \tilde{m}A_R$, where $\tilde{m}$ is a function of the separation h_r:

$$\tilde{m} = \frac{2}{\pi}\left[\cos^{-1}\frac{h_r}{2R} - \frac{h_r}{2R}\sqrt{1-\left(\frac{h_r}{2R}\right)^2}\,\right]$$

Then assume that the total induced velocity at the rear rotor is $v_{\text{rear}} = v_R + 2v_F\tilde{m}$. The total induced power becomes

$$P = (T_F^2 + T_R^2 + 2T_F T_R \tilde{m})/2\rho A V$$

and the interference factor is

$$\chi = \frac{\Delta P}{P_{\text{isolated}}} = \frac{2T_F T_R}{T_F^2 + T_R^2}\,\tilde{m}$$

For equal thrusts on the two rotors, $\chi = \tilde{m}$. For small separations, χ is slightly less than 1, and it falls to zero at $h_r = 2R$. Stepniewski found that this theory compares well with the measured losses of tandem rotor helicopters in forward flight. While it is a crude estimate of the interference, it is satisfactory for performance calculations when the induced power is small in forward flight.

Consider now the side-by-side configuration with lateral separation ℓ of the rotor shafts. Compared to the coaxial configuration, the span of the lifting system is increased by a factor $(1 + \ell/2R)$. Then, since coaxial rotors have twice the induced power loss of isolated rotors, the interference factor for the side-by-side configuration is

$$\chi = \frac{P}{P_{\text{isolated}}} - 1 = \frac{2}{(1 + \ell/2R)^2} - 1$$

This gives $\chi = -\frac{1}{2}$ for $\ell = 2R$, as above. The departure of the span loading from an elliptical distribution is ignored completely in this result, so χ approaches -1 rather than zero for very large separation. The value $\chi = -\frac{1}{2}$ for the disks just touching also overestimates the favorable interference; the above result should only be used out to about $\ell/R = 1.75$, beyond which the interference decreases to zero. Empirical data give $\chi \cong -0.2$ to -0.3 with the disks just touching, and the most favorable interference (at $\ell/2R \cong 1.75$) has $\chi \cong -0.25$ to -0.45. Thus the most favorable case still has 55% of the induced power of isolated rotors.

The literature on twin rotor performance in forward flight includes: Fail and Squire (1947), Stepniewski (1950), Dingeldein (1954), Huston (1963), Mil' (1966), Baskin, Vil'dgrube, Vozhdayev, and Maykapar (1976).

4–4 Ground Effect in Forward Flight

As discussed in section 3–6, the rotor induced power is decreased by the proximity of the ground, or equivalently the thrust is increased for a given power. In forward flight, where the wake is swept behind the rotor, the effect of the ground diminishes rapidly with forward speed. Ground effect is negligible for speeds above about $V = 2v_h$, or roughly $\mu = 0.15$. Figure 4-8 illustrates the influence of forward speed on ground effect. In hover the ground proximity significantly reduces the helicopter power required.

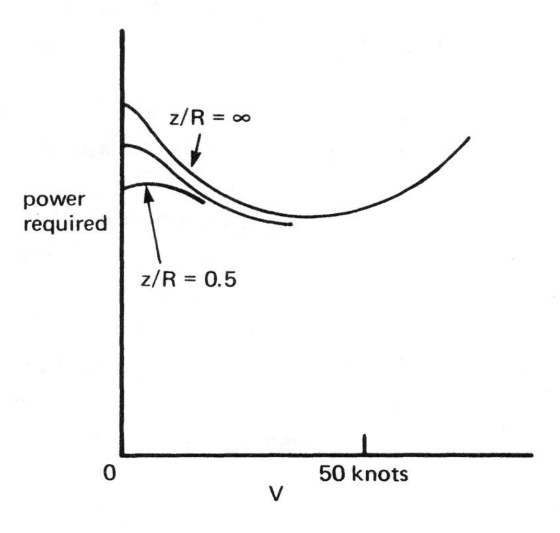

Figure 4-8 Sketch of the influence of forward speed on ground effect.

The effect continues at low speeds but decreases rapidly beyond transition until it is negligible at around 30 knots. The net effect of the ground is to reduce the sensitivity of the power required to changes in speed near hover. The sensitivity of ground effect to the helicopter velocity or, equivalently, to winds can be of considerable importance to helicopter operations.

Cheeseman and Bennett (1955) developed an approximate method for estimating the influence of the ground on the rotor lift in forward flight. Using the method of images, with the rotor modeled by a source a distance z above the ground, they obtained

$$\frac{T}{T_\infty} = \frac{1}{1 - \dfrac{(R/4z)^2}{1 + (\mu/\lambda)^2}}$$

which displays the correct behavior with height and forward speed as long as z/R is above about 0.5. Using blade element theory to incorporate the influence of the blade loading, they obtained

$$\frac{T}{T_\infty} = \frac{1}{1 - \dfrac{\sigma a \lambda}{4C_T}\dfrac{(R/4z)^2}{1 + (\mu/\lambda)^2}}$$

These expressions gave reasonable agreement with test data; the primary dependence is on z/R and μ, with only secondary effects due to the rotor loading.

Heyson (1960a) analyzed helicopter ground effect in forward flight, using an actuator disk model for the rotor and wake with an image vortex system below the ground plane. The ground was found to always reduce the power required, but in forward flight the effect decreases with height more rapidly than in hover. The influence of the ground decreased with forward speed, most of the effect of speed occurring below $V/v_h = 1.5$ to 2.0. It was also observed that for small heights the increase in power due to the decrease of ground effect with speed is more rapid than the decrease in power due to the normal reduction of induced velocity in forward flight. Thus the net power required in ground effect can actually increase with speed near hover (see Fig. 4-8).

Chapter 5

FORWARD FLIGHT II

5–1 The Helicopter Rotor in Forward Flight

Efficient hover capability is the fundamental characteristic of the helicopter, but without good forward flight performance the ability to hover would be of no value. During translational flight of the helicopter the rotor disk is moving edgewise through the air, remaining nearly horizontal (there is a small forward tilt to provide the propulsive force for the aircraft). Thus in forward flight the rotor blade sees a component of the helicopter forward velocity as well as the velocity due to its own rotation. On the advancing side of the disk the velocity of the blade is increased, while on the retreating side it is decreased, by the forward speed. Assuming a constant angle of attack of the blade, the varying dynamic pressure of the rotor aerodynamic environment in forward flight will tend to produce more lift on the advancing side than on the retreating side, that is, a rolling moment on the rotor. If nothing were done to counter this moment, the helicopter would respond by rolling toward the retreating side of the rotor until equilibrium was achieved with the rotor moment balanced by the gravitational force acting at the helicopter center of gravity. The rotor moment could possibly be so large that an equilibrium roll angle would not be achieved. A number of crashes of early helicopter designs as they attempted forward flight were due to this phenomenon. In addition, the rolling moment on the rotor disk corresponds to a large bending moment at the blade root that oscillates once per revolution, from maximum positive on the advancing side to maximum negative on the retreating side.

Since the rotor blade loading (T/A_{blade}) is limited by stall of the airfoil sections, for a given thrust (and tip speed) the rotary wing will tend to have about the same blade area regardless of the rotor diameter. It follows that the low disk loading helicopter rotor will have low solidity $\sigma = A_{\text{blade}}/A_{\text{rotor}}$

and thus high aspect ratio blades. The high aspect ratio, thin blades required for aerodynamic efficiency limit the structural load bearing capability at the root, and as a consequence the 1/rev loads due to forward flight are a severe problem. Some means is required to alleviate the root bending moments and reduce the blade stresses to an acceptable level. With stiff blades, such as on propellers, the structure must absorb all of the aerodynamic loads; but flexible blades respond to the aerodynamic forces with considerable bending motion, so the blade loads can be countered by the aerodynamic forces due to this motion rather than by structural forces. Hence in response to the lateral aerodynamic moment in forward flight there is a 1/rev motion of the blades out of the plane of the disk, called flapping motion. When the inertial and aerodynamic forces due to this flapping motion are accounted for, the net blade loads at the root and the rolling moment on the helicopter are small. The conventional approach has been to use a flap hinge at the blade root, about which the blade can rotate as a rigid body to produce the flap motion (see Fig. 1-4). Since the moment at the flap hinge must be zero, no hub moment at all can be transmitted to the helicopter (unless the hinge is offset from the center of rotation) and the bending moment throughout the blade root must be low. A rotor with mechanical flap hinges is called an articulated rotor. Recently there have been successful helicopter designs without flap hinges, which are called hingeless rotors. With modern materials, the blade root can be strong while still flexible enough to provide the flapping motion necessary to eliminate most of the root loads. Because of the large centrifugal forces on the blade, the flap motion of hingeless rotors is in fact very similar to that of articulated rotors. The root loads of a hingeless rotor are naturally higher than those of articulated rotors, and the increased hub moments have a significant effect on the helicopter handling qualities. In summary, the flap motion of the helicopter blades has the effect of reducing the asymmetry of the rotor lift distribution in forward flight. Thus the flap motion will be a principal concern of the analysis of the forward flight performance of the rotor.

Let us examine the velocity components seen by the rotating blades in forward flight (Fig. 5-1). The helicopter has forward velocity V and disk angle of attack α (positive for forward tilt of the rotor). The rotor rotates with speed Ω. Since the conventional rotation direction is counterclockwise when viewed from above, the advancing side of the disk is to the right (starboard). The fixed-frame coordinates x, y, and z are aft, right, and up,

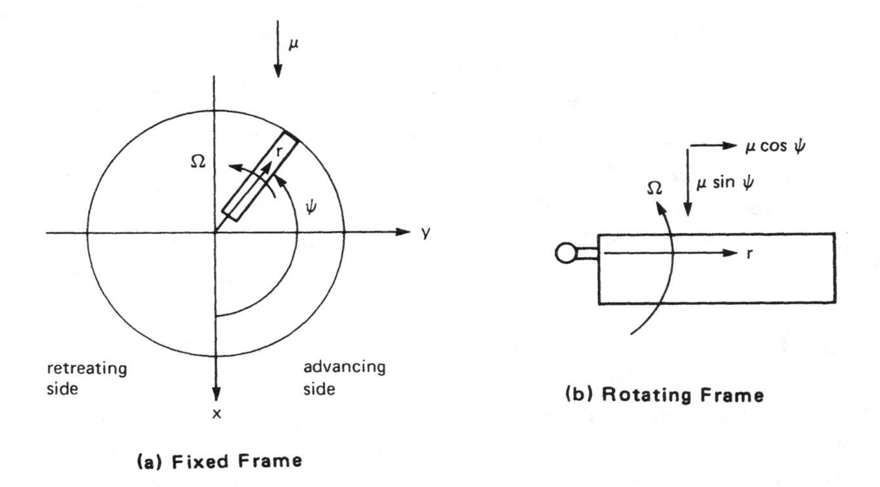

Figure 5-1 Rotor blade velocity in forward flight

respectively, with origin at the center of rotation of the rotor. The component of the helicopter velocity in the plane of the rotor disk is $V\cos\alpha$. Define the rotor advance ratio as the in-plane forward velocity component normalized by the rotor tip speed:

$$\mu = \frac{V\cos\alpha}{\Omega R}$$

Thus μ is the dimensionless forward speed of the rotor. The blade position is given by the azimuth angle $\psi = \Omega t$, measured from downstream. In a frame rotating with the rotor blade, the tangential component of the velocity seen by the blade is $\Omega r + V\cos\alpha\sin\psi$, and the radial component is $V\cos\alpha\cos\psi$. The dimensionless velocity components in the rotating frame are thus

$$u_T = r + \mu\sin\psi$$
$$u_R = \mu\cos\psi$$

The 1/rev variation of the tangential velocity u_T has a major influence on the aerodynamics of the rotor in forward flight. The advance ratio μ is small for typical helicopter cruise speeds. Early designs had a maximum speed corresponding to $\mu_{max} \cong 0.25$, while current helicopter designs have perhaps

$\mu_{max} = 0.35$ to 0.40. For a tip speed of $\Omega R \cong 200$ m/sec, an advance ratio of $\mu = 0.5$ corresponds to $V \cong 200$ knots.

A phenomenon introduced by forward flight is the reverse flow region, an area on the retreating side of the rotor disk where the velocity relative to the blade is directed from the trailing edge to the leading edge. The forward velocity component $\Omega R\mu \sin\psi$ is negative on the retreating side of the rotor ($\psi = 180°$ to $360°$), while the rotational velocity Ωr is positive and linearly increasing along the blade. Consequently, there will always be a region at the blade root where the rotational velocity is smaller in magnitude than the forward speed component, so that the flow is reversed. Specifically, for $\psi = 270°$ the total velocity is $\Omega R(r - \mu)$ and the flow is reversed for blade stations inboard of $r = \mu$. In general, the reverse flow region is defined as the area of the disk where $u_T < 0$, which has the boundary $r + \mu \sin\psi = 0$. The reverse flow boundary is thus a circle of diameter μ, centered at $r = \mu/2$ on the $\psi = 270°$ radial in the retreating side (Fig. 5-2). When $\mu \geqslant 1$, the reverse flow region includes the entire blade at $\psi = 270°$ and must have a significant impact on the rotor aerodynamics. An advance ratio of $\mu = 0.3$ to 0.4 is more typical of current helicopter forward speeds, however. For low advance ratio, the reverse flow region occupies only a small portion of the disk (The ratio of the reverse flow area to the total disk area is $\mu^2/4$.) Moreover, since by definition $u_T = 0$ at the boundary, the entire reverse flow region is characterized by low dynamic pressure until the advance ratio gets large. The root cutout, extending to typically 15% to 30% of the rotor radius, will cover much of the reverse flow region. Thus it is found that the effects of the reverse flow region are negligible up to an advance ratio of about $\mu = 0.5$.

The asymmetry of the aerodynamic environment in forward flight, which is due to the combination of the forward velocity and rotor rotation, means that the blade loads and motion will depend on the azimuthal position ψ. For steady-state conditions, the behavior of the blade as it revolves must always be the same at a given azimuth, which implies that the blade loads and motion are periodic around the azimuth, with period 2π. In dimensional terms, the rotor blade behavior is periodic with a fundamental frequency equal to the rotor speed Ω and a period $T = 2\pi/\Omega$. Periodic functions may be represented by a Fourier series. For example, the flap angle β may be written

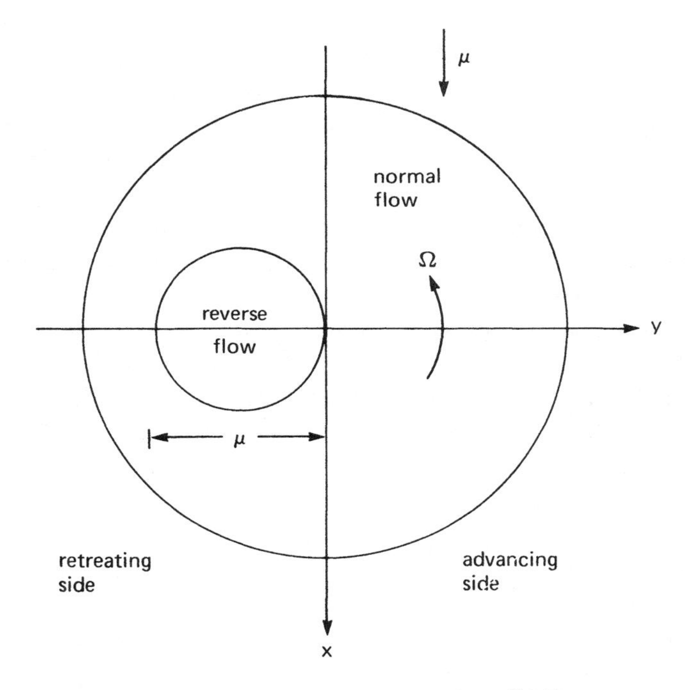

Figure 5-2 Reverse flow region (shown for $\mu \cong 0.7$).

$$\begin{aligned}\beta(\psi) &= \beta_0 + \beta_{1c}\cos\psi + \beta_{1s}\sin\psi + \beta_{2c}\cos 2\psi + \beta_{2s}\sin 2\psi + \dots \\ &= \beta_0 + \sum_{n=1}^{\infty}\left(\beta_{nc}\cos n\psi + \beta_{ns}\sin n\psi\right)\end{aligned}$$

The periodic function $\beta(\psi)$ is then defined by the harmonics β_0, β_{1c}, β_{1s}, etc. For rotors where it is generally found that only the lowest few harmonics are required to describe the motion adequately, the complete time behavior is described by a small number of parameters. The Fourier coefficients, or harmonics, are obtained from $\beta(\psi)$ as follows:

$$\beta_0 = \frac{1}{2\pi}\int_0^{2\pi}\beta d\psi$$

$$\beta_{nc} = \frac{1}{\pi}\int_0^{2\pi}\beta\cos n\psi\, d\psi$$

$$\beta_{ns} = \frac{1}{\pi}\int_0^{2\pi} \beta \sin n\psi \, d\psi$$

The motion of the blade degrees of freedom will be described by differential equations, which must be solved for the periodic motion in the rotating frame. One method of solution is the substitutional method, which consists of the following steps. The Fourier series representations of the degrees of freedom and their time derivatives are substituted into the equations of motion. Products of harmonics are reduced to sums of harmonics by means of trigonometric relations. All the terms in the equation for a given harmonic are then collected, and the coefficient of each harmonic (1, $\cos\psi$, $\sin\psi$, $\cos 2\psi$, $\sin 2\psi$, etc.) is set to zero. The result is a set of algebraic equations for the harmonics of the blade motion. An alternative approach is the operational method, in which each of the operators

$$\frac{1}{2\pi}\int_0^{2\pi} (\ldots) d\psi, \frac{1}{\pi}\int_0^{2\pi} (\ldots) \sin\psi \, d\psi, \frac{1}{\pi}\int_0^{2\pi} (\ldots) \cos\psi \, d\psi,$$

$$\frac{1}{\pi}\int_0^{2\pi} (\ldots) \sin 2\psi \, d\psi, \frac{1}{\pi}\int_0^{2\pi} (\ldots) \cos 2\psi \, d\psi, \text{etc.}$$

is applied to the differential equations of motion. Then the definitions of the harmonics are used to replace the integrals of the blade motion by the Fourier coefficients. The result is the same set of algebraic equations as obtained by the substitutional method, although the operational method obtains the equations one at a time. Linear differential equations reduce to linear algebraic equations for the harmonics. The solution for the blade motion is necessarily approximate, since the Fourier series must be truncated to obtain a finite set of equations. A finite number of harmonics is sufficient to describe the periodic functions encountered in the rotor analysis, however.

The flapping hinge is needed to alleviate the root stresses and hub moments by allowing out-of-plane motion of the blade. The flap motion also introduces aerodynamic and inertial forces, particularly Coriolis forces, in the plane of the rotor disk. Therefore, a lag hinge is frequently used as well, to alleviate the chordwise root loads by allowing in-plane motion of the blade. The lag hinge increases the mechanical complexity of the hub and introduces the possibility of a mechanical instability called ground resonance, which requires a mechanical lag damper to stabilize the motion. (Ground resonance involves the coupled motion of the blades about the lag hinges and the in-plane motion of the rotor hub, which is usually due to flexibility of the landing gear when the helicopter is on the ground; see section 12–4.) The alternative is to make the blade root strong and heavy enough to take the in-plane loads without a lag hinge. A pitch bearing or hinge is also required at the blade root to allow the blade pitch angle to be changed in response to control inputs. Thus the blade of a fully articulated rotor has three hinges at the root: flap, lag, and feather (sketched in Fig. 5-3). The

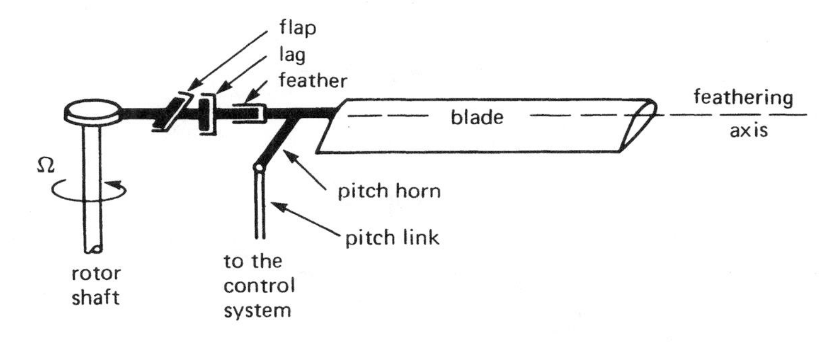

Figure 5-3 Schematic of the flap and lag hinges, and the pitch bearing at the hub of an articulated blade.

motion about the flap and lag hinges is restrained by centrifugal forces when the blade is rotating, while the motion about the feathering hinge is restrained by the control system. Note that when rotor blade hinges are spoken of, it is usually the flap and lag hinges that are being referred to; for example, a hingeless rotor often does have a pitch bearing. Mechanical considerations for fully articulated rotors require that the flap and lag hinges be offset slightly from the center of rotation. An offset of the lag hinge is necessary in any case, or it would not be possible to transmit torque from the shaft to the blades in order to turn the rotor; and a flap hinge offset improves the helicopter handling qualities by allowing some pitch and roll moments to be transmitted to the helicopter. In teetering or gimballed rotors the flap hinge is at the center of rotation, and these rotors are designed without lag hinges. With hingeless rotors the flap and lag motion is primarily due to bending at the blade root; such blades may be considered as roughly equivalent to articulated rotors with large hinge offsets (and probably effective hinge springs as well.)

The basic blade motion is represented by the flap, lag, and pitch degrees of freedom (Fig. 5-4). The out-of-plane or flap motion is due to rigid body

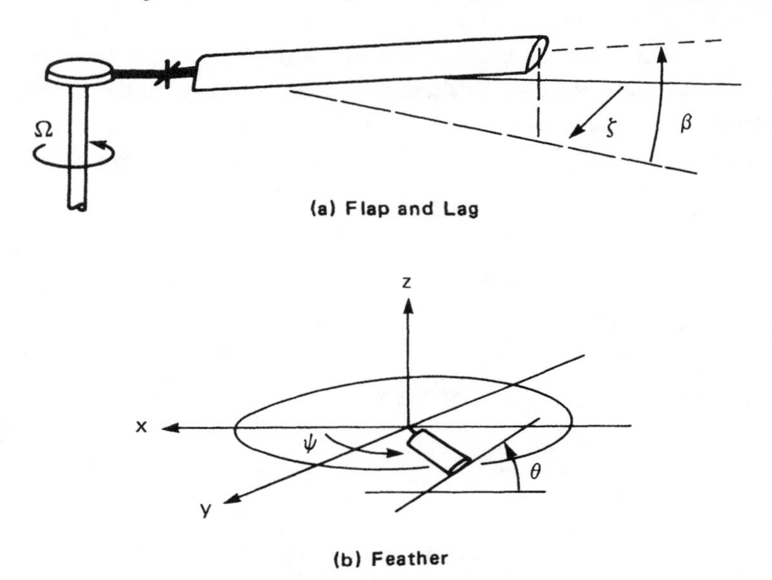

Figure 5-4 Rotor blade motion.

rotation about the flap hinge by the angle β (positive upward). The in-plane or lag motion is due to rotation about the lag hinge by the angle ζ (positive aft, opposite the direction of rotor rotation). Finally, the blade pitch or feather motion is due to rotation about the feathering axis by the angle θ (positive when nose up). The flap and pitch angles are measured from a disk reference plane (the various reference planes used in rotor analyses are discussed below). The steady-state flap motion is described by a Fourier series

$$\beta(\psi) = \beta_0 + \beta_{1c}\cos\psi + \beta_{1s}\sin\psi + \beta_{2c}\cos 2\psi + \beta_{2s}\sin 2\psi + \dots .$$

Let us examine what these harmonics imply in terns of the rotor motion as viewed in the fixed frame (Fig. 5-5). The zeroth harmonic or mean value β_0 is the coning angle. When $\beta = \beta_0$, the flap motion is independent of ψ and therefore the blades describe a cone as they rotate. The blade tips describe

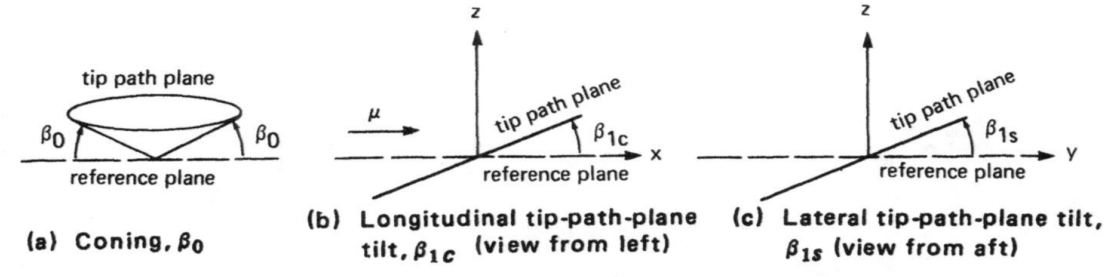

Figure 5-5 Interpretation of the blade flapping harmonics.

a circle that lies in a plane parallel to the reference plane. The first harmonic β_{1c} generates a once-per-revolution variation of the flap angle, $\beta = \beta_{1c} \cos \psi$. The blade out-of-plane deflection is then $z = r\beta = r\beta_{1c} \cos \psi = x\beta_{1c}$. Thus as they rotate the blades describe a plane tilted forward about the lateral (y) axis by the angle β_{1c} relative to the reference plane. Similarly, the first harmonic β_{1s} generates an out-of-plane deflection of $z = r\beta = r\beta_{1s} \sin \psi = y\beta_{1s}$, which corresponds to a plane tilted to the left (toward the retreating side) about the longitudinal axis by the angle β_{1s} relative to the reference plane. The combination of the harmonics β_0, β_{1c}, and β_{1s} forms a cone that has been tilted laterally and longitudinally. The circular path described by the blade tips still lies in a plane, which is called the tip-path plane. The orientation of the tip-path plane relative to the reference plane is given by β_{1c} and β_{1s}. The higher harmonics of the flap motion (β_{2c}, β_{2s}, etc.) produce a distortion of the tip-path plane. These harmonics are usually small, so the rotor flap motion is described primarily by β_0, β_{1c}, and β_{1s} for the helicopter in forward flight.

The lag motion may also be written as a Fourier series:

$$\zeta = \zeta_0 + \zeta_{1c} \cos \psi + \zeta_{1s} \sin \psi + \ldots$$

The zeroth harmonic ζ_0 is the mean lag angle of the blades relative to the rotor hub and shaft (Fig. 5-6). The first harmonic cyclic lag ζ_{1c} produces a lateral shift of the blades, to the left when $\zeta_{1c} > 0$ (see Fig. 5-6). Neglecting the hinge offset, the center of gravity of the blade is at $x_{CG} = r_{CG} \cos(\psi - \zeta) \cong r_{CG} (\cos \psi + \zeta \sin \psi)$ and $y_{CG} = r_{CG} \sin(\psi - \zeta) \cong r_{CG} (\sin \psi - \zeta \cos \psi)$, where r_{CG} is the radial location of the center of gravity. The mean center-of-gravity location, which is also the center of gravity for the entire rotor, is obtained by averaging over the rotor aximuth and multiplying by the number of blades. Using the definition of the lag harmonics, we obtain

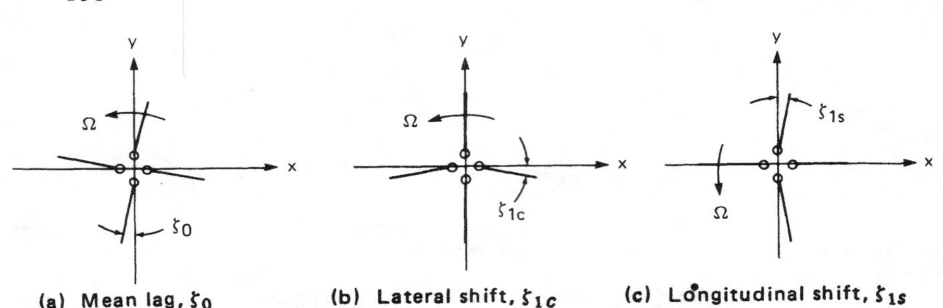

Figure 5-6 Interpretation of the blade lag harmonics.

$$(x_{CG})_{\text{rotor}} = \frac{N}{2\pi}\int_0^{2\pi} x_{CG}\, d\psi = \frac{N}{2} r_{CG}\zeta_{1s}$$

$$(y_{CG})_{\text{rotor}} = \frac{N}{2\pi}\int_0^{2\pi} y_{CG}\, d\psi = -\frac{N}{2} r_{CG}\zeta_{1c}$$

Thus the cyclic lag ζ_{1c} produces a lateral shift of the rotor center of gravity. Similarly, the cyclic lag ζ_{1s} produces a longitudinal shift of the blades in the plane of rotation (aft when $\zeta_{1s} > 0$) and a longitudinal shift in the rotor center of gravity. From the character of the rotor motion associated with the lowest harmonics of flap and lag, it may be inferred that the coning β_0 is the reaction to the mean blade lift, ζ_0 is the reaction to the mean rotor torque, the cyclic flap β_{1c} and β_{1s} are the response to moments on the rotor disk, and the cyclic lag ζ_{1c} and ζ_{1s} are the response to in-plane motion of the rotor hub.

The Fourier series representation for the blade pitch motion is

$$\theta = \theta_0 + \theta_{1c}\cos\psi + \theta_{1s}\sin\psi + \ldots$$

The zeroth harmonic θ_0 is the average blade pitch, while the first harmonics give a once-per-revolution variation of the pitch angle. Blade pitch or feathering motion has two sources. First, there is the elastic deformation of the control system and blade, described by dynamic degrees of freedom. Such motion is determined by the conditions for equilibrium of feathering moments on the blade, which give the equations of motion. The second source of blade pitch is the commanded input from the helicopter control aystem. It is by commanding the rotor blade pitch that the pilot controls the helicopter.

The feathering moments on the blade are low, and the lift changes due to pitch are large because the angle of attack is directly changed. Controlling the blade pitch is therefore a very effective means of controlling the forces on the rotor. For the present chapter we are concerned only with the blade pitch as a control variable; the torsion dynamics are considered in later chapters. The control inputs usually consist of just the mean and first harmonics: $\theta = \theta_0 + \theta_{1c} \cos \psi + \theta_{1s} \sin \psi$. The mean angle θ_0 is called the collective pitch, and the 1/rev harmonics θ_{1c} and θ_{1s} are called the cyclic pitch angles. Basically, collective pitch controls the average blade force, and hence the rotor thrust magnitude, while cyclic pitch controls the tip-path-plane tilt (that is, the 1/rev flapping) and hence the thrust vector orientation (θ_{1c} controls the lateral orientation, while θ_{1s} controls the longitudinal orientation).

The rotor must have a mechanical means of producing collective and cyclic pitch changes on the rotor blades. The blade pitch motion takes place about a pitch bearing or hinge (Fig. 5-3). A pitch horn is rigidly attached to the blade outboard of the pitch bearing, and a pitch link is attached to the pitch horn in such a way that vertical motion of the link produces blade pitch motion. Then what is required is a way to produce a steady and 1/rev sinusoidal vertical motion of the pitch link. This arrangement or its mechanical equivalent is fairly standard in rotor designs. There are other means of producing the blade lift control, such as the Kaman servo-flap; and there are many variations in the mechanical implementation of this arrangement. All means of controlling the rotor may be viewed along these general lines, however. A widely used method of providing the blade pitch control is by means of a swashplate. A swashplate is a mechanical device that transmits the pilot's control motion in the nonrotating frame to the blade cyclic pitch motion in the rotating frame. Figure 5-7 is a schematic of the swashplate arrangement; the actual mechanical arrangement varies widely, but this figure defines the principal components that must be present in some form. The swashplate has rotating and nonrotating rings concentric with the shaft, with bearings between the two rings. The rotating ring is gimballed to the shaft in an arrangement that allows an arbitrary orientation of the plane of the swashplate relative to the rotor shaft while one ring is stationary and the other rotates. The blade pitch links attach to the rotating ring, and links from the pilot's controls attach to the stationary ring. Vertical displacement of the swashplate provides a vertical motion of the pitch links

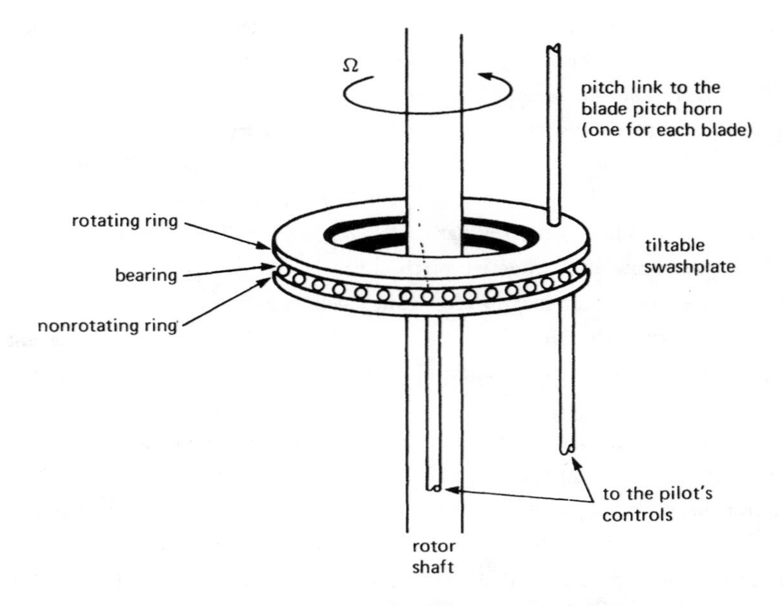

Figure 5-7 Schematic of the rotor swashplate

that is independent of azimuth, thereby changing the rotor collective pitch angle θ_0. If the swashplate is given a longitudinal tilt ϕ_{SP}, the vertical position at the pitch link exhibits a 1/rev sinusoidal variation: $z_{PL} = \phi_{SP} x_{PL} = \phi_{SP} r_{PL} \cos\psi$. Similarly, lateral tilt by ϕ_{SP} gives $z_{PL} = \phi_{SP} r_{PL} \sin\psi$. The swashplate tilt in response to the pilot's stick motion thus produces the cyclic blade pitch control, and vertical motion of the swashplate (or its equivalent, perhaps in an entirely separate mechanism) produces the collective pitch control. In general, the control system can be represented by a control plane, its tilt corresponding to cyclic control and its vertical position corresponding to collective control. Since there may be other sources of blade pitch motion, such as kinematic pitch-flap coupling, the control plane alone does not necessarily represent the blade pitch motion.

There always exists a reference plane relative to which the blade pitch has no 1/rev variation. Since the pitch angle θ as measured from this plane is constant, it is called the no-feathering plane. To locate the no-feathering plane, consider an arbitrary reference plane relative to which the cyclic pitch, θ_{1c} and θ_{1s}, is nonzero. The no-feathering plane then is obtained by rotating rearward about the lateral (y) axis by θ_{1s}, and to the left about the longitudinal (x) axis by θ_{1c}. Note that the component of these rotations about the feathering axis of the blade at azimuth ψ is $(\theta_{1c} \cos\psi + \theta_{1s} \sin\psi)$, which indeed cancels the cyclic pitch of the original reference plane. It follows that the longitudinal tilt of the no-feathering plane represents the sine cyclic θ_{1s}, while lateral tilt represents the cosine cyclic θ_{1c}. We shall find that in response to control the tip-path plane (and with it the thrust vector) tilts parallel to the no-feathering plane. Thus θ_{1s} provides longitudinal control of the helicopter and is called longitudinal cyclic, while θ_{1c} provides lateral control and is called lateral cyclic. The no-feathering plane is often used in rotor analyses, since the absence of cyclic pitch simplifies the calculations somewhat. Note that in general the no-feathering plane and control plane are not equivalent; the former represents the total blade pitch, whereas the latter represents the control system, that is, just the commanded pitch.

Now let us consider what follows from the observation that the flap and pitch angles (specifically, the 1/rev harmonics of β and θ) define the orientation of the plane of the rotor blade relative to the reference disk plane. We shall examine how β and θ transform as we change from one reference plane to another while maintaining the same orientation of the blade with respect to space. The orientation of the blade relative to space and the air has physical meaning, but the choice of a reference plane is arbitrary, although a particular reference plane may be more useful than the others for certain applications. Thus it is expected that there exist invariants of the transformation, which represent the orientation of the blade with respect to space and so must be independent of the reference plane chosen. Consider two reference planes, the second tilted forward by the angle ϕ_y relative to the first. This tilt decreases β by ϕ_y at $\psi = 0°$ and increases it by ϕ_y at $\psi = 180°$, while at $\psi = 90°$ the pitch is increased by ϕ_y and at $\psi = 270°$ it is decreased by ϕ_y. This suggests that the cyclic flap β_{1c} has been decreased by ϕ_y and the cyclic pitch θ_{1s} increased by ϕ_y as a result of the reference plane tilt. The flap and pitch angles transform in such a way that the 1/rev harmonics of β and θ as measured relative to the reference plane change by the same magnitude, but with a $90°$ shift in phase. Similarly, with a lateral tilt of the reference plane by ϕ_x, β_{1s} and θ_{1c} are decreased by the angle ϕ_x. The quantities $(\beta_{1c} + \theta_{1s})$ and $(\beta_{1s} - \theta_{1c})$ should therefore be independent of the reference plane. Now let us derive the reference plane transformation more rigorously.

The angles β and θ in the rotating frame define the orientation of the plane of the blade with respect to a particular reference plane. The components of the blade plane tilt in the nonrotating frame are ($\theta \cos\psi + \beta \sin\psi$) laterally and ($\theta \sin\psi - \beta \cos\psi$) longitudinally. Now tilt the reference plane by the angles ϕ_x laterally and ϕ_y longitudinally. Since the position of the blade in space is unchanged, the orientation of the blade relative to the first and second reference planes must be as follows:

$$(\theta)_2 \cos\psi + (\beta)_2 \sin\psi = (\theta)_1 \cos\psi + (\beta)_1 \sin\psi - \phi_x$$

$$(\theta)_2 \sin\psi - (\beta)_2 \cos\psi = (\theta)_1 \sin\psi - (\beta)_1 \cos\psi - \phi_y$$

or in the rotating frame:

$$(\theta)_2 = (\theta)_1 - \phi_x \cos\psi - \phi_y \sin\psi$$

$$(\beta)_2 = (\beta)_1 - \phi_x \sin\psi + \phi_y \cos\psi$$

which defines the transformation of the flap and pitch angles (only the first harmonics are affected). Writing β and θ as Fourier series gives

$$(\theta_{1c})_2 = (\theta_{1c})_1 - \phi_x$$

$$(\theta_{1s})_2 = (\theta_{1s})_1 - \phi_y$$

$$(\beta_{1c})_2 = (\beta_{1c})_1 + \phi_y$$

$$(\beta_{1s})_2 = (\beta_{1s})_1 - \phi_x$$

When ϕ_x and ϕ_y are eliminated, it becomes apparent that in the transformation from one reference plane to another the quantities ($\beta_{1s} - \theta_{1c}$) and ($\beta_{1c} + \theta_{1s}$) are constant. In terms of the no-feathering-plane (NFP) and tip-path-plane (TPP) variables,

$$\beta_{1s} - \theta_{1c} = (\beta_{1s})_{NFP} = -(\theta_{1c})_{TPP}$$

$$\beta_{1c} + \theta_{1s} = (\beta_{1c})_{NFP} = (\theta_{1s})_{TPP}$$

Relative to a general reference plane, θ_{1c} and θ_{1s} define the orientation of the no-feathering plane, while β_{1c} and β_{1s} define the orientation of the tip-path plane. Figure 5-8 shows that the quantities ($\beta_{1c} + \theta_{1s}$) and ($\beta_{1s} - \theta_{1c}$) are simply the longitudinal and lateral angles between the tip-path plane

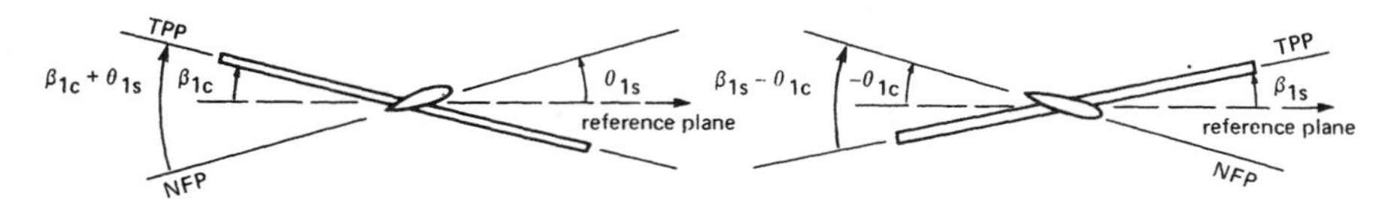

Figure 5-8 Equivalence of flapping and feathering.

and the no-feathering plane, which indeed must be independent of the choice of reference plane. The two planes of no flap motion and no pitch motion are physically relevant, hence their association with the invariants of flap and pitch in a reference plane transformation. The fact that by the reference plane transformation cyclic flapping may be exchanged for cyclic feathering, and vice versa, is referred to as the equivalence of flapping and feathering motion. It is useful to consider a blade with no offset of the flapping hinge or pitch bearing (while not mechanically practical, such a configuration is simple and well represents the basic behavior of an articulated rotor). In this case the flap and pitch hinges form a gimbal connecting the blade root to the hub, allowing an arbitrary orientation of the rotor shaft while the blade remains fixed with respect to space. In such a case the shaft has no influence on the blade aerodynamics or dynamics; only the relative orientation of the no-feathering plane and tip-path plane is significant. The analysis can therefore be conducted in the no-feathering plane or tip-path plane, ignoring the shaft orientation except to determine the actual cyclic pitch control required. Flapping-feathering equivalence then simply expresses the change in β and θ for the various possible shaft orientations. For a hingeless rotor or an articulated rotor with offset hinges, the shaft orientation relative to the no-feathering plane and tip-path plane has physical importance. The reference plane perpendicular to the rotor shaft is called the hub plane.

Figure 5-9 summarizes the various reference planes used for the helicopter rotor in forward flight. In vertical flight, the natural reference disk plane is the horizontal. With axial symmetry the tip-path plane and no-feathering plane are horizontal. The hub plane is not necessarily horizontal in vertical flight unless the helicopter center of gravity is on the rotor shaft

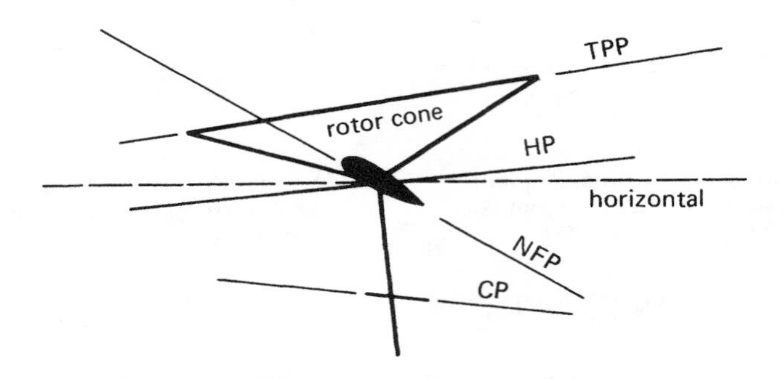

Figure 5-9 Rotor reference planes: tip-path plane (TPP), no-feathering plane (NFP), hub plane (HP), and control plane (CP).

axis, but it was not necessary to consider the hub plane in Chapters 2 and 3 since the hover analysis is primarily concerned with rotor aerodynamics. In forward flight, however, a number of reference planes have physical meaning, and owing to the asymmetry of the aerodynamics in forward flight these planes do not in general coincide with the horizontal plane or with each other. The tip-path plane (TPP) is parallel to the plane described by the blade tips, so there is no l/rev flapping motion. The orientation of the tip-path plane defines the cyclic flapping β_{1c} and β_{1s} relative to any other plane. The no-feathering plane (NFP) has no 1/rev pitch variation; its orientation thus defines the cyclic pitch θ_{1c} and θ_{1s} relative to any other plane. The control plane (CP) represents the commanded cyclic pitch from the rotor control system. It may be considered the swashplate plane. The hub plane (HP) is normal to the rotor shaft. The hub plane is the natural reference frame when there are important physical effects of the blade orientation relative to the hub, such as in the case of offset hinges or a hingeless rotor. In the hub plane there is both cyclic flapping and cyclic feathering motion. While in general no two of these planes coincide, there are special cases. For a flapping rotor with no cyclic pitch control (such as the tail rotor and some autogyros), the hub plane and control plane are equivalent; if there is no pitch-flap coupling or other pitch sources, then the control plane and no-feathering plane coincide as well. For a feathering rotor with no flapping (such as a propeller with cyclic pitch) the hub plane and tip-path plane are equivalent.

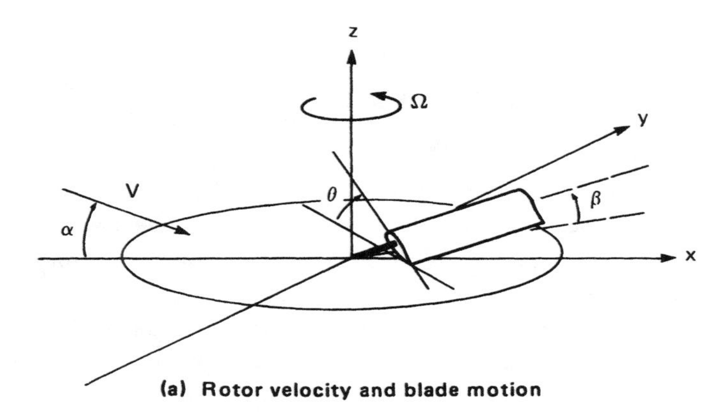

(a) Rotor velocity and blade motion

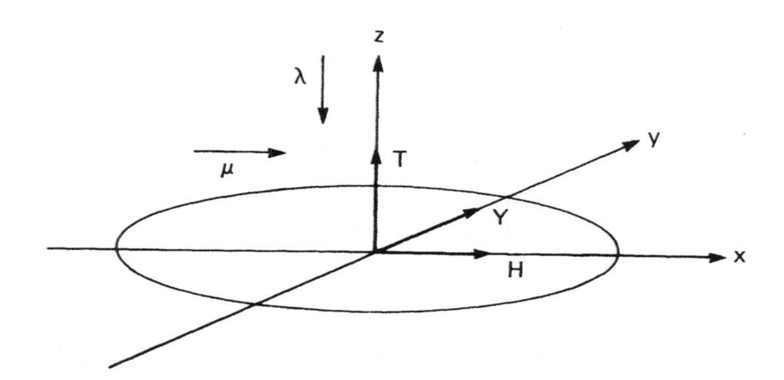

(b) Rotor forces and dimensionless velocity components

Figure 5-10 Definition of rotor motion, velocity, and forces relative to a given reference plane.

Figure 5-10 summarizes the quantities defining the rotor motion, velocity, and forces relative to a given reference plane. In the nonrotating axis system, x and y lie in the reference plane and z is normal to it. The flap and pitch angles are measured relative to the reference plane. The forward velocity has magnitude V and lies in the x-z plane at an angle of attack α (positive for forward tilt of the disk). The rotor induced velocity v is assumed to be normal to the reference plane. The advance ratio μ and inflow ratio λ are the respective dimensionless velocity components parallel to and normal to the reference plane:

$$\mu = \frac{V\cos\alpha}{\Omega R}$$

$$\lambda = \frac{V\sin\alpha + v}{\Omega R} = \mu\tan\alpha + \lambda_i$$

For small disk inclination, $\mu \cong V/\Omega R$ and $\lambda \cong \mu\alpha + \lambda_i$, so that while α (and hence λ also) depends on the reference plane orientation, the advance ratio μ is approximately independent of the reference plane used. The rotor force components are also defined relative to the reference plane chosen: the thrust T is normal to the disk, while the rotor drag force H and side force Y are in the reference plane. The coefficients are defined as usual:

$$C_T = T/\rho A(\Omega R)^2$$

$$C_H = H/\rho A(\Omega R)^2$$

$$C_Y = Y/\rho A(\Omega R)^2$$

Similarly, the hub moments C_{M_x} and C_{M_y} and the torque C_Q are defined relative to this reference plane. The resultant force of the rotor must be independent of the reference plane. Since the thrust is normally much greater than the drag or side force, it follows that the rotor thrust is itself approximately independent of the reference plane. In the preceding paragraphs, the behavior of the flap and pitch angles during a transformation of the reference plane was obtained. For a longitudinal rotation by ϕ_y and a lateral rotation by ϕ_x, we found that

$$(\theta_{1c})_2 = (\theta_{1c})_1 - \phi_x$$

$$(\theta_{1s})_2 = (\theta_{1s})_1 - \phi_y$$

$$(\beta_{1c})_2 = (\beta_{1c})_1 + \phi_y$$

$$(\beta_{1s})_2 = (\beta_{1s})_1 - \phi_x$$

Hence the quantities $(\beta_{1s} - \theta_{1c})$ and $(\beta_{1c} + \theta_{1s})$, which describe the orientation of the tip-path plane relative to the no-feathering plane, are invariant under the reference plane transformation. The invariants of the rotor forces and velocities are also of interest. For the above tilt of the reference plane, the velocity components, disk incidence, and force components transform as follows:

$$(\lambda)_2 = (\lambda)_1 - (\mu)_1 \phi_y$$

$$(\mu)_2 = (\mu)_1 + (\lambda)_1 \phi_y \cong (\mu)_1$$

$$(\alpha)_2 = (\alpha)_1 - \phi_y$$

$$(T)_2 = (T)_1 - \phi_x (Y)_1 + \phi_y (H)_1 \cong (T)_1$$

$$(H)_2 = (H)_1 - \phi_y (T)_1$$

$$(Y)_2 = (Y)_1 + \phi_x (T)_1$$

By eliminating ϕ_x and ϕ_y the quantities that are independent of the reference plane orientation can be obtained. Note that to the order considered here the advance ratio μ and rotor thrust T are themselves invariant. It is most useful to express the velocity and force invariants in forms relating quantities to the tip-path plane or the no-feathering plane:

$$\lambda = \lambda_{NFP} + \mu\theta_{1s} = \lambda_{TPP} - \mu\beta_{1c}$$

$$\alpha = \alpha_{NFP} + \theta_{1s} = \alpha_{TPP} - \beta_{1c}$$

$$H = H_{NFP} + \theta_{1s}T = H_{TPP} - \beta_{1c}T$$

$$Y = Y_{NFP} - \theta_{1c}T = Y_{TPP} - \beta_{1s}T$$

For the inflow ratio, $\lambda = \mu\alpha + \lambda_i = \mu(\alpha_{NFP} + \theta_{1s}) + \lambda_i = \mu(\alpha_{TPP} - \beta_{1c}) + \lambda_i$.

Now we shall begin the analysis of the aerodynamics and dynamics of the helicopter rotor in forward flight. At first only the simplest possible case is considered: a fully articulated rotor with no hinge offset, no hinge spring, and no pitch-flap coupling; rigid flapping is the only blade motion, with rigid pitch control; and the effects of reverse flow, tip loss, and root cutout are neglected. First the aerodynamics of the blade in forward flight will be derived, and the forces on the rotor obtained. Then the dynamics of the blade flapping motion will be investigated. The remaining sections of the chapter consider some of the factors neglected with this simple model.

5–2 Aerodynamics of Forward Flight

This section derives the aerodynamic forces on the rotor blade in forward flight. A fully articulated rotor is considered, with no hinge offset. The blade motion consists of rigid flapping β, and rigid pitch due to collective and cyclic control inputs. Elastic bending and torsion of the blade is neglected. Such a model is sufficient to determine the performance and control characteristics of an articulated rotor. Blade element theory is used to find the section aerodynamic forces. The effects of the reverse flow region are neglected for now. A general reference plane is considered.

Blade element theory assumes that each blade section acts as a two-dimensional airfoil for which the influence of the rotor wake consists entirely of an induced velocity at the section. Two-dimensional airfoil characteristics can then be used to evaluate the section loads in terms of the blade motion and aerodynamic environment at that section alone. The induced velocity may be obtained by various means: momentum theory, vortex theory, or nonuniform inflow calculations. Blade element theory requires that the aspect ratio be high, which is normally true for rotary wings. However, near the blade tip or in the large induced velocity gradients of a vortex-blade interaction, lifting-surface theory must be used for accurate results.

Consider the velocities seen by the blade section (Fig. 5-11). The blade section pitch θ is measured from the reference plane to the zero-lift line; it includes the collective and cyclic pitch control and the built-in twist of the

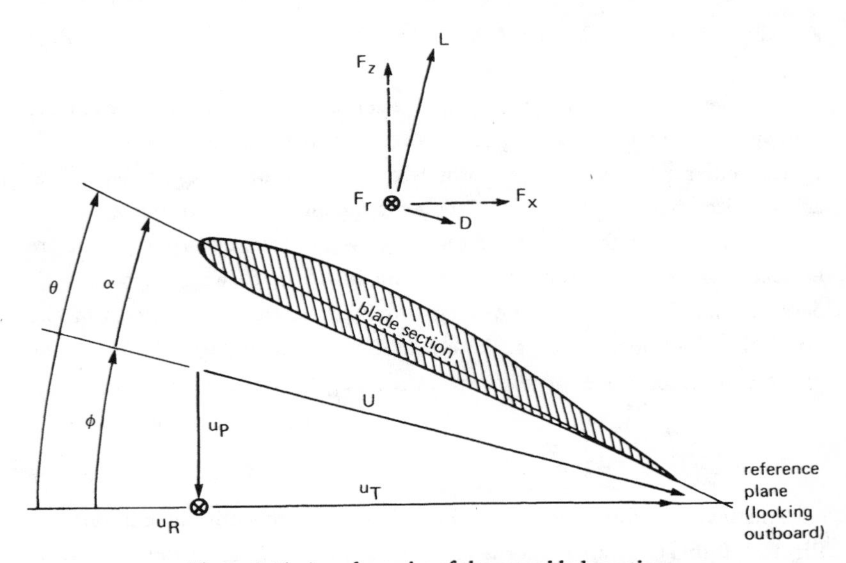

Figure 5-11 Aerodynamics of the rotor blade section.

blade. The components of the velocity of the air relative to the blade are u_T (tangential to the disk plane, positive toward the trailing edge), u_P (perpendicular to the disk plane, positive downward), and u_R (radial, positive outward). The resultant velocity and inflow angle of the section are $U = (u_T^2 + u_P^2)^{1/2}$ and $\phi = \tan^{-1} u_P/u_T$. The section angle of attack then is $\alpha = \theta - \phi$. The velocity seen by the blade is due to the rotor rotation, the helicopter forward speed and induced velocity, and the blade flap motion. To lowest order the tangential and radial components u_T and u_R are due solely to the rotor rotation and advance ratio (see Fig. 5-12):

$$u_T = r + \mu \sin \psi$$

$$u_R = \mu \cos \psi$$

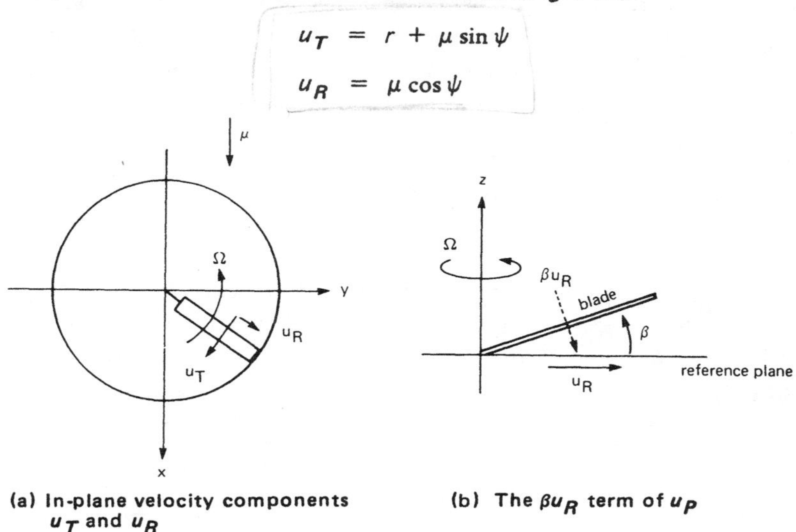

Figure 5-12 Air velocity relative to the blade in forward flight.

These components are then also independent of the reference plane used. The normal velocity u_P has three terms: $\Omega R \lambda$, which is the induced velocity plus the component of the free stream velocity normal to the rotor disk (recall $\lambda = \mu\alpha + \lambda_i$); $r d\beta/dt$, which is the angular velocity of the blade about the flap hinge; and $\Omega R \beta \mu \cos \psi$, which is a component of the radial velocity u_R normal to the blade when it is flapped up by the angle β (see Fig. 5-12). Thus the dimensionless normal velocity is

$$u_P = \lambda + r\dot{\beta} + \beta\mu \cos \psi$$

Note that each term in u_P depends on the reference plane. In deriving these expressions for the velocity components, it has been assumed that the flap angle β is small. Finally, while the pitch angle and velocity components depend on the reference plane, the section aerodynamic environment defined by the resultant velocity and angle of attack must be fixed. For small angles $U \cong u_T$, and $\alpha \cong \theta - u_P/u_T$ is easily shown to be invariant during a reference plane transformation.

Figure 5-11 also shows the aerodynamic forces on the blade section. The aerodynamic lift and drag (L and D) are respectively normal to and parallel to the resultant velocity U. The components of the section lift and drag resolved in the reference plane are F_z and F_x (normal and in-plane, respectively). The radial force on the section is F_r, defined to be positive when outward. The section forces can be expressed in terms of the lift and drag coefficients:

$$L = \tfrac{1}{2}\rho U^2 c c_l$$

$$D = \tfrac{1}{2}\rho U^2 c c_d$$

where c_l and c_d are in general functions of the section angle of attack α, and the Mach number $M = M_{\text{tip}} U$. Here ρ is the air density (which is omitted when dimensionless quantities are used), c is the blade chord, and $M_{\text{tip}} = \Omega R/c_s$ is the tip Mach number. The normal and in-plane forces are

$$F_z = L\cos\phi - D\sin\phi$$

$$F_x = L\sin\phi + D\cos\phi$$

(The first term in F_x is the induced drag, the second term the profile drag.) The blade radial force is

$$F_r = -\beta F_z + D_{\text{radial}}$$

The first term is the radial component of the normal force when the blade flaps up. The second term is a radial drag force due to radial flow along the blade, which will not be considered until section 5–12. Now substitute for L and D in terms of the section coefficients, divide by the chord c and the section two-dimensional lift curve slope a, and use dimensionless quantities. The resulting section forces in the reference axis system are

$$\frac{F_z}{ac} = U^2\left(\frac{c_l}{2a}\cos\phi - \frac{c_d}{2a}\sin\phi\right)$$

$$\frac{F_x}{ac} = U^2\left(\frac{c_l}{2a}\sin\phi + \frac{c_d}{2a}\cos\phi\right)$$

$$\frac{F_r}{ac} = -\beta\frac{F_z}{ac}$$

Next we make the small angle assumption and neglect stall and compressibility effects. Assuming λ, β, ϕ, and θ are all small angles, it follows that u_P/u_T and α are small; that $\phi \cong u_P/u_T$, $\sin\phi \cong \phi$, and $\cos\phi \cong 1$; that $U^2 \cong u_T^2$, and $\alpha \cong \theta - u_P/u_T$. Assuming a constant lift curve slope, then $c_\ell \cong a\alpha$; neglecting stall gives $c_d/c_\ell \ll 1$, so that with the small angle assumption $F_z \cong L$ and $F_x \cong L\phi + D$. Hence the section aerodynamic forces become

$$\frac{F_z}{ac} = \tfrac{1}{2}u_T^2\alpha = \tfrac{1}{2}(u_T^2\theta - u_P u_T)$$

$$\frac{F_x}{ac} = u_T^2\left(\frac{\alpha}{2}\phi + \frac{c_d}{2a}\right) = \tfrac{1}{2}(u_P u_T\theta - u_P^2) + \frac{c_d}{2a}u_T^2$$

$$\frac{F_r}{ac} = -\beta\frac{F_z}{ac}$$

5–3 Rotor Aerodynamic Forces

Now we shall derive the aerodynamic forces acting on the rotor. A general reference plane is used, although some of the results will be examined in the no-feathering plane and tip-path plane. The thrust T is normal to the rotor disk; the rotor drag force H is in the disk plane, positive aft; and the rotor side force Y is in the disk plane, positive toward the advancing side (see Fig. 5-10). The rotor drag and side forces are usually small in the tip-path plane, so that in general H/T and Y/T are of the order of the tip-path-plane tilt angles. In addition, there is a torque moment Q on the rotor, positive for a rotor absorbing power. For an articulated rotor with no flap hinge offset, there can be no net pitch or roll moment transmitted to the rotor hub. The rotor forces are obtained by integrating the blade section forces along the span. The rotor thrust is due to the normal force F_z, the drag and side forces are due to the in-plane forces F_x and F_r resolved in the nonrotating frame, and the torque is due to the in-plane force F_x. Multiplying by the number of blades N to obtain the forces on the entire rotor, the aerodynamic forces are

$$T = N \int_0^R F_z dr$$

$$H = N \int_0^R (F_x \sin\psi + F_r \cos\psi) dr$$

$$Y = N \int_0^R (-F_x \cos\psi + F_r \sin\psi) dr$$

$$Q = N \int_0^R r F_x dr$$

It is also necessary to average these expressions over the azimuth [by the operator $(1/2\pi) \int_0^{2\pi} (\ldots) d\psi$] to obtain the steady forces. Note that the rotor thrust coefficient is $C_T = T/[\rho A(\Omega R)^2] = (N/\pi) \int_0^R F_z/[\rho R(\Omega R)^2](dr/R) = (Nc/\pi R) \int_0^R F_z/[\rho c(\Omega R)^2](dr/R)$ or, using dimensionless quantities, $C_T = \int_0^1 \sigma(F_z/c) dr$. In general, the blade chord may be a function of r, but here only a constant chord will be considered. Then the solidity σ is a constant, and the rotor thrust coefficient is

$$\frac{C_T}{\sigma a} = \int_0^1 \frac{F_z}{ac} dr$$

Similarly,

$$\frac{C_H}{\sigma a} = \int_0^1 \frac{F_x}{ac} \sin\psi + \frac{F_r}{ac} \cos\psi \; dr$$

$$\frac{C_Y}{\sigma a} = \int_0^1 -\frac{F_x}{ac} \cos\psi + \frac{F_r}{ac} \sin\psi \; dr$$

$$\frac{C_Q}{\sigma a} = \int_0^1 r \frac{F_x}{ac} dr$$

Assuming small angles, and neglecting the tip loss and root cutout, we may substitute for the section forces to obtain

$$\frac{C_T}{\sigma a} = \int_0^1 \tfrac{1}{2}(u_T^2\theta - u_P u_T)dr$$

$$\frac{C_H}{\sigma a} = \int_0^1 \left\{ \sin\psi \left[\tfrac{1}{2}(u_P u_T \theta - u_P^2) + \frac{c_d}{2a} u_T^2 \right] - \beta\cos\psi \left[\tfrac{1}{2}(u_T^2\theta - u_P u_T) \right] \right\} dr$$

$$\frac{C_Y}{\sigma a} = \int_0^1 \left\{ -\cos\psi \left[\tfrac{1}{2}(u_P u_T \theta - u_P^2) + \frac{c_d}{2a} u_T^2 \right] - \beta\sin\psi \left[\tfrac{1}{2}(u_T^2\theta - u_P u_T) \right] \right\} dr$$

$$\frac{C_Q}{\sigma a} = \int_0^1 r \left[\tfrac{1}{2}(u_P u_T \theta - u_P^2) + \frac{c_d}{2a} u_T^2 \right] dr$$

where

$$u_T = r + \mu\sin\psi$$

$$u_P = \lambda + r\dot{\beta} + \beta\mu\cos\psi$$

$$\beta = \beta_0 + \beta_{1c}\cos\psi + \beta_{1s}\sin\psi + \beta_{2c}\cos 2\psi + \beta_{2s}\sin 2\psi + \ldots$$

$$\theta = \theta_0 + \theta_{tw} r + \theta_{1c}\cos\psi + \theta_{1s}\sin\psi$$

Linear twist has been assumed, and usually uniform inflow will be used. The flap motion has been written as a complete Fourier series, but in fact only the mean and first harmonics will be considered for most of this chapter.

The moments on the rotor hub may be obtained in a similar fashion. The pitch moment M_y and roll moment M_x (positive rearward and toward the retreating side, respectively) are

$$M_y = -N \int_0^R \cos\psi \, r F_z \, dr$$

$$M_x = N \int_0^R \sin\psi \, r F_z \, dr$$

Then in coefficient form

$$\frac{C_{M_y}}{\sigma a} = -\int_0^1 \cos\psi \frac{F_z}{ac} r \, dr$$

$$\frac{C_{M_x}}{\sigma a} = \int_0^1 \sin\psi \frac{F_z}{ac} r \, dr$$

Note that the root flapping moment on a single blade in the rotating frame is $M_F = \int_0^R r F_z \, dr$. From writing M_F as a Fourier series and remembering that the rotor forces and moments must be averaged over the azimuth, it follows that the pitch and roll moments are

$$M_y = -\frac{N}{2} M_{F_{1c}}$$

$$M_x = \frac{N}{2} M_{F_{1s}}$$

Hence the 1/rev flap moments at the center of rotation lead to the steady pitch and roll moments on the helicopter. In the case of an articulated rotor with the flap hinge at the center of rotation there is no moment on the hinge and for that reason there can be no hub moment acting on the helicopter. In general, it will be found that the pitch and roll moment can be related to the rotor tip-path-plane tilt, which is a measure of the 1/rev flapping moments.

It is convenient to separate the drag and side forces and the torque into two terms: a profile term due to the drag coefficient c_d, and an induced term due to the lift coefficient c_l. The former is denoted by the subscript o, and the latter by i. While such a separation is suggested by the division of induced and profile power, it is not quite consistent here because the induced terms will include the inflow ratio λ, part of which is due to the disk tilt necessary to encounter the rotor profile drag C_{H_o}. The division here is strictly formal, based on whether the source of the section force is the drag coefficient or the lift coefficient. In section 5-4 the rotor profile power and induced power are abtained from these expressions in accord with the definitions of the previous chapters. Thus $C_H = C_{H_i} + C_{H_o}$, $C_Y = C_{Y_i} + C_{Y_o}$, and $C_Q = C_{Q_i} + C_{Q_o}$ (the rotor thrust has no drag terms), where

$$C_{H_o} = \int_0^1 \frac{\sigma c_d}{2} \sin\psi\, u_T^2\, dr$$

$$C_{Y_o} = \int_0^1 \frac{\sigma c_d}{2} (-\cos\psi)\, u_T^2\, dr$$

$$C_{Q_o} = \int_0^1 \frac{\sigma c_d}{2}\, r u_T^2\, dr$$

and

$$C_{H_i} = \frac{\sigma a}{2} \int_0^1 (u_T\theta - u_P)(u_P \sin\psi - u_T\beta\cos\psi)\,dr$$

$$C_{Y_i} = \frac{\sigma a}{2} \int_0^1 (u_T\theta - u_P)(-u_P \cos\psi - u_T\beta\sin\psi)\,dr$$

$$C_{Q_i} = \frac{\sigma a}{2} \int_0^1 r(u_P u_T\theta - u_P^2)\,dr$$

Furthermore,

$$u_P \sin\psi - u_T\beta\cos\psi = \lambda\sin\psi + r\dot{\beta}\sin\psi - r\beta\cos\psi$$

$$u_P \cos\psi + u_T\beta\sin\psi = \lambda\cos\psi + r\dot{\beta}\cos\psi + r\beta\sin\psi + \mu\beta$$

The next step is to average over the rotor azimuth. By using the definitions of the Fourier coefficients, integrals of θ and β may be replaced by the appropriate harmonics. For example, the rotor thrust requires the term

$$\frac{1}{2\pi}\int_0^{2\pi} \theta\, u_T^2\, d\psi = \frac{1}{2\pi}\int_0^{2\pi} \theta\left[r^2 + 2r\mu\sin\psi + \tfrac{1}{2}\mu^2\left(1 - \cos 2\psi\right)\right] d\psi$$

$$= \left(\theta_0 + r\theta_{tw}\right)\left[r^2 + (\mu^2/2)\right] + \theta_{1s} r\mu - \theta_{2c}(\mu^2/4)$$

where the definitions

$$\theta_0 = \frac{1}{2\pi}\int_0^{2\pi} \theta\, d\psi, \quad \theta_{1s} = \frac{1}{\pi}\int_0^{2\pi} \theta \sin\psi\, d\psi, \quad \theta_{2c} = \frac{1}{\pi}\int_0^{2\pi} \theta\cos 2\psi\, d\psi$$

have been used. Also required is the term

$$\frac{1}{2\pi}\int_0^{2\pi} u_P u_T\, d\psi = \frac{1}{2\pi}\int_0^{2\pi} \left(\lambda + r\dot{\beta} + \mu\beta\cos\psi\right)\left(r + \mu\sin\psi\right) d\psi$$

$$= \lambda r + (\mu^2/4)\beta_{2s}$$

using the definition of β_{2s} and noting that

$$\int_0^{2\pi} \left(r^2\dot{\beta} + r\mu\dot{\beta}\sin\psi + r\mu\beta\cos\psi\right) d\psi = \int_0^{2\pi} \left(r^2\beta + r\mu\beta\sin\psi\right)^{\bullet} d\psi = 0$$

since the quantity $(r^2\beta + r\mu\beta\sin\psi)$ is periodic. There is no higher harmonic control, so $\theta_{2c} = 0$; and the higher harmonics of flapping are small, so β_{2s} is neglected. Thus the rotor thrust coefficient is

$$C_T = \frac{\sigma a}{2}\int_0^1 \left[\left(\theta_0 + r\theta_{tw}\right)\left(r^2 + (\mu^2/2)\right) + \theta_{1s} r\mu - \lambda r\right] dr$$

$$= \frac{\sigma a}{2}\left[\frac{\theta_0}{3}\left(1 + \frac{3}{2}\mu^2\right) + \frac{\theta_{tw}}{4}\left(1 + \mu^2\right) + \frac{\mu}{2}\theta_{1s} - \frac{\lambda}{2}\right]$$

Similarly, the induced drag and side force terms are

$$C_{H_i} = \frac{\sigma a}{2}\left[\theta_0\left(-\frac{1}{3}\beta_{1c} + \frac{1}{2}\mu\lambda\right) + \theta_{tw}\left(-\frac{1}{4}\beta_{1c} + \frac{1}{4}\mu\lambda\right) - \frac{1}{6}\theta_{1c}\beta_0 + \theta_{1s}\left(-\frac{1}{4}\mu\beta_{1c} + \frac{1}{4}\lambda\right) + \frac{3}{4}\lambda\beta_{1c} + \frac{1}{6}\beta_0\beta_{1s} + \frac{1}{4}\mu\left(\beta_0^2 + \beta_{1c}^2\right)\right]$$

$$C_{Y_i} = -\frac{\sigma a}{2}\left[\theta_0\left(\frac{3}{4}\mu\beta_0 + \frac{1}{3}\beta_{1s}\left(1 + \frac{3}{2}\mu^2\right)\right) + \theta_{tw}\left(\frac{1}{2}\mu\beta_0 + \frac{1}{4}\beta_{1s}\left(1 + \mu^2\right)\right) + \theta_{1c}\left(\frac{1}{4}\lambda + \frac{1}{4}\mu\beta_{1c}\right) + \theta_{1s}\left(\frac{1}{6}\beta_0\left(1 + 3\mu^2\right) + \frac{1}{2}\mu\beta_{1s}\right) - \frac{3}{4}\lambda\beta_{1s} + \beta_0\beta_{1c}\left(\frac{1}{6} - \mu^2\right) - \frac{3}{2}\mu\lambda\beta_0 - \frac{1}{4}\mu\beta_{1c}\beta_{1s}\right]$$

The induced torque is considered in section 5-4. For the profile terms, it is assumed that the section drag coefficient is constant over the entire rotor disk and has an appropriate mean value c_{d_0}. Then averaging over the azimuth gives

$$C_{H_o} = \int_0^1 \frac{\sigma c_d}{2} \sin\psi \, u_T^2 \, dr = \int_0^1 \frac{\sigma c_d}{2} r\mu dr = \frac{\sigma c_{d_0}}{4}\mu$$

$$C_{Y_o} = \int_0^1 \frac{\sigma c_d}{2}\left(-\cos\psi\right) u_T^2 \, dr = 0$$

$$C_{Q_o} = \int_0^1 \frac{\sigma c_d}{2} r u_T^2 \, dr = \int_0^1 \frac{\sigma c_d}{2} r\left(r^2 + \frac{\mu^2}{2}\right) dr = \frac{\sigma c_{d_0}}{8}\left(1 + \mu^2\right)$$

The profile side force is always zero because of the symmetry of the flow, as long as the variation of the drag coefficient is neglected. These results have been obtained neglecting reverse flow and radial flow effects. In section 5-12 the profile drag force, torque, and power will be extended to include reverse flow, radial flow, and the radial drag force. Note that since the radial drag cannot produce a torque on the rotor, and C_{Y_o} remains zero because of symmetry, the only influence of radial drag is on C_{H_o}.

In terms of the blade pitch at 75% radius ($\theta_{.75} = \theta_0 + \frac{3}{4}\theta_{tw}$) and the inflow in the no-feathering plane ($\lambda_{NFP} = \lambda - \mu\theta_{1s}$), the rotor thrust in forward flight is

$$C_T = \frac{\sigma a}{2}\left[\frac{\theta_{.75}}{3}\left(1 + \frac{3}{2}\mu^2\right) - \frac{\theta_{tw}}{8}\mu^2 - \frac{\lambda_{NFP}}{2}\right]$$

While this is the most compact form, it is the inflow relative to the tip-path plane [$\lambda - \mu\theta_{1s} = \lambda_{TPP} - \mu(\beta_{1c} + \theta_{1s})$] that has the most physical significance, since the angle of attack of the tip-path plane is determined directly by the drag of the helicopter and rotor. Thus the angle of the tip-path plane relative to the no-feathering plane ($\beta_{1c} + \theta_{1s}$) is needed to complete the evaluation of the thrust.

Relative to the tip-path plane, the rotor drag force is

$$C_{H_{TPP}} = \frac{\sigma c_{d_o}}{4}\mu + \frac{\sigma a}{2}\left[\frac{1}{2}\mu\lambda_{TPP}\left(\theta_0 + \frac{1}{2}\theta_{tw}\right) - \frac{1}{6}\theta_{1c}\beta_0 + \theta_{1s}\frac{1}{4}\lambda_{TPP} + \frac{1}{4}\mu\beta_0^2\right]$$

Then for a general reference plane the rotor drag is found by adding the term due to tilt of the thrust vector, $C_H = C_{H_{TPP}} - \beta_{1c}C_T$. Similarly, C_H in the no-feathering plane can be found by dropping the θ_{1c} and θ_{1s} terms from the general result, and then $C_H = C_{H_{NFP}} + \theta_{1s}C_T$. The rotor side force in the tip-path plane is

$$C_{Y_{TPP}} = -\frac{\sigma a}{2}\left[\frac{3}{4}\mu\beta_0\left(\theta_0 + \frac{2}{3}\theta_{tw}\right) + \theta_{1c}\frac{1}{4}\lambda_{TPP} + \theta_{1s}\frac{1}{6}\beta_0\left(1 + 3\mu^2\right) - \frac{3}{2}\mu\beta_0\lambda_{TPP}\right]$$

and then $C_Y = C_{Y_{TPP}} - \beta_{1s}C_T$ (or $C_Y = C_{Y_{NFP}} - \theta_{1c}C_T$). Since $C_{H_{TPP}}/C_T$ and $C_{Y_{TPP}}/C_T$ are usually small, the rotor thrust vector is tilted from the normal to the tip-path plane by only ½° to 1° or less in forward flight (the tilt is proportional to μ).

5–4 Power in Forward Flight

The expression for the rotor torque obtained in the preceding section is

$$\frac{C_Q}{\sigma a} = \int_0^1 r \frac{F_x}{ac} dr = \int_0^1 r \left[\frac{1}{2}\left(u_P u_T \theta - u_P^2\right) + \frac{c_d}{2a} u_T^2\right] dr$$

By integrating this expression over the disk, as for the rotor force coefficients, the torque may be evaluated. With all power transmitted to the rotor through the shaft, $P = \Omega Q$ or $C_P = C_Q$. An alternative procedure, which yields a simpler result, is the energy balance formulation of the rotor power. The energy expression is also more general, since it is not necessary to make many of the assumptions required in the force balance method.

Consider the general expressions for the rotor thrust, drag force, and torque in terms of the section forces:

$$\frac{C_T}{\sigma a} = \int_0^1 \frac{F_z}{ac} dr$$

$$\frac{C_H}{\sigma a} = \int_0^1 \left(\frac{F_x}{ac} \sin \psi + \frac{F_r}{ac} \cos \psi\right) dr$$

$$\frac{C_Q}{\sigma a} = \int_0^1 r \frac{F_x}{ac} dr$$

(as usual, an average over the azimuth is also required). In terms of the section lift and drag,

$$\frac{F_z}{ac} = U^2\left(\frac{c_\ell}{2a} \cos \phi - \frac{c_d}{2a} \sin \phi\right)$$

$$\begin{aligned}\frac{F_x}{ac} &= U^2\left(\frac{c_\ell}{2a} \sin \phi + \frac{c_d}{2a} \cos \phi\right) \\ &= \tan \phi \frac{F_z}{ac} + U^2 \frac{c_d}{2a} \frac{1}{\cos \phi} \\ &= \frac{u_P}{u_T} \frac{F_z}{ac} + \frac{U^3}{u_T} \frac{c_d}{2a}\end{aligned}$$

$$\frac{F_r}{ac} = -\eta'\beta \frac{F_z}{ac}$$

where $\tan\phi = u_P/u_T$ and $U^2 = u_T^2 + u_P^2$. No small angle assumptions have been made here. The radial drag force contribution to F_r has been neglected, however (see section 5-12). Then

$$\frac{C_H}{\sigma a} = \int_0^1 \left(\sin\psi \frac{u_P}{u_T} - \eta'\beta\cos\psi\right)\frac{F_z}{ac}\,dr + \int_0^1 \sin\psi \frac{U^3}{u_T}\frac{c_d}{2a}\,dr$$

$$= \frac{C_{H_i}}{\sigma a} + \frac{C_{H_o}}{\sigma a}$$

$$\frac{C_Q}{\sigma a} = \int_0^1 r\frac{u_P}{u_T}\frac{F_z}{ac}\,dr + \int_0^1 r\frac{U^3}{u_T}\frac{c_d}{2a}\,dr$$

$$= \frac{C_{Q_i}}{\sigma a} + \frac{C_{Q_o}}{\sigma a}$$

In C_{H_i} an important generalization has been introduced. The out-of-plane blade deflection considered here is $z = \beta\eta(r)$, where η is an arbitrary radial mode shape. For rigid flapping $\eta = r$, but with hinge offset or a hingeless rotor a more complicated mode shape is required. Then the normal velocity becomes $u_P = \lambda + \eta\dot{\beta} + \eta'\beta\mu\cos\psi$, and the radial force is $F_r = -\eta'\beta F_z$, from which the expression for C_{H_i} follows. Now consider the quantity $C_{Q_i} + \mu C_{H_i}$,

$$C_{Q_i} + \mu C_{H_i} = \sigma a \int_0^1 \left(r\frac{u_P}{u_T} + \mu\sin\psi\frac{u_P}{u_T} - \mu\cos\psi\,\eta'\beta\right)\frac{F_z}{ac}\,dr$$

$$= \sigma a \int_0^1 \left(u_P - \eta'\beta\mu\cos\psi\right)\frac{F_z}{ac}\,dr$$

$$= \sigma a \int_0^1 \left(\lambda + \eta\dot{\beta}\right)\frac{F_z}{ac}\,dr$$

The second term is zero, as will now be shown. Note that $\eta\dot{\beta} = \dot{z}$ is the velocity of the out-of-plane motion of the blade, so that $\int_0^{2\pi} \eta\dot{\beta}F_z d\psi = \int_0^{2\pi} \dot{z}F_z d\psi = \oint F_z dz$ is the work done on the blade section by the periodic aerodynamic force F_z during one revolution of the rotor. The total energy of the blade from one revolution to the next must be unchanged, however, since the steady-state rotor motion is periodic. Then the net work done on the blade during one revolution must be zero. This result can also be obtained by considering the dynamics of the blade flap motion. The following differential equation for the blade flapping motion is derived later in this chapter:

$$\ddot{\beta} + \nu^2\beta = \gamma \int_0^1 \eta \frac{F_z}{ac} dr$$

where ν is the natural frequency. Then averaging over the azimuth gives

$$\begin{aligned}
\frac{1}{2\pi}\int_0^{2\pi}\left(\int_0^1 \eta\dot{\beta}\frac{F_z}{ac}dr\right)d\psi &= \frac{1}{2\pi}\int_0^{2\pi}\dot{\beta}\left(\int_0^1 \eta\frac{F_z}{ac}dr\right)d\psi \\
&= \frac{1}{2\pi}\int_0^{2\pi}\frac{1}{\gamma}\dot{\beta}\left(\ddot{\beta} + \nu^2\beta\right)d\psi \\
&= \frac{1}{2\pi}\int_0^{2\pi}\frac{1}{2\gamma}\frac{d}{d\psi}\left(\dot{\beta}^2 + \nu^2\beta^2\right)d\psi \\
&= 0
\end{aligned}$$

since the energy of the out-of-plane motion $(\dot{\beta}^2 + \nu^2\beta^2)$ is periodic. This result is for an arbitrary flapping mode shape and frequency and hence applies to all rotors. The same result is obtained when more than one mode is used to represent the out-of-plane blade deflection. Thus

$$C_{Q_i} + \mu C_{H_i} = \sigma a \int_0^1 \lambda \frac{F_z}{ac} dr = \int \lambda dC_T$$

It follows that the rotor power may be written

$$C_P = C_Q = (C_{Q_i} + \mu C_{H_i}) - \mu C_H + (C_{Q_o} + \mu C_{H_o})$$
$$= \int \lambda dC_T - \mu C_H + (C_{Q_o} + \mu C_{H_o})$$

The expressions in section 5-3 for C_T and C_H could now be used to evaluate C_P. Here we are proceeding in another direction, which requires an expression for the induced velocity $\lambda = \mu \tan \alpha + \lambda_i$ and hence for the rotor disk angle of attack.

Consider force equilibrium for the helicopter in steady flight (Fig. 5-13).

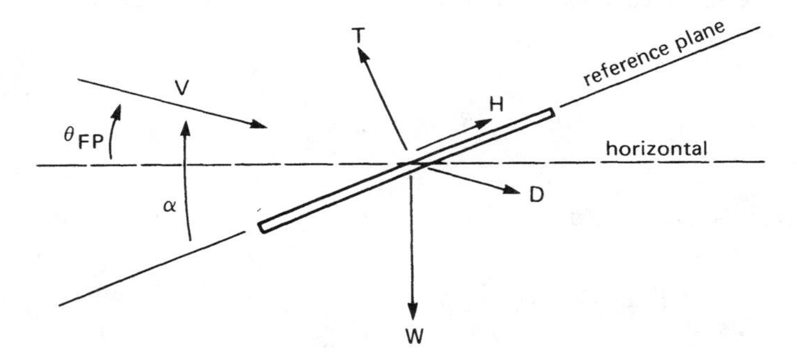

Figure 5-13 Longitudinal forces acting on the helicopter.

The rotor thrust T and drag H are defined relative to the reference plane used. The helicopter drag D is directed opposite to the free stream velocity V. The remaining force on the helicopter is the weight W, which is vertical. Auxiliary propulsion or lifting devices on the helicopter may be included by subtracting their forces from W and D. The helicopter flight path angle θ_{FP} gives a climb speed $V_c = V \sin \theta_{FP} (\lambda_c = V_c/\Omega R)$. For small angles $W = T$, and from the conditions for longitudinal force equilibrium $D + H = T(\alpha - \theta_{FP})$ or $\alpha = \theta_{FP} + (D + H)/T$. The inflow ratio then becomes

$$\lambda = \lambda_i + \mu\alpha = \lambda_i + \lambda_c + \mu \frac{D}{W} + \mu \frac{C_H}{C_T}$$

The small angle assumption is not really necessary, but it simplifies the analysis. A derivation valid for large angles is given in section 5-18.

Substituting for the inflow ratio λ in the power expression gives

$$C_P = \int \lambda dC_T - \mu C_H + (C_{Q_o} + \mu C_{H_o})$$

$$= \int \lambda_i dC_T + (C_{Q_o} + \mu C_{H_o}) + \mu \frac{D}{W} C_T + \lambda_c C_T$$

$$= C_{P_i} + C_{P_o} + C_{P_p} + C_{P_c}$$

C_{P_i} is the rotor induced power, which is required to produce the thrust; C_{P_o} is the rotor profile power, required to turn the rotor in a viscous fluid; C_{P_p} is the rotor parasite power, required to overcome the drag of the helicopter; and C_{P_c} is the rotor climb power, required to increase the gravitational potential energy. This is the energy balance expression for the helicopter performance in forward flight; it relates the power required to all the sources of energy loss. Note that the energy balance expression is independent of the reference plane used.

The rotor induced power is $C_{P_i} = \int \lambda_i dC_T$, where $dC_T = \sigma a(F_z/ac)dr$ (and an average over the azimuth is required as well). With uniform inflow this is simply $C_{P_i} = \lambda_i C_T$. For forward flight above about $\mu = 0.1$, $\lambda_i \cong \kappa C_T/2\mu$ is a good approximation, hence $C_{P_i} \cong \kappa C_T^2/2\mu$. The empirical factor κ accounts for tip loss, nonuniform inflow, and other losses.

The profile power is

$$C_{P_o} = C_{Q_o} + \mu C_{H_o} = \int_0^1 \frac{\sigma c_d}{2}\left(r\frac{U^3}{u_T} + \mu \sin\psi \frac{U^3}{u_T}\right) dr$$

$$= \int_0^1 \frac{\sigma c_d}{2} U^3 dr$$

where $U^2 = u_T^2 + u_P^2$. This can be written $C_{P_o} = \int_0^1 (\sigma/c) DU dr$, where DU is the power absorbed by the blade section. Assuming constant chord, a mean value for the drag coefficient, and low inflow so that $U^3 \cong u_T^3$, we obtain

$$C_{P_o} = \frac{\sigma c_{d_o}}{2}\int_0^1 u_T^3 dr = \frac{\sigma c_{d_o}}{2}\int_0^1 \left(r^3 + \frac{3}{2} r\mu^2\right) dr$$

$$= \frac{\sigma c_{d_o}}{8}\left(1 + 3\mu^2\right)$$

which could also have been obtained using $C_{Q_o} = (\sigma c_{d_o}/8)(1 + \mu^2)$ and

$C_{H_o} = (\sigma c_{d_o}/4)\mu$ from section 5-3. Two-thirds of the increase of the profile power with speed is thus due to C_{H_o}. When reverse flow and radial drag effects are included, a further increase of C_{P_o} with speed is found. In section 5-12 a reasonable approximation is shown to be

$$C_{P_o} = \frac{\sigma c_{do}}{8}\left(1 + 4.6\mu^2\right)$$

The parasite power is $C_{P_p} = \mu(D/W)C_T = VD/\rho A(\Omega R)^3$. If the helicopter drag is written in terms of an equivalent area f, so that $D = \frac{1}{2}\rho V^2 f$, then

$$C_{P_p} = \frac{1}{2}\frac{V^3 f}{(\Omega R)^3 A} \cong \frac{1}{2}\mu^3\frac{f}{A}$$

In summary, the rotor power in forward flight may be estimated from

$$\begin{aligned} C_P &= C_{P_i} + C_{P_o} + C_{P_p} + C_{P_c} \\ &\cong \frac{\kappa C_T^2}{2\mu} + \frac{\sigma c_{do}}{8}\left(1 + 4.6\mu^2\right) + \frac{1}{2}\frac{f}{A}\mu^3 + \lambda_c C_T \end{aligned}$$

which gives the power required as a function of gross weight or speed. The performance estimate may be improved by using a nonuniform induced velocity distribution; by considering the actual drag coefficient of the rotor blade, which requires the angle-of-attack distribution over the disk; and by refining the helicopter drag representation. Early formulations of the force balance method for helicopter performance calculations were essentially based on $C_P = \lambda C_T - \mu C_{H_i} + C_{Q_o}$, using the expressions for C_T and C_{H_i} in section 5-3 and often including the blade angle-of-attack distribution in the calculation of C_{Q_o}. Numerical calculations of the helicopter performance generally use the force balance method, obtaining the power from the rotor torque by $C_P = \int_0^1 \sigma a(F_x/ac)r\,dr$.

5–5 Rotor Flapping Motion

To complete the solution for the rotor behavior in forward flight, the harmonics of the blade flapping motion are required, particularly the coning and tip-path-plane tilt angles (β_0, β_{1c}, and β_{1s}). The angle of the tip-path plane relative to the no-feathering plane will be derived in this section.

From the tip-path-plane orientation (established by equilibrium of forces on the helicopter) the no-feathering-plane orientation can be found, and hence the cyclic pitch control required to fly the helicopter in the given operating condition. The blade flapping motion is determined by equilibrium of inertial and aerodynamic moments about the flap hinge. To introduce the analysis of the flapping motion, the simplest model is used: a rigid articulated rotor blade with no flap hinge offset or spring restraint.

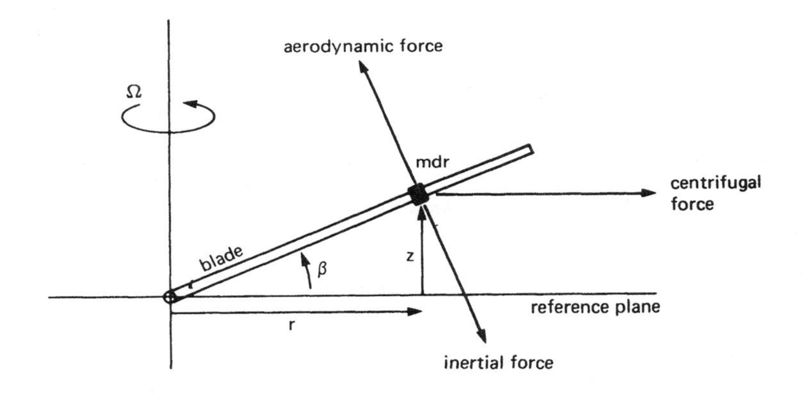

Figure 5-14 Rotor blade flapping moments.

Consider the the equilibrium of the inertial and aerodynamic moments about the flapping hinge (Fig. 5-14). The out-of-plane deflection is $z = \beta r$ for rigid motion with no hinge offset. Acting on a mass element $m\,dr$ (m is the blade mass per unit length) at radial station r are the following section forces:

i) an inertial force $m\ddot{z} = mr\ddot{\beta}$ opposing the flap motion, with moment arm r about the flap hinge;

ii) a centrifugal force $m\Omega^2 r$ directed radially outward, with moment arm $z = r\beta$; and

iii) an aerodynamic force F_z normal to the blade, with moment arm r.

For small angles recall that F_z is just the section lift L. The centrifugal force is the influence of the blade rotation. Since the centrifugal force always acts radially outward in a plane normal to the rotation axis, it acts as a spring force opposing the blade flap motion.

The moments about the flap hinge are given by integrals over the product of the span of the section forces and their corresponding moment arms. Since there is no flap hinge spring, the sum of the moments must be zero. Thus the equation of motion for the flap motion is

$$\int_0^R mr\ddot{\beta}\,r\,dr + \int_0^R m\Omega^2 r(r\beta)dr - \int_0^R F_z r\,dr = 0$$

or

$$\left(\int_0^R r^2 m\,dr\right)\left(\ddot{\beta} + \Omega^2\beta\right) = \int_0^R rF_z\,dr$$

Now define the moment of inertia about the flap hinge, $I_b = \int_0^R r^2 m\,dr$, and use dimensionless quantities based on ρ, Ω, and R. Then

$$\ddot{\beta} + \beta = \frac{1}{I_b}\int_0^1 rF_z\,dr$$

The dimensionless time variable is the rotor azimuth, $\psi = \Omega t$. Next define the Lock number γ:

$$\gamma = \frac{\rho acR^4}{I_b}$$

The Lock number is a dimensionless parameter of the blade representing the ratio of aerodynamic forces to inertial forces. Typically $\gamma = 8$ to 10 for articulated rotors and $\gamma = 5$ to 7 for hingeless rotors. Note that γ contains the sole influence of the air density on the flap motion. Assuming a constant chord and introducing the Lock number, the flapping equation becomes

$$\ddot{\beta} + \beta = \gamma\int_0^1 r\frac{F_z}{ac}\,dr = \gamma M_F$$

The left-hand side is a mass-spring system with a natural frequency of 1/rev (Ω dimensionally) due to the centrifugal spring. The right-hand side is the aerodynamic forcing moment. It follows that 1/rev aerodynamic forces will be exciting the blade flap motion at its resonant frequency. The amplitude of a system forced at resonance is determined by its damping alone, which in this case comes from the aerodynamic forces. The phase of the response is exactly a 90° lag, regardless of the magnitude of the damping.

The aerodynamic force normal to the blade is $F_z/ac = L/ac = u_T^2\alpha/2 = \frac{1}{2}(u_T^2\theta - u_P u_T)$. The aerodynamic flap moment is therefore

$$M_F = \int_0^1 r\frac{F_z}{ac}\,dr$$

$$= \int_0^1 r\,\frac{1}{2}\left[(r+\mu\sin\psi)^2\theta - (\lambda + r\dot{\beta} + \mu\beta\cos\psi)(r+\mu\sin\psi)\right]dr$$

Assuming uniform inflow and linear twist, the integration over the span can be performed:

$$\begin{aligned}
M_F &= M_\theta\theta_{con} + M_{\theta_{tw}}\theta_{tw} + M_\lambda\lambda + M_{\dot{\beta}}\dot{\beta} + M_\beta\beta \\
&= \theta_{con}\left(\frac{1}{8} + \frac{\mu}{3}\sin\psi + \frac{\mu^2}{4}\sin^2\psi\right) \\
&\quad + \theta_{tw}\left(\frac{1}{10} + \frac{\mu}{4}\sin\psi + \frac{\mu^2}{6}\sin^2\psi\right) \\
&\quad - \lambda\left(\frac{1}{6} + \frac{\mu}{4}\sin\psi\right) \\
&\quad - \dot{\beta}\left(\frac{1}{8} + \frac{\mu}{6}\sin\psi\right) \\
&\quad - \beta\mu\cos\psi\left(\frac{1}{6} + \frac{\mu}{4}\sin\psi\right)
\end{aligned}$$

where $\theta_{con} = \theta_0 + \theta_{1c}\cos\psi + \theta_{1s}\sin\psi$ is the collective and cyclic pitch control. The flapping motion is thus

$$\ddot{\beta} + \beta = \gamma(M_\theta\theta_{con} + M_{\theta_{tw}}\theta_{tw} + M_\lambda\lambda + M_{\dot{\beta}}\dot{\beta} + M_\beta\beta)$$

The aerodynamic coefficients are the flap moments due to angle-of-attack changes produced by the blade pitch, twist, inflow, flapping velocity, and flapping displacements, respectively. A flapping velocity produces an angle-of-attack perturbation that changes the blade lift to oppose the motion;

hence the blade has aerodynamic damping given by the coefficient $M_{\dot{\beta}}$.

The steady-state solution for the blade flapping is required, namely the harmonics of the periodic motion. Only the mean and first harmonics are sought here; without higher harmonic control (2/rev and above) the higher harmonics of flapping are small. The solution involves operating on the flapping equations with

$$\frac{1}{2\pi}\int_0^{2\pi}(\ldots)d\psi, \quad \frac{1}{\pi}\int_0^{2\pi}(\ldots)\cos\psi\, d\psi, \text{ and } \frac{1}{\pi}\int_0^{2\pi}(\ldots)\sin\psi\, d\psi$$

By using the definitions of the Fourier coefficients expressed in terms of the integrals of $\beta(\psi)$ and $\theta(\psi)$, linear algebraic equations are obtained for the harmonics of β. (An alternative approach is the substitutional method discussed in section 5-1.) Basically, the operational method evaluates the mean and 1/rev flap moments; the latter correspond to pitch and roll moments on the rotor disk (see section 5-3). For the inertial and centrifugal terms we obtain

$$\frac{1}{2\pi}\int_0^{2\pi}(\ddot{\beta}+\beta)d\psi = \frac{1}{2\pi}\int_0^{2\pi}\beta d\psi = \beta_0$$

and

$$\frac{1}{\pi}\int_0^{2\pi}(\ddot{\beta}+\beta)\cos\psi\, d\psi = \frac{1}{\pi}\int_0^{2\pi}(-\beta\cos\psi + \beta\cos\psi)d\psi = 0$$

by integrating the $\ddot{\beta}\cos\psi$ term by parts twice; similarly,

$$\frac{1}{\pi}\int_0^{2\pi}(\ddot{\beta}+\beta)\sin\psi\, d\psi = \frac{1}{\pi}\int_0^{2\pi}(-\beta\sin\psi + \beta\sin\psi)d\psi = 0$$

The centrifugal force gives an average flap moment when the rotor is coned by β_0. The 1/rev components of the inertial and centrifugal forces exactly cancel. It follows that the 1/rev components of the aerodynamic flap moments must also be zero. The requirement of zero aerodynamic pitch and roll moment on the rotor disk determines the tip-path-plane tilt angles β_{1c} and β_{1s}. The inertial and spring terms exactly cancel because the 1/rev aerodynamic forces are forcing the flap motion at its resonant frequency. With no aerodynamic forces, then, there would be no means of controlling the rotor, since the tip-path plane would be in equilibrium for any orientation.

Applying the operators to the flapping equation, using the aerodynamic coefficients given above, and neglecting second and higher harmonics of flap and pitch, we obtain the following equations:

$$\beta_0 = \gamma\left[\frac{\theta_0}{8}\left(1 + \mu^2\right) + \frac{\theta_{tw}}{10}\left(1 + \frac{5}{6}\mu^2\right) + \frac{\mu}{6}\theta_{1s} - \frac{\lambda}{6}\right]$$

$$0 = \frac{1}{8}\theta_{1c}\left(1 + \frac{1}{2}\mu^2\right) - \frac{1}{8}\beta_{1s} - \frac{\mu}{6}\beta_0 - \frac{\mu^2}{16}\beta_{1s}$$

$$0 = \frac{1}{8}\theta_{1s}\left(1 + \frac{3}{2}\mu^2\right) + \frac{\mu}{3}\theta_0 + \frac{\mu}{4}\theta_{tw} - \frac{\mu}{4}\lambda + \frac{1}{8}\beta_{1c} - \frac{\mu^2}{16}\beta_{1c}$$

Then with $\lambda - \mu\theta_{1s} = \lambda_{NFP}$, the solution for the rotor flapping motion is

$$\beta_0 = \gamma\left[\frac{\theta_{.8}}{8}\left(1 + \mu^2\right) - \frac{\mu^2}{60}\theta_{tw} - \frac{\lambda_{NFP}}{6}\right]$$

$$\beta_{1s} - \theta_{1c} = \frac{-(4/3)\mu\beta_0}{1 + (1/2)\mu^2}$$

$$\beta_{1c} + \theta_{1s} = \frac{-(8/3)\mu\,[\theta_{.75} - (3/4)\lambda_{NFP}]}{1 - (1/2)\mu^2}$$

Alternatively, in terms of $\lambda_{TPP} = \lambda_{NFP} + \mu(\beta_{1c} + \theta_{1s})$,

$$\beta_0 = \gamma\left[\frac{\theta_{.8}}{8}\left(1 + \mu^2\right) - \frac{\mu^2}{60}\theta_{tw} - \frac{\lambda_{TPP}}{6} + \frac{\mu}{6}\left(\beta_{1c} + \theta_{1s}\right)\right]$$

$$\beta_{1s} - \theta_{1c} = \frac{-(4/3)\mu\beta_0}{1 + (1/2)\mu^2}$$

$$\beta_{1c} + \theta_{1s} = \frac{-(8/3)\mu\,[\theta_{.75} - (3/4)\lambda_{TPP}]}{1 + (3/2)\mu^2}$$

While the expression for β_0 is simpler using λ_{NFP}, the expressions using λ_{TPP} are more appropriate since the tip-path-plane orientation has a direct physical meaning (essentially it specifies the thrust vector orientation, which is determined by the conditions for helicopter longitudinal force equilibrium). Note that the singularity of the $(\beta_{1c} + \theta_{1s})$ solution at $\mu = \sqrt{2}$

also disappears when the tip-path-plane inflow is used. (In any case, $\mu = \sqrt{2}$ is beyond the range of validity of these expressions; including reverse flow removes the singularity even for the expression involving λ_{NFP}.) The coning angle, roughly $\beta_0 \cong (3/4)\gamma(C_T/\sigma a)$, is directly proportional to the blade loading. The first harmonics β_{1c} and β_{1s} are proportional to the advance ratio μ and also to C_T/σ. Typically β_0 and β_{1c} have values of a few degrees, while the lateral flapping β_{1s} is somewhat smaller.

For hover the flapping solution reduces to $\beta_0 = \gamma(\theta_{.8}/8 - \lambda/6)$ and $\beta_{1c} + \theta_{1s} = 0$, $\beta_{1s} - \theta_{1c} = 0$. Recall that $(\beta_{1c} + \theta_{1s})$ and $(\beta_{1s} - \theta_{1c})$ give the orientation of the tip-path plane relative to the no-feathering plane. The hover solution for the flap motion is thus that the tip-path plane and no-feathering plane are always parallel. The flap motion in hover can be obtained as follows. The aerodynamic coefficients are constant when $\mu = 0$, and $M_\beta = 0$; there are no net 1/rev flap moments due to the inertial forces, the inflow, or the twist. Consequently, the equation of motion reduces to $M_\theta \theta + M_{\dot\beta}\dot\beta = 0$, which has components $M_\theta \theta_{1c} + M_{\dot\beta}\beta_{1s} = 0$ and $M_\theta \theta_{1s} - M_{\dot\beta}\beta_{1c} = 0$. Hence the solution is $\beta_{1s}/\theta_{1c} = -\beta_{1c}/\theta_{1s} = -M_\theta/M_{\dot\beta}$. As expected, the phase of the response of flapping to cyclic pitch is exactly 90°, with the magnitude determined by the ratio of the control moment to the damping. For the case considered here, $M_\theta = -M_{\dot\beta} = 1/8$; hence $-M_\theta/M_{\dot\beta} = 1$ and the tip-path plane is always parallel to the no-feathering plane. In forward flight, the rotor operating state uniquely determines the relative orientation of the tip-path plane and no-feathering plane, because $(\beta_{1c} + \theta_{1s})$ and $(\beta_{1s} - \theta_{1c})$ are obtained as functions of the helicopter speed and loading alone. As the no-feathering plane is tilted in response to pilot control inputs, then, the rotor tip-path plane and hence the thrust vector are tilted. By this means the pilot can control the attitude of the helicopter, using cyclic pitch (swashplate tilt) to produce moments about the helicopter center of gravity by tilting the thrust vector.

Let us examine further the role of inertial and aerodynamic forces in the rotor flap response. For the case of no aerodynamic forces, the rotor without flap hinge offset or restraint has the equation of motion $\ddot\beta + \beta = 0$. The solution of this equation is $\beta = \beta_{1c} \cos\psi + \beta_{1s} \sin\psi$, where β_{1c} and β_{1s} are arbitrary constants. The orientation of the rotor is thus arbitrary, but fixed in space since in the absence of aerodynamic forces or a hinge offset there is no means by which blade pitch or shaft tilt can produce a moment on the disk. The rotor behaves as a gyro. maintaining its orientation relative to inertial space in the absence of external moments. The rotor in air has the capability of producing an aerodynamic flap moment due to blade pitch (M_θ), which can be used to precess the rotor and hence control its orientation. If M_θ were the only moment, the rotor would respond to cyclic with a constant rate of tilt. However the rotor also has aerodynamic flap damping moments ($M_{\dot\beta}$). A tilt of the tip-path plane by β_{1c} or β_{1s} produces a flapping velocity in the rotating frame. Consequently, a moment due to control plane tilt precesses the rotor, tilting the tip-path plane until the flapping produces through $M_{\dot\beta}$ a moment just sufficient to counter the control moment on the disk. Because the moments due to θ and $\dot\beta$ balance, the rotor has achieved a new equilibrium position. Thus there are two ways to view the rotor flap dynamics. First, the rotor blade can be considered a system with natural frequency of 1/rev, so that aerodynamic moments due to cyclic pitch excite the system at resonance. The response has a phase lag of exactly 90° (i.e. one-quarter of a cycle, which at 1/rev means an azimuth angle of 90° also) and a magnitude determined by the damping. Alternatively, the rotor can be considered a gyro, with the flap hinges at the center of rotation forming the gyro gimbal. A control moment on the disk due to cyclic pitch then precesses the rotor with a 90° phase lag characteristic of a gyro until the flap damping produces a moment to stop the precession.

The rotor coning is proportional to the Lock number γ because it is determined by the balance of the centrifugal and aerodynamic flap moments. The coning is essentially proportional to the rotor thrust coefficient, the difference arising because of the extra factor r used in the integrand to obtain the flap moment rather than the total blade lift force. Because the rotor thrust produces a flap moment, the rotor cones upward until a centrifugal flap moment sufficient to cancel the aerodynamic moment is generated.

Since the longitudinal flap motion ($\beta_{1c} + \theta_{1s}$) is negative, in forward flight the tip-path plane tilts backward relative to the no-feathering plane. The lateral asymmetry of the blade velocity u_T in forward flight means that for constant pitch (i.e. relative to the no-feathering plane) the blade has a higher lift on the advancing side than on the retreating side of the disk. The result is a lateral flap moment on the rotor disk. In the rotating frame, where this flap moment is at the resonant frequency 1/rev, the blade responds with a 90° phase lag, the maximum flap displacement occurring on the front of the disk. The tip-path plane therefore tilts longitudinally (rearward) in response to the lateral moment. Now β_{1c} gives a flapping velocity $\dot{\beta} = -\beta_{1c} \sin \psi$ that has maximum amplitude at the sides of the disk. Through the flap damping, then, the tip-path-plane tilt produces a lateral flap moment. The rotor flaps back until this lateral moment due to the flap damping is just great enough to counter the lateral moment due to the aerodynamic asymmetry. With this balance of the aerodynamic forces the new equilibrium orientation is achieved.

Since the lateral flap motion $(\beta_{1s} - \theta_{1c})$ is negative, in forward flight the tip-path plane tilts toward the advancing side relative to the no-feathering plane. When the blade is at the coning angle β_0 there is a component of the forward velocity normal to the blade surface: $\beta_0 \mu \cos \psi$ (see Fig. 5-12). The angle of attack due to this normal velocity term is maximum positive at the front of the disk and maximum negative at the rear of the disk; hence it gives a longitudinal aerodynamic moment on the rotor. The blade responds to this 1/rev aerodynamic forcing in the rotating frame with a maximum flap amplitude 90° after the maximum moment, that is, with a lateral (to the right) tilt of the tip-path plane. Now β_{1s} gives a flapping velocity $\dot{\beta} = \beta_{1s} \cos \psi$, which through the flap damping also produces a longitudinal moment. The rotor flaps to the right until the longitudinal moment due to the flap damping balances the longitudinal moment due to the coning, and the rotor is in equilibrium with this orientation of the tip-path plane.

The tip-path-plane tilt is roughly proportional to the advance ratio μ. To keep the thrust vector orientation fixed as the speed increases it is necessary to tilt the no-feathering plane forward and toward the retreating side to counter the increased tip-path-plane tilt. Thus as speed increases a forward cyclic stick displacement is required in addition to the forward displacement to increase the propulsive force. An increasing cyclic stick displacement to the left is also required.

The control required to trim the helicopter is determined by the condition for helicopter force and moment equilibrium. As found in section 5-4, longitudinal force equilibrium determines the orientation of the tip-path plane relative to the horizontal (α_{TPP}, and also λ_{TPP}). The equilibrium of pitching moments on the helicopter determines the orientation of the hub plane relative to the horizontal (α_{HP}) as a function of the helicopter longitudinal center-of-gravity position and the aerodynamic forces on the aircraft (see section 5-18). The combination then determines the longitudinal flapping in the hub plane: $\beta_{1c_{HP}} = \alpha_{TPP} - \alpha_{HP}$. The conditions for rotor flapping equilibrium determine the orientation of the tip-path plane relative to the no-feathering plane, and from this the longitudinal cyclic control $\theta_{1s_{HP}}$ can be obtained. Similarly, the conditions for side force and roll moment equilibrium on the helicopter give the lateral flapping $\beta_{1s_{HP}}$ and therefore the lateral cyclic control $\theta_{1c_{HP}}$. The flapping solution given above is not in a form that may be directly used to calculate the cyclic required from the tip-path-plane tilt. It is necessary to simultaneously solve the C_T and β_{1c} equations for the collective pitch and longitudinal cyclic. Then the coning angle may be evaluated, and the lateral cyclic obtained from β_0. The result is:

$$\theta_{.75} = \frac{\left(1 + \frac{3}{2}\mu^2\right)\left(\frac{6C_T}{\sigma a} + \frac{3}{8}\mu^2\theta_{tw}\right) + \frac{3}{2}\lambda_{TPP}\left(1 - \frac{1}{2}\mu^2\right)}{1 - \mu^2 + \frac{9}{4}\mu^4}$$

$$\theta_{1s} = -\beta_{1c} - \frac{\frac{8}{3}\mu\left(\frac{6C_T}{\sigma a} + \frac{3}{8}\mu^2\theta_{tw}\right) + 2\mu\lambda_{TPP}\left(1 - \frac{3}{2}\mu^2\right)}{1 - \mu^2 + \frac{9}{4}\mu^4}$$

$$\beta_0 = \frac{\gamma/8}{1 - \mu^2 + \frac{9}{4}\mu^4}\left[\left(1 - \frac{19}{18}\mu^2 + \frac{3}{2}\mu^4\right)\frac{6C_T}{\sigma a} + \left(\frac{1}{20} + \frac{29}{120}\mu^2 - \frac{1}{5}\mu^4 + \frac{3}{8}\mu^6\right)\theta_{tw} + \left(\frac{1}{6} - \frac{7}{12}\mu^2 + \frac{1}{4}\mu^4\right)\lambda_{TPP}\right]$$

$$\theta_{1c} = \beta_{1s} + \frac{\frac{4}{3}\mu\beta_0}{1 + \frac{1}{2}\mu^2}$$

5–6 Examples of Performance and Flapping in Forward Flight

As an example, consider an articulated rotor with solidity $\sigma = 0.1$, Lock number $\gamma = 8$, twist $\theta_{tw} = -8^\circ$, and section lift curve slope $a = 5.7$; assume it operates at a blade loading of $C_T/\sigma = 0.12$ with $f/A = 0.015$ for the helicopter drag. This is a fairly representative example, except that the blade loading C_T/σ is much higher than that at which a rotor normally operates in forward flight. A high value is used so that the angle-of-attack distribution can be examined with the rotor near stall. The trends observed are similar with a more typical loading (say $C_T/\sigma = 0.08$). Cases with $C_T/\sigma = 0.04$, $f/A = 0$, or $\theta_{tw} = 0^\circ$ are also considered (changing just one parameter at a time). The expressions derived in the preceding sections were used to calculate the rotor loading and flap motion. These results are all for uniform induced velocity; the effects of nonuniform inflow are considered in section 13–2.

Fig. 5-15 shows the tip-path-plane inflow ratio as a function of μ, obtained from

$$\lambda_{TPP} = \frac{C_T}{2\sqrt{\mu^2 + \lambda^2}} + \frac{1}{2}\mu^3 \frac{f/A}{C_T}$$

At low speed λ_{TPP} is due to the induced velocity, while at high speed it is due to the propulsive force required to counter the helicopter parasite drag. Fig. 5-16 shows the solution for the collective pitch required. The variation with speed is primarily due to the change in inflow ratio; hence the collective stick position essentially follows the power required curve. Figs. 5-17 and 5-18 show the flapping relative to the no-feathering plane. Basically the flapping shows a linear variation with speed (neglecting nonuniform inflow) and rotor loading, with only secondary influences of the other parameters; the longitudinal flapping does show an effect of the induced velocity increase at high speed. Fig. 5-19 shows the coning angle, which depends mainly on the rotor thrust.

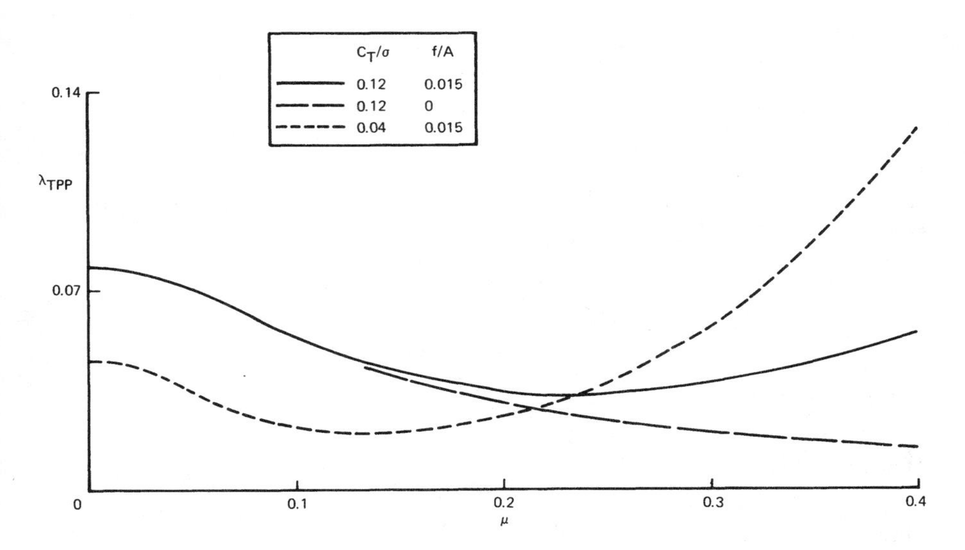

Figure 5-15 Variation of tip-path-plane inflow ratio with forward speed.

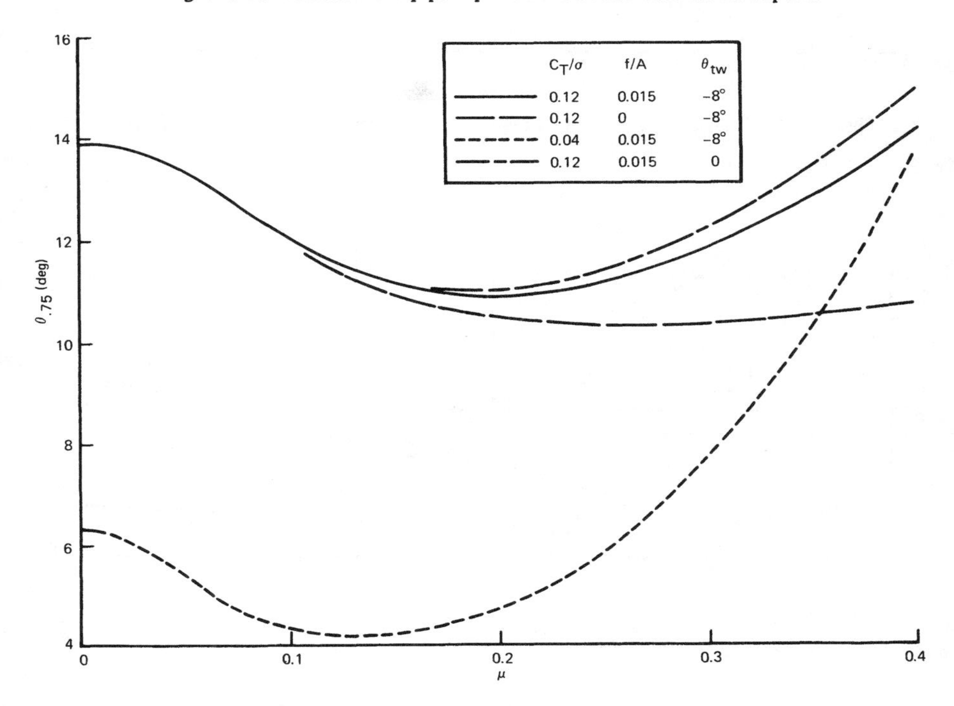

Figure 5-16 Variation of collective pitch with forward speed (uniform inflow).

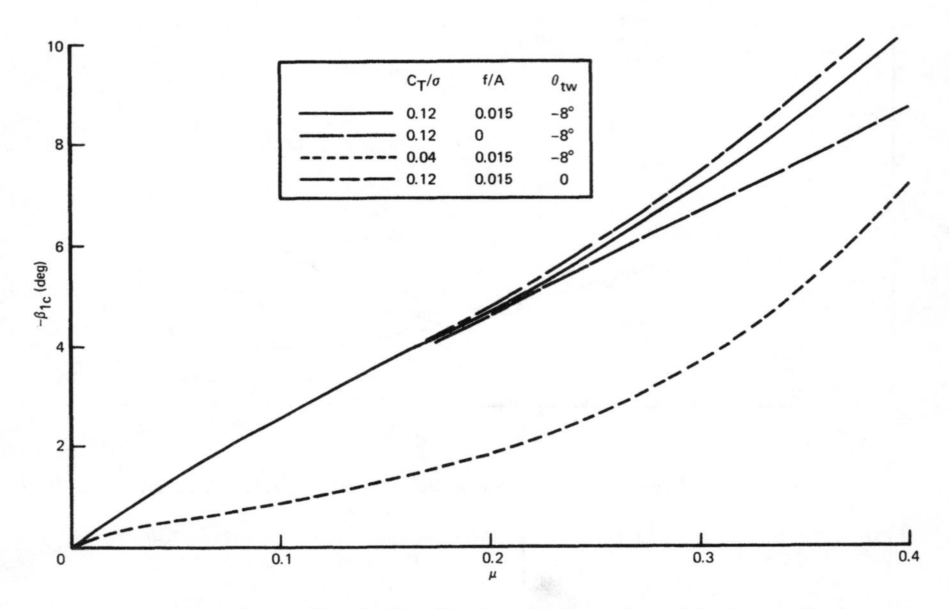

Figure 5-17 Variation of longitudinal flapping with forward speed (uniform inflow).

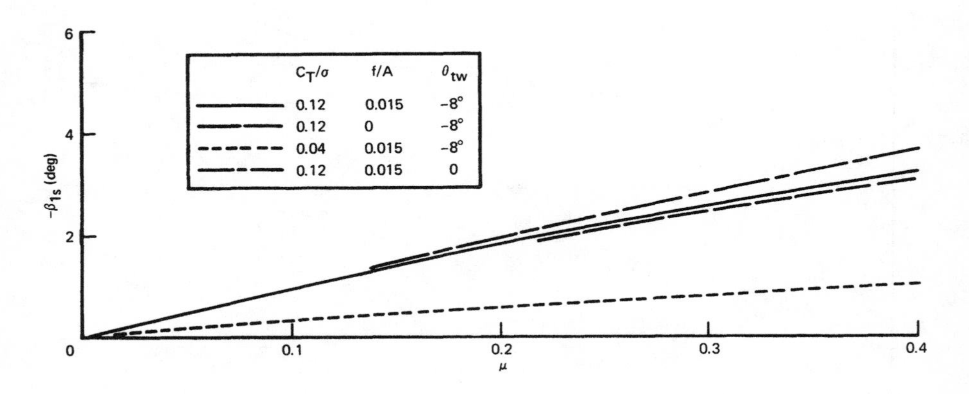

Figure 5-18 Variation of lateral flapping with forward speed (uniform inflow).

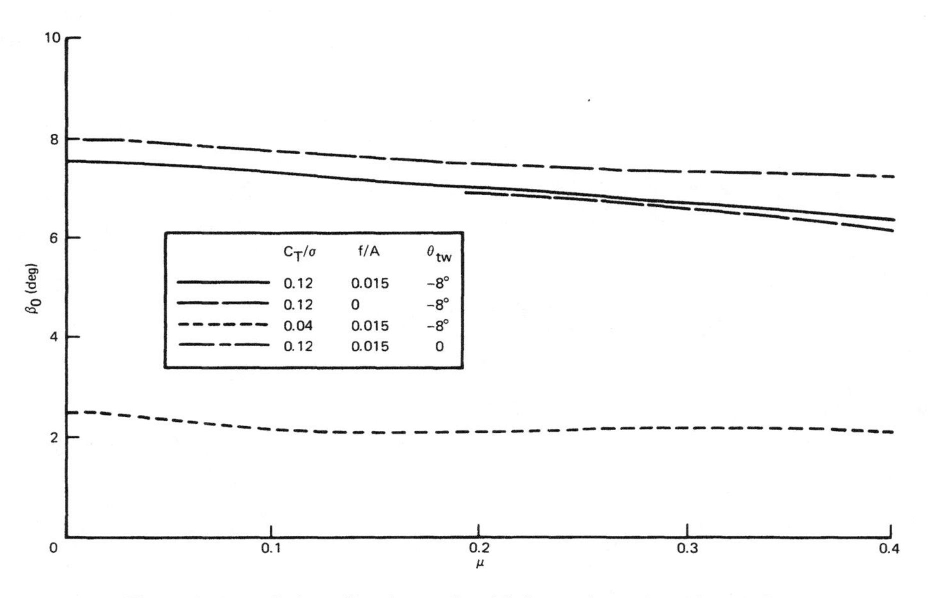

Figure 5-19 Variation of coning angle with forward speed (uniform inflow).

Fig. 5-20 shows the distribution over the rotor disk of the blade section angle of attack, calculated from

$$\begin{aligned}\alpha &= \theta - u_P/u_T \\ &= \theta_0 + r\theta_{tw} + (\beta_{1c} + \theta_{1s})\sin\psi - (\beta_{1s} - \theta_{1c})\cos\psi \\ &\quad - \frac{\lambda_{TPP} + \mu\beta_0\cos\psi}{r + \mu\sin\psi}\end{aligned}$$

Forward speed reduces the angle of attack on the advancing side and increases it on the retreating side as the blade flaps to maintain the same lift on both sides of the disk in the asymmetric aerodynamic environment of forward flight (really, to maintain zero 1/rev flapping moment). In forward flight the working angles of attack are concentrated on the front and rear of the disk. At this high loading, $C_T/\sigma = 0.12$, the angles of attack on the retreating side approach stall. The blade angle-of-attack distribution depends greatly on the flight condition (μ, C_T/σ, and f/A) and on the rotor twist. Figs. 5-21 and 5-22 show the angle of attack for rotors with no twist and

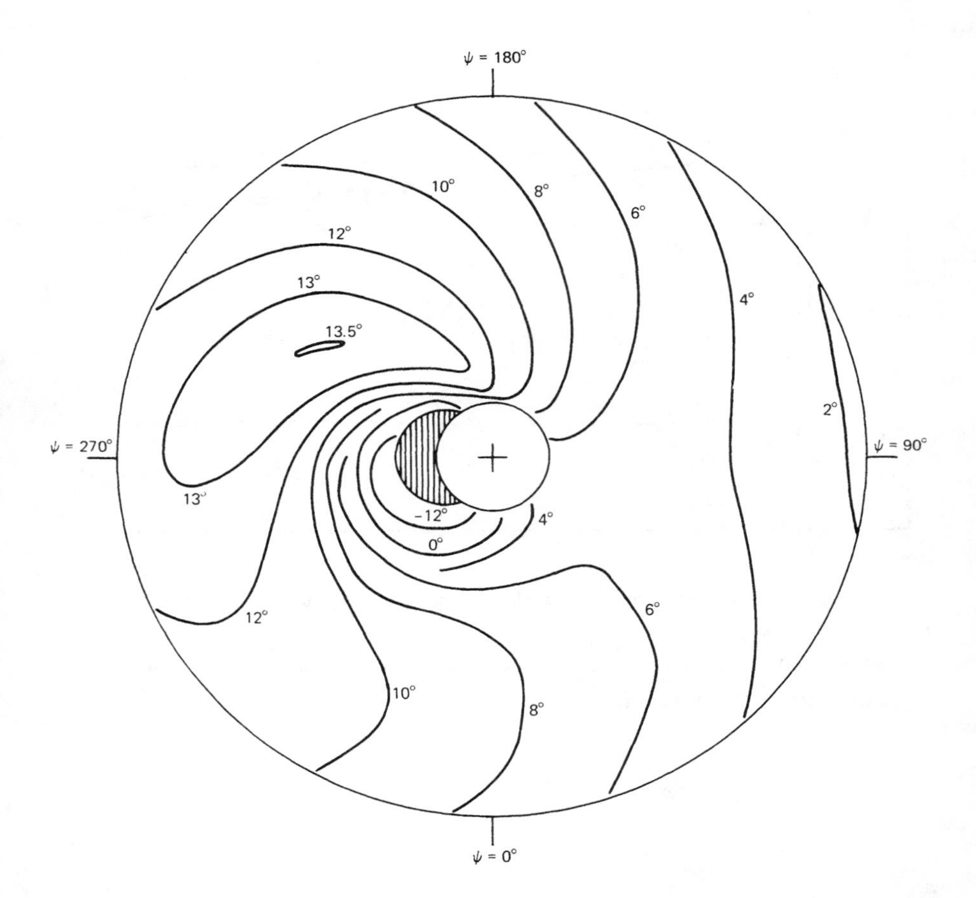

Figure 5.20 Blade angle-of-attack distribution (in degrees) at $\mu = 0.25$, for $C_T/\sigma = 0.12$, $f/A = 0.015$, and $\theta_{tw} = -8°$ (uniform inflow).

zero propulsive force respectively. Note that reducing the blade twist increases the maximum angle of attack on the blade and shifts the stall region toward the tips. It must be remembered, however, that these results are all for a uniform inflow distribution.

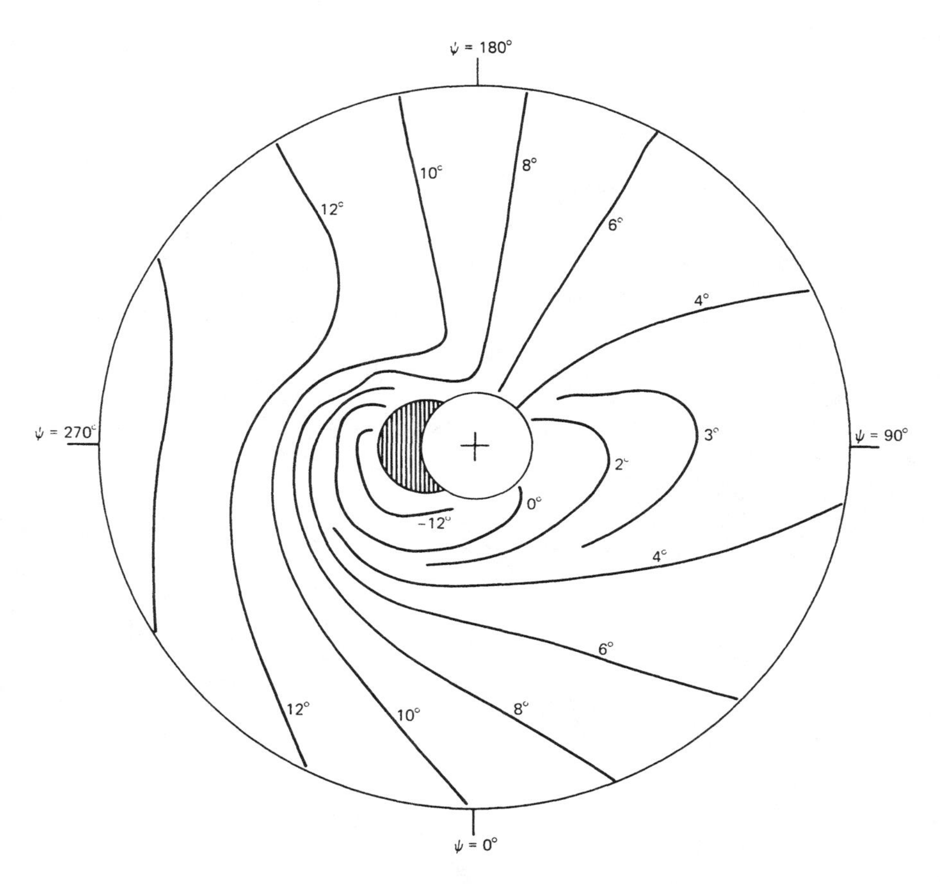

Figure 5.21 Blade angle-of-attack distribution (in degrees) at $\mu = 0.25$, for $C_T/\sigma = 0.12$, $f/A = 0.015$, and $\theta_{tw} = 0°$uniform inflow).

Fig. 5-23 shows the distribution of the dimensionless blade bound circulation $\Gamma/ac(\Omega R) = \frac{1}{2} u_T \alpha$ for $\mu = 0.025$, $C_T/\sigma = 0.12$, $f/A = 0.015$, and $\theta_{tw} = -8°$; Fig. 5-24 shows the distribution of the dimensionless blade section lift $L/\rho a c.(\Omega R)^2 = \frac{1}{2} u_T^2 \alpha$. Finally, Figs. 5-25 and 5-26 show the

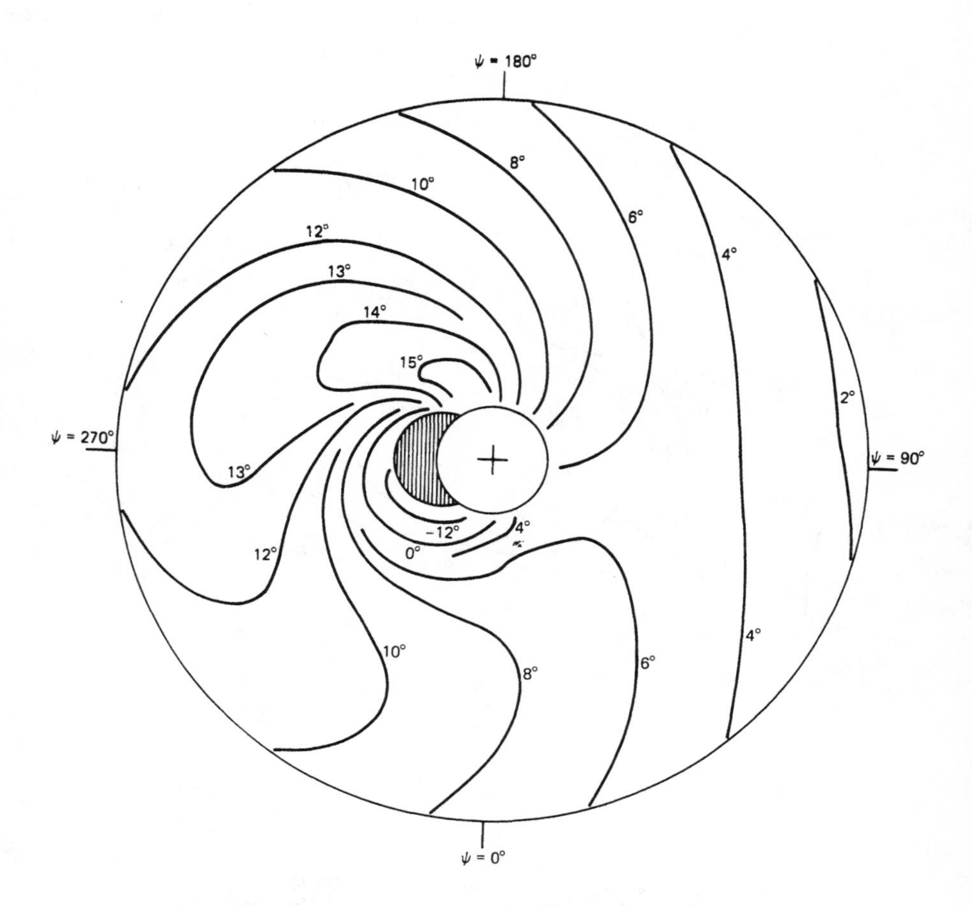

Figure 5-22 Blade angle-of-attack distribution (in degrees) at $\mu = 0.25$, for $C_T/\sigma = 0.12$, $f/A = 0$, and $\theta_{tw} = -8°$ (uniform inflow)

radial and azimuthal distribution of the blade section lift. Notice the reduction in lift on the tip of the advancing blade, which is needed to maintain a balance with the low lift on the retreating blade.

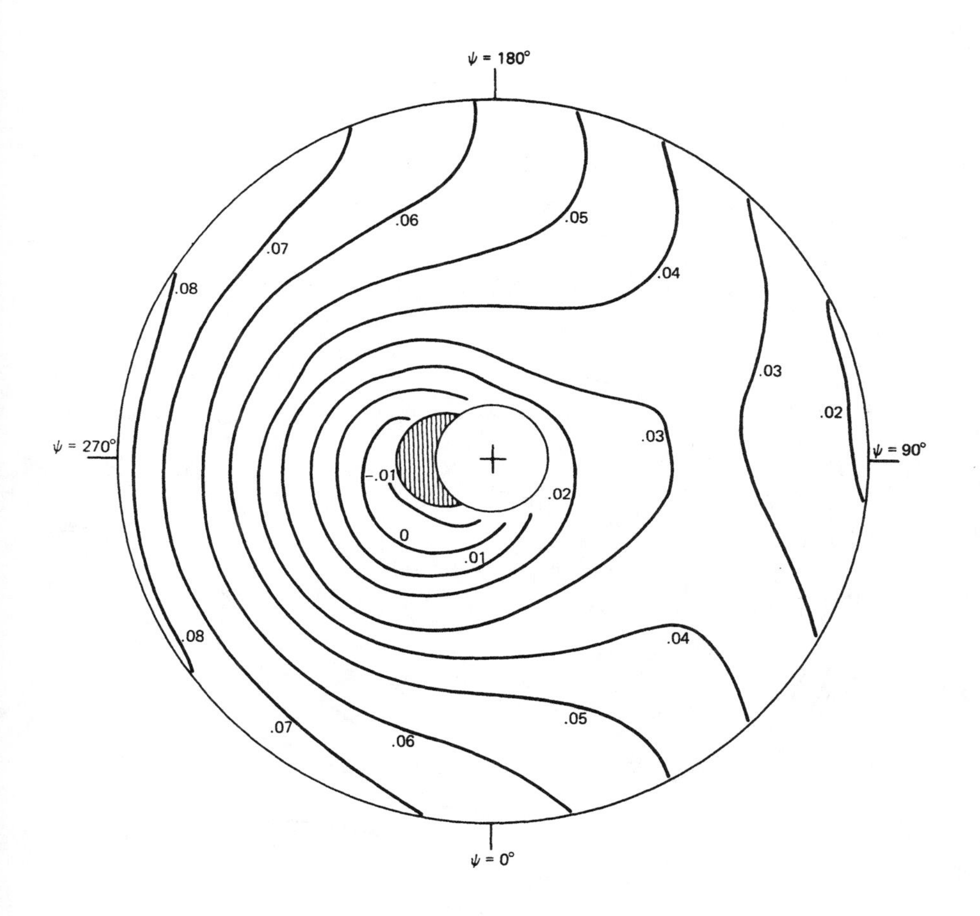

Figure 5-23 Blade bound circulation distribution $\Gamma/ac(\Omega R)$ at $\mu = 0.25$, for $C_T/\sigma = 0.12$, $f/A = 0.015$, and $\theta_{tw} = -8°$ (uniform inflow).

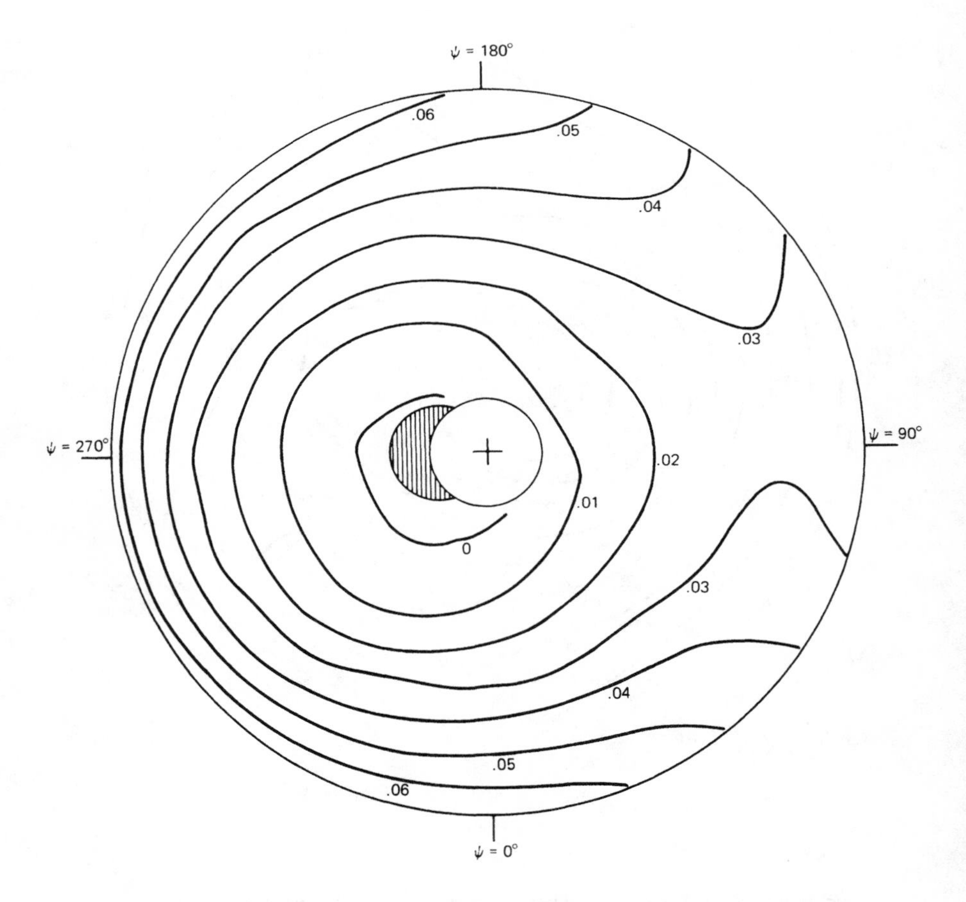

Figure 5-24 Blade section lift distribution $L/\rho ac(\Omega R)^2$ at $\mu = 0.25$, for $C_T/\sigma = 0.12$, $f/A = 0.015$, and $\theta_{tw} = -8°$ (uniform inflow).

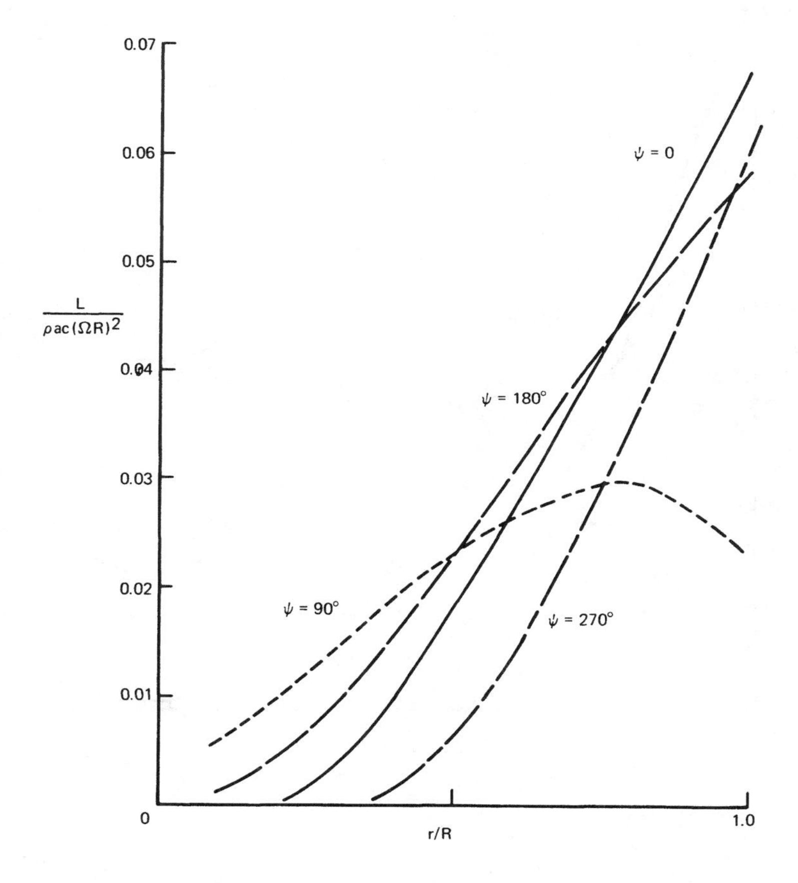

Figure 5-25 Radial distribution of the blade lift $L/\rho ac(\Omega R)^2$ at $\mu = 0.25$, for $C_T/\sigma = 0.12$, $f/A = 0.015$, and $\theta_{tw} = -8°$ (uniform inflow).

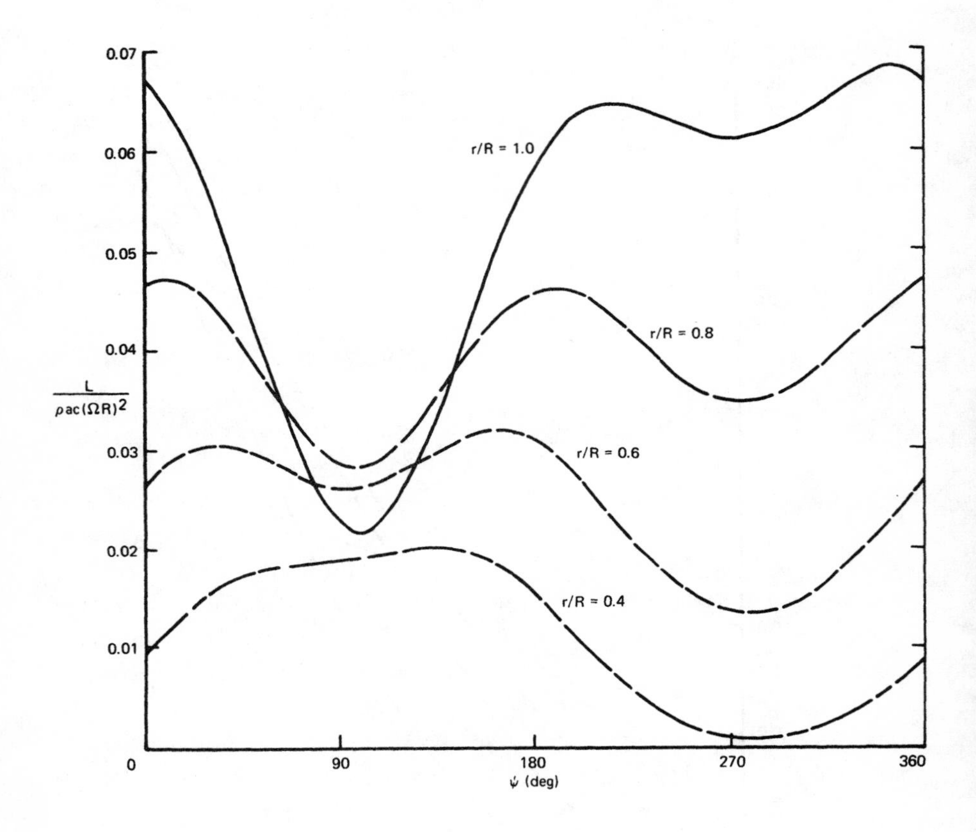

Figure 5-26 Azimuthal variation of the blade lift $L/\rho ac(\Omega R)^2$ at $\mu = 0.25$, for $C_T/\sigma = 0.12$, $f/A = 0.015$, and $\theta_{tw} = -8°$ (uniform inflow).

5–7 Review of Assumptions

The basic development of rotor behavior in forward flight has now been completed. The following sections consider a number of extensions of the analysis. Before proceeding, however, let us review the assumptions involved in the derivation so far.

Lifting-line theory has been used to determine the blade loading. Only rigid flap motion has been considered, and only collective and cyclic pitch control. There is no elastic flap motion, and no lag or pitch degrees of freedom. The rotor is articulated, with no flap hinge offset, spring restraint, or pitch-flap coupling. Reverse flow has been neglected. In general, small angles have been assumed. The section aerodynamic characteristics have been described by a constant lift curve slope and a mean profile drag coefficient. The effects of stall, compressibility, and radial flow have been neglected. A uniform induced velocity has been used. The blade has constant chord and linear twist. Root cutout and tip loss, the higher harmonics of the flap motion, and the blade weight have been neglected.

Lifting-line theory is a fundamental assumption of rotor aerodynamics, and it is valid except near the blade tip or in the vicinity of a close vortex, which, however, are important areas of the blade loading. The lag and pitch degrees of freedom and blade bending are important for vibration, blade loads, and aeroelastic stability, but may usually be neglected for helicopter performance and control. Similarly, the higher harmonic blade motion, important for vibration and loads, may be neglected. Reverse flow may be neglected up to about $\mu = 0.5$, which covers the speed range of most helicopters. Neglect of stall and compressibility limits the validity of the theory at extreme operating conditions (high μ or C_T/σ). Uniform inflow may be satisfactory for performance calculations at high speed, but it leads to significant errors in the calculation of flap motion, particularly β_{1s}. Nonuniform inflow is also important for rotor loads and vibration. The constant chord, linearly twisted blade is a typical rotor design. The rotor tip loss significantly influences both the performance and flapping motion.

Most of the assumptions introduced so far must be eliminated for a thorough analysis of the helicopter rotor in forward flight. While the basic features of the rotor behavior are contained in the solution derived above, the model is too limited for accurate quantitative results. The remaining sections of this chapter are devoted to extending the analysis of the rotor in forward flight in a number of these areas.

5–8 Tip Loss and Root Cutout

The decrease of the lift to zero over a finite distance at the blade tip may be accounted for by using a tip loss factor B such that the blade has drag but no lift when $r > BR$. In addition, the rotor has a root cutout, so that the blade airfoil starts at radial station r_R rather than at $r = 0$. When tip loss and root cutout are taken into account, the expression for the rotor thrust becomes

$$C_T = \frac{\sigma a}{2} \int_{r_R}^{B} \left[\left(\theta_0 + r\theta_{tw}\right)\left(r^2 + \tfrac{1}{2}\mu^2\right) - \lambda_{NFP} r \right] dr$$

$$= \frac{\sigma a}{2} \left\{ \frac{\theta_0}{3} \left[B^3 - r_R^{\,3} + \frac{3}{2}\left(B - r_R\right)\mu^2 \right] + \frac{\theta_{tw}}{4} \left[B^4 - r_R^{\,4} + \left(B^2 - r_R^{\,2}\right)\mu^2 \right] - \tfrac{1}{2}\lambda_{NFP}\left(B^2 - r_R^{\,2}\right) \right\}$$

The principal effect of the tip loss is to reduce the thrust for a given collective pitch, roughly by a factor B^3. The root cutout has only a minor influence on C_T. The tip loss similarly has a major effect on the flap moment, so it must be included in the flapping solution as well (see section 5-24). The tip loss also increases the rotor induced power loss for a given thrust, by a factor B^{-2} in forward flight (section 4-1.3).

5–9 Blade Weight Moment

The force of gravity, generally normal to the rotor disk, acts on the blade to produce a weight moment about the flap hinge. The blade weight opposes the lift force, and thereby reduces the coning angle. The gravitational force on a blade is mg, directed downward, with a moment arm r about the flap hinge. The additional flap moment is therefore

$$\int_0^R mgr\,dr = g \int_0^R rm\,dr = gS_b$$

where $S_b = \int_0^R rm\,dr = M_b r_{CG}$ is the first moment about the flap hinge (M_b is the blade mass and r_{CG} is the radial location of the center of gravity). This flap moment is added to the equation of motion; dimensionless quantities are used; and the equation is normalized by dividing by I_b. The result is

$$\ddot{\beta} + \beta = \gamma \int_0^1 r \frac{F_z}{ac}\,dr - S_b^* \frac{g}{\Omega^2 R}$$

where

$$S_b^* = \frac{RS_b}{I_b} = \frac{\int_0^1 rm\,dr}{\int_0^1 r^2 m\,dr} \cong \frac{3}{2}$$

(the last approximation is for a uniform mass distribution). The value of the dimensionless gravitational constant $g/\Omega^2 R$ is quite small, typically around 0.002 (assuming a constant tip speed, it scales with R).

The term $S_b^* g$ is a constant, so it only affects the solution for the coning angle. The coning is decreased by $\Delta\beta_0 = -S_b^* g/\Omega^2 R$, typically $-0.1°$ to $-0.2°$, which is small enough to be neglected for most purposes.

The dimensionless gravitational constant $g/\Omega^2 R$ can be viewed as the ratio of gravitational forces to centrifugal forces on the blade. Its small value implies that the rotor behavior is dominated by the centrifugal forces, and the blade weight generally has only a small influence.

5–10 Linear Inflow Variation

As a first approximation to the effects of nonuniform inflow in forward flight, consider an induced velocity of the form

$$\lambda_i = \lambda_0(1 + \kappa_x r \cos\psi + \kappa_y r \sin\psi)$$

This is a linear variation over the rotor disk, with λ_0 the mean induced velocity. The coefficients κ_x and κ_y will be functions of μ, since they must be zero in hover. For high speed, κ_x is around 1 and κ_y is somewhat smaller in magnitude and negative. Section 4-2.2 discussed a number of estimates for κ_x and κ_y. This linear inflow variation may be considered the first term in an expansion of the general nonuniform induced velocity $\lambda_i(r, \psi)$.

The lowest order terms are important for the rotor performance and flapping, while the higher order terms (which can be large in certain flight conditions) are important for the blade loads and vibration. So far a uniform inflow distribution has been used. Now it is necessary to find the additional contributions to the rotor forces and blade motion due to the inflow increment

$$\Delta\lambda = \lambda_0(\kappa_x r\cos\psi + \kappa_y r\sin\psi) = \lambda_x r\cos\psi + \lambda_y r\sin\psi$$

Here λ_x gives the longitudinal induced velocity vibration and λ_y the lateral vibration.

The rotor thrust is then

$$C_T = \frac{\sigma a}{2}\int_0^1 \left(-\Delta\lambda u_T\right) dr = \frac{\sigma a}{2}\int_0^1 \left(-\tfrac{1}{2}\lambda_y\right) r\mu\, dr$$

$$= \frac{\sigma a}{2}\left(-\tfrac{1}{4}\lambda_y\mu\right)$$

Hence

$$C_T = \frac{\sigma a}{2}\left[\frac{\theta_{.75}}{3}\left(1 + \frac{3}{2}\mu^2\right) - \frac{\theta_{tw}}{8}\mu^2 - \frac{1}{2}\left(\bar{\lambda}_{NFP} + \frac{\mu}{2}\lambda_y\right)\right]$$

where $\bar{\lambda}$ is the mean inflow. Thus there is a change of order μ^2 in the thrust for a given collective. The increments in the rotor drag and side forces are

$$C_{H_{TPP}} = \frac{\sigma a}{2}\left[\lambda_x\left(\theta_{1c}\frac{\mu}{16} + \beta_0\frac{1}{6}\right) + \lambda_y\left(\theta_{.75}\frac{1}{6} + \theta_{1s}\frac{3\mu}{16} - \bar{\lambda}_{TPP}\frac{1}{2}\right)\right]$$

$$C_{Y_{TPP}} = \frac{\sigma a}{2}\left[-\lambda_x\left(\theta_{.75}\frac{1}{6} + \theta_{1s}\frac{\mu}{16} - \bar{\lambda}_{TPP}\frac{1}{2}\right) - \lambda_y\left(\theta_{1c}\frac{\mu}{16} - \beta_0\frac{1}{6}\right)\right]$$

The flap moment increment is

$$\Delta M_F = \int_0^1 \left(-\Delta\lambda\right)\left(\frac{r^2}{2} + \frac{r\mu}{2}\sin\psi\right) dr$$

$$= -\left(\lambda_x\cos\psi + \lambda_y\sin\psi\right)\left(\frac{1}{8} + \frac{\mu}{6}\sin\psi\right)$$

and the flap equations become

$$\beta_0 = \gamma\left[\frac{\theta_{.8}}{8}\left(1+\mu^2\right) - \frac{\mu^2}{60}\theta_{tw} - \frac{1}{6}\left(\bar{\lambda}_{NFP} + \frac{\mu}{2}\lambda_y\right)\right]$$

$$0 = \frac{1}{8}\left(\theta_{1c} - \beta_{1s}\right)\left(1 + \frac{\mu^2}{2}\right) - \frac{\mu}{6}\beta_0 - \frac{\lambda_x}{8}$$

$$0 = \frac{1}{8}\left(\beta_{1c} + \theta_{1s}\right)\left(1 - \frac{\mu^2}{2}\right) + \frac{\mu}{3}\theta_{.75} - \frac{\mu}{4}\bar{\lambda}_{NFP} - \frac{\lambda_y}{8}$$

with solutions for the tip-path-plane tilt as follows:

$$\beta_{1s} - \theta_{1c} = \frac{-(4/3)\mu\beta_0 - \lambda_x}{1 + (\mu^2/2)}$$

$$\beta_{1c} + \theta_{1s} = \frac{-(8/3)\mu[\theta_{.75} - (3/4)\bar{\lambda}_{NFP}] + \lambda_y}{1 - (\mu^2/2)}$$

There is a change of order μ^2 in the coning angle as in the thrust, because the lateral inflow λ_y decreases (for $\lambda_y < 0$) the mean value of λu_T; this effect is small. The inflow variation has a significant influence on the tip-path-plane tilt, however. There are longitudinal and lateral angle-of-attack changes on the disk due respectively to λ_x and λ_y, which produce lateral and longitudinal flapping. The longitudinal flapping (and hence the longitudinal cyclic trim) change is small but not negligible, while the change in the lateral flapping and cyclic is large. Thus the rotor cyclic flapping and cyclic pitch trim are quite sensitive to nonuniform inflow. This is an important factor in the discrepancy between the calculations and measurements of the flapping.

Finally, nonuniform inflow gives an increment in the rotor induced power:

$$\Delta C_{P_i} = \int \Delta\lambda \, dC_T = \sigma a \int_0^1 \Delta\lambda \frac{F_z}{ac} dr$$

Here we shall consider an arbitrary induced velocity distribution $\lambda(r,\psi)$. To evaluate ΔC_{P_i}, expand the inflow as a Fourier series in azimuth and radially in terms of the orthogonal blade bending modes:

$$\Delta\lambda = \sum_{n=1}^{\infty}\sum_{i=1}^{\infty}\left(\lambda_{nc}^i \cos n\psi + \lambda_{ns}^i \sin n\psi\right)\eta_i(r)$$

The functions η_i are the mode shapes of the out-of-plane bending of the blade with corresponding natural frequencies ν_i (a general blade is considered, but for an articulated rotor with no hinge offset $\eta_1 = r$ and $\nu_1 = 1$). In section 9-2.2 the differential equation for the blade bending modes will be derived:

$$I_i^*(\ddot{q}_i + \nu_i^2 q_i) = \gamma \int_0^1 \eta_i \frac{F_z}{ac} dr$$

where I_i^* is the generalized mass of the ith mode. Then the integration over the span and azimuth in ΔC_{P_i} can be performed, giving

$$\Delta C_{P_i} = \sigma a \sum_{n=1}^{\infty} \sum_{i=1}^{\infty} \frac{1}{2\pi} \int_0^{2\pi} \left(\lambda_{nc}^i \cos n\psi + \lambda_{ns}^i \sin n\psi \right) \int_0^1 \eta_i \frac{F_z}{ac} dr\, d\psi$$

$$= \sigma a \sum_{n=1}^{\infty} \sum_{i=1}^{\infty} \frac{1}{2\pi} \int_0^{2\pi} \left(\lambda_{nc}^i \cos n\psi + \lambda_{ns}^i \sin n\psi \right) \frac{I_i^*}{\gamma} \left(\ddot{q}_i + \nu_i^2 q_i \right) d\psi$$

$$= \frac{\sigma a}{2} \sum_{n=1}^{\infty} \sum_{i=1}^{\infty} \frac{I_i^*(\nu_i^2 - n^2)}{\gamma} \left(\lambda_{nc}^i q_{nc}^i + \lambda_{ns}^i q_{ns}^i \right)$$

where q_{nc}^i and q_{ns}^i are the harmonics of the steady-state response of the ith bending mode. In general it will be simpler to just integrate $C_{P_i} = \int \lambda_i dC_T$ numerically. Note however that for a linear inflow variation, only the $n = 1$ terms are present; and that in addition considering just the first mode of an articulated blade with no hinge offset (so $\nu_1 = 1$) gives $\Delta C_{P_i} = 0$.

5–11 Higher Harmonic Flapping Motion

Consider next the solution of the flapping equation of motion for the second harmonics β_{2c} and β_{2s}. The higher harmonics of the blade motion are strongly influenced by nonuniform inflow and elastic blade bending, but the present solution will serve as a guide to the basic behavior of the higher harmonics. If it is still assumed that β_{2c} and β_{2s} are much smaller than β_{1c} and β_{1s}, the previous results for the first harmonic flapping will not be changed. The algebraic equations for β_{2c} and β_{2s} are obtained by operating on the flap differential equation with

$$\frac{1}{\pi}\int_0^{2\pi}(\ldots)\cos 2\psi\, d\psi \quad \text{and} \quad \frac{1}{\pi}\int_0^{2\pi}(\ldots)\sin 2\psi\, d\psi$$

By neglecting the influence of the second harmonics on the mean and first harmonics, it is only necessary to solve these two additional equations for β_{2c} and β_{2s} rather than five simultaneous equations for all five coefficients. The inertial and centrifugal terms now give

$$\frac{1}{\pi}\int_0^{2\pi}(\ddot{\beta}+\beta)\cos 2\psi\, d\psi = -3\beta_{2c}$$

$$\frac{1}{\pi}\int_0^{2\pi}(\ddot{\beta}+\beta)\sin 2\psi\, d\psi = -3\beta_{2s}$$

Since the 2/rev flap moments are acting above the system resonant frequency, the response is dominated by the blade inertia. In general the equations for β_{nc} and β_{ns} have the terms $(1-n^2)\beta_{nc}$ and $(1-n^2)\beta_{ns}$ from the inertial and centrifugal forces. As a result, the higher harmonic flapping in response to the aerodynamic flap moments decreases rapidly with harmonic order, roughly as n^{-2}. When the blade bending modes with natural frequencies above 1/rev are considered, there is again a possibility of large amplitude motion due to excitation near resonance.

With the aerodynamic terms as well, the equations of motion for β_{2c} and β_{2s} are

$$-3\beta_{2c} = \gamma\left(-\frac{\mu^2}{8}\theta_0 - \frac{\mu}{6}\theta_{1s} - \frac{\mu^2}{12}\theta_{tw} - \frac{1}{4}\beta_{2s} - \frac{\mu}{6}\beta_{1c} + \frac{\mu}{12}\lambda_y\right)$$

$$-3\beta_{2s} = \gamma\left(\frac{\mu}{6}\theta_{1c} + \frac{1}{4}\beta_{2c} - \frac{\mu}{6}\beta_{1s} - \frac{\mu^2}{8}\beta_0 - \frac{\mu}{12}\lambda_x\right)$$

A linear inflow variation has been included. The solution for the second harmonic flapping is

$$\beta_{2c} = \frac{\frac{\gamma}{24}\mu}{1+\left(\frac{\gamma}{12}\right)^2}\left\{\left[\mu\theta_{.67} + \frac{4}{3}\left(\beta_{1c} + \theta_{1s}\right) - \frac{2}{3}\lambda_y\right] + \frac{\gamma}{12}\left[\mu\beta_0 + \frac{4}{3}\left(\beta_{1s} - \theta_{1c}\right) + \frac{2}{3}\lambda_x\right]\right\}$$

$$\beta_{2s} = \frac{\frac{\gamma}{24}\mu}{1+\left(\frac{\gamma}{12}\right)^2}\left\{-\frac{\gamma}{12}\left[\mu\theta_{.67} + \frac{4}{3}\left(\beta_{1c} + \theta_{1s}\right) - \frac{2}{3}\lambda_y\right] + \left[\mu\beta_0 + \frac{4}{3}\left(\beta_{1s} - \theta_{1c}\right) + \frac{2}{3}\lambda_x\right]\right\}$$

Note that β_{2c} and β_{2s} are smaller than the first harmonic flapping by at least order μ. Typically the second harmonics have values of a few tenths of a degree, so they are indeed small, as has been assumed. In general it is found that the solution for β_{nc} and β_{ns} is of order μ^n/n^2.

The primary excitation of the higher harmonics of blade motion is provided by nonuniform inflow, which has not been considered except for the simple linear variation. With nonuniform inflow the higher harmonic blade motion has a significantly larger amplitude than that found here. In addition, the blade bending modes must also be included for a consistent and accurate calculation of the blade response at higher frequencies. The higher harmonic motion usually has little influence on the rotor performance and control, but it is of central importance to the helicopter vibration and blade loads.

Let us briefly examine the response to higher harmonic pitch control. Consider a hovering rotor only, so that there is no interharmonic coupling of the pitch control and flap response as occurs in forward flight because of the periodic aerodynamics of the blade. Then n/rev blade pitch gives just n/rev flapping. The flapping equation of motion in hover is

$$\ddot{\beta} + \beta = \frac{\gamma}{8}\left(-\dot{\beta} + \theta\right)$$

For an input of $\theta = \bar{\theta}\cos[n(\psi + \psi_0)]$ the flap response will be $\beta = \bar{\beta}\cos[n(\psi + \psi_0) - \Delta\psi]$. The equation of motion gives the magnitude and phase of the response:

$$\bar{\beta}/\bar{\theta} = \frac{\gamma/8}{\sqrt{(n\gamma/8)^2 + (n^2 - 1)^2}}$$

$$\Delta\psi = 90^\circ + \tan^{-1}\frac{n^2 - 1}{n\gamma/8}$$

For the first harmonic, $\bar{\beta}/\bar{\theta} = 1$ and $\Delta\psi = 90^\circ$ as expected. For large harmonic number, the amplitude decreases as $\bar{\beta}/\bar{\theta} \cong \gamma/8n^2$ as the blade inertia dominates the response, and the phase lag approaches $\Delta\psi = 180^\circ$. The effectiveness of 1/rev cyclic pitch in controlling the rotor thus lies in the fact the flap motion is being excited at resonance.

5–12 Profile Power and Radial Flow

The contributions of blade profile drag to the rotor forces, torque, and power were derived in sections 5–3 and 5–4:

$$C_{H_o} = \int_0^1 \sigma\left(\frac{F_x}{c}\sin\psi + \frac{F_r}{c}\cos\psi\right)dr$$

$$C_{Y_o} = \int_0^1 \sigma\left(-\frac{F_x}{c}\cos\psi + \frac{F_r}{c}\sin\psi\right)dr$$

$$C_{Q_o} = \int_0^1 \sigma r\frac{F_x}{c}\,dr$$

and

$$C_{P_o} = C_{Q_o} + \mu C_{H_o} = \int_0^1 \sigma\left(u_T\frac{F_x}{c} + u_R\frac{F_r}{c}\,dr\right)$$

The average over the azimuth is also required. The forces F_x and F_r are the normal and radial components of the section profile drag. Of particular interest is the rotor profile power loss C_{P_o}. Note that the terms $u_T F_x$ and $u_R F_r$ are the section power losses due to the normal and radial drag forces respectively. These coefficients have been evaluated already for a limited model. Now we shall examine the effects of reverse flow, radial flow, and the radial drag force. Because of the symmetry of the rotor flow field, $C_{Y_o} = 0$ for all of the cases considered here.

The radial flow along the blade ($u_R = \mu \cos \psi$) generates a radial component of the viscous drag force on the blade sections. We require an estimate of the normal and radial drag forces, preferably in terms of the two-dimensional section aerodynamic characteristics since little else is likely to be available. Consider the loading on an infinite wing of chord c, yawed at an angle Λ to the free stream velocity V. The loading must be the same at all stations of this infinite wing, but it will not be the same as the loading on an unyawed wing. The spanwise flow and spanwise pressure gradient on the yawed wing must influence the boundary layer and hence the drag. The spanwise flow has a great influence on the stall characteristics of the wing. The loading of the yawed wing can be viewed either in terms of a section normal to the span (unyawed sections), or a section aligned with the free stream velocity at yaw angle Λ (yawed sections). The geometric chord and angle of attack of the yawed and unyawed sections (the former denoted by the subscript y) are related by $c_y = c/\cos \Lambda$ and $\alpha_y = \alpha \cos \Lambda$. The lift and drag forces on the yawed section are L_y and D_y. It is now assumed that the total viscous drag force on the yawed wing section, D_y, has the same direction as the yawed free stream velocity. In fact, the drag force will have a greater yaw angle than the free stream velocity because of the action of the spanwise velocity on the boundary layer, but this approximation is sufficient for the present purposes. Resolving the drag forces normal and parallel to the span then gives the forces on the unyawed section: $L = L_y$, $D = D_y \cos \Lambda$, and $F_r = D_y \sin \Lambda = D \tan \Lambda$. The yawed section has a higher free stream velocity than the normal section, with the dynamic pressures related by $q_y = q/\cos^2 \Lambda$. Then in terms of section coefficients, the lift and drag forces are related by $c_\ell(\alpha) = c_{\ell_y}(\alpha_y)/\cos^2 \Lambda$ and $c_d(\alpha) = c_{d_y}(\alpha_y)/\cos \Lambda$. Since the chord increase is compensated by a corresponding decrease in section width for the yawed section, the loads act on the same differential area and the section coefficients differ only because of the dynamic pressure change.

Now the equivalence assumption for a swept wing is introduced: it is assumed that the yawed section drag coefficient $c_{d_y}(\alpha_y)$ is given by the two-dimensional unyawed airfoil characteristics, and that the normal section lift coefficient $c_\ell(\alpha)$ is not influenced by the yawed flow. The assumption for the lift is based on the fact that the wing when viewed in a frame moving spanwise at velocity $V \sin\Lambda$ is equivalent to an unyawed wing with free stream velocity $V \cos\Lambda$, except for changes in the boundary layer. Now below stall the lift of both the normal and yawed sections is proportional to the angle of attack, and the respective lift curve slopes are a and a_y; hence $c_\ell(\alpha) = a\alpha$ and $c_{\ell_y}(\alpha_y) = a_y\alpha_y$. But we already know that $c_\ell(\alpha) = c_{\ell_y}(\alpha_y)/\cos^2\Lambda$ and $\alpha_y = \alpha\cos\Lambda$, so it follows from the equivalence assumption that the yawed wing lift is $c_{\ell_y}(\alpha_y) = c_{\ell_{2D}}(\alpha_y \cos\Lambda)$. (Hence the lift curve slope of the yawed section is $a_y = a\cos\Lambda$.) For the yawed wing drag the equivalence assumption gives simply $c_{d_y}(\alpha_y) = c_{d_{2D}}(\alpha_y)$. The equivalence assumption allows the yawed wing forces to be obtained from unyawed two-dimensional data, although the smaller thickness ratio of the yawed section should be accounted for. The assumption is largely verified by experimental data for yawed wings. The use of unyawed data cannot always be valid, however. In particular, at high angle of attack or very large yaw angle the flow is radically altered by the spanwise velocity, and the equivalence assumption is no longer applicable.

The normal section characteristics for the yawed wing are now $c_\ell(\alpha) = c_{\ell_y}(\alpha_y)/\cos^2\Lambda = c_{\ell_{2D}}(\alpha\cos^2\Lambda)/\cos^2\Lambda$ and $c_d(\alpha) = c_{d_y}(\alpha_y)/\cos\Lambda = c_{d_{2D}}(\alpha\cos\Lambda)/\cos\Lambda$. For small angle of attack, the radial flow has no effect on the lift, while the drag increases by $(\cos\Lambda)^{-1}$, countered somewhat by the lower effective angle of attack. The longer effective chord of the yawed wing gives the boundary layer a longer time to grow, thus increasing the drag. At high angles of attack, the reduction of the effective angle of attack by $(\cos\Lambda)^{-1}$ for the drag and $(\cos\Lambda)^{-2}$ for the lift has the effect of delaying stall and compressible drag divergence of the wing. In terms of the rotor aerodynamics, the practice of neglecting the influence of radial flow on the lift has been verified. The radial flow increases the normal drag force for the blade and also introduces the radial drag force, both of which increase the profile power. To summarize, the lift, drag, and radial force on the normal blade section with radial flow are:

$$c_\ell(\alpha) = c_{\ell_{2D}}(\alpha\cos^2\Lambda)/\cos^2\Lambda$$

$$c_d(\alpha) = c_{d2D}(\alpha \cos\Lambda)/\cos\Lambda$$

$$F_r = D\tan\Lambda = (u_R/u_T)D,$$

where $\cos\Lambda = u_T/(u_T^2 + u_R^2)^{1/2}$. These results are based on the assumption that the resultant drag force is in the yawed free stream direction, and on the the equivalence assumption for swept wings.

The normal and radial drag forces required for the rotor profile power are thus $F_x = D\cos\phi \cong D$ and $F_r = D\tan\Lambda = D(u_R/u_T)$, where $D = \frac{1}{2}u_T|u_T|c\,c_d$ (the air density has been omitted because dimensionless quantities are used). The drag coefficient is given by $c_d = c_{d2D}(\alpha\cos\Lambda)/\cos\Lambda$, with $\cos\Lambda = |u_T|/(u_T^2 + u_R^2)^{1/2}$. The absolute value of u_T is used to account for the reverse flow region. Since D is defined to be positive when opposing the rotor rotation, it must change sign in the reverse flow region. Then with $u_T = r + \mu\sin\psi$ and $u_R = \mu\cos\psi$ as usual, the profile terms of the rotor drag force, torque, and power are:

$$C_{H_o} = \int_0^1 \sigma\left(r\sin\psi + \mu\right)\frac{D}{cu_T}\,dr$$

$$C_{Q_o} = \int_0^1 \sigma\left(ru_T\right)\frac{D}{cu_T}\,dr$$

$$C_{P_o} = \int_0^1 \sigma\left(u_T^2 + u_R^2\right)\frac{D}{cu_T}\,dr$$

where

$$\frac{D}{cu_T} = \frac{1}{2}\sqrt{u_T^2 + u_R^2}\;c_d(\alpha\cos\Lambda)$$

and the yawed resultant velocity is $u_T^2 + u_R^2 = r^2 + \mu^2 + 2r\mu\sin\psi$. Given the angle-of-attack distribution over the rotor disk and the appropriate section drag coefficient data, these expressions may be numerically integrated. To proceed further analytically, we shall assume that the drag coefficient is independent of the angle of attack: $c_{d2D}(\alpha\cos\Lambda) \cong c_{d_o}$. Also, a constant chord blade is considered.

It is useful to examine separately the effects of reverse flow, radial drag, and the yawed-flow drag coefficient. Without the radial drag force F_r, the rotor coefficients become

$$C_{H_o} = \int_0^1 \sigma\left(u_T \sin\psi\right) \frac{D}{cu_T}\, dr$$

$$C_{Q_o} = \int_0^1 \sigma\left(ru_T\right) \frac{D}{cu_T}\, dr$$

$$C_{P_o} = \int_0^1 \sigma\left(u_T^2\right) \frac{D}{cu_T}\, dr$$

Neglecting the yawed-flow increase of the drag coefficient gives $D/cu_T = \frac{1}{2} c_{d_o} |u_T|$, and then neglecting the reverse flow gives $D/cu_T = \frac{1}{2} c_{d_o} u_T$. Making all three approximations reduces the model to that considered in sections 5–3 and 5–4, with the solution:

$$C_{H_o} = \frac{\sigma c_{do}}{2} \int_0^1 \sin\psi\, u_T^2\, dr = \frac{\sigma c_{do}}{8}\, 2\mu$$

$$C_{Q_o} = \frac{\sigma c_{do}}{2} \int_0^1 r u_T^2\, dr = \frac{\sigma c_{do}}{8} \left(1 + \mu^2\right)$$

$$C_{P_o} = \frac{\sigma c_{do}}{2} \int_0^1 u_T^3\, dr = \frac{\sigma c_{do}}{8} \left(1 + 3\mu^2\right)$$

(where the average has been taken over the azimuth). If the radial drag force is now included:

$$C_{H_o} = \frac{\sigma c_{do}}{2} \int_0^1 \left(r \sin\psi + \mu\right) u_T dr = \frac{\sigma c_{do}}{8}\, 3\mu$$

$$C_{Q_o} = \frac{\sigma c_{do}}{2} \int_0^1 r\, u_T^2\, dr = \frac{\sigma c_{do}}{8} \left(1 + \mu^2\right)$$

$$C_{P_o} = \frac{\sigma c_{do}}{2} \int_0^1 \left(u_T^2 + u_R^2\right) u_T dr = \frac{\sigma c_{do}}{8}\left(1 + 4\mu^2\right)$$

The radial drag force thus increases the rotor drag coefficient by 50%, and hence it increases the profile power in forward flight. Including reverse flow only (a form frequently encountered in the literature) gives

$$C_{H_o} = \frac{\sigma c_{do}}{8}\left(2\mu + \tfrac{1}{2}\mu^3\right)$$

$$C_{Q_o} = \frac{\sigma c_{do}}{8}\left(1 + \mu^2 - \frac{1}{8}\mu^4\right)$$

$$C_{P_o} = \frac{\sigma c_{do}}{8}\left(1 + 3\mu^2 + \frac{3}{8}\mu^4\right)$$

Reverse flow simply replaces u_T by $|u_T|$ in the integrands, which may be treated as follows:

$$\frac{1}{2\pi}\int_0^{2\pi}\int_0^1 f(r,\psi)|u_T| dr d\psi = \frac{1}{2\pi}\int_0^{2\pi}\int_0^1 f u_T dr d\psi - \frac{1}{\pi}\int_\pi^{2\pi}\int_0^{-\mu\sin\psi} f u_T dr d\psi$$

The first integral is just the result neglecting reverse flow. With both the radial drag and reverse flow included:

$$C_{H_o} = \frac{\sigma c_{do}}{8}\left(3\mu + \frac{3}{4}\mu^3\right)$$

$$C_{Q_o} = \frac{\sigma c_{do}}{8}\left(1 + \mu^2 - \frac{1}{8}\mu^4\right)$$

$$C_{P_o} = \frac{\sigma c_{do}}{8}\left(1 + 4\mu^2 + \frac{5}{8}\mu^4\right)$$

The effect of reverse flow is secondary to the radial drag force effect therefore, because of the low dynamic pressure in the reverse flow region.

Finally, consider the rotor profile coefficients including the radial drag, reverse flow, and the effect of yawed flow on c_d:

$$C_{H_o} = \int_0^1 \frac{\sigma c_{do}}{2}\left(r\sin\psi + \mu\right)\left(u_T^2 + u_R^2\right)^{1/2} dr$$

$$C_{Q_o} = \int_0^1 \frac{\sigma c_{do}}{2}\, r u_T \left(u_T^2 + u_R^2\right)^{1/2} dr$$

$$C_{P_o} = \int_0^1 \frac{\sigma c_{do}}{2}\left(u_T^2 + u_R^2\right)^{3/2} dr$$

With the additional $(\cos\Lambda)^{-1}$ factor due to the yawed-flow increase of the drag coefficient, it is no longer possible to integrate analytically. Based on numerical evaluations of the integrals, approximate analytical expressions are

$$C_{H_o} = \frac{\sigma c_{do}}{8}\left(3\mu + 1.98\mu^{2.7}\right)$$

$$C_{Q_o} = \frac{\sigma c_{do}}{8}\left(1 + 1.5\mu^2 - 0.37\mu^{3.7}\right)$$

$$C_{P_o} = \frac{\sigma c_{do}}{8}\left(1 + 4.5\mu^2 + 1.61\mu^{3.7}\right)$$

These expressions are accurate to about 1% up to $\mu = 1$ (which means that they accurately evaluate the integrals, not necessarily that by using them an accurate estimate of the profile power will be obtained). A frequently used approximation is

$$C_{P_o} = \frac{\sigma c_{do}}{8}\left(1 + 4.6\mu^2\right)$$

which is accurate to about 1% for $\mu = 0$ to 0.3, and to about 5% up to $\mu = 0.5$. The factor $(1 + 4.6\mu^2)$, which gives the profile power increase with speed, has the following contributions: $(1 + \mu^2)$ from the rotor torque and $2\mu^2$ from the rotor drag C_{H_o} due to the normal blade drag force; μ^2 from C_{H_o} due to the radial drag force; $0.45\mu^2$ due to the yawed-flow increase of the drag coefficient; and $0.15\mu^2$ due to reverse flow. Fig. 5-27 shows the profile power as a function of advance ratio, comparing the exact result with the approximation $C_{P_o} = (\sigma c_{do}/8)(1 + 4.6\mu^2)$. Also shown are the

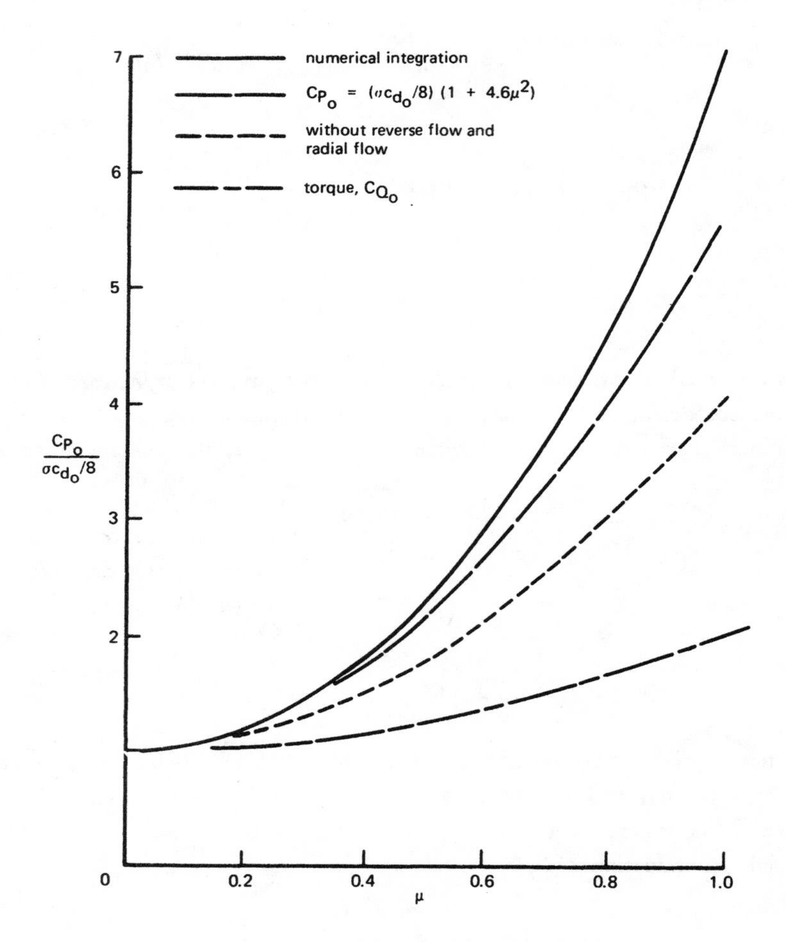

Figure 5-27 Profile power in forward flight.

profile power with no radial flow or reverse flow and the profile torque coefficient. The profile power increase is significant at moderate μ and is very large at high μ. At very high speeds, however, it is also necessary to include stall and compressibility effects in the evaluation of C_{P_o}.

Glauert (1926b) obtained

$$C_{P_o} = \int_0^1 \frac{\sigma c_{d_o}}{2} \left(u_T^2 + u_R^2\right)^{3/2} dr$$

by an energy analysis, and also

$$C_{P_o} = \int_0^1 \frac{\sigma c_{do}}{2} u_T^3\, dr = \frac{\sigma c_{do}}{8}\left(1 + 3\mu^2\right)$$

by blade element theory, neglecting reverse flow and radial flow. To evaluate the correct expression for C_{P_o}, he averaged the values at $\psi = 0°, 90°, 180°$, and $270°$ (where the radial integration can be performed analytically). Equating this average to $C_{P_o} = (\sigma c_{do}/8)(1 + n\mu^2)$, he obtained

$$1 + n\mu^2 \cong \frac{1}{2} + 3\mu^2 + \frac{1}{2}\mu^4 + \left(\frac{1}{2} + \frac{5}{4}\mu^2\right)\sqrt{1 + \mu^2} + \frac{3}{8}\mu^4 \ln \frac{\sqrt{1 + \mu^2} + 1}{\sqrt{1 + \mu^2} - 1}$$

Note that to order μ^2 this gives $n = 9/2$. Glauert used this result to evaluate the parameter n for a number of advance ratios. Bennett (1940) derived an expression for C_{P_o} by expanding the integral for small μ. He obtained

$$C_{P_o} = \frac{\sigma c_{do}}{8}\left(1 + \frac{9}{2}\mu^2 - \frac{3}{4}\mu^4 \ln\mu + \frac{3}{16}\mu^6 - \frac{3}{128}\mu^8 + \ldots\right)$$

The table below compares the results of Glauert, Bennett, and numerical integration in terms of the parameter n in the representation $C_{P_o} = (\sigma c_{do}/8)(1 + n\mu^2)$.

μ	0	0.3	0.4	0.5	0.6	0.75	1.0
numerical integration	4.50	4.69	4.83	4.99	5.18	5.49	6.11
Glauert	4.50	4.73	4.87	5.03	5.22	5.53	6.13
Bennett	4.50	4.58	4.61	4.64	4.66	4.67	4.67

Glauert's expression is evidently accurate. Bennett's expansion for small μ not unexpectedly is not good above about $\mu = 0.5$. Nevertheless, Bennett's results are clearly the source of the often used approximation $C_{P_o} = (\sigma c_{do}/8)(1 + 4.6\mu^2)$; Bennett suggested using $n = 4.65$. Other works concerned with the effect of radial flow on the profile power are Harris (1966a, 1966b) and Paglino (1969).

5–13 Flap Motion with a Hinge Spring

Consider an articulated rotor blade with no hinge offset from the center of rotation, but now with a spring ahout the flap hinge that produces a restoring moment on the blade (Fig. 5-28). Such a spring might be used to

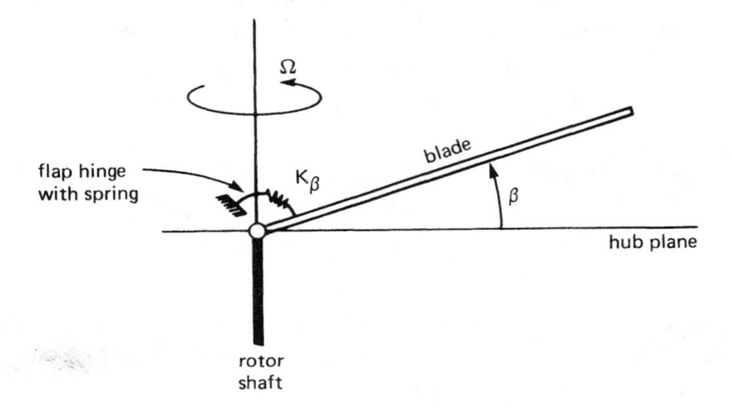

Figure 5.28 Blade flap motion, with a hinge spring.

augment the rotor control power, for with a spring the flap motion not only tilts the rotor thrust vector but also directly produces a moment on the hub. Since a hingeless rotor has a structural spring at the blade root, consideration of the blade with a flap hinge spring will serve as a guide to hingeless rotor behavior as well. It is assumed that the blade motion still consists of only rigid rotation about the flap hinge, so that the out-of-plane deflection is $z = r\beta$. For a very stiff spring the blade root restraint would approach that of a cantilevered blade, introducing considerable bending into the fundamental flapping mode shape. The spring stiffness which might be used on a rotor blade would be small compared to the centrifugal stiffening, however, so the rigid flapping assumption is reasonable. With rigid flapping motion, it follows that the equations for the rotor forces and power are unchanged. The hinge spring does change the rotor flapping motion, since it introduces an additional flap moment. Because the spring moment is proportional to the flapping displacement relative to the shaft, the hub plane is the appropriate reference plane in this case.

In the derivation of the flapping equation of motion, it is now only necessary to add the flap moment due to the hinge spring: $K_\beta(\beta - \beta_p)$, where K_β is the spring rate and β_p is the precone angle. With a spring at the flap hinge, the blade coning would produce a steady root moment except for the precone angle, which biases the hinge moment to zero for $\beta = \beta_p$. Then the flapping equation becomes

$$I_b(\ddot{\beta} + \Omega^2\beta) + K_\beta(\beta - \beta_p) = \int_0^R r F_z \, dr$$

or

$$\ddot{\beta} + \nu^2\beta = (\nu^2 - 1)\beta_p + \gamma \int_0^1 r \frac{F_z}{ac} \, dr,$$

where

$$\nu^2 = 1 + \frac{K_\beta}{I_b\Omega^2}$$

is the dimensionless natural frequency of the flap motion in the rotating frame. For practical flap springs, ν will be just slightly greater than 1. When $\nu > 1$, the aerodynamic forces acting at 1/rev are no longer forcing the flap motion exactly at resonance. Thus the rotor responds to this excitation with a reduced magnitude, and the lag is somewhat less than 90° in azimuth due to the spring quickening of the response. Flap hinge offset or cantilever root restraint will also increase the natural frequency of the flapping. By considering the articulated blade with a hinge spring it is possible to isolate the fundamental influence of the flap frequency, since the hinge spring changes nothing else. If the present problem is considered a model for an arbitrary rotor with flap frequency ν, the approximation lies in using rigid flapping for the blade mode shape.

The aerodynamic flap moments are unchanged by the hinge spring, but the inertial, centrifugal, and spring terms of the flapping equation now give

$$\frac{1}{2\pi}\int_0^{2\pi} \left[\ddot{\beta} + \nu^2\beta - (\nu^2 - 1)\beta_p\right] d\psi = \nu^2\beta_0 - (\nu^2 - 1)\beta_p$$

$$\frac{1}{\pi}\int_0^{2\pi}\left[\ddot{\beta} + \nu^2\beta - (\nu^2 - 1)\beta_p\right]\cos\psi\, d\psi = (\nu^2 - 1)\beta_{1c}$$

$$\frac{1}{\pi}\int_0^{2\pi}\left[\ddot{\beta} + \nu^2\beta - (\nu^2 - 1)\beta_p\right]\sin\psi\, d\psi = (\nu^2 - 1)\beta_{1s}$$

The flapping equations thus become

$$\nu^2\beta_0 = (\nu^2 - 1)\beta_p + \gamma\left[\frac{\theta_{.8}}{8}\left(1 + \mu^2\right) - \frac{\mu^2}{60}\theta_{tw} - \frac{\lambda_{NFP}}{6}\right]$$

$$(\nu^2 - 1)\beta_{1c} = \gamma\left[\frac{1}{8}\left(\theta_{1c} - \beta_{1s}\right)\left(1 + \frac{1}{2}\mu^2\right) - \frac{\mu}{6}\beta_0\right]$$

$$(\nu^2 - 1)\beta_{1s} = \gamma\left[\frac{1}{8}\left(\theta_{1s} + \beta_{1c}\right)\left(1 - \frac{1}{2}\mu^2\right) + \frac{\mu}{3}\theta_{.75} - \frac{\mu}{4}\lambda_{NFP}\right]$$

The solution for the coning is

$$\beta_0 = \frac{\nu^2 - 1}{\nu^2}\beta_p + \frac{\gamma}{\nu^2}\left[\frac{\theta_{.8}}{8}\left(1 + \mu^2\right) - \frac{\mu^2}{60}\theta_{tw} - \frac{\lambda_{NFP}}{6}\right]$$

The flap spring reduces the coning angle. Note that the solution can be written in terms of the coning with no spring,

$$\beta_0 = \frac{1}{\nu^2}\beta_{\text{ideal}} + \frac{\nu^2 - 1}{\nu^2}\beta_p$$

where β_{ideal} is the coning angle when $\nu = 1$. The role of the hub precone angle is to reduce the steady moments in the blade root. The mean hinge spring moment is

$$(\nu^2 - 1)(\beta_0 - \beta_p) = \frac{\nu^2 - 1}{\nu^2}\left(\beta_{\text{ideal}} - \beta_p\right)$$

The mean moment is thus nonzero when $\nu > 1$, unless the precone angle is selected to be $\beta_p = \beta_{\text{ideal}}$. Note that with ideal precone, the coning angle $\beta_0 = \beta_{\text{ideal}}$ is independent of the flap frequency. While the proper choice of precone will reduce the root loads, the ideal value is a function of the rotor loading and therefore the precone can be selected for a particular design operating condition only.

Now consider the tip-path-plane tilt. In hover the equations become

$$\beta_{1s} + \frac{\nu^2 - 1}{\gamma/8}\beta_{1c} = \theta_{1c}$$

$$\beta_{1c} - \frac{\nu^2 - 1}{\gamma/8}\beta_{1s} = -\theta_{1s}$$

which have the solution

$$\beta_{1s} = \left(\theta_{1c} + \frac{\nu^2 - 1}{\gamma/8}\theta_{1s}\right) \Big/ \left[1 + \left(\frac{\nu^2 - 1}{\gamma/8}\right)^2\right]$$

$$\beta_{1c} = \left(-\theta_{1s} + \frac{\nu^2 - 1}{\gamma/8}\theta_{1c}\right) \Big/ \left[1 + \left(\frac{\nu^2 - 1}{\gamma/8}\right)^2\right]$$

The parameter $(\nu^2 - 1)/(\gamma/8)$ is the ratio of the hinge spring to the aerodynamic flap damping. The inertia and centrifugal forces still exactly cancel, so it is the spring and damping forces that determine the response. The effect of a flap frequency ν greater than 1 is to introduce lateral flapping due to θ_{1s} and longitudinal flapping due to θ_{1c}. Write the cyclic flapping as $\bar{\beta}\cos(\psi + \psi_0 - \Delta\psi)$ and the cyclic pitch as $\bar{\theta}\cos(\psi + \psi_0)$. Then the magnitude and phase of the response are

$$\bar{\beta}/\bar{\theta} = \left[1 + \left(\frac{\nu^2 - 1}{\gamma/8}\right)^2\right]^{-1/2}$$

$$\Delta\psi = 90^\circ - \tan^{-1}\frac{\nu^2 - 1}{\gamma/8}$$

Increasing the flap frequency so that the system is excited below resonance slightly reduces the amplitude of the flap response to cyclic, and most importantly reduces the lag in the response. For example, when $\nu = 1.15$ and $\gamma = 8$, the amplitude is reduced only about 5%, but the lag becomes 72° instead of the 90° of an articulated rotor. This phase change constitutes a coupling of the lateral and longitudinal response of the tip-path plane to the no-feathering-plane control inputs. As far as control of the helicopter is concerned, such coupling can be eliminated by introducing a compensating phase shift between the control plane and no-feathering plane. That is, the control system geometry is changed so that the rotor still responds with purely longitudinal tip-path-plane tilt due to longitudinal cyclic stick displacements. In forward flight, the solution for the cyclic required to trim the helicopter is

$$\theta_{1c} = \beta_{1s} + \frac{\nu^2 - 1}{\gamma/8} \beta_{1c}\Big/\left(1 + \frac{1}{2}\mu^2\right) + \frac{4}{3}\mu\beta_0\Big/\left(1 + \frac{1}{2}\mu^2\right)$$

$$\theta_{1s} = -\beta_{1c} + \frac{\nu^2 - 1}{\gamma/8} \beta_{1s}\Big/\left(1 + \frac{3}{2}\mu^2\right)$$

$$-\frac{8}{3}\mu\left(\theta_{.75} - \frac{3}{4}\lambda_{TPP}\right)\Big/\left(1 + \frac{3}{2}\mu^2\right)$$

The tip-path-plane tilt relative to the hub plane (β_{1c} and β_{1s}) is determined by the equilibrium of helicopter forces and moments. The second term in these expressions gives the phase shift arising when $\nu > 1$. Note that forward speed has an influence on the phase shift, and moreover that the influence is not the same for both axes of cyclic. It follows that the ideal control rigging to compensate for the lateral-longitudinal coupling varies with speed (by typically 5% to 15% between hover and maximum speed) and is not the same for both lateral and longitudinal cyclic. The influence of forward flight is only of order μ^2, however, so it is possible to choose a single value for the control system phase that will in fact be satisfactory over the entire speed range of the helicopter.

The helicopter is controlled by using the rotor to produce moments about the center of gravity. An articulated rotor has no moment at the blade root and thus can produce moments on the helicopter only by tilting the rotor thrust vector. With a hinge spring, tilt of the rotor tip-path plane also produces a moment on the rotor hub. In the rotating frame, the hub moment due to the flap deflection of a single blade is

$$M = K_\beta(\beta - \beta_p) = (\nu^2 - 1)I_b\Omega^2(\beta - \beta_p)$$

The pitch and roll moments on the hub are obtained by resolving the flap moment in the nonrotating frame, multiplying by the number of blades, and averaging over the azimuth:

$$M_y = -(N/2\pi)\int_0^{2\pi} \cos\psi\, M\, d\psi \quad \text{and} \quad M_x = (N/2\pi)\int_0^{2\pi} \sin\psi\, M\, d\psi$$

(see section 5–3 also). In coefficient form, then, the hub pitch moment C_{M_y} and roll moment C_{M_x} are

$$\frac{2C_{My}}{\sigma a} = -\frac{\nu^2 - 1}{\gamma}\beta_{1c}$$

$$\frac{2C_{Mx}}{\sigma a} = \frac{\nu^2 - 1}{\gamma}\beta_{1s}$$

The inplane forces on the rotor may be written $H_{HP} = H_{TPP} - T\beta_{1c}$ and $Y_{HP} = Y_{TPP} - T\beta_{1s}$. Neglecting the tip-path-plane forces, the pitch and roll moments about the helicopter center of gravity a distance h below the hub are $M_y = h\,H_{HP} = -hT\beta_{1c}$ and $M_x = -h\,Y_{HP} = hT\beta_{1s}$. Upon combining the moments due to the thrust tilt and the hinge spring, the total moments about the helicopter center of gravity due to the rotor tip-path-plane tilt become

$$\begin{pmatrix} -\dfrac{2C_{My}}{\sigma a} \\ \\ \dfrac{2C_{Mx}}{\sigma a} \end{pmatrix} = \left(\frac{\nu^2 - 1}{\gamma} + h\,\frac{2C_T}{\sigma a}\right)\begin{pmatrix} \beta_{1c} \\ \\ \beta_{1s} \end{pmatrix}$$

The moment generating capability of the helicopter is increased greatly when $\nu > 1$. An articulated rotor normally obtains about half its moment from hinge offset and half from the thrust tilt. For a hingeless rotor the direct hub moment may be 2 to 4 times the thrust tilt term. Moreover, the direct hub moment term is independent of the helicopter load factor.

5–14 Flap Hinge Offset

Consider next an articulated rotor with the flap hinge offset from the center of rotation by a distance eR (Fig. 5-29). Such an arrangement is usually mechanically simpler than one with no offset, and in addition it has a favorable influence on the helicopter handling qualities, because it produces a flap frequency above 1/rev. Articulated rotors typically have an offset of $e = 0.03$ to 0.05. The analysis here will also consider a hinge spring. The blade radial coordinate r is still measured from the center of rotation. The blade motion is rigid rotation about the flap hinge, with

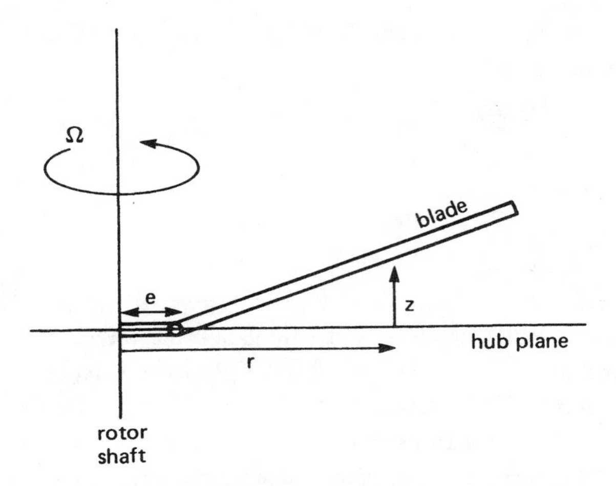

Figure 5-29 Flap motion with hinge offset.

degree of freedom β and mode shape $\eta(r)$, such that the out-of-plane deflection is $z = \beta\eta$.

Rigid rotation about a flap hinge offset by e corresponds to a mode shape

$$\eta = \begin{cases} k(r-e) & r > e \\ 0 & r < e \end{cases}$$

where k is a constant determined by the mode shape normalization. The normalization to be used here requires that the mode shape be equal to unity at the blade tip: $\eta(1) = 1$. Thus $k = (1 - e)^{-1}$, and the mode shape is $\eta = (r - e)/(1 - e)$. This reduces to $\eta = r$ for the case of no offset. Normalizing the mode shape to unity at the tip means that the degree of freedom β may be interpreted as the angle between the disk plane and a line extending from the center of rotation to the blade tip. This normalization is used because it is easily extended to the higher bending modes. An alternative mode shape is $\eta = (r - e)$, which makes β the actual angle of rotation about the flap hinge. The physically relevant quantities of the solution, such as the out-of-plane deflection $z = \beta\eta$, must of course be independent of the normalization chosen for the mode shape.

The normal velocity of the blade with an arbitrary flapping mode shape becomes

$$u_P = \lambda + \dot{z} + u_R \frac{dz}{dr} = \lambda + \eta\dot{\beta} + \eta'\beta\mu\cos\psi$$

There are no other changes to the blade aerodynamics. The rotor thrust is then

$$C_T = \sigma a \int_0^1 \tfrac{1}{2}(u_T{}^2\theta - u_P u_T)dr$$

$$= \frac{\sigma a}{2}\left[\frac{\theta_{.75}}{3}\left(1 + \frac{3}{2}\mu^2\right) - \frac{\theta_{tw}}{8}\mu^2 - \frac{1}{2}\left(\lambda - \mu\theta_{1s}\right) - \frac{\mu}{2}\beta_{1c}\frac{e}{1-e}\right]$$

The effect of the hinge offset on C_T is thus quite small. Similar results can be obtained for C_H and C_Y. Recall that the rotor power was derived for a general mode shape. The principal influence of the hinge offset is on the rotor flapping motion.

Consider again equilibrium of moments about the flap hinge. The forces acting on the blade section are:

(i) the inertial force $m\ddot{z} = m\eta\ddot{\beta}$, with moment arm $(r - e)$;

(ii) the centrifugal force $m\Omega^2 r$, with moment arm $z = \eta\beta$; and

(iii) the aerodynamic force F_z, with moment arm $(r - e)$.

There is also a spring moment at the flap hinge, $K_\beta(\beta - \beta_p)$, as in the previous section. For now a general mode shape $\eta = k(r - e)$ will be allowed. On integrating over the blade span, the conditions for equilibrium of the flap moments give

$$\int_e^R \eta(r-e)m\,dr\,\ddot{\beta} + \int_e^R \eta r m\,dr\,\beta\Omega^2 + K_\beta(\beta - \beta_p) = \int_e^R (r-e)F_z\,dr.$$

Now use dimensionless quantities and multiply by $k = \eta(1)/(1 - e)$:

$$\int_e^1 \eta^2 m\,dr\ddot{\beta} + k\int_e^1 \eta r m\,dr\beta + \frac{K_\beta k}{\Omega^2}\left(\beta - \beta_p\right) = \int_e^1 \eta F_z\,dr.$$

Let $I_b = \int_e^1 \eta^2 m\,dr$, and note that

$$k\int_e^1 \eta r m\,dr = \int_e^1 \eta^2 m\,dr + ke\int_e^1 \eta m\,dr = I_b + \frac{e}{1-e}\eta(1)\int_e^1 \eta m\,dr.$$

Then the flapping equation of motion is

$$\ddot{\beta} + \nu^2 \beta = \frac{K_\beta \eta(1)}{I_b \Omega^2 (1-e)} \beta_p + \gamma \int_e^1 \eta \frac{F_z}{ac} dr.$$

The Lock number is defined as $\gamma = \rho acR^4/I_b$ again, but note that here the definition of the characteristic moment of inertia I_b depends on the mode shape. If the definition $I_b = \int_0^1 r^2 m\, dr$ were retained, it would be necessary to introduce on the left-hand side of the flap equation the normalized flapping inertia $I_\beta{}^* = \int_e^1 \eta^2 m\, dr/I_b$. Such an approach is best when more degrees of freedom are considered, but here it is simplest to use the flapping generalized mass for I_b.

The natural frequency of the flap motion for the blade with hinge offset and spring is

$$\nu^2 = 1 + \frac{e}{1-e} \frac{\eta(1) \int_e^1 \eta m\, dr}{\int_e^1 \eta^2 m\, dr} + \frac{K_\beta}{I_b \Omega^2 (1-e)}$$

The first term is the centrifugal spring, the second term is the hinge offset effect (also due to the centrifugal forces, and the third term is the hinge spring. For a uniform mass distribution and no hinge spring, the result is

$$\nu^2 = 1 + \frac{3}{2} \frac{e}{1-e}$$

In general the flap frequency may be written $\nu^2 = 1 + er_{CG}M/I$, where M is the blade mass, I the moment of inertia about the flap hinge, and r_{CG} the radial center-of-gravity location relative to the hinge. Hinge offset thus raises the flap frequency above 1/rev. For the offsets of articulated rotors the increase is small, however, typically giving $\nu = 1.02$ to 1.04. This increase in flap frequency is the primary influence of the hinge offset. The flap dynamics of a rotor with $\nu > 1$ were examined in the last section. With a hinge offset there are also small changes in the aerodynamic flapping moments due to the mode shape change.

Now consider the aerodynamic forces. Again defining the aerodynamic coefficients by

$$M_F = \int_e^1 \eta \frac{F_z}{ac}\, dr = M_\theta \theta_{con} + M_{\theta_{tw}} \theta_{tw} + M_\lambda \lambda + M_{\dot\beta}\dot\beta + M_\beta \beta$$

we obtain

$$M_\theta = \frac{1}{8} c_2 + \frac{1}{3} c_1 \mu \sin\psi + \frac{1}{4} c_0 \mu^2 \sin^2\psi$$

$$M_{\theta_{tw}} = \frac{1}{10} c_3 + \frac{1}{4} c_2 \mu \sin\psi + \frac{1}{6} c_1 \mu^2 \sin^2\psi$$

$$M_\lambda = -\left(\frac{1}{6} c_1 + \frac{1}{4} c_0 \mu \sin\psi\right)$$

$$M_{\dot\beta} = -\left(\frac{1}{8} d_1 + \frac{1}{6} d_0 \mu \sin\psi\right)$$

$$M_\beta = -\left(\frac{1}{6} f_1 + \frac{1}{4} f_0 \mu \sin\psi\right) \mu \cos\psi$$

where $c_n = (n+2)\int_e^1 \eta r^n dr$, $d_n = (n+3)\int_e^1 \eta^2 r^n dr$, and $f_n = (n+2)\int_e^1 \eta\eta' r^n dr$.

With the mode shape $\eta = (r-e)/(1-e)$ the required constants are:

$$c_0 = 1 - e$$

$$c_1 = 1 - (e + e^2)/2$$

$$c_2 = 1 - (e + e^2 + e^3)/3$$

$$c_3 = 1 - (e + e^2 + e^3 + e^4)/4$$

$$d_0 = 1 - e$$

$$d_1 = 1 - (2e + e^2)/3$$

$$f_0 = 1$$

$$f_1 = 1 + e/2$$

Actually the constants c_n, d_n, and f_n should be evaluated by integrating from r_R to B, since the root cutout and especially the tip loss will have more effect than the hinge offset. The solution of the flapping equations is now:

$$\beta_0 = \frac{\gamma}{\nu^2}\left\{\frac{\theta_{.8}}{8}\left(c_2 + \mu^2 c_0\right) + \frac{\theta_{tw}}{10}\left[c_3 - c_2 + \mu^2\left(\frac{5}{6}c_1 - c_0\right)\right]\right.$$

$$\left. - \frac{\lambda_{NFP}}{6} c_1 + \frac{\mu}{12}\beta_{1c}\left(d_0 - f_1\right)\right\} + \frac{K_\beta \beta_p}{\nu^2 I_b \Omega^2 (1-e)}$$

$$\left(c_2 + \frac{1}{2}\mu^2 c_0\right)\theta_{1c} = \left(d_1 + \frac{1}{2}\mu^2 f_0\right)\beta_{1s} + \frac{\nu^2 - 1}{\gamma/8}\beta_{1c} + \frac{4}{3}\mu f_1 \beta_0$$

$$\left(c_2 - \frac{1}{2}\mu^2 c_0\right)\theta_{1s} = -\left(d_1 - \frac{1}{2}\mu^2 f_0\right)\beta_{1c} + \frac{\nu^2 - 1}{\gamma/8}\beta_{1s}$$

$$- \frac{8}{3}\mu\left[\theta_{0.75}c_1 + \frac{3}{4}\theta_{tw}\left(c_2 - c_1\right) - \frac{3}{4}c_0\lambda_{NFP}\right]$$

Thus the hinge offset produces small changes in the constants arising from the aerodynamic forces; to be consistent, however, the tip loss factor should also be included. The primary effect of the hinge offset on the flap response is the coupling of the lateral and longitudinal control that arises because $\nu > 1$. For hover, the phase lag between the flapping response and cyclic pitch input is reduced by

$$\Delta\psi = -\tan^{-1}\frac{\nu^2 - 1}{\gamma/8} \cong -\frac{12}{\gamma}e$$

which is small for articulated rotors.

Finally, consider the hub moment for a rotor with offset flapping hinges. The contributions to the moment about the hub ($r = 0$) are:

(1) the inertial force $m\eta\ddot{\beta}$, with moment arm r;

(ii) The centrifugal force $m\Omega^2 r$, with moment arm $\eta\beta$; and

(iii) the aerodynamic force F_z, with moment arm r.

Then the flapwise moment on the hub due to one blade is

$$M = -(\ddot{\beta} + \beta)\int_e^1 m\eta r\,dr + \int_e^1 r F_z\,dr$$

Substituting for $\ddot{\beta}$ from the flapping equation gives

$$M = -\left[\frac{K_\beta}{I_b\Omega^2(1-e)}\beta_p + \frac{1}{I_b}\int_e^1 \eta F_z dr + \beta\left(1-\nu^2\right)\right]\int_e^1 \eta r m dr + \int_e^1 r F_z dr$$

The precone term is constant, so it does not contribute to the pitch or roll moments on the hub. Using $r = (1-e)\eta + e$, note that

$$-\int_e^1 \eta F_z dr \int_e^1 \eta r m dr + \int_e^1 \eta^2 m dr \int_e^1 r F_z dr$$

$$= e\left[-\int_e^1 \eta F_z dr \int_e^1 \eta m dr + \int_e^1 \eta^2 m dr \int_e^1 F_z dr\right]$$

The factor in brackets is zero if the lift distribution is proportional to the mode shape, $F_z \sim (r-e)$. In general, then, this sum is second-order small and can be neglected. The hub moment thus reduces to

$$M = I_b(\nu^2 - 1)\beta$$

and the pitch and roll components from the N blades give

$$\begin{pmatrix} -\dfrac{2C_{M_Y}}{\sigma a} \\ \dfrac{2C_{M_X}}{\sigma a} \end{pmatrix} = \frac{\nu^2 - 1}{\gamma}\begin{pmatrix} \beta_{1c} \\ \beta_{1s} \end{pmatrix}$$

This is the same result as that obtained for the hinge spring alone. A more general derivation of the result is given in Chapter 9. While all the other effects of hinge offset examined have been only small refinements of the basic rotor behavior, the hub moment capability with offset hinges is of major importance. Articulated rotor helicopters generate about half the moment about the center of gravity by the thrust tilt and about half by the direct hub moment.

5–15 Hingeless Rotor

In hingeless rotors, which have no flap or lag hinges, the blades are attached to the hub with cantilever root restraint. Such a rotor has the advantages of a mechanically simple hub and generally improved handling qualities. The fundamental out-of-plane bending mode of a hingeless rotor blade is very similar to the rigid flapping mode of an articulated blade, because of the dominance of the centrifugal stiffening relative to the structural stiffening. The fundamental natural flap frequency of a hingeless blade is thus not far above 1/rev, although it is significantly greater than the frequency achieved with offset-hinged blades. Typically the flap frequency ν is 1.10 to 1.15 for hingeless rotors.

In the preceding section, the flapping equation was obtained for an arbitrary mode shape:

$$\ddot{\beta} + \nu^2 \beta = \gamma \int_0^1 \eta \frac{F_z}{ac} \, dr$$

With the proper value for the flap frequency ν, this equation can be used for the hingeless rotor blade as well. We have seen that the influence of the mode shape is secondary to that of the flap frequency. Thus a hingeless rotor can be modeled by using the correct flap frequency, but with a simple approximate mode shape. Since only the integrals of the mode shape must be accurate, such an approach should be reasonably accurate. The flap frequency will either be specified arbitrarily in the investigation or it must be obtained from a free vibration analysis of the blade. An appropriate mode shape is that of rigid rotation about an offset hinge, $\eta = (r - e)/(1 - e)$. The offset e can be chosen by matching the slope of the actual mode shape at an appropriate station such as 75% radius:

$$e = 1 - \frac{1}{\eta'(0.75)}$$

The effective offset is typically around $e = 0.10$ for hingeless rotors.

Although such an approximate model must be used with care to ensure that the assumptions are valid before the results are relied on too much, it does in general give correctly the fundamental behavior of the hingeless rotor which is determined primarily by the flap frequency ν. When other degrees of freedom (such as the lag or torsional motion) are involved, it is often necessary to use a more accurate model of the rotor motion that includes the correct mode shapes.

The literature concerned with the modeling of hingeless rotors includes: Allen (1946), Winson (1947), Payne (1955d), Young (1962a), Ward (1966a, 1966b), Bramwell (1969), Hohenemser and Yin (1973a). See also Chapter 9 and the literature on helicopter aeroelasticity and flap-lag dynamics.

5–16 Gimballed or Teetering Rotor

A gimballed rotor has three or more blades attached to the hub without flap or lag hinges (cantilever root restraint), and the hub is attached to the rotor shaft by a universal joint or gimbal. The motion of the gimballed hub relative to the shaft is described by two degrees of freedom, the longitudinal and lateral tilt angles β_{1c} and β_{1s}, which correspond to the tip-path-plane tilt of an articulated rotor by cyclic flapping. The hub may include spring restraint of the gimbal motion. During the coning motion of the blades the hub will not tilt, since there is no net pitch or roll moment on the rotor. Hence for the coning motion the blades behave as on a hingeless rotor. For the higher harmonics of the flap motions (β_{2c}, β_{2s}, etc) the hub also remains fixed.

The flapwise moment on the mth blade of a gimballed rotor is

$$\frac{M^{(m)}}{I_b\Omega^2} = -(\ddot{\beta} + \beta) + \gamma \int_0^1 r \frac{F_z}{ac}\, dr$$

(see section 5–14; the mode shape $\eta = r$ that has been used corresponds to rigid body motion of the rotor about the gimbal). The equations of motion for longitudinal and lateral tilt of the gimbal are obtained from equilibrium of moments on the entire rotor. Summing the pitch moments from all N blades, adding a hub spring moment, and averaging over the azimuth gives

$$\frac{M_{\text{spring}}}{I_b\Omega^2} + \frac{1}{2\pi}\int_0^{2\pi} \left[\sum_{m=1}^{N} \cos\psi_m \frac{M^{(m)}}{I_b\Omega^2} \right] d\psi = 0$$

where $\psi_m = \psi + m(2\pi/N)$ is the azimuth position of the mth blade. For the steady-state solution, since all the blades have the same periodic motion the sum over N blades followed by the average over ψ is equivalent to N times the average for one blade:

$$\frac{M_{\text{spring}}}{I_b\Omega^2} + \frac{1}{2\pi}\int_0^{2\pi}\left[N\cos\psi\,\frac{M}{I_b\Omega^2}\right]d\psi = 0$$

Now the longitudinal hub spring moment is

$$M_{\text{spring}} = -K_\beta\beta_{1c} = -K_\beta\frac{1}{\pi}\int_0^{2\pi}\beta\cos\psi\,d\psi$$

Thus the equation of motion is

$$\frac{1}{\pi}\int_0^{2\pi}\cos\psi\left[-\frac{K_\beta\beta}{\frac{1}{2}NI_b\Omega^2} + \frac{M}{I_b\Omega^2}\right]d\psi = 0$$

Similarly, for roll moments on the rotor we obtain

$$\frac{1}{\pi}\int_0^{2\pi}\sin\psi\left[-\frac{K_\beta\beta}{\frac{1}{2}NI_b\Omega^2} + \frac{M}{I_b\Omega^2}\right]d\psi = 0$$

The operators $\frac{1}{\pi}\int_0^{2\pi}(\ldots)\cos\psi d\psi$ and $\frac{1}{\pi}\int_0^{2\pi}(\ldots)\sin\psi\,d\psi$ are those used to obtain the equations for β_{1c} and β_{1s} of the articulated rotor. The equations of motion for the gimbal tilt are therefore the same as for the tip-path-plane tilt of an equivalent single blade with differential equation

$$\ddot{\beta} + \nu^2\beta = \gamma\int_0^1 r\frac{F_z}{ac}\,dr$$

and the solution is then the same as for the articulated rotor. Here the flap natural frequency is

$$\nu^2 = 1 + \frac{K_\beta}{\frac{1}{2}NI_b\Omega^2}$$

Unless there is a hub spring, the frequency is $\nu = 1$ as for an articulated blade with no hinge offset. Note that with a gimbal it is possible to put the hub spring in the nonrotating system so that it does not have to operate with a continual 1/rev motion. Moreover, different spring rates can then be used for longitudinal and lateral motions. For the coning and the second and higher harmonics of the flap motion the blade acts as a hingeless rotor. Again the solution can be obtained by considering an equivalent single blade and using the flap frequency corresponding to the cantilevered blade.

A teetering rotor has two blades attached to the hub without flap or lag hinges. The hub is attached to the shaft by a single flapping hinge, the two blades forming a single structure. The flapping motion is like that of a see-saw or teeter board, hence the name given this rotor. Such a hub configuration has the advantage of being mechanically very simple. As for the gimballed rotor, the coning motion gives no net moment about the teeter hinge and in effect the blades have cantilever root restraint. In general, the steady-state motion of the teetering rotor must be obtained by considering equilibrium of moments on the entire rotor. Since both blades must be executing the same periodic motion, the root flapping moment of the mth blade is a periodic function of ψ_m:

$$M^{(m)} = M_0 + \sum_{n=1}^{\infty} M_{nc} \cos n\psi_m + M_{ns} \sin n\psi_m$$

where $\psi_1 = \psi + \pi$ and $\psi_2 = \psi$. This may be written as

$$M^{(m)} = M_0 + \sum_{n=1}^{\infty} (-1)^{mn} (M_{nc} \cos n\psi + M_{ns} \sin n\psi)$$

The total flap moment about the teeter hinge is then

$$M = M^{(2)} - M^{(1)} = \sum_{n=1}^{\infty} [1 - (-1)^n] (M_{nc} \cos n\psi + M_{ns} \sin n\psi)$$

$$= 2 \sum_{n \text{ odd}} (M_{nc} \cos n\psi + M_{ns} \sin n\psi)$$

So for all even harmonics (including the coning motion) the flap moments from the two blades cancel each other. Only the odd harmonics, in particular the tip-path-plane tilt degrees of freedom β_{1c} and β_{1s}, produce a net moment about the hinge and hence teetering motion of the blade.

For the odd harmonics of the teetering rotor flap motion, the hinge moment consists of the root flapping moments from the two blades (which is equivalent to twice the moment of one of the blades) and a possible hub spring moment:

$$-\frac{K_\beta}{I_b\Omega^2} + 2\left[-(\ddot{\beta}+\beta) + \gamma\int_0^1 r\,\frac{F_z}{ac}\,dr\right] = 0$$

The equation of motion is therefore

$$\ddot{\beta} + \nu^2\beta = \gamma\int_0^1 r\frac{F_z}{ac}\,dr$$

where the natural frequency of the flapping is

$$\nu^2 = 1 + \frac{K_\beta}{2I_b\Omega^2}$$

Usually a teetering rotor does not have a hub spring, so $\nu = 1$. The tip-path-plane tilt motion of the teetering rotor is thus the same as that of an articulated rotor with no hinge offset.

To summarize the behavior of gimballed and teetering rotors, for those harmonics of the flap motion that give a net moment on the hub, including the tip-path-plane tilt, the blade acts as an articulated rotor with no hinge offset ($\eta = r$ and $\nu = 1$). For those harmonics (including the coning motion) where the flap moments are reacted internally in the hub, the blade acts as a hingeless rotor of very high stiffness. With these considerations, the solutions obtained for an articulated rotor are also applicable to gimballed and teetering rotors.

5–17 Pitch-Flap Coupling

Pitch-flap coupling in a kinematic feedback of the flapping displacement to the blade pitch motion, that may be described by $\Delta\theta = -K_p\beta$. For positive pitch-flap coupling ($K_p > 0$), flap up decreases the blade pitch and hence the blade angle of attack. The resulting lift reduction produces a change in flap moment that opposes the original flap motion. Thus positive pitch-flap coupling acts as an aerodynamic spring on the flap motion. Pitch-flap coupling may be obtained entirely by mechanical means. The simplest approach

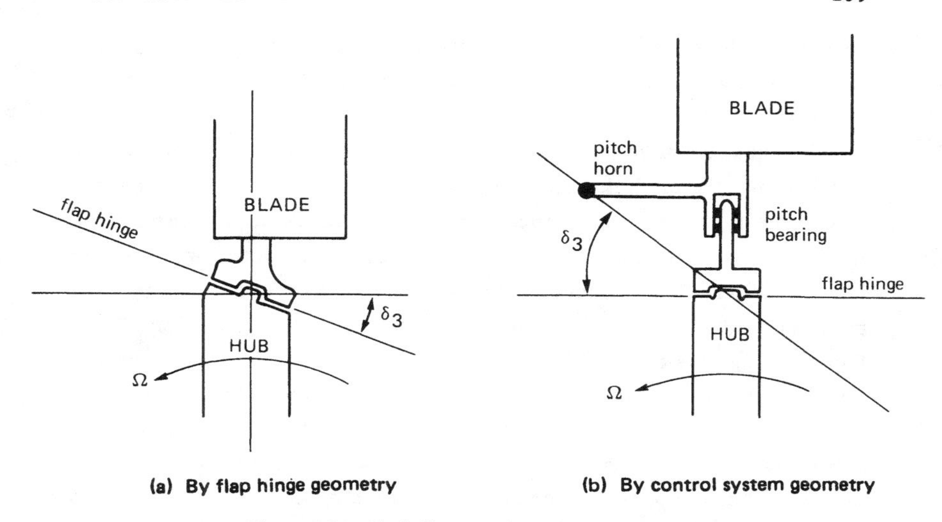

(a) By flap hinge geometry **(b) By control system geometry**

Figure 5-30 **Pitch-flap coupling of a rotor blade.**

is to skew the flap hinge by an angle δ_3 so that it is no longer perpendicular to the radial axis of the blade (see Fig. 5-30a). Then a rotation about the hinge with a flap angle β must also produce a pitch change of $-\beta \tan \delta_3$. The feedback gain for this arrangement is therefore $K_p = \tan \delta_3$. Pitch-flap coupling is usually defined in terms of the delta-three angle. Note that positive coupling $\delta_3 > 0$ represents negative feedback, decreasing the blade pitch for a flap increase. Pitch-flap coupling can also be introduced by the control system geometry (see Fig. 5-30b). When the pitch bearing is outboard of the flap hinge (the usual arrangement), the blade will experience a pitch change due to flapping if the pitch link is not in line with the axis of the flap hinge. For a fixed swashplate position, the flap motion can be viewed as occurring about a virtual hinge axis joining the end of the pitch horn and the actual flap hinge. The δ_3 angle then is the angle between this virtual hinge axis and the real flap hinge axis. Another source of pitch-flap coupling is the mean lag angle ζ_0 due to the rotor torque. If the flap hinge is outboard of the lag hinge, the mean lag angle is equivalent to a skew of the flap hinge; that is, $\delta_3 = \zeta_0$. There are similar coupling effects on hingeless rotors. Although pitch-flap and other coupling is determined for an articulated rotor by the hub, root, and control system geometry, for hingeless rotors it is also necessary to consider the structural and inertial characteristics of the blade. Often the δ_3 angle depends on the blade pitch because of changes in the control system geometry with collective, so that in general it is necessary to evaluate $K_p = -\partial\theta/\partial\beta$ for a given collective, coning, and mean lag angle of the blade.

The equation of motion for the blade flapping was derived above considering only the pitch due to the control system input, θ_{con}. The solution relates the flapping to the actual blade pitch. That solution remains valid with pitch-flap coupling, but the root pitch and the control input are no longer identical. The difference can be accounted for by noting that the root pitch is now $\theta - K_p\beta$ if θ retains its meaning as the control input only. Pitch-flap coupling thus changes the relative orientation of the control plane and no-feathering plane, while the solution for the orientation of the no-feathering plane relative to the tip-path plane is unchanged. Since pitch-flap coupling acts on the flapping with respect to the hub plane, $\theta_{HP} = \theta_{CP} - K_p\beta_{HP}$ is the actual blade root pitch. The flapping solution of section 5–5 determines θ_{HP} in terms of β_{HP}. There are two possible approaches to an analysis of the effects of pitch-flap coupling. The quantity $\theta_{CP} - K_p\beta_{HP}$ can be substituted for θ_{HP} in the differential equation of motion for the flapping, the solution of which will then give the control required θ_{CP} and show the other effects of K_p. Alternatively, the previous solutions may be used directly, with $\theta_{CP} = \theta_{HP} + K_p\beta_{HP}$ determining the control required.

Consider the differential equation obtained for the flap motion of a rotor with flap frequency ν. Here it is necessary to substitute $(\theta_{con} - K_p\beta)$ for θ_{con}, with the result

$$\ddot{\beta} + \nu^2\beta = \gamma\left[M_\theta(\theta_{con} - K_p\beta) + M_{\theta_{tw}}\theta_{tw} + M_\lambda\lambda + M_{\dot{\beta}}\dot{\beta} + M_\beta\beta\right]$$

For hover this becomes

$$\ddot{\beta} + \frac{\gamma}{8}\dot{\beta} + \left(\nu^2 + \frac{\gamma}{8}K_p\right)\beta = \frac{\gamma}{8}\theta_{con} + \frac{\gamma}{10}\theta_{tw} - \frac{\gamma}{6}\lambda$$

(the mode shape $\eta = r$ has been used to evaluate the aerodynamic coefficients). Thus pitch-flap coupling introduces an aerodynamic spring that increases the effective natural frequency of the flap motion to

$$\nu_e^2 = \nu^2 + \frac{\gamma}{8}K_P$$

The flapping response to cyclic depends on the effective spring ν_e. However, pitch-flap coupling does not produce a hub moment, which is still determined by $(\nu^2 - 1)$. The solution for the cyclic pitch control is

$$\theta_{1c} = \beta_{1s} + \left[\frac{\nu^2 - 1}{\frac{\gamma}{8}\left(1 + \frac{1}{2}\mu^2\right)} + K_P\right]\beta_{1c} + \frac{\frac{4}{3}\mu\beta_0}{1 + \frac{1}{2}\mu^2}$$

$$\theta_{1s} = -\beta_{1c} + \left[\frac{\nu^2 - 1}{\frac{\gamma}{8}\left(1 + \frac{3}{2}\mu^2\right)} + K_P\right]\beta_{1s}$$

$$- \frac{\frac{8}{3}\mu\left(\theta_{.75} - \frac{3}{4}\lambda_{TPP}\right)}{1 + \frac{3}{2}\mu^2} + \frac{\frac{8}{3}\mu K_P\beta_0}{1 + \frac{3}{2}\mu^2}$$

or in hover

$$\theta_{1c} = \beta_{1s} + \left(\frac{\nu^2 - 1}{\gamma/8} + K_P\right)\beta_{1c}$$

$$\theta_{1s} = -\beta_{1c} + \left(\frac{\nu^2 - 1}{\gamma/8} + K_P\right)\beta_{1s}$$

The magnitude and phase of the tip-path-plane response to cyclic becomes

$$\bar{\beta}/\bar{\theta} = \left[1 + \left(\frac{\nu^2 - 1}{\gamma/8} + K_P\right)^2\right]^{-1/2} = \left[1 + \left(\frac{\nu_e{}^2 - 1}{\gamma/8}\right)^2\right]^{-1/2}$$

$$\Delta\psi = 90° - \tan^{-1}\left(\frac{\nu^2 - 1}{\gamma/8} + K_P\right) = 90° - \tan^{-1}\frac{\nu_e{}^2 - 1}{\gamma/8}$$

For an articulated rotor $(\nu = 1)$, the result is

$$\bar{\beta}/\bar{\theta} = \frac{1}{\sqrt{1 + K_P{}^2}}$$

$$\Delta\psi = 90^\circ - \tan^{-1} K_P = 90^\circ - \delta_3.$$

Thus the swashplate phasing required is just equal to the δ_3 angle.

Now consider the effect of pitch-flap coupling in terms of the change of the control plane orientation relative to the no-feathering plane. The relation $\theta_{CP} = \theta_{HP} + K_P\beta_{HP}$ gives the collective and cyclic pitch required:

$$\begin{pmatrix} \theta_0 \\ \theta_{1c} \\ \theta_{1s} \end{pmatrix}_{CP} = \begin{pmatrix} \theta_0 \\ \theta_{1c} \\ \theta_{1s} \end{pmatrix}_{HP} + K_P \begin{pmatrix} \beta_0 \\ \beta_{1c} \\ \beta_{1s} \end{pmatrix}_{HP}$$

For a given thrust and positive pitch-flap coupling, the collective input must thus be increased to counter the feedback of the coning angle and keep the actual root collective at $(\theta_0)_{HP}$. Similarly, the cyclic pitch required may be determined from these relations. A special case is that of a rotor with no cyclic pitch control, an important example of which is the tail rotor. In that case the helicopter operating condition fixes the orientation of the control plane instead of the tip-path plane. With no cyclic in the control plane, $\theta_{CP} = \theta_{HP} + K_P\beta_{HP}$ gives

$$\theta_{1c_{HP}} + K_P\beta_{1c_{HP}} = 0$$

$$\theta_{1s_{HP}} + K_P\beta_{1s_{HP}} = 0$$

The orientation of the tip-path plane relative to the no-feathering plane,

$$\beta_{1c_{NFP}} = \beta_{1c_{HP}} + \theta_{1s_{HP}}$$

$$\beta_{1s_{NFP}} = \beta_{1s_{HP}} - \theta_{1c_{HP}},$$

is fixed by flap moment equilibrium. Eliminating θ_{HP} gives then

$$\beta_{1s_{HP}} = (\beta_{1s_{NFP}} - K_P\beta_{1c_{NFP}})/(1 + K_P^2)$$

$$\beta_{1c_{HP}} = (\beta_{1c_{NFP}} + K_P\beta_{1s_{NFP}})/(1 + K_P^2)$$

or

$$|\beta|_{HP} = |\beta|_{NFP}\Big/\sqrt{1 + K_P^2}$$

Thus pitch-flap coupling reduces the flapping magnitude relative to the rotor shaft. Note that negative coupling is as effective as positive coupling, because the effect of K_P is to remove flap motion from resonant excitation. The sign of the feedback influences the phase of the response, and large negative pitch-flap coupling does have an adverse effect on the flapping stability. It is common to use 45° of delta-three on tail rotors ($K_P = 1$) to reduce the transient and steady state flapping relative to the shaft.

5–18 Helicopter Force, Moment, and Power Equilibrium

The operating condition of the rotor is determined by force and moment equilibrium on the entire helicopter. In this section the longitudinal and lateral equilibrium for a helicopter in steady unaccelerated flight is examined. In the case of longitudinal force equilibrium the result for large angles will be obtained; this result can then be used to determine the rotor power required. While in numerical calculations the simultaneous equilibrium of all six components of force and moment on the helicopter can be found, the basic behavior may be determined by considering lateral and longitudinal equilibrium separately.

Longitudinal force equilibrium considers the forces in the vertical longitudinal plane of the helicopter (Fig. 5-31; see also section 5–4). The helicopter has speed V and a flight path angle θ_{FP}, so that a climb or descent velocity $V_c = V \sin \theta_{FP}$ is included. The forces on the rotor are the thrust T and rotor drag H, defined relative to the reference plane used. The reference plane has angle of attack α with respect to the forward speed (α is

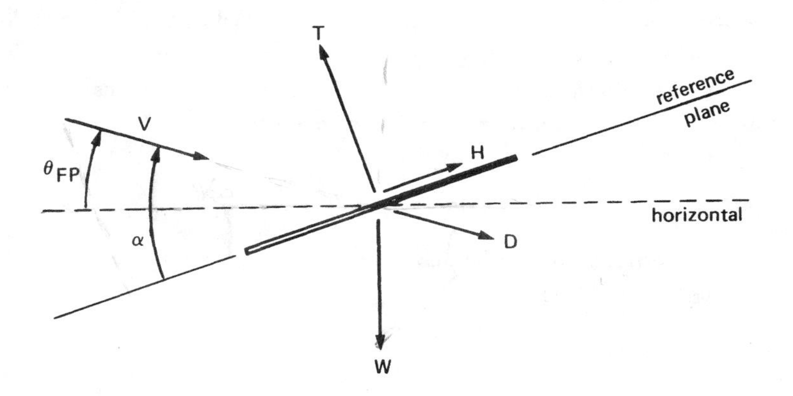

Figure 5-31 Longitudinal forces acting on the helicopter

positive for forward tilt of the rotor). The forces acting on the helicopter are the weight W (vertical) and the aerodynamic drag D (in the same direction as V). Auxiliary propulsion or lifting devices can be accounted for by including their forces in W and D. From the requirements for vertical and horizontal force equilibrium,

$$W = T\cos(\alpha - \theta_{FP}) - D\sin\theta_{FP} + H\sin(\alpha - \theta_{FP})$$

$$D\cos\theta_{FP} + H\cos(\alpha - \theta_{FP}) = T\sin(\alpha - \theta_{FP})$$

or for small angles $W = T$ and $D + H = T(\alpha - \theta_{FP})$. Therefore the rotor thrust equals the helicopter weight, and the conditions for horizontal force equilibrium give the angle of attack:

$$\alpha = \theta_{FP} + \frac{D}{W} + \frac{H}{T} = \frac{\lambda_c}{\mu} + \frac{D}{W} + \frac{C_H}{C_T}$$

where $\lambda_c = V_c/\Omega R \cong \mu\theta_{FP}$. Then with $H = H_{TPP} - \beta_{1c}T$,

$$\alpha = \frac{\lambda_c}{\mu} + \frac{D}{W} + \frac{C_{H\,TPP}}{C_T} - \beta_{1c}$$

and the inflow ratio is

$$\lambda = \lambda_i + \mu a \cong \lambda_i + \lambda_c + \mu\frac{D}{W} + \mu\frac{C_H}{C_T}$$

This is the same result as was obtained in section 5–4. Note that if H_{TPP} is neglected, the tip-path-plane inclination is determined by the helicopter drag and climb velocity alone: $\alpha_{TPP} = \theta_{FP} + D/W$.

For large angles, using the horizontal force equation to eliminate the drag force allows the vertical force equation to be written

$$W = \frac{T\cos\alpha}{\cos\theta_{FP}}\left(1 + \frac{H}{T}\tan\alpha\right)$$

Then the horizontal force equation can be written as

$$\frac{D}{T\cos\alpha} + \frac{H}{T}\left(1 + \tan\alpha\tan\theta_{FP}\right) = \tan\alpha - \tan\theta_{FP}$$

or

$$\frac{D}{W\cos\theta_{FP}}\left(1+\frac{H}{T}\tan\alpha\right)+\frac{H}{T}\left(1+\tan\alpha\tan\theta_{FP}\right) = \tan\alpha - \tan\theta_{FP}$$

Solving for $\tan\alpha$ gives

$$\tan\alpha = \frac{\tan\theta_{FP}+\dfrac{D}{W\cos\theta_{FP}}+\dfrac{H}{T}}{1-\dfrac{H}{T}\left(\tan\theta_{FP}+\dfrac{D}{W\cos\theta_{FP}}\right)}$$

from which the inflow ratio $\lambda = \mu\tan\alpha + \lambda_i$ may be obtained. Note that this result can be written

$$\begin{aligned}\alpha &= \tan^{-1}\left(\tan\theta_{FP}+\frac{D}{W\cos\theta_{FP}}\right)+\tan^{-1}\frac{H}{T}\\ &= \alpha|_{H=0} + \tan^{-1}\frac{H}{T}\end{aligned}$$

For small angles this reduces to the previous result. In summary, a forward tilt of the disk is required to produce the propulsive forces opposing the helicopter and rotor drag, and to provide the climb velocity.

The conditions for lateral force equilibrium (Fig. 5-32) determine the roll angle ϕ of the reference plane relative to the horizontal. The rotor thrust T and side force Y are defined relative to the reference plane used. The forces on the helicopter are the weight W and a side force Y_F (such as that due to the tail rotor). The conditions for horizontal and vertical force equilibrium

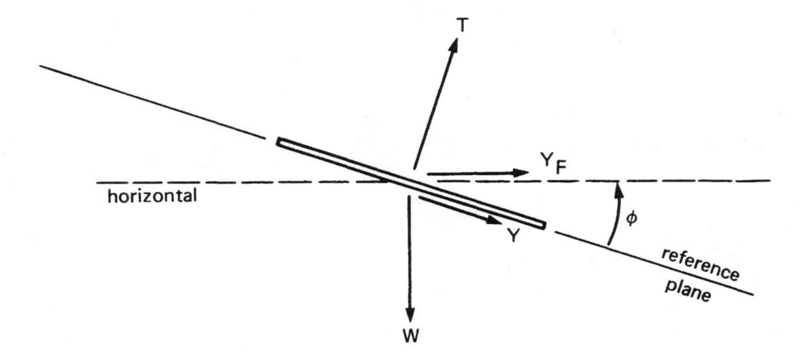

Figure 3-32 Lateral forces acting on the helicopter (view from aft).

give

$$Y_F + Y\cos\phi + T\sin\phi = 0$$

$$W = T\cos\phi - Y\sin\phi$$

with the solution

$$\tan\phi = \frac{-Y_F/W - Y/T}{1 - (Y_F/W)(Y/T)}$$

or $\phi = -\tan^{-1} Y_F/W - \tan^{-1} Y/T$. The rotor disk must roll to the left to provide a component of thrust to cancel the side forces of the helicopter and rotor. For small angles the result is $\phi = -Y_F/W - C_Y/C_T$, or using $Y = Y_{TPP} - \beta_{1s}T$,

$$\phi = -\frac{Y_F}{W} - \frac{C_{Y_{TPP}}}{C_T} + \beta_{1s}$$

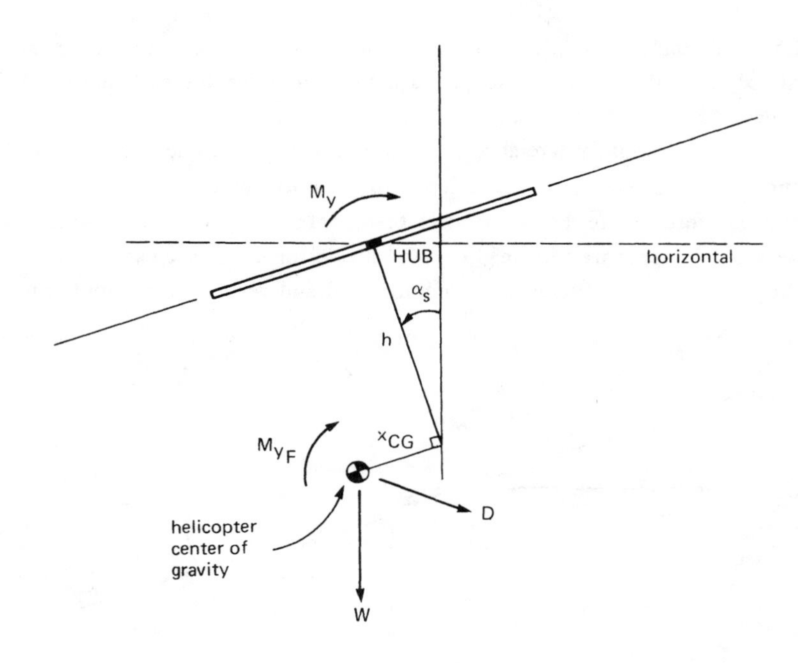

Figure 5-33 Longitudinal moments acting on the helicopter.

Next consider the equilibrium of pitch moments on the helicopter (Fig. 5-33), which determines the angle of attack of the rotor shaft relative to the vertical, α_s. Moments will be taken about the rotor hub so that the rotor forces will not be involved and the rotor reference plane will not enter the problem. The rotor hub moment M_y must be included, however. The forces at the helicopter center of gravity are the weight W, the aerodynamic drag D, and an aerodynamic pitch moment M_{yF}. The position of the helicopter center of gravity is defined relative to the rotor shaft (i.e. in the hub plane axis system). It is located a distance h below the hub and a distance x_{CG} forward of the shaft (x_{CG} is the longitudinal center-of-gravity position). For small angles, the requirements for moment equilibrium about the rotor hub give

$$M_y + M_{y_F} + W(h\alpha_s - x_{CG}) - hD = 0$$

which may be solved for the shaft angle of attack α_s,

$$\alpha_s = \alpha_{HP} - \theta_{FP} = \frac{x_{CG}}{h} + \frac{D}{W} - \frac{M_{yF}}{Wh} - \frac{M_y}{Wh}$$

Note that the shaft angle (the orientation of the hub plane relative to the horizontal) has also been written in terms of the hub plane tilt (relative to the aircraft velocity) and the flight path angle (between the velocity and the horizontal). Now the rotor hub moment is given by the tilt of the tip-path plane relative to the hub plane:

$$\frac{M_y}{Wh} = \frac{C_{My}}{hC_T} = -\frac{(\nu^2 - 1)/\gamma}{h2C_T/\sigma a}\beta_{1c_{HP}}$$

Next, recall that the requirements for longitudinal force equilibrium gave

$$\alpha_{HP} - \theta_{FP} - \frac{D}{W} = \frac{H_{HP}}{T} = \frac{H_{TPP}}{T} - \beta_{1c_{HP}}$$

After combining the force and moment equilibrium results to eliminate $(\alpha_{HP} - \theta_{FP} - D/W)$, solving for $\beta_{1c_{HP}}$ gives

$$\beta_{1c_{HP}} = \frac{-x_{CG}/h + M_{yF}/hW + C_{H_{TPP}}/C_T}{1 + \dfrac{(\nu^2 - 1)/\gamma}{h2C_T/\sigma a}}$$

and then the shaft angle

$$\alpha_s = \frac{x_{CG}/h - M_{y_F}/hW + \dfrac{(\nu^2 - 1)/\gamma}{h2C_T/\sigma a}\dfrac{C_{H_{TPP}}}{C_T}}{1 + \dfrac{(\nu^2 - 1)/\gamma}{h2C_T/\sigma a}} + \frac{D}{W}$$

Thus the rotor shaft angle and the tip-path-plane tilt relative to the shaft are determined by the requirements for moment equilibrium of the helicopter. From $\beta_{1c_{HP}}$ the flapping solution then gives the longitudinal cyclic control required, $\theta_{1s_{HP}}$. A forward center-of-gravity position requires a rearward tilt of the rotor and a forward tilt of the helicopter so that the center of gravity remains under the hub and the rotor thrust remains vertical. It is also observed that a flap frequency above 1/rev significantly reduces the tilt required for a given center-of-gravity offset and hence reduces the cyclic control travel.

Similarly, the requirements for roll moment equilibrium give the shaft roll angle ϕ_s (Fig. 5-34). The rotor hub roll moment is M_x, and the forces

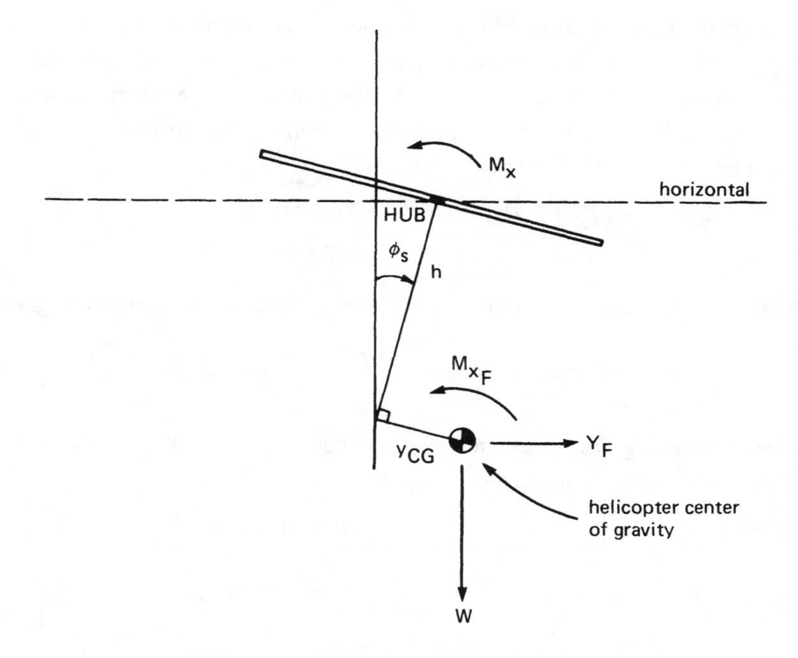

Figure 5-34 Lateral moments acting on the helicopter.

on the helicopter are the weight W, side force Y_F, and aerodynamic roll moment M_{x_F}. The helicopter center of gravity is offset to the right of the rotor shaft by the distance y_{CG}. Then for small angles, the requirement for roll moment equilibrium about the rotor hub gives

$$M_x + W(h\phi_s - y_{CG}) + Y_F h + M_{x_F} = 0$$

or

$$\phi_s = \phi_{HP} = \frac{y_{CG}}{h} - \frac{Y_F}{W} - \frac{M_{x_F}}{Wh} - \frac{M_x}{Wh}$$

Now the rotor hub moment is

$$\frac{M_x}{Wh} = \frac{C_{M_x}}{hC_T} = \frac{(\nu^2 - 1)/\gamma}{h2C_T/\sigma a}\beta_{1s_{HP}}$$

and from the conditions for lateral force equilibrium

$$\phi_{HP} + \frac{Y_F}{W} = -\frac{Y_{HP}}{T} = -\frac{Y_{TPP}}{T} + \beta_{1s_{HP}}$$

Solving for $\beta_{1s_{HP}}$ then:

$$\beta_{1s_{HP}} = \frac{y_{CG}/h - M_{x_F}/Wh + C_{Y_{TPP}}/C_T}{1 + \dfrac{(\nu^2 - 1)/\gamma}{h2C_T/\sigma a}}$$

and

$$\phi_s = \frac{y_{CG}/h - M_{x_F}/Wh - \dfrac{(\nu^2 - 1)/\gamma}{h2C_T/\sigma a}\dfrac{C_{Y_{TPP}}}{C_T}}{1 + \dfrac{(\nu^2 - 1)/\gamma}{h2C_T/\sigma a}} - \frac{Y_F}{W}$$

From the lateral tip-path-plane tilt $\beta_{1s_{HP}}$, the flapping solution gives the lateral cyclic control required, $\theta_{1c_{HP}}$.

Finally, consider the rotor power required. The expression

$$C_P = \int \lambda dC_T - \mu C_H + C_{P_o}$$

was derived in section 5–4 using no small angle approximations. The inflow ratio $\lambda = \lambda_i + \mu \tan\alpha$ is needed to complete the energy balance relation for the power required. Using the results above for horizontal and vertical force equilibrium,

$$\tan\alpha = \tan\theta_{FP} + \frac{D}{T\cos\alpha} + \frac{H}{T}\left(1 + \tan\alpha\tan\theta_{FP}\right)$$

$$W\cos\theta_{FP} = T\cos\alpha\left(1 + \frac{H}{T}\tan\alpha\right),$$

the quantity $(\tan\alpha - H/T)$ can be written

$$\begin{aligned}\tan\alpha - \frac{H}{T} &= \tan\theta_{FP}\left(1 + \frac{H}{T}\tan\alpha\right) + \frac{D}{T\cos\alpha}\\ &= \tan\theta_{FP}\frac{W\cos\theta_{FP}}{T\cos\alpha} + \frac{D}{T\cos\alpha}\\ &= \frac{D + W\sin\theta_{FP}}{T\cos\alpha}\end{aligned}$$

It follows that the rotor power is

$$\begin{aligned}C_P &= \int\lambda_i dC_T + C_{P_o} + \mu C_T(\tan\alpha - H/T)\\ &= \int\lambda_i dC_T + C_{P_o} + \frac{V\cos\alpha\, T}{\rho A(\Omega R)^3}\left(\frac{D + W\sin\theta_{FP}}{T\cos\alpha}\right)\\ &= \int\lambda_i dC_T + C_{P_o} + \frac{DV}{\rho A(\Omega R)^3} + \frac{V_c W}{\rho A(\Omega R)^3}\\ &= C_{P_i} + C_{P_o} + C_{P_p} + C_{P_c}\end{aligned}$$

Thus without any small angle assumptions, the helicopter parasite power is exactly $P_p = DV = \frac{1}{2}\rho V^3 f$, and the climb power is $P_c = V_c W$.

5–19 Lag Motion

The helicopter rotor blade has not only flap motion, but motion in the plane of the disk as well, called lag or lead-lag. An articulated rotor has a lag or drag hinge, so the lag motion consists of rigid body rotation about a vertical axis near the center of rotation. Generally the lag motion requires a more complicated analysis than does the flap motion. The flapping motion produces in-plane inertial forces that couple the flap and lag degrees of freedom of the blade. Also, for low inflow rotors the in-plane forces on the blade are small compared to the out-of-plane forces, and consequently more care is required in analyzing the motion resulting from lag moment balance. The present section is only an introduction to the topic; the rotor lag dynamics are covered in more detail in Chapters 9 and 12.

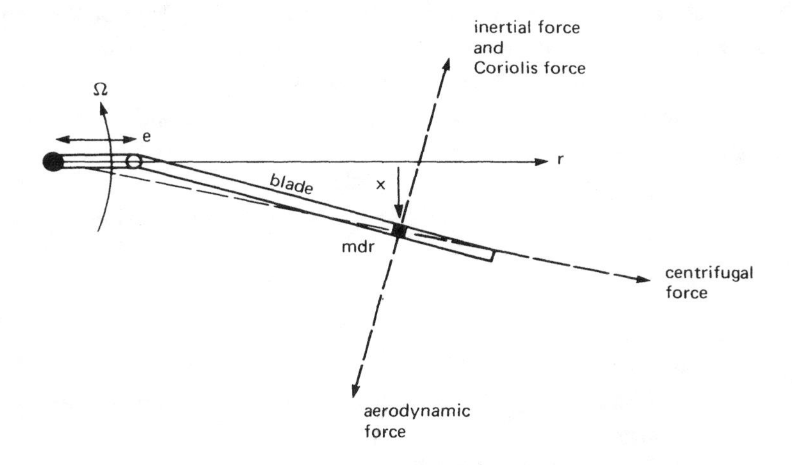

Figure 5-35 Rotor blade lag moments.

Consider the in-plane motion of a blade with a lag hinge offset by a distance eR from the center of rotation (Fig. 5-35). If there is no lag hinge spring the offset cannot be zero, or there would be no way to deliver torque to the rotor. Rigid body rotation about the lag hinge is represented by the lag degree of freedom ζ, defined to be positive for motion opposing the rotor rotation direction. With a mode shape $\eta = (r - e)/(1 - e)$, the in-plane deflection is $x = \eta\zeta$. A lag hinge spring with constant K_ζ is included in the analysis. The in-plane forces acting on the blade section at r, and their moment arms about the lag hinge at $r = e$, are:

(i) an inertial force $m\ddot{x} = m\eta\ddot{\zeta}$ opposing the lag motion, with moment arm $(r - e)$;

(ii) a centrifugal force $m\Omega^2 r$ directed radially outward from the center of rotation, hence with moment arm $x(e/r) = \eta\zeta(e/r)$ about the lag hinge;

(iii) an aerodynamic force F_x in the drag direction, with moment arm $(r - e)$; and

(iv) a Coriolis force $2\Omega \dot{z} z' m = 2\Omega \dot{\beta} \beta r m$ in the same direction as the inertial force, with moment arm $(r - e)$.

Note that if the lag hinge were at the center of rotation the centrifugal force would produce no lag moment. The Coriolis force is due to the product of the rotor angular velocity Ω and the radially inward section velocity $\dot{z} z'$. This radial velocity can be considered the in-plane component of the flap velocity $\dot{z} = r\dot{\beta}$, produced when the blade is coned upward at angle $z' = \beta$. The Coriolis force is in the blade lead direction when $\beta\dot{\beta} > 0$.

Equilibrium of moments about the lag hinge, including a spring moment $K_\zeta \zeta$, gives the equation of motion:

$$\int_e^R \left[(m\eta\ddot{\zeta})(r-e) + m\Omega^2 r\left(\frac{e}{r}\,\eta\zeta\right) + 2\Omega\beta\dot{\beta} r m (r-e) \right] dr$$

$$+ K_\zeta \zeta = \int_e^R F_x (r-e) dr$$

Expressed in terms of dimensionless quantities and divided by $(1 - e)$, the equation becomes

$$I_b \ddot{\zeta} + \left[\frac{e}{1-e} \int_e^1 \eta m dr + \frac{K_\zeta}{\Omega^2 (1-e)} \right] \zeta + 2 I_b \beta \dot{\beta} = \int_e^1 \eta F_x dr$$

with $I_b = \int_e^1 \eta^2 m dr$. If the blade Lock number is defined as $\gamma = \rho a c R^4 / I_b$, the differential equation for the blade lag motion is then

$$\ddot{\zeta} + \nu_\zeta^2 \zeta + 2\beta\dot{\beta} = \gamma \int_e^1 \eta \frac{F_x}{ac} dr$$

The lag dynamics are described by a mass and spring system excited by the in-plane aerodynamic forces (profile and induced drag) and a Coriolis force due to the blade flapping. The aerodynamic forces damp the lag motion, but much less effectively than out-of-plane motion; articulated rotors will have a mechanical lag damper, however. The natural frequency of the lag motion is

$$\nu_\zeta^2 = \frac{e}{1-e} \frac{\int_e^1 \eta m dr}{\int_e^1 \eta^2 m dr} + \frac{K_\zeta}{I_b \Omega^2 (1-e)}$$

The first term, the centrifugal spring on the lag motion, is zero if there is no hinge offset. For uniform mass distribution and no hinge spring, the result is simply

$$\nu_\zeta^2 = \frac{3}{2} \frac{e}{1-e}$$

Articulated rotors typically have a lag frequency of $\nu_\zeta = 0.2$ to 0.3/rev. With hingeless rotors (or with a lag-hinge spring) a higher lag frequency can be attained. Since the lag frequency must not be too near 1/rev to avoid excessive blade loads, hingeless rotors naturally fall into two classes: soft in-plane rotors, for which the lag frequency is below 1/rev (typically $\nu_\zeta = 0.65$ to 0.80/rev); and stiff in-plane rotors, for which the lag frequency is above 1/rev (typically $\nu_\zeta = 1.4$ to 1.6/rev). Gimballed and teetering rotors also fall into the stiff in-plane class. Soft in-plane rotors exhibit a mechanical instability called ground resonance (see Chapter 12) if the lag frequency or the lag damping is too low. For this reason an articulated rotor and even some soft in-plane hingeless rotors must have mechanical dampers.

The Coriolis force is a second order term, but because all the in-plane forces on the blade are small it is an important factor in the blade behavior. The in-plane loads generated by Coriolis forces, when the blade flaps, are the reason for equipping the articulated rotor with a lag hinge. For studies of transient lag dynamics (including aeroelastic stability) the Coriolis term is linearized about the blade position:

$$\beta\dot{\beta} \cong \beta_{\text{trim}}\delta\dot{\beta} + \dot{\beta}_{\text{trim}}\delta\beta$$

For hover, or when averaged trim values are used in forward flight, this becomes $\beta\dot{\beta} \cong \beta_0\delta\dot{\beta}$. The Coriolis force is therefore due primarily to the radial component of the flapping velocity of the blade coned at a trim angle β_0. For the steady-state solution, the Coriolis term acts as a forcing function and may be evaluated by considering the coning and first harmonics of the flap response:

$$\beta\dot{\beta} = (\beta_0 + \beta_{1c}\cos\psi + \beta_{1s}\sin\psi)(-\beta_{1c}\sin\psi + \beta_{1s}\cos\psi)$$

$$= \beta_0\beta_{1s}\cos\psi - \beta_0\beta_{1c}\sin\psi + \beta_{1c}\beta_{1s}\cos 2\psi + \tfrac{1}{2}(\beta_{1s}^2 - \beta_{1c}^2)\sin 2\psi$$

Consider the steady-state lag motion, which is periodic and therefore may be written as a Fourier series. Since the inertial and Coriolis forces have zero mean values, the mean lag angle is

$$\zeta_0 = \frac{\gamma}{\nu_\zeta^2}\frac{C_Q}{\sigma a}$$

(Note that the mean value of $\int_0^1 r(F_x/ac)\,dr$ is $C_Q/\sigma a$, the rotor torque coefficient.) The mean lag angle is typically a few degrees, varying from slightly negative in autorotation to perhaps 10° at maximum power.

The solution for the first harmonic lag motion due to the aerodynamic and Coriolis forces is

$$\zeta_{1c} = \frac{-(\gamma C_Q/\sigma a)_{1c} + 2\beta_0\beta_{1s}}{1 - \nu_\zeta^2}$$

$$\zeta_{1s} = \frac{-(\gamma C_Q/\sigma a)_{1s} - 2\beta_0\beta_{1c}}{1 - \nu_\zeta^2}$$

A lag frequency near 1/rev will give large 1/rev lag motion and hence high in-plane blade loads. The damping, which determines the response amplitude for $\nu_\zeta = 1$, will be low for the blade lag motion and therefore does not alter this conclusion. (An articulated blade with high damping from a mechanical damper also has a small lag frequency.) Thus the lag frequency of a soft in-plane rotor is generally a compromise between the requirements of low blade loads (low lag frequency) and ground resonance stability (high lag frequency). These expressions for ζ_{1c} and ζ_{1s} are somewhat misleading, because there are actually flap terms in the 1/rev aerodynamic lag moments that cancel part of the Coriolis excitation.

The solution for the 2/rev lag motion due to the Coriolis forces alone is

$$\zeta_{2c} = \frac{2\beta_{1c}\beta_{1s}}{4 - \nu_\zeta^2}$$

$$\zeta_{2s} = \frac{\beta_{1s}^2 - \beta_{1c}^2}{4 - \nu_\zeta^2}$$

or

$$|\zeta|_{2/\text{rev}} = \sqrt{\zeta_{2c}^2 + \zeta_{2s}^2} = \frac{\beta_{1c}^2 + \beta_{1s}^2}{4 - \nu_\zeta^2} = \frac{|\beta|^2_{1/\text{rev}}}{4 - \nu_\zeta^2}$$

The Coriolis forces thus produce a 2/rev lag motion proportional to the square of the 1/rev flap amplitude.

5–20 Reverse Flow

The reverse flow region is a circle of diameter μ on the retreating side of the rotor disc. For low advance ratio the influence of reverse flow is small, since it is confined to a small area where the dynamic pressure is low. Therefore, up to about $\mu = 0.5$ reverse flow effects may be neglected. At higher advance ratios, the reverse flow region occupies a large portion of the disk and must be accounted for in calculating the aerodynamic forces on the blade. An elementary model for the blade aerodynamics in the reverse flow region will be developed here. Near the reverse flow boundary at least, there will be significant separated and radial flow, which may require a better model.

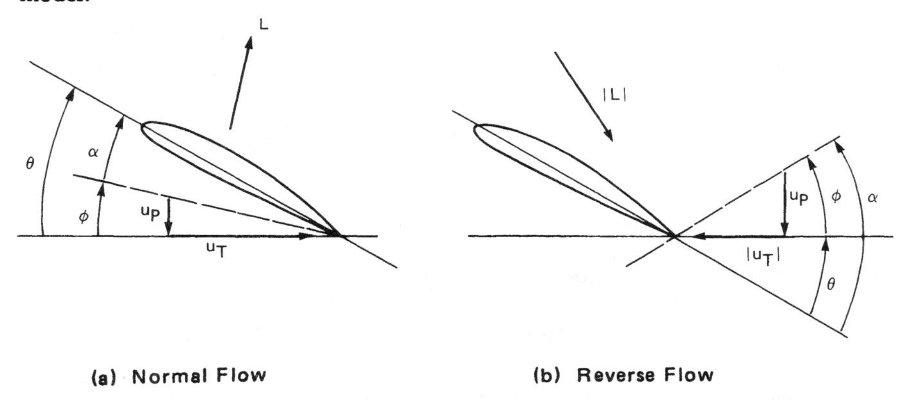

Figure 5-36 Rotor blade section aerodynamics in normal and reverse flow.

Figure 5-36 compares the section aerodynamics in the normal and reverse flow regions. Recall that in section 5–2 the normal aerodynamic force neglecting stall was given for small angles as

$$\frac{F_z}{ac} \cong \frac{L}{ac} \cong \tfrac{1}{2} u_T^2 \alpha = \tfrac{1}{2} u_T(\theta u_T - u_P).$$

This result neglects the reverse flow, however. The positive directions of the various quantities are as follows: F_z and L upward, θ nose up, u_P downward, and u_T from the leading edge to the trailing edge. Figure 5-36 shows that in the reverse flow region the angle of attack is

$$\alpha = \theta + \phi = \theta + \frac{u_P}{|u_T|} = \theta - \frac{u_P}{u_T}$$

just as in normal flow. However, in reverse flow a positive α gives a negative (downward) lift:

$$\frac{L}{ac} = -\tfrac{1}{2} u_T^2 \alpha = \tfrac{1}{2} |u_T| u_T \alpha$$

Thus an expression valid in both reverse and normal flow is

$$\frac{F_z}{ac} \cong \frac{L}{ac} \cong \tfrac{1}{2} |u_T| u_T \alpha = \tfrac{1}{2} |u_T| (\theta u_T - u_P)$$

Since the inertial and centrifugal flap moments are unaffected by reverse flow, the only change to the flap dynamics is in the aerodynamic moment:

$$M_F = \int_0^1 r \frac{F_z}{ac} dr = \int_0^1 \tfrac{1}{2} |u_T| (\theta u_T - u_P) r\,dr$$

$$= M_\theta \theta_{con} + M_{\theta_{tw}} \theta_{tw} + M_\lambda \lambda + M_{\dot\beta} \dot\beta + M_\beta \beta$$

The reverse flow introduces a change in sign, so that the aerodynamic coefficients now require integrations of the form $\int_0^1 \text{sign}\,(u_T)\, f(r,\psi)\, dr$. The evaluation of this integral depends on the azimuth angle. Assuming $\mu < 1$, it is only necessary to distinguish between the advancing and retreating sides of the disk:

$$\int_0^1 \text{sign}\,(u_T)\, f(r,\psi)\, dr = \begin{cases} \int_0^1 f\,dr & 0° < \psi < 180° \\ \int_0^1 f\,dr - 2 \int_0^{-\mu \sin\psi} f\,dr & 180° < \psi < 360° \end{cases}$$

Thus on the advancing side the aerodynamic coefficients are identical to the results already obtained, while on the retreating side a correction for the changed sign in the reverse flow region is required. If $\mu > 1$, there is also a region extending from $\psi = 270° - \cos^{-1} 1/\mu$ to $\psi = 270° + \cos^{-1} 1/\mu$ where the blade is entirely in the reverse flow region so that

$$\int_0^1 \text{sign}\,(u_T) f(r,\psi) dr = -\int_0^1 f\, dr$$

which is just the negative of the expression for the advancing side. Evaluating the flapping aerodynamic coefficients then gives for $\mu < 1$:

$$M_\theta = \begin{cases} \frac{1}{8} + \frac{1}{3}\mu\sin\psi + \frac{1}{4}\left(\mu\sin\psi\right)^2 \\ \\ \frac{1}{8} + \frac{1}{3}\mu\sin\psi + \frac{1}{4}\left(\mu\sin\psi\right)^2 - \frac{1}{12}\left(\mu\sin\psi\right)^4 \end{cases}$$

$$M_{\theta_{tw}} = \begin{cases} \frac{1}{10} + \frac{1}{4}\mu\sin\psi + \frac{1}{6}\left(\mu\sin\psi\right)^2 \\ \\ \frac{1}{10} + \frac{1}{4}\mu\sin\psi + \frac{1}{6}\left(\mu\sin\psi\right)^2 + \frac{1}{30}\left(\mu\sin\psi\right)^5 \end{cases}$$

$$M_\lambda = \begin{cases} -\frac{1}{6} - \frac{1}{4}\mu\sin\psi \\ \\ -\frac{1}{6} - \frac{1}{4}\mu\sin\psi + \frac{1}{6}\left(\mu\sin\psi\right)^3 \end{cases}$$

$$M_{\dot\beta} = \begin{cases} -\frac{1}{8} - \frac{1}{6}\mu\sin\psi \\ \\ -\frac{1}{8} - \frac{1}{6}\mu\sin\psi - \frac{1}{12}\left(\mu\sin\psi\right)^4 \end{cases}$$

$$M_\beta = \begin{cases} -\mu\cos\psi\left(\dfrac{1}{6}+\dfrac{1}{4}\mu\sin\psi\right) \\ -\mu\cos\psi\left(\dfrac{1}{6}+\dfrac{1}{4}\mu\sin\psi-\dfrac{1}{6}\left(\mu\sin\psi\right)^3\right) \end{cases}$$

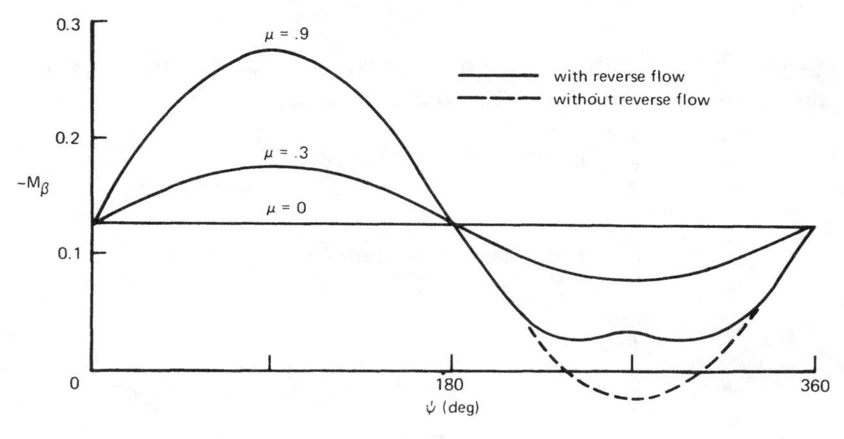

Figure 5-37 Flap damping coefficient $M_{\dot\beta}$.

(where the first expression is for the advancing side and the second is for the retreating side). Fig. 5-37 shows the flap damping coefficient $M_{\dot\beta}$ for several advance ratios. The flap damping is always positive ($M_{\dot\beta} < 0$). In hover, the damping is constant at $M_{\dot\beta} = -0.125$; when $\mu > 0$ it is higher on the advancing side and lower on the retreating side. When $\mu > 0.794$, the damping reaches a minimum value of $M_{\dot\beta} = -0.0258$ on the retreating side, with a local maximum at $\psi = 270°$. Figure 5-38 shows the pitch control coefficient M_θ, which has a value $M_\theta = 0.125$ for hover and is higher on the advancing side and lower on the retreating side when $\mu > 0$. When $\mu > 0.641$, M_θ is negative on the retreating side. The twist coefficient $M_{\theta_{tw}}$ behaves like M_θ, while the inflow coefficient M_λ and flap spring $M_\beta/(\mu\cos\psi)$ are similar to $M_{\dot\beta}$. The flap damping is always positive, even when $\mu > 1$, but the aerodynamic flap spring M_β is negative on the front of the disk because of the $\mu\cos\psi$ factor.

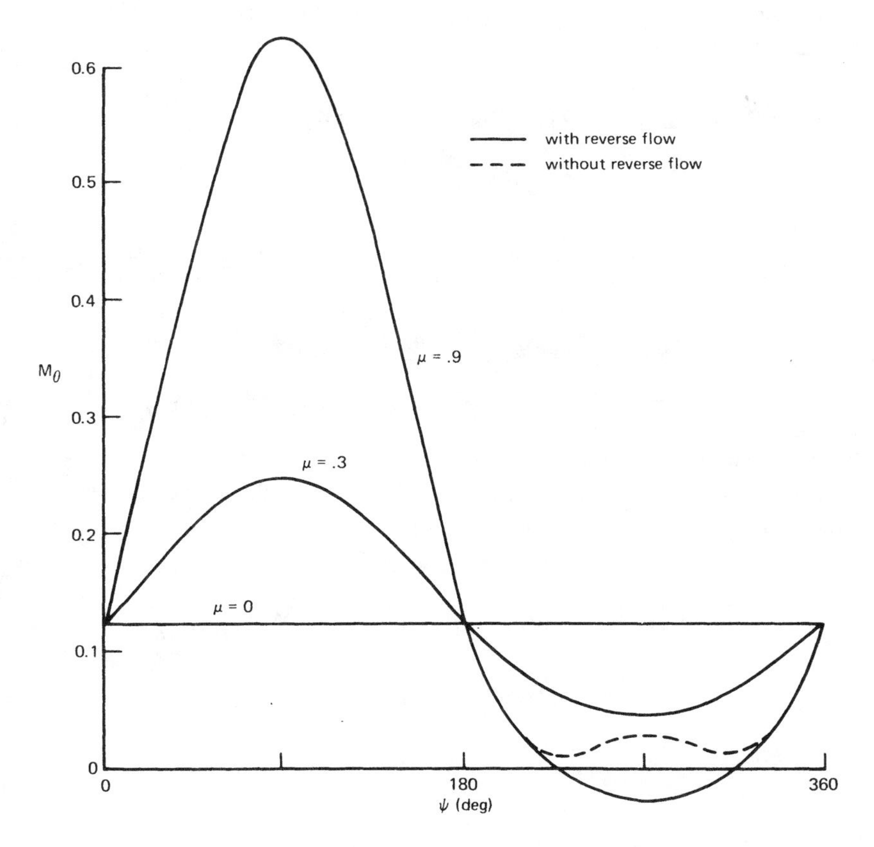

Figure 5-38 Pitch control coefficient M_θ.

To solve the flapping equations, it is convenient to express the aerodynamic coefficients as Fourier series. Because of the symmetry of the flow, half of the harmonics are found to be zero:

$$M_\theta = M_\theta^{0} + M_\theta^{1s} \sin\psi + M_\theta^{2c} \cos 2\psi + M_\theta^{3s} \sin 3\psi + \ldots$$

and similarly for $M_{\theta_{tw}}$, $M_{\dot\beta}$, M_λ; and

$$M_\beta = M_\beta^{1c} \cos\psi + M_\beta^{2s} \sin 2\psi + M_\beta^{3c} \cos 3\psi + \ldots$$

For an articulated rotor ($\nu = 1$), the equations for the flap harmonics β_0, β_{1c}, and β_{1s} then become

$$\frac{\beta_0}{\gamma} = \theta_0 M_\theta^{\;0} + \theta_{1s} \tfrac{1}{2} M_\theta^{\;1s} + \beta_{1c}\left(\tfrac{1}{2} M_\beta^{\;1c} - \tfrac{1}{2} M_{\dot\beta}^{\;1s}\right)$$

$$+ \theta_{tw} M_{\theta_{tw}}^{\;0} + \lambda M_\lambda^0$$

$$0 = \theta_{1c}\left(M_\theta^{\;0} + \tfrac{1}{2} M_\theta^{\;2c}\right) + \beta_{1s}\left(M_{\dot\beta}^{\;0} + \tfrac{1}{2} M_{\dot\beta}^{\;2c} + \tfrac{1}{2} M_\beta^{\;2s}\right) + \beta_0 M_\beta^{\;1c}$$

$$0 = \theta_0 M_\theta^{\;1s} + \theta_{tw} M_{\theta_{tw}}^{\;1s} + \theta_{1s}\left(M_\theta^{\;0} - \tfrac{1}{2} M_\theta^{\;2c}\right)$$

$$+ \beta_{1c}\left(- M_{\dot\beta}^{\;0} + \tfrac{1}{2} M_{\dot\beta}^{\;2c} + \tfrac{1}{2} M_\beta^{\;2s}\right) + \lambda M_\lambda^{\;1s}$$

To be consistent, the second harmonics of the flap motion should probably be considered also when the advance ratio is high enough to require reverse flows; such a calculation is best performed numerically, however. Using $\lambda = \lambda_{NFP} + \mu\theta_{1s} = \lambda_{TPP} - \mu\beta_{1c}$, the solution is:

$$\beta_0 = \gamma\Big[\theta_0 M_\theta^{\;0} + \theta_{tw} M_{\theta_{tw}}^{\;0} + \theta_{1s}\left(\tfrac{1}{2} M_\theta^{\;1s} + \mu M_\lambda^{0}\right)$$

$$+ \beta_{1s}\left(\tfrac{1}{2} M_\beta^{\;1c} - \tfrac{1}{2} M_{\dot\beta}^{\;1s}\right) + \lambda_{NFP} M_\lambda^{0}\Big]$$

$$\beta_{1s} = \frac{-\beta_0 M_\beta^{\;1c} - \left(\theta_{1c} M_\theta^{\;0} + \tfrac{1}{2} M_\theta^{\;2c}\right)}{M_{\dot\beta}^{0} + \tfrac{1}{2} M_{\dot\beta}^{\;2c} + \tfrac{1}{2} M_\beta^{\;2s}}$$

$$\beta_{1c} = \frac{\theta_0 M_\theta^{\;1s} + \theta_{tw} M_{\theta_{tw}}^{\;1s} + \theta_{1s}\left(M_\theta^{\;0} - \tfrac{1}{2} M_\theta^{\;2C} + \mu M_\lambda^{\;1s}\right) + \lambda_{NFP} M_\lambda^{1s}}{M_{\dot\beta}^{\;0} - \tfrac{1}{2} M_{\dot\beta}^{\;2c} - \tfrac{1}{2} M_\beta^{\;2s}}$$

$$= \frac{\theta_0 M_\theta^{\;1s} + \theta_{tw} M_{\theta_{tw}}^{\;1s} + \theta_{1s}\left(M_\theta^{\;0} - \tfrac{1}{2} M_\theta^{\;2C}\right) + \lambda_{TPP} M_\lambda^{1s}}{M_{\dot\beta}^{\;0} - \tfrac{1}{2} M_{\dot\beta}^{\;2c} - \tfrac{1}{2} M_\beta^{\;2S} + \mu M_\lambda^{\;1s}}$$

Evaluating the required harmonics then gives:

$$\beta_0 = \gamma\left[\frac{\theta_{.8}}{8}\left(1 + \mu^2 - \frac{\mu^4}{8}\right) - \frac{\theta_{tw}}{60}\left(\mu^2 - \frac{9}{8}\mu^4\right) - \frac{1}{6}\left(1 + \frac{2}{3\pi}\mu^3\right)\lambda_{NFP} - \frac{\mu^4}{15\pi}\left(\beta_{1c} + \theta_{1s}\right)\right]$$

$$\beta_{1s} - \theta_{1c} = \frac{-\frac{4}{3}\mu\left(1 + \frac{4}{15\pi}\mu^3\right)\beta_0}{1 + \frac{1}{2}\mu^2 - \frac{1}{24}\mu^4}$$

$$\beta_{1c} + \theta_{1s} = \frac{-\frac{8}{3}\mu\left[\theta_{.75}\left(1 + \frac{4}{15\pi}\mu^3\right) - \theta_{tw}\mu^3\frac{23}{75\pi} - \frac{3}{4}\lambda_{NFP}\left(1 - \frac{\mu^2}{4}\right)\right]}{1 - \frac{1}{2}\mu^2 + \frac{7}{24}\mu^4}$$

$$= \frac{-\frac{8}{3}\mu\left[\theta_{.75}\left(1 + \frac{4}{15\pi}\mu^3\right) - \theta_{tw}\mu^3\frac{23}{75\mu} - \frac{3}{4}\lambda_{TPP}\left(1 - \frac{\mu^2}{4}\right)\right]}{1 + \frac{3}{2}\mu^2 - \frac{5}{24}\mu^4}$$

Thus reverse flow introduces flap motion of order μ^4. When $\mu > 0.5$, it is necessary to consider such factors as stall and compressibility as well as reverse flow. Moreover, other blade degrees of freedom also become important. For example, the lift acting at the three-quarter chord in the reverse flow region induces a significant response in the blade pitch above $\mu \cong 0.7$, which alters the blade loading. In general, then, a numerical calculation of the blade loading and motion is required at high advance ratio in order that a consistent model may be considered.

Finally, the rotor thrust coefficient including the effects of reverse flow is

$$C_T = \sigma a \int_0^1 \tfrac{1}{2}|u_T|(\theta u_T - u_P)dr$$

$$= \frac{\sigma a}{2}\left[\theta_0\left(\frac{1}{3} + \frac{1}{2}\mu^2 - \frac{4}{9\pi}\mu^3\right) + \theta_{tw}\left(\frac{1}{4} + \frac{1}{4}\mu^2 - \frac{1}{32}\mu^4\right)\right.$$

$$\left. - \frac{1}{2}\lambda_{NFP}\left(1 + \frac{1}{2}\mu^2\right) - \frac{1}{8}\mu^3\beta_{1c_{NFP}}\right]$$

Reverse flow primarily introduces terms of higher order in μ; the $\lambda_{NFP}\mu^2$ term is found to be significant even at fairly low advance ratio, however.

For further discussion of reverse flow aerodynamics, particularly for rotor flap motion, see Perisho (1959), Sissingh (1968), and Sissingh and Kuczynski (1970).

5–21 Compressibility

Compressibility of the air influences the rotor performance and motion by its effects on the blade forces. Of particular importance are the increase in lift curve slope with Mach number and the sharp increase in drag and pitching moment above a certain critical Mach number. When the blade is operating at high, unsteady angles of attack, as on the retreating side with high rotor loading, compressibility effects are important even at low Mach number. The primary effect of compressibility on the rotor performance is a rapid increase in the profile power C_{P_o} when the tip Mach number exceeds the critical Mach number for drag divergence. Note that the critical Mach number depends on the angle of attack, and also that it is increased by the three-dimensional flow at the tip. The increased compressible lift curve slope has little effect on β_{1c} or β_{1s}/β_0 (which involve only a balance of aerodynamic moments), but it significantly increases the rotor thrust and coning angle at high tip speeds. The only practical means of accounting for the compressibility effects in detail is to use data for the airfoil aerodynamic characteristics expressed as a function of Mach number as well as angle of attack in a numerical calculation of the rotor loads and motion. The effects of three-dimensional flow must also be included, especially near the tip.

The blade normal Mach number in forward flight is

$$M = u_T/c_s = M_{\text{tip}}(r + \mu \sin\psi)$$

where $M_{\text{tip}} = \Omega R/c_s$ is the mach number based on the tip speed, and c_s is the speed of sound. Write $M_{r,\psi}$ for the Mach number at radial station r and azimuth angle ψ. The highest Mach number occurs on the tip of the advancing blade:

$$M_{1,90} = M_{tip}(1 + \mu)$$

M_{tip} is a good parameter for the average effects of compressibility, while $M_{1,90}$ is a measure of the extreme effects. Compressibility limits the maximum speed of the helicopter. On writing $M_{1,90} = (\Omega R + V)/c_s$ it becomes clear that the critical Mach number of the blade must constrain the forward speed of the helicopter, since the tip speed cannot be decreased too much without encountering other limits (see section 7–4).

The effect of the increased lift curve slope on the helicopter loads and flapping can be estimated by using the Prandtl-Glauert relation for a:

$$a = \frac{a_{incomp}}{\sqrt{1 - M^2}}$$

Because the Mach number varies over the disk, the compressible lift curve slope does also. Thus the Prandtl-Glauert factor $(1 - M^2)^{-1/2}$ must be included in the integrands, which then cannot be evaluated analytically. Alternatively, some averaged lift curve slope can be used for the entire rotor, for example, a slope based on the Mach number at an effective radius r_e:

$$a = \frac{a_{incomp}}{\sqrt{1 - (r_e M_{tip})^2}}$$

Payne (1959a) suggests using $r_e = 0.7$, which gives good results up to about $M_{tip} = 0.7$. Peters and Ormiston (1975) found that when the advancing tip Mach number $M_{1,90}$ is less than 0.9, a simple correction based on the 75% radius is sufficient. When $M_{1,90}$ is above about 0.9, it is necessary to include the radial and azimuthal variations of the compressibility effects.

Gessow and Crim (1956) calculated the effects of high tip Mach number on the flapping, thrust, and power of a helicopter rotor in forward flight. They found a minor increase in the rotor flapping and thrust due to compressibility. The largest effect found was an increase in profile power on the advancing side of the disk when the drag divergence Mach number was exceeded. The increase in profile power correlated well with ΔM_d, the amount by which the advancing tip Mach number $M_{1,90}$ exceeded the section drag-divergence Mach number, roughly according to the expression

$$\Delta C_{P_o}/\sigma = 0.007(\Delta M_d) + 0.052(\Delta M_d)^2$$

Norman and Sultany (1965) and Norman and Somsel (1967) give similar correlations. The work cited in sections 2–8, 5–24, and 6–6 includes experimental and theoretical investigations that consider the rotor operating at high Mach number in hover or forward flight.

5–22 Tail Rotor

The tail rotor of a single main rotor helicopter is a small diameter rotary wing with the function of balancing the main rotor torque and providing yaw control, which is achieved through the action of the tail rotor thrust on a longitudinal arm (usually somewhat longer than the main rotor radius) about the main rotor shaft. The tail rotor is usually a flapping rotor with a low disk loading, to which the analysis developed in this chapter is applicable. The special features of the tail rotor configuration make the use of the analysis somewhat different than for a main rotor, however. First, the tail rotor has no cyclic pitch control, just collective to control the thrust magnitude. Secondly, the tail rotor shaft angle is fixed by the geometry of the tail rotor installation and the helicopter yaw angle, instead of being determined by the conditions for force equilibrium of the rotor. The tail rotor drag or propulsive force is included in the airframe drag and is balanced by the main rotor.

When there is no cyclic pitch control, the expressions for blade flap moment equilibrium give the flapping produced rather than the cyclic required by the rotor operating state. The tail rotor usually has pitch-flap coupling, which gives the actual pitch in the hub plane in terms of the flapping: $\theta_{1c_{HP}} = -K_P\beta_{1c_{HP}}$ and $\theta_{1s_{HP}} = -K_P\beta_{1s_{HP}}$ (45° of delta-three is typical, hence $K_P = 1$). With the shaft orientation fixed, the hub-plane angle of attack α_{HP} is known and then the tip-path-plane incidence depends on the longitudinal flapping. Hence the inflow ratio is

$$\lambda_{TPP} = \lambda_{HP} + \mu\beta_{1c_{HP}} = \lambda_i + \mu(\alpha_{HP} + \beta_{1c_{HP}})$$

The tail rotor drag force, which must be reacted by the main rotor, is then

$$D_{tr} = H_{TPP} - T\alpha_{TPP} = H_{TPP} - T(\alpha_{HP} + \beta_{1c_{HP}})$$

See Chapters 6 and 7 for a further discussion of tail rotor performance and literature.

5–23 Numerical Solutions

To obtain an analytical solution for the rotor forces and flap motion, it has been necessary to simplify the model in a number of respects. In particular, the analysis has neglected stall and compressibility effects in the rotor aerodynamics; nonuniformities in the induced velocity distribution, beyond the simplest linear variation; the second and higher harmonics of blade flap motion; and all degrees of freedom except for the fundamental flap mode. The solution obtained under these approximations is useful for the information it provides about the rotor behavior and because it is in fact reasonably accurate over a wide range of helicopter operating conditions. For a helicopter operating in the more extreme flight conditions like high speed, high tip Mach number, or high gross weight, one or more of the assumptions is no longer valid and a better model is required. Moreover, even at operating conditions for which a simple model gives a good estimate of performance and flapping, a better model is required to calculate blade loads and vibration.

Thus it is frequently necessary to calculate the rotor behavior using an extended model of the rotor, in which as many of the assumptions about the blade motion and aerodynamics are removed as is practical and possible. Such calculations must be performed numerically, and really are practical only with a high-speed digital computer. Many numerical analyses of rotor behavior using the computer have been developed in recent years, and their application in helicopter design, testing, and evaluation is now routine. There is no question that these analyses have greatly expanded our knowledge of rotor behavior and improved our ability to predict it. It is also true that even with the most advanced model there remain great deficiencies in the predictive capability for rotary wings, which may be traced to the magnitude of the problem of calculating rotor behavior, and to a fundamental lack of understanding in a number of areas of helicopter aerodynamics and dynamics. The formulation of the numerical problem of the helicopter rotor, and its solution, is the subject of Chapter 14 and to some extent the literature discussed in the remainder of this chapter.

5–24 Literature

This chapter concludes with an examination of some of the work which forms the basis for the analysis of the rotor forces and flap motion in forward flight. The discussion here is primarily concerned with the analytical models and the flapping solution. The rotor performance solutions are discussed further in Chapter 6. The comprehensive numerical analyses are the subject of Chapter 14.

Glauert (1926b) developed the first analysis of the flapping rotor in forward flight in order to check the claims Cierva was making for his autogyro. Glauert considered a flapping rotor with no twist or taper, and with no cyclic pitch (i.e. a no-feathering-plane analysis). He solved for the coning and first harmonic flapping using blade element theory, with a momentum theory analysis for the induced velocity. The major restriction of the analysis was that only the terms of order μ were retained in forward flight. The small angle assumption was used in the blade aerodynamics together with a constant lift curve slope ($c_\ell = a\alpha$) and mean drag coefficient. Glauert developed the momentum theory result for the induced velocity in forward flight,

$$\lambda_i = \frac{C_T}{2\sqrt{\mu^2 + \lambda^2}}$$

which connects the induced velocity solutions for hover and high speed flight (see section 4-1.1). He also considered a linearly varying induced velocity of the form $\nu = \nu_0(1 + \kappa_x r \cos\psi)$ and suggested the approximation $\lambda_i \cong C_T/2\mu$ for high speed. Glauert also derived an expression for the profile power that includes the effects of reverse flow and radial drag in forward flight:

$$C_{P_o} = \int_0^1 \frac{\sigma c_d}{2}\left(u_T^2 + u_R^2\right)^{3/2} dr$$

This expression was approximated by $C_{P_o} = (\sigma c_{d_o}/8)(1 + n\mu^2)$, and the parameter n was evaluated for several advance ratios (see section 5–12). This result for the profile power was derived from energy conservation considerations as a check on the result obtained for C_Q by blade element theory. Since the connection between the two approaches was not obvious, the blade element theory form was viewed as the primary result by Glauert, and by those who later used this paper as the foundation for further investigations.

Lock (1927) extended Glauert's analysis by including the higher powers of μ, second harmonic flapping, and rotor cyclic pitch. He introduced the concept that a nonflapping rotor with cyclic pitch (a tip-path-plane analysis) is equivalent to a flapping rotor without cyclic (a no-feathering-plane analysis). He obtained the rotor forces, torque, and flapping in both reference planes, and verified that the two solutions are equivalent. Lock also demonstrated the equivalence of the force and energy expressions for the rotor torque, but he neglected reverse flow and radial drag and as a result obtained $C_{P_o} = (\sigma c_{d_o}/8)(1 + 3\mu^2)$. Lock introduced the parameter representing the ratio of aerodynamic and inertial forces on the blade $\gamma = \rho a c R^4/I_b$, which consequently bears his name. (Actually, Lock used a lift curve slope defined as $a = (dL/d\alpha)/(\rho V^2 c) = \frac{1}{2} dc_\ell/d\alpha$, which gives a Lock number with one-half the value according to the modern definition. This extra factor of 2 is still occasionally found in rotor analyses.)

Wheatley (1934d) extended the theory of Glauert and Lock and evaluated the accuracy of the theory by comparing it with test results. He considered a flapping rotor with no hinge offset; linearly twisted, constant chord blades; the tip loss factor B; linear induced velocity variation; the second harmonics of the flap motion; and reverse flow. (The reversed direction of the lift and drag in the reverse flow region was accounted for by introducing $|u_T|$ in place of u_T in the section forces.) His analysis assumed small angles, neglected terms of higher order than μ^4, used a linear lift curve slope ($c_\ell = a\alpha$) and mean drag coefficient; and neglected the radial flow and radial drag effects on the profile losses. The analysis was conducted in the no-feathering plane, in particular using the inflow velocity in the no-feathering plane as a parameter: $\lambda = \lambda_{NFP} = \lambda_i + \tan\alpha_{NFP}$. Wheatley obtained expressions for the forces, flapping, and performance of the rotor in forward flight based on this model. Comparing the calculations with data for the Pitcairn autogyro, Wheatley found reasonable agreement for most of the rotor characteristics. The prediction of α_{NFP}, C_T, and Ω out to $\mu = 0.4$ or 0.5 was good (the rotor speed was obtained from C_T and the autogyro gross weight). The theory was accurate for the rotor net forces and power. Longitudinal inflow variation had little effect on the rotor forces. The results for rotor flapping similarly showed a divergence from the experimental data at approximately $\mu = 0.4$ or 0.5. In general, however, the accuracy of the rotor flapping prediction was not as good as that of the rotor forces prediction, especially for the lateral flapping β_{1s}, which typically was about 1.5° low in magnitude. Longitudinal inflow variation (with $\kappa_x = 0.5$) reduced the lateral flapping discrepancy to about 1.0°. This error, although a significant fraction of β_{1s} (the measured values were around 3° to 4°), was a fairly consistent bias from $\mu = 0.1$ to $\mu = 0.6$. The accuracy of the longitudinal flapping was better; the calculations were perhaps 0.5° high, but generally within the data scatter. Thus it was concluded that prediction of the flap motion with great accuracy was not possible. Since most of the error was in β_{1s}, which is usually smaller than β_{1c}, the flap amplitude could be predicted fairly well, but not the azimuthal phase. This discrepancy between the theory and data is probably due to the simple induced velocity variation that was considered; the treatments of tip loss and reverse flow are also elementary, and there is no consideration of stall or compressibility effects.

Sissingh (1939) extended Wheatley's analysis, principally eliminating the assumption of a constant profile drag coefficient for the blade. He considered a general polar of the form $c_d = \delta_0 + \delta_1\alpha + \delta_2\alpha^2$ in calculating the rotor profile power. Sissingh also considered the effect of a flap hinge offset.

Bailey (1941) put Wheatley's analysis in practical form for routine use by expressing all quantities as direct functions of the blade collective pitch, twist, and no-feathering-plane inflow ratio. The coefficients of these expressions are functions only of the advance ratio, the Lock number, and the tip loss factor. A flapping rotor with no hinge offset, and a linearly twisted, constant chord blade were considered. Bailey divided the rotor torque into accelerating and decelerating terms, $C_Q = C_{Q_a} + C_{Q_d}$, corresponding to the induced and profile division used in this chapter:

$$C_{Q_a} = \frac{\sigma}{2}\int_0^1 |u_T| u_T c_\ell \phi r\,dr$$

$$C_{Q_d} = \frac{\sigma}{2}\int_0^1 |u_T| u_T c_d r\,dr$$

The rotor profile power was obtained as the ratio of profile drag to lift of the rotor:

$$C_{P_o} = \mu C_T\left(\frac{D}{L}\right)_o = \frac{\sigma}{2}\int_0^1 |u_T| u_T^2 c_d dr$$

Bailey extended Wheatley's analysis by using a drag coefficient of the form $c_d = \delta_0 + \delta_1\alpha + \delta_2\alpha^2$ in the calculation of C_{Q_d} and $(D/L)_o$. Thus the increase of the profile drag with lift was accounted for. The aerodynamic model still neglected the radial drag, and used $c_\ell = a\alpha$ for the lift. Bailey developed a method to find the coefficients δ_0, δ_1, and δ_2 from the aerodynamic characteristics of the airfoil section ($c_{\ell_{max}}$, $c_{d_{min}}$, $c_{\ell_{opt}}$, and c_{ℓ_a} at a given Reynolds number; see section 7–8). The expression

$$c_d = 0.0087 - 0.0216\alpha + 0.400\alpha^2$$

was obtained for the NACA 23012 airfoil at $Re = 2 \times 10^6$ and accurately gives the drag to about $\alpha = 12°$. Often this expression is used for the blade drag in rotor analyses, even for other airfoil sections. Bailey also considered the blade stall limit, in terms of when the quadratic expression for c_d is no longer valid. Thus Bailey obtained expressions for the flapping harmonics (β_0/γ, β_{1c}, β_{1s}/γ, β_{2c}/μ^2, and β_{2s}/μ^2), the thrust coefficient $2C_T/\sigma a$, the accelerating and decelerating torques ($2C_{Q_a}/\sigma a$ and $2C_{Q_d}/\sigma a$), and the rotor profile power $\mu(2C_T/\sigma)(D/L)_o$. The flapping harmonics and thrust are linear functions of θ_0, θ_{tw}, and λ_{NFP}, while the torque and profile power are quadratic functions (C_{Q_d} and $(D/L)_o$ also depend on the drag polar coefficients δ_0, δ_1, and δ_2). Bailey gives tables and analytical expressions for the coefficients of θ_0, θ_{tw}, and λ_{NFP}; Gessow and Myers (1952) reproduce the tables. The coefficients are functions of μ, B, and γ, although the coefficients for β_{2c} and β_{2s} are independent of advance ratio and the remaining coefficients are not very sensitive to Lock number up to about $\mu = 0.5$. Bailey developed the following procedure for calculating the helicopter performance in forward flight (for a full discussion see Chapter 6). The energy method is used to find the power. For the first iteration, a simple estimate of C_{P_o} is used. Then $C_P = C_{Q_a} + C_{Q_d}$ is a quadratic equation for λ_{NFP}. (If the thrust rather than the collective is the parameter given, it is necessary to substitute for θ_0 using the C_T equation to obtain this quadratic.) By solving for λ_{NFP}, a new estimate of the profile power can be obtained. This procedure is repeated until the solution converges. Then the C_T equation may be solved for the collective pitch, and the flapping harmonics may be evaluated. For an autogyro, $C_P = 0$ immediately gives the quadratic equation for λ_{NFP}, with no need to iterate. Bailey suggests presenting the rotor performance data in terms of $(D/L)_o$ vs. C_L/σ, with θ_0 and μ as parameters (this is equivalent to C_{P_o} vs. C_T/σ). Bailey was concerned with the autogyro case, but a chart such as his can be constructed for any value of C_P.

Based on the work of Wheatley and Bailey, it has been concluded that the tip loss factor significantly reduces β_0 and C_T (the quantities that depend directly on the blade lift) and has little influence on β_{1c} or β_{1s}/β_0. Also, the effect of reverse flow is negligible up to $\mu = 0.5$ except that the $\mu^2\lambda$ term in C_T is significant even at low speeds. Thus, when the principal effects of tip loss and reverse flow are accounted for, the results of sections 5–3 and 5–5 take the following form:

$$\frac{2C_T}{\sigma a} = \frac{\theta_0}{3}\left(1 + \frac{3}{2}\mu^2\right)B^3 + \frac{\theta_{tw}}{4}\left(1 + \mu^2\right)B^4 - \frac{1}{2}\left(\lambda_{NFP} + \frac{\mu}{2}\lambda_y\right)\left(1 + \frac{1}{2}\mu^2\right)B^2$$

$$\beta_0 = \gamma\left[\frac{\theta_0}{8}\left(1 + \mu^2\right)B^4 + \frac{\theta_{tw}}{10}\left(1 + \frac{5}{6}\mu^2\right)B^5 - \frac{1}{6}\left(\lambda_{NFP} + \frac{\mu}{2}\lambda_y\right)B^3\right] - S_b^* \frac{g}{\Omega^2 R}$$

$$\beta_{1c} = \frac{-\frac{8}{3}\mu\left(\theta_0 + \frac{3}{4}\theta_{tw} - \frac{3}{4}\lambda_{NFP}\right) + \lambda_y}{1 - \frac{1}{2}\mu^2}$$

$$\beta_{1s} = \frac{-\frac{4}{3}\mu\beta_0 - \lambda_x}{1 + \frac{1}{2}\mu^2}$$

(where β_{1c} and β_{1s} are the flapping relative to the no-feathering plane). These expressions are nearly as accurate as Bailey's up to $\mu = 0.5$; for higher speeds a more detailed, numerical analysis is usually required anyway. The corresponding results for the second harmonic flapping are:

$$\beta_{2c} = \frac{-\gamma\mu}{144+\gamma^2 B^8}\left[\mu\theta_0\left(\frac{46}{3}+\frac{7}{144}\gamma^2 B^8\right)B^2 + \mu\theta_{tw}\left(12+\frac{7}{180}\gamma^2 B^8\right)B^3 - \mu\lambda_{NFP}\left(16+\frac{7}{108}\lambda^2 B^8\right)B + \frac{1}{3}\gamma B^7\lambda_x - 4B^3\lambda_y\right]$$

$$\beta_{2s} = \frac{\gamma\mu}{144+\gamma^2 B^8}\left[\gamma\mu\left(\theta_0\frac{25}{36}B^6 + \theta_{tw}\frac{8}{15}B^7 - \lambda_{NFP}\frac{5}{9}B^5\right) - 4B^3\lambda_x - \frac{1}{3}\gamma B^7\lambda_y\right]$$

For the rotor profile power, Bailey included the variation of the section drag with angle of attack, but neglected the radial drag forces. The result may be more accurate at low speeds (although with uniform or linearly varying inflow the angle-of-attack distribution will not really be correct), but it will underestimate C_{P_o} at higher μ.

Castles and New (1952) extended the theory of Wheatley and Bailey to large angles of pitch and inflow. They represented the section aerodynamic coefficients as $c_\ell = a\sin\alpha$ and $c_d = \delta_0 + \delta_1\sin\alpha + \delta_2\cos\alpha$. These forms are convenient since $\sin\alpha$ and $\cos\alpha$ can be expanded as follows:

$$\sin\alpha = \sin(\theta-\phi) = \sin\theta\cos\phi - \sin\phi\cos\theta$$
$$\cos\alpha = \cos(\theta-\phi) = \cos\theta\cos\phi + \sin\phi\sin\theta$$

Because θ is known, $\sin\theta$ and $\cos\theta$ can be evaluated exactly; for large inflow angles $\sin\phi = u_P/(u_T^2+u_P^2)^{1/2}$ and $\cos\phi = u_T/(u_T^2+u_P^2)^{1/2}$. They considered an articulated blade with arbitrary twist and chord distributions, and root cutout. The tip loss, stall, compressibility, and reverse flow effects were neglected. A linear inflow variation was used, and the flap angle β was assumed to be small. They obtained expressions for the forces and moments on the rotor hub (C_T, C_H, C_Y, C_Q, C_{M_x}, and C_{M_y}) and for the lowest flapping harmonics (β_0, β_{1c}, and β_{1s}).

Gessow and Crim (1952) also extended the theory of Wheatley and Bailey to large θ and ϕ angles. They considered a linearly twisted, constant chord blade with the flap hinge at the center of rotation. The drag term was still neglected in finding the normal force on the blade, so that $F_z \cong L \cos\phi$. The angle of attack $\alpha = \theta - \phi$ was assumed to be small even though the pitch and inflow could be large. They observed that the flow at high speeds during powered operation is usually stalled in the reverse flow region. Thus the reverse flow aerodynamics were modeled by assuming a constant c_ℓ and c_d appropriate to the stalled state. For powered flight they suggested $c_\ell = 1.2$ and $c_d = 1.1$ in the reverse flow region, and for autorotation $c_\ell = 0.5$ and $c_d = 0.1$. With this model, Gessow and Crim obtained expressions for C_T, C_{Q_i}, C_{Q_o}, C_{P_o}, and the flapping harmonics (up to 2/rev). The results of these analytical expressions generally compared well with the results of numerical integration, but with significant differences at high μ or high C_T/σ.

Tapscott and Gessow (1956) give charts of the calculated blade flap motion (β_0, β_{1c}, β_{1s}, β_{2c}, and β_{2s}), based on the theory of Gessow and Crim (1952). Using this theory, Gessow and Tapscott (1956) prepared performance charts for a rectangular blade with linear twist ($\theta_{tw} = 0°$, $-8°$, and $-16°$) at advance ratios of $\mu = 0.05$ to 0.50. The performance solution is discussed in more detail in Chapter 6.

Gessow and Crim (1955) developed the equations and a solution procedure for the numerical calculation of the transient flap motion. They considered an articulated rotor with offset flapping hinge, and also a teetering rotor. General airfoil characteristics ($c_\ell(\alpha,M)$ and $c_d(\alpha,M)$) were used, and large angles of flapping, inflow, and pitch were considered. The equation of motion for flapping was derived by considering the balance of aerodynamic, inertial, centrifugal, and weight moments. The solution was obtained by numerical integration using a digital computer, with a Runge-Kutta method suggested. Gessow and Crim developed this theory for investigations of flapping dynamic stability (from the transient motion) and rotor performance (from the converged, periodic solution). By using numerical integration, general aerodynamic characteristics could be considered, including stall, compressibility, and reverse flow effects (assuming that the required airfoil characteristics are available).

Gessow (1956) further developed the equations for numerical calculation of the aerodynamic characteristics of rotors in a form intended for digital computer applications. He derived expressions for the rotor thrust, profile drag, power, pitch and roll moments, root shear force, and blade flapping. An articulated rotor with hinge offset was considered, and the pitch and roll rates of the helicopter were included. The model included arbitrary blade twist, chord, and mass distribution; general two-dimensional aerodynamic characteristics for the blade airfoil; and large angles of pitch and inflow. The blade flap angle β was assumed to be small, and radial flow effects were neglected. The flap equation of motion is solved for the harmonics of the blade motion, and then the rotor aerodynamic forces and moments are evaluated. The solution procedure involves calculating the aerodynamic flap moment over a single revolution from $\psi = 0°$ to $360°$. This flap moment is harmonically analyzed, and then the harmonics of the flap motion are obtained from the moment harmonics. These steps are repeated until the solution converges. This procedure obtains the steady-state periodic solution directly, as opposed to a method that simply numerically integrates the equations of motion. A fuller description of the procedure is given in section 14–2.

Tanner (1964a) developed a performance calculation method based on the analysis of Gessow and Crim (1955). The calculation assumed two-dimensional steady flow at each blade section, with uniform inflow; neglected radial flow effects; and considered only the rigid flap motion of an articulated blade with an offset hinge. No small angle assumptions were made. Stall and compressibility effects were included by using the appropriate airfoil characteristics. The flap equation was numerically integrated until the motion converged to the steady-state, periodic solution, and then the blade aerodynamic forces were integrated to obtain the rotor forces and power. Using this analysis, Tanner (1964b, 1964c) prepared performance charts and tables for the helicopter rotor at a given advance ratio ($\mu = 0.25$ to 1.4), twist ($\theta_{tw} = 0°$, $-4°$, and $-8°$), and advancing tip Mach number ($M_{1,90} = 0.7$ to 0.9). This performance solution is discussed in more detail in Chapter 6.

Harris (1972) evaluated the capability to predict the flap motion of an articulated rotor, in particular the lateral flapping β_{1s} at low advance ratio. Classical uniform inflow theory predicts a small negative β_{1s} that increases steadily with μ. The experimental results show instead a large lateral flap magnitude at low speed, with a peak around $\mu \doteq 0.1$. In the example considered, the maximum measured flapping was $\beta_{1s} = -3.4°$ at $\mu = 0.08$, while uniform inflow theory indicates only $\beta_{1s} = -0.4°$. The increased magnitude of β_{1s} is attributed to the longitudinal variation of the induced velocity over the rotor disk. Since the influence of the lateral inflow variation on the

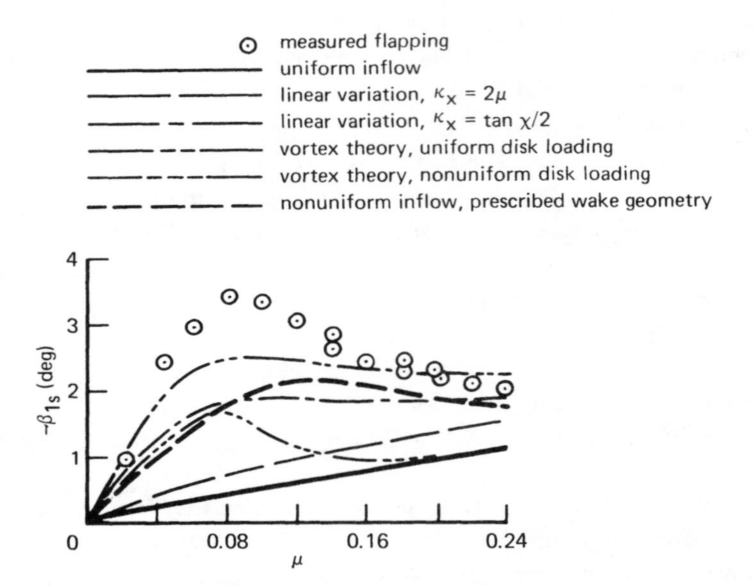

Figure 5-39 **Comparison of measured rotor lateral flapping with calculations using several induced velocity models, for $C_T/\sigma = 0.08$ and $\alpha_{TPP} = 1.0°$; from Harris (1972).**

longitudinal flapping β_{1c} is less significant, uniform inflow theory does estimate β_{1c} fairly well. The trends with collective and shaft angle are also predicted fairly well. Thus the major discrepancy is due to the nonuniform induced velocity in forward flight. Harris compared the lateral flapping measured at low forward speeds on a rotor operating at $C_T/\sigma = 0.08$ and $\alpha_{TPP} = 1.0°$ with the flapping calculated using several induced velocity models (Fig. 5-39). The induced velocity models considered included uniform inflow; a linear induced velocity variation with $\kappa_x = 2\mu$; a linear induced velocity variation with $\kappa_x = \tan \chi/2$, from Coleman, Feingold, and Stempin (1945); the vortex theory inflow distribution for a uniformly loaded disk, from Castles and De Leeuw (1954); the vortex theory inflow distribution for a nonuniformly loaded disk, from Heyson and Katzoff (1957); and finally, a numerically calculated nonuniform inflow distribution based on a discrete vortex wake model with a prescribed wake geometry. (The vortex theory results were discussed in Chapter 4.) It can be seen (Fig. 5-39) that nonuniform inflow indeed improves the estimation of the lateral flapping in both magnitude and trend, but still the flapping is substantially underpredicted. The implication is that the rotor theory still falls short of being able to accurately predict even the 1/rev variation of the the nonuniform induced velocity of the helicopter, which is responsible for the tip-path-plane tilt. Harris suggests that the primary source of the remaining discrepancies is the use of a prescribed wake geometry. The wake-induced velocity is very sensitive to the position of the tip vortex spirals, and there is a large self-induced distortion of the wake geometry near the disk at low speeds. Thus further advances in the predictive capability will require an accurate and efficient calculation of the helicopter wake geometry and induced velocity (see Chapter 13 for a further discussion of nonuniform inflow and wake geometry calculations).

Peters and Ormiston (1975) extended the calculation of the rotor steady-state flapping to hingeless rotors, considering the influence of various elements in the analytical model on the solution. Based on their work the following conclusions are reached regarding the modeling requirements for the rotor flapping and loads analysis. For an accurate calculation of the n/rev flapping harmonics (e.g. $n = 1$ for β_{1c} and β_{1s}), the analysis must include all the harmonics up to m/rev, where $m = n$ for $0 < \mu < 0.4$, and $m = n + 1$ for $0.4 < \mu < 1.0$. Reverse flow is required only for $\mu > 0.6$; the root cutout must be accounted for only above $\mu = 1.0$ and the tip loss factor is always important. The effects of compressibility are important, but when the advancing tip Mach number $M_{1,90} < 0.9$, a simple correction based on the 75% radius is sufficient. Above about $M_{1,90} = 0.9$, it is necessary to include the radial and azimuthal variations of the compressibility effects. The equivalent hinge spring and offset model is not really a very good representation of the hingeless rotor mode shape; the use of the actual elastic cantilever modes is greatly preferable. To accurately calculate the rotor loads and motion, it is necessary to include only a single flapwise mode for $0 < \mu < 0.6$; two modes for $0.6 < \mu < 1.2$; and three modes for $1.2 < \mu < 1.6$. These conclusions also apply to the articulated rotor analysis, since the hinged blade can be considered a limiting case of the cantilevered flexible blade.

Additional literature on the helicopter rotor in forward flight: Klemin (1925), Munk (1925), Glauert (1927a, 1928), Seiferth (1927), Glauert and Lock (1928), Wheatley (1932, 1933, 1934a, 1934c, 1935, 1936, 1937a, 1937b), Bennett (1933), Schrenk (1935), Squire (1935), Wheatley and

Bioletti (1935, 1936a, 1936b), Wheatley and Hood (1935), Wheatley and Windler (1935), Beavan and Lock (1936), Platt (1936), Brequet (1937), Bailey (1938, 1940), Hohenemser (1938), Knight (1938), Prewitt (1938), Hufton, Woodward-Nutt, Bigg, and Beavan (1939), Sissingh (1939, 1941), Bennett (1940), Pfluger (1940), Seibel (1944), Gustafson (1945a), Migotsky (1945, 1948), Allen (1946), Gustafson and Gessow (1946), Lichten (1946), Fail and Squire (1947), Myers (1947), Winson (1947), Dingeldein and Schaefer (1948), Carpenter and Peitzer (1949), Squire, Fail, and Eyre (1949), Stewart (1950, 1952a, 1952b), Dingeldein (1954), Payne (1954b, 1954c, 1955a, 1955b, 1955d, 1958, 1959b), Amer (1955a), Tapscott and Gustafson (1955), Bradley (1956), Castles and Durham (1956a), Gessow and Crim (1956), Jones (1957, 1964), Haffron, Bristow, Gass, and Brown (1959), McCarty, Brooks, and Maglieri (1959a), McKee and Naeseth (1959), Gessow and Gustafson (1960), Rabbott (1962), Young (1962a, 1966), Huston (1963), Jenney, Arcidiacono, and Smith (1963), Price (1963-1965), Arcidiacono (1964), McCloud and Biggers (1964), Sweet, Jenkins, and Winston (1964), Ekquist (1965), Jenkins (1965a), Norman and Sultany (1965), Harris (1966a, 1966b), Madden (1967), Norman and Somsel (1967), Clarke and Bramwell (1968), McCloud, Biggers, and Stroub (1968), Bramwell (1969), Paglino (1969), Yatsunovich (1969, 1970), La Forge and Rohtert (1970), Rohtert and La Forge (1970), Caradonna and Isom (1972, 1976), Johansson (1972, 1973, 1978), Le Nard (1972), Dietz (1973), Hohenemser and Yin (1973a), Le Nard and Boehler (1973), Isom (1974), Brotherhood (1975), van Holten (1975, 1977), Dooley (1976), Huber and Strehlow (1976), Blackwell (1977), Heyson (1977, 1978), Kawakami (1977), Sheridan and Wiesner (1977), Landgrebe, Moffitt, and Clark (1977), Balch (1978), Caradonna and Philippe (1978), Dooley and Ferguson (1978), Isogai (1978), Sheridan (1978), Smith (1978), Soohoo, Noll, Morino, and Ham (1978), Spangler and Smith (1978), Stroub (1978), Tran and Renaud (1978).

On the interference of the rotor wake with wind tunnel walls, and wind tunnel corrections: Ganzer and Rae (1960), Heyson (1960e, 1961a, 1969, 1970), Rae (1967), Rae and Shindo (1969), Lo and Binion (1970), Leyman and Besold (1971), Lo (1971), Rae and Shindo (1977).

The literature on rotor airloads data: Meyer and Falabella (1953), Rabbott (1956), Rabbott and Churchill (1956), Mayo (1959), Burpo and Lynn (1962), Duvivier (1962), Ham (1963, 1965), Huston (1963), Scheiman and

Kelley (1963, 1967), Scheiman and Ludi (1963), Scheiman (1964), Ham and Madden (1965), Du Waldt and Statler (1966), Rabbott, Lizak, and Paglino (1966a, 1966b), Rabbott and Paglino (1966), Pruyn and Alexander (1967), Pruyn, et al. (1967), Bartsch and Sweers (1968), Blaser and Velkoff (1968), Fenaughty and Beno (1969), Bowden and Shockey (1970), Tung and Du Waldt (1970), Ward (1971), Beno (1973), Dadone and Fukushima (1974), Riley and Brotherhood (1974), Van Gaasbeek (1975), Gray, et al. (1976), Shockey, Williamson, and Cox (1976), Shockey, Cox, and Williamson (1977), Knight, Haywood, and Williams (1978), Morris (1978), Shivananda, McMahon, and Gray (1978).

Chapter 6

PERFORMANCE

The calculation of helicopter performance is largely a matter of determining the power required and power available over a range of flight conditions. The power information may then be translated into quantities such as climb rate, ceiling, range, and maximum speed, which define the operational capabilities of the aircraft. The helicopter power required is divided into four parts: the induced power, required to produce the rotor thrust; the profile power, required to turn the rotor through the air; the parasite power, required to move the helicopter through the air; and the climb power, required to change the gravitational potential energy. In hover there is no parasite power, and the induced losses are 60% to 70% of the total. As the forward speed increases, the induced power decreases, the profile power increases slightly, and the parasite power increases until it is the dominant loss at high speed. Thus the total power required is high at hover, because of the induced loss with a low but reasonable disk loading; at first it decreases significantly with increasing speed, as the induced power decreases; and then it increases again at high speed, because of parasite power losses. Minimum power is required roughly in the middle of the helicopter speed range.

The major task in helicopter performance analysis is the calculation of the rotor forces and power. Procedures to perform these calculations have been developed in the preceding chapters. There are two basic approaches to the calculation of rotor performance, the force balance method and the energy balance method. In the force balance method, the blade section forces are integrated to obtain the net rotor forces and torque. The solution requires a knowledge of the rotor induced velocity and blade motion as well, to define the blade angle-of-attack distribution. The helicopter is trimmed to force and moment equilibrium to determine the rotor forces and attitude required to maintain the specified flight condition. Alternatively, the rotor performance may be calculated for a range of thrusts and attitudes and from it performance charts may be constructed. In Chapter 5, expressions were derived for the rotor thrust and torque in terms of the blade section forces. Even in its simplest form, the force balance method is complicated, so it is best suited for numerical calculations. It is also well suited for use with the most advanced models of the rotor and its aerodynamics. The force balance method can also be used to prepare performance charts for the rotor alone, which are then used to determine the performance of the entire helicopter.

The second approach to helicopter performance analysis is the energy balance method, in which the power required is expressed in terms of the individually identifiable sources of energy loss in the helicopter. In Chapter 5, the energy balance expression was derived from the force balance relations, showing that the two methods are equivalent and must give identical results when the same assumptions are used. The energy balance method is most useful for routine performance calculations for a number of reasons. First, the helicopter longitudinal force equilibrium has already been considered, so the power is obtained directly without the necessity of calculating the helicopter trim as well. Secondly, the parasite and climb power losses are given in simple yet exact forms; and with separate expressions given for the induced and profile losses it is easy to use approximate results for these terms. With the simplest approximations for the induced and profile losses, the energy balance method is fast and reasonably accurate, and hence it is well suited for preliminary design use. For more detailed performance analyses a better estimate of the induced and profile power is needed, requiring again a calculation of the blade angle-of-attack distribution. Thus with numerical methods the force balance and energy balance methods are even computationally equivalent, although it is still useful to break the total power into the induced, profile, parasite, and climb components in order to interpret the results.

In summary, the helicopter performance analysis generally takes one of three forms: the energy balance method, using fairly simple expressions for the induced and profile losses; rotor performance charts, based on analytical or numerical solutions, usually by the force balance method; or a numerical calculation, using as detailed a model as is possible and appropriate. With the digital computer, purely numerical performance calculations are practical for routine use in the detailed design of helicopters. The results for the rotor performance in hover and forward flight are summarized in this chapter. Detailed derivations and descriptions of the performance calculation methods are given elsewhere, particularly in Chapters 2 and 5, and in Chapter 14 for the numerical methods. Next, the quantities that describe the helicopter performance capabilities are discussed, including the power required, maximum speed, rate of climb and descent, ceilings, range, and endurance. In a broader sense, the helicopter performance requirement is the ability to complete a specified mission most efficiently, and the ability to calculate these specific performance quantities is required not only to define the operational limits of the aircraft, but also to perform a mission analysis. The chapter concludes with a discussion of the performance analyses and the results available in the literature.

6–1 Hover Performance

6-1.1 Power Required in Hover and Vertical Flight

The rotor power required in vertical flight has been obtained in the form $C_P = C_{P_i} + C_{P_o} + C_{P_c}$, where the induced, profile, and climb power coefficients are given by

$$C_{P_i} = \int_{r_R}^{B} \lambda_i \, dC_T$$

$$C_{P_o} = \int_{r_R}^{1} \frac{\sigma c_d}{2} r^3 \, dr$$

$$C_{P_c} = \lambda_c C_T$$

In dimensional form,

$$P = P_i + P_o + P_c = \int_{r_R}^{B} (V+v)\,dT + \rho A(\Omega R)^3 C_{P_o}$$

The induced loss C_{P_i} is the energy dissipated in the rotor wake by imparting a downward momentum to the air, from which the lift reaction on the rotor is obtained. Recall from Chapters 2 and 3 that momentum theory gives the simplest estimate of the induced power. Using a correlation of measured rotor performance data to define the law in regions where momentum theory is not applicable, the induced and climb power in vertical flight is given by a universal curve of $(V + v)/v_h = P/P_h$ as a function of V/v_h, where $v_h^2 = T/2\rho A$ (see Fig. 3-8). For hover, the induced velocity is

$$v = \kappa v_h = \kappa\sqrt{T/2\rho A}$$

where κ is an empirical factor correcting for the additional losses, principally tip losses and losses due to nonuniform inflow; typically $\kappa = 1.10$ to 1.20 (see section 3-1.3.1). For vertical flight, a somewhat better estimate of the induced power can be obtained from combined blade element and momentum theory (section 2-5).

The profile power C_{P_o} is the energy dissipated by the viscous drag of the blade. A rough estimate of C_{P_o} obtained by using a mean blade drag coefficient is $C_{P_o} = \sigma c_{d_o}/8$. A more accurate estimate requires an integration of the drag coefficient over the span of the blade, using the actual angle-of-attack distribution and Mach number of the blade. Such a calculation may be performed in combined blade element and momentum theory.

Hover performance may be expressed in terms of the polar of C_P as a function of C_T. The approximate expression for the hover polar is

$$C_P = \kappa C_T^{3/2}/\sqrt{2} + \sigma c_{d_o}/8$$

or

$$P = \kappa T\sqrt{T/2\rho A} + \rho A(\Omega R)^3(\sigma c_{d_o}/8)$$

In vertical flight the polar is $C_P = (\lambda_i + \lambda_c)C_T + \sigma c_{d_o}/8$, where for small climb or descent rates $\lambda_i + \lambda_c \cong \lambda_{\text{hover}} + \frac{1}{2}\lambda_c$.

Blade element theory may be used to obtain the collective pitch; for a linearly twisted, constant chord blade with uniform inflow,

$$\theta_{.75} = \frac{6C_T}{\sigma a} + \frac{3}{2}\lambda$$

In numerical calculations of the rotor performance, the collective pitch is in fact the parameter varied; the solution for C_P and C_T then gives the polar.

The rotor hovering efficiency is expressed in terms of the figure of merit:

$$M = \frac{T\sqrt{T/2\rho A}}{P} = \frac{C_T^{3/2}/\sqrt{2}}{C_P}$$

The figure of merit is a measure of the relative contributions of the induced and profile losses in hover. Typically the profile power is about 30% of the total, and the nonideal induced losses are around 10%, giving a figure of merit $M \cong 0.60$ (see section 2-1.4).

6-1.2 Climb and Descent

In section 3–3, the helicopter vertical climb speed for a given power increment was derived:

$$V = \frac{\Delta P}{v_h} \frac{2v_h + \Delta P/T}{v_h + \Delta P/T}$$

where $v_h^2 = T/2\rho A$, and ΔP is the power available less the power required to hover. Since this result comes from momentum theory, it is applicable to low rates of descent also. For small climb and descent rates (roughly when $V < v_h$), it gives

$$V \cong 2\frac{\Delta P}{T}$$

The induced velocity decrease due to the climb speed thus doubles the effectiveness of a given power change.

In Chapter 3, methods were discussed for deriving the descent rate in vertical autorotation from the universal induced power curve in the turbulent wake state. The autorotation of a real rotor is defined by $P = T(V + v) + P_o = 0$, or $(V + v)/v_h = -P_o/P_h$, from which the descent rate V/v_h may be determined. In section 3–2 an approximate solution is given, based on a straight line for the inflow curve in the turbulent wake state and assuming that the profile power loss is the same as in hover:

$$\frac{V}{v_h} \cong -\left[1.71 + 0.29\left(\frac{1}{M} - \kappa\right)\right]$$

6-1.3 Power Available

The helicopter power available is obtained from the performance data for the engine. The engine power usually decreases as the altitude or temperature increases, and there is some influence of speed. Thus the variations in the power available are important in calculating the helicopter performance. There are power losses in the engine and transmission that must also be accounted for, including gear train losses, any power required to cool the engine, and the power required to drive accessories such as the generator and oil pump. A frequent approach is to express these losses in terms of the overall efficiency factor η in such a way that the total power required is larger than the rotor power by the factor η^{-1}:

$$P_{\text{req, total}} = \frac{1}{\eta} P_{\text{req, rotor}}$$

Typically the engine and drive train losses correspond to $\eta \cong 0.91$ to 0.96.

The helicopter has power losses in addition to those of the isolated main rotor. Rotor-rotor and rotor-fuselage aerodynamic interference losses can be a significant fraction of the total power, particularly for a tandem helicopter configuration. For the single main rotor helicopter, the tail rotor power must be included also. The tail rotor performance calculation is complicated by the fact that the tail rotor operates in the wake of the main rotor and fuselage. The aerodynamic interference reduces the efficiency of the tail rotor, and in particular increases its loads and vibration. During yawing maneuvers, the tail rotor can even be operating in the vortex ring state, which reduces the control power and greatly increases the vibration. Since the tail rotor thrust is given by the main rotor torque— $T_{tr} = Q/\ell_{tr}$, where ℓ_{tr} is the tail rotor moment arm about the main rotor shaft—the tail rotor performance can be calculated. Because the power required for the tail rotor is a small fraction of the total, and the aerodynamic interference losses would have to be estimated anyway, a more approximate approach is often used. The helicopter aerodynamic interference losses and tail rotor power can also be included in the efficiency factor η. Then only the main rotor power need be estimated, from which the total power required is obtained by multiplying by the factor η^{-1}. The overall helicopter efficiency, including engine and transmission losses, aerodynamic interference, and the tail rotor power, typically gives $\eta \cong 0.80$ to 0.87 for hover. The efficiency usually improves in forward flight, as the aerodynamic interference and tail rotor losses decrease.

6–2 Forward Flight Performance

6-2.1 Power Required in Forward Flight

In Chapter 5 the rotor torque coefficient was expressed as an integral of the in-plane component of the blade section forces:

$$C_P = C_Q = \int_0^1 \tfrac{1}{2}\sigma r U^2 (c_\ell \sin\phi + c_d \cos\phi) dr$$

where $U^2 = u_T{}^2 + u_P{}^2$, and $\tan\phi = u_P/u_T$. The c_ℓ term is the accelerating torque, and the c_d term is the decelerating torque. From this, the energy balance relation for the helicopter power required in forward flight was derived:

$$C_P = C_{P_i} + C_{P_o} + C_{P_p} + C_{P_c}$$

where the induced, profile, parasite, and climb power terms are

$$C_{P_i} = \int_{r_R}^{B} \lambda_i dC_T$$

$$C_{P_o} = C_{Q_o} + \mu C_{H_o}$$

$$C_{P_p} = \frac{DV}{\rho A(\Omega R)^3}$$

$$C_{P_c} = \frac{V_c W}{\rho A(\Omega R)^3}$$

Recall that this result required no small angle assumptions (see sections 5–4 and 5–18). Forward flight introduces the helicopter parasite loss $P_p = DV$, which is the power required to move the helicopter through the air against the drag force D.

In Chapter 4 a solution was obtained for the ideal loss (no profile power) of the rotor in forward flight, $P = P_i + P_p + P_c = T(V \sin\alpha + v)$. Figure 4-4 presents the solution, based on a combination of momentum theory and experimental results, for $(V\sin\alpha + v)/v_h = P/P_h$ as a function of $V\cos\alpha/v_h$ and $V\sin\alpha/v_h$. The momentum theory result for the induced velocity in forward flight is

$$\lambda_i = \frac{C_T}{2\sqrt{\mu^2 + \lambda^2}}$$

where $\lambda = \lambda_i + \mu \tan \alpha$. For all but the lowest speeds, a good approximation in forward flight is $\lambda_i \cong C_T/2\mu$ (see section 4-1.1). This result is very useful because it is independent of the climb or descent velocity. Including the empirical correction factor κ, the induced power in forward flight is $C_{P_i} = \lambda_i C_T = \kappa C_T^2/2\mu$ or $P_i = \kappa T^2/2\rho A V$.

The rotor profile power loss was obtained in section 5–12 as

$$C_{P_o} = \int_0^1 \frac{\sigma c_d}{2} (u_T^2 + u_R^2)^{3/2} dr$$

which includes the effects of reverse flow, radial flow, and the radial drag force. Using a mean section drag coefficient, the following approximation for speeds up to $\mu = 0.5$ was obtained:

$$C_{P_o} = \frac{\sigma c_{d_o}}{8} (1 + 4.6\mu^2)$$

For high speeds or high loading, it is necessary to include the effects of stall and compressibility in calculating the profile power, and this requires a numerical solution including a determination of the blade angle-of-attack distribution in forward flight.

When the helicopter drag is written in terms of an equivalent parasite drag area, $D = \frac{1}{2}\rho V^2 f$, the parasite power is $P_p = DV = \frac{1}{2}\rho V^3 f$, or

$$C_{P_p} = \frac{1}{2}\left(\frac{V}{\Omega R}\right)^3 \frac{f}{A} \cong \frac{1}{2}\mu^3 \frac{f}{A}$$

Alternatively, in terms of the drag force we have $C_{P_p} \cong \mu(D/W)C_T$. The helicopter climb power is $P_c = V_c W$, where $V_c = V \sin\theta_{FP}$ is the climb velocity and W is the helicopter weight. In terms of $\lambda_c = V_c/\Omega R$ then

$$C_{P_c} = \lambda_c \frac{W}{\rho A(\Omega R)^3} \cong \lambda_c C_T.$$

Thus the energy balance method gives the following estimate of the rotor power required in forward flight:

$$C_P = \frac{\kappa C_T^2}{2\mu} + \frac{\sigma c_{d_o}}{8}(1 + 4.6\mu^2) + \lambda_c C_T + \frac{1}{2}\frac{f}{A}\mu^3$$

from which the power as a function of gross weight or speed can be found. At low speed, the induced power must be calculated instead from $C_{P_i} = \kappa C_T^2/2\sqrt{\mu^2 + \lambda^2}$, which is valid down to hover. At high speed, the neglect of stall and compressibility effects in the profile power becomes a significant consideration. At high speed, the small angle approximations made for the parasite and climb power in this result may not be correct, but the exact results can be easily used instead.

6-2.2 Climb and Descent in Forward Flight

In forward flight, the induced power loss is essentially independent of the disk inclination or climb speed: $C_{P_i} \cong \kappa C_T^2/2\mu$. This approximation is valid for $\mu > 0.1$ or so, that is, for speeds above $V = 25$ to 35 knots. The rotor profile power is also not very sensitive to climb or descent in forward flight, assuming that there is no great change in the blade angle-of-attack distribution; neither is the parasite power, if the change in the helicopter drag with angle of attack is neglected. Hence only the climb power $P_c = V_c W$ depends on the climb or descent rate in forward flight. The power required may thus be written as

$$P = P_i + P_o + P_p + P_c \cong (P_i + P_o + P_p)_{V_c=0} + P_c = P_{\text{level}} + P_c$$

which gives the climb rate

$$V_c = \frac{P - P_{\text{level}}}{W} = \frac{\Delta P}{W}$$

Here P_{level} is the power required for level flight at the given thrust and speed, and ΔP is the excess power available. It follows that the helicopter climb and descent characteristics in forward flight can be determined from the power required for level flight and the power available. At low forward speeds it is necessary to account for the change in induced power with climb speed (so the climb rate approaches $V_c \cong 2\Delta P/T$ for vertical flight).

6-2.3 D/L Formulation

The rotor power required can be written in terms of an equivalent drag force D by the definition $P = DV$. Hence $D = D_i + D_o + D_p + D_c$, or in terms of the drag-to-lift ratio,

$$\left(\frac{D}{L}\right)_{\text{total}} = \left(\frac{D}{L}\right)_i + \left(\frac{D}{L}\right)_o + \left(\frac{D}{L}\right)_p + \left(\frac{D}{L}\right)_c$$

where $L = T\cos\alpha$ is the rotor lift (for large angles, $L = T\cos\alpha + H\sin\alpha = W\cos\theta_{FP}$ should be used so that the definition of D/L is independent of the reference plane). The rotor drag-to-lift ratio is defined as

$$\left(\frac{D}{L}\right)_r = \left(\frac{D}{L}\right)_i + \left(\frac{D}{L}\right)_o$$

Note that the drag-to-lift ratio can also be written as

$$\frac{D}{L} = \frac{P}{VL} = \frac{P}{TV\cos\alpha} = \frac{C_P}{\mu C_T}$$

Now the induced, profile, parasite, and climb powers become

$$\left(\frac{D}{L}\right)_i = \frac{C_{P_i}}{\mu C_T} \cong \frac{\kappa C_T}{2\mu^2}$$

$$\left(\frac{D}{L}\right)_o = \frac{C_{P_o}}{\mu C_T} \cong \frac{\sigma c_{d_o}}{8}\frac{(1+4.6\mu^2)}{\mu C_T} = \frac{3}{4}\frac{c_{d_o}}{\bar{c}_\ell}\frac{(1+4.6\mu^2)}{\mu}$$

$$\left(\frac{D}{L}\right)_p = \frac{P_p}{VL} = \frac{D}{W\cos\theta_{FP}}$$

$$\left(\frac{D}{L}\right)_c = \frac{P_c}{VL} = \frac{V\sin\theta_{FP}W}{VW\cos\theta_{FP}} = \tan\theta_{FP}$$

where $\bar{c}_\ell = 6C_T/\sigma$ has been used in the expression for the parasite power. These results take a simpler form if the induced power and parasite power are written in terms of a helicopter lift coefficient C_L, defined as

$$C_L = \frac{L}{\frac{1}{2}\rho V^2 A}$$

Then

$$\left(\frac{D}{L}\right)_i = \frac{T^2/2\rho AV}{LV} \cong \frac{L}{2\rho AV^2} = \frac{C_L}{4}$$

$$\left(\frac{D}{L}\right)_p = \frac{DV}{LV} = \frac{\tfrac{1}{2}\rho V^2 f}{L} = \frac{f/A}{C_L}$$

The induced power result is simply the induced drag of a circular wing; when the aspect ratio $AR = 4/\pi$, the drag-to-lift ratio is $D_i/L = C_{D_i}/C_L = C_L/\pi AR = C_L/4$. Thus, in terms of C_L the helicopter power required is

$$\left(\frac{D}{L}\right)_{\text{total}} = \frac{C_L}{4} + \left(\frac{D}{L}\right)_o + \frac{f/A}{C_L} + \tan\theta_{FP}$$

For a given gross weight and speed, C_L can be evaluated. Then, using a simple expression like the one above (or some rotor performance charts), the profile losses $(D/L)_o$ can be found, completing the estimate of the helicopter power required. This formulation was developed for autogyro performance calculations. The lift coefficient C_L is used because the rotor on an autogyro functions like a fixed wing. Consequently, many of the earlier analyses express the results for improved profile power calculations in terms of $(D/L)_o$. For helicopter performance calculations this formulation is not very appropriate, however, since the drag-to-lift ratio $D/L = C_P/\mu C_T$ is singular at hover.

6-2.4 Rotor Lift and Drag

Theoretical and experimental rotor performance data are often expressed in terms of the rotor lift and drag, defined as the wind axis components of the total force on the rotor hub (Fig. 6-1). Thus, in terms of the rotor thrust and H-force, defined relative to some reference plane such as shaft axes, the coefficients C_L and C_X are

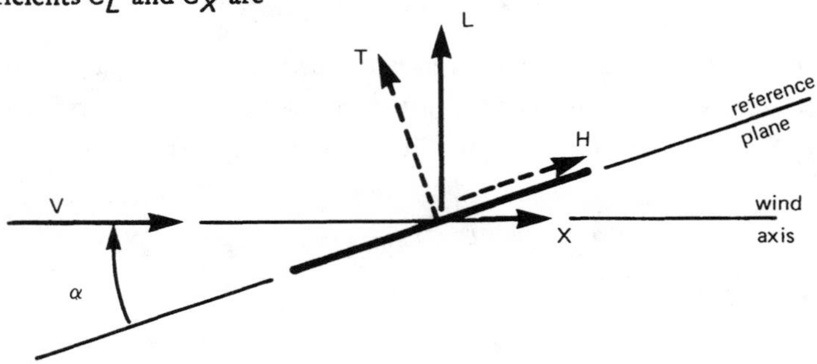

Figure 6-1 Rotor lift and drag forces (wind axes)

$$C_L = C_T \cos\alpha + C_H \sin\alpha$$

$$C_X = C_H \cos\alpha - C_T \sin\alpha$$

(note that here $C_L = L/\rho A(\Omega R)^2$, which is not the same quantity used in the preceding section). The calculated and measured results are then typically presented in terms of C_L/σ and C_X/σ. The rotor propulsive force (PF) is the negative of the X-force.

The rotor drag will be defined as

$$D_r = \frac{P}{V} - PF = \frac{P}{V} + X$$

The rotor lift-to-drag ratio $(L/D)_r$ is a useful expression of the rotor efficiency at high speed. Note that since the rotor propulsive force must equal the helicopter parasite drag, $PF = -X = D_p$, the rotor drag-to-lift ratio as defined here is

$$\left(\frac{D}{L}\right)_r = \frac{P/V - D_p}{L} = \left(\frac{D}{L}\right)_{\text{total}} - \left(\frac{D}{L}\right)_p$$

which is consistent with the definition of the preceding section,

$$\left(\frac{D}{L}\right)_r = \left(\frac{D}{L}\right)_i + \left(\frac{D}{L}\right)_o$$

By using a wind axes presentation of the data, performance charts can be directly interpreted in terms of the helicopter operating condition. The helicopter gross weight determines the rotor lift required, and the helicopter parasite drag determines the rotor propulsive force.

6-2.5 P/T Formulation

It is more useful for the helicopter to express the power required in terms of the power-to-thrust ratio P/T. Compared with the drag-to-lift formulation, $D/L = P/VL$, the principal difference is that P/T is not singular at hover. In coefficient form,

$$\frac{C_P}{C_T} = \frac{P}{\Omega R T}$$

so that

$$\left(\frac{C_P}{C_T}\right) = \left(\frac{C_P}{C_T}\right)_i + \left(\frac{C_P}{C_T}\right)_o + \left(\frac{C_P}{C_T}\right)_p + \left(\frac{C_P}{C_T}\right)_c$$

Then the induced, profile, parasite, and climb powers are

$$\left(\frac{C_P}{C_T}\right)_i = \lambda_i \cong \frac{\kappa C_T}{2\mu}$$

$$\left(\frac{C_P}{C_T}\right)_o \cong \frac{\sigma c_{d_o}}{8} \frac{(1 + 4.6\mu^2)}{C_T} = \frac{3}{4} \frac{c_{d_o}}{\bar{c}_\ell} (1 + 4.6\mu^2)$$

$$\left(\frac{C_P}{C_T}\right)_p = \frac{DV}{\Omega RT} = \mu \frac{D}{W}$$

$$\left(\frac{C_P}{C_T}\right)_c = \frac{V_c W}{\Omega RT} \cong \lambda_c$$

6–3 Helicopter Performance Factors

6-3.1 Hover Performance

The rotor hovering performance can be expressed in terms of C_P as a function of C_T, using collective pitch as the parameter (Fig. 6-2). At low thrust, the primary loss is the profile power; at moderate thrust levels C_P increases as $C_T^{3/2}$ because of the induced power rise; and at high thrust there is a steep increase in the profile power as a result of stall of the rotor blade. The maximum figure of merit occurs at minimum $C_P/C_T^{3/2}$, where the polar is tangent to the curve $C_P/C_T^{3/2}$ = constant. Without stall, the maximum figure of merit would be achieved at very high thrust; that is, at very high disk loading where M approaches 1 because of the induced power increase. With stall included in the rotor profile power, the maximum figure of merit is achieved at a value of C_T/σ just above the inception of stall. The minimum power per unit thrust is achieved at a point where a straight

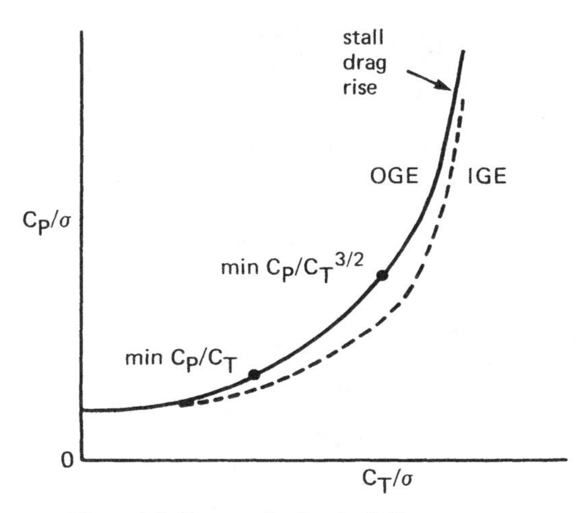

Figure 6-2 Hover polar for the helicopter rotor.

line through the origin is tangent to the polar.

The hover power required increases with gross weight, the induced power (which accounts for most of the hover losses) varying according to $P_i \sim W^{3/2}$. The air density decreases as the altitude or temperature increases, reducing the rotor profile power because of the smaller drag forces on the blades but increasing the induced power because of the higher effective disk loading. Except at very low disk loadings, the induced power increase dominates, and the total power required increases with altitude and temperature. The hover polar also depends on ground effect, which reduces the power required at a given gross weight for small distances above the ground (Fig. 6-2).

6-3.2 Minimum Power Loading in Hover

Consider now the disk loading for the best power loading of a hovering rotor. Without the profile losses, the solution is $T/A = 0$, which implies zero induced power. Including the profile power, the hover power per unit thrust can be written

$$\frac{P}{T} = \kappa \sqrt{T/2\rho A} + \frac{\sigma c_{d_0}}{8} \frac{(\Omega R)^3}{T/\rho A}$$

or

$$\frac{C_P}{C_T} = \kappa\sqrt{\frac{C_T}{2}} + \frac{\sigma c_{d_o}}{8C_T}$$

Minimizing C_P/C_T as a function of C_T (which for fixed tip speed is equivalent to minimizing P/T as a function of $T/\rho A$) gives the optimum solution

$$C_T = \frac{1}{2}\left(\frac{\sigma c_{d_o}}{\kappa}\right)^{2/3}$$

which occurs at the point where $P_i = 2P_o$, so

$$\frac{C_P}{C_T} = 3\frac{C_{P_o}}{C_T} = \frac{3}{4}(\kappa^2 \sigma c_{d_o})^{1/3}$$

Dimensionally, the solution is

$$\frac{T}{A} = \tfrac{1}{2}\rho(\Omega R)^2\left(\frac{\sigma c_{d_o}}{\kappa}\right)^{2/3}$$

which is the disk loading for minimum power loading. For a given gross weight, this disk loading determines the optimum radius of the rotor. As the profile power increases, the optimum disk loading increases and therefore the rotor radius decreases. This solution also gives a figure of merit of

$$M = \frac{T\sqrt{T/2\rho A}}{P} = \frac{C_T^{3/2}/\sqrt{2}}{C_P} = \frac{2}{3\kappa}$$

Hence the figure of merit for the rotor hovering at minimum power loading is a constant, depending only on the empirical induced power factor κ. For $\kappa \cong 1.15$, this figure of merit is $M \cong 0.58$. In practice, helicopters tend to be designed to a figure of merit near, but slightly above, this value at operational gross weight. It follows from the fixed value of the figure of merit that the basic relationship between size and power required is $P \sim W^{3/2}$. This optimum solution gives a disk loading somewhat lower than is normally used, for there are considerations besides power loading involved in selecting the disk loading. The variation of P/T, and hence the engine and fuel weight, with T/A is fairly flat near the optimum value, so the designer has some latitude in choosing the rotor radius. The weight of the rotor blade and transmission generally tends to decrease as the radius is reduced. Thus helicopters are usually designed to a higher disk loading than the optimum found here. The best disk loading considering system weight depends greatly on the specific weight of the engine (engine weight per unit power) and specific fuel consumption.

The disk loading for minimum power loading was found while assuming a fixed tip speed and solidity. If a constraint on C_T/σ is then introduced, the required solidity is

$$\sigma = \frac{1}{8} \frac{(c_d/\kappa)^2}{(C_T/\sigma)^3}$$

which typically is rather low. The same solution is obtained if the power loading is minimized as a function of C_T/σ, still assuming fixed solidity. Alternatively, consider the optimum P/T for a given disk loading. Then the induced power is fixed, and minimizing the profile power requires a low value of $\sigma(\Omega R)^3$. A constraint on C_T/σ further requires a constant value of $\sigma(\Omega R)^2 \doteq (T/\rho A)/(C_T/\sigma)$. We can then write

$$\sigma(\Omega R)^3 = \left(\frac{T/\rho A}{C_T/\sigma}\right)\Omega R = \left(\frac{T/\rho A}{C_T/\sigma}\right)^{3/2} \sigma^{-1/2}$$

There is no absolute minimum to this problem, unless system weight considerations are added. However, it follows that a low tip speed, and correspondingly a high solidity, are desired.

6-3.3 Power Required in Level Flight

Figure 6-3 sketches the variation with speed of the power required by the helicopter in level flight. The induced power is the largest component in hover, but it quickly decreases with speed. The profile power exhibits a slight increase with speed. The parasite power is negligible at low speeds but increases proportional to V^3 to dominate at high speed. Thus the total power required is high at hover, has a minimum value in the middle of the helicopter speed range, and then increases again at high speed because of the parasite power. At very high speeds, stall and compressibility effects will also increase the profile power. Ground effect significantly reduces the power required at hover and very low speeds, but it has little influence at

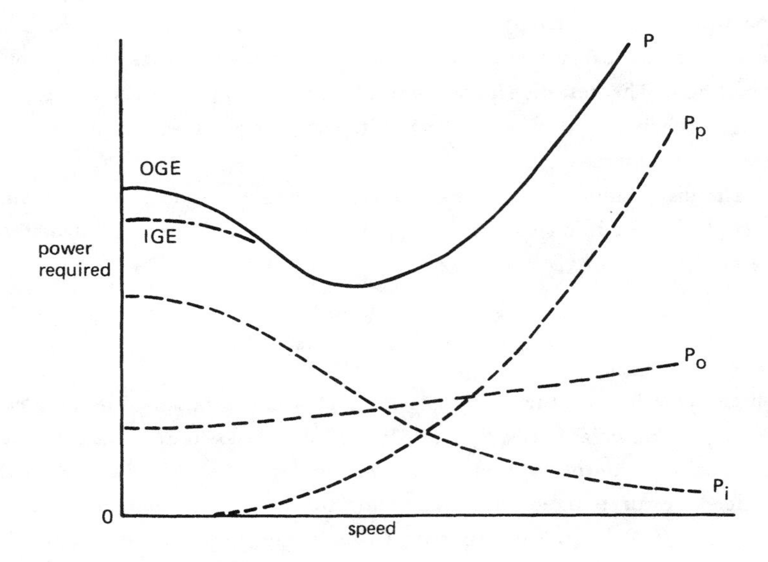

Figure 6-3 Helicopter power required for level flight at a given gross weight and altitude.

high speeds. The effect of gross weight is primarily on the induced power until the loading is high enough to increase C_{P_o} because of stall. For different aircraft the parasite drag increases with the gross weight, roughly as $f \sim GW^{2/3}$, so the parasite power increases with helicopter size.

For any given weight, there is a speed at which the helicopter power required is a minimum. The point at which the power required is a minimum is important, since it determines the best endurance, best climb rate, and minimum descent rate of the aircraft. The speed for minimum power is easily determined from the power required curve (Fig. 6-4). To estimate this speed, consider the power in forward flight:

$$C_P = \frac{\kappa C_T^2}{2\mu} + \frac{\sigma c_{d_0}}{8}(1 + 4.6\mu^2) + \frac{1}{2}\frac{f}{A}\mu^3$$

Since the profile power increase is small, the minimum power point is essentially determined by the changes in the induced power and parasite power. Neglecting the variation of C_{P_o} and minimizing C_P as a function of μ gives

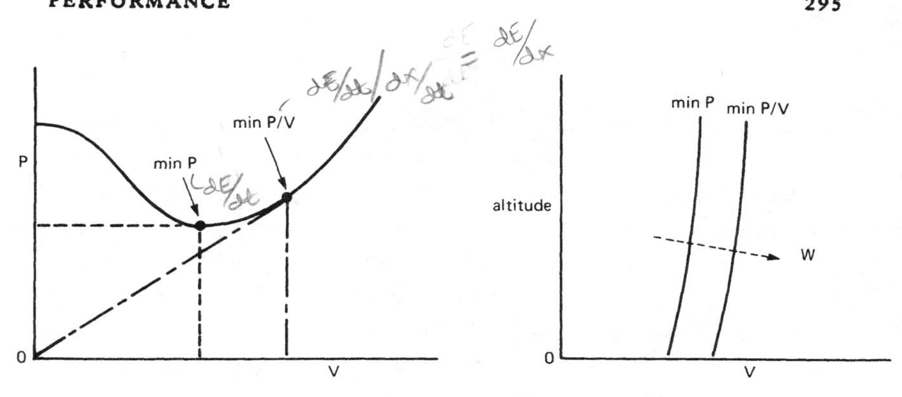

(a) Determining V from the power required curve

(b) Variation with altitude and gross weight

Figure 6-4 Speeds for minimum power and P/V.

$$\mu = \left(\frac{\kappa C_T^2}{3f/A}\right)^{1/4} = \lambda_h \left(\frac{4\kappa}{3f/A}\right)^{1/4}$$

or

$$V = v_h \left(\frac{4\kappa}{3\,f/A}\right)^{1/4}$$

where $v_h^2 = T/2\rho A$ as usual. This solution occurs where $P_i = 3P_p$. The speed for minimum power is typically $V = 60$ to 70 knots. The speed for minimum power increases with altitude and gross weight because it is proportional to v_h (Fig. 6-4).

We are also interested in the speed for minimum P/V, which is required for the best range and best descent angle. The point of minimum P/V is easily found on the power required curve as the point where a straight line through the origin is tangent to the curve (see Fig. 6-4).

6-3.4 Climb and Descent

The vertical climb rate can be calculated for a given excess power, using the procedure in section 6-1.2; for the low rates typical of helicopters, $V_c \cong 2\Delta P/T$. The climb rate at maximum power is reduced by gross weight, then, because of both the factor T^{-1} and the increase in hover power.

The climb rate slows with increasing altitude and temperature because of the hover power increase and the reduction in available engine power. The altitude at which the climb rate is zero defines the absolute hover ceiling.

The descent rate in power-off vertical autorotation can be estimated by the methods discussed in section 6-1.2 and Chapter 3. Since the descent speed is proportional to v_h, it increases with gross weight and altitude.

In forward flight, the climb or descent rate is expressed by $V_c = (P_{avail} - P_{level})/W = \Delta P/W$ (the influence of climb rate on the induced power is neglected in this approximation). The maximum climb rate is thus achieved at maximum ΔP or, neglecting the variation of the power available with speed, at the speed for minimum power required in level flight. The best angle of climb is achieved at maximum $V_c/V = \Delta P/WV$. If the helicopter can hover at the given gross weight and altitude, the best angle is vertical. Above the hover ceiling, the speed for the best angle of climb lies between the minimum speed and the speed for minimum power. The minimum power increases, and hence the best climb rate decreases, with gross weight; and the climb rate decreases with altitude. The point where the maximum climb rate reaches zero defines the absolute ceiling of the aircraft.

The descent rate in power-off autorotation in forward flight is given by simply $V_d = P_{level}/W$. The minimum descent rate thus occurs at the speed where minimum power is required. This descent rate is generally about one-half the rate in vertical autorotation. The best angle of descent, $V_d/V = P/WV$, is attained at the speed for minimum P/V in level flight. Usually this angle is between 30° and 45° from the horizontal. After power failure at high altitudes above the ground, the pilot will establish equilibrium autorotation at the forward speed giving the minimum descent rate. Near the ground the aircraft is flared to reduce both the vertical and forward speed to zero just before contacting the ground. When power failure occurs near the ground, however, there is not enough time to achieve a stabilized descent; for a power failure in hover, the optimum descent is purely vertical. Helicopter autorotation characteristics are discussed further in section 7–5.

6-3.5 Maximum Speed

The minimum and maximum velocities of a helicopter are determined by the intersection of the power required and power available curves for a given

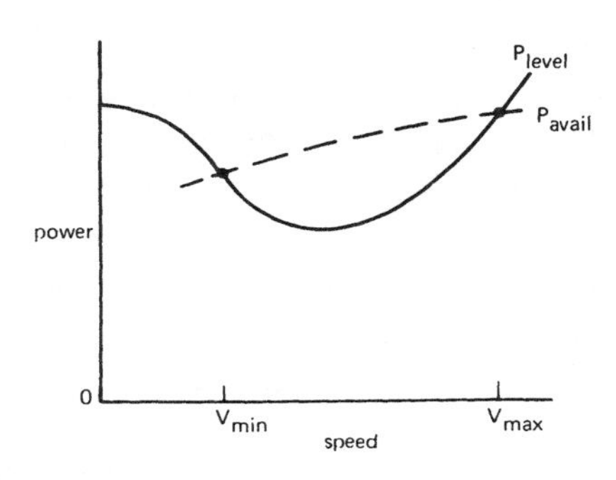

(a) V_{min} and V_{max} from power required curve

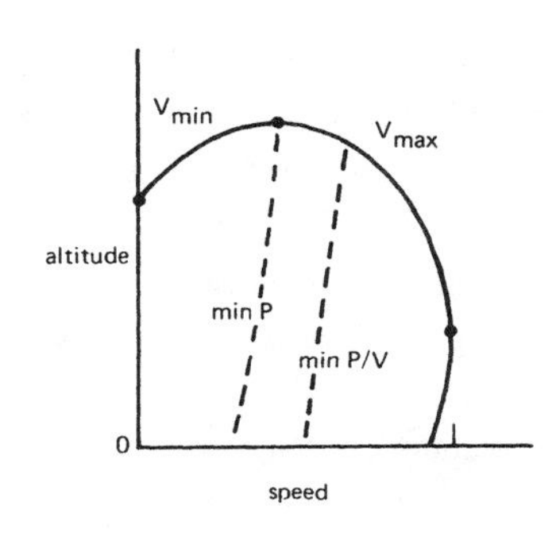

(b) Influence of altitude

Figure 6-5 Helicopter minimum and maximum speeds.

gross weight and altitude (Fig. 6-5). For $V > V_{max}$ there is insufficient power available to sustain level flight. If the helicopter can hover the minimum speed is zero, but at high altitude or high gross weight the power available may be insufficient to hover as well, so V_{min} is positive. The maximum speed of the helicopter may not be power limited, however. Rather, the maximum speed is often determined by retreating blade stall and advancing blade compressibility effects, which produce severe vibration and loads at high speed. This speed limitation is discussed in more detail in section 7–4. The power-limited maximum speed may be estimated by neglecting the variation of induced power and profile power with speed, compared to the parasite power increase. The result is

$$V_{max} = \left[\frac{2}{\rho f}\left(P_{\text{avail}} - P_i - P_o\right)\right]^{1/3}$$

or

$$\mu_{max} = \left[\frac{2}{f/A}\left(C_{P_{\text{avail}}} - C_{P_i} - C_{P_o}\right)\right]^{1/3}$$

Note that if it is assumed that the power required at maximum speed is about the same as that at hover (a balanced design), then $P_{\text{avail}} - P_i - P_o \cong P_{\text{hover}} - P_o \cong (P_i)_{\text{hover}} = T\sqrt{T/2\rho A}$, which gives

$$V_{max} \cong v_h\left(\frac{4}{f/A}\right)^{1/3}$$

Basically, the maximum speed is increased by increasing the installed power or by decreasing the helicopter drag. The parasite power rise is proportional to V^3, however, so a large change in drag or power is needed to achieve a significant maximum speed increment. The parasite power decreases with altitude, so initially the maximum speed may increase. Eventually the reduction in air density will reduce the power available, and then the maximum speed decreases with altitude. Above the hover ceiling there is a finite minimum speed also. At still higher altitude, the minimum and maximum speeds approach each other until they coincide (together with the speed for minimum power) at the absolute ceiling of the helicopter (Fig. 6-5).

6-3.6 Maximum Altitude

The helicopter ceiling is defined as the altitude at which the maximum power available is just equal to the power required; at a higher altitude, it is not possible to maintain level flight (see Fig. 6-5). This absolute ceiling is also defined as the altitude at which the climb rate becomes zero. Since the absolute ceiling can be approached from below only asymptotically, it is often more meaningful to work with the service ceiling, which is defined as the altitude where the climb rate is reduced to some small, finite value (typically around 0.5 m/sec). The principal factors defining the ceiling are the reduction of engine power with increasing altitude, the increase in power required with altitude and gross weight, and the variation of the power required with speed.

There are three ceilings of particular interest for the helicopter. First, there is the hover ceiling out of ground effect (OGE), determined by the point where the power available equals the power required to hover at a given gross weight. Secondly, there is the hover ceiling in ground effect (IGE). Since ground effect reduces the induced power required, the IGE ceiling is substantially higher than the OGE ceiling. The fact that ground effect increases the operational ceiling or weight of the helicopter can be used advantageously in operating the aircraft. The third ceiling of interest is the maximum ceiling, encountered in forward flight at the speed for minimum power. In both calculations and flight tests these ceilings are obtained by measurements of the helicopter climb rate at maximum power. Extrapolating the curves to zero climb rate gives the absolute ceilings.

6-3.7 Range and Endurance

The helicopter range is calculated by integrating the specific range dR/dW_F over the total fuel weight, for a given initial gross weight and flight condition:

$$R = \int \frac{dR}{dW_F}\, dW_F$$

Similarly, the endurance is obtained by integrating the specific endurance dE/dW_F:

$$E = \int \frac{dE}{dW_F}\, dW_F$$

The specific range and endurance are given by the specific fuel consumption of the engine (SFC, in kg/hp-hr or lb/hp-hr) as follows:

$$\frac{dR}{dW_F} = \frac{V}{P\ (SFC)}$$

$$\frac{dE}{dW_F} = \frac{1}{P\ (SFC)}$$

In general, dR/dW_F and dE/dW_F vary during a flight even if the helicopter is operated at the optimum conditions. Moreover, the power depends on the altitude and gross weight, and the specific fuel consumption depends on the power and altitude. Consequently, these expressions must be numerically integrated for an accurate determination of range and endurance. Since the total fuel weight is usually a small fraction of the gross weight, however, the integrals may be approximately evaluated using the specific range and endurance at the midpoint of the flight, where the weight is the initial gross weight less one-half the total fuel weight:

$$R = W_F \left(\frac{V}{P\ (SFC)} \right)_{W_G - \frac{1}{2} W_F}$$

$$E = W_F \left(\frac{1}{P\ (SFC)} \right)_{W_G - \frac{1}{2} W_F}.$$

The speeds for best range and endurance may be found by examining the specific range and endurance data as a function of velocity. Assuming that the specific fuel consumption is independent of velocity (which is not really true, because of the dependence of the *SFC* on the engine power), the minimum fuel consumption per unit distance and hence the maximum range are achieved at the speed for minimum P/V. Similarly, the maximum endurance is achieved at the speed for minimum P. The speeds for which fuel consumption is a minimum are more accurately obtained from a plot of $P(SFC)$ as a function of speed for a given altitude and gross weight. The speed for best endurance is at the minimum of $P(SFC)$, while the speed for best range lies at the point where a straight line through the origin is tangent to the curve (as in Fig. 6-4).

If it is assumed that P/T, the speed, and the specific fuel consumption are independent of the helicopter weight, then the range and endurance can be evaluated analytically. Write

$$\frac{dW_F}{dR} = \frac{P\ (SFC)}{V} = W \frac{P}{TV}\ (SFC) = \left(W_G - W_F\right)\frac{P}{TV}\ (SFC)$$

where W_G is the initial gross weight. Integrating over the total fuel weight then gives the Breguet range equation:

$$R = \frac{TV}{P\ (SFC)}\left[-\ln\left(1 - W_F/W_G\right)\right]$$

where W_F/W_G is the ratio of the fuel weight to the initial gross weight. Similarly, the endurance is

$$E = \frac{T}{P\ (SFC)}\left[-\ln\left(1 - W_F/W_G\right)\right]$$

These expressions account for the decrease in the gross weight as the fuel is used, a factor that reduces the fuel consumption since it has been assumed that $(P/T)SFC$ is constant.

Fig. 6-6 sketches the payload-range diagram for an aircraft. Point A is the maximum range of the helicopter at maximum gross weight and fuel capacity. Slightly higher payloads can be carried with the same gross weight by reducing the fuel carried, that is, by reducing the range. A slightly higher range can be achieved by reducing the payload with maximum fuel on board, since the reduced gross weight improves the fuel consumption.

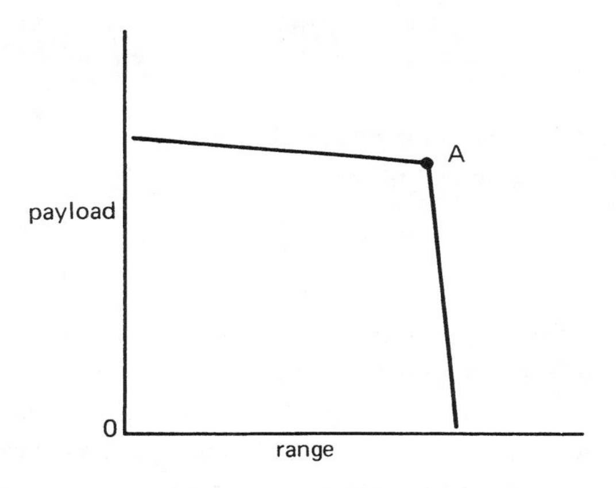

Figure 6-6 Helicopter payload-range diagram.

6–4 Other Performance Problems

6-4.1 Power Specified (Autogyro)

Consider the level flight of a rotary wing aircraft at a specified power P. The rotor induced power and profile power are determined by the speed and gross weight, so the parasite power must be given by $P_p = P - (P_i + P_o)$. Replacing P_p by DV gives

$$D = \frac{P - (P_i + P_o)}{V}$$

where D is the net drag of the helicopter. Alternatively, D is the propulsive force of the rotor at this operating condition. $D < 0$ implies that the rotor has a net drag force, which must then be balanced by an auxiliary propulsion device on the aircraft. In terms of the rotor inflow, $D < 0$ corresponds to a rearward tilt of the disk. Then there is a component of the forward velocity flowing upward through the disk, providing the additional energy required by the rotor when $P_i + P_o$ is greater than the supplied shaft power P.

The autogyro is a specific case with the shaft power fixed, namely at $P = 0$. Then the aircraft propulsive force required to balance the rotor drag is $D = -(P_i + P_o)/V$. In terms of the drag-to-lift formulation, the result is

$$-\frac{D}{W\cos\theta_{FP}} = \left(\frac{D}{L}\right)_i + \left(\frac{D}{L}\right)_o = \left(\frac{D}{L}\right)_r$$

So for the autogyro, the rotor acts much like a wing; aircraft lift is supplied at the cost of induced and profile drag.

6-4.2 Shaft Angle Specified (Tail Rotor)

If the rotor shaft angle is fixed, the performance solution gives the power required and the rotor propulsive force. The forces and moments required to balance the rotor must then be supplied by the rest of the aircraft. The most common case with the shaft angle (α_{HP}) specified is the tail rotor. The drag force of a tail rotor must be countered by the main rotor, adding to its parasite power.

The rotor drag force is $D = T\alpha_{HP} - H_{HP}$ (the sign convention of $D \geqslant 0$ for a rotor propulsive force has been retained). Since the rotor thrust is nearly perpendicular to the tip-path plane,

$$D = T(\alpha_{HP} + \beta_{1c_{HP}}) - H_{TPP} \cong T(\alpha_{HP} + \beta_{1c_{HP}})$$

Thus, finding the rotor drag or propulsive force with the shaft angle fixed (α_{HP}) requires knowing the longitudinal flapping relative to the shaft ($\beta_{1c_{HP}}$) as well, which gives the tip-path-plane angle. The performance solution then also requires the solution of the rotor flapping equations. In the case of the tail rotor there is no cyclic pitch, and usually a large pitch-flap coupling. These factors must be accounted for in the solution of the flapping equation for $\beta_{1c_{HP}}$. After the rotor propulsive force D is obtained, the power absorbed is calculated from $P = P_i + P_o + P_p$, where the parasite power $P_p = DV$.

The tail rotor has two contributions to the power required for the entire helicopter, the power absorbed directly through the tail rotor shaft and the main rotor parasite power required because of the tail rotor drag force. The total power attributed to the tail rotor is thus

$$P_{\text{total}} = P_{\text{shaft}} + P_{\text{drag}} = (P_i + P_o + P_p)_{tr} + (\Delta P_p)_{mr}$$

Now the tail rotor parasite power is $(P_p)_{tr} = DV$, and the increment of the main rotor parasite power due to the tail rotor drag force is $(\Delta P_p)_{mr} = -DV$. Hence

$$P_{\text{total}} = (P_i + P_o)_{tr}$$

The total power loss due to the tail rotor is independent of the tail rotor drag force, which simply determines the distribution of the total loss between the tail rotor and main rotor shaft powers. The helicopter performance can then be analyzed by ignoring the tail rotor drag or propulsive force. The result is a small change in the main rotor disk inclination, as determined by horizontal force equilibrium, but it is not necessary to consider the tail rotor flapping solution to find the tip-path-plane orientation.

6–5 Improved Performance Calculations

A comprehensive analysis of helicopter rotor performance must consider an arbitrary rotor, including general chord, twist, and profile distributions, and it must be applicable to extreme flight conditions, such as high loading or high speed. The climb and parasite power may be obtained exactly, assuming that the helicopter flight path angle and the parasite drag are known (that is, assuming that the rotor orientation can be accurately determined from the expressions for helicopter force and moment equilibrium). Thus efforts to improve the calculation of helicopter performance have been primarily concerned with the induced power and profile power,

$$C_{P_i} = \int \lambda_i dC_T$$

$$C_{P_o} = \int \frac{\sigma c_d}{2} \left(u_T^2 + u_R^2\right)^{3/2} dr$$

Improving the estimate of the induced power primarily requires a calculation of the nonuniform induced velocity distribution, although it also depends on an accurate loading distribution. The profile power estimate is improved by considering the actual angle of attack and Mach number distribution in calculating the section drag. Note that obtaining the blade angle of attack requires the nonuniform induced velocity calculation and also a solution for the blade motion. At extreme operating conditions it will be necessary to consider more blade degrees of freedom than the fundamental flapping mode. Thus an improved performance analysis is a complicated numerical problem requiring more attention to the details of the rotor and its aerodynamics. Moreover, it is important to be consistent in such an analysis, so an advance in one area of the problem is not really useful until equivalent assumptions in other areas can also be eliminated.

It is possible to retain an analytical formulation of the performance calculation when making some advances. For example, a drag polar of the form $c_d = \delta_0 + \delta_1\alpha + \delta_2\alpha^2$ improves the profile power calculations while still allowing the integrals to be evaluated analytically. Even such analytical solutions are fairly complicated, though, so the results are often used in the form of performance charts constructed for some representative rotor. Because of the complexity of the rotor aerodynamics, most rotor performance calculations beyond using the simplest expressions require extensive numerical computations. Again a convenient and economic presentation of such calculations is in the form of performance charts or tables. With a high-speed digital computer it is also practical to perform a numerical performance analysis for a specific rotor under consideration. Such an analysis is essential if the many details specific to individual rotors are to be included, such as planform and airfoil variations. Performance charts remain useful, however, particularly for preliminary design of helicopters.

6–6 Literature

We conclude this chapter with a discussion of the performance analyses available in literature. The models on which these analyses are based were discussed in section 5–24.

Bailey (1941) developed a performance analysis in which the rotor thrust, torque, and profile power are expressed as functions of θ_0 and λ_{NFP}. The coefficients of these expressions depend on the rotor twist, Lock number, tip loss factor, drag polar constants (δ_0, δ_1, and δ_2), and the advance ratio. The theory treats an articulated rotor with no hinge offset and a constant chord, linearly twisted blade. The aerodynamic model includes reverse flow (to order μ^4), and the blade section characteristics are represented by a constant lift curve slope and a drag polar ($c_\ell = a\alpha$ and $c_d = \delta_0 + \delta_1\alpha + \delta_2\alpha^2$). The theory assumes uniform inflow and neglects stall, compressibility, and radial flow effects. The analysis was developed for the autogyro, which is reflected in the formulation of the solution procedure and the presentation of the results. The performance problem specifies the rotor parameters, the flight speed, and either the helicopter drag or the rotor power required ($C_P = 0$ for the autogyro). Either the collective pitch or the rotor thrust can be used as the independent parameter because the relation between them in the C_T equation is linear. Consider the autogyro performance problem ($C_Q = 0$). For a given collective and advance ratio, the torque equation becomes a quadratic in λ_{NFP}. Solving for λ_{NFP}, the expressions for C_T and C_{P_o} can be evaluated. The induced power is obtained from $C_{P_i} = \kappa C_T^2/2\sqrt{\mu^2 + \lambda^2}$, and then the rotor drag force may be obtained from $(D/L)_r = (D/L)_i + (D/L)_o$. The shaft angle (really α_{NFP}) can be found from λ_{NFP} and λ_i. Finally, Bailey also gives expressions for the rotor coning and flapping in terms of θ_0 and λ_{NFP}. The helicopter performance problem may be solved with Bailey's analysis also, but an iterative procedure is required. For a given thrust, speed, and helicopter drag, the energy balance method gives the power required, C_P. For the first estimate, the simplest approximation for C_{P_o} can be used. Then the torque equation is again a quadratic equation which can be solved for λ_{NFP}. With θ_0 and λ_{NFP} now, the profile power can be recalculated from Bailey's expression, and a new estimate of the total power required can be obtained from the energy balance expression. These steps are repeated until the solution for the power (and λ_{NFP}) converges. Thus even Bailey's analysis requires many numerical calculations, since with C_T and C_P given two equations must be solved for θ_0 and λ_{NFP}, and for the helicopter problem it is necessary to iterate. The numerical problem may be avoided by using the theory to construct performance charts for a representative rotor (i.e. twist, Lock number, tip loss, and drag polar) over a wide range of operating conditions; specific performance problems may be quickly solved graphically using these charts. To construct a performance chart using Bailey's theory, an arbitrary total power and rotor twist are assumed. For a range of advance ratio and collectives, λ_{NFP} is obtained from the torque equation, and the thrust and profile power are evaluated. The result is a plot of C_{P_o}/σ as a function of C_T/σ for a given value of C_P/σ and θ_{tw}, using μ and $\theta_{.75}$ as parameters. Actually, Bailey was concerned with the autogyro problem, and for that reason suggested using the drag-to-lift formulation, a plot of $(D/L)_o$ as a function of C_L/σ, for a given value of $(D/L)_{total}$ (namely zero) and θ_{tw} (see Fig. 6-7). For the helicopter performance problem the total power is not known, so it is still necessary to iterate, but graphically rather than numerically. The greatest difficulty with this form of performance chart is the necessity to interpolate between graphs to find the total power. The performance charts

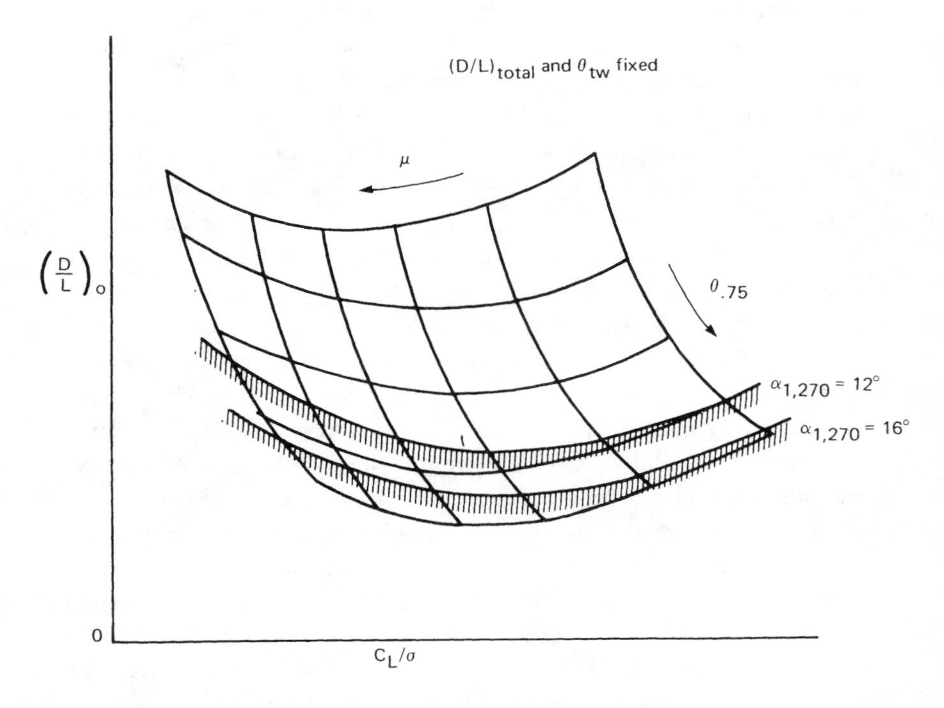

Figure 6-7 Bailey's rotor performance chart formulation: rotor profile drag as a function of lift, for a given total power and blade twist.

are constructed for a specific set of rotor parameters, but the influence of Lock number is found to be small and in the form given the influence of rotor solidity is small also. Separate charts must be constructed for different values of the blade twist, the most important remaining parameter. These performance charts (Fig. 6-7) also show lines corresponding to angles of attack on the retreating blade tip of $\alpha_{1,270} = 12°$ and $16°$ as an indication of the stall limits of the rotor (see Chapter 16).

Bailey and Gustafson (1944) and Gustafson (1953) present helicopter performance charts based on Bailey's theory. The charts are in the form shown in Fig 6-7, for blade twist of $\theta_{tw} = 0°$ and $-8°$. Note that in their notation, P/L is used where $(D/L)_{total} = (P/TV)_{total}$ is the notation in this book. Some of these charts are also given by Gessow and Myers (1952).

Gessow and Tapscott (1956) present performance charts based on the

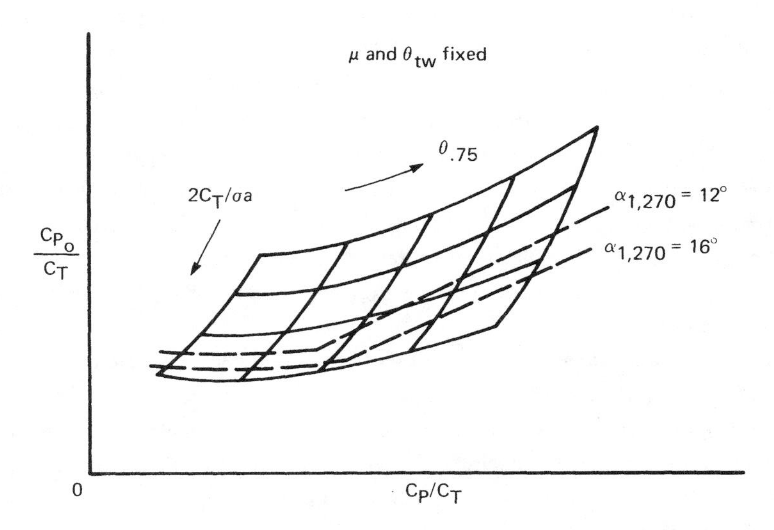

Figure 6-8 Gessow and Tapscott's rotor performance chart formulation: profile power as a function of total power, for a given speed and blade twist.

theory of Gessow and Crim (1952). The charts describe the calculated performance in the form of C_{P_o}/C_T as a function of C_P/C_T for given values of μ and θ_{tw}, using $2C_T/\sigma a$ and $\theta_{.75}$ as parameters (Fig. 6-8). The charts given cover $\mu = 0.05$ to 0.50 and are for linear twist of $\theta_{tw} = 0°, -8°$, and $-16°$. The calculations used $\gamma = 15$, but the results are applicable for $\gamma = 0$ to 25 with good accuracy; the theory considered rectangular blades, but the charts can be applied to tapered blades using an equivalent solidity. The airfoil characteristics were described by $a = 5.73$ and $c_d = 0.0087 - 0.0216\alpha + 0.400\alpha^2$. The helicopter performance problem is solved using the C_P/C_T formulation of the energy balance expression. For a given thrust and speed, the induced, parasite, and climb power can be evaluated. Then, on the performance chart a straight line with a 2:1 slope is drawn from $(C_{P_i} + C_{P_p} + C_{P_c})/C_T$ on the abscissa. (The slope would be 45° if the vertical and horizontal scales were the same.) The point where this line intersects the curve for the given C_T defines the performance solution, the profile power and total power, subject to the constraint $C_P = (C_{P_i} + C_{P_p} + C_{P_c}) + C_{P_o}$. The chart also gives the collective pitch $\theta_{.75}$. Gessow and Tapscott also give charts relating $2C_T/\sigma a$ to λ_{NFP} and $\theta_{.75}$, from which the disk inclination

α_{NFP} can be obtained. For a stall criterion, they use the retreating blade angle of attack $\alpha_{1,270}$ (for powered flight) and $\alpha_{\mu+0.4,270}$ (for autorotation). The angles $\alpha = 12°$ and $16°$ are considered to indicate incipient stall and excessive stall, respectively (see Chapter 16). The performance charts show lines where $\alpha = 12°$ or $16°$ (as in Fig. 6-8), and separate charts are given expressing these stall criteria in terms of limitations on the helicopter performance, particularly the speed, thrust, and propulsive force (i.e. the rotor disk inclination). The lines representing rotor stall also define the limits of validity of the calculations, since the theory does not include stall in the airfoil characteristics.

Gessow and Tapscott (1960) present tables and charts of calculated rotor performance, including flight conditions well into the stall range, based on the analyses of Gessow and Crim (1955) and Gessow (1956). The calculations were for a rectangular, articulated blade with $-8°$ of linear twist. Static, two-dimensional data (for a NACA 0015 section) were used, so that stall effects would be included. The rotor flapping, thrust, power, profile power, and H-force are given as functions of $\theta_{.75}$ and λ_{NFP} for $\mu = 0.1$ to 0.5.

Tanner (1964b, 1964c) presents tables and charts of numerically calculated rotor performance, including stall, compressibility, and large angle effects. The calculations were based on the theory of Tanner (1964a) and Gessow and Crim (1955), which uses two-dimensional, steady data for the blade section aerodynamic characteristics. The theory assumes uniform inflow and neglects radial flow effects, and it considers only the rigid flap motion of an articulated blade. The calculations were performed for a rectangular blade with solidity $\sigma = 0.1$, root cutout $r_R = 0.25$, tip loss factor $B = 0.97$, and Lock number $\gamma = 8$. Two-dimensional data for the lift and drag coefficients of a NACA 0012 airfoil were used. The performance charts present the results in terms of the rotor wind axis drag and lift coefficients (C_D/σ and C_L/σ, see section 6-2.4), as shown in Fig. 6-9. Charts are given for advance ratios $\mu = 0.25$ to 1.40, advancing-tip Mach numbers $M_{1,90} = 0.7$ to 0.9, and linear twist of $\theta_{tw} = 0°, -4°$, and $-8°$. The helicopter performance problem can be solved directly with these charts. The helicopter drag and gross weight determine C_D/σ and C_L/σ, from which the charts give the rotor power, collective pitch, longitudinal flapping β_{1c}, and disk inclination α_{NFP}. The results in the form presented are not very sensitive to the rotor solidity, but Tanner does specify a solidity

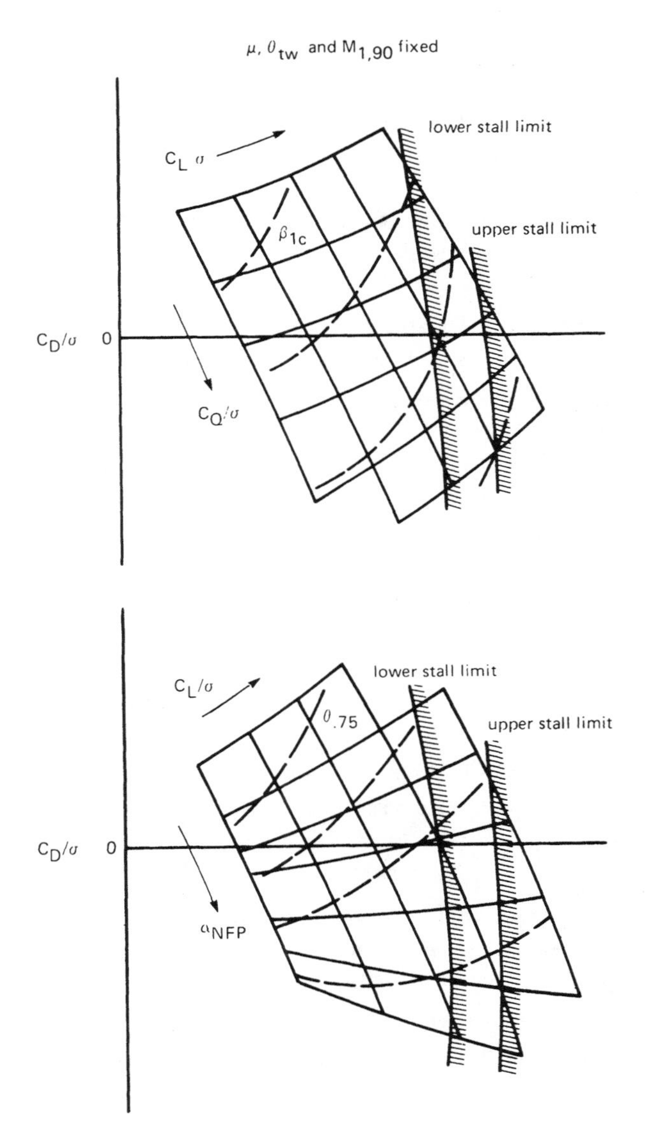

Figure 6-9 Tanner's rotor performance chart formulation: rotor wind axis lift and drag, for a given speed, twist, and advancing-tip Mach number.

correction for α_{NFP} and C_D/σ. The charts also show stall limits based on the maximum blade profile torque occurring around the rotor azimuth:

$$\frac{C_{Q_o}}{\sigma} = \left(\frac{1}{2}\int_0^1 c_d u_T^2 r\,dr\right)_{\text{max over }\psi}$$

Rotor stall is identified by a rapid increase in this parameter on the retreating side of the disk. The onset of significant stall effects, called the lower stall limit, is defined by $(C_{Q_o}/\sigma)_{max} = 0.004$ (see Fig. 6-9); the upper stall limit, beyond which operation is undesirable, is defined by $(C_{Q_o}/\sigma)_{max} = 0.008$. Tanner also gives charts for rotor hovering performance (C_T/σ vs. C_P/σ) calculated using combined blade element and momentum theory; the tables give the inflow ratio and flapping harmonics (up to 3/rev) as well as the parameters shown in the charts.

Kisielowski, Bumstead, Fissel, and Chinsky (1967) present performance charts for the helicopter in forward flight that are based on numerical calculations of the rotor forces and flap motion. Their analysis made no small angle assumptions; included stall, compressibility, and reverse flow effects; and used tabular data for the blade section static lift and drag coefficients (for a NACA 0012 airfoil). They did assume uniform inflow and neglected the effects of radial flow and dynamic stall. The calculations were for a rectangular, linearly twisted blade with solidity $\sigma = 0.062$ (solidity corrections are discussed), Lock number $\gamma = 7.6$, root cutout $r_R = 0.2$, tip loss factor $B = 0.97$, and flap hinge offset $e = 0.0226$. The performance charts give the results in terms of the wind axis rotor lift and drag forces (normalized using $\rho V^2 R^2 \sigma$) for a given helicopter speed, rotor tip speed, and twist (Fig. 6-10). The charts are for $V = 50$ to 300 knots and $\Omega R = 300$ to 800 fps ($\mu = 0.2$ to 1.5 and $M_{1,90} = 0.64$ to 0.98, with the high speed and Mach number cases predominating), and for blade twist of $\theta_{tw} = -4°, -8°$, and $-12°$. For a given helicopter speed and tip speed, the helicopter gross weight and drag define a point on the performance chart from which the rotor power and shaft angle are obtained. The performance charts use lines of the retreating blade tip angle of attack $\alpha_{1,270} = 12°$ and $14°$ as a stall criterion.

Literature on helicopter performance theory and measurements: Hohenemser (1938), Hufton, Woodward-Nutt, Bigg, and Beavan (1939), Sissingh (1941), Wald (1943), Gustafson (1944, 1945a), Castles (1945), Dingeldein and Schaefer (1945, 1948), Gustafson and Gessow (1945, 1946, 1948), Migotsky (1945), Talkin (1945, 1947), Autry (1946), Lichten (1946),

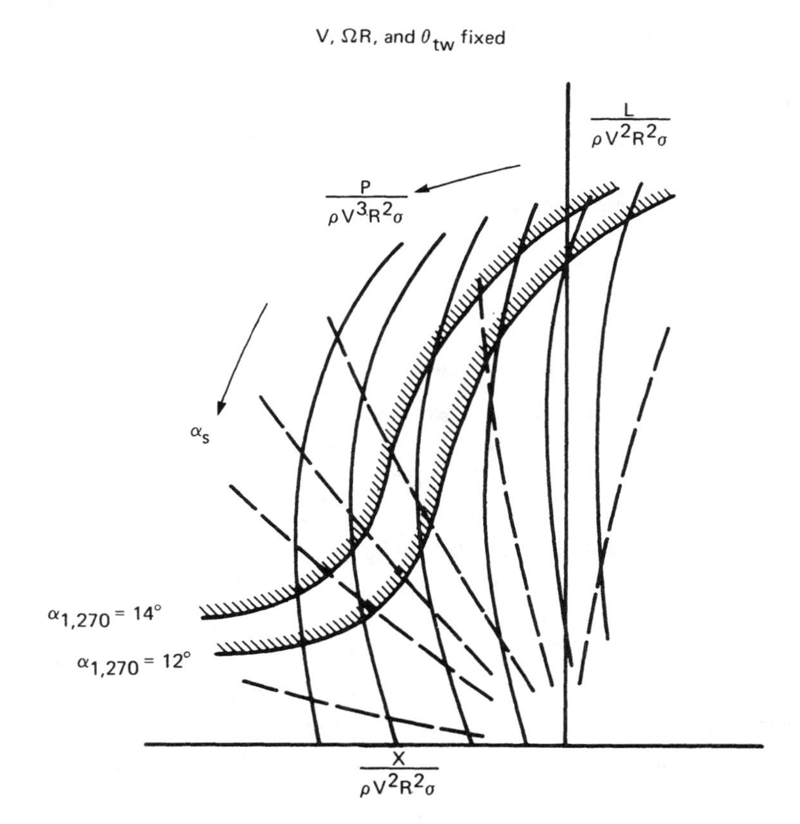

Figure 6-10 Kisielowski's rotor performance chart formulation: rotor wind axis lift and drag, for a given helicopter speed, rotor tip speed, and twist.

Lipson (1946), Fail and Squire (1947), Gessow and Myers (1947), Toms (1947), Carpenter (1948), Gessow (1948d, 1959), Squire, Fail, and Eyre (1949), Carpenter and Paulnock (1950), Harrington (1951, 1954), Carpenter (1952, 1958, 1959), Stepniewski (1952), Payne (1953), Dingeldein (1954, 1961), Powell (1954, 1957), Makofski (1956), Shivers and Carpenter (1956, 1958), Foster (1957), Jewel and Harrington (1958), McCloud and McCullough (1958), Powell and Carpenter (1958), Churchill and Harrington (1959), McKee and Naeseth (1959), Gessow and Gustafson (1960), Jewel (1960), Shivers (1960, 1961, 1967), Sikorsky (1960), Sweet (1960b), Rabbott

(1961, 1962), Biggers, McCloud, and Patterakis (1962), Jenkins, Winston, and Sweet (1962), Jepson (1962), Shivers and Monahan (1962), Sweet and Jenkins (1962, 1963), Huston (1963), McCloud, Biggers, and Maki (1963), Stutz and Price (1964), Sweet, Jenkins, and Winston (1964), Wood and Buffalano (1964), Ekquist (1965), Jenkins (1965a), Livingston (1965), Norman and Sultany (1965), Piper (1965), Schad (1965), Davenport and Front (1966), Harris (1966a, 1966b), Norman and Somsel (1967), Strand, Levinski, and Wei (1967), Clarke and Bramwell (1968), McCloud, Biggers, and Stroub (1968), Paglino and Logan (1968), Spivey (1968), Tanner and Van Wyckhouse (1968), Tanner, Van Wyckhouse, Cancro, and McCloud (1968), Charles and Tanner (1969), Paglino (1969, 1971), Cassarino (1970), LaForge and Rohtert (1970), Linville (1970, 1972), Spivey and Morehouse (1970), Charles (1971), Lee, Charles, and Kidd (1971), Putman and Traybar (1971), Sonneborn (1971), Bazov (1972), Bellinger (1972a), Fradenburgh (1972), Gilmore and Gartshore (1972), Landgrebe and Cheney (1972), Lewis (1972), Wells and Wood (1973), Davis and Stepniewski (1974), Gillespie (1974), Gillespie and Windsor (1974), Keys and Wiesner (1974), Niebanck (1974), Shipman (1974), Young (1974), Kerr (1975), Montana (1975, 1976a, 1976b), Paglino and Clark (1975), Schwartzberg (1975), Sheehy and Clark (1975, 1976), Smith (1975), Williams and Montana (1975), Wilson and Mineck (1975), Schmitz and Vause (1976), Sheehy (1976, 1977), Yeager and Mantay (1976), Landgrebe, Moffitt, and Clark (1977), Loiselle (1977), Mantay, Shidler, and Campbell (1977), Moffitt and Sheehy (1977), Schwartzberg, et al. (1977), Weller (1977), Weller and Lee (1977), Balch (1978), Beirun (1978), Keys and Rosenstein (1978), Morris (1978), Sheridan (1978), Stroub (1978).

Chapter 7

DESIGN

7–1 Rotor Types

The helicopter rotor type is largely determined by the construction of the blade root and its attachment to the hub. The blade root configuration has a fundamental influence on the blade flap and lag motion and hence on the helicopter handling qualities, vibration, loads, and aeroelastic stability. The basic distinction between rotor types is the presence or absence of flap and lag hinges, and thus whether the blade motion involves rigid body rotation or bending at the blade root.

An articulated rotor has its blades attached to the hub with both flap and lag hinges. The flap hinge is usually offset slightly from the center of rotation because of mechanical constraints and to improve the helicopter handling qualities. The lag hinge must be offset in order for the shaft to transmit torque to the rotor. The purpose of the flap and lag hinges is to reduce the root blade loads (since the moments must be zero at the hinge). With a lag hinge it is also necessary to have a mechanical lag damper to avoid a mechanical instability called ground resonance, involving the coupled motion of the rotor lag and hub in-plane displacement. The articulated rotor is the classical design solution to the problem of the blade root loads and hub moments. It is conceptually simple, and the analysis of the rigid body motion is straightforward. The articulated rotor is mechanicaslly complex, however, involving three hinges (flap, lag, and feather) and a lag damper for each blade. The flap and lag bearings are required to transmit both the blade thrust and centrifugal force to the hub, and so must operate in a highly stressed environment. The hub also has the swashplate, and the rotating and nonrotating links of the control system. The resulting hub requires a high level of maintenance and contributes substantially to the helicopter parasite drag. Recently, the use of elastomeric bearings has been introduced. By replacing the mechanical bearings, a major maintenance problem is eliminated.

The teetering rotor (also called a semi-articulated, semi-rigid, or see-saw rotor) has two blades attached rigidly to the hub without flap or lag hinges; the hub is attached to the rotor shaft with a single flap hinge. The two blades thus form a single structure that flaps as a whole relative to the shaft. The hub usually has a built-in precone angle to reduce the steady coning loads, and perhaps an undersling also to reduce Coriolis forces. The blades have feathering bearings. Without lag hinges, the blade in-plane loads must be reacted by the root structure. Similarly, the rotor coning produces structural loads, except at the design precone angle. To take these loads the rotor requires additional structure and weight relative to an articulated rotor. This factor is offset by the mechanical simplicity of the teetering configuration, which eliminates all the lag hinges and dampers and all but a single flap hinge. The flap hinge also does not have to carry the centrifugal loads of the blade, but only the rotor thrust, since the centrifugal forces cancel in the hub itself. The teetering configuration is perhaps the simplest and lightest for a small helicopter. It is not practical for large helicopters because a large chord is required to obtain the necessary blade area with only two blades. A gimballed rotor has three or more blades attached to the hub without flap or lag hinges (but with feathering hinges); the hub is attached to the shaft by a universal joint or gimbal. Basically, the gimballed rotor is the multi-blade counterpart of the teetering rotor, and like it has the advantage of a simpler hub than articulated rotors. The teetering and gimballed rotors are characterized by a flap hinge exactly at the center of rotation, giving a flap frequency of exactly 1/rev. The improvements in handling qualities due to offset hinges are not available. For example, flight at low or zero load factor is not possible with a teetering or gimballed rotor, since the control power and damping of the rotor are directly proportional to the thrust. However, a hub spring can be used to increase the flap frequency by as much as can be achieved in articulated rotors, although in the teetering rotor a hub spring leads to large 2/rev loads as well. The lag motion of teetering and gimballed rotors is usually stiff in-plane motion with a natural frequency above 1/rev.

The hingeless rotor (also called a rigid rotor) has its blades attached to the rotor hub and shaft with cantilever root constraint. While the rotor has no flap or lag hinges, there often are hinges or bearings for the feathering motion. The fundamental flap and lag motion involves bending at the blade root. The structural stiffness is still small compared to the centrifugal stiffening of the blade, so the mode shape is not too different from the rigid body rotation of articulated blades and the flap frequency is not far above 1/rev (typically $\nu = 1.10$ to 1.20 for hingeless rotors). Depending on the structural design of the root, the blade may be either soft in-plane (lag frequency below 1/rev) or stiff in-plane (lag frequency above 1/rev). Without hinges, there can be considerable coupling of the flap, lag, and pitch motions of the blade, which leads to significantly different aeroelastic characteristics than with articulated blades. The hingeless rotor is capable of producing a large moment on the hub due to the tip-path-plane tilt; this moment has a significant influence on the helicopter handling qualities, including increased control power and damping, but also increased gust response. The hingeless rotor is a simple design mechanically, with therefore a potentially low maintenance requirement and low hub drag. A stronger hub and blade root are required to take the hub moments, however. There are rotor designs that eliminate the blade pitch bearings as well (these are sometimes called bearingless rotors). The pitch motion in such designs takes place about torsionally soft structure at the blade root.

Most rotor designs have a hinge or bearing at the blade root to allow the feathering or pitch motion of the blade for collective and cyclic control inputs. While it is the most common design solution, the pitch bearing operates under very adverse conditions. It is required to transmit the centrifugal and thrust loads of the blade while undergoing a periodic motion due to the rotor cyclic pitch control. Thus there have been other approaches to achieving blade pitch control. A hinge can be used instead of a bearing, or an elastomeric bearing can be used instead of a mechanical one, to simplify the mechanical design. Another approach is to allow the pitch motion to take place about torsional flexibility at the root, or tension-torsion straps between the blade and hub. Kaman developed a rotor that uses a servo-flap on the outboard portion of a torsionally flexible blade. Servo-flap deflection causes the blade to twist, which can be used for the collective and cyclic control of the rotor in place of root pitch.

7–2 Helicopter Types

The helicopter configuration primarily involves the number and orientation of the main rotors, the means for torque balance and yaw control, and the fuselage arrangement. The basic rotor analysis is applicable to all helicopter types, but the configuration of the helicopter does have an influence on its behavior, notably on its stability and control characteristics.

A single main rotor and tail rotor is the most common configuration. The tail rotor is a small auxiliary rotor used for torque balance and yaw control. It is mounted vertically on a tail boom, with the thrust acting to the right for a counter-clockwise-rotating main rotor. The moment arm of the tail rotor thrust about the main rotor shaft is usually slightly greater than the main rotor radius. Pitch and roll control of this configuration is achieved by tilting the main rotor thrust using cyclic pitch; height control is achieved by changing the main rotor thrust magnitude using collective pitch; and yaw control is achieved by changing the tail rotor thrust magnitude using collective pitch. This configuration is simple, requiring only a single set of main rotor controls and a single main transmission. The tail rotor gives good yaw control, but it absorbs power in balancing the torque, which increases the helicopter power requirement by several percent. The single main rotor configuration typically has only a small center-of-gravity range, although it is increased with a hingeless rotor. The tail rotor is also some hazard to ground personnel unless it is located very high on the tail, and it is possible for the tail rotor to strike the ground during operation of the helicopter. The tail rotor operates in an adverse aerodynamic environment (as do the fixed vertical and horizontal tail surfaces) due to the wake of the main rotor and fuselage, which reduces the aerodynamic efficiency and increases the tail rotor loads and vibration. The single main rotor and tail rotor configuration is the simplest and lightest for small- and medium-size helicopters.

Many antitorque devices to replace the tail rotor have been considered. A successful alternative must have satisfactory stability, control power, autorotation capability, weight, and power loss. The tail rotor has satisfactory characteristics in all these areas, excellent characteristics in some. Most candidate replacements are seriously deficient in at least one area. The most likely alternative to the tail rotor appears to be the ducted fan. The primary deficiencies of the tail rotor are its hazard to personnel, noise, and vibration. The ducted fan offers some improvements, particularly regarding personnel hazard. Some development problems remain to be solved before the ducted fan can replace the tail rotor, however.

With two (or more) contrarotating main rotors torque balance is inherent in the helicopter configuration, and no specific antitorque device with its own power loss is required. There are aerodynamic losses from the interference between the main rotors and between the rotors and fuselage; these losses reduce the overall efficiency of twin main rotor configurations to about the same level as for the single main and tail rotor configuration. The mechanical complexity is greater with twin main rotors because of the duplication of control systems and transmissions. For large machines, the resulting increase in weight and maintenance is offset by the fact that rotors of smaller diameter than a single main rotor can be used for a given gross weight and disk loading, thereby reducing the rotor and transmission weights.

The tandem rotor helicopter has two contrarotating main rotors with longitudinal separation. The main rotor disks are usually overlapped, typically by around 30% to 50% (the shaft separation is thus around 1.7R to 1.5R). To minimize the aerodynamic interference created by the operation of the rear rotor in the wake of the front, the rear rotor is elevated on a pylon, typically 0.3 to 0.5R above the front rotor. Longitudinal control is achieved by differential change of the main rotor thrust magnitude, from differential collective; roll control is by lateral thrust tilt with cyclic pitch; and height control is by main rotor collective. Yaw control is achieved by differential lateral tilt of the thrust on the two main rotors using differential cyclic pitch. A large fuselage is inherent in the design, being required to support the two rotors. The tandem helicopter also has a large longitudinal center-of-gravity range because of the use of differential thrust to balance the helicopter in pitch. The operation of the rear rotor in the wake of the front rotor is a significant source of vibration, oscillatory loads, noise, and power loss. The high pitch and roll inertia, unstable fuselage aerodynamic moments, and low yaw control power adversely affect the helicopter handling qualities. There is a structural weight penalty for the rear rotor pylon. Generally the tandem rotor configuration is suitable for medium and large helicopters.

The side-by-side configuration has two contrarotating main rotors with lateral separation. The rotors are mounted on the tips of wings or pylons, with usually no overlap (so the shaft separation is at least 2R). Control is as for the tandem helicopter configuration, but with the pitch and roll axes reversed. Roll control is achieved by differential collective pitch, and helicopter pitch control by longitudinal cyclic pitch. The structure to support the rotors is only a source of drag and weight, unless the aircraft has a high enough speed to benefit from the lift of a fixed wing.

The coaxial rotor helicopter has two contrarotating main rotors with concentric shafts. Some vertical separation of the rotor disks is required to accommodate lateral flapping. Pitch and roll control is achieved by main rotor cyclic, and height control by collective pitch, as in the single main rotor configuration. Yaw control is achieved by differential torque of the two rotors. The concentric configuration complicates the rotor controls and transmission, but the extensive cross-shafting of other twin rotor configurations is not required. Yaw control by differential torque is somewhat sluggish. This helicopter configuration is compact, having small diameter main rotors and requiring no tail rotor. The synchropter is a helicopter with two contrarotating main rotors with very small lateral separation. It is therefore nearly a coaxial design, but is simpler mechanically because of the separate shafts.

In most helicopter designs the power is delivered to the rotor by a mechanical drive, that is, through the rotor shaft torque. Such designs require a transmission and a means for balancing the main rotor torque. An alternative is to supply the power by a jet reaction drive of the rotor, using cold or hot air ejected out of the blade tips or trailing edges. For example, helicopters have been designed with ram jets on the blade tips, or with jet flaps on the blade trailing edges that use compressed air generated in the fuselage. Since there is no torque reaction between the helicopter and rotor (except for the small bearing friction), no transmission or antitorque device is required, resulting in a considerable weight saving. With a jet reaction drive, the propulsion system is potentially lighter and simpler, although the aerodynamic and thermal efficiency will be lower. The helicopter must still have a mechanism for yaw control. Fixed aerodynamic surfaces (a rudder) may be used, but at low speeds they are not very effective, depending on the forces generated by the rotor wake velocities.

7–3 Preliminary Design

Preliminary design is the process of defining the basic parameters of the helicopter to meet a given set of performance or mission specifications. Basically, the preliminary design analysis involves sizing the helicopter, rotor, and powerplant, and thus it can be formulated as an iteration on gross weight. Basic parameters such as rotor radius, tip speed, and solidity are selected on the basis of a current estimate of the helicopter gross weight; fundamental limits such as those on disk loading, Mach number, advance ratio, and blade loading are considered. Next, the powerplant is sized by a performance analysis that consists primarily of a calculation of the power required for the specified mission. Typically, the energy balance method is used for the performance analysis. The simplest method that will accurately do the task is desired, assuming it is consistent with the preliminary definition of the aircraft that is available. The basic sizing of the helicopter is then complete, and the general layout can be sketched. The component weights can be estimated now from the size of the rotor and powerplant and from the fuel and payload required for the mission. The component weights are summed to obtain the gross weight of the helicopter, and the procedure is repeated until the gross weight converges. Design optimization is based on an examination of mission cost parameters (such as direct operating cost, or even gross weight, which controls first cost) or various performance indices (such as range, maximum speed, or noise) as a function of the basic rotor and helicopter parameters. Even rotor type and helicopter type can be considered in the optimization process if the performance analysis and weight estimation are detailed enough to be able to distinguish between the types.

The major rotor parameters to be selected in the preliminary design stage are the disk loading, tip speed, and solidity. For a given gross weight, the disk loading determines the rotor radius. The disk loading is a major factor in determining the power required, particularly the induced power in hover. The disk loading also influences the rotor downwash and the autorotation descent rate. The rotor tip speed is selected largely as a compromise between the effects of stall and compressibility. A high tip speed increases the advancing-tip Mach number, leading to high profile power, blade loads, vibration, and noise. A low tip speed increases the angle of attack on the retreating blade until limiting profile power, control loads, and vibration due to stall are encountered. Thus there will be only a limited range of acceptable tip speeds, which becomes smaller as the helicopter velocity increases (see section 7–4). For a given rotor radius, the tip speed also determines the rotational speed. The rotational speed should be high for good autorotation characteristics and for low torque (and hence low transmission weight). The blade area or solidity is determined by the stall limitations on the rotor blade loading. The limits placed by stall on the blade operating lift coefficient, and therefore on C_T/σ, require a minimum value of $(\Omega R)^2 A_{\text{blade}}$ for a given gross weight. The rotor weight and profile power increase with blade chord, however, so the smallest blade area that maintains an adequate stall margin is used. Parameters such as blade twist and planform, number of blades, and airfoil section are chosen to optimize the aerodynamic performance of the rotor. The choice will be a compromise for the various operating conditions that must be considered. With appropriate representations of their influence on the helicopter weight and performance, these and other parameters can be included in the preliminary design process. However, there are many factors influencing the basic design features of the helicopter that do not appear directly in the preliminary design analysis. For example, the rotor type is determined more by its influence on the helicopter handling qualities, aeroelastic stability, and maintenance than by its influence on performance and weight. Such considerations must be included by the engineer in the optimization process.

A key element in the preliminary design of aircraft is the estimation of the weights of the various components of the vehicle from the basic parameters of the design. For a new aircraft that has not reached the detailed design stage, the component weight estimates can only be obtained by interpolating and extrapolating the trends observed in the weight data for existing vehicles. Preliminary design analyses generally use analytical expressions based on correlation of such weight data. The fundamental difficulty with such an approach is the reliability of the trends, particularly when it is necessary to extrapolate far beyond existing designs. If this limitation is kept in mind, the formulas expressing empirical weight trends may be successfully employed in preliminary design.

Component weight formulas are typically obtained by correlating weight data from existing designs as a straight line with some parameter κ on a log-log scale, which leads to expressions of the form $W = c_1 \kappa^{c_2}$ (where c_1 and c_2 are empirical constants). The parameter κ will be a function of those quantities that have a primary influence on the component weight. As an example, for the helicopter rotor weight, κ would depend on at least the rotor radius, tip speed, and blade area. Determining the form of the parameter κ requires a combination of analysis, empirical correlation, and guesswork. There is no unique correlation expression, or even a best one. Consequently there are numerous component weight formulas in use for preliminary design analyses.

Detail design eompletes the specification of the construction of all components of the helicopter. All the individual components are designed to perform their required tasks in accordance with the results of the preliminary design analysis. The major task is the structural analysis of all components, which requires a detailed specification of the aerodynamic and inertial loads and a complete calculation of the helicopter performance. This stage in the helicopter design thus brings to bear the best developed and most complex analyses available to the engineer.

7–4 Helicopter Speed Limitations

As for fixed wing aircraft, the maximum speed of a helicopter in level flight is limited by the power available, but with a rotary wing there are a number of other speed limitations as well, among them stall, compressibility, and aeroelastic stability effects. The primary limitation with many current designs is retreating blade stall, which at high speed produces an increase in the rotor and control system loads and helicopter vibration, severe enough to limit the flight speed. The result of these limitations is that the design cruise speed of the pure helicopter is generally between 150 and 200 knots with current technology. To achieve a higher cruise speed requires either an improvement in rotor and fuselage aerodynamics or a significant change in the helicopter configuration.

The absolute maximum level flight speed is the speed at which the power required equals the maximum power available. At high speed the principal power loss is the parasite power. To increase the power-limited speed requires an increase in the installed power of the helicopter or a reduction in the hub and body drag. Because the parasite power is proportional to V^3, a substantial change in drag or installed power is required to noticeably influence the helicopter speed. The rotor profile power also shows a sharp increase at some high speed as a result of stall and compressibility effects.

A measure of the compressibility effects on the rotor blade is the Mach number of the advancing tip,

$$M_{1,90} = M_{tip}(1 + \mu) = \frac{V + \Omega R}{c_s}$$

where c_s is the speed of sound and $M_{tip} = \Omega R/c_s$. The significance of compressibility effects on the rotor speed and power depends primarily on whether $M_{1,90}$ is above or below the critical Mach number for the angle of attack of the advancing tip. Compressibility increases the rotor profile power due to drag divergence above the critical Mach number, and the high transient forces on the blade increase the helicopter vibration and rotor loads. It is also possible to encounter dynamic stability problems (flapping or flap-pitch flutter) due to compressibility. A limit on the rotor Mach number that is increasingly important is the rotor noise level. Power and vibration effects do not appear until a significant portion of the rotor disk is above the critical Mach number, so usually a value of $M_{1,90}$ five to ten percent above the section critical Mach number can be tolerated. If rotor noise is considered, a substantially lower rotor speed may well be required. An alternative to reducing the rotor speed to avoid compressibility effects is to increase the critical Mach number, for example by using thin airfoil sections at the blade tip. Since the compressibility limitation on the advancing-tip Mach number basically provides a maximum value for $\Omega R + V$, the designer must compromise between the rotor speed and flight speed.

A measure of stall effects on the rotor is the ratio of the thrust coefficient to solidity, C_T/σ, which represents the mean lift coefficient of the blade. In hover, quite high values of C_T/σ can be achieved before the profile power increase due to stall is encountered. In forward flight, however, the angle of attack increases on the retreating side of the disk to maintain the same loading as on the advancing side (see section 5–6), so that stall is encountered at significantly lower values of C_T/σ. The rotor profile power increases when a substantial portion of the disk is stalled, and more importantly there is a sharp increase in the rotor loads and vibration, particularly in the control system, as a result of the high transient pitch moments on the periodically stalling blade. Stall of the helicopter rotor is discussed fully in Chapter 16. The stall-limited C_T/σ in forward flight decreases as either forward speed or propulsive force increases, since both increase the nonuniformity of the blade angle-of-attack distribution. Alternatively, for a given C_T/σ severe rotor stall effects are encountered at some critical advance ratio, which increases as the blade loading is reduced. Since the lowest acceptable C_T/σ is limited by the amount the blade area can be increased (based on the weight and performance penalties), the advance ratio restriction due to stall is an important helicopter design criterion.

The maximum advance ratio at which the helicopter may be operated depends on several factors. As μ increases, the aeroelastic stability of the blade motion decreases, the blade and control loads increase because of the asymmetry of the flow, and the aerodynamic efficiency and propulsive force capability of the rotor decrease. Retreating blade stall often constitutes the primary restriction on μ. For a specified maximum advance ratio $\mu = V/\Omega R$, the designer must increase the rotor tip speed to obtain a high forward speed of the helicopter. However, compressibility limits the possible tip speed and thus limits the helicopter speed.

Compressibility effects on the advancing blade and stall effects on the retreating blade combine to restrict the maximum forward speed of the helicopter rotor. The advancing-tip Mach number and advance ratio specify the sum and ratio of the tip speed and velocity:

$$M_{1,90} = (V + \Omega R)/c_s$$

$$\mu = V/\Omega R$$

Solving for V and ΩR gives

$$V = c_s M_{1,90} \frac{\mu}{1 + \mu}$$

$$\Omega R = c_s M_{1,90} \frac{1}{1 + \mu}$$

A high helicopter speed thus requires a high tip Mach number and a high advance ratio. This relationship is shown graphically on the rotor speed vs. velocity diagram (Fig. 7-1), which plots ΩR as a function of V for constant advancing-tip Mach number and advance ratio. From this diagram the maximum helicopter speed for given limits on $M_{1,90}$ and μ can be determined. For example, a critical Mach number of $M_{1,90} = 0.9$ and a maximum advance ratio of $\mu = 0.5$ produce a tip speed $\Omega R = 200$ m/sec and a maximum velocity $V = 200$ knots.

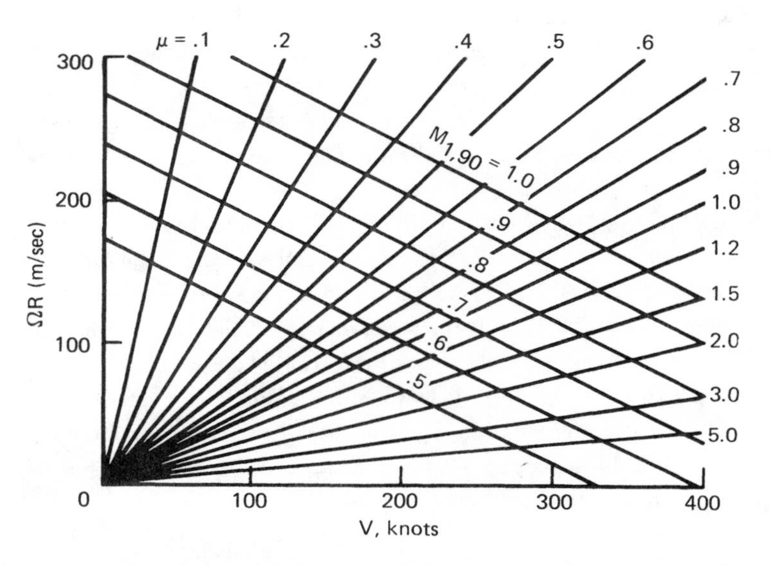

Figure 7-1 Rotor speed vs. velocity diagram (for speed of sound $c_s = 340$ m/sec).

There are many ideas for modifications to the basic helicopter configuration that are aimed principally at achieving higher speed in level flight. If a wing is added to the helicopter, its lift in forward flight will allow the rotor loading to be reduced, thus delaying stall effects. Since the rotor lift is also the source of the helicopter propulsive force, reducing the rotor loading to very low levels requires an auxiliary propulsion device as well. The result is the compound helicopter configuration. Unless a hingeless rotor is used, fixed aerodynamic control surfaces will also be required to maintain control with low rotor thrust. Avoiding compressibility limits at high speed will probably require that the rotor be slowed. The slowed and unloaded rotor might then be stopped completely and stowed, to minimize the aircraft drag at high speed. There are also suggestions for stopping the rotor and using it as a fixed wing in high-speed forward flight. An alternative approach is to tilt the rotors forward, so that they act as propellers in forward dlight. Then the many rotor problems due to the asymmetric aerodynamics of edgewise flight are eliminated. This is the tilting proprotor configuration. References dealing with rotary wing aircraft other than the pure helicopter configuration are given at the end of this chapter. None of these alternative configurations has yet achieved a success comparable to that of the helicopter, primarily because there has been no civil or military mission for which the higher speed is worth the penalties in performance, weight, and complexity.

7–5 Autorotational Landings after Power Failure

After loss of power due to engine failure, the helicopter has the capability of making an autorotation landing, in which the rotor lift is maintained while the aircraft descends at a steady rate. Because the equilibrium descent rate of the helicopter is fairly high, even in forward flight, autorotational descent is normally used only as an emergency procedure. Moreover, it is essential that the pilot take prompt and correct action to establish the optimum flight path both at the beginning and end of the maneuver.

After power failure, the rotor slows down as profile and induced losses absorb the rotor kinetic energy, which is the only power source available until the helicopter begins to descend. As the descent rate builds up, the inflow up through the rotor disk increases and therefore the blade angle of attack increases. Possibly the helicopter can then achieve an equilibrium descent rate, with the angle-of-attack increase countering the rotor speed loss to maintain the thrust equal to the gross weight. Stall places a limit on the angle of attack, however, and the rotor kinetic energy must be conserved for the end of the maneuver. If the rotor stalls, it will not be possible to establish equilibrium descent. Consequently, to keep the angle of attack in autorotation low and maintain the rotor speed, after a power failure it is necessary for the pilot to reduce the collective pitch. The transient lift capability of a rotor is higher than its static capability (see the discussion of dynamic stall in Chapter 16), which gives the pilot some additional time to react, but still the pilot must recognize the power loss and drop the collective within 2 or 3 seconds to prevent excessive rotor speed decay. The collective pitch required in autorotation is usually a small positive angle. On a single main rotor helicopter, the rotor torque loss will also require a pedal control change to reduce the tail rotor thrust. After the initial control actions, the pilot must establish equilibrium power-off descent at the minimum possible rate. The lowest autorotation descent rate is achieved in forward flight at the speed for the minimum power required in level flight (see section 6-3.4); the value is about one-half the descent rate in vertical autorotation. Consequently, the pilot must establish the proper vertical and forward speed after power failure and fly the helicopter to the ground. Near the ground the pilot must flare the helicopter, reducing the vertical and horizontal velocities for a gentle touchdown. Ideally, the helicopter has zero velocity just at the instant it contacts the ground. The flare maneuver requires that the collective be raised to increase the thrust and decelerate the helicopter, and that aft longitudinal cyclic be used to reduce the forward speed (producing a significant pitch-up motion as well). The power for the rotor during the flare maneuver is supplied by the rotational kinetic energy stored in the rotor. This is a limited power source, so the flare maneuver must be well timed by the pilot. Because the rotor slows down when the collective is increased, blade stall limits the flare capability of the helicopter. The total kinetic energy of the rotor is $KE = \frac{1}{2} N I_b \Omega^2$ (where $N I_b$ is the rotational moment of inertia of the entire rotor), but the fraction of the energy available before the rotor stalls and the thrust is lost is only $(1 - \Omega_f^2/\Omega_i^2)$. Here Ω_i and Ω_f are the rotor speeds at the beginning and end of the flare; assuming that the rotor thrust remains fairly constant,

$$\left(\frac{\Omega_f}{\Omega_i}\right)^2 = \frac{(C_T/\sigma)_i}{(C_T/\sigma)_f}$$

The rotor speed and C_T/σ at the beginning of the maneuver will be close to the normal operating values of the helicopter, and $(C_T/\sigma)_f$ is determined by the rotor stall limit (taking into account the lift overshoot possible in a transient maneuver).

If a power failure occurs when the helicopter is near the ground, it will not be possible to establish an equilibrium descent condition. Then the entire power-off landing is a transient maneuver, and the best flight path is somewhat different. If the power loss occurs in hover, the minimum contact velocity at the ground is achieved with a purely vertical flight path. Thus the pilot should not attempt to establish the forward velocity for lowest equilibrium descent rate, but only enough speed to avoid the vortex ring state and give a view of the landing spot.

The descent rate in autorotation is determined by the rotor disk loading, which should therefore be low. It follows that a low autorotation descent rate will be associated with low hover power. Helicopter flare capability is even more important for power-off landings than the steady-state descent rate, particularly since the choice of disk loading will be influenced primarily by performance considerations. The flare capability depends on the rotor kinetic energy, which requires a high rotor speed and a large blade moment of inertia. The stall margin should be high, both for good flare characteristics and for a minimal loss of rotor speed before the collective is reduced just after the power failure. Thus the helicopter operating C_T/σ should be low. The rotor inertia is the most effective parameter for improving helicopter autorotation characteristics. In dimensionless form, the relevant parameter is the blade Lock number, which should be low. A high inertia also implies heavy blades, however.

The helicopter must have a free-wheeling or over-riding clutch so that the engine can drive the rotor but not the other way around. Then upon a power failure, the engine automatically disengages from the rotor, and the rotor does not have the drag of the engine during autorotation. The tail rotor of a single main rotor helicopter must be geared directly to the main rotor, so that yaw control can be maintained in the event of power failure.

If the power loss occurs high above the ground, the pilot has ample time to establish equilibrium descent. The normal rotor speed can be recovered by a momentary increase in the descent rate, so that the flare can be initiated with the maximum possible stored energy in the rotor. If the power loss occurs near the ground, however, it will not be possible before the flare is started to make up for the rotor speed drop at the beginning of the maneuver, particularly when the pilot reaction time is accounted for. The result for most helicopters is that the flare can not be initiated with sufficient rotor energy and low enough descent rate to avoid an excessive vertical velocity at ground contact. Thus, on the helicopter height-velocity diagram (Fig. 7-2) there is a region at low speed in which the helicopter should not be operated, because a safe landing after power loss is not possible. The boundary of this region is called the deadman's curve. Above point A (typically 100 to 150 m in altitude) the rotor speed can be recovered sufficiently, and the descent rate kept low enough, to make a safe landing. With enough altitude, equilibrium autorotation descent can be established. For very low heights (point B in Fig. 7-2, typically 3 to 5 m above the ground), the ground will be reached before the helicopter has time to accelerate to an excessive velocity. With sufficient forward speed (point C, typically at 20 to 35 knots) a safe landing is again possible because of the reduction in

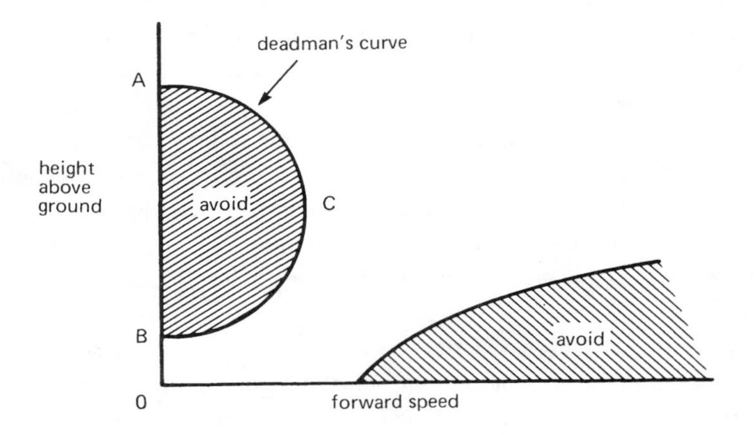

Figure 7-2 Helicopter height-velocity diagram

autorotative descent rate with forward flight. There is also usually a restriction on high speed flight near the ground, as shown in Fig. 7-2. If a power failure occurs at high speed and low altitude, there will not be time to reduce the horizontal velocity sufficiently to avoid damage to the landing gear, particularly for helicopters with skid-type gear. The two forbidden regions on the height-velocity diagram combine to constrain the helicopter takeoff and landing to a specific corridor. The limit on the operational use of the helicopter is not particularly restrictive, however. A purely vertical takeoff or landing is not usually made because of the deadman's region; rather, after a vertical climb to about 5 m altitude the pilot begins to accelerate the helicopter forward. With two or more engines, the deadman's region of the helicopter disappears, or at least becomes much smaller. The concern with multi-engine helicopters is more with the single-engine-out performance capability than with the consequences of a complete power failure.

Consider an analysis of the initial rate of descent and rotor speed decay following power failure but before the pilot reacts to the situation, so that the collective pitch is unchanged. This derivation is based on the work of McCormick (1956), but see also Katzenberger and Rich (1956), McIntyre (1970), and Wagner (1973). The equation of motion for the vertical acceleration of the helicopter is $M\ddot{h} = W - T$, where h is the helicopter height above the ground, W the gross weight, T the rotor thrust, and $M = W/g$ the helicopter mass. The equation of motion for the rotor speed is $NI_b\dot{\Omega} = -Q$, where NI_b is the total rotor moment of inertia and Q is the decelerating torque on the rotor. Before the power failure (at time $t = 0$), the rotor thrust equals the gross weight and the rotor speed is constant. After the power failure, the engine torque no longer balances the rotor decelerating torque, so the rotor slows down; Q then is just due to the rotor power requirement. Since it is assumed that the collective is unchanged from the hover value, and at least initially the descent rate has not built up sufficiently to change the inflow ratio, the rotor thrust and torque coefficients (C_T and C_Q, which are functions of $\theta_{0.75}$ and λ) must remain fixed at the same values as at the instant of power failure. The rotor thrust and torque are then changed only by variations in the rotor speed: $T = W(\Omega/\Omega_0)^2$ and $Q = Q_0(\Omega/\Omega_0)^2$. Here Ω_0 is the initial rotor speed and Q_0 the rotor torque required in level flight, so $P = \Omega_0 Q_0$ is the helicopter power required for level flight. The equation of motion of the rotor speed for $t > 0$ then becomes $NI_b\dot{\Omega} = -Q_0(\Omega/\Omega_0)^2$, which integrates to

$$\frac{\Omega}{\Omega_0} = \left(1 + \frac{tQ_0}{NI_b\Omega_0}\right)^{-1}$$

and the helicopter descent velocity has the solution

$$\dot{h} = gt^2\frac{Q_0}{NI_b\Omega_0}\left(1 + \frac{tQ_0}{NI_b\Omega_0}\right)^{-1}$$

These results may be written as

$$\frac{\Omega}{\Omega_0} = \frac{\tau}{t+\tau}$$

$$\dot{h} = \frac{gt^2}{t+\tau}$$

where the time constant is

$$\tau = \frac{2KE}{P} = \frac{NI_b\Omega_0}{Q_0}$$

P is the rotor power required for level flight, and $KE = \frac{1}{2}NI_b\Omega_0^{\,2}$ is the kinetic energy stored in the rotor. McCormick found that these expressions describe the helicopter behavior fairly well for the first few seconds after power failure.

The flare is a far more important part of the power-off landing, but the above analysis is useful because it introduces the parameter $\tau = 2KE/P$ as a measure of helicopter autorotation characteristics. A small decay of the rotor speed requires a large value of τ, hence a high rotor kinetic energy, and a low required power. The helicopter power enters as a measure of the torque acting to decelerate the rotor after engine power is lost. Typically, $KE/P \cong 4$ seconds, so the time for a significant decay of the rotor speed is around 1 to 2 seconds. The largest permissible reaction time can be estimated by setting the rotor speed decrease equal to the stall limit:

$$\left(\frac{\Omega}{\Omega_0}\right)^2 = \frac{C_T/\sigma}{(C_T/\sigma)_{stall}}$$

from which

$$t_{max} = 2\frac{KE}{P}\left[\left(\frac{(C_T/\sigma)_{stall}}{C_T/\sigma}\right)^{1/2} - 1\right]$$

Wood (1976) summarizes a number of autorotation performance indices: a time constant for the rotor speed decay, $t = (KE/P)(1 - T/0.8T_{max})$; the usable kinetic energy, $E = (KE/T)(1 - T/T_{max})$; an autorotation index, $AI = KE/P$; and an energy factor, $h = KE/T$. Here P is the installed power of the helicopter, T is the rotor thrust, T_{max} is the stall-limited thrust, and $KE = \frac{1}{2} N I_b \Omega^2$ is the rotor kinetic energy. These parameters are concerned with the overall autorotation characteristics of the helicopter, such as those represented by the deadman's curve. Wood discusses their origins and presents a correlation with autorotation characteristics.

The literature on helicopter power-off landing and rotor autorotation includes: Toussaint (1920), Wimperis (1926), Bennett (1932), Peck (1934), Gessow and Myers (1947), Gessow (1948d), Nikolsky and Seckel (1949a, 1949b), Slaymaker, Lynn, and Gray (1952), Slaymaker and Gray (1953), Katzenberger and Rich (1956), McCormick (1956), Jepson (1962), Davis, Kannon, Leone, and McCafferty (1965), Cooper, Hansen, and Kaplita (1966), Hansen (1966), Pegg (1968, 1969), Shapley, Kyker, and Ferrell (1970), McIntyre (1970), Wagner (1973), Wood (1976), Johnson (1977c), Benson, Bumstead, and Hutto (1978), Talbot and Schroers (1978), Young (1978). See also the references in section 3-2.

7–6 Helicopter Drag

The estimation of helicopter parasite drag is an important aspect of performance calculation because it establishes the propulsive force and power requirement at high speed. The helicopter drag is commonly expressed in terms of the parasite drag area f; specifically, $D = \frac{1}{2}\rho V^2 f$. Except for compressibility or Reynolds number effects, f will be independent of speed. The parasite drag area can be calculated from the drag coefficients of the various components of the airframe by

$$f = \sum_i C_{D_i} S_i$$

where S_i is the component wetted area or frontal area, on which C_{D_i} is based. A major contributor to the helicopter drag is the rotor hub, which typically accounts for 25% to 50% of the total parasite drag area. The drag of even a clean helicopter is significantly greater than that of an airplane of similar gross weight, partly because of the large rotor hub drag and partly because of higher fuselage drag. Early helicopter designs in particular tended to have high drag levels.

For a rough estimate of the helicopter drag, the parasite area can be correlated with rotor area for existing designs. It is found that $f/A \cong 0.025$ for old designs, $f/A \cong 0.010$ to 0.015 for helicopters in current production, and $f/A \cong 0.004$ to 0.008 for clean helicopter designs. The rotor hub accounts for a major fraction of the drag; $f_{\text{hub}}/A \cong 0.0025$ to 0.0050 for current rotor designs, and $f_{\text{hub}}/A \cong 0.0015$ for a very clean, faired hub. The parasite drag area is also often correlated with the helicopter gross weight, usually by expressions of the form $f/W^{2/3}$ = constant. For the purpose of estimating the drag, the results are equivalent to a square-cube scaling of the rotor area with gross weight; specifically, $A \cong 0.6W^{2/3}$ when the rotor area is in m^2 and the gross weight is in kg.*

*This relation between A and W is only approximately correct, but it is sufficiently accurate to correlate the helicopter drag with gross weight instead of rotor area. Note that the square-cube scaling law implies that the rotor disk loading tends to increase with helicopter size, which is true. For rotor areas in ft^2 and gross weights in lb, $A \cong 4W^{2/3}$.

7–7 Rotor Blade Airfoil Selection

The airfoil for a helicopter rotor blade is chosen to give the rotor good aerodynamic efficiency while allowing the structural requirements of the blade to be satisfied. The task of selecting an airfoil section, and even more so the task of designing airfoils specifically for rotor blades, are difficult because of the complex aerodynamic environment in which the rotary wing operates. The airfoil used is inevitably a compromise, balancing the many varied constraints imposed by the rotor flow.

The figure of merit is a useful measure of the aerodynamic efficiency of the hovering rotor. Recall from section 2-6.4 that the figure of merit can be written

$$M = \frac{1}{\kappa + \frac{3}{4}\frac{c_{d_0}/\bar{c}_l}{\lambda_h}}$$

Thus for a fixed disk loading it is essentially a measure of the ratio of the blade profile drag to lift. A high figure of merit requires that the airfoil section have a low drag for moderate to high lift coefficients.

Good stall characteristics are important for any wing, including the rotor blade. The rotor airfoil should have a high maximum lift coefficient, which allows the rotor to be designed to operate at a high C_T/σ value and hence have a low tip speed and blade area. The strictest limitation imposed by stall is on the retreating blade in forward flight; a high lift coefficient is therefore required at low to moderate Mach numbers. In forward flight, stall occurs periodically as the blade rotates, so really the airfoil must have good unsteady stall characteristics (see Chapter 16). Generally, though, it has been found that good static stall characteristics imply good dynamic stall characteristics, so the airfoil selection can reasonably be based on static data if unsteady measurements are not available.

At high forward speeds, the advancing-tip Mach number is high. The rotor airfoil should therefore have a high critical Mach number for drag divergence and shock formation at the low angle of attack characteristic of the advancing side of the disk.

Aerodynamic pitch moments on the blade are transmitted to the control system. A low moment about the aerodynamic center is required of the blade airfoil if excessive control system loads are to be avoided, particularly in forward flight, where there is a large periodic variation in angle of attack and dynamic pressure. If the control system is entirely mechanical, the aerodynamic pitch moments on the blade will also be transmitted to the pilot's cyclic and collective control sticks.

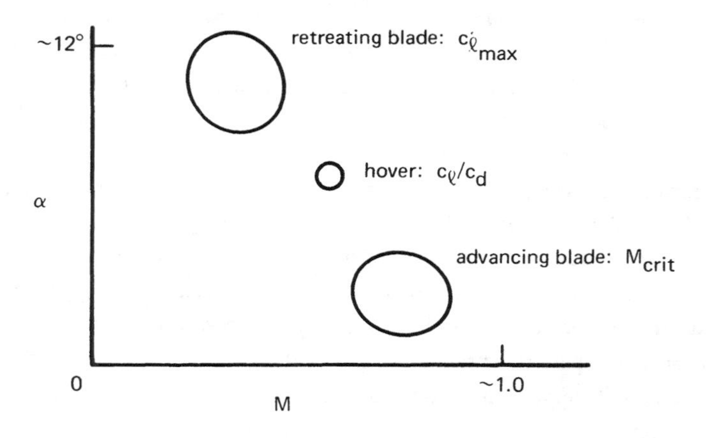

Figure 7-3 Rotor blade airfoil criteria.

Figure 7-3 illustrates the basic concerns in selecting or designing an airfoil for a helicopter rotor blade. The rotor blade section operates over a wide range of conditions. Low drag is required at the working conditions of the rotor in hover, namely moderately high angles of attack and Mach number. Good stall characteristics, including a high maximum lift coefficient, are required at the low to moderate Mach numbers of the retreating blade in forward flight. Finally, a high critical Mach number is required at the low angle of attack of the advancing blade in forward flight. The hover criterion is intended to give good lifting performance by the rotor, while the forward flight requirements are primarily based on achieving low vibration and loads at high speed. In addition, the airfoil should have a small pitching moment. A symmetrical, moderately thick airfoil section has frequently been the choice for rotor blades, with the same section over the entire span for simplicity of construction. The symmetrical section assures a zero pitching moment. The thickness ratio (typically 10% to 15%)

is a compromise between the thin section desired because of compressibility effects and the thick section desired for structural efficiency. Fortunately, extremely thick sections are required only at the blade root, where high aerodynamic efficiency is not required anyway. The NACA 0012 airfoil was a frequent selection for past rotor designs and has come to be considered the standard rotor airfoil. With improved aerodynamic, structural, and manufacturing technology more sophisticated blade designs are being used for current helicopters. A number of airfoils have been developed with characteristics optimized for the rotary wing environment, and it is fairly common to use thinner sections at the blade tip.

As a guide in the evaluation and selection of a rotor blade airfoil, both the section operating conditions and the airfoil characteristics may be plotted as a function of angle of attack and Mach number (Fig. 7-4). The airfoil characteristics shown as a function of Mach number for a hypothetical airfoil are the angles of attack for maximum lift coefficient (α_{max}) and for drag divergence or supercritical flow (α_{crit}). Also plotted is the operating condition for a particular radial station as the blade moves around the azimuth (a closed curve is generated in forward flight, converging to a single angle of attack and Mach number for hover). The tip sections of the blade

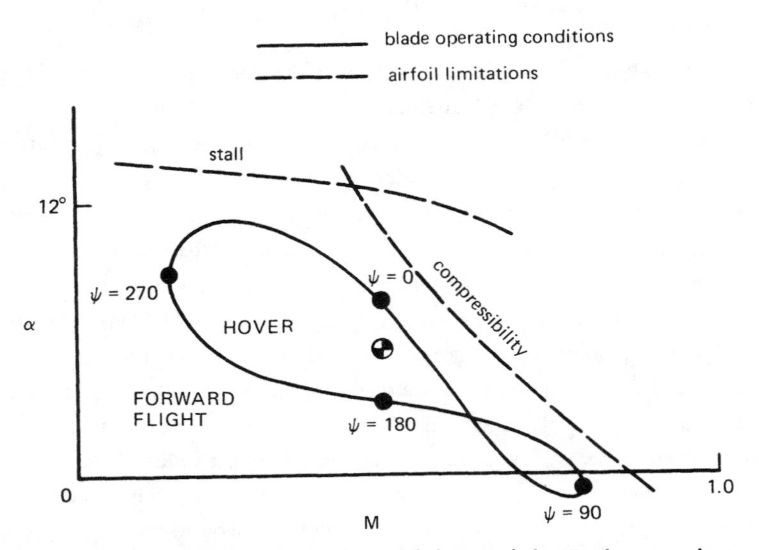

Figure 7-4 Rotor airfoil requirements and characteristics: section operating conditions as a function of azimuth angle and airfoil limitations.

will show the highest Mach numbers, while sections somewhat inboard (such as at 75% radius) will show the largest angle of attack. Thus the requirements dictated by the aerodynamic environment of the blade vary with the radial station. The requirements for a given rotor operating state can be compared with the stall and compressibility characteristics of a particular airfoil by a plot such as Fig. 7-4. This plot may also be used to graphically compare the characteristics of different airfoil sections; an improved airfoil should show increased angle-of-attack limits over the entire Mach number range.

Gustafson (1944) examined the influence of the airfoil section on rotor performance. As a measure of the section drag influence, he considered the profile power, which may be viewed as a weighted average of the drag coefficient c_d over the rotor disk. The profile power can be written equivalently as an integral over the angle of attack:

$$\frac{C_{P_o}}{\sigma} = \int f(\alpha) c_d(\alpha) d\alpha$$

Since the angle-of-attack distribution will not be changed much by using different airfoils, $f(\alpha)$ is a function that depends only on the operating state of the rotor and defines the relative contributions of the drag at various angles of attack to the profile power. Different airfoil sections may then be compared by plotting fc_d as a function of α.

Gustafson (1949) presents a bibliography and discussion of the literature on airfoil section characteristics and their application in helicopter rotor airfoil selection.

Davenport and Front (1966) give a brief historical review of the development of airfoil sections and their application to helicopter rotors. They define the objectives of improved rotor airfoils as reduced profile power and a postponement of the control load and vibration rise at high speed. The airfoil requirements that follow from these objectives are low drag at high and intermediate Mach number, high lift capability at moderate Mach numbers ($M = 0.3$ to 0.5), and low moment about the aerodynamic center under all conditions. They summarize the effects of airfoil thickness, leading edge radius, and camber on the airfoil characteristics important for rotor blades, and conclude that a thin or moderately thick (9% to 12%) section with a blunt leading edge and little leading edge camber should give the best characteristics. On the basis of these considerations airfoil sections were developed which produced modest improvements in rotor performance.

Benson, Dadone, Gormont, and Kohler (1973) describe the aerodynamic properties of several transonic airfoil sections designed specifically for the complex aerodynamic environment of rotor blades, with particular attention paid to the stall characteristics. They describe in detail the constraints imposed on the airfoils by rotor performance, noise, and loads considerations. It was found that the stall flutter boundary (see Chapter 16) correlated well with $c_{l_{max}}$ at $M = 0.40$, so they concluded that it is sufficient to consider the static stall characteristics. Airfoils which satisfy all the criteria well are not common, but they did find that airfoils can be developed which are clearly superior to the classical sections such as the NACA profiles.

Dadone and Fukushima (1975) discuss the airfoil requirements for rotor blades, specifying detailed objectives for advanced airfoil designs. Experience indicates that although the rotary wing aerodynamic environment is highly three dimensional and unsteady, significant improvements in rotor performance and loads can be achieved by considering the two-dimensional, static airfoil characteristics. In general, it is found that the stall and compressibility requirements (high maximum lift coefficient at moderate Mach numbers, and high critical Mach number at low lift) require a compromise. The best approach is to use different airfoils at the tip (where compressibility effects dominate) and at midspan (where stall effects dominate), rather than using a single airfoil. Dadone and Fukushima compare the aerodynamic characteristics of a number of airfoils developed for helicopter rotors, both standard sections and recent designs. The recent airfoils show definite improvements, particularly in maximum lift coefficients at Mach numbers around 0.6, and in the drag below the critical Mach number. Further improvements required include increased critical Mach number, increased maximum lift coefficient at low Mach number, and reduced pitch moments.

Additional literature on airfoil design and selection for rotor blades: Wheatley (1934b), Razak (1944), Lipson (1946), Gustafson (1948), Stewart (1948), Schaefer, Loftin, and Horton (1949), Powell (1954), Critzos, Heyson, and Boswinkle (1955), Spivey (1968), Wilby, Gregory, and Quincey (1969), Wortmann and Drees (1969), Spivey and Morehouse (1970), Pearcey, Wilby, Riley, and Brotherhood (1972), Reichert and Wagner (1972), Kemp (1973), Wilby (1973), Scarpati, Sandford, and Powell (1974), Bingham (1975), Brotherhood (1975b), Paglino and Clark (1975), Prouty (1975),

Dadone (1976, 1977, 1978), Noonan and Bingham (1977), Thibert and Gallot (1977), Morris and Yeager (1978). See also Chapter 16 for further discussion of rotor stall.

7–8 Rotor Blade Profile Drag

The calculation of rotor performance requires a knowledge of the blade section profile drag coefficient, preferably including its dependence on angle of attack and Mach number. There are other factors that influence the drag coefficient in the three-dimensional, unsteady aerodynamic environment of the rotor blade in forward flight. In particular, it may be necessary to account for the radial flow, the time-varying angle of attack, and three-dimensional flow effects at the tip. Roughness and blade construction quality also influence the section drag, often increasing the drag coefficient by 20% to 50% compared to its value for smooth, ideally shaped airfoils. The general practice in numerical work is to rely on tabular data for c_l, c_d, and c_m as a function of α and M for the particular profile used, with semi-empirical corrections to account for the other factors that are considered important. Often it is difficult to obtain a complete and reliable set of even static, two-dimensional airfoil data, however. The measured aerodynamic characteristics can be sensitive to small variations in the airfoil or test facility, leading to different properties for airfoils that are nominally identical.

At the other extreme, the rotor analysis can use a mean profile drag coefficient to represent the overall effects of the blade drag on the rotor. The mean drag coefficient can be evaluated using the mean lift coefficient of the rotor, and the Mach number and Reynolds number at some representative radial station (say 75% radius). The use of a mean drag coefficient greatly simplifies the analysis and has been frequently used in the previous chapters to obtain elementary expressions for the rotor profile losses. Such an analysis is sufficiently accurate for some purposes, such as preliminary design, or when detailed aerodynamic characteristics for the blade section are not available. A mean drag coefficient is not appropriate when localized aerodynamic phenomena are important, such as stall and compressibility effects in forward flight. Additional corrections or a more detailed analysis is thus required for rotors at extreme operating conditions.

Frequently, helicopter performance calculations use a drag polar of the form $c_d = \delta_0 + \delta_1\alpha + \delta_2\alpha^2$ (see section 5-24). This is a better representation than a mean value, but it is simple enough to allow an analytical treatment if desired. The constants δ_0, δ_1, and δ_2 depend on the airfoil section. Hoerner (1965) suggests the following procedure for estimating the profile drag polar. A basic skin friction coefficient is obtained for the appropriate section Reynolds number. As an example, for a turbulent boundary layer in the Reynolds number range $10^6 < Re < 10^8$. Hoerner suggests

$$c_f = 0.44\, Re^{-1/6}$$

The minimum profile drag coefficient for the airfoil is then twice c_f, multiplied by a factor accounting for the airfoil thickness. For the NACA 4- or 5-digit airfoil series the result is

$$c_{d_{min}} = 2c_f[1 + 2(t/c) + 60(t/c)^4]$$

where t/c is the section thickness ratio. The term $2(t/c)$ accounts for the velocity increase due to thickness, and the term $60(t/c)^4$ is due to the pressure drag. Hoerner then gives the effect of lift on the profile drag as

$$c_d \cong c_{d_{min}}(1 + c_\ell{}^2)$$

which completes the construction of the drag polar.

Bailey (1941) developed a procedure for identifying the constants in the drag polar $c_d = \delta_0 + \delta_1\alpha + \delta_2\alpha^2$, given the basic section characteristics; see also Bailey and Gustafson (1944). Using this method, the polar $c_d = 0.0087 - 0.0216\alpha + 0.400\alpha^2$ was obtained for a NACA 23012 airfoil at $Re = 2 \times 10^6$. This particular result is quoted and used so often in the helicopter literature that the method by which it was obtained deserves some attention. Bailey started with the result that the profile drag can be written as $c_d = c_{d_{min}} + \Delta c_d$, where the minimum drag depends on the Reynolds number and Δc_d depends on the angle of attack. It is found that for all profiles Δc_d is approximately a unique function of

$$\ell = \frac{c_\ell - c_{\ell opt}}{c_{\ell max} - c_{\ell opt}}$$

where $c_{\ell max}$ is the maximum lift coefficient of the airfoil and $c_{\ell opt}$ is the lift coefficient at minimum drag (at the appropriate Reynolds number). Bailey wrote the profile drag function as $c_d = K_0 + K_1\ell + K_2\ell^2$ and determined the constants $K_0 = 0.0003$, $K_1 = -0.0025$, and $K_2 = 0.0229$ by matching the function to the empirical curve at $\ell = 0.125$, 0.4, and 0.675. This expression is a good approximation to about $\ell = 0.8$. At higher lift the stall effects are large and therefore the drag is significantly underestimated by this expression. Using $c_\ell = a\alpha$, the constants in the drag polar $c_d = \delta_0 + \delta_1\alpha + \delta_2\alpha^2$ may then be evaluated:

$$\delta_0 = c_{d_{min}} + K_0 - \frac{K_1 c_{\ell opt}}{c_{\ell max} - c_{\ell opt}} + \frac{K_2 c_{\ell opt}^2}{(c_{\ell max} - c_{\ell opt})^2}$$

$$\delta_1 = \frac{aK_1}{c_{\ell max} - c_{\ell opt}} - \frac{2aK_2 c_{\ell opt}}{(c_{\ell max} - c_{\ell opt})^2}$$

$$\delta_2 = \frac{a^2 K_2}{(c_{\ell max} - c_{\ell opt})^2}$$

Thus, given $c_{d_{min}}$, $c_{\ell max}$, $c_{\ell opt}$, and $c_{\ell\alpha} = a$ at the required Reynolds number, the profile drag polar can be constructed. Note however that Bailey's expression for Δc_d does not reduce to zero at $c_{\ell_{opt}}$, where $\Delta c_d = 0.0003$ instead; in fact, the minimum occurs at $(c_\ell - c_{\ell_{opt}})/(c_{\ell_{max}} - c_{\ell_{opt}}) = 0.055$, where $\Delta c_d = 0.0002$. An alternative choice of constants that is nearly as accurate and gives a minimum $\Delta c_d = 0$ at $c_\ell = c_{\ell_{opt}}$ is $K_0 = K_1 = 0$ and $K_2 = 0.0200$. Bailey's expression is more accurate in the working range for the blade angle of attack, however. As an example, Bailey considers the NACA 23012 airfoil at $Re = 2 \times 10^6$. For this airfoil, $c_{\ell_{max}} = 1.45$, $c_{d_{min}} = 0.0066$, $c_{\ell_{opt}} = 0.08$, and $a = 5.73$. The minimum drag was increased by 25% to $c_{d_{min}} = 0.0082$ to account for roughness. The resulting polar is $c_d = 0.0087 - 0.0216\alpha + 0.400\alpha^2$. As another example, consider the NACA 0012 airfoil at $Re = 2 \times 10^6$. From $c_{\ell_{max}} = 1.40$, $c_{\ell_{opt}} = 0$, $a = 5.73$, and $c_{d_{min}} = 0.0065$ (increased to $c_{d_{min}} = 0.0081$ for roughness), one obtains $c_d = 0.0084 - 0.0102\alpha + 0.384\alpha^2$. The limit on the validity of these expressions is $\ell < 0.8$, or

$$\alpha < \alpha_{\text{limit}} = \frac{0.8c_{\ell max} + 0.2c_{\ell opt}}{a}$$

This limit is the drag rise due to stall at high angle of attack. For the NACA 23012 airfoil the limit is $\alpha < 11.8°$, and for the NACA 0012 airfoil, $\alpha < 11.2°$.

7–9 Literature

On helicopter design, development, and construction: Focke (1938), Prewitt (1942), Taylor (1942), Gustafson (1945b), Stewart (1948, 1962b), Maloy (1949), Hafner (1954), Douglas (1954, 1958), Lightfoot (1958, 1966), Kee (1959), Sibley and Jones (1959), Cresap (1960, 1962), Sikorsky (1960), Dingeldein (1961), Donovan and Leoni (1963), Gillmore and Schneider (1963), Statler, Heppe, and Cruz (1963), Campbell (1964), Fradenburgh and Kiely (1964), Lockheed-California Company (1964a, 1964b), Wachs and Rabbott (1964), Bossler (1965), Carter (1965), Foulke (1965), Hiller Aircraft Company (1965), Bilezikjian et al. (1966), Gallant, Scully, and Lange (1966), Mil'(1966), Wax and Tocci (1966), Dutton (1967), Simpson (1967), Brown and Fischer (1968), Ellis, Acurio, and Schneider (1968), Fagan (1968, 1971), Seibel (1968), Wood (1968), Cardinale (1969), Lightfoot and Immenschuh (1969), Twelvetrees (1969), Weiland (1969), Yatsunovich (1969, 1970), Jones (1970, 1973), Schneider (1970), Duke and Hooper (1971), Eastman (1971), Johnston and Cook (1971), Rao (1971), Scully and Faulkner (1971a, 1971b), Sonneborn (1971), Albrecht (1972), Bazov (1972), Dumond and Simon (1972), Hunt (1972), Lewis (1972), McCutcheon (1972), Tencer and Cosgrove (1972), Davis and Wisniewski (1973), Huber (1973a), McCall, Field, and Reddick (1973), Wagner (1973), Austin, Rogers, and Smith (1974), Berrington (1974), Foster, Kidwell, and Wells (1974), Gorenberg and Harvick (1974), Hansen, McKeown, and Gerdes (1974), Metzger, Plaks, Meier, and Berman (1974), Scarpati, Sandford, and Powell (1974), Asher, Donelson, and Higgins (1975), Dajani, Warner, Epstein, and O'Brien (1975), Grina (1975), Kefford and Munch (1975), Magee, Clark, and Widdison (1975), Mouille (1975), Sonneborn and Drees (1975), Teleki (1975), Unger (1975), Arcidiacono and Zincone (1976), Hoffstedt and Swatton (1976), Maloney and Porterfield (1976), Nazarov (1976), Porterfield and Clark (1976), Rich (1976), Thompson, Kulkerni, and Lee (1976), Wood (1976), Blackwell (1977), Boeing Vertol Company (1977a, 1977b), Culhane (1977), Fenaughty and Noehren (1977), Gormont and Wolfe (1977), Harris, Cancro, and Dixon (1977), Mack (1977), Niven, Sanders, and McManus (1977), Stratton, Scarpati, and Feenan (1977), Blackwell and Merkley (1978), Cresap, Myers, and Viswanathan (1978), Fradenburgh (1978), Jonda and Frommlet (1978), Niebanck and Girvan (1978), Richardson and

Alwang (1978), Shaw and Edwards (1978), Stepniewski (1978), Twomey and Ham (1978).

On helicopter operating conditions, particularly load factors: Crim and Hazen (1952), Gustafson and Crim (1953), Hazen (1955), Connor (1960), Connor and Ludi (1960), Barun and Giessler (1966), Clay, Braun, Chestnutt, and Bartek (1966), Di Carlo (1967, 1975), Porterfield and Maloney (1968), Porterfield and Alexander (1970), Gissler, Clay, and Nash (1971), Herskovitz and Steinmann (1973), Spreuer, Snackenburg, and Roeck (1974), Cox, Johnson, and Russell (1975).

On the tail rotor and other antitorque devices: Sowyrda (1968), Huston and Morris (1970, 1971), Lynn et al. (1970), Mouille (1970), Robinson (1970), Tung, Erickson, and Du Waldt (1970), Grumm and Herrick (1971), Lehman (1971), Velazquez (1971), Davidson, Harvey, and Sharrieb (1972), Akeley and Carson (1974), Empey and Ormiston (1974), Wiesner and Kohler (1974a, 1974b), Yeager, Young, and Mantay (1974), Clark (1975), Meier, Groth, Clark, and Verzella (1975), Potash (1975), Raitch (1975), Simon and Savage (1976), Mineck (1977a), Logan (1978).

On the compound helicopter configuration: Michel (1960), Fradenburgh (1961), Harris (1961), Kuhn (1961), Fradenburgh and Rabbott (1962), Cresap and Lynn (1963), Drees (1963), Robinson (1963), Van Wyckhouse and Cresap (1963-1964), Stepniewski and Schneider (1964), Tanner and Bergquist (1964), Wachs and Rabbott (1964), Foulke (1965), Fradenburgh and Segel (1965), Klingloff, Sardanowsky, and Baker (1965), Wyrick (1965), Lynn (1966), Segel and Bain (1966), Van Wyckhouse (1966), Bain and Landgrebe (1967), Lynn and Drees (1967), Blackburn (1968), Blackburn and Rita (1968), Brown and Fischer (1968), Fradenburgh and Chuga (1968), Hohenemser (1968), Lentine, Groth, and Oglesby (1968), Ludi (1968), Meyers, Tompkins, and Goldberg (1968), Spreuer (1968), Cruz, Gorenberg, and Kerr (1969), Jenkins and Deal (1970), Putman and Traybar (1971), Sonneborn (1971), Sonneborn and Hartwig (1971), Dumond and Simon (1973), Condon (1974), Kefford and Munch (1975), McHugh and Harris (1976), Torres (1976), Huston, Jenkins, and Shipley (1977).

On the tilting proprotor aircraft: Lichten (1949, 1959), Koenig, Greif, and Kelly (1959), Deckert and Ferry (1960), Quigley and Koenig (1961), Houbolt and Reed (1962), Liu (1962), Hafner (1964, 1971), Hall (1966), Reed (1966, 1967), Young and Lytwyn (1967), Brown and Fischer (1968),

Edenborough (1968), Wernicke (1968, 1969), Bell Helicopter Company (1969, 1971, 1972a, 1972b, 1972c, 1973), De Larm (1969), Gaffey, Yen, and Kvaternik (1969), Magee, Maisel, and Davenport (1969), De Tore and Gaffey (1970), Johnston and Kefford (1970), Magee and Pruyn (1970), Richardson, Liiva, et al. (1970), Pruyn and Taylor (1971), Fry (1971), Richardson (1971), Tiller and Nicholson (1971), Baird, Bauer, and Kohn (1972), Boeing Vertol Company (1972a, 1972b), Cook and Poisson-Quinton (1972), De Tore and Brown (1972), Edenborough, Gaffey, and Weiberg (1972), Gupta and Bryson (1972), Johnston (1972), Marr and Neal (1972), Sambell (1972), Wernicke and Edenborough (1972), Alexander, Eason, Gillmore, Morris, and Spittle (1973), Harendra, Joglekar, Gaffey, and Marr (1973), Kaza (1973, 1974), Kvaternik (1973, 1974), McHugh, Eason, Alexander, and Mutter (1973), Magee and Alexander (1973, 1976), Marr, Sambell, and Neal (1973), Matthys, Joglekar, and Ksieh (1973), Rosenstein, McVeigh, and Molienkof (1973), Soule (1973a, 1973b), Soule and Clark (1973), Alexander, Hengen, and Weiberg (1974), Faulkner (1974b, 1974c), Frick and Johnson (1974), Gaffey and Maisel (1974), Johnson (1974a, 1974b, 1974d, 1975a, 1975b, 1976a, 1977a, 1977b), Magee, Clark, and Alexander (1974), Marr, Ford, and Ferguson (1974), Marr and Roderick (1974), Widdison, Magee, and Alexander (1974), Yasue (1974), Detore and Sambell (1975), Kiessling (1975), Magee, Clark, and Widdison (1975), Ungar (1975), Faulkner and Swan (1976), McVeigh (1976, 1977), McVeigh and Widdison (1976), Marr (1976), Marr, Willis, and Churchill (1976), Radford et al. (1976), Wilson, Mineck, and Freeman (1976), Amos and Alexander (1977), Kvaternik and Kohn (1977), Soohoo, Morino, Noll, and Ham (1977), Hofman et al. (1978), Shovlin and Gambucci (1978).

On other rotary wing configurations, including stopped rotor designs and jet reaction rotors: Miller (1946), Brightwell, Peters, and Sanders (1948), Barnaby, Berkowitz, and Colcord (1949), Gessow (1950), Hickey (1956), Hohenemser (1957a), Marks (1957, 1960), Hohenemser and Perisho (1958), MacNeal (1958), Bolanovich and Marks (1959), Doblhoff (1959), Dorand and Boehler (1959), Sissingh (1962), Young (1962b), Pruyn and Swales (1964), Evans and McCloud (1965), Cheeseman and Seed (1967), Donham and Harvick (1967), Segel (1967), Deckert and McCloud (1968), Smith (1968), Winston (1968), Cheeseman (1969), Cheney (1969), Donham, Watts, and Cardinale (1969), Huston and Shivers (1969), LTV Aerospace

Corporation (1969), Lockheed-California Company (1969), White (1969a, 1969b), Bailey and Hammer (1970), Briardy, La Forge, and Neff (1970), Deckert and Hickey (1970), Shivers (1970), Yuan (1970), Paglino (1971), Paglino and Beno (1970), Rose, Hammer, and Kizilos (1971), Stroub, Falarski, McCloud, and Soderman (1971), Trenka (1971), Watts, London, and Snoddy (1971), Bell Helicopter Company (1972d), Cheney (1972), Lemnios and Smith (1972), Lemnios, Smith, and Nettles (1972), Linden et al. (1972a, 1972b), McCloud (1972), Nichols (1972), Sullivan, La Forge, and Halchin (1972), Carlson and Cassarino (1973), Fradenburgh, Murrill, and Kiely (1973), Halley (1973), Kretz, Aubrun, and Lerche (1973), Landgrebe and Bellinger (1973, 1974a), Hughes and Wernicke (1974), McNeill, Plaks, and Blackburn (1974), Piziali and Trenka (1974), Fradenburgh (1975), Robinson, Nettles, and Howes (1975), Wilkerson and Linck (1975), Fradenburgh, Hager, and Keffort (1976), Gangwani (1976), Rorke (1976), Williams (1976), Arents (1977), Lemnios and Howes (1977), Putman and Curtiss (1977), Ruddell (1977), Wilkerson (1977b, 1977c), Young and Simon (1977), MacNeal and Hedgepeth (1978), Phelps and Mineck (1978).

Chapter 8

MATHEMATICS OF ROTATING SYSTEMS

This chapter presents some mathematics that are useful in the analysis of periodic dynamic systems, specifically an N-bladed helicopter rotor rotating at speed Ω. The period for a single blade is $T = 2\pi/\Omega$. In terms of the dimensionless time, measured by the azimuth angle ψ, the period is 2π. For the entire rotor, viewed in the nonrotating frame, the period is $T = 2\pi/N\Omega$. We are interested in the steady-state behavior of a rotating system, which in the rotating frame must be periodic, and thus a Fourier series analysis is appropriate (as in Chapter 5). We are also concerned with the transient behavior of a rotating system, particularly the dynamic stability.

8–1 Fourier Series

A Fourier series is a representation of a periodic function $\beta(\psi)$ as a linear combination of sine and cosine functions with fundamental period 2π:

$$\begin{aligned}\beta(\psi) &= \beta_0 + \beta_{1c}\cos\psi + \beta_{1s}\sin\psi + \beta_{2c}\cos 2\psi + \beta_{2s}\sin 2\psi + \dots \\ &= \beta_0 + \sum_{n=1}^{\infty}\left(\beta_{nc}\cos n\psi + \beta_{ns}\sin n\psi\right)\end{aligned}$$

(it is assumed that the time scale has been normalized so that the dimensionless period is 2π). The Fourier coefficients or harmonics are constants, which may be evaluated from integrals of $\beta(\psi)$ as follows:

$$\beta_0 = \frac{1}{2\pi}\int_0^{2\pi}\beta\,d\psi$$

$$\beta_{nc} = \frac{1}{\pi} \int_0^{2\pi} \beta \cos n\psi \, d\psi$$

$$\beta_{ns} = \frac{1}{\pi} \int_0^{2\pi} \beta \sin n\psi \, d\psi$$

A more concise representation is given by the complex form of the Fourier series:

$$\beta(\psi) = \sum_{n=-\infty}^{\infty} \beta_n e^{in\psi}$$

where

$$\beta_n = \frac{1}{2\pi} \int_0^{2\pi} \beta e^{-in\psi} \, d\psi$$

Since β is real, it follows that β_n and β_{-n} must be complex conjugates. The real and complex harmonics are related by

$$\beta_n = \tfrac{1}{2}(\beta_{nc} - i\beta_{ns})$$

for $n \geqslant 1$ (β_0 has the same definition in both forms). The complex form can be useful in manipulating the equations of a periodic system, since a single expression defines all the harmonics. To interpret the results it is still necessary to consider the real form.

The Fourier series is a linear transformation between a representation of a periodic motion by a continuous function $\beta(\psi)$ over one period, and a representation by an infinite set of constants (β_0, β_{1c}, β_{1s}, ...). The Fourier coefficients represent the motion in the nonrotating frame, as for the flap and lag motion discussed in section 5–1. The usefulness of the Fourier series description of steady-state rotor motion is based on the fact that only the lowest few harmonics have significant magnitude, so that the complete periodic motion is described by a small set of numbers.

The Fourier coefficients describing the blade motion are the steady-state solution of the linear differential equation for the motion. For example, recall the flapping equation as derived in Chapter 5:

$$\ddot{\beta} + \nu^2\beta = \gamma[M_\theta\theta + M_\lambda\lambda + M_{\dot{\beta}}\dot{\beta} + M_\beta\beta]$$

In general, the coefficients of the equations of motion (in this case the aerodynamic flap moments M_θ, M_λ, $M_{\dot{\beta}}$, and M_β) are periodic functions of ψ. There are two approaches to solving the equations of motion for the Fourier coefficients, the substitutional method and the operational method. In the substitutional method, the degrees of freedom (and their time derivatives) as well as the equation coefficients are written as Fourier series. Then products of sines and cosines are reduced to sums of sines and cosines by using trigonometric relations. Next, all the coefficients of like harmonics in the equation (that is, the coefficients of 1, $\cos\psi$, $\sin\psi$, $\cos 2\psi$, $\sin 2\psi$, etc.) are collected. Finally, the collected coefficients of $\sin n\psi$ and $\cos n\psi$ are individually set to zero. The result is an infinite set of linear algebraic equations for the harmonics (β_0, β_{1c}, β_{1s}, etc.). It is necessary to truncate the Fourier series representation to obtain a finite set of algebraic equations, which may then be solved for the required harmonics.

In the operational method, the following operators are applied to the differential equation of motion:

$$\frac{1}{2\pi}\int_0^{2\pi}(\ldots)d\psi, \quad \frac{1}{\pi}\int_0^{2\pi}(\ldots)\cos n\psi\, d\psi, \quad \frac{1}{\pi}\int_0^{2\pi}(\ldots)\sin n\psi\, d\psi$$

The periodic coefficients are again written as Fourier series, and products of harmonics reduced to sums of harmonics. The operation here is simpler than in the substitutional method since Fourier series have not been introduced for the degrees of freedom. The integral operators only act on the product of the degrees of freedom and a cosine or sine harmonic, i.e. on terms of the form $\beta\cos k\psi$ or $\beta\cos k\psi$. The definitions of the Fourier coefficients may then be used to replace these integrals by the appropriate harmonics of the blade motion. The result is the set of linear algebraic equations, which can be solved for the required harmonics. The substitutional and operational methods produce the same algebraic equations. The operational method has the advantage of obtaining the equations one at a time; it can be interpreted as resolving into the nonrotating frame the moment equilibrium that produces the equation of motion.

8–2 Sum of Harmonics

In order to determine the total influence of a rotor with N blades undergoing identical periodic motion, it is necessary to evaluate sums of harmonics of the form $\sum_{m=1}^{N} \cos n\psi_m$ or $\sum_{m=1}^{N} \sin n\psi_m$. Here the azimuth angle of each blade is $\psi_m = \psi + m\Delta\psi$, with ψ the dimensionless time variable (and the azimuth angle of the reference blade) and $\Delta\psi = 2\pi/N$ the azimuthal spacing between the blades. The summation is over all the blades: $m = 1$ to N. The result for the sum of these harmonics is

$$\frac{1}{N}\sum_{m=1}^{N} \cos n\psi_m = f_n \cos n\psi$$

$$\frac{1}{N}\sum_{m=1}^{N} \sin n\psi_m = f_n \sin n\psi$$

$$\frac{1}{N}\sum_{m=1}^{N} e^{in\psi_m} = f_n e^{in\psi}$$

where $f_n = 1$ only if n is a multiple of the number of blades (i.e. $n = pN$, where p is an integer); $f_n = 0$ otherwise. Hence the sum is zero unless the harmonic number is a multiple of the number of blades.

To prove this result, consider the sum

$$S = \sum_{m=1}^{N} e^{inm\Delta\psi} = \sum_{m=1}^{N} e^{2\pi imn/N}$$

After factoring $e^{in\psi}$ from $\sum_{m=1}^{N} e^{in\psi_m}$, what must be proved is that $S = Nf_n$.

If n/N is an integer, then

$$(e^{2\pi i})^{(n/N)m} = (1)^{(n/N)m} = 1$$

for all m, and so $S = \sum_{m=1}^{N} 1 = N$. For the case of n/M not an integer, note that multiplying the series S by $e^{2\pi in/N}$ is equivalent to subtracting the first ($m = 1$) term and adding an $m = N+1$ term:

$$\begin{aligned} Se^{2\pi in/N} &= S + e^{2\pi i(n/N)(N+1)} - e^{2\pi i(n/N)} \\ &= S + e^{2\pi in} e^{2\pi i(n/N)} - e^{2\pi i(n/N)} \\ &= S \end{aligned}$$

since $e^{2\pi in} = 1$. But $e^{2\pi i(n/N)} \neq 1$ if n/N is not an integer, so necessarily $S = 0$. Hence $S = Nf_n$ as required.

In rotor dynamics, sums of the following form are also encountered:

$$\frac{1}{N}\sum_{m=1}^{N} (-1)^m \cos n\psi_m = g_n \cos n\psi$$

$$\frac{1}{N}\sum_{m=1}^{N} (-1)^m \sin n\psi_m = g_n \sin n\psi$$

$$\frac{1}{N}\sum_{m=1}^{N} (-1)^m e^{in\psi m} = g_n e^{in\psi}$$

where $g_n = 1$ if $n = N/2 + pN$ (p some integer), and $g_n = 0$ otherwise. Thus the sums are zero unless the harmonic number equals an odd multiple of $N/2$, which also requires that the rotor have an even number of blades. The proof of this result is similar to that given above; note that $(-1)^m = e^{i(N/2)m\Delta\psi}$

8–3 Harmonic Analysis

In numerical work, a periodic function $f(\psi)$ will usually be evaluated at J equally spaced points around the azimuth: $f_j = f(\psi_j)$, where $\psi_j = j2\pi/J$ for $j = 1$ to J. The function f may be estimated at points between the known values using the Fourier interpolation formula:

$$\tilde{f}(\psi) = \sum_{\ell=-L}^{L} F_\ell e^{i\ell\psi}$$

$(L \leqslant (J-1)/2)$, where

$$F_\ell = \frac{1}{J}\sum_{j=1}^{J} f_j e^{-i\ell\psi_j}$$

is a numerical evaluation of the harmonics of a Fourier series representation of $f(\psi)$. If $L < (J-1)/2$, this expression is a least-squared-error representation of f. If $L = (J-1)/2$, it gives $\tilde{f}(\psi_j) = f_j$ exactly.

While it matches the periodic function exactly at the known values, the Fourier interpolation formula usually is a poor representation elsewhere. It gives large excursions because of the higher harmonics and does not estimate derivatives of the function well. For numerical harmonic analysis it is therefore better to use a linear interpolation:

$$\tilde{f}(\psi) = f(\psi_j) + \frac{\psi - \psi_j}{\psi_{j+1} - \psi_j}\left[f(\psi_{j+1}) - f(\psi_j)\right]$$

for $\psi_j \leqslant \psi \leqslant \psi_{j+1}$. This interpolation is equivalent to

$$\tilde{f}(\psi) = \sum_{\ell=-\infty}^{\infty} F_\ell e^{i\ell\psi}$$

with the harmonics

$$F_\ell = \left(\frac{J}{\pi\ell}\sin\frac{\pi\ell}{J}\right)^2 \frac{1}{J}\sum_{j=1}^{J} f_j e^{-i\ell\psi_j}$$

The factor $[(J/\pi\ell)\sin(\pi\ell/J)]^2$ reduces the magnitude of the higher harmonics, but now an infinite number of harmonics are required. By truncating the Fourier series ($\ell = -L$ to L) the corners of the linear interpolation will be rounded off. Usually $L \cong J/3$ is satisfactory.

8–4 Fourier Coordinate Transformation

Generally, the rotor equations of motion are derived in the rotating frame, with degrees of freedom describing the motion of each blade separately. An example is the flapping equation as derived in Chapter 5. In fact, however, the rotor responds as a whole to excitation (such as aerodynamic gusts, control inputs, or shaft motion) from the nonrotating frame. It is desirable to work with degrees of freedom that reflect this behavior. Such a representation of the rotor motion simplifies both the analysis and the understanding of the behavior. For the steady state solution, the appropriate representation of the blade motion is a Fourier series, the harmonics of which describe the motion of the rotor as a whole. The equations of motion in the nonrotating frame are simply algebraic equations for the harmonics. Now, however, instead of the steady-state solution we are concerned with the general dynamic behavior, including the transient response of the rotor.

The appropriate transformation of the degrees of freedom and the equations of motion to the nonrotating frame is of the Fourier type. There are many similarities between this coordinate change and Fourier series, Fourier interpolation, and discrete Fourier transforms. The common factor is the periodic nature of the system. The Fourier coordinate transformation has been widely used in the classical literature, although often with only a heuristic basic. For example, it was used by Coleman and Feingold (1958) to represent the rotor lag motion in ground resonance analyses, and by Miller (1948) to represent the flap motion for a helicopter stability and control analysis. More recently there have been applications of the Fourier coordinate transformation with a sounder mathematical basis, for example by Hohenemser and Yin (1972b).

8-4.1 Transformation of the Degrees of Freedom

Consider a rotor with N blades equally spaced around the azimuth at $\psi_m = \psi + m\Delta\psi$, where ψ is the dimensionless time variable ($\psi = \Omega t$ for constant rotational speed) and $\Delta\psi = 2\pi/N$ is the azimuthal spacing between blades. The blade index m ranges from 1 to N. Let $\beta^{(m)}$ be the degree of freedom in the rotating frame for the mth blade. The Fourier coordinate transformation is a linear transformation of the degrees of freedom from the rotating to the nonrotating frame. The following new degrees of freedom are introduced:

$$\beta_0 = \frac{1}{N}\sum_{m=1}^{N} \beta^{(m)}$$

$$\beta_{nc} = \frac{2}{N} \sum_{m=1}^{N} \beta^{(m)} \cos n\psi_m$$

$$\beta_{ns} = \frac{2}{N} \sum_{m=1}^{N} \beta^{(m)} \sin n\psi_m$$

$$\beta_{N/2} = \frac{1}{N} \sum_{m=1}^{N} \beta^{(m)} (-1)^m$$

These degrees of freedom describe the motion of the rotor in the nonrotating frame. As an example, for the rotor flap motion, β_0 is the coning degree of freedom, while β_{1c} and β_{1s} are the tip-path-plane-tilt degrees of freedom. The remaining degrees of freedom are called reactionless modes, since they involve no net force or moment on the rotor hub. The corresponding inverse transformation is

$$\beta^{(m)} = \beta_0 + \sum_n (\beta_{nc} \cos n\psi_m + \beta_{ns} \sin n\psi_m) + \beta_{N/2}(-1)^m$$

which gives the motion of the individual blades again. The summation over the harmonic index n goes from $n = 1$ to $(N - 1)/2$ for N odd, and from $n = 1$ to $(N - 2)/2$ for N even. The reactionless degree of freedom $\beta_{N/2}$ appears in the transformation only if N is even.

The variables β_0, β_{nc}, β_{ns}, and $\beta_{N/2}$ are degrees of freedom, i.e. functions of time, just as the variables $\beta^{(m)}$ are. These degrees of freedom describe the motion of the rotor as a whole, in the nonrotating frame, while $\beta^{(m)}$ describes the motion of an individual blade in the rotating frame. Thus we have a linear, reversible transformation between the N degrees of freedom $\beta^{(m)}$ in the rotating frame ($m = 1 \ldots N$) and the N degrees of freedom β_0, β_{nc}, β_{ns}, $\beta_{N/2}$ in the nonrotating frame. Compare this coordinate transformation with a Fourier series representation of the steady-state solution. In the latter case, where $\beta^{(m)}$ is a periodic function of ψ_m, the motions of all the blades are identical. It follows that the motion in the rotating frame may be represented by a Fourier series, the coefficients of which are constant in time but infinite in number. Thus there are similarities between the

Fourier coordinate transformation and the Fourier series, but they are by no means identical.

The collective and cyclic modes (β_0, β_{1c}, and β_{1s}, where in general β may be any degree of freedom of the blade) are of particular importance because of their fundamental role in the coupled motion of the rotor and the nonrotating system. It will be seen in later chapters that for axial flow only the collective and cyclic modes of the rotor degrees of freedom couple with the fixed system, while the reactionless modes (β_{2c}, β_{2s}, ... β_{nc}, β_{ns}, and $\beta_{N/2}$) correspond to purely internal rotor motion. Nonaxial flow to some extent couples all the rotor degrees of freedom and the fixed system variables, but still the collective and cyclic motions dominate the rotor dynamic behavior. The reactionless mode $\beta_{N/2}$, present for rotors with an even number of blades, generally introduces some special behavior in the analysis. In this mode all the blades have identical motion, but the displacement alternates the sign from one blade to the next around the azimuth. Note that for a two-bladed rotor the nonrotating degrees of freedom are the coning and teetering modes,

$$\beta_0 = \tfrac{1}{2}(\beta^{(2)} + \beta^{(1)})$$

$$\beta_1 = \tfrac{1}{2}(\beta^{(2)} - \beta^{(1)})$$

In this case β_1 replaces the cyclic modes β_{1c} and β_{1s} and therefore couples with the fixed system. Because of the absence of the cyclic modes, two-bladed rotor dynamics are fundamentally different from the dynamics of rotors with three or more blades.

The proof that the rotating and nonrotating degrees of freedom describe the same motion proceeds as follows. Consider the case of N odd, and for this purpose use a complex representation of the nonrotating degrees of freedom:

$$\beta^{(m)} = \sum_{n=-(N-1)/2}^{(N-1)/2} \beta_n e^{in\psi_m}$$

$$\beta_n = \frac{1}{N}\sum_{m=1}^{N} \beta^{(m)} e^{-in\psi_m}.$$

It is necessary to show that these transformations are equivalent and reversible. Substituting for $\beta^{(m)}$ gives

$$\beta_\ell = \frac{1}{N}\sum_{m=1}^{N}\left[\sum_{n=-(N-1)/2}^{(N-1)/2} \beta_n e^{in\psi_m}\right] e^{-i\ell\psi_m}$$

$$= \sum_n \beta_n \left[\frac{1}{N}\sum_{m=1}^{N} e^{i(n-\ell)\psi_m}\right]$$

$$= \sum_n \beta_n S_{n\ell}$$

Using the results of section 8–2 for the sums of harmonics, $S_{n\ell} = e^{i(n-\ell)\psi}$ if $(n-\ell)$ is a multiple of N, and $S_{n\ell} = 0$ otherwise. Since both n and ℓ have a magnitude less than or equal to $(N-1)/2$, $(n-\ell)$ is a multiple of N only for $(n-\ell) = 0$. Hence $S_{n\ell} = 1$ for $n = \ell$ and zero otherwise, and $\beta_\ell = \beta_\ell$ is obtained as required. Substituting for β_n in the reverse transformation:

$$\beta^{(k)} = \sum_{n=-(N-1)/2}^{(N-1)/2}\left[\frac{1}{N}\sum_{m=1}^{N}\beta^{(m)} e^{-in\psi_m}\right] e^{in\psi_k}$$

$$= \sum_{m=1}^{N}\beta^{(m)}\left[\frac{1}{N}\sum_{n=-(N-1)/2}^{(N-1)/2} e^{in(k-m)\Delta\psi}\right]$$

$$= \sum_{m=1}^{N}\beta^{(m)}\left[\frac{1}{N}\sum_{n=1}^{N} e^{i[n-(N+1)/2](k-m)\Delta\psi}\right]$$

$$= \sum_{m=1}^{N}\beta^{(m)} S_{mk}$$

Using the results for the sum of harmonics, $S_{mk} = 1$ only if $(k-m)$ is a multiple of N, which requires $k-m=0$. Hence $\beta^{(k)} = \beta^{(k)}$ is obtained as required. The proof of the transformation for N even follows in a similar manner, although it is complicated somewhat by the presence of the $\beta_{N/2}$ mode.

Finally, consider the transformation of time derivatives of the motion. From

$$\beta^{(m)} = \beta_0 + \sum_n (\beta_{nc} \cos n\psi_m + \beta_{ns} \sin n\psi_m) + \beta_{N/2}(-1)^m$$

it follows that

$$\dot\beta^{(m)} = \dot\beta_0 + \sum_n [(\dot\beta_{nc} + n\Omega\beta_{ns}) \cos n\psi_m + (\dot\beta_{ns} - n\Omega\beta_{nc}) \sin n\psi_m] + \dot\beta_{N/2}(-1)^m$$

$$\ddot\beta^{(m)} = \ddot\beta_0 + \sum_n [(\ddot\beta_{nc} + 2n\Omega\dot\beta_{ns} + n\dot\Omega\beta_{ns} - n^2\Omega^2\beta_{nc}) \cos n\psi_m + (\ddot\beta_{ns} - 2n\Omega\dot\beta_{nc} - n\dot\Omega\beta_{nc} - n^2\Omega^2\beta_{ns}) \sin n\psi_m] + \ddot\beta_{N/2}(-1)^m$$

where $\Omega = \dot\psi$. The Ω's are omitted for dimensionless equations, and usually the trim rotor speed is constant (or its perturbations are represented by a separate degree of freedom), so $\dot\Omega = 0$. Then the harmonics of the time derivatives are defined as follows:

$$\frac{1}{N}\sum_{m=1}^{N} \dot\beta^{(m)} = \dot\beta_0$$

$$\frac{2}{N}\sum_{m=1}^{N} \dot\beta^{(m)} \cos n\psi_m = \dot\beta_{nc} + n\beta_{ns}$$

$$\frac{2}{N}\sum_{m=1}^{N} \dot\beta^{(m)} \sin n\psi_m = \dot\beta_{ns} - n\beta_{nc}$$

$$\frac{1}{N}\sum_{m=1}^{N} \dot\beta^{(m)} (-1)^m = \dot\beta_{N/2}$$

and

$$\frac{1}{N}\sum_{m=1}^{N} \ddot{\beta}^{(m)} = \ddot{\beta}_0$$

$$\frac{2}{N}\sum_{m=1}^{N} \ddot{\beta}^{(m)} \cos n\psi_m = \ddot{\beta}_{nc} + 2n\dot{\beta}_{ns} - n^2\beta_{nc}$$

$$\frac{2}{N}\sum_{m=1}^{N} \ddot{\beta}^{(m)} \sin n\psi_m = \ddot{\beta}_{ns} - 2n\dot{\beta}_{nc} - n^2\beta_{ns}$$

$$\frac{1}{N}\sum_{m=1}^{N} \ddot{\beta}^{(m)} (-1)^m = \ddot{\beta}_{N/2}$$

The transformation of the velocity and acceleration from the rotating frame introduces Coriolis and centrifugal terms in the nonrotating frame.

8-4.2 Conversion of the Equations of Motion

The Fourier coordinate transformation must be accompanied by a conversion of the differential equations of motion from the rotating to the nonrotating frame. This conversion is accomplished by operating on the rotating-frame equation of motion with the following summation operators:

$$\frac{1}{N}\sum_{m=1}^{N}(\ldots),\ \frac{2}{N}\sum_{m=1}^{N}(\ldots)\cos n\psi_m,\ \frac{2}{N}\sum_{m=1}^{N}(\ldots)\sin n\psi_m,\ \frac{1}{N}\sum_{m=1}^{N}(\ldots)(-1)^m$$

The result is N differential equations in the nonrotating frame, obtained by summing the rotating equation over all N blades. Note that these same operators are involved in transforming the degrees of freedom. The conversion of the equations is not complete, however, until the summation operator is eliminated by using it to transform to the nonrotating degrees of freedom.

A procedure analogous to the substitutional method for Fourier series consists of the following steps. The periodic coefficients in the rotating-frame equation of motion are written as Fourier series, and the Fourier coordinate transformation is introduced for the degrees of freedom and their time derivatives. Then products of harmonics are written as sums of harmonics using trigonometric relations. Next, all coefficients of 1, $\cos \psi_m$, $\sin \psi_m$, ...$\cos n\psi_m$, $\sin n\psi_m$, $(-1)^m$ are collected and individually set to zero, producing the required differential equations. There is a difficulty with this approach that arises because, unlike the Fourier series case, only N equations are to be obtained. Thus any harmonics $\cos \ell\psi_m$ and $\sin \ell\psi_m$ with $\ell > N/2$ must be rewritten as products of harmonics in the proper range ($\ell < N/2$) and harmonics of N/rev. For example, consider a second harmonic appearing in the equations for a three-bladed rotor. By writing

$$\cos 2\psi_m = \cos 3\psi_m \cos \psi_m + \sin 3\psi_m \sin \psi_m$$

$$\sin 2\psi_m = \sin 3\psi_m \cos \psi_m - \cos 3\psi_m \sin \psi_m$$

it follows that the second harmonics contribute 3/rev terms to the $\cos \psi_m$ and $\sin \psi_m$ equations. A better approach is to apply the summation operators given above instead of trying to collect coefficients of like harmonics. Then, since the summation over all blades acts only on the harmonics, it is only necessary to evaluate terms of the form

$$\sum_{m=1}^{N} \cos \ell\psi_m, \quad \sum_{m=1}^{N} \sin \ell\psi_m, \quad \sum_{m=1}^{N} (-1)^m \cos \ell\psi_m, \quad \sum_{m=1}^{N} (-1)^m \sin \ell\psi_m$$

to complete the equations. These sums may be evaluated using the results of section 8–2 for sums of harmonics. Recall that the first two sums give harmonics of N/rev if ℓ is a multiple of N, and the sums involving $(-1)^m$ give harmonics of $\frac{1}{2}N$/rev if ℓ is an odd multiple of $N/2$.

An operational method, which requires less manipulation of the harmonics, proceeds as follows. Again the periodic coefficients of the rotating equations are written as Fourier series, and the summation operators are applied to the equations. Products of harmonics are reduced to sums of harmonics as usual. Since the rotating degrees of freedom are still present, it is necessary to evaluate terms of the form

$$\frac{2}{N}\sum_{m=1}^{N} \beta^{(m)} \cos \ell\psi_m, \quad \frac{2}{N}\sum_{m=1}^{N} \beta^{(m)} \sin \ell\psi_m,$$

$$\frac{2}{N}\sum_{m=1}^{N}\beta^{(m)}(-1)^m \cos \ell\psi_m, \quad \frac{2}{N}\sum_{m=1}^{N}\beta^{(m)}(-1)^m \sin \ell\psi_m$$

If $\ell < N/2$, the first two sums are simply the definitions of the nonrotating degrees of freedom $\beta_{\ell c}$ and $\beta_{\ell s}$. For the general case, write $\ell = n + pN$, where p is an integer and n is the principal value of the harmonic, such that $n < N/2$. Then if the complex form is used and the definition of the nonrotating degrees of freedom (for n, which now has the proper range) is applied,

$$\begin{aligned}
\frac{1}{N}\sum_{m=1}^{N}\beta^{(m)}e^{-i\ell\psi_m} &= \frac{1}{N}\sum_{m=1}^{N}e^{-in\psi_m}\, e^{-ipN\psi_m}\, \beta^{(m)} \\
&= e^{-ipN\psi}\,\frac{1}{N}\sum_{m=1}^{N}\beta^{(m)}\, e^{-in\psi_m} \\
&= e^{-ipN\psi}\,\beta_n
\end{aligned}$$

since $e^{-ipNm\Delta\psi} = e^{-2\pi ipm} = 1$. If N is even it is also necessary to consider the case $\ell = n + pN$ with $n = N/2$, for which

$$\frac{1}{N}\sum_{m=1}^{N}\beta^{(m)}\, e^{-i\ell\psi_m} = e^{-i(p+\frac{1}{2})N\psi}\,\beta_{N/2}$$

The real form is as follows. Writing $\ell = n + pN$ where $n < N/2$, then

$$\frac{2}{N}\sum_{m=1}^{N}\beta^{(m)}\cos \ell\psi_m = \beta_{nc}\cos pN\psi - \beta_{ns}\sin pN\psi$$

$$\frac{2}{N}\sum_{m=1}^{N}\beta^{(m)}\sin \ell\psi_m = \beta_{nc}\sin pN\psi + \beta_{ns}\cos pN\psi$$

or if $n = N/2$,

$$\frac{1}{N}\sum_{m=1}^{N} \beta^{(m)} \cos \ell\psi_m = \beta_{N/2} \cos (p + ½)N\psi$$

$$\frac{1}{N}\sum_{m=1}^{N} \beta^{(m)} \sin \ell\psi_m = \beta_{N/2} \sin (p + ½)N\psi$$

Similarly, for the summations involving $(-1)^m$, write $\ell = n + (p - ½)N$, where $n < N/2$, then

$$\frac{2}{N}\sum_{m=1}^{N} \beta^{(m)} (-1)^m \cos \ell\psi_m = \beta_{nc} \cos (p - ½)N\psi - \beta_{ns} \sin (p - ½)N\psi$$

$$\frac{2}{N}\sum_{m=1}^{N} \beta^{(m)} (-1)^m \sin \ell\psi_m = \beta_{nc} \sin (p - ½)N\psi + \beta_{ns} \cos (p - ½)N\psi$$

or if $n = N/2$,

$$\frac{1}{N}\sum_{m=1}^{N} \beta^{(m)} (-1)^m \cos \ell\psi_m = \beta_{N/2} \cos pN\psi$$

$$\frac{1}{N}\sum_{m=1}^{N} \beta^{(m)} (-1)^m \sin \ell\psi_m = \beta_{N/2} \sin pN\psi$$

With these results the construction of the differential equations in the nonrotating frame is straightforward.

Two assumptions have been made in outlining these procedures for the conversion of the equations to the nonrotating frame: first that the number of degrees of freedom involved is small enough for an analytical construction to be practical; and secondly that analytical expressions are available for the periodic coefficients as Fourier series. For comprehensive dynamics analyses neither assumption is valid, and a procedure better suited to numerical work is required. The rotating degrees of freedom are written in terms of the Fourier coordinate transformation:

$$\beta^{(m)} = \beta_0 + \sum_n (\beta_{nc} \cos n\psi_m + \beta_{ns} \sin n\psi_m) + \beta_{N/2} (-1)^m$$

(with similar expressions for the time derivatives), and the summation operators are applied to the rotating equation. Then it is necessary to evaluate summations over all N blades that involve the periodic coefficient multiplied by one of the factors 1, $\cos n\psi_m$, $\sin n\psi_m$, or $(-1)^m$ from the Fourier coordinate transformation, and are multiplied by one of the factors 1, $\cos k\psi_m$, $\sin k\psi_m$, or $(-1)^m$ from the summation operators. The construction of the nonrotating equations in this manner is simple, and the evaluation easily performed numerically. The value of an analytical approach is that with simple periodic coefficients many of these summations are exactly zero, which greatly simplifies the nonrotating equations of motion.

With a constant coefficient differential equation, the conversion to the nonrotating frame is elementary. The summation operators then act only on the degrees of freedom, not on the equation coefficients, and the definitions of the nonrotating degrees of freedom (and their derivatives) can be applied directly. Consider for example a mass-spring-damper system of the form

$$\ddot{\beta}^{(n)} + \frac{\gamma}{8} \dot{\beta}^{(m)} + \nu^2 \beta^{(m)} = \frac{\gamma}{8} \theta^{(m)}$$

(the flapping equation in hover). The resulting nonrotating equations are

$$\ddot{\beta}_0 + \frac{\gamma}{8} \dot{\beta}_0 + \nu^2 \beta_0 = \frac{\gamma}{8} \theta_0$$

for β_0,

$$\begin{pmatrix} \beta_{nc} \\ \beta_{ns} \end{pmatrix}^{\cdot\cdot} + \begin{bmatrix} \frac{\gamma}{8} & 2n \\ -2n & \frac{\gamma}{8} \end{bmatrix} \begin{pmatrix} \beta_{nc} \\ \beta_{ns} \end{pmatrix}^{\cdot} + \begin{bmatrix} \nu^2 - n^2 & n\frac{\gamma}{8} \\ -n\frac{\gamma}{8} & \nu^2 - n^2 \end{bmatrix} \begin{pmatrix} \beta_{nc} \\ \beta_{ns} \end{pmatrix} = \frac{\gamma}{8} \begin{pmatrix} \theta_{nc} \\ \theta_{ns} \end{pmatrix}$$

for β_{nc} and β_{ns}, and

$$\ddot{\beta}_{N/2} + \frac{\gamma}{8} \dot{\beta}_{N/2} + \nu^2 \beta_{N/2} = \frac{\gamma}{8} \theta_{N/2}$$

for $\beta_{N/2}$. These equations show the basic manner in which inertia, damping, and spring terms transform to the nonrotating frame. The conversion introduces Coriolis and centrifugal terms to the β_{nc} and β_{ns} equations. Note that the only coupling of the nonrotating degrees of freedom occurs in the β_{nc} and β_{ns} equations as a result of the Coriolis terms. The number of blades influences only the number of degrees of freedom and equations that must be analyzed. The equations, and the procedure by which they are obtained, are much more complicated with periodic coefficients. Consider the differential equation for the flapping motion in forward flight:

$$\ddot{\beta}^{(m)} + \left(\frac{\gamma}{8} + \frac{\gamma}{6}\mu \sin\psi_m\right)\dot{\beta}^{(m)} + \left(\nu^2 + \frac{\gamma}{6}\mu\cos\psi_m + \frac{\gamma}{8}\mu^2 \sin 2\psi_m\right)\beta^{(m)}$$

$$= \left(\frac{\gamma}{8}(1+\mu^2) + \frac{\gamma}{3}\mu\sin\psi_m - \frac{\gamma}{8}\mu^2\cos 2\psi_m\right)\theta^{(m)}$$

$$-\left(\frac{\gamma}{6} + \frac{\gamma}{4}\mu\sin\psi_m\right)\lambda$$

(see section 5–5). The inertia and centrifugal-structural spring terms ($\ddot{\beta}^{(m)} + \nu^2\beta^{(m)}$) transform as above for hover. The transformation of the aerodynamic terms to the nonrotating frame is given in section 11–4 for the cases of two, three, and four blades. It is found that as the number of blades increases, the periodic coefficients are cleared from the lower degrees of freedom and equations. There are always periodic coefficients in the complete set of equations, though, regardless of the number of blades. It is also noted (see section 11–4) that the higher harmonics of the coefficients in the rotating frame contribute to the mean values of the coefficients in the nonrotating frame. Only N/rev harmonics (in general multiples of N/rev) appear in the equations for the N-bladed rotor. This follows from the results for

$$\sum_{m=1}^{N} \beta^{(m)} \cos \ell\psi_m \text{ and } \sum_{m=1}^{N} \beta^{(m)} \sin \ell\psi_m$$

given above. Hence while the rotating equations have period 2π, the equations in the nonrotating frame have a period $T = 2\pi/N$, as expected with identical blades. The exception is that ½N/rev harmonics (in general, odd multiples of ½N/rev) appear in the matrix elements coupling the $\beta_{N/2}$ mode with the other degrees of freedom. So, for a rotor with an even number of blades, when the $\beta_{N/2}$ degree of freedom is included in the analysis, the period is $T = 4\pi/N$. The period is twice the expected result because the blades are no longer identical; the $\beta_{N/2}$ mode identifies alternate blades with a plus or minus amplitude by the $(-1)^m$ factor. Thus the period of

$4\pi/N$ follows from the mathematical description of the rotor motion; the solution must still correspond to a physical system with period $2\pi/N$.

The equations of motion for the rotor in forward flight must always have periodic coefficients, whether they are written in the rotating or nonrotating frame. The solutions of such equations have very distinctive behavior, but are also more difficult to obtain than the solutions of constant coefficient equations (see section 8–6). When the equations are only weakly periodic, it may be hoped that there is some constant coefficient system which closely represents the behavior of the true system. An example is the periodic coefficients arising from the aerodynamics of forward flight, which have higher harmonics only of order μ and smaller. It is necessary to establish the best means to construct such a constant coefficient approximation and to determine its range of validity. The constant coefficient system can be constructed by retaining only the mean values of the original periodic coefficients. Clearly a better approximation will result if the coefficients are averaged in the nonrotating frame, for the higher harmonics in the rotating equation contribute to the mean values of the coefficients in the nonrotating equations. By working in the nonrotating frame it is necessary to solve more equations, however. The constant coefficient approximation is an important tool in the analysis of rotor dynamics, and it will be discussed further in the chapters to follow.

8–5 Eigenvalues and Eigenvectors of the Rotor Motion

We shall now examine the characteristics of the rotor motion, in particular the eigenvalues and eigenvectors of the system described by the nonrotating-frame degrees of freedom and equations. Consider a constant-coefficient, mass-spring-damper system in the rotating frame; for example, the rotor flapping equation in hover:

$$\ddot{\beta}^{(m)} + \frac{\gamma}{8}\dot{\beta}^{(m)} + \nu^2 \beta^{(m)} = 0$$

(The homogeneous equation is sufficient, since only the roots and mode shapes are required here.) The uncoupled motion of all the rotor degrees of freedom (lag, pitch, elastic bending, etc.) will be described by similar equations. To be general, an arbitrary level of damping ($\gamma/8$) is allowed, and a natural frequency (ν) is considered that unlike the flap motion is not necessarily near 1/rev. The eigenvalues of the rotating equation are the solution of the quadratic equation $s_R^2 + (\gamma/8)s_R + \nu^2 = 0$, or

$$s_R = -\frac{\gamma}{16} + i\sqrt{\nu^2 - \left(\frac{\gamma}{16}\right)^2}$$

and its conjugate.

In the nonrotating frame, the equations for β_0 and $\beta_{N/2}$ are identical to the rotating equation:

$$\ddot{\beta}_0 + \frac{\gamma}{8}\dot{\beta}_0 + \nu^2\beta_0 = 0$$

$$\ddot{\beta}_{N/2} + \frac{\gamma}{8}\dot{\beta}_{N/2} + \nu^2\beta_{N/2} = 0$$

The roots of both equations are then the same as the rotating roots: $s = s_R$ and its conjugate. The differential equations for β_{nc} and β_{ns} are

$$\ddot{\beta}_{nc} + 2n\dot{\beta}_{ns} + \frac{\gamma}{8}\dot{\beta}_{nc} + (\nu^2 - n^2)\beta_{nc} + \frac{\gamma}{8}n\beta_{ns} = 0$$

$$\ddot{\beta}_{ns} - 2n\dot{\beta}_{nc} + \frac{\gamma}{8}\dot{\beta}_{ns} + (\nu^2 - n^2)\beta_{ns} - \frac{\gamma}{8}n\beta_{nc} = 0$$

or

$$\begin{bmatrix} s^2 + \frac{\gamma}{8}s + \nu^2 - n^2 & 2ns + \frac{\gamma}{8}n \\ -\left(2ns + \frac{\gamma}{8}n\right) & s^2 + \frac{\gamma}{8}s + \nu^2 - n^2 \end{bmatrix}\begin{pmatrix}\beta_{nc} \\ \beta_{ns}\end{pmatrix} = 0$$

(see section 8-4.2). The transformation to the nonrotating frame introduces centrifugal and Coriolis terms that couple the β_{nc} and β_{ns} equations. The roots are the solution of the characteristic equation:

$$\left(s^2 + \frac{\gamma}{8}s + \nu^2 - n^2\right)^2 + \left(2ns + \frac{\gamma}{8}n\right)^2 = 0$$

or

$$s = -\frac{\gamma}{16} \pm in + i\sqrt{\nu^2 - \left(\frac{\gamma}{16}\right)^2}$$

and their conjugates. Hence the nonrotating eigenvalues for the β_{nc} and β_{ns} degrees of freedom are simply the rotating roots shifted in frequency by n/rev: $s = s_R \pm in$. The corresponding eigenvectors are found to be $\beta_{nc}/\beta_{ns} = i$ for $s = s_R + in$, and $\beta_{nc}/\beta_{ns} = -i$ for $s = s_R - in$.

The eigenvalues $s = s_R \pm in$ correspond to a coupled motion of β_{nc} and β_{ns} that is a damped oscillation at frequency $\text{Im}\, s = \text{Im}\, s_R \pm n$/rev. The exponential decay rate, $\text{Re}\, s = \text{Re}\, s_R$, is the same as for the rotating roots. The $s = s_R + in$ root has frequency $\text{Im}\, s_R + n$/rev, and $\beta_{nc} = i\beta_{ns}$ implies that the motion of β_{nc} leads that of β_{ns} by a phase of $90°$ [i.e. by one-quarter of the oscillation period, which is $2\pi/(\text{Im}\, s_R + n)$]. Thus $s = s_R + in$ is a high frequency, progressive mode (note that the frequency $\text{Im}\, s_R + n$/rev is always greater than the rotor speed). The $s = s_R - in$ root has frequency $|\text{Im}\, s_R - n\text{/rev}|$. If $\text{Im}\, s_R > n$/rev, then $\beta_{nc} = -i\beta_{ns}$ implies that the motion of β_{nc} lags β_{ns} by $90°$. If however $\text{Im}\, s_R < n$/rev, so that the frequency $\text{Im}\, s_R - n$/rev is negative, then $\beta_{nc} = -i\beta_{ns}$ implies that β_{nc} leads β_{ns} by $90°$. Thus $s = s_R - in$ is a low frequency mode (the frequency can be below 1/rev if the rotating frequency is near n/rev), regressive if $\text{Im}\, s_R > n$/rev and progressive if $\text{Im}\, s_R < n$/rev.

Consider the important case of the cyclic modes ($n = 1$) for the flap and lag motion of the rotor. For the flap motion, the rotating natural frequency $\text{Im}\, s_R$ will usually be slightly below 1/rev for articulated rotors, and perhaps slightly above 1/rev for hingeless rotors. Then for the high frequency mode $s = s_R + i$, β_{1c} leading β_{1s} means that the tip-path plane is wobbling in the same direction as the rotor rotation, at a speed around 2/rev. For the low frequency mode $s = s_R - i$, the tip-path plane wobbles at a low rate, again in the same direction as the rotor rotation if the rotating frequency is below 1/rev, but in the opposite direction if $\text{Im}\, s_R$ is above 1/rev. For the lag motion, articulated and soft in-plane hingeless rotors will have a rotating frequency below 1/rev. The high frequency lag mode is a progressive mode in which the rotor center of gravity whirls in the same direction as the rotor rotation at a speed above 1/rev; the low frequency lag mode is also a progressive whirling, but with a low frequency. For stiff in-plane rotors the rotating lag frequency is above 1/rev and the low frequency lag mode is a regressive mode in which the rotor center of gravity whirls in the opposite direction to the rotor rotation.

Fig. 8-1 summarizes the transformation of the eigenvalues describing

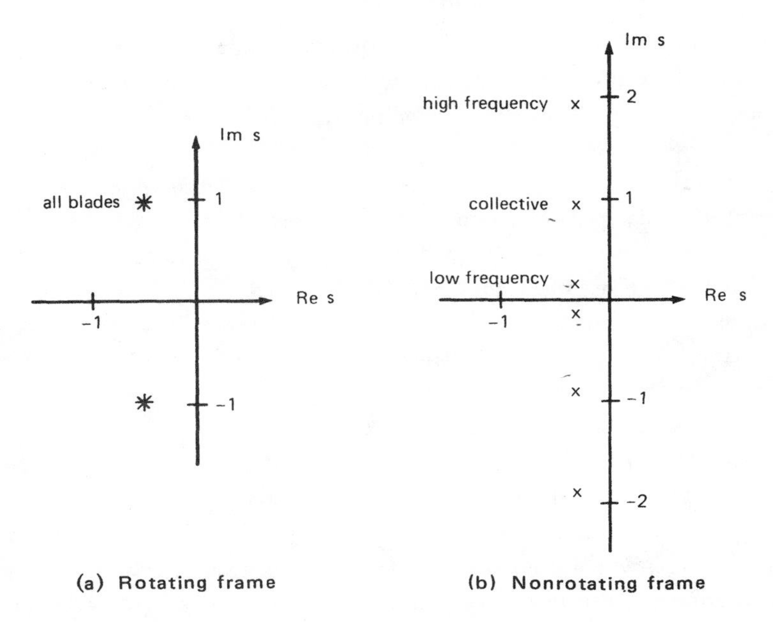

Figure 8-1 Transformation of eigenvalues from the rotating to the nonrotating frame (shown for $N = 3$).

the rotor dynamics from the rotating to the nonrotating frame. The case of a three-bladed rotor with a rotating frequency just below 1/rev is shown. In the rotating frame there are triple roots at s_R and its conjugate corresponding to the three independent blades, or in general there are N pairs of roots for an N-bladed rotor. In the nonrotating frame there are still N pairs of roots, at s_R and its conjugate again for the β_0 and $\beta_{N/2}$ modes; and at $s_R \pm in$ and their conjugates for the coupled β_{nc} and β_{ns} motion. Thus the transformation leaves the real part of the roots unchanged and shifts the frequency by $\pm\, n$/rev. Fig. 8-1 shows the collective, high frequency, and low frequency modes for the three-bladed rotor case. When the individual blades of the rotor are not independent, but rather are coupled through the fixed system (such as by the control system or shaft motion), the nonrotating modes are not all influenced in the same manner, and the real parts of the roots will not necessarily be identical nor will the frequencies be separated by exactly n/rev. The basic character illustrated by Fig. 8-1 still dominates the roots in the nonrotating frame, however.

8–6 Analysis of Linear, Periodic Systems

The aeroelastic behavior of the rotor or helicopter is described in many cases by linear differential equations with periodic coefficients. The periodic coefficients arise as a result of the aerodynamic forces in forward flight or a basic asymmetry in the rotor system (such as with a two-bladed rotor). Helicopter analysis therefore requires a means for obtaining the dynamic behavior of periodic systems, in particular the eigenvalues describing the stability.

Consider a physical system described by linear ordinary differential equations of second order,

$$A_2\ddot{\vec{x}}_1 + A_1\dot{\vec{x}}_1 + A_0\vec{x}_1 = B_0\vec{v}$$

Here $\vec{x}_1$ is the vector of degrees of freedom; $\vec{v}$ is the vector of input variables; and A_2, A_1, A_0, and B_0 are matrices of the coefficients of the equations of motion. For a time-invariant system, the coefficient matrices are constant. We are also interested in the more general case of time-varying coefficients, especially periodic types. It is convenient to deal with these equations in a standard first-order form, both in the mathematical development of the theory and the actual computation of the dynamic response. Thus we define $\vec{x}_2 = \dot{\vec{x}}_1$, so that

$$\dot{\vec{x}}_2 = \ddot{\vec{x}}_1 = -A_2^{-1}(A_1\dot{\vec{x}}_1 + A_0\vec{x}_1 - B_0\vec{v})$$

Then the equations of motion become

$$\begin{pmatrix}\vec{x}_2\\ \vec{x}_1\end{pmatrix}^{\bullet} = \left[\begin{array}{c|c} -A_2^{-1}A_1 & -A_2^{-1}A_0 \\ \hline I & 0 \end{array}\right]\begin{pmatrix}\vec{x}_2\\ \vec{x}_1\end{pmatrix} + \left[\begin{array}{c} A_2^{-1}B_0 \\ \hline 0 \end{array}\right]\vec{v}$$

or

$$\dot{\vec{x}} = A\vec{x} + B\vec{v}$$

where

$$\vec{x} = \begin{pmatrix}\vec{x}_2\\ \vec{x}_1\end{pmatrix} = \begin{pmatrix}\dot{\vec{x}}_1\\ \vec{x}_1\end{pmatrix}$$

is the state variable vector, consisting of the displacement and velocity of all second-order degrees of freedom. By transforming from second order to first order, the degree of the system (i.e. the dimension of $\vec{x}$) has been doubled. It is not unusual for the spring terms to be absent from a particular degree of freedom (i.e. to find a zero column in A_0), so that it is actually first order. Such lower order variables should be represented as a single state to avoid spurious zero eigenvalues. The degrees of freedom are reordered so that the second order variables $\vec{x}_1$ appear first and the second order variables $\vec{x}_0$ appear last in the vector. Then, since the last columns of A_0 corresponding to $\vec{x}_0$ are zero, it can be written $A_0 = [\tilde{A}_0 | 0]$. The differential equations then are

$$\begin{pmatrix} \dot{\vec{x}}_1 \\ \dot{\vec{x}}_0 \\ \vec{x}_1 \end{pmatrix}^{\bullet} = \left[\begin{array}{c|c} -A_2^{-1} A_1 & -A_2^{-1} \tilde{A}_0 \\ \hline I \quad 0 & 0 \end{array}\right] \begin{pmatrix} \dot{\vec{x}}_1 \\ \dot{\vec{x}}_0 \\ \vec{x}_1 \end{pmatrix} + \left[\begin{array}{c} A_2^{-1} B_0 \\ \hline 0 \end{array}\right] \vec{v}$$

or $\dot{\vec{x}} = A\vec{x} + B\vec{v}$, which is again the standard first-order form. The theory of linear, periodic systems will be developed using this matrix form for the equations.

8-6.1 Linear, Constant Coefficient Equations

The analysis of a linear, time-invariant system will be developed first, as a background for the periodic system analysis. The emphasis in this chapter is on the periodic system and its special behavior, but the time-invariant case is more practical to solve and hence more widely used. Consider the system described by ordinary differential equations of the form $\dot{\vec{x}} = A\vec{x} + B\vec{v}$, where A and B are constant matrices. The state vector $\vec{x}$ has dimension n. The dynamic behavior of this system is determined by the eigenvalues and eigenvectors of the matrix A. For a system of order n, there are n eigenvalues λ_i ($i = 1 \ldots n$) and corresponding eigenvectors $\vec{u}_i$, which are the solution of the algebraic equations $(A - \lambda_i I)\vec{u}_i = 0$. These homogeneous equations have a nonzero solution for $\vec{u}$ only if the determinant of the coefficients is zero: $\det(A - \lambda I) = 0$. This determinant defines a polynomial of order n in λ called the characteristic equation, which may be solved for the eigenvalues. Define Λ as the diagonal eigenvalue matrix, and define the modal matrix M as the matrix with the eigenvectors as columns (ordered to correspond to the eigenvalues in Λ). Then the eigenvalue equation becomes $AM - M\Lambda = 0$, or $A = M\Lambda M^{-1}$.

To show the relation between the eigenvalues and the linear differential equation, consider the homogeneous equation $\dot{\vec{x}} = A\vec{x}$. Expand the state vector $\vec{x}$ in terms of the eigenvectors of A:

$$\vec{x}(t) = \sum_{i=1}^{n} \alpha_i(t)\vec{u}_i$$

where the α_i are scalar functions of time. This expansion is possible because the eigenvectors $\vec{u}_i$ form a complete linearly independent set, so the constants α_i can be found for any $\vec{x}$. Substituting for $\vec{x}$ in $\dot{\vec{x}} = A\vec{x}$, and using $A\vec{u}_i = \vec{u}_i\lambda_i$, gives $\dot{\alpha}_i = \lambda_i\alpha_i$. The solution is $\alpha_i = c_i e^{\lambda_i t}$, where the c_i are scalar constants. The solution of the differential equation has now been obtained in terms of the eigenvalues and eigenvectors:

$$\vec{x}(t) = \sum_{i=1}^{n} c_i e^{\lambda_i t}\vec{u}_i$$

The constants c_i are obtained from the initial conditions $\vec{x}(0) = \sum_{i=1}^{n} c_i\vec{u}_i$. This expression is called the normal mode expansion of the response. In matrix form, the expansion of the state vector in terms of the eigenvectors is accomplished by the linear transformation $\vec{x} = M\vec{q}$, where M is the modal matrix and $\vec{q}$ is the vector of the normal coordinates (which is equivalent to $\alpha_i(t)$ above). Then using $AM = M\Lambda$, the differential equation $\dot{\vec{x}} = A\vec{x}$ becomes $\dot{\vec{q}} = \Lambda\vec{q}$. Since the eigenvalue matrix Λ is diagonal, the differential equations for q_i are decoupled and easily integrated, giving

$$\vec{q} = e^{\Lambda t}\vec{q}(0)$$

or

$$\vec{x} = M e^{\Lambda t}\vec{q}(0)$$

Solving the initial conditions $\vec{x}(0) = M\vec{q}(0)$ for $\vec{q}(0)$ and substituting into the above expression gives

$$\vec{x} = M e^{\Lambda t} M^{-1}\vec{x}(0)$$

or

$$\vec{x} = e^{At}\vec{x}(0)$$

which is the required solution of the homogeneous differential equation.

The analysis of a linear time-invariant system thus basically requires an evaluation of the eigenvalues and eigenvectors of A. The normal mode expansion shows that the solution is unstable if $\text{Re}(\lambda_i) > 0$ for any mode, since in that case $e^{\lambda_i t}$ increases without bound as time increases. Thus the eigenvalues determine the stability of the system, a fact that is often shown graphically as the variation of the roots with some parameter in the plane of $\text{Im}\lambda$ vs. $\text{Re}\lambda$. The system is stable if all the roots are in the left half-plane of the root locus diagram. The eigenvectors $\vec{u}_i$ describe the mode shape of the state variable x corresponding to each eigenvalue. The eigenvalues of the real matrix A must be either real or occur in complex conjugate pairs. Complex roots are usually described in terms of the frequency $\omega = \text{Im}\lambda$ and the damping ratio $\zeta = -\text{Re}\lambda/|\lambda|$ (the natural frequency $\omega_n = |\lambda|$ is also used). The motion is a decaying oscillation at frequency ω. The roots are on the imaginary axis (neutrally damped) when $\zeta = 0$, and on the real axis when $\zeta = 1$; ζ is thus the fraction of critical damping. For an unstable oscillation, $\zeta < 0$. Real roots are described by the time constant $\tau = -1/\lambda$, or by the time to decay to one-half amplitude $t_{1/2} = 0.693\tau$. The eigenvectors corresponding to complex eigenvalues must be complex conjugates also. It follows that the corresponding initial values of the normal coordinates $[\vec{q}(0) = M^{-1}\vec{x}(0)]$ are conjugates. Hence the total contribution of the pair of complex roots to the state vector,

$$\Delta\vec{x} = \vec{u}_1 e^{\lambda_1 t} q_1(0) + \vec{u}_2 e^{\lambda_2 t} q_2(0) = 2\text{Re}[\vec{u}_1 e^{\lambda_1 t} q_1(0)],$$

is real, as required of a physical system.

Now consider the solution including the response to the input vector $\vec{v}$. Using $\vec{x} = M\vec{q}$, the normal form of the differential equation $\dot{\vec{x}} = A\vec{x} + B\vec{v}$ becomes $\dot{\vec{q}} = \Lambda\vec{q} + M^{-1}B\vec{v}$. Since Λ is diagonal, these equations are readily integrated to obtain

$$\vec{q}(t) = e^{\Lambda(t-t_0)}\vec{q}(t_0) + \int_{t_0}^{t} e^{\Lambda(t-\tau)} M^{-1} B\vec{v}\, d\tau$$

The first term is the transient response and depends on the initial conditions, while the second term is the forced response to the input v occurring after t_0. In a stable system, the transient dies out as t increases. In terms of the state vector, the solution is then

$$\vec{x}(t) = e^{A(t-t_0)}\vec{x}(t_0) + \int_{t_0}^{t} e^{A(t-\tau)}B\vec{v}\,d\tau$$

The matrix $\phi(t,t_0) = e^{A(t-t_0)}$ is called the state transition matrix, which relates the state at t to the state at t_0. As an example of forced response of the system, consider a sinusoidal input $\vec{v} = \vec{v}_0 e^{i\omega t}$. It is a property of a linear, time-invariant system that the forced response must also be a sinusoid at frequency ω; hence $x = x_0 e^{i\omega t}$. Integrating the above expression (with $t_0 = -\infty$), or substituting into the differential equation and solving directly, gives

$$\vec{x}_0 = (i\omega - A)^{-1} B\vec{v}_0$$

$$= -(A + i\omega)(A^2 + \omega^2 I)^{-1} B\vec{v}_0$$

This is in the form $\vec{x}_0 = H\vec{v}_0$, where $H(\omega)$ is the matrix of the transfer functions of the system response. Also useful is the step response, obtained by integrating $v = 0$ for $t < 0$ and $v = \vec{v}_0$ for $t > 0$:

$$\vec{x} = A^{-1}(e^{At} - I)B\vec{v}_0$$

The limit as time approaches infinity is the steady-state response, $\vec{x} = -A^{-1}B\vec{v}_0$.

8-6.2 Linear, Periodic Coefficient Equations

Now let us consider a linear, time-varying dynamic system described by ordinary differential equations of the form $\dot{\vec{x}} = A(t)\vec{x} + B(t)\vec{v}$. The coefficient matrices A and B are functions of time. We are particularly interested in periodic systems, for which $A(t + T) = A(t)$, where T is the period. The analysis of periodic coefficient equations is called Floquet-Liapunov theory.

The solution of $\dot{\vec{x}} = A(t)\vec{x}$ must be of the form $\vec{x}(t) = \phi(t,t_0)\vec{x}(t_0)$, since for a linear system the degrees of freedom at t must always be a linear combination of the degrees of freedom at t_0. The matrix $\phi(t,t_0)$ is called the state transition matrix. By definition, $\phi(t_0,t_0) = I$ and $\phi(t_2,t_0) = \phi(t_2,t_1)\phi(t_1,t_0)$, from which it also follows (letting $t_2 = t_0$) that $\phi(t_1,t_0) = \phi^{-1}(t_0,t_1)$. By substituting $\vec{x}(t) = \phi\vec{x}(t_0)$ into $\dot{\vec{x}} = A\vec{x}$, the differential equation for ϕ is obtained: $\dot{\phi} = A\phi$, with initial conditions $\phi(t_0,t_0) = I$. When the response to the input $\vec{v}$ is included, the state transition matrix gives the complete solution:

$$\vec{x}(t) = \phi(t,t_0)\vec{x}(t_0) + \int_{t_0}^{t} \phi(t,\tau)B(\tau)\vec{v}(\tau)\,d\tau$$

Thus the analysis of a linear system response involves finding the state transition matrix. For a time-invariant system, ϕ must have the further property of depending only on the difference $t - t_0$. The results of section 8-6.1 become $\phi(t - t_0) = e^{A(t-t_0)}$ for constant coefficient equations.

Now we shall restrict the system to the periodic coefficient case, $A(t + T) = A(t)$. The differential equation for ϕ becomes

$$\frac{d}{dt}\phi(t,t_0) = A(t)\phi(t,t_0)$$

and

$$\frac{d}{dt}\phi(t+T,t_0) = A(t+T)\phi(t+T,t_0) = A(t)\phi(t+T,t_0)$$

So $\phi(t+T,t_0)$ must be a linear combination of $\phi(t,t_0)$, since both are solutions of the same equation. That is,

$$\phi(t+T,t_0) = \phi(t,t_0)\alpha$$

where α is a constant matrix, depending on the system. Write the state transition matrix as

$$\phi(t,0) = P(t)e^{\beta t}$$

or more generally

$$\phi(t,t_0) = P(t)e^{\beta(t-t_0)}P^{-1}(t_0)$$

where β is a constant matrix defined by $\alpha = e^{\beta T}$. Now

$$\begin{aligned} P(t+T) &= \phi(t+T,0)e^{-\beta(t+T)} \\ &= \phi(t,0)\alpha e^{-\beta T}e^{-\beta t} \\ &= \phi(t,0)e^{-\beta t} \\ &= P(t) \end{aligned}$$

Hence the matrix P is periodic, with initial conditions $P(0) = I$. It has thus been established that the solution of a periodic system must take the form of an exponential factor with decay or growth determined by the constant matrix β, multiplied by a purely periodic factor P. This is the principal result of Floquet theory.

From $\phi(t + T,t_0) = \phi(t,t_0)\alpha$, it follows that $\phi(t + NT,t_0) = \phi(t,t_0)\alpha^N$ Consequently, all the information about the solution is contained in the state transition matrix for a single period. Since by definition $\alpha = \phi(t_0 + T,t_0)$, the solution for all other times can be constructed from that data. Let Θ be the eigenvalue matrix of α, and S the corresponding modal matrix, so that $\alpha = S\Theta S^{-1}$. Then $\alpha^N = S\Theta^N S^{-1}$, from which it follows that the system is unstable, the state transition matrix increasing without bound as time increases, if $|\theta_i| > 1$ for any eigenvalue of α. The more conventional roots of the system are the eigenvalues of β. Let Λ be the eigenvalue matrix of β, and let S be the modal matrix, so that $\beta = S\Lambda S^{-1}$ (α and β have the same eigenvectors). From the definition $\alpha = e^{\beta T}$ it follows the eigenvalues are related by $\Theta = e^{\Lambda T}$, or

$$\Lambda = \frac{1}{T} \ln \Theta$$

The solution is thus unstable if $\mathrm{Re}\lambda_i > 0$ for any eigenvalue. Note that the logarithm of a complex function has many branches, giving values for λ_i that differ in frequency by multiples of $2\pi i/T$. The principal value of λ_i can be used, or the value with the frequency expected from physical considerations.

The state transition matrix for a periodic system can be written in a normal form analogous to that of a time-invariant system. Using the eigenvalues of β gives

$$\begin{aligned} \phi(t,t_0) &= P(t)\, e^{\beta(t-t_0)}\, P^{-1}(t_0) \\ &= [P(t)S]\; e^{\Lambda(t-t_0)}\; [P(t_0)S]^{-1} \end{aligned}$$

which may be compared with the result for a time-invariant system,

$$\phi(t,t_0) = e^{A(t-t_0)} = M e^{\Lambda(t-t_0)} M^{-1}$$

The periodic matrix PS may therefore be considered the modal matrix (i.e. the eigenvectors) of the periodic system, with the eigenvalues Λ determining the principal frequency and damping of the modes. Thus the expansion in normal coordinates $\vec{q}$ is defined as $\vec{x} = PS\vec{q}$. The transient solution $x(t) =$

$\phi(t,t_0)\vec{x}(t_0)$ gives the solution for the normal coordinates, simply $\vec{q}(t) = e^{\Lambda(t-t_0)}\vec{q}(t_0)$ as for the time-invariant case. When $\vec{u}$ is written for the columns of PS, the normal form of the solution is

$$\begin{aligned} \vec{x}(t) &= P(t)S\, e^{\Lambda t}\, \vec{q}(0) \\ &= \sum_i \vec{u}_i(t)\, e^{\lambda_i t}\, q_i(0) \end{aligned}$$

where the initial conditions are obtained from $\vec{q}(0) = S^{-1}\vec{x}(0)$. Compared to the time-invariant system, the periodic system is also described by normal modes u_i and roots λ_i, but now the eigenvectors are periodic functions rather than constants; $\vec{u}_i(t + T) = \vec{u}_i(T)$ follows from the periodicity of P. If the substitution $\phi = Pe^{\beta t}$ is made, the differential equation for ϕ gives

$$\dot{P} = AP - P\beta$$

From this there follows a differential equation for the eigenvectors $\vec{u}_i$, which are the columns of PS:

$$\dot{\vec{u}}_i = (A - \lambda_i I)\vec{u}_i$$

The requirement that $\vec{u}_i$ be periodic is then sufficient to determine the eigenvalues λ_i. For the time-invariant case (A constant) the only "periodic" solution is $\vec{u}_i$ = constant, and the problem reduces to $(A - \lambda_i I)\vec{u}_i = 0$ as above.

The analysis of the dynamic behavior of a system of linear, periodic coefficient equations therefore requires that the state transition matrix ϕ be obtained over one period, $t = 0$ to T, by integrating $\dot{\phi} = A\phi$ with $\phi(0) = I$. The eigenvalues and eigenvectors of the matrix $\alpha = \phi(T)$ are then obtained, and the roots of the system are $\Lambda = (1/T) \ln \Theta$. The mode shapes are given by $PS = \phi S e^{-\Lambda t}$, or $\vec{u}_i = e^{-\lambda_i t}\phi\vec{v}_i$ (where the $\vec{v}_i$ are the eigenvectors of α). The system is unstable if $|\theta_i| > 1$ or Re $\lambda_i > 0$ for any mode. Because the time-varying eigenvectors of a periodic system involve a great deal of information, often the analysis is only concerned with the eigenvalues. For a second-order system with a single degree of freedom, it is possible to construct the characteristic equation directly. The equation $a_2\ddot{x} + a_1\dot{x} + a_0 x = 0$ is integrated over one period, from $t = 0$ to T, with two independent initial conditions. Let x_R be the solution with initial conditions $\dot{x}(0) = 1$ and $x(0) = 0$, and x_P be the solution with initial conditions $\dot{x}(0) = 0$ and $x(0) = 1$. Then the eigenvalues of

$$\alpha = \begin{bmatrix} \dot{x}_R(T) & \dot{x}_P(T) \\ x_R(T) & x_P(T) \end{bmatrix}$$

are the solution of the quadratic equation

$$\theta^2 - [\dot{x}_R(T) + x_P(T)]\theta + [\dot{x}_R(T)x_P(T) - x_R(T)\dot{x}_P(T)] = 0$$

Including the forced response to the input $\vec{v}$, the solution for $\vec{x}$ can be obtained from the state transition matrix as above. Alternatively, by using $\vec{x} = PS\vec{q}$, the normal equations can be integrated to obtain

$$\vec{q}(t) = e^{\Lambda(t-t_0)}\vec{q}(t_0) + \int_{t_0}^{t} e^{\Lambda(t-\tau)}(PS)^{-1}B(\tau)\vec{v}(\tau)d\tau$$

While this is formally similar to the solution for the time-invariant case, here PS and very likely B are periodic matrices. Besides making it difficult to evaluate the response, this periodicity has a fundamental influence on its character. For example, the response to sinusoidal excitation at frequency ω will not be at that same frequency alone, but rather it will be composed of harmonics at frequencies $\omega \pm n2\pi/T$ for all intergers n, where $2\pi/T$ is the fundamental frequency of the system. Thus the frequency response of a periodic system is not described by a single transfer-function matrix, but rather by a transfer function $H_n(\omega)$ for each of the harmonics $\omega + n2\pi/T$, or equivalently by a periodic function of time

$$\sum_{n=-\infty}^{\infty} H_n(\omega)^{i2\pi nt/T}$$

Finally, let us examine in more detail the behavior of the eigenvalues of a periodic system. The eigenvalues θ_i of the matrix $\alpha = \phi(T)$ are either real or occur in complex conjugate pairs. Then the roots λ_i are obtained from $\lambda = (1/T)\ln\theta$, or

$$\lambda = \frac{1}{T}(\ln|\theta| + i\angle\theta) + n\frac{2\pi}{T}i$$

(where $\angle\theta$ is the argument or phase angle of θ). The principal part of the eigenvalue is

$$\lambda_P = \frac{I}{T} (\ln|\theta| + i\angle\theta)$$

and a multiple of the fundamental frequency $2\pi/T$ may be added, depending on the branch of the logarithm that the root is on. A complex conjugate pair for θ gives a conjugate pair for the roots λ_P also. A real, positive θ gives a principal root λ_P with zero imaginary part, so that the frequency of λ is a multiple of the fundamental frequency of the system (i.e. n/rev). For a real and negative θ, the frequency of the principal root λ_P is π/T, one-half the fundamental frequency; the frequency of λ will be $(n+½)$/rev. To interpret these roots, two questions must be answered: how is the branch of the logarithm selected (that is, what multiple of the fundamental frequency is added to the frequency of λ_P?); and what is the meaning of the λ roots associated with real θ? (The λ roots are complex but do not have a corresponding conjugate when θ is real.) As for the interpretation of complex roots of a time-invariant system, these concerns are resolved by considering the actual physical response $\vec{x}(t)$ rather than the eigenvalues and eigenvectors separately. The principal value λ_P is uniquely determined from θ, and there is a corresponding principal value of the mode shape $\vec{u}$. The physical response of the system depends on the product $\vec{u}e^{\lambda t}$. Hence adding a multiple of the fundamental, $in2\pi/T$, to the frequency of the root corresponds to multiplying the mode by the periodic function $e^{-i2\pi nt/T}$. Since the theory only requires that the mode shape $\vec{u}(t)$ be periodic, it offers no guidance on apportioning this periodicity between the eigenvalue and eigenvector. If the system being analyzed is time invariant for some limit, then the frequencies of the roots are determined by the requirement that the roots be continuous as the periodicity is introduced. For example, the periodic coefficients due to the aerodynamic forces in forward flight drop out in the hover limit, $\mu = 0$. One way to mechanize this choice of frequencies is to require that the mean value of the eigenvector have the largest magnitude; then the harmonic of largest magnitude in the eigenvector corresponding to the principal value of the eigenvalue gives the frequency $n2\pi/T$. This criterion gives the correct results for the time-invariant case. The frequencies of the roots may also be established by using a knowledge of the uncoupled natural frequencies of the system, or other considerations of the physical characteristics of the response.

For a real and positive θ root, there is a single complex λ root with a frequency equal to a multiple of the system fundamental frequency. The principal value λ_p is on the real axis, however, so requiring the contribution to $\vec{x}(t)$ be real means that the corresponding principal value of the eigenvector is also real. Giving λ a frequency $in2\pi/T$ then corresponds to multiplying the mode shape by $e^{-i2\pi nt/T}$ without changing the product $\vec{u}e^{\lambda t}$. For a real and negative θ root, the principal value λ_p has a frequency of one-half the system fundamental, $\lambda_p = (1/T)(\ln|\theta| + i\pi)$. Requiring $ue^{\lambda t}$ be real implies that the function $\vec{w}(t) = \vec{u}(t)e^{i\pi t/T}$ is real, and since $\vec{u}$ is periodic it follows that $\vec{w}$ is antiperiodic: $\vec{w}(t + T) = -\vec{w}(t)$. Thus the implication of the ½/rev frequency of λ is that the contribution to the response is of the form

$$\Delta\vec{x} = c_i \vec{w}(t) e^{(t/T)\ln|\theta|}$$

where $\vec{w}(t)$ is a real, antiperiodic function. Therefore, while as eigenvalues of the real matrix α the roots θ must appear as real numbers or complex conjugate pairs, the λ roots are under no such restriction. A real θ gives a single λ root with a frequency equal to a multiple of one-half the fundamental frequency of the system. The property of the solution that allows such behavior is the corresponding periodicity of the eigenvectors.

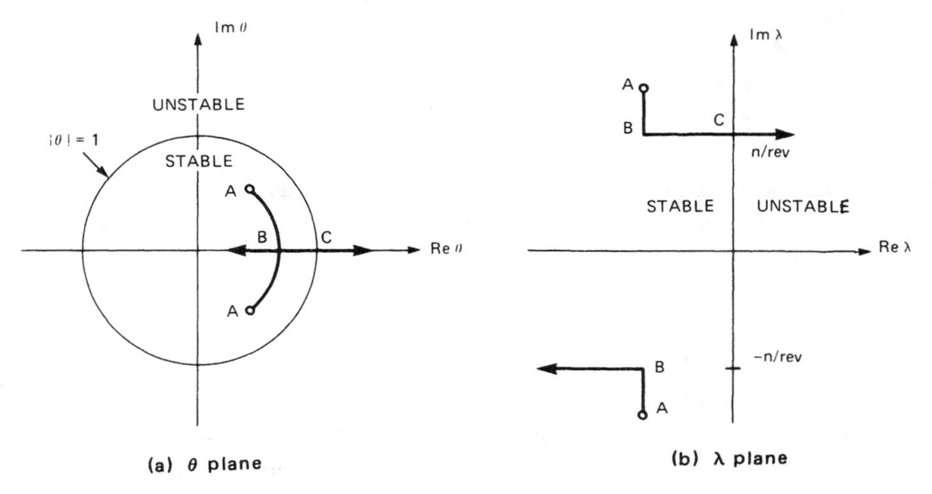

Figure 8-2 Sketch of a possible root locus for a periodic system.

Fig. 8-2 sketches a root locus that might therefore be encountered with a periodic system. The behavior illustrated is typical of systems with strongly periodic coefficients. If the parameter being varied, for example the advance ratio μ, is such that at $\mu = 0$ the system is time invariant, the roots start out as complex conjugates on both the θ and λ planes (point A). As μ increases, the system periodicity increases, and the roots change. The λ roots remain complex conjugates, though, as long as the θ roots are complex. If the θ roots reach the real axis (point B), one will increase along the real axis while the other decreases. On the λ plane the roots have reached an n/rev frequency for some critical μ (or $n + \frac{1}{2}$/rev for negative, real θ), and as μ increases further the real part of one root increases and that of the other decreases while the frequency remains fixed at n/rev. The criterion for instability is $|\theta| > 1$ or Re $\lambda > 0$, so a stability boundary is crossed when the locus moves outside the $|\theta| = 1$ circle on the θ plane, or into the right half-plane on the λ plane. With a time-invariant system, two types of instabilities are possible: a complex conjugate pair of roots may cross the Im λ axis at a positive frequency, or a single root on the real axis may go through the origin into the right half-plane. With periodic systems a third type of instability is introduced, and in fact dominates the behavior for strong periodicity. Fig. 8-2 illustrates this instability of periodic systems. After the θ roots reach the real axis, one becomes less stable and the other more stable. Often the root being destabilized will eventually cross over the stability boundary. For a time-invariant system, such a splitting of the branches of the root loci on the λ plane can only occur at the real axis. With periodic systems this behavior is generalized so that it can occur at any frequency that is a multiple of one-half the fundamental frequency of the system. The interpretation of this behavior is that the instability occurs with the oscillatory motion locked to the frequency of the system.

The Fourier coordinate transformation described in section 8—4 is often associated with the generalized Floquet analysis of linear, periodic coefficient differential equations. Indeed, there is a fundamental link between these topics, because both are associated with the rotation of the system. However, since either one can be required in the rotor analysis without the other, they are truly separate subjects. For example, the Fourier coordinate transformation is needed to represent the blade motion of a rotor in axial flow when coupling with the nonrotating system is involved (such as shaft motion or control inputs), but the rotor is then a constant coefficient system. Alternatively, for the shaft-fixed dynamics of a rotor in forward flight, a single blade representation in the rotating frame is appropriate, but there are periodic coefficients due to the forward flight aerodynamics, and as a consequence the Floquet analysis is needed to determine the system stability.

Chapter 9

ROTARY WING DYNAMICS I

The differential equations of motion for the rotor blade are derived in this chapter. The principal concern here is with the inertial and structural forces on the blade. The rotor aerodynamics are considered only in terms of the net forces and moments on the blade section. In Chapter 11 the equations are completed by analyzing the aerodynamic forces in more detail, and in Chapter 12 the equations are solved for a number of fundamental rotor problems. In Chapter 5 the flap and lag dynamics of an articulated rotor were analyzed for only the rigid motion of the blade, perhaps with a hinge spring or offset. The present chapter extends the derivation of the equations of motion to include a hingeless rotor, higher blade bending modes, and the blade torsion and pitch degrees of freedom. The corresponding hub reactions and blade loads will be derived, and the rotor shaft motion will be included in the analysis.

The rotor blade equations of motion are derived using the Newtonian approach, with a normal mode representation of the blade motion. The chapter concludes with a discussion of the other approaches by which the dynamics may be analyzed. The solution for the blade bending mode shapes and frequencies is also covered in this chapter. Engineering beam theory is almost universally used in helicopter blade analyses. The blade is assumed to be rigid chordwise, so its motion is represented by the bending and rotation of a slender beam. This is normally a very good model for the rotor blade, although a more detailed structural analysis may be required to obtain the effective beam parameters for some portions of the blade, such as at the root.

9–1 Sturm-Liouville Theory

The results of Sturm-Liouville theory will be required in dealing with the normal modes of the blade bending and torsion motion. Consider an ordinary differential equation of the form $\mathcal{L}y + \lambda Ry = 0$, where $\mathcal{L}$ is a linear differential operator of the form

$$\mathcal{L} = \frac{d^2}{dx^2} S \frac{d^2}{dx^2} + \frac{d}{dx} P \frac{d}{dx} + Q$$

Here S, P, Q, and R are symmetric operators. (An operator S is symmetric if $\phi_1 S\phi_2 = \phi_2 S\phi_1$ for all functions ϕ_1 and ϕ_2.) With the appropriate boundary conditions at the end points $x = a$ and $x = b$, this is an eigenvalue problem for λ.

Consider any two distinct eigenvalues λ_1 and λ_2 and their corresponding eigenfunctions ϕ_1 and ϕ_2. Using the differential equations satisfied by these functions, and integrating twice by parts, we obtain

$$\begin{aligned}(\lambda_2 - \lambda_1) \int_a^b \phi_1 R\phi_2 dx &= \int_a^b (\phi_2 \mathcal{L}\phi_1 - \phi_1 \mathcal{L}\phi_2) dx \\ &= \left[\phi_2 \left(\frac{d}{dx} S \frac{d^2\phi_1}{dx^2} + P \frac{d\phi_1}{dx}\right)\right. \\ &\quad \left.- \phi_1 \left(\frac{d}{dx} S \frac{d^2\phi_2}{dx^2} + P \frac{d\phi_2}{dx}\right)\right]\Bigg|_a^b \\ &\quad - \left[\frac{d\phi_2}{dx} S \frac{d^2\phi_1}{dx^2} - \frac{d\phi_1}{dx} S \frac{d^2\phi_2}{dx^2}\right]\Bigg|_a^b\end{aligned}$$

The right-hand side is zero for boundary conditions of the following form:

$$\frac{d}{dx} S \frac{d^2y}{dx^2} = K_1 y \quad \text{and} \quad S \frac{d^2y}{dx^2} = K_2 \frac{dy}{dx}$$

or $S = 0$; and

$$P \frac{dy}{dx} = K_3 y$$

or $P = 0$ (where K_1, K_2, and K_3 are constants). With such boundary conditions

$$\int_a^b \phi_1 R\phi_2 dx = 0$$

so the eigensolutions are orthogonal over the interval from a to b, with weighting function R. For a beam in bending, the following end restraints will satisfy these boundary conditions: (a) a free end, for which $d^2y/dx^2 = d^3y/dx^3 = 0$ and $P = 0$; (b) a hinged end, for which $y = 0$ and $Sd^2y/dx^2 = Kdy/dx$, where K is the hinge spring constant ($d^2y/dx^2 = 0$ with no spring); or (c) a cantilever end, for which $y = 0$ and $dy/dx = 0$ (which is also the limit of $K \rightarrow \infty$ with a spring). For a rod in torsion (so $S = 0$), the following end restraints satisfy the boundary conditions: (a) a free end with $dy/dx = 0$; (b) a fixed end with $y = 0$; or (c) a restrained end with $Pdy/dx = Ky$, where K is the spring constant.

A proper Sturm-Liouville problem has boundary conditions of the form given above, and R and P of opposite sign to S and Q. These conditions are satisfied for the blade bending and torsion problems encountered in this chapter. It follows then that the eigensolutions are orthogonal, the eigenvalues λ are real and positive; and that an expansion of an arbitrary function over the interval $x = a$ to $x = b$ as a series in the eigensolutions will converge.

The eigenvalue λ may be obtained from the eigensolutions as follows:

$$-\lambda \int_a^b \phi R \phi dx = \int_a^b \phi \, \mathcal{L} \phi dx$$

$$= \left[\phi \frac{d}{dx} S \frac{d^2\phi}{dx^2} - \frac{d\phi}{dx} S \frac{d^2\phi}{dx^2} + \phi P \frac{d\phi}{dx} \right] \Bigg|_a^b$$

$$+ \int_a^b \left[\frac{d^2\phi}{dx^2} S \frac{d^2\phi}{dx^2} - \frac{d\phi}{dx} P \frac{d\phi}{dx} + \phi Q \phi \right] dx$$

For example, for a beam with a free end at $x = b$ and a general restrained end at $x = a$,

$$-\lambda \int_a^b \phi R \phi dx = \frac{d\phi}{dx} K \frac{d\phi}{dx} \Bigg|_{x=a}$$

$$+ \int_a^b \left[\frac{d^2\phi}{dx^2} S \frac{d^2\phi}{dx^2} - \frac{d\phi}{dx} P \frac{d\phi}{dx} + \phi Q \phi \right] dx$$

and for a rod with a free end at $x = b$ and a restrained end at $x = a$,

$$\lambda \int_a^b \phi R \phi dx = \phi K \phi \Big|_{x=a} + \int_a^b \left[\frac{d\phi}{dx} P \frac{d\phi}{dx} - \phi Q \phi \right] dx$$

The exact value for λ is obtained if the exact eigenfunction is used, but these expressions are also useful to estimate λ by using mode shapes that are only approximately correct.

9–2 Out-of-Plane Motion

9-2.1 Rigid Flapping

As a guide to the analyses that follow, let us review the derivation of the equation of motion for rigid flapping of an articulated blade, which is given in detail in Chapter 5. The degree of freedom β is the angle of rigid rotation about the flap hinge (Fig. 9-1), so the out-of-plane deflection is $z = \beta r$. There is no flap hinge offset or spring restraint. The equation of motion is obtained from equilibrium of moments about the flap hinge. (Based on the results of section 5-9, the gravitational moments will be neglected in this chapter.) The section forces and their moment arms about the flap hinge are as follows:

(i) an inertial force $m\ddot{z} = mr\ddot{\beta}$ opposing the flap motion, with moment r;

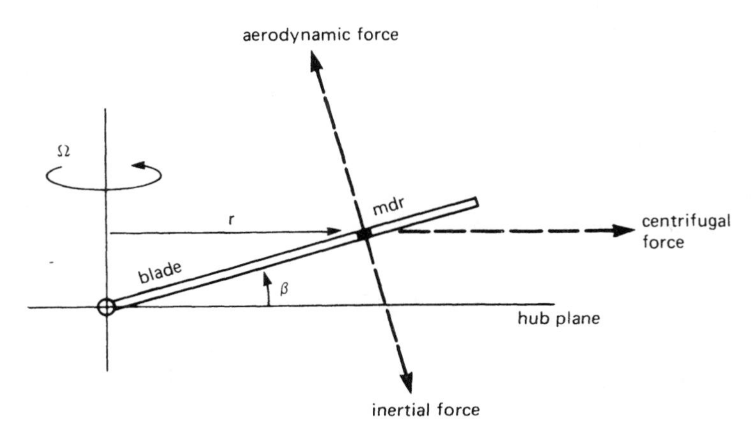

Figure 9-1 Rotor blade flapping moments.

(ii) a centrifugal force $m\Omega^2 r$ directed radially outward, with moment arm $z = r\beta$; and

(iii) an aerodynamic force F_z, with moment arm r.

Here m is the blade section mass per unit length. From the conditions for equilibrium of moments about the flap hinge,

$$\left(\int_0^R r^2 m dr\right)\left(\ddot{\beta} + \Omega^2\beta\right) = \int_0^R r F_z dr$$

On dividing by the flap moment of inertia $I_b = \int_0^R r^2 m dr$ and using dimensionless quantities,

$$\ddot{\beta} + \beta = \frac{1}{I_b}\int_0^1 r F_z dr$$

$$= \gamma \int_0^1 r \frac{F_z}{ac} dr$$

where $\gamma = \rho ac R^4 / I_b$ is the blade Lock number. This is the required equation of motion for the rigid flapping of an articulated rotor blade. The centrifugal spring gives a natural frequency of $\nu = 1$/rev in the rotating frame.

With an offset flap hinge, the blade out-of-plane deflection due to rigid rotation about the hinge becomes $z = \beta\eta$, where β is the degree of freedom and the mode shape is $\eta = (r - e)/(1 - e)$ (e is the flap hinge offset). Since the mode shape has been normalized to $\eta = 1$ at the tip, β is the angle a line from the center of rotation to the blade tip makes with the hub plane. The section forces are now:

(i) an inertial force $m\ddot{z} = m\eta\ddot{\beta}$, with moment arm $(r - e)$;

(ii) a centrifugal force $m\Omega^2 r$, with moment arm $z = \eta\beta$; and

(iii) an aerodynamic force F_z, with moment arm $(r - e)$.

Including a hinge spring (with precone angle β_p), the expression for moment equilibrium about the flap hinge becomes

$$\int_e^R m\eta\ddot{\beta}(r-e)dr + \int_e^R m\Omega^2 r\eta\beta dr + K_\beta(\beta - \beta_p) = \int_e^R (r-e)F_z dr$$

Divide by $(1 - e)$ and write $I_\beta = \int_e^R \eta^2 m\,dr$ for the generalized mass of the flap mode. Then

$$I_\beta(\ddot{\beta} + \nu^2\beta) = \frac{K_\beta}{\Omega^2(1-e)}\beta_p + \int_e^1 \eta F_z\,dr$$

where the natural frequency of the flap motion is

$$\nu^2 = 1 + \frac{e}{1-e}\frac{\int_e^1 \eta m\,dr}{\int_e^1 \eta^2 m\,dr} + \frac{K_\beta}{I_\beta\Omega^2(1-e)}$$

Finally, divide by the characteristic inertia I_b to obtain

$$I_\beta^*(\ddot{\beta} + \nu^2\beta) = \frac{K_\beta}{I_b\Omega^2(1-e)}\beta_p + \gamma\int_e^1 \eta\frac{F_z}{ac}\,dr$$

where $I_\beta^* = I_\beta/I_b$, and the Lock number is defined again as $\gamma = \rho acR^4/I_b$. The flap frequency $\nu = 1$/rev if there is no hinge offset or spring. For a uniform mass distribution,

$$\nu^2 = 1 + \frac{3}{2}\frac{e}{1-e} + \frac{K_\beta}{I_\beta\Omega^2(1-e)}$$

so in general $\nu > 1$/rev.

We shall follow the practice in this chapter of using I_b for the characteristic inertia of the rotor blade. The function of this parameter is to normalize the generalized masses of the blade motion (such as $I_\beta^* = I_\beta/I_b$) and to represent the blade inertial forces in the Lock number, $\gamma = \rho acR^4/I_b$. The normalization is desired since the dimensionless moments of inertia are divided by ρR^5, and thus vary with the air density. Note that the actual value of I_b has no influence on the numerical solution, since the entire equation of motion is divided by I_b. A good choice for the blade inertia is $I_b = \int_0^R r^2 m\,dr$, even for blades with hinge offset or no flap hinge at all. This is a well-defined parameter of the blade that may be obtained from the rotary inertia about the shaft and avoids any dependence on the flap mode shape.

9-2.2 Out-of-Plane Bending

Consider now the out-of-plane bending of a rotor blade with arbitrary root constraint. This model will include the higher bending modes of articulated blades and cover the case of a hingeless rotor blade with cantilever root restraint. In Chapter 5, hingeless rotor dynamics were discussed in terms of the fundamental flapping mode; the present analysis adds the equation by which the frequency and mode shape may be calculated, as well as a rigorous derivation of the differential equation of motion. The equation of motion is obtained by considering the equilibrium of aerodynamic, inertial, and structural bending moments on the blade portion outboard of radial station r. Let $z(r)$ be the out-of-plane deflection of the blade. The forces acting on the blade section at radial station ρ, with their moment arm about the radial station at r, are as follows:

(i) the inertial force $m\ddot{z}(\rho)$, with moment arm $(\rho - r)$;

(ii) the centrifugal force $m\Omega^2\rho$, with moment arm $z(\rho) - z(r)$; and

(iii) the aerodynamic force F_z, with moment arm $(\rho - r)$.

The moment on the blade section at r due to the forces acting on the blade outboard of r is then

$$M(r) = \int_r^R \left[(F_z - m\ddot{z})(\rho - r) - m\,\Omega^2\rho\left(z(\rho) - z(r)\right)\right] d\rho$$

Now engineering beam theory relates the structural moment to the bending curvature of the blade:

$$M(r) = EI\frac{d^2z}{dr^2}$$

where E is the modulus of elasticity of the blade section and I is the modulus-weighted area moment about the chordwise principal axis. The equilibrium distribution of the structural, inertial, and aerodynamic moments on the section gives

$$EI\frac{d^2z}{dr^2} + \int_r^R m\Omega^2\rho\left(z(\rho) - z(r)\right)d\rho + \int_r^R m\ddot{z}(\rho - r)d\rho$$

$$= \int_r^R F_z(\rho - r)d\rho$$

The second derivative then gives the partial differential equation for the out-of-plane bending of a rotor blade:

$$\frac{d^2}{dr^2}\,EI\frac{d^2z}{dr^2} - \frac{d}{dr}\left[\int_r^R m\Omega^2\rho d\rho\,\frac{dz}{dr}\right] + m\ddot{z} = F_z$$

The boundary conditions are as follows. The blade tip is a free end, with zero moment and shear force, so $d^2z/dr^2 = d^3z/dr^3 = 0$ at $r = R$. The root of an articulated blade has a hinge, with zero displacement and moment, so $z = d^2z/dr^2 = 0$ at $r = e$ (allowing for a hinge offset). The root of a hingeless rotor has cantilever restraint, with zero displacement and slope, so $z = dz/dr = 0$ at $r = e$ (allowing for an extremely rigid hub). The root restraint can be generalized by considering a hinge with spring constant K_β, so that $EI(d^2z/dr^2) = K_\beta(dz/dr)$ at $r = e$. For $K_\beta = 0$ this reduces to the articulated rotor case, and for $K_\beta = \infty$ it reduces to the hingeless rotor case.

The partial differential equation for the blade bending will be solved by the method of separation of variables, which leads to ordinary differential equations (in time) for the degrees of freedom, as for rigid flapping. Thus the out-of-plane deflection $z(r,t)$ will be expanded as a series in mode shapes describing the spanwise deformation. A single equation of motion is obtained for the degree of freedom corresponding to each mode. First, it is necessary to obtain an appropriate series of mode shapes for the rotating blades. When the mode shapes are chosen such that the forced response of the blade is well described by the first few modes, the rotor dynamics problems can be solved by considering the smallest number of degrees of freedom. Consider the free vibration of the rotating blade at frequency ν. That is, in the homogeneous partial differential equation for the bending (without the aerodynamic force F_z in this case), write $z = \eta(r)e^{i\nu t}$, where η is the spanwise mode shape. The result is

$$\frac{d^2}{dr^2} EI \frac{d^2\eta}{dr^2} - \frac{d}{dr}\left[\int_r^R m\Omega^2 \rho\, d\rho \frac{d\eta}{dr}\right] - \nu^2 m\eta = 0$$

with the same boundary conditions on η as given above for z, for a hinged or cantilevered blade as appropriate. Since the aerodynamic force has been dropped, this can be viewed as the equation for vibration in a vacuum, involving the equilibrium of structural, centrifugal, and inertial moments alone. This modal equation and its boundary conditions constitute an eigenvalue problem for the natural frequencies ν and mode shapes $\eta(r)$. According to section 9–1 it is a proper Sturm-Liouville problem, from which it follows that there exists a series of eigensolutions $\eta_k(r)$ and corresponding eigenvalues $\nu_k{}^2$. The mode shapes are orthogonal with weighting function m; if $i \neq k$,

$$\int_0^R \eta_k \eta_i m\, dr = 0$$

Moreover, an expansion of an arbitrary function of r (such as the actual blade bending deflection) as a series in these modes will converge. The modal equation is linear, so the solutions are only defined to within a multiplicative factor. The mode shapes will be normalized to unit deflection at the blade tip: $\eta(1) = 1$ (or $\eta(R) = R$ with dimensional quantities). The series of natural frequencies ν_1, ν_2, ν_3, etc. will be ordered by magnitude, such that the fundamental mode ν_1 has the lowest frequency. When the modes are ordered in this fashion, it can be shown that the kth mode shape has $k-1$ modes where $\eta(r) = 0$ (not counting the root, where $\eta(0) = 0$ always).

Now the bending deflection z is expanded as a series in the rotating natural mode shapes:

$$z(r,t) = \sum_{k=1}^{\infty} \eta_k(r) q_k(t)$$

The degrees of freedom of the bending motion are $q_k(t)$. With the modes normalized to unit deflection at the tip, q_k represents the angle from the hub plane made by a line from the center of rotation to the tip for the kth mode. Since orthogonal modes are being used, a simple set of equations for q_k will be obtained. Substitute this expansion for z into the partial differential equation for bending:

$$\sum_k \left\{ (EI\eta_k'')'' - \left[\int_r^R m\Omega^2 \rho\, d\rho\, \eta_k' \right]' \right\} q_k + \sum_k m\eta_k \ddot{q}_k = F_z$$

Now the differential equation satisfied by the mode shape η_k states that the terms in brackets equal just $\nu_k^2 m\eta_k$, giving

$$\sum_{k=1}^{\infty} m\eta_k(\ddot{q}_k + \nu_k^2 q_k) = F_z$$

Next, operate on this equation with $\int_0^R (...)\eta_k\, dr$. Define

$$I_{q_k} = \int_0^R \eta_k^2 m\, dr$$

as the generalized mass of the kth bending mode, and recall that $\int_0^R \eta_k\eta_i m dr = 0$ if $i \neq k$. Then the bending equation becomes

$$I_{q_k}(\ddot{q}_k + \nu_k^2 q_k) = \int_0^R \eta_k F_z dr$$

Using the free vibration modes of the rotating blade has allowed the structural and centrifugal terms to be replaced by the natural frequencies ν_k, and because these modes are orthogonal the differential equation for the kth mode is not coupled with other bending modes (except through the aerodynamic force). Dividing by I_b and using dimensionless quantities gives

$$I_{q_k}^*(\ddot{q}_k + \nu_k^2 q_k) = \gamma \int_0^1 \eta_k \frac{F_z}{ac} dr$$

where $I_{q_k}^* = I_{q_k}/I_b$. This is the differential equation of motion for the kth out-of-plane bending mode of the elastic rotor blade.

A further result of Sturm-Liouville theory (section 9–1) is that the natural frequencies can be obtained from the mode shapes by

$$\nu^2 = \frac{K_\beta[\eta'(e)]^2 + \int_0^R \left[EI\eta''^2 + \int_r^R m\Omega^2 \rho\, d\rho\, \eta'^2\right] dr}{\int_0^R \eta^2 m\, dr}$$

(The dimensionless frequency is obtained by dividing by Ω^2.) This relation can be interpreted as an energy balance: $\nu^2 \int \eta^2 m\, dr$ is the maximum kinetic energy of the vibrating blade, $\int EI\eta''^2 dr$ is the maximum potential energy of bending, $K_\beta(\eta'(e))^2$ is the potential energy in the hinge spring, and $\int\int m\Omega^2 \rho\, d\rho\, \eta'^2 dr$ is the potential energy in the centrifugal spring, Note that this relation can be written as $\nu^2 = K_1 + K_2\Omega^2$, which is the Southwell form. The Southwell coefficients K_1 and K_2 (representing the structural and centrifugal stiffening respectively) are constants involving integrals of the blade mode shape, which in fact will also be somewhat sensitive to the rotor speed Ω, but the Southwell form gives the basic dependence of the blade bending frequencies on the rotor speed (for a further discussion see section 9-8.3). This energy relation gives the exact frequency when the correct mode shape (which must be obtained by solving the modal equation) is used. It is also the basis for obtaining estimates of the natural frequencies by using approximate mode shapes. Since the modes are integrated, the accuracy of the frequency estimate will be good as long as the modes are fairly close to the correct shape.

The fundamental flapping mode is the lowest frequency solution of the modal equation. For an articulated rotor with no hinge offset or spring it is easily verified that $\eta = r$ satisfies the differential equation with dimensionless natural frequency $\nu = 1$/rev, which is also obtained from

$$\nu^2 = \frac{\int_0^1 \int_r^1 m\rho\, d\rho\, dr}{\int_0^1 r^2 m\, dr} = \frac{\int_0^1 \rho m \int_0^\rho dr\, d\rho}{\int_0^1 r^2 m\, dr} = 1$$

The equation thus reduces to that for rigid flapping, as is required. With a hinge offset and spring, the mode shape $\eta = (r - e)/(1 - e)$ gives the same equation of motion and the natural frequency

$$\nu^2 = \frac{\dfrac{1}{(1-e)^2}\displaystyle\int_e^1\int_r^1 m\rho\, d\rho\, dr}{\displaystyle\int_e^1 \eta^2 m\, dr} + \frac{K_\beta}{I_\beta\Omega^2(1-e)^2}$$

$$= \frac{\displaystyle\int_e^1 m\rho(\rho-e)d\rho}{(1-e)^2\displaystyle\int_e^1 \eta^2 m\, dr} + \frac{K_\beta}{I_\beta\Omega^2(1-e)^2}$$

$$= 1 + \frac{e}{1-e}\,\frac{\displaystyle\int_e^1 \eta m\, dr}{\displaystyle\int_e^1 \eta^2 m\, dr} + \frac{K_\beta}{I_\beta\Omega^2(1-e)^2}$$

as in section 9-2.1. [The additional factor of $(1 - e)$ in the spring term is due to a different definition of the spring constant K_β.] Note that the modal equation is actually not quite satisfied if $\eta = (r - e)/(1 - e)$, but the bending involved in the fundamental mode of an articulated blade will be very small. For a hingeless rotor, there must be bending at the blade root, where the cantilever restraint requires zero slope. However, the centrifugal stiffening dominates the fundamental mode of even the hingeless blade, as indicated by the fact that the natural frequency is only slightly above 1/rev (typically $\nu = 1.10$ to 1.20). Except in the root region, therefore, the mode shape of the hingeless blade will not differ substantially from that of the articulated rotor. It is the natural frequency that is the dominant parameter of the blade bending mode, not the mode shape. Thus even the small increase of the fundamental frequency above 1/rev for the hingeless blade has a major impact on the root loads of the blade, and on the behavior of the rotor in general.

The second flapwise bending mode has a rotating natural frequency typically around 2.6 to 2.8/rev. As the modal number increases, so do the number of nodes and the curvature of the mode shape. The higher modes thus play an important role in the blade bending loads and their calculation. For an articulated blade, the second out-of-plane mode is often called the first bending mode, since the fundamental flap mode does not involve elastic motion of the blade. If no better estimate is available, $\eta = 4r^2 - 3r$ can be used as an approximation to the second out-of-plane mode shape of an articulated rotor blade. This expression is orthogonal to the first mode, $\eta = r$; however, the boundary conditions of zero moment at the root and tip are not satisfied. The expression $\eta = r - (\pi/3)\sin \pi r$ has also been suggested; it satisfies all the conditions except for zero shear at the tip. Such approximate mode shapes are useful for evaluating the inertial and aerodynamic coefficients in a dynamics analysis, and particularly for estimating the natural frequency of the second mode from the energy relation.

The utility of the normal mode representation of the blade motion depends on being able to use only a small number of modes to solve most rotor problems. The frequency content of the forces exciting the blade provides a good guide to the number of modes that must be included. In many cases, the fundamental flap mode is a sufficient representation of the blade for both articulated and hingeless rotors. For problems such as the calculation of oscillatory rotor loads or helicopter vibration, up to 3 or 5 out-of-plane modes may be required.

9-2.3 Nonrotating Frame

The degrees of freedom and equations of motion in the nonrotating frame are obtained using the Fourier coordinate transformation, as discussed in section 8–4. The equations derived for the out-of-plane bending are for each blade of an N-bladed rotor in the rotating frame. The Fourier coordinate transformation introduces N degrees of freedom ($\beta_0, \beta_{1c}, \beta_{1s}, \ldots, \beta_{nc}, \beta_{ns}, \beta_{N/2}$) to describe the rotor motion in the nonrotating frame. The corresponding N equations of motion are obtained by operating on the rotating equation with

$$\frac{1}{N}\sum_{m=1}^{N}(\ldots),\ \frac{2}{N}\sum_{m=1}^{N}(\ldots)\cos n\psi_m,\ \frac{2}{N}\sum_{m=1}^{N}(\ldots)\sin n\psi_m,\ \frac{1}{N}\sum_{m=1}^{N}(\ldots)(-1)^m$$

as appropriate. The inertial and structural terms of the equation that will be encountered in this chapter have constant coefficients. These summation operators therefore act only on the rotating degrees of freedom and their time derivatives. By using the definitions of the nonrotating degrees of freedom (and the corresponding transformations of the time derivatives, given in section 8-4.1), the conversion of the equations of motion to the nonrotating frame is straightforward.

With independent blades the equations of motion in the rotating frame can be used directly. Unless there is some coupling of the blades through the fixed frame, there is no reason to use the Fourier coordinate transformation (except that the constant coefficient approximation for the aerodynamics in forward flight is better made in the nonrotating frame). The usefulness of the transformation will be more apparent later in this chapter, when rotor shaft motion is involved.

Consider the fundamental flap mode of an articulated or hingeless rotor blade. The equation of motion for the mth blade ($m = 1$ to N) in the rotating frame is

$$I_\beta^* (\ddot{\beta}^{(m)} + \nu^2 \beta^{(m)}) = \gamma \int_0^1 \eta \frac{F_z}{ac} dr = \gamma M_F^{(m)}$$

Applying the summation operators, which act only on $\ddot{\beta}^{(m)}$ and $\beta^{(m)}$, then gives directly

$$I_\beta^* (\ddot{\beta}_0 + \nu^2 \beta_0) = \frac{1}{N} \sum_{m=1}^{N} \gamma M_F^{(m)} = \gamma M_{F_0}$$

$$I_\beta^* \left(\ddot{\beta}_{nc} + 2n \dot{\beta}_{ns} + (\nu^2 - n^2) \beta_{nc} \right) = \frac{2}{N} \sum_{m=1}^{N} \gamma M_F^{(m)} \cos n\psi_m = \gamma M_{F_{nc}}$$

$$I_\beta^* \left(\ddot{\beta}_{ns} - 2n \dot{\beta}_{nc} + (\nu^2 - n^2) \beta_{ns} \right) = \frac{2}{N} \sum_{m=1}^{N} \gamma M_F^{(m)} \sin n\psi_m = \gamma M_{F_{ns}}$$

$$I_\beta^* (\ddot{\beta}_{N/2} + \nu^2 \beta_{N/2}) = \frac{1}{N} \sum_{m=1}^{N} \gamma M_F^{(m)} (-1)^m = \gamma M_{F_{N/2}}$$

The influence of this transformation on the eigenvalues and eigenvectors of the rotor dynamics is discussed in section 8–5. We will not consider the solution further until Chapter 12, when the aerodynamic forces can be included.

9-2.4 Bending Moments

The flapwise bending moment on the blade has been obtained in section 9-2.2 as

$$M(r) = \int_r^R \left[(F_z - m\ddot{z})(\rho - r) - m\Omega^2 \rho \left(z(\rho) - z(r) \right) \right] d\rho$$

Substituting for the modal expansion of z and using dimensionless quantities gives

$$M(r) = \int_r^1 F_z(\rho - r)\, d\rho$$

$$- \sum_k \left[\ddot{q}_k \int_r^1 m\eta_k(\rho - r) d\rho + q_k \int_r^1 m\rho \left(\eta_k(\rho) - \eta_k(r) \right) d\rho \right]$$

Now expand the aerodynamic loading as a series in the bending mode shapes: $F_z = \sum_k F_{z_k} m\eta_k(r)$. It is easily shown that the constants are $F_{z_k} = \int_0^1 \eta_k F_z dr / I_{q_k}$. On substituting for the expansion of F_z into the bending moment, and noting that the equation of motion for the kth bending mode gives $F_{z_k} = \ddot{q}_k + \nu_k{}^2 q_k$, the bending moment becomes

$$M(r) = \sum_k q_k \left[\nu_k^2 \int_r^1 m\eta_k(\rho - r) d\rho - \int_r^1 m\rho \left(\eta_k(\rho) - \eta_k(r) \right) d\rho \right]$$

Thus the bending moment can be evaluated from the response of the blade modes and from the corresponding mode shapes and frequencies. The bending moment can also be obtained from the blade curvature:

$$M(r) = EI \frac{d^2 z}{dr^2} = \sum_k q_k (EI \eta_k'')$$

which is equivalent to the previous expression, as may be shown by integrating the differential equation for η_k twice.

When the number of blade modes is large, all of these expressions for the bending moment must give the same results. With a small number of modes, however, the best accuracy is obtained by directly integrating the section aerodynamic forces and acceleration over the blade (the first expression given here). In the other methods it is assumed that the aerodynamic loading distribution can be well represented by the response of the bending modes (i.e. the truncated series $\sum_k F_{z_k} m \eta_k$), which may not be true with a small number of modes. The greatest difficulty is encountered using $M = EI d^2 z/dr^2$. Not only are a large number of modes required as a result of the greater relative contribution of the higher modes to the curvature, but also there are numerical problems because the second derivative of the deflection (and hence of the mode shapes) is required.

9–3 In-Plane Motion

9-3.1 Rigid Flap and Lag

In Chapter 5 an introduction was provided to the lag dynamics of an articulated rotor. Here the coupled equations for rigid flap and lag motion will be derived in more detail. Consider an articulated rotor with both flap and lag hinges. Hinge offsets and springs will be included, and the flap and lag offsets are not necessarily equal. The degree of freedom for rigid rotation about the flap hinge is again β, with mode shape $\eta_\beta = (r - e)/(1 - e)$. The in-plane motion consists of rigid rotation about the lag hinge, generating an in-plane displacement $x = \zeta \eta_\zeta$, where ζ is the lag degree of freedom with mode shape $\eta_\zeta = (r - e)/(1 - e)$. The flap motion is positive when upward, and the lag motion is positive in the direction opposite the rotor rotation. The equations of motion are obtained from the conditions for equilibrium of moments about the hinges.

The section forces producing flap moments remain the same as in section 9-2.1, with the addition of a Coriolis force due to the lag motion. The Coriolis acceleration is twice the cross product of the angular velocity vector and the velocity vector relative to the rotating frame. The inertial force in the d'Alembert sense is then in the opposite direction. The product of the rotor rotational velocity Ω and the in-plane velocity of the section $\dot{x}$ gives a Coriolis force $2\Omega\dot{x}m = 2\Omega\dot{\zeta}\eta_\zeta m$, directed radially inward. This force has a moment arm $z = \eta_\beta\beta$ about the flap hinge, producing a total moment of

$$-\int_e^R (2\Omega\eta_\zeta\dot{\zeta}m)(\eta_\beta\beta)dr$$

Including this term in the moment equilibrium gives for the flap equation of motion

$$I_\beta^*(\ddot{\beta} + \nu_\beta^2\beta) - I_{\beta\zeta}^*\,2\beta\dot{\zeta} = \frac{K_\beta}{I_\beta\Omega^2(1-e)}\beta_p + \gamma\int_e^1 \eta_\beta\frac{F_z}{ac}dr$$

where $I_{\beta\zeta}^* = \int_e^1 \eta_\beta\eta_\zeta m\,dr/[(1-e)I_b]$.

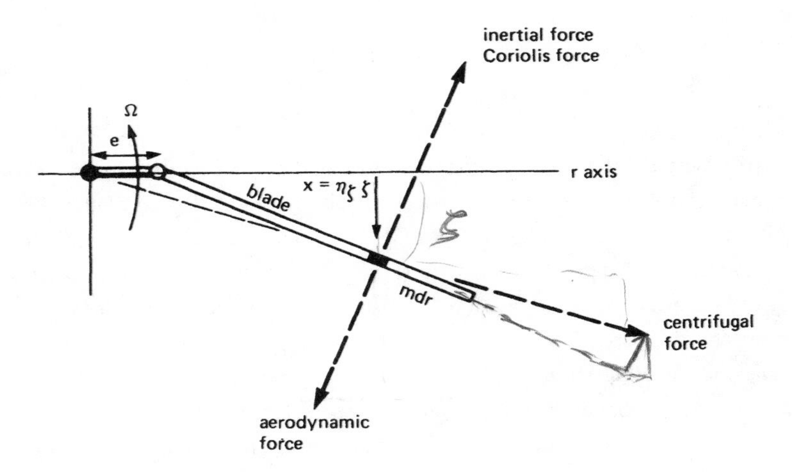

Figure 9-2 Rotor blade lagging moments

The in-plane forces acting on the blade section (Fig. 9-2) and their moment arms about the offset lag hinge are as follows:

(i) an inertial force $m\ddot{x} = m\,\eta_\zeta\ddot{\zeta}$ opposing the lag motion, with moment arm $(r-e)$;

(ii) a centrifugal force $m\Omega^2 r$ directed radially outward, with a moment arm about the lag hinge of $(e/r)x = (e/r)\eta_\zeta\zeta$;

(iii) an aerodynamic force F_x, with moment arm $(r - e)$; and

(iv) a Coriolis force $2\Omega\dot{z}z'm = 2\Omega\beta\dot{\beta}\eta_\beta\eta'_\beta m$, in the same direction as the inertial force and with moment arm $(r - e)$.

Since the out-of-plane velocity $\dot{z}$ has a radially inward component $\dot{z}(dz/dr)$ when the blade is flapped up, the Coriolis force arises from the product of the rotational speed of the rotor and this radial velocity of the blade. The conditions for equilibrium of moments about the lag hinge gives

$$\left(\int_e^R (r-e)\eta_\zeta m\,dr\right)\ddot{\zeta} + \left(\int_e^R e\eta_\zeta m\,dr\right)\Omega^2\zeta$$

$$+\left(\int_e^R \eta_\beta\eta_\zeta m\,dr\right)2\Omega\beta\dot{\beta} + K_\zeta\zeta = \int_e^R (r-e)F_x\,dr$$

Dividing by $(1 - e)$ and using dimensionless quantities gives

$$\left(\int_e^1 \eta_\zeta^2 m\,dr\right)\ddot{\zeta} + \left(\frac{e}{1-e}\int_e^1 \eta_\zeta m\,dr\right)\zeta + \frac{K_\zeta}{\Omega^2(1-e)}\zeta$$

$$+\left(\frac{1}{1-e}\int_e^1 \eta_\beta\eta_\zeta m\,dr\right)2\beta\dot{\beta} = \int_e^1 \eta_\zeta F_x\,dr$$

Next, define the lag inertia as $I_\zeta = \int_e^R \eta_\zeta^2 m\,dr$ and divide by I_b to obtain

$$I_\zeta^*(\ddot{\zeta} + \nu_\zeta^2\zeta) + I_{\beta\zeta}^* 2\beta\dot{\beta} = \gamma\int_0^1 \eta_\zeta \frac{F_x}{ac}\,dr$$

which is the equation of motion for the rigid lag of an articulated blade.

The rotating natural frequency of the lag motion is

$$\nu_\zeta^2 = \frac{e}{1-e}\,\frac{\int_e^1 \eta_\zeta m\,dr}{\int_e^1 \eta_\zeta^2 m\,dr} + \frac{K_\zeta}{I_\zeta\Omega^2(1-e)}$$

As discussed in section 5–19, the lag hinge must have an offset or spring to obtain a nonzero lag frequency. For a uniform mass distribution and no spring,

$$\nu_\zeta^2 = \frac{3}{2}\,\frac{e}{1-e}$$

More generally, the lag frequency is given by $\nu_\zeta^2 = eS_\zeta/I_\zeta$, where I_ζ is the second moment of inertia about the lag hinge and S_ζ is the first moment. (S_ζ equals the product of the blade mass and the radial distance of the center of gravity from the lag hinge.) Assuming the same mode shapes and spring constants for the flap and lag motion, the expressions for the natural frequencies here and in section 9-2.1 give

$$\nu_\beta^2 = 1 + \nu_\zeta^2$$

For an articulated blade with coincident flap and lag hinges the mode shapes are in fact identical and this result is correct. In general, this relation is an expression of the fundamentally different role of centrifugal forces in flap and lag dynamics. The centrifugal force always acts as a spring on the flap motion to produce a natural frequency of at least 1/rev. However, since the centrifugal force goes through the center of rotation, the lag motion must depend on the hinge offset to obtain a centrifugal spring.

The flap and lag equations of motion are coupled by nonlinear terms due to the blade Coriolis forces: $-I_{\beta\zeta}^* 2\beta\dot\zeta$ in the flap equation and $I_{\beta\zeta}^* 2\beta\dot\beta$ in the lag equation. For a linear stability analysis, these terms are linearized about the trim motion:

$$\Delta(\beta\dot\zeta) = \beta_{trim}\Delta\dot\zeta + \zeta_{trim}\Delta\beta \cong \beta_0\Delta\dot\zeta$$

$$\Delta(\beta\dot\beta) = \beta_{trim}\Delta\dot\beta + \dot\beta_{trim}\Delta\beta \cong \beta_0\Delta\dot\beta$$

where β_0 is the trim coning angle. The last approximation is based on using the mean values of the periodic trim lag and flap motion; it is exact for hover. Thus a blade with a finite coning angle has a flap moment due to lag velocity, and a lag moment due to flap velocity. Since the Coriolis terms are nonlinear, these coupling moments are small. However, all the lag moments are small compared to the flap moments, so the Coriolis force due to flapping velocity is an important factor in the lag dynamics.

Since an articulated rotor will have a mechanical lag damper, the term

$C_\zeta^* \dot{\zeta}$ should be added to the lag equation of motion. Here $C_\zeta^* = C_\zeta / I_b \Omega$, and C_ζ is the lag moment due to angular velocity about the lag hinge. For hingeless rotors the structural damping of the blade should be included in the analysis by adding the term $I_\zeta^* g_s \nu_\zeta \dot{\zeta}$ to the lag equation, where g_s is the structural damping coefficient (typically $g_s = 0.01$ to 0.03). The structural damping is low, but it can be important to the lag dynamics because in general the in-plane forces are small.

9-3.2 In-Plane Bending

Consider the pure in-plane motion of a rotating blade, including now the blade bending and an arbitrary root restraint. The in-plane forces due to out-of-plane motion are important, but are neglected for now in order to concentrate on the in-plane natural frequencies and mode shapes. The section forces and their moment arms about the blade section at r are as follows:

(i) an inertial force $m\ddot{x}(\rho)$, with moment arm $(\rho - r)$;

(ii) the centrifugal force $m\Omega^2 \rho$, with moment arm $(r/\rho)x(\rho) - x(r)$; and

(iii) the aerodynamic force F_x, with moment arm $(\rho - r)$.

The lag moment at r, due to the inertial and aerodynamic forces on the section outboard of r, is thus

$$M_z(r) = \int_r^R \left[(F_x - m\ddot{x})(\rho - r) - m\Omega^2 \rho \left(x(\rho) \frac{r}{\rho} - x(r) \right) \right] d\rho$$

Engineering beam theory gives the structural bending moment as $M_z(r) = EI_{xx} d^2x/dr^2$, where E is the modulus of elasticity and I_{xx} the modulus-weighted area moment about the vertical principal axis of the section. Equating the structural moment with the inertial and aerodynamic moments, and taking the second derivative, gives the partial differential equation for the in-plane bending motion of the rotating blade:

$$\frac{d^2}{dr^2} EI_{xx} \frac{d^2 x}{dr^2} - \frac{d}{dr}\left[\int_e^R m\Omega^2 \rho \, d\rho \frac{dx}{dr} \right] - \Omega^2 m x + m\ddot{x} = F_x$$

The boundary conditions for articulated and hingeless blades are as discussed for out-of-plane bending in section 9-2.2.

The modal equation is obtained by assuming free vibration of the rotating blade. Substituting $x = \eta e^{i\nu t}$ in the homogeneous equation then gives

$$\frac{d^2}{dr^2} EI_{xx} \frac{d^2\eta}{dr^2} - \frac{d}{dr}\left[\int_r^R m\Omega^2 \rho d\rho \frac{d\eta}{dr}\right] - \Omega^2 m\eta - \nu^2 m\eta = 0$$

This is again a proper Sturm-Liouville eigenvalue problem, for which there exists a series of orthogonal eigensolutions η_{x_k} and corresponding eigenvalues $\nu_{x_k}^2$.

The in-plane displacement may now be expanded as a series in the normal modes:

$$x(r,t) = \sum_{k=1}^{\infty} \eta_{x_k}(r) q_{x_k}(t)$$

where q_{x_k} are the in-plane bending degrees of freedom. This modal expansion is substituted into the partial differential equation, and the modal equation is used to replace the structural and centrifugal spring terms by the natural frequency ν_{x_k}. Operating with $\int_0^R (\ldots)\eta_{x_k} dr$ and using the orthogonality of the modes then gives

$$I_{\zeta_k}^* (\ddot{q}_{x_k} + \nu_{x_k}^2 q_{x_k}) = \gamma \int_0^1 \eta_{x_k} \frac{F_x}{ac} dr$$

where $I_{\zeta_k}^* = \int_0^R \eta_{x_k}^2 dr/I_b$. This is the equation of motion for pure in-plane bending of the blade.

The natural frequency can be obtained from the mode shape by using the energy relation from Sturm-Liouville theory:

$$\nu^2 = \frac{K_\zeta\left(\eta'(e)\right)^2 + \int_0^R \left[EI\eta''^2 + \int_r^R m\Omega^2 \rho d\rho \eta'^2 - \Omega^2 m\eta^2\right] dr}{\int_0^R \eta^2 m\, dr}$$

Note that assuming the same mass and stiffness distributions this is formally equivalent to $\nu^2_{flap} = 1 + \nu^2_{lag}$, which can also be deduced by directly comparing the modal equations for in-plane and out-of-plane bending (see section 9-2.2). However the chordwise bending stiffness (EI_{xx}) is much greater than the flapwise bending stiffness (EI_{zz}), typically by a factor of 20 to 40. Moreover, in general the in-plane and out-of-plane mode shapes are not the same. Thus the relation $\nu^2_{flap} = 1 + \nu^2_{lag}$ really is only applicable to the fundamental modes of an articulated blade with coincident hinges. The similarity between the out-of-plane and in-plane modal problems can be used to advantage in numerical solutions for the modes, however.

9-3.3 In-Plane and Out-of-Plane Bending

Now the equations of motion for in-plane and out of plane bending will be derived. This is a generalization of the rigid flap and lag results. It is assumed that there is no structural coupling of the bending motion, so that the displacement z is still purely out of plane, and the displacement x is purely in plane. The only coupling of the equations of motion is due to the Coriolis forces. Therefore, it is only necessary to add the Coriolis terms to the results of sections 9-2.2 and 9-3.2.

For out-of-plane bending, there is a Coriolis force $2\Omega\dot{x}m$ directed radially inward, with moment arm $z(\rho) - z(r)$ about the blade station at r. The flapwise bending moment at r then becomes

$$M_x(r) = \int_r^R \left[(F_z - m\ddot{z})(\rho - r) - (m\Omega^2\rho - 2\Omega\dot{x}m)\left(z(\rho) - z(r)\right)\right] d\rho$$

and the partial differential equation for out-of-plane bending is

$$(EI_{zz}z'')'' - \left[\int_r^R m\Omega^2 \rho d\rho z'\right]' + m\ddot{z} + \left[z' \int_r^R 2\Omega\dot{x}m d\rho\right]' = F_z$$

When the aerodynamic force and the Coriolis term due to the in-plane velocity are dropped, the same modal equation as in section 9-2.2 is obtained. The out-of-plane deflection is now expanded as a series in the modes η_{z_k}:

$$z = \sum_k \eta_{z_k}(r) q_{z_k}(t)$$

where q_{z_k} are the degrees of freedom. This expansion is substituted into the partial differential equation, and the modal equation is used to replace the structural and centrifugal spring terms with the natural frequency ν_{z_k}. Then the operation $\int_0^R (\ldots)\eta_{z_k}\,dr$ produces the ordinary differential equation for the kth out-of-plane bending mode of the rotating blade:

$$I_{\beta_k}(\ddot{q}_{z_k} + \nu_{z_k}^2 q_{z_k}) + \int_0^1 \eta_{z_k}\left[z'\int_r^1 2\dot{x}m\,d\rho\right]' dr = \int_0^1 \eta_{z_k} F_z\,dr$$

Integrating by parts and changing the order of integration converts the Coriolis term to

$$\int_0^1 \eta_{z_k}\left[z'\int_r^1 2\dot{x}\,m\,d\rho\right]' dr = -2\int_0^1 \dot{x}m\int_0^r \eta'_{z_k} z'\,d\rho\,dr$$

$$\cong -2\beta_0\int_0^1 \dot{x}\,\eta_{z_k}\,m\,dr$$

The last approximation follows from linearizing $\dot{x}z'$ about the trim condition, using the mean trim values of $\dot{x}$ and z', and assuming that the trim blade slope is principally due to the coning angle β_0. For rigid flap and lag this Coriolis force reduces to the previous result, $-2\beta_0\dot{\zeta}I_{\beta\zeta}$. (The extra factor of $(1-e)$ in $I_{\beta\zeta}$ was lost when we assumed $z' \cong \beta_0$ instead of $z' = \eta'\beta = \beta/(1-e)$.)

There are two Coriolis forces that must be considered for in-plane bending. The lag velocity $\dot{x}$ and rotor speed Ω give a radially inward Coriolis force $2\Omega\dot{x}m$. This is the same force that produces a flapwise bending moment. It also produces a chordwise moment, with moment arm $x(\rho) - x(r)$ about the blade station at r. Secondly, the in-plane and out-of-plane deflection produces a nonlinear radial shortening of the blade by

$$-\frac{1}{2}\int_0^\rho (x'^2 + z'^2)\,d\rho^*$$

and thus there is a radially inward velocity of the blade section equal to

$$-\int_0^{\rho} (x'\dot{x}' + z'\dot{z}')d\rho^*$$

The cross product of this velocity and the rotor rotational speed gives an in-plane Coriolis force with moment arm $(\rho - r)$ about the blade station at r (see Fig. 9-2). The total lag bending moment is thus

$$M_z(r) = \int_r^R \left[(F_x - m\ddot{x})(\rho - r) - m\Omega^2\rho\left(x(\rho)\frac{r}{\rho} - x(r)\right) + 2\Omega\dot{x}m\left(x(\rho) - x(r)\right) - 2\Omega m \int_0^{\rho} (x'\dot{x}' + z'\dot{z}')d\rho^*(\rho - r)\right] d\rho$$

and the partial differential equation for in-plane bending becomes

$$(EI_{xx}x'')'' - \left[\int_r^R m\Omega^2\rho\, d\rho\, x'\right]' - \Omega^2 mx + m\ddot{x} + \left[2\Omega x' \int_r^R \dot{x}m\, d\rho\right]' + 2\Omega m \int_0^r (x'\dot{x}' + z'\dot{z}')d\rho = F_x$$

After the in-plane deflection is expressed in terms of the normal modes, $x = \sum_k \eta_{x_k} q_{x_k}$, the ordinary differential equation for the kth in-plane bending mode can be obtained by the usual steps:

$$I_{\zeta_k}(\ddot{q}_{x_k} + \nu_{x_k}^2 q_{x_k}) + \int_0^1 \eta_{x_k} 2m \int_0^r (x'\dot{x}' + z'\dot{z}')d\rho\, dr + \int_0^1 \eta_{x_k}\left[x' \int_r^1 2\dot{x}m\, d\rho\right]' dr = \int_0^1 \eta_{x_k} F_x\, dr$$

The two Coriolis terms can then be written

$$2\int_0^1 \eta_{x_k} m \int_0^r (x'\dot{x}' + z'\dot{z}')d\rho\, dr - 2\int_0^1 \dot{x} m \int_0^r \eta'_{x_k} x' d\rho\, dr$$

$$\cong 2\int_0^1 \eta_{x_k} m \int_0^r z'\dot{z}'\, d\rho\, dr$$

$$\cong 2\beta_0 \int_0^1 \dot{z}\eta_{x_k} m\, dr$$

For rigid lag motion, where x is independent of r, the two in-plane velocity terms cancel exactly. Similarly, they cancel if the trim lag displacement is primarily due to the rigid mode. Therefore these two terms have been neglected for the general case.

Substituting the modal expansions for $\dot{x}$ in the flap equation and for $\dot{z}$ in the lag equation then completes the coupled equations of motion for out-of-plane and in-plane bendings

$$I^*_{\beta_k}(\ddot{q}_{z_k} + \nu_{z_k}^2 q_{z_k}) - \sum_{i=1}^{\infty} \dot{q}_{x_i} 2\beta_0 I^*_{\beta_k \zeta_i} = \gamma \int_0^1 \eta_{z_k} \frac{F_z}{ac}\, dr$$

$$I^*_{\zeta_k}(\ddot{q}_{x_k} + \nu_{x_k}^2 q_{x_k}) + \sum_{i=1}^{\infty} \dot{q}_{z_i} 2\beta_0 I^*_{\beta_i \zeta_k} = \gamma \int_0^1 \eta_{x_k} \frac{F_x}{ac}\, dr$$

where

$$I^*_{\beta_k} = \int_0^R \eta_{z_k}^2\, dr/I_b,\quad I^*_{\zeta_k} = \int_0^R \eta_{x_k}^2\, dr/I_b,\ \text{and}\ \beta_0 I^*_{\beta_k\zeta_i} = z'_{\text{trim}} \int_0^R \eta_{z_k}\eta_{x_i}\, dr/I_b.$$

This set of equations is not a good model for the out-of-plane and in-plane bending of a rotor blade, however. Unless the blade is untwisted and operating at zero pitch, there will be considerable structural coupling of the in-plane and out-of-plane deflections. The structural principal axes are rotated by the blade pitch angle, while the centrifugal forces always act relative to the shaft axes. Therefore, when the blade pitch is nonzero the axes of structural and centrifugal stiffening do not coincide, and the free vibration mode of the blade will not be purely out of plane or purely in plane as was assumed above. A better analysis would use a single series of coupled flap-lag bending modes to represent the blade deflection. The blade torsional motion must be included in such an analysis, since the coupling between bending and pitch has a major influence on the dynamics. The structural coupling is most important at the blade root, so these considerations apply to a hingeless rotor in particular. For an articulated rotor the equations given here may be satisfactory, although often a major reason a simpler analysis must be used is the great complexity of deriving the equations for the fully coupled flap-lag-torsion motion of a rotor blade.

9–4 Torsional Motion

9-4.1 Rigid Pitch and Flap

The rotor dynamics analysis will now be extended to include the blade pitch degree of freedom. Consider an articulated rotor, with no flap hinge offset (Fig. 9-3). A general flap frequency can be obtained by using a hinge

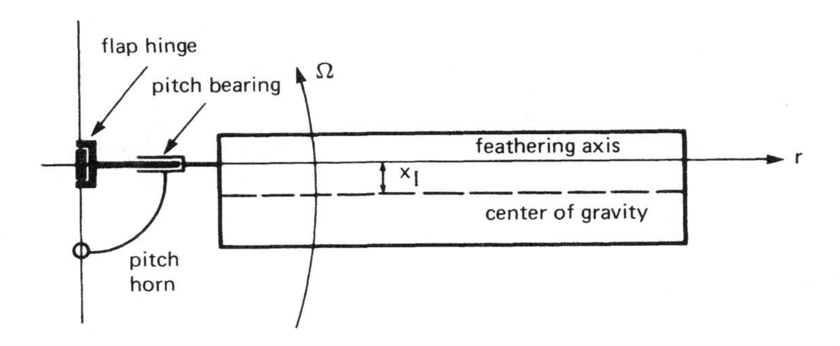

Figure 9-3 Articulated rotor blade with flapping and feathering motion.

spring. In addition, we now consider the blade pitch motion, consisting of rigid rotation about the feathering axis, restrained by the rotor control system. If there is flexibility in the control system, the blade rigid pitch motion is a degree of freedom, not a control input (as in Chapter 5). Since in most rotors the control system stiffness is less than the elastic torsional stiffness of the blades, the rigid flap and rigid pitch motion is in fact a good model for an articulated blade. It is assumed that the pitch bearing is outboard of the flap hinge, and that there is no δ_3 pitch-flap coupling. The chordwise position of the blade section of gravity is a distance x_I behind the feathering axis (Fig. 9-3).

The flap degree of freedom β is the angle of rigid rotation about the flap hinge. The out-of-plane deflection of the blade is thus $z = r\beta$. Let θ be the degree of freedom for the pitch motion, defined as the nose-up angle of rigid rotation about the feathering axis. The built-in twist of the blade is not considered here, since it is only involved in the trim forces on the blade. The rotor control system commands a pitch angle θ_{con}, while the actual blade pitch angle is θ. The difference $\theta - \theta_{\text{con}}$ is due to control system flexibility and produces a restoring moment about the feathering axis equal to $K_\theta(\theta - \theta_{\text{con}})$, where K_θ is the control system spring constant.

The flapping equation of motion is obtained as usual from the conditions for equilibrium of moments about the flap hinge. The forces acting on the blade section center of gravity are now:

(i) the inertial force $m(\ddot{z} - x_I\ddot{\theta}) = m(r\ddot{\beta} - x_I\ddot{\theta})$, with moment arm r;

(ii) the centrifugal force $m\Omega^2 r$, with moment arm $z - x_I\theta = r\beta - x_I\theta$; and

(iii) the aerodynamic force F_z, with moment arm r.

Including the hinge spring moment, the equation of motion becomes

$$\int_0^R m(r\ddot{\beta} - x_I\ddot{\theta})r\,dr + \int_0^R m\Omega^2 r(r\beta - x_I\theta)\,dr + K_\beta\beta = \int_0^R rF_z\,dr$$

or

$$\left(\int_0^R r^2 m\,dr\right)(\ddot{\beta} + \nu^2\beta) - \left(\int_0^R x_I r m\,dr\right)(\ddot{\theta} + \Omega^2\theta) = \int_0^R rF_z\,dr$$

where ν is the rotating natural frequency of the flap motion. Define $I_b = \int_0^R r^2 m\,dr$, and $I_x = \int_0^R x_I r m\,dr$. Dividing by I_b and using dimensionless quantities then gives

$$\ddot{\beta} + \nu^2\beta - I_x^*(\ddot{\theta} + \theta) = \gamma\int_0^1 r\frac{F_z}{ac}\,dr$$

where $I_x^* = I_x/I_b$. Thus the pitch motion introduces inertial and centrifugal flap moments when the center of gravity is offset from the feathering axis.

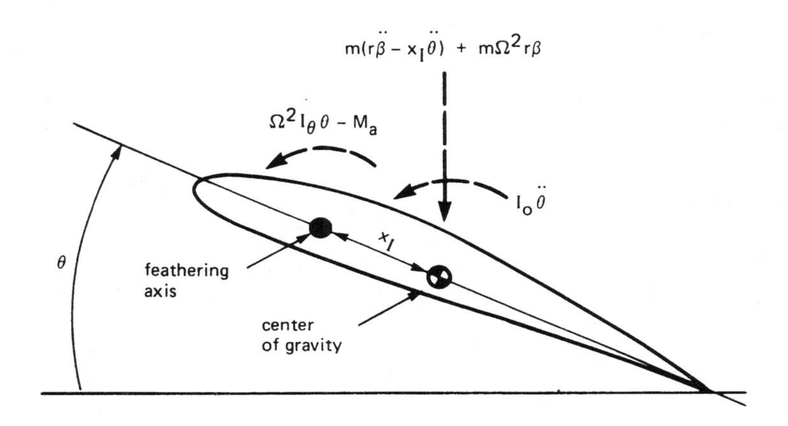

Figure 9-4 Blade section pitch moments.

The pitch equation of motion is obtained from the conditions for equilibrium of moments about the feathering axis (Fig. 9-4). The forces acting on the blade section and their moment arms about the feathering axis are as follows:

(i) an inertial moment $I_0\ddot{\theta}$ about the section center of gravity, and the inertial force $m(r\ddot{\beta} - x_I\ddot{\theta})$ acting on the center of gravity with moment arm x_I about the feathering axis;

(ii) A propeller moment $I_\theta\Omega^2\theta$ about the feathering axis, acting to oppose the pitch motion, and the flapping centrifugal spring force $m\Omega^2 r\beta$ acting at the center of gravity with moment arm x_I about the feathering axis; and

(iii) a nose-up aerodynamic moment M_a about the feathering axis.

Here I_0 is the pitch moment of inertia of the section, about the center of gravity, and $I_\theta = I_0 + x_I^2 m$ is the section moment of inertia about the feathering axis. When the blade flaps up, the centrifugal force has a component $m\Omega^2 r\beta$ normal to the blade. This force is responsible for the centrifugal flap moment, and when the center of gravity is offset from the feathering axis it also produces a pitch moment. The propeller moment is also due to centrifugal forces. The centrifugal force on a blade mass element dm acts on a line through the center of rotation (Fig. 9-5). For an element a distance x behind the feathering axis there is then a chordwise component of this centrifugal force equal to

$$\left(\sqrt{r^2 + x^2}\,\Omega^2 dm\right) \frac{x}{\sqrt{r^2 + x^2}} = x\Omega^2 dm$$

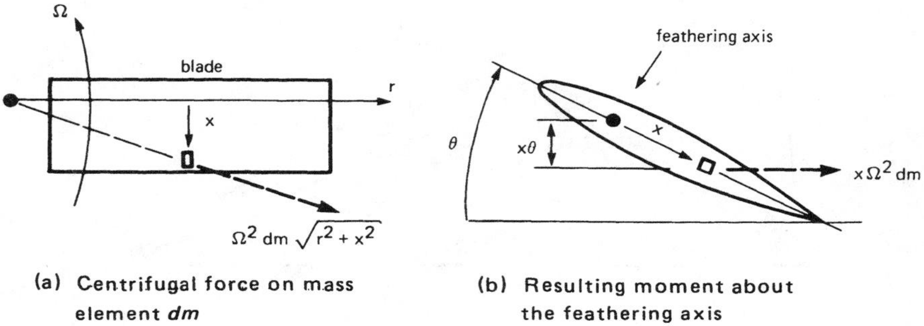

(a) Centrifugal force on mass element *dm*

(b) Resulting moment about the feathering axis

Figure 9-5 Origin of the propeller moment.

When the blade is pitched up by the angle θ, this chordwise force will be acting on a line a distance $x\theta$ below the feathering axis (see Fig. 9-5). For mass elements forward of the feathering axis, the centrifugal force component is directed forward and acts on a line above the feathering axis. Thus there is a centrifugal feathering moment opposing the pitch motion. The propeller moment is obtained by integrating over the blade section:

$$\int_{\text{section}} (x\theta)(x\Omega^2 dm) = \theta\Omega^2 \int_{\text{section}} x^2 dm = \theta\,\Omega^2 I_\theta$$

where I_θ is the section moment of inertia about the feathering axis.

Equilibrium of moments about the feathering axis implies that

$$\int_0^R \left[I_0\ddot{\theta} - (r\ddot{\beta} - x_I\ddot{\theta})x_I m + I_\theta\Omega^2\theta - m\Omega^2 r\beta x_I \right] dr$$

$$+ K_\theta(\theta - \theta_{\text{con}}) = \int_0^R M_a dr$$

or

$$\left(\int_0^R I_\theta \, dr\right)(\ddot{\theta} + \Omega^2\theta) - \left(\int_0^R x_I r m \, dr\right)(\ddot{\beta} + \Omega^2\beta)$$

$$+ \quad K_\theta(\theta - \theta_{con}) = \int_0^R M_a \, dr$$

The restoring moment from the control system, $K_\theta(\theta - \theta_{con})$, has been included: θ_{con} is the pitch angle commanded by the control system, and K_θ is the effective spring constant of the flexible control system. Now define the total moment of inertia about the feathering axis as $I_f = \int_0^R I_\theta \, dr$, and write

$$\omega^2 = \frac{K_\theta}{I_f \Omega^2}$$

where ω is the dimensionless natural frequency of the blade pitch motion. Dividing by I_b and using dimensionless quantities gives the pitch equation of motion,

$$I_f^* \left(\ddot{\theta} + (\omega^2 + 1)\theta\right) - I_x^*(\ddot{\beta} + \beta) = \gamma \int_0^1 \frac{M_a}{ac} \, dr + I_f^* \omega^2 \theta_{con}$$

where $I_f^* = I_f/I_b$. Note that ω is the nonrotating natural frequency of the pitch motion, and that the propeller moment gives a spring equivalent to a 1/rev natural frequency. The rotating pitch natural frequency is therefore $(\omega^2 + 1)^{1/2}$. Typically the control system stiffness gives $\omega = 3$ to 5/rev, so the propeller moment is small compared to the structural spring.

To summarize, the equations of motion for rigid flapping and rigid pitch about the feathering axis are as follows:

$$\ddot{\beta} + \nu^2\beta - I_x^*(\ddot{\theta} + \theta) = \gamma \int_0^1 r \frac{F_z}{ac} \, dr$$

$$I_f^* \left(\ddot{\theta} + (\omega^2 + 1)\theta\right) - I_x^*(\ddot{\beta} + \beta) = \gamma \int_0^1 \frac{M_a}{ac} \, dr + I_f^* \omega^2 \theta_{con}$$

where $I_x^* = \int_0^R x_I r m\,dr/I_b$ and $I_f^* = \int_0^R I_\theta\,dr/I_b$. The flap and pitch motions are coupled by inertial and centrifugal forces when the blade center of gravity is offset from the feathering axis. Here ν is the rotating natural frequency of the flap motion and ω is the nonrotating pitch natural frequency due to the control system stiffness. I_θ is the pitch moment of inertia of the section, about the feathering axis, and x_I is the offset of the section center of gravity behind the feathering axis. For constant x_I we have $I_x^* = x_I \int_0^R r m\,dr/I_b = x_I S_b^* \cong (3/2)x_I$. Since the center of gravity offset will be a small fraction of the chord, x_I (which is normalized using the rotor radius) is a second-order-small quantity. The ratio of the pitch inertia to the flap inertia, I_f^*, is roughly equal to $0.1(c/R)^2$. In general, all the pitch moments are two orders smaller than the flap moments.

In the limit of a very stiff control system, the restoring moment $K_\theta(\theta - \theta_{\text{con}})$ must remain finite while $K_\theta \to \infty$, since the restoring moment is equal to the sum of the inertial and aerodynamic pitch moments on the blade. Then $\theta \to \theta_{\text{con}}$ in this limit, and the pitch motion is just the input commanded by the control system. Alternatively, in the limit of $\omega \to \infty$ the equation for the pitch motion reduces to $I_f^* \omega^2 \theta = I_f^* \omega^2 \theta_{\text{con}}$, or $\theta = \theta_{\text{con}}$ again. Kinematic pitch-flap coupling due to the control system geometry is a feedback of the flap angle to the commanded pitch of the form $\Delta\theta_{\text{con}} = -K_P\beta$. The equation of motion becomes

$$I_f^*\left(\ddot{\theta} + (\omega^2 + 1)\theta\right) - I_x^*(\ddot{\beta} + \beta) + K_P I_f^* \omega^2 \beta = \gamma \int_0^1 \frac{M_a}{ac}\,dr + I_f^* \omega^2 \theta_{\text{con}}$$

and in the limit of infinite control system stiffness it reduces to $\theta = \theta_{\text{con}} - K_P\beta$ as required.

9-4.2 Structural Pitch-Flap and Pitch-Lag Coupling

The order of the flap and pitch hinges, or for a hingeless rotor the distribution of bending inboard and outboard of the pitch bearing, has an important influence on the blade dynamics. The preceding analysis assumed that the pitch bearing was outboard of the flap hinge, so that flap motion tilts the feathering axis along with the blade. If the pitch bearing is inboard of the flap hinge, the feathering axis remains in the hub plane when the blade flaps, resulting in different moment arms of the section forces about the feathering axis.

Consider the rigid flap and rigid pitch of an articulated rotor blade, now with the pitch bearing inboard of the flap and lag hinges. The flapping equation is not changed, at least for small angles of flap and pitch. However, there is a change in the manner in which the centrifugal forces produce pitch moments. The centrifugal force $m\Omega^2 r$ does not now have a component about the feathering axis when the blade flaps, because the centrifugal force and feathering axis are both parallel to the hub plane. However, the chordwise component of the centrifugal force has a moment arm of $x\theta - r\beta$ about the feathering axis, so the propeller moment becomes

$$\int_{\text{section}} (x\theta - r\beta)(x\Omega^2 dm) = \theta\Omega^2 I_\theta - (m\Omega^2 r\beta)x_I$$

Thus there is no net change in the pitch moment due to the centrifugal forces, but there are a number of nonlinear effects of the flap and lag motion that must be considered when the pitch bearing is inboard. The trim flap and lag motion displaces the blade section from the feathering axis, so that all in-plane and out-of-plane forces have a moment arm to produce pitch moments. In particular, the pitching motion produces an in-plane acceleration of the blade when it is flapped up, and an out-of-plane acceleration when it is lagged back. Hence the effective pitch inertia with the flap and lag hinges outboard is increased to

$$\begin{aligned} I &= \int_0^R (I_\theta + z^2_{\text{trim}} m + x^2_{\text{trim}} m)dr \\ &= \int_0^R I_\theta dr + \beta^2_{\text{trim}} \int_0^R \eta_\beta^2 m\, dr + \zeta^2_{\text{trim}} \int_0^R \eta_\zeta^2\, m\, dr \\ &\cong I_f + (\beta_0^2 + \zeta_0^2) I_b \end{aligned}$$

The resulting increase in pitch inertia and decrease in the effective pitch natural frequency can be quite substantial.

If the flap and lag motion occur outboard of the pitch bearing, there is a coupling of the pitch moment with the flap and lag moments that is particularly important for hingeless rotors. Consider the pitch moment resulting from the flap and lag motion of an articulated blade, with hinge springs to obtain general frequencies. The forces on the blade section and their moment arms about the feathering axis are:

(i) the out-of-plane force $F_z - mr\ddot{\beta}$, with moment arm $r\zeta$ due to the lag motion; and

(ii) the in-plane force $F_x - mr\ddot{\zeta} - 2\Omega mr\beta\dot{\beta} + mr\zeta\Omega^2$, with moment arm $r\beta$ due to the flap motion.

Then the nose-down moment about the feathering axis is

$$\Delta M_\theta = \zeta\left[\int_0^1 rF_z dr - \ddot{\beta}\int_0^1 r^2 m\,dr\right]$$

$$- \beta\left[\int_0^1 rF_x dr - (\ddot{\zeta} + 2\beta\dot{\beta} - \zeta)\int_0^1 r^2 m\,dr\right]$$

Substituting for $\ddot{\beta}$ and $\ddot{\zeta}$ in the flap and lag equations of motion (section 9-3.1) gives

$$\Delta M_\theta = \zeta(I_b\nu_\beta^2\beta - K_\beta\beta_p) - \beta(I_b\nu_\zeta^2\zeta + I_b\zeta - K_\zeta\zeta_p)$$

$$= \beta\zeta I_b(\nu_\beta^2 - 1 - \nu_\zeta^2) - \zeta K_\beta\beta_p + \beta K_\zeta\zeta_p$$

where K_β and K_ζ are the hinge spring constants, β_p is the precone angle, and ζ_p is the prelag angle. (Note that here β and ζ are the total flap and lag angles. The pitch moment in terms of the flap and lag relative to the feathering axis at the precone angle β_p and the prelag angle ζ_p can be obtained by substituting $\beta = \tilde{\beta} + \beta_p$ and $\zeta = \tilde{\zeta} + \zeta_p$.) This result may be interpreted as follows. The net flap moment at the root $M_\beta = I_b(\nu_\beta^2 - 1)\beta - K_\beta\beta_p$ has a nose-down component about the feathering axis when the blade is lagged by ζ. Similarly, the lag moment $M_\zeta = I_b\nu_\zeta^2\zeta - K_\zeta\zeta_p$ has a nose-up pitch component when the blade is flapped by β. Then the total pitch moment is $\Delta M_\theta = M_\beta\zeta - M_\zeta\beta$, as above. Although nonlinear in the flap and lag motion, this pitch moment can be very significant. The principal effect of this moment is to produce a static pitch deflection due to the control system flexibility, $\Delta\theta = -\Delta M_\theta/K_\theta$. Thus the effect on the linearized dynamics is to introduce an effective pitch-flap and pitch-lag coupling. The pitch-flap coupling is

$$K_{P_\beta} = -\frac{\partial\theta}{\partial\beta} = \frac{1}{K_\theta}\left[I_b(\nu_\beta^2 - 1 - \nu_\zeta^2)\zeta + K_\zeta\zeta_p\right]$$

for a given trim lag deflection; and the pitch-lag coupling is

$$K_{P_\zeta} = -\frac{\partial\theta}{\partial\zeta} = \frac{1}{K_\theta}\left[I_b(\nu_\beta^2 - 1 - \nu_\zeta^2)\beta - K_\beta\beta_p\right]$$

for a given trim flap deflection. To evaluate these coupling factors it is necessary to solve for the trim coning and lag angles, which depend on the rotor thrust and torque, and also on the precone and prelag angles (see Chapter 5). For an articulated rotor with no springs but with coincident flap and lag hinges, $\nu_\beta^2 = 1 + \nu_\zeta^2$, so the pitch moment is zero and this coupling disappears.

A similar result can be derived for the torsional moment at an arbitrary blade section. Consider bending of a blade with out-of-plane deflection $z(r)$ and in-plane deflection $x(r)$. The forces acting on the blade outboard of r will thus produce a torsional moment on the section at r:

$$\Delta M_r = \int_r^R \left\{\left[z(\rho) - z(r) - (\rho - r)z'(r)\right]G_x - \left[x(\rho) - x(r) - (\rho - r)x'(r)\right]G_z\right\}d\rho$$

where G_x is the total section in-plane force, including both inertial and aerodynamic contributions, and G_z is the total section out-of-plane force. Then the nose-down torsional loading on the section is

$$\Delta T = \frac{\partial M_r}{\partial r} = x''\int_r^R (\rho - r)G_z\,d\rho - z''\int_r^R (\rho - r)G_x\,d\rho$$

Now $M_x = \int_r^R (\rho - r)G_z d\rho$ and $M_z = \int_r^R (\rho - r)G_x d\rho$ are respectively the flapwise and chordwise bending moments on the section at r, so

$$\Delta T = M_x x'' - M_z z''$$

In terms of the flapwise and chordwise bending stiffnesses, then, the torsional loading is

$$\Delta T = M_x M_z \left(\frac{1}{EI_{xx}} - \frac{1}{EI_{zz}} \right) = x'' z'' (EI_{zz} - EI_{xx})$$

That is, the coupling is proportional to the product of the in-plane and out-of-plane deflections and the difference between the flapwise and chordwise bending stiffnesses. For a blade with $EI_{zz} = EI_{xx}$ the torsion-bending coupling disappears. This is called the matched-stiffness case, and it corresponds to the condition $\nu_\beta^2 = 1 + \nu_\zeta^2$ obtained for a rigid blade. Note that the matched-stiffness blade will have equal nonrotating flap and lag frequencies. Usually the chordwise stiffness of a rotor blade is much greater than the flapwise stiffness. However, it is possible to achieve the matched-stiffness condition with a soft in-plane hingeless rotor, at least at the root, where it is most important for the fundamental modes.

The effects that have been discussed in this section are primarily important for a hingeless rotor, which requires a more complete model of the bending and torsion dynamics for an accurate analysis. It may be concluded, however, that flap or lag bending outboard of the feathering axis will produce substantial torsional moments. The resulting effective pitch-lag and pitch-flap coupling is an important factor in hingeless rotor dynamics.

For a further discussion, particularly of the pitch-flap and pitch-lag coupling for hingeless rotors, see Mil′ (1966), Hansford and Simon (1973), Hodges and Ormiston (1973), Huber (1973); and also the references in Chapter 12 on flap-lag dynamics.

9-4.3 Torsion and Out-of-Plane Bending

Consider now the torsion and out-of-plane bending motion of an elastic blade. It is not entirely consistent to exclude the in-plane motion from such an analysis. For example, the in-plane forces on the blade produce torsional moments when there is out-of-plane bending, as was seen from the preceding section. These forces are relieved by the blade lag motion, however, so they should not be considered unless the model includes the in-plane motion as well. For hingeless rotors in particular, a fully coupled flap-lag-torsion analysis is required to adequately represent the dynamics. Thus we are primarily concerned here with extending the rigid flap and rigid pitch analysis of section 9-4.1 to include the higher bending modes and elastic torsion, and with laying the foundation for the development of more complete models.

It is assumed that the blade . .as a straight elastic axis coincident with the feathering axis. The blade pitch now consists of the rigid pitch angle p_0 due to control system flexibility, and a deflection θ_e due to elastic torsion of the blade: $\theta = p_0 + \theta_e$. (Again the built-in twist only influences the trim forces, and so it can be ignored. The notation for the rigid pitch angle is chosen to be consistent with the modal expansion introduced below for the elastic torsion θ_e.)

The equation of motion for bending is obtained from the conditions for equilibrium of moments on the blade outboard of r. The section forces at radial station ρ and their moment arms about the elastic axis at r are as follows:

(i) an inertial force $m(\ddot{z} - x_I\ddot{\theta})$, with moment arm $(\rho - r)$;

(ii) the centrifugal force $m\Omega^2\rho$ acting on the center of gravity, with moment arm $(z - \theta x_I - z(r))$ about the elastic axis at r;

(iii) a centrifugal moment $(m\Omega^2 rx_I)\theta(r)$; and

(iv) an aerodynamic force F_z, with moment arm $(\rho - r)$.

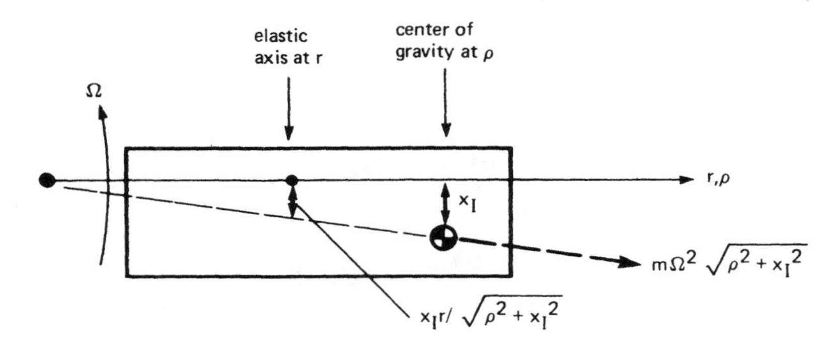

Figure 9-6 Origin of the centrifugal bending moment $m\Omega^2 rx_I\theta(r)$.

The centrifugal moment $m\Omega^2 rx_I\theta(r)$ at r due to the forces at ρ arises as follows. Fig. 9-6 shows that the centrifugal force $m\Omega^2(\rho^2 + x_I^2)^{1/2}$ acting on the section center of gravity has a moment arm $x_I r/(\rho^2 + x_I^2)^{1/2}$ about the elastic axis at r, producing an in-plane bending moment $m\Omega^2 r x_I$. The section at r has a pitch angle $\theta(r)$, however, so the flapwise component of this bending moment on the blade is $m\Omega^2 r x_I \theta(r)$. Thus the total moment on the blade section at r is

$$M(r) = \int_r^R \Big[\big(F_z - m(\ddot{z} - x_I\ddot{\theta})\big)(\rho - r) - m\Omega^2 \rho \big(z - \theta x_I - z(r)\big) - \theta(r) m\Omega^2 r x_I \Big] d\rho$$

By equating this expression to $M(r) = EI\, d^2z/dr^2$ and taking the second derivative the partial differential equation for bending of the elastic axis is obtained:

$$\frac{d^2}{dr^2} EI \frac{d^2 z}{dr^2} - \frac{d}{dr}\left[\int_r^R m\Omega^2 \rho\, d\rho \frac{dz}{dr} \right] + m\ddot{z} - m x_I \ddot{\theta} + \frac{d}{dr}\left[\frac{d(r\theta)}{dr} \int_r^R \Omega^2 x_I m\, d\rho \right] = F_z$$

The out-of-plane deflection can be expanded as a series in the normal modes, $z(r,t) = \sum_k \eta_k(r) q_k(t)$, where q_k are the bending degrees of freedom. The modal equation is the same as in section 9-2.2. By substituting for z and operating with $\int_0^R (\ldots)\,\eta_k\, dr$, the equation of motion for the kth bending mode can be obtained:

$$I_{q_k}(\ddot{q}_k + \nu_k^2 q_k) - \int_0^R \eta_k x_I \ddot{\theta} m\, dr + \int_0^R \eta_k \frac{d}{dr}\left[\frac{d(r\theta)}{dr} \int_r^R x_I \Omega^2 m\, d\rho \right] dr = \int_0^R \eta_k F_z\, dr$$

The θ term can be written as

$$\int_0^R \eta_k \frac{d}{dr}\left[\frac{d(r\theta)}{dr} \int_r^R x_I \Omega^2 m\, d\rho \right] dr = -\int_0^R x_I \Omega^2 m \int_0^r \eta'_k (\rho\theta)' d\rho\, dr$$

For rigid pitch and flap ($\eta = r$ and θ independent of r) this equation of motion reduces to the result of section 9-4.1.

The equations of motion for rigid pitch and elastic torsion are obtained from the conditions for equilibrium of torsion moments about the elastic axis. The forces acting on the blade section at ρ and their moment arms about the elastic axis at r are as follows:

(i) an inertial moment $I_0\ddot{\theta}$ about the section center of gravity, and the inertial force $m(\ddot{z} - x_I\ddot{\theta})$ acting at the center of gravity with moment arm x_I about the elastic axis;

(ii) a nose-down propeller moment $I_\theta \theta \Omega^2 - x_I \Omega^2 m(z - z(r))$ about the elastic axis, and a nose-up centrifugal moment $(m\Omega^2 x_I r)z'(r)$; and

(iii) the nose-up aerodynamic moment M_a about the elastic axis.

The propeller moment is due to the in-plane centrifugal force component $x\Omega^2 dm$ (see Fig. 9-5) acting with moment arm $(z(r) - (z - x\theta))$ about the elastic axis at r, so that

$$\int_{\text{section}} \Big(z(r) - (z - x\theta)\Big) x\Omega^2 \, dm = I_\theta \theta \Omega^2 - x_I \Omega^2 m \Big(z - z(r)\Big)$$

The centrifugal moment $(m\Omega^2 x_I r)z'(r)$ is due to the in-plane bending moment $m\Omega^2 x_I r$ discussed for the flapping equation (see Fig. 9-6). When the blade is flapped up by the angle $z'(r)$, this moment has a torsional component $(m\Omega^2 x_I r)z'(r)$ about the elastic axis at r. The total nose-up torsional moment on the blade section at r is thus

$$M_r = \int_r^R \Big[M_a - I_\theta \ddot{\theta} - I_\theta \theta \Omega^2 + m x_I \ddot{z} + x_I \Omega^2 m \Big(z - z(r) + r z'(r)\Big) \Big] d\rho$$

The equation of motion for rigid pitch is obtained from the conditions for equilibrium of moments about the pitch bearing at $r = 0$. The inertial and aerodynamic pitch moments of the blade are reacted by a moment from the control system:

$$M_r(0) = K_\theta \Big(\theta(0) - \theta_{\text{con}}\Big)$$

where K_θ is the control system stiffness, θ_{con} is the root pitch angle commanded by the control system, and $\theta(0)$ is the actual root pitch. We shall define the elastic torsion of the blade to be zero at the pitch bearing, $\theta_\theta(0) = 0$, so that the root pitch equals the rigid pitch degree of freedom; that is, $\theta(0) = p_0$. The differential equation of motion for rigid pitch is thus

$$\int_0^R \left[I_\theta \ddot{\theta} + I_\theta \theta \Omega^2 - m x_I \ddot{z} - m x_I z \Omega^2 \right] dr + K_\theta (p_0 - \theta_{con}) = \int_0^R M_a \, dr$$

For rigid flap and pitch this reduces to the result of section 9-4.1.

Engineering beam theory relates the torsional moment to the elastic torsion deflection by

$$M_r = GJ \frac{d\theta_e}{dr}$$

where GJ is the torsional rigidity of the blade section. If the expressions for the structural moment on the section and the total inertial and aerodynamic moment are equated and the derivative taken with respect to r,

$$-\frac{d}{dr} GJ \frac{d\theta_e}{dr} + I_\theta \ddot{\theta} + I_\theta \theta \Omega^2 - m x_I \ddot{z} + r \cdot \frac{d}{dr}\left[\frac{dz}{dr} \int_r^R m \Omega^2 x_I \, d\rho \right] = M_a$$

This is the partial differential equation for elastic torsion of the rotating blade. The boundary conditions are $d\theta_e/dr = 0$ at $r = R$ (a free end at the tip) and $\theta_e = 0$ at $r = 0$ (a fixed end at the root). Consider free torsional vibration of the nonrotating blade, i.e. the homogeneous equation

$$-\frac{d}{dr} GJ \frac{d\theta_e}{dr} + I_\theta \ddot{\theta}_e = 0$$

Solving this equation by separation of variables, we write $\theta_e = \xi(r) e^{i\omega t}$, which gives

$$\frac{d}{dr} GJ \frac{d\xi}{dr} + \omega^2 I_\theta \xi = 0$$

with the boundary conditions $\xi(0) = 0$ and $\xi'(R) = 0$. This is a proper Sturm-Liouville eigenvalue problem, for which there exists a series of eigensolutions $\xi_k(r)$ and corresponding eigenvalues ω_k^2. Since the mode shapes are orthogonal with weighting function I_θ,

$$\int_0^R \xi_k \xi_i I_\theta \, dr = 0$$

if $k \neq i$. The eigenvalues will be ordered according to size (ω_1 is the smallest torsion frequency) and the mode shapes will be normalized to unit deflection at the tip, $\xi(R) = 1$. According to Sturm-Liouville theory (section 9-1), the natural frequencies can be obtained from the mode shapes by

$$\omega^2 = \frac{\int_0^R GJ \xi'^2 \, dr}{\int_0^R I_\theta \xi^2 \, dr}$$

The free vibration of a nonrotating blade with uniform GJ and I_θ distributions has the exact solution

$$\xi_k = \sin\left[(k - \tfrac{1}{2})\pi(r/R)\right]$$

with the corresponding natural frequencies

$$\omega_k = (k - \tfrac{1}{2})\pi \sqrt{\frac{GJ}{I_\theta R^2}}$$

for $k = 1$ to ∞. These functions are useful in solving the modal equation for the true mode shapes, such as by the Galerkin method, and serve as approximate mode shapes when better estimates are not available. The simple function $\xi_1 = r/R$ can also be used as an approximation to the first mode shape. Note that for torsion the nonrotating free vibration modes are being used. Rotating modes could be used instead by retaining the centrifugal spring term (the propeller moment) in the modal equation. For the torsional stiffness typical of rotor blades the rotation has little effect on the free vibration frequencies and mode shapes, however. In fact, for the present case the rotating and nonrotating torsion modes are identical. Thus it is appropriate to use the nonrotating modes, which are generally simpler to calculate.

Now the torsional deflection is expanded as a series in the normal modes:

$$\theta_e(r,t) = \sum_{k=1}^{\infty} \xi_k(r) p_k(t)$$

where p_k are the degrees of freedom of elastic torsion. With the mode shapes normalized to $\xi_k = 1$ at the tip, p_k is the pitch angle at the tip for the kth mode. Using a mode shape $\xi_0 = 1$ for rigid pitch, the total blade pitch can be written as

$$\theta = p_0 + \theta_e = \sum_{k=0}^{\infty} \xi_k(r) p_k(t)$$

Next, the expansion for θ_e is substituted into the partial differential equation for torsion, the modal equation satisfied by ξ_k is used to replace the torsional stiffness term by the natural frequency $\omega_k{}^2$, and the equation is operated on with $\int_0^R (...)\xi_k dr$. Using the orthogonality of the elastic torsion modes (note that the rigid pitch and elastic torsion modes are not orthogonal), the following differential equation is obtained for the kth mode:

$$I_{p_k}\left(\ddot{p}_k + (\omega_k{}^2 + \Omega^2)p_k\right) + \left(\int_0^R I_\theta \xi_k dr\right)(\ddot{p}_0 + \Omega^2 p_0)$$

$$- \int_0^R \xi_k m x_I \ddot{z} dr + \int_0^R \xi_k r \frac{d}{dr}\left[\frac{dz}{dr}\int_r^R \Omega^2 m x_i d\rho\right] dr$$

$$= \int_0^R \xi_k M_a dr$$

where $I_{p_k} = \int_0^R \xi_k{}^2 I_\theta dr$ is the generalized mass of the mode. The bending term can be written as

$$\int_0^R \xi_k r \frac{d}{dr}\left[\frac{dz}{dr}\int_r^R \Omega^2 m x_I d\rho\right] dr = -\int_0^R x_I \Omega^2 m \int_0^r z'(\rho\xi_k)' d\rho\, dr$$

Now substituting the expansion $\theta = \sum_{k=0}^{\infty} \xi_k p_k$ into the equation of motion for rigid pitch gives

$$I_{p_0}\left(\ddot{p}_0 + (\omega_0^2 + 1)\Omega^2 p_0\right) + \sum_{j=1}^{\infty}\left(\int_0^R I_\theta \xi_j dr\right)(\ddot{p}_j + \Omega^2 p_j)$$

$$- \int_0^R m x_I \ddot{z} dr - \int_0^R m x_I z \Omega^2 dr$$

$$= \int_0^R M_a dr + I_{p_0} \omega_0^2 \Omega^2 \theta_{\text{con}}$$

where $I_{p_0} = \int_0^R I_\theta dr$ is the pitch moment of inertia of the blade and ω_0 is the natural frequency of the rigid pitch motion due to control system flexibility: $\omega_0^2 = K_\theta/(I_{p_0}\Omega^2)$.

Finally, substitute the modal expansion for z into the torsion and pitch equations, and the expansion for θ into the bending equation; divide by I_b and use dimensionless quantities. The result is the equations of motion for out-of-plane bending, rigid pitch, and elastic torsion of the rotating blade:

$$I^*_{q_k}(\ddot{q}_k + \nu_k^2 q_k) - \sum_{j=0}^{\infty}(I^*_{q_k\ddot{p}_j}\ddot{p}_j + I^*_{q_k p_j} p_j)$$

$$= \gamma \int_0^1 \eta_k \frac{F_z}{ac} dr$$

$$I^*_{p_0}\left(\ddot{p}_0 + (\omega_0^2 + 1)p_0\right) + \sum_{j=1}^{\infty} I^*_{p_0 p_j}(\ddot{p}_j + p_j)$$

$$- \sum_{j=1}^{\infty}(I^*_{q_j\ddot{p}_0}\ddot{q}_j + I^*_{q_j p_0} q_j)$$

$$= \gamma \int_0^1 \frac{M_a}{ac} dr + I^*_{p_0} \omega_0^2 \theta_{\text{con}}$$

$$- \sum_{j=1}^{\infty} I^*_{p_0} \omega_0^2 K_{p_j} q_j$$

$$I^*_{p_k}\left(\ddot{p}_k + (\omega_k^2 + 1)p_k\right) + I^*_{\ddot{p}_0 p_k}(\ddot{p}_0 + p_0)$$
$$- \sum_{j=1}^{\infty} (I^*_{q_j \ddot{p}_k}\ddot{q}_j + I^*_{q_j p_k} q_j)$$
$$= \gamma \int_0^1 \xi_k \frac{M_a}{ac}\, dr$$

where the inertial coefficients are

$$I^*_{q_k} = \frac{1}{I_b}\int_0^1 \eta_k^2\, m\, dr$$

$$I^*_{q_k \ddot{p}_j} = \frac{1}{I_b}\int_0^1 \eta_k \xi_j x_I\, m\, dr$$

$$I^*_{q_k p_j} = \frac{1}{I_b}\int_0^1 x_I m \int_0^r \eta'_k (\rho \xi_j)'\, d\rho\, dr$$

$$I^*_{p_k} = \frac{1}{I_b}\int_0^1 \xi_k^2 I_\theta\, dr$$

$$I^*_{p_0 p_k} = \frac{1}{I_b}\int_0^1 \xi_k I_\theta\, dr$$

The bending and torsion equations are coupled by inertial and centrifugal forces if the section center of gravity is offset from the elastic axis. Note that kinematic pitch-bending coupling of the form $\Delta\theta_{con} = -\sum_j K_{p_j} q_j$ has been included. For rigid flap and pitch, these equations reduce to those obtained in section 9-4.1.

Because the rigid pitch mode is not orthogonal to the elastic bending modes, the equations for p_0 and p_k $(k \geqslant 1)$ are coupled by inertial and centrifugal forces. The problem can also be formulated without the separate rigid pitch degree of freedom. Then the p_0 degree of freedom and equation of motion are dropped, and θ_e represents the complete pitch motion, including that due to control system flexibility. The boundary condition for the torsion equation becomes

$$GJ\frac{d\theta_e}{dr} = K_\theta(\theta_e - \theta_{\text{con}} + \sum_k K_{p_k} q_k)$$

at $r = 0$. The modal equation for free vibration can be solved with the boundary condition

$$GJ\frac{d\xi}{dr} = K_\theta \xi$$

for a general restrained end. The solution is a single series of orthogonal modes including both control system flexibility and blade torsional flexibility. However, this series of modes always gives $GJ\theta_e' = K_\theta \theta_e$ at the pitch bearing, which implies that the commanded pitch control and the pitch-bending feedback are zero. This is a typical result for normal modes, and it implies that point forces and moments applied at the end points cannot be handled. The problem also arises in treating the lag damper of an articulated blade, where the normal modes imply that the moment at the hinge is always zero. For this reason the rigid pitch and elastic torsion motion were separated in the present normal modes analysis. This is a rigorous approach and is easily implemented in a numerical solution. Moreover, the rigid pitch alone is a sufficient model of the blade pitch motion for many rotors. The coupled rigid pitch/elastic torsion modes can be used in the rotor analysis, including a proper representation of the end conditions, with the Rayleigh-Ritz or Galerkin methods (see section 9-9).

9-4.4 Nonrotating Frame

The rotor control system couples the pitch motion of the individual blades. Each nonrotating mode of pitch motion will have a different load path in the fixed control system, and hence a different effective stiffness. This coupling may be accounted for by using a separate natural frequency for each nonrotating degree of freedom. Consider the pitch equation of motion for the mth blade in the rotating frame:

$$I_f^*\left(\ddot{\theta}^{(m)} + (\omega^2 + 1)\theta^{(m)}\right) = \gamma \int_0^1 \frac{M_a}{ac}\, dr = \gamma M_f^{(m)}$$

The corresponding equations of motion in the nonrotating frame are

$$I_f^*\left(\ddot{\theta}_0 + (\omega_0^2 + 1)\theta_0\right) = \gamma M_{f_0}$$

$$I_f^*\left(\ddot{\theta}_{nc} + 2n\dot{\theta}_{ns} + (\omega_{nc}^2 + 1 - n^2)\theta_{nc}\right) = \gamma M_{f_{nc}}$$

$$I_f^*\left(\ddot{\theta}_{ns} - 2n\dot{\theta}_{nc} + (\omega_{ns}^2 + 1 - n^2)\theta_{ns}\right) = \gamma M_{f_{ns}}$$

$$I_f^*\left(\ddot{\theta}_{N/2} + (\omega_{N/2}^2 + 1)\theta_{N/2}\right) = \gamma M_{f_{N/2}}$$

where a separate natural frequency has been introduced for each equation. This is equivalent to assuming that the restoring moment provided by the control system responds to the nonrotating modes in such a way that

$$M_\theta^{(m)} = K_0(\theta_0 - \theta_0^{\text{con}}) + \sum_n \left(K_{nc}(\theta_{nc} - \theta_{nc}^{\text{con}})\cos n\psi_m + K_{ns}(\theta_{ns} - \theta_{ns}^{\text{con}})\sin n\psi_m\right) + K_{N/2}(\theta_{N/2} - \theta_{N/2}^{\text{con}})(-1)^m$$

instead of $M_\theta = K_\theta(\theta - \theta_{\text{con}})$ as in section 9-4.1. Thus ω_0 is here the stiffness of the collective control system, while ω_{1c} and ω_{1s} are the stiffnesses of the cyclic control system. The higher modes produce no net force in the nonrotating control system and hence are only due to flexibility in the pitch horn and pitch link, and swashplate bending. Thus for the reactionless modes ($\omega_{2c}, \omega_{2s}, \ldots, \omega_{N/2}$) the frequencies are usually much higher than for the collective and cyclic modes.

This technique of using different natural frequencies in the nonrotating frame is also useful for the flap and lag motion. A gimballed rotor can be modelled by using $\nu = 1$ for the β_{1c} and β_{1s} rigid flap degrees of freedom, and the appropriate cantilever frequency and mode shape for the coning and other degrees of freedom. Similarly, the collective lag degree of freedom ζ_0 can be used for the rotor rotational speed perturbation by setting the lag frequency to zero for this mode.

9–5 Hub Reactions

The net forces and moments at the root of the rotating blade are transmitted to the helicopter airframe. The steady components of these hub reactions in the nonrotating frame are the forces and moments required to trim the aircraft. The higher frequency components are responsible for helicopter vibration. When the shaft motion is included in the model, these rotor forces and moments determine the helicopter stability and control characteristics. Fig. 9-7 shows the definition of the root shears and moments on the rotating blade, and the rotor forces and moments acting on

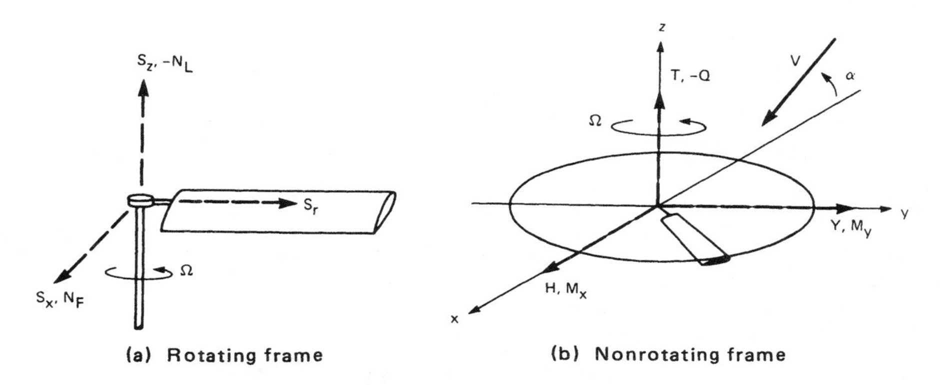

Figure 9-7 Definitions of the rotor forces and moments acting on the hub.

the hub in the nonrotating frame. The vertical shear force S_z produces the rotor thrust, and the in-plane shear forces S_x and S_r produce the rotor side and drag forces. The flapwise root moment N_F produces the rotor pitch and roll moments, and the lagwise moment N_L produces the rotor shaft torque. Note that positive rotor hub reactions are acting on the helicopter, with the exception of the rotor torque Q, which is defined as the moment on the rotor (the torque reaction of the rotor on the hub is positive in the direction opposing the rotor rotation). Fig. 9-7 indicates the positive directions of the rotor thrust T, drag force H, side force Y, pitch moment M_y, and roll moment M_x.

9-5.1 Rotating Loads

The net root forces and moments on the rotating blade can be obtained by integrating the section inertial and aerodynamic forces, as in the derivation of the blade equations of motion. Consider an articulated rotor with no hinge offset, as in section 9-2.1. The vertical forces acting on the blade section are the inertial force $m\ddot{z} = mr\ddot{\beta}$ and the aerodynamic force F_z. (The centrifugal force is always parallel to the hub plane; see Fig. 9-8.) The

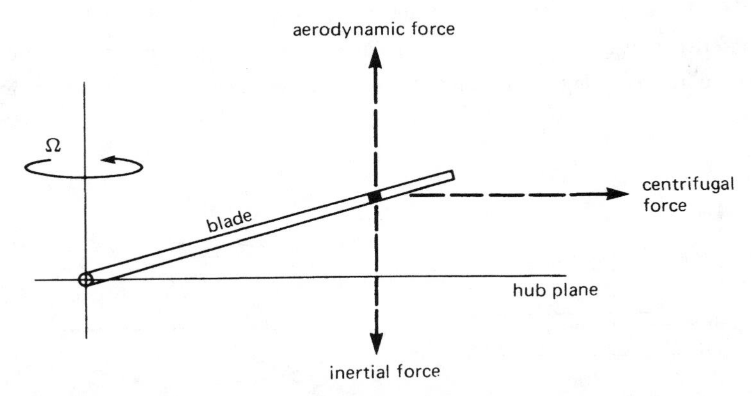

Figure 9-8 Blade section forces producing the vertical shear and flapwise moment at the root.

vertical shear force at the blade root is therefore

$$S_z = \int_0^R F_z dr - \ddot{\beta} \int_0^R rm\, dr$$

The root flap moment was obtained in section 9-2.1 for the equation of flapping motion due to the inertial, centrifugal, and aerodynamic forces on the section:

$$N_F = \int_0^R r F_z dr - (\ddot{\beta} + \Omega^2 \beta) \int_0^R r^2 m\, dr$$

The root moment is simply the hinge moment, since there is no offset of the flap hinge; the moment can be nonzero only with a hinge spring. The moment transmitted to the hub through the hinge spring is $N_F = K_\beta(\beta - \beta_p)$, or since $\nu_\beta^2 = 1 + K_\beta/(I_b\Omega^2)$,

$$N_F = I_b \Omega^2 (\nu_\beta^2 - 1)(\beta - \beta_p)$$

It was found in section 5–14 that this relation applies when there is a hinge offset as well.

Next consider the case of general out-of-plane bending motion, including both articulated and hingeless rotors. The forces acting on the blade section are described in section 9-2.2, and the equation of motion for the normal bending modes is derived. The vertical shear force at the root is obtained by integrating the aerodynamic and inertial forces on the blade:

$$S_z = \int_0^R (F_z - m\ddot{z})dr$$

Substituting for the modal expansion $z = \sum_k \eta_k q_k$ gives

$$S_z = \int_0^R F_z dr - \sum_k \ddot{q}_k \int_0^R \eta_k m\, dr$$

The root moment is obtained from the flap moments due to the aerodynamic, inertial, and centrifugal forces on the blade section (see Fig. 9-8), or simply by evaluating the flapwise bending moment expression given in section 9-2.2 at the root:

$$N_F = \int_0^R [(F_z - m\ddot{z})r - m\Omega^2 rz]\, dr$$

$$= \int_0^R rF_z dr - \sum_k (\ddot{q}_k + \Omega^2 q_k) \int_0^R r\eta_k m\, dr$$

Recall that the differential equation of motion for q_k is

$$I_{q_k}(\ddot{q}_k + \nu_k^2 q_k) = \int_0^R \eta_k F_z dr$$

The aerodynamic loading F_z therefore contributes directly to the root shear and moment, but it also excites the blade bending motion, which then cancels part of the hub reaction. Indeed, the flap hinge was introduced so that the blade motion rather than the structure would absorb the root moments. Since the mode shapes η_k form a complete series, it is possible to expand the aerodynamic loading as $F_z = \sum_k F_{z_k} \eta_k m$. It is easily verified that $F_{z_k} = \int_0^R \eta_k F_z dr / \int_0^R \eta_k^2 m\, dr$. When the expansion for F_z is substituted, the root moment becomes

$$N_F = \sum_k (F_{z_k} - \ddot{q}_k - \Omega^2 q_k) \int_0^R r\eta_k m\, dr$$

The equation of motion for q_k gives $F_{z_k} = \ddot{q}_k + \nu_k^2 q_k$, however. Hence

$$N_F = \sum_k q_k \Omega^2 (\nu_k^2 - 1) \int_0^R r\eta_k m\, dr$$

Note that for an articulated rotor with no hinge offset, $\nu_1 = 1$ and $\eta_1 = r$ for the first mode, and all the higher mode shapes are orthogonal to $r = \eta_1$; hence $N_F = 0$, as required. If only a single flap mode is used, and the mode shape is approximated by $\eta \cong r$, the expression reduces to

$$N_F = I_b \Omega^2 (\nu_\beta^2 - 1)\beta$$

as above. Thus the hub moment may be obtained from the flap deflection and the natural frequency of the fundamental flap mode. The simplicity of this result makes it very useful. In a similar fashion the vertical shear force of the root can be expressed as

$$S_z = \sum_k (F_{z_k} - \ddot{q}_k) \int_0^R \eta_k m\, dr = \sum_k q_k \Omega^2 \nu_k^2 \int_0^R \eta_k m\, dr$$

It will be more convenient, though, to relate the vertical shear, and hence the rotor thrust, directly to the aerodynamic force.

If the number of modes is large, the same result should be obtained for the hub moment regardless of whether the forces are integrated along the blade or the expression

$$N_F = \sum_k q_k \Omega^2 (\nu_k^2 - 1) \int_0^R r\eta_k m\, dr$$

is used. With the latter approach, using a finite number of modes is equivalent to truncating the expansion $F_z = \sum_k F_{z_k}\eta_k m$, which may not be an adequate representation of the loading if only a small number of modes are used. Thus better results are generally to be expected from using the integrals of the blade section forces to obtain the hub reactions, although in some cases the improved accuracy may not be as valuable as a simple equation.

Next let us examine the in-plane shear forces and torque moment at the blade root, including the in-plane blade motion. Consider an articulated blade with an in-plane displacement given by $x = \eta_\zeta \zeta$. There are three forces acting in the radial direction: the radial aerodynamic force F_r, due to the radial drag and an in-plane component of the lift when the blade flaps; the centrifugal force $m\Omega^2 r$; and a radially inward Coriolis force $2\Omega \dot{x} m = 2\Omega \eta_\zeta \dot{\zeta} m$. The Coriolis force is due to the product of the rotor rotational speed Ω and the in-plane velocity $\dot{x}$; this is the force responsible for the $\beta\dot{\zeta}$ flap moment (see section 9-3.1). Thus the radial shear force at the root is

$$S_r = \int_0^R (F_r + m\Omega^2 r - 2\Omega \eta_\zeta \dot{\zeta} m)\, dr$$

$$= \int_0^R F_r\, dr + \Omega^2 \int_0^R rm\, dr - 2\Omega \dot{\zeta} \int_0^R \eta_\zeta m\, dr$$

The centrifugal force is constant and is reacted by the identical centrifugal forces on the other blades. Hence only the aerodynamic and Coriolis forces contribute to the hub reactions in the nonrotating frame.

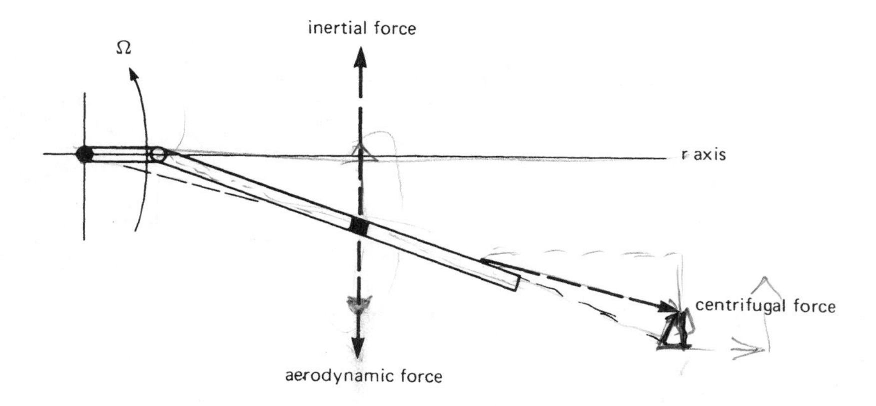

Figure 9-9 Blade section forces producing the in-plane shear at the root.

Fig. 9-9 shows the in-plane forces normal to the r axis that act on the blade section: the aerodynamic force F_x, consisting of profile and induced drag terms; the inertial force $m\ddot{x} = m\eta_\zeta \ddot{\zeta}$; and the centrifugal force $m\Omega^2 x = m\Omega^2 \eta_\zeta \zeta$. The last force arises because the centrifugal force $m\Omega^2 r$ has a component $(m\Omega^2 r)(x/r)$ normal to the r axis that acts in the same direction as the lag motion (Fig. 9-9). The in-plane Coriolis force due to the flapping motion is small compared to the centrifugal force and has been neglected. Thus the total in-plane shear force at the blade root is

$$S_x = \int_0^R (F_x - m\eta_\zeta \ddot{\zeta} + m\Omega^2 \eta_\zeta \zeta)dr$$

$$= \int_0^R F_x dr - (\ddot{\zeta} - \Omega^2 \zeta) \int_0^R \eta_\zeta m\, dr$$

The torque moment acting on the rotor hub is due to the section in-plane forces as discussed in deriving the lag equation of motion: the aerodynamic force F_x, the inertial force $m\ddot{x}$, and a Coriolis force $2\Omega \dot{z} z' m$. These forces have moment arm r about the center of rotation. The centrifugal force always passes through the shaft axis and so does not contribute to the torque. The moment at the root is thus

$$N_L = \int_0^R r(F_x - m\eta_\zeta \ddot{\zeta} - 2\Omega m \beta \dot{\beta} \eta_\beta \eta_\beta')dr$$

$$= \int_0^R rF_x dr - \ddot{\zeta} \int_0^R r\eta_\zeta m\, dr - 2\Omega \beta \dot{\beta} \int_0^R \eta_\beta \eta_\beta' rm\, dr$$

These results are readily extended to the case of general in-plane bending. As in section 9-3.2, expand the in-plane deflection as a series in the normal modes: $x = \sum_k \eta_{x_k} q_{x_k}$. Then the radial and in-plane shear forces are

$$S_r = \int_0^R (F_r + m\Omega^2 r - 2\Omega \dot{x} m)dr$$

$$= \int_0^R F_r dr + \Omega^2 \int_0^R rm\, dr - 2\Omega \sum_k \dot{q}_{x_k} \int_0^R \eta_{x_k} m\, dr$$

$$S_x = \int_0^R (F_x - m\ddot{x} + m\Omega^2 x)dr$$

$$= \int_0^R F_x\,dr - \sum_k (\ddot{q}_{x_k} - \Omega^2 q_{x_k}) \int_0^R \eta_{x_k} m\,dr$$

Neglecting the Coriolis terms, the torque moment is

$$N_L = \int_0^R r(F_x - m\ddot{x})\,dr$$

$$= \int_0^R rF_x\,dr - \sum_k \ddot{q}_{x_k} \int_0^R r\eta_{x_k} m\,dr$$

By expanding the aerodynamic loading F_x as a series in the bending mode shapes as for the hub moment, the in-plane shear force and the torque moment can be written as

$$S_x = \sum_k q_{x_k} \Omega^2 (\nu_{x_k}^2 + 1) \int_0^R \eta_{x_k} m\,dr$$

$$N_L = \sum_k q_{x_k} \Omega^2 \nu_{x_k}^2 \int_0^R r\eta_{x_k} m\,dr$$

These results are not as useful as the corresponding expression for the flap moment, however, since the blade lag motion q_{x_k} must be found in order to evaluate S_x and N_L in this manner. If the in-plane shear and torque are left in terms of the integrated aerodynamic forces, they can be evaluated even if the analysis neglects the lag motion.

9-5.2 Nonrotating Loads

The total forces and moments acting on the rotor hub are obtained by resolving the rotating forces in the nonrotating frame and summing over all N blades:

$$T = \sum_{m=1}^{N} S_z$$

$$H = \sum_{m=1}^{N} (S_r \cos\psi_m + S_x \sin\psi_m)$$

$$Y = \sum_{m=1}^{N} (S_r \sin\psi_m - S_x \cos\psi_m)$$

$$M_x = \sum_{m=1}^{N} N_F \sin\psi_m$$

$$M_y = -\sum_{m=1}^{N} N_F \cos\psi_m$$

$$Q = \sum_{m=1}^{N} N_L$$

where ψ_m is the azimuth angle of the mth blade. It is convenient to work with the hub reactions in rotor coefficient form. Note that in terms of dimensionless quantities

$$\frac{T}{NI_b} = \frac{T/\rho R^4 \Omega^2}{NI_b/\rho R^5} = \frac{\rho a c R^4}{I_b} \frac{T/\rho A(\Omega R)^2}{(Nc/\pi R)a} = \gamma \frac{C_T}{\sigma a}$$

Thus the equations for the hub reactions are divided by NI_b, giving

$$\gamma \frac{C_T}{\sigma a} = \frac{1}{N} \sum_{m=1}^{N} \frac{S_z}{I_b} = \frac{\gamma}{N} \sum_{m=1}^{N} \frac{S_z}{ac}$$

and similar results for the other forces and moments.

Consider first the rotor thrust. The following result (using dimensionless quantities) for the vertical shear force of the mth blade was obtained in the previous section:

$$\frac{S_z}{I_b} = \gamma \int_0^1 \frac{F_z}{ac}\, dr - \sum_k S_{q_k}^* \ddot{q}_k^{(m)}$$

where $S_{q_k}^* = \int_0^1 \eta_k m\, dr/I_b$. So

$$\gamma \frac{C_T}{\sigma a} = \frac{\gamma}{N} \sum_{m=1}^{N} \int_0^1 \frac{F_z}{ac}\, dr - \sum_k S_{q_k}^* \left(\frac{1}{N} \sum_{m=1}^{N} \ddot{q}_k^{(m)} \right)$$

Now from the definition of the nonrotating degrees of freedom in the Fourier coordinate transformation (section 8-4.1) it follows that the acceleration of the coning degree of freedom of the kth bending mode is

$$\ddot{\beta}_0^{(k)} = \frac{1}{N} \sum_{m=1}^{N} \ddot{q}_k^{(m)}$$

Hence the rotor thrust becomes

$$\gamma \frac{C_T}{\sigma a} = \frac{\gamma}{N} \sum_{m=1}^{N} \int_0^1 \frac{F_z}{ac}\, dr - \sum_k S^*_{q_k} \ddot{\beta}_0^{(k)}$$

The first term is the net aerodynamic lift on the rotor, and the second term is the vertical acceleration due to the coning motion of the blades.

The root flapwise moment on the rotating blade was found to be

$$\frac{N_F}{I_b} = \sum_k q_k^{(m)} (\nu_k^2 - 1) I^*_{q_k a}$$

where $I^*_{q_k a} = \int_0^1 r \eta_k\, m\, dr / I_b$. The pitch and roll moments on the rotor hub are then

$$-\gamma \frac{2C_{M_y}}{\sigma a} = \sum_k (\nu_k^2 - 1) I^*_{q_k a} \frac{2}{N} \sum_{m=1}^{N} q_k^{(m)} \cos\psi_m$$

$$\gamma \frac{2C_{M_x}}{\sigma a} = \sum_k (\nu_k^2 - 1) I^*_{q_k a} \frac{2}{N} \sum_{m=1}^{N} q_k^{(m)} \sin\psi_m$$

For a rotor with three or more blades, the definitions of the cyclic degrees of freedom $\beta_{1c}^{(k)}$ and $\beta_{1s}^{(k)}$ then give

$$\begin{pmatrix} -\gamma \dfrac{2C_{M_y}}{\sigma a} \\ \gamma \dfrac{2C_{M_x}}{\sigma a} \end{pmatrix} = \sum_k (\nu_k^2 - 1) I^*_{q_k a} \begin{pmatrix} \beta_{1c}^{(k)} \\ \beta_{1s}^{(k)} \end{pmatrix}$$

Using just a single flap mode, the hub moments are then simply proportional to the rotor tip-path-plane tilt:

$$\begin{pmatrix} -\gamma \dfrac{2C_{M_y}}{\sigma a} \\ \gamma \dfrac{2C_{M_x}}{\sigma a} \end{pmatrix} = I^*_\beta(\nu_\beta^2 - 1) \begin{pmatrix} \beta_{1c} \\ \beta_{1s} \end{pmatrix}$$

(a result that was also obtained in section 5-13 for a more limited model of the blade motion). If instead

$$\frac{N_F}{I_b} = \gamma \int_0^1 r \frac{F_z}{ac}\, dr - \sum_k (\ddot{q}_k^{(m)} + q_k^{(m)}) I^*_{q_k a}$$

is used as the expression for the moment, then

$$-\gamma \frac{2C_{M_y}}{\sigma a} = \frac{2\gamma}{N} \sum_{m=1}^{N} \cos\psi_m \int_0^1 r \frac{F_z}{ac}\, dr - \sum_k (\ddot{\beta}_{1c}^{(k)} + 2\dot{\beta}_{1s}^{(k)}) I^*_{q_k a}$$

$$\gamma \frac{2C_{M_x}}{\sigma a} = \frac{2\gamma}{N} \sum_{m=1}^{N} \sin\psi_m \int_0^1 r \frac{F_z}{ac}\, dr - \sum_k (\ddot{\beta}_{1s}^{(k)} - 2\dot{\beta}_{1c}^{(k)}) I^*_{q_k a}$$

In the steady-state case, the tip-path-plane tilt is constant. Then only the aerodynamic forces contribute to the hub pitch and roll moment, which is the result derived in section 5—3.

The rotor drag and side forces are obtained by resolving into the non-rotating frame the in-plane and radial shear forces on the root of the rotating blade:

$$\frac{S_x}{I_b} = \gamma \int_0^1 \frac{F_x}{ac}\, dr - S^*_\zeta(\ddot{\zeta}^{(m)} - \zeta^{(m)})$$

$$\frac{S_r}{I_b} = \gamma \int_0^1 \frac{F_r}{ac}\, dr - 2S^*_\zeta \zeta^{(m)}$$

where $S^*_\zeta = \int_0^1 \eta_\zeta m\, dr/I_b$. Only the fundamental lag mode has been considered for simplicity, and the centrifugal force has been dropped from S_r since it does not contribute to the total hub forces. Now the definitions of the cyclic lag degrees of freedom in the Fourier coordinate transformation give

$$\frac{2}{N}\sum_{m=1}^{N}\left[(\ddot{\zeta}-\zeta)\sin\psi_m+2\dot{\zeta}\cos\psi_m\right]$$

$$=(\ddot{\zeta}_{1s}-2\dot{\zeta}_{1c}-\zeta_{1s}-\zeta_{1s})+2(\dot{\zeta}_{1c}+\zeta_{1s})$$

$$=\ddot{\zeta}_{1s}$$

and

$$\frac{2}{N}\sum_{m=1}^{N}\left[(\ddot{\zeta}-\zeta)\cos\psi_m-2\dot{\zeta}\sin\psi_m\right]$$

$$=(\ddot{\zeta}_{1c}+2\dot{\zeta}_{1s}-\zeta_{1c}-\zeta_{1c})-2(\dot{\zeta}_{1s}-\zeta_{1c})$$

$$=\ddot{\zeta}_{1c}$$

(again assuming the rotor has three or more blades). Hence the rotor drag and side forces are

$$\gamma\frac{2C_H}{\sigma a}=\frac{2\gamma}{N}\sum_{m=1}^{N}\left[\cos\psi_m\int_0^1\frac{F_r}{ac}\,dr+\sin\psi_m\int_0^1\frac{F_x}{ac}\,dr\right]-S^*_\zeta\ddot{\zeta}_{1s}$$

$$\gamma\frac{2C_Y}{\sigma a}=\frac{2\gamma}{N}\sum_{m=1}^{N}\left[\sin\psi_m\int_0^1\frac{F_r}{ac}\,dr-\cos\psi_m\int_0^1\frac{F_x}{ac}\,dr\right]+S^*_\zeta\ddot{\zeta}_{1c}$$

The in-plane forces on the rotor have inertial reactions due to the longitudinal and lateral shifts of the rotor center of gravity that are associated with the cyclic lag degrees of freedom. Recall that in Chapter 5 the steady-state rotor forces were expressed as $H=\beta_{1c}T+H_{TPP}$ and $Y=-\beta_{1s}T+Y_{TPP}$. To write the present results in terms of the tilt of the thrust vector with the tip-path plane requires a detailed consideration of the aerodynamic forces F_x and F_r, which will be given in Chapter 11. Finally, if the Coriolis term is neglected, the torque on a single blade is

$$\frac{N_L}{I_b}=\gamma\int_0^1 r\frac{F_x}{ac}\,dr-I^*_{\zeta a}\ddot{\zeta}^{(m)}$$

where $I^*_{\zeta a}=\int_0^1 r\eta_\zeta m\,dr/I_b$. Then the total rotor torque is

$$\gamma \frac{C_Q}{\sigma a} = \frac{\gamma}{N} \sum_{m=1}^{N} \int_0^1 r \frac{F_x}{ac} dr - I^*_{\zeta a} \ddot{\zeta}_0$$

since by the definition of the collective lag degree of freedom, $(1/N) \sum_m \ddot{\zeta}^{(m)} = \ddot{\zeta}_0$.

For a two-bladed rotor the cyclic flap and lag degrees of freedom in the nonrotating frame do not exist, so different results are obtained. Instead of the cyclic degrees of freedom like β_{1c} and β_{1s}, the two-bladed rotor has a single teetering degree of freedom, β_1. Determining the hub moment requires an evaluation of the sums:

$$\frac{2}{N} \sum_{m=1}^{N} \beta^{(m)} \sin \psi_m = 2 \sin \psi \frac{1}{N} \sum_{m=1}^{N} \beta^{(m)} (-1)^m = 2\beta_1 \sin \psi$$

$$\frac{2}{N} \sum_{m=1}^{N} \beta^{(m)} \cos \psi_m = 2 \cos \psi \frac{1}{N} \sum_{m=1}^{N} \beta^{(m)} (-1)^m = 2\beta_1 \cos \psi$$

where $\beta_1 = \frac{1}{2}(\beta^{(2)} - \beta^{(1)})$ is the teetering degree of freedom. Then the hub moments are

$$-\gamma \frac{2C_{M_Y}}{\sigma a} = I^*_\beta(\nu_\beta^2 - 1) 2\beta_1 \cos \psi$$

$$\gamma \frac{2C_{M_X}}{\sigma a} = I^*_\beta(\nu_\beta^2 - 1) 2\beta_1 \sin \psi$$

Similarly, the rotor drag and side force depend on the differential lag degree of freedom $\zeta_1 = \frac{1}{2}(\zeta^{(2)} - \zeta^{(1)})$:

$$\gamma \frac{2C_H}{\sigma a} = \frac{2\gamma}{N} \sum_{m=1}^{N} \left[\cos \psi_m \int_0^1 \frac{F_r}{ac} dr + \sin \psi_m \int_0^1 \frac{F_x}{ac} dr \right]$$
$$- 2S^*_\zeta\left((\ddot{\zeta}_1 - \zeta_1) \sin \psi + 2\dot{\zeta}_1 \cos \psi \right)$$

$$\gamma \frac{2C_Y}{\sigma a} = \frac{2\gamma}{N} \sum_{m=1}^{N} \left[\sin \psi_m \int_0^1 \frac{F_r}{ac} dr - \cos \psi_m \int_0^1 \frac{F_x}{ac} dr \right]$$
$$+ 2S^*_\zeta\left((\ddot{\zeta}_1 - \zeta_1) \cos \psi - 2\dot{\zeta}_1 \sin \psi \right)$$

Therefore, while the steady-state, periodic motion of the two-bladed rotor is the same as for rotors with three or more blades, the transient dynamics are fundamentally different because of the absence of the cyclic degrees of freedom.

9–6 Shaft Motion

So far only the motion of the rotor itself has been considered in the dynamic analysis. The shaft motion is also an important factor, both for helicopter stability and control problems involving the rigid body degrees of freedom, and for aeroelastic problems involving the coupled motion of the airframe and rotor. Fig. 9-10 defines the linear and angular hub motion considered. The perturbation of the hub position from the equilibrium flight path is given by the displacements x_h, y_h, and z_h. The perturbed orientation is given by the hub rotations α_x, α_y, and α_z. For now an inertial reference frame is used, so that the coordinate frame remains fixed in space during the perturbed motion of the hub.

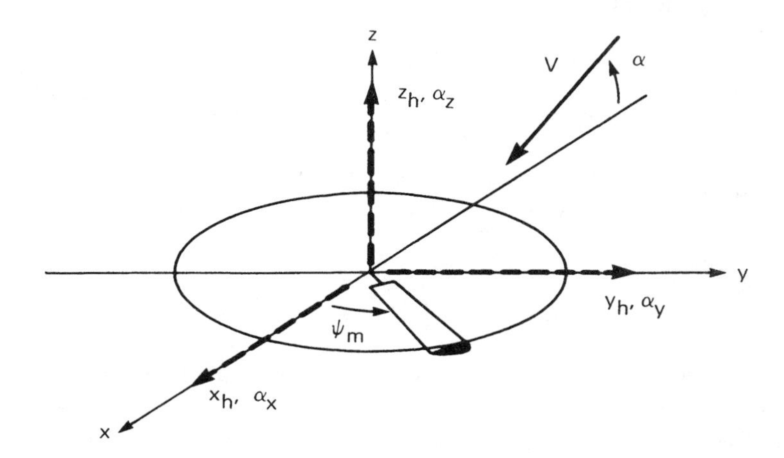

Figure 9-10 Definition of the linear and angular shaft motion.

The shaft motion introduces additional out-of-plane and in-plane acceleration terms that must be included in the bending equations of motion. Consider the rigid flap and lag model developed in sections 9-2.1 and 9-3.1.

The additional section accelerations producing flap moments are as follows:

(i) the angular acceleration $r(\ddot{\alpha}_x \sin\psi_m - \ddot{\alpha}_y \cos\psi_m)$;

(ii) the Coriolis acceleration $2\Omega r(\dot{\alpha}_x \cos\psi_m + \dot{\alpha}_y \sin\psi_m)$; and

(iii) the vertical acceleration $\ddot{z}_h$.

Each of these terms gives a downward inertial force on the section, with moment arm $(r - e)$ about the offset flap hinge. The angular acceleration $(\ddot{\alpha}_x \sin\psi_m - \ddot{\alpha}_y \cos\psi_m)$ is the flapwise component of the pitch and roll acceleration of the hub. The Coriolis acceleration arises from the cross product of the angular velocity $(\dot{\alpha}_x \cos\psi_m + \dot{\alpha}_y \sin\psi_m)$ of the rotor disk about the blade radial axis, and the rotational velocity of the section Ωr. Integrating these forces over the blade span gives the additional flap moment

$$\left(\int_0^R \eta_\beta r m\, dr\right)\left[(\ddot{\alpha}_x + 2\Omega\dot{\alpha}_y)\sin\psi_m - (\ddot{\alpha}_y - 2\Omega\dot{\alpha}_x)\cos\psi_m\right] + \left(\int_0^R \eta_\beta m\, dr\right)\ddot{z}_h$$

The flap equation of motion becomes

$$I_\beta^*(\ddot{\beta} + \nu_\beta^2 \beta) - I_{\beta\zeta}^* 2\beta\dot{\zeta} + I_{\beta\alpha}^*\left((\ddot{\alpha}_x + 2\dot{\alpha}_y)\sin\psi_m - (\ddot{\alpha}_y - 2\dot{\alpha}_x)\cos\psi_m\right) + S_\beta^*\ddot{z}_h = \gamma\int_0^1 \eta_\beta \frac{F_z}{ac}\, dr = \gamma M_F$$

where $I_{\beta\alpha}^* = \int_0^1 r\eta_\beta m\, dr/I_b$ and $S_\beta^* = \int_0^1 \eta_\beta m\, dr/I_b$. Note that the shaft motion appears in the blade equation of motion with periodic coefficients, because it is defined in the nonrotating frame.

The additional in-plane accelerations producing lag moments on the blade are:

(i) the hub angular acceleration $r\ddot{\alpha}_z$; and

(ii) the hub in-plane linear acceleration $(\ddot{x}_h \sin\psi_m - \ddot{y}_h \cos\psi_m)$.

The angular acceleration term gives an inertial force in the lag direction, and the linear acceleration gives a force opposing the lag motion; both have moment arms $(r - e)$ about the lag hinge. Integrating over the span gives the lag moments

$$-\left(\int_0^R r\eta_\zeta m\,dr\right)\ddot{\alpha}_z + \left(\int_0^R \eta_\zeta m\,dr\right)\left(\ddot{x}_h \sin\psi_m - \ddot{y}_h \cos\psi_m\right)$$

so that the lag equation of motion becomes

$$I^*_\zeta(\ddot{\zeta} + \nu_\zeta^2 \zeta) + I^*_{\beta\zeta} 2\beta\dot{\beta} - I^*_{\zeta a}\ddot{\alpha}_z$$

$$+ S^*_\zeta(\ddot{x}_h \sin\psi_m - \ddot{y}_h \cos\psi_m) = \gamma \int_0^1 \eta_\zeta \frac{F_x}{ac}\,dr = \gamma M_L$$

where $I^*_{\zeta a} = \int_0^1 r\eta_\zeta m\,dr/I_b$ and $S^*_\zeta = \int_0^1 \eta_\zeta m\,dr/I_b$.

Next, transform the flap and lag equations of motion to the nonrotating frame. The hub acceleration and velocity are independent of the blade index, so the summation operators act only on the $\sin\psi_m$ and $\cos\psi_m$ factors. It follows that the hub motion contributes only to the collective and cyclic equations in the nonrotating frame (at least for the inertial forces). The result for the flap motion is:

$$I^*_\beta(\ddot{\beta}_0 + \nu_\beta^2 \beta_0) - I^*_{\beta\zeta} 2\beta_{\text{trim}}\dot{\zeta}_0 + S^*_\beta \ddot{z}_h = \gamma M_{F_0}$$

$$I^*_\beta\left(\ddot{\beta}_{1c} + 2\dot{\beta}_{1s} + (\nu_\beta^2 - 1)\beta_{1c}\right) - I^*_{\beta\zeta} 2\beta_{\text{trim}}(\dot{\zeta}_{1c} + \zeta_{1s})$$

$$- I^*_{\beta a}(\ddot{\alpha}_y - 2\dot{\alpha}_x) = \gamma M_{F_{1c}}$$

$$I^*_\beta\left(\ddot{\beta}_{1s} - 2\dot{\beta}_{1c} + (\nu_\beta^2 - 1)\beta_{1s}\right) - I^*_{\beta\zeta} 2\beta_{\text{trim}}(\dot{\zeta}_{1s} - \zeta_{1c})$$

$$+ I^*_{\beta a}(\ddot{\alpha}_x + 2\dot{\alpha}_y) = \gamma M_{F_{1s}}$$

and for the lag motion:

$$I^*_\zeta(\ddot{\zeta}_0 + \nu_\zeta^2 \zeta_0) + I^*_{\beta\zeta} 2\beta_{\text{trim}}\dot{\beta}_0 - I^*_{\zeta a}\ddot{\alpha}_z = \gamma M_{L_0}$$

$$I_\zeta^*\left(\ddot{\zeta}_{1c} + 2\dot{\zeta}_{1s} + (\nu_\zeta^2 - 1)\zeta_{1c}\right) + I_{\beta\zeta}^* 2\beta_{\text{trim}}(\dot{\beta}_{1c} + \beta_{1s})$$
$$- S_\zeta^* \ddot{y}_h = \gamma M_{L_{1c}}$$

$$I_\zeta^*\left(\ddot{\zeta}_{1s} - 2\dot{\zeta}_{1c} + (\nu_\zeta^2 - 1)\zeta_{1s}\right) + I_{\beta\zeta}^* 2\beta_{\text{trim}}(\dot{\beta}_{1s} - \beta_{1c})$$
$$+ S_\zeta^* \ddot{x}_h = \gamma M_{L_{1s}}$$

In the nonrotating frame the inertial coupling between the rotor and shaft motion is thus quite limited. The coning mode responds to vertical acceleration, the cyclic flap modes respond to the pitch and roll motion, the collective lag responds to yaw acceleration of the shaft, and the cyclic lag modes respond to the longitudinal and lateral hub acceleration. There is no coupling at all of the shaft motion with the equations of motion for the reactionless degrees of freedom ($2c$, $2s$, ..., nc, ns, $N/2$).

All three of the vertical inertial forces due to the shaft motion that produce flap moments must also be included in the root shear force, which becomes

$$S_z = \int_0^R F_z\,dr - \ddot{\beta}\int_0^R \eta_\beta m\,dr - \ddot{z}_h \int_0^R m\,dr$$
$$-\left((\ddot{\alpha}_x + 2\dot{\alpha}_y)\sin\psi_m - (\ddot{\alpha}_y - 2\dot{\alpha}_x)\cos\psi_m\right)\int_0^R rm\,dr$$

Similarly, the root hub moment is

$$N_F = \int_0^R rF_z\,dr - (\ddot{\beta} + \Omega^2\beta)\int_0^R r\eta_\beta m\,dr - \ddot{z}_h \int_0^R rm\,dr$$
$$-\left((\ddot{\alpha}_x + 2\dot{\alpha}_y)\sin\psi_m - (\ddot{\alpha}_y - 2\dot{\alpha}_x)\cos\psi_m\right)\int_0^R r^2\,m\,dr$$

Alternatively, the expression $N_F = I_b\Omega^2(\nu_\beta^2 - 1)\beta$ can be used here. The radial acceleration $(\ddot{x}_h \cos\psi_m + \ddot{y}_h \sin\psi_m)$ due to the in-plane hub motion must be added to the radial shear force:

$$S_r = \int_0^R F_r\,dr - 2\Omega\dot{\zeta}\int_0^R \eta_\zeta m\,dr - (\ddot{x}_h \cos\psi_m + \ddot{y}_h \sin\psi_m)\int_0^R m\,dr$$

The in-plane inertial forces due to the shaft motion that produce lag moments must be included in the in-plane root shear force and the blade torque:

$$S_x = \int_0^R F_x\,dr - (\ddot{\zeta} - \Omega^2\zeta)\int_0^R \eta_\zeta m\,dr + \ddot{\alpha}_z\int_0^R rm\,dr - (\ddot{x}_h \sin\psi_m - \ddot{y}_h \cos\psi_m)\int_0^R m\,dr$$

$$N_L = \int_0^R rF_x\,dr - \ddot{\zeta}\int_0^R r\eta_\zeta m\,dr - 2\Omega\beta\dot{\beta}I_{\beta\zeta} + \ddot{\alpha}_z\int_0^R r^2 m\,dr - (\ddot{x}_h \sin\psi_m - \ddot{y}_h \cos\psi_m)\int_0^R rm\,dr$$

The results for the rotor forces and moments, obtained by summing the root reaction over all N blades, are simpler because many of the new terms cancel. The rotor thrust is

$$\gamma\frac{C_T}{\sigma a} = \frac{1}{N}\sum_{m=1}^{N}\frac{S_z}{I_b} = \left(\gamma\frac{C_T}{\sigma a}\right)_{\text{aero}} - S^*_\beta\ddot{\beta}_0 - M^*_b\ddot{z}_h$$

where $S^*_\beta = \int_0^R \eta_\beta m\,dr/I_b$ and $M^*_b = \int_0^R m\,dr/I_b$ (see section 9-5.2 for the aerodynamic term). Note that M^*_b is the normalized mass of a single blade. For the pitch and roll moments, we can still use

$$\begin{pmatrix} -\gamma\dfrac{2C_{M_y}}{\sigma a} \\ \gamma\dfrac{2C_{M_x}}{\sigma a}\end{pmatrix} = I^*_\beta(\nu_\beta^2 - 1)\begin{pmatrix}\beta_{1c} \\ \beta_{1s}\end{pmatrix}$$

or

$$-\gamma\frac{2C_{M_Y}}{\sigma a} = -\left(\gamma\frac{2C_{M_Y}}{\sigma a}\right)_{\text{aero}} - I^*_{\beta a}(\ddot{\beta}_{1c} + 2\dot{\beta}_{1s}) + I^*_0(\ddot{\alpha}_y - 2\dot{\alpha}_x)$$

$$\gamma\frac{2C_{M_X}}{\sigma a} = \left(\gamma\frac{2C_{M_X}}{\sigma a}\right)_{\text{aero}} - I^*_{\beta a}(\ddot{\beta}_{1s} - 2\dot{\beta}_{1c}) - I^*_0(\ddot{\alpha}_x + 2\dot{\alpha}_y)$$

where $I^*_{\beta a} = \int_0^R r\eta_\beta m\, dr/I_b$ and $I^*_0 = \int_0^R r^2 m dr/I_b$. The rotor drag and side forces become

$$\gamma\frac{2C_H}{\sigma a} = \frac{2}{N}\sum_{m=1}^{N}\left(\frac{S_r}{I_b}\cos\psi_m + \frac{S_x}{I_b}\sin\psi_m\right)$$

$$= \left(\gamma\frac{2C_H}{\sigma a}\right)_{\text{aero}} - S^*_\zeta\ddot{\zeta}_{1s} - 2M^*_b\ddot{x}_h$$

$$\gamma\frac{2C_Y}{\sigma a} = \frac{2}{N}\sum_{m=1}^{N}\left(\frac{S_r}{I_b}\sin\psi_m - \frac{S_x}{I_b}\cos\psi_m\right)$$

$$= \left(\gamma\frac{2C_Y}{\sigma a}\right)_{\text{aero}} + S^*_\zeta\ddot{\zeta}_{1c} - 2M^*_b\ddot{y}_h$$

Finally, the rotor torque becomes

$$\gamma\frac{C_Q}{\sigma a} = \frac{1}{N}\sum_{m=1}^{N}\frac{N_L}{I_b}$$

$$= \left(\gamma\frac{C_Q}{\sigma a}\right)_{\text{aero}} - I^*_{\zeta a}\ddot{\zeta}_0 + I^*_0\ddot{\alpha}_z$$

The only inertial contributions to the thrust, drag, and side forces of the rotor are the reactions to the linear acceleration of the total rotor mass. The angular acceleration reactions of the entire rotor produce moments on the hub.

The utility of the Fourier coordinate transformation lies in its association with the interaction between the rotor and the nonrotating system. The shaft motion appears in the rotating equations of motion with periodic coefficients, which are eliminated by conversion to the nonrotating frame; summing the blade root forces to obtain the total rotor hub reactions naturally leads to the nonrotating degrees of freedom for the rotor motion. Moreover, the coupling between the rotor and the fixed system is limited, because the nonrotating degrees of freedom define the motion of the rotor as a whole, in certain patterns that naturally lead to an association with only certain components of the shaft motion and hub forces. Specifically, the rotor tip-path-plane tilt appears with the shaft pitch and roll motion, and with the hub pitch and roll moments. The cyclic lag degrees of freedom, which produce an in-plane shift of the rotor center of gravity, are associated with the hub in-plane displacements and forces. The rotor coning motion appears with the vertical shaft displacement and the rotor thrust, while the collective lag motion appears with the shaft yaw and rotor torque. Finally, the reactionless rotor modes do not couple with the shaft motion and hub forces at all. In axial flow, there is some additional coupling of the dynamics by the aerodynamic forces, but the motion still separates into a vertical system (z_h and α_z), a lateral-longitudinal system (x_h, y_h, α_x, and α_y), and the reactionless modes. In forward flight the aerodynamic forces tend to couple all the degrees of freedom of the helicopter, but this basic decomposition remains a dominant characteristic of the behavior.

The influence of the shaft motion is different for the case of a two-bladed rotor because of the absence of the cyclic modes. The equations for β_{1c} and β_{1s} are replaced by the teetering equation, which including the shaft motion is

$$I_\beta^*(\ddot{\beta}_1 + \nu_\beta^2 \beta_1) - I_{\beta\zeta}^* 2\beta_{\text{trim}} \dot{\zeta}_1 + I_{\beta\alpha}^*\Big((\ddot{\alpha}_x + 2\dot{\alpha}_y)\sin\psi - (\ddot{\alpha}_y - 2\dot{\alpha}_x)\cos\psi\Big) = \gamma M_{F_1}$$

Similarly, an equation for the differential lag motion ζ_1 replaces the ζ_{1c} and ζ_{1s} equations:

$$I_\zeta^*(\ddot{\zeta}_1 + \nu_\zeta^2 \zeta_1) + I_{\beta\zeta}^* 2\beta_{\text{trim}} \dot{\beta}_1 + S_\zeta^*(\ddot{x}_h \sin\psi - \ddot{y}_h \cos\psi) = \gamma M_{L_1}$$

The equations of motion for β_0 and ζ_0 given above are valid for both $N \geqslant 3$ and the two-bladed rotor case, as are the results for the rotor thrust and torque. For the in-plane hub forces, the inertial reaction to the rotor acceleration must be added to the expressions given in section 9-5.2:

$$\Delta\left(\gamma\frac{2C_H}{\sigma a}\right) = -2M_b^*\ddot{x}_h$$

$$\Delta\left(\gamma\frac{2C_Y}{\sigma a}\right) = -2M_b^*\ddot{y}_h$$

The pitch and roll moments of the two-bladed rotor were obtained in terms of the tip-path-plane motion, and so are unchanged; the shaft motion influences the hub moments through the solution for β_1. The most notable feature of the coupling between the fixed frame and the two-bladed rotor in both the hub forces and the shaft motion is the periodic coefficients due to the lack of axisymmetry of the rotor. As a result, the analysis of the dynamics of a two-bladed rotor is distinctly different from that for rotors with three or more blades.

In stability and control analyses, for helicopters as well as for airplanes, a body axis system is most frequently used. With a body axis system the coordinate axes remain fixed in the body during its perturbed motion, whereas an inertial axis system remains fixed relative to space. Since the trim velocity of aircraft is defined relative to the reference axes, the angular velocity of the body axes must rotate the velocity vector as well, which implies a centrifugal acceleration relative to inertial space:

$$\left(\frac{d\vec{v}}{dt}\right)_{\substack{\text{inertial}\\\text{axes}}} = \left(\frac{d\vec{v}}{dt}\right)_{\substack{\text{body}\\\text{axes}}} + \dot{\vec{\alpha}}\times\vec{v}$$

The results that have been derived in this section for the shaft motion require knowing the hub acceleration in inertial space. The rotor in steady flight has trim velocity components μ in the disk plane and $\mu\tan\alpha$ normal to the disk plane. Then the inertial accelerations in terms of the body axis motion are

$$(\ddot{x}_h)_{\substack{\text{inertial}\\\text{space}}} = (\ddot{x}_h + \dot{\alpha}_y\mu\tan\alpha)_{\substack{\text{body}\\\text{axes}}}$$

$$(\ddot{y}_h)_{\substack{\text{inertial}\\\text{space}}} = (\ddot{y}_h - \dot{\alpha}_z\mu - \dot{\alpha}_x\mu\tan\alpha)_{\substack{\text{body}\\\text{axes}}}$$

$$(\ddot{z}_h)_{\substack{\text{inertial}\\\text{space}}} = (\ddot{z}_h + \dot{\alpha}_y\mu)_{\substack{\text{body}\\\text{axes}}}$$

In hover, where the trim velocity of the helicopter is zero, there is no difference between the inertial axes and body axes for a linear analysis of the inertial forces. The use of body axes affects the formulation of the rotor aerodynamic forces due to shaft motion as well (see section 11-6).

9–7 Coupled Flap-Lag-Torsion Motion

The intent of this chapter has been to explore the basic physics of the rotating blade, and the models considered have been no more complicated than necessary to accomplish that. There are rotors and problems for which the equations developed here are not sufficient to describe the dynamic behavior. A number of comments have already been made in the text concerning the need for more advanced analyses, and more will follow. A particular extension that is often required is the consideration of the fully coupled flap-lag bending, rigid pitch, and elastic torsion motion of a blade. There are other degrees of freedom that may be also required, such as gimbal or teeter hinge motion, and the rotational speed perturbation. Often there are details of the geometric, inertial, and structural definition of the rotor that must be added or extended, such as a consideration of bending flexibility inboard or droop and sweep outboard of the pitch bearing. Most often the requirement for an advanced dynamics analysis is encountered when dealing with hingeless rotors.

The derivation of the equation of motion for the fully coupled flap-lag-torsion dynamics of a rotor blade is a lengthy and complicated process. The number of interactions that must be considered increases at least as the square of the number of degrees of freedom, and many of the couplings required are fundamentally nonlinear. Moreover, helicopter dynamics is still a subject of research, with a complete definition of what forces must be included and what approximations may be made yet to be established. For these reasons the development of more general equations of motion for the rotor is left to the literature. See for example Arcidiacono (1969), Piziali (1970), Hodges and Dowell (1974), Johnson (1977f); and the references of Chapters 12 and 14.

9–8 Rotor Blade Bending Modes

9-8.1 Engineering Beam Theory for a Twisted Blade

The blade pitch and twist introduce structural coupling of the out-of-plane and in-plane bending deflection. The free vibration modes in the centrifugal force field of the rotating blade hence involve coupled flap and lag motion, which has a major impact on the rotor dynamics. It is therefore necessary to re-examine engineering beam theory for a rotor blade, including the effects of the blade pitch and twist. The objective is to relate the bending moments at the section to the blade bending deflection. Elastic torsion will also be included in the model. This analysis is based on the work of Houbolt and Brooks (1958).

It is assumed that the undeformed elastic axis is a straight line, and that the blade has a high structural aspect ratio so that engineering beam theory is applicable. Fig. 9-11 shows the blade geometry considered. The spanwise variable r is measured from the center of rotation along the elastic axis. The coordinates x and z are the principal axes of the section, with origin at the elastic axis. Then by definition $\int xzEdA = 0$. Note that this integral is weighted with the modulus of elasticity. The modulus-weighted centroid, or tension center, is assumed to be located on the x-axis at a distance x_C aft of the elastic axis; hence $\int zEdA = 0$ and $\int xEdA = x_C \int EdA$. The angle of the major principal axis (the x-axis) with respect to the hub plane is the blade pitch θ. The existence of the eleastic axis means that torsion about the elastic axis occurs without bending of the blade. Thus the pitch angle consists of the root pitch θ_0, the built-in twist θ_{tw}, and an elastic torsion deflection θ_e: $\theta = \theta_0 + \theta_{tw} + \theta_e$. The twist θ_{tw} is a function of r and is defined to be zero at the root. Shear stress in the blade is due to θ_e only. It is assumed that θ_e is small, but the trim pitch angles θ_0 and θ_{tw} may be large.

The unit vectors of the rotating hub plane axis system are $\vec{i}_B$, $\vec{j}_B$, and $\vec{k}_B$ (see Fig. 9-11). The unit vectors of the principal axes of the section are $\vec{i}$, $\vec{j}$, and $\vec{k}$, which are rotated by the angle θ from the hub plane:

$$\vec{i} = \vec{i}_B \cos\theta - \vec{k}_B \sin\theta$$

$$\vec{j} = \vec{j}_B$$

$$\vec{k} = \vec{i}_B \sin\theta + \vec{k}_B \cos\theta$$

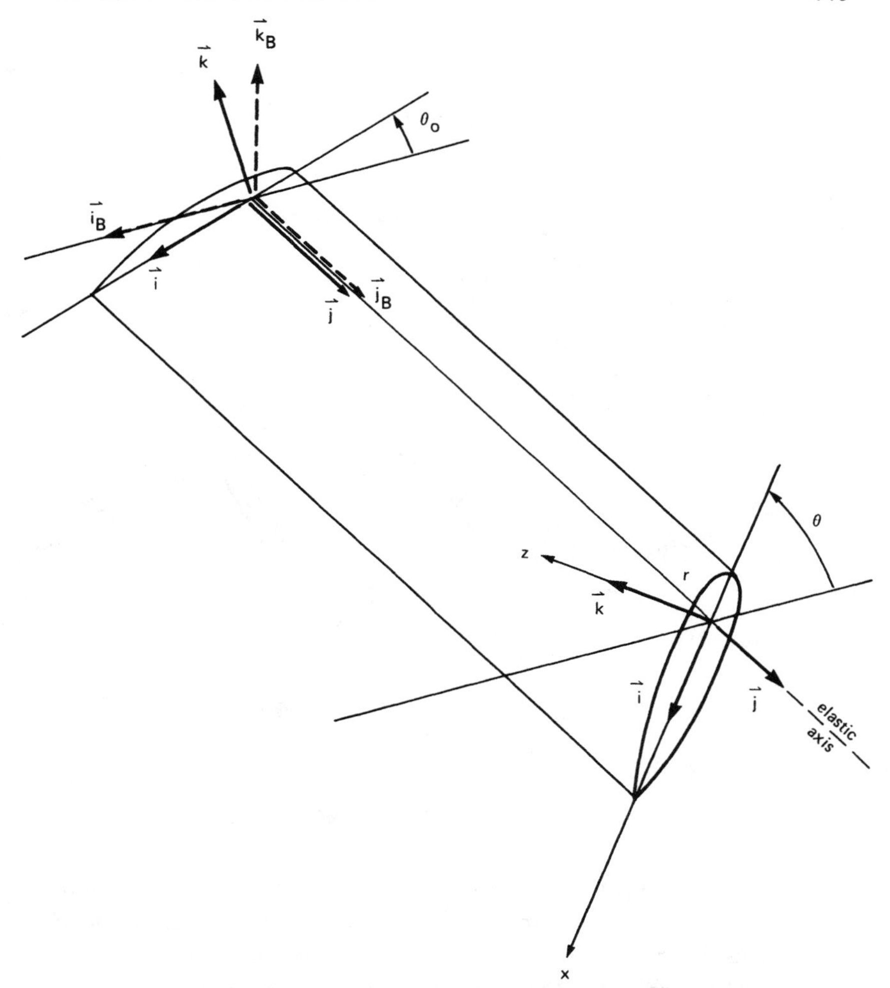

Figure 9-11 Geometry of the twisted rotor blade (before the bending distortion).

Note that the elastic torsion is included in the definition of $\vec{i}$ and $\vec{k}$, but not the blade bending. It also follows from this definition that $\partial\vec{i}/\partial r = -\theta'\vec{k}$ and $\partial\vec{k}/\partial r = \theta'\vec{i}$.

The distortion of the blade will be described by a deflection of the elastic axis with components x_0, r_0, and z_0 (Fig. 9-12). The bending of the elastic axis produces a rotation of the section by the angles ϕ_x and ϕ_z. The elastic torsion θ_e has already been included in θ. Engineering beam theory assumes

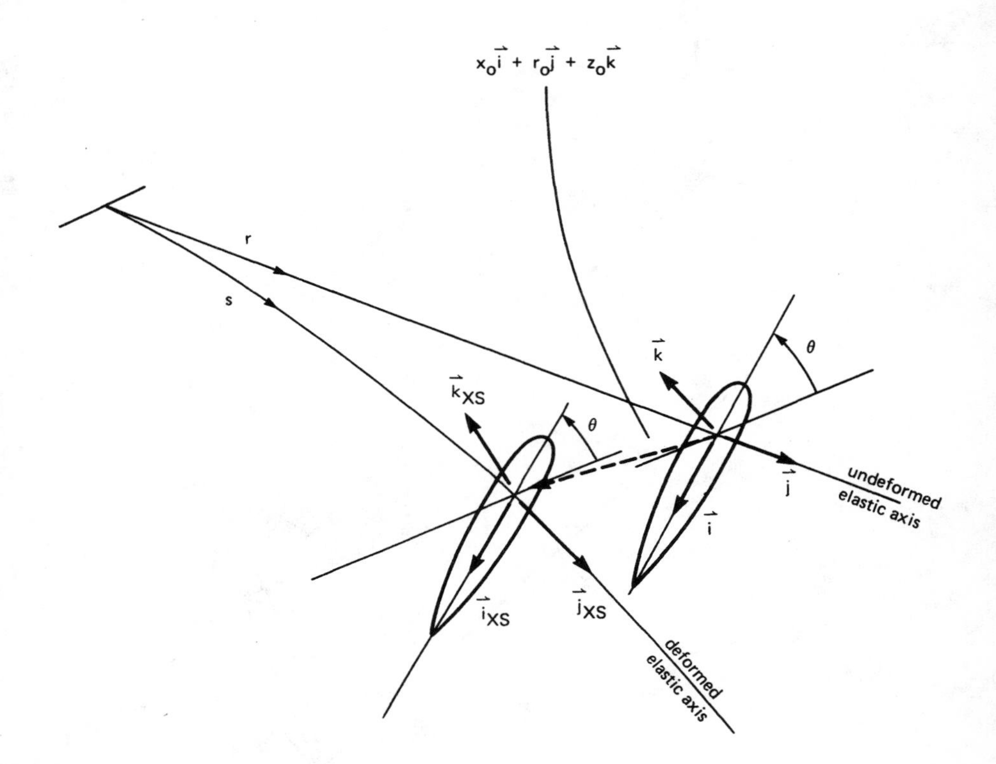

Figure 9-12 Definition of the blade deformation.

that plane sections perpendicular to the elastic axis remain so after bending. This description of the blade motion is sufficient to define the distortion of all elements of the section. It is assumed that the quantities x_0, r_0, z_0, ϕ_x, ϕ_z, and θ_e are small. The unit vectors of the deformed cross section ($\vec{i}_{XS}$, $\vec{j}_{XS}$, and $\vec{k}_{XS}$ as shown in Fig. 9-12) are rotated by ϕ_x and ϕ_z from the undeformed cross section:

$$\vec{i}_{XS} = \vec{i} + \phi_z\vec{j}$$

$$\vec{j}_{XS} = \vec{j} - \phi_z\vec{i} + \phi_x\vec{k}$$

$$\vec{k}_{XS} = \vec{k} - \phi_x\vec{j}$$

The vector $\vec{j}_{XS}$ is the tangent to the deformed elastic axis. Hence by definition $\vec{j}_{XS} = d\vec{r}/ds$, where $\vec{r} = x_0\vec{i} + (r + r_0)\vec{j} + z_0\vec{k}$ is the deflected position, and s is the arc length along the deformed elastic axis. To first order, then,

$$\begin{aligned} \vec{j}_{XS} &= \vec{j} + (x_0\vec{i} + z_0\vec{k})' \\ &= \vec{j} + (x'_0 + z_0\theta')\vec{i} + (z'_0 - x_0\theta')\vec{k} \end{aligned}$$

Comparing the two definitions of $\vec{j}_{XS}$ shows that the rotation of the section is $-\phi_z = x'_0 + z_0\theta'$ and $\phi_x = z'_0 - x_0\theta'$, or

$$\phi_x\vec{i} + \phi_z\vec{k} = (z_0\vec{i} - x_0\vec{k})'$$

The undeflected position of a blade element is $\vec{r} = r\vec{j} + x\vec{i} + z\vec{k}$, and the deflected position is

$$\begin{aligned} \vec{r} &= (r + r_0)\vec{j} + x_0\vec{i} + z_0\vec{k} + x\vec{i}_{XS} + z\vec{k}_{XS} \\ &= r\vec{j} + x_0\vec{i} + r_0\vec{j} + z_0\vec{k} + (x\phi_z - z\phi_x)\vec{j} + x\vec{i} + z\vec{k} \end{aligned}$$

For now the elastic extension r_0 will be neglected. The strain analysis is then simplified since to first order $s = r$. The extension r_0 just produces a uniform strain over the section, which is easily introduced later.

The fundamental metric tensor g_{mn} of the undistorted blade is defined by

$$(ds)^2 = d\vec{r}\cdot d\vec{r} = \left(\frac{\partial\vec{r}}{\partial x_m}dx_m\right)\cdot\left(\frac{\partial\vec{r}}{\partial x_n}dx_n\right) = g_{mn}\,dx_m\,dx_n$$

where ds is the differential length in the material and x_m are general curvilinear coordinates. Similarly, the metric tensor G_{mn} of the deformed blade is

$$(dS)^2 = d\vec{r}\cdot d\vec{r} = \left(\frac{\partial\vec{r}}{\partial x_m}dx_m\right)\cdot\left(\frac{\partial\vec{r}}{\partial x_n}dx_n\right) = G_{mn}\,dx_m\,dx_n$$

Then the strain tensor γ_{mn} is defined by the differential length increment

$$2\gamma_{mn}\,dx_m\,dx_n = (dS)^2 - (ds)^2$$

or

$$\gamma_{mn} = \tfrac{1}{2}(G_{mn} - g_{mn})$$

For engineering beam theory, only the axial component of the strain is required. For the specific case of the twisted rotor blade, the metric g_{mn} is obtained from the undistorted position vector, $r = x\vec{i} + r\vec{j} + z\vec{k}$. The axial component is

$$g_{rr} = \frac{\partial \vec{r}}{\partial r} \cdot \frac{\partial \vec{r}}{\partial r} = 1 + \theta'^{\,2}_{tw}(x^2 + z^2)$$

The metric G_{mn} of the deformed blade is obtained from the position vector $r = (x + x_0)\vec{i} + (r + x\phi_z - z\phi_x)\vec{j} + (z + z_0)\vec{k}$:

$$G_{rr} = \frac{\partial \vec{r}}{\partial r} \cdot \frac{\partial \vec{r}}{\partial r} = (1 + x\phi'_z - z\phi'_x)^2$$

$$+ \left(x'_0 + \theta'(z + z_0)\right)^2 + \left(z'_0 - \theta'(x + x_0)\right)^2$$

Hence the axial component of the strain tensor is

$$\gamma_{rr} = \tfrac{1}{2}(G_{rr} - g_{rr})$$

$$= \tfrac{1}{2}\Big[(1 + x\phi'_z - z\phi'_x)^2 - 1 + \left(x'_0 + \theta'(z + z_0)\right)^2 - \theta'^{\,2}_{tw} z^2$$

$$+ \left(z'_0 - \theta'(x + x_0)\right)^2 - \theta'^{\,2}_{tw} x^2\Big]$$

The linear strain is then

$$\gamma_{rr} \cong \epsilon_{rr} = x\phi'_z - z\phi'_x + \theta'^{\,2}_{tw}(xx_0 + zz_0)$$

$$+ \theta'_{tw}\left(zx'_0 - xz'_0 + \theta'_e(x^2 + z^2)\right)$$

since x_0, z_0, ϕ_x, ϕ_z, and θ_e are small.

The strain due to the blade tension, ϵ_T, is a constant defined by $T = \int E\epsilon_{rr} dA = \epsilon_T \int E dA$, where T is the tension force on the section. Substituting for ϵ_{rr} and including the strain due to the blade extension r_0 again, we obtain

$$\epsilon_T = \phi'_z x_C + \theta'^{\,2}_{tw} x_0 x_C - \theta'_{tw} z'_0 x_C + \theta'_{tw}\theta'_e k_p^{\,2} + r'_0$$

where $\int zEdA = 0$, $\int xEdA = x_C \int EdA$, and $\int (x^2 + z^2)EdA = k_p^{\,2} \int EdA$ (k_p is the modulus-weighted radius of gyration about the elastic axis). It follows the strain can be written as

$$\epsilon_{rr} = \epsilon_T + (x - x_C)(\phi'_z - \theta'_{tw}\phi_x) - z(\phi'_x + \theta'_{tw}\phi_z) + \theta'_{tw}\theta'_e(x^2 + z^2 - k_p^2)$$

where ϵ_T is obtained from the tension force.

Engineering beam theory assumes that all the stresses due to bending are negligible except for the axial component, $\sigma_{rr} = E\epsilon_{rr}$. The direction of σ_{rr} is given by the unit vector

$$\hat{e} = \frac{\partial \vec{r}/\partial r}{|\partial \vec{r}/\partial r|}$$

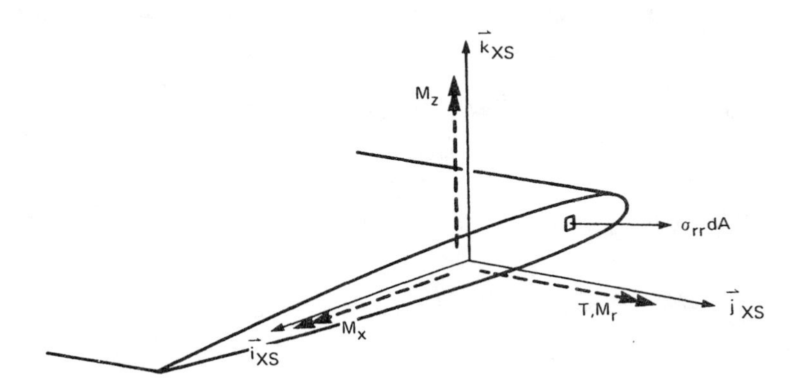

Figure 9-13 Bending and torsion moments on the blade section.

where $\vec{r}$ is the deformed position of the section (see Fig. 9-13). The moment about the elastic axis due to the elemental force $\sigma_{rr}dA$ on the cross section is

$$d\vec{M} = (x\vec{i}_{XS} + z\vec{k}_{XS}) x (\sigma_{rr}\hat{e})DA$$

$$= \left(-z\vec{i}_{XS} + x\vec{k}_{XS} + \theta'_{tw}(x^2 + z^2)\vec{j}_{XS}\right)\sigma_{rr}dA$$

Then the total moment, whose components M_x, M_z, and M_r are defined in Fig. 9-13, is obtained by integrating over the section:

$$M_x\vec{i}_{XS} + M_r\vec{j}_{XS} + M_z\vec{k}_{XS} = \int_{\text{section}} d\vec{M}$$

or

$$(M_x)_{EA} = -\int_{\text{section}} z\,\sigma_{rr}\,dA$$

$$(M_z)_{EA} = \int_{\text{section}} x\,\sigma_{rr}\,dA$$

$$M_r = GJ\theta'_e + \int_{\text{section}} (x^2 + z^2)\theta'\,\sigma_{rr}\,dA$$

The torsion moment $GJ\theta'_e$ due to the shear forces of elastic torsion has been added to M_r. For bending it is more convenient to work with moments about the tension center at x_C:

$$M_x = -\int z\,\sigma_{rr}\,dA$$

$$M_z = \int (x - x_C)\,\sigma_{rr}\,dA$$

After substituting for σ_{rr} and integrating, the bending and torsion moments are as follows:

$$M_x = EI_{zz}(\phi'_x + \theta'\phi_z) - \theta'_{tw}\theta'_e EI_{ZP}$$

$$M_z = EI_{xx}(\phi'_z - \theta'\phi_x) + \theta'_{tw}\theta'_e EI_{XP}$$

$$M_r = (GJ + k_P^2 T + \theta'^2_{tw} EI_{PP})\theta'_e + \theta'_{tw} k_P^2 T$$
$$+ \theta'_{tw}\left(EI_{XP}(\phi'_z - \theta'\phi_x) - EI_{ZP}(\phi'_x + \theta'\phi_z)\right)$$

where

$$EI_{zz} = \int z^2\,E\,dA$$

$$EI_{xx} = \int (x - x_C)^2\,E\,dA$$

$$EI_P = k_P^2 \int E\,dA = \int (x^2 + z^2)E\,dA$$

$$EI_{XP} = \int (x - x_C)(x^2 + z^2)E\,dA$$

$$EI_{ZP} = \int z(x^2 + z^2)E\,dA$$

$$EI_{PP} = \int (x^2 + z^2 - k_P^2)^2\,E\,dA$$

Since the tension T acts at the tension center x_C, the bending moments about the elastic axis can be obtained from those about the tension center by $(M_z)_{EA} = M_z + x_C T$ and $(M_x)_{EA} = M_x$.

This result may be more conveniently expressed using a vector representation of the blade bending. Define the section bending moment vector $\vec{M}$ and the flap-lag deflection $\vec{u}$ as follows:

$$\vec{M} = M_x\vec{i} + M_z\vec{k}$$
$$\vec{u} = z_0\vec{i} - x_0\vec{k}$$

The derivatives of $\vec{u}$ are

$$\vec{u}' = (z_0\vec{i} - x_0\vec{k})' = (z_0' - x_0\theta')\vec{i} - (x_0' + z_0\theta')\vec{k}$$
$$= \phi_x\vec{i} + \phi_z\vec{k}$$

$$\vec{u}'' = (z_0\vec{i} - x_0\vec{k})'' = (\phi_x' + \theta'\phi_z)\vec{i} + (\phi_z' - \theta'\phi_x)\vec{k}$$

Then the bending and torsion moments can be written as

$$\vec{M} = (EI_{zz}\vec{i}\,\vec{i} + EI_{xx}\vec{k}\,\vec{k})\cdot(z_0\vec{i} - x_0\vec{k})''$$
$$+ \theta'_{tw}\theta'_e(EI_{XP}\vec{k} - EI_{ZP}\vec{i})$$

$$M_r = (GJ + k_P^2 T + \theta'^2_{tw} EI_{PP})\theta'_e + \theta'_{tw}k_P^2 T$$
$$+ \theta'_{tw}(EI_{XP}\vec{k} - EI_{ZP}\vec{i})\cdot(z_0\vec{i} - x_0\vec{k})''$$

This is the result sought in engineering beam theory to relate the structural moments and deflections of the rotor blade section. For a blade with zero pitch, it reduces to the usual results for uncoupled in-plane and out-of-plane bending of a beam:

$$M_x = EI_{zz}z_0''$$
$$M_z = EI_{xx}x_0''$$
$$M_r = (GJ + k_P^2 T)\theta'_e$$

Writing $EI = EI_{zz}\vec{i}\,\vec{i} + EI_{xx}\vec{k}\,\vec{k}$ for the bending stiffness dyadic gives $\vec{M} = EI\,\vec{u}''$ for bending alone, which appears as an elementary extension of the zero-pitch case. The vector form allows a simultaneous treatment of the coupled in-plane and out-of-plane bending of the blade, with considerable simplification of the analysis as a consequence.

Houbolt and Brooks (1958) considered the bending displacement defined in terms of the hub plane axes. Their result can be obtained from the present relations by writing $\vec{u} = w\vec{i}_B - v\vec{k}_B$. Then

$$\vec{u}'' = w''\vec{i}_B - v''\vec{k}_B = (w''\cos\theta + v''\sin\theta)\vec{i} + (x''\sin\theta - v''\cos\theta)\vec{k}$$

The bending moment can also be obtained in terms of the hub plane axes by writing $\vec{M} = M_x\vec{i}_B + M_z\vec{k}_B$, and for the bending stiffnes dyadic

$$\begin{aligned} EI &= EI_{zz}\vec{i}\,\vec{i} + EI_{xx}\vec{k}\,\vec{k} \\ &= (EI_{zz}\cos^2\theta + EI_{xx}\sin^2\theta)\vec{i}_B\,\vec{i}_B \\ &\quad + (EI_{zz}\sin^2\theta + EI_{xx}\cos^2\theta)\vec{k}_B\,\vec{k}_B \\ &\quad + (EI_{xx} - EI_{zz})\cos\theta\sin\theta(\vec{i}_B\,\vec{k}_B + \vec{k}_B\,\vec{i}_B) \end{aligned}$$

The partial differential equations for blade bending and torsion are often derived from the conditions for equilibrium of forces and moments on a differential element of the beam extending from r to $r + dr$. Consider the shear forces, bending moments, tension, and torsion moment on the blade section, as shown in Fig. 9-14, defined relative to the hub plane axes (so that the axes are the same at r and at $r + dr$). The blade also has distributed forces

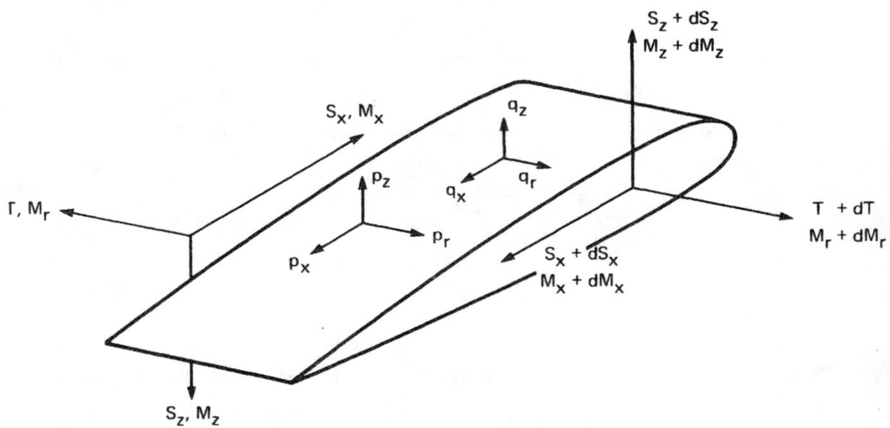

Figure 9-14 Forces and moments acting on the blade element from r to $r+dr$.

(components p_x, p_z, and p_r) and moments (components q_x, q_z, and q_r) acting on the section element. The conditions for equilibrium of forces and moments on this blade element give

$$S'_x + p_x = 0$$

$$S'_z + p_z = 0$$

$$T' + p_r = 0$$

$$M'_x - Tw' + S_z + q_x = 0$$

$$M'_z + Tv' - S_x + q_z = 0$$

$$M'_r + S_x w' - S_z v' + q_r = 0$$

where w and v are the bending deflections in hub plane axes. Eliminating the shears, these equations become

$$M''_x - (Tw')' + q'_x - p_z = 0$$

$$M''_z + (Tv')' + q'_z + p_x = 0$$

$$M'_r - w'\int^r p_x dr + v'\int^r p_z dr + q_r = 0$$

where the tension is $T = -\int^r p_r dr$. The structural analysis gives the section moments M_x, M_z, and M_r. The distributed forces and moments are due to the inertial and aerodynamic forces for the blade. To combine the in-plane and out-of-plane equations again, define the following two-dimensional vectors:

$$\vec{M} = M_x \vec{i}_B + M_z \vec{k}_B$$

$$\vec{u} = w\vec{i}_B - v\vec{k}_B$$

$$\vec{q} = q_x \vec{i}_B + q_z \vec{k}_B$$

$$\vec{p} = p_z \vec{i}_B - p_x \vec{k}_B$$

Then the equilibrium equation for bending is

$$\vec{M}'' - (T\vec{u}')' + \vec{q}' - \vec{p} = 0$$

Recall that the engineering beam theory analysis gave $\vec{M} = EI\vec{u}''$ (with no torsion), so

$$(EI\vec{u}'')'' - (T\vec{u}')' + \vec{q}' - \vec{p} = 0$$

is the partial differential equation for the blade bending.

9-8.2 Modal Equations

The modal equation for coupled flap-lag bending of the twisted rotating blade is obtained from the conditions for equilibrium of the elastic, inertial, and centrifugal bending moments. In the preceding paragraph the following partial differential equation for bending was obtained:

$$(EI\vec{u}'')'' - (T\vec{u}')' + \vec{q}' - \vec{p} = 0$$

where EI is the bending stiffness dyadic and $\vec{u} = (z_0\vec{i} - x_0\vec{k})$ is the displacement vector. Now it is necessary to identify the centrifugal and inertial contributions to the tension force and distributed loadings on the blade. Only the terms due to bending deflection are required for the modal equation. The tension is due to the centrifugal force: $T = \Omega^2 \int_r^R \rho m d\rho$. The out-of-plane and in-plane accelerations give an inertial force on the blade section: $\vec{p} = -m\ddot{\vec{u}}$. The section centrifugal force $m\Omega^2\rho$ has an in-plane component $(m\Omega^2\rho)(v/\rho)$ in the lag direction due to the in-plane deflection v (see Fig. 9-15); thus the total lag moment produced on the section at r is $q_z = -\int_r^R m\Omega^2 v d\rho$. Hence $\vec{q}' = -\vec{k}_B\Omega^2 m\vec{k}_B\cdot\vec{w} = -\vec{\Omega}m\vec{\Omega}\cdot\vec{w}$, where $\vec{\Omega} = \Omega\vec{k}_B$.

The partial differential equation of the rotating blade is then

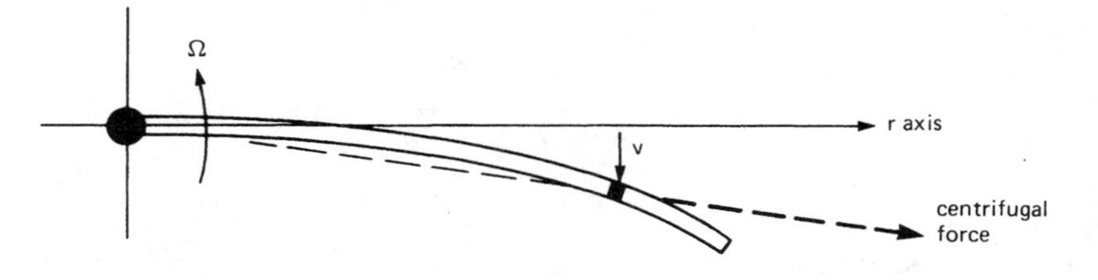

Figure 9-15 In-plane component of the centrifugal force responsible for the lag moment on a blade section.

$$\left[(EI_{zz}\vec{i}\vec{i} + EI_{xx}\vec{k}\vec{k})(z_0\vec{i} - x_0\vec{k})''\right]'' - \Omega^2\left[\int_r^R \rho m d\rho (z_0\vec{i} - x_0\vec{k})'\right]'$$

$$- \vec{\Omega} m \vec{\Omega} \cdot (z_0\vec{i} - x_0\vec{k}) + m(z_0\vec{i} - x_0\vec{k})^{\cdot\cdot} = 0$$

For more details of the derivation see Houbolt and Brooks (1958), Hodges and Dowell (1974), or Johnson (1977f).

Using the method of separation of variables, write the bending deflection as $(z_0\vec{i} - x_0\vec{k}) = \vec{\eta}(r)e^{i\nu t}$. Then the equation for the vector mode shape $\vec{\eta}$ is

$$(EI\vec{\eta}'')'' - \Omega^2\left(\int_r^R \rho m d\rho \vec{\eta}'\right)' - \vec{\Omega} m \vec{\Omega} \cdot \vec{\eta} - m\nu^2\vec{\eta} = 0$$

This is the modal equation for coupled flap-lag bending of the rotating blade. For a blade with no pitch or twist, this vector equation separates into two equations for the purely out-of-plane deflection η_z and the purely in-plane deflection η_x:

$$(EI_{zz}\eta''_z)'' - \Omega^2\left(\int_r^R \rho m d\rho \eta'_z\right)' - m\nu_z^2 \eta_z = 0$$

$$(EI_{xx}\eta''_x)'' - \Omega^2\left(\int_r^R \rho m d\rho \eta'_x\right)' - m(\Omega^2 + \nu_x^2)\eta_x = 0$$

which were obtained in sections 9-2.2 and 9-3.2. The boundary conditions are as follows:

(i) at the tip ($r = R$) there is zero shear force and moment, so $EI\vec{\eta}'' = (EI\vec{\eta}'')' = 0$;

(ii) at the root of an articulated blade ($r = e$, allowing for hinge offset) there is zero deflection, $\vec{\eta} = 0$; and moment balance with a hinge spring requires $EI\vec{\eta}'' = K_s\vec{\eta}'$ (where K_s is the hinge spring dyadic);

(iii) at the root of a hingeless blade ($r = e$, allowing for a very stiff hub) there is zero deflection and slope, so $\vec{\eta} = \vec{\eta}' = 0$.

The modal equation and its boundary conditions thus form a proper Sturm-Liouville eigenvalue problem (see section 9—1) for which there is a series of eigensolutions $\vec{\eta}_k(r)$ and corresponding eigenvalues ν_k^2. The modes are orthogonal with weighting function m, so if $i \neq k$,

$$\int_e^R \vec{\eta}_i \cdot \vec{\eta}_k \, m \, dr = 0$$

The natural frequencies ν_k may be obtained from the mode shapes using the energy relation

$$\nu^2 = \frac{\vec{\eta}\,'(e) K_s \vec{\eta}\,'(e) + \int_e^R \left[\vec{\eta}\,'' EI \vec{\eta}\,'' + \Omega^2 \int_r^R \rho m \, d\rho \, \eta'^2 - m(\vec{\Omega} \cdot \vec{\eta})^2 \right] dr}{\int_e^R \eta^2 m \, dr}$$

which can also be used to estimate the frequencies from approximate mode shapes. Note that the influence of the hinge offset e can be obtained by transforming the radial variable r so that it is measured from the hinge rather than the center of rotation. Then the centrifugal tension force becomes $T = \Omega^2 \int_0^{R-e} (\rho + e) m \, d\rho$, which gives the lowest order effect of e directly.

The Holzer-Myklestad method is probably most commonly used to calculate the mode shapes and natural frequencies of the rotating blade. Alternatively, the modal equation can be solved by the Galerkin method, or by a finite element analysis.

9-8.3 Bending Natural Frequencies

The result for the natural frequency of coupled flap-lag bending of a rotating blade can be written as

$$\nu^2 = K_1 + K_2 \Omega^2$$

where K_1 and K_2 are the Southwell coefficients, corresponding to the structural and centrifugal stiffening of the blade respectively:

$$K_1 = \left[\vec{\eta}\,'(e) K_s \vec{\eta}\,'(e) + \int_e^R \vec{\eta}\,'' EI \vec{\eta}\,'' \, dr \right] \bigg/ \left(\int_e^R \eta^2 m \, dr \right)$$

$$K_2 = \left[\int_e^R \eta'^2 \int_r^R \rho m \, d\rho \, dr - \int_e^R (\vec{\eta} \cdot k_B)^2 \, m \, dr \right] \bigg/ \left(\int_e^R \eta^2 m \, dr \right)$$

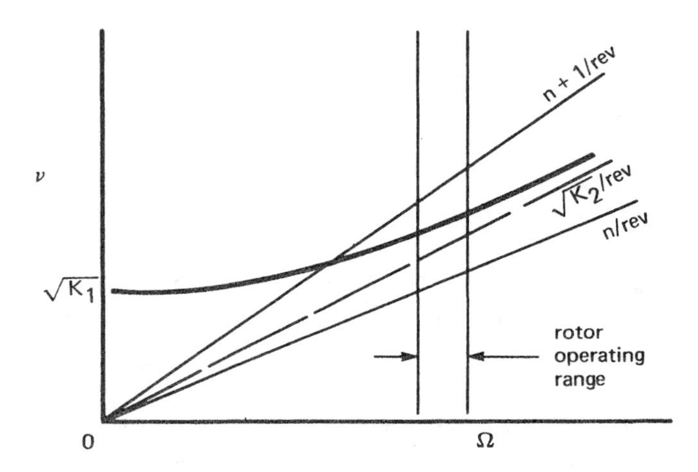

Figure 9-16 Blade bending frequency diagram.

Southwell and Gough (1921) obtained this expression from Rayleigh energy considerations. Fig. 9-16 shows the variation of the bending frequency with rotor speed that the Southwell form implies. The natural frequencies of the blade must be kept away from resonance with harmonics of the rotor speed, so the n/rev lines are also usually shown on the frequency diagram. In the limit $\Omega = 0$, the blade frequency is $\nu^2 = K_1$. Therefore $\sqrt{K_1}$ is the non-rotating natural frequency of bending, due to the structural stiffnesss. In the limit of large Ω, the blade frequency approaches $(\nu/\Omega)^2 = K_2$. Therefore $\sqrt{K_2}$ is the dimensionless (per-revolution) natural frequency at high speed, due to the centrifugal forces. Actually, since the mode shapes will themselves depend on the rotor speed, the Southwell coefficients are not exactly constants. Nevertheless, the Southwell form for the frequency emphasizes the relative strengths of structural and centrifugal stiffening.

The zero stiffness limit is a useful case, for it gives the minimum possible bending frequency, which is due to centrifugal stiffening alone. When $EI = 0$, the modal equation becomes

$$\left(\int_r^1 \rho m \, d\rho \, \vec{\eta}\,' \right)' + m \vec{k}_B \vec{k}_B \cdot \vec{\eta} \,.+ m \nu^2 \vec{\eta} = 0$$

which separates into purely out-of-plane and purely in-plane equations:

$$\left(\int_r^1 \rho m \, d\rho \, \eta_z'\right)' + m\nu_z^2 \eta_z = 0$$

$$\left(\int_r^1 \rho m \, d\rho \, \eta_x'\right)' + m(\nu_x^2 + 1)\eta_x = 0$$

Without the structural terms, there is no coupling of the flap and lag modes. Moreover, the in-plane and out-of-plane motion have identical mode shapes, with corresponding frequencies given by $\nu_{\text{flap}}^2 = 1 + \nu_{\text{lag}}^2$. This is a singular limit, however, since dropping the structural terms reduces the order of the equations. For small EI, the boundary conditions are satisfied in small regions near the ends of the blade. For $EI = 0$ it is necessary to drop the two boundary conditions at the tip. For a uniform mass distribution, the equation for η reduces to Legendre's equation,

$$\left((1 - r^2)\eta'\right)' + 2\nu^2 \eta = 0$$

The solutions that satisfy the boundary conditions and are finite at the tip are the odd Legendre polynomials, $\eta_k(r) = P_{2k-1}(r)$, with eigenvalues $\nu_k^2 = k(2k - 1)$. These polynomials may be obtained from

$$P_k = \frac{1}{2^k k!} \frac{d^k}{dr^k}\left(r^2 - 1\right)^k$$

which gives $\eta_1 = P_1 = r$, $\eta_2 = P_3 = \frac{1}{2}(5r^3 - 3r)$, etc. The corresponding frequencies are $\nu_1 = 1$, $\nu_2 = 2.45$, and $\nu_3 = 3.87$ for flap, and $\nu_1 = 0$, $\nu_2 = 2.24$, and $\nu_3 = 3.74$ for lag. For the third modes and above, the curvature is large enough that the structural stiffening begins to dominate the solution; beginning with ν_3 (or even ν_2 for lag), this lower bound on the frequency is therefore very conservative.

For articulated rotors, the fundamental flap and lag modes are rigid rotation about the hinges. Moreover, these modes are pure in-plane and pure out-of-plane deflection. The energy relation gives a good estimate of the natural frequencies, using $\eta = (r - e)/(1 - e)$. For the second and higher modes the structural stiffness becomes increasingly important, and twist of the blade couples the in-plane and out-of-plane deflections. Generally twist does not influence the natural frequencies much, unless it happens to couple two modes with close frequencies. If the correct mode shape is not available, $\eta = 4r^2 - 3r$ is a fairly good approximation for the first elastic mode.

For a hingeless rotor blade, bending at the root is important even for the fundamental modes. The principal axes for the centrifugal stiffening are always the hub plane axes, while the principal axes for the structural stiffening are determined by the blade pitch. Only if these axes coincide will the free vibration modes of the blade be purely in-plane and purely out-of-plane. Pitch of the blade, particularly at the root, introduces significant coupling of the flap and lag motion in the fundamental modes. For many hingeless rotors, particularly stiff in-plane designs, the centrifugal stiffening dominates the out-of-plane motion, while structural stiffening dominates the in-plane motion. Even a small root pitch (5° to 10°) will then greatly influence the modes. Soft in-plane blades tend to have matched stiffnesses at the root, which reduces the coupling caused by collective pitch. For the outboard portions of the blade the centrifugal forces dominate the fundamental flap and lag modes. Consequently, there is little bending outboard, and also the blade twist has little influence compared to the root pitch. For the higher bending modes, the structural stiffening becomes increasingly important, and consequently twist rather than root pitch has the most influence on the mode shape.

9-8.4 Literature

On the free vibration modes and frequencies of the rotating blade: Southwell and Gough (1921), Prewitt and Wagner (1940), Kelley (1945), Horvay (1948), Simpkinson, Eatherton, and Millenson (1948), Morduchow (1950), de Guillenchmidt (1951), Yntema (1955), Houbolt and Brooks (1958), Taylor (1958), Brooks and Leonard (1960), Isakson and Eisley (1960, 1964), Craig (1963), Jones and Bhuta (1963), Wilde and Price (1965), Lipelas (1960), Wadsworth and Wilde (1968), Piziali (1970), Giansante (1971), Young (1971, 1972), Rao and Carnegie (1972), Sadler (1972a), Peters (1973), Rudy (1973), Bratanow and Ecer (1974a, 1974b), Dowell (1974), Johnson (1974d, 1977f), Kaza (1974), Tram, Twomey, and Dat (1974), Bennett (1975), Dowell and Traybar (1975), Hodges and Peters (1975), White and Malatino (1975), Kiessling (1976), Laurenson (1976),

Murthy (1976, 1977), Dowell, Traybar, and Hodges (1977), Giurgiutiu and Stafford (1977), McDaniel and Murthy (1977), Wilkerson (1977a), Carlson and Wong (1978), Gaukroger and Hassal (1978), Weller and Mineck (1978).

9–9 Derivation of the Equations of Motion

The equations of motion for the rotor blade have been derived in this chapter using an integral Newtonian method to obtain the partial differential equations for bending or torsion, and a normal mode expansion to obtain ordinary differential equations for the normal coordinates. The choice of this approach was based on the physical insight that is gained by working directly with the forces and accelerations of the blade. Other methods are also commonly used to derive the equations of motion for analyses of rotor dynamics. As a guide to what may be encountered in the literature, this chapter concludes with brief outlines of a number of these alternative approaches.

Rotor dynamics analyses are frequently based on Lagrange's equations,

$$\frac{d}{dt}\frac{\partial T}{\partial \dot{q}_i} - \frac{\partial T}{\partial q_i} + \frac{\partial U}{\partial q_i} = Q_i$$

where T and U are the kinetic and potential energies of the entire system, q_i are the generalized coordinates (degrees of freedom), and Q_i are generalized forces. Usually T gives the inertial terms, U the structural terms, and Q_i the aerodynamic terms of the equations of motion. The derivation of the equations of motion by Lagrange's equations is simply formulated and routinely (if somewhat laborously) executed. Consequently, Lagrangian methods are often used in the development of the most comprehensive models found in the literature.

To illustrate the various derivation methods, we shall consider the bending of a cantilever beam (Fig. 9-17). A distributed loading $p(r)$, tip force F_T, and moment M_T are included. The spanwise variable is r and the bending deflection is z. The beam is not rotating, since the present purpose is to examine the methods of analysis rather than the rotor blade behavior. The objective is to obtain a set of ordinary differential equations (in time) describing the bending motion $z(r,t)$ of the cantilever beam.

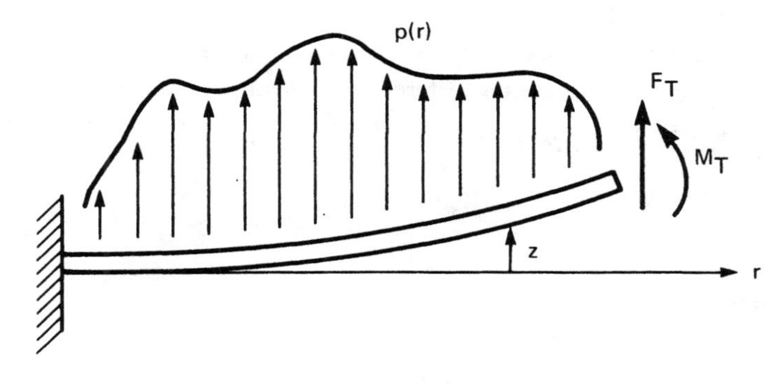

Figure 9-17 Bending of a cantilever beam (nonrotating).

9-9.1 Integral Newtonian Method

Newtonian methods derive the equations of motion from the conditions for equilibrium of forces on the body. This chapter has used an integral Newtonian method. In the case of the cantilever beam, the moment on the section at r consists of the tip moment M_T, the tip force F_T with moment arm $(R - r)$, and the section loading $(p - m\ddot{z})$ integrated over the portion outboard of r with moment arm $(\rho - r)$:

$$M(r) = \int_r^R (p - m\ddot{z})(\rho - r)d\rho + M_T + (R - r)F_T$$

Equating this to $M(r) = EIz''$ from engineering beam theory and taking the second derivative gives the required partial differential equation for bending:

$$(EIz'')'' + m\ddot{z} = p$$

The boundary conditions are needed to complete the problem formulation. Evaluating $M(r)$ and $M'(r)$ at the tip gives $EIz'' = M_T$ and $(EIz'')' = -F_T$ at $r = R$. The boundary conditions for the cantilever root are just $z = z' = 0$ at $r = 0$.

9-9.2 Differential Newtonian Method

The equation of motion can also be derived from the conditions for equilibrium of forces and moments on the differential beam element extending from r to $r+dr$. Let S and M be the shear and moment on the section at r; and $S+dS=S+S'dr$ and $M+dM=M+M'dr$ the reactions on the section at $r+dr$. Force balance on the differential element implies

$$p\,dr + (S+S'dr) - S = m\ddot{z}dr$$

or

$$p + S' = m\ddot{z}$$

Moment balance implies

$$(M+M'dr) - M + S\,dr = 0$$

or

$$M' + S = 0$$

Eliminating S then gives the partial differential equation

$$M'' + m\ddot{z} = p$$

which with $M=EIz''$ becomes

$$(EIz'')'' + m\ddot{z} = p$$

9-9.3 Normal Mode Method

An expansion of the bending deflection as a series in the normal modes of free vibration has been used in this chapter to obtain the ordinary differential equations of motion. Consider the cantilever beam, but with no tip force or moment. The differential equation for free vibration is

$$(EIz'')'' + m\ddot{z} = 0$$

If it is assumed that $z=\eta(r)e^{i\nu t}$, the modal equation

$$(EI\eta'')'' - \nu^2 m\eta = 0$$

with boundary conditions $\eta=\eta'=0$ at $r=0$, and $\eta''=\eta'''=0$ at $r=R$ is obtained. This is a proper Sturm-Liouville eigenvalue problem, with a series of eigensolutions η_k and eigenvalues ν_k^2. The mode shapes are orthogonal with weighting function m, so if $i\neq k$,

$$\int_0^R \eta_i\eta_k m\,dr = 0$$

Sturm-Liouville theory also gives for the eigenvalues

$$\nu^2 = \int_0^R EI\eta''^2\, dr \Big/ \int_0^R \eta^2 m\, dr$$

Finally, Sturm-Liouville theory shows that an expansion of a function as a series in η_k will converge.

The bending deflection is next expanded as a series in the free vibration modes

$$z(r,t) = \sum_{k=1}^{\infty} \eta_k(r)\, q_k(t)$$

where q_k is the degree of freedom for the kth mode. Substituting this expansion in the partial differential equation gives

$$\sum_{k=1}^{\infty} \left[(EI\eta_k'')'' q_k + m\eta_k \ddot{q}_k \right] = p$$

Using the equation satisfied by η_k, the structural term is replaced by the natural frequency:

$$\sum_{k=1}^{\infty} \left[m\eta_k (\ddot{q}_k + \nu_k^2 q_k) \right] = p$$

Operating with $\int_0^R (\ldots)\eta_k\, dr$ and using the orthogonality of the modes gives

$$\left(\int_0^R \eta_k^2 m\, dr \right) (\ddot{q}_k + \nu_k^2 q_k) = \int_0^R \eta_k p\, dr,$$

the ordinary differential equation of motion for the kth bending modes. Note that using the energy expression for $\nu_k{}^2$, this equation can be written as

$$\left(\int_0^R \eta_k^2 m\, dr \right) \ddot{q}_k + \left(\int_0^R EI\eta_k''^2\, dr \right) q_k = \int_0^R \eta_k p\, dr$$

By using the orthogonal modes of free vibration, the structural and inertial terms of the equations of motion are uncoupled.

The limitation of using normal modes is that since each of the modes satisfies a homogeneous boundary condition, the solution z must also. Thus it is not possible to include the tip force and moment in this approach directly.

9-9.4 Galerkin Method

The Galerkin method also uses a modal expansion of the bending deflection to obtain the ordinary differential equations of motion, but not necessarily the normal modes of free vibration. Let $z = \sum_k \eta_k(r) q_k(t)$, where q_k are the generalized coordinates describing the motion and η_k are a series of modes. It will be required that each of the modes η_k satisfy the boundary conditions at the root, and that the total deflection z satisfy the boundary conditions at the tip. The true solution satisfies the differential equation

$$(EIz'')'' + m\ddot{z} = p$$

The Galerkin solution will not in general satisfy this equation exactly, since a finite number of modes are used. Therefore, define an error function

$$\epsilon = p - (EIz'')'' - m\ddot{z}$$

The equations of motion are obtained from the requirement that the equation error ϵ be small, and specifically that

$$\int_0^R \eta_i \epsilon \, dr = 0$$

Substituting for ϵ and employing the modal expansion for z gives

$$\sum_k \int_0^R m \eta_i \eta_k \, dr \, \ddot{q}_k + \sum_k \int_0^R \eta_i (EI \eta_k'')'' \, dr \, q_k = \int_0^R \eta_i p \, dr$$

Now integrating twice by parts and using the boundary conditions, the structural term can be written

$$\sum_k \int_0^R \eta_i (EI\eta_k'')'' dr\, q_k = \sum_k \Big[\eta_i (EI\eta_k'')' q_k - \eta_i'(EI\eta_k'')q_k\Big]_0^R$$

$$+ \sum_k \int_0^R EI\, \eta_i'' \eta_k'' \, dr\, q_k$$

$$= \Big[\eta_i (EIz'')' - \eta_i'(EIz'')\Big]_0^R$$

$$+ \sum_k \int_0^R EI\eta_i'' \eta_k'' \, dr\, q_k$$

$$= -\eta_i(R)F_T - \eta_i'(R)M_T + \sum_k \int_0^R EI\eta_i''\eta_k'' \, dr q_k$$

Thus the equation of motion for the ith mode is

$$\sum_k M_{ik}\ddot{q}_k + \sum_k K_{ik} q_k = \int_0^R \eta_i p\, dr + \eta_i(R)F_T + \eta_i'(R)M_T$$

where $M_{ik} = \int_0^R \eta_i \eta_k m\, dr$ and $K_{ik} = \int_0^R EI\eta_i''\eta_k''\, dr$. The normal mode expansion gave a similar result. With the Galerkin method the mass and spring matrices are not diagonal because the free vibration modes are not necessarily used; but the excitation by the tip force and moment are now included. The Galerkin method is equivalent to the Rayleigh-Ritz method (discussed below) when the proper weighting function is used for the equation error integral (η_i in this case). The Rayleigh-Ritz procedure has a stronger physical and mathematical basis, but the Galerkin procedure allows the use of a Newtonian approach to derive the equation of motion.

Note that if the free vibration modes are used in the Galerkin method, the mass and spring matrices will be diagonal and the equations of motion become

$$\left(\int_0^R \eta_k^2 m\, dr\right)(\ddot{q}_k + \nu_k^2 q_k) = \int_0^R \eta_k p\, dr + \eta_k(R)F_T + \eta_k'(R)M_T$$

This is just the normal mode result, but now with the excitation due to the tip loading included. This suggests that the equations of motion for the rotor can be derived using the normal mode procedure, and then the influence of point loads on the blade can be accounted for by adding terms according to the Galerkin procedure. Such an approach is useful for example in adding a lag damper to the normal mode analysis, or for the control system force and moment at the pitch bearing.

9-9.5 Lagrangian Method

Lagrangian methods derive the equation of motion from energy considerations instead of from equilibrium of forces. Hamilton's principle states that the motion of a dynamic system is determined by the condition

$$\int_{t_1}^{t_2} (\delta T - \delta U + \delta W) dt = 0$$

where T is the system kinetic energy, U the potential energy, and δW the virtual work of nonconservative forces. Since for a conservative system the criterion is that $\int_{t_1}^{t_2} (T - U)dt$ have a minimum value, it is also called the principle of least action.

For the cantilever beam, the kinetic and potential energy are

$$T = \int_0^R \tfrac{1}{2} m \dot{z}^2 \, dr$$

$$U = \int_0^R \tfrac{1}{2} EI z''^2 \, dr$$

and the virtual work by the distributed force and tip loads is

$$\delta W = M_T \delta z'(R) + F_T \delta z(R) + \int_0^R p \, \delta z \, dr$$

Hamilton's principle thus requires

$$\int_{t_1}^{t_2}\left[\int_0^R \left(p - m\ddot{z} - (EIz'')''\right)\delta z\, dr + (-EIz'' + M_T)\delta z'(R) + \left((EIz'')' + F_T\right)\delta z(R)\right] dt = 0$$

(It is necessary to integrate the kinetic energy term by parts with respect to t, and the potential energy term twice by parts with respect to r.) Since the variational δz is arbitrary, it follows that

$$(EIz'')'' + m\ddot{z} = p$$

$$(EIz'')|_{r=R} = M_T$$

$$(EIz'')'|_{r=R} = -F_T$$

These expressions are the partial differential equation for bending of the beam, and the boundary conditions at the tip. From this point the normal mode method or Galerkin method can be used as desired.

9-9.6 Rayleigh-Ritz Method

If the energy and virtual work are expressed in terms of the generalized coordinates q_i:

$$T = T(q_i, \dot{q}_i)$$

$$U = U(q_i)$$

$$\delta W = \sum_i Q_i \delta q_i$$

then application of Hamilton's principle leads to Lagrange's equations,

$$\frac{d}{dt}\frac{\partial T}{\partial \dot{q}_i} - \frac{\partial T}{\partial q_i} + \frac{\partial U}{\partial q_i} = Q_i$$

By means of Lagrange's equations the ordinary differential equations of motion for the generalized coordinates describing the motion are obtained directly from the expressions for the system energy, without going through the partial differential equation.

Consider again a modal expansion for the bending deflection, as in the Galerkin method: $z = \sum_k \eta_k q_k$. Substituting for z gives the energy and virtual work in terms of the generalized coordinates:

$$T = \sum_i \sum_k \tfrac{1}{2} \int_0^R \eta_i \eta_k m\, dr\, \dot{q}_i \dot{q}_k = \sum_i \sum_k \tfrac{1}{2} M_{ik} \dot{q}_i \dot{q}_k$$

$$U = \sum_i \sum_k \tfrac{1}{2} \int_0^R EI \eta_i'' \eta_k'' dr\, q_i q_k = \sum_i \sum_k \tfrac{1}{2} K_{ik} q_i q_k$$

$$\delta W = \sum_i Q_i \delta q_i = \left[\int_0^R \eta_i p\, dr + \eta_i(R) F_T + \eta_i'(R) M_T \right] \delta q_i$$

Application of Lagrange's equations then gives directly

$$\sum_k M_{ik} \ddot{q}_k + \sum_k K_{ik} q_k = \int_0^R \eta_i p\, dr + \eta_i(R) F_T + \eta_i'(R) M_T$$

which is identical to the result obtained by the Galerkin method.

9-9.7 Lumped Parameter Methods

In lumped parameter or finite element methods the continuous physical system is modeled by a series of discrete elements. For example, the cantilever beam considered here might be represented by finite masses located at a series of points, and connected by massless elastic elements with uniform properties. The equations of motion are usually derived by Lagrangian methods. The greatest advantage of finite element methods is that they generally have the flexibility to treat complex structures. The problem for a new system is to specify its geometry and properties in the manner required by the lumped parameter method to be used, rather than developing an entirely new analysis.

Chapter 10

ROTARY WING AERODYNAMICS I

The principal topic of this chapter is the unsteady aerodynamics of a rotary wing. Generally, helicopter rotor analyses use a combination of lifting-line theory and two-dimensional airfoil theory to calculate the aerodynamic loads. The rotation of the wing introduces a number of features that require special attention, notably the returning vortex wake, the time-varying free stream and radial flow, and a fundamentally transcendental geometry that requires either approximate or numerical solutions.

10–1 Lifting-Line Theory

The lift on a wing is due to its bound circulation. Conservation of vorticity in three-dimensional flow requires that there be trailed and shed vorticity in the wake behind the wing. The spanwise variation of bound circulation results in trailed vorticity parallel to the free stream direction. Time variation of the bound circulation leads to shed vorticity parallel to the wing span. The wake is composed of sheets of vorticity convected downstream from the trailing edge by the free stream velocity. Classical lifting-line theory treats the case of a high-aspect-ratio, planar, fixed wing in steady flow. In the linearized model both the wing and wake are represented by thin planar sheets of vorticity. The assumption of high aspect ratio separates the lifting-line problem into two parts. The first is an inner problem that considers the aerodynamics of a wing section. The flow over the section behaves as if it is locally two-dimensional, with the influence of the wake and the rest of the wing represented entirely by a uniform induced downwash at the section (an induced angle-of-attack change). Two-dimensional airfoil theory or experimental section characteristics are used to obtain the blade section aerodynamic loads (lift, drag, and pitching moment). The second part is an outer problem in which the induced velocity is obtained. The wing is modeled by just a bound vortex line, and the trailed wake is a sheet extending behind it. The induced velocity is calculated along the bound vortex line. The inner problem relates the section loading to the induced velocity, while the outer problem derives the induced velocity from the wing loading, which determines the strength of the wake vorticity. Combining the two parts of the lifting-line problem then gives a solution for the spanwise loading of the three-dimensional wing.

A three-dimensional wing is characterized by a drag due to lift, called the induced drag, which arises because of the energy convected downstream in the vortex wake of the wing. For high-aspect-ratio wings, the induced drag can be associated with an induced velocity at the wing, so lifting-line theory can obtain the wing loading and induced velocity for a particular geometric angle-of-attack and chord distribution.

Blade element theory is essentially lifting-line theory for the rotary wing. The linearized wake model consists of helical vortex sheets trailed behind each blade. For the fixed wing the distortion of the wake geometry and the roll-up of the tip vortices can generally be neglected, because the wake is convected downstream away from the wing. In contrast, the rotary wing encounters the wake from preceding blades of the rotor. Consequently, a more detailed and more accurate model of the wake is required to obtain an estimate of the induced velocity in lifting-line theory for the rotary wing. The blade vorticity quickly rolls up into concentrated tip vortices, which are best represented by line vortex elements instead of sheets. In many flight conditions the self-induced distortion of the tip vortex helices must also be accounted for to accurately obtain the blade loads. So far in this text only the simplest possible calculations of the induced velocity have been considered, most often an estimate of the mean induced velocity obtained by momentum theory for an actuator disk model of the rotor and wake. An entirely analytical solution analogous to that of the fixed wing is not possible for the rotary wing because of the helical geometry of the wake, except in the case of the continuous wake of an actuator disk model. To obtain a tractable mathematical problem for calculating the induced velocity, the vorticity in the wake is usually modeled by a series of discrete line vortex elements. This is equivalent to considering a stepped bound circulation distribution, both radially and azimuthally. The rolled-up tip vortex is well represented by such line vortex elements. A vortex-line representation of the inboard trailed and shed vorticity introduces singularities in the induced velocity near each line element, but good numerical results can be obtained if care is taken in the analysis, particularly in the choice of points at which the induced velocity is calculated. As the rotor blade encounters the wake from the preceding blades, substantial time-varying loads are produced. Thus even in steady-state forward flight the blade loading is periodic in the rotating frame, and the rotor analysis requires unsteady aerodynamic theory. A further discussion is given in Chapter 13 for the subject of nonuniform inflow calculation.

The basic assumption of lifting-line theory, that the wing has a high aspect ratio, is almost always satisfied with a rotor blade. However, a more accurate specification of the requirement for validity of lifting-line theory is that there be no rapid spanwise change in the aerodynamic environment of the blade; this requirement is frequently not satisfied with the helicopter rotor blade, even though the geometric aspect ratio is large. There are two important cases where this requirement is not satisfied for the rotary wing: at the blade tip, and near a vortex-blade interaction. The loading near a wing tip must drop to zero in a finite distance. On the rotor blade, where the loading is concentrated at the blade tip because of the higher velocities there, the gradient of the lift at the tip is particularly high, and any small distortion of the loading due to three-dimensional flow effects is very important. In certain flight conditions the rotor blade passes quite close to a tip vortex from a preceding blade. The vortex-induced velocity gradients at the blade will be large for such close passages, and lifting-line theory significantly overestimates the loading produced. Hence lifting-line theory must be corrected or modified in order to handle some of the aerodynamic problems of the rotary wing. The correction may be as simple as the use of a tip loss factor, or as complex as a complete lifting-surface-theory analysis of the rotor aerodynamics.

10–2 Two-Dimensional Unsteady Airfoil Theory

Since the aerodynamic environment of the rotor blade in forward flight or during transient motion is unsteady, lifting-line theory requires an analysis of the unsteady aerodynamics of a two-dimensional airfoil. Let us therefore consider the problem of a two-dimensional airfoil undergoing unsteady motion in a uniform free stream. Linear, incompressible aerodynamic theory represents the airfoil and its wake by thin surfaces of vorticity (two-dimensional vortex sheets) in a straight line parallel to the free stream velocity. For the linear problem the solution for the thickness and camber loads can be separated from the loads due to angle of attack and unsteady motion. Only the latter solution is considered here. In the development of unsteady thin-airfoil theory, we will also be constructing the foundation for a number of extensions of the analysis for rotary wings, which will be presented in later sections of this chapter.

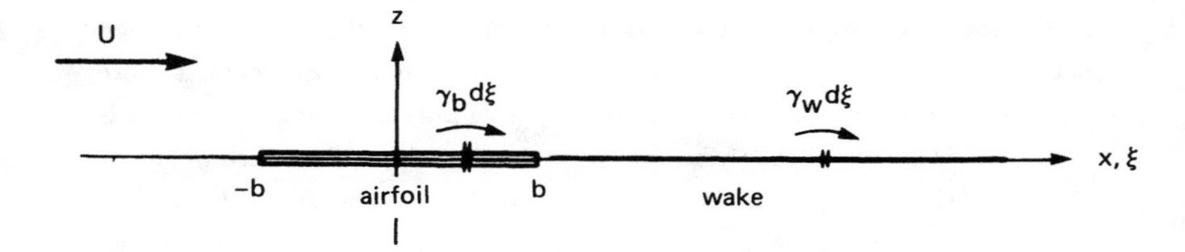

Figure 10-1 Unsteady thin airfoil theory model of the two-dimensional wing and wake.

The airfoil and shed wake in unsteady thin-airfoil theory are modeled by planar sheets of vorticity, as shown in Fig. 10-1. An airfoil of chord $2b$ is in a uniform free stream with velocity U. Since the bound circulation of the section varies with time, there is shed vorticity in the wake downstream of the airfoil. The vorticity strength on the airfoil is γ_b, and in the wake γ_w. The blade motion will be described by a heaving motion h (positive downward) and by a pitch angle α about an axis at $x = ab$ (positive for nose upward; see Fig. 10-2). The aerodynamic pitch moment will also be evaluated

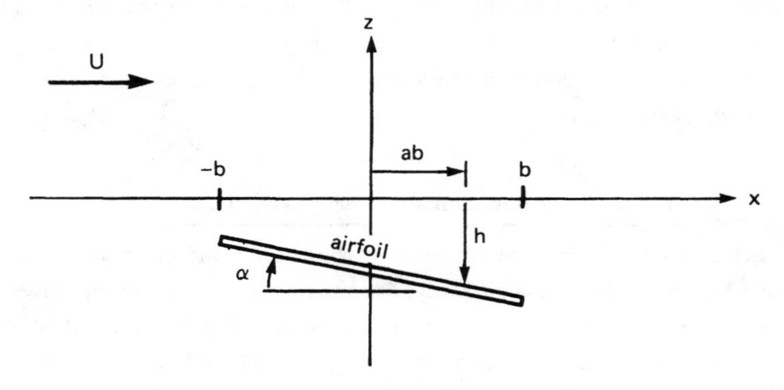

Figure 10-2 Unsteady pitching and heaving motion of the airfoil.

about the axis at $x = ab$. The airfoil motion produces an upwash velocity relative to the blade of

$$w_a = U\alpha + \dot{h} + \dot{\alpha}(x - ab)$$

Besides the velocity w_a, there is at the blade section also a downwash velocity λ due to the shed wake, and w_b due to the vorticity representing the blade surface. From the strength of the vortex sheets representing the airfoil and shed wake, it follows that these induced velocities are

$$w_b(x) = \frac{1}{2\pi} \int_{-b}^{b} \frac{\gamma_b \, d\xi}{x - \xi}$$

$$\lambda(x) = \frac{1}{2\pi} \int_{b}^{\infty} \frac{\gamma_w \, d\xi}{x - \xi}$$

The boundary condition of no flow through the wing surface, $w_b + \lambda - w_a = 0$, thus gives an integral equation for the bound vorticity γ_b:

$$\frac{1}{2\pi} \int_{-b}^{b} \frac{\gamma_b \, d\xi}{x - \xi} = w_a - \lambda$$

From the bound circulation γ_b the chordwise pressure loading can be found. The shed wake vorticity is given by the time rate of change in the total bound circulation $\Gamma = \int_{-b}^{b} \gamma_b \, dx$:

$$\gamma_w = -\frac{1}{U} \frac{d\Gamma}{dt}$$

evaluated at the time the element was shed, $t - (x - b)/U$. So the wake-induced velocity λ is also defined by the blade vorticity γ_b. The boundary condition of no pressure difference across the wake requires that the shed vorticity be convected with the free stream, so $\gamma_w = \gamma_w(x - Ut)$. Finally, the Kutta condition of finite velocity at the blade trailing edge requires $\gamma_b = 0$ there.

With the Kutta condition, the integral equation inverts to

$$\gamma_b = -\frac{2}{\pi} \sqrt{\frac{b - x}{b + x}} \int_{-b}^{b} \sqrt{\frac{b + \xi}{b - \xi}} \frac{w_a - \lambda}{x - \xi} \, d\xi$$

Now write for the wake-induced velocity and the upwash due to the airfoil motion:

$$\lambda = \sum_{n=0}^{\infty} \lambda_n \cos n\theta$$

$$w_a = \sum_{n=0}^{\infty} w_n \cos n\theta$$

where $x = b \cos\theta$. Then the solution for γ_b reduces to

$$\gamma_b = 2 \sum_{n=0}^{\infty} (w_n - \lambda_n) f_n(\theta)$$

where f_n is the Glauert series:

$$f_n(\theta) = \begin{cases} \tan(\theta/2) & n = 0 \\ \sin n\theta & n \geqslant 1 \end{cases}$$

In terms of x rather than θ, the expansion of the normal velocity is

$$w_a = w_0 + w_1(x/b) + w_2(2x^2/b^2 - 1) + \ldots .$$

For the blade motion considered then $w_0 = U\alpha + \dot{h} - ab\dot{\alpha}$ (w_a at the mid-chord), $w_1 = b\dot{\alpha}$, and $w_n = 0$ for $n \geqslant 2$. The first terms in the Glauert series are

$$f_0 = \sqrt{\frac{b-x}{b+x}}$$

$$f_1 = \sqrt{1 - (x/b)^2}$$

$$f_2 = 2(x/b)\sqrt{1 - (x/b)^2}$$

The coefficients w_n can be evaluated for a particular blade motion. To complete the solution, the wake-induced velocity λ is required.

On substituting for γ_b, the airfoil bound circulation becomes

$$\Gamma = \int_{-b}^{b} \gamma_b \, dx = 2\pi b[(w_0 + \tfrac{1}{2}w_1) - (\lambda_0 + \tfrac{1}{2}\lambda_1)]$$

Now let us divide γ_b into two parts: the circulatory vorticity γ_{bC}, which gives Γ but corresponds to $w_b = 0$ and so has no effect of the boundary conditions; and the noncirculatory vorticity γ_{bNC}, which satisfies the boundary conditions but gives $\Gamma = 0$. Hence $\gamma_b = \gamma_{bC} + \gamma_{bNC}$, and it may be verified that the expressions

$$\gamma_{bC} = \frac{2b}{\sqrt{b^2 - x^2}} \left[(w_0 + \tfrac{1}{2} w_1) - (\lambda_0 + \tfrac{1}{2}\lambda_1) \right]$$

$$\gamma_{bNC} = -\frac{2}{\sin\theta} \left[(w_0 - \lambda_0)\cos\theta + \tfrac{1}{2}(w_1 - \lambda_1)\cos 2\theta \right. \\ \left. + 2\sum_{n=2}^{\infty} (w_n - \lambda_n) f_n(\theta) \right]$$

give

$$\int_{-b}^{b} \gamma_{bC}\, dx = \Gamma$$

$$\int_{-b}^{b} \gamma_{bNC}\, dx = 0$$

$$\frac{1}{2\pi} \int_{-b}^{b} \frac{\gamma_{bC}}{x - \xi}\, d\xi = 0$$

$$\frac{1}{2\pi} \int_{-b}^{b} \frac{\gamma_{bNC}}{x - \xi}\, d\xi = w_a - \lambda$$

as required. The relation

$$\frac{1}{\pi} \int_{0}^{\pi} \frac{\cos n\theta\, d\theta}{\cos\theta - \cos\phi} = \frac{\sin n\phi}{\sin\phi}$$

is used to establish the last two results.

The pressure is obtained by linearizing the unsteady Bernoulli equation:

$$p = -\rho \left(U \frac{\partial \phi}{\partial x} + \frac{\partial \phi}{\partial t} \right)$$

where ϕ is the velocity potential. The differential pressure on the airfoil surface is then

$$-\Delta p = \rho\left(U\frac{\partial\Delta\phi}{\partial x} + \frac{\partial\Delta\phi}{\partial t}\right)$$

where Δp is the upper surface pressure minus the lower surface pressure. Now the velocity parallel to the blade surface is $u = \partial\phi/\partial x$ and the blade vorticity strength is $\gamma_b = \Delta u$. Then

$$\frac{\partial\Delta\phi}{\partial x} = \Delta u = \gamma_b$$

$$\frac{\partial\Delta\phi}{\partial t} = \frac{\partial}{\partial t}\int_{-\infty}^{x}\Delta u\,dx = \frac{\partial}{\partial t}\int_{-b}^{x}\gamma_b\,dx$$

The differential pressure is thus

$$-\Delta p = \rho\left(U\gamma_b + \frac{\partial}{\partial t}\int_{-b}^{x}\gamma_{b_{NC}}\,dx\right)$$

Note that only the noncirculatory vorticity contributes pressure through the $\partial\phi/\partial t$ term. The unsteady circulatory vorticity produces pressure through the shed-wake induced velocity λ, and hence is already accounted for.

The net aerodynamic forces on the airfoil are the lift L (positive upward) and moment M about the axis at $x = ab$ (positive nose upward):

$$L = \int_{-b}^{b}(-\Delta p)\,dx$$

$$M = \int_{-b}^{b}(-\Delta p)(-x + ab)\,dx$$

Substituting for Δp gives:

$$L = \rho\left(U\Gamma - \frac{\partial}{\partial t}\Gamma_{NC}^{(1)}\right)$$

$$M = -\rho\left(U\Gamma^{(1)} - \tfrac{1}{2}\frac{\partial}{\partial t}\Gamma_{NC}^{(2)}\right)$$

where

$$\Gamma^{(n)} = \int_{-b}^{b}x^n\gamma_b\,dx$$

$$\Gamma_{NC}^{(n)} = \int_{-b}^{b} x^n \gamma_{b_{NC}} dx$$

The required circulations can be evaluated by substituting for γ_b:

$$\Gamma = 2\pi b\left[(w_0 + \tfrac{1}{2}w_1) - (\lambda_0 + \tfrac{1}{2}\lambda_1)\right]$$

$$\Gamma^{(1)} = 2\pi b^2\left[-(\tfrac{1}{2}+a)\Big((w_0 + \tfrac{1}{2}w_1) - (\lambda_0 + \tfrac{1}{2}\lambda_1)\Big) + \tfrac{1}{4}\Big((w_1 + w_2) - (\lambda_1 + \lambda_2)\Big)\right]$$

$$\Gamma_{NC}^{(1)} = 2\pi b^2\left[-\tfrac{1}{2}(w_0 - \tfrac{1}{2}w_2) + \tfrac{1}{2}(\lambda_0 - \tfrac{1}{2}\lambda_2)\right]$$

$$\Gamma_{NC}^{(2)} = 2\pi b^3\left[a\Big((w_0 - \tfrac{1}{2}w_2) - (\lambda_0 - \tfrac{1}{2}\lambda_2)\Big) - \tfrac{1}{8}\Big((w_1 - w_3) - (\lambda_1 - \lambda_3)\Big)\right]$$

For the blade motion considered here,

$$w_0 + \tfrac{1}{2}w_1 = U\alpha + \dot{h} + (\tfrac{1}{2} - a)b\dot{\alpha} = w_a \text{ at the three-quarter chord}$$

$$w_0 - \tfrac{1}{2}w_2 = U\alpha + \dot{h} - ab\dot{\alpha} = w_a \text{ at the midchord}$$

$$w_1 + w_2 = b\dot{\alpha}$$

$$w_1 - w_3 = b\dot{\alpha}$$

and the coefficients λ_n in the expansion of the induced velocity over the chord can be written in terms of the wake vorticity γ_w as follows:

$$\begin{aligned}
\lambda_n &= \frac{2}{\pi}\int_0^{\pi} \lambda \cos n\theta \, d\theta \\
&= \frac{2}{\pi}\int_0^{\pi}\left[\frac{1}{2\pi}\int_b^{\infty} \frac{\gamma_w \, d\xi}{x - \xi}\right]\cos n\theta \, d\theta \\
&= -\frac{1}{\pi}\int_b^{\infty} \gamma_w \left[\frac{1}{\pi}\int_0^{\pi} \frac{\cos n\theta}{\xi - b\cos\theta} d\theta\right] d\xi \\
&= -\frac{1}{\pi}\int_b^{\infty} \gamma_w \left[\frac{(\xi - \sqrt{\xi^2 - b^2})^n}{b^n\sqrt{\xi^2 - b^2}}\right] d\xi
\end{aligned}$$

So

$$\lambda_0 + \tfrac{1}{2}\lambda_1 = -\frac{1}{2\pi b}\int_b^\infty \gamma_W \left[\sqrt{\frac{\xi+b}{\xi-b}} - 1\right] d\xi$$

$$\lambda_0 - \tfrac{1}{2}\lambda_2 = -\frac{1}{\pi b^2}\int_b^\infty \gamma_W \left[\xi - \sqrt{\xi^2 - b^2}\right] d\xi$$

$$\lambda_1 + \lambda_2 = -\frac{1}{\pi b^2}\int_b^\infty \gamma_W \left(\xi - \sqrt{\xi^2 - b^2}\right)\left(\sqrt{\frac{\xi+b}{\xi-b}} - 1\right) d\xi$$

$$\lambda_1 - \lambda_3 = -\frac{2}{\pi b^3}\int_b^\infty \gamma_W \left(\xi - \sqrt{\xi^2 - b^2}\right)^2 d\xi$$

The circulations required for the airfoil lift are then

$$\Gamma = 2\pi b\left(U\alpha + \dot{h} + b\dot{\alpha}(\tfrac{1}{2} - a)\right) + \int_b^\infty \left(\sqrt{\frac{\xi+b}{\xi-b}} - 1\right)\gamma_W \, d\xi$$

and

$$\frac{\partial}{\partial t}\Gamma^{(1)}_{NC} = \frac{\partial}{\partial t}\left[-\pi b^2(U\alpha + \dot{h} - ab\dot{\alpha}) - \int_b^\infty \left(\xi - \sqrt{\xi^2 - b^2}\right)\gamma_W \, d\xi\right]$$

$$= -\pi b^2(U\dot{\alpha} + \ddot{h} - ab\ddot{\alpha}) - U\int_b^\infty \frac{\partial}{\partial \xi}\left(\xi - \sqrt{\xi^2 - b^2}\right)\gamma_W \, d\xi$$

$$= -\pi b^2(U\dot{\alpha} + \ddot{h} - ab\ddot{\alpha}) - U\int_b^\infty \left(1 - \frac{\xi}{\sqrt{\xi^2 - b^2}}\right)\gamma_W \, d\xi$$

The airfoil lift now is

$$L = \rho U 2\pi b\left(U\alpha + \dot{h} + b\dot{\alpha}(\tfrac{1}{2} - a)\right) + \rho\pi b^2(U\dot{\alpha} + \ddot{h} - ab\ddot{\alpha})$$

$$+ \rho U\int_b^\infty \frac{b}{\sqrt{\xi^2 - b^2}}\,\gamma_W \, d\xi$$

$$= L_Q + L_{NC} + L_W$$

L_Q is the quasistatic lift, which is the only term present for the steady case ($L = 2\pi\rho U^2 b\alpha$); L_{NC} is the noncirculatory lift, which is due to $\partial\Gamma_{NC}^{(1)}/\partial t$; and L_W is the lift due to the shed-wake induced velocity. Note that for the unsteady case L_Q is due to the angle of attack at the three-quarter chord point. Now the bound circulation is

$$\Gamma = \frac{L_Q}{\rho U} + \int_b^\infty \left(\sqrt{\frac{\xi + b}{\xi - b}} - 1\right) \gamma_W \, d\xi$$

and conservation of vorticity requires $\Gamma = -\int_b^\infty \gamma_W \, d\xi$; hence

$$L_Q = -\rho U \int_b^\infty \sqrt{\frac{\xi + b}{\xi - b}} \, \gamma_W \, d\xi$$

and

$$L_C = L_Q + L_W = -\rho U \int_b^\infty \frac{\xi}{\sqrt{\xi^2 - b^2}} \gamma_W \, d\xi$$

The lift can therefore be written as

$$L = \frac{\displaystyle\int_b^\infty \frac{\xi}{\sqrt{\xi^2 - b^2}} \gamma_W \, d\xi}{\displaystyle\int_b^\infty \sqrt{\frac{\xi + b}{\xi - b}} \, \gamma_W \, d\xi} L_Q + L_{NC}$$

The effect of the shed wake is to multiply the quasistatic lift L_Q by a factor depending on γ_W, and hence on the airfoil motion. To evaluate this factor it is necessary to consider a specific time history of motion. Assume that the airfoil has purely harmonic motion at frequency ω: $\alpha = \bar{\alpha} e^{i\omega t}$ and $h = \bar{h} e^{i\omega t}$ It follows then that the wake vorticity γ_W must also be periodic in time, and thus it has the form $\gamma_W = \bar{\gamma}_W e^{i\omega(t - \xi/U)}$ when the requirement of convection with the free stream velocity is applied as well. Then $\gamma_W e^{i\omega t}$ factors out of the integrals over the wake, giving

$$\begin{aligned} L &= C(k) L_Q + L_{NC} \\ &= 2\pi\rho U b C(k)\left(U\alpha + \dot{h} + b\dot{\alpha}(\tfrac{1}{2} - a)\right) + \rho\pi b^2 (U\dot{\alpha} + \ddot{h} - ab\ddot{\alpha}) \end{aligned}$$

where $C(k)$ is a function depending only on the dimensionless frequency $k = \omega b/U$. $C(k)$ is the Theodorsen lift deficiency function. Since the magnitude of C varies from 1 at low frequency to 0.5 at high frequency, the effect of the shed wake is to reduce the circulatory lift below the quasi-static value.

The circulations required for the aerodynamic moment about the axis $x = ab$ are obtained by similar manipulations:

$$\Gamma^{(1)} = -b(\tfrac{1}{2}+a)\Gamma + \tfrac{1}{2}\pi b^3\dot{\alpha} + \tfrac{1}{2}\int_b^\infty \gamma_w\left(\xi - \sqrt{\xi^2-b^2}\right)\left(\sqrt{\frac{\xi+b}{\xi-b}} - 1\right)d\xi$$

$$\tfrac{1}{2}\frac{\partial}{\partial t}\Gamma^{(2)}_{NC} = -b(\tfrac{1}{2}+a)\frac{\partial}{\partial t}\Gamma^{(1)}_{NC} - \tfrac{1}{2}\pi b^3\left(U\dot{\alpha} + \ddot{h} + b(\tfrac{1}{4}-a)\ddot{\alpha}\right) + \frac{U}{2}\int_b^\infty \gamma_w\left(\xi - \sqrt{\xi^2-b^2}\right)\left(\sqrt{\frac{\xi+b}{\xi-b}} - 1\right)d\xi$$

Thus the moment is

$$\begin{aligned} M &= b(\tfrac{1}{2}+a)L + M_{QC} \\ &= b(\tfrac{1}{2}+a)L - \tfrac{1}{2}\rho\pi b^3\left(2U\dot{\alpha} + \ddot{h} + b(\tfrac{1}{4}-a)\ddot{\alpha}\right) \\ &= b(\tfrac{1}{2}+a)C(k)L_Q + \rho\pi b^3\left(a\ddot{h} - (\tfrac{1}{2}-a)U\dot{\alpha} - b(\tfrac{1}{8}+a^2)\ddot{\alpha}\right) \end{aligned}$$

M_{QC} is the moment about the quarter chord, which is the aerodynamic center predicted by thin-airfoil theory. With the pitch axis at the quarter chord ($a = -\tfrac{1}{2}$) there is no moment due to the lift. The virtual mass terms ($\ddot{h}$ and $\ddot{\alpha}$) arise both from M_{QC} and from the noncirculatory lift L_{NC}. The noncirculatory pitch damping moment is due to lift acting at the three-quarter chord; for $a = \tfrac{1}{2}$ this moment is zero.

Let us now examine in detail the Theodorsen lift deficiency function $C(k)$, which defines the influence of the shed wake on the aerodynamic loads during unsteady motion. Recall that to evaluate the wake influence, harmonic motion at frequency ω was assumed, giving $\gamma_w = \bar{\gamma}_w e^{i\omega(t-\xi/U)}$. Hence

$$C(k) = \frac{\int_b^\infty \frac{\xi}{\sqrt{\xi^2 - b^2}} \gamma_w d\xi}{\int_b^\infty \sqrt{\frac{\xi + b}{\xi - b}} \gamma_w d\xi}$$

$$= \frac{\int_1^\infty \frac{\xi}{\sqrt{\xi^2 - 1}} e^{-ik\xi} d\xi}{\int_1^\infty \sqrt{\frac{\xi + 1}{\xi - 1}} e^{-ik\xi} d\xi}$$

$$= \frac{H_1^{(2)}(k)}{H_1^{(2)}(k) + iH_0^{(2)}(k)}$$

where $H_n^{(2)} = J_n - iY_n$ is the Hankel function, and the reduced frequency is $k = \omega b/U$. Fig. 10-3 shows the magnitude and phase of the lift deficiency function for reduced frequencies up to $k = 1$. For $k = 0$, $C = 1$ as is required of the static limit; and for large frequencies the magnitude approaches $|C| = 0.5$, so the shed wake reduces the circulatory lift to one-half the quasistatic values. There is a moderate phase shift that has a maximum just above 15° around $k = 0.3$ and approaches zero again at high frequencies. For small frequencies, the lift deficiency function is approximately

$$C(k) \cong \left(1 - \frac{\pi}{2} k\right) + ik\left(\ln\frac{k}{2} + \gamma\right)$$

where $\gamma = 0.5772156...$ is Euler's constant. For a rotor, the frequency of the blade motion can be expressed in terms of the rotational speed Ω. Consider n/rev motion, where $\omega = n\Omega$. The free stream velocity is Ωr, and the semichord is $c/2$, so the reduced frequency becomes $k = nc/2r$. For the high-aspect-ratio blades of rotors, $k \cong 0.05n$ typically. For the lower harmonics, then, the reduced frequency is small, and the lift deficiency function is near unity. For 1/rev motion there is perhaps a 5% reduction in the lift due to the shed wake. Thus the neglect of the shed wake and other unsteady

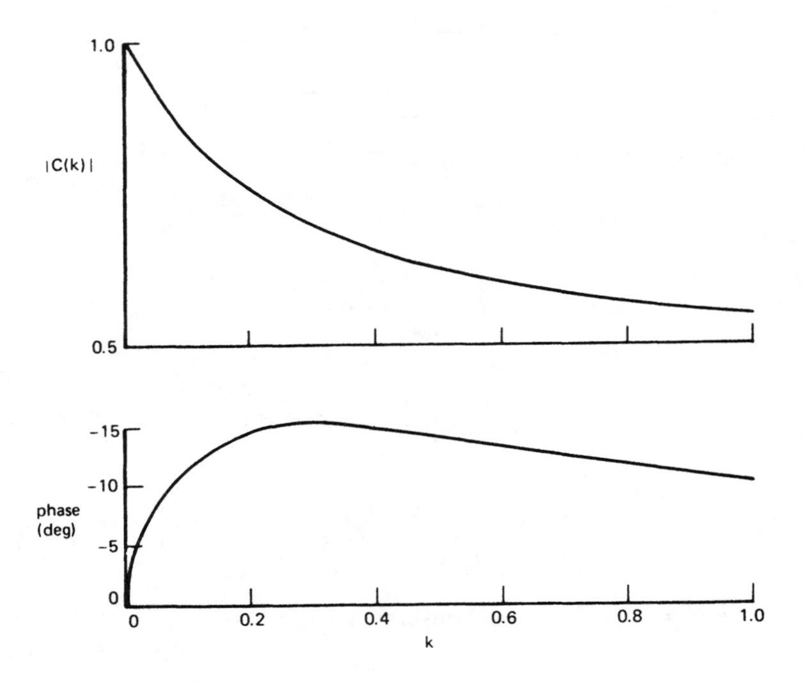

Figure 10-3 Theodorsen lift deficiency function.

aerodynamic effects in the analysis of the rotor performance and flap motion of the earlier chapters is justified. For the higher harmonics, however, the reduced frequency is large enough that the shed wake effects must be accounted for to obtain an accurate estimate of the loads.

An alternative form of the unsteady thin-airfoil result is a Glauert series for the pressure developed by Cicala:

$$-\Delta p = \rho U^2 e^{i\omega t} \sum_{n=0}^{\infty} a_n f_n(\theta)$$

where $x = b \cos \theta$. Expand the upwash due to the blade motion as a cosine series:

$$w_a = U e^{i\omega t}\left(A_0 + 2 \sum_{n=1}^{\infty} A_n \cos n\theta\right)$$

Then the solution can be written as

$$a_0 = 2(A_0 + A_1)C(k) - 2A_1$$

$$a_n = -\frac{2ik}{n}\left(A_{n+1} - A_{n-1}\right) + 4A_n$$

with the lift and moment given by

$$L = \rho U^2 b\pi\left(a_0 + \tfrac{1}{2}a_1\right)e^{i\omega t}$$

$$M_{QC} = -\rho U^2 b^2 \frac{\pi}{4}\left(a_1 + a_2\right)e^{i\omega t}$$

For example, consider an encounter with a sinusoidal gust of wavelength $2\pi b/k$, so that the airfoil sees the upwash velocity

$$w_a = w_0\, e^{i\omega(t - x/U)}$$

$$= w_0\, e^{i\omega t} e^{-ikx/b}$$

$$= w_0\, e^{i\omega t} e^{-ik\cos\theta}$$

Expanding $e^{-ik\cos\theta}$ as a cosine series in θ gives

$$A_n = \frac{w_0}{U}(-i)^n J_n(k)$$

where J_n is the Bessel function. It follows that

$$a_0 = \frac{w_0}{U}\, 2\left[\left(J_0(k) - iJ_1(k)\right)C(k) + iJ_1(k)\right]$$

$$a_n = \frac{w_0}{U}\frac{2ik}{n}(-1)^{n-1}\left(J_{n+1} + J_{n-1} - \frac{2n}{k}J_n\right) = 0$$

The pressure then has only the first term in the Glauert series:

$$-\frac{\Delta p}{\rho U^2} = e^{i\omega t}\frac{w_0}{U}\, 2S(k)\sqrt{\frac{b-x}{b+x}}$$

The lift is

$$\frac{L}{\rho U^2 b} = e^{i\omega t}\frac{w_0}{U}\, 2\pi S(k)$$

and $M_{QC} = 0$. Here $S(k)$ is the Sears function,

$$S(k) = \left(J_0(k) - iJ_1(k)\right)C(k) + iJ_1(k),$$

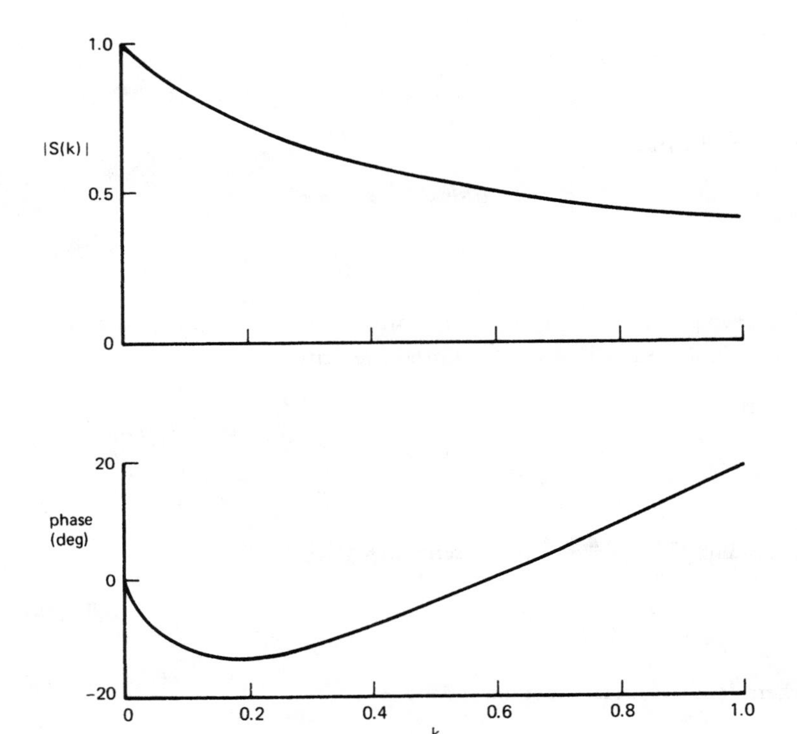

Figure 10-4 Sears function for sinusoidal gust loading.

which is shown in Figure 10-4. Since any gust can be Fourier analyzed, the resulting aerodynamic lift always acts at the quarter chord. At $k = 0$, the Sears function is $S = 1$. For large frequency, it is approximately

$$S(k) \sim \frac{1}{\sqrt{2\pi k}} e^{i(k-\pi/4)}$$

so the magnitude approaches zero (in contrast to the Theodorsen function), while the phase is linear with k.

10–3 Near Shed Wake

The two-dimensional airfoil analysis indicates that the shed wake is an important factor in determining the unsteady aerodynamic loading at frequencies characteristic of rotor blade motion. Unlike the two-dimensional model considered, the rotary wing shed wake is in a helical sheet behind the blade. The major effects will be due to the shed wake nearest the trailing edge of the blade, however. The question that concerns us then is how the near shed wake (extending 15° to 45° in azimuth behind the blade) should be modeled in the numerical calculation of the induced velocity and airloading of a rotary wing. The two-dimensional model of the last section will be used to explore this problem.

Helicopter airloads analyses generally use lifting-line theory to calculate the wake-induced velocity at the bound vortex. The inflow is evaluated only at a single point on the chord. Because of the complex geometry of the rotor wake, extensive numerical calculations are required to obtain the induced velocity. However, unsteady airfoil theory requires a knowledge of the distribution of the shed-wake induced velocity over the chord. For example, in section 10–2 the coefficients λ_0, λ_1, and λ_2 in the cosine series representation of the inflow were needed to obtain the unsteady lift and moment. A numerical calculation of the shed-wake induced velocity at several points along the chord is to be avoided if possible, to minimize the computation required. It must be established then how the near shed wake should be modeled so that calculation of the induced velocity at a single chordwise point in lifting-line theory will give the proper unsteady loads.

Miller (1964a, 1964b) considered a lifting-line theory approximation for the near shed wake. Note that since the lifting-line assumption of high aspect ratio also implies low reduced frequency, the result is expected to be equivalent to a low frequency approximation. The approach is to determine what treatment of the shed wake in the lifting-line evaluation of the induced velocity will correctly give the unsteady loads on the two-dimensional airfoil, specifically the lift deficiency function. Write the airfoil unsteady lift as $L = L_C + L_{NC}$, where

$$L_{NC} = \rho\pi b^2(U\dot{\alpha} + \ddot{h} - ab\ddot{\alpha})$$

$$L_C = \rho U\Gamma = \rho U2\pi b\left[(w_0 + \tfrac{1}{2}w_1) - (\lambda_0 + \tfrac{1}{2}\lambda_1)\right]$$

The quasistatic lift is defined as before:

$$L_Q = \rho U2\pi b(w_0 + \tfrac{1}{2}w_1) = \rho U2\pi b\left(U\alpha + \dot{h} + b\dot{\alpha}(\tfrac{1}{2} - a)\right)$$

Therefore the circulatory lift can be written as

$$L_C = L_Q - \rho U 2\pi b \lambda$$

where the induced velocity is obtained from the shed wake vorticity:

$$\lambda = \frac{1}{2\pi} \int_b^\infty \frac{\gamma_w}{x - \xi} d\xi$$

In the lifting-line approximation, the induced velocity is evaluated at a single point, namely at the quarter chord where the bound vortex is located ($x = -b/2$). Also the wake vorticity is extended up to the quarter chord. Then

$$\lambda = \frac{1}{2\pi} \int_{-b/2}^\infty \frac{\gamma_w}{-\left(\xi + \frac{b}{2}\right)} d\xi$$

Now the wake vorticity is given by the time variation of the bound circulation:

$$\gamma_w = -\frac{1}{U} \frac{d\Gamma}{dt}$$

at $t - (\xi - b)/U$. Assuming harmonic motion so that $\Gamma = \overline{\Gamma} e^{i\omega t}$, γ_w is

$$\gamma_w = -\frac{i\omega}{U} \Gamma e^{-i\omega(\xi - b)/U}$$

and the induced velocity becomes

$$\lambda = -\frac{i\omega}{U} \Gamma \frac{1}{2\pi} \int_{-b/2}^\infty \frac{e^{-i\omega(\xi - b)/U}}{-\left(\xi + \frac{b}{2}\right)} d\xi$$

$$= \Gamma \frac{ike^{i(3/2)k}}{2\pi b} \int_0^\infty \frac{e^{-ik\xi}}{\xi} d\xi$$

$$= \Gamma \frac{ke^{i(3/2)k}}{2\pi b} \left(\int_0^\infty \frac{\sin k\xi}{\xi} d\xi + i \int_0^\infty \frac{\cos k\xi}{\xi} d\xi \right)$$

(the factor $e^{i(3/2)k} \cong 1$ to this order of approximation). The cosine integral is not finite, so it must be omitted now. The remaining integral (the real part) becomes

$$\lambda = \Gamma \frac{k}{2\pi b} \int_0^\infty \frac{\sin k\xi}{\xi} d\xi = \Gamma \frac{k}{4b}$$

$$= \frac{L_C}{\rho U 2\pi b} \frac{\pi}{2} k$$

Then the unsteady lift is $L_C = L_Q - \rho U 2\pi b\lambda = L_Q - L_C(\pi/2)k$, or

$$L_C = \frac{L_Q}{1 + \frac{\pi}{2} k}$$

Thus an approximate lift deficiency function has been obtained:

$$C(k) = \frac{1}{1 + \frac{\pi}{2} k}$$

which is in fact a correct approximation to order k for the Theodorsen lift deficiency function (see section 10–2). Fig. 10-5 compares this result with Theodorsen's function; the approximation is good even for fairly large values of reduced frequency. At high k the correct value for $C(k)$ is significantly underestimated, however.

The lifting-line approximation therefore gives the proper results, except that actually the integral over the wake vorticity was divergent. The difficulty is due to the singularity in the induced velocity at the edge of the vortex sheet, which was extended up to the quarter chord. To correct the model, consider stopping the shed wake a distance ϵb behind the quarter chord (where the induced velocity is evaluated). It follows that λ is:

$$\lambda = \frac{i\omega}{U} \Gamma \frac{1}{2\pi} \int_{-(b/2)+b\epsilon}^{\infty} \frac{e^{-i\omega(\xi - b)/U}}{\xi + \frac{b}{2}} d\xi$$

$$= \Gamma \frac{ik}{2\pi b} e^{i(3/2)k} \int_{\epsilon}^{\infty} \frac{e^{-ik\xi}}{\xi} d\xi$$

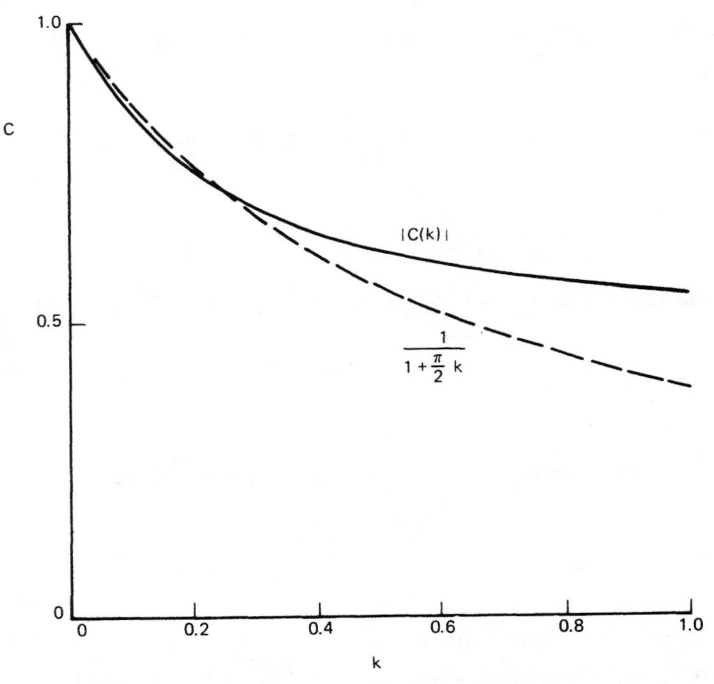

Figure 10-5 Comparison of the approximate lift deficiency function with the Theodorsen function $C(k)$.

$$= \Gamma \frac{k}{2\pi b} I$$

$$= \frac{L_C}{\rho U 2\pi b} kI$$

where

$$I = e^{i(3/2)k}\left[\int_\epsilon^\infty \frac{\sin k\xi}{\xi}\,d\xi + i\int_\epsilon^\infty \frac{\cos k\xi}{\xi}\,d\xi\right]$$

The lift deficiency function obtained by this model is then

$$C(k) = \frac{1}{1 + kI}$$

Requiring that this approximation give exactly the Theodorsen function determines the parameter ϵ. Actually there are two values, ϵ_s and ϵ_c (for the sine and cosine integrals respectively), which are evaluated from the real and imaginary parts of the equation. The important parameter is ϵ_c, which prevents the divergence of the cosine integral. The limit for low frequency is $\epsilon_c \rightarrow \frac{1}{2}$. Fig. 10-6 shows the results obtained for ϵ_c and ϵ_s over a range of frequencies. For $k = 0$ to 1, $\epsilon \cong \frac{1}{2}$ is a good approximation, particularly for the cosine integral. Thus from Miller's results it has been concluded that the near shed wake in the lifting-line model should be extended to a quarter chord ($\epsilon b \cong b/2 = c/4$) behind the point where the induced velocity is being calculated.

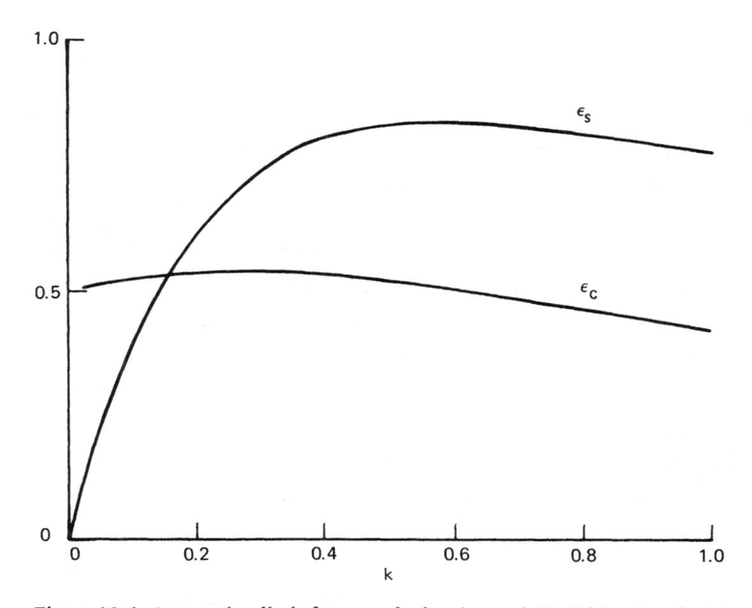

Figure 10-6 Integration limit for near shed wake model in lifting-line theory.

Piziali (1966b) considered the effect of a discrete vortex representation of the rotor wake on the shed wake influence. The spirals of the rotary-wing wake are most easily represented by a lattice of finite-strength line vortex segments. In two-dimensional airfoil theory, the corresponding shed wake model is a series of point vortices (Fig. 10-7). The distance between the vortices in the wake is $d = 2\pi U/N\omega$, for N vortices per cycle of oscillation. The induced velocity is calculated N times per cycle. The discrete shed vortices correspond to a step change in the airfoil bound circulation. The

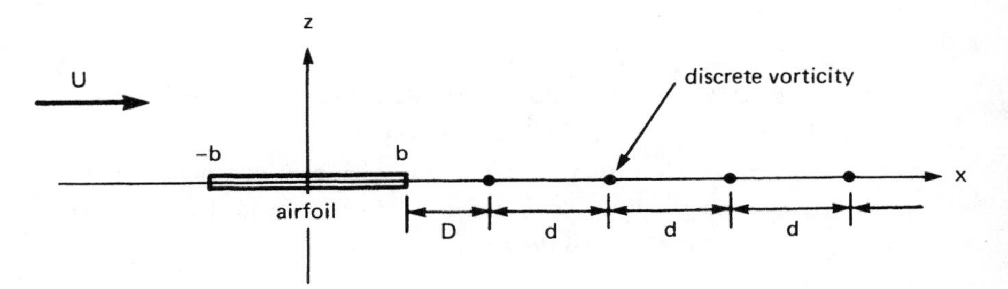

Figure 10-7 Discrete vortex model for the shed wake of a two-dimensional airfoil.

distance of the first vortex behind the trailing edge is D. Piziali calculated the ratio of the unsteady lift and moment to their quasisteady counterparts for this model, and compared it with the Theodorsen function for pitch and heaving of the airfoil at various frequencies. It was found that $D = d$ does not give good results even with a large number of points per cycle. However, if the entire discrete wake model is advanced closer to the trailing edge so that $D = d/3$, reasonable results are obtained over the frequency range of interest. It was concluded that with a vortex lattice wake model in a rotary wing airloads analysis, the shed wake elements should be advanced by about 70% of the azimuthal spacing, so that the first elements are closer to the blade trailing edge. The bound circulation is being evaluated at intervals of d (in terms of the distance traveled by the airfoil). The discrete vortices in the wake should be positioned midway between the points where the bound circulation is evaluated. By this reasoning, the distance between the bound vortex and the first wake element should be $d/2$, which gives $D = d/2 - 3b$.

Daughaday and Piziali (1966) found that the calculation of the lift and moment at high frequency can be improved by representing the shed wake just behind the blade as a continuous distribution of vorticity, replacing the first few discrete elements in the vortex lattice (Fig. 10-8). The loads were calculated for a two-dimensional airfoil with this wake model and compared with the Theodorsen results. Note that the lifting-line approximation is not being used, but rather the chordwise distribution of the induced velocity is still being evaluated. The only approximation is made in the wake vorticity representation. The first two discrete vortices were replaced by a continuous sheet as shown in Fig. 10-8. The vorticity distribution γ_w in the sheet was

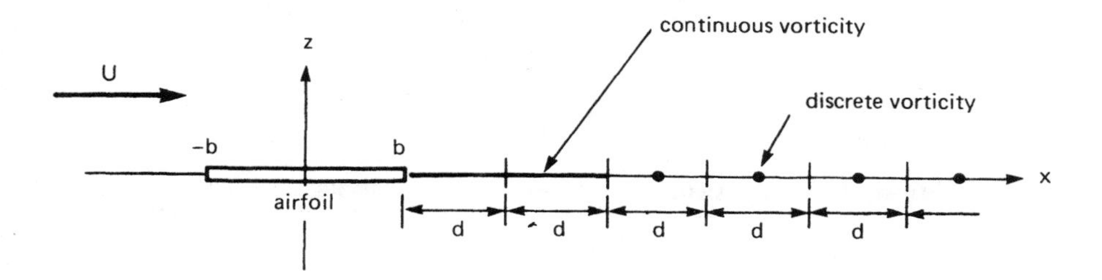

Figure 10-8 Combined continuous and discrete shed wake vorticity model.

given by a polynomial obtained from the loading variation. It was found that Theodorsen's results for the lift and moment could be obtained with 5 to 8 points per oscillation cycle. The improvement over using only discrete vortices in the wake (even advanced toward the trailing edge) was substantial, an order of magnitude fewer points per cycle being required with the continuous near shed wake for the same accuracy.

In summary, lifting-line theory calculates the blade loads from the velocity induced at the section by the shed and trailed vorticity in the wake. For the inflow calculation, the blade is modeled by the bound vortex at the quarter chord, and the trailed vorticity (due to the spanwise lift variation) is extended up to the bound vortex. The induced velocity is then evaluated along the bound vortex. The simplest and most economical representation of the complex wake structure is a lattice of finite-strength vortex-line elements. A line vortex is in fact a good representation of the rolled-up tip vortices of the blades. Based on the two-dimensional airfoil analyses discussed in this section, it has been concluded that such a lifting-line model can be used for the shed wake as well. With a proper representation of the near shed wake, the unsteady aerodynamic effects will be correctly calculated. The shed-wake induced velocity can be calculated at the bound vortex point alone. To correct for the neglect of the chordwise variation of the induced velocity, the shed wake is not extended all the way to the bound vortex, however, but is stopped a quarter chord behind it. Furthermore, a vortex sheet with a continuous vorticity distribution should be used for the near shed wake and extend for one or two azimuth increments before the discrete shed vortices begin. In general, for the trailed wake and the far shed wake a vortex lattice model is satisfactory. The use of vortex sheets may also be required to model the returning wake (except for the tip vortices) where it passes very close to a blade, and in other areas where the wake and blade interaction is significant.

10–4 Unsteady Airfoil Theory with a Time-Varying Free Stream

The rotating blade of a helicopter rotor in forward flight sees a periodically varying free stream velocity: $u_T = r + \mu \sin\psi = r(1 + (\mu/r)\sin\psi)$. For either high advance ratio or the inboard sections, the 1/rev variation of the velocity is a significant fraction of the mean. In such cases the time-varying free stream must be included in the unsteady airfoil theory, both for its direct effects and for its influence through the shed wake. Only the case $\mu/r < 1$ will be considered. If $\mu/r > 1$, the blade section passes through the reverse flow region, and a simple wake model is not applicable.

Consider the two-dimensional airfoil and wake model described in section 10–2. Only a few modifications to that analysis are required to account for a time-varying free stream velocity U. The time derivative acts on the velocity now, so

$$\frac{\partial}{\partial t}\Gamma_{NC}^{(1)} = -\pi b^2\left((U\alpha)^{\cdot} + \ddot{h} - ab\ddot{\alpha}\right) - U\int_b^\infty\left(1 - \frac{\xi}{\sqrt{\xi^2 - b^2}}\right)\gamma_w\, d\xi$$

$$\tfrac{1}{2}\frac{\partial}{\partial t}\Gamma_{NC}^{(2)} = -b(\tfrac{1}{2}+a)\frac{\partial}{\partial t}\Gamma_{NC}^{(1)} - \tfrac{1}{2}\pi b^3\left((U\alpha)^{\cdot} + \ddot{h} + b(\tfrac{1}{4}-a)\ddot{\alpha}\right) + \frac{U}{2}\int_b^\infty \gamma_w\left(\xi - \sqrt{\xi^2 - b^2}\right)\left(\sqrt{\frac{\xi+b}{\xi-b}} - 1\right)d\xi$$

Then the lift and moment become

$$L = L_C + L_{NC} = L_Q + \rho U\int_b^\infty \frac{b}{\sqrt{\xi^2 - b^2}}\gamma_w\, d\xi + \rho\pi b^2\left((U\alpha)^{\cdot} + \ddot{h} - ab\ddot{\alpha}\right)$$

$$M = b(\tfrac{1}{2}+a)L + M_{QC}$$

$$= b(\tfrac{1}{2}+a)L - \tfrac{1}{2}\rho\pi b^3\left(U\dot{\alpha} + (U\alpha)^{\cdot} + \ddot{h} + b(\tfrac{1}{4}-a)\ddot{\alpha}\right)$$

where $L_Q = \rho U 2\pi b\left(U\alpha + \dot{h} + b\dot{\alpha}(\tfrac{1}{2}-a)\right)$ again. The only changes here are $\dot{U}\alpha$ terms added to the noncirculatory lift and moment. The quasistatic lift and circulatory lift are still given in terms of the wake vorticity as follows:

$$L_Q = -\rho U \int_b^\infty \sqrt{\frac{\xi+b}{\xi-b}}\,\gamma_w\,d\xi$$

$$L_C = -\rho U \int_b^\infty \frac{\xi}{\sqrt{\xi^2-b^2}}\,\gamma_w\,d\xi$$

Relating L_Q and L_C in terms of a lift deficiency function requires a knowledge of the dependence of γ_w on ξ. The criterion that there be no pressure difference across the vortex sheet gives

$$-\Delta p = \rho\left(\frac{\partial}{\partial t} + U\frac{\partial}{\partial \xi}\right)\Delta\phi = 0$$

which implies

$$\left(\frac{\partial}{\partial t} + U\frac{\partial}{\partial \xi}\right)\frac{\partial \Delta\phi}{\partial \xi} = \left(\frac{\partial}{\partial t} + U\frac{\partial}{\partial \xi}\right)\gamma_w = 0$$

the solution of which is

$$\gamma_w = \gamma_w\left(\xi - \int^t U\,dt\right)$$

If the free stream velocity is constant, the wake vorticity is convected at a constant rate and γ_w is a function of $(\xi - Ut)$ as before. Considering the rotor blade in forward flight, the dimensionless free stream velocity is $U = r + \mu \sin\psi$, so

$$\gamma_w = \gamma_w(\xi - r\psi + \mu\cos\psi)$$

($\psi = \Omega t$ is the dimensionless time variable). Now assume periodic motion of the blade. For the flow field to be entirely periodic, the blade motion can consist only of harmonics of the fundamental frequency Ω of the free stream variation. The period of the flow is then $2\pi/\Omega$. The wake vorticity must be a periodic function of ξ with a wavelength equal to the distance the wake is convected during the period:

$$\int_0^{2\pi} U d\psi = 2\pi r$$

Next, write the periodic function γ_w as a Fourier series in ξ with period $2\pi r$:

$$\gamma_w = \sum_{m=-\infty}^{\infty} \gamma_{w_m}(\psi) e^{-im\xi/r}$$

Since γ_w must be a function of the quantity $(\xi - r\psi + \mu \cos \psi)$ alone, it follows that

$$\gamma_w = \sum_{m=-\infty}^{\infty} \bar{\gamma}_m e^{im(\psi - (\mu/r)\cos\psi) - im\xi/r}$$

where $\bar{\gamma}_m$ are constants. For $\mu = 0$ this reduces to $\gamma_w = \bar{\gamma} e^{i\omega(t - \xi/U)}$ as before.

With the structure of the wake vorticity established, the relation between the quasistatic and circulatory lift can be constructed. Substituting for γ_w gives

$$L_Q = -\rho U \sum_{m=-\infty}^{\infty} \bar{\gamma}_m e^{im(\psi - (\mu/r)\cos\psi)} \int_b^{\infty} \sqrt{\frac{\xi + b}{\xi - b}} e^{-im\xi/r} d\xi$$

$$L_C = -\rho U \sum_{m=-\infty}^{\infty} \bar{\gamma}_m e^{im(\psi - (\mu/r)\cos\psi)} \int_b^{\infty} \frac{\xi}{\sqrt{\xi^2 - b^2}} e^{-im\xi/r} d\xi$$

Noting that

$$\frac{1}{2\pi} \int_0^{2\pi} e^{in(\psi - (\mu/r)\cos\psi)} (1 + (\mu/r)\sin\psi) d\psi = 1$$

if $n = 0$ and is zero otherwise, the harmonics $\bar{\gamma}_m$ can be evaluated:

$$\bar{\gamma}_m = \frac{-\int_0^{2\pi} (1 + (\mu/r)\sin\psi) e^{-im(\psi - (\mu/r)\cos\psi)} L_Q d\psi}{2\pi\rho U \int_b^{\infty} \sqrt{\frac{\xi + b}{\xi - b}} e^{-im\xi/r} d\xi}$$

Thus the circulatory lift is

$$L_C = 2\pi\rho Ub \sum_{m=-\infty}^{\infty} e^{im(\psi - (\mu/r)\cos\psi)} C(mb/r)$$

$$\frac{1}{2\pi} \int_0^{2\pi} (1 + (\mu/r)\sin\psi)\, e^{-im(\psi - (\mu/r)\cos\psi)} Q\, d\psi$$

where $C(mb/r)$ is the Theodorsen lift deficiency function at reduced frequency $k = mb/r$ (i.e. $\omega = m\Omega$ and an average velocity $\bar{U} = \Omega r$), and

$$Q = \frac{L_Q}{2\pi\rho Ub} = U\alpha + \dot{h} + b\dot{\alpha}(\tfrac{1}{2} - a)$$

If now the quasistatic circulation Q is written as a Fourier series

$$Q = \sum_{n=-\infty}^{\infty} Q_n e^{in\psi}$$

then the lift can be written in a form analogous to the constant velocity result:

$$L_C = 2\pi\rho Ub \sum_{n=-\infty}^{\infty} Q_n e^{in\psi} C_\mu(n,\psi)$$

where $C_\mu(n,\psi)$ is the modified lift deficiency function for the nth harmonic of the blade motion with free stream velocity $U = r + \mu\sin\psi$:

$$C_\mu(n,\psi) = \sum_{m=-\infty}^{\infty} e^{im(\psi - (\mu/r)\cos\psi) - in\psi} C(mb/r)$$

$$\frac{1}{2\pi} \int_0^{2\pi} (1 + (\mu/r)\sin\psi)\, e^{-im(\psi - (\mu/r)\cos\psi) + in\psi} d\psi$$

For a constant free stream velocity ($\mu = 0$), the integral is nonzero only for $m = n$, and hence $C_{\mu=0}(n,\psi) = C(nb/r)$ as required. An alternative form is

$$L_C = 2\pi\rho Ub \sum_{n=-\infty}^{\infty} Q_n \left\{ \sum_{\ell=-\infty}^{\infty} C_{\ell n} e^{i\ell\psi} \right\}$$

where

$$C_{\ell n} = \sum_{m=-\infty}^{\infty} \left[\frac{1}{2\pi} \int_0^{2\pi} e^{im(\psi - (\mu/r)\cos\psi) - i\ell\psi} \, d\psi \right] C(mb/r)$$

$$\left[\frac{1}{2\pi} \int_0^{2\pi} (1 + (\mu/r)\sin\psi)\, e^{-im(\psi - (\mu/r)\cos\psi) + in\psi} \, d\psi \right]$$

The coefficients C_{ℓ_n} are the harmonics in a Fourier series expansion of $e^{in\psi} C_\mu(n,\psi)$. This form shows that the time-varying free stream couples the harmonics of the circulation and lift through the influence of the shed wake.

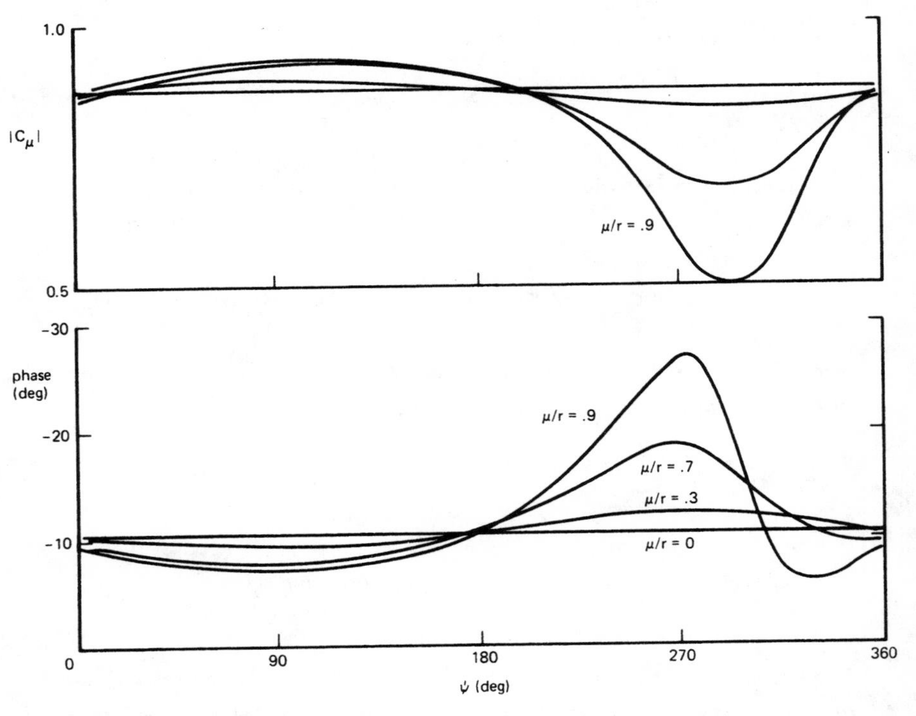

Figure 10-9 Lift deficiency function with a time-varying free stream; for the second harmonic ($n = 2$) and $b/r = 0.04$.

The integrals appearing in the lift deficiency function for a time-varying free stream can be evaluated in terms of Bessel functions. Fig. 10-9 shows typical results for $C_\mu(n,\psi)$, with $n = 2$ and $b/r = 0.04$. The 1/rev variation in the free stream velocity produces a basic 1/rev variation of C_μ with ψ. The largest influence occurs nearest the reverse flow boundary, at $\psi = 270°$. Most of the range of velocities and radial stations of interest are covered by $0 < \mu/r < 0.7$. The model breaks down for $\mu/r > 1$, when the section passes through the reverse flow region. For small μ/r, the lift deficiency function is approximately

$$C_\mu(n,\psi) \cong \sum_{m=-\infty}^{\infty} e^{i(m-n)\psi} \, [1 - im(\mu/r)\cos\psi] \; C(mb/r)$$

$$\frac{1}{2\pi}\int_0^{2\pi} e^{i(n-m)\psi} \, [1 + (\mu/r)(\sin\psi + im\cos\psi)] \, d\psi$$

$$= C_n + (\mu/r)(in/2)\,[\cos\psi(C_{n+1} + C_{n-1} - 2C_n) + i\sin\psi(C_{n+1} - C_{n-1})]$$

where C_n means $C(nb/r)$. Assuming small values of b/r then gives

$$C_\mu(n,\psi) \cong C(nb/r) - [(nb/r)(\mu/r)\sin\psi]\, C'(nb/r)$$

$$= C[nb/(r + \mu\sin\psi)]$$

Thus for small variations in the free stream velocity (small $nb\mu/r^2$) the lift deficiency function is nearly the same as the Theodorsen function $C(k)$; the reduced frequency is based on the local velocity, $k = \omega b/U$. This approximation works well for moderate n. Fig. 10-9 shows the basic dependence on the local reduced frequency. On the advancing side, the increased velocity lowers the reduced frequency and hence the lift deficiency function is nearer unity. On the retreating side there is the greatest accumulation of shed vorticity in the wake near the trailing edge, and thus the greatest reduction in the lift.

In summary, a time-varying free stream has the following influence on the unsteady aerodynamics of a two-dimensional airfoil: there are additional noncirculatory lift and moment terms due to $d(U\alpha)/dt$; there is coupling by the wake of all the harmonics of the quasistatic and unsteady circulation; and there is a significant influence on the lift deficiency function due to stretching and compressing of the vorticity in the shed wake. For the free stream variation of the rotor blade in forward flight, all these effects basically produce 1/rev variations of the loads. The noncirculatory lift and moment terms are valid for a general time variation of U. The simple approximation $C_\mu(n,\psi) \cong C(k)$ using the local reduced frequency is quite good up to $\mu/r = 0.7$. For small enough μ/r the cruder approximation $C_\mu(n,\psi) \cong C(nb/r)$ using the mean reduced frequency can be chosen (i.e. the influence of a time-varying free stream on the shed wake can be neglected entirely).

Other work on unsteady airfoil theory with a time-varying free stream: Isaacs (1945, 1946), and Greenberg (1947).

10-5 Two-Dimensional Model for Rotary Wing Unsteady Aerodynamics

The wake of a rotor in hover or vertical flight consists of helical vortex sheets below the disk, one from each blade. Unsteady motion of the rotor blade will produce shed vorticity in the wake spirals. With low disk loading the wake remains near the rotor disk and therefore passes close to the following blades. Thus the wake vorticity is not convected downstream of the airfoil as with fixed wings, and the shed vorticity sheets below the rotor disk must be accounted for to correctly estimate the unsteady loads. For high inflow or forward flight the rotor wake will be convected away from the blades; the returning shed wake influence is primarily a concern of vertical flight. Assuming a high aspect ratio of the blade, lifting-line theory requires a knowledge of the loads on the blade section, and the returning shed wake of the rotor must be incorporated into the two-dimensional unsteady airfoil theory. The wake far from the blade section will have little influence, so the emphasis is on modeling the wake near the blade, which for low inflow consists of vortex sheets that are nearly planar surfaces parallel to the disk plane. Based on these considerations, a two-dimensional model for the unsteady aerodynamics of the rotor can be constructed.

Loewy (1957) developed a two-dimensional model for the unsteady aerodynamics of the blade of a hovering rotor, including the effect of the returning shed wake. Fig. 10-10 shows the two-dimensional model of the helical wake assumed. Consider first a single-bladed rotor, so that all the vorticity comes from the same blade. There is a two-dimensional thin airfoil, with a shed wake vortex sheet extending downstream from the trailing edge to infinity. The surfaces of the wake spiral below the blade are modeled by a series of planar, two-dimensional vortex sheets with vertical separation h that extend from infinity upstream to infinity downstream. All the wake

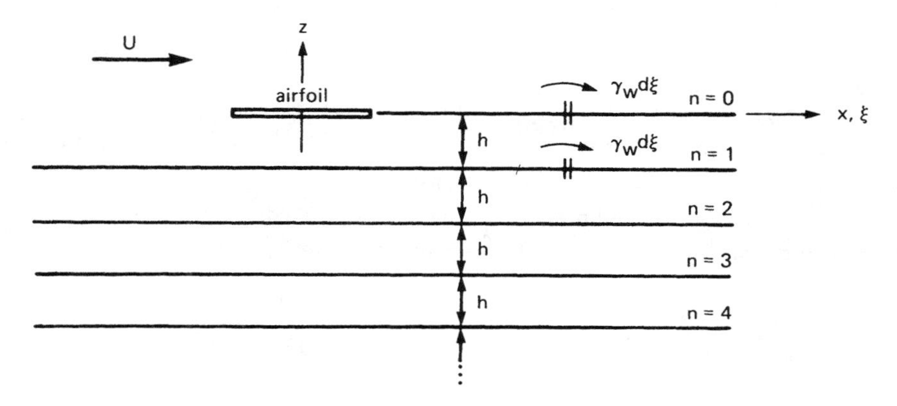

Figure 10-10 Two-dimensional model of rotary wing unsteady aerodynamics (single blade case).

vortex sheets are parallel to the free stream velocity. Except for the wake induced velocity, the model and its analysis are the same as in section 10–2. The free stream velocity U is constant here.

Assuming harmonic blade motion at frequency ω, the strength of the vortex sheet directly behind the blade ($n = 0$, see Fig. 10-10) is as before $\gamma_w = \bar{\gamma}_w e^{i\omega(t - x/U)}$. Since the sheets below the blade represent the successive spirals of a single vortex helix, the vorticity strength must be related to that of the first sheet. Consider a point x on a sheet. Moving one rotor revolution downstream must be equivalent to going directly downward to the next sheet at the same x. It follows that the vortex strength must be a function of the quantity $(x + n\Delta x)$, where n is the index of the wake sheets and Δx is the distance the wake is convected in a single revolution of the rotor:

$$\Delta x = 2\pi r = 2\pi(\Omega r)/\Omega = 2\pi U/\Omega$$

Then increasing x by Δx is indeed equivalent to increasing n by 1. Furthermore, since all the vorticity is convected downstream at the velocity U, the vortex strength in the sheets must also be a function of $t - x/U$. Assuming harmonic motion requires the time dependence of the wake strength to be $e^{i\omega t}$, and the wake structure is completely specified:

$$\gamma_{w_n} = \bar{\gamma}_w e^{i\omega(t - x/U - n2\pi/\Omega)}$$

$$= \bar{\gamma}_w e^{i\omega t} e^{-ikx/b} e^{-in2\pi\omega/\Omega}$$

where γ_{w_n} is the strength of the nth sheet. If ω/Ω is an integer, the loading is at a harmonic of the rotor speed. Since $e^{-in2\pi\omega/\Omega} = 1$ in that case, the vorticity in all the sheets is exactly in phase.

The analysis of the unsteady aerodynamics proceeds as in section 10–2. With the present wake model, the wake-induced velocity is

$$\lambda = \frac{1}{2\pi}\int_b^\infty \frac{\gamma_w d\xi}{x-\xi} + \sum_{n=1}^\infty \frac{1}{2\pi}\int_{-\infty}^\infty \frac{\gamma_{w_n}(x-\xi)}{(x-\xi)^2 + h^2n^2}\, d\xi$$

The second term is the additional contribution of the returning shed wake. Substituting for γ_{w_n} gives

$$\Delta\lambda = \sum_{n=1}^\infty \frac{1}{2\pi}\int_{-\infty}^\infty \frac{\gamma_{w_n}(x-\xi)}{(x-\xi)^2 + h^2n^2}\, d\xi$$

$$= \bar{\gamma}_w e^{i\omega t}\sum_{n=1}^\infty e^{-in2\pi\omega/\Omega}\frac{1}{2\pi}\int_{-\infty}^\infty \frac{e^{-ik\xi/b}(x-\xi)}{(x-\xi)^2 + h^2n^2}\, d\xi$$

$$= \bar{\gamma}_w e^{i\omega t}\sum_{n=1}^\infty \frac{i}{2} e^{-ikx/b} e^{-n[(kh/b) + i2\pi(\omega/\Omega)]}$$

$$= \frac{i}{2}\bar{\gamma}_w e^{i\omega t} e^{-ikx/b}\, W$$

where

$$W = \sum_{n=1}^\infty e^{-n[(kh/b) + i2\pi(\omega/\Omega)]} = \frac{1}{e^{kh/b} e^{i2\pi\omega/\Omega} - 1}$$

To evaluate the unsteady loads on the airfoil, the wake-induced velocity must be expanded as a series:

$$\lambda = \sum_{n=0}^\infty \lambda_n \cos n\theta$$

where $x = b\cos\theta$. Using the following expression for the Bessel function,

$$J_n(k) = \frac{i^n}{\pi}\int_0^{\pi} e^{-ik\cos\theta}\cos n\theta\, d\theta,$$

we obtain

$$\Delta\lambda_0 = \frac{1}{\pi}\int_0^{\pi}\Delta\lambda\, d\theta = \frac{i}{2}\bar{\gamma}_w e^{i\omega t} W J_0(k)$$

$$\Delta\lambda_1 = \frac{2}{\pi}\int_0^{\pi}\Delta\lambda\cos\theta\, d\theta = \bar{\gamma}_w e^{i\omega t} W J_1(k)$$

$$\Delta\lambda_2 = \frac{2}{\pi}\int_0^{\pi}\Delta\lambda\cos 2\theta\, d\theta = -i\bar{\gamma}_w e^{i\omega t} W J_2(k)$$

Then

$$\Delta(\lambda_0 + \tfrac{1}{2}\lambda_1) = \bar{\gamma}_w e^{i\omega t} W \tfrac{1}{2}\big(J_1(k) + iJ_0(k)\big)$$

$$\frac{b}{U}\frac{\partial}{\partial t}\Delta(\lambda_0 - \tfrac{1}{2}\lambda_2) = -\bar{\gamma}_w e^{i\omega t} W\frac{k}{2}\big(J_2(k) + J_0(k)\big)$$

$$= -\bar{\gamma}_w e^{i\omega t} W J_1(k)$$

The changes in the bound circulation and lift are

$$\Delta\Gamma = -2\pi b\,\Delta(\lambda_0 + \tfrac{1}{2}\lambda_1)$$

$$= -2\pi b\bar{\gamma}_w e^{i\omega t} W\tfrac{1}{2}\big(J_1(k) + iJ_0(k)\big)$$

$$\Delta L = -\rho U 2\pi b\left[\Delta(\lambda_0 + \tfrac{1}{2}\lambda_1) + \tfrac{1}{2}\frac{b}{U}\frac{\partial}{\partial t}\Delta(\lambda_0 - \tfrac{1}{2}\lambda_2)\right]$$

$$= -\rho U 2\pi b\bar{\gamma}_w e^{i\omega t} W (i/2) J_0(k)$$

Thus the total lift is now

$$L = L_Q + L_{NC} + \rho U\int_b^{\infty}\frac{b}{\sqrt{\xi^2 - b^2}}\gamma_w\, d\xi$$

$$- \rho U 2\pi b\bar{\gamma}_w e^{i\omega t} W (i/2) J_0$$

with L_Q and L_{NC} defined as in section 10–2. The total circulation is

$$\Gamma = \frac{L_Q}{\rho U} + \int_b^\infty \left(\sqrt{\frac{\xi + b}{\xi - b}} - 1 \right) \gamma_w \, d\xi - 2\pi b \bar{\gamma}_w e^{i\omega t} W \tfrac{1}{2}(J_1 + iJ_0)$$

$$= - \int_b^\infty \gamma_w \, d\xi$$

which gives

$$L_Q = -\rho U \int_b^\infty \sqrt{\frac{\xi + b}{\xi - b}} \gamma_w \, d\xi + \rho U 2\pi b \gamma_w e^{i\omega t} W \tfrac{1}{2}(J_1 + iJ_0)$$

On substituting $\gamma_w = \bar{\gamma}_w e^{i\omega(t - \xi/b)}$, the last expression gives $\bar{\gamma}_w$ in terms of L_Q, which may then be used to evaluate the circulatory lift in terms of L_Q. The result is

$$L = C' L_Q + L_{NC}$$

where

$$C'(k, \omega/\Omega, h) = 1 + \frac{-\int_1^\infty \frac{1}{\sqrt{\xi^2 - 1}} e^{-ik\xi} d\xi + \pi i J_0 W}{\int_1^\infty \sqrt{\frac{\xi + 1}{\xi - 1}} e^{-ik\xi} d\xi - \pi(J_1 + iJ_0)W}$$

$$= \frac{\int_1^\infty \frac{\xi}{\sqrt{\xi^2 - 1}} e^{-ik\xi} d\xi - \pi J_1 W}{\int_1^\infty \sqrt{\frac{\xi + 1}{\xi - 1}} e^{-ik\xi} d\xi - \pi(J_1 + iJ_0)W}$$

$$= \frac{H_1^{(2)}(k) + 2J_1(k)W}{H_1^{(2)}(k) + iH_0^{(2)}(k) + 2\left(J_1(k) + iJ_0(k)\right) W}$$

is Loewy's lift deficiency function. The only influence of the returning shed wake on the two-dimensional unsteady loads of an airfoil is in the lift deficiency function, with Loewy's function replacing Theodorsen's. The modification of the lift deficiency function by the returning shed wake is determined by the quantity W. For a single blade it is

$$W(kh/b,\ \omega/\Omega) = \frac{1}{e^{kh/b}\, e^{i2\pi(\omega/\Omega)} - 1}$$

As h approaches infinity, W approaches zero and hence Loewy's function C' reduces to the Theodorsen function $C(k)$. Besides the reduced frequency k, the rotor model also introduces the parameters h/b and ω/Ω. The wake spacing is given by h/b, and ω/Ω determines the relative phase of the vorticity in the successive wake sheets. When ω/Ω is equal to an integer, the strengths of all the sheets are exactly in phase. Note that it is only the fractional part of ω/Ω that is important.

Now consider the case of an N-bladed rotor. Again the two-dimensional model of the rotor wake is a series of parallel vortex sheets with vertical spacing h arrayed below the blade. Here only every Nth sheet is due to a given blade, however. Let n be the index of the rotor revolutions as above, and let $m = 0, 1, 2, \ldots, N\text{-}1$ be the blade index (see Fig. 10-11). Note that when $n = 0$ all the wake sheets should extend upstream to a blade. It is consistent with the two-dimensional model to extend these sheets upstream to infinity. The shed vorticity of sheets from a given blade (fixed m) must again be a function of $(x + n\Delta x) = (x + n2\pi U/\Omega)$. To determine the relation between the vorticity strength in sheets from different blades, assume that all the blades have the same motion, but that the motion of each blade leads to the following one by the time $\Delta t = \Delta\psi/\Omega$. Then moving directly down to the next sheet must be equivalent to moving downstream a distance $(\Delta x/N - U\Delta t)$, where $\Delta x/N$ is the spacing between the blades. Therefore the shed vorticity must also be a function of $x + m(\Delta x/N - U\Delta t)$. For harmonic blade motion at frequency ω, it follows that

$$\begin{aligned}\gamma_{w_{nm}} &= \bar{\gamma}_w\, e^{i\omega(t - x/U - n2\pi/\Omega + m\Delta\psi/\Omega\ -\ 2\pi m/N\Omega)} \\ &= \bar{\gamma}_w\, e^{i\omega t}\, e^{-ikx/b}\, e^{-i2\pi(\omega/\Omega)[n + (m/N)(1 - N\Delta\psi/2\pi)]}\end{aligned}$$

The wake-induced velocity for the N-bladed rotor is then

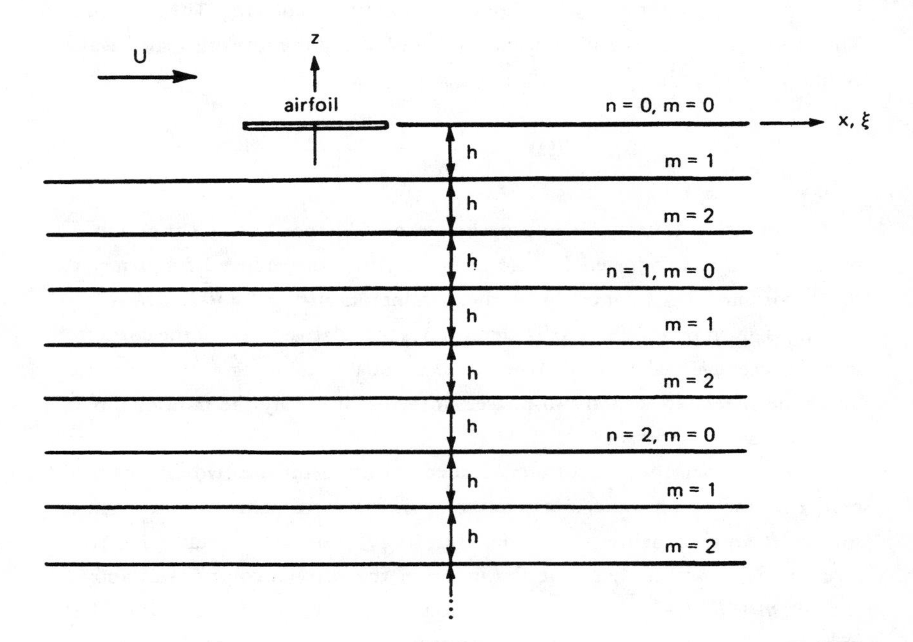

Figure 10-11 Two-dimensional wake model of an N-bladed rotor (three-bladed case shown).

$$\lambda = \frac{1}{2\pi}\int_b^\infty \frac{\gamma_w\, d\xi}{x-\xi} + \sum_{m=1}^{N-1} \frac{1}{2\pi}\int_{-\infty}^{\infty} \frac{\gamma_{w_{0m}}(x-\xi)}{(x-\xi)^2 + h^2 m^2}\, d\xi$$

$$+ \sum_{m=1}^{\infty}\sum_{m=0}^{N-1} \frac{1}{2\pi}\int_{-\infty}^{\infty} \frac{\gamma_{w_{nm}}(x-\xi)}{(x-\xi)^2 + h^2(Nn+m)^2}\, d\xi$$

Substituting for $\gamma_{w_{nm}}$ gives

$$\Delta\lambda = \sum_{n=0}^{\infty}\sum_{m=0}^{N-1}\left\{\frac{1}{2\pi}\,\bar{\gamma}_w\, e^{i\omega t}\, e^{-i2\pi(\omega/\Omega)[n + (m/N)(1 - N\Delta\psi/2\pi)]} \int_{-\infty}^{\infty} \frac{e^{-ik\xi/b}(x-\xi)\,d\xi}{(x-\xi)^2 + h^2(Nn+m)^2}\right\}$$

$$-\frac{1}{2\pi}\,\bar{\gamma}_w\, e^{i\omega t}\int_{-\infty}^{\infty} \frac{e^{-ik\xi/b}}{x-\xi}\, d\xi$$

or

$$\Delta\lambda = i\tfrac{1}{2}\bar{\gamma}_w e^{i\omega t} e^{-ikx/b} W$$

where

$$W = -1 + \sum_{n=0}^{\infty} \sum_{m=0}^{N-1} e^{-i2\pi(\omega/\Omega)[n + (m/N)(1 - N\Delta\psi/2\pi)]} e^{-kh(Nn+m)/b}$$

$$= \frac{1 + \sum_{m=1}^{N-1} \left(e^{kNh/b} e^{i2\pi(\omega/\Omega)}\right)^{1-m/N} e^{im\Delta\psi\omega/\Omega}}{e^{kNh/b} e^{i2\pi(\omega/\Omega)} - 1}$$

This is the same form for $\Delta\lambda$ as was obtained for the single blade. Then the same result for the unsteady loads, and in particular for the lift deficiency function C' is obtained. The number of blades influences the solution only through the function W appearing in C'.

The Fourier coordinate transformation (see Chapter 8) introduces degrees of freedom that describe the motion of the rotor as a whole. Each of the nonrotating modes (collective, cyclic, and reactionless) defines the relative motion of the N blades of the rotor, and hence the relationship between the shed vorticity in successive sheets of the wake. So for each nonrotating mode the function W, and then the lift deficiency function C', can be evaluated. In the collective mode the motions of all the blades are exactly in phase, and the only phase shift in the wake vorticity is due to the spacing between the blades. Then the time phase shift $\Delta\psi = 0$, for which

$$W = \frac{1}{e^{kh/b} e^{i2\pi\omega/N\Omega} - 1}$$

The collective mode thus is equivalent to a single-bladed rotor with wake spacing $h_e = h$ and

$$(\omega/\Omega)_e = \omega/N\Omega$$

The relative phase between the vorticity in successive wake sheets is determined by the two parameters $\Delta\psi$ and ω/Ω for the N-bladed rotor, but only by ω/Ω with a single blade. Therefore, all the wake sheets are in phase [$(\omega/\Omega)_e$ is an integer] only if the collective mode oscillation occurs at a frequency that is a multiple of N/rev. For the reactionless mode (the $N/2$

mode, which is present only with an even number of blades, as discussed in section 8-4.1) successive blades have identical motion except for opposite signs. The blades are then 180° out of phase, giving $\Delta\psi = \pi(\Omega/\omega)$ and

$$W = \frac{1}{e^{kh/b}\, e^{i2\pi(\omega/N\Omega + \frac{1}{2})} - 1}$$

Thus the reactionless mode is also equivalent to a single-bladed rotor, with $h_e = h$ and

$$(\omega/\Omega)_e = (\omega/N\Omega) + \tfrac{1}{2}$$

In the cyclic modes (the nc and ns degrees of freedom in general) the blade motion is the same at a given rotor azimuth angle, so $\Delta\psi = 2\pi/N$. Then

$$W = \frac{1 + e^{kNh/b}\, e^{i2\pi\omega/\Omega} \displaystyle\sum_{m=1}^{N-1} \left(e^{-kh/b} \right)^m}{e^{kNh/b}\, e^{i2\pi\omega/\Omega} - 1}$$

For the case of ω/Ω equal to an integer, this reduces to

$$W = \frac{1}{e^{kh/b} - 1}$$

So for the important case of oscillation at a harmonic of the rotor speed, the cyclic modes are equivalent to a single blade (with $h_e = h$) that also undergoes harmonic oscillation [$(\omega/\Omega)_e$ equal to an integer].

Next let us examine the behavior of Loewy's lift deficiency function,

$$C' = F' + iG' = \frac{H_1^{(2)} + 2J_1 W}{H_1^{(2)} + iH_0^{(2)} + 2(J_1 + iJ_0)W}$$

It has been shown that the N-bladed rotor case is equivalent to a single-bladed rotor, using the same wake spacing and a value of ω/Ω such that the successive wake sheets have the proper relative phase. It is sufficient therefore to consider the single-bladed rotor case, for which

$$W = \frac{1}{e^{kh/b}\, e^{i2\pi\,\omega/\Omega} - 1}$$

Figs. 10-12 to 10-14 show the magnitude and phase of C' for the cases of

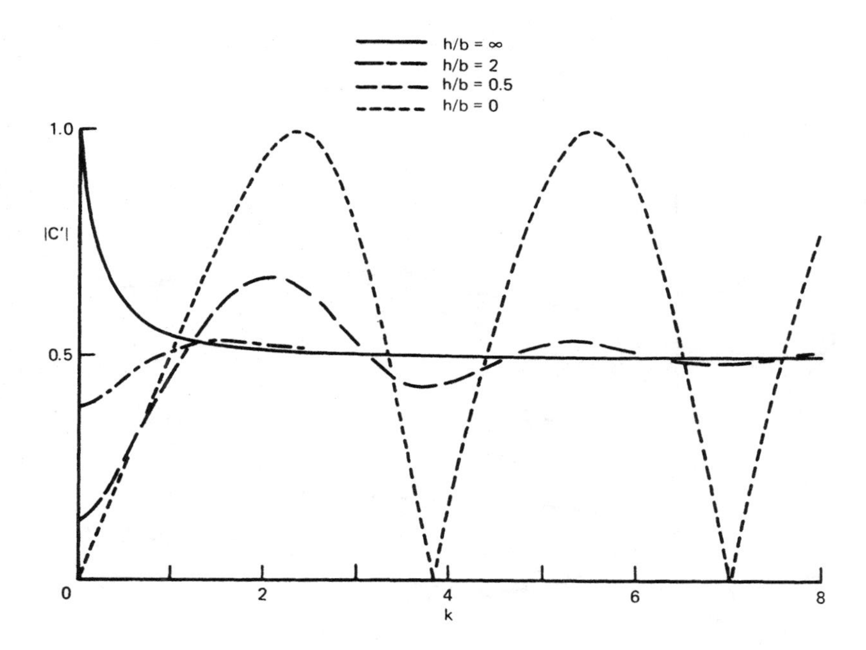

Figure 10-12 Magnitude of Loewy's lift deficiency function as a function of reduced frequency and wake spacing, for ω/Ω = integer.

ω/Ω = integer and ω/Ω = integer + ½. The limit as h approaches infinity is $W = 0$, and thus $C' = C(k)$, the Theodorsen function. The case $h = 0$ is not a physically realistic limit, but it does show the behavior of C' for small wake separation. With $h = 0$,

$$W = \frac{1}{e^{i2\pi\omega/\Omega} - 1}$$

Then when ω/Ω = integer,

$$C' = \frac{J_1}{J_1 + iJ_0}$$

and when ω/Ω = integer + ½,

$$C' = \frac{Y_1}{Y_1 + iY_0}$$

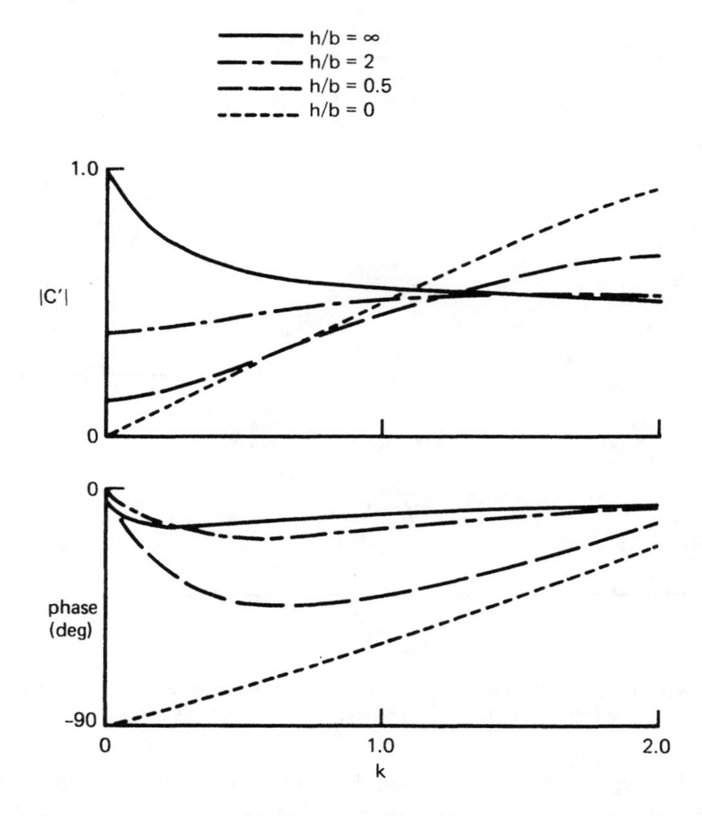

Figure 10-13 Magnitude and phase of Loewy's lift deficiency function as a function of reduced frequency and wake spacing, for ω/Ω = integer.

These Bessel functions give C' an oscillatory behavior at large k (see Fig. 10-12). The real part F' oscillates between 0 and 1, and the imaginary part between -0.5 and 0.5, with a period of π. Note that $|C'| = \sqrt{F'}$ for these cases, so the magnitude of C' goes to zero at certain values of k. In general, for large reduced frequency the lift deficiency function is

$$C' \sim \tfrac{1}{2}(1 + i e^{-kh/b} e^{-i2\pi\omega/\Omega} e^{i2k})$$

Hence the oscillatory behavior observed in Fig. 10-12 is a general result for large k. The period is π and the amplitude of the oscillation diminishes with increasing wake spacing h. For small reduced frequency the lift deficiency function is approximately

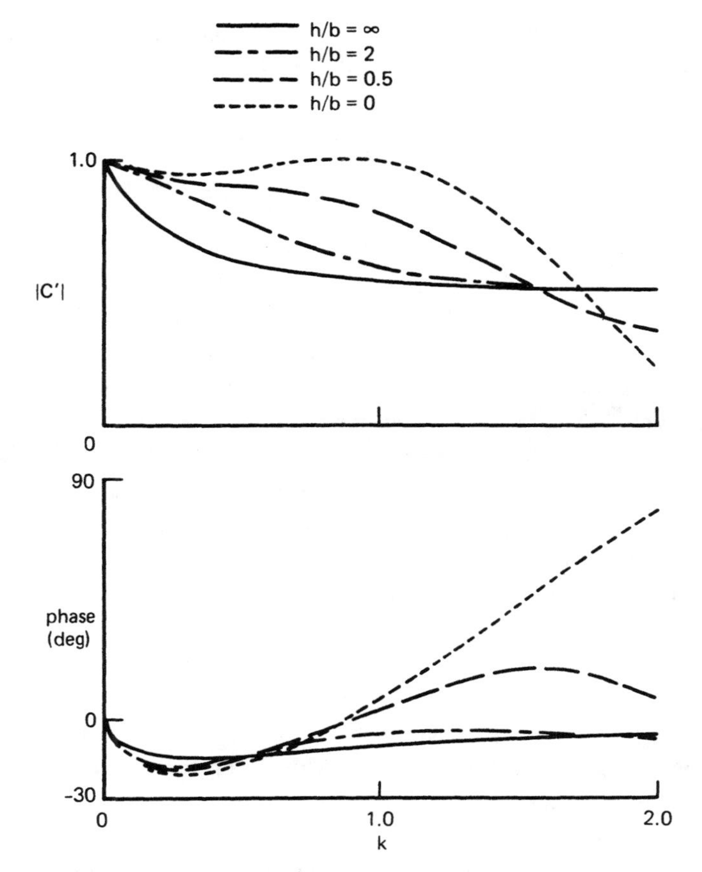

Figure 10-14 Magnitude and phase of Loewy's lift deficiency function as a function of reduced frequency and wake spacing, for ω/Ω - integer + ½.

$$C' \cong \frac{1 - i\frac{\pi}{2}k^2 W}{1 + \frac{\pi}{2}k - ik\left(\ln\frac{k}{2} + \gamma\right) + \left(1 - \frac{i}{2}k\right)\pi k W}$$

where γ is Euler's constant. If ω/Ω is not equal to an integer, W is of order 1 or smaller for all h; then to order k

$$C' \cong \frac{1}{1 + \frac{\pi}{2}k - ik\left(\ln\frac{k}{2} + \gamma\right) + \pi k W}$$

Note that this gives $C' = 1$ at $k = 0$, independent of the wake spacing h (see Fig. 10-14). For ω/Ω = integer, however, $W \cong b/kh$; so to order 1

$$C' \cong \frac{1}{1 + \pi k W} = \frac{1}{1 + \dfrac{\pi}{h/b}}$$

These results for small reduced frequency are of most interest for helicopter rotors. If ω/Ω is not an integer, there is only an order k correction of the Theodorsen function due to the returning shed wake. For oscillations at harmonics of the rotor speed (ω/Ω = integer), however, the shed wake reduces the lift deficiency function at low frequency to $C' = h/(h + \pi b)$. Fig. 10-13 shows that this low frequency result of order 1 is a good approximation even to $k = 0.5$ or so. Note that $C'(0) \neq 1$ now; in fact, $C'(0) = 0$ when $h = 0$, because of the returning shed wake. There is a substantial reduction of the unsteady loads in the case of oscillations at harmonics of the rotor speed, because the vorticity in successive wake sheets is exactly in phase. Important examples of such motion in rotor dynamics are cyclic pitch control and flapping, which give 1/rev motion in the rotating frame; and flutter instabilities with a natural frequency at n/rev. In summary, there are two major effects of the returning shed wake on the lift deficiency function. At high reduced frequencies C' has an oscillatory behavior, with minima near zero for small wake spacing. This large-k behavior is not very important for helicopter rotor aerodynamics. The other major effect occurs at low reduced frequency, with oscillation at a harmonic of the rotor speed. For that case there is a substantial reduction of C', and hence of the unsteady circulatory loads, at small wake spacing.

The wake spacing h/b is determined by the rate at which the helical vortex sheets of the rotor are convected downward. Using the mean induced velocity at the rotor disk for the convection velocity of the wake near the rotor, the distance the wake moves in a single rotor revolution is $Nh = v(2\pi/\Omega)$, or

$$\frac{h}{b} = \frac{v 2\pi}{\Omega N b} = \frac{4\lambda}{\sigma}$$

where λ is the inflow ratio and σ is the rotor solidity. The lift deficiency function for low-k, harmonic oscillations is then

$$C' \cong \frac{1}{1 + \dfrac{\pi\sigma}{4\lambda}}$$

Typically $\lambda \cong 0.07$ for the hovering helicopter rotor, which gives $h/b \cong 3$ or 4, and so $C' \cong 0.5$. Thus the reduction of the unsteady loads due to the returning shed wake can indeed be large, with serious consequences for rotor control, loads, and stability in the critical circumstances of low inflow and harmonic oscillations. The reduction of the circulatory lift will decrease the rotor response to collective and cyclic pitch control. It will also decrease the damping of the flap motion and the flapwise bending modes, therefore increasing the response of the blade vibration to harmonic airloads. The blade pitch damping moments are also influenced by the returning shed wake if the pitch axis is off the aerodynamic center.

Miller (1964a) considered the higher harmonic loading of the rotor in forward flight. He concluded that the nonuniform inflow in forward flight is primarily due to the geometry of the discrete tip vortices, with strength given by the mean rotor lift. It is therefore the trailed rather than the shed vorticity that is the dominant factor in the higher harmonic rotor blade loads in forward flight. Next in importance is the inflow variation due to the near shed wake. Miller found that for very low advance ratio the returning shed wake effects persist. However, by about $\mu = 0.2$ only the near shed wake effects, as given by Theodorsen's lift deficiency function, are observed.

Jones (1958) developed a theory for the unsteady aerodynamics of rotary wings that is essentially the same as Loewy's two-dimensional model for the case of a single blade. Timmen and van de Vooren (1957) analyzed the critical limiting case of zero wake spacing ($h = 0$). Piziali (1964) gives tables of Loewy's lift deficiency function C'.

Daughaday, Du Waldt, and Gates (1957) conducted an experimental investigation of the bending load amplification on model helicopter rotor blades with particular attention to the effect of the returning shed wake on the aerodynamic damping of the bending modes. The damping of a single flapping rotor blade in hover was obtained from forced response (to flap moments) and transient decay data. They found good correlation with the results of Loewy's theory. The predicted reduction in the damping at frequencies that are harmonics of the rotor speed, due to the reduction in the unsteady lift by C', was verified.

Ham, Moser, and Zvara (1958) measured the flap response of a two-bladed articulated rotor to oscillations of the collective pitch and to vertical oscillations of the hub. For low collective a significant influence of the returning shed wake was observed near harmonics of 2/rev (as expected for collective modes). At high collective the wake influence was negligible. Good agreement was obtained between the measured response and that calculated using Loewy's lift deficiency function. The magnitude of the flap response to collective showed the reduction near 2/rev as indicated by the theory. (The reduction of the lift reduces the flap moment due to control. Since the response of the blade at 2/rev is dominated by the inertia, the corresponding reduction in the flap damping has little effect.) The magnitude of the rotor response to 2/rev control is small, however, so the influence of the wake is most apparent in the phase, which was also verified by the experiment. The flap response to vertical oscillations of the hub showed a large increase in magnitude at harmonics of 2/rev, increasing with frequency. This was the direct result of the reduced flap damping. Excellent correlation was obtained with calculations using Loewy's theory. The effect of the wake decreased as collective increased, until the response was down to the quasistatic levels at $\theta_0 = 10^\circ$. It was also found that the increased response due to the shed wake at low collective persisted in forward flight (measurements were made to about $\mu = 0.2$).

Silveira and Brooks (1959) conducted an experimental study of the damping of the flapwise bending modes of a two-bladed hovering rotor at low pitch (and hence low inflow). They measured the response to flap moments and vertical oscillation of the hub, and obtained the damping from decay of the transient occurring when the external force was removed. The data show the decrease in the aerodynamic damping due to the shed wake, down to about zero at harmonics of 2/rev. Reasonable quantitative correlation was obtained with the results of Loewy's theory.

Additional literature concerned with the returning shed wake of the rotor blade: Ashley, Brunelle, and Moser (1958), Ashley, Moser, and Dugundji (1958), Brunelle (1958), Du Waldt, Gates, and Piziali (1959), Jones (1960, 1966), Jones and Rao (1970, 1971), Hammond and Pierce (1971), Anderson and Watts (1973, 1976), Bass, Johnson, and Unruh (1974), Murthy and Pierce (1976), Kunz (1977).

10–6 Approximate Solutions for Rotary Wing Unsteady Aerodynamics

10-6.1 Lifting-Line Approximation

The lifting-line approximation was examined in section 10–3 for the near shed wake in unsteady two-dimensional airfoil theory. Following Miller (1964a, 1964b), this analysis will now be extended to include the returning shed wake as well. Recall that the lifting-line approximation involves calculating the induced velocity at a single point on the chord, rather than using the distribution over the chord. Because the treatment required for the near shed wake in this approximation has already been derived, here the near shed wake will be treated correctly and the lifting-line approximation will be used only for the wake sheets below the airfoil. In section 10–5 an expression was obtained for the additional induced velocity due to the returning shed wake:

$$\Delta\lambda = i\tfrac{1}{2}\overline{\gamma}_w e^{i\omega t} e^{-ikx/b} W$$

Here $\Delta\lambda$ is evaluated at the quarter chord ($x = -b/2$) only:

$$\Delta\lambda = i\tfrac{1}{2}\overline{\gamma}_w e^{i\omega t} e^{i\frac{1}{2}k} W$$

Since basically the lifting-line approximation corresponds to low reduced frequency in the unsteady aerodynamics, the approximation $e^{i\frac{1}{2}k} \cong 1$ can also be used. The lift and bound circulation increments are then

$$\Delta L = -\rho U 2\pi b \overline{\gamma}_w e^{i\omega t} W \tfrac{1}{2} i$$

$$\Delta\Gamma = -2\pi b \overline{\gamma}_w e^{i\omega t} W \tfrac{1}{2} i$$

from which the lift deficiency function is obtained as

$$C' = \frac{H_1^{(2)}(k)}{H_1^{(2)}(k) + iH_0^{(2)}(k) + 2iW}$$

This result can be obtained from Loewy's function by using the order 1 approximations of the Bessel functions. Miller shows that this is a good approximation to Loewy's function at least up to $k = 0.5$ for the complete range of wake spacing. The greatest error is in the imaginary part of C' (hence the phase shift) at low h/b. Miller concludes that the lifting-line analysis is a satisfactory treatment for the returning shed wake as well as for the trailed wake. Only the near shed wake requires special treatment.

10-6.2 Two-Dimensional, Continuous Wake Approximation

Miller (1962a, 1964b) also considers a further approximation to Loewy's analysis, modeling the wake sheets below the blade by a continuous vorticity distribution. This model is analogous to the actuator disk model of the rotor (to be considered in section 10-6.3), and thus it serves to illustrate the connection between the continuous vorticity analysis for the rotor and the discrete wake model of Loewy's two-dimensional theory. Only the case of harmonic blade motion is considered, ω/Ω = integer. Since the loading on the blade is periodic, the vorticity strength in the wake is independent of the vertical position. Also assuming low reduced frequency, the variation of the induced velocity over the chord is neglected. The wake induced velocity at $x = 0$ on the airfoil is then

$$\lambda = -\frac{1}{2\pi}\int_{-\infty}^{0}\int_{-\infty}^{\infty}\frac{\gamma\xi}{\xi^2 + z^2}\,d\xi\,dz$$

where $\gamma(\xi,t)$ is the strength of the continuous vorticity distribution below the airfoil. Since each discrete wake sheet with strength γ_w has now been spread over the distance h, it follows that $\gamma = \gamma_w/h$. The shed wake strength is obtained from the time derivative of the bound circulation Γ. If sinusoidal motion at frequency ω is assumed,

$$\gamma = -\frac{1}{Uh}\frac{d}{dt}\Gamma(t - \xi/U) = -\frac{i\omega}{Uh}\overline{\Gamma}\,e^{i\omega(t-\xi/U)}$$

$$= -\frac{ik\overline{\Gamma}/b^2}{h/b}\,e^{-i\omega\xi/U}$$

With the wake convected downward by the mean inflow ratio λ_0, the wake spacing was found in section 10–5 to be $h/b = 4\lambda_0/\sigma$; so

$$\gamma = -\frac{ik\overline{\Gamma}/b^2}{4\lambda_0/\sigma}\,e^{-i\omega\xi/U}$$

After substituting for γ, the velocity induced by the shed wake is

$$\lambda = \frac{k\Gamma/b^2}{4\lambda_0/\sigma}\frac{1}{2\pi}\int_{-\infty}^{0}\int_{-\infty}^{\infty}\frac{\xi\, e^{-ik\xi/b}}{\xi^2+z^2}\,d\xi\, dz$$

$$= \frac{k\Gamma/b^2}{4\lambda_0/\sigma}\;\frac{1}{2k/b}$$

$$= \frac{\pi\sigma}{4\lambda_0}\;\frac{L}{\rho U 2\pi b}$$

Then from $L = L_Q - \rho U 2\pi b\lambda = L_Q - (\pi\sigma/4\lambda_0)L$, it follows that $L = C' L_Q$, where the lift deficiency function

$$C' = \frac{1}{1 + \dfrac{\pi\sigma}{4\lambda_0}}$$

This is exactly the low frequency limit of Loewy's function for the case of harmonic oscillation (see section 10–5). Miller shows that it is a good approximation to F' up to $k = 0.5$ at least, although the imaginary part is neglected entirely.

10-6.3 Rotary Wing Actuator Disk Model

Analyzing the unsteady aerodynamics of the rotor blade is difficult because of the complex structure of the rotor wake. So far only two-dimensional models have been considered. Following Miller (1962a, 1964b), this section develops an analysis for the rotating blade. Two assumptions are made to obtain a tractable mathematical problem. First, the actuator disk model is used, so that the distribution of vorticity in the wake is continuous rather than discrete. Secondly, only hover or vertical flight is considered, so that the wake geometry is axisymmetric. The analysis is an extension of classical vortex theory to the case of unsteady loading, and it produces an approximate lift deficiency function for the rotary wing.

We begin with the case of uniform disk loading and hence constant bound circulation over the blade span. Then the trailing vorticity is concentrated in root and tip vortices (see the discussion of vortex theory in section 2-7.2). Neglecting the contraction, the wake is a circular cylinder below the rotor disk. The tip vortex spirals form a continuous distribution of ring vortices on the surface of this cylinder. Continuity of vorticity requires a root vortex on the axis of the wake, and axial vorticity on the surface of the cylinder as well. Such axial vorticity does not contribute to the axial induced velocity at the rotor disk. Thus in the steady case only the ring vortices of strength γ contribute to the induced velocity. Integrating the Biot-Savart law over the surface of the cylinder gives the steady induced velocity at the point (r,ψ) on the rotor disk:

$$\begin{aligned}
\nu &= \frac{1}{4\pi}\int_{-\infty}^{0}\int_{0}^{2\pi} \frac{\gamma\left(R^2 - rR\cos(\psi-\psi^*)\right)d\psi^*\,dz}{\left(R^2 - 2Rr\cos(\psi-\psi^*) + r^2 + z^2\right)^{3/2}} \\
&= \frac{\gamma}{4\pi}\int_{0}^{2\pi} \frac{R^2 - rR\cos(\psi-\psi^*)}{R^2 - 2Rr\cos(\psi-\psi^*) + r^2}\,d\psi^* \\
&= \frac{\gamma}{4\pi}\int_{0}^{2\pi} \sum_{\ell=0}^{\infty}\left(\frac{r}{R}\right)^{\ell} \cos\ell(\psi-\psi^*)\,d\psi^* \\
&= \frac{\gamma}{2}
\end{aligned}$$

The strength of the ring vortex γ is obtained from the bound circulation Γ of N blades, spread over the vertical distance the wake is convected by ν in one revolution:

$$\gamma = \frac{N\Gamma}{\nu 2\pi/\Omega}$$

or in terms of the rotor thrust, $T = ½N\rho\Omega R^2\Gamma$, so $\gamma = T/\rho A\nu$. The induced velocity is then $\nu = \gamma/2 = T/2\rho A\nu$, or

$$\nu = \sqrt{T/2\rho A}$$

This is simply another vortex theory derivation of the actuator disk result for the induced velocity of a uniformly loaded rotor. For the present purposes, the form

$$\nu = \frac{N\Omega}{4\pi\nu}\Gamma$$

is more interesting. This result for steady loading and induced velocity will now be extended to the unsteady case.

Consider a rotor for which the bound circulation varies azimuthally but is still constant radially. Because only the case of harmonic loading is considered, the bound circulation is periodic and can be expressed as a Fourier series in ψ:

$$\Gamma = \sum_{n=0}^{\infty}\left(\Gamma_{nc}\cos n\psi + \Gamma_{ns}\sin n\psi\right)$$

The trailing vorticity again consists of vortex rings on the surface of the wake cylinder, with strength

$$\gamma = \sum_{n=0}^{\infty}\left(\gamma_{nc}\cos n\psi + \gamma_{ns}\sin n\psi\right)$$

The wake vorticity strength is periodic in ψ due to the bound circulation variation. The vorticity in all the wake spirals is in phase because harmonic loading is assumed, and therefore γ is independent of the axial distance z. The strength of the nth harmonic of γ is obtained from the nth harmonic of the bound circulation of all N blades, spread over the vertical distance that the wake is convected in one revolution:

$$\gamma_n = \frac{N\Omega}{2\pi v_0}\Gamma_n = \frac{T_n}{\rho A v_0}$$

Here v_0 is the mean induced velocity of the rotor. The azimuthal variation of the bound circulation means that the wake cylinder also contains radial shed vorticity. The strength of the shed vorticity is given by the time rate of change of the bound circulation of all N blades, again spread vertically over the distance the wake is convected in one revolution. The strength of the shed vorticity is thus $d\gamma/d\psi$, where γ is the strength of the trailing vortex rings; this relation is also required to account for the azimuthal change in the strength of the rings. The Fourier series for γ gives

$$\frac{d\gamma}{d\psi} = \sum_{n=1}^{\infty} n(-\gamma_{nc}\sin n\psi + \gamma_{ns}\cos n\psi)$$

The induced velocity at the rotor disk due to the nth harmonic of the trailing vorticity is then

$$v_{n_T} = \frac{1}{4\pi} \int_{-\infty}^{0} \int_{0}^{2\pi} \frac{(\gamma_{nc} \cos n\psi^* + \gamma_{ns} \sin n\psi^*)[R^2 - rR \cos(\psi - \psi^*)]\, d\psi^*\, dz}{[R^2 - 2Rr \cos(\psi - \psi^*) + r^2 + z^2]^{3/2}}$$

$$= \frac{1}{4\pi} \int_0^{2\pi} \left(\gamma_{nc} \cos n\psi^* + \gamma_{ns} \sin n\psi^* \right) \sum_{\ell=0}^{\infty} \left(\frac{r}{R}\right)^{\ell} \cos \ell(\psi - \psi^*)\, d\psi^*$$

$$= \frac{1}{4}\left(\frac{r}{R}\right)^n \left(\gamma_{nc} \cos n\psi + \gamma_{ns} \sin n\psi\right)$$

The induced velocity due to the nth harmonic of the shed vorticity is obtained by integrating the Biot-Savart law over the volume of the wake cylinder:

$$v_{n_S} = \frac{1}{4\pi} \int_0^{\infty} \int_0^{2\pi} \int_0^{R} \frac{d\gamma_n}{d\psi^*} \frac{r \sin(\psi - \psi^*) dr^*\, d\psi^*\, dz}{[r^2 + r^{*2} + z^2 - 2rr^* \cos(\psi - \psi^*)]^{3/2}}$$

$$= \frac{1}{4\pi} \int_0^{2\pi} \int_0^{R} \frac{d\gamma_n}{d\psi^*} \frac{r \sin(\psi - \psi^*)\, dr^*\, d\psi^*}{r^2 + r^{*2} - 2rr^* \cos(\psi - \psi^*)}$$

$$= \frac{1}{4\pi} \int_0^{2\pi} \frac{d\gamma_n}{d\psi^*} \sum_{\ell=1}^{\infty} \sin \ell(\psi - \psi^*) \left[\frac{1}{r} \int_0^{r} \left(\frac{r^*}{r}\right)^{\ell - 1} dr^* + \frac{1}{r} \int_r^{R} \left(\frac{r}{r^*}\right)^{\ell+1} dr^* \right] d\psi^*$$

$$= \frac{1}{4\pi} \int_0^{2\pi} n\left(-\gamma_{nc} \sin n\psi^* + \gamma_{ns} \cos n\psi^*\right) \sum_{\ell=1}^{\infty} \frac{1}{\ell}\left[2 - \left(\frac{r}{R}\right)^{\ell}\right] \sin \ell(\psi - \psi^*)\, d\psi^*$$

$$= \frac{1}{4}\left[2 - \left(\frac{r}{R}\right)^n\right] \left(\gamma_{nc} \cos n\psi + \gamma_{ns} \sin n\psi\right)$$

The total induced velocity is thus

$$v_n = v_{n_T} + v_{n_S} = \frac{\gamma_n}{2} = \frac{N\Omega}{4\pi v_0} \Gamma_n = \frac{T_n}{2\rho A v_0}$$

So the nth harmonic of the loading produces the nth harmonic of the induced velocity. Moreover, the constant spanwise bound circulation produces an induced velocity independent of r. This expression is a simple extension of the steady inflow result. Now the rotor thrust can be written $T_n = T_{nQ} + T_{nW}$, where T_{nQ} is the quasistatic thrust and T_{nW} is the thrust due to the harmonic inflow ν_n:

$$\begin{aligned} T_{nW} &= -N \int_0^R \tfrac{1}{2}\rho(\Omega r)c\, 2\pi \nu_n\, dr \\ &= -N\rho\Omega c \nu_n R^2 \frac{\pi}{2} \\ &= -\rho A \Omega R \frac{\pi}{2} \sigma \nu_n \\ &= -\frac{\sigma\pi}{4} \frac{\Omega R}{\nu_0} T_n \end{aligned}$$

So $T_n = T_{nQ} - (\sigma\pi/4\lambda_0)T_n$, which gives $T_n = C'T_{nQ}$ with the lift deficiency function

$$C' = \frac{1}{1 + \dfrac{\sigma\pi}{4\lambda_0}}$$

for the rotary wing. Note that C' is independent of the harmonic number. Remarkably, this lift deficiency function is identical to the function obtained using a two-dimensional, continuous wake model and is the low frequency approximation to Loewy's function for harmonic loading.

Miller (1964b) considered the case of a rotor with bound circulation that varies radially as well as azimuthally. In this case, vorticity is trailed into the wake from all along the blade span, not just at the tip and root. The result for the nth harmonic of the induced velocity is

$$\nu_n(r,\psi) = \frac{N\Omega}{4\pi\nu_0} \Gamma_n(r,\psi) = \frac{N}{\rho 4\pi \nu_0 r} L_n(r,\psi)$$

The induced velocity now varies radially, but it depends only on the local bound circulation of the blade. The total section lift is $L_n = L_{nQ} + L_{nW} = L_{nQ} - \tfrac{1}{2}\rho\Omega r c 2\pi\nu_n = L_{nQ} - (\sigma\pi/4\lambda_0)L_n$, or $L_n = C'L_{nQ}$ with

$$C' = \frac{1}{1 + \dfrac{\sigma\pi}{4\lambda_0}}$$

again. Since the lift deficiency function is independent of r as well as frequency, the loading can be integrated over the span to obtain $T_n = C'T_{n_Q}$ as for the case of constant bound circulation.

The harmonic blade loading can also be expressed in terms of the local disk loading of the actuator disk:

$$L_n = \frac{2\pi r}{N} \frac{dT_n}{dA}$$

Then the inflow perturbation due to the unsteady loading can be written

$$\delta\nu = \frac{dT/dA}{2\rho\nu_0}$$

This may be viewed as a differential form of momentum theory ($dT = 2\dot{m}\,\delta\nu$ where $\dot{m} = \rho\nu_0 dA$), applied to the time-varying load as well.

10-6.4 Perturbation Inflow Model for Rotor Unsteady Aerodynamics

The wake-induced velocities at the rotor disk play an important role in rotor unsteady aerodynamics, and therefore must be accounted for in determining both the periodic and transient blade loading. The relationship between the induced velocity and the unsteady loading is very complex, however. An elementary representation of the rotor unsteady aerodynamics useful for aeroelastic analyses can be obtained from the vortex theory results of the preceding section. For the hovering rotor the perturbation inflow $\delta\nu(r,\psi)$ at a point on the rotor disk was related to the perturbation of the local disk loading dT/dA by

$$\delta\nu = \frac{dT/dA}{2\rho\nu_0}$$

where ν_0 is the mean induced velocity. This result was derived for harmonic variation of the blade loading at frequency n/rev in the rotating frame (where n is a nonzero integer). Recall that this relation gives the low frequency approximation for the rotary wing lift deficiency function. When the result is viewed as the differential form of the induced velocity solution of vortex or momentum theory, the basic assumption is that the rotor forces vary slowly enough (compared to the wake response) for the actuator disk results to be applicable to the perturbation as well as the trim inflow velocities. A perturbation to the induced velocity of the following form will be considered:

$$\delta\lambda = \lambda + \lambda_x r\cos\psi + \lambda_y r\sin\psi$$

This inflow perturbation consists of a uniform component λ and components due to λ_x and λ_y that vary linearly over the rotor disk. These inflow components will be related to the net unsteady aerodynamic forces on the rotor, specifically the thrust C_T, pitching moment C_{M_y}, and rolling moment C_{M_x}.

Assuming a linear variation of the loading over the rotor disk, the pitch and roll moments give

$$\frac{dT}{dA} = -4\frac{\delta M_y}{RA}r\cos\psi + 4\frac{\delta M_x}{RA}r\sin\psi$$

The moments therefore correspond to 1/rev loading variations, and the unsteady aerodynamics result for $\delta\nu$ is applicable. Substituting for dT/dA gives a linearly varying inflow perturbation:

$$\delta\lambda = -\frac{2\delta C_{M_y}}{\lambda_0}r\cos\psi + \frac{2\delta C_{M_x}}{\lambda_0}r\sin\psi$$

This relation can be extended to forward flight following the usual approach of momentum theory (as in section 4-1.1). The mass flow through the differential disk area dA is determined by the resultant velocity through the disk: $\dot{m} = \rho(V^2 + \nu_0^2)^{1/2}dA$. Then $dT = 2\dot{m}\delta\nu$ gives

$$\delta\nu = \frac{dT/dA}{2\rho\sqrt{V^2 + \nu_0^2}}$$

The inflow perturbation due to the aerodynamic moments becomes

$$\delta\lambda = -\frac{2\delta C_{M_y}}{\sqrt{\mu^2 + \lambda_0^2}}r\cos\psi + \frac{2\delta C_{M_x}}{\sqrt{\mu^2 + \lambda_0^2}}r\sin\psi$$

For speeds above transition ($\mu > 0.10$ or 0.15) this is approximately

$$\delta\lambda = -\frac{2\delta C_{M_y}}{\mu} r\cos\psi + \frac{2\delta C_{M_x}}{\mu} r\sin\psi$$

which may also be obtained directly from the differential form of the induced velocity in forward flight ($\lambda_i \cong C_T/2\mu$).

Next consider the induced velocity perturbation due to transient changes of the rotor thrust. The relation $\delta v = (dT/dA)/2\rho v_0$ was found in section 10-6.3 to be equivalent to the small reduced frequency limit ($k = \omega b/\Omega r \ll 1$) for loading oscillations at harmonics of the rotor speed (ω/Ω equal to a nonzero integer). It is thus applicable to low frequency variations of the harmonic loading of a rotor blade, such as the 1/rev loading due to moment perturbations. The thrust changes, however, correspond to low frequency variation of the mean blade loading, so a different approach is required. For thrust variations it is possible to simply consider a perturbation form of the hover momentum theory result for the trim inflow, $\lambda_0 = (C_T/2)^{1/2}$. For low frequency thrust changes then,

$$\delta\lambda = \frac{\partial\lambda}{\partial C_T}\delta C_T = \frac{\delta C_T}{4\lambda_0}$$

Note that the harmonic loading result would give an inflow change twice as large, $\delta\lambda = \delta C_T/2\lambda_0$. The difference is due to the shed wake. The derivation in section 10-6.3 shows that for harmonic loading (such as the 1/rev variations due to moments) part of the inflow perturbation comes from the shed vorticity in the wake, and part from the trailed vorticity. The rotor thrust produces trailed wake vorticity only (the tip vortices), and hence only one-half the wake influence of the rotor hub moments. The extension to forward flight is based on the momentum theory result for the trim inflow,

$$\lambda_0 = \mu\tan\alpha + \frac{C_T}{2\sqrt{\mu^2+\lambda_0^2}}$$

(see section 4-1.1). Then

$$\delta\lambda = \frac{\partial\lambda}{\partial C_T}\delta C_T = \frac{\delta C_T}{2\sqrt{\mu^2+\lambda_0^2} + C_T\lambda_0/(\mu^2+\lambda_0^2)}$$

$$\cong \frac{\delta C_T}{2(\lambda_0 + \sqrt{\mu^2+\lambda_0^2})}$$

where the last approximation is valid for low inflow ratio.

In summary, the inflow perturbation due to the unsteady thrust and moment changes is

$$\delta\lambda = \frac{\delta C_T}{2(\lambda_0 + \sqrt{\mu^2 + \lambda_0^2})} - \frac{2\delta C_{M_y}}{\sqrt{\mu^2 + \lambda_0^2}} r \cos\psi + \frac{2\delta C_{M_x}}{\sqrt{\mu^2 + \lambda_0^2}} r \sin\psi$$

which may be written as a linear relation between the inflow perturbations and the unsteady aerodynamic hub reactions of the rotor:

$$\begin{pmatrix} \lambda \\ \lambda_x \\ \lambda_y \end{pmatrix} = \begin{bmatrix} \frac{1}{2(\lambda_0 + \sqrt{\mu^2 + \lambda_0^2})} & 0 & 0 \\ 0 & \frac{2}{\sqrt{\mu^2 + \lambda_0^2}} & 0 \\ 0 & 0 & \frac{2}{\sqrt{\mu^2 + \lambda_0^2}} \end{bmatrix} \begin{pmatrix} C_T \\ -C_{M_y} \\ C_{M_x} \end{pmatrix}$$

A more general relationship is $\vec{\lambda} = (\partial\lambda/\partial L)\vec{L}$, where $\partial\lambda/\partial L$ can be a full nine-element matrix if a means is available to evaluate all nine elements. In addition, a time lag can be included in this model for the rotor unsteady aerodynamics:

$$\tau\dot{\vec{\lambda}} + \vec{\lambda} = \frac{\partial\lambda}{\partial L}\vec{L}$$

The wake-induced velocity depends not only on the rotor loading, but also on the rotor velocity, which determines the mass flow through the disk. Hence there will also be inflow changes due to the rotor velocity changes. Consider again the momentum theory result for the trim induced velocity:

$$\lambda_i = \frac{C_T}{2\sqrt{\mu^2 + (\mu_z + \lambda_i)^2}}$$

where $\mu_z = \mu \tan\alpha$ is the component of the free stream velocity normal to the rotor disk. Then the perturbation of the uniform inflow due to changes in the in-plane and normal velocity components of the rotor is

$$\delta\lambda = -\frac{C_T/2}{(\mu^2 + \lambda_0^2)^{3/2} + \lambda_0 C_T/2}\left(\mu\delta\mu + \lambda_0\delta\mu_z\right)$$

In forward flight this result is approximately

$$\delta\lambda = -\frac{C_T}{2\mu^2}\,\delta\mu - \frac{C_T\lambda_0}{2\mu^3}\,\delta\mu_z$$

while in hover ($\mu = 0$ and $\lambda_0 = \sqrt{C_T/2}$) it reduces to

$$\delta\lambda = -\tfrac{1}{2}\delta\mu_z$$

Recall that in section 3—3 it was also found that for low climb speed the induced velocity is reduced by an amount equal to one-half the climb speed, because of the increased mass flow through the disk.

Finally, let us examine the lift deficiency function that is implied by the above results. Recall from section 5—3 that the aerodynamic pitch and roll moments on the rotor are

$$-\frac{2C_{M_y}}{\sigma a} = \frac{1}{\pi}\int_0^{2\pi}\int_0^1 \cos\psi\,\frac{F_z}{ac}\,r\,dr\,d\psi$$

$$\frac{2C_{M_x}}{\sigma a} = \frac{1}{\pi}\int_0^{2\pi}\int_0^1 \sin\psi\,\frac{F_z}{ac}\,r\,dr\,d\psi$$

The lift force due to the inflow perturbation is $\delta(F_z/ac) = -\tfrac{1}{2}u_T\delta\lambda = -\tfrac{1}{2}r^2(\lambda_x\cos\psi + \lambda_y\sin\psi)$, which gives

$$\begin{pmatrix} -\dfrac{2C_{M_y}}{\sigma a} \\[2ex] \dfrac{2C_{M_x}}{\sigma a}\end{pmatrix} = \begin{pmatrix} -\dfrac{2C_{M_y}}{\sigma a} \\[2ex] \dfrac{2C_{M_x}}{\sigma a}\end{pmatrix}_{QS} - \frac{1}{8}\begin{pmatrix}\lambda_x \\ \lambda_y\end{pmatrix}$$

(See also the derivation in Chapter 11, which shows that the factor $1/8$ is valid for the constant coefficient approximation in forward flight as well as for hover.) Here the subscript QS means the quasistatic loading, namely all the moments except those due to the wake-induced velocities. The present model for the rotor unsteady aerodynamics relates the induced velocity to the moment perturbation as follows:

$$\begin{pmatrix}\lambda_x \\ \lambda_y\end{pmatrix} = \frac{2}{\sqrt{\mu^2 + \lambda_0^2}}\begin{pmatrix}-C_{M_y} \\ C_{M_x}\end{pmatrix}$$

Eliminating the inflow from these two equations produces

$$\begin{pmatrix} -\dfrac{2C_{M_y}}{\sigma a} \\[2ex] \dfrac{2C_{M_x}}{\sigma a} \end{pmatrix} = C' \begin{pmatrix} -\dfrac{2C_{M_y}}{\sigma a} \\[2ex] \dfrac{2C_{M_x}}{\sigma a} \end{pmatrix}_{QS}$$

where the lift deficiency function C' is

$$C' = \frac{1}{1 + \dfrac{\sigma a}{8\sqrt{\mu^2 + \lambda_0^2}}}$$

The rotor wake reduces all the rotor aerodynamic hub moments by the factor C', which can significantly affect the dynamic behavior. For forward flight the lift deficiency function is

$$C' = \frac{1}{1 + \dfrac{\sigma a}{8\mu}}$$

and for hover

$$C' = \frac{1}{1 + \dfrac{\sigma a}{8\lambda_0}}$$

The hover result is again the low reduced frequency, harmonic loading limit of Loewy's lift deficiency function. Similarly, from section 5–3 the aerodynamic thrust of the rotor is

$$\frac{C_T}{\sigma a} = \frac{1}{2\pi} \int_0^{2\pi} \int_0^1 \frac{F_z}{ac}\, dr\, d\psi$$

Substituting $\delta(F_z/ac) = -\frac{1}{2} r \delta\lambda$ gives

$$\frac{C_T}{\sigma a} = \left(\frac{C_T}{\sigma a}\right)_{QS} - \frac{1}{4}\delta\lambda$$

(See also Chapter 11, which gives $T_\lambda = -\frac{1}{4}$ for hover and for the constant coefficient approximation in forward flight.) The uniform inflow perturbation is here related to the unsteady thrust by

$$\delta\lambda = \frac{C_T}{2\left(\lambda_0 + \sqrt{\mu^2 + \lambda_0^2}\right)}$$

Eliminating $\delta\lambda$ gives

$$\frac{C_T}{\sigma a} = C'\left(\frac{C_T}{\sigma a}\right)_{QS}$$

with the lift deficiency function

$$C' = \frac{1}{1 + \dfrac{\sigma a}{8(\lambda_0 + \sqrt{\mu^2 + \lambda_0^2})}}$$

For forward flight the same expression is obtained for C' as for the moments, but for thrust changes in hover,

$$C' = \frac{1}{1 + \dfrac{\sigma a}{16\lambda_0}}$$

The aerodynamic thrust is reduced by the wake effects, but the influence of the wake in hover is greater for moments because of the shed vorticity. Typical values of these lift deficiency functions are $C' \cong 0.8$ for forward flight; $C' \cong 0.7$ for thrust changes in hover; and $C' \cong 0.5$ for moment changes in hover.

Development of a model for the rotor unsteady aerodynamics that is more accurate than the elementary one derived here, and yet practical for routine applications, is still the subject of research. The literature available on this topic is listed in section 12-1.6.

10–7 Unsteady Airfoil Theory for the Rotary Wing

We shall now develop a theory of airfoil unsteady aerodynamics in a form suitable for rotary wing analyses. The two-dimensional airfoil theories of the preceding sections are the basis for this development, but there are a number of additional factors that must be considered for the rotor blade. Lifting-line theory separates the problem of calculating the unsteady loads on a three-dimensional wing into two parts: an inner problem involving the aerodynamic behavior of a two-dimensional airfoil, and an outer problem involving the calculation of the velocity induced by the rotor vortex wake at the blade section. The steady two-dimensional airfoil loads may be obtained as a function of angle of attack and Mach number from measured airfoil data in tabular form. For the unsteady section loads (below stall), thin airfoil theory results can be used. The rotary wing wake has a very complex structure. The rotation of the wing produces interlocking helical vortex sheets behind each blade that are distorted by the self-induced velocities in the wake and exhibit concentrated vorticity in the rolled-up tip vortices. An analytical solution of the outer problem for the induced velocity at the rotor blade section is not possible without a drastic simplification of the wake model, such as the use of an actuator disk representation of the rotor. A practical, accurate evaluation of the nonuniform induced velocity thus requires a numerical calculation (for a detailed discussion, see Chapter 13). Therefore, it is not the intention here to obtain a relation between induced velocity and the blade loading, which would lead to a lift deficiency function. Instead, the unsteady airfoil theory must be left in terms of the induced velocity. For the lifting-line analysis it is also desirable to calculate the induced velocity only at a single chordwise point. The preceding investigations of the lifting-line approximation for the two-dimensional airfoil establish the manner in which this must be done to maintain an accurate representation of the wake effects. A special treatment is required only for the near shed wake. Furthermore, the unsteady aerodynamic theory for a rotary wing must account for reverse flow and radial flow effects. Corrections will also be included for the real flow effects on the blade lift curve slope and aerodynamic center.

Consider the rotor blade shown in Fig. 10-15. The blade chord is c (semichord b), with the leading and trailing edges at $x = x_L$ and $x = x_T$ respectively. In thin wing theory the blade is defined by a surface a distance $z_b(r, x, \psi)$ above the disk plane. Since forward flight is considered, there is a time-varying free stream velocity normal to the blade (u_T), and also a radial velocity (u_R). The rotor azimuth ψ is the dimensionless time variable. The boundary condition that determines the wing loading is that there be no flow through the surface of the blade: $w_a + w_z = 0$, where w_a is the

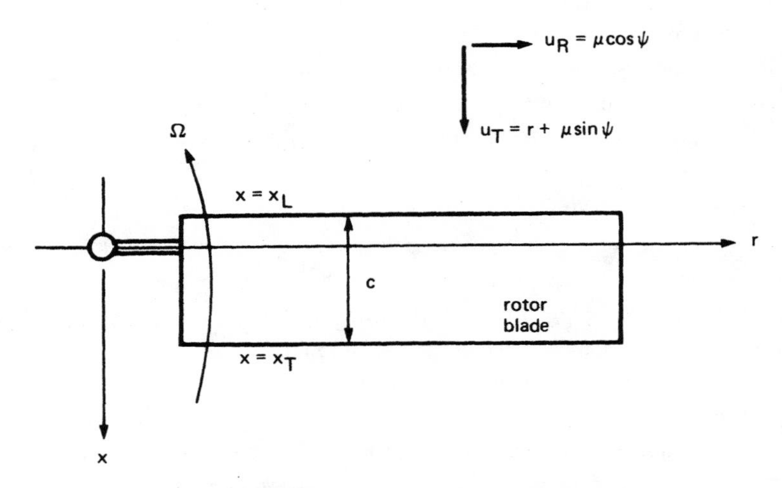

Figure 10-15 Blade geometry and velocity in the rotating frame.

vertical velocity of the airfoil surface and w_z is the air velocity induced by the wing and wake vorticity. The blade velocity is given by the total derivation of the surface deflection:

$$w_a = \frac{D}{D\psi} z_b = \left[\frac{\partial}{\partial\psi} + \left(r + \mu \sin\psi\right)\frac{\partial}{\partial x} + \left(-x + \mu\cos\psi\right)\frac{\partial}{\partial r}\right] z_b$$

The induced velocity can be written $w_z = w_{z_b} + \lambda$, where w_{z_b} is the downwash due to the vorticity representing the wing surface, and λ is the wake-induced downwash.

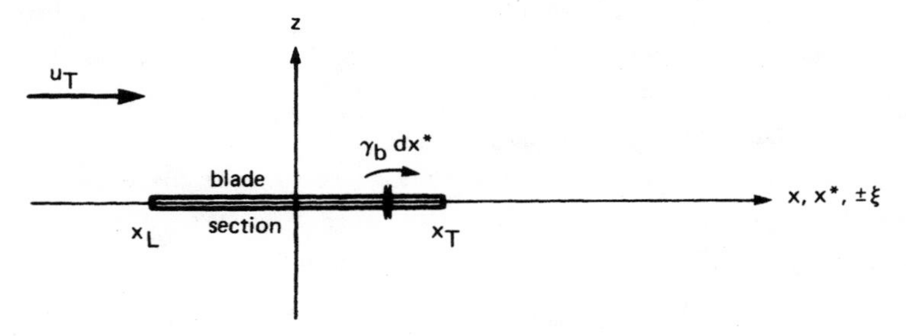

Figure 10-16 Thin airfoil representation of a rotor blade section.

In the lifting-line analysis, the velocity w_{z_b} induced by the wing loading is found by considering only the two-dimensional blade section (Fig. 10-16).

Let γ_b be the strength of the vorticity representing the airfoil. Integrating over the chord gives

$$w_{z_b} = \frac{1}{2\pi} \int_{x_L}^{x_T} \frac{\gamma_b}{x - x^*} dx^*$$

The boundary condition thus becomes

$$-\frac{1}{2\pi} \int_{x_L}^{x_T} \frac{\gamma_b}{x - x^*} dx^* = w_a + \lambda$$

which is an integral equation for γ_b and hence for the aerodynamic loading, given the blade motion w_a and the wake-induced velocity λ. The Kutta condition of finite velocity at the trailing edge ($\gamma_b = 0$) must also be satisfied. In forward flight, the inboard sections of the rotary wing pass through the reverse flow region. In the reverse flow region it is necessary to apply the Kutta condition at the geometric leading edge of the blade, $x = x_L$. To account for reversed flow, the chordwise variable ξ will be defined by $x = (\pm\xi - a)b$, where the upper sign is used for normal flow and the lower sign for reverse flow ($\pm = \text{sign}(u_T)$). Then $\xi = -1$ at the aerodynamic leading edge and $\xi = 1$ at the aerodynamic trailing edge always. Here b is the blade semichord and ab is the distance the r-axis is behind the midchord (for example, $a = -\frac{1}{2}$ with the r-axis at the quarter chord). The integral equation for γ_b is then

$$-\frac{1}{2\pi} \int_{-1}^{1} \frac{\gamma_b}{\xi - \xi^*} d\xi^* = w_a + \lambda$$

which with the requirement $\gamma_b = 0$ at $\xi = 1$ inverts to

$$\gamma_b = \pm \frac{2}{\pi} \sqrt{\frac{1-\xi}{1+\xi}} \int_{-1}^{1} \sqrt{\frac{1+\xi^*}{1-\xi^*}} \frac{w_a - \lambda}{\xi - \xi^*} d\xi^*$$

Next, the induced velocity λ is expanded as a Fourier series over the blade chord:

$$\lambda = \sum_{n=0}^{\infty} \lambda_n \cos n\theta$$

where $\xi = \cos\theta$. The blade velocity is written as a mean value plus a term linear in x: $w_a = -(A + Bx)$. For example, assume that the blade motion is described by vertical deflection of the elastic axis z_0, and nose-up pitch about the elastic axis θ, so that

$$z_b = z_0(r,\psi) - x\theta(r,\psi)$$

It follows from the definition of w_a that for this case

$$A = -\dot{z}_0 + (r + \mu\sin\psi)\theta - \mu\cos\psi\, z_0'$$

$$B = \dot{\theta} + z_0' + \mu\cos\psi\,\theta'.$$

Hence A gives the angle of attack of the section times the free stream velocity, $(A - \lambda) = u_T\alpha$, while B is the effective pitch rate of the blade. After substituting for w_a and λ, the solution for γ_b integrates to

$$\gamma_b = \pm 2\sqrt{\frac{1-\xi}{1+\xi}}\left[A \pm Bb\left(\xi + 1 \mp a\right)\right] \mp 2\sum_{n=0}^{\infty}\lambda_n f_n(\theta)$$

where f_n is the Glauert series:

$$f_n = \begin{cases} \tan\theta/2 & n = 0 \\ \sin n\theta & n \geqslant 1 \end{cases}$$

This solution can be written as a term $\gamma_{b_{NC}}$, which satisfies the boundary condition but gives no bound circulation, and a term γ_{b_C}, which does not affect the boundary condition but makes the sum γ_b satisfy the Kutta condition: $\gamma_b = \gamma_{b_{NC}} + \gamma_{b_C}$, where

$$\gamma_{b_C} = \pm\frac{2}{\sqrt{1-\xi^2}}\left[A - \left(\lambda_0 + \tfrac{1}{2}\lambda_1\right) \pm bB\left(\tfrac{1}{2} \mp a\right)\right]$$

The total bound circulation of the airfoil is then

$$\Gamma = \int_{x_L}^{x_T}\gamma_b\,dx$$

$$= \pm 2\pi b\left[A - \left(\lambda_0 + \tfrac{1}{2}\lambda_1\right) \pm bB\left(\tfrac{1}{2} \mp a\right)\right]$$

The other integrals of γ_b that will be required for the section loading are:

$$\Gamma^{(1)} = \int_{x_L}^{x_T} x\gamma_b\,dx = 2\pi b^2\Big[-(\tfrac{1}{2} \pm a)\big(A - (\lambda_0 + \tfrac{1}{2}\lambda_1)\big) \pm Ba^2 b - \tfrac{1}{4}(\lambda_1 + \lambda_2)\Big]$$

$$\Gamma^{(1)}_{NC} = \int_{x_L}^{x_T} x\gamma_{b_{NC}}\,dx = 2\pi b^2\Big[-\tfrac{1}{2}\big(A - (\lambda_0 + \tfrac{1}{2}\lambda_1)\big) + \tfrac{1}{2}abB - \tfrac{1}{4}(\lambda_1 + \lambda_2)\Big]$$

$$\Gamma^{(2)}_{NC} = \int_{x_L}^{x_T} x^2\gamma_{b_{NC}}\,dx = 2\pi b^3\Big[a\big(A - (\lambda_0 + \tfrac{1}{2}\lambda_1)\big) - bB\Big(\frac{1}{8} + a^2\Big) + \frac{a}{2}\big(\lambda_1 + \lambda_2\big) \pm \frac{1}{8}\big(\lambda_1 - \lambda_3\big)\Big]$$

$$\Gamma^{(3)}_{NC} = \int_{x_L}^{x_T} x^3\gamma_{b_{NC}}\,dx = 2\pi b^4\Big[-\Big(\frac{3}{8} + \frac{3}{2}a^2\Big)\big(A - (\lambda_0 + \tfrac{1}{2}\lambda_1)\big) + abB\Big(\frac{3}{4} + \frac{3}{2}a^2\Big) - \frac{3}{4}\big(a \pm \tfrac{1}{2}\big)^2\big(\lambda_1 + \lambda_2\big) \pm \frac{3}{8}a\lambda_2 - \frac{1}{8}\big(1 \mp a\big)\lambda_3 - \frac{1}{16}\lambda_4\Big]$$

The linearized form of Kelvin's equation leads to the following result for the differential pressure on the blade surface:

$$-\frac{\Delta p}{\rho} = \frac{D}{D\psi}\Delta\phi = \Big[\frac{\partial}{\partial\psi} + \big(r + \mu\sin\psi\big)\frac{\partial}{\partial x} + \big(-x + \mu\cos\psi\big)\frac{\partial}{\partial r}\Big]\Delta\phi$$

The velocity potential ϕ is related to the airfoil vorticity strength by $\Delta\phi = \int_{x_\ell}^{x} \gamma_b\,dx$ (see section 10–2); hence

$$-\frac{\Delta p}{\rho} = \big(r + \mu\sin\psi\big)\gamma_b + \frac{\partial}{\partial\psi}\int_{x_\ell}^{x}\gamma_{b_{NC}}\,dx + \big(-x + \mu\cos\psi\big)\frac{\partial}{\partial r}\int_{x_\ell}^{x}\gamma_{b_{NC}}\,dx$$

Integrating the pressure difference over the chord gives the section lift and moment, $L = \int_{x_L}^{x_T} (-\Delta p)dx$ and $M = \int_{x_L}^{x_T} (-\Delta p)x\,dx$, where L is positive upward and M is the nose-up moment about the r-axis. Substituting for Δp and using dimensionless quantities (so that the air density ρ can be omitted), we obtain

$$L = (r+\mu\sin\psi)\Gamma - \frac{\partial}{\partial\psi}\Gamma_{NC}^{(1)} - \mu\cos\psi\frac{\partial}{\partial r}\Gamma_{NC}^{(1)} + \frac{1}{2}\frac{\partial}{\partial r}\Gamma_{NC}^{(2)}$$

$$M = -(r+\mu\sin\psi)\Gamma^{(1)} + \frac{1}{2}\frac{\partial}{\partial\psi}\Gamma_{NC}^{(2)} + \frac{1}{2}\mu\cos\psi\frac{\partial}{\partial r}\Gamma_{NC}^{(2)} - \frac{1}{3}\frac{\partial}{\partial r}\Gamma_{NC}^{(3)}$$

With the expressions for $\Gamma^{(n)}$ and $\Gamma_{NC}^{(n)}$ given above, this is the required solution for the unsteady loads on the rotary wing. The first two terms in L and M are the circulatory and noncirculatory loads of thin airfoil theory. The last two terms are due to the radial flow. To lowest order the radial flow contribution to the lift is

$$\Delta L = \mu\cos\psi\frac{\partial}{\partial r}\left[\pi b^2(A-\lambda)\right]$$

which is just the slender body result for a wing of span $2b$. Thus the radial flow terms arise from the blade acting as a low aspect ratio wing with velocity $u_R = \mu\cos\psi$. These slender body terms are zero in hover, and in general are not an important factor in the rotor unsteady aerodynamics; they are, however, of the same order (in the chord c) as some of the noncirculatory lift and moment terms.

For lifting-line theory the wake-induced velocity is calculated only at a single point on the chord. Thus only the term λ_0 will be retained in the analysis (and written simply as λ). The chord c is a small quantity for high aspect ratio blades (in dimensional terms, the chord-to-radius ratio c/R is small). Therefore we shall simplify the results by retaining terms in the lift only up to order c^2, and terms in the moment up to order c^3. The chord c rather then the semichord b will be used now, and the tangential and radial blade velocities will be written as $(r+\mu\sin\psi) = u_T$ and $\mu\cos\psi = u_R$. Finally, write $w = A - \lambda + Bx$ for the upwash at the point x on the blade, including the contributions from both the blade motion and the wake-induced velocity. It follows that

$$\frac{\Gamma}{2\pi c} = \pm \frac{1}{2} w|_{\frac{3}{4}c}$$

$$\frac{\Gamma^{(1)}}{2\pi c} = -\frac{c}{2}\left(\pm \tfrac{1}{2} + a\right)\frac{\Gamma}{2\pi c} \pm \frac{c^2}{32} B$$

$$\frac{\Gamma^{(1)}_{NC}}{2\pi c} = -\frac{c}{8} w|_{\frac{1}{2}c}$$

$$\frac{\Gamma^{(2)}_{NC}}{2\pi c} = -ac\frac{\Gamma^{(1)}_{NC}}{2\pi c} - \frac{c^3}{128} B$$

and so the section loads are

$$\frac{L}{2\pi c} = \frac{1}{2}|u_T| w|_{\frac{3}{4}c} + \frac{c}{8}\left(\dot{w} + u_R w'\right)$$

$$\frac{M}{2\pi c} = \frac{c}{2}\left(\pm \tfrac{1}{2} + a\right)\frac{L}{2\pi c} \mp \frac{c^2}{32}\left(u_T B + \dot{w} + u_R w'\right)$$

Here the loads have been normalized using the chord and the theoretical lift curve slope $c_{\ell\alpha} = 2\pi$. Note that $w|_{\frac{3}{4}c}$ is the downwash at the three-quarter chord relative to the actual flow, and hence at the geometric quarter chord in reverse flow. Frequently only the lowest order term is retained for the lift, but it is still necessary to account for the noncirculatory lift terms in the moment. Thus the above result can also be written as

$$\frac{L}{2\pi c} = \frac{1}{2}|u_T| w + \frac{c}{8}\left(\dot{w} + u_R w'\right) + \frac{c}{4}\left(\tfrac{1}{2} \mp a\right) u_T B$$

$$\frac{M}{2\pi c} = \frac{c}{2}\left(\pm \tfrac{1}{2} + a\right)\frac{1}{2}|u_T| w \mp \frac{c^2}{8} a^2 u_T B + \frac{c^2}{16} a\left(\dot{w} + u_R w'\right)$$

where here w is the upwash velocity along the blade chord, with no terms of order c; and B is the gradient of the upwash along the chord. To lowest order the lift is $L/2\pi c = \tfrac{1}{2}|u_T| u_T \alpha$, as derived in section 5–20. An important term in the aerodynamic moment is the pitch damping. Since B has a term $\dot{\theta}$ and w has a term $u_T\theta$, the total moment about the pitch axis (at $x = ab$) is

$$\frac{\partial M}{\partial \dot{\theta}} = 2\pi \frac{c^3}{8} u_T a\left(\tfrac{1}{2} \mp a\right)$$

Note that with the pitch axis at the quarter chord ($a = -\tfrac{1}{2}$) the pitch damping is zero in the reverse flow region. The $u_R w'$ terms in the lift and moment are the slender body forces due to radial flow along the blade.

Thin airfoil theory predicts that the section lift curve slope is $c_{\ell_\alpha} = 2\pi$, and that the aerodynamic center is at the quarter chord. It is therefore necessary to correct the unsteady aerodynamic theory results to account for the lift curve slope and location of the aerodynamic center of the real airfoil. The lift curve slope correction requires multiplying the lift and moment by the ratio $a/2\pi$, where a is the two-dimensional lift curve slope of the real airfoil section. Typically $a = 5.7$ is used for helicopter blades in incompressible flow. (The parameter a has also been used for the pitch axis location in thin airfoil theory, but we are about to replace that notation.) Thin airfoil theory places the aerodynamic center at a distance $-b(\tfrac{1}{2} + a)$ behind the pitch axis in normal flow, and a distance $-b(-\tfrac{1}{2} + a)$ behind the pitch axis in reversed flow. Let x_A be the distance the real aerodynamic center of the airfoil is behind the pitch axis. It is still assumed that the aerodynamic center is shifted by $c/2$ in reverse flow, so the effective aerodynamic center offset is defined as

$$x_{A_e} = \begin{cases} x_A & \text{in normal flow} \\ x_A + (c/2) & \text{in reverse flow} \end{cases}$$

Thus we may substitute x_{A_e} for $-b(\pm\tfrac{1}{2} + a)$ in the thin airfoil theory results and then use the real airfoil data for x_A. Incorporating these corrections for the lift curve slope and aerodynamic center gives the unsteady aerodynamic lift and moment about the pitch axis for the rotary wing:

$$\frac{L}{ac} = \frac{1}{2}|u_T|w + \frac{c}{4}u_T B\left(1 \pm 2\frac{x_{A_e}}{c}\right) + \frac{c}{8}\left(\dot{w} + u_R w'\right)$$

$$\frac{M}{ac} = -x_{A_e}\frac{1}{2}|u_T|w \mp \frac{c^2}{32}u_T B\left(1 \pm 4\frac{x_{A_e}}{c}\right)^2$$
$$\mp \frac{c^2}{32}\left(\dot{w} + u_R w'\right)\left(1 \pm 4\frac{x_{A_e}}{c}\right).$$

In these expressions u_T and u_R are respectively the tangential and radial velocities seen by the blade; w is the upwash at the section, for example $w = u_T\theta - u_P$ (with no terms of order c); and B is the gradient of the upwash along the chord, such as produced by pitch rate. The upper part of the double sign is for normal flow and the lower is for reverse flow. Radial flow effects are included in the slender body pressure terms (from the radial derivative w') and in the contributions to the upwash w. The time derivative $\dot{w}$ includes terms due to the time-varying free stream. Corrections for real flow effects on the lift curve slope and aerodynamic center have been included. Finally, the shed and trailed wake effects are included, by the induced velocity contribution to w. The induced velocity is calculated at a single point, with the near shed wake treated as discussed in section 10–3.

10–8 Vortex-Induced Velocity

The direct consequence of the lift on a wing in three-dimensional or unsteady flow is a wake of trailed and shed vorticity. This wake vorticity in turn induces a velocity at the wing surface that has a major influence on the loading. The calculation of the wake-induced velocity is therefore an important part of rotary wing aerodynamic analyses. Considering both accuracy and efficiency, the calculation of the wake-induced velocity is best accomplished for the helicopter rotor by modeling the wake as a series of discrete elements. For each vortex element in the wake the induced velocity is evaluated by analytical expressions, and then the total induced velocity is obtained by summing the contributions from all elements. The tip vortex, which is the dominant portion of the rotor wake, is well represented by a connected series of straight-line vortex segments. The inboard shed and trailed vorticity is much less important to the induced velocity calculation, and as a result a more approximate model may be used that includes simplifications ranging from neglecting the inboard wake entirely to using a vortex lattice representation or vortex sheets.

Calculations of the rotor nonuniform inflow are thus based on the induced velocity due to a discrete element of the vortex wake. We conclude this chapter with a derivation of the velocity induced by line and sheet vortex elements. The case of a straight, infinite line vortex is considered first, in order to examine the general characteristics of the vortex flow field.

The finite-strength line vortex is an idealization in which a finite amount of vorticity is concentrated into a line of infinitesimal cross-section. There is a singularity at such a vortex line, with the induced velocity increasing as the inverse of the distance from the line. In the real fluid, viscosity eliminates this singularity by diffusing the vorticity into a tube of small but finite cross-section radius, called the vortex core. The maximum induced velocity occurs at some distance from the center of the line vortex, which will be defined as the core radius. Because the rotor blade often passes very close to the tip vortices from preceding blades, the vortex core has a significant role in the wake-induced velocity of the rotor, and it must be included in the representation of the wake vorticity. The tip vortex core radius is typically around 10% of the blade chord. Measurements of the core size are not very extensive, however, particularly for rotating wings.

10-8.1 Straight, Infinite Line Vortex

Consider a straight line vortex of inifnite extent and strength Γ (Fig. 10-17). The induced velocity is required at the point P. The axis system is

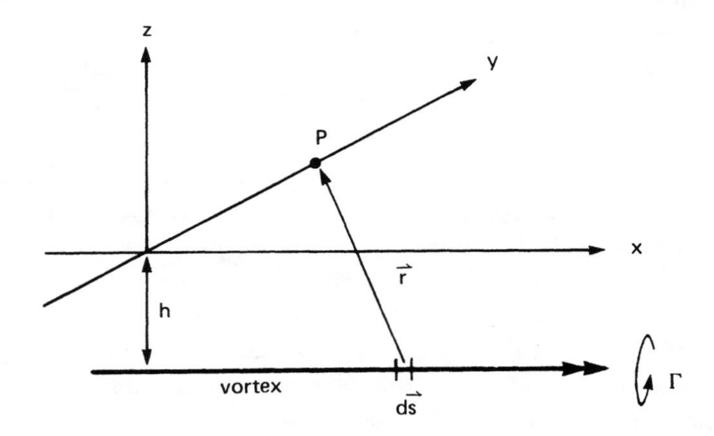

Figure 10-17 Geometry of a straight, infinite line vortex.

chosen so that the x-axis is parallel to the vortex and the point P is on the y-axis. Since the vortex is a distance h below the x-y plane, the point P is at a distance $(y^2 + h^2)^{1/2}$ from the vortex. We shall examine the behavior of the induced velocities along the y-axis, which is representative of the position of a rotor blade encountering a tip vortex from a preceding blade on the advancing side of the disk in forward flight.

The Biot-Savart law gives the induced velocity as

$$\vec{v} = -\frac{\Gamma}{4\pi}\int \frac{\vec{r} \times d\vec{s}}{r^3}$$

where $\vec{r}$ is the vector from the vortex element $\Gamma d\vec{s}$ to the point P:

$$\vec{r} = -x\vec{i} + y\vec{j} + h\vec{k}$$

and $d\vec{s} = \vec{i}\,dx$. The integral is over the length of the vortex. Thus

$$\vec{v} = -\frac{\Gamma}{4\pi}\int_{-\infty}^{\infty} \frac{-y\vec{k} + x\vec{j}}{(x^2 + y^2 + h^2)^{3/2}}\,dx$$

$$= \frac{\Gamma}{2\pi}\,\frac{-h\vec{j} + y\vec{k}}{y^2 + h^2}$$

When the y-axis is viewed as the blade span and the z-axis as normal to the blade, the vortex-induced upwash is

$$w = \vec{k}\cdot\vec{v} = \frac{\Gamma}{2\pi}\,\frac{y}{y^2 + h^2}$$

which is characterized by peaks at $y = \pm h$ of magnitude $w_{max} = \Gamma/4\pi h$ (see Fig. 10-18). This normal velocity will induce a loading on the blade with a similar radial variation. The radial component of the induced velocity is

$$u = \vec{j}\cdot\vec{v} = \frac{\Gamma}{2\pi}\,\frac{-h}{y^2 + h^2}$$

which has a peak at $y = 0$ (Fig. 10-18).

For $h = 0$ the induced velocity is $\vec{v} = \vec{k}\,\Gamma/2\pi y$, which is singular at the vortex line ($y = 0$). Thus for small h and y it is necessary to include a vortex core in the wake model in order to obtain physically realistic calculations of the induced velocity. A simple model for the core is solid body rotation inside the core of radius r_c, and potential flow (the above result) outside the core. Then for $h = 0$

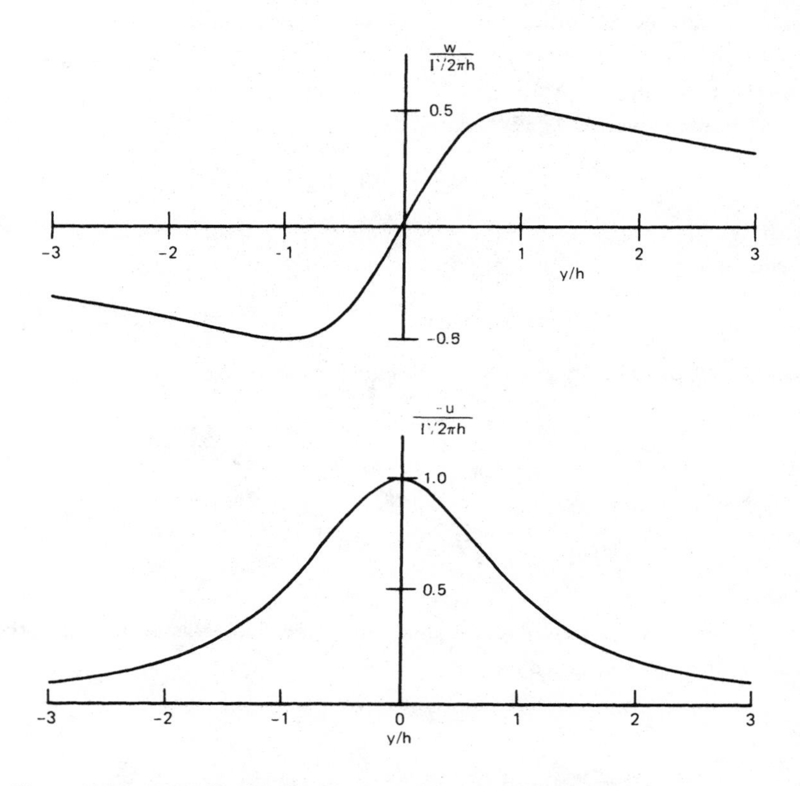

Figure 10-18 Normal and radial velocity induced by a straight infinite line vortex.

$$w = \begin{cases} \dfrac{\Gamma}{2\pi y} & y > r_c \\[2ex] \dfrac{\Gamma y}{2\pi r_c^{\,2}} & y < r_c \end{cases}$$

or more generally

$$\vec{v} = \begin{cases} \dfrac{\Gamma}{2\pi}\,\dfrac{-h\vec{j} + y\vec{k}}{y^2 + h^2} & y^2 + h^2 > r_c^2 \\[2ex] \dfrac{\Gamma}{2\pi}\,\dfrac{-h\vec{j} + y\vec{k}}{r_c^{\,2}} & y^2 + h^2 < r_c^2 \end{cases}$$

The vorticity distribution for solid body rotation inside the core is

$$\zeta = \frac{1}{y}\frac{d}{dy}yw = \begin{cases} \Gamma/\pi r_c^2 & y < r_c \\ 0 & y > r_c \end{cases}$$

So the vorticity is uniform inside the core, and zero outside. The peaks in the normal component of the induced velocity are now

$$w_{max} = \begin{cases} \dfrac{\Gamma}{4\pi h} \quad \text{at } y = \pm h, \quad \text{for } h > r_c/\sqrt{2} \\ \dfrac{\Gamma}{2\pi r_c}\sqrt{1 - (h/r_c)^2} \text{ at } y = \pm\sqrt{r_c^2 - h^2}, \text{ for } h < r_c/\sqrt{2} \end{cases}$$

Outside the core the peaks are still at $y = \pm h$ with the potential flow values (the edges of the core reach $y = \pm h$ for $h = r_c/\sqrt{2}$). Inside the core the peak velocity occurs on the core boundary. The vortex core limits the maximum induced velocity to $\Gamma/2\pi r_c$, which is achieved at $h = 0$.

Confining all the vorticity within the vortex core radius (defined as the point of maximum tangential velocity), as with the solid body rotation, is not a very good model for real vortices, however. Since the tangential velocity at a given radius from the vortex is determined only by the vorticity inside that radius, having some of the vorticity outside the core radius r_c means that the peak induced velocity will be less than $\Gamma/2\pi r_c$. It is found that the maximum tangential velocity of real vortices is much less than $\Gamma/2\pi r_c$ (where r_c is the location of the peak), implying that a substantial fraction of the vorticity is outside of the core radius. Scully (1975) suggests the following model for the circulation distribution of a real vortex:

$$\gamma = \Gamma \frac{r^2}{r^2 + r_c^2}$$

based on measured velocity distributions of vortices from nonrotating wings. The corresponding vorticity distribution is

$$\zeta = \frac{1}{2\pi r}\frac{d\gamma}{dr} = \frac{\Gamma}{\pi r_c^2}\frac{1}{[1 + (r/r_c)^2]^2}$$

(here r is the distance from the vortex line). Hence half the vorticity is inside the core radius, and half outside the core. This vorticity distribution gives the induced velocity for the straight infinite line vortex as

$$\vec{v} = \frac{\Gamma}{2\pi} \frac{-h\vec{j} + y\vec{k}}{y^2 + h^2 + r_c^2}$$

Note that the core effect here is equivalent to moving the vortex farther away, to an equivalent distance $h_{eq} = (h^2 + r_c^2)^{1/2}$; such a simple correction can be useful for analytical work. The peak downwash is now

$$w_{max} = \frac{\Gamma}{4\pi\sqrt{h^2 + r_c^2}}$$

at $$y = \pm\sqrt{h^2 + r_c^2}\,.$$

The maximum tangential velocity (at $r = r_c$) is now $\Gamma/4\pi r_c$, which is one-half the value obtained using solid body rotation inside the vortex core. This reduction of the peak velocity due to the distributed vorticity is based on observation of the measured vortex velocity distributions. Fig. 10-19 compares the magnitude and position of the peak downwash velocity as a function of the vortex position h for the three cases considered: the Biot-Savart law (no core), concentrated vorticity inside the core radius (solid body rotation), and distributed vorticity (Scully's expression). The effect of the distributed vorticity is quite significant. For application to other vortex elements, Scully suggests using the factor $r^2/(r^2 + r_c^2)$ to correct for effects of the vortex core on the induced velocity. He also suggests a value of $r_c = 0.0025R$ for the core radius of rotor tip vortices. It should be noted that the experimental data for vortex core properties is limited, particularly for rotary wings. Improvements to the core model may well be necessary when more data is available.

10-8.2 Finite-Length Vortex Line Element

A finite-length vortex line segment is the most useful element in modeling the rotor wake for nonuniform inflow calculations. A connected series of straight line segments can represent the tip vortex spirals, and even the inboard shed and trailed vorticity can be modeled with line elements (using

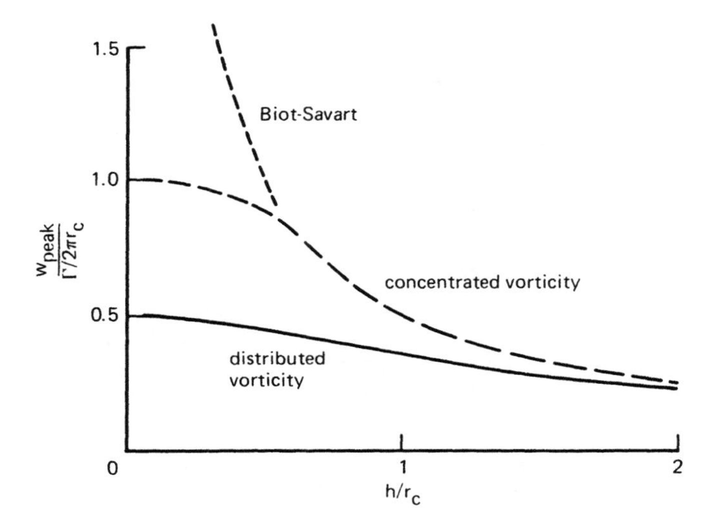

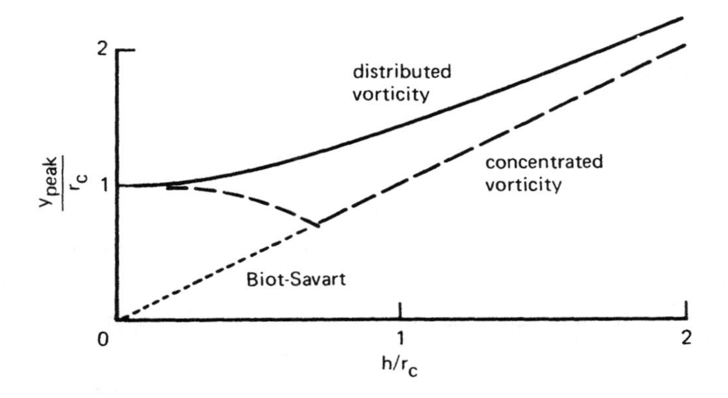

Figure 10-19 Magnitude and position of peak downwash velocity for three vortex core models: Biot-Savart (no core); concentrated vorticity; and distributed vorticity.

a large core to smooth out the singularity of the line). An expression is required for the velocity induced by a finite-length, constant-strength vortex line element at an arbitrary point in the flow field, including the effect of the vortex core.

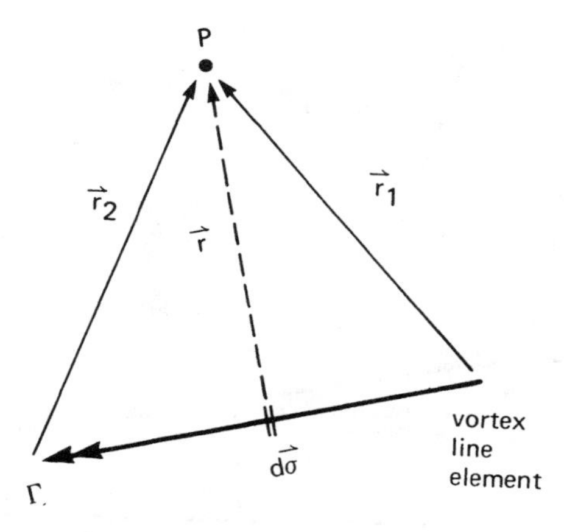

Figure 10-20 Finite-length vortex line element.

Consider the straight vortex line segment of length s and strength Γ shown in Fig. 10-20. The induced velocity is required at the point P, defined by the position vectors $\vec{r}_1$ and $\vec{r}_2$ from the ends of the segment. The vectors $\vec{r}_1$ and $\vec{r}_2$ can be in any convenient coordinate frame. The Biot-Savart law gives the induced velocity due to this line segment:

$$\Delta\vec{v}_{.} = -\frac{1}{4\pi}\int\frac{\Gamma\vec{r}\times d\vec{\sigma}}{r^3}$$

where $\vec{r}$ is the vector from the element $d\vec{\sigma}$ on the segment to the point P, and $r = |\vec{r}|$. Now write $\vec{r} = \vec{r}_m - \sigma\hat{e}$, where $\vec{r}_m$ is the minimum distance from the vortex line (including its extension beyond the end points of the segment) to the point P, and $\hat{e}$ is the unit vector in the direction of the vortex:

$$\vec{r}_m = \frac{\vec{r}_1(r_2^2 - \vec{r}_1\cdot\vec{r}_2) + \vec{r}_2(r_1^2 - \vec{r}_1\cdot\vec{r}_2)}{s^2}$$

$$\hat{e} = \frac{\vec{r}_1 - \vec{r}_2}{s}$$

with the length of the vortex element given by

$$s^2 = |\vec{r}_1 - \vec{r}_2|^2 = r_1^2 + r_2^2 - 2\vec{r}_1\cdot\vec{r}_2$$

The coordinate σ is measured from s_1 to s_2 along the vortex segment, where

$$s_1 = \frac{\vec{r}_1 \cdot \vec{r}_2 - r_1^2}{s}$$

$$s_2 = \frac{r_2^2 - \vec{r}_1 \cdot \vec{r}_2}{s}$$

Note that $\vec{r}_m$ and $\hat{e}$ are perpendicular. It follows that

$$\Delta\vec{v} = \frac{\Gamma}{4\pi} \vec{r}_1 \times \vec{r}_2 \int_{s_1}^{s_2} \frac{d\sigma}{s(r_m^2 + \sigma^2)^{3/2}}$$

$$= \frac{\Gamma}{4\pi} \vec{r}_1 \times \vec{r}_2 \frac{s_2 r_1 - s_1 r_2}{s r_m^2 r_1 r_2}$$

Substituting for s_1, s_2, s, and r_m gives the velocity induced by this vortex element:

$$\Delta\vec{v} = \frac{\Gamma}{4\pi} \vec{r}_1 \times \vec{r}_2 \frac{(r_1 + r_2)(r_1 r_2 - \vec{r}_1 \cdot \vec{r}_2)}{r_1 r_2 [r_1^2 r_2^2 - (\vec{r}_1 \cdot \vec{r}_2)^2]}$$

$$= \frac{\Gamma}{4\pi} \vec{r}_1 \times \vec{r}_2 \left(\frac{1}{r_1} + \frac{1}{r_2}\right) \frac{1}{r_1 r_2 + \vec{r}_1 \cdot \vec{r}_2}$$

Without a vortex core, this result is singular as the vortex segment is approached. Following Scully (1975), the influence of the vortex core is accounted for by multiplying the induced velocity by the factor

$$\frac{r_m^2}{r_m^2 + r_c^2} = \frac{r_1^2 r_2^2 - (\vec{r}_1 \cdot \vec{r}_2)^2}{r_1^2 r_2^2 - (\vec{r}_1 \cdot \vec{r}_2)^2 + r_c^2 s^2}$$

Here r_m is the minimum distance from the vortex line to the point P, and r_c is the vortex core radius. Then the velocity induced by the finite-length vortex line element is

$$\Delta\vec{v} = \frac{\Gamma}{4\pi} \vec{r}_1 \times \vec{r}_2 \frac{(r_1 + r_2)(1 - \vec{r}_1 \cdot \vec{r}_2 / r_1 r_2)}{r_1^2 r_2^2 - (\vec{r}_1 \cdot \vec{r}_2)^2 + r_c^2 (r_1^2 + r_2^2 - 2\vec{r}_1 \cdot \vec{r}_2)}$$

10-8.3 Rectangular Vortex Sheet

The inboard trailed and shed vorticity of the rotor is distributed on helical surfaces behind each blade. For the far wake a vortex lattice representation of this vorticity is satisfactory, and perhaps even for near the rotor blades if a large effective core is used to limit the induced velocities close to the line elements. The vortex lattice representation of the wake is computationally the most economical, but there are cases where a vortex sheet element must be used for accurate calculation of the inflow. For example, a vortex sheet is needed for the near shed wake of the blade, and sometimes for close passages of the inboard wake from preceding blades. One sheet element for which it is practical to integrate the Biot-Savart law is the planar, rectangular sheet.

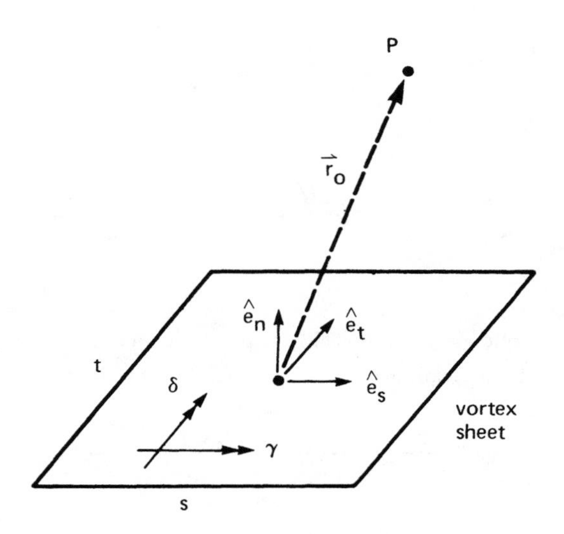

Figure 10-21 Rectangular vortex sheet.

Consider the rectangular vortex sheet with sides s and t shown in Fig. 10-21. The induced velocity is required at an arbitrary point P in the flow field, defined by the vector $\vec{r}_0$ from the center of the sheet. The orientation of the sheet is defined by the orthogonal unit vectors $\hat{e}_s$ and $\hat{e}_t$ parallel to the sides of the sheet, and the normal $\hat{e}_n = \hat{e}_s \times \hat{e}_t$. The vorticity strength is δ in the $\hat{e}_t$ direction and γ in the $\hat{e}_s$ direction. Linear variation of δ and γ along the directions $\hat{e}_t$ and $\hat{e}_s$ respectively is allowed. The minimum distance from P to the sheet or its extension is r_m:

$$\vec{r}_m = \vec{r}_0 \cdot \hat{e}_n \hat{e}_n.$$

The vector $\vec{r}_m$ is perpendicular to the plane of the sheet and intersects it at a point M. A coordinate system (σ,τ) will be used on the sheet plane, with origin at M so that the center of the sheet is at

$$\sigma = s_0 = -\vec{r}_0 \cdot \hat{e}_s$$

$$\tau = t_0 = -\vec{r}_0 \cdot \hat{e}_t$$

The edges of the rectangle are then given by $\sigma = s_0 \pm s/2$ and $\tau = t_0 \pm t/2$. The distance from a point on the sheet to P is

$$\vec{r} = \vec{r}_m - \sigma \hat{e}_s - \tau \hat{e}_t$$

The linearly varying vorticity of the vortex sheet is

$$\vec{\omega} = \gamma \hat{e}_s + \delta \hat{e}_t = (\gamma_m + \sigma \gamma_s)\hat{e}_s + (\delta_m + \tau \delta_t)\hat{e}_t$$

Note that conservation of vorticity requires that $\partial\gamma/\partial\sigma = -\partial\delta/\partial\tau$ or $\gamma_s = -\delta_t$.

Now the Biot-Savart law gives the induced velocity of the vortex sheet:

$$\vec{v} = -\frac{1}{4\pi}\int \frac{\vec{r} \times \vec{\omega}}{r^3}\, dA$$

$$= -\frac{1}{4\pi}\int_{t_0-(t/2)}^{t_0+(t/2)} \int_{s_0-(s/2)}^{s_0+(s/2)} \frac{r_m \gamma \hat{e}_t - r_m \delta \hat{e}_s + (\tau\gamma - \sigma\delta)\hat{e}_n}{(r_m^2 + \sigma^2 + \tau^2)^{3/2}}\, d\sigma\, d\tau$$

$$= -\frac{1}{4\pi}\Big[(\gamma_m \hat{e}_t - \delta_m \hat{e}_s)I_1 + (-r_m \delta_t \hat{e}_s + \gamma_m \hat{e}_n)I_2$$

$$+ (r_m \gamma_s \hat{e}_t - \delta_m \hat{e}_n)I_3 + \hat{e}_n(\gamma_s - \delta_t)I_4\Big] \Bigg|_{t_0-(t/2)}^{t_0+(t/2)} \Bigg|_{s_0-(s/2)}^{s_0+(s/2)}$$

where

$$I_1 = \iint \frac{r_m}{r^3}\, d\sigma\, d\tau = \tan^{-1} \frac{\sigma\tau}{r_m r}$$

$$I_2 = \iint \frac{\tau}{r^3}\, d\sigma d\tau = \ln(r - \sigma)$$

$$I_3 = \iint \frac{\sigma}{r^3}\, d\sigma d\tau = \ln(r - \tau)$$

$$I_4 = \iint \frac{\sigma\tau}{r^3}\, d\sigma d\tau = -r\,.$$

Evaluating the integrals at the four corners of the rectangle determines $\Delta\vec{v}$. This expression is more complicated than the one for the vortex line segment, and hence its use significantly increases the computation required to calculate the nonuniform inflow.

The basic induced velocity of a vortex sheet is given by the arc-tangent term, which produces for the point P approaching the surface

$$\Delta\vec{v} \to \pm \tfrac{1}{2}(\delta\,\hat{e}_s - \gamma\hat{e}_t)$$

where the plus sign is for just above the sheet, and the minus sign is for just below. There is a velocity jump across the sheet that is equal to the vorticity strength. For a point approaching an edge, such as $s_0 = 0$ and $t_0 = t/2$, the velocity is

$$\Delta\vec{v} \to \frac{1}{2\pi}\,\gamma_m\,\hat{e}_n \ln r_m$$

Thus the logarithmic terms give the singularity at the edge of a vortex sheet. At the side edges of the rotor wake there is a large velocity normal to the sheet and due to the trailed vorticity; this velocity is responsible for the roll-up of the tip vortices. Elsewhere in the wake the logarithmic singularity is due to the discreteness of the wake model. Representing a curved sheet by a series of flat panels introduces infinite curvature where the edges join. Moreover, whenever rectangular elements are used to model the wake helix, there will always be gaps or overlap between the trailing edge of one sheet and the leading edge of the next. It is these approximations to the actual wake structure that are responsible for the singular behavior of the induced velocity. If the sheets joined smoothly with no gaps and finite curvature the logarithmic terms of matching edges would cancel. Thus it is not appropriate to handle the side edge singularity of the rectangular vortex sheet by introducing a vortex core. Probably the best procedure is to simply avoid the singularity by arranging the geometry of the sheets in the wake so that the induced velocity is not calculated near an edge.

Chapter 11

ROTARY WING AERODYNAMICS II

The present chapter derives the aerodynamic forces required to complete the differential equations describing the helicopter rotor blade motion. In Chapter 9 the inertial and structural terms of the equations of motion were derived, and the net aerodynamic forces required were defined in terms of integrals of the section forces and pitch moment over the blade span. Here these aerodynamic forcing terms will be expressed as functions of the blade and shaft motion and the rotor loading. Solutions of the resulting differential equations for a number of fundamental rotor problems are given in Chapter 12. Blade element theory is the basis of the rotor aerodynamic model, so the section loads depend on the aerodynamic environment at that radial station alone, with the wake influence given by the induced velocity. For the blade torsion dynamics in particular it is necessary to consider the unsteady aerodynamic effects discussed in Chapter 10. The forces on the blade section will be derived for the general case of large pitch and inflow, but it will be necessary to assume small angles in order to obtain analytical expressions for the aerodynamic coefficients. For the same reason the effects of stall, compressibility, and reverse flow will be neglected. Such approximations are satisfactory except for extreme operating conditions of the rotor. The small angle assumption is usually very good for the low disk loading helicopter rotor, and compressibility can be accounted for in a rough fashion by using the lift curve slope at the Mach number of a representative blade radius. The neglect of reverse flow restricts the model to advance ratios below about $\mu = 0.5$, which covers the speed range of most current designs. When the aerodynamic coefficients are to be evaluated numerically, a more complex aerodynamic model is easily implemented. The linear differential equations describing the rotor dynamics are completed by expressing the aerodynamic forces in terms of the perturbation blade motion. The analysis begins with a derivation of the forces acting on the blade section.

11–1 Section Aerodynamics

Consider the air velocity and aerodynamic forces at the rotor blade section, as shown in Fig. 11-1. A hub plane reference axis system is used for the aerodynamic analysis. The hub plane is fixed relative to the shaft,

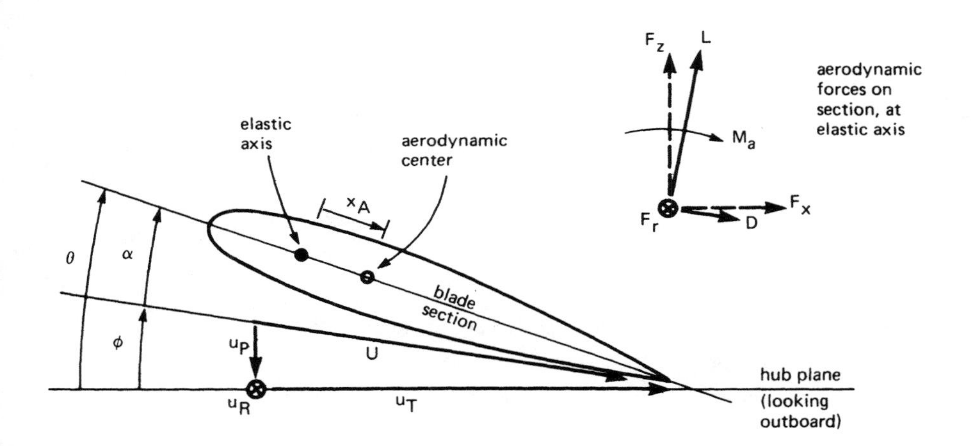

Figure 11-1 Rotor blade section aerodynamics.

and thus is tilted and displaced by the shaft motion. The pitch angle θ is measured from the reference plane. The velocities u_T, u_P, and u_R are the components of the air velocity seen by the blade, resolved in the hub plane axis system. The tangential velocity u_T is in the hub plane, positive in the blade drag direction; the radial velocity u_R is positive when directed radially outward; and the perpendicular velocity u_P is normal to the reference plane, positive when directed down through the disk. The resultant velocity in the section is $U = (u_T^2 + u_P^2)^{1/2}$ and the inflow ratio is $\phi = \tan^{-1} u_P/u_T$. Then the blade section angle of attack is $\alpha = \theta - \phi$. The aerodynamic lift and drag forces (L and D) are respectively normal to and parallel to the resultant velocity U. F_x and F_z are the components of the section lift and drag resolved into the hub plane axes. The radial force F_r is positive outward (in the same direction as u_R); it consists of a radial drag force, and an in-plane component of the blade lift due to flapwise bending of the blade. The section aerodynamic moment at the elastic axis is M_a, defined positive in the nose-upward direction. The aerodynamic center of the section is a distance x_A behind the elastic axis.

The blade lift and drag forces may be written in terms of the section coefficients:

$$L = \tfrac{1}{2}\rho U^2 c c_\ell$$

$$D = \tfrac{1}{2}\rho U^2 c c_d$$

where ρ is the air density and c is the rotor chord. Dimensionless quantities will be used from this point on in the analysis, so the air density ρ is omitted. The section lift and drag coefficients, $c_\ell = c_\ell(\alpha, M)$ and $c_d = c_d(\alpha, M)$, are functions of the angle of attack and Mach number:

$$\alpha = \theta - \phi$$

$$M = M_{\text{tip}} U$$

where M_{tip} is the tip Mach number (the tip speed ΩR divided by the speed of sound) in hover. In fact, the lift and drag of the rotor blade depend on other parameters as well, such as the local yaw angle of the flow and unsteady angle-of-attack changes. Such effects can be included in a numerical analysis, but are neglected here. The radial force on the section is

$$F_r = \frac{u_R}{U} D - z' F_z = \tfrac{1}{2} U u_R c c_d - z' F_z$$

The first term is the radial drag force, obtained by assuming that the viscous drag force on the section has the same yaw angle as the local velocity. (See section 5–12 for a derivation and discussion of this result.) The second term in F_r is the radial component of the normal force F_z due to the local flapwise bending slope z'. The nose-up moment about the elastic axis is

$$\begin{aligned} M_a &= -x_A L + M_{AC} + M_{US} \\ &= -x_A \tfrac{1}{2} U^2 c c_\ell + \tfrac{1}{2} U^2 c^2 c_{m_{ac}} + M_{US} \end{aligned}$$

where x_A is the distance the aerodynamic center is behind the elastic axis, M_{AC} is the moment about the aerodynamic center, and M_{US} is the unsteady aerodynamic moment.

The section forces resolved relative to the hub plane axes are now

$$F_z = L\cos\phi - D\sin\phi = (Lu_T - Du_P)/U$$

$$F_x = L\sin\phi + D\cos\phi = (Lu_P + Du_T)/U$$

Next, substitute for L and D in terms of the section coefficients, and divide by the two-dimensional lift curve slope a and the section chord c. The result is

$$\frac{F_z}{ac} = U\left(u_T\frac{c_\ell}{2a} - u_P\frac{c_d}{2a}\right)$$

$$\frac{F_x}{ac} = U\left(u_P\frac{c_\ell}{2a} + u_T\frac{c_d}{2a}\right)$$

$$\frac{F_r}{ac} = Uu_R\frac{c_d}{2a} - z'\frac{F_z}{ac}$$

$$\frac{M_a}{ac} = -x_A U^2\frac{c_\ell}{2a} + U^2 c\frac{c_m}{2a} + \frac{M_{US}}{2a}$$

The integrals of these forces over the blade span are required in the rotor equations of motion. The section pitch moment M_a will not be considered further until section 11–8.

The objective of this chapter is to obtain the aerodynamic forces in the rotor blade equations in terms of the perturbed motion of the blade. Thus the perturbation section forces must be expressed in terms of the perturbations of the velocities and pitch angle. Each component of the velocity seen by the blade has a trim term, due to operation of the rotor in its equilibrium state, and a perturbation due to the perturbed motion of the system. The latter term is due to the system degrees of freedom, and is assumed to be small in deriving linear differential equations describing the rotor dynamics. Thus the blade pitch and section velocities are written as trim plus perturbation terms:

$$\theta = (\theta)_{\text{trim}} + \delta\theta$$

$$u_T = (u_T)_{\text{trim}} + \delta u_T$$

$$u_P = (u_P)_{\text{trim}} + \delta u_P$$

$$u_R = (u_R)_{\text{trim}} + \delta u_R$$

(After this substitution has been made, the subscript "trim" can be dropped.) The perturbation of the angle of attack, resultant velocity, and Mach number are then

$$\delta\alpha = \delta\theta - (u_T \delta u_P - u_P \delta u_T)/U^2$$

$$\delta U = (u_T \delta u_T + u_P \delta u_P)/U$$

$$\delta M = M_{\text{tip}}\, \delta U$$

and the perturbations of the section coefficients are

$$\delta c_\ell = \frac{\partial c_\ell}{\partial \alpha}\delta\alpha + \frac{\partial c_\ell}{\partial M}\delta M = c_{\ell_\alpha}\delta\alpha + c_{\ell_M}\delta M$$

$$\delta c_d = \frac{\partial c_d}{\partial \alpha}\delta\alpha + \frac{\partial c_d}{\partial M}\delta M = c_{d_\alpha}\delta\alpha + c_{d_M}\delta M$$

The perturbations of the section aerodynamic forces are obtained by carrying out the differential operation on the relations for F_x, F_z, and F_r, using the above results to express the perturbation forces in terms of $\delta\theta$, δu_T, δu_P, and δu_R. The coefficients of the perturbation quantities are evaluated at the trim state. The result for the section forces is as follows:

$$\begin{aligned}
\delta\frac{F_z}{ac} &= \left(Uu_T\frac{c_{\ell_\alpha}}{2a} - Uu_P\frac{c_{d_\alpha}}{2a}\right)\delta\theta \\
&+ \left[-\frac{u_T}{U}\left(u_T\frac{c_{\ell_\alpha}}{2a} - u_P\frac{c_{d_\alpha}}{2a}\right) + \left(\frac{c_\ell}{2a} + M\frac{c_{\ell_M}}{2a}\right)\frac{u_T u_P}{U}\right. \\
&\qquad \left. - \left(\frac{c_d}{2a} + M\frac{c_{d_M}}{2a}\right)\frac{u_P^2}{U} - \frac{c_d}{2a}U\right]\delta u_P \\
&+ \left[\frac{u_P}{U}\left(u_T\frac{c_{\ell_\alpha}}{2a} - u_P\frac{c_{d_\alpha}}{2a}\right) + \left(\frac{c_\ell}{2a} + M\frac{c_{\ell_M}}{2a}\right)\frac{u_T^2}{U}\right. \\
&\qquad \left. + \frac{c_\ell}{2a}U - \left(\frac{c_d}{2a} + M\frac{c_{d_M}}{2a}\right)\frac{u_T u_P}{U}\right]\delta u_T \\
&= F_{z_\theta}\delta\theta + F_{z_P}\delta u_P + F_{z_T}\delta u_T
\end{aligned}$$

$$\delta \frac{F_x}{ac} = \left(Uu_P \frac{c_{\ell_\alpha}}{2a} + Uu_T \frac{c_{d_\alpha}}{2a}\right) \delta\theta$$

$$+ \left[-\frac{u_T}{U}\left(u_P \frac{c_{\ell_\alpha}}{2a} + u_T \frac{c_{d_\alpha}}{2a}\right) + \left(\frac{c_\ell}{2a} + M\frac{c_{\ell_M}}{2a}\right)\frac{u_P^2}{U}\right.$$

$$\left. + \frac{c_\ell}{2a} U + \left(\frac{c_d}{2a} + M\frac{c_{d_M}}{2a}\right)\frac{u_T u_P}{U}\right]\delta u_P$$

$$+ \left[\frac{u_P}{U}\left(u_P \frac{c_{\ell_\alpha}}{2a} + u_T \frac{c_{d_\alpha}}{2a}\right) + \left(\frac{c_\ell}{2a} + M\frac{c_{\ell_M}}{2a}\right)\frac{u_P u_T}{U}\right.$$

$$\left. + \left(\frac{c_d}{2a} + M\frac{c_{d_M}}{2a}\right)\frac{u_T^2}{U} + \frac{c_d}{2a} U\right]\delta u_T$$

$$= F_{x_\theta}\delta\theta + F_{x_P}\delta u_P + F_{x_T}\delta u_T$$

$$\delta \frac{F_r}{ac} = \left(Uu_R \frac{c_{d_\alpha}}{2a}\right)\delta\theta$$

$$+ \left[-\frac{u_T u_R}{U}\frac{c_{d_\alpha}}{2a} + \left(\frac{c_d}{2a} + M\frac{c_{d_M}}{2a}\right)\frac{u_P u_R}{U}\right]\delta u_P$$

$$+ \left[\frac{u_P u_R}{U}\frac{c_{d_\alpha}}{2a} + \left(\frac{c_d}{2a} + M\frac{c_{d_M}}{2a}\right)\frac{u_T u_R}{U}\right]\delta u_T$$

$$+ \left[U\frac{c_d}{2a}\right]\delta u_R$$

$$- \left(\frac{F_z}{ac}\right)\delta z' - \left(z'\right)\delta\frac{F_z}{ac}$$

Since the low disk loading helicopter rotor is characterized by low inflow, it is possible to assume small angles in the aerodynamic analysis. Specifically, it is assumed that θ, ϕ, and c_d/c_ℓ are all small compared to unity. It follows that α and u_P/u_T are also small, that $U \cong u_T$, and that $\phi \cong u_P/u_T$. The section forces are then

$$\frac{F_z}{ac} \cong \frac{L}{ac} \cong u_T^2 \frac{c_\ell}{2a}$$

$$\frac{F_x}{ac} \cong u_T u_P \frac{c_\ell}{2a} + u_T^2 \frac{c_d}{2a}$$

$$\frac{F_r}{ac} \cong u_T u_R \frac{c_d}{2a} - z' \frac{F_z}{ac} .$$

Furthermore, since it is usually consistent with the small angle approximation to also assume a constant lift curve slope and neglect stall, the lift coefficient is simply $c_\ell = a\alpha = a(\theta - u_P/u_T)$. Then

$$\frac{F_z}{ac} = \tfrac{1}{2} u_T^2 \alpha = \tfrac{1}{2}(u_T^2 \theta - u_T u_P)$$

$$\frac{F_x}{ac} = \tfrac{1}{2} u_T u_P \alpha + u_T^2 \frac{c_d}{2a} = \tfrac{1}{2}(u_T u_P \theta - u_P^2) + u_T^2 \frac{c_d}{2a}$$

With the small angle assumption, the perturbation forces become

$$\delta \frac{F_z}{ac} = \tfrac{1}{2} u_T^2 \delta\theta - \tfrac{1}{2} u_T \delta u_P + \tfrac{1}{2}(u_P + 2u_T \alpha)\delta u_T$$

$$\delta \frac{F_x}{ac} = \tfrac{1}{2} u_T u_P \delta\theta + \tfrac{1}{2}(u_T \alpha - u_P)\delta u_P + \left(\tfrac{1}{2} u_P \theta + 2u_T \frac{c_d}{2a}\right)\delta u_T$$

$$\delta \frac{F_z}{ac} = \frac{c_d}{2a} u_R \delta u_T + \frac{c_d}{2a} u_T \delta u_R - \left(\frac{F_z}{ac}\right) \delta z' - \left(z'\right) \delta \frac{F_z}{ac}$$

In the process of making the small angle approximation we have also neglected reverse flow; and the terms due to c_{d_α}, c_{ℓ_M}, and c_{d_M} have been dropped. In this form, the section forces can be integrated analytically, while retaining the basic characteristics of the rotor aerodynamics.

To complete the specification of the perturbation forces, the trim and perturbation velocities are required. The trim terms are as follows. The components of the helicopter trim forward speed in the hub plane axes are the advance ratio and inflow ratio,

$$\mu = \frac{V \cos\alpha_{HP}}{\Omega R}$$

$$\lambda_{HP} = \lambda_i + \mu \tan \alpha_{HP}$$

The in-plane trim velocity seen by the rotor blade is due to the rotor rotation and the advance ratio, giving $u_T = r + \mu \sin \psi$ and $u_R = \mu \cos \psi$ (as obtained in Chapter 5). It follows that any aerodynamic forces involving u_T and u_R alone depend only on the advance ratio μ. The blade pitch θ and normal velocity u_P depend on the operating condition of the rotor, in particular on the thrust coefficient, as well as on the advance ratio. Consequently, those aerodynamic forces involving the trim values of θ or u_P require the solution for the blade angle of attack and loading at the given operating state. The velocity and loading of the rotor blade in forward flight are periodic because of the rotation of the blade relative to the forward speed of the helicopter. Hence the aerodynamics of forward flight introduce periodic coefficients in the equations of motion describing the dynamics. For hover or vertical flight the aerodynamic environment is axisymmetric, so the differential equations have constant coefficients.

The perturbation velocities depend on the degrees of freedom considered in the model of the blade motion. The following sections of this chapter are concerned with deriving the perturbation velocities corresponding to the various models of the rotor motion considered in Chapter 9, and integrating the resulting perturbation forces to obtain the required aerodynamic terms in the equations of motion. The rotor shaft motion will also be included in the blade velocity perturbations. The definitions of the displacements and rotations of the rotor hub are shown in Fig. 11-2 (see also section 9–6).

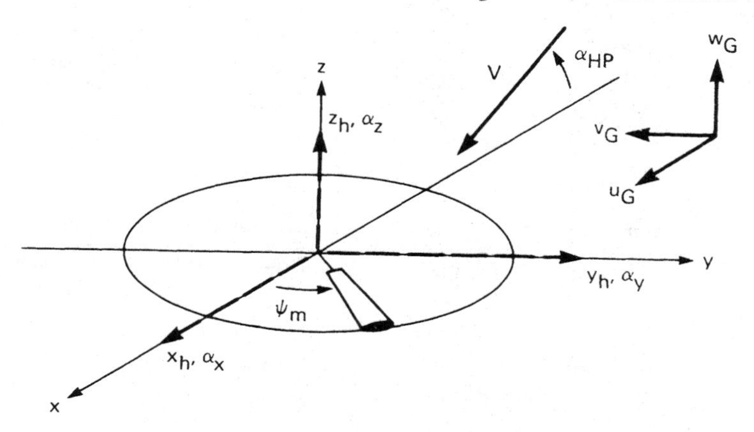

Figure 11-2 Definition of rotor hub motion and aerodynamic gust components.

The dimensionless shaft motion variables are assumed to be small. In addition, the excitation of the rotor by aerodynamic turbulence is considered. The gust velocity components defined relative to the nonrotating hub plane axes are u_G, v_G, and w_G (longitudinal, lateral, and vertical, respectively, as shown in Fig. 11-2). It is assumed that the gust velocity components are uniform throughout space, and that the dimensionless velocities (based on the rotor tip speed ΩR) are small quantities. The aerodynamic forces due to a uniform perturbation of the wake-induced inflow velocity will also be obtained. Such an inflow perturbation can be used to model the unsteady aerodynamics of the rotor, as discussed in section 10-6.4, and in section 11–7 below.

11–2 Flap Motion

Consider the rigid flap motion of a blade with no hinge offset (section 9-2.1). The aerodynamic flap moment is

$$M_F = \int_0^1 r \frac{F_z}{ac} dr$$

For small angles $F_z/ac \cong L/ac \cong \frac{1}{2} u_T^2 \alpha$, so the perturbation force is

$$\delta \frac{F_z}{ac} = (\tfrac{1}{2} u_T^2)\delta\theta + (-\tfrac{1}{2} u_T)\delta u_P$$

The only degree of freedom for the blade is the rigid flap motion β. The blade pitch control is included, and also a uniform perturbation of the inflow velocity. Hence $\delta\theta$ and δu_P are here

$$\delta\theta = \theta - K_P \beta$$

$$\delta u_P = \lambda + r\dot{\beta} + \beta u_R$$

The perturbation pitch angle consists of the control input and the kinematic pitch-flap coupling. The perturbation normal velocity consists of the inflow perturbation, flapping velocity, and the normal component of the radial velocity u_R when the blade is flapped up. Here θ, λ, and β are small perturbations of the trim quantities. Recall from section 5–2 that the pitch angle and normal velocity for the steady-state flapping solution are

$$\theta = \theta_{con} + \theta_{tw} r - K_P \beta$$

$$u_P = \lambda_{HP} + r\dot{\beta} + \beta\mu\cos\psi$$

Since θ and u_P are linear functions, the perturbation quantities have the same form.

Substituting for δF_z in the flap moment gives

$$M_F = M_\theta(\theta - K_P\beta) + M_\lambda\lambda + M_{\dot{\beta}}\dot{\beta} + M_\beta\beta$$

where the aerodynamic coefficients are

$$M_\theta = \int_0^1 \frac{1}{2} r u_T^2\, dr = \frac{1}{8} + \frac{\mu}{3}\sin\psi + \frac{\mu^2}{4}\sin^2\psi$$

$$M_\lambda = -\int_0^1 \frac{1}{2} r u_T\, dr = -\left(\frac{1}{6} + \frac{\mu}{4}\sin\psi\right)$$

$$M_{\dot{\beta}} = -\int_0^1 \frac{1}{2} r^2 u_T\, dr = -\left(\frac{1}{8} + \frac{\mu}{6}\sin\psi\right)$$

$$M_\beta = -\int_0^1 \frac{1}{2} r u_T u_R\, dr = u_R M_\lambda$$

Including the tip loss factor, the upper limit of integration should be $r = B$ rather than $r = 1$. All of these flap moments are due to the lift increment produced by the blade angle-of-attack change. The identical coefficients were derived in section 5–5 (where the steady-state solution for the flap motion was obtained by assuming a Fourier series for the flap response). Here we have a linear differential equation for the perturbed flap motion. In hover ($\mu = 0$) this differential equation has constant coefficients. In forward flight the aerodynamic coefficients in the equation of motion are periodic functions of the azimuth angle ψ.

For a fundamental flapping mode that is applicable generally, to offset-hinge articulated rotors and hingeless rotors, the required flap moment is

$$M_F = \int_0^1 \eta \frac{F_z}{ac} dr$$

where $\eta(r)$ is the out-of-plane mode shape. With the deflection $z = \eta\beta$ now, the normal velocity is

$$\delta u_P = \lambda + \dot{z} + z' u_R = \lambda + \eta\dot{\beta} + \eta'\beta u_R$$

The flap moment is again written as

$$M_F = M_\theta(\theta - K_P\beta) + M_\lambda\lambda + M_{\dot{\beta}}\dot{\beta} + M_\beta\beta$$

where the aerodynamic coefficients are

$$M_\theta = \int_0^1 \frac{1}{2}\eta u_T^2 dr = \frac{1}{8} c_2 + \frac{\mu}{3} c_1 \sin\psi + \frac{\mu^2}{4} c_0 \sin^2\psi$$

$$M_\lambda = -\int_0^1 \frac{1}{2}\eta u_T dr = -\left(\frac{1}{6} c_1 + \frac{\mu}{4} c_0 \sin\psi\right)$$

$$M_{\dot{\beta}} = -\int_0^1 \frac{1}{2}\eta^2 u_T dr = -\left(\frac{1}{8} d_1 + \frac{\mu}{6} d_0 \sin\psi\right)$$

$$M_\beta = -\int_0^1 \frac{1}{2} u_T \eta\eta' u_R dr = -\mu\cos\psi\left(\frac{1}{6} f_1 + \frac{\mu}{4} f_0 \sin\psi\right)$$

with $c_n = (n+2)\int_0^1 \eta r^n dr$, $d_n = (n+3)\int_0^1 \eta^2 r^n dr$, and $f_n = (n+2)\int_0^1 \eta\eta' r^n dr$ (see also section 5–14).

For the case of out-of-plane bending of the blade (section 9-2.2) the aerodynamic force is

$$M_{F_k} = \int_0^1 \eta_k \frac{F_z}{ac} dr$$

and $z = \sum_i \eta_i q_i$ gives

$$\delta u_P = \lambda + \sum_i \eta_i \dot{q}_i + \sum_i \eta_i' q_i u_R$$

$$\delta\theta = \theta - \sum_i K_{P_i} q_i$$

Then

$$M_{F_k} = M_{q_k\theta}\left(\theta - \sum_i K_{P_i} q_i\right) + M_{q_k\lambda}\lambda + \sum_i M_{q_k\dot{q}_i}\dot{q}_i + \sum_i M_{q_k q_i} q_i$$

where the aerodynamic coefficients are

$$M_{q_k\theta} = \int_0^1 \tfrac{1}{2}\eta_k u_T^2\, dr$$

$$M_{q_k\lambda} = -\int_0^1 \tfrac{1}{2}\eta_k u_T\, dr$$

$$M_{q_k\dot{q}_i} = -\int_0^1 \tfrac{1}{2}\eta_k \eta_i u_T\, dr$$

$$M_{q_k q_i} = -u_R \int_0^1 \tfrac{1}{2}\eta_k \eta_i' u_T\, dr$$

The influence of the rotor wake on the unsteady aerodynamic forces of the blade can be accounted for by using an appropriate model for the inflow perturbation λ. Alternatively, the quasistatic lift can be multiplied by a lift deficiency function $C'(k)$. Then the factor C' must be included in the integrands of the aerodynamic coefficients, for example

$$M_\theta = \int_0^1 C'(k)\tfrac{1}{2} r u_T^2\, dr.$$

The reduced frequency and therefore C' vary along the blade span, but it is usually satisfactory to evaluate the lift deficiency function just at an effective radius (typically $0.75R$), in which case

$$M_\theta \cong C'(k_e) \int_0^1 \tfrac{1}{2} r u_T^2 \, dr$$

The lift deficiency function is based on harmonic motion and hence is applicable to the frequency response or flutter boundary calculation. In forward flight Theodorsen's function should be used for $C'(k)$. If the lift deficiency function is to be integrated numerically, the reduced frequency should be based on the local free stream velocity: $k = \omega b/u_T$. For the low harmonics of flap motion the reduced frequency is small, and the near shed wake effects are small (Theodorsen's function $C \cong 1$). In hover at low thrust, the returning shed wake effects can be significant, and Loewy's lift deficiency function should be used for C' (see section 10–5). When the spacing between the wake spirals is small and oscillation occurs at a harmonic of the rotor speed so that the layers of wake vorticity are in phase, Loewy's function will substantially reduce the blade lift.

11–3 Flap and Lag Motion

The aerodynamic forces required for the fundamental flap and lag motions of a rotor blade are

$$M_F = \int_0^1 \eta_\beta \frac{F_z}{ac} \, dr$$

$$M_L = \int_0^1 \eta_\zeta \frac{F_x}{ac} \, dr$$

(see section 9-3.1). Here η_β and η_ζ are the mode shapes for the flap (purely out-of-plane) and lag (purely in-plane) motion, respectively. From section 11–1, the perturbations of the section forces are

$$\delta \frac{F_z}{ac} = (\tfrac{1}{2} u_T^2)\delta\theta + (-\tfrac{1}{2} u_T)\delta u_P + \tfrac{1}{2}(u_P + 2u_T\alpha)\delta u_T$$

$$\delta \frac{F_x}{ac} = (\tfrac{1}{2} u_T u_P)\delta\theta + \tfrac{1}{2}(u_T\alpha - u_P)\delta u_P + \left(\tfrac{1}{2} u_P\theta + 2u_T \frac{c_d}{2a}\right)\delta u_T$$

In the case of the lag degree of freedom it is necessary to consider the in-plane force and velocity. The in-plane velocity perturbation δu_T produces dynamic pressure and small angle-of-attack changes. The resulting lift perturbation is much smaller than the lift due to $\delta\theta$ and δu_P, which directly produce angle-of-attack changes. The in-plane forces are due to the induced drag perturbations, and hence are much smaller than the out-of-plane forces. The δu_T term of δF_x also has a profile drag change due to the dynamic pressure perturbation. The blade pitch and normal velocity perturbations are

$$\delta\theta = \theta - K_P\beta$$

$$\delta u_P = \lambda + \eta_\beta\dot{\beta} + \eta'_\beta\beta u_R$$

as above. The lag motion produces a perturbation of the in-plane blade velocity:

$$\delta u_T = -\eta_\zeta\dot{\zeta} - \eta'_\zeta\zeta u_R$$

Substituting for the force and velocity perturbations then gives the flap and lag moments:

$$M_F = M_\theta(\theta - K_P\beta) + M_\lambda\lambda + M_{\dot{\beta}}\dot{\beta} + M_\beta\beta + M_{\dot{\zeta}}\dot{\zeta} + M_\zeta\zeta$$

$$M_L = Q_\theta(\theta - K_P\beta) + Q_\lambda\lambda + Q_{\dot{\beta}}\dot{\beta} + Q_\beta\beta + Q_{\dot{\zeta}}\dot{\zeta} + Q_\zeta\zeta$$

where the aerodynamic coefficients are

$$M_\theta = \int_0^1 \tfrac{1}{2}\eta_\beta u_T^2\,dr$$

$$M_\lambda = -\int_0^1 \tfrac{1}{2}\eta_\beta u_T\,dr$$

$$M_{\dot{\beta}} = -\int_0^1 \tfrac{1}{2}\eta_\beta^2 u_T\,dr$$

$$M_\beta = -u_R\int_0^1 \tfrac{1}{2}\eta_\beta\eta'_\beta u_T\,dr$$

$$M_{\dot{\zeta}} = -\int_0^1 \tfrac{1}{2}\eta_\beta\eta_\zeta(u_P + 2u_T\alpha)\,dr$$

$$M_\zeta = -u_R\int_0^1 \tfrac{1}{2}\eta_\beta\eta'_\zeta(u_P + 2u_T\alpha)\,dr$$

$$Q_\theta = \int_0^1 \tfrac{1}{2}\eta_\zeta u_T u_P \, dr$$

$$Q_\lambda = \int_0^1 \tfrac{1}{2}\eta_\zeta (u_T\alpha - u_P) dr$$

$$Q_{\dot\beta} = \int_0^1 \tfrac{1}{2}\eta_\zeta \eta_\beta (u_T\alpha - u_P) dr$$

$$Q_\beta = u_R \int_0^1 \tfrac{1}{2}\eta_\zeta \eta'_\beta (u_T\alpha - u_P) dr$$

$$Q_{\dot\zeta} = -\int_0^1 \eta_\zeta^2 \left(\tfrac{1}{2}u_P\theta + 2u_T \frac{c_d}{2a}\right) dr$$

$$Q_\zeta = -u_R \int_0^1 \eta_\zeta \eta'_\zeta \left(\tfrac{1}{2}u_P\theta + 2u_T \frac{c_d}{2a}\right) dr$$

The coefficients M_θ, M_λ, $M_{\dot\beta}$, and M_β are the flap moments produced by the lift changes due to angle-of-attack perturbations. These coefficients are therefore defined entirely by the rotor advance ratio (and the flapping mode shape of the blade; see the expressions given in the preceding section). The remaining coefficients involve either in-plane velocity or in-plane force, or both, and therefore require the solution for the trim blade motion and loading (i.e. θ, u_P, and α as well as u_T and u_R) in order to be evaluated. These coefficients then depend on the rotor operating state, particularly the thrust coefficient.

In hover or vertical flight, where the trim solution for the blade loading is axisymmetric, the aerodynamic coefficients are more readily evaluated than in forward flight. The trim velocities in vertical flight are $u_T = r$, $u_P = \lambda_{HP}$, and $u_R = 0$; the blade angle of attack becomes $\alpha = \theta - \lambda_{HP}/r$. In order to evaluate the integrals analytically, it is assumed that the induced velocity λ_{HP} is uniform over the rotor disk and that the blade mode shapes are $\eta_\beta = \eta_\zeta = r$. If the coefficients are to be evaluated numerically, the actual mode shapes can be used, and perhaps also the combined blade element and momentum theory result for the nonuniform inflow. It follows that for hover or vertical flight

$$M_\theta = \int_0^1 \tfrac{1}{2} r^3\, dr = \frac{1}{8}$$

$$M_\lambda = -\int_0^1 \tfrac{1}{2} r^2\, dr = -\frac{1}{6}$$

$$M_{\dot\beta} = -\int_0^1 \tfrac{1}{2} r^3\, dr = -\frac{1}{8}$$

$$M_{\dot\zeta} = -\int_0^1 \tfrac{1}{2} r^2 (\lambda_{HP} + 2r\alpha) dr = -\left(\frac{\lambda_{HP}}{6} + 2\int_0^1 \tfrac{1}{2} r^3\, \alpha\, dr\right)$$

$$Q_\theta = \int_0^1 \tfrac{1}{2} r^2\, \lambda_{HP}\, dr = \frac{\lambda_{HP}}{6}$$

$$Q_\lambda = \int_0^1 \tfrac{1}{2} r (r\alpha - \lambda_{HP}) dr = \int_0^1 \tfrac{1}{2} r^2 \alpha\, dr - \frac{\lambda_{HP}}{4}$$

$$Q_{\dot\beta} = \int_0^1 \tfrac{1}{2} r^2 (r\alpha - \lambda_{HP}) dr = \int_0^1 \tfrac{1}{2} r^3 \alpha\, dr - \frac{\lambda_{HP}}{6}$$

$$Q_{\dot\zeta} = -\int_0^1 r^2 \left(\tfrac{1}{2}\lambda_{HP}\theta + 2r\frac{c_d}{2a}\right) dr = -\left(\lambda_{HP}\int_0^1 \tfrac{1}{2} r^2\, \theta\, dr + \frac{c_d}{4a}\right)$$

and $M_\beta = M_\zeta = Q_\beta = Q_\zeta = 0$. These coefficients can be expressed directly in terms of the rotor thrust. Momentum theory relates the induced velocity and thrust; for example, in hover $\lambda_{HP} = \kappa\sqrt{C_T/2}$. Recall that the definition of the thrust coefficient is

$$\frac{C_T}{\sigma a} = \int_0^1 \frac{F_z}{ac}\, dr = \int_0^1 \tfrac{1}{2} r^2\, \alpha\, dr$$

So integrals of the blade angle-of-attack distribution can be related to the thrust also. With $\alpha = \theta - \lambda_{HP}/r = \theta_0 + \theta_{tw} r - \lambda_{HP}/r$:

$$\frac{C_T}{\sigma a} = \frac{\theta_{.75}}{6} - \frac{\lambda_{HP}}{4}$$

It follows that

$$\int_0^1 \tfrac{1}{2} r^3 \alpha\, dr = \frac{\theta_{.8}}{8} - \frac{\lambda_{HP}}{6} = \frac{3}{4}\frac{C_T}{\sigma a} + \frac{\lambda_{HP}}{48} + \frac{\theta_{tw}}{160}$$

$$\int_0^1 \tfrac{1}{2} r^2 \theta\, dr = \frac{C_T}{\sigma a} + \frac{\lambda_{HP}}{4}$$

The results for the aerodynamic coefficients are

$$M_{\dot{\zeta}} = -\left(\frac{3}{2}\frac{C_T}{\sigma a} + \frac{5}{24}\lambda_{HP} + \frac{1}{80}\theta_{tw}\right)$$

$$Q_\lambda = \frac{C_T}{\sigma a} - \frac{\lambda_{HP}}{4}$$

$$Q_{\dot{\beta}} = \frac{3}{4}\frac{C_T}{\sigma a} - \frac{7}{48}\lambda_{HP} + \frac{1}{160}\theta_{tw}$$

$$Q_{\dot{\zeta}} = -\left[\lambda_{HP}\left(\frac{C_T}{\sigma a} + \frac{\lambda_{HP}}{4}\right) + \frac{c_d}{4a}\right]$$

The rotor thrust coefficient thus defines these aerodynamic coefficients.

11–4 Nonrotating Frame

The aerodynamic flap moment on the mth blade of an N-bladed rotor has been obtained in the following form:

$$M_F^{(m)} = M_\theta\left(\theta^{(m)} - K_P\beta^{(m)}\right) + M_{\dot{\beta}}\dot{\beta}^{(m)} + M_\beta\beta^{(m)} + M_\lambda\lambda$$

The aerodynamic forces in the nonrotating frame are obtained by introducing the Fourier coordinate transformation, and evaluating the summations

$$M_{F_0} = \frac{1}{N}\sum_{m=1}^{N} M_F^{(m)}$$

$$M_{F_{nc}} = \frac{2}{N}\sum_{m=1}^{N} M_F^{(m)} \cos n\psi_m$$

$$M_{F_{ns}} = \frac{2}{N} \sum_{m=1}^{N} M_F^{(m)} \sin n\psi_m$$

$$M_{F_{N/2}} = \frac{1}{N} \sum_{m=1}^{N} M_F^{(m)} (-1)^m$$

where $\psi_m = \psi + m\Delta\psi$ is the azimuth angle of the mth blade ($\Delta\psi = 2\pi/N$). When the aerodynamic coefficients are independent of ψ_m, as they are for the hovering rotor, the summation operators act only on the degrees of freedom, not on the aerodynamic coefficients. The summations are easily evaluated using the definitions of the nonrotating degrees of freedom and the corresponding results for their time derivatives (see section 8-4.1). The result for constant coefficients is

$$M_{F_0} = M_\theta(\theta_0 - K_P\beta_0) + M_{\dot\beta}\dot\beta_0 + M_\lambda\lambda$$

$$M_{F_{nc}} = M_\theta(\theta_{nc} - K_P\beta_{nc}) + M_{\dot\beta}(\dot\beta_{nc} + n\beta_{ns})$$

$$M_{F_{ns}} = M_\theta(\theta_{ns} - K_P\beta_{ns}) + M_{\dot\beta}(\dot\beta_{ns} - n\beta_{nc})$$

$$M_{F_{N/2}} = M_\theta(\theta_{N/2} - K_P\beta_{N/2}) + M_{\dot\beta}\dot\beta_{N/2}$$

(recall that $M_\beta = 0$ in hover). In forward flight the aerodynamic coefficients are periodic functions of ψ_m, and therefore the evaluation of the summations is more complicated. In section 8-4.2 the techniques for converting the equations of motion to the nonrotating frame were discussed. We have seen that the solutions of differential equations with periodic coefficients have a number of unique characteristics. A transformation of the degrees of freedom cannot change the physical behavior of a system, although it may make the analysis much easier. Therefore, periodic coefficients must appear in the rotor equations in the nonrotating frame if they appear in the rotating frame. Moreover, with periodic coefficients the differential equations in the nonrotating frame depend on the number of blades. Using the forward flight aerodynamic coefficients given in section 11–2, the flap moments in the nonrotating frame are as follows, for $N = 2$:

$$\begin{pmatrix} M_{F_0} \\ M_{F_1} \end{pmatrix} = \begin{bmatrix} \frac{1}{8}(1+\mu^2) - \frac{\mu^2}{8}\cos 2\psi & \frac{\mu}{3}\sin\psi \\ \frac{\mu}{3}\sin\psi & \frac{1}{8}(1+\mu^2) - \frac{\mu^2}{8}\cos 2\psi \end{bmatrix} \begin{pmatrix} \theta_0 - K_P\beta_0 \\ \theta_1 - K_P\beta_1 \end{pmatrix}$$

$$- \begin{bmatrix} \frac{1}{8} & \frac{\mu}{6}\sin\psi \\ \frac{\mu}{6}\sin\psi & \frac{1}{8} \end{bmatrix} \begin{pmatrix} \beta_0 \\ \beta_1 \end{pmatrix}^{\bullet} - \begin{bmatrix} \frac{\mu^2}{8}\sin 2\psi & \frac{\mu}{6}\cos\psi \\ \frac{\mu}{6}\cos\psi & \frac{\mu^2}{8}\sin 2\psi \end{bmatrix} \begin{pmatrix} \beta_0 \\ \beta_1 \end{pmatrix}$$

$$+ \begin{bmatrix} -\frac{1}{6} \\ -\frac{\mu}{4}\sin\psi \end{bmatrix} \lambda$$

for $N = 3$:

$$\begin{pmatrix} M_{F_0} \\ M_{F_{1c}} \\ M_{F_{1s}} \end{pmatrix} = \begin{bmatrix} \frac{1}{8}(1+\mu^2) & -\frac{\mu^2}{16}\cos 3\psi & \frac{\mu}{6} - \frac{\mu^2}{16}\sin 3\psi \\ -\frac{\mu^2}{8}\cos 3\psi & \frac{1}{8}\left(1+\frac{\mu^2}{2}\right) + \frac{\mu}{6}\sin 3\psi & -\frac{\mu}{6}\cos 3\psi \\ \frac{\mu}{3} - \frac{\mu^2}{8}\sin 3\psi & -\frac{\mu}{6}\cos 3\psi & \frac{1}{8}\left(1+\frac{3}{2}\mu^2\right) - \frac{\mu}{6}\sin 3\psi \end{bmatrix} \begin{pmatrix} \theta_0 - K_\beta\beta_0 \\ \theta_{1c} - K_P\beta_{1c} \\ \theta_{1s} - K_P\beta_{1s} \end{pmatrix}$$

$$- \begin{bmatrix} \frac{1}{8} & 0 & \frac{\mu}{12} \\ 0 & \frac{1}{8} + \frac{\mu}{12}\sin 3\psi & -\frac{\mu}{12}\cos 3\psi \\ \frac{\mu}{6} & -\frac{\mu}{12}\cos 3\psi & \frac{1}{8} - \frac{\mu}{12}\cos 3\psi \end{bmatrix} \begin{pmatrix} \beta_0 \\ \beta_{1c} \\ \beta_{1s} \end{pmatrix}^{\bullet}$$

$$- \begin{bmatrix} 0 & \frac{\mu^2}{16}\sin 3\psi & -\frac{\mu^2}{16}\cos 3\psi \\ \frac{\mu}{6} + \frac{\mu^2}{8}\sin 3\psi & \frac{\mu}{6}\cos 3\psi & \frac{1}{8}\left(1+\frac{\mu^2}{2}\right) + \frac{\mu}{6}\sin 3\psi \\ -\frac{\mu^2}{8}\cos 3\psi & -\frac{1}{8}\left(1-\frac{\mu^2}{2}\right) + \frac{\mu}{6}\sin 3\psi & -\frac{\mu}{6}\cos 3\psi \end{bmatrix} \begin{pmatrix} \beta_0 \\ \beta_{1c} \\ \beta_{1s} \end{pmatrix}$$

$$+\begin{bmatrix} -\dfrac{1}{6} \\ 0 \\ -\dfrac{\mu}{4} \end{bmatrix}\lambda$$

and for $N = 4$:

$$\begin{pmatrix} M_{F_0} \\ M_{F_{1c}} \\ M_{F_{1s}} \\ M_{F_2} \end{pmatrix} \begin{bmatrix} \dfrac{1}{8}\left(1+\mu^2\right) & 0 & \dfrac{\mu}{6} & -\dfrac{\mu^2}{8}\cos 2\psi \\ 0 & \dfrac{1}{8}\left(1+\dfrac{\mu^2}{2}-\dfrac{\mu^2}{2}\cos 4\psi\right) & -\dfrac{\mu^2}{16}\sin 4\psi & \dfrac{\mu}{3}\sin 2\psi \\ \dfrac{\mu}{3} & -\dfrac{\mu^2}{16}\sin 4\psi & \dfrac{1}{8}\left(1+\dfrac{3}{2}\mu^2+\dfrac{\mu^2}{2}\cos 4\psi\right) & -\dfrac{\mu}{3}\cos 2\psi \\ -\dfrac{\mu^2}{8}\cos 2\psi & \dfrac{\mu}{6}\sin 2\psi & -\dfrac{\mu}{6}\cos 2\psi & \dfrac{1}{8}\left(1+\mu^2\right) \end{bmatrix} \begin{pmatrix} \theta_0 - K_P\beta_0 \\ \theta_{1c} - K_P\beta_{1c} \\ \theta_{1s} - K_P\beta_{1s} \\ \theta_2 - K_P\beta_2 \end{pmatrix}$$

$$-\begin{bmatrix} \dfrac{1}{8} & 0 & \dfrac{\mu}{12} & 0 \\ 0 & \dfrac{1}{8} & 0 & \dfrac{\mu}{6}\sin 2\psi \\ \dfrac{\mu}{6} & 0 & \dfrac{1}{8} & -\dfrac{\mu}{6}\cos 2\psi \\ 0 & \dfrac{\mu}{12}\sin 2\psi & -\dfrac{\mu}{12}\cos 2\psi & \dfrac{1}{8} \end{bmatrix} \begin{pmatrix} \beta_0 \\ \beta_{1c} \\ \beta_{1s} \\ \beta_2 \end{pmatrix}^{\bullet}$$

$$-\begin{bmatrix} 0 & 0 & 0 & \dfrac{\mu^2}{8}\sin 2\psi \\ \dfrac{\mu}{6} & \dfrac{\mu^2}{16}\sin 4\psi & \dfrac{1}{8}\left(1+\dfrac{\mu^2}{2}-\dfrac{\mu^2}{2}\cos 4\psi\right) & \dfrac{\mu}{6}\cos 2\psi \\ 0 & -\dfrac{1}{8}\left(1-\dfrac{\mu^2}{2}+\dfrac{\mu^2}{2}\cos 4\psi\right) & -\dfrac{\mu^2}{16}\sin 4\psi & \dfrac{\mu}{6}\sin 2\psi \\ \dfrac{\mu^2}{8}\sin 2\psi & \dfrac{\mu}{6}\cos 2\psi & \dfrac{\mu}{6}\sin 2\psi & 0 \end{bmatrix} \begin{pmatrix} \beta_0 \\ \beta_{1c} \\ \beta_{1s} \\ \beta_2 \end{pmatrix}$$

$$+\begin{bmatrix} -\frac{1}{6} \\ 0 \\ -\frac{\mu}{4} \\ 0 \end{bmatrix} \lambda$$

Note that for the two-bladed rotor the pitch control variables are collective $\theta_0 = \frac{1}{2}(\theta^{(2)} + \theta^{(1)})$ and differential pitch $\theta_1 = \frac{1}{2}(\theta^{(2)} - \theta^{(1)})$. The usual swashplate control gives $\theta_1 = \theta_{1c} \cos \psi + \theta_{1s} \sin \psi$. Also observe that increasing the number of blades has the effect of clearing the periodic coefficients from the lower degrees of freedom and equations, although the periodicity always remains in the higher elements of the matrices.

The analysis of time-invariant system response is much simpler than the analysis of periodic system response, and more powerful tools are available. Thus we are interested in the possibility of an accurate constant coefficient representation of the rotor dynamics. Such a representation can only be approximate, since it can never correctly model all the behavior of a periodic system. From the above equations for the flap moments it may be concluded that the constant coefficient approximation should be introduced in the nonrotating frame. If the mean values of the aerodynamic coefficients in the rotating frame are used, all the influence of forward flight is lost except for an increase of order μ^2 in M_θ. The mean values of the coefficients in the nonrotating frame include some of the higher harmonics of the coefficients in the rotating frame. From the results for three- and four-bladed rotors above, the constant coefficient approximation for the flap moments is:

$$\begin{pmatrix} M_{F_0} \\ M_{F_{1c}} \\ M_{F_{1s}} \end{pmatrix} = \begin{bmatrix} \frac{1}{8}\left(1+\mu^2\right) & 0 & \frac{\mu}{6} \\ 0 & \frac{1}{8}\left(1+\frac{\mu^2}{2}\right) & 0 \\ \frac{\mu}{3} & 0 & \frac{1}{8}\left(1+\frac{3}{2}\mu^2\right) \end{bmatrix} \begin{pmatrix} \theta_0 - K_P\beta_0 \\ \theta_{1c} - K_P\beta_{1c} \\ \theta_{1s} - K_P\beta_{1s} \end{pmatrix}$$

$$-\begin{bmatrix} \frac{1}{8} & 0 & \frac{\mu}{12} \\ 0 & \frac{1}{8} & 0 \\ \frac{\mu}{6} & 0 & \frac{1}{8} \end{bmatrix} \begin{pmatrix} \beta_0 \\ \beta_{1c} \\ \beta_{1s} \end{pmatrix}^{\bullet}$$

$$-\begin{bmatrix} 0 & 0 & 0 \\ \frac{\mu}{6} & 0 & \frac{1}{8}\left(1+\frac{\mu^2}{2}\right) \\ 0 & -\frac{1}{8}\left(1-\frac{\mu^2}{2}\right) & 0 \end{bmatrix} \begin{pmatrix} \beta_0 \\ \beta_{1c} \\ \beta_{1s} \end{pmatrix}$$

$$+\begin{bmatrix} -\frac{1}{6} \\ 0 \\ -\frac{\mu}{4} \end{bmatrix} \lambda$$

(there are additional degrees of freedom and equations for $N \geq 4$). Since increasing the number of blades tends to sweep the periodic coefficients to the higher modes, it is expected that for a rotor with a large number of blades the constant coefficient approximation will be a good representation of the dynamics involving primarily the collective and cyclic degrees of freedom (β_0, β_{1c}, and β_{1s} here).

Alternatively, the constant coefficient approximation is readily obtained directly from the rotating equation. Consider a typical term in the non-rotating equation, of the form

$$\frac{2}{N} \sum_{m=1}^{N} \left(M_\theta^{(m)} \theta^{(m)}\right) \cos \psi_m$$

Substituting the Fourier coordinate transformation for $\theta^{(m)}$ gives

$$\frac{2}{N} \sum_{m=1}^{N} M_\theta^{(m)} \left[\theta_0 \cos \psi_m + \theta_{1c} \tfrac{1}{2}(1 + \cos 2\psi_m) + \theta_{1s} \tfrac{1}{2} \sin 2\psi_m\right]$$

If the complete periodic coefficient is now desired, the evaluation of the summation over $M_\theta^{(m)}$ is rather complicated and also depends on N. If only the mean value of the coefficient is required, however, the summation operator simply picks out the corresponding harmonic in the Fourier series expansion of $M_\theta^{(m)}$. Hence in the present example we obtain

$$\theta_0 M_\theta^{1c} + \theta_{1c}\left(M_\theta^0 + \tfrac{1}{2} M_\theta^{2c}\right) + \theta_{1s} \tfrac{1}{2} M_\theta^{2s}$$

where M_θ^{nc} and M_θ^{ns} are the Fourier series harmonics of M_θ. The complete result for the β_0, β_{1c}, and β_{1s} flap moments is:

$$\begin{pmatrix} M_{F_0} \\ M_{F_{1c}} \\ M_{F_{1s}} \end{pmatrix} = \begin{bmatrix} M_\theta^0 & 0 & \frac{1}{2} M_\theta^{1s} \\ 0 & M_\theta^0 + \frac{1}{2} M_\theta^{2c} & 0 \\ M_\theta^{1s} & 0 & M_\theta^0 - \frac{1}{2} M_\theta^{2c} \end{bmatrix} \begin{pmatrix} \theta_0 - K_P\beta_0 \\ \theta_{1c} - K_P\beta_{1c} \\ \theta_{1s} - K_P\beta_{1s} \end{pmatrix}$$

$$+ \begin{bmatrix} M_{\dot\beta}^0 & 0 & \frac{1}{2} M_{\dot\beta}^{1s} \\ 0 & M_{\dot\beta}^0 + \frac{1}{2} M_{\dot\beta}^{2c} & 0 \\ M_{\dot\beta}^{1s} & 0 & M_{\dot\beta}^0 - \frac{1}{2} M_{\dot\beta}^{2c} \end{bmatrix} \begin{pmatrix} \beta_0 \\ \beta_{1c} \\ \beta_{1s} \end{pmatrix}^{\bullet}$$

$$+ \begin{bmatrix} 0 & \frac{1}{2}\left(M_\beta^{1c} - M_{\dot\beta}^{1s}\right) & 0 \\ M_\beta^{1c} & 0 & M_{\dot\beta}^0 + \frac{1}{2} M_{\dot\beta}^{2c} + \frac{1}{2} M_\beta^{2s} \\ 0 & -M_{\dot\beta}^0 + \frac{1}{2} M_{\dot\beta}^{2c} + \frac{1}{2} M_\beta^{2s} & 0 \end{bmatrix} \begin{pmatrix} \beta_0 \\ \beta_{1c} \\ \beta_{1s} \end{pmatrix}$$

$$+ \begin{bmatrix} M_\lambda^0 \\ 0 \\ M_\lambda^{1s} \end{bmatrix} \lambda$$

(use has been made of the fact that all the odd-cosine and even-sine harmonics of M_θ, $M_{\dot\beta}$, and M_λ are zero, as are the even-cosine and odd-sine harmonics of M_β). Expressions are available for the harmonics of the coefficients M_θ, $M_{\dot\beta}$, M_β, and M_λ in forward flight. Substituting the results for the harmonics calculated with $\eta_\beta = r$ gives the previous result for the constant coefficient approximation.

Often it is neither necessary nor possible to obtain explicit expressions for the periodic coefficients in the nonrotating frame, as was done above for the flap moments. Because the conversion of the equations to the nonrotating frame is rather tedious and must be repeated for every value of N, such an approach is justified only for analytical investigations with a small number of degrees of freedom. Moreover, with any but the simplest models the harmonics of the coefficients in the rotating frame must themselves be evaluated numerically. Thus a general procedure is desired for converting the rotor blade equations of motion to the nonrotating frame, and one that can easily be implemented in numerical investigations. Consider again the flap moment

$$M_F^{(m)} = M_\theta\left(\theta^{(m)} - K_P\beta^{(m)}\right) + M_{\dot\beta}\dot\beta^{(m)} + M_\beta\beta^{(m)} + M_\lambda\lambda$$

where the aerodynamic coefficients are periodic functions of ψ_m. Substituting the Fourier coordinate transformation for $\theta^{(m)}$ and $\beta^{(m)}$, and applying the summation operators to convert the equations to the nonrotating frame, there follows directly:

$$\begin{pmatrix} M_{F_0} \\ M_{F_{1c}} \\ M_{F_{1s}} \end{pmatrix} = \frac{1}{N}\sum_{m=1}^{N}\begin{bmatrix} M_\theta & M_\theta C & M_\theta S \\ M_\theta 2C & M_\theta 2C^2 & M_\theta 2CS \\ M_\theta 2S & M_\theta 2CS & M_\theta 2S^2 \end{bmatrix}\begin{pmatrix} \theta_0 - K_P\beta_0 \\ \theta_{1c} - K_P\beta_{1c} \\ \theta_{1s} - K_P\beta_{1s} \end{pmatrix}$$

$$+ \frac{1}{N}\sum_{m=1}^{N}\begin{bmatrix} M_{\dot\beta} & M_{\dot\beta} C & M_{\dot\beta} S \\ M_{\dot\beta} 2C & M_{\dot\beta} 2C^2 & M_{\dot\beta} 2CS \\ M_{\dot\beta} 2S & M_{\dot\beta} 2CS & M_{\dot\beta} 2S^2 \end{bmatrix}\begin{pmatrix} \beta_0 \\ \beta_{1c} \\ \beta_{1s} \end{pmatrix}^{\bullet}$$

$$+ \frac{1}{N}\sum_{m=1}^{N}\begin{bmatrix} M_\beta & M_\beta C - M_{\dot\beta} S & M_\beta S + M_{\dot\beta} C \\ M_\beta 2C & M_\beta 2C^2 - M_{\dot\beta} 2CS & M_\beta 2CS + M_{\dot\beta} 2C^2 \\ M_\beta 2S & M_\beta 2CS - M_{\dot\beta} 2S^2 & M_\beta 2S^2 + M_{\dot\beta} 2CS \end{bmatrix}\begin{pmatrix} \beta_0 \\ \beta_{1c} \\ \beta_{1s} \end{pmatrix}$$

$$+\frac{1}{N}\sum_{m=1}^{N}\begin{bmatrix} M_\lambda \\ M_\lambda 2C \\ M_\lambda 2S \end{bmatrix}\lambda$$

where $C = \cos\psi_m$ and $S = \sin\psi_m$. The summation of the coefficients over all the blades ($m = 1 \ldots N$) is to be performed numerically now. The corresponding rows and columns of these matrices are easily obtained for the β_{nc}, β_{ns}, and $\beta_{N/2}$ degrees of freedom as required, depending on the number of blades. Each row of a matrix has one of the factors 1, $2\cos k\psi_m$, $2\sin k\psi_m$, or $(-1)^m$ from the summation operator. Each column has one of the factors 1, $\cos n\psi_m$, $\sin n\psi_m$, or $(-1)^m$ from the Fourier coordinate transformation (or 0, $-n\sin n\psi_m$, $n\cos n\psi_m$, 0 for the Coriolis terms resulting from the transformation of time derivatives). Note that the form of this result does not depend on the number of blades, except for the size of the matrices. It is the result of the summations over all the blades that depends on N. (See also the discussion in section 8-4.2.)

The constant coefficient approximation requires the mean values of the equation coefficients in the nonrotating frame, which are obtained by applying the operator

$$\frac{1}{2\pi}\int_0^{2\pi}(\ldots)\,d\psi$$

to the matrices above. The result is terms of the form:

$$\frac{1}{2\pi}\int_0^{2\pi}\left(\frac{1}{N}\sum_{m=1}^{N}\begin{Bmatrix} 1 \\ \cos\psi_m \\ \sin\psi_m \\ 2\cos^2\psi_m \\ 2\sin^2\psi_m \\ 2\cos\psi_m\sin\psi_m \end{Bmatrix} M(\psi_m)\right)d\psi$$

$$= \frac{1}{N} \sum_{m=1}^{N} \frac{1}{2\pi} \int_0^{2\pi} \begin{Bmatrix} 1 \\ \cos\psi_m \\ \sin\psi_m \\ 2\cos^2\psi_m \\ 2\sin^2\psi_m \\ 2\cos\psi_m \sin\psi_m \end{Bmatrix} M(\psi_m)\,d\psi_m$$

$$= \begin{Bmatrix} M^0 \\ \tfrac{1}{2}M^{1c} \\ \tfrac{1}{2}M^{1s} \\ M^0 + \tfrac{1}{2}M^{2c} \\ M^0 - \tfrac{1}{2}M^{2c} \\ \tfrac{1}{2}M^{2s} \end{Bmatrix}$$

Here M^{nc} and M^{ns} are the harmonics of a Fourier series representation of the rotating blade aerodynamic coefficient M:

$$M(\psi_m) = M^0 + \sum_{n=1}^{\infty} (M^{nc} \cos n\psi_m + M^{ns} \sin n\psi_m)$$

In the present case these harmonics must be evaluated numerically, using the Fourier interpolation formulas (see section 8–3). The aerodynamic coefficient M is evaluated at J points equally spaced around the azimuth. Then the harmonics are obtained from

$$M^0 = \frac{1}{J} \sum_{j=1}^{J} M(\psi_j)$$

$$M^{nc} = \frac{2}{J} \sum_{j=1}^{J} M(\psi_j) \cos n\psi_j$$

$$M^{ns} = \frac{2}{J} \sum_{j=1}^{J} M(\psi_j) \sin n\psi_j$$

Where $\psi_j = j2\pi/J$. The number of harmonics that will be required is $n = N - 1$ for an odd number of blades, and $n = N - 2$ for N even. Good accuracy from the Fourier interpolation requires at least $J = 6n$. Using these expressions, the required harmonics are:

$$\begin{Bmatrix} M^0 \\ \tfrac{1}{2}M^{1c} \\ \tfrac{1}{2}M^{1s} \\ M^0 + \tfrac{1}{2}M^{2c} \\ M^0 - \tfrac{1}{2}M^{2c} \\ \tfrac{1}{2}M^{2s} \end{Bmatrix} = \frac{1}{J}\sum_{j=1}^{J} \begin{Bmatrix} 1 \\ \cos\psi_j \\ \sin\psi_j \\ 2\cos^2\psi_j \\ 2\sin^2\psi_j \\ 2\cos\psi_j\sin\psi_j \end{Bmatrix} M(\psi_j)$$

It follows that the constant coefficient approximation is obtained from the periodic coefficient expressions by the simple transformation

$$\frac{1}{N}\sum_{m=1}^{N}\left\{\ldots\right\} M(\psi_m) \Rightarrow \frac{1}{J}\sum_{j=1}^{J}\left\{\ldots\right\} M(\psi_j)$$

The summation over N blades ($m = 1...N$, $\Delta\psi = 2\pi/N$) for the periodic coefficient case is replaced by a summation over the rotor azimuth ($j = 1...J$, $\Delta\psi = 2\pi/J$) for the constant coefficient approximation. Thus the same procedure may be used to evaluate the coefficients for the two cases, with simply a change in the azimuth increment. The periodic coefficients must be evaluated throughout the period, of course, while the mean values for the constant coefficient approximation are evaluated only once.

11–5 Hub Reactions

11-5.1 Rotating Frame

To evaluate the net rotor forces acting on the hub, it is first necessary to find the shears and moments at the root of an individual blade (see section 9–5). The blade motion considered is the fundamental flap and lag modes. The perturbation forces and velocities for this case were given in section 11–3. The vertical shear forces at the root is

$$\int_0^1 \frac{F_z}{ac}\,dr = T_\theta(\theta - K_P\beta) + T_\lambda\lambda + T_{\dot\beta}\dot\beta + T_\beta\beta + T_{\dot\zeta}\dot\zeta + T_\zeta\zeta$$

where the aerodynamic coefficients are the same as for the flap moments, but without the factor η_β in the integrands:

$$T_\theta = \int_0^1 \tfrac{1}{2}u_T^2\,dr = \frac{1}{6} + \frac{\mu}{2}\sin\psi + \frac{\mu^2}{2}\sin^2\psi$$

$$T_\lambda = -\int_0^1 \tfrac{1}{2}u_T\,dr = -\left(\frac{1}{4} + \frac{\mu}{2}\sin\psi\right)$$

$$T_{\dot\beta} = -\int_0^1 \tfrac{1}{2}\eta_\beta u_T\,dr = -\left(\frac{1}{6} + \frac{\mu}{4}\sin\psi\right)$$

$$T_\beta = -u_R\int_0^1 \tfrac{1}{2}\eta_\beta' u_T\,dr = -u_R\left(\frac{1}{4} + \frac{\mu}{2}\sin\psi\right)$$

$$T_{\dot\zeta} = -\int_0^1 \tfrac{1}{2}\eta_\zeta(u_P + 2u_T\alpha)\,dr$$

$$T_\zeta = -u_R\int_0^1 \tfrac{1}{2}\eta_\zeta'(u_P + 2u_T\alpha)\,dr$$

Here $T_{\dot\beta}$ and T_β have been evaluated using $\eta_\beta = r$. For hover, $T_\zeta = 0$ and

$$T_{\dot\zeta} = -\left[\frac{\lambda_{HP}}{4} + 2\int_0^1 \tfrac{1}{2}r^2\,\alpha\,dr\right] = -\left[\frac{\lambda_{HP}}{4} + \frac{2C_T}{\sigma a}\right]$$

assuming that the induced velocity is uniform and $\eta_\zeta = r$. The behavior of these aerodynamic coefficients for thrust is similar to the behavior of the flap moments, with just a change in the numerical constants because the factor η_β has been removed from the integrands.

The flapwise aerodynamic moment at the rotor hub is $\int_0^1 r(F_z/ac)dr$, which can be evaluated from the flap moment M_F by simply replacing the mode shape η_β by r. (The aerodynamic coefficients of M_F are given in section 11–3). Alternatively, the hub moment can be obtained directly from the flap response using $N_F = I_b\Omega^2(\nu_\beta^2 - 1)\beta$ as discussed in Chapter 9.

The perturbation of the section radial force is

$$\delta\frac{F_r}{ac} = \frac{c_d}{2a}u_R\,\delta u_T + \frac{c_d}{2a}u_T\,\delta u_R - \left(\frac{F_z}{ac}\right)\delta z' - (z')\delta\frac{F_z}{ac}$$

$$= (-z'\tfrac{1}{2}u_T^2)\delta\theta + (z'\tfrac{1}{2}u_T)\delta u_P$$

$$+ \left(\frac{c_d}{2a}u_R - z'\tfrac{1}{2}(u_P + 2u_T\alpha)\right)\delta u_T$$

$$+ \left(\frac{c_d}{2a}u_T\right)\delta u_R - \left(\frac{F_z}{ac}\right)\delta z'$$

The perturbation velocities are given in section 11–3. Here we also require $\delta u_R = \eta_\zeta'\zeta\mu\sin\psi$ and $\delta z' = \eta_\beta'\beta$. Then the radial aerodynamic shear force at the blade root is

$$\int_0^1 \frac{F_r}{ac}\,dr = R_\theta(\theta - K_P\beta) + R_\lambda\lambda + R_{\dot\beta}\dot\beta + R_\beta\beta + R_{\dot\zeta}\dot\zeta + R_\zeta\zeta$$

where

$$R_\theta = -\int_0^1 z'\tfrac{1}{2}u_T^2\,dr$$

$$R_\lambda = \int_0^1 z'\tfrac{1}{2}u_T\,dr$$

$$R_{\dot\beta} = \int_0^1 z'\tfrac{1}{2}\eta_\beta u_T\,dr$$

$$R_\beta = -\int_0^1 \frac{F_z}{ac}\eta_\beta'\,dr + u_R\int_0^1 z'\tfrac{1}{2}\eta_\beta' u_T\,dr$$

$$R_{\dot\zeta} = -\int_0^1 \eta_\zeta\left(\frac{c_d}{2a}u_R - z'\tfrac{1}{2}(u_P + 2u_T\alpha)\right)dr$$

$$R_\zeta = \int_0^1 \eta_\zeta'\left[-u_R\left(\frac{c_d}{2a}u_R - z'\tfrac{1}{2}(u_P + 2u_T\alpha)\right) + \mu\sin\psi\,\frac{c_d}{2a}u_T\right]dr$$

Most of the terms in these aerodynamic coefficients are due to the radial tilt of the thrust vector. The coefficient R_β, which gives the radial force due to flap displacement, is particularly important. Assuming $\eta_\beta = r$, the first term in R_β is

$$-\int_0^1 \frac{F_z}{ac}\,dr = -\frac{S_z}{ac} = -\frac{C_T}{\sigma a}$$

where $S_z = T/N$ is the thrust of a single blade. Therefore, this term gives a radial force $\Delta R = -(T/N)\beta$, which is just the in-plane component of the blade thrust when it is tilted by flap deflection of the blade. Because of the importance of this term, we write

$$R_\beta = -\frac{C_T}{\sigma a} + R_\beta^*$$

where

$$R_\beta^* = \int_0^1 \left(\frac{C_T}{\sigma a} - \tfrac{1}{2} u_T^2\, \alpha \eta_\beta'\right) dr + u_R \int_0^1 z' \tfrac{1}{2}\eta_\beta' u_T\, dr$$

R_β^* will be nonzero if $\eta_\beta' \neq 1$, or in forward flight where the radial velocity u_R is nonzero and the blade lift varies around the azimuth.

If it is assumed that the trim slope of the blade z' is independent of r, that is, $z' = \eta_\beta'\beta = \beta_{trim}$, then the radial force aerodynamic coefficients can be related to the corresponding vertical force coefficients. For example,

$$R_\theta \cong -z' \int_0^1 \tfrac{1}{2} u_T^2\, dr = -z' T_\theta$$

The coefficients can be readily evaluated for the hover case. Assuming uniform inflow, $\eta_\beta = \eta_\zeta = r$, and constant z', we obtain

$$R_\theta = -\frac{z'}{6}$$

$$R_\lambda = \frac{z'}{4}$$

$$R_{\dot\beta} = \frac{z'}{6}$$

$$R_{\dot{\zeta}} = z'\left(\frac{\lambda_{HP}}{4} + \frac{2C_T}{\sigma a}\right)$$

and $R_\beta^* = R_\zeta = 0$.

The in-plane aerodynamic shear force at the blade root is obtained from the section drag force F_x as was the lag moment, giving

$$\int_0^1 \frac{F_x}{ac}\,dr = H_\theta(\theta - K_P\beta) + H_\lambda\lambda + H_{\dot{\beta}}\dot{\beta} + H_\beta\beta + H_{\dot{\zeta}}\dot{\zeta} + H_\zeta\zeta$$

with the aerodynamic coefficients as follows:

$$H_\theta = \int_0^1 \tfrac{1}{2} u_T u_P\,dr$$

$$H_\lambda = \int_0^1 \tfrac{1}{2}(u_T\alpha - u_P)dr$$

$$H_{\dot{\beta}} = \int_0^1 \tfrac{1}{2}\eta_\beta(u_T\alpha - u_P)dr$$

$$H_\beta = u_R\int_0^1 \tfrac{1}{2}\eta'_\beta(u_T\alpha - u_P)dr$$

$$H_{\dot{\zeta}} = -\int_0^1 \eta_\zeta\left(\tfrac{1}{2}u_P\theta + 2u_T\frac{c_d}{2a}\right)dr$$

$$H_\zeta = -u_R\int_0^1 \eta'_\zeta\left(\tfrac{1}{2}u_P\theta + 2u_T\frac{c_d}{2a}\right)dr$$

The coefficient $H_{\dot{\beta}}$, which gives the in-plane force due to flapping velocity, is of particular importance. For hover it becomes

$$H_{\dot{\beta}} = \int_0^1 (\tfrac{1}{2}r^2\alpha - \tfrac{1}{2}r\lambda_{HP})dr = \frac{C_T}{\sigma a} - \frac{\lambda_{HP}}{4}$$

assuming $\eta_\beta = r$. The first term is an in-plane force $\Delta H = (T/N)\dot{\beta}$. The flapping velocity gives an angle-of-attack change $\delta\alpha = -\eta_\beta\dot{\beta}/r = -\dot{\beta}$ that tilts the thrust vector backward and thus produces an in-plane shear force on the blade. This coefficient will therefore be written as

$$H_{\dot{\beta}} = \frac{C_T}{\sigma a} + H_{\dot{\beta}}^*$$

where

$$H_{\dot{\beta}}^* = \int_0^1 \left(\tfrac{1}{2}\eta_\beta u_T \alpha - \frac{C_T}{\sigma a} - \tfrac{1}{2}\eta_\beta u_P\right) dr$$

Even for hover there is a nonzero inflow contribution, $H_{\dot{\beta}}^* = -\lambda_{HP}/4$, which can be a significant fraction of the thrust term in $H_{\dot{\beta}}$. The aerodynamic coefficients for hover are

$$H_\theta = \frac{\lambda_{HP}}{4}$$

$$H_\lambda = \frac{3}{2}\frac{C_T}{\sigma a} - \frac{5}{8}\lambda_{HP} - \frac{\theta_{tw}}{48}$$

$$H_{\dot{\beta}}^* = -\frac{\lambda_{HP}}{4}$$

$$H_{\dot{\zeta}} = -\left[\lambda_{HP}\left(\frac{3}{2}\frac{C_T}{\sigma a} + \frac{3}{8}\lambda_{HP} - \frac{\theta_{tw}}{48}\right) + \frac{c_d}{3a}\right]$$

with $H_\beta = H_\zeta = 0$.

The aerodynamic torque moment acting on the rotor hub at the center of rotation can be obtained from the lag moment M_L (derived in section 11–3) by replacing the mode shape η_ζ by r:

$$\int_0^1 r\frac{F_x}{ac}\,dr = Q_\theta(\theta - K_P\beta) + Q_\lambda\lambda + Q_{\dot{\beta}}\dot{\beta} + Q_\beta\beta + Q_{\dot{\zeta}}\dot{\zeta} + Q_\zeta\zeta$$

Since we have been using the approximation $\eta_\zeta \cong r$ to evaluate the aerodynamic coefficients, there is no need to distinguish in the notation between the lag and torque moments. For numerical work the proper mode shapes for the flap and lag motion can be used.

11-5.2 Nonrotating Frame

The total aerodynamic forces and moments acting on the rotor hub were derived in section 9-5.2. The aerodynamic thrust, torque, drag force, and side force of the rotor are obtained by summing the root reactions of all N blades:

$$\left(\frac{C_T}{\sigma a}\right)_{\text{aero}} = \frac{1}{N} \sum_{m=1}^{N} \int_0^1 \frac{F_z}{ac}\, dr$$

$$\left(\frac{C_Q}{\sigma a}\right)_{\text{aero}} = \frac{1}{N} \sum_{m=1}^{N} \int_0^1 r \frac{F_x}{ac}\, dr$$

$$\left(\frac{2C_H}{\sigma a}\right)_{\text{aero}} = \frac{2}{N} \sum_{m=1}^{N} \left[\cos\psi_m \int_0^1 \frac{F_r}{ac}\, dr + \sin\psi_m \int_0^1 \frac{F_x}{ac}\, dr\right]$$

$$\left(\frac{2C_Y}{\sigma a}\right)_{\text{aero}} = \frac{2}{N} \sum_{m=1}^{N} \left[\sin\psi_m \int_0^1 \frac{F_r}{ac}\, dr - \cos\psi_m \int_0^1 \frac{F_x}{ac}\, dr\right]$$

Expressions for the root forces and moments were derived in the preceding paragraphs as linear functions of the rotating degrees of freedom of the blade. In hover, since the aerodynamic coefficients in these expressions are constants, the summation operators in the total hub reactions act only on the blade degrees of freedom. Hence for this constant coefficient case the summations are easily evaluated using the definitions of the degrees of freedom in the nonrotating frame. It follows that in hover the thrust and torque only involve the rotor collective degrees of freedom (here the coning and collective lag modes). The result is

$$\frac{C_T}{\sigma a} = T_\theta(\theta_0 - K_P\beta_0) + T_\lambda\lambda + T_{\dot\beta}\dot\beta_0 + T_{\dot\zeta}\dot\zeta_0$$

$$\frac{C_Q}{\sigma a} = Q_\theta(\theta_0 - K_P\beta_0) + Q_\lambda\lambda + Q_{\dot\beta}\dot\beta_0 + Q_{\dot\zeta}\dot\zeta_0$$

For hover, the rotor in-plane hub forces involve only the cyclic degrees of freedom in the nonrotating frame. Neglecting the forces due to the blade lag motion, which are much smaller than those due to the flap motion, the rotor drag and side forces in hover are:

$$\begin{pmatrix} \dfrac{2C_H}{\sigma a} \\ \\ \dfrac{2C_Y}{\sigma a} \end{pmatrix} = -\frac{2C_T}{\sigma a}\begin{pmatrix} \beta_{1c} \\ \\ \beta_{1s} \end{pmatrix} + \begin{bmatrix} R_{\dot\beta} & H_{\dot\beta} \\ \\ -H_{\dot\beta} & R_{\dot\beta} \end{bmatrix}\begin{pmatrix} \beta_{1c} \\ \\ \beta_{1s} \end{pmatrix}^{\bullet}$$

$$+\begin{bmatrix} -H_{\dot\beta}^{*} & R_{\dot\beta} \\ -R_{\dot\beta} & -H_{\dot\beta}^{*} \end{bmatrix}\begin{pmatrix} \beta_{1c} \\ \beta_{1s} \end{pmatrix} + \begin{bmatrix} R_\theta & H_\theta \\ -H_\theta & R_\theta \end{bmatrix}\begin{pmatrix} \theta_{1c} - K_P\beta_{1c} \\ \theta_{1s} - K_P\beta_{1s} \end{pmatrix}$$

using $R_\beta = -C_T/\sigma a$ and $H_{\dot\beta} = H_{\dot\beta}^{*} + C_T/\sigma a$. Substituting for the aerodynamic coefficients gives:

$$\begin{pmatrix} \dfrac{2C_H}{\sigma a} \\ \\ \dfrac{2C_Y}{\sigma a} \end{pmatrix} = -\frac{2C_T}{\sigma a}\begin{pmatrix} \beta_{1c} \\ \\ \beta_{1s} \end{pmatrix} + \begin{bmatrix} 0 & \dfrac{C_T}{\sigma a} \\ \\ -\dfrac{C_T}{\sigma a} & 0 \end{bmatrix}\begin{pmatrix} \beta_{1c} \\ \\ \beta_{1s} \end{pmatrix}^{\bullet}$$

$$+\begin{bmatrix} -\dfrac{z'}{6} & \dfrac{\lambda_{HP}}{4} \\ \\ -\dfrac{\lambda_{HP}}{4} & -\dfrac{z'}{6} \end{bmatrix}\begin{pmatrix} -\dot\beta_{1c} - \beta_{1s} + \theta_{1c} - K_P\beta_{1c} \\ \\ -\dot\beta_{1s} + \beta_{1c} + \theta_{1s} - K_P\beta_{1s} \end{pmatrix}$$

Note that this result is of the form

$$\begin{pmatrix} H \\ Y \end{pmatrix} = -T\begin{pmatrix} \beta_{1c} \\ \beta_{1s} \end{pmatrix} + \begin{pmatrix} H_{TPP} \\ Y_{TPP} \end{pmatrix}$$

Thus the rotor drag and side hub forces consist of the in-plane component of the thrust vector tilted with the tip-path plane, plus the in-plane forces relative to the tip-path plane. Consequently, the flapping response is a principal factor in the rotor hub reactions. Recall that the rotor hub moments can also be related to the tip-path-plane tilt. The total moment about the helicopter center of gravity a distance h below the hub is then

$$\begin{pmatrix} -\dfrac{2C_{My}}{\sigma a} - h\dfrac{2C_H}{\sigma a} \\ \dfrac{2C_{Mx}}{\sigma a} - h\dfrac{2C_Y}{\sigma a} \end{pmatrix} = \left[\frac{I_\beta^*(\nu_\beta^2 - 1)}{\gamma} + h\frac{2C_T}{\sigma a}\right]\begin{pmatrix} \beta_{1c} \\ \beta_{1s} \end{pmatrix}$$

plus the moments due to the in-plane forces relative to the tip-path plane.

The in-plane force due to the tilt of the rotor thrust vector with the tip-path plane arises from two sources, one-half from R_β and one-half from $H_{\dot\beta}$. The slope of the blade due to flap deflection tilts the blade lift radially, producing an in-plane component of the thrust (R_β). The rotating-frame flap velocity due to tip-path-plane tilt changes the blade angle of attack, which tilts the blade lift chordwise and thereby produces an in-plane component of the thrust ($H_{\dot\beta}$). While R_β acts only on the flap displacement, the $H_{\dot\beta}$ coefficient produces forces due to tip-path-plane tilt rate ($\dot\beta_{1c}$ and $\dot\beta_{1s}$) as well. Also, any blade pitch change, flap displacement, or flap velocity changes the blade lift magnitude. Since the lift has an in-plane component due to the trim induced velocity, in-plane hub forces are produced by these lift magnitude changes (through $-H_{\dot\beta}^* = H_\theta = \lambda_{HP}/4$).

For a two-bladed rotor the cyclic flap degrees of freedom β_{1c} and β_{1s} are replaced by the teetering mode β_1, so the above results are applicable only when $N \geqslant 3$. When $N = 2$ the in-plane hub forces become

$$\begin{pmatrix} \dfrac{2C_H}{\sigma a} \\ \dfrac{2C_Y}{\sigma a} \end{pmatrix} = R_\beta\begin{pmatrix} 2\cos\psi \\ 2\sin\psi \end{pmatrix}\beta_1 + \begin{pmatrix} R_{\dot\beta}2\cos\psi + H_{\dot\beta}2\sin\psi \\ R_{\dot\beta}2\sin\psi - H_{\dot\beta}2\cos\psi \end{pmatrix}\dot\beta_1$$

$$+ \begin{pmatrix} R_\theta 2\cos\psi + H_\theta 2\sin\psi \\ R_\theta 2\sin\psi - H_\theta 2\cos\psi \end{pmatrix}\left(\theta_1 - K_P\beta_1\right)$$

Thus even in hover the two-bladed rotor dynamics are described by periodic coefficient differential equations.

The derivation of the hub reactions in forward flight follows the derivation for the flap moment in the nonrotating frame (section 11–4). For the constant coefficient approximation we obtain:

$$\begin{pmatrix} \dfrac{C_T}{\sigma a} \\ \dfrac{2C_H}{\sigma a} \\ \dfrac{2C_Y}{\sigma a} \end{pmatrix} = \begin{bmatrix} T_{\dot\beta}^0 & \tfrac{1}{2}T_{\dot\beta}^{1c} & \tfrac{1}{2}T_{\dot\beta}^{1s} \\ R_{\dot\beta}^{1c} + H_{\dot\beta}^{1s} & R_{\dot\beta}^0 + \tfrac{1}{2}R_{\dot\beta}^{2c} + \tfrac{1}{2}H_{\dot\beta}^{2s} & \tfrac{1}{2}R_{\dot\beta}^{2s} + H_{\dot\beta}^0 - \tfrac{1}{2}H_{\dot\beta}^{2c} \\ R_{\dot\beta}^{1s} - H_{\dot\beta}^{1c} & \tfrac{1}{2}R_{\dot\beta}^{2s} - H_{\dot\beta}^0 - \tfrac{1}{2}H_{\dot\beta}^{2c} & R_{\dot\beta}^0 - \tfrac{1}{2}R_{\dot\beta}^{2c} - \tfrac{1}{2}H_{\dot\beta}^{2s} \end{bmatrix} \begin{pmatrix} \dot\beta_0 \\ \dot\beta_{1c} + \beta_{1s} \\ \dot\beta_{1s} - \beta_{1c} \end{pmatrix}$$

$$+ \begin{bmatrix} T_\beta^0 & \tfrac{1}{2}T_\beta^{1c} & \tfrac{1}{2}T_\beta^{1s} \\ R_\beta^{1c} + H_\beta^{1s} & R_\beta^0 + \tfrac{1}{2}R_\beta^{2c} + \tfrac{1}{2}H_\beta^{2s} & \tfrac{1}{2}R_\beta^{2s} + H_\beta^0 - \tfrac{1}{2}H_\beta^{2c} \\ R_\beta^{1s} - H_\beta^{1c} & \tfrac{1}{2}R_\beta^{2s} - H_\beta^0 - \tfrac{1}{2}H_\beta^{2c} & R_\beta^0 - \tfrac{1}{2}R_\beta^{2c} - \tfrac{1}{2}H_\beta^{2s} \end{bmatrix} \begin{pmatrix} \beta_0 \\ \beta_{1c} \\ \beta_{1s} \end{pmatrix}$$

$$+ \begin{bmatrix} T_\theta^0 & \tfrac{1}{2}T_\theta^{1c} & \tfrac{1}{2}T_\theta^{1s} \\ R_\theta^{1c} + H_\theta^{1s} & R_\theta^0 + \tfrac{1}{2}R_\theta^{2c} + \tfrac{1}{2}H_\theta^{2s} & \tfrac{1}{2}R_\theta^{2s} + H_\theta^0 - \tfrac{1}{2}H_\theta^{2c} \\ R_\theta^{1s} - H_\theta^{1c} & \tfrac{1}{2}R_\theta^{2s} - H_\theta^0 - \tfrac{1}{2}H_\theta^{2c} & R_\theta^0 - \tfrac{1}{2}R_\theta^{2c} - \tfrac{1}{2}H_\theta^{2s} \end{bmatrix} \begin{pmatrix} \theta_0 - K_P\beta_0 \\ \theta_{1c} - K_P\beta_{1c} \\ \theta_{1s} - K_P\beta_{1s} \end{pmatrix}$$

$$+ \begin{bmatrix} T_\lambda^0 \\ R_\lambda^{1c} + H_\lambda^{1s} \\ R_\lambda^{1s} - H_\lambda^{1c} \end{bmatrix} \lambda$$

The superscripts denote the harmonics of the Fourier series expansions of the aerodynamic coefficients for forward flight. With more than three blades there are additional degrees of freedom and equations, but the coupled dynamics of the fixed frame and rotor are dominated by these collective and cyclic modes. The forces due to the lag motion have also been neglected here. For hover only the mean terms remain in the matrices, and these equations reduce to the previous results. Perhaps the most important effect of forward flight is the coupling of the vertical and the lateral-longitudinal dynamics.

11–6 Shaft Motion

The linear and angular shaft motions were defined in Fig. 11-2. The perturbation linear velocity of the hub has components $\dot{x}_h$, $\dot{y}_h$, and $\dot{z}_h$, while the orientation of the shaft relative to the inertial reference frame is given by the perturbation angles α_x, α_y, and α_z. We also consider aerodynamic turbulence with velocity components u_G, v_G, and w_G (normalized by the rotor tip speed ΩR). Including the shaft motion and gust, the perturbation velocities of the rotor blade section become

$$\begin{aligned}
\delta u_P &= (\lambda + \dot{z}_h - w_G - \mu\alpha_y) \\
&\quad + r(\dot{\beta} + \dot{\alpha}_x \sin\psi - \dot{\alpha}_y \cos\psi) \\
&\quad + \mu \cos\psi\,\beta \\
\delta u_T &= -r(\dot{\zeta} - \dot{\alpha}_z) - \mu\cos\psi(\zeta - \alpha_z) \\
&\quad - (\dot{x}_h - u_G - \lambda_{HP}\alpha_y)\sin\psi \\
&\quad + (\dot{y}_h + v_G + \lambda_{HP}\alpha_x)\cos\psi \\
\delta u_R &= \mu\sin\psi(\zeta - \alpha_z) \\
&\quad - (\dot{x}_h - u_G - \lambda_{HP}\alpha_y)\cos\psi \\
&\quad - (\dot{y}_h + v_G + \lambda_{HP}\alpha_x)\sin\psi
\end{aligned}$$

The vertical velocity of the hub contributes to δu_P, and the in-plane velocity resolved in the rotating frame contributes to δu_T and δu_R. The influence of the aerodynamic gust components is analogous to the hub velocities. The angular rates of pitch and roll of the rotor disk give a normal velocity of the blade section (δu_P), while the yaw motion of the hub produces velocity perturbations in a manner similar to the blade lag motion. Finally, the trim velocity of the rotor (with components μ and λ_{HP}) is defined relative to the unperturbed inertial reference frame. Pitch and roll rotations of the shaft (α_y and α_x) therefore produce perturbation components of these velocities relative to the hub plane. Since the resulting $\lambda_{HP}\alpha_x$ and $\lambda_{HP}\alpha_y$ terms in the velocity perturbations are an order smaller than the other terms, they can usually be neglected for low inflow helicopter rotors. The blade pitch is measured from the hub plane, so $\delta\theta = \theta - K_P\beta$ still. Only first mode flap and lag motion has been considered for the rotor blade. Since the equivalent mode shape for the angular motion of the hub is exactly $\eta = r$, the blade mode shapes have been approximated by $\eta_\beta = \eta_\zeta = r$ also. Then the same aerodynamic coefficients can be used in many cases for both the blade and shaft motion, simplifying the analysis. For numerical work the actual blade mode shapes can be used, which will modify the aerodynamic coefficients of the rotor degrees of freedom slightly, but will not greatly influence the basic behavior of the rotor.

With these velocity perturbations, the aerodynamic flap and lag moments now become:

$$\begin{aligned} M_F = {} & M_\theta(\theta - K_P\beta) + M_\lambda(\lambda + \dot{z}_h - w_G - \mu\alpha_y) \\ & + M_{\dot\beta}(\dot\beta + \dot\alpha_x \sin\psi - \dot\alpha_y \cos\psi) + M_\beta\beta \\ & + M_{\dot\zeta}(\dot\zeta - \dot\alpha_z) + M_\zeta(\zeta - \alpha_z) \\ & + M_\mu[(-\dot{x}_h + u_G)\sin\psi + (\dot{y}_h + v_G)\cos\psi] \end{aligned}$$

$$\begin{aligned} M_L = {} & Q_\theta(\theta - K_P\beta) + Q_\lambda(\lambda + \dot{z}_h - w_G - \mu\alpha_y) \\ & + Q_{\dot\beta}(\dot\beta + \dot\alpha_x \sin\psi - \dot\alpha_y \cos\psi) + Q_\beta\beta \\ & + Q_{\dot\zeta}(\dot\zeta - \dot\alpha_z) + Q_\zeta(\zeta - \alpha_z) \\ & + Q_\mu[(-\dot{x}_h + u_G)\sin\psi + (\dot{y}_h + v_G)\cos\psi] \end{aligned}$$

(see section 11–3). Since the velocity produced by the shaft motion is similar to that produced by the blade motion already considered, only two new aerodynamic coefficients appear:

$$M_\mu = \int_0^1 \tfrac{1}{2} r(u_P + 2u_T\alpha)dr$$

$$Q_\mu = \int_0^1 r\left(\tfrac{1}{2}u_P\theta + 2u_T \frac{c_d}{2a}\right)dr$$

which are the flap and lag moments due to in-plane velocity of the blade. For hover these coefficients are

$$M_\mu = \frac{2C_T}{\sigma a} + \frac{\lambda_{HP}}{4}$$

$$Q_\mu = \lambda_{HP}\left(\frac{3}{2}\frac{C_T}{\sigma a} + \frac{3}{8}\lambda_{HP} - \frac{\theta_{tw}}{48}\right) + \frac{c_d}{3a}$$

On transforming to the nonrotating frame, the flap moments in hover become

$$M_{F_0} = M_\theta(\theta_0 - K_P\beta_0) + M_\lambda(\lambda + \dot{z}_h - w_G) + M_{\dot\beta}\dot\beta_0 - M_{\dot\zeta}\dot\alpha_z$$

$$M_{F_{1c}} = M_\theta(\theta_{1c} - K_P\beta_{1c}) + M_{\dot\beta}(\dot\beta_{1c} + \beta_{1s} - \dot\alpha_y) + M_\mu(\dot y_h + v_G)$$

$$M_{F_{1s}} = M_\theta(\theta_{1s} - K_P\beta_{1s}) + M_{\dot\beta}(\dot\beta_{1s} - \beta_{1c} + \dot\alpha_x) + M_\mu(-\dot x_h + u_G)$$

for a rotor with three or more blades. If $N \geqslant 4$ there are additional degrees of freedom and equations, but in hover these are not influenced by the shaft motion. The characteristic pattern of limited interaction between the shaft motion and nonrotating degrees of freedom, already found in the inertial terms, is also observed in the hover aerodynamics. There are coning moments due to the vertical velocity and yaw rate of the hub and due to the vertical gusts. There is a longitudinal flap moment due to the lateral in-plane velocity and pitch rate of the hub and due to the lateral gusts. Finally there are lateral flap moments due to the longutudinal in-plane velocity and roll rate of the hub and due to longitudinal gusts. For a two-bladed rotor the aerodynamic flap moment for the teetering mode is instead

$$M_{F_1} = M_\theta(\theta_1 - K_P\beta_1) + M_{\dot\beta}(\dot\beta_1 + \dot\alpha_x \sin\psi - \dot\alpha_y \cos\psi)$$
$$+ M_\mu(-\dot x_h \sin\psi + \dot y_h \cos\psi + u_G \sin\psi + v_G \cos\psi)$$

Thus there are periodic coefficients coupling the rotor and shaft motion, even in hover.

The vertical and in-plane aerodynamic shear forces at the blade root due to the shaft motion are

$$\Delta\int_0^1 \frac{F_z}{ac}\,dr = T_\lambda(\dot z_h - w_G - \mu\alpha_y) + T_{\dot\beta}(\dot\alpha_x \sin\psi - \dot\alpha_y \cos\psi)$$
$$- T_{\dot\zeta}\dot\alpha_z - T_\zeta\alpha_z + T_\mu[(-\dot x_h + u_G)\sin\psi + (\dot y_h + v_G)\cos\psi]$$

$$\Delta\int_0^1 \frac{F_x}{ac}\,dr = H_\lambda(\dot z_h - w_G - \mu\alpha_y) + H_{\dot\beta}(\dot\alpha_x \sin\psi - \dot\alpha_y \cos\psi)$$
$$-H_{\dot\zeta}\dot\alpha_z - H_\zeta\alpha_z + H_\mu[(-\dot x_h + u_G)\sin\psi + (\dot y_h + v_G)\cos\psi]$$

As for the flap and lag moments, there are only two new aerodynamic coefficients, which are due to the in-plane velocity perturbations:

$$T_\mu = \int_0^1 ½(u_P + 2u_T\alpha)\,dr$$

$$H_\mu = \int_0^1 \left(½u_P\theta + 2u_T\frac{c_d}{2a}\right)dr$$

and which for hover are

$$T_\mu = \frac{3C_T}{\sigma a} + \frac{\lambda_{HP}}{4} - \frac{\theta_{tw}}{24}$$

$$H_\mu = \lambda_{HP}\left(\frac{3C_T}{\sigma a} + \frac{3}{4}\lambda_{HP} - \frac{\theta_{tw}}{8}\right) + \frac{c_d}{2a}$$

The radial aerodynamic force due to the shaft motion is

$$\Delta\int_0^1 \frac{F_r}{ac}\,dr = R_\lambda(\dot{z}_h - w_G - \mu\alpha_y) + R_{\dot\beta}(\dot\alpha_x \sin\psi - \dot\alpha_y \cos\psi)$$

$$- R_{\dot\zeta}\dot\alpha_z - R_\zeta\alpha_z + R_\mu[(-\dot{x}_h + u_G)\cos\psi - (\dot{y}_h + v_G)\sin\psi]$$

$$+ R_r[(-\dot{x}_h + u_G)\sin\psi + (\dot{y}_h + v_G)\cos\psi]$$

The two new aerodynamic coefficients in the radial force are due to the in-plane velocity perturbations resolved in the radial direction (R_μ) and in the chordwise direction (R_r):

$$R_\mu = \int_0^1 \frac{c_d}{2a}u_T\,dr = \frac{c_d}{2a}\left(\frac{1}{2} + \mu\sin\psi\right)$$

$$R_r = \int_0^1 \left[\frac{c_d}{2a}u_R - z'½(u_P + 2u_T\alpha)\right]dr = \frac{c_d}{2a}\mu\cos\psi - z'T_\mu$$

Note that in hover R_μ is the single contribution of the radial drag force to the rotor dynamics. All the other radial forces are due to the tilt of the thrust vector by the blade flap deflection. The torque moment at the hub center of rotation is here identical to the lag moment (since $\eta_\zeta = r$ has been assumed), giving

$$\Delta\int_0^1 r\frac{F_x}{ac}\,dr = Q_\lambda(\dot{z}_h - w_G - \mu\alpha_y) + Q_{\dot\beta}(\dot\alpha_x \sin\psi - \dot\alpha_y \cos\psi)$$

$$- Q_{\dot{\zeta}}\dot{\alpha}_z - Q_{\zeta}\alpha_z + Q_{\mu}[(-\dot{x}_h + u_G)\sin\psi + (\dot{y}_h + v_G)\cos\psi]$$

Summing the root forces for all N blades gives the total rotor hub reactions in the nonrotating frame. For hover, the thrust and torque perturbations including the shaft motion are

$$\frac{C_T}{\sigma a} = T_\theta(\theta_0 - K_P\beta_0) + T_\lambda(\lambda + \dot{z}_h - w_G) + T_{\dot{\beta}}\dot{\beta}_0 + T_{\dot{\zeta}}(\dot{\zeta}_0 - \dot{\alpha}_z)$$

$$\frac{C_Q}{\sigma a} = Q_\theta(\theta_0 - K_P\beta_0) + Q_\lambda(\lambda + \dot{z}_h - w_G) + Q_{\dot{\beta}}\dot{\beta}_0 + Q_{\dot{\zeta}}(\dot{\zeta}_0 - \dot{\alpha}_z)$$

The rotor drag and side forces due to the shaft motion are:

$$\Delta\begin{pmatrix} \dfrac{2C_H}{\sigma a} \\[2ex] \dfrac{2C_Y}{\sigma a} \end{pmatrix} = \begin{bmatrix} -R_{\dot{\beta}} & H_{\dot{\beta}} \\[2ex] H_{\dot{\beta}} & R_{\dot{\beta}} \end{bmatrix} \begin{pmatrix} \dot{\alpha}_y \\[2ex] \dot{\alpha}_x \end{pmatrix} + \begin{bmatrix} -(H_\mu + R_\mu) & R_r \\ -R_r & -(H_\mu + R_\mu) \end{bmatrix} \begin{pmatrix} \dot{x}_h - u_G \\ \dot{y}_h + v_G \end{pmatrix}$$

(for three or more blades). The flap response to the rotor shaft motion tilts the rotor thrust vector, and by this means also contributes to the hub in-plane forces.

With a two-bladed rotor, the summation of the root shears over both blades to obtain the hub in-plane forces does not eliminate the sinusoidal variation of the coefficients. The contributions of the shaft motion to the hub forces are for this case:

$$\Delta\begin{pmatrix} \dfrac{2C_H}{\sigma a} \\[2ex] \dfrac{2C_Y}{\sigma a} \end{pmatrix} = \begin{bmatrix} -R_{\dot{\beta}}2C^2 - H_{\dot{\beta}}2CS & R_{\dot{\beta}}2CS + H_{\dot{\beta}}2S^2 \\[2ex] -R_{\dot{\beta}}2CS + H_{\dot{\beta}}2C^2 & R_{\dot{\beta}}2S^2 - H_{\dot{\beta}}2CS \end{bmatrix} \begin{pmatrix} \dot{\alpha}_y \\[2ex] \dot{\alpha}_x \end{pmatrix}$$

$$+ \begin{bmatrix} -H_\mu 2S^2 - R_\mu 2C^2 - R_r 2CS & H_\mu 2CS - R_\mu 2CS + R_r 2C^2 \\ -H_\mu 2CS - R_\mu 2CS - R_r 2S^2 & -H_\mu 2C^2 - R_\mu 2S^2 + R_r 2CS \end{bmatrix} \begin{pmatrix} \dot{x}_h - u_G \\ \dot{y}_h + v_G \end{pmatrix}$$

where $C = \cos\psi$ and $S = \sin\psi$.

For aircraft stability and control analyses, a body axis reference frame is most frequently used. With the inertial axis system considered so far, angular motion of the shaft tilts the axes relative to the trim velocity components μ and λ_{HP}, which are fixed in space, producing perturbations of the air velocity as seen in the reference frame. With body axes, however, the helicopter trim velocity vector remains fixed relative to the reference axes when the shaft is tilted. Thus for body axes the velocity perturbations are

$$\begin{aligned} \delta u_P &= (\lambda + \dot{z}_h - w_G) + r(\dot{\beta} + \dot{\alpha}_z \sin\psi - \dot{\alpha}_y \cos\psi) + \mu\cos\psi\beta \\ \delta u_T &= -r(\dot{\zeta} - \dot{\alpha}_z) - \mu\cos\psi\zeta - (\dot{x}_h - u_G)\sin\psi + (\dot{y}_h + v_G)\cos\psi \\ \delta u_R &= \mu\sin\psi\zeta - (\dot{x}_h - u_G)\cos\psi - (\dot{y}_h + v_G)\sin\psi \end{aligned}$$

So the $\mu\alpha_y$, $\mu\alpha_z$, $\lambda_{HP}\alpha_y$, and $\lambda_{HP}\alpha_x$ terms are dropped from the rotor equations of motion and the hub reactions. The use of body axes adds corresponding terms to the inertial forces, as discussed in section 9–6.

The shaft motion contributions to the nonrotating equations of motion in forward flight can be derived following section 11–4. The constant coefficient approximation for the aerodynamic flap moments in forward flight is:

$$\Delta \begin{pmatrix} M_{F_0} \\ M_{F_{1c}} \\ M_{F_{1s}} \end{pmatrix} = \begin{bmatrix} 0 & \tfrac{1}{2}M_{\dot{\beta}}^{1s} & M_{\dot{\zeta}}^{0} \\ M_{\dot{\beta}}^{0} + \tfrac{1}{2}M_{\dot{\beta}}^{2c} & 0 & M_{\dot{\zeta}}^{1c} \\ 0 & M_{\dot{\beta}}^{0} - \tfrac{1}{2}M_{\dot{\beta}}^{2c} & M_{\dot{\zeta}}^{1s} \end{bmatrix} \begin{pmatrix} -\dot{\alpha}_y \\ \dot{\alpha}_x \\ -\dot{\alpha}_z \end{pmatrix}$$

$$+ \begin{bmatrix} M_\lambda^0 & \tfrac{1}{2}M_\mu^{1s} & \tfrac{1}{2}M_\mu^{1c} \\ 0 & \tfrac{1}{2}M_\mu^{2s} & M_\mu^0 + \tfrac{1}{2}M_\mu^{2c} \\ M_\lambda^{1s} & M_\mu^0 - \tfrac{1}{2}M_\mu^{2c} & \tfrac{1}{2}M_\mu^{2s} \end{bmatrix} \begin{pmatrix} \dot{z}_h - w_G \\ -\dot{x}_h + u_G \\ \dot{y}_h + v_G \end{pmatrix}$$

(this result is for body axes, since the problem considered in this text involving the shaft motion is the helicopter stability and control analysis). Similarly, the constant coefficient approximation for the hub forces is:

$$\Delta \begin{pmatrix} \dfrac{C_T}{\sigma a} \\ \dfrac{2C_H}{\sigma a} \\ \dfrac{2C_Y}{\sigma a} \end{pmatrix} = \begin{bmatrix} \tfrac{1}{2}T_{\dot\beta}^{1c} & \tfrac{1}{2}T_{\dot\beta}^{1s} & T_{\dot\zeta}^{0} \\ R_{\dot\beta}^{0} + \tfrac{1}{2}R_{\dot\beta}^{2c} + \tfrac{1}{2}H_{\dot\beta}^{2s} & \tfrac{1}{2}R_{\dot\beta}^{2s} + H_{\dot\beta}^{0} - \tfrac{1}{2}H_{\dot\beta}^{2c} & R_{\dot\zeta}^{1c} + H_{\dot\zeta}^{1s} \\ \tfrac{1}{2}R_{\dot\beta}^{2s} - H_{\dot\beta}^{0} - \tfrac{1}{2}H_{\dot\beta}^{2c} & R_{\dot\beta}^{0} - \tfrac{1}{2}R_{\dot\beta}^{2c} - \tfrac{1}{2}H_{\dot\beta}^{2s} & R_{\dot\zeta}^{1s} - H_{\dot\zeta}^{1c} \end{bmatrix} \begin{pmatrix} -\dot\alpha_y \\ \dot\alpha_x \\ -\dot\alpha_z \end{pmatrix}$$

$$+ \begin{bmatrix} T_{\lambda}^{0} & \tfrac{1}{2}T_{\mu}^{1s} & \tfrac{1}{2}T_{\mu}^{1c} \\ R_{\lambda}^{1c} + H_{\lambda}^{1s} & \begin{matrix} R_{\mu}^{0} + \tfrac{1}{2}R_{\mu}^{2c} + \tfrac{1}{2}R_{r}^{2s} \\ + H_{\mu}^{0} - \tfrac{1}{2}H_{\mu}^{2c} \end{matrix} & \begin{matrix} -\tfrac{1}{2}R_{\mu}^{2s} + R_{r}^{0} + \tfrac{1}{2}R_{r}^{2c} \\ + \tfrac{1}{2}H_{\mu}^{2s} \end{matrix} \\ R_{\lambda}^{1s} - H_{\lambda}^{1c} & \begin{matrix} \tfrac{1}{2}R_{\mu}^{2s} + R_{r}^{0} - \tfrac{1}{2}R_{r}^{2c} \\ - \tfrac{1}{2}H_{\mu}^{2s} \end{matrix} & \begin{matrix} -R_{\mu}^{0} + \tfrac{1}{2}R_{\mu}^{2c} + \tfrac{1}{2}R_{r}^{2s} \\ - H_{\mu}^{0} - \tfrac{1}{2}H_{\mu}^{2c} \end{matrix} \end{bmatrix} \begin{pmatrix} \dot z_h - w_G \\ -\dot x_h + u_G \\ \dot y_h + v_G \end{pmatrix}$$

Once again observe that forward flight fully couples the dynamics of the rotor and shaft motion.

11–7 Summary

Let us summarize the results derived for the hover aerodynamics, including the hub reactions and shaft motion. For simplicity, the cyclic lag degrees of freedom are dropped, and the special case of a two-bladed rotor is not considered. The axisymmetry of the aerodynamics in vertical flight separates the dynamics into a vertical group, consisting of the coning moment and the rotor thrust and torque:

$$M_{F_0} = M_\theta(\theta_0 - K_P\beta_0) + M_{\dot\beta}\dot\beta_0 + M_\lambda(\lambda + \dot z_h - w_G) + M_{\dot\zeta}(\dot\zeta_0 - \dot\alpha_z)$$

$$\left(\frac{C_T}{\sigma a}\right)_{\text{aero}} = T_\theta(\theta_0 - K_P\beta_0) + T_{\dot\beta}\dot\beta_0 + T_\lambda(\lambda + \dot z_h - w_G) + T_{\dot\zeta}(\dot\zeta_0 - \dot\alpha_z)$$

$$\left(\frac{C_Q}{\sigma a}\right)_{\text{aero}} = Q_\theta(\theta_0 - K_P\beta_0) + Q_{\dot\beta}\dot\beta_0 + Q_\lambda(\lambda + \dot z_h - w_G) + Q_{\dot\zeta}(\dot\zeta_0 - \dot\alpha_z)$$

and a lateral-longitudinal group, consisting of the pitch and roll flap moments and the rotor in-plane hub forces:

$$\begin{pmatrix} M_{F_{1c}} \\ M_{F_{1s}} \end{pmatrix} = M_\theta \begin{pmatrix} \theta_{1c} - K_P\beta_{1c} \\ \theta_{1s} - K_P\beta_{1s} \end{pmatrix} + M_{\dot\beta}\begin{pmatrix} \dot\beta_{1c} + \beta_{1s} - \dot\alpha_y \\ \dot\beta_{1s} - \beta_{1c} + \dot\alpha_x \end{pmatrix} + M_\mu \begin{pmatrix} \dot y_h + v_G \\ -\dot x_h + u_G \end{pmatrix}$$

$$\begin{pmatrix} \dfrac{2C_H}{\sigma a} \\ \dfrac{2C_Y}{\sigma a} \end{pmatrix}_{\text{aero}} = R_\beta \begin{pmatrix} \beta_{1c} \\ \beta_{1s} \end{pmatrix} + \begin{bmatrix} R_{\dot\beta} & H_{\dot\beta} \\ -H_{\dot\beta} & R_{\dot\beta} \end{bmatrix}\begin{pmatrix} \dot\beta_{1c} + \beta_{1s} - \dot\alpha_y \\ \dot\beta_{1s} - \beta_{1c} + \dot\alpha_x \end{pmatrix} + \begin{bmatrix} R_\theta & H_\theta \\ -H_\theta & R_\theta \end{bmatrix}\begin{pmatrix} \theta_{1c} - K_P\beta_{1c} \\ \theta_{1s} - K_P\beta_{1s} \end{pmatrix} + \begin{bmatrix} -(H_\mu + R_\mu) & R_r \\ -R_r & -(H_\mu + R_\mu) \end{bmatrix}\begin{pmatrix} \dot x_h - u_G \\ \dot y_h + v_G \end{pmatrix}$$

The aerodynamic coefficients for hover can be evaluated analytically assuming uniform induced velocity, $\eta_\beta = \eta_\zeta = r$, and neglecting the tip losses:

$$M_\theta = -M_{\dot\beta} = \frac{1}{8}$$

$$M_\lambda = -\frac{1}{6}$$

$$M_\mu = \frac{2C_T}{\sigma a} + \frac{\lambda_{HP}}{4}$$

$$M_{\dot{\xi}} = -\left(\frac{3}{2}\frac{C_T}{\sigma a} + \frac{5}{24}\lambda_{HP} + \frac{\theta_{tw}}{80}\right)$$

$$T_\theta = -T_{\dot{\beta}} = \frac{1}{6}$$

$$T_\lambda = -\frac{1}{4}$$

$$T_{\dot{\xi}} = -\left(\frac{2C_T}{\sigma a} + \frac{\lambda_{HP}}{4}\right)$$

$$Q_\theta = \frac{\lambda_{HP}}{6}$$

$$Q_\lambda = \frac{C_T}{\sigma a} - \frac{\lambda_{HP}}{4}$$

$$Q_{\dot{\beta}} = \frac{3}{4}\frac{C_T}{\sigma a} - \frac{7}{48}\lambda_{HP} + \frac{\theta_{tw}}{160}$$

$$Q_{\dot{\xi}} = -\left[\lambda_{HP}\left(\frac{C_T}{\sigma a} + \frac{\lambda_{HP}}{4}\right) + \frac{c_d}{4a}\right]$$

$$H_\theta = \frac{\lambda_{HP}}{4}$$

$$H_{\dot{\beta}} = \frac{C_T}{\sigma a} - \frac{\lambda_{HP}}{4}$$

$$H_\mu + R_\mu = \lambda_{HP}\left(\frac{3C_T}{\sigma a} + \frac{3}{4}\lambda_{HP} - \frac{\theta_{tw}}{8}\right) + \frac{3c_d}{4a}$$

$$R_\theta = -R_{\dot{\beta}} = -\frac{\beta_{\text{trim}}}{6}$$

$$R_r = -\beta_{\text{trim}}\left(\frac{3C_T}{\sigma a} + \frac{\lambda_{HP}}{4} - \frac{\theta_{tw}}{24}\right)$$

$$R_\beta = -\frac{C_T}{\sigma a}$$

The behavior of a particular aerodynamic coefficient depends primarily on whether it is an out-of-plane or in-plane force, and whether it is due to the blade pitch or an out-of-plane or in-plane velocity. Hence a set of six aerodynamic coefficients is sufficient to establish the basic behavior of the forces, for example:

$$M_\theta = \int_0^1 r F_{z_\theta} dr = \int_0^1 \tfrac{1}{2} r u_T^2 dr$$

$$M_{\dot\beta} = \int_0^1 r^2 F_{z_P} dr = -\int_0^1 \tfrac{1}{2} r^2 u_T dr$$

$$M_\mu = \int_0^1 r F_{z_T} dr = \int_0^1 \tfrac{1}{2} r (u_P + 2u_T\alpha) dr$$

$$H_\theta = \int_0^1 F_{x_\theta} dr = \int_0^1 \tfrac{1}{2} u_T u_P dr$$

$$H_{\dot\beta} = \int_0^1 r F_{x_P} dr = \int_0^1 \tfrac{1}{2} r(u_T\alpha - u_P) dr$$

$$H_\mu = \int_0^1 F_{x_T} dr = \int_0^1 \left(\tfrac{1}{2} u_P\theta + 2u_T\frac{c_d}{2a}\right) dr$$

The remaining coefficients are all similar to one from this set, as may be seen in the expressions for hover given above.

Evaluating the aerodynamic coefficients in forward flight is a more involved task than in hover. The trim pitch and velocities are then periodic functions of the rotor azimuth:

$$u_T = r + \mu \sin\psi$$

$$u_R = \mu \cos\psi$$

$$u_P = \lambda_{HP} + r\dot\beta + \beta\mu\cos\psi$$

$$= \lambda_{HP} + r(\beta_{1s} \cos\psi - \beta_{1c} \sin\psi) + \mu \cos\psi(\beta_0 + \beta_{1c} \cos\psi + \beta_{1s} \sin\psi)$$

$$\theta = \theta_0 + r\theta_{tw} + \theta_{1c} \cos\psi + \theta_{1s} \sin\psi$$

(see Chapter 5). A complete trim solution is thus required, not just a specification of the rotor thrust coefficient. The conditions for helicopter force and moment equilibrium give the tip-path-plane tilt, and the hub-plane inflow ratio $\lambda_{HP} = \lambda_{TPP} - \mu\beta_{1c}$. The thrust coefficient and flapping equations can then be solved for the collective and cyclic pitch control and the coning angle. As for hover, the coefficients can be integrated analytically over the span assuming uniform inflow and $\eta_\beta = r$. It is simplest to leave the expressions in terms of the harmonics of the trim pitch and flap motion, rather than trying to obtain the explicit dependence on the parameters of the operating condition (such as thrust coefficient) as for hover. The results for the basic set of six aerodynamic coefficients are as follows:

$$M_\theta = \frac{1}{8} + \frac{\mu}{3} \sin\psi + \frac{\mu^2}{4} \sin^2\psi$$

$$M_{\dot\beta} = -\left(\frac{1}{8} + \frac{\mu}{6} \sin\psi\right)$$

$$M_\mu = \theta\left(\frac{1}{3} + \frac{\mu}{2} \sin\psi\right) + \theta_{tw}\left(\frac{1}{4} + \frac{\mu}{3} \sin\psi\right) - \left(\frac{1}{4} \lambda_{HP} + \frac{1}{6}\dot\beta + \frac{1}{4}\beta u_R\right)$$

$$H_\theta = \lambda_{HP}\left(\frac{1}{4} + \frac{\mu}{2} \sin\psi\right) + \dot\beta\left(\frac{1}{6} + \frac{\mu}{4} \sin\psi\right) + \beta u_R\left(\frac{1}{4} + \frac{\mu}{2} \sin\psi\right)$$

$$H_{\dot\beta} = \theta\left(\frac{1}{6} + \frac{\mu}{4} \sin\psi\right) + \theta_{tw}\left(\frac{1}{8} + \frac{\mu}{6} \sin\psi\right) - \left(\frac{1}{2} \lambda_{HP} + \frac{1}{3}\dot\beta + \frac{1}{2}\beta u_R\right)$$

$$H_\mu = \theta\left(\frac{1}{2} \lambda_{HP} + \frac{1}{4}\dot\beta + \frac{1}{2}\beta u_R\right) + \theta_{tw}\left(\frac{1}{4} \lambda_{HP} + \frac{1}{6}\dot\beta + \frac{1}{4}\beta u_R\right)$$
$$+ \frac{c_d}{2a}\left(1 + 2\mu \sin\psi\right)$$

In forward flight all the aerodynamic coefficients are periodic functions of the rotor azimuth.

A uniform perturbation of the wake-induced velocity has been included in the aerodynamic analysis for use in a model of the rotor unsteady aerodynamics. In section 10-6.4 such a model was derived, relating the uniform and linear inflow perturbations to the transient changes in the aerodynamic thrust and hub moments on the rotor:

$$\begin{pmatrix} \lambda \\ \lambda_x \\ \lambda_y \end{pmatrix} = \begin{bmatrix} \dfrac{1}{2(\lambda_0 + \sqrt{\mu^2 + \lambda_0^2})} & 0 & 0 \\ 0 & \dfrac{2}{\sqrt{\mu^2 + \lambda_0^2}} & 0 \\ 0 & 0 & \dfrac{2}{\sqrt{\mu^2 + \lambda_0^2}} \end{bmatrix} \begin{pmatrix} C_T \\ -C_{M_y} \\ C_{M_x} \end{pmatrix}_{\text{aero}}$$

or more generally $\tau\dot{\vec{\lambda}} + \vec{\lambda} = (\partial\lambda/\partial L)\vec{L}$, where τ is the time lag of the inflow response and $\partial\lambda/\partial L$ may be a full nine-element matrix. Section 10-6.4 also related the uniform inflow perturbation to the rotor velocity changes:

$$\delta\lambda = -\frac{C_T/2}{(\mu^2 + \lambda_0^2)^{3/2} + \lambda_0 C_T/2}\left(\mu\,\delta\mu + \lambda_0\,\delta\mu_z\right)$$

where here $\delta\mu = -\dot{x}_h + u_G$ and $\delta\mu_z = \dot{z}_h - w_G$. For hover this relation reduces to $\delta\lambda = -\frac{1}{2}\delta\mu_z = -\frac{1}{2}(\dot{z}_h - w_G)$. Hence the coning moment, thrust, and torque due to the rotor vertical velocity perturbations $(\dot{z}_h - w_G)$ in hover are reduced by a factor of one-half by the effect of this inflow perturbation, i.e. $\lambda + \dot{z}_h - w_G = \frac{1}{2}(\dot{z}_h - w_G)$. Without the time lag, these inflow equations reduce to linear algebraic equations for the induced velocity perturbations in terms of the system degrees of freedom. Eliminating λ, λ_x, and λ_y from the model leads to a lift deficiency function representation of the wake effects, as shown in Chapter 10. With large order systems it is more practical to accomplish this substitution numerically, and if the time lag is included the inflow perturbations are actually degrees of freedom. While the model derived in section 10-6.4 is quite elementary, the unsteady aerodynamics are often important to the dynamic behavior of the rotor, so some representation is necessary. The development of a more sophisticated and accurate model is still the subject of research.

This chapter has found the aerodynamic force required for the rotor dynamics investigations that follow, specifically the flap and flap-lag dynamics in Chapter 12, and the helicopter stability and control analysis in Chapter 15. With the present analysis and the rotary wing literature as guides, the aerodynamic forces for other models of the blade motion can be derived as needed. The chapter concludes with a derivation of the aerodynamic forces involved in the pitch and flap motion of the rotor blade.

11–8 Pitch and Flap Motion

Consider the rigid flap and rigid pitch motion of a rotor blade, as in section 9-4.1. For the feathering moments on the blade it is necessary to include the effects of unsteady aerodynamics. The aerodynamic forces required for the equations of motion are the flap moment M_F and the pitch moment about the feathering axis M_f:

$$M_F = \int_0^1 r \frac{F_z}{ac} dr \cong \int_0^1 r \frac{L}{ac} dr$$

$$M_f = \int_0^1 \frac{M_a}{ac} dr$$

The section lift and pitch moment are obtained from the unsteady aerodynamic theory developed for rotary wings in section 10–7:

$$\frac{L}{ac} = C'(k)\tfrac{1}{2}u_T w + \frac{c}{4} C'(k) u_T(\dot{\theta} + \beta)\left(1 + 2\frac{x_A}{c}\right) + \frac{c}{8}\left(w + u_R w'\right)$$

$$\frac{M_a}{ac} = -x_A C'(k)\tfrac{1}{2}u_T w - \frac{c^2}{32} u_T(\dot{\theta} + \beta)\left[1 + 8\frac{x_A}{c} C'(k)\left(1 + 2\frac{x_A}{c}\right)\right]$$
$$- \frac{c^2}{32}\left(\dot{w} + u_R w'\right)\left(1 + 4\frac{x_A}{c}\right)$$

The lift deficiency function $C'(k)$ has been included to account for shed wake effects if necessary, and the effects of reverse flow have been neglected. Here c is the blade chord and x_A is the distance the aerodynamic center is behind the elastic axis. The upwash velocity seen by the blade section is $w = u_T\alpha = u_T\theta - u_P$. With the rigid flap and rigid pitch degrees of freedom this becomes

$$w = u_T\theta - (\lambda + r\dot{\beta} + \beta u_R)$$

The inflow perturbation λ has been included as an alternative to the lift deficiency function in modeling the wake effects. Recall that the degree of freedom θ is the actual blade pitch angle, whereas in the preceding sections of this chapter it has been the pitch control variable. Here the commanded pitch and kinematic pitch-flap coupling enter the solution through the pitch equation of motion (see section 9-4.1). Neglecting the virtual mass terms we have

$$\dot{w} + u_R w' = u_T\dot{\theta} + 2u_R\theta - 2u_R\dot{\beta} + \mu\sin\psi\beta$$

which in hover is just $r\dot{\theta}$ and hence a source of pitch damping. Substituting for the perturbation forces and velocities therefore gives

$$M_F = M_\theta\theta + M_{\dot{\theta}}\dot{\theta} + M_\lambda\lambda + M_{\dot{\beta}}\dot{\beta} + M_\beta\beta$$

$$M_f = m_\theta\theta + m_{\dot{\theta}}\dot{\theta} + m_\lambda\lambda + m_{\dot{\beta}}\dot{\beta} + m_\beta\beta$$

where the aerodynamic coefficients are:

$$M_\theta = \int_0^1 r\left[C'(k)\tfrac{1}{2}u_T^2 + \frac{c}{4}u_R\right]dr$$

$$M_{\dot{\theta}} = \int_0^1 ru_Tc\left[\frac{1}{8} + \frac{1}{4}C'(k)\left(1 + 2\frac{x_A}{c}\right)\right]dr$$

$$M_\lambda = -\int_0^1 r\,[C'(k)\tfrac{1}{2}u_T]\,dr$$

$$M_{\dot{\beta}} = -\int_0^1 r\left[C'(k)\tfrac{1}{2}ru_T + \frac{c}{4}u_R\right]dr$$

$$M_\beta = -\int_0^1 r\left[C'(k)\tfrac{1}{2}u_Ru_T - \frac{c}{4}C'(k)u_T\left(1 + 2\frac{x_A}{c}\right) - \frac{c}{8}\mu\sin\psi\right]dr$$

$$m_\theta = -\int_0^1\left[x_AC'(k)\tfrac{1}{2}u_T^2 + \frac{c^2}{16}u_R\left(1 + 4\frac{x_A}{c}\right)\right]dr$$

$$m_{\dot{\theta}} = -\int_0^1 \frac{c^2}{16} u_T\left(1 + 2\frac{x_A}{c}\right)\left(1 + 4\frac{x_A}{c}\,C'(k)\right) dr$$

$$m_\lambda = \int_0^1 x_A\,C'(k)\,½u_T\,dr$$

$$m_{\dot{\beta}} = \int_0^1 \left[x_A\,C'(k)\,½u_T r + \frac{c^2}{16} u_R\left(1 + 4\frac{x_A}{c}\right)\right] dr$$

$$m_\beta = \int_0^1 \left[x_A\,C'(k)\,½u_R u_T - \frac{c^2}{32} u_T\left(1 + 8\frac{x_A}{c}\,C'(k)\left(1 + 2\frac{x_A}{c}\right)\right)\right.$$

$$\left. - \frac{c^2}{32}\,\mu \sin\psi \left(1 + 4\frac{x_A}{c}\right)\right] dr$$

Virtual mass terms ($\ddot{\theta}$ and $\ddot{\beta}$) have been neglected. The aerodynamic coefficients can be integrated analytically assuming constant chord and aerodynamic center offset, and evaluating the lift deficiency function at an effective radius (such as $r_e = 0.75$) so that the reduced frequency is $k_e = \omega b/(r_e + \mu\ \sin\psi)$. The results are:

$$M_\theta = C'(k_e)\left(\frac{1}{8} + \frac{\mu}{3}\sin\psi + \frac{\mu^2}{4}\sin^2\psi\right) + \frac{c}{8}\,\mu\cos\psi$$

$$M_{\dot{\theta}} = \frac{c}{4}\left[\frac{1}{2} + C'(k_e)\left(1 + 2\frac{x_A}{c}\right)\right]\left(\frac{1}{3} + \frac{\mu}{2}\sin\psi\right)$$

$$M_\lambda = -C'(k_e)\left(\frac{1}{6} + \frac{\mu}{4}\sin\psi\right)$$

$$M_{\dot{\beta}} = -C'(k_e)\left(\frac{1}{8} + \frac{\mu}{6}\sin\psi\right) - \frac{c}{8}\,\mu\cos\psi$$

$$M_\beta = -\mu\cos\psi\,C'(k_e)\left(\frac{1}{6} + \frac{\mu}{4}\sin\psi\right)$$

$$+ \frac{c}{12}\,C'(k_e)\left(1 + 2\frac{x_A}{c}\right)\left(1 + \frac{3}{2}\mu\sin\psi\right) + \frac{c}{16}\,\mu\sin\psi$$

$$m_\theta = -x_A C'(k_e)\left(\frac{1}{6} + \frac{\mu}{2}\sin\psi + \frac{\mu^2}{2}\sin^2\psi\right) - \frac{c^2}{16}\mu\cos\psi\left(1 + 4\frac{x_A}{c}\right)$$

$$m_{\dot\theta} = -\frac{c^2}{32}\left(1 + 2\frac{x_A}{c}\right)\left(1 + 4\frac{x_A}{c}C'(k_e)\right)\left(1 + 2\mu\sin\psi\right)$$

$$m_\lambda = x_A C'(k_e)\left(\frac{1}{4} + \frac{\mu}{2}\sin\psi\right)$$

$$m_{\dot\beta} = x_A C'(k_e)\left(\frac{1}{6} + \frac{\mu}{4}\sin\psi\right) + \frac{c^2}{16}\mu\cos\psi\left(1 + 4\frac{x_A}{c}\right)$$

$$m_\beta = x_A C'(k_e)\mu\cos\psi\left(\frac{1}{4} + \frac{\mu}{2}\sin\psi\right) - \frac{c^2}{64}\left[1 + 8\frac{x_A}{c}C'(k_e)\left(1 + 2\frac{x_A}{c}\right)\right]\left(1 + 2\mu\sin\psi\right) - \frac{c^2}{32}\mu\sin\psi\left(1 + 4\frac{x_A}{c}\right)$$

The noncirculatory lift terms are an order c/R smaller than the flap moments due to the circulatory lift. The rotor wake can significantly reduce the circulatory lift forces through the lift deficiency function, however. The circulatory lift also produces feathering moments, through the pitch axis-aerodynamic center offset x_A. The noncirculatory forces are the source of the aerodynamic pitch damping moment of the blade ($m_{\dot\theta}$). In hover, the aerodynamic coefficients reduce to:

$$M_\theta = \frac{1}{8}C'(k_e)$$

$$M_{\dot\theta} = \frac{c}{12}\left[\tfrac{1}{2} + C'(k_e)\left(1 + 2\frac{x_A}{c}\right)\right]$$

$$M_\lambda = -\frac{1}{6}C'(k_e)$$

$$M_{\dot\beta} = -\frac{1}{8}C'(k_e)$$

$$M_{\beta} = \frac{c}{12} C'(k_e)\left(1 + 2\frac{x_A}{c}\right)$$

$$m_{\theta} = -\frac{x_A}{6} C'(k_e)$$

$$m_{\dot{\theta}} = -\frac{c^2}{32}\left(1 + 2\frac{x_A}{c}\right)\left(1 + 4\frac{x_A}{c} C'(k_e)\right)$$

$$m_{\lambda} = \frac{x_A}{4} C'(k_e)$$

$$m_{\dot{\beta}} = \frac{x_A}{6} C'(k_e)$$

$$m_{\beta} = -\frac{c^2}{64}\left[1 + 8\frac{x_A}{c} C'(k_e)\left(1 + 2\frac{x_A}{c}\right)\right]$$

The circulatory lift produces flap moments due to θ, $\dot{\beta}$, and λ; and the corresponding pitch moments through x_A. The noncirculatory forces produce flap and pitch moments due to $\dot{\theta}$ and β.

Chapter 12

ROTARY WING DYNAMICS II

The aeroelastic equations of motion for the helicopter rotor have been derived in Chapters 9 and 11. The present chapter examines the solutions of these equations for a number of fundamental problems in rotor dynamics. To obtain analytical solutions it is generally necessary to restrict the problem to a small number of degrees of freedom, and to only the fundamental blade motion. Helicopter engineering currently has the capability to routinely calculate the dynamic behavior for much more detailed and complex models of the rotor and airframe, using the high-speed digital computer. Thus elementary analyses such as those presented here are less necessary for actual numerical solutions, but are even more important as the basis for understanding the rotor dynamics.

12–1 Flapping Dynamics

The flapping motion of the rotor blade has a dominant role in almost every aspect of the helicopter behavior. Chapter 5 was largely devoted to finding the steady-state solution for the flap response in forward flight. Here we are concerned with the dynamic behavior of the flap motion. Thus we shall consider the eigenvalues in the rotating and nonrotating frame, and the flap response to control, gust, and shaft motion inputs. The hub reactions in response to shaft motion, including the effects of the flapping dynamics, will also be examined. The equations derived will be of use in Chapter 15 in the investigation of helicopter stability and control characteristics. For the shaft-fixed problems, a single independent blade in the rotating frame, which is a single degree of freedom system, can be considered. When the motion of the entire rotor is considered there are N degrees of freedom, one for each blade.

12-1.1 Rotating Frame

The equation of motion for the fundamental flapping mode of a rotor blade in the rotating frame has been derived in section 9-2.1:

$$I_\beta^*(\ddot{\beta} + \nu_\beta^2 \beta) = \gamma \int_0^1 \eta_\beta \frac{F_z}{ac} dr = \gamma M_F$$

As usual, β is the flapping degree of freedom. An arbitrary rotor blade is considered, described by the rotating natural frequency of the flap motion ν_β and by the out-of-plane mode shape $\eta_\beta(r)$. The normalized flap inertia I_β^* will have a value of approximately 1. The Lock number $\gamma = \rho acR^4/I_b$ characterizes the relative magnitudes of the aerodynamic and inertial forces acting on the blade. In section 11–6 the aerodynamic flap moment

$$M_F = M_\theta(\theta - K_P\beta) + M_{\dot{\beta}}\dot{\beta} + M_\beta\beta + M_\lambda(\lambda - w_G)$$

was obtained. Besides the aerodynamic forces caused by the flap motion, those due to the blade pitch control and a vertical gust velocity are included, as is the pitch-flap kinematic coupling (K_P). Assuming $\eta_\beta = r$, and neglecting reverse flow and tip losses, the aerodynamic coefficients are

$$M_\theta = \frac{1}{8} + \frac{\mu}{3}\sin\psi + \frac{\mu^2}{4}\sin^2\psi$$

$$M_\lambda = -\left(\frac{1}{6} + \frac{\mu}{4}\sin\psi\right)$$

$$M_{\dot{\beta}} = -\left(\frac{1}{8} + \frac{\mu}{6}\sin\psi\right)$$

$$M_\beta = -\mu\cos\psi\left(\frac{1}{6} + \frac{\mu}{4}\sin\psi\right)$$

Thus the flap equation of motion is

$$I_\beta^*\ddot{\beta} - \gamma M_{\dot{\beta}}\dot{\beta} + (I_\beta^*\nu_\beta^2 + K_P\gamma M_\theta - \gamma M_\beta)\beta = \gamma M_\theta\theta + \gamma M_\lambda(\lambda - w_G)$$

which is a linear ordinary differential equation that has periodic coefficients in forward flight. The aerodynamic forces provide the flap damping, flap springs in forward flight (M_β) and through the pitch-flap coupling, and the moments due to the control and gust inputs.

12-1.1.1 Hover Roots

In hover ($\mu = 0$) the aerodynamic environment is axisymmetric, and hence the aerodynamic coefficients are constants. In addition, $M_\beta = 0$ in hover. Thus the homogeneous equation is

$$I_\beta^* \ddot{\beta} - \gamma M_{\dot{\beta}} \dot{\beta} + (I_\beta^* \nu_\beta^2 + K_P \gamma M_\theta)\beta = 0$$

The characteristic equation is

$$I_\beta^* s^2 - \gamma M_{\dot{\beta}} s + (I_\beta^* \nu_\beta^2 + K_P \gamma M_\theta) = 0$$

which can be solved for the eigenvalues or roots of the flap dynamics in hover:

$$s = \frac{\gamma M_{\dot{\beta}}}{2 I_\beta^*} \pm i \sqrt{\nu_\beta^2 + K_P \frac{\gamma M_\theta}{I_\beta^*} - \left(\frac{\gamma M_{\dot{\beta}}}{2 I_\beta^*}\right)^2}$$

Substituting $I_\beta^* = 1$, $-M_{\dot{\beta}} = M_\theta = 1/8$ gives

$$s = -\frac{\gamma}{16} \pm i \sqrt{\nu_\beta^2 + K_P \frac{\gamma}{8} - \left(\frac{\gamma}{16}\right)^2}$$

Unless the Lock number γ is very large, the transient flap motion in the rotating frame is a damped oscillation, with frequency, natural frequency, and damping ratio as follows:

$$\omega = \operatorname{Im} s = \sqrt{\nu_\beta^2 + K_P \frac{\gamma}{8} - \left(\frac{\gamma}{16}\right)^2}$$

$$\omega_n = |s| = \sqrt{\nu_\beta^2 + K_P \frac{\gamma}{8}} = \nu_{\beta_e}$$

$$\zeta = -\operatorname{Re} s/|s| = \frac{\gamma}{16 \nu_{\beta_e}}$$

The pitch-flap coupling K_P introduces an aerodynamic spring on the flap motion through M_θ, giving the effective flapping natural frequency ν_{β_e}. The damping ratio is typically around 50% critical damping, so the flap motion is highly damped. The source of this damping is the aerodynamic lift forces on the blade due to the angle-of-attack change produced by a flapping velocity. Fig. 12-1 shows the hover eigenvalues, with typical roots for articulated and hingeless rotors. Note that for the articulated rotor the

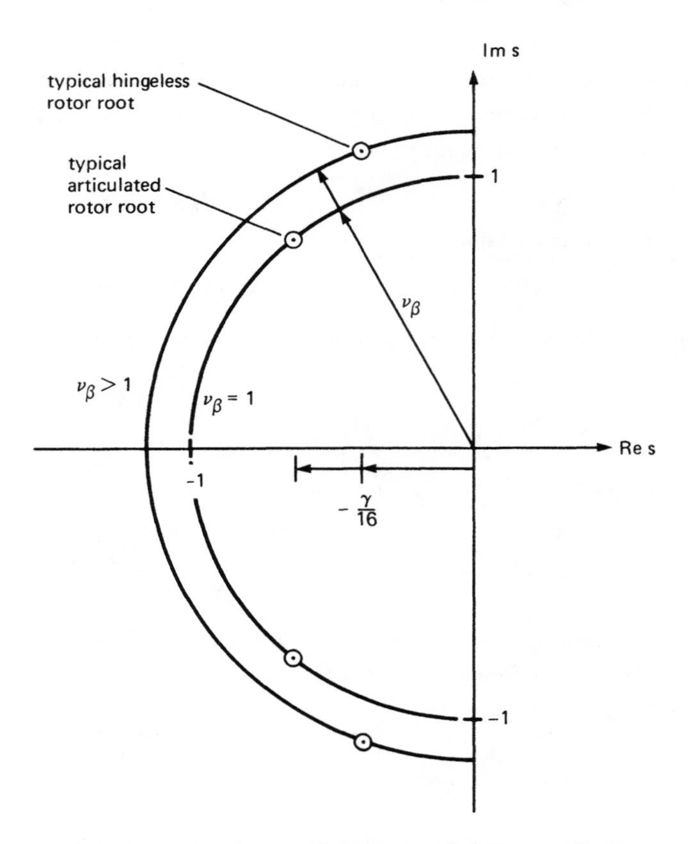

Figure 12-1 Flapping roots in hover (no pitch-flap coupling).

frequency is below 1/rev, while for the hingeless rotor ($\nu_\beta > 1$ and small γ) the frequency is likely to be above 1/rev. The location of the roots is determined by the natural frequency ν_{β_e} (which gives the distance from the origin) and the damping $\mathrm{Re}\, s = -\gamma/16$ (which gives the distance from the imaginary axis).

For $\gamma = 0$, the roots are $s = \pm i\nu_\beta$. Since there are no aerodynamic forces in this case, the motion is an undamped oscillation at the frequency ν_β determined by the centrifugal and structural stiffness of the blade. For $\gamma > 0$, the locus describes a circle with radius $(\nu_\beta^2 + K_P^2)^{1/2}$ and center at $s = -K_P$ on the real axis. The location of the two complex conjugate roots on this circle can be determined from the real part $\mathrm{Re}\, s = -\gamma/16$, which depends on the Lock number alone. For large enough γ, namely at $\gamma/16 = K_P + (\nu_\beta^2 + K_P^2)^{1/2}$ with $s = -\gamma/16$, the loci intercept the real axis. In the absence of pitch-flap coupling the intercept occurs at $\gamma = 16\nu_\beta$, which is quite large for modern rotors. Consequently, unless the pitch-flap coupling ($K_P = \tan\delta_3$) is significantly negative the flap roots will be a complex conjugate pair, implying oscillatory transient motion. As the Lock number becomes still larger ($\gamma \to \infty$), one of the roots approaches $s = -\infty$ on the real axis and the other goes to $s = -K_P$. If $K_P < 0$, one of the roots on the real axis goes through the origin into the right half-plane as γ increases. The criterion for stable motion ($\mathrm{Re}\, s < 0$) is then

$$\frac{\gamma}{16} < -\frac{\nu_\beta^2}{2K_P}$$

The main rotors of the helicopter are generally well away from this boundary, but it may be a concern for rotors with large negative pitch-flap coupling and small Lock number. Since this instability is a static divergence, the boundary is simply determined by the spring terms in the flap equation. For stable motion a net positive flap spring is required, or $\nu_{\beta_e}^2 > 0$, which gives the above boundary. This flap divergence is primarily a limit on the allowable negative pitch-flap coupling, so the stability requirement can be written instead as $K_P > -8\nu_\beta^2/\gamma$.

12-1.1.2 Forward Flight Roots

The homogeneous equation of the flap motion in forward flight is

$$I_\beta^* \ddot{\beta} - \gamma M_{\dot{\beta}} \dot{\beta} + (I_\beta^* \nu_\beta^2 + K_P \gamma M_\theta - \gamma M_\beta)\beta = 0$$

The flap inertia I_β can just be included in the Lock number here, so with no loss of generality $I_\beta^* = 1$ can be assumed. The aerodynamic coefficients are

$$M_{\dot{\beta}} = -\left(\frac{1}{8} + \frac{\mu}{6}\sin\psi\right)$$

$$M_\beta = -\mu\cos\psi\left(\frac{1}{6} + \frac{\mu}{4}\sin\psi\right)$$

$$M_\theta = \frac{1}{8} + \frac{\mu}{3}\sin\psi + \frac{\mu^2}{4}\sin^2\psi$$

assuming $\eta_\beta = r$ and neglecting reverse flow. Above about $\mu = 0.5$ it is necessary to include reverse flow in the aerodynamic coefficients, as in section 5–20. The hover solution was found in the last section:

$$s = -\frac{\gamma}{16} \pm i\sqrt{\nu_\beta^2 + K_P\frac{\gamma}{8} - \left(\frac{\gamma}{16}\right)^2}$$

Forward flight ($\mu > 0$) introduces periodic coefficients due to the rotation of the blade relative to the helicopter forward velocity; these coefficients radically influence the behavior of the root loci and also the analysis techniques required. The root loci of a time-invariant system may exhibit behavior in which two roots start as complex conjugates, meet at the real axis, and then proceed in opposite directions on the real axis. With periodic coefficients this behavior is generalized so that it can occur at any frequency that is a multiple of ½/rev, not just on the real axis. The property of the solution that allows this behavior is the fact that the eigenvectors are themselves periodic, instead of constant as for a time-invariant system. In Chapter 8 the behavior of roots of periodic coefficient differential equations was discussed and procedures for calculating the roots were developed.

The stability of the rotor flapping motion in forward flight has been examined in a number of investigations (see section 12-1.6). For small μ it is possible to obtain analytical solutions for the roots, but numerical methods are required at moderate and high advance ratios. At $\mu = 0$ the roots are complex conjugate pairs (or perhaps two real roots) determined by ν_β, γ, and K_P as described in the last section. The frequency of the hover roots has an important influence on the behavior at low μ. For values of γ and ν_β such that the hover root frequency is not too close to a multiple of ½/rev, the roots for low advance ratio only exhibit an order μ^2 change in frequency, which is quite small even at $\mu = 0.5$; there is no change in the real part of the root. When the hover root frequency is near a multiple of one-half the fundamental frequency of the system, there can occur a degradation of the stability, perhaps even an instability, which is characteristic of periodic coefficient equations. If the hover root frequency is near n/rev, then as μ increases the roots approach n/rev while remaining a complex conjugate pair. The roots reach $\text{Im}\, s = n$/rev for some critical μ, and then for still larger μ the frequency remains fixed while the real part of one root is increased and that of the other root decreased. The root being destabilized may cross into the right half-plane for large enough μ, indicating that the system has become unstable because of the periodic coefficients. Similar behavior can occur if the hover root frequency is near $n + \frac{1}{2}$/rev. For the hover root frequency near $\frac{1}{2}$/rev there are order μ influences of the periodic coefficients. There is thus initially an order μ change in the frequency, with the real part of the root remaining at the hover value $\text{Re}\, s = -\gamma/16$. The roots reach $\text{Im}\, s = \frac{1}{2}$ rev for a value of the advance ratio that decreases as the hover root moves closer to $\frac{1}{2}$/rev. For larger μ there are order μ changes in the real part of the roots while the frequency is fixed at $\frac{1}{2}$/rev. The order μ reduction in damping is small compared to the large aerodynamic damping in hover, so the flapping stability remains high for small advance ratio, even with the influence of the periodic coefficients. The roots exhibit a similar behavior when the hover frequency is near 1/rev, except that all the changes are of order μ^2, and hence are much smaller than those near $\frac{1}{2}$/rev. At $\mu = 2.25$ or so (there is some dependence on ν_β, γ, and K_P) a flapping instability is encountered because of the periodic forces on the rotor blade in forward flight. This instability usually occurs in a region where the frequency is fixed at 1/rev and the real part of one root has been decreased enough for it to go into the right half-plane. For advance ratios high enough to encounter this instability it is of course necessary to include the reverse flow effects in the aerodynamic coefficients. It is also found that other degrees of freedom (such as elastic bending, lag motion, and torsion motion) significantly reduce the advance ratio at the stability boundary. A representation of the rotor blade motion by just the fundamental flap mode is not adequate at very high advance ratio.

Fig. 12-2 shows typical root loci of the rotor blade flap motion from hover ($\mu = 0$) up to about $\mu = 0.5$. Three cases are shown: (a) a typical articulated rotor with $\gamma = 12$ and $\nu_\beta = 1.0$, for which the hover frequency is near $\frac{1}{2}$/rev; (b) a typical hingeless rotor with $\gamma = 6$ and $\nu_\beta = 1.15$, for which the hover frequency is near 1/rev; and (c) an intermediate case with $\gamma = 6$ and $\nu_\beta = 1.0$, for which the hover frequency is not near a multiple of $\frac{1}{2}$/rev. There is a pair of roots for each case. The articulated rotor (case a) illustrates the order μ behavior near $\frac{1}{2}$/rev, and the hingeless rotor (case b) illustrates the order μ^2 behavior near 1/rev. Case (c) shows just the small frequency change of order μ^2 for roots away from a multiple of $\frac{1}{2}$/rev. The eigenvalues depend primarily on the Lock number and advance ratio, so the above results can be presented as contours of constant real and imaginary parts of the roots on the

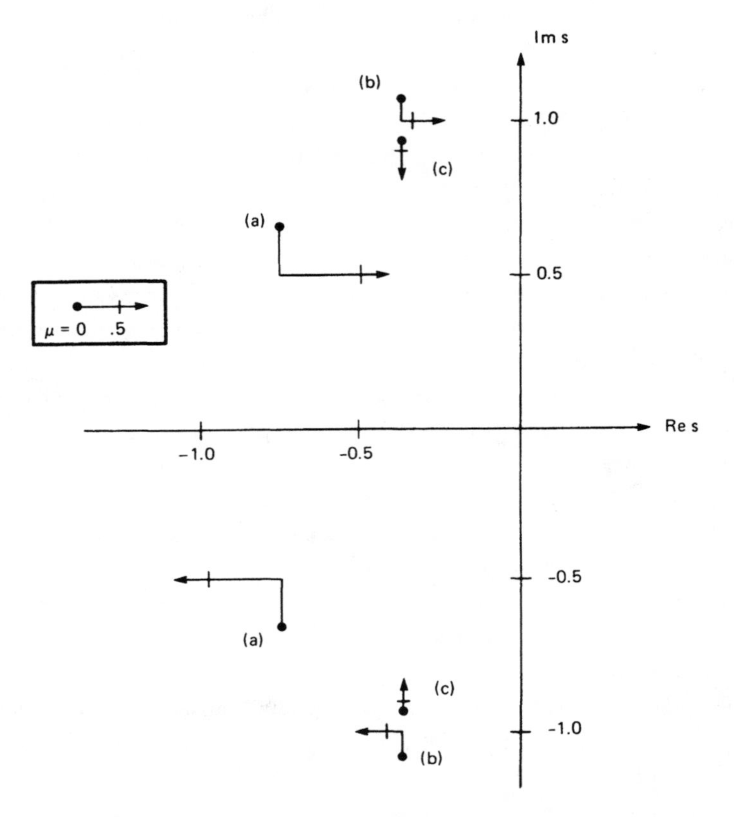

Figure 12-2 Influence of forward flight on the flapping roots for (a) $\nu = 1$ and $\gamma = 12$; (b) $\nu = 1.15$ and $\gamma = 6$; and (c) $\nu = 1$ and $\gamma = 6$.

γ–μ plane. Fig. 12-3 is such a plot for the case $\nu_\beta = 1$ and $K_P = 0$. The regions in which the frequency is fixed at ½/rev or 1/rev are due to the periodic coefficients. Since a horizontal line in Fig. 12-3 corresponds to constant γ, the variation of Im s and Re s as such a line is traversed gives the root locus for varying μ. For example, consider the articulated rotor with $\gamma = 12$. As μ increases the ½/rev region comes closer, indicating that the frequency is approaching ½/rev. When the $\gamma = 12$ line goes into the ½/rev region, the frequency of the root remains fixed while for each point in the region there are two values of Re s, one more stable and one less stable than the hover root.

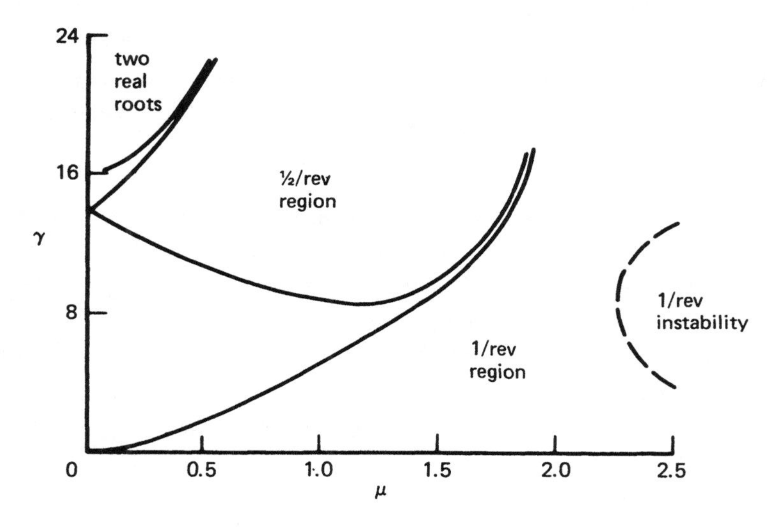

Figure 12-3 Flap roots in forward flight ($\nu_\beta = 1$ and $K_P = 0$).

When μ is of order 1 the real parts of the roots approach each other again, and at about $\mu = 1.7$ the roots move quickly from ½/rev to 1/rev as complex conjugates, hence with the same real parts. After the 1/rev region is entered, the damping of one root decreases again while that of the other increases. The branch being destabilized finally crosses into the right half-plane at about $\mu = 2.3$. Fig. 12-3 shows that the critical regions where the frequency is fixed at a multiple of ½/rev increase in importance as the periodic forces increase with μ, eventually dominating the behavior of the root loci at high advance ratio.

Because of the greater ease and scope of the analysis of constant coefficient differential equations, it is most desirable to have a time-invariant model of the rotor dynamics in forward flight. Such a model must be approximate, since periodic systems have unique behavior, but the approximation may be satisfactory for certain applications. If the mean values of the coefficients in the rotating frame are used, the only influence of forward flight on the flap moments that remains is an order μ^2 change in M_θ. If there is no pitch-flap coupling, forward flight has no influence at all on the eigenvalues. Unless μ is very small this will not be a satisfactory approximation. However, it was found in section 11–4 that by using the mean values of the coefficients in the nonrotating frame much more of the influence of forward flight is retained. The constant coefficient approximation in the nonrotating frame has been examined for the flap motion in a number of investigations, including Hohenemser and Yin (1972b, 1974a), Biggers (1974), and Johnson (1974c).

Consider for example the case of the articulated rotor with $\gamma = 12$ and $\nu_\beta = 1.0$. In the rotating frame the roots encounter the ½/rev critical region as μ increases. Recall from the discussion in section 8–5 that in the transformation to the nonrotating frame, the coning roots are unchanged while the low frequency and high frequency flap mode roots are shifted in frequency by 1/rev from the rotating roots, as shown in Fig. 12-4. (There are additional roots for rotors with more than three blades.) Fig. 12-4 also shows the results of the constant coefficient approximation in the nonrotating frame, which

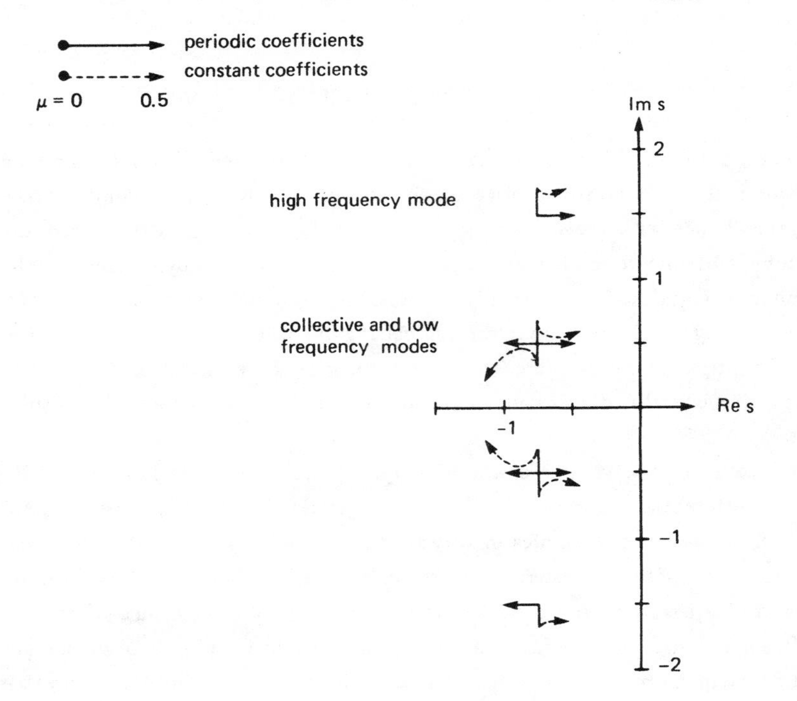

Figure 12-4 Comparison of the flapping roots of a three-bladed rotor from hover to about $\mu = 0.5$, for the periodic coefficient solution and the constant coefficient approximation; from Biggers (1974).

gives the influence of forward flight on the flap roots remarkably well. The approximation does not work in the rotating frame because without periodic coefficients the two flap roots must always remain complex conjugates. The transformation to the nonrotating frame places four roots near ½/rev, however, two from the coning mode and two from the low frequency flap mode. These four roots can behave in a fashion similar to the roots of a periodic system in a critical region (i.e. the frequency is fixed at a multiple of ½/rev while the real parts decrease for one root and increase for the other), while remaining complex conjugates as required of a constant coefficient system. For the hingeless rotor example ($\gamma = 6$ and $\nu_\beta = 1.15$, which places the rotating frequency near 1/rev as shown in Fig. 12-2), the transformation to the nonrotating frame shifts the roots of the low frequency flap mode to the real axis, where the constant coefficient approximation can and does model the correct behavior of the roots. In general, the characteristic behavior of the roots of a periodic system can be exhibited by the roots of a time-invariant system only in the cases of two roots on the real axis, or four complex roots (two at positive frequency and two at negative frequency). The transformation to the nonrotating frame produces such loci because it shifts the frequency of the rotating roots (by $\pm n$/rev for the β_{nc} and β_{ns} modes). The high frequency modes in the nonrotating frame will always be isolated pairs (one at positive frequency and one at negative frequency), which must remain complex conjugates in the constant coefficient approximation. It is expected therefore that the constant coefficient approximation will be least satisfactory when the rotor high frequency dynamics must be modeled. Increasing the number of blades will improve the approximation by increasing the number of coupled degrees of freedom of the model in the nonrotating frame. In summary, the constant coefficient approximation to the rotor dynamics in forward flight produces differential equations that can be more easily and more thoroughly analyzed, but no longer describe the real rotor. The results from the time-invariant model must always be approximate at best. However, the constant coefficient approximation in the nonrotating frame generally gives results that are remarkably close to the correct solution, particularly for the behavior involving the lower frequency modes, as long as the advance ratio is not too large. The validity of the constant coefficient approximation should always be checked, by comparing the exact and approximate solutions for the particular type of problem being considered.

12-1.1.3 Hover Transfer Function

The response of the blade flapping motion to pitch and gust imputs can be defined by the transfer function. Only the hovering case will be considered, since the periodic coefficients in forward flight introduce interharmonic coupling. In forward flight a sinusoidal input at a single frequency ω does not produce an output at that frequency alone, but rather at all frequencies $\omega \pm n$/rev. For a small advance ratio at least, the dominant response will still be at the input frequency ω, however.

For hover, the flapping equation of motion in the rotating frame is

$$I_\beta^* \ddot{\beta} - \gamma M_{\dot{\beta}} \dot{\beta} + (I_\beta^* \nu_\beta^2 + K_P \gamma M_\theta)\beta = \gamma M_\theta \theta - \tfrac{1}{2}\gamma M_\lambda w_G$$

Recall from section 11–7 that in hover the inflow perturbation reduces the effect of the vertical velocity by a factor of one-half, $\lambda - w_G = -\tfrac{1}{2} w_G$, which has been used in the above equation. The transfer function of the flap motion in hover is then

$$\beta = \frac{\gamma M_\theta \theta - \tfrac{1}{2}\gamma M_\lambda w_G}{I_\beta^* \left(s^2 - \dfrac{\gamma M_{\dot{\beta}}}{I_\beta^*} s + \nu_{\beta_e}^2 \right)}$$

Here s is the Laplace variable, and $\nu_{\beta_e}^2 = \nu_\beta^2 + K_P \gamma M_\theta / I_\beta^*$. The poles (the roots of the denominator polynomial) are the hover eigenvalues, and there are no zeros in this case. Substituting for the coefficients gives

$$\beta = \frac{\dfrac{\gamma}{8}\theta + \dfrac{\gamma}{12} w_G}{s^2 + \dfrac{\gamma}{8} s + \nu_{\beta_e}^2}$$

The frequency response is obtained from $s = i\omega$:

$$\beta = \frac{\dfrac{\gamma}{8}\theta + \dfrac{\gamma}{12} w_G}{\nu_{\beta_e}^2 - \omega^2 + \dfrac{\gamma}{8} i\omega}$$

Fig. 12-5 shows the magnitude and phase of the frequency response of the blade flap motion to pitch control inputs for the typical case $\nu_\beta = 1$, $K_P = 0$, and $\gamma = 8$. The frequency response is that of a highly damped second-order

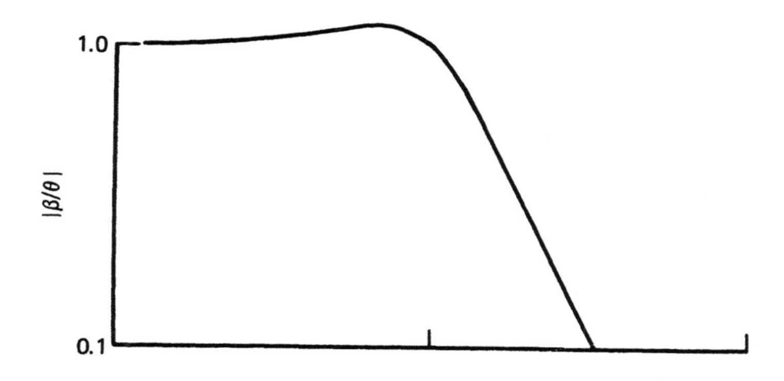

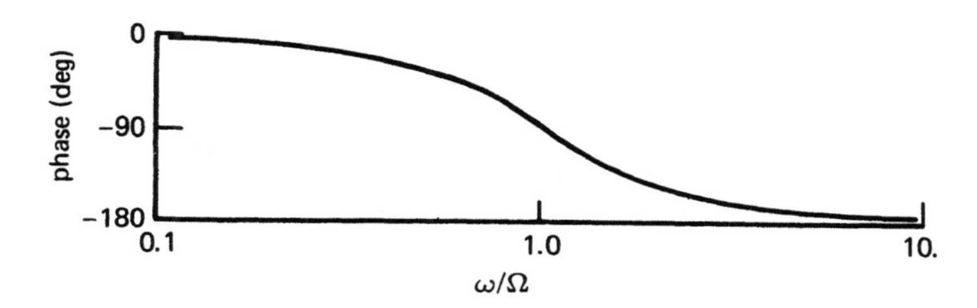

Figure 12-5 Frequency response of the flap motion to pitch control imputs ($\nu_\beta = 1, K_P = 0, \gamma = 8$).

system with natural frequency ν_{β_e}. The static response ($\omega = 0$) is $\beta/\theta = \gamma/8\nu_{\beta_e}^2$, and at high frequency the magnitude decreases and the phase shifts to $-180°$ as the inertia dominates the system.

12-1.2 Nonrotating Frame

The equations for the flap motion in the nonrotating frame are obtained by application of the Fourier coordinate transformation to the rotating equation. Here only the case of hover and three or more blades will be considered, so the equations have constant coefficients. Excitation by blade pitch control, shaft motion, and aerodynamic gusts is included. The results of the derivations in sections 9–6 and 11–7 for this case are as follows:

$$I_\beta^*\ddot{\beta}_0 - \gamma M_{\dot\beta}\dot{\beta}_0 + (I_\beta^*\nu_\beta^2 + K_P\gamma M_\theta)\beta_0$$
$$= \gamma M_\theta\,\theta_0 + M_\lambda(\lambda + \dot{z}_h - w_G) - S_\beta^*\ddot{z}_h$$

$$I_\beta^*\begin{pmatrix}\beta_{1c}\\ \beta_{1s}\end{pmatrix}^{\cdot\cdot} + \begin{bmatrix}-\gamma M_{\dot\beta} & 2I_\beta^*\\ -2I_\beta^* & -\gamma M_{\dot\beta}\end{bmatrix}\begin{pmatrix}\beta_{1c}\\ \beta_{1s}\end{pmatrix}^{\cdot}$$
$$+ \begin{bmatrix}I_\beta^*(\nu_\beta^2 - 1) + K_P\gamma M_\theta & -\gamma M_{\dot\beta}\\ \gamma M_{\dot\beta} & I_\beta^*(\nu_\beta^2 - 1) + K_P\gamma M_\theta\end{bmatrix}\begin{pmatrix}\beta_{1c}\\ \beta_{1s}\end{pmatrix}$$
$$= \gamma M_\theta\begin{pmatrix}\theta_{1c}\\ \theta_{1s}\end{pmatrix} + \gamma M_\mu\begin{pmatrix}\dot{y}_h + v_G\\ -\dot{x}_h + u_G\end{pmatrix} + I_{\beta\alpha}^*\begin{pmatrix}\ddot{\alpha}_y\\ -\ddot{\alpha}_x\end{pmatrix}$$
$$+ \begin{bmatrix}-\gamma M_{\dot\beta} & 2I_{\beta\alpha}^*\\ -2I_{\beta\alpha}^* & -\gamma M_{\dot\beta}\end{bmatrix}\begin{pmatrix}\dot{\alpha}_y\\ -\dot{\alpha}_x\end{pmatrix}$$

$$I_\beta^*\begin{pmatrix}\beta_{nc}\\ \beta_{ns}\end{pmatrix}^{\cdot\cdot} + \begin{bmatrix}-\gamma M_{\dot\beta} & 2nI_\beta^*\\ -2nI_\beta^* & -\gamma M_{\dot\beta}\end{bmatrix}\begin{pmatrix}\beta_{nc}\\ \beta_{ns}\end{pmatrix}^{\cdot}$$
$$+ \begin{bmatrix}I_\beta^*(\nu_\beta^2 - n^2) + K_P\gamma M_\theta & -n\gamma M_{\dot\beta}\\ n\gamma M_{\dot\beta} & I_\beta^*(\nu_\beta^2 - n^2) + K_P\gamma M_\theta\end{bmatrix}\begin{pmatrix}\beta_{nc}\\ \beta_{ns}\end{pmatrix}$$
$$= \gamma M_\theta\begin{pmatrix}\theta_{nc}\\ \theta_{ns}\end{pmatrix}$$

$$I_\beta^*\ddot{\beta}_{N/2} - \gamma M_{\dot\beta}\dot{\beta}_{N/2} + (I_\beta^*\nu_\beta^2 + K_P\gamma M_\theta)\beta_{N/2} = \gamma M_\theta\,\theta_{N/2}$$

The flapping degrees of freedom in the nonrotating frame are the coning mode β_0, the tip-path-plane tilt modes β_{1c} and β_{1s}, and the reactionless modes (β_{nc}, β_{ns}, $\beta_{N/2}$) as required to give a total of N degrees of freedom for an N-bladed rotor. Note that the only coupling of the flap degrees of freedom is between β_{1c} and β_{1s}, and between β_{nc} and β_{ns}. Moreover, only the coning and tip-path-plane tilt degrees of freedom respond to the shaft motion and gusts; consequently, these three degrees of freedom are of the most interest. For the shaft-fixed case, these equations also apply to the two-bladed rotor in hover, where the degrees of freedom are the coning mode β_0 and the teetering mode β_1.

12-1.2.1 Hover Roots and Modes

The homogeneous equations give the following characteristic equations for the flap motion in the nonrotating frame:

$$(I_\beta^* s^2 - \gamma M_{\dot\beta} s + I_\beta^* \nu_\beta^2 + K_P \gamma M_\theta)\beta_0 = 0$$

$$\begin{bmatrix} I_\beta^* s^2 - \gamma M_{\dot\beta} s + I_\beta^*(\nu_\beta^2 - n^2) + K_P\gamma M_\theta & n(2I_\beta^* s - \gamma M_{\dot\beta}) \\ -n(2I_\beta^* s - \gamma M_{\dot\beta}) & I_\beta^* s^2 - \gamma M_{\dot\beta} s + I_\beta^*(\nu_\beta^2 - n^2) + K_P\gamma M_\theta \end{bmatrix} \begin{pmatrix} \beta_{nc} \\ \beta_{ns} \end{pmatrix} = 0$$

$$(I_\beta^* s^2 - \gamma M_{\dot\beta} s + I_\beta^* \nu_\beta^2 + K_P \gamma M_\theta)\beta_{N/2} = 0$$

Since the characteristic equations for β_0 and $\beta_{N/2}$ are the same as for the single blade in the rotating frame (see section 12-1.1.1), the nonrotating eigenvalues s_{NR} for the coning and reactionless modes are equal to the rotating eigenvalues:

$$s_{NR} = s_R = \frac{\gamma M_{\dot\beta}}{2I_\beta^*} \pm i\sqrt{\nu_\beta^2 + K_P \frac{\gamma M_\theta}{I_\beta^*} - \left(\frac{\gamma M_{\dot\beta}}{2I_\beta^*}\right)^2}$$

For the β_{nc} and β_{ns} degrees of freedom, the determinant of the matrix gives

$$(I_\beta^* s^2 - \gamma M_{\dot\beta} s + I_\beta^*(\nu_\beta^2 - n^2) + K_P\gamma M_\theta)^2 + n^2(2I_\beta^* s - \gamma M_{\dot\beta})^2 = 0$$

which has the solution $s_{NR} = s_R \pm in$. Thus the transformation to the nonrotating frame simply shifts the frequency of the β_{nc} and β_{ns} roots by $\pm n$/rev, while the real part remains unchanged. The corresponding eigenvectors (using the characteristic equation to replace the numerator) are

$$\frac{\beta_{nc}}{\beta_{ns}} = \frac{I_\beta^* s^2 - \gamma M_{\dot\beta} s + I_\beta^*(\nu_\beta^2 - n^2) + K_P \gamma M_\theta}{n(2I_\beta^* s - \gamma M_{\dot\beta})} = \pm i$$

Thus the high frequency mode $s_{NR} = s_R + in$ has the eigenvector $\beta_{nc}/\beta_{ns} = i$, and the low frequency mode $s_{NR} = s_R - in$ has the eigenvector $\beta_{nc}/\beta_{ns} = -i$. For both modes, then, β_{nc} and β_{ns} have equal magnitude but are 90° apart in phase. For the high frequency mode β_{nc} leads β_{ns} by one-quarter of an oscillation cycle, while for the low frequency mode β_{ns} leads β_{nc}. (See section 8–5 for a further discussion of the engenvalues and eigenvectors in the nonrotating frame.)

The normal modes of the flap motion in hover are thus as follows. The only coupling is between the β_{nc} and β_{ns} degrees of freedom. The coning (β_0) and reactionless ($\beta_{N/2}$) modes are highly damped oscillations, with the same eigenvalues as the rotating flap response. The β_{nc} and β_{ns} degrees of freedom have two modes, each a damped oscillation with a frequency equal to the rotating flap frequency plus or minus n/rev. The high frequency mode with eigenvalue $s_{NR} = s_R + in$ (assuming that the frequency of s_R is positive) is a whirling or wobbling motion of β_{nc} and β_{ns} at a frequency $\text{Im}\, s_R + n$/rev in the same direction as the rotor rotation. The low frequency mode with eigenvalue $s_{NR} = s_R - in$ is a motion of β_{nc} and β_{ns} at a frequency $|\text{Im}\, s_R - n/\text{rev}|$, in the same direction as the rotor rotation if the rotating frequency is below n/rev, and in the opposite direction if $\text{Im}\, s_R > n$/rev. The high frequency flap motion can thus be called a progressive mode, and the low frequency motion a regressive mode.

Since the β_{1c} and β_{1s} degrees of freedom represent tilt of the tip-path plane, their coupled motion is a wobble of the tip-path plane. The high frequency flap mode is a wobble in the same direction as the rotor rotation but at frequency $\text{Im}\, s_R + 1$/rev (around 2/rev). The low frequency mode is a wobble at frequency $|\text{Im}\, s_R - 1/\text{rev}|$ (around zero), in a direction depending on whether the rotating flap frequency is above or below 1/rev. For an articulated rotor the rotating frequency is below 1/rev, so the low frequency mode is a wobble in the same direction as the rotor. For a hingeless rotor $\text{Im}\, s_R$ is likely above 1/rev, in which case the low frequency motion is truly a regressive mode, the tip-path plane wobbling opposite the direction of rotor rotation.

Similar behavior of the roots in the nonrotating frame is found for the other degrees of freedom of the rotor. The differences are in the frequency of the motion in the rotating frame and the interpretation of the motion implied by the nonrotating degrees of freedom. For example, since the cyclic lag degrees of freedom ζ_{1c} and ζ_{1s} produce a net shift in the rotor center of gravity, the coupled modes correspond to a whirling motion of the rotor center of gravity.

12-1.2.2 Hover Transfer Functions

The equation of motion for the coning mode β_0 of a hovering rotor gives the following transfer function:

$$\beta_0 = \frac{\gamma M_\theta \theta_0 - \frac{1}{2}\gamma M_\lambda w_G + (-S_\beta^* s + \frac{1}{2}\gamma M_\lambda)\dot{z}_h}{I_\beta^*\left(s^2 - \frac{\gamma M_{\dot\beta}}{I_\beta^*} s + \nu_{\beta_e}^2\right)}$$

using for the hover inflow perturbation $\lambda + \dot{z}_h - w_G = \frac{1}{2}(\dot{z}_h - w_G)$. The response to collective pitch θ_0 and to vertical gusts is the same as the flap response of the blade in the rotating frame. The coning response to vertical shaft motion involves the rotor inertia as well as aerodynamic forces and introduces a zero at $s = \frac{1}{2}\gamma M_\lambda / S_\beta^*$ on the negative real axis. For low frequency shaft motion, the aerodynamic forces dominate and the response is like that to the vertical gusts. For high frequency shaft motion the inertia dominates, so the coning response approaches $\beta_0/z_h = -S_\beta^*/I_\beta^*$.

The response of the tip-path-plane tilt to cyclic pitch inputs is defined by the transfer function:

$$\begin{bmatrix} I_\beta^* s^2 - \gamma M_{\dot\beta} s + I_\beta^*(\nu_\beta^2 - 1) + K_P\gamma M_\theta & 2I_\beta^* s - \gamma M_{\dot\beta} \\ -2I_\beta^* s + \gamma M_{\dot\beta} & I_\beta^* s^2 - \gamma M_{\dot\beta} s + I_\beta^*(\nu_\beta^2 - 1) + K_P\gamma M_\theta \end{bmatrix} \begin{pmatrix} \beta_{1c} \\ \beta_{1s} \end{pmatrix}$$

$$= \gamma M_\theta \begin{pmatrix} \theta_{1c} \\ \theta_{1s} \end{pmatrix}$$

Inverting the matrix gives

$$\begin{pmatrix} \beta_{1c} \\ \beta_{1s} \end{pmatrix} = \frac{1}{\Delta} \begin{bmatrix} I_\beta^*\left(s^2 - \frac{\gamma M_{\dot\beta}}{I_\beta^*} s + \nu_{\beta e}^2 - 1\right) & -2I_\beta^* s + \gamma M_{\dot\beta} \\ 2I_\beta^* s - \gamma M_{\dot\beta} & I_\beta^*\left(s^2 - \frac{\gamma M_{\dot\beta}}{I_\beta^*} s + \nu_{\beta e}^2 - 1\right) \end{bmatrix} \gamma M_\theta \begin{pmatrix} \theta_{1c} \\ \theta_{1s} \end{pmatrix}$$

where

$$\Delta = I_\beta^{*2}\left(s^2 - \frac{\gamma M_{\dot\beta}}{I_\beta^*} s + \nu_{\beta e}^2 - 1\right)^2 + I_\beta^{*2}\left(2s - \frac{\gamma M_{\dot\beta}}{I_\beta^*}\right)^2$$

is the characteristic equation. It is convenient to introduce the parameter

$$N_* = \frac{I_\beta^*(\nu_{\beta e}^2 - 1)}{-\gamma M_{\dot\beta}} = \frac{I_\beta^*(\nu_\beta^2 - 1)}{-\gamma M_{\dot\beta}} + K_P \frac{M_\theta}{-M_{\dot\beta}} \cong \frac{\nu_\beta^2 - 1}{\gamma/8} + K_P$$

which defines the phase shift of the flap response due to the structural stiffening of the blade ($\nu_\beta > 1$) and the pitch-flap coupling. Then the transfer function can be written as:

$$\begin{pmatrix} \beta_{1c} \\ \beta_{1s} \end{pmatrix} = \frac{\begin{bmatrix} \frac{2I_\beta^*}{-\gamma M_{\dot\beta}} s + 1 & \frac{I_\beta^*}{-\gamma M_{\dot\beta}} s^2 + s + N_* \\ -\left(\frac{I_\beta^*}{-\gamma M_{\dot\beta}} s^2 + s + N_*\right) & \frac{2I_\beta^*}{-\gamma M_{\dot\beta}} s + 1 \end{bmatrix}}{\left(\frac{2I_\beta^*}{-\gamma M_{\dot\beta}} s + 1\right)^2 + \left(\frac{I_\beta^2}{-\gamma M_{\dot\beta}} s^2 + s + N_*\right)^2} \left(\frac{M_\theta}{-M_{\dot\beta}}\right) \begin{pmatrix} -\theta_{1s} \\ \theta_{1c} \end{pmatrix}$$

The static response ($s = 0$) is

$$\begin{pmatrix} \beta_{1c} \\ \beta_{1s} \end{pmatrix} = \frac{1}{1 + N_*^2} \begin{bmatrix} 1 & N_* \\ -N_* & 1 \end{bmatrix} \left(\frac{M_\theta}{-M_{\dot\beta}}\right) \begin{pmatrix} -\theta_{1s} \\ \theta_{1c} \end{pmatrix}$$

For an articulated rotor ($\nu_\beta = 1$, $K_P = 0$, and $-M_\theta/M_{\dot\beta} = 1$) this reduces to

$$\begin{pmatrix} \beta_{1c} \\ \\ \beta_{1s} \end{pmatrix} = \begin{pmatrix} -\theta_{1s} \\ \\ \theta_{1c} \end{pmatrix}$$

which states that the tip-path plane remains exactly parallel to the control plane. In general, $-M_\theta/M_{\dot\beta}$ is the static gain of the flap response to cyclic, and N_* defines the phase shift between the tip-path-plane and control-plane tilt (see also sections 5–13 and 5–17).

After substituting for the coefficients, the transfer functions of the direct and cross response of the tip-path-plane tilt to cyclic pitch are as follows:

$$\frac{\beta_{1c}}{-\theta_{1s}} = \frac{\beta_{1s}}{\theta_{1c}} = \frac{\dfrac{16}{\gamma}s + 1}{\left(\dfrac{16}{\gamma}s + 1\right)^2 + \left(\dfrac{8}{\gamma}s^2 + s + N_*\right)^2}$$

$$\frac{\beta_{1c}}{\theta_{1c}} = \frac{\beta_{1s}}{\theta_{1s}} = \frac{\dfrac{8}{\gamma}s^2 + s + N_*}{\left(\dfrac{16}{\gamma}s + 1\right)^2 + \left(\dfrac{8}{\gamma}s^2 + s + N_*\right)^2}$$

The poles (roots of the denominator polynomial) are the eigenvalues of the β_{1c} and β_{1s} motion in the nonrotating frame: $s_{NR} = s_R \pm i$. The direct transfer function has a single zero at

$$s = \frac{\gamma M_{\dot\beta}}{2I_\beta{}^*} \cong -\frac{\gamma}{16}$$

This zero is on the negative real axis, with the same real part as the poles. The cross transfer function has two zeros, which are the solution of the quadratic

$$\frac{I_\beta^*}{-\gamma M_{\dot\beta}} s^2 + s + N_* = 0$$

namely

$$s = \frac{\gamma M_{\dot\beta}}{2I_\beta^*} \pm \sqrt{1 - \nu_{\beta_e}{}^2 + \left(\frac{\gamma M_{\dot\beta}}{2I_\beta^*}\right)^2}$$

$$= \mathrm{Re}\, s_R \pm \sqrt{1 - (\mathrm{Im}\, s_R)^2}$$

If the rotating flap root frequency is below 1/rev, there are two real zeros, at equal distances on either side of the real part of the poles. If $\text{Im}\, s_R >$ 1/rev there are two complex conjugate zeros, with the same real part as the poles. For an articulated rotor ($\nu_\beta = 1$, $K_P = 0$) the two real zeros are $s = 0$ and $\gamma M_{\dot\beta}/I_\beta^*$, which are at the origin and twice the real part of the poles. The zero at the origin is responsible for the static response of β_{1c}/θ_{1c} and β_{1s}/θ_{1s} being zero in this case. With negative pitch-flap coupling such that $\nu_{\beta_e} < 1$, the zeros are shifted farther away from the real part of the poles, the zero at the origin moving into the right half-plane. For $\nu_{\beta_e} > 1$, the two zeros move instead toward $\text{Re}\, s_R$. At

$$\nu_{\beta_e}^2 = 1 + \left(\frac{\gamma M_{\dot\beta}}{2I_\beta^*}\right)^2$$

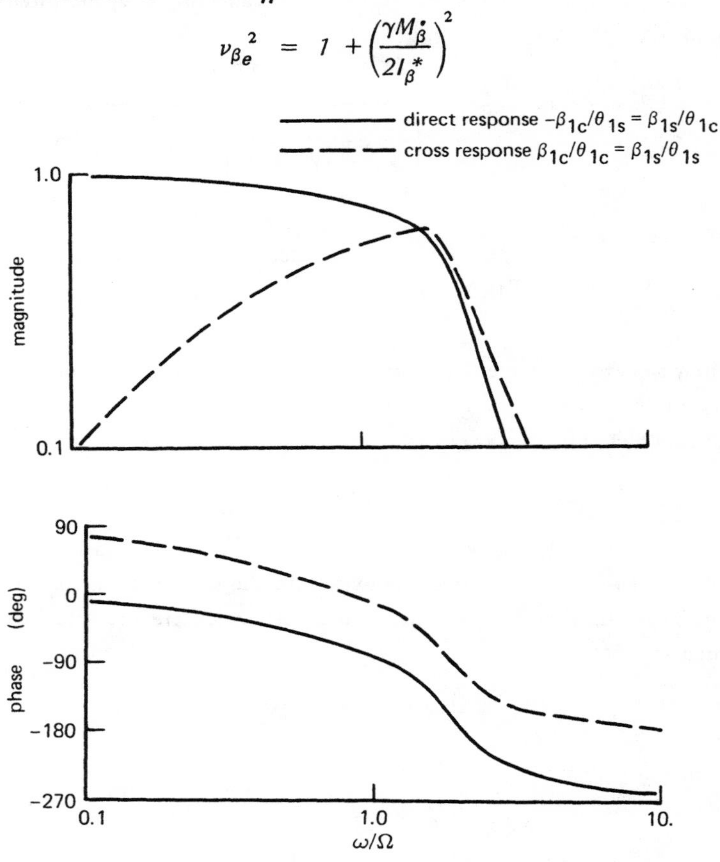

Figure 12-6 Frequency response of the tip-path-plane tilt to cyclic pitch inputs for an articulated rotor ($\nu_\beta = 1.0$ and $\gamma = 10$) in hover.

where the rotating frequency $\operatorname{Im} s_R = 1$/rev, the two zeros coincide at $s = \operatorname{Re} s_R$. Moreover, the two poles of the low frequency flap mode are also at $s = \operatorname{Re} s_R$ on the real axis for this case. For still larger ν_{β_e}, so that $\operatorname{Im} s_R >$ 1/rev as is likely with a hingeless rotor, there are two complex conjugate zeros. The zeros have the same real part as the poles and a larger frequency than the low frequency flap mode poles.

Substituting $s = i\omega$ gives the frequency response. Figs. 12-6 and 12-7 present the direct ($-\beta_{1c}/\theta_{1s} = \beta_{1s}/\theta_{1c}$) and cross ($\beta_{1c}/\theta_{1c} = \beta_{1s}/\theta_{1s}$) response of the tip-path plane to cyclic control inputs for typical articulated and hingeless rotors respectively. The frequency response shows the resonance

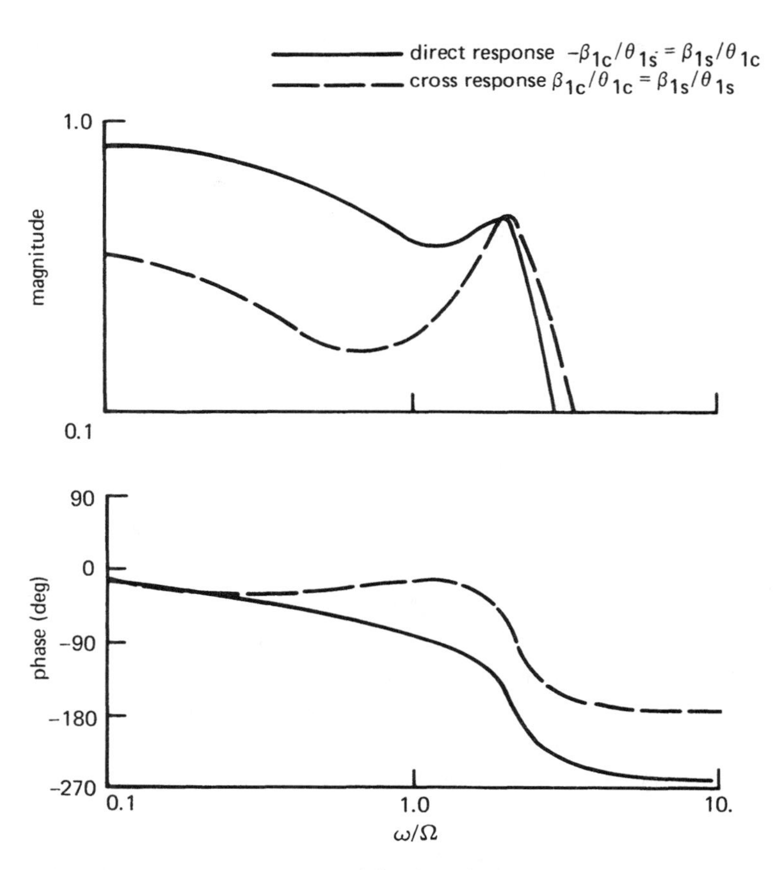

Figure 12-7 Frequency response of the tip-path-plane tilt to cyclic pitch inputs for a hingeless rotor ($\nu_\beta = 1.15$ and $\gamma = 6$) in hover.

with the high frequency flap mode around 2/rev. The low frequency mode has a very large damping ratio, so it is more evident in the phase than in the magnitude of the response.

The response of the tip-path-plane tilt to the in-plane shaft motion and gusts is similar to the response to cyclic pitch, but with a static gain of $-M_\mu/M_{\dot\beta} \cong 8(2C_T/\sigma a + \lambda_{HP}/4)$ instead of $-M_\theta/M_{\dot\beta} \cong 1$. The response to the shaft pitch and roll rate is more complicated, involving inertial and Coriolis forces as well as the aerodynamic forces.

The reactionless modes β_{nc} and β_{ns} respond only to the pitch control inputs θ_{nc} and θ_{ns}, which are not likely to be present in fact. In hover the reactionless rotor modes are therefore not coupled with the nonrotating system, either through shaft motion, gusts, or control inputs. These degrees of freedom (β_{nc}, β_{ns}, and $\beta_{N/2}$) then represent purely internal rotor motion. In forward flight all the rotor degrees of freedom are coupled, but the coning and tip-path-plane tilt response still dominate the flap dynamics.

12-1.3 Low Frequency Response

The transient flap motion is a damped oscillation that decays proportionally to $e^{(\mathrm{Re}\, s_R)\psi}$, where

$$\mathrm{Re}\, s_R = \frac{\gamma M_{\dot\beta}}{2I_\beta^*} \cong -\frac{\gamma}{16}$$

The time constant of the response is thus

$$\tau = \frac{1}{-\mathrm{Re}\, s_R} = \frac{2I_\beta^*}{-\gamma M_{\dot\beta}} \cong \frac{16}{\gamma}$$

The time to half amplitude, $t_{1/2} = 0.693\tau$, is then typically 90° of the rotor azimuth; because of the high flap damping, the flapping transients die out in less than one revolution of the rotor. In dimensional terms, the time to half amplitude is of the order of 0.05 sec. Hence the rotor flap motion responds on a much shorter time scale than the inputs from the pilot, from shaft motion due to the helicopter rigid body degrees of freedom, or from aerodynamic gusts. For problems such as the helicopter stability and control analysis, it should be sufficient therefore to consider only the steady-state response of the rotor, neglecting the transient flapping dynamics. Alternatively, it can be assumed that since the inputs to the rotor are slow compared to the flapping response, it is sufficient to consider only the low frequency response of the rotor, in fact just the steady state response. This approach, introduced by Hohenemser (1939) in an investigation of the helicopter flight stability, is called the quasistatic representation of the rotor dynamics.

Consider the response of the tip-path-plane tilt for a hovering rotor with three or more blades. To lowest order in the Laplace variable s, the equations of motion for β_{1c} and β_{1s} become:

$$\begin{bmatrix} I_\beta^*(\nu_\beta^2 - 1) + K_P\gamma M_\theta & -\gamma M_{\dot\beta} \\ \gamma M_{\dot\beta} & I_\beta^*(\nu_\beta^2 - 1) + K_P\gamma M_\theta \end{bmatrix} \begin{pmatrix} \beta_{1c} \\ \beta_{1s} \end{pmatrix}$$

$$= \gamma M_\theta \begin{pmatrix} \theta_{1c} \\ \theta_{1s} \end{pmatrix} + \gamma M_\mu \begin{pmatrix} \dot y_h + v_G \\ -\dot x_h + u_G \end{pmatrix} + \begin{bmatrix} -\gamma M_{\dot\beta} & -2I_{\beta a}^* \\ -2I_{\beta a}^* & \gamma M_{\dot\beta} \end{bmatrix} \begin{pmatrix} \dot\alpha_y \\ \dot\alpha_x \end{pmatrix}$$

(see section 12-1.2). Assuming uniform induced velocity, $\eta_\beta = r$, and neglecting tip losses, the coefficients are

$$M_\theta = -M_{\dot\beta} = \frac{1}{8}$$

$$M_\mu = \frac{2C_T}{\sigma a} + \frac{\lambda_{HP}}{4}$$

$$I_\beta^* = I_{\beta a}^* = 1$$

Defining again

$$N_* = \frac{I_\beta^*(\nu_{\beta_e}^2 - 1)}{-\gamma M_{\dot\beta}} = \frac{I_\beta^*(\nu_\beta^2 - 1)}{-\gamma M_{\dot\beta}} + K_P \frac{M_\theta}{-M_{\dot\beta}}$$

the low frequency tip-path plane tilt response is

$$\begin{pmatrix} \beta_{1c} \\ \beta_{1s} \end{pmatrix} = \frac{1}{1 + N_*^2} \begin{bmatrix} 1 & N_* \\ -N_* & 1 \end{bmatrix} \left\{ \frac{M_\theta}{-M_{\dot\beta}} \begin{pmatrix} -\theta_{1s} \\ \theta_{1c} \end{pmatrix} + \frac{M_\mu}{-M_{\dot\beta}} \begin{pmatrix} \dot x_h - u_G \\ \dot y_h + v_G \end{pmatrix} + \frac{2I_{\beta a}^*}{-\gamma M_{\dot\beta}} \begin{pmatrix} \dot\alpha_y \\ -\dot\alpha_x \end{pmatrix} + \begin{pmatrix} \dot\alpha_x \\ \dot\alpha_y \end{pmatrix} \right\}$$

The parameter N_* determines the lateral-longitudinal coupling of the rotor flap response. In this result most of the Lock number factors have cancelled, indicating that the flap response is primarily a balance of aerodynamic forces. The exception is the third term, which is a balance of Coriolis inertial forces due to the shaft angular velocity, and the aerodynamic forces due to the flap motion. Blade pitch control produces an aerodynamic flap moment through M_θ. A 1/rev pitch input from longitudinal cyclic θ_{1s} produces a lateral aerodynamic moment on the disk. The rotor responds with a 90° lag (less if $N_* > 0$), hence with longitudinal tilt of the tip-path plane. The flapping velocity in the rotating frame due to longitudinal tip-path-plane tilt β_{1c} produces a lateral aerodynamic moment on the disk through $M_{\dot\beta}$, which opposes the moment due to the cyclic pitch input. The tip-path plane tilts until equilibrium of moments is achieved, which gives the steady-state response. The effectiveness of cyclic pitch in producing flapping is thus governed by $-M_\theta/M_{\dot\beta}$; assuming $\eta_\beta = r$, this quantity has the value 1, implying that the tip-path plane remains parallel to the control plane. Longitudinal shaft velocity $\dot x_h$ and gust velocity u_G produce a lateral moment on the rotor disk through M_μ. This moment is due to the lateral asymmetry in the air velocity seen by the blades, which is similar to the effect of advance ratio in forward flight. Thus the steady-state response to longitudinal velocity is also longitudinal tip-path-plane tilt, but with effectiveness

$$\frac{M_\mu}{-M_{\dot\beta}} = 8\left(\frac{2C_T}{\sigma a} + \frac{\lambda_{HP}}{4}\right)$$

which typically has a value 0.4 in hover. A shaft roll rate $\dot\alpha_x$ also produces a lateral aerodynamic moment on the disk, through $M_{\dot\beta}$. The rotor responds with longitudinal tip-path-plane tilt until the flapping velocity in the rotating frame just cancels the velocity due to shaft roll. Thus in this case the effectiveness is given by $M_{\dot\beta}/M_{\dot\beta} = 1$ (again assuming $\eta_\beta = r$, so that the mode shapes of flapping and shaft tilt are identical). Similarly, lateral cyclic pitch θ_{1c}, lateral shaft velocity $\dot y_h$ and gust velocity v_G, and shaft pitch rate $\dot\alpha_y$ produce longitudinal aerodynamic moments on the rotor disk, and hence lateral tip-path-plane tilt.

Angular velocity of the shaft also produces tip-path-plane tilt, in this case proportional to

$$\frac{2I^*_{\beta a}}{-\gamma M_{\dot\beta}} \cong \frac{16}{\gamma}$$

This is the lag of the tip-path plane required to precess the rotor to follow the shaft motion. For the rotor to follow the shaft with a pitch rate $\dot\alpha_y$, there must be an angular roll acceleration $2\Omega I^*_{\beta a}\dot\alpha_y$ on the disk due to the Coriolis forces on the rotating blades. There must then be a lateral moment on the disk to provide this acceleration to precess the rotor. This moment is supplied by the aerodynamic forces on the blade. The rotor tip-path plane tilts back, lagging the shaft tilt, until the lateral moment due to the flap velocity in the rotating frame ($\gamma M_{\dot\beta}\beta_{1c}$) is sufficient to provide the required moment. Similarly, the rotor disk follows shaft roll angular velocity $\dot\alpha_x$, but with a lateral tip-path-plane tilt lag to provide the aerodynamic moment to precess the disk.

The parameter N_* is a measure of the structural and aerodynamic springs on the flap motion. When $N_* = 0$, the rotor disk responds to applied moments with exactly a 90° phase shift (which is 90° in azimuth because the flap natural frequency is at 1/rev in this case). When $N_* > 0$ because $\nu_\beta > 1$ or because of pitch-flap coupling, the flap response is quickened and hence the lag in the response is less than 90°. Consider the tip-path-plane tilt due to cyclic pitch alone,

$$\begin{pmatrix}\beta_{1c}\\ \beta_{1s}\end{pmatrix} = \frac{1}{1+N_*^2}\begin{bmatrix}1 & N_*\\ -N_* & 1\end{bmatrix}\frac{M_\theta}{-M_{\dot\beta}}\begin{pmatrix}-\theta_{1s}\\ \theta_{1c}\end{pmatrix}$$

The off-diagonal terms represent the lateral-longitudinal coupling, which is zero only if $N_* = 0$. The magnitude of this response is

$$\frac{|\beta|}{|\theta|} = \frac{M_\theta}{-M_{\dot\beta}}\frac{1}{\sqrt{1+N_*^2}}$$

and the azimuthal phase shift is $\Delta\psi = \tan^{-1}N_*$. Thus when $N_* > 0$, the magnitude of the flap response is reduced slightly, and the phase lag has been reduced from 90° to 90° $-\Delta\psi$. Note that when $N_* < 0$ (because of negative pitch-flap coupling), the magnitude of the flap response is again reduced, but the phase lag is increased. This behavior can also be viewed as due to removing the natural frequency of the flap motion from resonance with the 1/rev exciting forces, as discussed in section 5–13. There is a similar phase shift and magnitude reduction in the flap response to shaft motion and gusts. When $N_* = 0$, the response to lateral or longitudinal shaft motion is purely lateral or purely longitudinal tip-path-plane tilt respectively. The phase shift when $N_* \neq 0$ couples the lateral and longitudinal motions of the helicopter. The coupling of the response to cyclic control can be cancelled by a corresponding phase shift in the swashplate rigging, but the lateral-longitudinal coupling of the rotor response to shaft motion remains and can be troublesome if large. As an example, consider an articulated main rotor with offset hinges and no pitch-flap coupling. Using $\nu_\beta = 1.03$, $\gamma = 10$, and $K_P = 0$ gives $N_* = 0.05$ and thus a negligible change in response magnitude and phase shift of only $\Delta\psi = \tan^{-1}N_* = 3^\circ$. For a hingeless rotor with $\nu_\beta = 1.15$, $\gamma = 6$, and $K_P = 0$ we obtain $N_* = 0.43$. Then the response magnitude is reduced by 8%, and the phase shift is an appreciable $\Delta\psi = 23^\circ$. For an articulated tail rotor with $\nu_\beta = 1$ and $K_P = 1$, the phase shift of $\Delta\psi = 45^\circ$ due to $N_* = 1$ is large but not important; the pitch-flap coupling reduces the flap response magnitude by 29%, however.

The rotor dynamics in forward flight are described by periodic coefficient differential equations, but we have seen that the constant coefficient approximation in the nonrotating frame does provide a good representation of the flap dynamics as long as the advance ratio is not too large. The constant coefficient approximation is particularly good for the low frequency modes of the rotor. Consider a rotor with three or more blades in forward flight, with the flap motion described by the coning and tip-path-plane tilt modes. The inertial terms in the equations of motion are the same as in hover, and the constant coefficient approximation for the aerodynamic forces can be found in sections 11–4 and 11–6. Body axes are used for the aerodynamics here, since this result is intended for the helicopter stability and control analysis. When only the lowest order terms in the Laplace variable s are retained, the low frequency response of the rotor flap motion in forward flight is:

$$\begin{bmatrix} I_\beta^*\nu_\beta^2 + K_P\gamma M_\theta^0 & -\gamma\tfrac{1}{2}(M_\beta^{1c} - M_{\dot\beta}^{1s}) & K_P\gamma\tfrac{1}{2}M_\theta^{1s} \\ -K_P\gamma M_\theta^{1s} & -\gamma(M_{\dot\beta}^0 - \tfrac{1}{2}M_\beta^{2s} - \tfrac{1}{2}M_{\dot\beta}^{2c}) & -I_\beta^*(\nu_\beta^2 - 1) - K_P\gamma(M_\theta^0 - \tfrac{1}{2}M_\theta^{2c} \\ -\gamma M_\beta^{1c} & I_\beta^*(\nu_\beta^2 - 1) + K_P\gamma(M_\theta^0 + \tfrac{1}{2}M_\theta^{2c}) & -\gamma(M_{\dot\beta}^0 + \tfrac{1}{2}M_{\dot\beta}^{2c} + \tfrac{1}{2}M_\beta^{2s}) \end{bmatrix} \begin{pmatrix} \beta_0 \\ \beta_{1c} \\ \beta_{1s} \end{pmatrix}$$

$$= \begin{bmatrix} \gamma M_\theta^0 & -\gamma \tfrac{1}{2} M_\theta^{1s} & 0 \\ -\gamma M_\theta^{1s} & \gamma(M_\theta^0 - \tfrac{1}{2} M_\theta^{2c}) & 0 \\ 0 & 0 & \gamma(M_\theta^0 + \tfrac{1}{2} M_\theta^{2c}) \end{bmatrix} \begin{pmatrix} \theta_0 \\ -\theta_{1s} \\ \theta_{1c} \end{pmatrix}$$

$$+ \begin{bmatrix} \gamma M_\lambda^0 & -\gamma \tfrac{1}{2} M_\mu^{1s} & \gamma \tfrac{1}{2} M_\mu^{1c} \\ -\gamma M_\lambda^{1s} & \gamma(M_\mu^0 - \tfrac{1}{2} M_\mu^{2c}) & -\gamma \tfrac{1}{2} M_\mu^{2s} \\ 0 & -\gamma \tfrac{1}{2} M_\mu^{2s} & \gamma(M_\mu^0 + \tfrac{1}{2} M_\mu^{2c}) \end{bmatrix} \begin{pmatrix} \dot{z}_h - w_G \\ \dot{x}_h - u_G \\ \dot{y}_h + v_G \end{pmatrix}$$

$$+ \begin{bmatrix} -\gamma M_{\dot{\xi}}^0 & 0 & -\gamma \tfrac{1}{2} M_{\dot{\beta}}^{1s} \\ \gamma M_{\dot{\xi}}^{1s} & 2I_{\beta a}^* & \gamma(M_{\dot{\beta}}^0 - \tfrac{1}{2} M_{\dot{\beta}}^{2c}) \\ -\gamma M_{\dot{\xi}}^{1c} & -\gamma(M_{\dot{\beta}}^0 + \tfrac{1}{2} M_{\dot{\beta}}^{2c}) & 2I_{\beta a}^* \end{bmatrix} \begin{pmatrix} \dot{\alpha}_z \\ \dot{\alpha}_y \\ -\dot{\alpha}_x \end{pmatrix}$$

Forward speed of the helicopter has the effect of coupling the vertical and the lateral-longitudinal motions by aerodynamic forces of order μ, resulting in more complex behavior than for hover. Moreover, the task of calculating the dynamic behavior is more difficult because of the higher order of the system that must be considered and the additional aerodynamic coefficients that must be obtained. Of particular significance is the lateral aerodynamic moment due to vertical velocity of the helicopter:

$$-\gamma M_\lambda^{1s} \left(\dot{z}_h - w_G\right) = \gamma \frac{\mu}{4} \left(\dot{z}_h - w_G\right)$$

recalling that $M_\lambda = -(1/6 + \mu/4 \sin \psi)$. Ignoring various other effects, this lateral moment on the disk primarily produces a longitudinal tip-path-plane tilt of

$$\Delta\beta_{1c} = \frac{M_\lambda^{1s}}{M_{\dot{\beta}}^0} \left(\dot{z}_h - w_G\right) = 2\mu\left(\dot{z}_h - w_G\right)$$

The downward velocity through the disk of $(\dot{z}_h - w_G)$ decreases the angle of attack of the blades. In forward flight the resulting lift decrease is largest on the advancing side and smallest on the retreating side, so there is a lateral moment of the rotor disk, in response to which the rotor flaps forward.

12-1.4 Hub Reactions

Next let us examine the forces and moments acting on the rotor hub, including the effects of the flap response. Since a knowledge of the hub reactions is required in this text mainly for the helicopter stability and control investigation (Chapter 15), we are primarily interested in the low frequency response again. Consider first the hovering rotor, for which the analysis is simplified not only by the constant coefficients, but also by a complete decoupling of the vertical and longitudinal-lateral dynamics because of the axisymmetry.

From sections 9–6 and 11–7, the perturbation thrust due to the rotor is

$$\gamma \frac{C_T}{\sigma a} = \gamma T_\theta \theta_0 + \tfrac{1}{2}\gamma T_\lambda(\dot{z}_h - w_G) - M_b{}^* \ddot{z}_h - S_\beta{}^* \ddot{\beta}_0 + \gamma T_{\dot{\beta}} \dot{\beta}_0 - K_P \gamma T_\theta \beta_0$$

and the coning equation is

$$I_\beta^* \ddot{\beta}_0 - \gamma M_{\dot{\beta}} \dot{\beta}_0 + (I_\beta^* \nu_\beta^2 + K_P \gamma M_\theta)\beta_0$$
$$= \gamma M_\theta \theta_0 + \tfrac{1}{2}\gamma M_\lambda(\dot{z}_h - w_G) - S_\beta^* \ddot{z}_h$$

again using $\lambda + \dot{z}_h - w_G = \tfrac{1}{2}(\dot{z}_h - w_G)$. Assuming $\eta_{\dot{\beta}} = r$ and neglecting tip losses, the hover aerodynamic coefficients are $M_\theta = -M_{\dot{\beta}} = 1/8$, $-M_\lambda = T_\theta = -T_{\dot{\beta}} = 1/6$, and $T_\lambda = -1/4$. To the lowest order in s, the low frequency thrust and coning equations are then

$$\frac{C_T}{\sigma a} = T_\theta \theta_0 + \tfrac{1}{2} T_\lambda(\dot{z}_h - w_G) - K_P T_\theta \beta_0$$

$$\beta_0 = \frac{1}{I_\beta^* \nu_{\beta_e}^2}\left[\gamma M_\theta \theta_0 + \tfrac{1}{2}\gamma M_\lambda(\dot{z}_h - w_G)\right]$$

It is assumed that the rotor mass has been included in the helicopter gross weight, so the $M_b{}^* \ddot{z}_h$ term in $C_T/\sigma a$ can be dropped. Note that the coning motion influences the rotor low frequency thrust only through the change in collective with pitch-flap coupling. Eliminating β_0 gives

$$\frac{C_T}{\sigma a} = \left(T_\theta - K_P T_\theta \frac{\gamma M_\theta}{I_\beta^* \nu_{\beta_e}^2}\right)\theta_0 + \tfrac{1}{2}\left(T_\lambda - K_P T_\theta \frac{\gamma M_\lambda}{I_\beta^* \nu_{\beta_e}^2}\right)\left(\dot{z}_h - w_G\right)$$

$$\cong \left[T_\theta \theta_0 + \tfrac{1}{2} T_\lambda\left(\dot{z}_h - w_G\right)\right]\frac{\nu_\beta^2}{\nu_{\beta_e}^2}$$

where

$$\frac{\nu_\beta^2}{\nu_{\beta_e}^2} = \frac{1}{1 + \dfrac{K_P \gamma M_\theta}{I_\beta^* \nu_\beta^2}}$$

Basically, the thrust is given by the direct response to collective and vertical velocity perturbations. Pitch-flap coupling reduces the thrust response by the factor $\nu_\beta^2/\nu_{\beta_e}^2$ due to the collective pitch reduction with coning. A lift deficiency factor C' should also be included in the thrust to account for the effects of the rotor wake in the unsteady aerodynamics.

The in-plane forces on the hub, C_H and C_Y, and the hub moments C_{M_x} and C_{M_y} are closely related to the rotor tip-path-plane tilt. The low frequency flap response was discussed in the last section. From sections 9–6 and 11–7 the in-plane hub forces of the hovering rotor are

$$\begin{pmatrix} \dfrac{2C_H}{\sigma a} \\ \dfrac{2C_Y}{\sigma a} \end{pmatrix} = -\frac{2M_b^*}{\gamma}\begin{pmatrix} \ddot{x}_h \\ \ddot{y}_h \end{pmatrix} - \frac{C_T}{\sigma a}\begin{pmatrix} \beta_{1c} \\ \beta_{1s} \end{pmatrix}$$

$$+ \begin{bmatrix} R_{\dot\beta} & H_{\dot\beta} \\ -H_{\dot\beta} & R_{\dot\beta} \end{bmatrix}\begin{pmatrix} \dot\beta_{1c} + \beta_{1s} - \dot\alpha_y \\ \dot\beta_{1s} - \beta_{1c} + \dot\alpha_x \end{pmatrix}$$

$$+ \begin{bmatrix} R_\theta & H_\theta \\ -H_\theta & R_\theta \end{bmatrix}\begin{pmatrix} \theta_{1c} - K_P\beta_{1c} \\ \theta_{1s} - K_P\beta_{1s} \end{pmatrix}$$

$$+ \begin{bmatrix} -(H_\mu + R_\mu) & R_r \\ -R_r & -(H_\mu + R_\mu) \end{bmatrix}\begin{pmatrix} \dot{x}_h - u_G \\ \dot{y}_h + v_G \end{pmatrix}$$

Assuming that the rotor mass M_b^* is included in the helicopter gross weight, and writing $H_{\dot\beta} = H_{\dot\beta}^* + C_T/\sigma a$, the low frequency response of the hub forces is

$$\begin{pmatrix} \dfrac{2C_H}{\sigma a} \\ \dfrac{2C_Y}{\sigma a} \end{pmatrix} = -\frac{2C_T}{\sigma a}\begin{pmatrix} \beta_{1c} \\ \beta_{1s} \end{pmatrix} + \begin{bmatrix} -H_{\dot\beta}^* - K_P R_\theta & R_{\dot\beta} - K_P H_\theta \\ -R_{\dot\beta} + K_P H_\theta & -H_{\dot\beta}^* - K_P R_\theta \end{bmatrix}\begin{pmatrix} \beta_{1c} \\ \beta_{1s} \end{pmatrix}$$

$$+ \begin{bmatrix} -H_\theta & R_\theta \\ -R_\theta & -H_\theta \end{bmatrix}\begin{pmatrix} -\theta_{1s} \\ \theta_{1c} \end{pmatrix} + \begin{bmatrix} -R_{\dot\beta} & -H_{\dot\beta} \\ H_{\dot\beta} & -R_{\dot\beta} \end{bmatrix}\begin{pmatrix} \dot\alpha_y \\ -\dot\alpha_x \end{pmatrix}$$

$$+ \begin{bmatrix} -(H_\mu + R_\mu) & R_r \\ -R_r & -(H_\mu + R_\mu) \end{bmatrix}\begin{pmatrix} \dot x_h - u_G \\ \dot y_h + v_G \end{pmatrix}$$

The first term is the in-plane force due to tilt of the thrust vector with the tip-path plane. The rotor hub moments can be obtained directly from the tip-path-plane tilt:

$$\begin{pmatrix} -\dfrac{2C_{M_y}}{\sigma a} \\ \dfrac{2C_{M_x}}{\sigma a} \end{pmatrix} = \frac{I_\beta^*(\nu_\beta^2 - 1)}{\gamma}\begin{pmatrix} \beta_{1c} \\ \beta_{1s} \end{pmatrix}$$

The total moment about the helicopter center of gravity a distance h below the rotor hub is then

$$\begin{pmatrix} -\dfrac{2C_{M_y}}{\sigma a} \\ \dfrac{2C_{M_x}}{\sigma a} \end{pmatrix}_{CG} = \left(\frac{I_\beta^*(\nu_\beta^2 - 1)}{\gamma} + h\,\frac{2C_T}{\sigma a}\right)\begin{pmatrix} \beta_{1c} \\ \beta_{1s} \end{pmatrix}$$

plus some in-plane forces due to tilt of the thrust vector relative to the tip-path plane. For an articulated rotor without hinge offset there are no hub moments, so all moments about the center of gravity must come from the thrust vector tilt. With no hub moment capability a helicopter must avoid flight at low load levels, where the control and damping from the rotor will be lost because they are proportional to the rotor thrust.The moment-producing capability of an articulated rotor can be roughly doubled by using flap hinge offset, and the hub moment term does not depend on the thrust magnitude. With a hingeless rotor the hub moment term dominates, being typically three or four times the thrust tilt term. Thus a hingeless rotor has much greater control power and damping than the articulated rotor rotor, but also higher gust response. (See also the discussion in section 5—13.)

The rotor forces and moments acting on the helicopter are thus basically proportional to the tip-path-plane tilt. Longitudinal flapping β_{1c} produces a drag force C_H and pitch moment C_{M_Y} on the helicopter; lateral flapping β_{1s} produces a side force C_Y and roll moment C_{M_X}. Cyclic pitch tilts the tip-path plane and consequently produces pitch and roll moments about the helicopter center of gravity. Thus the pilot can use cyclic pitch inputs to control the helicopter. Hub in-plane velocity ($\dot{x}_h$ or $\dot{y}_h$) produces flapping and thus an in-plane component of the thrust vector that opposes the motion. Hence there is a damping force on the helicopter speed perturbations. The corresponding moments due to $\dot{x}_h$ and $\dot{y}_h$ couple the linear and angular motion of the helicopter and are a major factor in the flight dynamics. Longitudinal and lateral aerodynamic gusts produce forces and moments on the helicopter by the same means. The rotor responds to shaft angular velocity with a tip-path-plane lag in order to precess the rotor. The result of this flapping response is a moment opposing the shaft motion; hence the rotor provides angular damping of the helicopter pitch and roll motion. The hub forces due to shaft angular velocity are the one case where there is an important contribution relative to the tip-path plane. For an articulated rotor we find that

$$\Delta\begin{pmatrix}\dfrac{2C_H}{\sigma a}\\[2ex]\dfrac{2C_Y}{\sigma a}\end{pmatrix} = -\left(\frac{2C_T}{\sigma a} + H_{\dot{\beta}}^{*}\right)\frac{16}{\gamma}\begin{pmatrix}\dot{\alpha}_y\\[2ex]-\dot{\alpha}_x\end{pmatrix}$$

Thus the effectiveness of the shaft angular velocity is not given by the thrust vector tilt alone, but instead by

$$H_{\dot{\beta}} - R_{\beta} = \frac{2C_T}{\sigma a} + H_{\dot{\beta}}^{*} = \frac{2C_T}{\sigma a} - \frac{\lambda_{HP}}{4}$$

The $H_{\dot\beta}^*$ contribution reduces the pitch and roll damping from the rotor. (For the hub forces due to cyclic or in-plane velocity, the $H_{\dot\beta}^*$ terms are cancelled by other terms in C_H and C_Y.)

The constant coefficient approximation for the hub forces in forward flight is given in sections 11-5.2 and 11–6. As in hover, the low frequency response requires only the aerodynamic terms. The hub in-plane forces and moments are still dominated by the flap response. Recall from the last section that forward flight introduces a longitudinal tip-path-plane tilt due to vertical velocity of the helicopter:

$$\Delta\beta_{1c} \cong 2\mu(\dot z_h - w_G)$$

Now a downward velocity perturbation ($\dot z_h < 0$) is an increase in the aircraft angle of attack in forward flight. The rotor flaps back in response, thereby producing a pitch-up moment on the helicopter, which tends to increase the angle of attack still further. Thus in forward flight the rotor is a source of an angle-of-attack instability that has an important role in the helicopter flight dynamics.

12-1.5 Two-Bladed Rotor

The case of a two-bladed rotor requires a special consideration because, unlike rotors with three or more blades, it does not have the tip-path-plane tilt degrees of freedom β_{1c} and β_{1s}. Instead, the dynamic behavior of the two-bladed rotor is described by the motion of the teetering degree of freedom β_1, which leads to periodic coefficient equations coupling the rotor and the nonrotating frame. For helicopter flight dynamics, however, the primary concern is with the low frequency response of the flap motion and hub reactions to control inputs, shaft motion, and gusts. It will be found that the low frequency response of the two-bladed rotor is nearly identical to that for the $N \geqslant 3$ case.

The frequency response of a linear, time-invariant dynamic system is described by a transfer function $H(\omega)$ relating the magnitude and phase of the input and output at frequency ω: $\bar F = H(\omega)\bar\alpha$. The implication of the periodic coefficients in the equations of motion for the two-bladed rotor is that such a transfer does not exist, for an input at frequency ω produces in general a response at all frequencies $\omega \pm n\Omega$ (n an integer).

Then the input-output relation for sinusoidal excitation takes the form

$$F = \left(\sum_{n=-\infty}^{\infty} H_n(\omega) e^{in\Omega t} \right) \bar{\alpha} e^{i\omega t}$$

Specifically, consider the flapping equation of motion for the two-bladed rotor as derived in sections 9–6 and 11–6:

$$I_\beta^* \ddot{\beta}_1 - \gamma M_{\dot\beta} \dot{\beta}_1 + (I_\beta^* \nu_\beta^2 + K_P \gamma M_\theta) \beta_1$$

$$= [\gamma M_\theta \theta_{1c} + I_{\beta a}^* (\ddot{\alpha}_y - 2\dot{\alpha}_x) - \gamma M_{\dot\beta} \dot{\alpha}_y + \gamma M_\mu (\dot{y}_h + v_G)] \cos \psi$$

$$+ [\gamma M_\theta \theta_{1s} - I_{\beta a}^* (\ddot{\alpha}_x + 2\dot{\alpha}_y) + \gamma M_{\dot\beta} \dot{\alpha}_x + \gamma M_\mu (-\dot{x}_h + u_G)] \sin \psi$$

(Assuming that the pitch control input is from the swashplate, we have written $\theta_1 = \theta_{1c} \cos \psi + \theta_{1s} \sin \psi$.) For this equation, only the transfer functions $H_1(\omega)$ and $H_{-1}(\omega)$ are nonzero, meaning that the teetering response to sinusoidal inputs occurs only at the frequencies $\omega \pm \Omega$. It follows that since the response to low frequency inputs occurs at the frequencies $\pm\Omega$, the low frequency flap motion can be written $\beta_1 = \beta_{1c} \cos \psi + \beta_{1s} \sin \psi$. The low frequency response of the two-bladed rotor can thus be described by the steady state motion of the tip-path plane. On substituting for β_1, the solution found for β_{1c} and β_{1s} here becomes identical to the low frequency flap response obtained in section 12-1.3 for rotors with three or more blades.

The in-plane hub forces for the rotor with two blades are derived in sections 9–6, 11-5.2, and 11–6:

$$\begin{pmatrix} \dfrac{2C_H}{\sigma a} \\ \dfrac{2C_Y}{\sigma a} \end{pmatrix} = -\frac{2M_b^*}{\gamma} \begin{pmatrix} \ddot{x}_h \\ \ddot{y}_h \end{pmatrix} + R_\beta \begin{pmatrix} 2C \\ 2S \end{pmatrix} \beta_1 + \begin{pmatrix} H_{\dot\beta} 2S + R_{\dot\beta} 2C \\ -H_{\dot\beta} 2C + R_{\dot\beta} 2S \end{pmatrix} \dot{\beta}_1$$

$$+ \begin{pmatrix} R_\theta 2C + H_\theta 2S \\ R_\theta 2S - H_\theta 2C \end{pmatrix} \left(\theta_1 - K_P \beta_1 \right)$$

$$+ \begin{bmatrix} R_{\dot\beta} 2CS + H_{\dot\beta} 2S^2 & -R_{\dot\beta} 2C^2 - H_{\dot\beta} 2CS \\ R_{\dot\beta} 2S^2 - H_{\dot\beta} 2CS & -R_{\dot\beta} 2CS + H_{\dot\beta} 2C^2 \end{bmatrix} \begin{pmatrix} \dot{\alpha}_x \\ \dot{\alpha}_y \end{pmatrix}$$

$$+ \begin{bmatrix} -H_\mu 2S^2 - R_\mu 2C^2 - R_r 2CS & H_\mu 2CS - R_\mu 2CS + R_r 2C^2 \\ -H_\mu 2CS - R_\mu 2CS - R_r 2S^2 & -H_\mu 2C^2 - R_\mu 2S^2 + R_r 2CS \end{bmatrix} \begin{pmatrix} \dot{x}_h - u_G \\ \dot{y}_h + v_G \end{pmatrix}$$

and the hub moment in section 9–5:

$$\begin{pmatrix} \dfrac{-2C_{M_y}}{\sigma a} \\ \dfrac{2C_{M_x}}{\sigma a} \end{pmatrix} = \frac{I_\beta^*(\nu_\beta^2 - 1)}{\gamma} \begin{pmatrix} 2C \\ 2S \end{pmatrix} \beta_1$$

where $C = \cos\psi$ and $S = \sin\psi$. Substituting $\beta_1 = \beta_{1c}\cos\psi + \beta_{1s}\sin\psi$, the low frequency response of the hub reactions is then:

$$\begin{pmatrix} \dfrac{2C_H}{\sigma a} \\ \dfrac{2C_Y}{\sigma a} \end{pmatrix} = -\frac{2C_T}{\sigma a}\begin{pmatrix} \beta_{1c} \\ \beta_{1s} \end{pmatrix}$$

$$+ \begin{bmatrix} -H_{\dot\beta}^* 2S^2 - R_{\dot\beta} 2CS & H_{\dot\beta}^* 2CS + R_{\dot\beta} 2C^2 \\ -K_P R_\theta 2C^2 - K_P H_\theta 2CS & -K_P R_\theta 2CS - K_P H_\theta 2S^2 \\ H_{\dot\beta}^* 2CS - R_{\dot\beta} 2S^2 & -H_{\dot\beta}^* 2C^2 + R_{\dot\beta} 2CS \\ -K_P R_\theta 2CS + K_P H_\theta 2C^2 & -K_P R_\theta 2S^2 + K_P H_\theta 2CS \end{bmatrix} \begin{pmatrix} \beta_{1c} \\ \beta_{1s} \end{pmatrix}$$

$$+ \begin{bmatrix} -H_\theta 2S^2 - R_\theta 2CS & H_\theta 2CS + R_\theta 2C^2 \\ H_\theta 2CS - R_\theta 2S^2 & -H_\theta 2C^2 + R_\theta 2CS \end{bmatrix} \begin{pmatrix} -\theta_{1s} \\ \theta_{1c} \end{pmatrix}$$

$$+ \begin{bmatrix} -H_{\dot\beta} 2CS - R_{\dot\beta} 2C^2 & -H_{\dot\beta} 2S^2 - R_{\dot\beta} 2CS \\ H_{\dot\beta} 2C^2 - R_{\dot\beta} 2CS & H_{\dot\beta} 2CS - R_{\dot\beta} 2S^2 \end{bmatrix} \begin{pmatrix} \dot\alpha_y \\ -\dot\alpha_x \end{pmatrix}$$

$$+ \begin{bmatrix} -H_\mu 2S^2 - R_\mu 2C^2 - R_r 2CS & H_\mu 2CS - R_\mu 2CS + R_r 2C^2 \\ -H_\mu 2CS - R_\mu 2CS - R_r 2S^2 & -H_\mu 2C^2 - R_\mu 2S^2 + R_r 2CS \end{bmatrix} \begin{pmatrix} \dot x_h - u_G \\ \dot y_h + v_G \end{pmatrix}$$

and

$$\begin{pmatrix} -\dfrac{2C_{M_y}}{\sigma a} \\ \dfrac{2C_{M_x}}{\sigma a} \end{pmatrix} = \frac{I_\beta^*(\nu_\beta^2 - 1)}{\gamma} \begin{bmatrix} 2C^2 & 2CS \\ 2CS & 2S^2 \end{bmatrix} \begin{pmatrix} \beta_{1c} \\ \beta_{1s} \end{pmatrix}$$

The average of these coefficients gives exactly the same expressions for the hub reactions as those obtained in section 12-1.4 for the low frequency response of rotors with three or more blades. However, while this constant coefficient result is exact for the rotor with three or more blades in hover, because of the inertial and aerodynamic axisymmetry of the rotor, for the two-bladed rotor there are periodic variations of the coefficients in the hub reactions. The asymmetry of the rotor with two blades leads to large 2/rev variations of the coefficients even in hover. The thrust tilt term is obtained without periodic coefficients even with $N = 2$. Recall that the thrust dominates the hub in-plane forces except for the response to shaft angular velocity, where the $H_{\dot{\beta}}^{*}$ term due to tip-path-plane tilt is also important. Thus the primary influence of the periodic coefficients on the in-plane hub forces will be in the rotor pitch and roll damping. There is also a large 2/rev variation in the hub moment if $\nu_\beta > 1$, which is why two-bladed rotors are not often designed with a hub spring.

In summary, the two-bladed rotor is indeed a special case. The description of the dynamics is unique, involving the teetering degree of freedom β_1, which is fundamentally in the rotating frame, rather than the tip-path-plane tilt degrees of freedom. The frequency response of the two-bladed rotor motion is not given by the usual transfer function relation because the system is not time-invariant. The low frequency flap response does reduce to a tip-path-plane representation identical to the result for $N \geqslant 3$, but the steady-state limit ($\omega = 0$), which allows writing $\beta_1 = \beta_{1c} \cos \psi + \beta_{1s} \sin \psi$, is a special case. The equations for the rotor low frequency response that are used in the analysis of the helicopter flight dynamics are the same as for $N \geqslant 3$ if the averaged coefficients are used, but in fact the hub reactions involve large amplitude periodic coefficients even in hover. The 2/rev variation of the coefficients due to the lack of axisymmetry with a two-bladed rotor is expected to influence primarily the rotor pitch and roll damping, and the hub moments if $\nu_\beta > 1$.

The special characteristics of the two-bladed rotor influence several aspects of the analysis of the aeroelastic behavior. In general it is necessary to analyze periodic coefficient equations more often than for a rotor with three or more blades. A special procedure may also be required to implement the quasistatic approximations (i.e. to derive the rotor low frequency response). For a rotor with three or more blades the low frequency response can be obtained by dropping the flapping acceleration and velocity terms from the equations in the nonrotating frame (see section 12-1.3). Such a procedure does not work with a two-bladed rotor because the equation of motion for β_1 is still in the rotating frame, so the β_1 response to low frequency inputs from the nonrotating frame is not at low frequency also, but rather at 1/rev. Furthermore, a constant coefficient approximation cannot be used directly for the helicopter flight dynamics, since averaging the periodic coefficients of the two-bladed rotor equations of motion eliminates the coupling between the rotor and the shaft motion.

12-1.6 Literature

On the helicopter rotor flapping response: Prewitt and Wagner (1940), Sissingh (1946, 1950, 1961, 1964a, 1964b, 1965), Reissner and Morduchow (1948), Carpenter and Peitzer (1949), Morduchow (1949), Zbrozek (1949a, 1949b), Payne (1954a), Maglieri and Reisert (1955), Daughaday, Du Waldt, and Gates (1957), Ham, Moser, and Zvara (1958), Du Waldt, Gates and Piziali (1959), McCarty, Brooks, and Maglieri (1959b), McKee and Naeseth (1959), Silveira and Brooks (1959), Jones (1960, 1964), Zvara and Ham (1960), Liu (1962), Young (1962a), Gates and DuWaldt (1963), Jenney, Arcidiacono, and Smith (1963), Marinescu (1965), Ward (1966b, 1969), Azuma (1967), Bramwell (1969), Loewy (1969), Niebanck (1969), Skingle, Gaukroger, and Taylor (1969), Ward and Snyder (1969), Niebanck and Bain (1970), Curtiss and Shupe (1971), Gaonkar (1971), Hohenemser and Yin (1971, 1972a, 1972b, 1973a, 1973b, 1974a, 1974b, 1974c, 1977), Kuczynski and Sissingh (1971, 1972), Kuczynski (1972), Kuczynski, Sharp, and Sissingh (1972), Curtiss (1973), Hohenemser and Prelewicz (1973), London, Watts, and Sissingh (1973), Takasawa (1973), Arcidiacono, Bergquist, and Alexander (1974), Azuma and Nakamura (1974), Bass, Johnson, and Unruh (1974), Hohenemser (1974), Hohenemser and Crews (1974), Kanning and Biggers (1974), Hohenemser, Banerjee, and Yin (1975, 1976), Banerjee and Hohenemser (1976), Dooley (1976), Gaukroger and Cansdale (1976), Banerjee, Crews, Hohenemser, and Yin (1977), Hohenemser and Banerjee (1977), Dooley and Ferguson (1978).

On the stability of the flap motion in forward flight: Horvay (1947), Horvay and Yuan (1947), Parkus (1948), Gessow and Crim (1955), Shulman (1956), Shutler and Jones (1958), Perisho (1959), Jenney Arcidiacono, and Smith (1963), Lowis (1963), Drees and McGuigan (1965), Wilde, Bramwell, and Summerscales (1965), Jenkins (1967, 1968), Sissingh (1968), Arcidiacono (1969), Elman, Niebanck, and Bain (1969), Sissingh and Kuczynski (1970), Hohenemser and Yin (1971, 1972a, 1972b), Peters and Hohenemser (1971), Johnson (1972a, 1972b, 1973, 1974c), Biggers (1974). Other literature on the treatment of periodic coefficient differential equations: Perisho and Garcia (1962), Grant (1966), Gladwell and Stammers (1968), Crimi (1969, 1970), Piarulli and White (1970), Gaonkar (1971), Gockel (1972), Hsu (1974), Von Kerczek and Davis (1975), Friedmann, Hammond, and Woo (1977), Ibrahim and Barr (1978).

On the response of the rotor to random turbulence, particularly in forward flight: Grant (1966), Hohenemser and Gaonkar (1967-1970, 1971), Gaonkar and Hohenemser (1969, 1971, 1972), Gaonkar (1971, 1972, 1974a, 1974b, 1977), Gaonkar, Hohenemser, and Yin (1972), Kana and Chu (1972), Lakshmikantham and Rao (1972a, 1972b), Wan and Lakshmikantham (1973, 1974), Gaonkar and Subramanian (1977).

On the influence of the wake induced velocity in rotor dynamics: Slaymaker, Lynn, and Gray (1952), Carpenter and Fridovich (1953), Rebont, Soulez-Lariviere, and Valensi (1960), Rebont, Valensi, and Soulez-Lariviere (1960, 1961), Shupe (1970), Curtiss and Shupe (1971), Hohenemser and Crews (1971, 1972, 1973a, 1973b, 1974, 1975, 1976, 1977), Ormiston and Peters (1972), Crews, Hohenemser, and Ormiston (1973), Azuma and Nakamura (1974), Peters (1974), Ormiston (1976), Banerjee, Crews, Hohenemser, and Yin (1977), Hohenemser and Banerjee (1977).

12–2 Flutter

Traditionally, the term flutter refers to an aeroelastic instability involving the coupled bending and torsion motion of a wing. For the rotary wing, flutter refers to the pitch-flap motion of the blade. Often the term is generalized to include any aeroelastic instability of the rotor or helicopter, but the subject of this section is the blade pitch-flap stability. The classical problem considers two degrees of freedom, the rigid flap and rigid pitch motion of an articulated rotor blade. Since the control system is usually the softest element in the torsion motion, the rigid pitch degree of freedom is a good representation of the blade dynamics. A general fundamental flap mode with natural frequency ν_β is considered here. A thorough analysis of the flutter of a hingeless rotor blade usually requires that the in-plane motion be modeled as well, however. The rotation of the wing introduces a number of effects that make helicopter blade flutter much different from the fixed-wing phenomenon. The centrifugal forces couple the flap and pitch motion if the center of gravity is offset from the feathering axis. Moreover, the returning shed wake has an important influence on the blade aerodynamic forces, as does the periodic aerodynamic environment of the blade in forward flight.

12-2.1 Pitch-Flap Equations

The differential equations for the rigid flap and rigid pitch motion of a rotor blade were derived in section 9-4.1:

$$I_\beta^*(\ddot{\beta} + \nu_\beta^2\beta) - I_x^*(\ddot{\theta} + \theta) = \gamma \int_0^1 r \frac{F_z}{ac}\, dr = \gamma M_F$$

$$I_f^*[\ddot{\theta} + (\omega_\theta^2 + 1)\theta] - I_x^*(\ddot{\beta} + \beta) + K_P I_f^* \omega_\theta^2 \beta = \gamma \int_0^1 \frac{M_a}{ac}\, dr = \gamma M_f$$

Here β is the degree of freedom of the perturbation flap motion, with rotating natural frequency ν_β; and θ is the pitch degree of freedom, with nonrotating natural frequency ω_θ. The inertial coefficients are

$$I_\beta^* = \int_0^1 r^2 m\, dr/I_b,\ I_f^* = \int_0^1 I_\theta\, dr/I_b, \text{ and } I_x^* = \int_0^1 x_I r m\, dr/I_b,$$

where x_I is the distance the blade center of gravity is behind the feathering axis. In section 11–8 the aerodynamic flap and pitch moments:

$$M_F = M_\theta \theta + M_{\dot{\theta}}\dot{\theta} + M_{\dot{\beta}}\dot{\beta} + M_\beta \beta$$

$$M_f = m_\theta \theta + m_{\dot{\theta}}\dot{\theta} + m_{\dot{\beta}}\dot{\beta} + m_\beta \beta$$

are obtained. For hover, the aerodynamic coefficients are:

$$M_\theta = \frac{1}{8} C'(k_e)$$

$$M_{\dot{\theta}} = \frac{c}{12}\left[\tfrac{1}{2} + C'(k_e)\left(1 + 2\frac{x_A}{c}\right)\right]$$

$$M_{\dot\beta} = -\frac{1}{8} C'(k_e)$$

$$M_\beta = \frac{c}{12} C'(k_e)\left(1 + 2\frac{x_A}{c}\right)$$

$$m_\theta = -\frac{x_A}{6} C'(k_e)$$

$$m_{\dot\theta} = -\frac{c^2}{32}\left[1 + 2\frac{x_A}{c}\right]\left[1 + 4\frac{x_A}{c} C'(k_e)\right]$$

$$m_{\dot\beta} = \frac{x_A}{6} C'(k_e)$$

$$m_\beta = -\frac{c^2}{64}\left[1 + 8\frac{x_A}{c} C'(k_e)\left(1 + 2\frac{x_A}{c}\right)\right]$$

where x_A is the distance the aerodynamic center is behind the feathering axis, and the lift deficiency function C' is evaluated at an effective radius, so that $k_e = \omega c/2r_e$ (typically with $r_e = 0.75R$). In section 11–8 the aerodynamic coefficients in forward flight are also given.

The coupled differential equations for the flap and pitch motion are thus:

$$\begin{bmatrix} I_\beta^* & -I_x^* \\ -I_x^* & I_f^* \end{bmatrix}\begin{pmatrix}\beta \\ \theta\end{pmatrix}^{\cdot\cdot} + \begin{bmatrix} -\gamma M_{\dot\beta} & -\gamma M_{\dot\theta} \\ -\gamma m_{\dot\beta} & -\gamma m_{\dot\theta} \end{bmatrix}\begin{pmatrix}\beta \\ \theta\end{pmatrix}^{\cdot}$$

$$+ \begin{bmatrix} I_\beta^*\nu_\beta^2 - \gamma M_\beta & -I_x^* - \gamma M_\theta \\ -I_x^* + K_p I_f^* \omega_\theta^2 - \gamma m_\beta & I_f^*(\omega_\theta^2 + 1) - \gamma m_\theta \end{bmatrix}\begin{pmatrix}\beta \\ \theta\end{pmatrix} = 0$$

No forcing terms have been included since only the stability of the motion is of interest here. In terms of the Laplace variable s (for hover, so that the aerodynamic coefficients are constant), the equations of motion are:

$$\begin{bmatrix} I_\beta^* s^2 - \gamma M_{\dot\beta} s + I_\beta^*\nu_\beta^2 - \gamma M_\beta & -I_x^* s^2 - \gamma M_{\dot\theta} s - I_x^* - \gamma M_\theta \\ -I_x^* s^2 - \gamma m_{\dot\beta} s - I_x^* + K_p I_f^* \omega_\theta^2 - \gamma m_\beta & I_f^* s^2 - \gamma m_{\dot\theta} s + I_f^*(\omega_\theta^2 + 1) - \gamma m_\theta \end{bmatrix}\begin{pmatrix}\beta \\ \theta\end{pmatrix} = 0$$

The eigenvalues are then the roots of the following characteristic equation:

$$\begin{aligned}(I_\beta^* s^2 - \gamma M_{\dot\beta} s + I_\beta^* \nu_\beta^2 - \gamma M_\beta)[I_f^* s^2 - \gamma m_{\dot\theta} s + I_f^*(\omega_\theta^2 + 1) - \gamma m_\theta] \\ - (I_x^* s^2 + \gamma M_{\dot\theta} s + I_x^* + \gamma M_\theta)(I_x^* s^2 + \gamma m_{\dot\beta} s + I_x^* - K_P I_f^* \omega_\theta^2 + \gamma m_\beta) = 0\end{aligned}$$

While the four roots of this equation must in general be found numerically, the stability boundary can be determined analytically. A plane of the system parameters has regions in which all the roots have negative real parts, so that the motion is stable; and regions in which one or more roots have positive real parts, so that the motion is unstable. On the stability boundary one root must be on the imaginary axis of the s—plane, crossing from the left half-plane into the right half-plane. There are two ways a root can cross the imaginary axis into the right half-plane, producing an unstable system: as a real root along the real axis, and as a complex conjugate pair at finite frequency. The instability associated with a real root going through the origin into the right half-plane is called divergence. It is a static instability, since with zero frequency no velocity or acceleration forces are involved. The instability associated with a complex conjugate pair of roots crossing the imaginary axis is called flutter. This instability involves an oscillatory motion of the system.

The most significant parameters for the rotor blade flutter stability are the pitch natural frequency ω_θ, determined by the control system stiffness; and the offsets of the center of gravity and aerodynamic center from the feathering axis. The separation of the center of gravity and aerodynamic center $(x_I - x_A)$ is more important than their distance from the feathering axis, but x_A must usually be kept small to avoid large oscillatory control loads in forward flight. Thus the principal parameters controlling the blade flutter stability are the control stiffness (ω_θ) and the chordwise mass balance (x_I).

12-2.2 Divergence Instability

A divergence instability occurs when a real root goes through the origin of the s—plane into the right half-plane. The divergence stability boundary is thus defined by the requirement that one root be $s = 0$, for which the characteristic equation becomes:

$$(I_\beta^* \nu_\beta^2 - \gamma M_\beta)[I_f^*(\omega_\theta^2 + 1) - \gamma m_\theta] - (I_x^* + \gamma M_\theta)(I_x^* - K_P I_f^* \omega_\theta^2 + \gamma m_\beta) = 0$$

which is a balance of the spring terms alone. Since increasing the flap or pitch springs should produce static stability, the criterion for a stable system is that this quantity be positive. Neglecting I_x^* relative to γM_θ, and γM_β relative to $I_\beta^* \nu_\beta^2$ in the flap equation, the stability criterion can be written as

$$(I_\beta^* \nu_\beta^2 + K_P \gamma M_\theta)[I_f^*(\omega_\theta^2 + 1) - \gamma m_\theta] > \gamma M_\theta [I_x^* + K_P(I_f^* - \gamma m_\theta) + \gamma m_\beta]$$

The left-hand side is the product of the net flap and pitch springs, while the right-hand side is the product of the moments coupling the flap and pitch motion (primarily the aerodynamic flap moment due to pitch, M_θ, and the centrifugal pitch moment due to flapping, I_x^*). The flap and pitch springs are certainly positive. Negative pitch-flap coupling ($K_P < 0$) or a forward aerodynamic center ($x_A < 0$) will contribute a negative spring, but these terms are unlikely to be larger than even the centrifugal springs alone. Divergence stability thus requires that the quantity $[I_x^* + K_P(I_f^* - \gamma m_\theta) + \gamma m_\beta]$ be small or negative; hence a forward center-of-gravity position is desired. Divergence stability can also be ensured by a large enough pitch spring (ω_θ).

We are primarily interested in the pitch-flap stability as a function of the parameters ω_θ and I_x^*, representing the control system stiffness and chordwise mass balance. The divergence stability criterion can be written

$$\omega_\theta^2 > \frac{1}{I_\beta^* \nu_{\beta_e}^2}\left[\gamma M_\theta \frac{I_x}{I_f} - I_\beta^* \nu_\beta^2\left(1 - \frac{\gamma m_\theta}{I_f^*}\right) + \gamma M_\theta \frac{\gamma m_\beta}{I_f^*}\right]$$

which is a straight line on the plane of ω_θ^2 vs. I_x/I_f, or a parabola for ω_θ vs. I_x/I_f. In terms of the mass balance required, the criterion is

$$I_x^* < \frac{I_\beta^* \nu_\beta^2 I_f^*(\omega_\theta^2 + 1)}{\gamma M_\theta} - \frac{m_\theta}{M_\theta} I_\beta^* \nu_\beta^2 - \gamma m_\beta + K_p I_f^* \omega_\theta^2$$

Using $I_x^* \cong (3/2)x_I$ and the results in section 12-2.1 for the aerodynamic coefficients, this becomes

$$x_I - \frac{8\nu_\beta^2}{9} x_A < \frac{16}{3\gamma} \nu_\beta^2 I_f^*\left(\omega_\theta^2 + 1\right) + \frac{\gamma_c^2}{96} + \frac{2}{3} K_p I_f^* \omega_\theta^2$$

which shows that divergence depends on the distance the center of gravity is aft of the aerodynamic center $(x_I - x_A)$. The boundary is relatively insensitive to the pitch axis location for a fixed $x_I - x_A$. Since the right-hand side of this criterion is always positive, divergence stability is assured regardless of the pitch spring if the blade is mass balanced in such a way that the center of gravity is ahead of the aerodynamic center $(x_I - x_A < 0)$.

12-2.3 Flutter Instability

A flutter instability occurs when a pair of complex conjugate roots crosses the imaginary axis into the right half-plane. The flutter stability boundary is thus defined by the requirement that one root be on the imaginary axis, $s = i\omega$, where ω is a real and positive frequency. On substituting $s = i\omega$, the real and imaginary parts of the characteristic equation are

$$\begin{aligned}(-I_\beta^* \omega^2 + I_\beta^* \nu_\beta^2 - \gamma M_\beta)(-I_f^* \omega^2 + I_f^*(\omega_\theta^2 + 1) - \gamma m_\theta) \\ - \omega^2 \gamma M_{\dot\beta} \gamma m_{\dot\theta} + \omega^2 \gamma M_{\dot\theta} \gamma m_{\dot\beta} \\ - (-I_x^* \omega^2 + I_x^* + \gamma M_\theta)(-I_x^* \omega^2 + I_x^* - K_p I_f^* \omega_\theta^2 + \gamma m_\beta) = 0\end{aligned}$$

and

$$\begin{aligned}M_{\dot\beta}[-I_f^* \omega^2 + I_f^*(\omega_\theta^2 + 1) - \gamma m_\theta] \\ + m_{\dot\theta}(-I_\beta^* \omega^2 + I_\beta^* \nu_\beta^2 - \gamma M_\beta) \\ + M_{\dot\theta}(-I_x^* \omega^2 + I_x^* - K_p I_f^* \omega_\theta^2 + \gamma m_\beta) \\ + m_{\dot\beta}(-I_x^* \omega^2 + I_x^* + \gamma M_\theta) = 0\end{aligned}$$

Eliminating ω^2 from these two equations gives a single relation defining the flutter boundary in terms of the blade parameters. The imaginary part of the characteristic equation can be solved for ω^2:

$$\omega^2 = \frac{M_{\dot\beta} I_f^*(\omega_\theta^2 + 1) + m_{\dot\theta} I_\beta^* \nu_\beta^2}{M_{\dot\beta} I_f^* + m_{\dot\theta} I_\beta^*}$$

(using $M_{\dot\beta} m_\theta = m_{\dot\beta} M_\theta$). After substituting for ω^2 in the real part of the characteristic equation and dropping higher order terms (such as I_x^* relative to γM_θ again), the criterion for flutter stability can be written as

$$\omega_\theta^4 + A\omega_\theta^2 \frac{I_x}{I_f} + B\omega_\theta^2 + C\frac{I_x}{I_f} + D > 0$$

where the coefficients are

$$A = -\frac{\gamma M_\theta}{I_\beta^*}\alpha$$

$$B = \left[-\frac{\gamma m_\theta}{I_f^*} + \gamma^2 \frac{M_{\dot\beta} m_{\dot\theta}}{I_\beta^* I_f^*}\right]\alpha - 2\left(\nu_\beta^2 - 1\right) - K_p \frac{\gamma M_\theta m_{\dot\theta}}{M_{\dot\beta} I_f^*}\alpha^2$$

$$C = -\frac{\gamma M_\theta m_{\dot\theta}}{M_{\dot\beta} I_f^*}\left(\nu_\beta^2 - 1\right)\alpha$$

$$D = \gamma^2\left[\left(\frac{m_{\dot\theta}}{I_f^*}\right)^2 + \frac{M_\theta m_{\dot\theta} m_\beta}{M_{\dot\beta} I_f^{*2}}\right]\alpha^2 + (\nu_\beta^2 - 1)^2$$

$$+ (\nu_\beta^2 - 1)\left[\frac{\gamma m_\theta}{I_f^*} + \gamma^2\left(\frac{m_{\dot\theta}}{I_f^*}\right)^2\right]\alpha$$

with $\alpha = 1 + (M_{\dot\beta} I_f^*)/(I_\beta^* m_{\dot\theta})$. Thus on the plane of ω_θ^2 vs. I_x/I_f the flutter stability boundary is a hyperbola, as sketched in Fig. 12-8. Increasing ω_θ^2 must stabilize both flutter and divergence. The asymptotes of the hyperbola have slopes

$$\frac{\partial\omega_\theta^2}{\partial I_x/I_f} = 0 \text{ and } -A$$

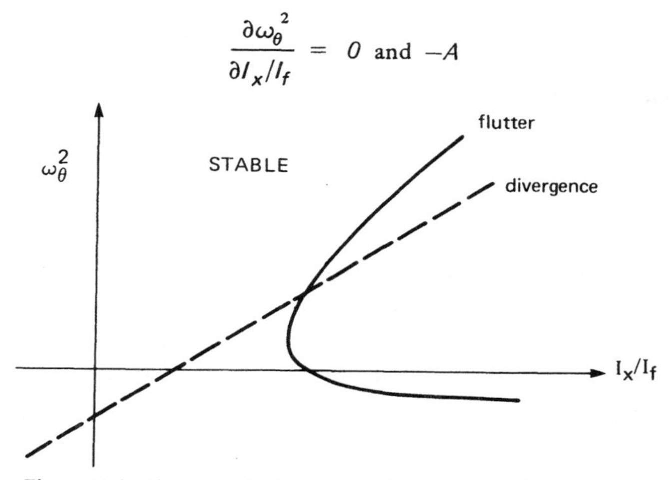

Figure 12-8 Sketch of the flutter and divergence stability boundaries.

The ratio of the slope of the upper asymptote of the flutter boundary to the slope of the divergence boundary is $\nu_{\beta_e}^2 \alpha$, which is always greater than 1. Hence for large enough I_x/I_f, flutter will always be the critical instability. The lower asymptote of the flutter boundary is a line of zero slope at $\omega_\theta^2 = -C/A$, which is always negative. Since the condition $\omega_\theta^2 < 0$ is not physically realizable with structural stiffness, only the upper branch of the flutter hyperbola is of practical interest. The minimum I_x^* of the flutter hyperbola occurs at

$$\frac{I_x}{I_f} = \frac{-2\sqrt{D} - B}{A}$$

for the articulated rotor ($\nu_\beta = 1$ and $K_p = 0$), which gives

$$I_x^* + \frac{I_\beta^* m_\theta}{M_\theta} = \frac{2I_\beta^*}{M_\theta}\sqrt{m_{\dot\theta}^2 + \frac{M_\theta}{M_{\dot\beta}} m_{\dot\theta} m_\beta + \frac{\gamma M_{\dot\beta} m_{\dot\theta}}{M_\theta}}$$

With $I_x^* \cong (3/2)x_I$ and the aerodynamic coefficients given in section 12-2.1, this becomes approximately

$$x_I - \frac{8}{9}x_A = c^2\left(\frac{1}{3\sqrt{2}} + \frac{\gamma}{48}\right)$$

Consequently, if the blade is mass balanced in such a way that the center of gravity is no farther aft than this distance, flutter stability is assured regardless of the pitch stiffness ω_θ. Note that again it is the distance between the center of gravity and aerodynamic center that is the primary parameter in the pitch-flap dynamics. In terms of the required chordwise mass balance, the criterion for flutter stability can be written as

$$\frac{I_x}{I_f} < -\frac{\omega_\theta^4 + B\omega_\theta^2 + D}{A\omega_\theta^2 + C}$$

From this expression the flutter boundary can be constructed for the desired range of ω_θ. Alternatively, to avoid the approximations involved in the derivation of this equation for the flutter hyperbola, the imaginary part of the characteristic equation can be used directly to obtain the flutter frequency ω for a given ω_θ. Then ω_θ and ω are substituted into the real part of the characteristic equation, which can be solved for I_x^*.

Fig. 12-9 is an example of the flutter boundary obtained for an articulated

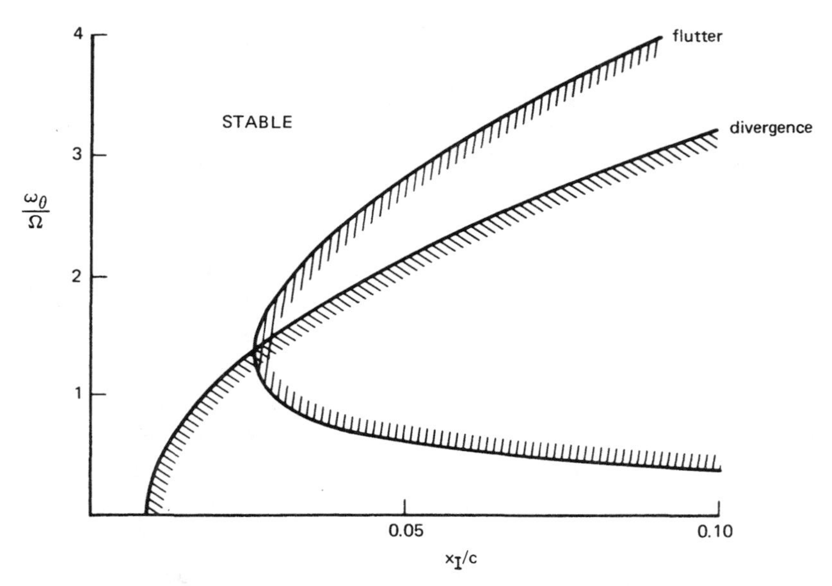

Figure 12-9 Flutter and divergence boundaries for an articulated rotor ($\nu_\beta = 1, \gamma = 12, c/R = 0.05, x_A = 0$, and $I_f^* = 0.001$).

rotor ($\nu_\beta = 1$ and $K_P = 0$) with uniform properties and the parameters $\gamma = 12$, $c/R = 0.05$, $I_f^* = 0.001$, and $x_A = 0$. The shed wake effects are neglected in this case, so $C'(k_e) = 1$ was used.

In summary, both flutter and divergence stability are increased by increasing the control system stiffness ω_θ, or decreasing I_x by moving the blade center of gravity toward the leading edge. A conservative approach is to place the center of gravity at the aerodynamic center or just aft of it. Rotor blades are therefore generally mass balanced about the quarter chord, for loads considerations as well as stability. Most blade designs require a leading edge balance weight to accomplish this. While the control system stiffness is an important flutter parameter, it is difficult to calculate because of the complicated geometry of the control system. Reliable measurements of the stiffness as it appears at the pitch bearing of the rotating blade are not easily obtained either. Mechanical or frictional damping in the control system and pitch bearing can also significantly influence the flutter stability. Usually such damping is nonlinear, and thus to include it in the analysis requires a numerical integration of the equations of motion.

Alternatively, a nonlinear damper can be represented in a linear stability analysis by using an equivalent viscous damping coefficient, with a value such that the same amount of energy is dissipated during a cycle of motion. Except for possible anomalous nonlinear effects, it is generally conservative to neglect such contributions to the pitch damping in the flutter analysis.

The present analysis has used dimensionless parameters, so the system stiffness is represented by $(\omega_\theta/\Omega)^2$. In terms of dimensional quantities, a given rotor will have a fixed value of ω_θ. The minimum allowable ω_θ/Ω at the flutter boundary then becomes a restriction on the maximum rotor speed Ω. For designing the rotor the dimensionless parameter ω_θ/Ω is most useful, but when the rotor has been built flutter places a limit on Ω. Flutter testing of rotors is generally conducted on this basis also; the rotor speed is increased until the flutter or divergence boundary is encountered as a result of decreasing ω_θ/Ω. The best indication of flutter when testing rotors is in the control loads, which are a measure of the pitch motion. Interpreting the frequencies in the cyclic control system requires accounting for the frequency shift Ω between the rotating and nonrotating frames.

12-2.4 Other Factors Influencing Pitch-Flap Stability

12-2.4.1 Shed Wake Influence

The effect of the rotor returning shed wake on the unsteady aerodynamic loads can be accounted for by using the lift deficiency function $C'(k_e)$. For certain operating conditions the wake can have a significant impact on the flutter stability. In Chapter 10, Theodorsen's, Loewy's, and a number of approximate lift deficiency functions were discussed. However, solving the characteristic equation for the stability boundary including the wake effects is not as simple as for the quasistatic case ($C' = 1$). Since C' is a complex number depending on the flutter frequency ω, the procedure described in the last section for constructing the flutter boundary will not work.

Consider the problem of finding the stability boundary for a given value of $I_x{}^*$. The characteristic equation at the flutter boundary ($s = i\omega$) can be solved for the control system stiffness to yield an equation of the form $\omega_\theta{}^2 = f(\omega)$, where f is a complex function of the flutter frequency ω that includes the dependence of the aerodynamic coefficients on $C'(k_e)$. The solution is defined by the requirement that ω_θ^2, and hence f, be real. The function $f(\omega)$ is evaluated for a series of frequencies ω, and the zeros of Im f are located either graphically or numerically. The control system stiffness at the flutter boundary is then given by the real part of $f(\omega)$ at the flutter frequencies corresponding to the zeros of Im f; that is, $\omega_\theta^2 = \text{Re} f$. Following this procedure for a series of values of I_x^*, the flutter stability boundary can be constructed. For the quasistatic case, as in the last section, there will be two solutions for ω_θ^2 for a given I_x, or no solution if I_x is small enough or negative. If Loewy's lift deficiency function is used, there are often more than two solutions, indicating that there are several unstable regions when the wake influence is included, instead of a single region as in the quasistatic case.

The results in the literature indicate that a flutter calculation based on quasistatic aerodynamics (i.e. $C' = 1$) is at least slightly conservative. The quasistatic flutter boundary tends to form an envelope around the boundaries calculated including the wake influence through the lift deficiency function. The effect of the wake is to divide the flutter instability region into several regions because of an increased stability in narrow ranges about certain critical values of ω_θ corresponding to harmonic excitation. Such a modification of the flutter boundary is of little practical significance.

12-2.4.2 Wake-Excited Flutter

In certain operating conditions the returning wake of the rotor can also produce a single-degree-of-freedom instability, called wake-excited flutter. With Loewy's lift deficiency function the aerodynamic damping of the pitch and flap motions can be significantly reduced by the shed wake. Thus in practice wake-excited flutter usually occurs under the operating conditions for which the returning wake has the strongest influence; low collective pitch as in run-up on the ground or autorotation, low forward speed or hover, and a pitch natural frequency near a harmonic of the rotor speed. Under these conditions the wake remains close to the rotor disk, and successive spirals are acting in phase. With higher collective, a climb velocity, or forward speed the influence of the wake and hence the possibility of wake-excited flutter is reduced. Single-degree-of-freedom instabilities are included in the coupled flap-pitch flutter solution, and may be identified by the dominance of one mode in the eigenvector corresponding to the unstable root.

12-2.4.3 *Influence of Forward Flight*

The aerodynamics of the rotor in forward flight introduce periodic coefficients in the equations of motion for flap and pitch of the blade. The eigenvalues of these periodic coefficient linear differential equations can be obtained by the methods discussed in section 8-6.2. At high advance ratio ($\mu > 0.5$, say) such an analysis is essential to propefly evaluate the stability including the influence of the periodic coefficients; it is also necessary to include reverse flow in the aerodynamic model at such high speeds. At low to moderate advance ratio, a constant coefficient approximation may be sufficiently accurate to be useful. If the mean values of the coefficients in the rotating frame are used, the only effect of forward flight retained is an order μ^2 increase in M_θ and m_θ. As for the flapping dynamics, then, the averaging of the coefficients should be performed in the nonrotating frame. The constant coefficient approximation in forward flight best models the lower frequency behavior of rotors with a large number of blades (see section 12-1.1.2). Since the pitch natural frequency tends to be relatively high, it may be expected that the exact solution of the periodic coefficient equations will be required for pitch-flap flutter more often than when only the flapping motion is involved.

Alternatively, the dynamic stability can be assessed from the results of a direct numerical integration of the equations of motion. Such an approach is also necessary if nonlinear effects are to be included in the analysis, such as those due to blade stall or compressibility. The evaluation of the stability of periodic systems from the transient motions is not an elementary matter, however. Another approach is the frozen coefficient method, in which the eigenvalues are obtained for a time-invariant system constructed using the coefficients evaluated at a particular azimuth. A number of critical azimuth locations are normally used, such as $\psi = 90°$ and $\psi = 270°$. Basically, this approach assumes that the variation of the aerodynamic coefficients due to forward flight (which is primarily at 1/rev, at least for small μ) is much slower than the blade motion involved in the flutter dynamics (which is usually at a frequency somewhat lower than ω_θ). The frozen coefficient method should be used with caution however, for that assumption is often not justified.

12-2.4.4 Coupled Blades

The flutter analysis developed here has so far considered a single independent blade. Even in the shaft-fixed case, all the blades are coupled through the rotor control system, however. The load path through the control system, which determines the stiffness of the restraint at the pitch bearing, depends on the pitch motion of all the blades. Thus the rigid pitch natural frequency ω_θ, which is a primary parameter of the flutter stability, cannot in general be defined for the motion of an individual blade alone; the flutter analysis must consider the entire rotor.

In the nonrotating frame the coupling of the blades through the control system appears as a different stiffness or natural frequency for each of the nonrotating equations of motion (i.e. for the θ_0, θ_{nc}, θ_{ns}, and $\theta_{N/2}$ equations; see section 9-4.4). With three or more blades the rotating control system is axisymmetric, so the coupling is due to the nonrotating control system. The primary load paths are through the collective and cyclic pitch control systems (θ_0, θ_{1c}, and θ_{1s} equations). The higher modes involve flexibility of the pitch horn and pitch link, and bending of the swashplate (θ_{2c}, θ_{2s}, ..., $\theta_{N/2}$ equations as required). In hover, the only coupling of the nonrotating equations is between θ_{nc} and θ_{ns} (and also the corresponding β_{nc} and β_{ns} equations). The equation for the collective mode is identical to that of a single blade; hence the flutter solution can be obtained by considering an independent blade with the appropriate collective natural frequency ω_{θ_0}. Similarly, the equation of motion for $\theta_{N/2}$ is not coupled with the other pitch degrees of freedom and is the same as an independent blade with pitch frequency $\omega_{\theta N/2}$. If $\omega_{\theta nc} = \omega_{\theta ns}$, then the coupled equations for θ_{nc} and θ_{ns} are also equivalent to a single independent blade. For the reactionless modes this will be the case because of the axisymmetry of the rotating control system. The effective natural frequencies of the lateral and longitudinal cyclic control are not likely to be equal, however ($\omega_{\theta_{1c}} \neq \omega_{\theta 1s}$). If the difference in cyclic control system stiffnesses is significant, the flutter analysis must consider a four-degree-of-freedom problem for θ_{1c}, θ_{1s}, β_{1c}, and β_{1s}. The coupled motion of the θ_{1c} and θ_{1s} degrees of freedom will be a progressive mode at a frequency near $\omega_\theta + 1$/rev and a regressive mode at a frequency near $\omega_\theta - 1$/rev.

The natural frequencies of the reactionless modes (θ_{2c}, θ_{2s}, ..., $\theta_{N/2}$) are generally higher than those of the collective and cyclic modes. The stiffness of the cyclic control system is usually less than that of the collective control system, in which case the critical flutter problem involves the cyclic pitch degrees of freedom (θ_{1c} and θ_{1s}).

It may also be necessary to consider aerodynamic effects that differ for each nonrotating mode of the rotor. For example, the cyclic modes involve identical motion of each blade at a given azimuth and thus are more susceptable to excitation by a disturbance at a particular point in the fixed frame, such as aerodynamic interference due to the fuselage or tail rotor. When Loewy's lift deficiency function is used to account for the returning shed wake effects, there is a separate function C' for each nonrotating mode (see section 10–5), although an equivalent single blade model is again often possible.

12-2.4.5 Additional Degrees of Freedom

In some cases additional degrees of freedom may be required to properly model even the flutter of an articulated rotor blade. The equations of motion for the coupled flapwise bending and elastic torsion of the blade were derived in section 9-4.3. Numerical methods are required to obtain the eigenvalues of such high-order systems. In the case of hingeless rotors it is usually necessary to include the in-plane motion to accurately calculate the aeroelastic stability of a blade, so a fully coupled flap-lag-torsion analysis is required.

Flutter of a two-bladed teetering rotor is called weaving. With a teetering rotor the precone can be outboard of the feather bearing. The increased effective torsional inertia of such a configuration (see section 9-4.2) has an unfavorable influence on the pitch-flap stability, decreasing the pitch natural frequency for a given control stiffness. In hover the divergence and flutter analyses developed in the preceding sections are applicable to the two-bladed rotor (with $\nu_\beta = 1$ for the teetering mode and the appropriate cantilever flap frequency for the coning mode). In forward flight the moment carry-through in the teetering hub cancels the odd harmonics in the periodic coefficients of the β_1 and θ_1 equations of motion.

12-2.5 Literature

Miller and Ellis (1956) conducted a theoretical investigation of helicopter rotor divergence and flutter in hover. They derived the equations of motion for rigid flap and rigid pitch, and also with the first flapwise elastic bending mode. Examples were given for the flutter and divergence boundaries. They examined the effect of Loewy's lift deficiency function, and concluded that quasistatic aerodynamics ($C' = 1$) give a slightly conservative stability boundary. The elastic bending mode was found to have only a small effect on the flutter of an articulated blade.

Daughaday, Du Waldt, and Gates (1957) conducted an experimental and theoretical investigation of the flutter of helicopter rotor blades, deriving the equations of motion for rigid flap, rigid pitch, and flapwise bending. Good correlation of the theory with measured flutter speed (ω_θ/Ω) and flutter frequency was found, using quasistatic aerodynamics ($C' = 1$). Using Loewy's lift deficiency function, nearly the same or slightly better prediction of the flutter frequency was obtained, but the predicted flutter speed was somewhat low. Wake-excited flutter was investigated for the case of zero collective, where the unsteady aerodynamics are most important. Using Loewy's function C', good correlation with the measured flutter frequency and flutter speed as a function of chordwise mass balance (x_I) was obtained. The shed wake had a stabilizing effect in a narrow range of flutter frequencies near 2/rev, which was predicted using Loewy's function but not with quasistatic aerodynamics. It was also found that the blade bending or torsion flexibility can have a considerable effect on flutter, especially if the rotor speed is such that a natural frequency of the blade is at resonance with n/rev.

Jones (1958) discusses the influence of the rotor wake on flutter and vibration. He compared the flutter boundaries calculated using quasistatic ($C' = 1$) and unsteady aerodynamics. The boundaries obtained using $C' = 1$ tended to be conservative, forming an envelope around the boundaries obtained when the shed wake influence was included. The increased flutter stability for certain critical ranges of ω_θ is of little practical importance, but a considerable influence of the lift deficiency function on the flutter frequency was found.

Brooks and Baker (1958) conducted an experimental investigation of the flutter of a hovering model rotor, examining the effects of tip Mach number, structural damping, and chordwise center of gravity position. The flutter speed $\Omega R/\omega_\theta c$ was found to be roughly constant for collective pitch settings below stall, and the flutter frequency was considerably below the pitch natural frequency ($\omega \cong 0.7\omega_\theta$). A forward center-of-gravity shift generally raised the flutter speed at low collective. Some wake-excited flutter was observed at collective settings near zero, occurring at a flutter speed about 85% that of the classical flutter instability, with a flutter frequency $\omega \cong 0.8\omega_\theta$. Data for stall flutter at high collective were also obtained. A large beneficial effect of compressibility was observed near the section critical Mach number: if flutter did not occur when $M_{tip} < 0.73$, it did not occur at all. Below stall the flutter speed decreased first as the tip Mach number was increaased, and then at a certain point the flutter speed rapidly increased. This stabilizing effect of compressibility was attributed to the rearward shift of the aerodynamic center above the critical Mach number. A tentative design criterion was formulated, as follows: for torsion structural damping above 1.5% critical ($g_s > 0.03$), the rotor should be flutter free if

$$\frac{\omega_\theta}{\Omega} > \frac{0.6M_{tip}}{c/R}$$

according to which ω_θ/Ω is typically around 5 to 6/rev.

Gates, Piziali, and Du Waldt (1963) conducted an experimental and theoretical investigation of the flutter of a model rotor in forward flight. The theory included rigid flap, rigid pitch, and flapwise bending motion, with the equations solved using an analog computer. A conservative prediction of flutter was found. Little effect of advance ratio on the measured flutter speed (Ω/ω_θ) was found if the blade was nearly mass balanced, at least up to $\mu = 0.5$. As the mass unbalance was increased, the flutter speed became more sensitive to forward speed, decreasing with μ. Generally, the theory and experiments showed that the flutter speed tends to decrease with μ, although there is not much effect below about $\mu = 0.2$ or 0.3.

On the flutter and divergence stability of helicopter rotor blades: Rosenberg (1944), Coleman (1946), Coleman and Stempin (1947), Reissner and Morduchow (1948), Turner and Duke (1949), Morduchow (1950), Brooks and Sylvester (1955), Goland and Perlmutter (1957), Leone (1957), Timman and van de Vooren (1957), Du Waldt, Gates, and Piziali (1959), Perisho (1959), Du Waldt and Piziali (1960), Zvara and Ham (1960), White and Crimi (1961), Crimi and White (1962), Gates and Du Waldt (1962, 1963), Trenka and White (1963), Mil′ (1966), Stammers (1968a, 1968b, 1970), Arcidiacono (1969), Astill and Niebanck (1969), Du Waldt (1969), Loewy

(1969), Niebanck (1969), Niebanck and Elman (1969), Paul (1969a), Niebanck and Bain (1970), Sissingh and Kuczynski (1970), Brunelle and Robertson (1971), Piarulli (1971), Shipman and Wood (1971), Hodges (1972), Pierce and White (1972), Anderson and Watts (1973, 1976), Ham (1973), Hodges and Ormiston (1973a, 1976, 1977), Dowell (1974), Hodges and Dowell (1974), Huber and Strehlow (1976), Murthy and Pierce (1976), Chopra (1977), Johnson (1977d), Kunz (1977), Yasue (1977).

12–3 Flap-Lag Dynamics

Let us examine next the stability of the coupled flap and lag motion of a rotor blade. The flap or lag motion alone has positive aerodynamic damping, although it is low for the lag mode. The blade in-plane and out-of-plane motions are coupled by Coriolis and aerodynamic forces, however, which can produce an instability. The problem considered here has only two degrees of freedom, the fundamental flap and lag modes.

12-3.1 Flap-Lag Equations

The equations of motion for the first out-of-plane and first in-plane modes of the rotor blade were derived in section 9-3.1. Assuming purely out-of-plane deflection for the flap mode, and purely in-plane deflection for the lag mode, the equations are as follows:

$$I_\beta^*(\ddot{\beta} + \nu_\beta^2 \beta) - I_{\beta\zeta}^* 2\beta_0 \dot{\zeta} = \gamma \int_0^1 \eta_\beta \frac{F_z}{ac} \, dr = \gamma M_F$$

$$I_\zeta^*(\ddot{\zeta} + C_\zeta^* \dot{\zeta} + \nu_\zeta^2 \zeta) + I_{\beta\zeta}^* 2\beta_0 \dot{\beta} = \gamma \int_0^1 \eta_\zeta \frac{F_x}{ac} \, dr = \gamma M_L$$

Here β is the flap degree of freedom with rotating natural frequency ν_β, and ζ the lag degree of freedom with natural frequency ν_ζ. While arbitrary blade frequencies will be considered, often it is necessary to account for the coupled flap and lag deflection in the fundamental bending modes in order to accurately model the dynamics of a hingeless rotor blade. The inertial constants $I_\beta^* = I_\zeta^* = I_{\beta\zeta}^* = 1$ exactly if it is assumed that the mode shapes are $\eta_\beta = \eta_\zeta = r$. Mechanical or structural damping of the lag motion has been included as a viscous damping coefficient C_ζ^*. The flap and lag motion of the blade are coupled inertially by Coriolis forces, which have been linearized about the trim coning angle β_0. Lag velocity $\dot{\zeta}$ aft produces a downward flap acceleration of the blade, or in the d'Alembert view an upward flap moment. Similarly, upward flap velocity $\dot{\beta}$ produces a lead Coriolis moment on the blade.

The aerodynamic flap and lag moments were derived in section 11–3:

$$M_F = M_\theta(-K_{P_\beta}\beta - K_{P_\zeta}\zeta) + M_{\dot{\beta}}\dot{\beta} + M_\beta\beta + M_{\dot{\zeta}}\dot{\zeta} + M_\zeta\zeta$$

$$M_L = Q_\theta(-K_{P_\beta}\beta - K_{P_\zeta}\zeta) + Q_{\dot{\beta}}\dot{\beta} + Q_\beta\beta + Q_{\dot{\zeta}}\dot{\zeta} + Q_\zeta\zeta$$

Since only the stability is of interest here, the pitch control input has not been included. There is, however, a pitch change due to kinematic pitch-flap and pitch-lag coupling. $\Delta\theta = -K_{P_\beta}\beta - K_{P_\zeta}\zeta$, which produces flap and lag moments. The sign convention for positive coupling is that a nose-down pitch change is produced by flap-up or lag-back deflection of the blade.

For hover or vertical flight the aerodynamic coefficients are constant, and $M_\beta = M_\zeta = Q_\beta = Q_\zeta = 0$. Assuming uniform inflow, $\eta_\beta = \eta_\zeta = r$, and neglecting tip losses, the aerodynamic coefficients are then:

$$M_\theta = \frac{1}{8}$$

$$M_{\dot{\beta}} = -\frac{1}{8}$$

$$M_{\dot{\zeta}} = -\left[\frac{3}{2}\frac{C_T}{\sigma a} + \frac{5}{24}\lambda_{HP} + \frac{\theta_{tw}}{80}\right]$$

$$Q_\theta = \frac{\lambda_{HP}}{6}$$

$$Q_{\dot{\beta}} = \frac{3}{4}\frac{C_T}{\sigma a} - \frac{7}{48}\lambda_{HP} + \frac{\theta_{tw}}{160}$$

$$Q_{\dot{\zeta}} = -\left[\lambda_{HP}\left(\frac{C_T}{\sigma a} + \frac{\lambda_{HP}}{4}\right) + \frac{c_d}{4a}\right]$$

For hover the trim induced velocity is $\lambda_{HP} = \kappa\sqrt{C_T/2}$, so the aerodynamic forces are determined by a single parameter of the rotor operating state, namely the thrust coefficient C_T. Alternatively, the collective pitch of the rotor can be used as the parameter, since it is related to the rotor thrust by

$$\theta_{.75} = \frac{6C_T}{\sigma a} + \frac{3}{2}\lambda_{HP}$$

The Coriolis forces also require the rotor coning angle, which for hover is

$$\beta_0 = \frac{1}{I_\beta^* \nu_{\beta_e}^2}\left[\gamma \int_0^1 \frac{1}{2} r^3 \alpha dr + K_\beta \beta_p\right]$$

$$= \frac{1}{I_\beta^* \nu_{\beta_e}^2}\left[\gamma\left(\frac{3}{4}\frac{C_T}{\sigma a} + \frac{\lambda_{HP}}{48} + \frac{\theta_{tw}}{160}\right) + K_\beta \beta_p\right]$$

where β_p is the precone angle, K_β is the flap hinge spring constant, and $I_\beta^* \nu_{\beta_e}^2 = I_\beta^* \nu_\beta^2 + K_{P_\beta} \gamma M_\theta$. The precone is used to reduce the mean flap bending moments at the hinge spring, which are proportional to $(\beta_0 - \beta_p)$. The ideal precone is the value of β_p that gives $\beta_0 = \beta_p$, and hence exactly zero moment. The ideal precone (and also β_0 then) is the value obtained by setting the flap hinge spring K_β to zero (in ν_β as well; see also section 5–13).

The differential equations for the rotor flap and lag motion in hover are thus

$$\begin{bmatrix} I_\beta^* & 0 \\ 0 & I_\zeta^* \end{bmatrix}\begin{pmatrix}\beta \\ \zeta\end{pmatrix}^{\cdot\cdot} + \begin{bmatrix} -\gamma M_{\dot\beta} & -\gamma M_{\dot\zeta} - 2I_{\beta\zeta}^* \beta_0 \\ -\gamma Q_{\dot\beta} + 2I_{\beta\zeta}^* \beta_0 & -\gamma Q_{\dot\zeta} + I_\zeta^* C_\zeta^* \end{bmatrix}\begin{pmatrix}\beta \\ \zeta\end{pmatrix}^{\cdot}$$

$$+ \begin{bmatrix} I_\beta^* \nu_\beta^2 + K_{P_\beta}\gamma M_\theta & K_{P_\zeta}\gamma M_\theta \\ K_{P_\beta}\gamma Q_\theta & I_\zeta^* \nu_\zeta^2 + K_{P_\zeta}\gamma Q_\theta \end{bmatrix}\begin{pmatrix}\beta \\ \zeta\end{pmatrix} = 0$$

and the characteristic equation is

$$[I_\beta^* s^2 - \gamma M_{\dot\beta} s + I_\beta^* \nu_\beta^2 + K_{P_\beta}\gamma M_\theta][I_\zeta^* s^2 + (-\gamma Q_{\dot\zeta} + I_\zeta^* C_\zeta^*)s + I_\zeta^* \nu_\zeta^2 + K_{P_\zeta}\gamma Q_\theta]$$

$$- [(\gamma M_{\dot\zeta} + 2I_{\beta\zeta}^* \beta_0)s - K_{P_\zeta}\gamma M_\theta][(\gamma Q_{\dot\beta} - 2I_{\beta\zeta}^* \beta_0)s - K_{P_\beta}\gamma Q_\theta] = 0$$

which can be solved for the four roots of the system (a complex conjugate pair for the flap mode, and for the lag mode). The uncoupled motion has positive damping for both flap and lag, so an instability can only be encountered because of the coupling terms. Divergence is seldom a factor in flap-lag dynamics, assuming that K_{P_β} and K_{P_ζ} are not so negative that there is a net negative flap or lag spring.

When there is zero pitch-lag coupling ($K_{P_\zeta} = 0$), the only lag influence on the flap equation is the velocity term. The flap moment due to $\dot{\zeta}$ consists of aerodynamic and Coriolis terms and can be written

$$\gamma M_{\dot{\zeta}} + 2I_{\beta\zeta}^{*}\beta_0 = -\gamma\left(\frac{\lambda_{HP}}{6} + 2\int_0^1 \frac{1}{2} r^3 \alpha dr\right) + 2\frac{I_{\beta\alpha}^{*}}{I_\beta^{*}\nu_{\beta_e}^2}\left(\gamma\int_0^1 \frac{1}{2} r^3 \alpha dr + K_\beta\beta_p\right)$$

$$\cong -\frac{2\gamma(\nu_{\beta_e}^2 - 1)}{\nu_{\beta_e}^2}\int_0^1 \frac{1}{2} r^3 \alpha dr + \frac{2K_\beta\beta_p}{\nu_{\beta_e}^2}$$

Thus for an articulated rotor with no flap hinge spring or offset and no pitch-flap coupling ($\nu_{\beta_e} = 1$ and $K_\beta = 0$) the aerodynamic and Coriolis flap moments due to lag velocity nearly cancel, and the flap equation is decoupled from the lag motion. The flap and lag motion are stable in this case. A lag motion produced by the Coriolis forces due to flapping is important for the blade loads and vibration, but not for stability. Note that when there is no flap hinge offset and no pitch-flap coupling, $\nu_{\beta_e}^2 = 1 + K_\beta$. In addition, the ideal precone is $\beta_{\text{ideal}} = \gamma \int_0^1 \frac{1}{2} r^2 \alpha dr$. Thus, when $\nu_{\beta_e} > 1$ because of a hinge spring, but with ideal precone, the total flap moment due to lag velocity is still zero. With ideal precone the hinge spring does not contribute to the balance of flap moments determining the coning, and the solution for β_0 is consequently the same as for the articulated rotor. It follows that for an articulated rotor with a flap frequency near 1/rev and small pitch-flap and pitch-lag coupling, in hover or low forward speed, the flap-lag motion of the blade is expected to remain stable.

Consider next the case of zero thrust, so that all the aerodynamic coefficients except M_θ, $M_{\dot{\beta}}$, and $Q_{\dot{\zeta}}$ are zero or nearly so. Assume also zero precone, so $\beta_0 = 0$. Then the only remaining coupling of the flap and lag equations is a flap moment due to ζ, acting through M_θ when there is kinematic pitch-lag coupling ($K_{P_\zeta} \neq 0$). The lag equation is decoupled from the flap motion, so the system is stable. It follows that if a flap-lag instability is encountered, it should be a high thrust or high collective phenomenon. The flap-lag stability boundary should give a critical thrust level, or equivalently a collective pitch or coning angle limit.

12-3.2 Articulated Rotors

For an articulated rotor the lag frequency is small, typically around $\nu_\zeta = 0.25$ to 0.30/rev. (Recall that $\nu_\zeta^2 = (3/2)e$, where e is the lag hinge offset.) An approximate solution for the flap-lag stability can be obtained in this case to show the influence of pitch-lag and pitch-flap coupling. When $K_{P_\zeta} \neq 0$, the flap moment produced by ζ through M_θ dominates the small flap moments due to the lag velocity $\dot{\zeta}$, and the latter will be neglected. All the aerodynamic lag moments due to flapping are neglected compared to the Coriolis term. For convenience, the lag moments $Q_{\dot{\zeta}}$ and $K_{P_\zeta} Q_\theta$ are also dropped, since they can be considered included in the lag damping C_ζ^* and lag spring $I_\zeta^* \nu_\zeta^2$. With these approximations, the equations of motion in Laplace form are

$$\begin{bmatrix} I_\beta^* s^2 - \gamma M_{\dot{\beta}} s + I_\beta^* \nu_\beta^2 + K_{P_\beta} \gamma M_\theta & K_{P_\zeta} \gamma M_\theta \\ 2I_{\beta\zeta}^* \beta_0 s & I_\zeta^* s^2 + I_\zeta^* C_\zeta^* s + I_\zeta^* \nu_\zeta^2 \end{bmatrix} \begin{pmatrix} \beta \\ \zeta \end{pmatrix} = 0$$

The equations are now coupled only because of the Coriolis lag moment, and the flap moment produced by pitch-lag coupling.

Since the flap damping is high, it is the lag mode that is most likely to go unstable. The frequency of that mode is near ν_ζ, which is small for an articulated rotor. It follows that the eigenvalue at the flutter boundary will be small, and hence the flap equation can be approximated by a quasi-static balance of the flap moments due to β and ζ:

$$(I_\beta^* \nu_\beta^2 + K_{P_\beta} \gamma M_\theta)\beta + (K_{P_\zeta} \gamma M_\theta)\zeta = 0$$

or

$$\beta = -\frac{K_{P_\zeta} \gamma M_\theta}{I_\beta^* \nu_\beta^2 + K_{P_\beta} \gamma M_\theta} \zeta$$

This is the flap motion that accompanies the lag oscillation in the flutter mode because of the pitch-lag coupling. On substituting for this flap motion in the Coriolis lag moment, the lag equation becomes

$$\left[I_\zeta^* s^2 + \left(I_\zeta^* C_\zeta^* - 2I_{\beta\zeta}^* \beta_0 \frac{K_{P_\zeta} \gamma M_\theta}{I_\beta^* \nu_{\beta_e}^2} \right) s + I_\zeta^* \nu_\zeta^2 \right] \zeta = 0$$

For an articulated rotor, the flap response to the moments produced by the pitch-lag coupling is in phase with the low frequency lag motion. Hence the Coriolis lag moment due to flap velocity results in a lag damping term, which determines the stability of the lag motion.

The criterion for stability follows directly from the requirement of positive net lag damping:

$$I_\zeta^* C_\zeta^* - 2I_{\beta\zeta}^* \beta_0 \frac{K_{P_\zeta} \gamma M_\theta}{I_\beta^* \nu_\beta^2 + K_{P_\beta} \gamma M_\theta} > 0$$

which was first obtained by Chou (1958). The aerodynamic drag damping can be included as well as C_ζ^*, and the aerodynamic moment $Q_{\dot\beta}$ in addition to the Coriolis moment. Since $2I_{\beta\zeta}^*\beta_0 - \gamma Q_{\dot\beta} \cong I_{\beta\zeta}^*\beta_0$, the criterion becomes

$$I_\zeta^* C_\zeta^* - \gamma Q_{\dot\zeta} - I_{\beta\zeta}^* \beta_0 \frac{K_{P_\zeta} \gamma M_\theta}{I_\beta^* \nu_\beta^2 + K_{P_\beta} \gamma M_\theta} > 0$$

This expression gives the lag damping required for stability, or alternatively the maximum pitch-lag coupling allowed. For a given rotor, the stability decreases as the coning angle β_0 increases with thrust. Positive pitch-lag coupling (lag back, pitch down) is destabilizing for articulated rotors.

12-3.3 Hingeless Rotors

Consider now the stability of the flap-lag motion of a hingeless rotor blade, as modeled by purely out-of-plane and purely in-plane modes with arbitrary natural frequencies ν_β and ν_ζ. Assuming no pitch-flap or pitch-lag coupling ($K_{P_\beta} = K_{P_\zeta} = 0$), the characteristic equation becomes

$$(I_\beta^* s^2 - \gamma M_{\dot\beta} s + I_\beta^* \nu_\beta^2)(I_\zeta^* s^2 + C_\zeta s + I_\zeta^* \nu_\zeta^2)$$

$$- (\gamma M_{\dot\zeta} + 2I_{\beta\zeta}^* \beta_0)(\gamma Q_{\dot\beta} - 2I_{\beta\zeta}^* \beta_0) s^2 = 0$$

where $C_\zeta = I_\zeta^* C_\zeta^* - \gamma Q_{\dot\zeta}$ is here the total lag damping. The flutter stability boundary is obtained by substituting $s = i\omega$ in the characteristic equation. The imaginary part of the characteristic equation can then be solved for the flutter frequency:

$$\omega^2 = \frac{-\gamma M_{\dot\beta} I_\zeta^* \nu_\zeta^2 + C_\zeta I_\beta^* \nu_\beta^2}{-\gamma M_{\dot\beta} I_\zeta^* + C_\zeta I_\beta^*}$$

If the flap damping is much higher than the lag damping, the flutter frequency will be just slightly above ν_ζ, implying that it is the lag mode which is unstable. Substituting for ω^2 in the real part of the characteristic equation gives the equation of the flutter boundary,

$$(\gamma M_{\dot\zeta} + 2I_{\beta\zeta}^*\beta_0)(\gamma Q_{\dot\beta} - 2I_{\beta\zeta}^*\beta_0)$$
$$= -\gamma M_{\dot\beta} C_\zeta \left[1 + \frac{I_\beta^{*2} I_\zeta^{*2} (\nu_\beta^2 - \nu_\zeta^2)^2}{(-\gamma M_{\dot\beta} I_\zeta^* + C_\zeta I_\beta^*)(-\gamma M_{\dot\beta} I_\zeta^* \nu_\zeta^2 + C_\zeta I_\beta^* \nu_\beta^2)}\right]$$

The left-hand side is the product of the coupling terms, which for stability must be less than the product of the damping terms on the right-hand side. This equation can be considered as a criterion for the minimum lag damping required for stability, or for the maximum allowable rotor thrust, which determines the aerodynamic forces and coning angle in the coupling terms. Alternatively, it may be viewed as defining a stability boundary on the ν_β vs. ν_ζ plane, or as a function of some other parameters.

Now in the flap equation, the aerodynamic and Coriolis moments due to $\dot\zeta$ are nearly equal in magnitude but opposite in sign, so

$$\gamma M_{\dot\zeta} + 2I_{\beta\zeta}^*\beta_0 \cong -\gamma \frac{\lambda_{HP}}{6} - \frac{2\gamma(\nu_\beta^2 - 1)}{\nu_\beta^2} \frac{3}{4} \frac{C_T}{\sigma a} + \frac{2K_\beta \beta_p}{\nu_\beta^2}$$

When $\nu_\beta > 1$, the aerodynamic term is larger. In the lag equation, the aerodynamic moment due to $\dot\beta$ has about one-half the magnitude of the Coriolis terms and opposite sign, so

$$\gamma Q_{\dot\beta} - 2I_{\beta\zeta}^*\beta_0 \cong -\gamma \frac{3}{16} \lambda_{HP} + \frac{\gamma(\nu_\beta^2 - 2)}{\nu_\beta^2} \frac{3}{4} \frac{C_T}{\sigma a} - \frac{2K_\beta \beta_p}{\nu_\beta^2}$$

Hence

$$(\gamma M_{\dot\zeta} + 2I_{\beta\zeta}^*\beta_0)(\gamma Q_{\dot\beta} - 2I_{\beta\zeta}^*\beta_0)$$
$$\cong \frac{2(\nu_\beta^2 - 1)(2 - \nu_\beta^2)}{\nu_\beta^4} \left[\gamma \frac{3}{4} \frac{C_T}{\sigma a} - \frac{K_\beta \beta_p}{\nu_\beta^2 - 1}\right] \left[\gamma \frac{3}{4} \frac{C_T}{\sigma a} + \frac{2K_\beta \beta_p}{2 - \nu_\beta^2}\right]$$
$$+ \frac{\gamma^2}{8} \lambda_{HP} \left(\frac{C_T}{\sigma a} + \frac{1}{4} \lambda_{HP}\right)$$

The last term can be combined with $-\gamma M_{\dot\beta} C_\zeta$ on the right-hand side of the equation to give

$$-\gamma M_{\dot\beta} C_\zeta - \frac{\gamma^2}{8} \lambda_{HP}\left(\frac{C_T}{\sigma a} + \frac{1}{4}\lambda_{HP}\right) = -\gamma M_{\dot\beta}\left(\frac{\gamma c_d}{4a} + I_\zeta^* C_\zeta^*\right)$$

(The aerodynamic damping $Q_{\dot\zeta}$ in C_ζ cancels, except for the viscous drag term.) The criterion for flap-lag stability of the hovering rotor with no pitch-flap or pitch-lag coupling is thus:

$$\left(\frac{6C_T}{\sigma a} - \frac{8}{\gamma}\frac{K_\beta \beta_p}{\nu_\beta^2 - 1}\right)\left(\frac{6C_T}{\sigma a} + \frac{16}{\gamma}\frac{K_\beta \beta_p}{2 - \nu_\beta^2}\right)\frac{2(\gamma/8)^2(\nu_\beta^2 - 1)(2 - \nu_\beta^2)}{\nu_\beta^4}$$

$$< -\gamma M_{\dot\beta}\left[\frac{\gamma c_d}{4a} + I_\zeta^* C_\zeta^* + \left(\frac{8}{\gamma}\right)^2 \frac{(\nu_\beta^2 - \nu_\zeta^2)^2 C_\zeta}{[1 + (8C_\zeta/\gamma)][\nu_\zeta^2 + (8C_\zeta/\gamma)\nu_\beta^2]}\right]$$

Since the flap damping ($-\gamma M_{\dot\beta}$) and lag damping (C_ζ) are positive, the right-hand side of this equation is always positive. It follows that the flap-lag motion is stable if the left-hand side is zero or negative. One such case is the articulated rotor, for which $\nu_\beta = 1$ and the left-hand side is zero. As discussed in section 12-3.1, this result is due to the decoupling of the flap equation from the lag motion when $\nu_\beta = 1$ and there is no pitch-lag coupling. In general, the flap-lag motion will be stable unless $1 < \nu_\beta^2 < 2$ ($1 < \nu_\beta < 1.414$), but that covers the usual range of flap frequencies for articulated and hingeless rotors. The left-hand side will be positive for high enough rotor thrust or collective pitch, and the flap-lag motion will be unstable at some critical C_T depending on the lag damping. The term in brackets on the right-hand side is of order C_ζ, so it follows that the dimensionless lag damping C_ζ required for stability is of the order $\bar\alpha^2 = (6C_T/\sigma a)^2$, which is small. Hence an articulated rotor with ν_β slightly above 1/rev and a mechanical damper giving a high level of lag damping will almost certainly be stable (assuming $K_{P_\zeta} = 0$). For a hingeless rotor, however, ν_β is significantly above 1/rev and the structural lag damping is small, so a flap-lag instability is a possibility.

Let us consider the case for which the flap-lag motion is least stable. The second term on the right-hand side of the stability criterion has a minimum value (zero) when $\nu_\beta = \nu_\zeta$. Furthermore, the factor $2(\nu_\beta^2 - 1)(2 - \nu_\beta^2)/\nu_\beta^4$ on the left-hand side has a maximum value of ¼ at $\nu_\beta^2 = 4/3$, or $\nu_\beta = 1.15$.

Thus the stiff in-plane hingeless rotor with $\nu_\beta = \nu_\zeta = \sqrt{4/3}$ has the minimum flap-lag stability margin. For this case the stability criterion becomes

$$\left(\frac{\gamma}{8}\right)^2 \left(\frac{6C_T}{\sigma a}\right)^2 - \left(3K_\beta \beta_p\right)^2 < \frac{\gamma}{2}\left(\frac{\gamma c_d}{4a} + I_\zeta^* C_\zeta^*\right)$$

Recall that the ideal precone is $\beta_p \cong \gamma\tfrac{3}{4}(C_T/\sigma a)$, and $K_\beta = \nu_\beta^2 - 1 = 1/3$ here. Then the left-hand side is zero with ideal precone, in which case the flap-lag motion is stable. Neglecting the precone term, and writing the lag damping in terms of a structural damping coefficient g_s ($C_\zeta^* = g_s \nu_\zeta = g_s\sqrt{4/3}$), we obtain for the stability criterion

$$\frac{6C_T}{\sigma a} < 4\sqrt{\frac{c_d}{2a} + \frac{4}{\gamma\sqrt{3}} g_s}$$

Since the blade viscous drag damping alone gives roughly $C_T/\sigma < 0.10$, any reasonable level of structural damping should be sufficient to stabilize the flap-lag motion. Fig. 12-10 shows typical flap-lag stability boundaries in terms of the natural frequencies of the blade flap and lag motion. For a given rotor thrust (C_T or $\theta_{.75}$) the motion is unstable inside a roughly elliptical region centered on the worst case $\nu_\beta = \nu_\zeta = 1.15$. These results are for a rotor with no precone, no pitch-flap or pitch-lag coupling, and purely in-plane and out-of-plane degrees of freedom.

To summarize the results for the case of no pitch-flap or pitch-lag coupling, the articulated rotor with flap frequency near 1/rev, small lag frequency, and large lag damping will likely be stable. The worst case for flap-lag stability is a stiff in-plane hingeless rotor with equal rotating flap and lag frequencies: $\nu_\beta = \nu_\zeta = \sqrt{4/3}$. With small precone, little structural damping, and high thrust a flap-lag instability is possible. The motion is stabilized by separating the flap and lag frequencies (moving away from the line $\nu_\beta = \nu_\zeta$) and by keeping the flap frequency away from the critical value $\nu_\beta = 1.15$ if possible.

The principal effect of pitch-flap coupling on the stability criterion is that the flap frequency ν_β^2 is replaced by the effective frequency $\nu_{\beta_e}^2 = \nu_\beta^2 + (\gamma/8)K_{P_\beta}$. The flap-lag motion is then stable unless the effective frequency is in the range $1 < \nu_{\beta_e}^2 < 2$. By using negative pitch-flap coupling ($K_{P_\beta} < 0$) the effective flap frequency ν_{β_e} can be placed below 1/rev, thus assuring flap-lag stability regardless of the lag frequency. This case is discussed by Gaffey (1969).

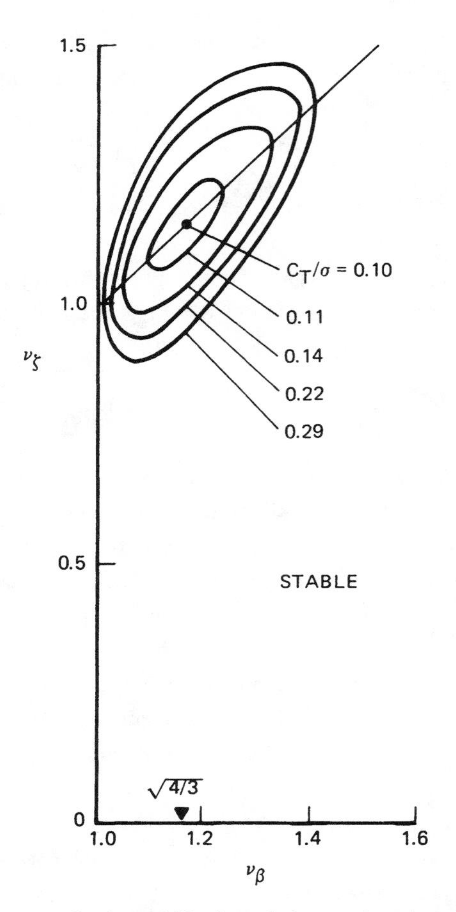

Figure 12-10 Flap-lag stability boundaries as a function of rotor thrust; from Ormiston and Hodges (1972).

Finally, consider a rotor with large pitch-lag coupling K_{P_ζ}. Assuming that the flap moments due to ζ can be neglected compared to the K_{P_ζ} term, the characteristic equation becomes

$$(I_\beta^* s^2 - \gamma M_{\dot\beta} s + I_\beta^* \nu_{\beta_e}^{\ 2})(I_\zeta^* s^2 + C_\zeta s + I_\zeta^* \nu_{\zeta_e}^{\ 2})$$

$$- K_{P_\zeta} \gamma M_\theta [-(\gamma Q_{\dot\beta} - 2 I_{\beta\zeta}^* \beta_0)s + K_{P_\beta} \gamma Q_\theta] = 0$$

where $C_\zeta = I_\zeta^* C_\zeta^* - \gamma Q_{\dot\zeta}$, $I_\beta^* \nu_{\beta_e}^2 = I_\beta^* \nu_\beta^2 + K_{P_\beta}\gamma M_\theta$, and $I_\zeta^* \nu_{\zeta_e}^2 = I_\zeta^* \nu_\zeta^2 + K_{P_\zeta}\gamma Q_\theta$. On substituting $s = i\omega$ to obtain the flutter boundary, the imaginary part of the characteristic equation gives the flutter frequency:

$$\omega^2 = \nu_{\zeta_e}^2 + \frac{C_\zeta I_\beta^*(\nu_{\beta_e}^2 - \nu_{\zeta_e}^2) - K_{P_\zeta}\gamma M_\theta(2I_{\beta\zeta}^*\beta_0 - \gamma Q_{\dot\beta})}{-\gamma M_{\dot\beta} I_\zeta^* + C_\zeta I_\beta^*}$$

and the real part is

$$I_\beta^* I_\zeta^*(\omega^2 - \nu_{\beta_e}^2)(\omega^2 - \nu_{\zeta_e}^2) + \gamma M_{\dot\beta} C_\zeta \omega^2 - K_{P_\zeta}\gamma M_\theta K_{P_\beta}\gamma Q_\theta = 0$$

The flutter boundary is obtained then by eliminating ω^2. Using the fact that the damping required at the stability boundary is order-β_0 small, the following approximate result is obtained:

$$I_\beta^* I_\zeta^*(\nu_{\beta_e}^2 - \nu_{\zeta_e}^2)\left[C_\zeta I_\beta^*(\nu_{\beta_e}^2 - \nu_{\zeta_e}^2) - K_{P_\zeta}\gamma M_\theta(2I_{\beta\zeta}^*\beta_0 - \gamma Q_{\dot\beta})\right] + (\gamma M_{\dot\beta})^2 C_\zeta \nu_{\zeta_e}^2 = 0$$

or for stability

$$C_\zeta > \frac{(\nu_{\beta_e}^2 - \nu_{\zeta_e}^2) K_{P_\zeta}\gamma M_\theta(2I_{\beta\zeta}^*\beta_0 - \gamma Q_{\dot\beta})}{I_\beta^*(\nu_{\beta_e}^2 - \nu_{\zeta_e}^2)^2 + (\gamma/8)^2 \nu_{\zeta_e}^2}$$

With small ν_ζ this reduces to the result obtained in section 12-3.2 for articulated rotors. Note that the right-hand side changes sign when $\nu_{\zeta_e} > \nu_{\beta_e}$. Hence for articulated and soft in-plane hingeless rotors, positive pitch-lag coupling is destabilizing; while for stiff in-plane hingeless rotors (with no structural coupling of the flap and lag motions) negative pitch-lag coupling (lag back, pitch up) is destabilizing.

12-3.4 Improved Analytical Models

The stability of the flap-lag motion of a rotor blade is a sensitive dynamics problem that frequently requires a better model than has been used here if accurate numerical results are to be obtained. The structural and inertial coupling of the in-plane and out-of-plane bending motion is a significant factor in the stability of hingeless rotors. Even a small amount of out-of-plane motion in the lag mode greatly increases its aerodynamic damping, and thus is very stabilizing. It is also usually necessary to include the blade pitch motion when analyzing hingeless rotor dynamics. It has been seen that the pitch-flap and pitch-lag coupling play important roles in the blade dynamics. For an articulated rotor these couplings are determined by the geometry of the root hinges and control system, but for a hingeless rotor they depend on the bending and torsion loads acting on the blade. Thus a fully coupled flap-lag-torsion model of the blade motion is usually required for an accurate analysis of the aeroelastic stability of a hingeless rotor. A consistent, general derivation of the nonlinear equations of motion for such a model is still the subject of research. The present analysis has only considered hover, but the aerodynamics in forward flight also strongly influence the blade flap-lag dynamics.

12-3.5 Literature

Morduchow and Hinchey (1950) investigated the flap-lag stability of an articulated rotor in hover, deriving the equations of motion for rigid flap and lag. They found that positive pitch-lag coupling ($K_{P_\zeta} > 0$) destabilizes the lag mode and increases its frequency. Pitch-lag coupling has little effect on the flap mode frequency or damping, but it introduces substantial flap motion into the lag mode. Pitch-flap coupling was found to have little effect on the lag frequency, but considerable influence on the damping of the lag mode. Positive pitch-flap coupling increases the flap frequency.

Chou (1958) investigated a lag instability of a fully articulated rotor blade due to pitch-lag coupling, which was encountered in a rotor test at high collective and low rotor speed. A lag oscillation of about 30° amplitude and 0.32/rev frequency was observed, with the flap motion at the same frequency. When the control linkage was examined, it was found to give positive pitch-lag coupling. Chou obtained a criterion for stability by considering the Coriolis damping of the lag motion arising from the flapping produced by the pitch-lag coupling, as derived here in section 12-3.2. He also obtained the stability criterion directly by means of Routh's discriminant for the equations given in section 12-3.2, and demonstrated that for articulated rotors the result is equivalent to the approximate criterion.

Blake, Burkam, and Loewy (1961) extended Chou's analysis of the articulated rotor to include all the aerodynamic terms ($M_{\dot{\zeta}}$, $Q_{\dot{\beta}}$, Q_θ, and $Q_{\dot{\zeta}}$); the equations were solved with an analog computer. Chou's criterion was found to be generally conservative (apparently because of the neglect of the aerodynamic lag damping $Q_{\dot{\zeta}}$), except in certain cases of large pitch-flap coupling. They concluded that the aerodynamic lag damping and hinge-offset effects should be included to accurately calculate the flap-lag stability.

Hohenemser and Heaton (1967) conducted a theoretical investigation of the flap-lag stability of a rotor in hover and forward flight. They considered a rigid blade with no hinge offset and no pitch-flap or pitch-lag coupling, but with hinge springs to obtain arbitrary flap and lag frequencies. They showed that for hover the flap-lag coupling is proportional to $\beta_0^2(\nu_\beta^2 - 1)$ (see section 12-3.1), from which it may be concluded that the stability decreases with coning angle, but that an articulated rotor will always be stable. If $\nu_\beta > 1$, the coning angle β_0 and thus the Coriolis forces are reduced, so that there is a net flap moment due to $\dot{\zeta}$ and an instability is possible. They point out that with ideal precone, the same coning angle β_0 is obtained as when $\nu_\beta = 1$, so the coupling flap moment is zero again. Flap-lag stability is thus assured when $\nu_\beta > 1$ if ideal precone is used. In general, it was found that a small amount of lag damping, such as is usually available from the aerodynamic and structural damping, is sufficient to stabilize the flap-lag motion. In forward flight (say $\mu = 0.4$) an instability can occur even for an articulated rotor. A moderate amount of lag damping is still sufficient to stabilize the motion, however. They compared the results of a numerical integration of the nonlinear equations of motion with the stability predictions from the linearized model, and concluded that the linear model is very good. Thus the Coriolis and aerodynamic coupling is determined primarily by the trim coning angle β_0, and the true nonlinear effects are small.

Ormiston and Hodges (1972) conducted a theoretical investigation of the flap-lag stability of a hovering rotor. By means of Routh's discriminant they obtained the stability boundary for the case of no precone and no pitch-flap or pitch-lag coupling, as given in section 12-3.3, and showed that the case $\nu_\beta = \nu_\zeta = \sqrt{4/3}$ has the minimum stability margin. They found that with ideal precone the motion is always stable, but with other values the precone can be either stabilizing or destabilizing. They examined the influence of pitch-lag coupling, and found that the condition $K_{P_\zeta} > 0$ is destabilizing for soft in-plane rotors, but that negative coupling ($K_{P_\zeta} < 0$) is destabilizing for stiff in-plane rotors. Ormiston and Hodges also investigated the influence of structural flap-lag coupling. They considered a model of the blade with hinge springs outboard of the pitch bearing as well as inboard, so that the principal axes of part of the flexibility rotated with the blade collective pitch. For the uncoupled case (all the flexibility inboard of the pitch bearing) the blade modes are purely in-plane or purely out of plane. For the fully coupled case (all the flexibility outboard of the pitch bearing) the lag mode has a significant amount of out-of-plane motion. It was found that such coupling is an extremely important factor in the flap-lag stability. Flexibility outboard of the pitch bearing is highly stabilizing, because then increasing the collective pitch not only introduces the Coriolis and aerodynamic coupling of the flap and lag modes, but also increases the out-of-plane deflection in the lag mode, and hence increases its aerodynamic damping. Such structural flap-lag coupling greatly increases the lag damping and virtually eliminates any flap-lag stability problems. It was also found that for the fully coupled case positive pitch-lag coupling is destabilizing with a stiff in-plane rotor.

Burkam and Miao (1972) conducted a theoretical and experimental investigation of the flap-lag stability of a soft in-plane hingeless rotor. They found that positive pitch-flap coupling, negative pitch-lag coupling, elastic flap-lag coupling, and precone were stabilizing. Lag damping was also stabilizing, with only a small amount needed to avoid an instability. They pointed out the importance of whether the pitch bearing is located inboard or outboard of the bending flexibility. For their case of a hingeless rotor with all the bending outboard of the pitch bearing, a coning value different from the precone resulted in pitch-flap and pitch-lag coupling (see section 9-4.2), which has a significant influence on the flap-lag stability.

Hodges and Ormiston (1973a) conducted a theoretical investigation of the flap-lag bending stability of a uniform rotor blade in hover, including the effects of elastic torsion. For the matched stiffness case ($EI_{zz} = EI_{xx}$) the bending and torsion are not coupled structurally and then for a blade with typical torsion stiffness ($\omega_\theta = 5$ to 8/rev) there is little influence of torsion on the flap-lag dynamics (for the matched stiffness case). Precone can introduce an instability at low collective, because of elastic pitch-lag coupling. As discussed in section 9-4.2, when the coning angle is below the precone angle (at low thrust) a lag deflection produces a moment about the feathering axis, and hence pitch-lag coupling. For the matched stiffness case this coupling is absent, however. Hodges and Ormiston demonstrated that this consequence of the torsion dynamics is basically a quasistatic effect by also solving the equations without the torsion inertia and damping terms. The quasistatic torsion model gave nearly the same results as the complete model, except for torsionally soft blades. Neglecting the torsion deflection does not give a good model of the hingeless rotor dynamics even for ω_θ around 10/rev, because the pitch-bending coupling effects are very important.

Huber (1973b) examined the flap-lag dynamics of a soft in-plane hingeless rotor, particularly the influence of the effective pitch-lag and pitch-flap coupling of a torsionally flexible blade with bending outboard of the pitch bearing. Such coupling was found to have a significant effect on the helicopter air resonance stability and flying qualities, as well as on the flap-lag stability. The analysis and flight tests also showed a flap-lag instability at high collective due to the loss of flap damping when the rotor blade stalls. This phenomenon only occurred in hover, since in forward flight stall occurs only on part of the rotor disk.

On the flap-lag dynamics of helicopter rotor blades: Reissner and Morduchow (1948), Carpenter and Peitzer (1949), Morduchow (1949, 1950), Goodier (1950), Sanders (1953), Hohenemser (1957b, 1974), Hohenemser and Perisho (1958), McKee and Naeseth (1959), Young (1962b, 1964, 1971), Jenkins (1967, 1968), Arcidiacono (1969), Elman, Niebanck, and Bain (1969), Gaffey (1969), Loewy (1969), Johnson and Hohenemser (1970), Tong (1971a, 1971b, 1972, 1974), Friedmann (1972, 1973a, 1973b, 1976, 1977a, 1977b), Friedmann and Tong (1972, 1973), Hodges (1972, 1976), Ormiston and Bousman (1972), Anderson (1973), Donham and Cardinale (1973), Hansford and Simons (1973), Hodges and Ormiston (1973b, 1976, 1977), Johnston and Conner (1973), Peters and Ormiston (1973), Dowell (1974), Friedmann and Silverthorn (1974a, 1974b, 1975), Hodges and Dowell (1974), Ormiston (1974b), White (1974), Curtiss (1975), Johnson (1975b, 1977d, 1977e), Ohtsuka (1975), Ormiston and Bousman (1975), Peters (1975a, 1975b), Bousman, Sharpe, and Ormiston (1976), Curtiss and Putman (1976), Kaza and Kvaternik (1976, 1977a, 1977b), Kvaternik and Kaza (1976), Chopra (1977), Edwards and Miao (1977), Fertis (1977), Friedmann and Shamie (1977a, 1977b), Friedmann and Yuan (1977), Kunz (1977), Rosen and Friedmann (1977), Shamie and Friedmann

(1977), Yasue (1977), Bousman (1978), Chopra and Dugundji (1978), Chopra and Johnson (1978), Hodges (1978a, 1978b), Shaw and Edwards (1978), Wei and Peters (1978), White, Sutton, and Nettles (1978).

12–4 Ground Resonance

Ground resonance is a dynamic instability involving the coupling of the blade lag motion with the in-plane motion of the rotor hub. This instability is characterized by a resonance of the frequency of the rotor lag motion (specifically the low frequency lag mode in the nonrotating frame) and a natural frequency of the structure supporting the rotor. Since the lag frequency depends on the rotor rotational speed, such resonances define certain critical speed ranges for the rotor. An instability is possible at a resonance if the rotating lag frequency ν_ζ is below 1/rev, as for articulated and soft in-plane hingeless rotors. With articulated rotors, the critical mode is usually an oscillation of the helicopter on the landing gear when in contact with the ground, hence the name ground resonance. Sometimes the phenomenon can occur in flight as well, particularly with a hingeless rotor, and then it is called air resonance.

The hub in-plane motions are coupled with the cyclic lag modes ζ_{1c} and ζ_{1s}, which correspond to lateral and longitudinal shifts of the net rotor center of gravity from the center of rotation. Since the low frequency lag mode involves whirling of the rotor center of gravity about the shaft, ground resonance is potentially very destructive, and avoiding this instability is an important consideration in helicopter design. The basic requirement is that resonances of the support structure with the lag mode be kept out of the operating range of the helicopter. Generally, resonances above 120% normal operating speed or below 40% normal speed are acceptable. The rotor has little energy at low speed, so it is possible to run up through very low frequency resonances without a large amplitude motion building up. For a fairly large range about the normal rotor speed range it is therefore necessary to either avoid resonances or provide sufficient damping in the system to prevent any instability.

The classical ground resonance analysis considers four degrees of freedom: longitudinal and lateral in-plane motion of the rotor hub, and the two cyclic lag degrees of freedom. The actual vibration modes of the rotor support, such as the motion of the helicopter on its landing gear, will probably involve tilt of the shaft as well, but it is the in-plane motion of the hub that is the dominant factor in ground resonance. The rotor aerodynamic forces have little influence on ground resonance, compared to the structural and inertial forces. Damping of the rotor and support comes almost entirely from mechanical dampers or structural damping. Thus the aerodynamic forces will be neglected in the ground resonance analysis. Such a model provides a good description of the fundamental characteristics of ground resonance, and even gives good numerical results, particularly for articulated rotors. In some cases, especially with hingeless rotors, a more complete model is required, including the rotor aerodynamics and flap motion and a better description of the support motion. The basic analysis of ground resonance is the work of Coleman and Feingold (1958).

12-4.1 Ground Resonance Equations

The degrees of freedom involved in ground resonance are the cyclic rotor lag modes ζ_{1c} and ζ_{1s}, which produce a shift of the net rotor center of gravity; and the hub longitudinal and lateral displacements, x_h and y_h. For now, only the case of a rotor with three or more blades is considered. The equations of motion for the rotor lag degrees of freedom, including the influence of the hub motion, were derived in section 9–6. Dropping the aerodynamic forces, but including a lag damping term, the equations of motion in the nonrotating frame are

$$I_\zeta[\ddot{\zeta}_{1c} + C_\zeta^*(\dot{\zeta}_{1c} + \zeta_{1s}) + 2\dot{\zeta}_{1s} + (\nu_\zeta^2 - 1)\zeta_{1c}] - S_\zeta \ddot{y}_h = 0$$

$$I_\zeta[\ddot{\zeta}_{1s} + C_\zeta^*(\dot{\zeta}_{1s} - \zeta_{1c}) - 2\dot{\zeta}_{1c} + (\nu_\zeta^2 - 1)\zeta_{1s}] + S_\zeta \ddot{x}_h = 0$$

Here ν_ζ is the rotating natural frequency of the lag motion and C_ζ^* is the lag damping coefficient (in the rotating frame, due to aerodynamic, structural, or mechanical damping); the first and second moments of the blade lag inertia are $S_\zeta = \int_0^1 \eta_\zeta m\, dr$ and $I_\zeta = \int_0^1 \eta_\zeta^2 m\, dr$. Since the hub in-plane motion is coupled by the inertial forces with ζ_{1c} and ζ_{1s} only, the other nonrotating lag degrees of freedom are not involved in ground resonance. The in-plane forces acting on the rotor hub were also derived in section 9–6. Retaining only the inertial terms, the rotor drag and side force are

$$H = -\frac{N}{2} S_\zeta \ddot{\zeta}_{1s} - NM_b \ddot{x}_h$$

$$Y = \frac{N}{2} S_\zeta \ddot{\zeta}_{1c} - NM_b \ddot{y}_h$$

where N is the number of blades, and the blade mass is $M_b = \int_0^1 m\,dr$. The rotor support structure will be represented by a mass-spring-damper system in the longitudinal and lateral directions, excited by the rotor hub forces:

$$M_x \ddot{x}_h + C_x \dot{x}_h + K_x x_h = H$$

$$M_y \ddot{y}_h + C_y \dot{y}_h + K_y y_h = Y$$

These equations are often a good model of the actual helicopter or wind tunnel support dynamics if the generalized mass and damping of the appropriate free vibration modes are used, as determined from the hub impedance. Substituting for the rotor hub forces gives

$$(M_x + NM_b)\ddot{x}_h + C_x \dot{x}_h + K_x x_h + \frac{N}{2} S_\zeta \ddot{\zeta}_{1s} = 0$$

$$(M_y + NM_b)\ddot{y}_h + C_y \dot{y}_h + K_y y_h - \frac{N}{2} S_\zeta \ddot{\zeta}_{1c} = 0$$

Now write $K_x = (M_x + NM_b)\omega_x^2$ and $C_x = (M_x + NM_b)C_x^*$, where $M_x + NM_b$ is the total mass including the rotor; and similarly define the natural frequency ω_y and damping coefficient C_y^* of the lateral mode.

The equations are normalized by dividing the lag equations by the characteristic rotor inertia I_b, and the support equations by $(N/2)I_b$. Define the normalized inertias as follows: $S_\zeta^* = S_\zeta/I_b$, $I_\zeta^* = I_\zeta/I_b$, $M_x^* = (M_x + NM_b)/(N/2)I_b$, and $M_y^* = (M_y + NM_b)/(N/2)I_b$. (The equations are already dimensionless.) The coupled lag and support equations of motion describing the ground resonance dynamics are then

$$\begin{bmatrix} I_\zeta^* & 0 & -S_\zeta^* & 0 \\ 0 & I_\zeta^* & 0 & S_\zeta^* \\ -S_\zeta^* & 0 & M_y^* & 0 \\ 0 & S_\zeta^* & 0 & M_x^* \end{bmatrix} \begin{pmatrix} \zeta_{1c} \\ \zeta_{1s} \\ y_h \\ x_h \end{pmatrix}^{\cdot\cdot} +$$

$$+\begin{bmatrix} I_\zeta^* C_\zeta^* & 2I_\zeta^* & 0 & 0 \\ -2I_\zeta^* & I_\zeta^* C_\zeta^* & 0 & 0 \\ 0 & 0 & M_y^* C_y^* & 0 \\ 0 & 0 & 0 & M_x^* C_x^* \end{bmatrix} \begin{pmatrix} \zeta_{1c} \\ \zeta_{1s} \\ y_h \\ x_h \end{pmatrix}^{\bullet}$$

$$+\begin{bmatrix} I_\zeta^*(\nu_\zeta^2 - 1) & I_\zeta^* C_\zeta^* & 0 & 0 \\ -I_\zeta^* C_\zeta^* & I_\zeta^*(\nu_\zeta^2 - 1) & 0 & 0 \\ 0 & 0 & M_y^* \omega_y^2 & 0 \\ 0 & 0 & 0 & M_x^* \omega_x^2 \end{bmatrix} \begin{pmatrix} \zeta_{1c} \\ \zeta_{1s} \\ y_h \\ x_h \end{pmatrix} = 0$$

Note that the only coupling of the rotor and support motion is due to the inertial terms. In terms of the Laplace variable s, the characteristic equation of this system is

$$I_\zeta^{*2}\,[(s^2 + C_\zeta^* s + \nu_\zeta^2 - 1)^2 + (2s + C_\zeta^*)^2\,]\,M_y^*(s^2 + C_y^* s + \omega_y^2)M_x^*(s^2 + C_x^* s + \omega_x^2)$$
$$-I_\zeta^*(s^2 + C_\zeta^* s + \nu_\zeta^2 - 1)S_\zeta^{*2} s^4\,[M_y^*(s^2 + C_y^* s + \omega_y^2) + M_x^*(s^2 + C_x^* s + \omega_x^2)]$$
$$+\ S_\zeta^{*4} s^8 \ = \ 0$$

The solution of this eighth-order polynomial gives the four eigenvalues of the system (and their complex conjugates) and hence the ground resonance stability. Note that a divergence type instability is not possible for for this system. On setting s equal to zero, the characteristic equation gives the divergence stability criterion

$$I_\zeta^{*2}\,[(\nu_\zeta^2 - 1)^2 \ + \ C_\zeta^{*2}\,]\,M_y^* \omega_y^2\, M_x^* \omega_x^2 \ > \ 0$$

which is always satisfied (assuming that either $\nu_\zeta \neq 1$ or $C_\zeta^* \neq 0$).

Consider the uncoupled dynamics, obtained by setting S_ζ^* equal to zero. The uncoupled hub motion consists of damped oscillations with natural frequencies ω_x and ω_y. The uncoupled (i.e. shaft-fixed) lag motion is a damped oscillation with eigenvalue

$$s_R = \frac{C_\zeta^*}{2} +_{\cdot} i \sqrt{\nu_\zeta^2 - \left(\frac{C_\zeta^*}{2}\right)^2}$$

in the rotating frame. In the nonrotating frame there are two cyclic lag modes, with eigenvalues $s_{NR} = s_R \pm i$. The high frequency lag mode ($s_{NR} = s_R + i$) corresponds to a progressive whirling motion of the rotor center of gravity at frequency $\mathrm{Im}\, s_R + 1$/rev. The low frequency lag mode ($s_{NR} = s_R - i$) is a regressive whirling motion of the rotor center of gravity at frequency $|\mathrm{Im}\, s_R - 1/\mathrm{rev}|$, if $\mathrm{Im}\, s_R > 1$/rev (such as for a stiff in-plane hingeless rotor); or a progressive whirling mode if $\mathrm{Im}\, s_R < 1$/rev, as for an articulated rotor. Thus for $S_\zeta^* = 0$, the characteristic equation factors into a product of the rotor and the support characteristic equations, with the solutions

$$s = \begin{cases} \dfrac{C_\zeta^*}{2} + i \sqrt{\nu_\zeta^2 - \left(\dfrac{C_\zeta^*}{2}\right)^2} \pm i \\[2ex] \dfrac{C_y^*}{2} + i \sqrt{\omega_y^2 - \left(\dfrac{C_y^*}{2}\right)^2} \\[2ex] \dfrac{C_x^*}{2} + i \sqrt{\omega_x^2 - \left(\dfrac{C_x^*}{2}\right)^2} \end{cases}$$

and their conjugates. The uncoupled rotor and support motion is stable, and a ground resonance instability can only be due to the inertial coupling when $S_\zeta^* \neq 0$. When there is no damping and $S_\zeta^* = 0$ as well, the solution is $s = \pm i\omega$ for frequencies $\omega = \nu_\zeta \pm 1$, ω_x, and ω_y.

The coupling of the rotor and support is determined by the parameter $S_\zeta = \int_0^1 \eta_\zeta m\, dr$. For an articulated rotor, S_ζ is the product of the blade mass and the radial distance of the blade center of gravity from the lag hinge. The following ratio is a measure of the coupling between the rotor and support terms in the characteristic equation:

$$\frac{S_\zeta^{*2}}{I_\zeta^* M_y^*} = \frac{N}{2} \frac{S_\zeta^2}{I_\zeta M_y} \cong \frac{3}{8} \frac{NM_b}{M_y} = \frac{3}{8} \frac{M_{\mathrm{rotor}}}{M_{\mathrm{support}}}$$

The support mass is usually much larger than the rotor mass, so this parameter is quite small. For example, the generalized mass in the case of rigid body motion of the aircraft on its landing gear is roughly equal to the helicopter gross weight. While an exact analytical solution of the eighth-order ground resonance characteristic equation is not possible, useful results can be obtained on the basis of the assumption that S_ζ^* is small (really that $S_\zeta^{*2}/I_\zeta^* M_y^* \ll 1$). Since this parameter is indeed very small for most practical cases, the solution for small S_ζ^* generally gives accurate numerical results as well.

The fundamental character of the ground resonance dynamics is determined by the rotating natural frequency of the lag motion, ν_ζ. For a general rotor, the dimensionless blade lag frequency varies with the rotor speed according to $\nu_\zeta^2 = K_1/\Omega^2 + K_2$, where K_1 and K_2 are the Southwell coefficients (see section 9-8.3). For an articulated rotor, $K_1 = 0$ and $\nu_\zeta^2 = K_2 = (3/2)e/(1 - e)$, where e is the lag hinge offset. For a hingeless rotor K_1 is not zero; it specifies the dimensional nonrotating lag frequency $\nu_{NR} = \sqrt{K_1}$. The lag frequency at high rotor speed approaches $\nu_\zeta = \sqrt{K_2}$ (per rev). For a soft in-plane rotor (hingeless or articulated), $K_2 < 1$; for a stiff in-plane rotor, probably $K_2 > 1$, although $\nu_\zeta > 1$/rev can always be obtained at low rotor speed even if $K_2 < 1$ (assuming $K_1 \neq 0$).

For a given rotor and support, it is preferable to present the ground resonance solution in terms of dimensional frequencies, using the rotor speed Ω as a parameter. The dimensional support frequencies ω_x and ω_y are constants, and the dimensional lag frequency depends on the rotor speed according to $\nu_\zeta^2 = K_1 + K_2\Omega^2$. Hence at low speed $\nu_\zeta \cong \nu_{NR} = \sqrt{K_1}$, and at high speed ν_ζ/Ω approaches $\sqrt{K_2}$/rev. The variation of ν_ζ with Ω determines the resonances with the various natural frequencies of the supports. In the ground resonance analysis, dimensionless parameters will still be used, however. The dimensionless support natural frequencies ω_x and ω_y vary inversely with the rotor speed for a fixed dimensional frequency.

12-4.2 No-Damping Case

Consider first the case of no damping of the lag or support motion. On setting $C_\zeta^* = C_x^* = C_y^*$ equal to zero, the characteristic equation becomes

$$
\begin{aligned}
&[(s^2 + \nu_\zeta^2 - 1)^2 + 4s^2](s^2 + \omega_y^2)(s^2 + \omega_x^2) \\
&\quad - \left(s^2 + \nu_\zeta^2 - 1\right) \frac{S_\zeta^{*2}}{I_\zeta^* M_y^* M_x^*} s^4 \left[M_y^*(s^2 + \omega_y^2) + M_x^*(s^2 + \omega_x^2)\right] \\
&\quad + \frac{S_\zeta^{*4}}{I_\zeta^{*2} \, M_y^* M_x^*} \, s^8 = 0
\end{aligned}
$$

The eigenvalues of this polynomial with real coefficients must appear as complex conjugate pairs. If there is no damping, however, the characteristic equation is actually a polynomial in s^2. The substitution $s = i\omega$, or $s^2 = -\omega^2$, therefore gives a polynomial in ω^2 that also has real coefficients. Complex conjugate solutions for ω correspond to solutions for s that are symmetric about the imaginary s-axis. It follows that complex solutions of the characteristic equation with zero damping will occur in groups of four roots, symmetric about both the real and imaginary axes. There will be one root in each quadrant of the s-plane. Since two of these roots will be in the right half-plane ($\mathrm{Re}\, s > 0$), such complex solutions correspond to an unstable system. Moreover, there is no way that the system can be stable, with all the roots in the left half-plane. The requirement of symmetry about the imaginary axis will be satisfied, however, if all the roots are on the imaginary axis, which corresponds to neutral stability. Neutral stability is the best that can be achieved when there is no damping in the rotor or support to extract energy from the system.

Therefore, for the case of no damping, there are boundaries not between stable and unstable conditions, but rather between neutrally stable and unstable conditions. Inside a neutral stability region, all the roots are on the imaginary axis. At the stability boundary, two roots meet at positive frequency and two at negative frequency, and break off from the imaginary axis. Then inside an unstable region there will be four complex roots, corresponding to the support mode in resonance, and the low frequency lag mode. Substituting $s = i\omega$ (where ω is a real number) now defines the entire neutral stability region, not just the flutter boundary. The simplest means of defining the stability boundary in this case is to solve the characteristic equation assuming $s = i\omega$. Where it is not possible to obtain all eight solutions with ω real must be an unstable condition. The uncoupled solution ($S_\zeta = 0$) is exactly $s = \pm i\omega$, with $\omega = \nu_\zeta \pm 1$, ω_y, and ω_x. Since an instability involves four roots, it requires a resonance of a support mode and a rotor mode. At such a resonance, the coupling due to S_ζ will produce the instability under certain conditions.

On substituting $s = i\omega$, the characteristic equation for the case of no damping becomes

$$[(\nu_\zeta^2 - 1 - \omega^2)^2 - 4\omega^2](\omega_y^2 - \omega^2)(\omega_x^2 - \omega^2)$$

$$- (\nu^2 - 1 - \omega^2)\omega^4 \frac{S_\zeta^{*2}}{I_\zeta^* M_y^* M_x^*} [M_y^* (\omega_y - \omega^2) + M_x^*(\omega_x^2 - \omega^2)]$$

$$+ \frac{S_\zeta^{*4}}{I_\zeta^{*2} M_y^* M_x^*} \omega^8 = 0$$

This polynomial can be solved numerically for ω^2. Alternatively values of ω^2 can be assumed, so that the characteristic equation becomes a quadratic for the dimensional frequency $(\nu_\zeta^2 - \Omega^2)$. Thus the solution for ω^2 as a function of Ω or ν_ζ can be constructed. Where less than four values of ω^2 are obtaind for a given Ω, the motion is unstable.

While an exact solution of the characteristic equation can be obtained numerically, it is useful to obtain an analytical solution based on the assumption that the coupling parameter S_ζ is small. The roots for small coupling should be near the exact solution for $S_\zeta = 0$: $\omega = \nu_\zeta \pm 1$, ω_x, or ω_y. Therefore, we shall find a correction of order S_ζ^2 to the uncoupled roots. Considering the solution near $\omega = \omega_x$, write the eigenvalues as $\omega^2 = \omega_x^2 + S_\zeta^{*2} s_1$. Then to lowest order in S_ζ^2, the characteristic equation is

$$[(\nu_\zeta^2 - 1 - \omega_x^2)^2 - 4\omega_x^2](\omega_y^2 - \omega_x^2)(-s_1)$$

$$- (\nu_\zeta^2 - 1 - \omega_x^2)\omega_x^4 \frac{1}{I_\zeta^* M_y^* M_x^*} M_y^*(\omega_y^2 - \omega_x^2) = 0$$

which gives

$$s_1 = - \frac{1}{I_\zeta^* M_x^*} \frac{(\nu_\zeta^2 - 1 - \omega_x^2)\omega_x^4}{(\nu_\zeta^2 - 1 - \omega_x^2)^2 - 4\omega_x^2}$$

or

$$\omega^2 = \omega_x^2 \left[1 - \frac{(\nu_\zeta^2 - 1 - \omega_x^2)\omega_x^2}{(\nu_\zeta^2 - 1 - \omega_x^2)^2 - 4\omega_x^2} \frac{S_\zeta^{*2}}{I_\zeta^* M_x^*}\right]$$

The order S_ζ^{*2} solution near $\omega = \omega_y$ is similar. For the solution near $\omega = \nu_\zeta \pm 1$, write $\omega^2 = (\nu_\zeta \pm 1)^2 + S_\zeta^{*2} s_1$. Then to lowest order in S_ζ^2, the characteristic equation is

$$s_1 2\nu_\zeta [\omega_y^2 - (\nu_\zeta \pm 1)^2][\omega_x^2 - (\nu_\zeta \pm 1)^2]$$

$$+ (\nu_\zeta \pm 1)^5 \frac{1}{I_\zeta^* M_y^* M_x^*} \left\{ M_y^*[\omega_y^2 - (\nu_\zeta \pm 1)^2] + M_x^*[\omega_x^2 - (\nu_\zeta \pm 1)^2] \right\} = 0$$

so

$$\omega^2 = (\nu_\zeta \pm 1)^2 \left[1 - \frac{(\nu_\zeta \pm 1)^3}{2\nu_\zeta} \frac{M_y^*[\omega_y^2 - (\nu_\zeta \pm 1)^2] + M_x^*[\omega_x^2 - (\nu_\zeta \pm 1)^2]}{[\omega_y^2 - (\nu_\zeta \pm 1)^2][\omega_x^2 - (\nu_\zeta \pm 1)^2]} \frac{S_\zeta^{*2}}{I_\zeta^* M_x^* M_y^*} \right]$$

Note that near a resonance of ω_x or ω_y with a rotor root $(\nu_\zeta \pm 1)$, these expansions diverge; this singular behavior does not necessarily indicate an instability, however. In the limit $\Omega = 0$, the dimensional lag frequency approaches the nonrotating lag frequency ν_{NR}. In terms of the dimensionless frequencies, ω_x and ω_y become infinite proportionally to Ω^{-1}, and $\nu_\zeta \cong \nu_{NR}/\Omega$. Then the solution near $\omega = \omega_x$ becomes

$$\omega^2 = \omega_x^2 \left[1 + \frac{\omega_x^2}{\omega_x^2 - \nu_{NR}^2} \frac{S_\zeta^{*2}}{I_\zeta^* M_x^*} \right]$$

in terms of the dimensional frequencies. Hence the solution ω^2 is increased at low rotor speed if $\omega_x > \nu_{NR}$, and decreased if $\omega_x < \nu_{NR}$. The solution near $\omega = \nu_\zeta \pm 1$ for $\Omega = 0$ is

$$\omega^2 = \nu_{NR}^2 \left[1 - \frac{\nu_{NR}^2}{2} \frac{M_y^*(\omega_y^2 - \nu_{NR}^2) + M_x^*(\omega_x^2 - \nu_{NR}^2)}{(\omega_y^* - \nu_{NR}^2)(\omega_x^2 - \nu_{NR}^2)} \frac{S_\zeta^{*2}}{I_\zeta^* M_y^* M_x^*} \right]$$

The direction of the shift in ω^2 depends on the magnitude of the nonrotating lag frequency ν_{NR} relative to the support frequencies ω_x and ω_y. In the limit of large rotor speed, the dimensional lag frequency becomes infinite proportionally to Ω. In terms of the dimensionless frequencies, ω_x and ω_y approach zero proportionally to Ω^{-1} while the lag frequency ν_ζ approaches a constant per-rev value. Then the solution near $\omega = \omega_x$ becomes

$$\omega^2 = \omega_x^2 \left[1 - \frac{\omega_x^2}{\nu_\zeta^2 - 1} \frac{S_\zeta^{*2}}{I_\zeta^* M_x^*} \right]$$

Thus the solution ω^2 is increased if $\nu_\zeta < 1$/rev. The solution near $\omega = \nu_\zeta \pm 1$ for Ω approaching infinity is

$$\omega^2 = (\nu_\zeta \pm 1)^2 \left[1 + \frac{\nu_\zeta \pm 1}{2\nu_\zeta} \left(M_y^* + M_x^* \right) \frac{S_\zeta^{*2}}{I_\zeta^* M_y^* M_x^*} \right]$$

so ω^2 is increased for the $\omega = \nu_\zeta + 1$ solution, and also for the $\omega = \nu_\zeta - 1$ solution if $\nu_\zeta > 1$/rev. Finally consider the limit $\omega = 0$, for which the characteristic equation reduces to $(\nu_\zeta^2 - 1)^2 = 0$, which gives $\nu_\zeta - 1 = 0$. With the rotor speed as a parameter, the $\omega = 0$ solution defines where the roots intercept the Ω-axis, namely at the rotor speed for which $\nu_\zeta = 1$/rev. In the uncoupled case, it is the solution $\omega = \nu_\zeta - 1$ that intercepts the Ω-axis at this point. The above result shows that the low frequency lag mode root intercepts the Ω-axis at the same point for all values of S_ζ.

The ground resonance solution for the case of no damping can be presented graphically in a form known as a Coleman diagram, which is a plot of the dimensional frequencies ω (the roots of the characteristic equation) as a function of the rotor speed Ω. The dimensional solution for the uncoupled case ($S_\zeta = 0$) is $\omega = \omega_x, \omega_y, \Omega \pm \nu_\zeta$, plus the corresponding negative frequencies, for a total of eight roots. (The negative solutions for ω are just mirror images of the positive solutions, and so need not be plotted.) The solution for $S_\zeta > 0$ can easily be sketched using the above results for the influence of small S_ζ in the limits $\Omega = 0$ and $\Omega = \infty$, plus the knowledge that for a coupled system the loci of roots never cross. The character of the ground resonance solution depends primarily on the lag frequency $\nu_\zeta^2 = K_1 + K_2\Omega^2$ (dimensionally). Figs. 12-11 to 12-13 present the Coleman diagrams for three types of rotors: an articulated rotor ($K_1 = 0$ and $K_2 < 1$), a soft in-plane hingeless rotor ($K_1 > 0$ and $K_2 < 1$), and a stiff in-plane hingeless rotor ($K_1 > 0$ and $K_2 > 1$). These are sketches of typical results for small coupling ($S_\zeta^2/I_\zeta M_y \ll 1$); they assume that the nonrotating lag frequency ν_{NR} is less than ω_x and ω_y for the hingeless rotors. The uncoupled roots for the support modes are horizontal lines at $\omega = \omega_x$ and $\omega = \omega_y$, and the uncoupled roots for the rotor are the low and high frequency rotor modes at $\omega = \nu_\zeta \pm \Omega$, which approach $\nu_{NR} = \sqrt{K_1}$ at low

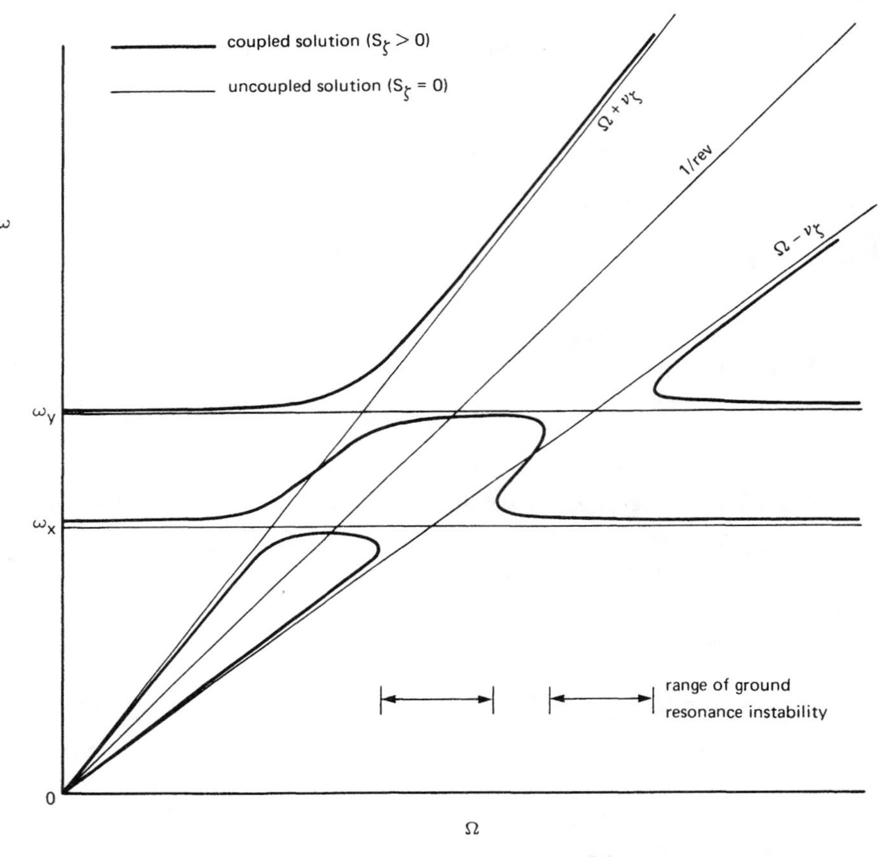

Figure 12-11 Coleman diagram of the ground resonance solution for an articulated rotor.

rotor speed and are asymptotic to constant per-rev values ($\sqrt{K_2} \pm 1$/rev) at high rotor speed. Thus the lag mode frequencies are in resonance with the support mode frequencies at some rotor speed.

For $S_\zeta > 0$, the solution is displaced from the uncoupled frequencies, as indicated by the results for small coupling. If there are four positive solutions for ω at a given rotor speed, then the system is stable (neutrally stable for this case of zero damping). For the articulated and soft in-plane hingeless rotors (Figs. 12-11 and 12-12), however, there are ranges of Ω where only two positive real solutions for ω exist, occurring around the

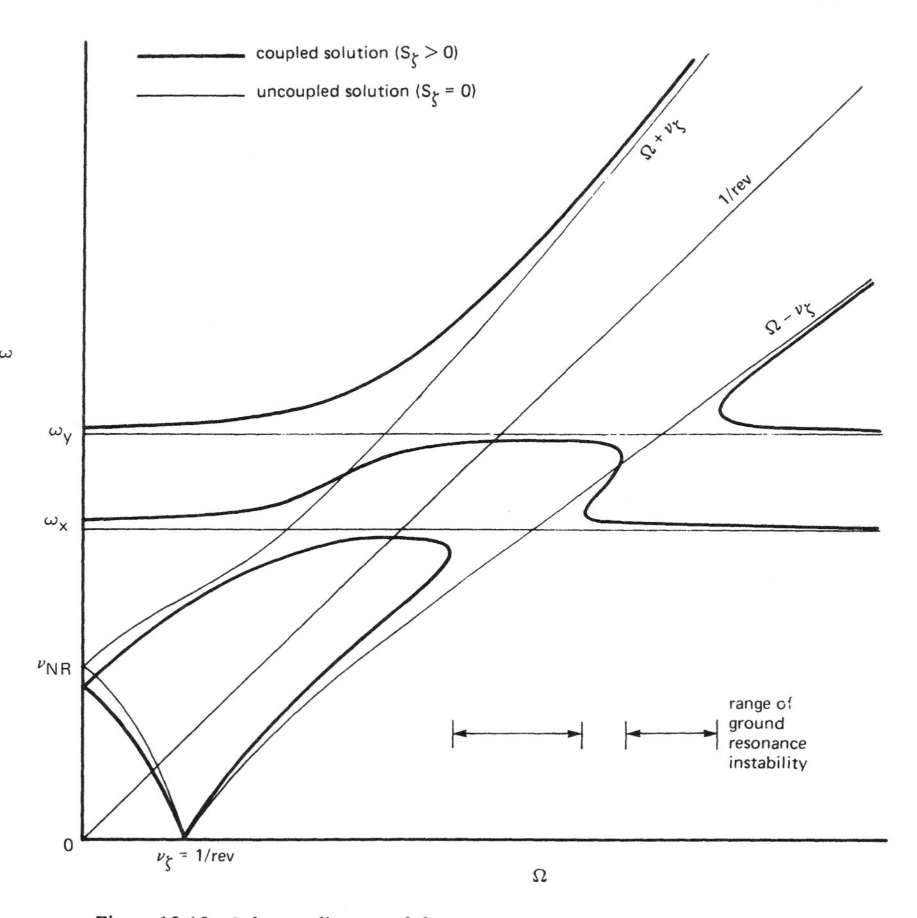

Figure 12-12 Coleman diagram of the ground resonance solution for a soft in-plane hingeless rotor.

resonances of the low frequency lag mode ($\Omega - \nu_\zeta$) with a support mode (ω_x or ω_y). The characteristic equation has four complex solutions in these ranges, so the system is unstable. For the stiff in-plane hingeless rotor (Fig. 12-13), four positive solutions for ω exist at all rotor speeds, and a ground resonance instability does not occur. This behavior of the ground resonance solution is determined by the direction the roots are shifted when $S_\zeta > 0$, which depends on whether $\nu_\zeta < 1$/rev or $\nu_\zeta > 1$/rev at the resonance of the low frequency lag mode with a support mode.

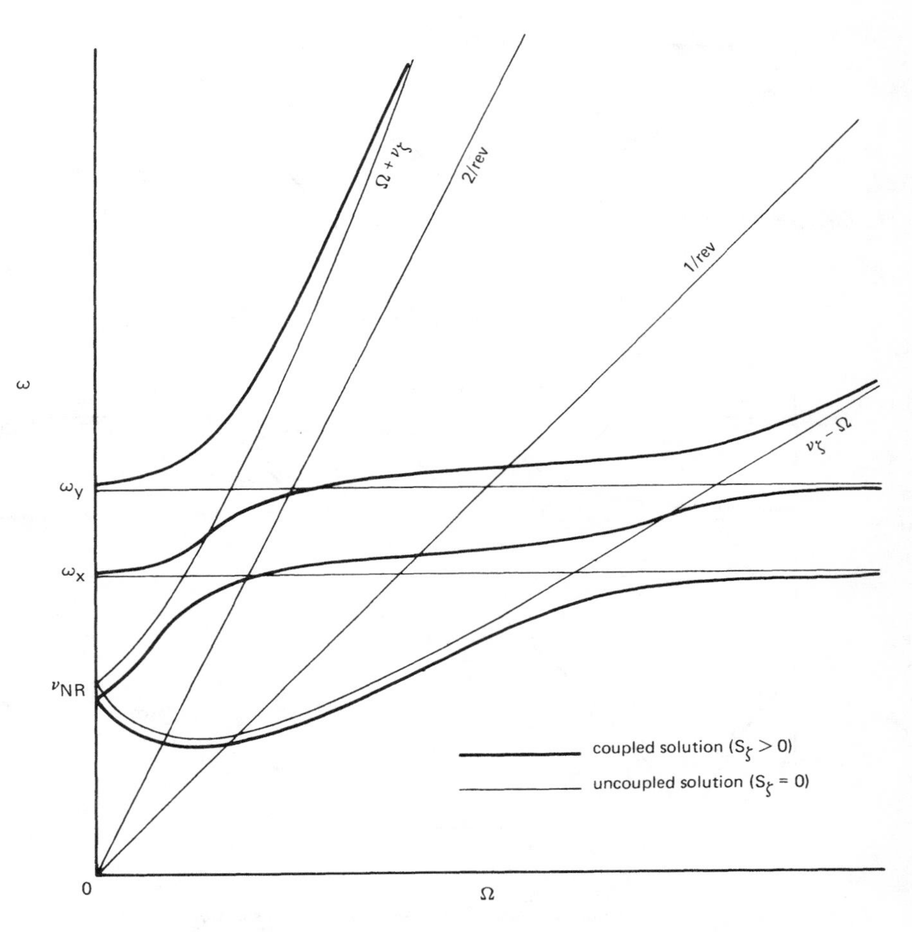

Figure 12-13 Coleman diagram of the ground resonance solution for a stiff in-plane hingeless rotor.

In conclusion, a ground resonance instability can occur at a resonance of a rotor mode and a support mode. The resonances of the high frequency lag mode ($\omega = 1 + \nu_\zeta$) are always stable, but resonances of the low frequency lag mode ($\omega = 1 - \nu_\zeta$) will be unstable if the rotating natural frequency ν_ζ is below 1/rev, as for articulated and soft in-plane hingeless rotors. Thus the placement of the rotor lag frequency determines whether or not a ground resonance instability can occur.

12-4.3 Damping Required for Ground Resonance Stability

For the case with damping of the lag and support motion, the stability boundary is obtained by setting s equal to $i\omega$ in the characteristic equation:

$$
\begin{aligned}
&I_\zeta^{*2}\,[(-\omega^2 + C_\zeta^* i\omega + \nu_\zeta^2 - 1)^2 \\
&\quad + (2i\omega + C_\zeta^*)^2]\,M_y^*(-\omega^2 + C_y^* i\omega + \omega_y^2)M_x^*(-\omega^2 + C_x^* i\omega + \omega_x^2) \\
&\quad - I_\zeta^*(-\omega^2 + C_\zeta^* i\omega + \nu_\zeta^2 - 1)S_\zeta^{*2}\omega^4\,[M_y^*(-\omega^2 + C_y^* i\omega + \omega_y^2) \\
&\quad + M_x^*(-\omega^2 + C_x^* i\omega + \omega_x^2)] + S_\zeta^{*4}\omega^8 = 0
\end{aligned}
$$

It is not possible to obtain an analytical solution for the stability boundary by eliminating ω^2 from the real and imaginary parts of this equation. However, an inverse solution is possible in which both the real and imaginary equations are solved for ω, or more likely solved for the rotor speed Ω for a range of flutter frequencies. The real solutions for ω are then plotted as a function of Ω, as in the Coleman diagram. A point where the solutions of the real and imaginary equations cross identifies a flutter frequency for which $s = i\omega$ satisfies the characteristic equation, and hence it defines the stability boundary. Note that the Coleman diagrams shown in the last section give the solution of the real part for the case of no damping. If a numerical solution is to be obtained, a more direct procedure is to solve the characteristic equation for all four roots, and plot the frequency and damping ratio of the modes as a function of Ω or some other parameter, for a given level of damping. An analytical solution for an approximate stability criterion can be obtained by again assuming S_ζ to be small.

It has been seen that an instability can occur at a resonance of the low frequency mode with a support mode. When there is no damping and $S_\zeta > 0$, such a resonance is unstable if $\nu_\zeta < 1$/rev. It is desired to find the damping required to stabilize this motion. Since the point of exact resonance of the uncoupled frequencies is always roughly in the center of the instability region, that point is expected to be the most critical case, requiring the most damping to stabilize. Therefore we consider the stability boundary exactly at resonance, $\omega_x = 1 - \nu_\zeta$. The solution will be expanded for small S_ζ^2, so $\omega^2 \cong \omega_x^2$. Since the instability is due to the inertial coupling S_ζ^*, the damping (C_ζ^*, C_x^*, and C_y^*) at the stability boundary must also be of order S_ζ^*. It is assumed for now that $\omega_x \neq \omega_y$.

Then to lowest order in S_ζ^2, the characteristic equation gives the stability boundary:

$$I_\zeta^{*2}(4C_\zeta^* \nu_\zeta i\omega_x)M_y^*(\omega_y^2 - \omega_x^2)M_x^*(C_x^* i\omega_x) + 2I_\zeta^*(1-\nu_\zeta)S_\zeta^{*2}\omega_x^4 M_y^*(\omega_y^2 - \omega_x^2) = 0$$

Hence the criterion for stability is

$$C_\zeta^* C_x^* > \frac{(1-\nu_\zeta)\omega_x^2}{2\nu_\zeta} \frac{S_\zeta^{*2}}{I_\zeta^* M_x^*}$$

which was first obtained by Deutsch (1946). For stiff in-plane rotors ($\nu_\zeta > 1$/rev), the right-hand side is negative and the motion is always stable. For a soft in-plane rotor ($\nu_\zeta < 1$/rev), the product of the lag and support damping must be greater than this critical value for stability. Similarly, the criterion for the lateral mode resonance at $\omega_y = 1 - \nu_\zeta$ is

$$C_\zeta^* C_y^* > \frac{(1-\nu_\zeta)\omega_y^2}{2\nu_\zeta} \frac{S_\zeta^{*2}}{I_\zeta^* M_y^*}$$

The stability boundary for a resonance with the high frequency lag mode, $\omega_x = 1 + \nu_\zeta$, gives the criterion

$$C_\zeta^* C_x^* > -\frac{(1+\nu_\zeta)\omega_x^2}{2\nu_\zeta} \frac{S_\zeta^{*2}}{I_\zeta^* M_x^*}$$

which is always satisfied, even for zero damping.

Thus it has been verified that the resonance of the low frequency lag mode with a support mode will be unstable if the lag frequency is below 1/rev, and the product of the lag and support damping is below a critical level. The other resonances of the lag and support modes are stable even with no damping. The damping required for ground resonance stability is proportional to the inertial coupling parameter $S_\zeta^{*2}/I_\zeta^* M_x^*$ and hence to the ratio of the rotor mass to the support mass. The damping required is also proportional to $(1 - \nu_\zeta)/\nu_\zeta$. For the small lag frequency typical of articulated rotors a large amount of lag damping is thus required. Mechanical lag dampers are generally needed to insure ground resonance stability. For typical soft in-plane hingeless rotors, however, the factor $(1 - \nu_\zeta)/\nu_\zeta$ is an order of magnitude smaller than for articulated rotors, so the blade structural damping may provide a sufficient level of C_ζ. For ground resonance stability as high a lag frequency as possible is desired, but if ν_ζ is too close to 1/rev the blade loads and vibration will be excessive. Thus even a hingeless rotor may require mechanical lag dampers for stability.

The dimensional form of the Deutsch criterion is generally most useful. Recalling the definitions of the normalized inertia and damping coefficients, we obtain for the required damping

$$\frac{C_\zeta C_x}{\omega_x^2} > \frac{N}{4} \frac{1-\nu_\zeta}{\nu_\zeta} S_\zeta^2$$

where N is the number of blades, ν_ζ is the lag frequency (still per-rev), and $S_\zeta = \int_0^R \eta_\zeta m dr$ is the first moment of inertia of the lag mode. The support mode is defined by the natural frequency ω_x (rad/sec) and the linear damping coefficient C_x (force per unit velocity). These parameters can be obtained for each vibration mode of the rotor support from the measured frequency response of the hub to excitation by in-plane forces. The lag damping coefficient C_ζ is the lag moment per unit angular velocity of the lag degree of freedom. This criterion defines the lag damping required at the resonance of the low frequency lag mode with ω_x, which occurs at the rotor speed $\Omega = \omega_x/(1-\nu_\zeta)$. Then for each lateral and longitudinal support mode a critical ground resonance rotor speed is obtained, as well as the lag damping required at that Ω to stabilize the motion. By comparing the lag damping required with the damping available as a function of rotor speed, the ground resonance stability can be assessed for a given rotor and helicopter.

The above result required the assumption $\omega_x \neq \omega_y$. For the case of an isotropic support ($\omega_x = \omega_y$), the characteristic equation to lowest order in S_ζ^2 gives instead the following stability criterion:

$$C_\zeta^* \frac{M_y^* C_y^* M_x^* C_x^*}{M_y^* C_y^* + M_x^* C_x^*} > \frac{(1-\nu_\zeta)\omega_x^2}{2\nu_\zeta} \frac{S_\zeta^{*2}}{I_\zeta^*}$$

For isotropic support damping as well ($M_y^* C_y^* = M_x^* C_x^*$), this becomes

$$C_\zeta^* C_x^* > \frac{(1-\nu_\zeta)\omega_x^2}{\nu_\zeta} \frac{S_\zeta^{*2}}{I_\zeta^* M_x^*}$$

The isotropic case requires twice the damping as the anisotropic support because equal lateral and longitudinal support frequencies allow a whirling motion of the hub that couples best with the whirling motion of the low frequency lag mode. The definition of an isotropic support requires that the frequencies ω_x and ω_y be of order $S_\zeta^{*2}/I_\zeta^* M_x^*$ apart, which is an extremely small frequency difference. In practice, then, the isotropic case is not important except when the rotor support structure is truly axisymmetric.

Ground resonance stability with articulated and soft in-plane hingeless rotors is achieved by providing a sufficient level of damping of the rotor lag motion and of the support motion. Instabilities can also be avoided by a proper placement of the natural frequencies of the airframe to avoid resonances, but usually there are too many other constraints on the structural design for this to be a practical means of handling the ground resonance problem. With a stiff in-plane rotor (for example, two-bladed teetering rotors and some hingeless rotor designs) the resonances are all stable. With articulated rotors, mechanical dampers on the landing gear and at the lag hinges are standard features of the helicopter design. The linear analysis developed here has assumed viscous damping, in which the force opposing the motion is proportional to the velocity of the motion. The actual damping of the rotor and support will almost certainly be nonlinear, however, particularly if mechanical dampers are used. It is possible to determine an equivalent viscous damping coefficient to describe nonlinear lag dampers, based on the energy dissipated during a cycle of motion. By this means the linear analysis can be applied to the real rotor. The equivalent viscous damping will depend on the frequency and amplitude of the lag motion. For example, frictional damping [restoring force proportional to sign($\dot{\zeta}$)] will give an equivalent viscous damping coefficient equal to a constant divided by $\omega\zeta_{amp}$, while hydraulic damping (restoring force proportional to $\dot{\zeta}|\dot{\zeta}|$) gives an equivalent coefficient equal to a constant multiplied by $\omega\zeta_{amp}$. The frequency of the ground resonance mode can be assumed to be near the lag frequency, $\omega \cong \nu_\zeta \Omega$ in the rotating frame, so that the rotor speed defines the frequency for the lag dampers. Then the lag damping level required for stability can be interpreted as a limitation on the lag amplitude. The damping of the support is also likely to be nonlinear, because of the complex structure of the helicopter and the presence of nonlinear elements such as oleo struts and tires. The analysis should use the lowest equivalent viscous damping that is likely to be encountered. Since calculation of the support characteristics is difficult at best, the ground resonance analysis should rely on the measured hub frequencies and damping. The ground resonance instability is a simple phenomenon physically, and therefore with good measurements of the rotor and support damping the stability can generally be accurately predicted.

12-4.4 Two-Bladed Rotor

Now let us consider the ground resonance stability of a two-bladed rotor. Because the rotor inertial properties are not axisymmetric as they are when $N \geqslant 3$, the cyclic lag degrees of freedom are not applicable to the two-bladed rotor. Instead, the lag motion is described by the differential lag degree of freedom ζ_1. The equation of motion for ζ_1 was obtained in section 9–6:

$$I_\zeta(\ddot{\zeta}_1 + C_\zeta^*\dot{\zeta}_1 + \nu_\zeta^2\zeta_1) + S_\zeta(\ddot{x}_h \sin\psi - \ddot{y}_h \cos\psi) = 0$$

and the hub forces in sections 9-5.2 and 9–6:

$$H = -NM_b\ddot{x}_h + NS_\zeta[-(\ddot{\zeta}_1 - \zeta_1)\sin\psi - 2\dot{\zeta}_1 \cos\psi]$$

$$Y = -NM_b\ddot{y}_h + NS_\zeta[(\ddot{\zeta}_1 - \zeta_1)\cos\psi - 2\dot{\zeta}_1 \sin\psi]$$

The aerodynamic forces have been dropped, and a lag damper included in the equation for ζ_1. Again the hub longitudinal and lateral equations of motion are

$$M_x\ddot{x}_h + C_x\dot{x}_h + K_x x_h = H$$

$$M_y\ddot{y}_h + C_y\dot{y}_h + K_y y_h = Y$$

The natural frequency and damping coefficient of the support modes are defined by $K_x = (M_x + NM_b)\omega_x^2$ and $C_x = (M_x + NM_b)C_x^*$, and similarly for ω_y and C_y^*. Finally the rotor equation is normalized by dividing by I_b and the support equations by dividing by NI_b, using the definitions $M_x^* = (M_x + NM_b)/NI_b$ and $M_y^* = (M_y + NM_b)/NI_b$. Then the equations of motion describing the ground resonance of a two-bladed rotor are as follows:

$$
\begin{bmatrix} I_\zeta^* & -S_\zeta^* \cos\psi & S_\zeta^* \sin\psi \\ -S_\zeta^* \cos\psi & M_y^* & 0 \\ S_\zeta^* \sin\psi & 0 & M_x^* \end{bmatrix} \begin{pmatrix} \zeta_1 \\ y_h \\ x_h \end{pmatrix}^{\cdot\cdot}
+ \begin{bmatrix} I_\zeta^* C_\zeta^* & 0 & 0 \\ 2S_\zeta^* \sin\psi & M_y^* C_y^* & 0 \\ 2S_\zeta^* \cos\psi & 0 & M_x^* C_x^* \end{bmatrix} \begin{pmatrix} \zeta_1 \\ y_h \\ x_h \end{pmatrix}^{\cdot}
+ \begin{bmatrix} I_\zeta^* \nu_\zeta^2 & 0 & 0 \\ S_\zeta^* \cos\psi & M_y^* \omega_y^2 & 0 \\ -S_\zeta^* \sin\psi & 0 & M_x^* \omega_x^2 \end{bmatrix} \begin{pmatrix} \zeta_1 \\ y_h \\ x_h \end{pmatrix} = 0
$$

These equations have periodic coefficients because of the inertial asymmetry of the rotor when $N = 2$, and the fact that the lag degree of freedom ζ_1 is really still in the rotating frame. The methods for analyzing the stability of such equations are discussed in section 8-6.2.

For the case of a completely isotropic support, constant coefficient differential equations can be obtained in the rotating frame. These equations can be analyzed as for the $N \geqslant 3$ case covered in the last section. Therefore let us assume $\omega_x = \omega_y$, $M_x^* = M_y^*$, and $C_x^* = C_y^*$; and define the hub deflections in the rotating frame:

$$
\begin{aligned} y_r &= y_h \cos\psi - x_h \sin\psi \\ x_r &= y_h \sin\psi + x_h \cos\psi \end{aligned}
$$

A similar transformation of the hub in-plane forces generates the differential equations for x_r and y_r in the rotating frame. Then the constant coefficient equations describing the ground resonance dynamics with an isotropic support are:

$$
\begin{bmatrix} I_\zeta^* & -S^* & 0 \\ -S_\zeta^* & M_y^* & 0 \\ 0 & 0 & M_y^* \end{bmatrix} \begin{pmatrix} \zeta_1 \\ y_r \\ x_r \end{pmatrix}^{\cdot\cdot}
+ \begin{bmatrix} I_\zeta^* C_\zeta^* & 0 & -2S_\zeta^* \\ 0 & M_y^* C_y^* & 2M_y^* \\ 2S_\zeta^* & -2M_y^* & M_y^* C_y^* \end{bmatrix} \begin{pmatrix} \zeta_1 \\ y_r \\ x_r \end{pmatrix}^{\cdot}
$$

$$+ \begin{bmatrix} I_\zeta^* \nu_\zeta^2 & S_\zeta^* & 0 \\ S_\zeta^* & M_y^*(\omega_y^2 - 1) & M_y^* C_y^* \\ 0 & -M_y^* C_y^* & M_y^*(\omega_y^2 - 1) \end{bmatrix} \begin{pmatrix} \zeta_1 \\ y_r \\ x_r \end{pmatrix} = 0$$

Note that in the rotating frame there are Coriolis and centrifugal forces coupling the equations for the support motion. The characteristic equation is

$$M_y^{*2}\left[(s^2 + C_y^* s + \omega_y^2 - 1)^2 + (2s + C_y^*)^2\right] I_\zeta^*(s^2 + C_\zeta^* s + \nu_\zeta^2)$$
$$+ M_y^* S_\zeta^{*2}\left[(s^2 + C_y^* s + \omega_y^2 - 1)(4s^2 - (s^2 - 1)^2) - 4s(s^2 - 1)(2s + C_y^*)\right] = 0$$

For the case of no damping, the characteristic equation reduces to

$$M_y^{*2}\left[(s^2 + \omega_y^2 - 1)^2 + 4s^2\right] I_\zeta^*(s^2 + \nu_\zeta^2)$$
$$+ M_y^* S_\zeta^{*2}\left[(s^2 + \omega_y^2 - 1)(4s^2 - (s^2 - 1)^2) - 8s^2(s^2 - 1)\right] = 0$$

The uncoupled solution ($S_\zeta = 0$) is just $s = \pm i\omega$, where $\omega = \nu_\zeta$ and $\omega = \omega_y \pm 1$/rev. Hence the support mode frequencies in the rotating frame are shifted by $\pm\Omega$ from the frequencies in the nonrotating frame, and the rotor mode is at the rotating lag natural frequency ν_ζ.

As when $N \geqslant 3$, the solution of the characteristic equation for the case of no damping can be expanded in S_ζ^2 about the uncoupled solution. Writing $\omega^2 = \nu_\zeta^2 + S_\zeta^{*2} s_1$ for the solution near $\omega = \nu_\zeta$, we obtain

$$\omega^2 = \nu_\zeta^2 + \frac{(\nu_\zeta^2 - \omega_y^2 + 1)\left[4\nu_\zeta^2 + (\nu_\zeta^2 + 1)^2\right] - 8\nu_\zeta^2(\nu_\zeta^2 + 1)}{(\nu_\zeta^2 - \omega_y^2 + 1)^2 - 4\nu_\zeta^2} \frac{S_\zeta^{*2}}{I_\zeta^* M_y^*}$$

and near the solution $\omega = \omega_y \pm 1$:

$$\omega^2 = (\omega_y \pm 1)^2 - \frac{\omega_y^3(\omega_y \pm 1)}{2\left[\nu_\zeta^2 - (\omega_y \pm 1)^2\right]} \frac{S_\zeta^{*2}}{I_\zeta^* M_y^*}$$

With these expressions, the directions the solutions shift when $S_\zeta > 0$ can be established for the limits $\Omega = 0$ and Ω approaching infinity. At $\Omega = 0$, the uncoupled roots are $\omega = \nu_{NR}$ and ω_y; it is found that the larger root is increased while the smaller root is decreased. When Ω is large, of the three roots corresponding to $\omega = \nu_\zeta$ and $\omega_y \pm 1$ the largest and smallest are increased, while the middle root is decreased. From this behavior the solution for $S_\zeta > 0$ can be sketched. Figs. 12-14 and 12-15 present typical Coleman diagrams for articulated (soft in-plane) and stiff in-plane two-bladed rotors. As in the case of three or more blades, a ground resonance instability appears with soft in-plane rotors ($\nu_\zeta < 1$/rev) at the resonance of the support and the low frequency lag mode — which in the rotating frame means $\nu_\zeta = \Omega - \omega_y$.

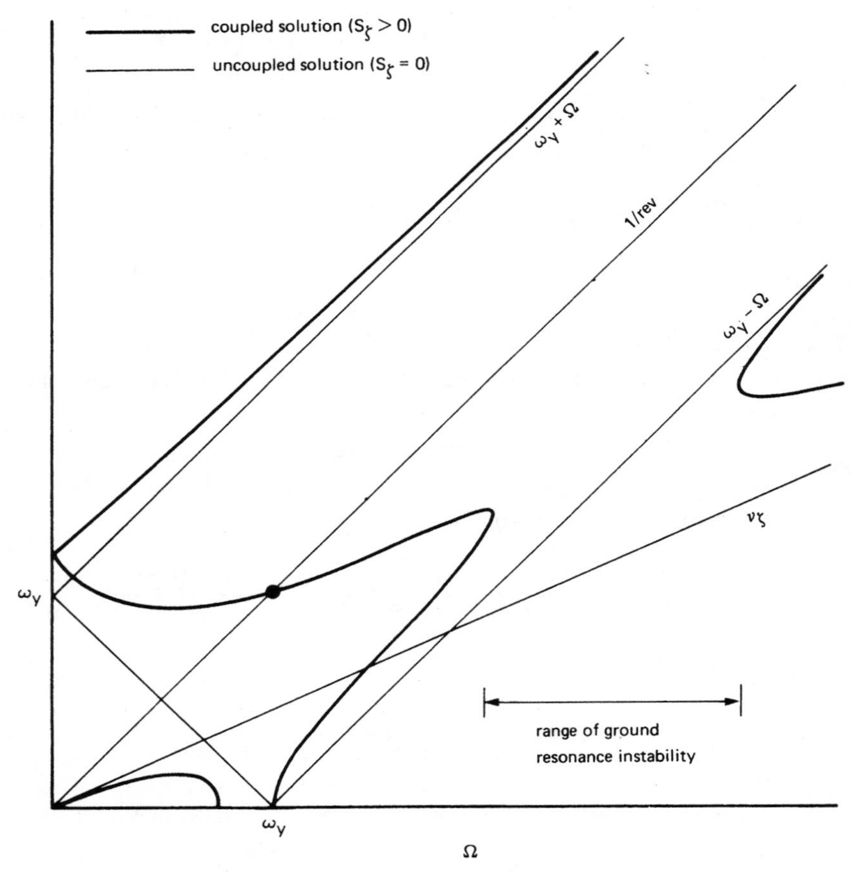

Figure 12-14 Coleman diagram of the ground resonance solution for a two-bladed articulated rotor ($\nu_\zeta < 1$/rev) on an isotropic support.

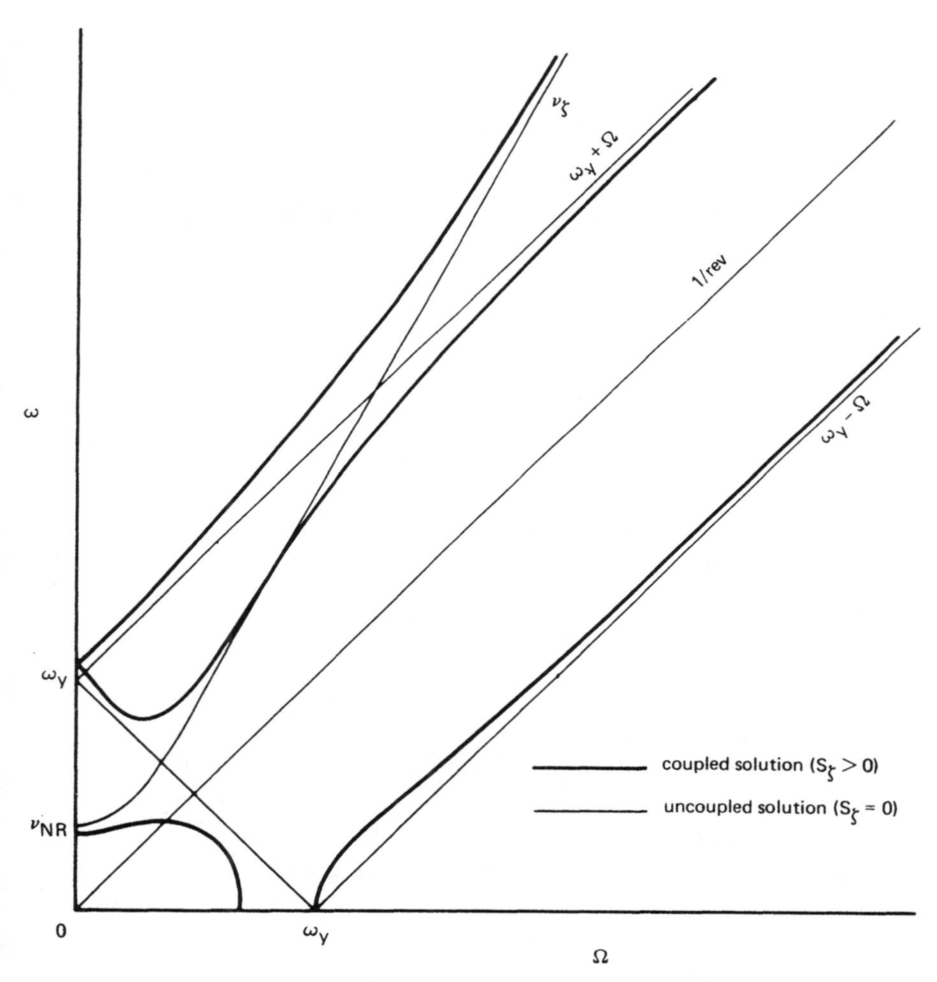

Figure 12-15 Coleman diagram of the ground resonance solution for a two-bladed stiff in-plane rotor ($\nu_\zeta > 1$/rev) on an isotropic support.

Note that for $N = 2$ the center of the gound resonance instability range is shifted to a rotor speed above the uncoupled resonance, in contrast to the $N \geqslant 3$ case, for which the instability range remains centered about the resonance. This suggests that for large enough coupling the instability region might be shifted above the rotor operating range. To examine this possibility further, consider the intersection of the locus with the 1/rev line, as indicated in Fig. 12-14. On substituting $s^2 = -1$, the characteristic equation becomes

$$M_y^* \omega_y^2 [(\omega_y^2 - 4) M_y^* I_\zeta^* (\nu_\zeta^2 - 1) - 8 S_\zeta^{*2}] = 0$$

Now since $\omega = 1$/rev in the rotating frame corresponds to $\omega = 0$ and $\omega = 2$/rev in the nonrotating frame, the uncoupled solutions are just $\omega_y = 0$ and $\omega_y = 2$/rev, and $\nu_\zeta = 1$/rev. It is the $\omega_y = 2$/rev solution that is of interest here. For the coupled case ($S_\zeta > 0$), this resonance occurs at the rotor speed

$$\Omega^2 = \frac{(\omega_y/2)^2}{1 - \dfrac{2}{1-\nu_\zeta^2} \dfrac{S_\zeta^{*2}}{I_\zeta^* M_y^*}}$$

which increases with S_ζ for the soft in-plane rotor. The ground resonance instability always occurs at a rotor speed above this value and thus it provides a conservative criterion for avoiding ground resonance. Note that if

$$\frac{S_\zeta^{*2}}{I_\zeta^* M_y^*} > \frac{1 - \nu_\zeta^2}{2}$$

then both the 1/rev resonance (2/rev in the nonrotating frame) and the instability region are swept to $\Omega = \infty$. The inertial coupling required is rather large, however, even when $\nu_\zeta = 0.85$ or so, which is about the upper limit for soft in-plane hingeless rotors. For a further discussion of this effect, see Leone (1956b).

In Fig. 12-14 there is also a region just below $\omega = \omega_y$ where there are only two real solutions for ω and hence the motion is unstable. This phenomenon occurs also for the stiff in-plane rotor (see Fig. 12-15) and has not been observed for rotors with three or more blades. In this instability region two roots of the characteristic equation are on the real axis, with zero frequency, one positive and one negative. The frequency $\omega = 0$ in the rotating frame corresponds to $\omega = 1$/rev in the nonrotating frame. Hence this instability is associated with the shaft critical speeds, where the rotor speed passes through the support natural frequency ω_y. On setting s equal to zero, the characteristic equation becomes

$$(\omega_y^2 - 1)[M_y^* I_\zeta^* \nu_\zeta^2 (\omega_y^2 - 1) - S_\zeta^{*2}] = 0$$

which has the two solutions $\omega_y^2 = 1$ and

$$\omega_y^2 = 1 + \frac{1}{\nu_\zeta^2}\frac{S_\zeta^{*2}}{I_\zeta^* M_y^*}$$

These equations are valid for large as well as small S_ζ and give the two points where the coupled roots intercept the Ω-axis. The shaft critical instability thus occurs in the range

$$\frac{\omega_y^2}{1 + \dfrac{1}{\nu_\zeta^2}\dfrac{S_\zeta^{*2}}{I_\zeta^* M_y^*}} < \Omega^2 < \omega_y^2$$

For an articulated rotor (with small ν_ζ^2) this rotor speed range can be large even though the inertial coupling is small.

Next, consider the damping required to stabilize the ground resonance motion of a two-bladed rotor. With lag and support damping, $s = i\omega$ defines the stability boundary. As in the $N \geqslant 3$ case, the equation for the stability boundary is expanded in S_ζ^2 about the resonance $\nu_\zeta = 1 - \omega_y$. To lowest order in S_ζ^2, the characteristic equation gives the stability criterion

$$C_\zeta^* C_y^* > -\frac{\omega_y^3(\omega_y - 1)}{2\nu_\zeta^2}\frac{S_\zeta^{*2}}{I_\zeta^* M_y^*}$$

or since $\omega_y(\omega_y - 1)/\nu_\zeta = \nu_\zeta - 1$,

$$C_\zeta^* C_y^* > \frac{(1-\nu_\zeta)\omega_y^2}{2\nu_\zeta}\frac{S_\zeta^{*2}}{I_\zeta^* M_y^*}$$

For a stiff in-plane rotor ($\nu_\zeta > 1$/rev) this resonance is always stable. In dinensional terms, the damping required for stability is

$$\frac{C_\zeta C_y}{\omega_y^2} > \frac{N}{2}\frac{1-\nu_\zeta}{\nu_\zeta} S_\zeta^2$$

Note that this result is exactly the same criterion as that obtained for a rotor with three or more blades on an isotropic support.

The damping required to stabilize the shaft critical mode (a divergence instability in the rotating frame) is obtained by substituting $s = 0$ in the characteristic equation:

$$M_y^{*2}[(\omega_y^2 - 1)^2 + C_y^{*2}]I_\zeta^*\nu_\zeta^2 - M_y^* S_\zeta^{*2}(\omega_y^2 - 1) = 0$$

The motion is stable if the support damping satisfies the criterion

$$C_y^{*2} > (\omega_y^2 - 1)\left[\frac{S_\zeta^{*2}}{\nu_\zeta^2 I_\zeta^* M_y^*} - (\omega_y^2 - 1)\right]$$

The damping required is zero at the end points of the shaft critical speed region, and has a maximum midway between them at

$$\omega_y^2 = 1 + \frac{1}{2\nu_\zeta^2}\frac{S_\zeta^{*2}}{I_\zeta^* M_y^*}$$

or at a rotor speed of

$$\Omega^2 = \frac{\omega_y^2}{1 + \dfrac{1}{2\nu_\zeta^2}\dfrac{S_\zeta^{*2}}{I_\zeta^* M_y^*}}$$

The damping required to stabilize the entire range is

$$C_y^* > \frac{1}{2\nu_\zeta^2}\frac{S_\zeta^{*2}}{I_\zeta^* M_y^*}$$

or dimensionally,

$$C_y > \frac{N}{2\nu_\zeta^2}\frac{S_\zeta^2}{I_\zeta}\Omega \cong \frac{N}{2\nu_\zeta^2}\frac{S_\zeta^2}{I_\zeta}\omega_y$$

This result is exact for all values of the inertial coupling S_ζ. In contrast to the ground resonance instability, only support damping is required to stabilize this motion. The level of support damping specified by this criterion is usually not very large.

Coleman and Feingold (1958) investigated the general case of a two-bladed rotor on an anisotropic support and obtained the stability by the infinite determinant method for analyzing periodic coefficient differential equations. They found that the dynamic behavior, and specifically the possible instabilities, are much the same as for the case with isotropic support. However, the periodic coefficients introduce additional resonances. A ground resonance instability can occur for soft in-plane rotors at frequencies near $\omega_y = 1 - \nu_\zeta + 2n$/rev, or at $\Omega = \omega_y/(1 - \nu_\zeta + 2n)$, where n is a positive integer. The shaft critical speed now occurs at $\omega_y = 1 + 2n$/rev, or at $\Omega = \omega_y/(1 + 2n)$. Thus for given rotor lag and support frequencies, additional resonances occur at rotor speeds lower than the fundamental. These resonances due to the periodic coefficients tend to occur at low Ω and have a narrower instability range than the fundamental resonance. Hence much less damping is required to stabilize the motion in these regions.

Finally, it should be noted that the most common two-bladed rotor design is the stiff in-plane teetering configuration. Ground resonance is not a concern since the lag frequency is above 1/rev, and only a low level of support damping is required to handle the shaft critical speed instability.

12-4.5 Literature

Coleman and Feingold (1958) developed the classical analysis of ground resonance. This report is actually a republication of work by Coleman in 1943 on the case of three or more blades; by Feingold in 1943 on the case of a two-bladed rotor with an isotropic support; and by Coleman and Feingold in 1947 on the undamped two-bladed rotor with an anisotropic support. Their notation is often seen in the literature on ground resonance. Coleman and Feingold used the notation ω for the rotor speed, for which Ω is used here. They used $\Lambda_2 + \Lambda_1\omega^2$ for the lag frequency $\nu_\zeta^2 = K_1 + K_2\Omega^2$; ω_f and ω_a for the flutter frequency ω; λ_β for the lag damping C_ζ^*; λ_f for the support damping C_x^* or C_y^*; and Λ_3 for the inertial coupling $S_\zeta^{*2}/I_\zeta^* M_y^*$. To normalize the frequencies they used the support natural frequency ω_x or ω_y, since the rotor speed Ω is a major parameter of the problem.

On the ground resonance instability of helicopters: Kelley (1945), Deutsch (1946), Horvay (1946), Howarth and Jones (1954), Leone (1956a, 1956b), Warming (1956), Hooper (1959), Sibley and Jones (1959), Price (1960, 1962), Gabel and Capurso (1962), Mil′ (1966), Gladwell and Stammers (1968), Loewy (1969), Done (1974), Hammond (1974), Metzger (1974), Schroder (1974), Young and Bailey (1974), Hohenemser and Yin (1977), Ormiston (1977).

On the air resonance instability of helicopters: Cardinale (1969), Donham, Cardinale, and Sachs (1969), Lytwyn, Miao, and Woitsch (1971), Baldock (1972), Bramwell (1972), Burkam and Miao (1972), Huber (1973b), Miao and Huber (1974), Johnson (1977e), Ormiston (1977), Weller (1977), Bousman (1978), Hodges (1978a, 1978b), White, Sutton, and Nettles (1978).

12–5 Vibration and Loads

12-5.1 Vibration

Vibration is the oscillatory response of the helicopter airframe (and other components in the nonrotating frame) to the rotor hub forces and moments. There are other important sources of helicopter vibration, notably the engine and transmission, and aerodynamic forces on the fuselage; but it is the rotor influence that is of interest in this text. In steady-state forward flight, the periodic forces at the root of the blade are transmitted to the helicopter, producing a periodic vibratory response. Thus helicopter vibration is characterized by harmonic excitation in the nonrotating frame, primarily at 1/rev and N/rev (where N is the number of blades). The vibration is generally low in hover and increases with forward flight to high levels at the maximum speed of the aircraft. There is also a high level of vibration in transition ($\mu \cong 0.1$) because of the rotor wake influence on the blade loading.

Let us examine how the periodic rotor forces are transmitted through the hub to the aircraft. It is assumed that the root reaction of the mth blade ($m = 1$ to N) is a periodic function of $\psi_m = \psi + m\Delta\psi$ ($\Delta\psi = 2\pi/N$). Therefore, all the blades have identical loading and motion. Consider first the vertical shear force $S_z^{(m)}$ at the root of the mth blade (see section 9–5 and Fig. 9-7). Write $S_z^{(m)}$ as a complex Fourier series in ψ_m:

$$S_z^{(m)} = \sum_{n=-\infty}^{\infty} S_{z_n} e^{in\psi_m}$$

The total thrust force of the rotor is obtained by summing the root vertical shears over all N blades:

$$T = \sum_{m=1}^{N} S_z^{(m)}$$

Using the results of section 8–2 for the summation of harmonics, it follows that

$$T = \sum_{n=-\infty}^{\infty} S_{z_n} \left(\sum_{m=1}^{N} e^{in\psi_m} \right)$$

$$= \sum_{p=-\infty}^{\infty} N S_{z_{pN}} e^{ipN\psi}$$

The forces from all the blades exactly cancel at the hub, except for those harmonics at multiples of N/rev, which are transmitted to the aircraft. The in-plane shear forces on the rotating blade are $S_x^{(m)}$ in the blade drag direction, and $S_r^{(m)}$ radially. The in-plane hub forces in the nonrotating frame, the rotor drag force H and side force Y, are given by

$$H = \sum_{m=1}^{N} (S_r^{(m)} \cos\psi_m + S_x^{(m)} \sin\psi_m)$$

$$Y = \sum_{m=1}^{N} (S_r^{(m)} \sin\psi_m - S_x^{(m)} \cos\psi_m)$$

Writing the rotating shear forces as Fourier series in ψ_m, we obtain

$$H = \sum_{n=-\infty}^{\infty} \left[S_{r_n} \left(\sum_{m=1}^{N} e^{in\psi_m} \cos\psi_m \right) + S_{x_n} \left(\sum_{m=1}^{N} e^{in\psi_m} \sin\psi_m \right) \right]$$

$$= \sum_{n=-\infty}^{\infty} \left[S_{r_n} \left(\frac{1}{2} \sum_{m=1}^{N} e^{i(n+1)\psi_m} + e^{i(n-1)\psi_m} \right) \right.$$

$$\left. + S_{x_n} \left(\frac{1}{2i} \sum_{m=1}^{N} e^{i(n+1)\psi_m} - e^{i(n-1)\psi_m} \right) \right]$$

$$= \sum_{p=-\infty}^{\infty} \left(\frac{N}{2} S_{r_{pN-1}} + \frac{N}{2} S_{r_{pN+1}} + \frac{N}{2i} S_{x_{pN-1}} - \frac{N}{2i} S_{x_{pN+1}} \right) e^{ipN\psi}$$

and similarly

$$Y = \sum_{p=-\infty}^{\infty} \left(\frac{N}{2i} S_{r_{pN-1}} - \frac{N}{2i} S_{r_{pN+1}} - \frac{N}{2} S_{x_{pN-1}} - \frac{N}{2} S_{x_{pN+1}} \right) e^{ipN\psi}$$

Thus for the in-plane hub forcces as well only the harmonics of N/rev appear in the nonrotating frame, produced by the $pN \pm 1$/rev harmonics of the rotating shear forces. The rotor torque transmitted through the hub is obtained from the root lagwise moment $N_L^{(m)}$ in a fashion similar to the rotor thrust, giving

$$Q = \sum_{m=1}^{N} N_L^{(m)} = N \sum_{p=-\infty}^{\infty} N_{L_{pN}} e^{ipN\psi}$$

Finally, the hub pitch and roll moments are obtained from the flapwise moment $N_F^{(m)}$ at the root of the rotating blade:

$$M_y = -\sum_{m=1}^{N} N_F^{(m)} \cos\psi_m = -\frac{N}{2}\sum_{p=-\infty}^{\infty}\left(N_{F_{pN-1}} + N_{F_{pn+1}}\right)e^{ipN\psi}$$

$$M_x = \sum_{m=1}^{N} N_F^{(m)} \sin\psi_m = \frac{N}{2i}\sum_{p=-\infty}^{\infty}\left(N_{F_{pN-1}} - N_{F_{pN+1}}\right)e^{ipN\psi}$$

So the rotor transmits forces and moments to the nonrotating frame only at harmonics of N/rev, as summarized in Table 12-1. The transmission of the

Table 12-1. Transmission of helicopter vibration through the rotor hub

Nonrotating Frame		Rotating Frame
thrust at pN/rev	from	vertical shear at pN/rev
torque at pN/rev	from	lagwise moment at pN/rev
rotor drag and side forces at pN/rev	from	in-plane shears at $pN \pm 1$/rev
pitch and roll moments at pN/rev	from	flapwise moment at $pN \pm 1$/rev
collective control system forces at pN/rev	from	feathering moments at pN/rev
cyclic control system forces at pN/rev	from	feathering moments at $pN \pm 1$/rev

blade feathering moments to the collective and cyclic control systems has also been included in this table. If the control system is entirely mechanical, these control loads will produce vibration in the pilot's collective and cyclic sticks. Basically, the rotor hub acts as a filter, transmitting to the helicopter only harmonics of the rotor forces at multiples of N/rev. This result is based on the assumption that all the blades are identical and have the same periodic motion. While this will not be perfectly true, still the N/rev harmonics dominate the vibration produced by real rotors. The filtering of the blade vibratory forces by the rotor hub makes the task of vibration reduction or avoidance easier, because only a few frequencies need be considered, and because the low harmonics with the largest magnitude are not transmitted to the helicopter (except in the case of a two-bladed rotor).

Helicopter rotors generally produce a significant 1/rev vibration as well, because of the large 1/rev variation of the loading in forward flight and the fact that any aerodynamic or inertial dissimilarity between the blades primarily generates 1/rev vibration. A major effort is made with every rotor to eliminate the differences between the blades in the tracking and balancing operations. The inertial properties of the blades can be adjusted using small balance weights, particularly at the tips; and the aerodynamic properties can be matched using aerodynamic trim tabs and by adjusting the pitch links. However, enough 1/rev vibration often remains that it must be considered in the helicopter design.

The helicopter N/rev vibration is due to the higher harmonic loading of the rotor. The sources of this loading are the rotor wake and the effects of stall and compressibility at high speed. The helicopter vibration is low in hover where the aerodynamic environment is nearly axisymmetric. The only sources of higher harmonic loading are the small asymmetries such as those due to aerodynamic interference with the fuselage and other rotors. In transition, at advance ratios around $\mu = 0.1$, there is normally a peak in the vibration level due to the wake-induced loads on the rotor. Since the helicopter drag is small at low speeds, the tip-path-plane incidence remains small and the tip vortices in the wake remain close to the disk plane. The advance ratio is high enough, though, so that the blades sweep past the tip vortices from preceding blades. Such close blade-vortex encounters produce significant higher harmonic airloading at the harmonics transmitted through the hub as vibration. This vibration is increased by operations that keep the wake near the plane of the disk, such as decelerating or descending flight. As the speed increases, the tip-path plane tilts forward to provide the propulsive force, which means that the wake will be convected away from the disk plane and the wake-induced vibration will decrease. At still higher speeds the vibration increases again, primarily as a result of the higher harmonic loading produced by stall and compressibility effects. Such vibration often in fact limits the maximum speed of the aircraft.

The basic principle in designing an aircraft to minimize vibration is to avoid structural resonances with the frequencies of the exciting forces. The helicopter airframe must be designed to avoid resonances with the harmonics of the rotor speed, particularly near 1/rev and N/rev. (Resonances must be avoided as well with the speeds of other rotating components, including the engine, transmission, and tail rotor.) The analysis of the vibration modes of a helicopter is a difficult task because of the complexity of the structure, but reasonable accuracy is possible with modern infinite element techniques. A shake test of the actual structure is necessary to determine the true natural frequencies, however. Adjusting the airframe frequencies to avoid resonances is also complicated by the large number of exciting frequencies that must be considered. Resonances in the rotor itself will amplify the root loads, and hence the transmitted vibration. Therefore, it is also necessary to design the blades to avoid resonances with N/rev and $N \pm 1$/rev. (If the distinction is relevant, namely for teetering and gimballed rotors, the collective modes of the rotor should avoid N/rev resonances while the cyclic modes should avoid $N \pm 1$/rev.) Considering the blade loads and the fact that the rotor hub is not a perfect filter of the root forces, it is generally necessary to avoid resonances of the rotating natural frequencies of the blade with all harmonics of the rotor speed. The blade manufacturing process should be chosen to minimize the structural and aerodynamic differences between the blades, and thus to minimize the helicopter 1/rev vibration.

Passive vibration isolation is sometimes used, including approaches such as a soft mounting of the rotor and transmission to the airframe. For articulated and soft in-plane hingeless rotors, ground resonance considerations will likely require a stiff mounting, however. A dynamic vibration isolation system, consisting of a mass and spring system attached between the rotor blades and the airframe, can be used in either the rotating or nonrotating frame. Such an isolator is tuned so that a particular frequency of vibration, usually N/rev, is highly attenuated. Then energy of the blade root loads at this frequency goes into the isolator rather than into airframe motion. It is possible to actually use the blade itself as a vibration isolator of this sort, although it is a simpler task to design an entirely separate device. For example, with a torsionally soft blade, the torsion motion can be coupled with the first flapwise bending mode to reduce the vibratory loads at the root. It is a frequent practice also to take advantage of nodes (points of zero motion) in the structural vibration modes of the helicopter airframe to minimize the vibration at critical points.

12-5.2 Loads

The blade, hub, and control loads produced by the aerodynamic and inertial forces acting on the rotor are needed in order to design the helicopter structural components to the specified strength and fatigue criteria. The structural design actually requires a knowledge of the stresses in the blade, but with engineering beam theory it is consistent to consider just the bending and torsional moments acting on the blade section. With articulated blades the critical bending moment is usually the flapwise load somewhere around the blade midspan. For hingeless rotors the highest bending moments will be at the blade root. The net reactions at the blade root are needed to determine the loads in the rotor hub. The feathering moments on the blades lead to loads in the rotor control system, which are often a limiting factor in extreme operating conditions of the helicopter. The designer is usually concerned with the periodic or nearly periodic loads occurring in steady-state or maneuvering flight. Since the periodic aerodynamic environment of the helicopter rotor produces high oscillatory loads in the blades, hub, and control system, the fatigue analysis is a major part of rotor structural design. Because it depends critically on the details of the stress distribution, the fatigue life must normally be verified by tests. This is particularly true for helicopter rotors since many components are designed for finite fatigue life because of the high load levels.

The bending and torsion moments acting on the rotor blade section were derived in Chapter 9. A notable feature of the blade loads calculation is that it is essential to consider the elastic bending and torsion modes as well as the fundamental blade motion. The aerodynamic forces are well determined by just the fundamental modes, which dominate the deflection (for example, the rigid flapping motion of an articulated blade). However, the excitation of the elastic modes by the higher harmonics of these aerodynamic forces can produce large loads on the section even though the deflections due to these modes are small. Usually at least four to six coupled flap-lag bending modes are required for reasonably accurate calculations of the blade loads.

In section 9-2.4, three expressions were obtained for the flapwise bending moment on the rotor blade section. The conditions for equilibrium of moments on the blade outboard of the radial station r give the bending moment $M(r)$:

$$M(r) = \int_r^1 F_z(\rho - r)\, d\rho$$

$$- \sum_k \left[\ddot{q}_k \int_r^1 m\eta_k(\rho - r) d\rho + q_k \int_r^1 m\rho(\eta_k(\rho) - \eta_k(r)) d\rho \right]$$

Here F_z is the aerodynamic loading (the lift) and q_k is the degree of freedom for the kth bending mode of the rotating blade. Eliminating $\ddot{q}_k$ by using the equation of motion gives

$$M(r) = \sum_k q_k \left[\nu_k^2 \int_r^1 m\eta_k(\rho - r) d\rho - \int_r^1 m\rho(\eta_k(\rho) - \eta_k(r)) d\rho \right]$$

Alternatively, engineering beam theory can be used to relate the section moment to the local curvature:

$$M(r) = \sum_k q_k \left(EI d^2\eta_k / dr^2 \right)$$

Similar expressions can be derived for the other blade loads, and with additional degrees of freedom. In general the first approach, integrating the aerodynamic and inertial loading along the span to obtain the section moment, gives the best accuracy in numerical calculations with a finite number of modes. The last expression $M = EId^2z/dr^2$, frequently does not give satisfactory results. It requires a large number of modes because of the greater relative contribution of the higher modes to the curvature, and also there may be numerical problems because the second derivatives of the mode shapes are required. Moreover, if the equations of motion are obtained by the Galerkin or Rayleigh-Ritz approach, this expression may not even be applicable, since the boundary conditions of the modes need not be consistent with the loads applied at the blade root (such as those due to a lag damper, or control system inputs). If the section moment is required in terms of the modal deflection alone, the second expression given above is preferable since it deals with integrals of the mode shapes.

Consider now the solution of the equations of motion for the elastic response of the blade, which is needed to evaluate the loads. The equation of motion for the kth flapwise bending mode was derived in section 9-2.2:

$$I_{q_k}^*(\ddot{q}_k + \nu_k^2 q_k) = \gamma \int_0^1 \eta_k \frac{F_z}{ac} dr = \gamma M_{q_k}$$

where ν_k is the natural frequency and I_{q_k} is the generalized mass of the kth mode. The higher harmonics of the loading F_z that excite these degrees of freedom are not influenced much by the blade bending, however, but are due primarily to the basic aerodynamics of the rotor in forward flight, particularly the effects of the wake-induced velocity, stall, and compressibility. This suggests a separation of the loads analysis into two parts: first, obtaining the air loading distribution with a small number of degrees of freedom; second, using that airloading to calculate the excitation of the higher bending modes, and then the blade loads. Therefore, the aerodynamic forcing of the kth bending mode is written as

$$M_{q_k} = \bar{M}_{q_k} + M_{q_k \dot{q}_k} \dot{q}_k$$

where the aerodynamic damping coefficient $M_{q_k \dot{q}_k} = -\int_0^1 \tfrac{1}{2}\eta_k{}^2 r dr$ is given in section 11–2. Here $\bar{M}_{q_k}$ is just M_{q_k} evaluated using the airloading F_z, which was calculated neglecting the higher bending modes. It is necessary to include the aerodynamic damping of the q_k mode, which has the function of eliminating the singular response at resonances. Since resonant excitation is to be avoided anyway, the mean value of the damping has been used. (Including the periodic variation of the damping in forward flight would couple the harmonics of the forced response.) The equation of motion for q_k is now

$$I_{q_k}^* \ddot{q}_k - \gamma M_{q_k \dot{q}_k} \dot{q}_k + I_{q_k}^* \nu_k^2 q_k = \gamma \bar{M}_{q_k}$$

Since the steady state solution in forward flight is periodic, the forcing and the response can be expanded as Fourier series:

$$\bar{M}_{q_k} = \sum_{n=-\infty}^{\infty} F_n e^{in\psi}$$

$$q_k = \sum_{n=-\infty}^{\infty} q_{kn} e^{in\psi}$$

The solution for the harmonics of the bending motion is then

$$q_{kn} = \frac{\gamma F_n}{I_{q_k}^*(\nu_k^2 - n^2) - in\gamma M_{q_k \dot{q}_k}}$$

Generally, the dominant response of a bending mode is due to the harmonics of excitation near its natural frequency ν_k because of resonant amplification there. Exactly at a resonance, $\nu_k = n$, the response amplitude is determined by the damping alone. If such a resonance occurs it is necessary to take more care in evaluating the damping; structural damping should be included, and in certain flight conditions the returning shed wake (represented by Loewy's lift deficiency function) can significantly reduce the aerodynamic forces.

It follows from this solution that a fundamental design requirement for minimizing rotor blade loads is to avoid excessive resonant amplification of the modal response to the periodic aerodynamic exciting forces. Thus the natural frequencies of the blade bending and torsion modes must be kept away from harmonics of the rotor speed Ω.

Flax and Goland (1951) developed an approximate amplification factor method to estimate the bending moments on a blade from its static loads. The solution for the harmonics of the modal deflection can be written

$$q_{kn} = \left(\frac{\gamma F_n}{I_{q_k}^* \nu_k^2}\right)\left(\frac{1}{1 - n^2/\nu_k^2 - in\gamma M_{q_k \dot{q}_k}/I_{q_k}^* \nu_k^2}\right)$$

or in general $q_{kn} = (q_{kn})_{\text{static}} A_{kn}$. Here $(q_{kn})_{\text{static}}$ is the modal deflection calculated neglecting the inertia and damping of the mode, and A_{kn} is a dynamic amplification factor. The advantages of this formulation are that A_{kn} does not depend on the airloading of the blade, only on the modal properties; and calculating the static deflection is a standard structural analysis task. Flax and Goland suggest the following approximate method for calculating the blade loads. The static bending moment on the blade due to the aerodynamic forces is calculated by an accurate method, perhaps one that does not require the blade bending modes or frequencies. This static loading is harmonically analyzed, and the harmonics are multiplied by the amplification factor A_{kn} to account for the blade inertia and damping. Assuming that the blade response at a particular frequency is dominated by a single bending mode, the amplification factor A_{kn} is applied to the total loading instead of the loading due to the kth mode alone. The mode chosen for the amplification factor is the one with its natural frequency ν_k nearest the n/rev harmonic considered. This approximate method was found to give the magnitude of the bending moment harmonics very well, although the phase is not estimated as accurately.

It is not necessary to consider the blade loads in terms of the modal response. The solution for the mode shapes and frequencies may not be available, and better accuracy may be possible by calculating the bending moments directly from the aerodynamic loading. In section 9-2.2 the partial differential equation was derived for the out-of-plane bending deflection of a rotating blade:

$$\frac{d^2}{dr^2} EI \frac{d^2 z}{dr^2} - \frac{d}{dr}\left[\int_r^R m\Omega^2 \rho d\rho \frac{dz}{dr}\right] + m\ddot{z} = F_z$$

For a given loading F_z this differential equation can be integrated along the span. Integrating a fourth-order equation and then differentiating the deflection twice to obtain the moment is not a good approach for numerical work, however. It is preferable to work directly in terms of the bending moment. Equilibrium of forces outboard of r implies that

$$M(r) = \int_r^1 F_z(\rho - r)d\rho - \int_r^1 \left[\ddot{z}(\rho - r) + \rho\big(z(\rho) - z(r)\big)\right] m\, d\rho$$

where z is obtained from the integral of $M = EIz''$. If periodic loading is considered, the exciting force F_z and the blade response are expanded as Fourier series:

$$F_z = \sum_{n=-\infty}^{\infty} F_{z_n} e^{in\psi}$$

$$M = \sum_{n=-\infty}^{\infty} M_n e^{in\psi}$$

$$z = \sum_{n=-\infty}^{\infty} z_n e^{in\psi}$$

Then the equations for the nth harmonic of the bending moment are

$$M_n = \int_r^1 F_{z_n}(\rho - r)d\rho + \int_r^1 \left[n^2 z_n(\rho - r) - \rho\big(z_n(\rho) - z_n(r)\big)\right] m\, d\rho$$

$$z_n = z_n(1) - (1 - r)z_n'(1) + \int_r^1 \frac{M_n}{EI}(\rho - r)d\rho$$

These equations are numerically integrated, starting at the tip where the boundary conditions are $M_n(1) = M_n'(1) = 0$ (which are automatically satisfied by the equation for M_n). The values of $z_n(1)$ and $z_n'(1)$ must be chosen to satisfy the two boundary conditions at the root. These equations can be linear or nonlinear, depending on the aerodynamic model (F_z). The linear problem can be solved by superposition, while the nonlinear problem can be solved using some search algorithm. Note that the bending moment equation used here is the same as that given at the beginning of this section, but here it appears directly in terms of the deflection $z = \sum_k q_k \eta_k$ rather than the modal response. The present approach does not involve the approximation of truncating the modal representation of the deflection. Since the aerodynamic damping of the modes is important for the high frequency response, the lift due to $\dot{z}$ should be included in F_z.

Finally, let us examine an approximate method for calculating blade bending moments, using the airloading and motion obtained when rigid flapping alone is considered. Elastic bending of the blade significantly reduces the loads, however, and so must be accounted for. Consider the limit of a rigid blade. For an articulated rotor the blade motion then is just rigid flapping, $z = \beta r$, and the bending moment on the rigid blade is

$$M_R = \int_r^1 F_z(\rho - r)d\rho - (\ddot{\beta} + \beta)\int_r^1 \rho(\rho - r)m\,d\rho$$

which implies a radius of curvature

$$r_R = \frac{1}{d^2z/dr^2} = \frac{EI}{M_R}$$

In the limit of zero structural stiffness ($EI = 0$) the blade has only centrifugal stiffness, and the partial differential equation of bending reduces to

$$-\frac{d}{dr}\left[\int_r^1 \rho m d\rho \frac{dz}{dr}\right] = F_z - m\ddot{z}$$

or

$$-\frac{d}{dr}\left[\int_r^1 \rho m\, d\rho \frac{dz_e}{dr}\right] = F_z - (\ddot{\beta} + \beta)m r$$

where z_e is the blade elastic deflection ($z = \beta r + z_e$), and the inertial force $m\ddot{z}_e$ has been neglected. This equation integrates to

$$\int_r^1 \rho m \, d\rho \, \frac{dz_e}{dr} = \int_r^1 F_z \, dr - (\ddot{\beta} + \beta) \int_r^1 \rho m \, d\rho$$

or $T(dz_e/dr) = S_R$, where S_R is the vertical shear force calculated for the rigid blade, and T is the centrifugal tension force. The radius of curvature is then

$$r_F = \frac{1}{d^2 z/dr^2} = \frac{1}{d^2 z_e/dr^2} = \frac{1}{\dfrac{d}{dr}\dfrac{S_R}{T}}$$

Define M_F as the moment on the blade with stiffness EI, but with the curvature of the zero stiffness solutions: $M_F = EI/r_F$. Now construct a composite solution for the radius of curvature of the blade with stiffness EI, valid for both the limits EI approaching infinity and $EI = 0$:

$$r_C = r_R + r_F = \frac{EI}{M_R} + \frac{1}{\dfrac{d}{dr}\dfrac{S_R}{T}}$$

Then the bending moment on the actual blade is

$$M = \frac{EI}{r_C} = \frac{EI}{\dfrac{EI}{M_R} + \dfrac{1}{\dfrac{d}{dr}\dfrac{S_R}{T}}} = \frac{M_R M_F}{M_R + M_F}$$

This result can be written

$$M = M_R \left(\frac{1}{1 + \dfrac{M_R}{EI \dfrac{d}{dr}\dfrac{S_R}{T}}} \right)$$

which is a correction of the rigid blade moment for the effects of bending. Thus the bending moment on the blade can be obtained from the moment and shear force calculated considering rigid flap motion alone. This approximation is due to Cierva; Flax (1947) discusses its analytical basis. Duberg and Luecker (1945) found that the results of this method compare well with exact calculations.

12-5.3 Calculation of Vibration and Loads

The prediction of helicopter vibration and rotor loads is a difficult task, and is not entirely successful even with the most sophisticated analytical models currently available. Basically, it is necessary to calculate the periodic aerodynamic and inertial forces of the blade, and thus the resulting motion of the rotor and airframe. Since the higher harmonic blade loading is the principal source of high vibration and loads, an accurate analysis of the rotor aerodynamics is required, including the effects of the rotor wake, stall, and compressibility. The high frequencies involved and the importance of resonant excitation mean that good inertial and structural models are required as well. The calculation of helicopter aeroelastic response, which includes vibration and loads, is discussed further in Chapter 14.

12-5.4 Blade Frequencies

A basic requirement for minimum vibration and loads is that the natural frequencies of the blade bending and torsion modes avoid resonances with harmonics of the rotor speed. The rotating natural frequencies of the blade can be written as $\nu^2 = K_1 + K_2\Omega^2$, where K_1 and K_2 are the Southwell coefficients (this is the dimensional form; see section 9-8.3). Thus the blade frequencies vary with rotor speed as shown in Fig. 12-16. The n/rev lines are also shown in this figure. The design criterion is that any intersections of the blade frequencies and the n/rev lines should not occur within the normal rotor speed operating range. Typically it is necessary to consider resonances up to 4 or 5/rev for a single main rotor helicopter, and up to 6 or 8/rev for a tandem rotor helicopter (because of the greater aerodynamic interference). Except for the rigid modes of an articulated rotor, the structural stiffness dominates at low rotor speed, and therefore the natural frequencies cross the n/rev lines. Such resonances will produce load amplification during the rotor run-up, but are not a major problem as long as they occur at low rotor speed.

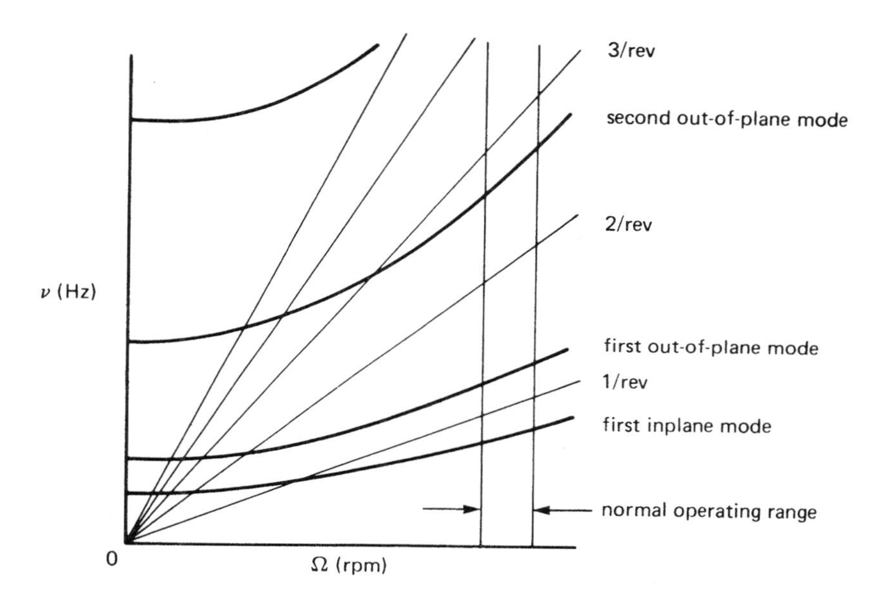

Figure 12-16 Blade frequency diagram (for a soft in-plane hingeless rotor in this case).

12-5.5 Literature

On helicopter vibration, and vibration reduction: Bailey (1940), Seibel (1944), Kelley (1945), Gillmore (1955), Payne (1955b), Miller and Ellis (1956), American Helicopter Society (1957), Morduchow and Muzyka (1958), Yeates (1958), Burkam, Capurso, and Yntema (1959), Ham and Zvara (1959), Laufer (1960), Ludi and Yeates (1960), Lynn (1960), Zvara and Ham (1960), Brooks and Silveira (1961), Capurso, Yntema, and Gabel (1961), Duvivier (1962), Ricks (1962), Silveira (1962), Smollen, Marshal, and Gabel (1962), Ellis and Jones (1963), Gabel (1963, 1975), Lockheed-California Company (1964a), Foulke (1965), Ham (1965), Brandt (1966), Flannelly (1966), Hooper (1966), Bain (1967), Kidd, Spivey, and Lawrence (1967), Pruyn, et al. (1967), Bies (1968), Fradenburgh and Chuga (1968), Jones (1968, 1970a, 1970b, 1971), Jones and Flannelly (1968), McIntyre (1968), Schuett (1968, 1969), Calcaterra and Schubert (1969), Fenaughty and Beno (1969), Ketchel, Danaher, and Morrissey (1969), Paul (1969b, 1969c), Malone, Schweichert, and Ketchel (1970), Balke (1971), Gabel,

Henderson, and Reed (1971), Piarulli (1971), von Hardenberg and Saltanis (1971), Allen and Calcaterra (1972), Griffin (1972), Hutchins (1972), Shipman, White, and Cronkhite (1972), Watts and Biggers (1972), Watts and London (1972), Beno (1973), Stave (1973), Amer and Neff (1974), Anderson and Gaidelis (1974), Anderson and Wood (1974), Bartlett and Flannelly (1974), Laing (1974), Taylor and Teare (1975), Weiss (1975), van der Harten (1976), Balmford (1977), Blackwell (1977), Griffin (1977), Jones and McGarvey (1977), Schrage and Peskar (1977), Bowes (1978), Desjardins and Hooper (1978a, 1978b), Needham and Banerjee (1978), Richardson and Alwang (1978), Rita, McGarvey, and Jones (1978), Twomey and Ham (1978).

On the structural dynamics of the helicopter airframes: Brooks (1956), Yeates (1956), Yeates, Brooks, and Houbolt (1957), Silveira and Brooks (1958), Ricks (1962), Flannelly, Berman, and Barnsby (1970), Flannelly, Berman, and Giansante (1972), Kenigsberg (1973), Flannelly and Giansante (1974), Lee and White (1974), Cronkhite (1976), Done and Hughes (1976), White (1976), Flannelly, Bartlett, and Forsberg (1977).

On helicopter vibration and loads reduction by higher harmonic control: Stewart (1952b), Payne (1958), Arcidiacono (1961), Wernicke and Drees (1963), Daughaday (1967, 1968), Shaw (1968), White and DuWaldt (1968), Balcerack and Erickson (1969), Balcerack, Erickson, and McGarvey (1969), McCloud (1972, 1975), London, Watts, and Sissingh (1973), McCloud and Kretz (1974), Piziali and Trenka (1974), Trenka (1975), Kretz (1976), Lemnios and Dunn (1976), McHugh and Shaw (1976, 1978), Schulz (1977), Weisbrich, Perley, and Howes (1977), Biggers and Mc Cloud (1978), McCloud and Weisbrich (1978).

Work concerned specifically with the fatigue life calculation and measurement of helicopter rotors includes: Winson (1948), Soulez-Lariviere (1961), Bott (1964), Rich, Israel, Kenigsberg, and Cook (1968), Murthy and Swartz (1972), Prinz (1972), Schumacher (1972), Ryan, Berens, Coy, and Roth (1975).

On helicopter rotor loads, particularly analysis methods: Klemin, Walling, and Wiesner (1942), Duberg and Luecker (1945), Dommasch (1946), Flax (1947), Horvay (1947), Yuan (1947), de Guilenschmidt (1951), Flax and Goland (1951), HIrsch (1953), Leone (1955), Berman (1956a, 1956b), Daughaday and Du Waldt (1956), Hirsch, Hutton, and Rasumoff (1956),

Mayo (1958), Rhodes and Gaidelis (1963), Landgrebe (1969a, 1971b), Kawakami (1977), Bielawa (1978), Mirandy (1978). See also the references in Chapter 14.

On helicopter rotor loads, including test data: Hufton, Woodward-Nutt, Bigg, and Beavan (1939), Meyer (1952), Jewel and Carpenter (1954), McCarty and Brooks (1955), Maglieri and Reisert (1955), Ludi (1958a, 1959, 1961), McCarty, Brooks and Maglieri (1959a), Mayo (1959), Head (1962), Rabbott (1962), Schelman and Ludi (1963), Wood, Hilzinger, and Buffalano (1963), Lockheed-California Company (1964a, 1964b), Scheiman (1964), Ward and Huston (1964), Wood and Buffalano (1964), Foulke (1965), Livingston (1965), DuWaldt and Statler (1966), Rabbott, Lizak, and Paglino (1966b), Ward (1966a, 1966b, 1969, 1971), Kidd, Spivey, and Lawrence (1967), Pruyn et al. (1967), Bartsch and Sweers (1968), Paglino and Logan (1968), Snyder (1968), Fenaughty and Beno (1969), Paul (1969c), Ward and Snyder (1969), La Forge and Rohtert (1970), Lee, Charles, and Kidd (1971), Watts and Biggers (1972), Watts and London (1972), Adams (1973), Anderson, Conner, and Kerr (1973), Beno (1973), Donham and Cardinale (1973), Taylor, Fries, and MacDonald (1973), Anderson and Gaidelis (1974), Lee and White (1974), Niebanck (1974), Tarzanin and Mirick (1974), Tarzanin and Ranieri (1974), Kerr (1975), Smith (1975), Huber and Strehlow (1976), Edwards and Miao (1977), Mantay, Shidler, and Campbell (1977), Shockey, Cox, and Williamson (1977), Vehlow (1977), Weller (1977), Weller and Lee (1977), Yasue (1977), Rabbott and Niebanck (1978).

Chapter 13

ROTARY WING AERODYNAMICS III

13–1 Rotor Vortex Wake

Associated with the lift of a rotor blade is a bound circulation. On the three-dimensional blade, conservation of vorticity requires that the bound circulation be trailed into the wake from the blade tip and root (see Fig. 13-1). Vorticity is also left in the rotor wake as a consequence of radial

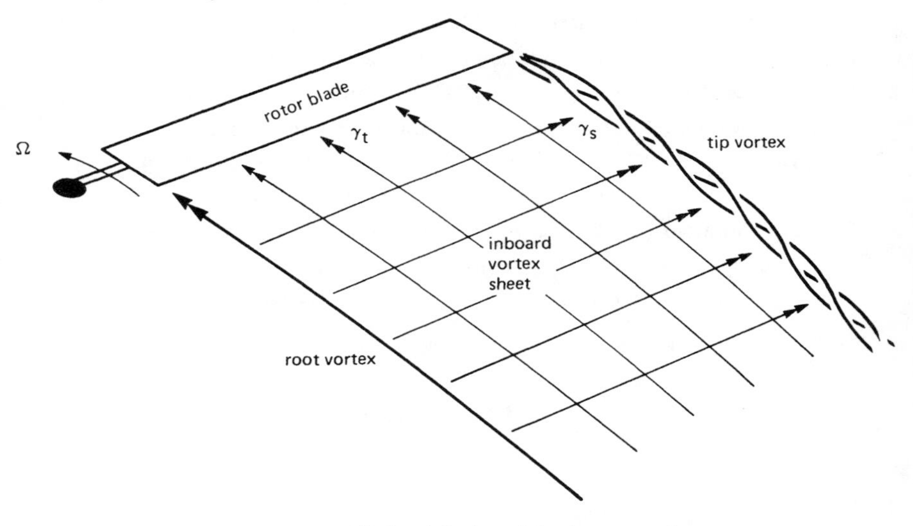

Figure 13-1 Trailed and shed vorticity in rotor wake.

and azimuthal changes in the bound circulation. The trailed vorticity γ_t, which is due to the radial variation of the bound circulation, is parallel to the local free stream at the instant it leaves the blade. The shed vorticity γ_s, due to the azimuthal variation of the bound circulation, is oriented radially in the wake. The strength of the rotor trailed and shed vorticity is

$$\gamma_t = \left.\frac{\partial \Gamma}{\partial r}\right|_{\psi-\phi}$$

$$\gamma_s = \left.-\frac{1}{u_T}\frac{\partial \Gamma}{\partial \psi}\right|_{\psi-\phi}$$

where the derivatives of the bound circulation are evaluated at the time the wake element left the blade, namely the current blade azimuth angle ψ less the wake age ϕ (in dimensionless time). Note that the existence of shed vorticity implies that the trailed vorticity strength varies along the length of the vortex filaments. Similarly, the trailed vorticity implies that the shed vorticity strength varies radially. Specifically,

$$-\frac{\partial}{\partial r} u_T \gamma_s = \frac{\partial}{\partial \psi} \gamma_t .$$

Because of the rotation of the blade, the lift and circulation are concentrated at the tip, as sketched in Fig. 13-2. Thus even in hover the ideal of constant

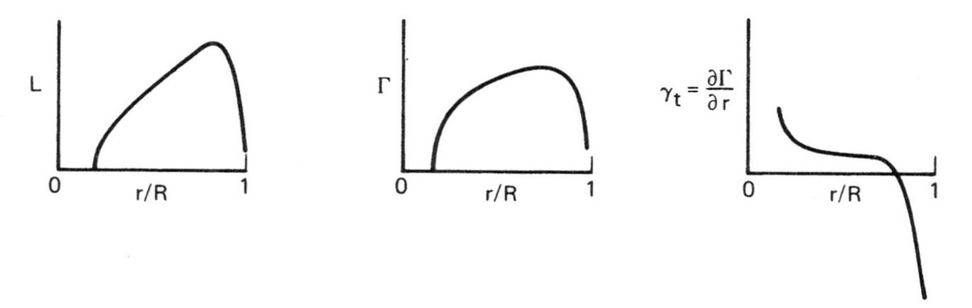

Figure 13-2 Sketch of the radial distribution of the blade lift L, bound circulation Γ, and trailed vorticity γ_t.

bound circulation is not achieved, but rather there is a peak in Γ typically at 80% to 95% radius. While the circulation drops to zero at the tip over a finite distance, the rate of decrease is still very high. The result is a large trailing vorticity strength at the outer edge of the wake, causing the vortex sheet to quickly roll up into a concentrated tip vortex. The formation of this tip vortex is also influenced by the blade tip geometry. With square tips most of the roll-up has occurred by the time the vortex leaves the trailing edge. The rolled-up tip vortex quickly reaches a strength nearly equal to the maximum bound circulation of the blade. The tip vortex has a small core radius that depends on the blade geometry and loading. Recall from section 10–8 that the vorticity is distributed over a small but finite region because of the viscosity of the fluid, rather than concentrated in an infinitesimal line. The vortex core radius is defined at the maximum tangential velocity.

On the inboard portion of the blade, the bound circulation drops off gradually to zero at the root. Hence there is an inboard sheet of trailed vorticity in the wake, with opposite sign to the tip vortex. Since the gradient of the bound circulation is low, the root vortex is generally much weaker and more diffuse than the tip vortex. When the bound circulation varies azimuthally, either as a result of the periodic loading in forward flight or because of transient loads, there is also an inboard sheet of shed vorticity in the wake.

The trailed and shed vorticity of the rotor wake is deposited in the flow field as the blades rotate, and then convected with the local velocity in the fluid. This local velocity consists of the free stream velocity and the wake self-induced velocity. The wake is transported downward, normal to the disk plane, by a combination of the mean wake induced velocity and the free stream velocity. (The component of the free stream velocity that is normal to the disk is a result of tilt of the disk plane in forward flight or axial velocity in vertical flight.) The wake is transported aft of the rotor disk by the in-plane component of the free stream velocity. The self-induced velocity of the wake produces substantial distortion of the vortex filaments as they are convected with the local flow. Thus the wake geometry basically consists of distorted interlocking helices, one behind each blade, skewed aft in forward flight. For further discussion of the vortex wake in hover and forward flight, see sections 2-7.1 and 4–2.

The strong concentrated tip vortices are by far the dominant feature of the rotor wake. Because of its rotation, a rotor blade encounters the tip vortex from the preceding blade in both hover and forward flight. When a vortex passes close to a blade it induces a large velocity and hence a large change in loading on the wing. Such vortex-induced loading is a principal source of the rotor higher harmonic airloading, which is an important factor in helicopter vibration, noise, loads, and even performance. The following sections of this chapter are concerned with the calculation of the nonuniform inflow due to the wake, the self-induced distortion of the wake, and the vortex-induced loads.

13–2 Nonuniform Inflow

The analyses developed so far in this text have generally assumed a uniform distribution of the wake-induced velocity over the rotor disk, or at most a simple linear variation. In fact, however, the induced velocity is highly nonuniform, since the requirements for uniform inflow (constant bound circulation and a very large number of blades) are far from satisfied with real rotors. The dominant factor in the rotor induced velocity is the discrete tip vortices trailed in helices from each blade. Because of the rotation of the wing, the wake is laid down in spirals close to the rotor disk, to be encountered again by the blades. In particular, a blade passes close to the tip vortex from the preceding blade, both in hover and forward flight. It was found in section 10-8.1 that a line vortex induces a tangential velocity that varies inversely with the distance from the vortex and has a maximum value at the vortex core radius. Thus the tip vortices in the wake induce a highly nonuniform flow field through which the blades must pass.

In hover, the tip vortex is convected downward only slightly until after it encounters the next blade, and there is little radial contraction. The vortex-blade encounter takes place near the tip, with the vortex a small distance from the blade. The vortex produces a large variation in the loading at the tip and has a substantial influence on the rotor hover performance (see the discussion in section 2-7.4 also). In forward flight the rotor wake is convected downstream, so the tip vortices are swept past the entire rotor disk instead of remaining in the tip region. The close vortex-blade encounters occur primarily on the advancing and retreating sides of the disk, where the blades sweep over the vortices. Thus in forward flight there is a large azimuthal variation of the induced velocity, which produces a large higher harmonic content of the loading. Nonuniform inflow thus is an important factor in helicopter vibration, loads, and noise during forward flight. Even the influence on the 1/rev loading, and hence on the cyclic control of the rotor, is large. The effects of nonuniform inflow in forward flight vary greatly with the flight condition, being largest in states such as transition, where the wake is closest to the rotor. In a tandem helicopter, the rear rotor also encounters the wake of the front rotor in forward flight, resulting in large vortex-induced loads on the forward portion of the rear rotor disk in particular. Fig. 13-3 is an example of the rotor blade loading in forward flight ($\mu = 0.15$ and $C_T/\sigma = 0.089$). The section lift at three radial stations is shown as a function of azimuth. The vortex-induced loading is apparent around $\psi = 90°$ and $270°$, particularly at the tip ($r/R = 0.95$).

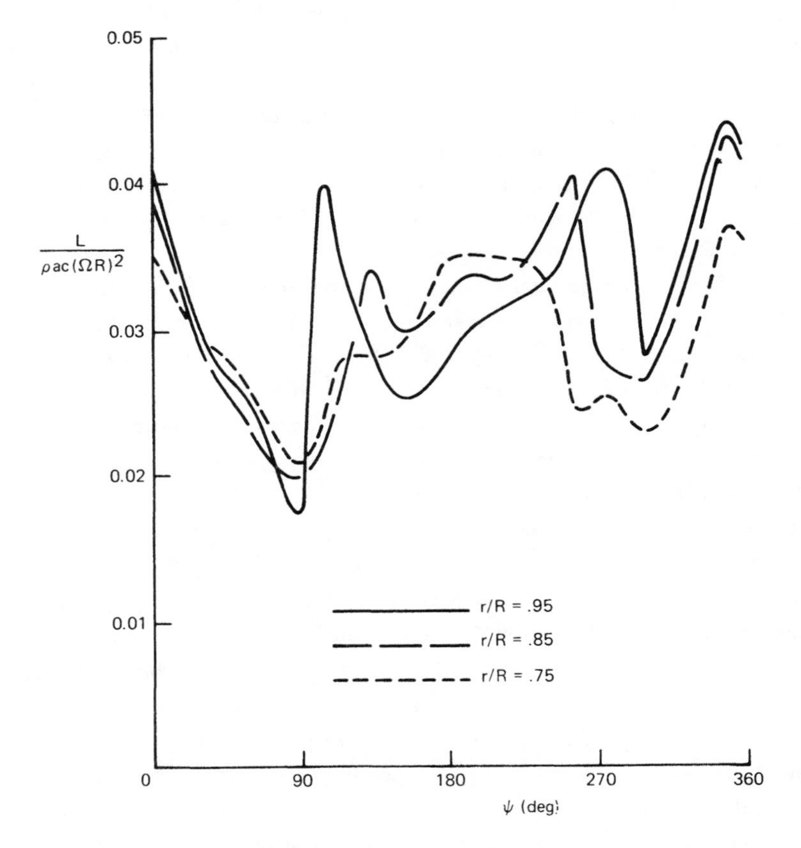

Figure 13-3 Rotor blade lift measured on a helicopter in flight at $\mu = 0.15$ and $C_T/\sigma = 0.089$, showing the wake influence; from Scheiman (1964), flight no. 7.

The accurate calculation of the wake-induced nonuniform velocity, and the resulting airloads and blade motion, is thus a prerequisite for the satisfactory prediction of rotor blade loads, rotor noise, helicopter vibration, and even the rotor performance and cyclic control. A complicated, detailed model of the rotor aerodynamics and dynamics is clearly required for such a calculation, and only a numerical solution is practical. The general problem of calculating the rotor aeroelastic behavior is discussed in Chapter 14; here we are concerned with just the problem of calculating the nonuniform inflow. Basically, the wake-induced velocity is obtained by integrating the Biot-Savart law over the vortex wake elements in the rotor wake. The wake strength is determined by the radial and azimuthal variation of the bound circulation. For the wake geometry, a simple assumed model, experimental measurements, or a calculated geometry can be used. Given the vorticity strength and geometry, the induced velocity can be evaluated; as an example, for line elements the Biot-Savart law is

$$\vec{u}(\vec{r}) = -\frac{1}{4\pi}\int \gamma(\vec{r}^{\,*})\,\frac{\vec{s} \times d\vec{\ell}\,(\vec{r}^{\,*})}{s^3}$$

where $\vec{s} = \vec{r} - \vec{r}^{\,*}$ is the distance between the vortex element $\gamma d\vec{\ell}$ and the point where the induced velocity $\vec{u}$ is required. With the helical geometry of the rotary-wing wake it is not possible to evaluate such integrals analytically, even if the self-induced distortion of the wake is neglected. Moreover, a direct numerical integration is not satisfactory in general, because the large variations of the induced velocity at close vortex-blade encounters would require a very small step size for accurate results. It is most accurate and most efficient to calculate the nonuniform inflow with the wake modeled using a set of discrete vortex elements. The vortex elements used are chosen such that the Biot-Savart law can be integrated analytically; a finite-length line segment or a rectangular vortex sheet (see sections 10-8.2 and 10-8.3) are typical choices. Then the induced velocity at any point in the flow field is obtained by summing the contributions from all the elements in the rotor wake. The vortex line segment is particularly important since the helical tip vortices can be modeled very well by a connected series of such straight-line elements. The approximations involved in such models of the rotor wake include replacement of the curvilinear geometry by a series of straight-line or planar segments; simplified distribution of vorticity over the discrete wake elements (usually constant, or linear at most); and perhaps physical approximations, such as the use of line elements to represent the inboard vortex sheet. The development of a practical model involves a balance between the accuracy and efficiency resulting from such approximations.

There is a tremendous amount of computation involved in calculating the rotor induced velocity using any realistic representation of the rotor wake. The induced velocity is typically required at 5 to 10 radial stations and 20 to 30 azimuthal stations, for a total of around 100 to 200 points on the rotor disk (for steady-state flight, so that the solution is periodic). Considering only the tip vortices, the contributions from roughly 300 to 800 elements in the wake must be evaluated to find the induced velocity at each point (for two to six blades; and two to six wake spirals, depending on the advance ratio). If the distorted wake geometry is to be calculated, the induced velocity is also required for a large number of points in the rotor wake. Hence routine calculation of the rotor nonuniform inflow became practical only with the development of high-speed digital computers. The application of such machines to helicopter airloads calculations was a significant step in the development of rotary wing analyses.

A number of models for the rotor vortex wake have been used. The strong, concentrated tip vortices are well represented by line elements with an appropriate viscous core radius (see section 10–8), and the helical geometry can be reasonably approximated using a connected series of finite-length straight-line segments (typically the azimuth increment required is 15° to 30°). Such a model for the rolled-up tip vortices is almost universally used, the primary differences being in the representation of the vortex core needed to avoid the velocity singularity near a line vortex. There is considerable variation in the models developed for the inboard shed and trailing vorticity, however. The influence of this vorticity is much less than that of the tip vortices, allowing greater latitude in selecting a model while retaining the required accuracy. A common wake model consists of a vortex lattice representation, so finite-length straight-line vortex elements are used for the inboard vortex sheets as well as for the tip vortices (see Fig. 13-4). Such a wake model corresponds to a variation of the blade bound circulation that is stepped both radially and azimuthally. The bound circulation is represented by a line vortex at the blade quarter chord. Typically 5 to 10 trailed vortex lines are used, normally concentrated at the tip where the greatest loading changes occur; the azimuthal increment for the shed wake elements is 15° to 30°. Replacing the inboard vortex sheets by line elements introduces singularities in the induced velocity near

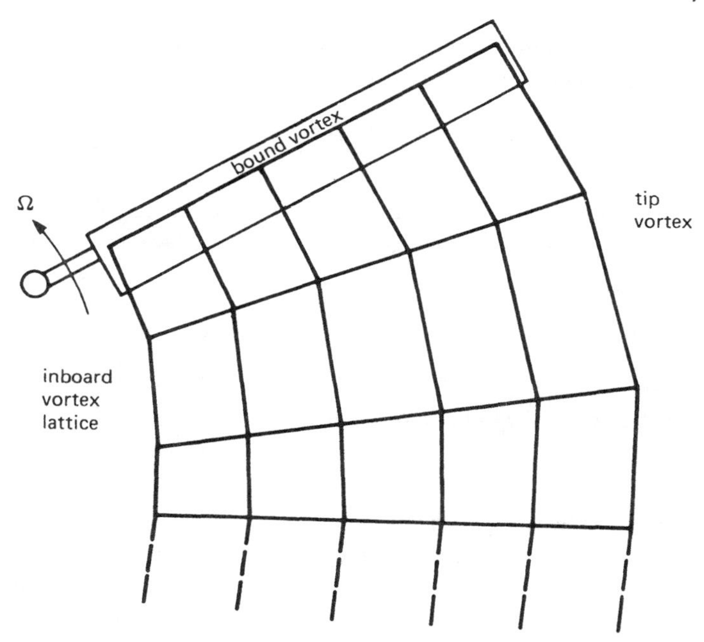

Figure 13-4 Vortex lattice model of the rotor wake.

the lines, which are not physically correct. The vortex lattice is the most efficient and flexible representation of a complicated wake structure, however, and with proper care to avoid these artificial singularities accurate results can be obtained. The induced velocity on the blade should be calculated at radial stations midway between the trailing vortex lines. When the blade passes over the inboard wake from preceding blades, a large "vortex core radius" can be used to improve the model. In this case the core radius is just a convenient mathematical means of limiting the induced velocity of the individual lines so that the induced velocity seen by the blade better approximates that of a vortex sheet. Typically a core radius of 0.3R is used, or a radius such that the cores of adjacent line elements touch. The tip vortex still will have a small core radius, since the large velocities induced near it are physically correct. The strength of a line element in the wake is determined from the blade bound circulation $\Gamma(\psi_i, r_j)$, which is evaluated at a finite number of azimuthal and radial points on the disk. Continuity of vorticity requires that the trailed vortex strength be given by the difference in bound circulation of adjacent radial stations, evaluated at the time the vortex element was left in the wake:

$$\gamma_t = \Gamma(\psi - \phi, r_{j+1}) - \Gamma(\psi - \phi, r_j)$$

Here ψ is the current azimuth angle (dimensionless time) and ϕ is the age of the vortex element. Similarly, the shed vorticity strength is obtained from the difference in bound circulation at successive azimuthal stations, which are $\Delta\psi$ apart:

$$\gamma_s = -\Gamma(\psi + \Delta\psi - \phi, r_j) + \Gamma(\psi - \phi, r_j)$$

The near shed wake usually requires some special treatment in order to obtain the effects of unsteady aerodynamics correctly. The models that can be used for the near shed wake in rotor induced velocity calculations have been discussed in section 10–3.

Sometimes when a vortice lattice model is used, the inboard shed and trailed vorticity is retained only for part of a revolution behind the blade (typically 45° to 90°); see Fig. 13-5), on the basis that the near wake and the tip vortices dominate the rotor induced velocity in forward flight. Such an approximation greatly reduces the number of wake elements, and hence it simplifies the computation required to evaluate the induced velocity. Also, the singularities when the blade passes over the returning inboard wake lattice are avoided. There are cases (such as the returning shed wake influence in hover) where such a model will not be correct. As a further simplification, the inboard shed and trailed vorticity can be neglected entirely (Fig. 13-6). Then the wake model consists of the trailed vortices alone. There are numerous variations on these models also, such as neglecting all the shed vorticity in the vortex lattice, but using trailed vortex lines from a number of radial stations (that is, retaining the azimuthal variation of the trailed vortex strength).

For a better model than the vortex lattice representation, rectangular vortex sheet elements can be used for the inboard shed and trailed vorticity (see Fig. 13-7). The tip vortices are still represented by line elements. Using vortex sheets for the inboard wake greatly improves the physical model of the wake, since it eliminates the singularity of the line element and since it corresponds to a linear variation of the bound circulation. The computational effort required to evaluate the induced velocity is significantly greater then when the vortex lattice model is used. The vortex sheets need

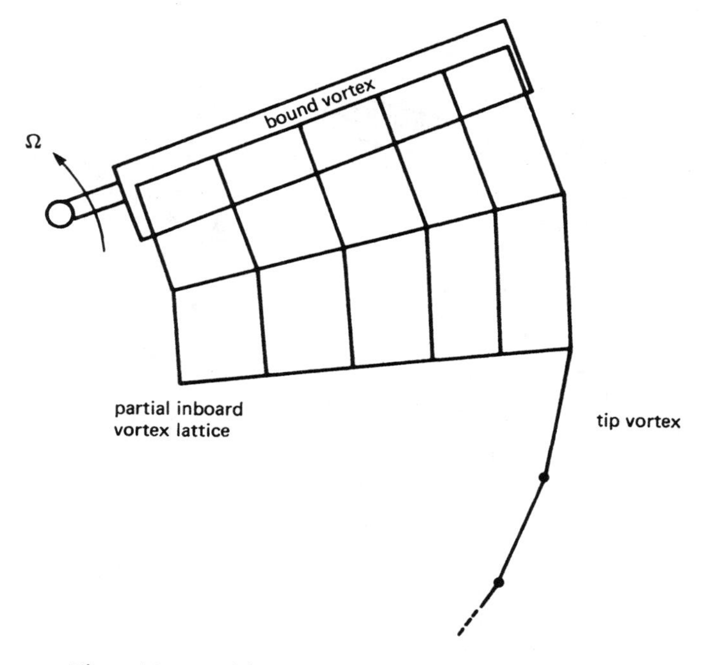

Figure 13-5 Partial vortex lattice model of the rotor wake

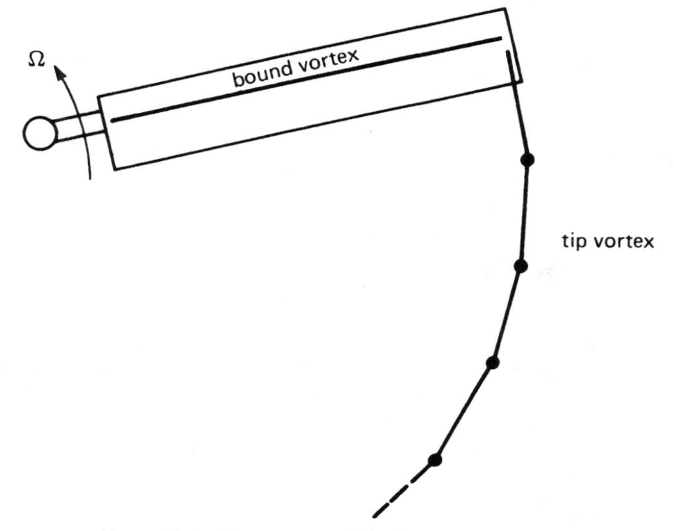

Figure 13-6 Tip vortex model of the rotor wake

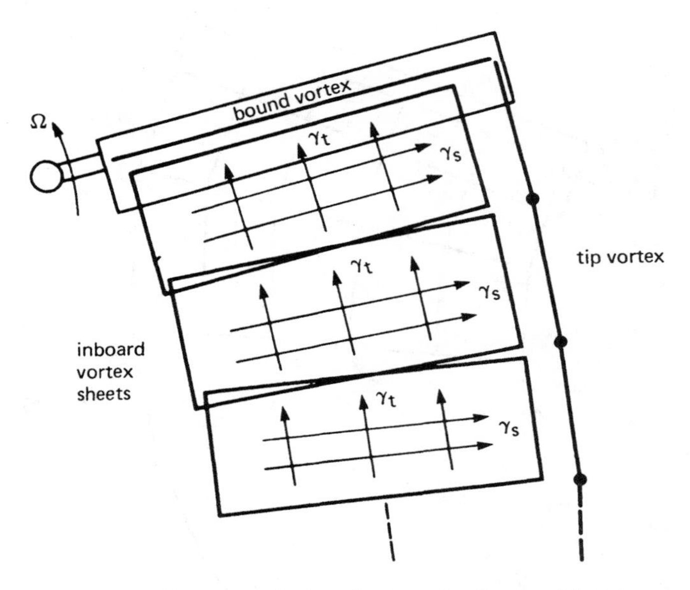

Figure 13-7 Line tip vortex and inboard rectangular sheet model of the rotor wake.

only be used for the inboard wake near the blade, however, so if the vortex lattice model is retained for the far wake the improvement in the induced velocity calculation can be obtained for a modest increase in computation.

Figs. 13-8 to 13-14 present the results of an airloads calculation including nonuniform inflow. An articulated rotor with solidity $\sigma = 0.1$, Lock number $\gamma = 8$, twist $\theta_{tw} = -8°$, and three blades is considered. The same example was examined in section 5–6, there with uniform inflow. The operating condition is again a blade loading of $C_T/\sigma = 0.12$ and a helicopter drag of $f/A = 0.015$. Here only a single speed is considered, $\mu = 0.25$. The induced velocity calculation is based on an undistorted wake geometry. Also, the blade motion calculation does not include elastic torsion or control system deflection, which would be large for the high loading considered here and would therefore significantly influence the loading distribution (see Chapter 16). Fig. 13-8 shows the tip-path-plane inflow ratio as a function of azimuth angle for a number of radial stations; Fig. 13-9 shows the distribution of λ_{TPP} over the rotor disk. For comparison, momentum theory gives a mean inflow ratio of $\lambda_{TPP} = 0.034$ (of which 0.024 is the induced velocity λ_i, and

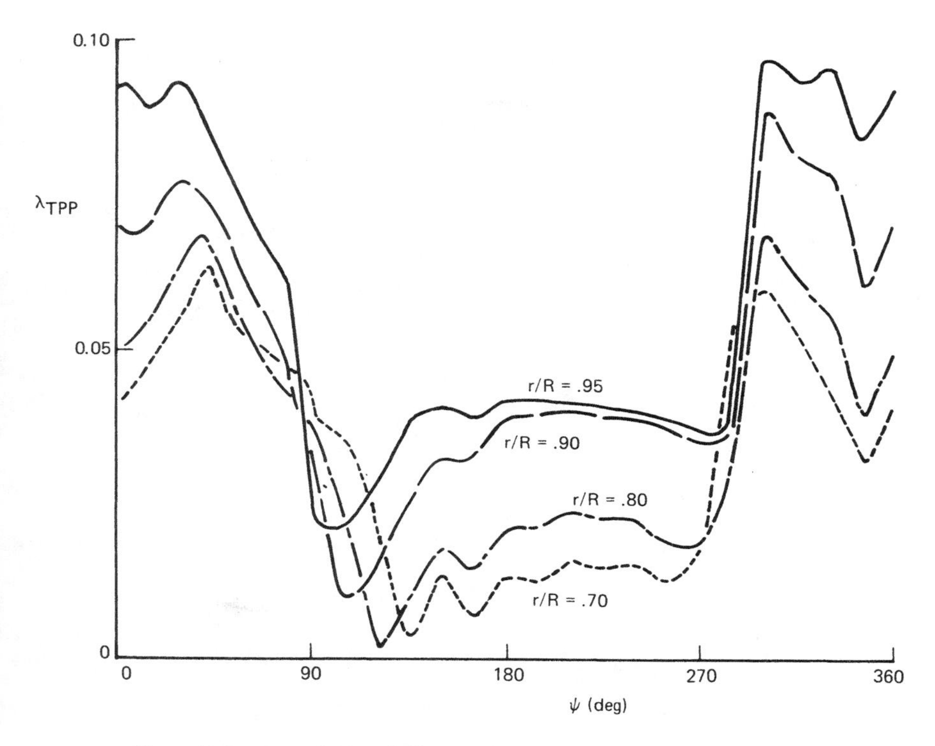

Figure 13-8 Tip-path plane inflow ratio at several radial stations for $\mu = 0.25$, $C_T/\sigma = 0.12$, $\theta_{tw} = -8^\circ$, and $f/A = 0.015$.

0.010 is the inflow due to the disk tilt, $\mu\alpha_{TPP}$). Generally, the nonuniform inflow is larger at the rear of the disk and smaller at the front. Sharp variations in the induced velocity occur around $\psi = 90^\circ$ and 270°, where the blade encounters the tip vortex from the preceding blade. The wake-induced velocity is indeed highly nonuniform. Fig. 13-10 shows the distribution over the rotor disk of the blade section angle of attack, which should be compared with Fig. 5-20 for uniform inflow. The nonuniform inflow substantially alters the angle-of-attack distribution, particularly on the advancing and retreating sides where the blade-vortex encounters occur. Note that the stall region on the retreating side of the disk is modified somewhat. The maximum angle of attack is significantly increased, and also the rate of change of α near stall is increased. Hence nonuniform inflow is

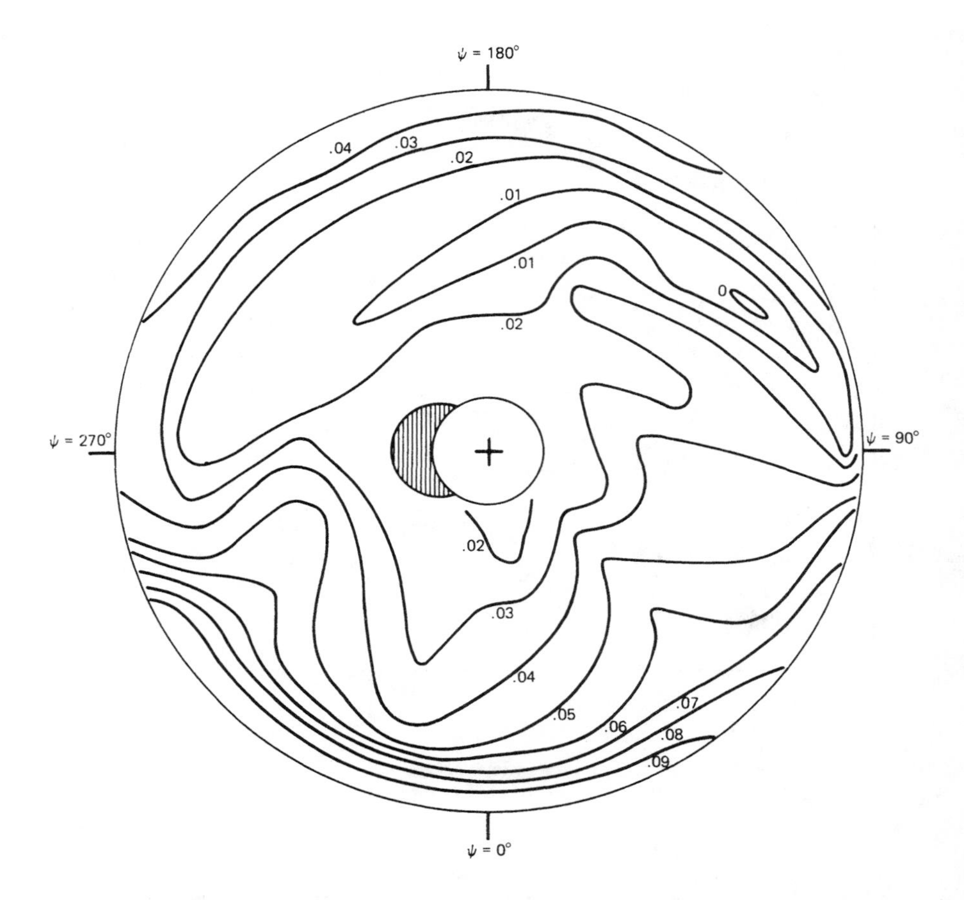

Figure 13-9 Distribution of the tip-path plane inflow ratio λ_{TPP} over the rotor disk at $\mu = 0.25$, for $C_T/\sigma = 0.12$, $\theta_{tw} = -8°$, and $f/A = 0.015$.

a significant factor in the rotor stall characteristics (see Chapter 16). Fig. 13-11 shows the angle-of-attack distribution for a rotor with untwisted blades, and Fig. 13-12 shows the distribution for the case of no propulsive force (compare with Figs. 5-21 and 5-22 respectively). When $\theta_{tw} = 0°$, there is a substantial shift of the high angle-of-attack region due to nonuniform inflow. With uniform inflow the maximum angle of attack is predicted to occur on the tip of the retreating blade ($\psi = 270°$ and $r/R = 1$; see Fig. 5-21); nonuniform inflow shifts the stall region inboard and

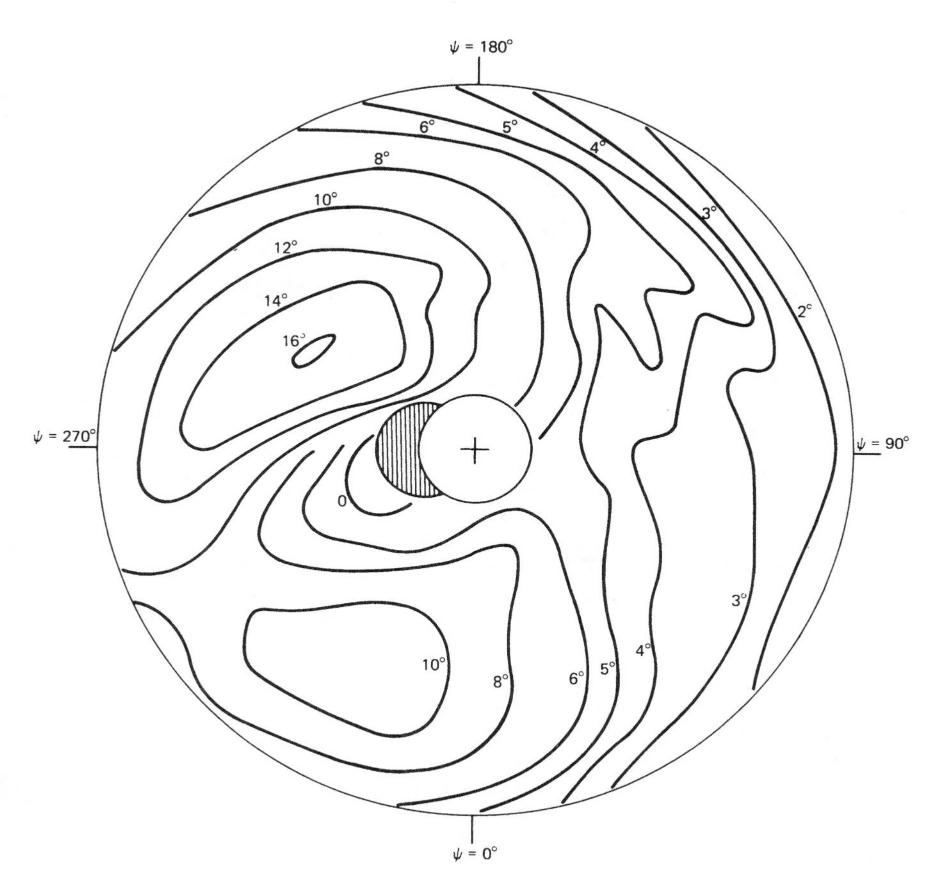

Figure 13-10 Blade angle-of-attack distribution (in degrees) at $\mu = 0.25$, for $C_T/\sigma = 0.12$, $f/A = 0.015$, and $\theta_{tw} = -8^\circ$ (nonuniform inflow).

into the third quadrant, so the maximum α occurs in this case at about $\psi = 255^\circ$ and $r/R = .65$. Note that with nonuniform inflow there is much less difference between the twisted and untwisted blades than with uniform inflow.Fig. 13-13 shows the distribution of the dimensionless blade section lift, for $\theta_{tw} = -8^\circ$ and $f/A = 0.015$. The corresponding results for uniform inflow were given in Fig. 5-24. Fig. 13-14 gives the variation of the section loading with azimuth at a number of radial stations (compare with Fig. 5-26).

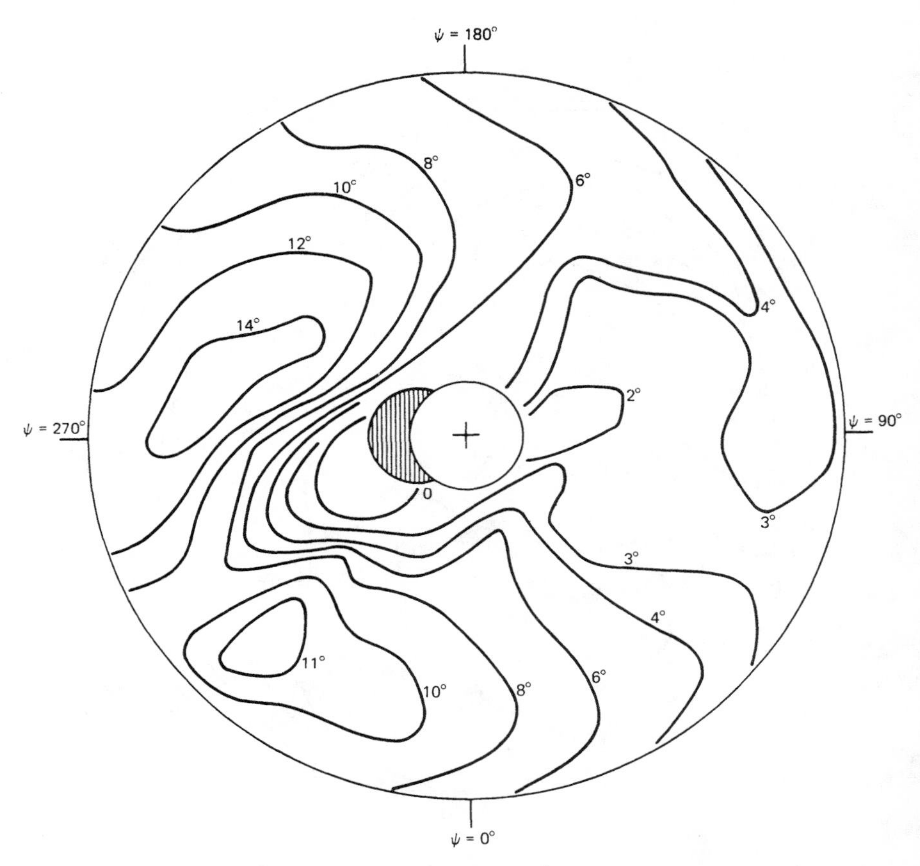

Figure 13-11 Blade angle-of-attack distribution (in degrees) at $\mu = 0.25$, for $C_T/\sigma = 0.12$, $f/A = 0.015$, and $\theta_{tw} = 0°$ (nonuniform inflow).

The influence of the tip vortices on the advancing and retreating sides of the disk is evident. There is also a significant effect of the wake-induced velocities on the blade motion, even on the mean and 1/rev flapping and hence on the collective and cyclic pitch control angles required for a given flight state. The flapping (relative to the no-feathering plane) and collective pitch angles calculated at $\mu = 0.25$ and $C_T/\sigma = 0.12$ with uniform and nonuniform inflow are tabulated below:

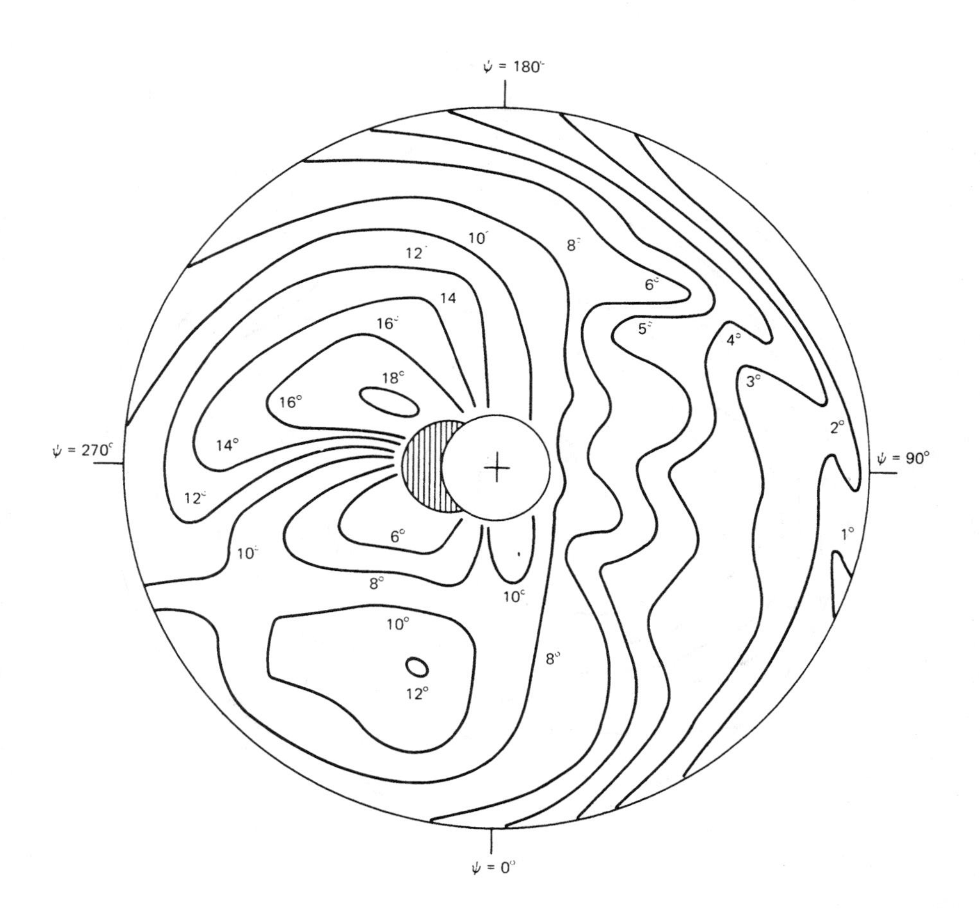

Figure 13-12 Blade angle-of-attack distribution (in degrees at $\mu = 0.25$, for $C_T/\sigma = 0.12$, $f/A = 0$, and $\theta_{tw} = -8°$ (nonuniform inflow).

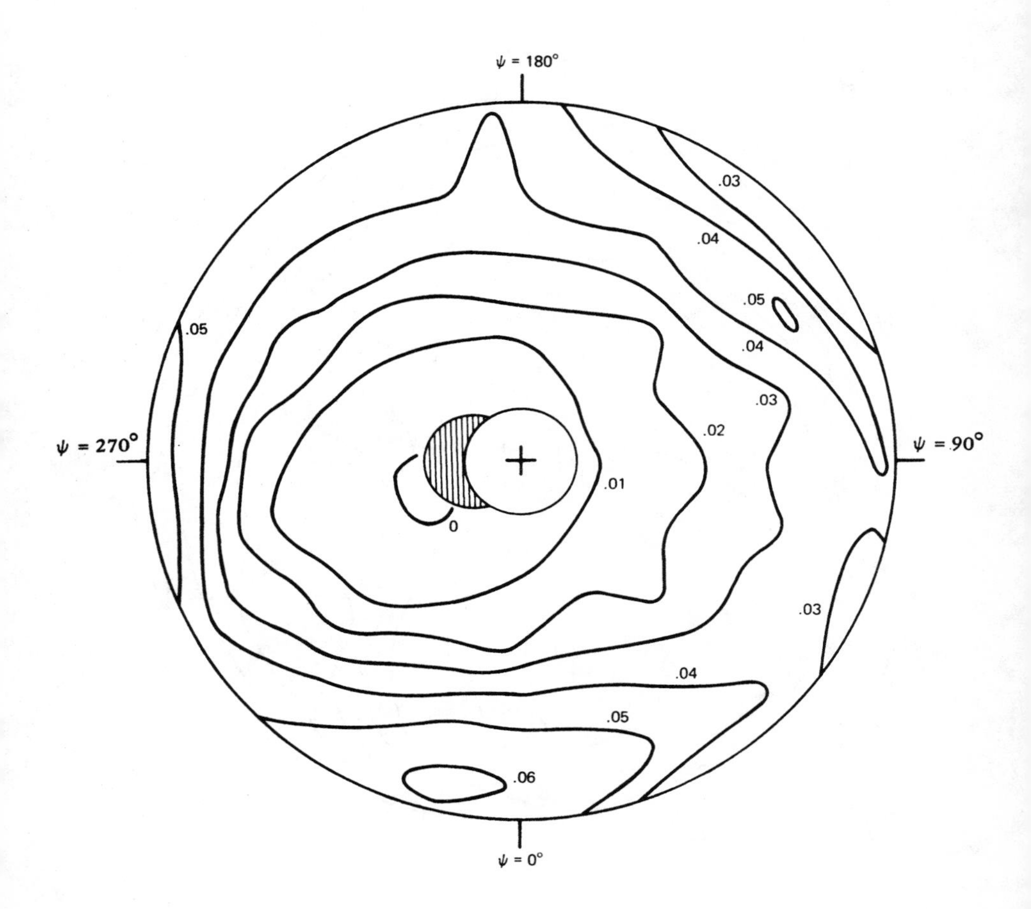

Figure 13-13 Blade section lift distribution $L/\rho ac(\Omega R)^2$ at $\mu = 0.25$, for $C_T/\sigma = 0.12$, $f/A = 0.015$, and $\theta_{tw} = -8°$ (nonuniform inflow).

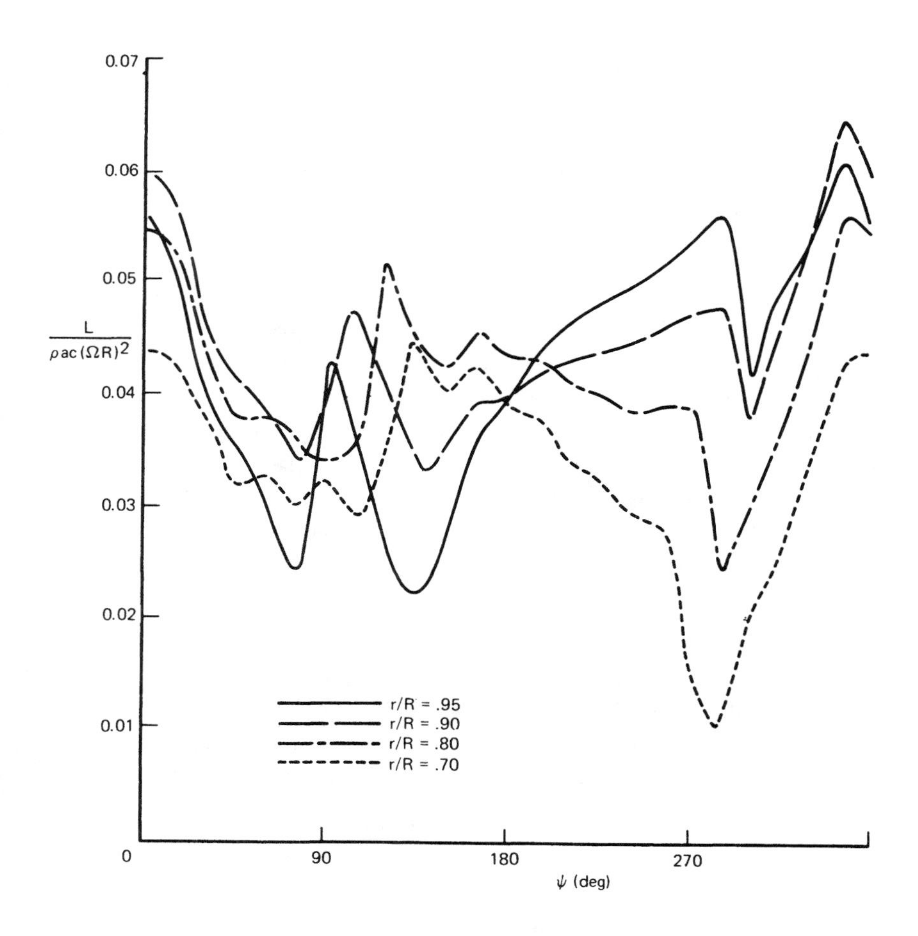

Figure 13-14 Azimuthal variation of the blade lift $L/\rho ac(\Omega R)^2$ at $\mu = 0.25$, for $C_T/\sigma = 0.12$, $f/A = 0.015$, and $\theta_{tw} = -8^\circ$ (nonuniform inflow).

	β_0	$-\beta_{1c}$	$-\beta_{1s}$	$\theta_{.75}$ (deg)
Uniform inflow results				
$\theta_{tw} = -8°$, $f/A = 0.015$	6.9	5.9	2.2	11.1
$\theta_{tw} = 0°$, $f/A = 0.015$	7.5	6.0	2.4	11.3
$\theta_{tw} = -8°$, $f/A = 0$	6.8	5.6	2.2	10.3
Nonuniform inflow results				
$\theta_{tw} = -8°$, $f/A = 0.015$	6.9	6.0	3.9	11.5
$\theta_{tw} = 0°$, $f/A = 0.015$	7.4	6.1	4.3	11.8
$\theta_{tw} = -8°$, $f/A = 0$	6.8	5.5	4.0	10.6

The principal influence of nonuniform inflow is to increase the lateral flapping and hence the lateral cyclic required to trim the rotor.

Miller (1962a, 1964b) developed a method for calculating the rotor nonuniform inflow and higher harmonic loads. With a uniform or linear inflow distribution, the predicted harmonic blade loading is of the order μ^n (where n is the harmonic number), in contrast to the large fifth or sixth harmonics that are measured in certain flight states such as transition or flare. This large harmonic loading is the source of the roughness and noise associated with such flight states, and is due primarily to the wake-induced velocities. Since Miller considered the case of steady-state flight only, the inflow, loads, and blade motion are all periodic and a Fourier analysis can be used. The blade aerodynamic forces were calculated using two-dimensional unsteady airfoil theory, and the downwash at the blade was obtained by integrating over the wake vorticity. The influence of the near shed wake (extending roughly 90° in azimuth behind the blade) was accounted for by using Theodorsen's lift deficiency function $C(k)$. For each harmonic the blade circulation due to the near shed wake is then just $(1 - C(k))$ times the total bound circulation. The higher harmonic inflow is almost entirely due to the far wake, particularly the returning spirals passing under the blades. For the far wake and the near trailing wake, the downwash is calculated by integrating the Biot-Savart law over the wake vorticity. The variation of the downwash over the blade chord is neglected. The radial integration over the wake helix can be performed analytically, and the wake strength can be expressed in terms of the bound circulation Γ. The result is that the wake-induced circulation is of the form $\int \Gamma f(\psi,\phi) d\phi$, where the integration is over the wake age ϕ (i.e. the helix azimuth angle). The Fourier series for Γ is now substituted, and the integrals over ϕ are evaluated numerically to obtain the coefficients of the harmonics of Γ. For the shed wake the integration starts about $\psi = 90°$ behind the blade (little sensitivity was found using $45°$ to $270°$), since the near shed wake has already been accounted for. Now the total bound circulation is $\Gamma = \Gamma_{wake} + \Gamma_{QS}$, where the wake-induced circulation has been evaluated as described above from the near and far wake, and the quasistatic circulation is given by the blade motion. The wake-induced circulation is in the form of a Fourier series, the coefficients a linear combination of the harmonics of Γ. The blade motion is also periodic. Setting to zero the $\sin n\psi$ and $\cos n\psi$ coefficients of the equation $\Gamma = \Gamma_{wake} + \Gamma_{QS}$ gives a set of linear equations relating the harmonics of Γ and the harmonics of the blade motion. The flap motion is forced by the blade lift, which two-dimensional unsteady airfoil theory relates to the circulation and blade motion. Thus the flapping equation of motion gives the harmonics of the flap response in terms of the harmonics of the circulation, completing the set of equations for the harmonics of Γ and β. An iterative procedure is proposed in which the downwash is calculated given the circulation, and then the harmonics of the blade motion and loads are calculated, including the new estimate of the circulation. The procedure is repeated until the harmonics converge to the steady-state solution. Since the higher harmonic downwash is determined mainly by the geometry of the wake, the iteration can be started using just the mean circulation for the required rotor thrust. Miller emphasizes that the harmonic airloads are sensitive to the vertical spacing of the wake, and points out the probable importance of a nonrigid wake geometry and lifting-surface theory for close tip vortex passages. He also introduces the concept of a "semirigid" wake, in which each vorticity element has a vertical velocity in the fluid equal to the inflow at the disk at the time it left the blade.

Miller (1962b) revised the treatment of the near shed wake in the nonuniform inflow calculation. The near shed wake was calculated only at a single point on the blade chord, namely at the bound vortex (see section 10–3). The unsteady aerodynamic effects are obtained correctly by extending the near shed wake to a quarter-chord behind this point. Thus in evaluating the induced velocity, the integral over ϕ extends from the bound vortex for the trailed wake, and from $\phi = c/4r$ behind the bound vortex for the shed wake (the r integration can still be performed analytically).

Miller (1964a) discussed the higher harmonic loading due to nonuniform inflow, concluding that the primary factor is the bound circulation associated with the mean blade lift acting together with the complicated wake geometry in forward flight to produce the higher harmonic downwash. This downwash is mainly due to the tip vortices. In addition, an nth harmonic of the downwash is induced by the wake vorticity due to the nth harmonic of the bound circulation, which is just the lift deficiency function effect. Miller found that in forward flight above $\mu = 0.2$ or 0.3, this latter component of the induced velocity is due to the near shed wake only (i.e. Theodorsen's lift deficiency function), but at lower advance ratio there is a significant influence of the returning shed wake as well. Miller again points out that the inflow calculation depends on the accuracy of the wake geometry. At transition speeds a nonrigid wake geometry (including the self-induced distortion) is probably required.

Scully (1965) modified the induced velocity calculation developed by Miller, replacing the numerical integration over the wake helices by a wake model consisting of a connected series of straight-line vortex elements. The numerical integration requires a very small integration step size for accuracy, while Scully found that vortex elements 30° or 40° in length were satisfactory. The result was a reduction in the computation required by about a factor of six.

Piziali and Du Waldt (1962) developed a method for calculating the wake-induced nonuniform inflow. Their wake model consisted of a vortex lattice with undistorted geometry. At this stage they used measured blade motion, calculating only the airloads. The only unknown quantities then were the bound circulation at a finite set of radial and azimuthal points on the rotor disk. Airfoil theory gives the bound circulation in terms of the angle of attack, and therefore in terms of the blade motion and induced velocity. The induced velocity is given by the Biot-Savart law in terms of the strength of the vortex elements, and hence the bound circulation. Thus the problem can be formulated as a set of simultaneous linear algebraic equations for the bound circulation at points distributed over the rotor disk. This set of equations is of very large order, since typically the loading must be calculated at 100 to 200 points on the disk.

Piziali, Daughaday, and Du Waldt (1963) developed further this procedure for calculating the nonuniform inflow and higher harmonic airloads. Their solution included the blade aeroelastic response, for example the flap motion $\beta(\psi)$, which must be found at the same time as the bound circulation. The blade equations of motion involve acceleration and velocity terms. Since the blade motion is periodic, however, these time derivatives can be expressed in terms of the Fourier coefficients of the displacement. Moreover, these harmonics can be obtained from the displacement at a finite number of points around the azimuth. Consequently, the blade equations of motion reduce to a set of linear algebraic equations for the displacement at a finite number of azimuthal points. Since the algebraic equations for the bound circulation and blade motion are coupled, a simultaneous solution for $\Gamma(r_j,\psi_i)$ and $\beta(\psi_i)$ is required. The importance of the wake geometry to an accurate airloads calculation is emphasized, as is the difficulty in calculating it. They discuss an extension of the blade model to include unsteady airfoil theory, including the variation of the induced velocity over the chord and the near shed wake effects. They point out that the shed wake probably requires more care in its treatment.

Piziali (1966a, 1966b) developed a solution for the rotor airloads and aeroelastic response, including the wake-induced velocity and blade unsteady aerodynamics. He considered a prescribed wake geometry: undistorted (all elements convected with the same velocity), semirigid (all wake elements having the velocity of the point on the disk they came from), or semirigid for the first part of a helix and then undistorted. The wake model consisted of a vortex lattice for the first part of the helix, extending about 45° behind the blade; and then just the tip and root trailed vortices. (It was found that the root vortex has little influence on the loading.) The treatment of the near shed wake was based on a comparison of two-dimensional unsteady airfoil theory results with continuous and discrete models (see section 10–3). To correctly obtain the unsteady aerodynamic effects, the entire shed wake was advanced by $0.7\Delta\psi$, so that the first element was $0.3\Delta\psi$ behind the trailing edge. Two-dimensional airfoil theory was used to obtain the blade section circulation, lift, and moment from the blade motion and the induced velocity over the blade chord. The Biot-Savart law gives the induced velocity in terms of the bound circulation at discrete points on the rotor disk $\Gamma_j = \Gamma(r,\psi)$, in the form $w_k(x) = \sum_j C_{kj}(x)\Gamma_j$, where C_{kj} is the contribution of Γ_j to the induced velocity w_k at the kth point on the rotor disk. The induced velocity $w_k(x)$ and therefore the coefficients $C_{kj}(x)$ can be written as a cosine series in $\theta = \cos^{-1}x/b$. Two-dimensional airfoil theory relates the Glauert coefficients γ_n of the blade bound circulation distribution $\gamma_b(x)$ to the blade motion and induced velocity (see section 10–2); hence γ_n can be written as a linear combination of the bound circulation Γ_j. By then writing the bound circulation in terms of the Glauert coefficients γ_n, a set of linear algebraic equations for Γ_j is obtained, which have nonhomogeneous terms that are determined by the blade motion and the operating state. Blade stall is included in this analysis by limiting the bound circulation to the value at the stall angle of attack. Piziali considered a modal representation of the blade out-of-plane motion, with the degrees of freedom excited by integrals of the loading over the span. The harmonics of the airloads then gave the harmonics of the blade motion. An iterative solution for the circulation and blade motion was used. The algebraic equations for the bound circulation are solved using the current estimate of the blade motion. (Note that the coefficients of the equations for Γ_j need only be calculated once, since they are independent of the blade motion and a prescribed wake geometry is used.) Then the Glauert coefficients for $\gamma_b(x)$ are evaluated using the solution for the induced velocity, and from γ_n the section lift and moment are evaluated. Next the blade motion is obtained from this aerodynamic loading. The procedure is repeated until the solution converges.

Segel (1966) extended the analysis of Piziali and Du Waldt to the case of transient rather than periodic loading, such as loading due to transient control inputs, Balcerak (1967) extended Piziali's analysis to the case of a tandem rotor helicopter. Chang (1967) extended Piziali's analysis to include blade in-plane and torsional motions in the solution. The correlation between predicted and measured blade airloads and bending moments was not substantially improved, however. It was concluded that further improvement in the predictive capability will require a better wake model, particularly regarding the distorted wake geometry, and perhaps also a better representation of the blade motion.

Davenport (1964) developed a method for calculating nonuniform inflow, applicable to single and tandem rotors. The wake model consisted of a number of finite-strength trailed vortices, each composed of a connected series of straight-line segments. The shed wake was neglected entirely. A rigid, undistorted wake geometry was used. Davenport (1965) found a significant influence of nonuniform inflow on rotor performance predictions, due to the radical change in the blade angle-of-attack distribution over the rotor disk.

Scully (1975) developed a method for calculating the nonuniform inflow and harmonic airloading of a helicopter rotor. His analysis also included a free wake geometry calculation, discussed in the next section. In the wake model, the tip vortices were represented by a series of straight-line segments. Directly behind the blade tip, a vortex sheet was also used to simulate the wake before the roll-up is complete. The width of the sheet and distribution of vorticity between the sheet and the line vortex at the tip depends on the blade circulation distribution. The sheet strength is decreased linearly to zero (so all the vorticity is in the line elements) at a point about 75° behind the blade. A vortex core radius of 0.0025R was used for the tip vortices. The inboard wake was found to have a small, but not negligible, effect on the loads. The inboard trailed wake was represented by a single vortex line typically at $r/R = 0.475$, with a large vortex core (the core radius was one-half the width of the inboard trailed wake, typically 0.35R). The shed wake was represented by radial vortex lines with a large core radius (one-half the distance between adjacent shed lines, roughly $0.4R\Delta\psi$; $\Delta\psi$ is the azimuth increment, so $R\Delta\psi$ is the spacing at the tip). An extensive investigation of the inboard trailed and shed wake representation was conducted, including the use of vortex sheets or a large number of trailers. It was concluded that this model using line vortex elements, with a large core to better simulate the sheet vorticity, is the best in terms of both accuracy and economy. For the near shed wake a rectangular vortex sheet was used, extended to a quarter chord behind the bound vortex to correctly obtain the unsteady aerodynamic effects of the wake (see section 10–3). For close vortex-blade encounters, the lifting surface theory developed by Johnson (1971a, 1971b) was used to calculate the vortex-induced loads. Scully's wake model also included the effect of "bursting" of the tip vortex core induced by a blade-vortex interaction. A larger core radius (0.1R) was used for a tip vortex element after it first encountered a blade. Only the rigid flap and first flapwise bending motions of the blade were considered, calculated by numerically integrating the modal equations of motion.

The wake-induced velocity $\lambda_j = \lambda(r,\psi)$ at the jth point on the rotor disk is the sum of contributions from all elements in the wake. Since the strength of each wake element is determined by the blade bound circulation $\Gamma_i = \Gamma(\psi)$ at the ith azimuth position, the induced velocity can be written as a summation over the bound circulation: $\lambda_j = \sum_i C_{ji}\Gamma_i$. The coefficient matrix C_{ji} depends only on the wake geometry, so the calculation procedure begins with the evaluation of this matrix. Then an iterative calculation of the airloading proceeds as follows. With an initial estimate of the bound circulation, the nonuniform inflow distribution over the rotor disk is calculated. Then the angle of attack, aerodynamic loading, and blade motion are calculated. This procedure is repeated using the new circulation estimate until the solution converges. Usually 5 iterations are sufficient when the inflow and loads are calculated at 24 azimuthal and 6 radial stations. When the free wake calculation was included, the airloads were first obtained using a rigid wake model. Next, using this estimate of the bound circulation, the distorted geometry of the tip vortices was calculated. Finally, the induced velocity and aerodynamic loading were recalculated using the distorted wake geometry. Since the free wake geometry calculation is not very sensitive to changes in the bound circulation, further iterations are not usually required. An examination of measured rotor airloads indicated that the vortex-induced loads on the advancing side are generally high when the blade first encounters the vortex from the preceding blade, but decrease at larger ψ as the blade sweeps over the vortex. There is evidently some phenomenon limiting the loads, as observed by Johnson (1970c). It is indeed likely that a close blade-vortex encounter will influence the vortex properties. One possibility is that bursting of the vortex core is induced by the blade. Another possibility is that the vortex interacts with the trailed wake it induces behind the blade, with the effect of diffusing the circulation in the vortex. Local-flow separation due to the high radial pressure gradients on the blade could also be responsible for limiting the vortex-induced loads. Scully modeled such effects by increasing the core radius when a vortex element encountered a blade and including upstream propagation of this "bursting". Such blade-induced bursting, with upstream propagation, was found to be essential for an adequate prediction of the rotor aerodynamic loads. Lifting-surface theory substantially reduced the predicted vortex-induced loads, but was not sufficient alone to account for the behavior of the measured loads. The exact physical mechanism involved in this phenomenon remains speculative; the increase in vortex core radius is simply a convenient way to reduce the influence of the vortex. Scully concluded that the wake geometry, and also lifting-surface theory and the details of the wake model, are important factors in the calculation of helicopter nonuniform inflow and airloads. The loads induced on the blade during close encounters with the tip vortices in the wake are the primary concern. The current understanding of the aerodynamic phenomena involved in the interaction of a tip vortex and a rotating wing is far from complete, however. The inability to completely model such phenomena limits the accuracy of present predictions of helicopter airloads.

Other work on helicopter nonuniform inflow calculation includes: Ham (1963), Ghareeb (1964), Miller (1964c), Carlson and Hilzinger (1965), Segel (1965), Harrison and Ollerhead (1966), White (1966), Madden (1967), Clarke and Bramwell (1968), Jenney, Olson, and Landgrebe (1968), Clark and Leiper (1970), Dat (1970), Erickson and Hough (1970), Tung and Du Waldt (1970), Woodley (1971), Isay (1972a, 1972b), Johnson and Scully (1972), Sadler (1972b), Costes (1973).

13–3 Wake Geometry

In the close encounters of the rotating helicopter rotor blade with the wake from preceding blades, the induced loads are sensitive to the relative position of the tip vortex and blade. Thus the geometry of the rotor vortex wake is a major factor in determining the nonuniform inflow and aerodynamic loading. As the blade rotates through the air, trailed and shed vorticity is left in the wake. After each vortex element is created, it is convected with the local velocity of the flow field, which consists of the forward or climb speed of the helicopter, plus a velocity component induced by the wake itself. If uniform induced velocity is assumed, together with the rotation of the wing it creates the basic helical geometry of the wake. However, as at the rotor disk, the induced velocity throughout the wake is highly nonuniform. The actual position of the vortex elements, determined by the integral of the local convection velocity, is highly distorted from the basic helical form. Overall, the rotor wake tends to roll up so that far downstream it is similar to the two tip vortices of a circular wing. It is the detailed geometry near the rotor disk that is the most important for the blade loads, however, particularly the position of the tip vortices when they are first encountered by the following blades. The effects of a vortex-blade encounter are not limited to the loading induced on the blade. There are also influences of the blade on the vortex, including a strong contribution of the bound circulation to the convection velocity of the vortex, a local distortion of the geometry, and perhaps a blade-induced bursting phenomenon of some sort. Another factor in the wake geometry is the stability of the vortex elements. A vortex line is susceptible to short wavelength instabilities, and a helical vortex also exhibits long wavelength instabilities due to the interaction of successive turns of the helix. Such instabilities are generally of secondary importance for the loading, though, since they occur in the wake some distance from the rotor. It must also be recognized that the concept of a unique geometry is an idealization, since in the real flow, turbulence and vortex instabilities can produce significant variations in the wake geometry with time, even for a nominally steady operating state.

From the dominant role of the tip vortices in determining the inflow and loading, it follows that their position is the most important part of the wake geometry. Because the inboard trailed and shed vorticity has less influence on the rotor, a knowledge of its position less accurate than for the tip vortices is generally acceptable. Frequently, therefore, it is only the tip vortex position that is specified in a calculated or measured wake geometry. With the tip vortices modeled by a connected series of straight-line segments, the location of all the node points where two segments join is sufficient to define the geometry. The wake geometry is required for each azimuth position of the blade for which the induced velocity is to be calculated.

The wake geometry is most conveniently described in a nonrotating, tip-path-plane axis system. Relative to the tip-path plane there is no 1/rev flapping of the rotor to consider, and it is the tip-path-plane orientation that is determined by the helicopter operating condition. Consider the position of a wake element of age ϕ (see Fig. 13-15). The blade is now at azimuth angle ψ (which is also the dimensionless time variable). Since the age of the wake element is ϕ, it was created when the blade was at azimuth $\psi - \phi$, and the position of the blade at that time was $x = r\cos(\psi - \phi)$, $y = r\sin(\psi - \phi)$, and $z = r\beta_0$ (where r is the radial station considered, and

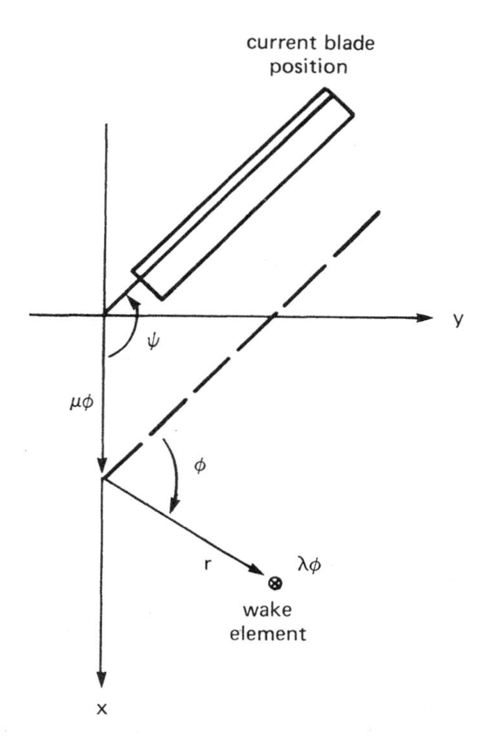

Figure 13-15 Description of rotor wake geometry.

β_0 is the coning angle). Since it left the blade, the wake element has been convected with the local flow. Consider just the mean convection velocity, which has components λ and μ in the tip-path plane. The inflow ratio λ includes the mean induced velocity. Then the current position of the wake element is

$$x = r\cos(\psi - \phi) + \mu\phi$$

$$y = r\sin(\psi - \phi)$$

$$z = r\beta_0 - \lambda\phi$$

which defines the basic skewed helix of the rotor wake in forward flight. Actually, there are N such helices, one behind each blade, obtained by substituting for ψ the azimuth angle of the mth blade: $\psi_m = \psi + m\Delta\psi$

($m = 1$ to N, $\Delta\psi = 2\pi/N$). When the nonuniform induced velocity is included in the convection of the wake element, the vertical position becomes

$$z = r\beta_0 - \int_0^{\phi} \lambda d\phi$$

which defines the distorted wake geometry (the distortion in the x and y directions is required as well).

There are a number of wake geometries used in rotary wing analyses. The rigid wake model is the undisturbed helical geometry, in which all the wake elements are convected with the same mean velocity. An elementary extension is the semirigid wake model, in which each element is convected downward with the induced velocity of the point on the rotor disk where it was created. It is probably best to use the mean induced velocity for the convection after a vortex element encounters the following blade (roughly, for ages $\phi > 2\pi/N$). The free wake or nonrigid wake model includes the distortion from the basic helix, as each wake element is convected with the local flow, including the velocity induced by the wake itself. The distorted geometry may be calculated, or it may be measured experimentally. When measured wake geometry information is used, it is often called a prescribed wake model.

The rigid wake model is the simplest, requiring a negligible amount of computation. It is also the farthest from the true rotor wake geometry, which can involve significant distortion. If the flight condition is such that the wake is convected away from the disk (large tip-path-plane incidence α_{TPP} at high speed, or high climb rates), and hence there is no significant vortex-blade interaction, then a rigid wake geometry is a satisfactory model. The semirigid wake geometry does not require additional computation, since it uses only the nonuniform inflow at the rotor disk. The assumption that the wake elements are convected with the velocity at the disk should be good for small age, but not very accurate by the time the vortex encounters the following blade. Thus the semirigid wake model generally offers little improvement over the rigid wake model. When the helicopter operating state is such that the wake remains close to the rotor disk, the distortion of the wake geometry can have a large effect on the loading, and the free wake model may be required. Calculating the distorted wake geometry requires evaluating the induced velocity throughout the wake rather than just at the rotor disk, and therefore it involves a very large amount of computation. The use of a prescribed wake model is limited by the necessity of performing measurements for the required rotor and flight condition. The choice of the wake geometry model is usually a balance between accuracy and economy. For many problems an economical free wake calculation is not presently possible, so a rigid wake model is used. Moreover, the increased accuracy possible with a distorted wake model will be wasted unless consistent advances are also made in the other parts of the analysis.

The major problem in a free wake analysis is developing an efficient yet accurate procedure for performing the calculations. Conceptually, the free wake calculation is simple. At each time step, the induced velocity is evaluated at every element in the wake, by summing the contributions from all elements in the wake as in calculating the nonuniform inflow at the rotor disk. Then the convection velocities are numerically integrated to obtain the positions of the wake elements at the next time step. In the extreme case, the rotor is started impulsively from rest (no wake at all) and the integration continues until the steady state geometry is obtained. Such a direct approach is sometimes used, but it is inefficient, requiring orders of magnitude more computation than just finding the nonuniform inflow at the disk. It is important to develop special procedures for economic calculation of the distorted wake, in particular taking advantage of the periodicity of the solution in steady forward flight.

Figs. 13-16 to 13-19 present a comparison of the rotor aerodynamic loads obtained using a rigid wake geometry and a calculated free wake geometry. An articulated rotor with solidity $\sigma = 0.07$, Lock number $\gamma = 8$, and three blades is considered. The flight state involves low speed, $\mu = 0.1$; moderate thrust, $C_T/\sigma = 0.09$; and a small tip-path-plane incidence, $\alpha_{TPP} = 0.6°$. The wake therefore remains close to the rotor disk. The results of using lifting-line theory or lifting-surface theory for the vortex-induced loads will also be compared. The lifting-surface theory used was that developed by Johnson (1971a, 1971b), which is also discussed in section 13–4. Fig. 13-16 shows the blade loading calculated at several radial stations using a rigid wake geometry and lifting-line theory. Fig. 13-17 shows the blade loading obtained using a calculated free wake geometry. For the rigid wake model, the tip vortices are always convected downward, and

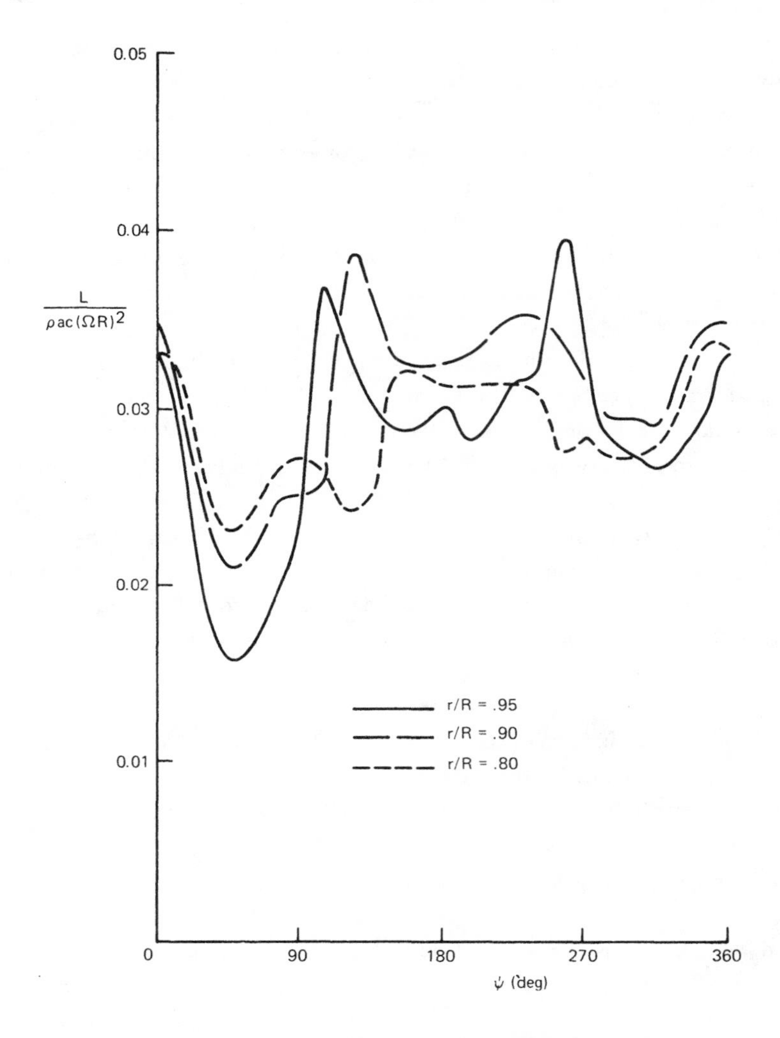

Figure 13-16 Blade section lift calculated using a rigid wake model, and lifting-line theory for the vortex-induced loads.

hence are a considerable distance below the disk plane by the time they reach the following blade. For the free wake model, however, the tip vortices are much closer to the disk plane when they encounter the blade. In fact, for this case the free wake calculation predicts an extremely close encounter

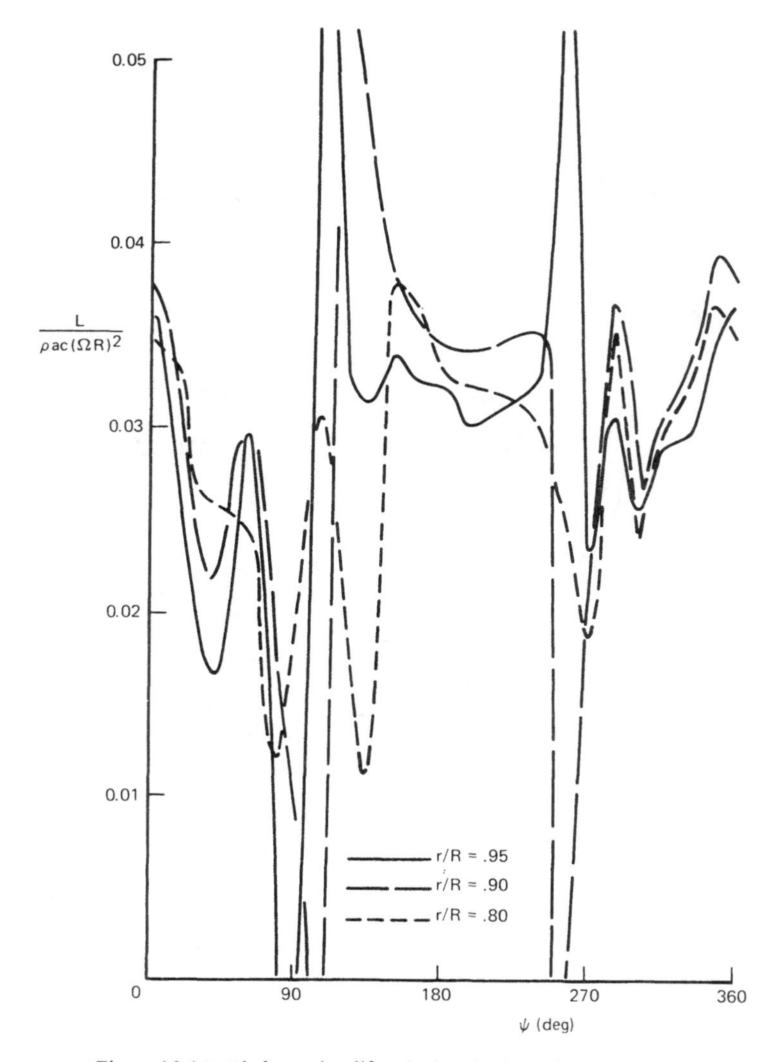

Figure 13-17 Blade section lift calculated using a free wake model, and lifting-line theory for the vortex-induced loads.

on the advancing side, perhaps even with the blade cutting through the tip vortex. The result is a large increase in the predicted loading with the

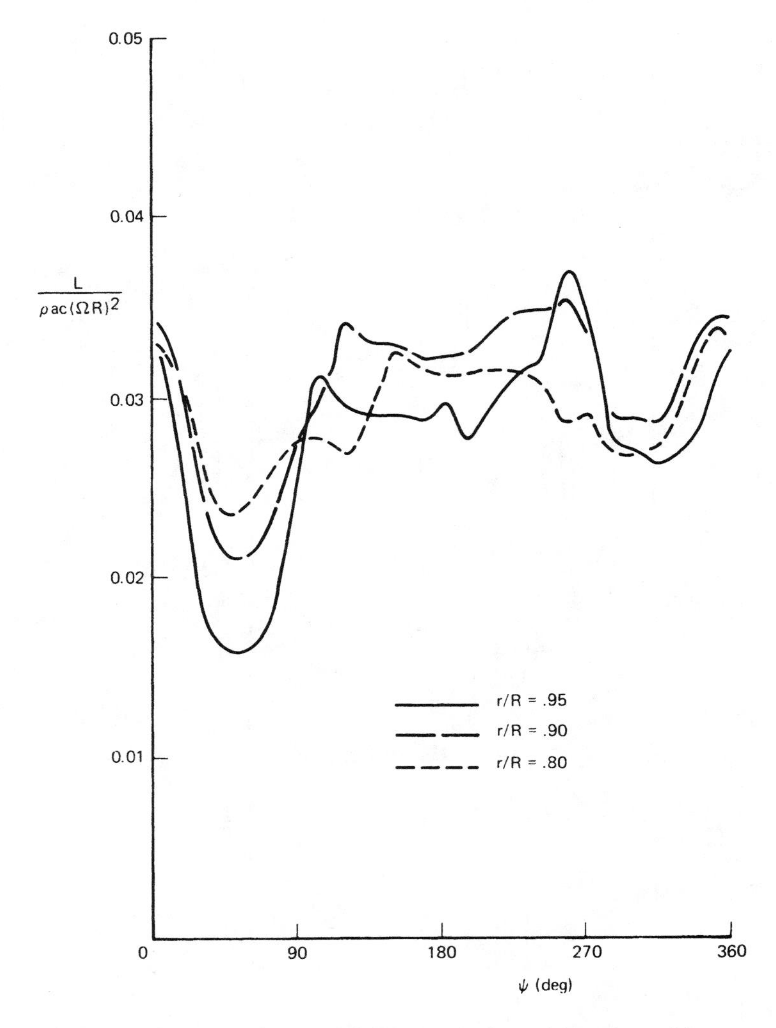

Figure 13-18 Blade section lift calculated using a rigid wake model, and lifting-surface theory for the vortex-induced loads.

free wake, as shown in Fig. 13-17. (The loading peaks are far off the scale of this figure, the maximum being $L/\rho ac(\Omega R)^2 = 0.099$ for $r/R = 0.90$ on the advancing side, and the minimum being $L/\rho ac(\Omega R)^2 = -0.077$.) Such

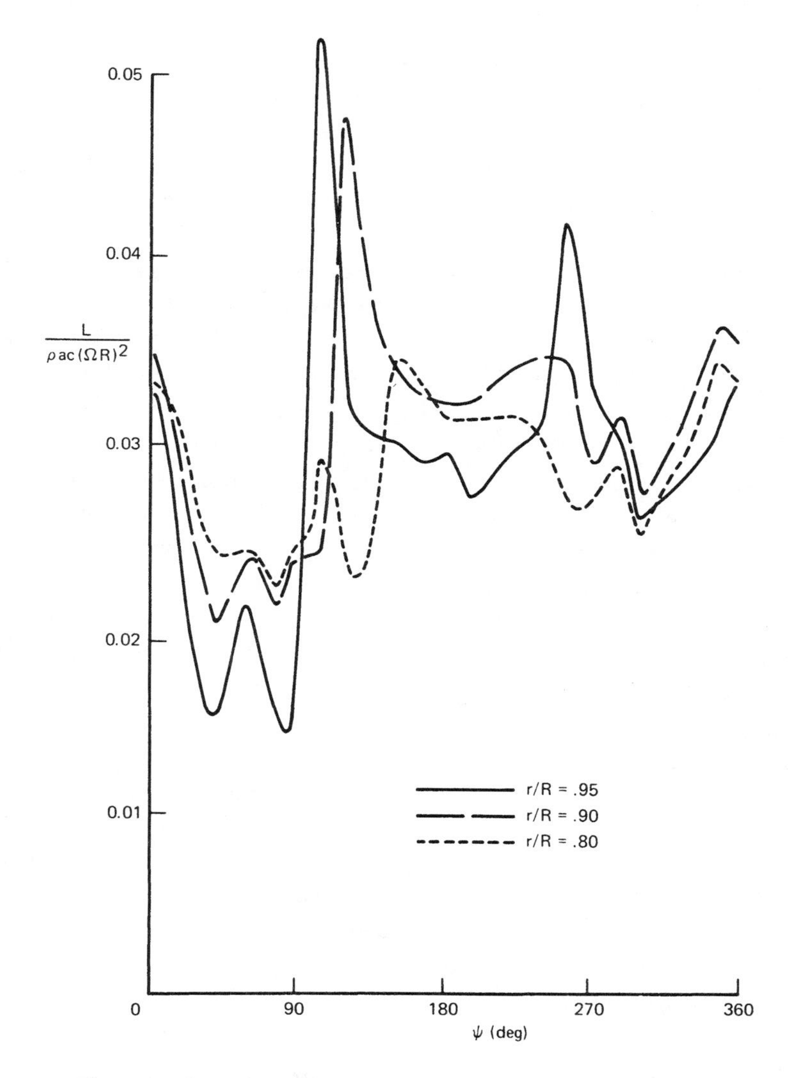

Figure 13-19 Blade section lift calculated using a free wake model, and lifting surface theory for the vortex-induced loads.

high loading is unrealistic. Lifting-surface theory is required if the vortex-induced loading on the blade in such close encounters is to be calculated accurately. Figs. 13-18 and 13-19 show the substantial reduction in predicted loads obtained using lifting-surface theory, particularly with the free wake geometry. There are two effects involved in the loads reduction. The first is the direct three-dimensional flow effect, which reduces the loads by a factor of 0.26 in this case (for the peak at $\psi \cong 105°$, $r/R = 0.95$).There is a further reduction due to the returning wake of the rotor. Lifting-surface theory reduces the lift and bound circulation of the blade where it encounters the vortex on the advancing side, hence it reduces the strength of the tip vortex trailed from the blade at that point. At this low speed, the tip vortex element left in the wake at $\psi \cong 105°$ is convected downstream very little, and as a result it is encountered by the following blade, still at nearly $\psi = 105°$. In that interaction, the reduced tip vortex strength further reduces the loading induced on the blade. The net effect of this feedback through the returning wake of the rotor is that the predicted vortex-induced loads are reduced by a factor of 0.15 when lifting-surface theory is used. This result demonstrates the importance of being consistent in developing an aerodynamic theory of the helicopter rotor. The wake geometry distortions in the free wake model greatly influence the loads, but the loading is overpredicted unless lifting-surface theory is also introduced for the close vortex-blade interactions.

Crimi (1965, 1966) developed a method for calculating the induced velocity at any point in the flow field, including a calculation of the distorted wake geometry and the effects of a fuselage. The wake model consisted of just the tip vortices; the inboard shed and trailed wake was neglected entirely. The blade loading and circulation were assumed to be known, so that only the wake geometry had to be calculated. Crimi used an azimuth increment of $\Delta\psi = 30°$, with two revolutions of the wake at $\mu = 0.25$, four revolutions at $\mu = 0.15$, and about eight at hover (for a two-bladed rotor). Some evidence of an instability in the wake structure was found in terms of a lack of convergence of the calculated geometry. The instability occurred at low speed ($\mu < 0.07$) and started 2 or 3 revolutions below the disk.

Simons, Pacifico, and Jones (1966) conducted a flow visualization experiment with a model rotor. They found that at low μ the trailing vortices from the leading edge of the rotor disk tend to pass upward through the disk first, and then downward. At the rear of the disk the vortices tend to be convected downward at a rate higher than the mean inflow.

Scully (1967) developed a method for calculating the distorted tip vortex geometry of a helicopter rotor in forward flight. His wake model was a vortex lattice, with only two trailers (the tip vortex, and an inboard trailer typically at $r/R = 0.50$). The effects of the tip vortices dominated the solution, but the effects of the inboard wake were not negligible. This investigation was particularly concerned with developing techniques for improving the convergence of the solution and hence reducing computation time. Little sensitivity of the wake geometry to the vortex core radius was found.

Landgrebe (1969b) developed a method for calculating the rotor distorted wake geometry. The wake model consisted of up to 10 trailed vortex lines; the shed wake vorticity was neglected. Only the geometry of the tip vortices was calculated. An azimuth increment of $\Delta\psi = 15$ to $30°$ was used, with about five revolutions of the wake. The wake geometry was not sensitive to the vortex core radius used. To reduce the computation required, Landgrebe divided the wake elements into far wake and near wake regions. The near wake elements were those that were found in the first iteration to contribute significantly to the induced velocity at a given point in the wake. For successive iterations, then, only the induced velocity contributions of the near wake elements were updated. The result is a reduction of about an order of magnitude in the computation required to obtain the free wake geometry.

Clark and Leiper (1970) developed a method for calculating the distorted wake geometry of a hovering rotor. Their wake model consisted of a number of constant-strength trailed vortex lines; in hover there is no shed vorticity in the wake. The far wake was approximated by segments of ring vortices. Two revolutions of the free wake were used, followed by 30 revolutions of the far wake. The distorted geometry of all the trailers was calculated. A substantial influence of the distorted geometry on the loading distribution was found, particularly near the tip, and hence on the hover performance of the rotor.

Landgrebe (1971a, 1972) conducted an experimental investigation of the performance and wake geometry of a model hovering rotor. The wake geometry was measured by flow visualization, and the data were used to develop expressions for the axial convection and radial contraction of the tip vortices and inboard vortex sheets. The tip vortex elements were found to have a roughly constant rate of descent before and after passing beneath the following blade. Prior to the encounter with the blade the descent rate is proportional to the blade loading C_T/σ. After the encounter the axial convection rate is higher; it is proportional to the mean inflow ratio $\sqrt{C_T/2}$, but about 40% larger than the momentum theory value. This generalized wake geometry information was used in calculations of the rotor performance, and it provided a significant improvement in correlation with measured performance, compared to the results based on an uncontracted, rigid wake model. The uncontracted wake model gives too large a thrust at a given power, especially for high loading and a large number of blades. The rotor performance was quite sensitive to the tip vortex geometry, particularly at the first blade-vortex interaction, because of the large vortex influence on the tip loading. It was found that an increase in the distance of the vortex from the blade by $0.01R$ could produce as much as a 10% increase in C_T. Landgrebe also calculated the distorted tip vortex geometry for the hovering rotor. The wake geometry calculations gave a lower-than-measured initial axial velocity, and hence a closer passage of the vortex to the following blade. This error was probably due to the neglect of the bound circulation, which was necessary to avoid unrealistic distortions arising from the blade-vortex interaction. The predicted contraction rate was also smaller than measured. The final rate of axial convection of the tip vortices after the blade passage was predicted well, and the general dependence of the wake geometry on the rotor parameters (such as C_T, C_T/σ, and N) was obtained correctly. It was concluded that the wake geometry calculation is qualitatively very good, but that the important first blade-vortex interaction was still not well predicted. Landgrebe found a reduction in the stability of the wake vortices with increasing distance from the rotor disk in both the measurements and the calculations. An instability in which successive coils of the helices rolled around one another was observed in many of the model rotor flow visualizations. Shortly beyond this instability, further observation of the tip vortices was difficult. In no case was a smoothly contracting tip vortex observed for a large enough distance below the rotor disk to definitely preclude the possibility of an instability. Usually three or four turns of the helices were clearly evident, and then nothing of the wake structure could be seen. The hovering wake geometry calculations showed a similar instability, which did not appear to be a numerical problem. The wake near the disk, which is important for the rotor loads and performance, was stable in both experiment and theory, however.

Scully (1975) developed a method for calculating the free wake geometry of a helicopter rotor, for use in an analysis of rotor nonuniform inflow and aerodynamic loading (discussed in the last section). The emphasis in this work was on developing efficient yet accurate computation techniques for the wake geometry and inflow calculations. Since the case of steady-state flight was considered, the solution was periodic. The wake model was essentially the same as for the inflow calculation (see section 13–2). Only the geometry of the tip vortices was obtained. The distortion of the tip vortex geometry from the basic helix was described by the displacement vector $\vec{D}(\psi,\delta)$ of the wake element with current age δ that was created when the blade was at azimuth angle ψ. The procedure for calculating the wake geometry consists basically of integrating the induced velocity at each wake element. The outer loop in the calculation was on the wake age δ. The induced velocities $\vec{q}(\psi)$ were calculated at all wake elements for a given age δ, and at all azimuth angles ψ. Then the increment in the distortion as the wake increased by $\Delta\psi$ was

$$\vec{D}(\psi,\delta) = \vec{D}(\psi,\delta-\Delta\psi) + \Delta\psi\vec{q}(\psi)$$

An efficient calculation of the wake geometry requires many variations on this basic procedure. Scully adopted Landgrebe's near wake and far wake scheme for reducing the computation. The first time the induced velocity is evaluated at a point in the wake, the contributions from all wake elements must be found. For subsequent evaluations of the induced velocity at that point, only the induced velocity due to the near wake elements is recalculated. The other major consideration for minimizing the computation was the matter of updating the induced velocity calculation. At a given point in the wake geometry calculation, there is a boundary in the wake between the distorted geometry and the initial, rigid geometry. The distortion has been calculated between the rotor disk and the boundary; downstream of the boundary, the wake is undistorted. As time increases by $\Delta\psi$, the entire wake is convected downstream, and the rotor blades move forward by $\Delta\psi$, adding new trailed and shed vorticity to the beginning of the wake. If there were no distortion of the wake during the time $\Delta\psi$, the induced velocity at a given wake element would not change except for the contributions from the newly created wake vorticity just behind the blade. Thus the normal calculation procedure consisted of calculating the induced velocity at the boundary by just adding at each step the contribution from the new wake just behind the blade. Of course, the wake does distort as it is convected and as the estimate of the distortion improves. Thus it was necessary to update the calculation of the induced velocity in the wake. In boundary updating, the induced velocity was still calculated at the boundary, by summing the contributions from all elements in the wake. In general updating, the induced velocity was recalculated at all points in the wake upstream of the boundary. Boundary updating was typically done every 90° on the front and rear portions of the helices, and every 45° along the sides where the distortion is greater. General updating was typically done every 180°. General updating cannot be done often if the amount of computation is to be kept low, but it does improve the accuracy and convergence considerably. Numerous techniques for secondary improvements in the efficiency and accuracy was also developed. For example, the latest calculation of the distortion was averaged with the last distortion estimate in order to improve stability and convergence. In calculating the downwash at a point in the wake it was necessary to consider at least two revolutions downstream of the point. The distorted wake geometry must be calculated for m revolutions, where m decreases with forward speed approximately as $m = 0.4/\mu$. A single iteration consisted of calculating the distortion $\vec{D}(\psi,\delta)$ for $\psi = 0$ to 2π, and $\delta = 0$ to $2\pi m$. Usually two iterations were sufficient to obtain the converged solution for the wake geometry. The computation required for this procedure depends on the number of blades and the advance ratio, which are the primary parameters determining how much wake must be analyzed. Typically the wake geometry calculation required about the same computation time as the calculation of the nonuniform inflow and airloads. Scully found that the wake geometry has indeed a significant influence on the predicted rotor aerodynamic loading, because the distorted wake tends to be much closer to the blades than the rigid wake model would indicate.

Landgrebe and Egolf (1976a, 1976b) conducted extensive comparisons of calculated wake geometry and induced velocities with the experimental data from several sources. They considered the time-averaged and instantaneous velocities within and outside the wake of rotors in hover and forward flight. The overall correlation was fairly good. The sensitivity of the data to the detailed characteristics of the flow, particularly the fine structure of the wake geometry, can lead to large local discrepancies between calculations and measurements, however.

Other work on helicopter rotor wake geometry, either calculated or measured: Levy and Forsdyke (1928), Gray (1960), Du Waldt (1966), Jenney, Olson, and Landgrebe (1968), Lehman (1968a, 1968b, 1971a), Erickson (1969), Rorke and Wells (1969), Joglekar and Loewy (1970), Levinsky and Strand (1970), Landgrebe and Bellinger (1971, 1973, 1974a), Sadler (1971a, 1971b, 1972b), Boatwright (1972), Gilmore and Gartshore (1972), Johnson and Scully (1972), Landgrebe and Cheney (1972), Scully and Sullivan (1972), Walters and Skujins (1972), Widnall (1972), Gupta and Loewy (1973, 1974), Skujins and Walters (1973), Sullivan (1973), Tangler, Wohlfeld, and Miley (1973), Clark (1974), Shipman (1974), Young (1974), Chou and Fanucci (1975, 1976), Gupta and Lessen (1975), Hall (1975a), Landgrebe and Egolf (1975), Gray et al. (1976), Summa (1976), Kocurek and Tangler (1977), Landgrebe and Bennett (1977), Landgrebe, Moffitt, and Clark (1977).

13–4 Vortex-Induced Loads

A tip vortex encountering the following blade induces a large aerodynamic loading on that blade for the small vortex-blade separations characteristic of the helicopter rotor in both hover and forward flight. In view of the fact that such vortex-induced loading is a principal source of rotor higher harmonic loads and vibration, its calculation is an important part of rotary wing analysis. It was found in section 10-8.1 that a vortex a distance h below the blade induces a downwash velocity (the component normal to the blade surface) which is zero directly over the vortex and has positive and negative peaks a distance h on either side of the intersection. The induced bound circulation and loading has the same general form as the induced velocity distribution (see Fig. 13-20), although the peaks are somewhat further apart than $2h$ because of lifting-surface effects. The spanwise gradient of the bound circulation indicates that there is trailed vorticity in the wake behind the blade, induced by the tip vortex from the preceding

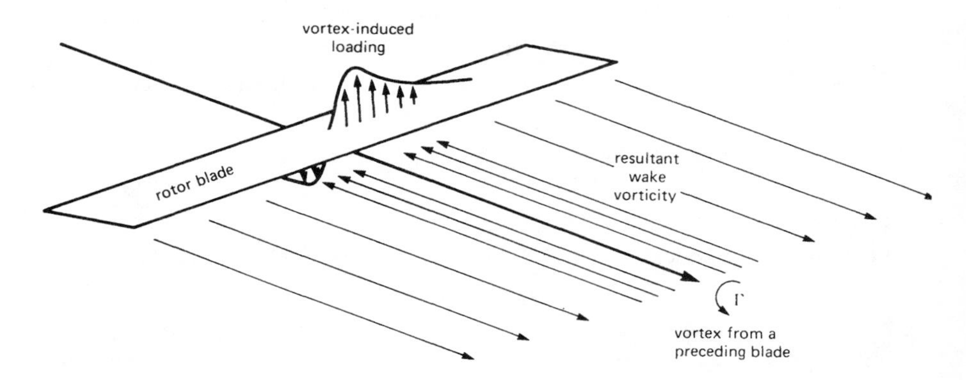

Figure 13-20 Vortex-induced blade loading.

blade. Since this induced wake vorticity has a direction parallel to the free tip vortex, if the free vortex is not perpendicular to the blade span there will actually be a radial component of the trailed vorticity (i.e. shed wake). If the vortex is not perpendicular to the span, the vortex-blade intersection sweeps radially along the blade as the vortex is convected with the free stream, and the problem is unsteady. For the extremely small vortex-blade separations typical of the rotary wing, the loading varies greatly in a short distance along the span of the blade. Lifting-line theory does not give an accurate prediction of the loading in such cases with a small effective aspect ratio. Therefore, a lifting-surface analysis is required for an accurate treatment of the vortex-induced loads on a rotor blade. The treatment of the strong concentrated trailed vorticity behind the blade which is induced by a close encounter with a free vortex further complicates the rotary wing analysis. It would not be efficient to use a highly detailed model for the entire wake to automatically account for such induced wake vorticity. Some special treatment of the wake model will therefore be required near vortex-blade interactions, whether lifting-line or lifting-surface theory is used. Note that the free vortex and the induced wake vorticity just above it have opposite signs, and will likely interact downstream of the blade. The result of such an interaction would be a diffusion of the vorticity, which could be important if the vortex encounters yet another blade of the rotor.

The direct application of lifting-surface theory to the rotary wing is discussed in section 13–6. An alternative approach is to develop a lifting-surface solution for a model problem representing blade-vortex interaction, and then to apply that solution in the calculation of vortex-induced loads in a helicopter airloads analysis. Such an approach avoids the large amount of computation required for a direct application of lifting-surface theory. Moreover, since the effects of the wake induced by the free vortex are included in the solution for the model problem, a special representation of that part of the rotor wake is not required in the rotary wing applications.

Johnson (1971a, 1971b) developed a lifting-surface theory solution for vortex-induced loads that is applicable to the calculation of helicopter airloads. The model problem considered is an infinite aspect ratio, non-rotating wing in a subsonic free stream, encountering a straight, infinite vortex at an angle Λ with the wing (Fig. 13-21). The wing has chord c; the vortex lies in a plane parallel to the wing, a distance h below it. The vortex is convected past the wing by the free stream. The distortion of the vortex-line geometry by the interaction with the wing were not considered. In linear lifting-surface theory, the blade and wake are represented by a planar

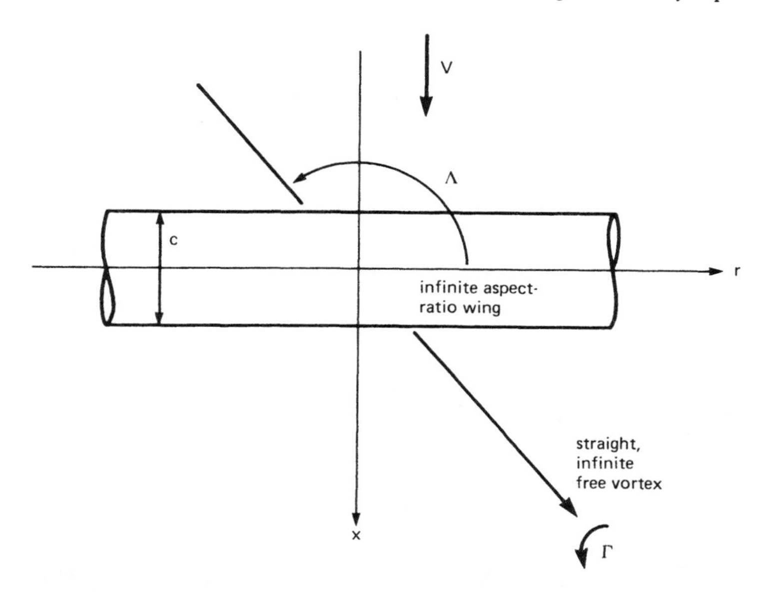

Figure 13-21 Model problem for lifting-surface theory analysis of vortex-induced loads.

distribution of vorticity. This model problem was solved for the case of a sinusoidal induced velocity distribution with wave fronts parallel to the vortex line. The vortex-induced velocity distribution can be obtained by a suitable combination of sinusoidal waves of various wavelengths (i.e. by a Fourier transform), so for this linear problem the same superposition gives the vortex-induced loading from the sinusoidal loading solution. The lifting-surface solution was obtained numerically, and then an analytical approximation was constructed for the loading due to sinusoidal induced velocity. The form of this approximation was chosen such that an analytical expression could be obtained for the vortex-induced loading. The approximate solution is not valid for extremely small wavelengths, but the range of validity is sufficient to handle the cases arising in rotary wing applications. For the velocity induced by a vortex of strength Γ:

$$\frac{w}{V} = \frac{\Gamma}{2\pi V} \frac{-r \sin \Lambda}{(r \sin \Lambda)^2 + h^2}$$

the approximate lifting surface solution for the section lift was obtained in the form

$$L_{ls} = \rho V \Gamma f(r/c, h/c, \Lambda, M)$$

by fitting analytical expressions to the numerical solution for sinusoidal loading. The corresponding lifting-line theory solution for the vortex-induced loading can be obtained in the form:

$$L_{ll} = \rho V \Gamma g(r/c, h/c, \Lambda, M)$$

Such a lifting-surface solution can be used in helicopter airloads calculations in the following manner. For each finite-length line segment in the wake model for the tip vortices, it is determined whether the segment is close enough to the blade for lifting-surface effects to be important. If so, the induced velocity contribution of that segment is reduced by the factor L_{ls}/L_{ll}. Then the loading on the blade is determined from the section angle of attack, including the wake-induced velocity, by using lifting-line theory as usual. The use of lifting-surface theory was found to substantially reduce the predicted vortex-induced loads. It was concluded that lifting-surface theory is required to accurately treat the vortex-blade interaction. Lifting-line theory is satisfactory only in those cases where the vortex-induced loading is so small that it is not important anyway. A consistently accurate analysis of the vortex-blade interaction requires a consideration of much more than just lifting-surface theory, however. The exact nature of the aerodynamic phenomena involved in such interactions, particularly on the rotating wing, is still the subject of research.

Literature on vortex-induced loads includes: Scheiman and Ludi (1963), Jones (1965), Kfoury (1966), Simons (1966), Johnson (1970a, 1970c, 1970d, 1971a, 1971b), Lucassen and Vodegel (1970), McCormick and Surendraiah (1970), Rudhman (1970), Hancock (1971), Padakannaya (1971, 1974), Filotas (1972, 1973a, 1973b), Isay (1972a, 1972b), Johnson and Scully (1972), Monnerie and Toguet (1972), Adamczyk (1974), Chu and Widnall (1974), Ham (1974, 1975), Patel and Hancock (1974), Paterson, Amiet, and Munch (1975), Scully (1975), Selic (1975), Wong (1975).

13–5 Vortices and Wakes

Clearly the fluid dynamic characteristics of the vorticity that makes up the rotor wake are important in the analysis of helicopter aerodynamics. The tip vortex and the blade-vortex interaction are of primary concern. The factors that must be considered are the vortex core, bursting, and other viscous effects; self-induced distortion and the stability of the wake filaments; vortex-induced loads, including stall and other viscous effects, and local distortion of the vortex; and tip vortex roll-up. These aerodynamic phenomena are complex, and many are not yet thoroughly understood even for nonrotating wings. There is considerable literature concerned with the vortices and wakes of fixed wings. The literature concerned specifically with rotary wing vortices and wakes includes: Simons, Pacifico, and Jones (1966), Spencer, Sternfeld, and McCormick (1966), Spivey (1968), McCormick and Surendraiah (1970), Spivey and Morehouse (1970), Chigier and Corsiglia (1971), Hoffman and Velkoff (1971), Padakannaya (1971), Rinehart (1971), Rinehart, Balcerak, and White (1971), Schumacher (1971), Boatwright (1972, 1974), Cook (1972), Rorke, Moffitt, and Ward (1972), Velkoff, Hoffman, and Blaser (1972), White and Balcerak (1972a, 1972b), Widnall (1972), Balcerak and Feller (1973a, 1973b), Landgrebe and Bellinger (1973, 1974a), Tangler, Wohlfeld, and Miley (1973), White (1973), Hall, et al. (1974), Ham (1974), Shrager (1974), Pegg, Hosier, Balcerak, and Johnson (1975), Selic (1975), Tangler (1975), Wong (1975), Biggers, Lee, Orloff, and Lemmer (1977a, 1977b), Rorke and Moffitt (1977), White (1977).

13–6 Lifting-Surface Theory

Lifting-line theory is generally used in rotary wing analyses to obtain the blade aerodynamic loading. The basic assumption of high aspect ratio is usually satisfied with rotor blades, but the rotary wing aerodynamic environment can produce situations where the effective aspect ratio is small because of large spanwise variation of the loading. There are two major cases where lifting-line theory does not give sufficiently accurate predictions of the loads: at the blade tip, where the load must drop quickly to zero; and when a blade passes close to a tip vortex in the wake (which usually occurs near the tip as well). The rotor performance is sensitive to the tip loading, and the vortex-induced airloads are a principal factor in helicopter vibration, loads, and noise. Thus it is necessary to develop accurate analyses for such cases.

Lifting-surface theory retains the effects of mutual interaction of all the elements of the wing and wake by representing the wing by vortex surfaces and satisfying the boundary conditions over the entire surface. As a result, lifting-surface theory can handle the large variations of induced velocity and loading that occur at the blade tip or in an encounter with a wake vortex. Considerable progress has been made in developing lifting-surface theory for fixed wing aircraft. The development of lifting-surface theory for the rotary wing is a more difficult task, however. Because the rotor blade encounters its own wake, the wake model must be very detailed and accurate if it is to be consistent with the use of lifting-surface theory. A free wake geometry calculation is required, and also a calculation of the tip vortex roll-up and other fine structures of the wake. Only for the hovering rotor is the problem steady; in forward flight an unsteady lifting-surface theory is required for the rotor. Although the solution is periodic rather than a general transient, all the harmonics are coupled. Moreover, the high tip speed of most rotors requires that compressibility effects be included as well.

For the complex geometry of the rotor and its wake, a finite-element lifting-surface theory may be the only practical approach. In the simplest case, a vortex lattice is used to represent the wing surface as well as the wake. Such an analysis would be similar to the nonuniform inflow calculation described above, except that the induced velocity must be calculated at orders of magnitude more points on the blade surface. Even without a free wake calculation, a lifting-surface theory analysis for the rotary wing would require many times the computation needed for the nonuniform inflow calculation. At the current stage of development of lifting-surface theory for rotary wings, such a large amount of computation is seldom justified for routine applications.

On lifting-surface theory for the rotary wing: Ichikawa (1967), Sopher (1969), Dat (1970, 1974, 1976), Woodley (1971), Caradonna and Isom (1972, 1976), Csencsitz, Fanucci, and Chou (1973), Isom (1974), Chou and Fanucci (1975, 1976), Hall (1975a), Costes (1976), Summa (1976), Isay (1977), Kocurek and Tangler (1977), Rao and Schatzle (1978). See also the literature concerning lifting-surface theory for fixed wing aircraft.

13–7 Boundary Layers

The literature of both experimental and theoretical investigations concerned specifically with the boundary layers of rotating wings includes: Sears (1948, 1950), Fogarty and Sears (1950), Fogarty (1951), Mager (1952), Tan (1953), Rott and Smith (1956), Banks and Gadd (1963), Tanner and Buettiker (1966), Tanner and Yaggy (1966), Velkoff (1966), Tanner (1967), Tanner and Wohlfeld (1967), McCroskey and Yaggy (1968), Velkoff, Blaser, Shumaker, and Jones (1969), Bowden and Shockey (1970), Williams and Young (1970), Blaser and Velkoff (1971, 1973), Clark and Arnoldi (1971), Dwyer and McCroskey (1971), Hicks and Nash (1971), McCroskey (1971), McCroskey, Nash, and Hicks (1971), Velkoff, Blaser, and Jones (1971), Clark and Lawton (1972), Velkoff, Hoffman, and Blaser (1972), Young and Williams (1972), Warsi (1974). See also Chapter 16, and the literature on the boundary layers of nonrotating wings.

Chapter 14

HELICOPTER AEROELASTICITY

14–1 Aeroelastic Analyses

In the broadest sense, helicopter aeroelasticity encompasses all of rotary wing analysis, for there are few problems where it is possible to ignore either the inertial, structural, or aerodynamic forces on the blade. This chapter is specifically concerned with comprehensive analyses of the aeroelastic behavior that bring together the most advanced models of the geometry, structure, inertia, and aerodynamics available in rotary wing technology, subject to the conflicting constraints of accuracy and economy. Such analyses of the aeroelastic behavior can be applied to the calculation of the helicopter performance and trim, handling qualities, blade motion and airloading, dynamic stability, blade and control loads, and vibration. Often the range of application of an analysis is restricted in order to improve its efficiency, since the same degree of sophistication is not required in all elements of the model for all problems. Usually the primary focus is on the rotor, but a general model will include a representation of the entire aircraft. Invariably such analyses are implemented in the form of a program for a high-speed digital computer. It was the development of such machines that made attempts to calculate the general aeroelastic behavior of helicopters practical.

The construction of an aeroelastic analysis begins with the specification of the scope of the problem to be solved (performance, blade loads, etc.) and the extent of the aircraft to be modeled (a single blade, the rotor, or the entire helicopter). This specification usually depends on the stage of development of the analysis as well as on the problems of interest. Then the basic elements of the analysis are derived: the detailed description of the system, the structural and inertial model (i.e. the equations of motion), and the aerodynamic model. Many alternative models are available for the wake structure and geometry, the induced velocity calculation, the rotor and airframe degrees of freedom, the blade aerodynamics, and the other elements in the analysis. It is important that the models used for various elements have a consistent level of sophistication. Using an advanced model in one area alone often results in either a misleading sense of accuracy or a loss of efficiency. Concentrating on a single element of the analysis is appropriate for developing more advanced models, however. There are two basic formulations of the helicopter aeroelastic analysis, linearized eigenvalue solutions and nonlinear time history calculations. The basic task is to solve the equations of motion for the aircraft response, which must be accomplished numerically because of the complexity of the models.

A nonlinear aeroelastic analysis of the helicopter typically involves the following sequence of calculations. The inputs include the description of the rotor, the helicopter, and the operating state. The output depends on the problem considered (rotor performance, blade loads, helicopter transient motion, etc.), At each time step the analysis successively calculates the wake geometry, wake induced velocity, and the aerodynamic forces on the rotor and helicopter, and the resulting motion of the rotor blades and airframe, using as simple or complex model for each element as is appropriate to the problem involved. After the equations of motion have been integrated to obtain the rotor and airframe response, the time is incremented and the calculation process repeated. The iteration on time continues until the converged, periodic solution for steady-state flight is obtained, or for as long as is required to establish the transient behavior of the aircraft. Such a direct approach requires a tremendous amount of computation for the more complex models. Considerable attention is given therefore to developing more efficient variations of this procedure, according to the problem being investigated and the resources available.

An important special case is the aeroelastic behavior in steady flight, which includes the helicopter performance, aerodynamic loading, blade and control loads, and vibration. Since the solution is periodic in this case and the motion of all the blades is identical, a direct time history calculation is not mandatory. Thus the basic iteration sequence of the aeroelastic analysis can be modified to improve the efficiency of the calculation. The fundamental principle is to keep to a minimum the number of times the computationally intensive steps must be performed to obtain a converged solution. As an example, consider a problem involving the calculation of the wake-induced nonuniform inflow. In the direct approach, the downwash would be evaluated at each time step until the airloads and blade motion converged to the periodic solution. However, the inflow calculation is not very sensitive to small changes in the loading and motion of the rotor. Thus the inflow calculation can be separated from the solution for the periodic airloads and blade motion. The induced velocity distribution over the entire rotor disk is evaluated, and then the equations of motion are integrated for as many revolutions as is required to obtain a converged solution. This basic cycle is repeated, with only two or three calculations of the induced velocity distribution being required to obtain a converged solution for both the inflow and the blade motion. The result is a significant reduction in the computation required compared to the most direct approach. Other elements of the aeroelastic analysis, such as a distorted wake geometry calculation, can be handled in a similar fashion. There are many variations of the solution for even the helicopter steady flight response, but the successful approaches are those where considerable effort has been made to obtain efficient calculations.

The rotor aeroelastic response can be calculated for a given control setting. However, the helicopter operating state is usually specified in terms of parameters such as speed and gross weight, not in terms of the control positions. Thus to the analysis there must be added a trim calculation procedure, involving an iteration on the controls to achieve equilibrium of the net forces and moments on the rotor or helicopter. If just the rotor is considered, there are three control variables: collective pitch, longitudinal cyclic pitch, and lateral cyclic pitch. These controls may be adjusted to trim three quantities, typically the rotor thrust and tip-path-plane tilt (e.g. for zero flapping relative to the shaft), or the thrust, propulsive force, and side force. If the entire helicopter is considered there are six control variables to trim the six forces and moments on the aircraft: the pilot's collective stick, longitudinal cyclic stick, lateral cyclic stick, and pedal positions; and the helicopter pitch and roll angles relative to the flight path. The basic trim procedure consists of comparing the current solution for the forces and moments on the helicopter with the target values, and incrementing the controls in the manner required to approach the targets in the next cycle. These steps are repeated until the desired values for the forces and moments on the helicopter are achieved, within a specified tolerance. To increment the controls in the proper direction and magnitude requires knowing the derivatives of the helicopter forces with respect to the control variables. These derivatives can either be obtained from a simple analysis, or calculated prior to the trim iteration by individually incrementing the control variables by specified amounts and noting the resulting changes in the forces. The latter procedure is usually required, particularly in extreme operating conditions. Trim of a single quantity, such as the rotor thrust with collective, is simple and easily accomplished. Trimming the six forces and moments of the complete helicopter is a more difficult problem, with convergence to the desired state by no means assured. It helps to start close to the trim state, to take small control increments, and to update the control derivatives occasionally—all of which increase the required computation.

The aeroelastic stability of the rotor can be evaluated by using a nonlinear, open-form analysis to calculate the transient response of the system. The disadvantages of this approach are that much more computation is needed to obtain the transient response than to obtain the periodic solution (which in fact must be obtained first as a starting point for the transient motion), and that it is not a simple matter to obtain quantitative information about the complete dynamics from the transient response. An alternative approach is to calculate the aeroelastic stability using the methods of linear system theory (see section 8–6). Linear differential equations are derived for the perturbed motion of the rotor and helicopter from the trim state. Then the stability is found directly from the eigenvalues of these equations. In this approach more effort is required to derive the equations of motion describing the system, so that the efficient analysis techniques of linear system theory can be used. When the entire helicopter is considered, the aeroelastic stability calculation includes the rigid body motions, so the analysis also encompasses the aircraft flight dynamics.

Using the best helicopter analyses currently available, good correlation between the measured and predicted behavior is found for the general, overall quantities. The prediction of the detailed, specific quantities is often quite poor, however. Predictions of quantities such as rotor performance or the mean and alternating loads are generally reliable provided a theoretical model appropriate to the problem is used, although this capability has been achieved only with considerable use of empirical models (for dynamic stall, three-dimensional flow effects, aerodynamic interference, and so on). However, such use of empiricism and approximations often leads to inaccurate prediction of detailed characteristics. The structural characteristics, the inertial characteristics, and the aerodynamic environment of the rotary wing are complex, and evidently considerable further development of the theoretical models is required before consistently reliable prediction of the aeroelastic behavior is possible.

14–2 Integration of the Equations of Motion

One element of the helicopter aeroelastic analysis that is not covered at all elsewhere in this text is the numerical integration of the equations of motion. The differential equations to be solved may be written in the form $\ddot{\beta} = f(\dot{\beta}, \beta, \psi)$, where β represents the degrees of freedom of the system and ψ is the dimensionless time variable. In general since many degrees of freedom are involved, this is really a set of equations. With linear equations and a small number of degrees of freedom analytical solutions are possible. The aeroelastic analysis often involves nonlinear aerodynamic, structural, and inertial forces, so a numerical solution is required. Given the values of β and $\dot{\beta}$ at $\psi = \psi_n$ (from which $\ddot{\beta} = f$ can be evaluated), the problem is to integrate the equations over the time step $\Delta\psi$ to obtain the motion $\dot{\beta}$ and β at $\psi_{n+1} = \psi_n + \Delta\psi$.

For the problem of the steady-state behavior of the rotor, the solution for the motion will be periodic. It is possible then to solve the equations of motion directly for the harmonics of a Fourier series representation of the motion. By making use of the periodicity of the solution in this fashion, the convergence of the integration is greatly improved. Gessow (1956) developed a harmonic analysis method for integrating the differential equation for the blade flap motion. The equation of motion for flapping in the rotating frame is

$$\ddot{\beta} + \nu_\beta^2 \beta = \gamma M_F$$

where M_F is the aerodynamic flap moment and the flap natural frequency ν_β is near 1/rev. Gessow's procedure is to calculate M_F at a finite number of points around the azimuth from the current estimate of the blade motion.

Then the harmonics of a Fourier expansion of M_F can be evaluated:

$$M_F = \sum_{n=0}^{\infty} (M_{F_{nc}} \cos n\psi + M_{F_{ns}} \sin n\psi)$$

Assuming periodic motion, the solution of the flap equation is then

$$\beta_{nc,ns} = \frac{\gamma M_{F_{nc,ns}}}{\nu_\beta^2 - n^2}$$

where β_{nc} and β_{ns} are the harmonics of the flap motion. With this new estimate of the blade motion the flap moments can be recalculated. The successive calculations of the flap moments and blade motion are repeated until the solution converges, which is indicated when the change in blade motion from one iteration to the next falls below a specified tolerance level. With the converged solution for the blade motion, the rotor forces and performance can then be calculated. The only difficulty lies with the first harmonics of the flap motion, β_{1c} and β_{1s}. For $n = 1$ the flap equation gives

$$(\nu_\beta^2 - 1)\beta_{1c} = \gamma M_{F_{1c}}$$

$$(\nu_\beta^2 - 1)\beta_{1s} = \gamma M_{F_{1s}}$$

For an articulated rotor ($\nu_\beta = 1$) the left-hand side vanishes, and in general a different approach is required because the tip-path-plane tilt is primarily determined by the balance of aerodynamic moments on the blade (see Chapter 5). Expand the lateral flap moment as

$$\left(M_{F_{1s}}\right)_{\substack{\text{correct}\\ \text{solution}}} = \left(M_{F_{1s}}\right)_{\substack{\text{current}\\ \text{estimate}}} + \frac{\partial M_{F_{1s}}}{\partial \beta_{1c}}\left(\beta_{1c_{\text{correct}}} - \beta_{1c_{\text{current}}}\right)$$

(recall that the balance of lateral moments on the disk determines the longitudinal tip-path-plane tilt β_{1c}). Now

$$\left(\gamma M_{F_{1s}}\right)_{\substack{\text{correct}\\ \text{solution}}} = (\nu_\beta^2 - 1)\beta_{1s}$$

so

$$(\beta_{1c})_{n+1} = \left[\beta_{1c} - \frac{M_{F_{1s}} - \beta_{1s}(\nu_\beta^2 - 1)/\gamma}{\partial M_{F_{1s}}/\partial \beta_{1c}}\right]_n$$

and similarly,

$$(\beta_{1s})_{n+1} = \left[\beta_{1s} - \frac{M_{F_{1c}} - \beta_{1c}(\nu_\beta^2 - 1)/\gamma}{\partial M_{F_{1c}}/\partial \beta_{1s}}\right]_n$$

The derivatives of the flap moment can be estimated from a simple analysis, since they do not affect the final solution, but only the convergence to it. Gessow gives

$$\frac{\partial M_{F_{1s}}}{\partial \beta_{1c}} = \frac{1}{8}\left(1 - \frac{1}{2}\mu^2\right)$$

$$\frac{\partial M_{F_{1c}}}{\partial \beta_{1s}} = -\frac{1}{8}\left(1 + \frac{1}{2}\mu^2\right)$$

With these expressions the 1/rev flap motion can be updated from the current calculation of the flap moments.

Now let us develop a more general harmonic analysis method for integrating the rotor equations of motion. Consider equations of the form

$$\ddot{\beta} + \nu^2\beta = g$$

where β is the degree of freedom, ν is the appropriate natural frequency in the rotating frame, and g is a forcing function (usually nonlinear). To avoid the singularity of the resonant response at harmonics near the natural frequency, it is necessary to include the damping terms on the left-hand side of this equation. Thus the term $C\dot{\beta}$ is added to both sides, giving

$$\ddot{\beta} + C\dot{\beta} + \nu^2\beta = g + C\dot{\beta} = F$$

where C is the damping coefficient. As an example, $C = \gamma/8$ can be used for the fundamental flap mode of the blade. For good convergence the damping coefficient used should be close to the actual damping of the particular degree of freedom, including structural, mechanical, and aerodynamic damping sources. The damping estimate does not have to be exact, however, since it is added to both sides of the equation. In fact, since the actual damping in the forcing function g will often be time-varying and even nonlinear, the viscous damping coefficient has to be an approximation. The sole function of this damping term is to avoid divergence of the solution near resonance; the value of C has no influence on the final converged solution. Now the function F is evaluated at J points around the rotor azimuth:

$$F_j = F(\psi_j) = (g + C\dot{\beta})\Big|_{\psi_j}$$

where $\psi_j = j\Delta\psi$ ($j = 1$ to J, and $\Delta\psi = 2\pi/J$). Then the harmonics of a complex Fourier series representation of F are

$$F_n = \frac{1}{J}\sum_{j=1}^{J} F_j e^{-in\psi_j} K_n$$

where

$$K_n = \left(\frac{J}{\pi n}\sin\frac{\pi n}{J}\right)^2$$

The factor K_n is introduced to smooth out the Fourier interpolation for the function F between the known points at ψ_j (see section 8–3). For harmonics up to about $n = J/12$, $K_n = 1$ can be used. The solution of the equation of motion for the harmonics of β is then

$$\beta_n = \frac{F_n}{\nu^2 - n^2 + inC}$$

The iterative solution proceeds as follows. At a given azimuth ψ_j, the blade motion is calculated using the current estimates of the harmonics:

$$\beta = \sum_n \beta_n e^{in\psi_j}$$

$$\dot{\beta} = \sum_n \beta_n in e^{in\psi_j}$$

The forcing function F_j is evaluated next. The estimates of the flapping harmonics are then updated to account for the difference between the current value of F_j and that found in the last revolution:

$$\Delta\beta_n = \left[F_j - (F_j)_{\substack{\text{last}\\\text{rev}}}\right] \frac{K_n}{J} \frac{e^{-in\psi_j}}{\nu^2 - n^2 + inC}$$

After $\Delta\beta_n$ is added to the flap harmonic β_n, the azimuth angle is incremented to ψ_{j+1}. The calculation proceeds around the azimuth in this fashion until the solution converges. A test for convergence is performed once each revolution and may, for example, require that the change in the root-mean-square level of the blade motion from one revolution to the next be below a specified tolerance.

The standard techniques of numerical analysis can also be used to integrate the equations of motion for the rotary wing. For the periodic case a harmonic analysis method is preferable, but for the general transient motion an open-form integration method is required. The simplest numerical integration technique is Euler's method, based on the expansion

$$\beta_{n+1} \cong \beta_n + \dot{\beta}_n \Delta\psi$$

For the second order equation $\ddot{\beta} = f$ considered here, the Taylor series expansion gives the iteration procedure:

$$\begin{aligned}
\beta_{n+1} &= \beta_n + \dot{\beta}_n \Delta\psi + \ddot{\beta}_n \tfrac{1}{2} \Delta\psi^2 \\
\dot{\beta}_{n+1} &= \dot{\beta}_n + \ddot{\beta}_n \Delta\psi \\
\ddot{\beta}_{n+1} &= f(\beta_{n+1}, \dot{\beta}_{n+1}, \psi_{n+1})
\end{aligned}$$

The accuracy and convergence of these two techniques is poor, however, even with a very small time step. The Runge-Kutta numerical integration techniques give much better results. The fourth-order Runge-Kutta solution of $\ddot{\beta} = f(\beta, \dot{\beta}, \psi)$ is

$$\beta_{n+1} = \beta_n + \Delta\psi \left(\dot{\beta}_n + \frac{\Delta\psi}{6}\left(f_1 + f_2 + f_3\right)\right)$$

$$\dot{\beta}_{n+1} = \dot{\beta}_n + \frac{\Delta\psi}{6}\left(f_1 + 2f_2 + 2f_3 + f_4\right)$$

where

$$f_1 = f(\beta_n, \dot{\beta}_n, \psi_n)$$

$$f_2 = f\left(\beta_n + \frac{1}{2}\Delta\psi\dot{\beta}_n + \frac{1}{8}\Delta\psi^2 f_1, \dot{\beta}_n + \frac{1}{2}\Delta\psi f_1, \psi_n + \frac{1}{2}\Delta\psi\right)$$

$$f_3 = f\left(\beta_n + \frac{1}{2}\Delta\psi\dot{\beta}_n + \frac{1}{8}\Delta\psi^2 f_1, \dot{\beta}_n + \frac{1}{2}\Delta\psi f_2, \psi_n + \frac{1}{2}\Delta\psi\right)$$

$$f_4 = f\left(\beta_n + \Delta\psi\dot{\beta}_n + \frac{1}{2}\Delta\psi^2 f_3, \dot{\beta}_n + \Delta\psi f_3, \psi_n + \Delta\psi\right)$$

Davis, Bennett, and Blankenship (1974) recommend the use of a Runge-Kutta method, after considering a number of numerical integration techniques. Mil′ (1966) suggests using a predictor-corrector technique based on the Taylor series expansion.

$$\dot{\beta}^p_{n+1} = \dot{\beta}_n + \Delta\psi\ddot{\beta}_n$$

$$\beta^p_{n+1} = \beta_n + \Delta\psi\dot{\beta}_n + \tfrac{1}{2}\Delta\psi^2\ddot{\beta}_n$$

$$\ddot{\beta}^p_{n+1} = f(\beta^p_{n+1}, \dot{\beta}^p_{n+1}, \psi_{n+1})$$

$$\dot{\beta}_{n+1} = \dot{\beta}_n + \tfrac{1}{2}\Delta\psi(\ddot{\beta}_n + \ddot{\beta}^p_{n+1})$$

$$\beta_{n+1} = \beta_n + \Delta\psi\dot{\beta}_n + \tfrac{1}{4}\Delta\psi^2(\ddot{\beta}_n + \ddot{\beta}^p_{n+1})$$

$$\ddot{\beta}_{n+1} = f(\beta_{n+1}, \dot{\beta}_{n+1}, \psi_{n+1})$$

With a Runge-Kutta or predictor-corrector technique good accuracy and convergence can be obtained with a reasonable step size. Note, however, that these techniques require that the forcing function f be evaluated more than once per cycle, which complicates the analysis besides increasing the computation. Frequently the equations of motion are integrated using a modification of the Taylor series expansion:

$$\dot{\beta}_{n+1} = \dot{\beta}_n + \ddot{\beta}_n\Delta\psi$$

$$\beta_{n+1} = \beta_n + \dot{\beta}_{n+1}\Delta\psi = \beta_n + \dot{\beta}_n\Delta\psi + \ddot{\beta}_n\Delta\psi^2$$

$$\ddot{\beta}_{n+1} = f(\beta_{n+1}, \dot{\beta}_{n+1}, \psi_{n+1})$$

This method seems to be about as accurate as the Runge-Kutta or predictor-corrector techniques for many applications, and is simpler to implement.

Tanner (1964a, 1964b) and others have used a numerical integration method designed specifically for the steady state case, in which the converged solution must be periodic. If the 2/rev and higher harmonics are neglected so that $\beta = \beta_0 + \beta_{1c} \cos\psi + \beta_{1s} \sin\psi$, it follows exactly that

$$\beta_{n+1} = \beta_n + \dot{\beta}_n \sin\Delta\psi + \ddot{\beta}_n(1 - \cos\Delta\psi)$$

$$\dot{\beta}_{n+1} = \dot{\beta}_n \cos\Delta\psi + \ddot{\beta}_n \sin\Delta\psi$$

For small $\Delta\psi$ this reduces to the Taylor series expansion. This technique can be extended to the case of transient motion of a general degree of freedom as follows. Consider the equation of motion in the form

$$\ddot{\beta} + \nu^2\beta = g(\beta, \dot{\beta}, \psi)$$

It is assumed that the forcing function is constant over the interval from ψ_n to ψ_{n+1}; that is, $g \cong g_n$. Then the linear differential equation $\ddot{\beta} + \nu^2\beta = g_n$ can be integrated analytically, with initial conditions β_n and $\dot{\beta}_n$ at ψ_n. Evaluating the solution at $\psi_{n+1} = \psi_n + \Delta\psi$ gives

$$\beta_{n+1} = \beta_n \cos\nu\Delta\psi + \dot{\beta}_n \frac{\sin\nu\Delta\psi}{\nu} + g_n \frac{1 - \cos\nu\Delta\psi}{\nu^2}$$

$$= \beta_n + \dot{\beta}_n \frac{\sin\nu\Delta\psi}{\nu} + \ddot{\beta}_n \frac{1 - \cos\nu\Delta\psi}{\nu^2}$$

$$\dot{\beta}_{n+1} = \dot{\beta}_n \cos\nu\Delta\psi - \beta_n \nu \sin\nu\Delta\psi + g_n \frac{\sin\nu\Delta\psi}{\nu^2}$$

$$= \dot{\beta}_n \cos\nu\Delta\psi + \ddot{\beta}_n \frac{\sin\nu\Delta\psi}{\nu}$$

For small $(\nu\Delta\psi)$ these equations reduce to the Taylor series expansion. For best results, the damping estimate should be included here also. Then the equation

$$\ddot{\beta} + C\dot{\beta} + \nu^2\beta \cong (g + C\dot{\beta})_n$$

must be integrated from ψ_n to ψ_{n+1}. Berman (1965) developed a variation of this numerical integration technique that is applicable to a system of nonlinear equations.

The basic difference between the harmonic analysis and numerical integration methods is that the former takes advantage of the periodicity of the solution to obtain some knowledge of the motion at times after as well as before ψ_n; while the latter only has available information about the motion up to the present time ψ_n. It follows that integrating the equations to obtain the transient motion entails more problems with accuracy and convergence than obtaining the periodic motion by a harmonic analysis method. The superiority of the Runge-Kutta and predictor-corrector techniques is due to their use of an estimate of the motion at ψ_{n+1} as well as at ψ_n. The computation required to integrate the equations of motion can often be reduced by updating certain calculations (such as the nonuniform inflow) as infrequently as possible while retaining the required accuracy.

14–3 Literature

On the aeroelastic analyses of the rotor and helicopter: Burkam, Capurso, and Yntema (1959), Capurso, Yntema, and Gabel (1961), Blankenship and Harvey (1962), Miller (1962a, 1962b, 1964a, 1964b, 1964c), Perlmutter, Yackle, Chou, and Miller (1962), Gerstenberger and Wood (1963), Piziali, Daughaday, and Du Waldt (1963), Wood and Hilzinger (1963), Drees and McGuigan (1965), Duhon, Harvey, and Blankenship (1965), Harrison and Ollerhead (1966), Piziali (1966a, 1966b), Balcerak (1967), Chang (1967), Jones (1967), Alexander et al. (1969), Arcidiacono (1969), Harvey, Blankenship, and Drees (1969), Livingston, et al. (1970), Yen, Weber, and Gaffey (1971), Bobo (1972), Livingston (1972), Anderson, Conner, and Kerr (1973), Arcidiacono and Carlson (1973), Austin and Vann (1973), Bennett (1973), Carlson and Kerr (1973), Gabel (1973), Gallott (1973), Lemnios (1973), McKenzie and Howell (1973), Reichert (1973a), Vann, Mirick, and Austin (1973), Davis, Bennett, and Blankenship (1974), Freeman and Bennett (1974), Johnson (1974d, 1976b, 1977e, 1977f, 1978). Niebanck (1974), Ormiston (1974), Shipman (1974), Hall (1975b), Sopher (1975), Sutton (1975), Sutton and Gangwani (1975), Anderson, Conner, Kretsinger, and Reaser (1976), Bielawa (1976, 1977, 1978), Bielawa, Cheney, and Novak (1976), Briczinski (1976), Hodges (1976), Johnston (1976), Johnston and Cassarino (1976), Meirovitch, Kraige, and Hale (1976), Staley (1976), Sutton, Gangwani, and Nettles (1976), Houck et al. (1977), McLarty, Van Gaasbeek, and Hsieh (1977), Sutton (1977), Costes (1978), Hodges (1978a), Meirovitch and Hale (1978), Reaser (1978), Reaser and Kretsinger (1978), Tran and Renaud (1978), Van Gaasbeek and Austin (1978), Warmbrodt (1978).

Chapter 15

STABILITY AND CONTROL

Stability and control are among the most important aspects of the analysis and design of rotary wing aircraft. As with the airplane, the problem of controlling the vehicle was one of the major obstacles in the development of a successful helicopter. Designing for satisfactory flying qualities remains a major concern in the development of a helicopter, with new applications of the vehicle always demanding improved behavior. Stability and control analysis basically involves solving the equations of motion for force and moment equilibrium on the entire aircraft.

15–1 Control

Helicopter control requires the ability to produce moments and forces on the vehicle for two purposes: first, to produce equilibrium and thereby hold the helicopter in a desired trim state; and secondly, to produce accelerations and thereby change the helicopter velocity, position, and orientation. Like airplane control, helicopter control is accomplished primarily by producing moments about all three aircraft axes: pitch, roll, and yaw. The engine power is controlled by the throttle. The helicopter has in addition direct control over the vertical force on the aircraft, corresponding to its VTOL capability. This additional control variable is part of the versatility of the helicopter, but it also makes the piloting task more difficult. Usually the control task is eased by the use of a rotor speed governor on the engine throttle to automatically manage the power.

Direct control over moments on the aircraft is satisfactory for trajectory control in forward flight. In hover and at low speed, direct control over the forces would be more desirable, to obtain direct command of the helicopter velocity and displacement. Such control is available only for the vertical force, however. The lateral and longitudinal velocities of the helicopter in hover must be controlled using pitch and roll moments about the aircraft center of gravity, which is a more difficult task. The pilot directly commands a change in pitch or roll attitude that then produces a longitudinal or lateral force and finally the desired velocity of the helicopter. There usually is considerable coupling of the forces and moments produced by the helicopter controls, so that any control application to produce a particular moment will require some compensating control inputs on the other axes as well. Moreover, without an automatic stability augmentation system, the helicopter is not dynamically or statically stable, particularly in hover. Consequently, the pilot is required to provide the feedback control to stabilize the vehicle, an operation that demands constant attention. The use of an automatic control system to augment the helicopter stability and control characteristics is desirable, and for some applications essential, but such systems increase the cost and complexity of the aircraft.

The rotor is almost universally used to control the helicopter. In forward flight, fixed aerodynamic surfaces such as a horizontal stabilizer and elevator may be used as well. There have been designs using nonrotating aerodynamic surfaces operating in the rotor wake to provide control in both hover and forward flight, but none has been successful. The rotor controls consist of cyclic and collective pitch. A collective pitch change gives a change in the mean blade angle of attack, which produces a change in the thrust magnitude. Cyclic pitch control gives a 1/rev pitch motion in the rotating frame, which produces a tip-path-plane tilt. The thrust vector tilts with the tip-path plane, producing a moment about the helicopter center of gravity below the rotor hub. With an offset-hinge articulated rotor or a hingeless rotor, the tip-path-plane tilt also produces a moment at the rotor hub directly. Thus command of the rotor collective and cyclic pitch gives efficient control over the magnitude and direction of the rotor thrust vector. When the rotor is operated at constant speed, the blade pitch bearings and collective pitch control are required in any case for thrust management. Then using cyclic control as well does not particularly increase the mechanical complexity of the rotor. The 1/rev pitch change of the blades required for cyclic control is obtained using a swashplate mechanism of some kind (see section 5–1).

The pilot's controls for the helicopter consist of a cyclic stick for control of longitudinal and lateral moments; a collective stick for control of the vertical force; foot pedals for control of the yaw moment; and a throttle for control of the rotor speed and torque (power management). These controls are similar in function to those of the airplane, with the addition of the collective stick, which is used for direct height control in hover and low speed flight. In forward flight, the collective control is used mainly for thrust trim. The cyclic stick is in the pilot's right hand and has fore-aft and lateral motions similar to an airplane control stick. The collective stick in the pilot's left hand has basically a vertical motion. The throttle is generally on the collective grip. The pedals follow airplane convention for directional control. It is important to maintain the rotor speed at the proper value. Since the rotor power required varies with both thrust and forward speed, it is necessary to coordinate the throttle with the collective and cyclic stick motions. A speed governor on the engine to automatically handle the power management is desirable, since it greatly reduces the pilot's work load. Basically, the cyclic stick controls the longitudinal and lateral motions in hover, but the helicopter is characterized by considerable coupling between the controls. The manner in which the pilot's cyclic and collective control sticks are connected to the cyclic and collective pitch of each rotor depends on the helicopter configuration. The connection of the pilot's controls to the rotor can be by a direct mechanical linkage (at least for small helicopters), or electro-hydraulic actuators can be used to produce or augment the rotor control inputs commanded by displacements of the pilot's sticks.

The manner in which the actions of the main and auxiliary rotors are combined to produce the required control moments and forces depends on the helicopter configuration. Table 15-1 summarizes how the control is accomplished for the major shaft-driven helicopter configurations. The rotor cyclic tilts the tip-path plane, and thereby tilts the thrust vector and produces a hub moment. The rotor collective changes the thrust magnitude. For all configurations, vertical control is accomplished by changing the main rotor thrust magnitude using collective pitch. Longitudinal and lateral control is generally obtained using cyclic pitch to tilt the tip-path plane of the main rotor. When there are two, nonaxial main rotors, however, one axis of the helicopter can be controlled by means of differential main rotor collective pitch changes. The means of producing a yaw moment for directional control is closely related to the manner in which balance of the main rotor torque is achieved, and as a result it varies with the helicopter configuration.

Table 15-1 Helicopter Control

Helicopter Configuration	Torque Balance	Longitudinal Control	Lateral Control	Height Control	Directional Control
		Pitch Moment	Roll Moment	Vertical Force	Yaw Moment
Single main rotor and tail rotor	tr thrust	mr cyc	mr cyc	mr coll	tr coll
Coaxial	mr diff torque	mr cyc	mr cyc	mr coll	mr diff coll
Tandem	mr diff torque	mr diff coll	mr cyc	mr coll	mr diff cyc
Side-by-side	mr diff torque	mr cyc	mr diff coll	mr coll	mr diff cyc

Note: mr = main rotor, tr = tail rotor, cyc = cyclic, coll = collective, diff = differential

In systems with a direct mechanical linkage between the pilot's controls and the rotor, blade feathering moments are transmitted through the control system to the pilot's sticks. The collective stick force comes from the mean blade pitch moment, and the cyclic stick forces come from the 1/rev pitch moments. The proper behavior of these control forces is important for good handling qualities. The general requirements for good control forces are low friction, low vibration, and logical control force transients. A means for trimming the control forces in steady flight is also required. It is desirable to have a moderate but increasing force gradient opposing any stick motion. In a helicopter with a mechanical control system, the feathering moments and hence the stick forces are sensitive to the blade dynamics and geometry. Moreover, the vibratory blade pitch moments as well as the steady loads are transmitted through the control system. There will be N/rev stick forces, which in the collective control system are due to the N/rev feathering moments and in the cyclic control system are due to $N \pm 1$/rev moments. There may also be 1/rev stick vibration due to unbalanced aerodynamic and inertial forces of the blades. For low steady and vibratory control forces, first of all the proper airfoil section must be chosen; in particular, it should have a low moment coefficient about the aerodynamic center. The chordwise offsets of the aerodynamic center and center of gravity from the feathering axis must be small also. The blades must be well balanced and tracked to eliminate the differences that lead to 1/rev vibratory control loads. Because stick forces increase with the helicopter size, larger helicopters generally use servo-actuators between the pilot's sticks and the rotor controls. With irreversible actuators, the blade pitch moments will not be transmitted back to the pilot. The control forces must be provided by the actuators or an automatic force-feel system in this case.

The control required to trim the helicopter in a desired equilibrium flight state can be calculated by an aeroelastic analysis such as described in Chapter 14. To design the control system and insure that the helicopter will have the proper control characteristics, the positions of the pilot's sticks must be calculated over the operating range of the aircraft, particularly as a function of speed, gross weight, and center-of-gravity position. In the trim solution the pilot's collective, longitudinal cyclic, lateral cyclic, and pedal positions and the aircraft pitch and roll attitude are iteratively determined so that the six force and moment components on the helicopter are simultaneously zero. It is necessary to solve the rotor equations of motion for the fundamental flap mode at least, and for some quantities such as lateral cyclic a very good rotor aerodynamic model is required. A complete analysis of the helicopter trim is thus fairly complex. While an entirely accurate prediction of the helicopter control is not currently possible with such an analysis, significant improvements over the simple analytical solutions have been made. For example, the lateral flapping prediction is much improved when the rotor nonuniform inflow is included. The basic analysis developed in chapter 5 gives the principal characteristics of the control required to trim the helicopter. The conditions for force and moment equilibrium of the helicopter determine the orientation of the rotor hub plane and tip-path plane required for a given flight state. The flapping relative to the shaft, β_{1c} and β_{1s}, is therefore determined as a function of the helicopter center-of-gravity position and speed. The cyclic and collective control required to obtain this rotor thrust and tip-path-plane orientation was derived in section 5–5 for an articulated rotor by solving the flapping equation of motion. For a rotor with a general flap frequency ($\nu \geqslant 1$), the result is:

$$\theta_{.75} = \frac{\left(1+\frac{3}{2}\mu^2\right)\left(\frac{6C_T}{\sigma a}+\frac{3}{8}\mu^2\theta_{tw}\right)+\frac{3}{2}\lambda_{TPP}\left(1-\frac{1}{2}\mu^2\right)-\frac{3}{2}\mu\frac{\nu^2-1}{\gamma/8}\beta_{1s}}{1-\mu^2+\frac{9}{4}\mu^4}$$

$$\theta_{1s} = -\beta_{1c} - \frac{\frac{8}{3}\mu\left(\frac{6C_T}{\sigma a}+\frac{3}{8}\mu^2\theta_{tw}\right)+2\mu\lambda_{TPP}\left(1-\frac{3}{2}\mu^2\right)-\left(1+\frac{3}{2}\mu^2\right)\frac{\nu^2-1}{\gamma/8}\beta_{1s}}{1-\mu^2+\frac{9}{4}\mu^4}$$

$$\theta_{1c} = \beta_{1s} + \frac{\frac{4}{3}\mu\beta_0 + \frac{\nu^2-1}{\gamma/8}\beta_{1c}}{1+\frac{1}{2}\mu^2}$$

$$\begin{aligned}\beta_0 = \frac{\nu^2-1}{\nu^2}\beta_p + \frac{\gamma/8\nu^2}{1-\mu^2+\frac{9}{4}\mu^4}\Bigg\{&\left(1-\frac{19}{18}\mu^2+\frac{3}{2}\mu^4\right)\frac{6C_T}{\sigma a}\\ &+\left(\frac{1}{20}+\frac{29}{120}\mu^2-\frac{1}{5}\mu^4+\frac{3}{8}\mu^6\right)\theta_{tw}\\ &+\left(\frac{1}{6}-\frac{7}{12}\mu^2+\frac{1}{4}\mu^4\right)\lambda_{TPP}\\ &-\frac{\mu}{6}\left(1-3\mu^2\right)\frac{\nu^2-1}{\gamma/8}\beta_{1s}\Bigg\}\end{aligned}$$

This result neglects tip losses and reverse flow, assumes uniform inflow, and uses $\eta = r$ for the flap mode shape. The basic behavior of the rotor control is contained in these expressions. The rotor collective pitch varies directly with the helicopter gross weight. The collective pitch varies significantly with speed also, mainly as a result of the variation of the inflow ratio with μ. The collective required first decreases as the helicopter speed increases from hover, because of the decrease in the induced velocity; then at high speed it increases again, because of the helicopter parasite drag contribution to $\lambda_{TPP} = \mu\alpha_{TPP} + \lambda_i$. The results for the cyclic pitch were discussed in section 5–5 in terms of the flapping response. They can also be interpreted in terms of the rotor disk moments that must be supplied by the rotor control. A lateral moment on the disk is required to sustain a longitudinal tip-path-plane tilt β_{1c}, due to the flap velocity in the rotating frame. This moment is supplied by the longitudinal cyclic θ_{1s}, which cancels the blade angle-of-attack change due to the flapping. There is also a lateral moment on the disk due to the higher velocity on the advancing side in forward flight and requiring a longitudinal cyclic pitch proportional to the advance ratio μ. Finally, lateral tip-path-plane tilt β_{1s} requires a lateral hub moment proportional to $(\nu^2 - 1)$, which is provided by longitudinal cyclic. The terms in the expression for the lateral cyclic θ_{1c} have a similar origin. A forward shift of the helicopter center of gravity requires an aft tilt of the tip-path plane ($\Delta\beta_{1c} < 0$) to tilt the thrust vector and maintain pitch moment equilibrium. The longitudinal cyclic control system is rigged so that rearward stick displacement produces a pitch-up moment on the helicopter by tilting the rotor thrust vector aft. The aft tilt of the control plane required by the center of gravity shift ($\Delta\theta_{1s} > 0$, from the above expressions) then corresponds to a rearward displacement of the cyclic stick. Similarly, a shift of the center of gravity to the right requires a leftward tilt of the tip-path plane and control plane, and hence a left displacement of the cyclic stick. The tip-path plane tilts rearward and to the advancing side relative to the control plane, roughly in proportion to the advance ratio (see Chapter 5). Thus a forward tilt of the control plane is required to maintain the tip-path-plane orientation as speed increases, and therefore a forward displacement of the cyclic stick, which is the desired control motion for increased speed. However, a lateral cyclic stick displacement is also required as speed increases, to counter the lateral flapping. The lateral flapping is sensitive to nonuniform inflow, which generally increases this undesirable coupling, particularly at low speeds.

15–2 Stability

Now let us examine helicopter flying qualities in terms of the dynamic stability and response to control of the aircraft rigid body motions. The six rigid body degrees of freedom are the basic motion involved in the flight dynamics analysis. Naturally, the rotor is a major factor in the analysis, and therefore it is necessary to consider the rotor degrees of freedom as well, particularly the flap motion. Two major simplifications will be considered for the analysis developed here. First, it will usually be assumed that the lateral and longitudinal dynamics of the helicopter are separable, which reduces by a factor of two the number of degrees of freedom that must be dealt with at one time. Secondly, only the low frequency dynamics of the rotor will be used, so that the rotor dynamics do not add separate degrees of freedom to the system. Instead, the rotor is just a source of hub forces and moments due to the shaft motion and controls. The resulting model of the helicopter is not entirely correct. Because of the rotation of the rotor the helicopter does not have lateral symmetry, except for certain configurations (such as side-by-side or coaxial), so there is in fact considerable coupling of the longitudinal and lateral motions. There are also circumstances in which the low frequency rotor response is not an adequate representation of the rotor influence on the flight dynamics. Generally, however, the helicopter model with these two approximations does retain the basic characteristics of the flight dynamics, and the model is simple enough for analytical work. Indeed, the low frequency rotor response is usually an acceptable approximation even with more complex analyses.

Since flying qualities of the helicopter have different characteristics in hover and in forward flight, these two regimes will be analyzed separately. The hovering analysis is simpler, because of the axisymmetry of the rotor aerodynamics in vertical flight. The analysis will be developed primarily for the single main rotor and tail rotor configuration, followed by a consideration of the special characteristics of twin main rotor configurations, particularly the tandem helicopter. Another basic assumption will be that the rotor speed is constant, which implies intervention by the pilot or an automatic governor system during the helicopter motion

15–3 Flying Qualities in Hover

15-3.1 Equations of Motion

Consider the rigid body motions of a single main rotor and tail rotor helicopter in hover. It is assumed that the aircraft has complete axisymmetry, so that the vertical and the longitudinal-lateral dynamics are completely separated. Such separation is a basic feature of the rotor in hover, and generally holds for the hover flying qualities even though the entire helicopter is not exactly axisymmetric. A body axis system is used, with the origin at the helicopter center of gravity. The z-axis is vertical with the center of gravity directly under the rotor hub. The x-axis is positive forward, the y-axis is positive to the right, and the z-axis is positive downward (the usual convention for stability and control analyses). It is assumed that these axes are the principal axes of the aircraft, so inertial cross-coupling of the helicopter yaw and roll is neglected. Because the effects of the tail rotor are neglected except on the helicopter yaw motion, the yaw dynamics are decoupled from the other degrees of freedom. The direction of rotor rotation discriminates between the left and right sides of the rotor, so the helicopter does not have a lateral symmetry plane. However, the analysis will assume at least initially that the longitudinal and lateral dynamics are decoupled also.

Fig. 15-1 shows the rigid body degrees of freedom of the aircraft. The perturbed displacement of the center of gravity has the components x_B, y_B, and z_B; the roll, pitch, and yaw rotations of the helicopter about the center of gravity are given by the Euler angles ϕ_B, θ_B, and ψ_B. The main rotor hub is a distance h above the center of gravity, and the tail rotor a distance ℓ_{tr} aft. The linear and angular displacements of the main rotor hub are thus:

$$
\begin{aligned}
x_h &= -x_B + h\theta_B \\
y_h &= y_B + h\phi_B \\
z_h &= -z_B \\
\alpha_x &= -\phi_B \\
\alpha_y &= \theta_B \\
\alpha_z &= -\psi_B
\end{aligned}
$$

(Figures 9-10 and 11-2 show the definition of rotor hub motion; note that the directions of the x and z-axes are reversed.) Neglecting the trim Euler angle terms, the helicopter angular velocity has components $p = \dot{\phi}_B$, $q = \dot{\theta}_B$,

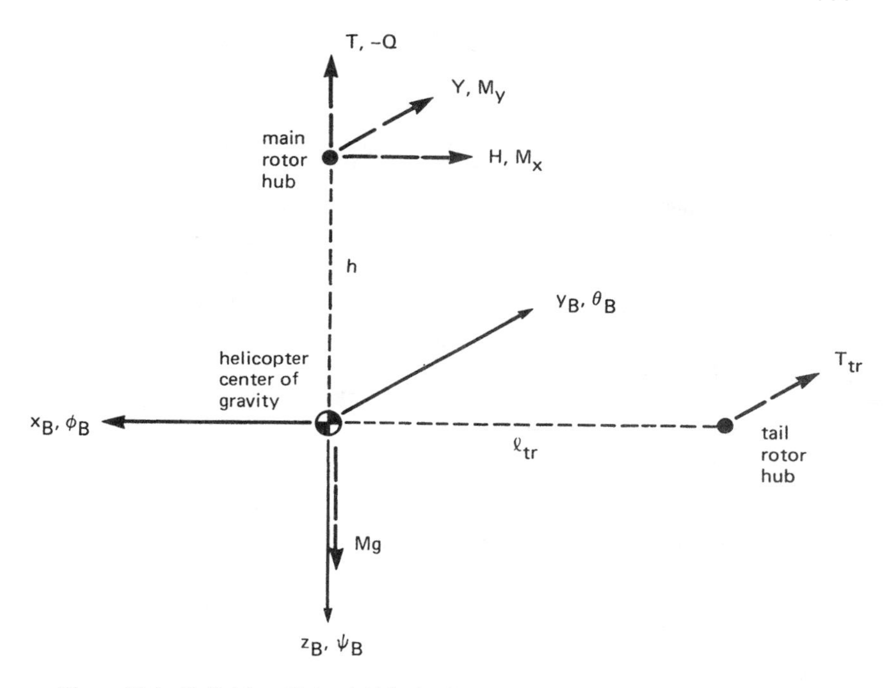

Figure 15-1 Definition of the rigid-body degrees of freedom of a helicopter, and the forces and moments due to the main rotor, tail rotor, and gravity.

and $r = \dot{\psi}_B$. The components of the linear velocity perturbation are $u = \dot{x}_B$, $v = \dot{y}_B$, and $w = \dot{z}_B$. Dimensionless quantities are used, with the linear displacement, velocity, and acceleration based on R, ΩR, and $\Omega^2 R$, respectively, and the angular velocity and acceleration based on Ω and Ω^2.

The helicopter has mass M and moments of inertia I_x, I_y, and I_z about the roll, pitch, and yaw axes respectively. The rotor mass is included in the aircraft inertia. Fig. 15-1 also shows the forces and moments acting on the helicopter in hover: the main rotor hub reactions, the tail rotor thrust, and the gravitational force. The differential equations for the rigid body degrees of freedom are thus:

$$
\begin{aligned}
M\ddot{x}_B &= -H - Mg\theta_B \\
M\ddot{y}_B &= Y + Mg\phi_B \\
M\ddot{z}_B &= -T
\end{aligned}
$$

$$I_x \ddot{\phi}_B = -M_x + hY$$
$$I_Y \ddot{\theta}_B = M_y + hH$$
$$I_z \ddot{\psi}_B = Q - \ell_{tr} T_{tr}$$

Dimensionless quantities (based on ρ, Ω, and R) are used, and the equations will be normalized by dividing by $\frac{1}{2}NI_b$ (where N is the number of blades and I_b is the characteristic inertia of the blade) to place the rotor forces in helicopter coefficient form. Define the normalized inertias as follows: $M^* = M/(\frac{1}{2}NI_b)$, $I_x^* = I_x/(\frac{1}{2}NI_b)$, and similarly for the pitch and yaw inertias. Then the normalized equations of motion are:

$$M^* \ddot{x}_B = -\gamma \frac{2C_H}{\sigma a} - M^* g \theta_B$$

$$M^* \ddot{y}_B = \gamma \frac{2C_Y}{\sigma a} + M^* g \phi_B$$

$$M^* \ddot{z}_B = -\gamma \frac{2C_T}{\sigma a}$$

$$I_x^* \ddot{\phi}_B = \gamma \left(-\frac{2C_{M_X}}{\sigma a} + h \frac{2C_Y}{\sigma a} \right)$$

$$I_y^* \ddot{\theta}_B = \gamma \left(\frac{2C_{M_Y}}{\sigma a} + h \frac{2C_H}{\sigma a} \right)$$

$$I_z^* \ddot{\psi}_B = \gamma \left(\frac{2C_Q}{\sigma a} - \ell_{tr} \frac{(\sigma a A(\Omega R)^2)_{tr}}{(\sigma a A(\Omega R)^2)_{mr}} \left(\frac{2C_T}{\sigma a} \right)_{tr} \right)$$

Note that since vertical force trim gives $Mg = T$ dimensionally, M^* can be evaluated from the trim rotor thrust coefficient: $M^* g = \gamma(2C_T/\sigma a)_{trim}$. The moments of inertia can be written in terms of radii of gyration, for example $I_x^* = M^* k_x^2$. The dimensionless gravitational constant g has been divided by $\Omega^2 R$, so it tends to increase with rotor radius (assuming that the tip speed is fixed). It follows that the normalized helicopter mass M^* tends to decrease with the helicopter size roughly in proportion to R^{-1}.

In the sections to follow, the rotor forces and moments will be linearly expanded in terms of the body motion perturbations (after first dividing by the helicopter mass M^* or the appropriate moment of inertia). The coefficients of these expansions are the stability derivatives. The notation for the longitudinal, lateral, and vertical force derivatives will be X, Y, and Z respectively; and for the roll, pitch, and moment derivatives L, M, and N. The directions of these force and moment components follow the definition of the body axes (Fig. 15-1). Derivatives with respect to the linear velocity of the helicopter will be designated by a subscript u, v, or w; and with respect to the angular velocity by a subscript p, q, or r. These stability derivatives are dimensionless, based on the rotor radius and the rotor rotational speed. The dimensional derivatives are readily obtained by inserting factors of R and Ω as required. Note that the forces divided by the helicopter mass (for example $Z = -\gamma(2C_T/\sigma a)/M^*$) have dimensions of linear acceleration ($\Omega^2 R$), and the moments divided by the moment of inertia have dimensions of angular acceleration (Ω^2). Also, derivatives with respect to the helicopter linear velocity should be divided by ΩR, and derivatives with respect to angular velocity should be divided by Ω.

Assuming that the motion involved in helicopter flight dynamics is slow as far as the rotor is concerned, the low frequency or quasistatic response is a sufficient model of the rotor. The low frequency rotor response including the effects of the flap motion was derived in section 12–1, where the hub reactions due to shaft motion, pitch control, and aerodynamic gusts are given. The low frequency flap response is just the static solution, obtained from algebraic rather than differential equations of motion, and therefore the flap motion does not add degrees of freedom to the system. For hover the rotor dynamics decouple into vertical and longitudinal-lateral systems. The low frequency response:

$$\frac{C_T}{\sigma a} = C' \frac{\nu^2}{\nu_e^2} \left[T_\theta \theta_0 - \tfrac{1}{2} T_\lambda (\dot{z}_B + w_G) \right]$$

was derived in section 12-1.4; the aerodynamic coefficients are given in section 11–7. The factor of one-half in the $(\dot{z}_B + w_G)$ term is the result of the inflow perturbation due to the vertical velocity of the rotor. The lift deficiency function C' accounts for the influence of the rotor wake-induced velocity; the factor ν^2/ν_e^2 is the sole effect of the blade coning motion (and is due to the pitch-cone coupling K_P). The inertial reactions of the rotor have been accounted for by including the rotor mass in the helicopter mass. The thrust perturbations due to collective pitch and vertical velocity of the helicopter are produced by direct changes in the blade angle of attack. Since vertical velocity $\dot{z}_B$ increases the angle of attack, T_λ is negative.

The low frequency response of the rotor tip-path-plane tilt was derived in section 12-1.3. Substituting for the hub motion in terms of the body degrees of freedom give

$$\begin{pmatrix}\beta_{1c}\\ \beta_{1s}\end{pmatrix} = \frac{1}{1+N_*^2}\begin{bmatrix}1 & N_*\\ -N_* & 1\end{bmatrix}\left\{\frac{M_\theta}{-M_{\dot\beta}}\begin{pmatrix}-\theta_{1s}\\ \theta_{1c}\end{pmatrix} + \frac{M_\mu}{-M_{\dot\beta}}\begin{pmatrix}-\dot{x}_B + h\dot\theta_B - u_G\\ \dot{y}_B + h\dot\phi_B + v_G\end{pmatrix} + \frac{2I^*_{\beta a}}{-\gamma M_{\dot\beta}}\begin{pmatrix}\dot\theta_B\\ \dot\phi_B\end{pmatrix} + \begin{pmatrix}-\dot\phi_B\\ \dot\theta_B\end{pmatrix}\right\}$$

where the aerodynamic coefficients are

$$\frac{M_\theta}{-M_{\dot\beta}} = 1$$

$$\frac{M_\mu}{-M_{\dot\beta}} = 8\left(\frac{2C_T}{\sigma a} + \frac{\lambda_{HP}}{4}\right)$$

$$\frac{2I^*_{\beta a}}{-\gamma M_{\dot\beta}} = \frac{16}{\gamma}$$

(assuming uniform inflow, no tip loss, and $\eta = r$), and

$$N_* = \frac{\nu_e^2 - 1}{-\gamma M_{\dot\beta}} = \frac{\nu^2 - 1}{\gamma/8} + K_P$$

The parameter N_* defines the lateral-longitudinal coupling of the rotor response due to a flap frequency $\nu > 1$, or to pitch-flap coupling. This coupling produces a decrease in the magnitude of the tip-path-plane response by $(1 + N_*^2)^{-1/2}$ and a phase shift by $\Delta\psi = \tan^{-1}N_*$, as discussed in section 12-1.3. Longitudinal cyclic θ_{1s}, longitudinal hub velocity $(\dot{x}_B - h\dot\theta_B + u_G)$, and shaft rolling velocity $\dot\phi_B$ produce a lateral aerodynamic moment on the rotor disk. The rotor flap motion responds with maximum amplitude 90° after the maximum excitation ($90° - \tan^{-1}N_*$ in general), and hence with longitudinal tip-path-plane tilt β_{1c}. The resulting flapping velocity in the rotating frame then produces a lateral aerodynamic moment on the disk. The tip-path plane tilts until the aerodynamic flap moments are in equilibrium.

Since this new equilibrium position is reached quickly, the static response can be used for the low frequency rotor dynamics. The transients before equilibrium is achieved are not important for the slow inputs likely to be involved in helicopter flight dynamics. The lateral tip-path-plane tilt response β_{1s} has a similar origin. The remaining term in the flap response is the lag of the tip-path-plane tilt required to precess the rotor disk to follow the shaft angular velocity. A pitching rate $\dot{\theta}_B$ requires a roll moment on the rotor disk in order for the tip-path plane to follow the shaft; this roll moment is provided by the longitudinal flapping β_{1c}. Similarly the tip-path plane follows a shaft rolling velocity $\dot{\phi}_B$, with a steady lateral tilt β_{1s} relative to the shaft. The low frequency response of the rotor hub forces and moments was obtained in section 12-1.4:

$$\begin{pmatrix} \dfrac{2C_H}{\sigma a} \\ \dfrac{2C_Y}{\sigma a} \end{pmatrix} = \begin{bmatrix} -\left(\dfrac{2C_T}{\sigma a} + H_\beta^* + K_P R_\theta\right) & R_{\dot\beta} - K_P H_\theta \\ -R_{\dot\beta} + K_P H_\theta & -\left(\dfrac{2C_T}{\sigma a} + H_\beta^* + K_P R_\theta\right) \end{bmatrix} \begin{pmatrix} \beta_{1c} \\ \beta_{1s} \end{pmatrix}$$

$$+ \begin{bmatrix} -H_\theta & R_\theta \\ -R_\theta & -H_\theta \end{bmatrix} \begin{pmatrix} -\theta_{1s} \\ \theta_{1c} \end{pmatrix} + \begin{bmatrix} -R_{\dot\beta} & -H_{\dot\beta} \\ H_{\dot\beta} & -R_{\dot\beta} \end{bmatrix} \begin{pmatrix} \dot\theta_\beta \\ \dot\phi_\beta \end{pmatrix}$$

$$+ \begin{bmatrix} -(H_\mu + R_\mu) & R_r \\ -R_r & -(H_\mu + R_\mu) \end{bmatrix} \begin{pmatrix} -\dot x_B + h\dot\theta_B - u_G \\ \dot y_B + h\dot\phi_B + v_G \end{pmatrix}$$

$$\begin{pmatrix} -\dfrac{2C_{M_y}}{\sigma a} \\ \dfrac{2C_{M_x}}{\sigma a} \end{pmatrix} = \frac{(\nu^2 - 1)}{\gamma} \begin{pmatrix} \beta_{1c} \\ \beta_{1s} \end{pmatrix}$$

The aerodynamic coefficients are given in section 11–7; in particular,

$$H_{\dot\beta} = \frac{C_T}{\sigma a} + H_{\dot\beta}^* = \frac{C_T}{\sigma a} - \frac{\lambda_{HP}}{4}$$

The principal hub force is the in-plane component of the thrust vector when it is tilted with the tip-path plane. Combined with the hub moment, it gives a total moment about the helicopter center of gravity equal to

$$\begin{pmatrix} -\dfrac{2C_{M_y}}{\sigma a} - h\dfrac{2C_H}{\sigma a} \\ \\ \dfrac{2C_{M_x}}{\sigma a} - h\dfrac{2C_Y}{\sigma a} \end{pmatrix} = \left[\frac{(\nu^2-1)}{\gamma} + h\frac{2C_T}{\sigma a}\right]\begin{pmatrix} \beta_{1c} \\ \\ \beta_{1s} \end{pmatrix} - h\begin{pmatrix} \dfrac{2C_H}{\sigma a} \\ \\ \dfrac{2C_Y}{\sigma a} \end{pmatrix}_{TPP}$$

where the last term is due to the in-plane hub forces relative to the tip-path plane. The pitch and roll moments have a greater role in the helicopter flight dynamics than the in-plane hub forces. Note that for an articulated rotor the moments and forces are always directly proportional. The hub forces and moments are determined primarily by the rotor flap response. Longitudinal and lateral control plane tilt produce respectively pitch and roll moments about the center of gravity, which the pilot can use to control the aircraft. Longitudinal hub velocity $\dot{x}_B$ produces an in-plane force C_H opposing the motion, and a corresponding pitch moment responsible for the speed stability of the helicopter. Similarly, there is a side force C_Y due to lateral velocity $\dot{y}_B$, and a corresponding roll moment similar to the dihedral effect of an airplane wing. By the same mechanism, rotor hub reactions due to longitudinal and lateral gusts are produced. Pitching rate $\dot{\theta}_B$ of the helicopter produces a pitch moment C_{M_y} due to the lag of the tip-path plane required to precess the rotor; similarly, rolling rate $\dot{\phi}_B$ produces a roll moment C_{M_x}. Since these moments oppose the helicopter motion, the rotor provides damping of the helicopter angular motion. There are two major effects that occur when $\nu > 1$, as is the case with an offset flap hinge or a hingeless rotor. First, the moments are increased as a result of the hub moment capability, particularly for hingeless rotors. Secondly, the lateral and longitudinal responses of the rotor are coupled (since $N_* \neq 0$).

15-3.2 Vertical Dynamics

The conditions for vertical force equilibrium give the equation of motion for the helicopter vertical velocity. Substituting for the main rotor thrust perturbations due to collective pitch control and the net vertical velocity $(\dot{z}_B + w_G)$ gives

$$M^*\ddot{z}_B = -\gamma\frac{2C_T}{\sigma a} = M^*(Z_{\theta_0}\theta_0 + Z_w(\dot{z}_B + w_G))$$

or

$$\ddot{z}_B - Z_w\dot{z}_B = Z_{\theta_0}\theta_0 + Z_w w_G$$

where the stability derivatives are

$$Z_w = \frac{\gamma C' T_\lambda}{M^*} = -\frac{\gamma C'/4}{M^*}$$

$$Z_{\theta_0} = -\frac{2\gamma C' T_\theta}{M^*} = -\frac{\gamma C'/3}{M^*}$$

and

$$M^* = \frac{1}{g}\gamma\left(\frac{2C_T}{\sigma a}\right)_{\text{trim}}$$

It has been assumed that there is no pitch-flap coupling, which is usually true for main rotors. The vertical dynamics of the helicopter are described by a first order differential equation with time constant

$$\tau_z = -\frac{1}{Z_w} = \frac{8C_T/\sigma a}{gC'}$$

Dimensionally, the time to half amplitude is typically $t_{1/2} = 28(C_T/\sigma)$ sec, or around 2 sec. Because the dimensional time constant increases with the rotor tip speed ΩR and the blade loading C_T/σ, it tends to vary little from one design to another, but for a given aircraft it increases with gross weight.

The expressions describing the vertical dynamics of the helicopter have no zeros, and there is a single pole typically at $s = Z_w = -0.01$ to -0.02. This dimensionless root is quite small, justifying the use of the rotor low frequency response. The static (long time) response to control is $\dot{z}_B/\theta_0 = -Z_{\theta_0}/Z_w = -4/3$, or dimensionally $\dot{z}_B/\theta_0 = -(4/3)\Omega R$. The static response is determined by the balance of the rotor aerodynamic forces and hence does not depend on the Lock number or the lift deficiency function. However, the inflow perturbation due to vertical velocity of the rotor reduces the vertical damping and increases the vertical control sensitivity of the helicopter by a factor of one-half in hover, since the larger mass flow through the rotor in climb decreases the induced velocity (see section 10-6.4).

Recall also that section 3–3 obtained the collective pitch change required to produce a small steady-state climb rate, including the effect of the smaller induced velocity: $\Delta\theta_0 = (3/4)\lambda_c$. This result corresponds to a control sensitivity of $\dot{z}_B/\theta_0 = -(4/3)$, as above. The short time response is

$$\ddot{z}_B = Z_{\theta_0}\theta_0 + Z_w w_G$$

or dimensionally (with the vertical acceleration in g's)

$$\frac{\ddot{z}_B}{g} = -\frac{C'}{6C_T/\sigma a}\left(\theta_0 + \frac{3}{4}\frac{w_G}{\Omega R}\right)$$

The response to both collective control and vertical gusts is large.

The rotor rotational speed and ground effect are also important factors in the vertical dynamics. If the throttle is fixed so that the rotor speed can vary during the vertical motions, the vertical damping and control power of the helicopter will be reduced. The additional degree of freedom (Ω) also introduces an overshoot since the vertical dynamics are then second order. The helicopter height control is therefore more difficult if the rotor speed is not fixed. An automatic governor to control the rotor speed eases the piloting task. There is no spring on vertical displacements of the helicopter, so for height control the pilot must provide the feedback of height to collective. Near the ground, however, ground effect provides a vertical spring term, due to the rotor induced velocity changes with height. Ground effect thus helps the tasks of precision height control in hover and the helicopter flare in landing.

15-3.3 Yaw Dynamics

The conditions for equilibrium of yaw moments about the center of gravity give the differential equation for the helicopter yaw motion. Only the yaw moments due to the tail rotor thrust will be considered. The control variable is the tail rotor collective pitch θ_{0tr}, and the axial damping of the tail rotor gives thrust perturbations due to the yaw rate and the lateral gusts. The thrust perturbations due to the helicopter side velocity will also be included. It is assumed, however, that the influence of ψ_B on the lateral dynamics is small, and therefore that the yaw motion is independent of the other degrees of freedom, with $\dot{y}_B$ appearing as just another input. Using the low frequency response of the tail rotor thrust, the equation of motion becomes

$$I_z^* \ddot{\psi}_B = -\gamma \ell_{tr} \frac{(\sigma a A(\Omega R)^2)_{tr}}{(\sigma a A(\Omega R)^2)_{mr}} \left(\frac{2C_T}{\sigma a}\right)_{tr}$$

$$= I_z^*(N_{\theta_{tr}} \theta_{0_{tr}} + N_v(\dot{y}_B + v_G) + N_r \dot{\psi}_B)$$

or

$$\ddot{\psi}_B - N_r \dot{\psi}_B = N_{\theta_{tr}} \theta_{0_{tr}} + N_v(\dot{y}_B + v_G)$$

where the stability derivatives are

$$N_r = -\frac{\gamma \ell_{tr}^2}{4I_z^*} \frac{(\sigma a A \Omega R)_{tr}}{(\sigma a A \Omega R)_{mr}} \left(C' \frac{\nu^2}{\nu_e^2}\right)_{tr}$$

$$N_v = -N_r/\ell_{tr}$$

$$N_{\theta_{tr}} = \frac{4}{3} \frac{(\Omega R)_{tr}}{(\Omega R)_{mr}} \frac{N_r}{\ell_{tr}}$$

and

$$I_z^* = \frac{k_z^2}{g} \gamma \left(\frac{2C_T}{\sigma a}\right)_{\text{trim}}$$

Thus the yaw dynamics are described by a first-order differential equation for the yaw rate, with time constant $\tau_r = -1/N_r$. The yaw time constant is generally about the same as the vertical time constant τ_z; typically, $\tau_r = 2$ sec. The steady-state response to directional control is

$$\frac{\dot{\psi}_B}{\theta_{0_{tr}}} = -\frac{N_{\theta_{tr}}}{N_r} = -\frac{4}{3\ell_{tr}} \frac{(\Omega R)_{tr}}{(\Omega R)_{mr}}$$

or dimensionally

$$\frac{\dot{\psi}_B}{\theta_{0_{tr}}} = -\frac{4}{3\ell_{tr}} (\Omega R)_{tr}$$

The tail length ℓ_{tr} is usually just slightly greater than the rotor radius, so the yaw rate commanded by the directional control tends to decrease with the helicopter size. After a small first-order lag, the tail rotor collective produces a high steady-state yaw rate even for large helicopters. The yaw time constant τ_r will be increased with positive pitch-cone coupling, by the factor $\nu_e^2/\nu^2 = (1 + \gamma K_P/8\nu^2)$. An increased time constant implies a reduction of the short time response. In particular, the response to lateral gusts can be reduced by 30% to 50% with large pitch-cone coupling. The steady-state yaw response to control is not influenced by pitch-cone coupling. As for the vertical motion, the tail rotor induced velocity perturbations due to its axial velocity have reduced the yaw damping and increased the control sensitivity.

Sideward velocity of the helicopter produces a change in tail rotor thrust, and hence a yaw of the helicopter. The steady-state response to $\dot{y}_B$ is $\dot{\psi}_B/\dot{y}_B = -N_v/N_r = 1/\ell_{tr}$. Thus the lateral and yaw motions of the helicopter in hover are coupled. Sideward velocity is produced by lateral cyclic control, but to maintain the helicopter heading a pedal control input is required as well. The tail rotor control required to prevent yaw during the lateral motions of the helicopter is $\theta_{0tr}/\dot{y}_B = -N_v/N_{\theta_{tr}} = 3/(4\Omega R)_{tr}$ (dimensionally). Neglecting the influence of the tail rotor thrust on the lateral dynamics, the poles of the yaw and lateral dynamics are still uncoupled, but a coordination of the pedal control with lateral cyclic is required. In forward flight this yaw moment due to sideslip is still present, and it provides the directional stability of the helicopter.

Assuming that the rotor speed is constant, the main rotor rotational damping torque produces a small increase in the helicopter yaw damping. The thrust changes in response to vertical motion and inputs also produce torque changes that couple the vertical and yaw control in such a way that a coordination of the pedal control with collective stick inputs is needed to maintain the helicopter heading during vertical motions. If the rotor speed is not fixed, the torque changes will be absorbed by a rotor speed perturbation instead of by the helicopter yaw motion.

Tail rotor design and operation are complicated. When the yaw rate or translational velocity is large, the tail rotor can be operating in the vortex ring state; it regularly operates in an adverse aerodynamic environment due to the wake of the main rotor, fuselage, and vertical tail. The directional control and yaw damping are greatly influenced by these factors. In general, however, the tail rotor is a powerful and efficient design solution for the torque balance, directional stability, and control of single main rotor helicopters.

15-3.4 Longitudinal Dynamics

15-3.4.1 Equations of Motion

The dynamics of a helicopter in hover separate into vertical and lateral-longitudinal motions. It is assumed for now that the lateral and longitudinal dynamics can also be analyzed separately. Such decoupling actually exists for the coaxial rotor configuration, and the uncoupled analysis is also applicable to the lateral dynamics of a tandem helicopter or the longitudinal dynamics of the side-by-side configuration. For the single main rotor and tail rotor configuration the principal characteristics of the helicopter flying qualities are contained in the separate analyses of the longitudinal and lateral dynamics, although the coupling must be considered for a complete model (see section 15-3.6).

Neglecting the influence of vertical and lateral motions, the longitudinal dynamics consist of just two degrees of freedom, pitch θ_B (positive nose upward) and longitudinal velocity $\dot{x}_B$ (positive forward). The inputs considered are longitudinal cyclic θ_{1s} and longitudinal aerodynamic gust velocity u_G. The conditions for equilibrium of longitudinal forces and pitch moments give the differential equations for the longitudinal motions of the helicopter:

$$M^*\ddot{x}_B = -\gamma\frac{2C_H}{\sigma a} - M^*g\theta_B$$

$$I_y^*\ddot{\theta}_B = \gamma\left(\frac{2C_{M_y}}{\sigma a} + h\frac{2C_H}{\sigma a}\right)$$

Using the low frequency response (section 15-3.1), the rotor reactions can be expanded in terms of the stability derivatives as follows:

$$-\gamma\frac{2C_H}{\sigma a} = M^*(X_\theta\theta_{1s} + X_u(\dot{x}_B + u_G) + X_q\dot{\theta}_B)$$

$$\gamma\left(\frac{2C_{M_y}}{\sigma a} + h\frac{2C_H}{\sigma a}\right) = I_y^*(M_\theta\theta_{1s} + M_u(\dot{x}_B + u_G) + M_q\dot{\theta}_B)$$

Then in Laplace form the equations of motion are

$$\begin{bmatrix} s - X_u & -X_q s + g \\ -M_u & s^2 - M_q s \end{bmatrix} \begin{pmatrix} \dot{x}_B \\ \theta_B \end{pmatrix} = \begin{pmatrix} X_\theta \\ M_\theta \end{pmatrix} \theta_{1s} + \begin{pmatrix} X_u \\ M_u \end{pmatrix} u_G$$

The characteristic equation is

$$\Delta = s^3 - (X_u + M_q)s^2 + (X_u M_q - X_q M_u)s + gM_u = 0$$

The three solutions of this polynomial are the poles of the helicopter longitudinal dynamics. The equations of motion invert to

$$\begin{pmatrix} \dot{x}_B \\ \theta_B \end{pmatrix} = \frac{1}{\Delta} \begin{pmatrix} X_\theta s^2 + (X_q M_\theta - X_\theta M_q)s - M_\theta g \\ M_\theta s + (X_\theta M_u - X_u M_\theta) \end{pmatrix} \theta_{1s}$$

$$+ \frac{1}{\Delta} \begin{pmatrix} X_u s^2 + (X_q M_u - X_u M_q)s - M_u g \\ M_u s \end{pmatrix} u_G$$

The longitudinal velocity response to cyclic and gusts has two zeros, while the pitch response has one zero.

15-3.4.2 Poles and Zeros

Consider now the case of an articulated rotor with no flap hinge offset (so the flap frequency $\nu = 1$) and no pitch-flap coupling. Because there is then no hub moment, the pitching moments are only due to the in-plane hub force:

$$I_y^* \ddot{\theta}_B = \gamma h \frac{2C_H}{\sigma a}$$

It follows that each pitch moment stability derivative is given by the product of the mast height h and the longitudinal force derivative, divided by $-I_y^*/M^* = -k_y^2$: $M_\theta = -(h/k_y^2)X_\theta$, $M_u = -(h/k_y^2)X_u$, and $M_q = -(h/k_y^2)X_q$. The characteristic equation then reduces to

$$\Delta = s^3 - (X_u + M_q)s^2 + gM_u = 0$$

From section 15-3.1, the stability derivatives for an articulated rotor are as follows:

$$X_\theta = -\frac{\gamma}{M^*}\frac{2C_T}{\sigma a} = -g$$

$$X_u = -\frac{\gamma}{M^*}\frac{2C_T}{\sigma a} 8M_\mu + \frac{\gamma}{M^*}\left[H_\theta\, 8M_\mu - (H_\mu + R_\mu)\right]$$

$$= -g8M_\mu + \frac{\gamma}{M^*}\left[H_\theta\, 8M_\mu - (H_\mu + R_\mu)\right]$$

$$X_q = \frac{\gamma}{M^*}\left(\frac{2C_T}{\sigma a} + H_{\dot\beta}^*\right)\frac{16}{\gamma} - hX_u$$

$$= g\frac{16}{\gamma}\left(1 + \frac{H_{\dot\beta}^*}{2C_T/\sigma a}\right) - hX_u$$

where the rotor aerodynamic coefficients are $M_\mu = 2C_T/\sigma a + \lambda_{HP}/4$ and $H_{\dot\beta}^* = -\lambda_{HP}/4$. Since $H_{\dot\beta}M_\mu/M_{\dot\beta}H_\mu \cong 1$, the speed stability is given primarily by the first term: $X_u \cong -g8M_\mu$. Thus for the articulated rotor the hub forces are primarily due to the tilt of the thrust vector with the tip-path plane, and hence they are proportional to $\gamma 2C_T/\sigma a$; the moments are entirely due to the hub forces acting a distance h above the center of gravity. The tip-path-plane tilt due to cyclic pitch is given by $-M_\theta/M_{\dot\beta} = 1$, the tilt due to helicopter speed perturbations is given by $8M_\mu$, and the tilt due to the helicopter angular velocity is given by $16/\gamma$. For pitch motions of the rotor the thrust vector does not remain perpendicular to the tip-path plane; rather, it lags the tip-path plane by a significant amount. The $2C_T/\sigma a$ term in X_q is the thrust tilt with the tip-path plane, and the $H_{\dot\beta}^*$ term is the additional direct hub force. Since $H_{\dot\beta}^*$ is negative,

$$\frac{H_{\dot\beta}^*}{2C_T/\sigma a} = -\frac{\sigma a}{16\lambda_{HP}}$$

[for hover, where $\lambda_{HP} = (C_T/2)^{1/2}$], it reduces the rotor pitch damping, perhaps by 30% at normal operating conditions and more at low loading.

Miller (1948) noted the existence of this term reducing the helicopter damping. Amer (1950) derived expressions for the helicopter rotor pitch and roll damping, including the effects of the direct hub forces since the resultant rotor force is not perpendicular to the tip-path plane for pitching or rolling motions. Amer found little influence of forward speed on this damping, up to about $\mu = 0.5$. These results were confirmed by flight tests, which indicated that the helicopter tends to have low damping at high speeds and in climb (where the inflow ratio is highest), and that unstable damping can occur during maneuvers at load factors well below $1\ g$. Using the assumption that the rotor force is perpendicular to the tip-path plane in order to calculate the pitch and roll damping can therefore give misleading results. The direct hub force term ($H_{\dot{\beta}}^{*}$) is always important. Amer and Gustafson (1951) showed that assuming the resultant rotor force is perpendicular to the tip-path plane leads to grossly incorrect stability derivatives.

In the characteristic equation for the articulated rotor helicopter, the s^2 coefficient is dominated by the pitch damping term M_q. If the $X_u s^2$ term is neglected for now, the characteristic equation becomes

$$\Delta \cong s^3 - M_q s^2 + g M_u = 0$$

To establish the behavior of the solution of this cubic equation for the hover roots, it can be viewed as the characteristic equation for a classical feedback system with gain $M_u > 0$ (for positive speed stability). The "open loop poles" are the solution when $M_u = 0$: two roots at the origin, $s = 0$; and a root on the negative real axis due to the pitch damping, $s = M_q = -(h/k_y^2)X_q$. The usual rules for constructing the root loci of systems with feedback can then be used to find the solutions of the characteristic equation when $M_u > 0$. Fig. 15-2 sketches the roots of the longitudinal dynamics obtained in this manner. The influence of the speed stability is to increase the pitch root on the real axis, and to produce a low frequency, slightly unstable oscillatory mode. If the X_u term is retained, the characteristic equation can be written as

$$s^3 - M_q s^2 + M_u[(k_y^2/h)s^2 + g] = 0$$

Hence the system with gain M_u really has two "open loop zeros" on the imaginary axis at $s = \pm i(gh/k_y^2)^{1/2}$. These "zeros" are quite large, and have no influence on the locus shown in Fig. 15-2 until the speed stability is well beyond the value appropriate to the helicopter rotor. Alternatively, the characteristic equation may be written so that the pitch damping M_q can be considered the feedback gain:

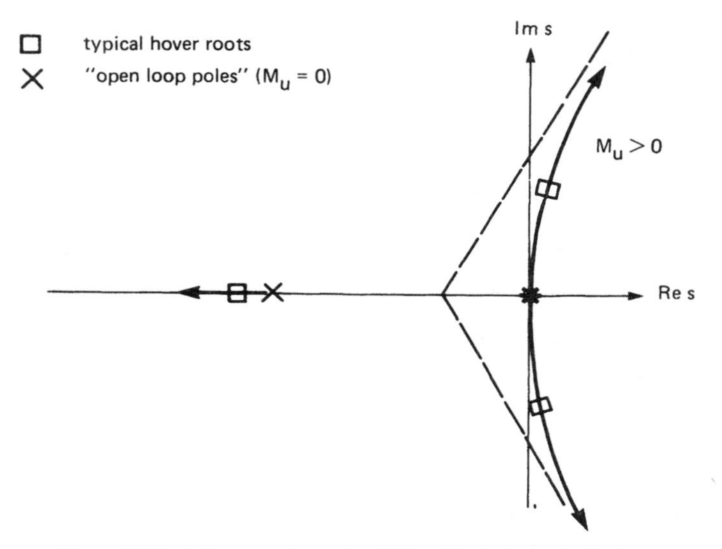

Figure 15-2 Influence of speed stability ($M_u > 0$) on the helicopter longitudinal roots.

$$s^3 + gM_u - M_q s^2 = 0$$

(again neglecting the $X_u s^2$ term). The three "open loop poles" are

$$s = -(gM_u)^{1/3},\ (gM_u)^{1/3}\tfrac{1}{2}(1 + i\sqrt{3}),\ \text{and}\ (gM_u)^{1/3}\tfrac{1}{2}(1 - i\sqrt{3})$$

and there are two "open loop zeros" at the origin. The influence of positive damping ($M_q < 0$) is sketched in Fig. 15-3. Again it is found that the hover roots of the longitudinal dynamics are a stable real root and an unstable complex pair.

The uncoupled pitch and longitudinal motions of the hovering helicopter both have positive damping; the former has the pitch damping M_q and the latter the drag damping X_u. The instability of the longitudinal dynamics for hover must therefore be a result of the coupling of the motion by the pitch moments due to longitudinal velocity (the speed stability M_u) and the longitudinal component of the gravitational force due to pitch. The approximate characteristic equation

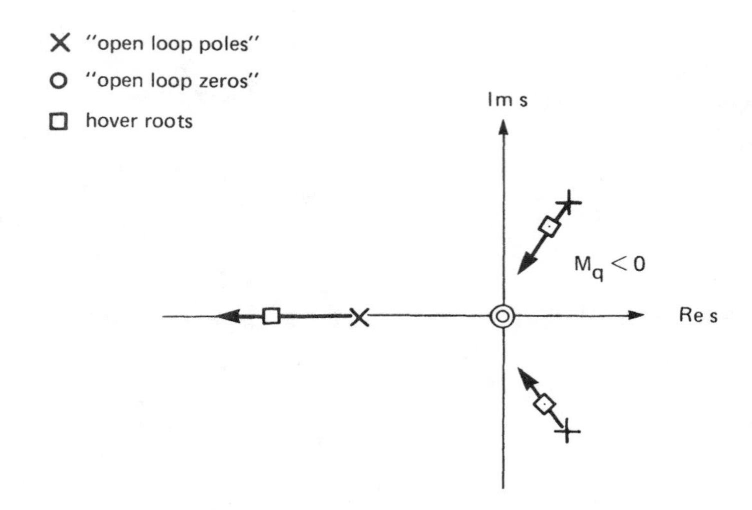

Figure 15-3 Influence of pitch damping ($M_q < 0$) on the helicopter longitudinal roots.

$$s^3 - M_q s^2 + gM_u = 0$$

is equivalent to the following differential equations:

$$\begin{bmatrix} s & g \\ -M_u & s^2 - M_q s \end{bmatrix} \begin{pmatrix} \dot{x}_B \\ \theta_B \end{pmatrix} = 0$$

These equations contain the essence of the longitudinal dynamics in hover. [Compare them with the complete equations given in section 15-3.4.1; the $(h^2/k_y^2)X_u$ term in M_q can also be neglected here.] The longitudinal dynamics may thus be interpreted as follows. Consider an oscillatory motion at frequency ω. Because of the rotor speed stability M_u, longitudinal velocity $\dot{x}_B$ produces a nose-up pitch moment. This moment is balanced by the pitch inertia, so the pitch response is 180° out of phase with the longitudinal velocity. The pitch equation gives $\theta_B = -(M_u/\omega^2)\dot{x}_B$, so forward velocity $\dot{x}_B$ produces a nose-down pitch motion. The pitch motion rotates the body axes relative to the vertical, so there is a forward component of the gravitational force. Thus in the longitudinal velocity equation there is a force $g\theta_B = -(g/\omega^2)M_u\dot{x}_B$, which is negative damping, and therefore the coupled motion is unstable. The pitch moments are also reacted by the pitch damping

M_q, which strongly influences the phase and frequency of the coupled motion.

For a more general view of the role of the speed stability, consider the divergence and flutter stability given by the characteristic equation. The requirement for static or divergence stability is that the constant term in the characteristic equation be positivc, which is satisfied since $M_u > 0$. The requirement for dynamic or flutter stability can be obtained by applying Routh's criterion. The coefficients of the characteristic equation are positive, so all the roots are in the left-hand plane (stable) if

$$gM_u + (X_u + M_q)(X_u M_q - X_q M_u) < 0$$

The second term is zero for an articulated rotor, so the criterion reduces to $M_u < 0$, and the motion is unstable. The speed stability thus has a dominant role in determining the stability of the helicopter dynamics in hover. Because the criteria for static and dynamic stability are conflicting, the helicopter motion will be unstable regardless of the sign or magnitude of M_u (see Fig. 15-2).

For an articulated rotor, the response becomes

$$\begin{pmatrix} \dot{x}_B \\ \theta_B \end{pmatrix} = \frac{1}{\Delta}\left(X_\theta \theta_{1s} + X_u u_G\right)\begin{pmatrix} s^2 + gh/k_y^2 \\ -(h/k_y^2)s \end{pmatrix}$$

The pitch response to cyclic control and gusts has a single zero at the origin, $s = 0$. The longitudinal velocity response has two zeros on the imaginary axis at $s = \pm i(gh/k_y^2)^{1/2}$. These zeros usually have several times the magnitude of the poles; hence they only influence the response for large gain.

In summary, the helicopter longitudinal dynamics are characterized by three roots: a negative real root (stable convergence) principally due to the pitch damping of the rotor; and a complex conjugate pair in the right half-plane (a mildly unstable oscillation) due to the coupling of the pitch and longitudinal velocity by the speed stability derivative M_u. Typically for an articulated rotor the real root or short period mode has a time to half amplitude of $t_{1/2} = 1$ to 2 sec. The oscillatory root or long period mode typically has a period of $T = 10$ to 20 sec (a frequency of 0.05 to 0.10 Hz), and a time to double amplitude of $t_2 = 3$ to 4 sec. All three of these roots are small compared to the rotor speed, justifying the use of the low frequency rotor response. The real root is roughly the same as the root of the vertical motion. The instability is not too objectionable, since the period and time to double amplitude are long enough for the motion to be controllable by the pilot. However, the control characteristics of the helicopter require the pilot to adopt a fairly complex compensation scheme in order to successfully stabilize this mode.

15-3.4.3 Loop Closures

In order to achieve stable flight, the longitudinal dynamics of the hovering helicopter require feedback control, either from the pilot or from an automatic control system (perhaps a mechanical system, often using a gyro). The longitudinal velocity and pitch attitude must be sensed and, after appropriate compensation, fed back to the longitudinal cyclic pitch. From the open loop poles and zeros of the longitudinal dynamics the root loci for various loop closures can be constructed.

Fig. 15-4 shows the root loci for feedback of the helicopter longitudinal displacement, longitudinal velocity, and a combination of rate and translation (a lead network). Since the open loop zeros of the $\dot{x}_B$ response are very large compared to the poles, they do not influence the loci except

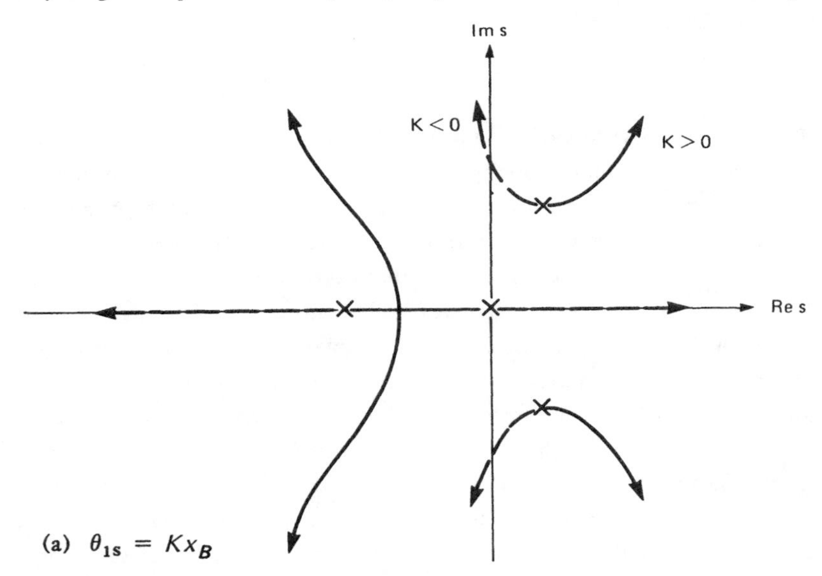

Figure 15-4 Longitudinal velocity feedback to cyclic for an articulated rotor helicopter in hover.

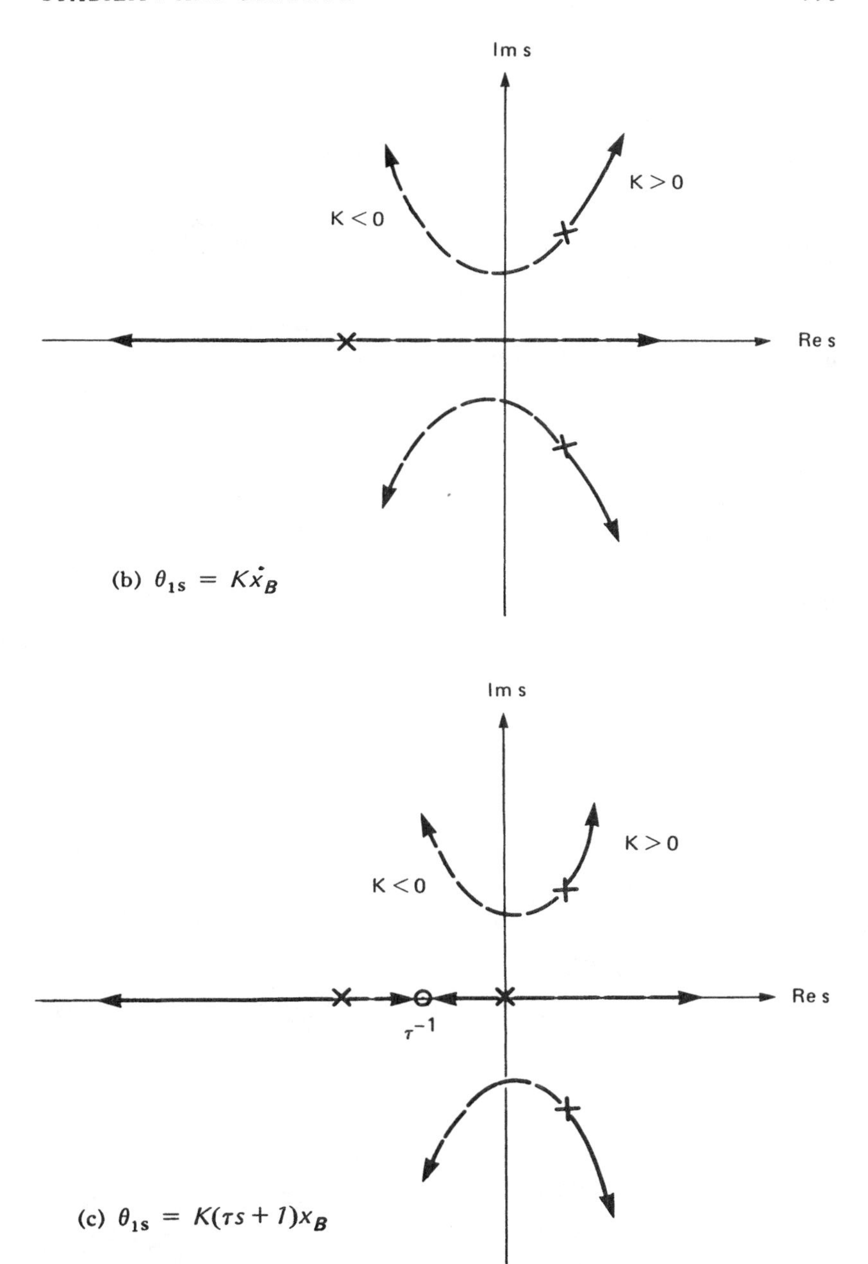

Figure 15-4 concluded.

for very high gains. None of these networks based on measuring x_B or $\dot{x}_B$ is satisfactory. Negative feedback ($K > 0$) of x_B or $\dot{x}_B$ generally destabilizes the oscillatory mode, while positive feedback produces a static instability. Longitudinal velocity feedback $\theta_{1s} = K\dot{x}_B$ is equivalent to changing the speed stability, so this lack of success is not surprising.

Fig. 15-5 shows the root loci for feedback of the helicopter pitch attitude, pitch rate, and a combination of attitude and rate. The pitch response has a single open loop zero at the origin. Pitch attitude feedback can stabilize the oscillatory mode with positive gain, although it is limited to rather low damping for an articulated rotor. Attitude feedback, however, decreases the damping of the real root, which is undesirable. Pitch rate feedback both increases the real root damping and stabilizes the oscillatory mode. The period and time to double amplitude of the oscillatory root are both increased, but this mode remains unstable. Pitch rate feedback is equivalent to increasing the rotor pitch damping derivatives X_q and M_q. These results suggest the use of a combination of pitch attitude feedback to stabilize the oscillation, and pitch rate feedback to keep the pitch damping high.

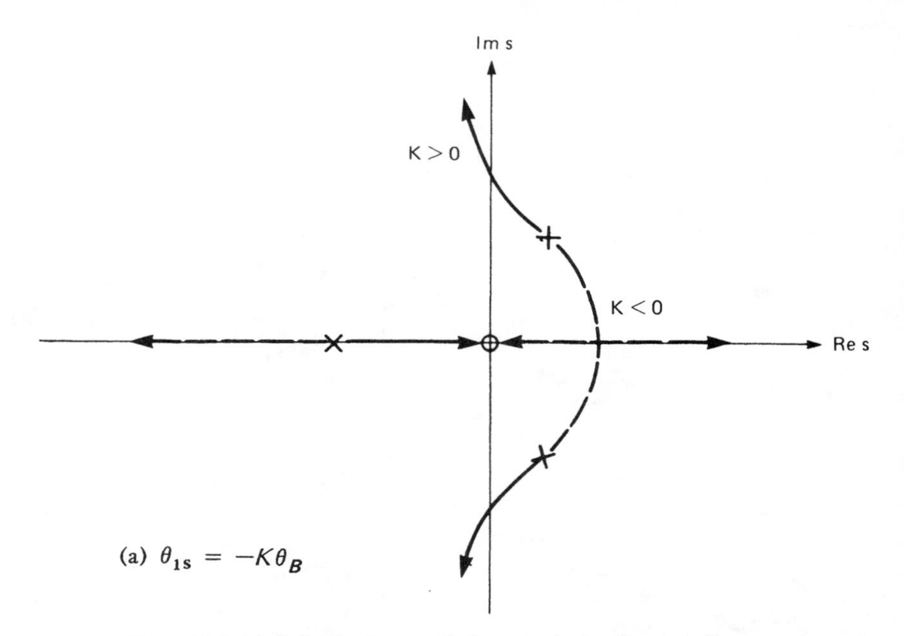

Figure 15-5 Pitch feedback to cyclic for an articulated rotor helicopter in hover.

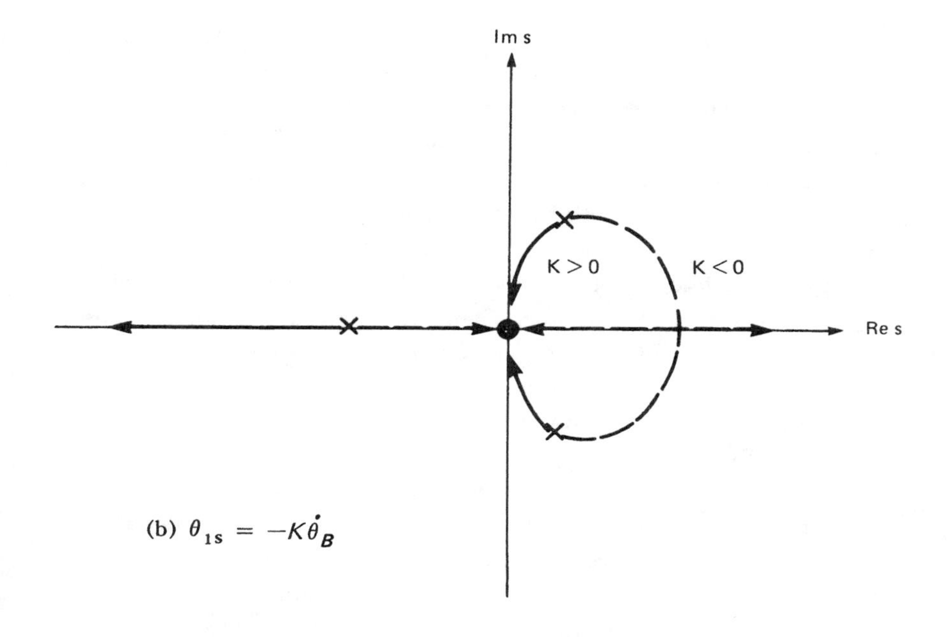

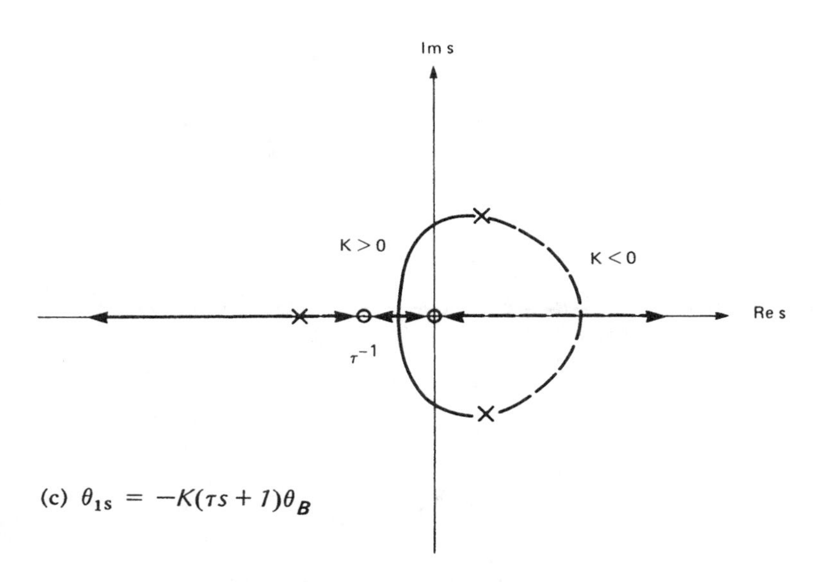

Figure 15-5 concluded.

Introducing some attitude feedback into the rate network pulls one of the zeros off the origin into the left half-plane, bringing the oscillatory branches of the loci with it, while maintaining the real root damping increase of the rate feedback. The lead τ must be large enough for the zero to remain close to the origin. If the zero gets to the left of the open loop pole, the network will operate more like attitude feedback, and the real root damping will decrease. Typically a lead of $\tau = 2$ to 4 sec is required for articulated rotor helicopters. Thus rate plus attitude feedback of pitch (or attitude feedback with lead) produces the desired handling qualities of high pitch damping and a stable oscillatory mode. Such feedback is not the most desirable for the pilot, however. The lead required is a bit higher than can be easily handled, so the hovering task is tiring for the pilot and requires some training. Thus an automatic stability augmentation system is frequently desirable.

Fig. 15-6 shows the root loci for feedback of the helicopter pitch attitude or rate with some lag in the network. Mechanical systems in particular are likely to introduce such lag, with typically $\ell \cong 1$ sec so there is an additional pole somewhat to the left of the real root of the hover dynamics. Generally the flying qualities deteriorate with lag in the system. With large enough lag

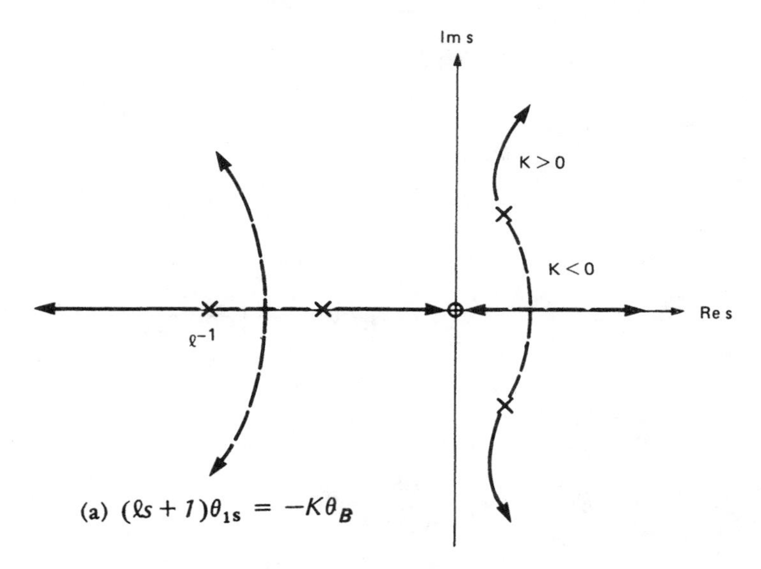

Figure 15-6 Lagged pitch feedback to cyclic for an articulated rotor helicopter in hover.

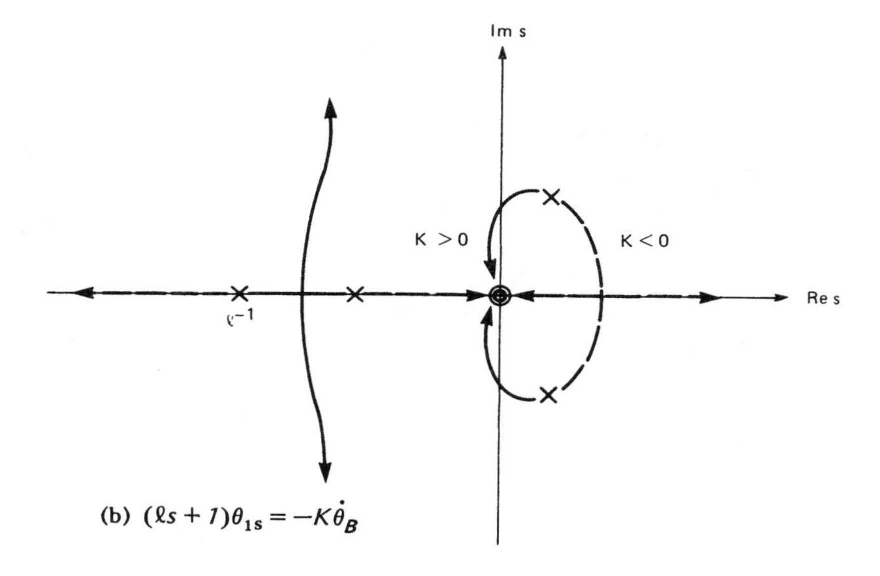

(b) $(\ell s + 1)\theta_{1s} = -K\dot{\theta}_B$

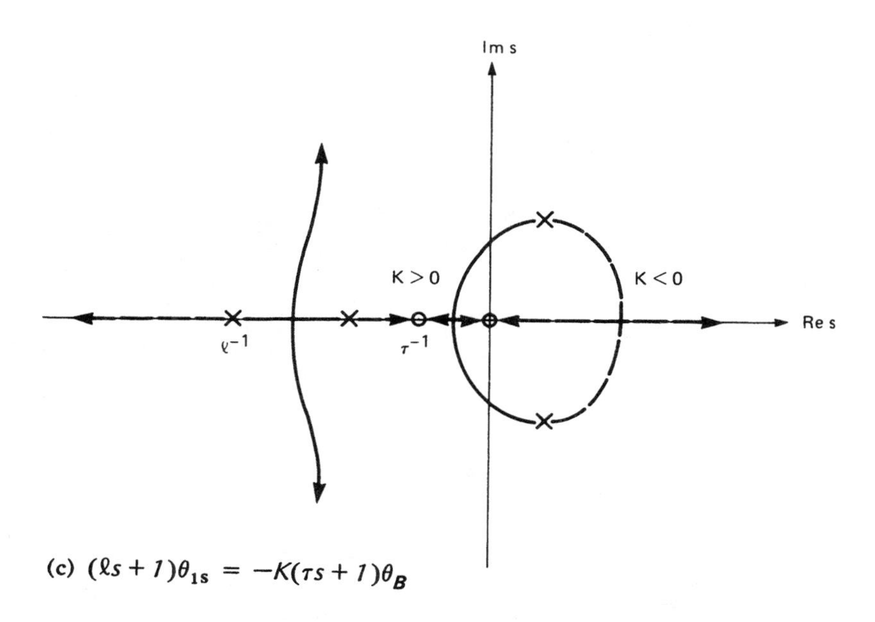

(c) $(\ell s + 1)\theta_{1s} = -K(\tau s + 1)\theta_B$

Figure 15-6 concluded.

the attitude feedback no longer stabilizes the oscillatory mode, and for rate feedback the lag introduces a limit on the possible amount of increase in the real root damping. As long as the lag is significantly smaller than the time constant of the aircraft pitch root, the root loci will not be greatly influenced. In particular, the lagged rate plus attitude feedback remains satisfactory as long as the lead character of the network dominates (the lag pole must be to the left of the lead zero, and preferably also to the left of the pitch root). Lagged rate feedback $(\ell s + 1)\theta_{1s} = -K\dot{\theta}_B$ is of interest since there are mechanical systems that produce such control (see section 15–6). This system is generally similar to pure rate feedback. While rate or lagged rate feedback does not produce a stable system, it definitely improves the helicopter flight dynamics. (At very high gain the oscillatory mode may even be slightly stable with lagged rate feedback, but this is not a practical consideration.)

15-3.4.4 Hingeless Rotors

Consider now the case of an offset-hinge articulated rotor or a hingeless rotor, so that the flap frequency is above 1/rev. The major factor introduced by $\nu > 1$ is the hub moment produced by tip-path-plane tilt, which greatly increases the capability of the rotor to produce moments about the helicopter center of gravity. There is also increased coupling of the lateral and longitudinal motions, but here only the longitudinal dynamics are considered. Flap hinge offset of an articulated rotor does not radically alter the character of the helicopter flight dynamics, although there is an important quantitative improvement of the handling qualities due to the hub moment capability. For a hingeless rotor the flap frequency is large enough to have a major impact on the dynamics.

From section 15-3.1, the stability derivatives for the general case of $\nu \geqslant 1$ are as follows:

$$X_\theta = -\frac{\gamma}{M^*(1+N_*^2)}\left[\frac{2C_T}{\sigma a} - \frac{\nu^2-1}{\gamma/8}(R_\theta - H_\theta N_*)\right]$$

$$X_u = -\frac{\gamma}{M^*(1+N_*^2)}\left[\frac{2C_T}{\sigma a} - \frac{\nu^2-1}{\gamma/8}(R_\theta - H_\theta N_*)\right]8M_\mu + \frac{\gamma}{M^*}\left[H_\theta 8M_\mu - (H_\mu + R_\mu)\right]$$

$$X_q = \frac{\gamma}{M^*(1+N_*^2)}\left[\left(\frac{2C_T}{\sigma a} + H_\beta^* - \frac{\nu^2-1}{\gamma/8} R_\theta\right)\left(\frac{16}{\gamma} + N_*\right) - H_\theta K_P\left(\frac{16}{\gamma} N_* - 1\right)\right] - hX_u$$

$$M_\theta = \frac{\nu^2-1}{k_y^2 M^*(1+N_*^2)} - \frac{h}{k_y^2} X_\theta$$

$$M_u = \frac{(\nu^2-1)8M_\mu}{k_y^2 M^*(1+N_*^2)} - \frac{h}{k_y^2} X_u$$

$$M_q = -\frac{(\nu^2-1)\left(\frac{16}{\gamma} + N_* + h\,8M_\mu\right)}{k_y^2 M^*(1+N_*^2)} - \frac{h}{k_y^2} X_q$$

where

$$N_* = \frac{\nu^2-1}{\gamma/8} + K_P$$

$$M^* = \frac{1}{g}\gamma\left(\frac{2C_T}{\sigma a}\right)_{\text{trim}}$$

and the rotor aerodynamic coefficients are given in section 11–7. When $\nu = 1$ the moments on the helicopter are due to tilt of the rotor thrust vector with the tip-path plane, but with flap hinge offset or hingeless blades there is also a direct moment acting on the rotor hub. The ratio of the pitch moment derivatives in the two cases is roughly

$$\frac{M_{\nu>1}}{M_{\nu=1}} \cong 1 + \frac{\nu^2-1}{h\gamma 2C_T/\sigma a}$$

The force derivatives vary little with the flap frequency, but it is the pitch moments that dominate the longitudinal dynamics. The moment derivatives can be roughly doubled by using flap hinge offset. For a typical hingeless rotor the control derivative M_θ and speed stability M_u are increased by a factor of three or four compared to the articulated rotor case (for no flap hinge offset). The pitch damping M_q is increased even more, because the

$H_{\dot\beta}^*$ term reduces the damping produced by the thrust vector tilt but does not influence the hub moment contribution.

As for the articulated rotor with no flap hinge offset, the longitudinal motion has three poles: a real negative root due to the pitch damping, and an unstable oscillation due to the speed stability. The high pitch damping of the hingeless rotor greatly increases the magnitude of the real root, and it even counters the increased speed stability to increase the period and time to double amplitude of the oscillatory mode. Typically for hingeless rotors the pitch mode has a time to half amplitude of 0.2 to 0.5 sec; the oscillatory mode has a period of 10 to 20 sec, with a time to double amplitude of 10 to 15 sec.

The pitch response has a single zero (the equations of motion are given in section 15-3.4.1), always exactly at the origin for θ_B/u_G, and very slightly negative for θ_B/θ_{1s} if $\nu > 1$. The longitudinal velocity response has a complex conjugate pair of zeros. When $\nu > 1$ the magnitude of these zeros is increased and they are shifted off the imaginary axis into the left half-plane.

The hingeless rotor helicopter has larger pitch damping and a less unstable oscillatory mode than the articulated rotor helicopter. In view of its larger control power as well, the task of controlling the helicopter is easier. Still, it is necessary for the pilot or an automatic system to provide closed loop stability. The root loci for various loop closures can be readily obtained from the open loop poles and zeros. The root loci for an offset-hinge articulated rotor or a hingeless rotor are similar to the loci given in the last section, but the quantitative differences in the roots have important influences on the gain, lead, and lag requirements of the feedback network. With the greatly increased pitch damping, pure attitude feedback might provide the required stability of the oscillatory mode, but such a network is unsatisfactory if there is any significant lag. Thus rate plus attitude feedback is again required for satisfactory dynamics, but the increased damping and control power means that less lead and a lower gain are required, which eases the piloting task. Generally, the lead zero should be to the right of the open loop real root, so that negative feedback still increases its damping; hence the lead should be greater than the time constant of the pitch mode. Typically for a hingeless rotor a lead of about $\tau = 1$ sec is required, which is within the range the pilot can comfortably adopt.

15-3.4.5 Response to Control

The steady-state response to control is given by the equations of motion (section 15-3.4.1) in the limit $s = 0$:

$$\begin{pmatrix} \dot{x}_B \\ \theta_B \end{pmatrix} = \begin{pmatrix} -M_\theta / M_u \\ (X_\theta M_u - X_u M_\theta)/(gM_u) \end{pmatrix} \theta_{1s} \cong \begin{pmatrix} -1/(8M_\mu) \\ 0 \end{pmatrix} \theta_{1s}$$

The last approximation is quite good in general, and is exact for an articulated rotor with no hinge offset. Cyclic pitch control thus produces longitudinal velocity $\dot{x}_B$ but no attitude change in the steady-state perturbation from hover. Longitudinal cyclic produces a pitch moment $M_\theta \theta_{1s}$, and equilibrium is achieved when the moment $M_u \dot{x}_B$ due to the speed stability is sufficient to cancel the control moment. At this equilibrium state there is no net force or moment on the helicopter (since $X_\theta M_u - X_u M_\theta \cong 0$) and hence no pitch attitude change. The zero steady-state pitch response to cyclic implies neutral static stability of the hovering helicopter relative to pitch attitude changes. The gradient of the cyclic control with the steady-state velocity perturbation $(\theta_{1s}/\dot{x}_B)$ is a measure of the rotor speed stability. Because the longitudinal motion in hover is unstable, a finite steady-state response will be achieved only if the pilot or an automatic control system intervenes to ensure that the transients die out. This solution for the steady-state response is thus best interpreted as the gradient of the control to trim the helicopter for small velocity and pitch changes from hover.

The short time response to cyclic control and longitudinal gusts (the limit of s approaching infinity) is

$$\begin{pmatrix} \ddot{x}_B \\ \ddot{\theta}_B \end{pmatrix} = \begin{pmatrix} X_\theta \\ M_\theta \end{pmatrix} \theta_{1s} + \begin{pmatrix} X_u \\ M_u \end{pmatrix} u_G$$

For an articulated rotor the longitudinal acceleration response to control is $(\ddot{x}_B/g)/\theta_{1s} = X_\theta/g = -1$ g/rad, which is quite small. Note that this result is independent of any parameters of the rotor; the response varies little with the flap frequency also. The primary response to longitudinal cyclic is the pitch acceleration. For an articulated rotor $\ddot{\theta}_B/\theta_{1s} = M_\theta = hg/k_y^2$, and for a hingeless rotor the pitch acceleration is three or four times larger.

Similarly, the longitudinal acceleration response to gusts is small while the pitch acceleration response is large, especially with a hingeless rotor. The lack of direct command of the helicopter velocity makes more difficult the tasks of precisely controlling the longitudinal or lateral position in hover and providing the feedback required to stabilize the oscillatory modes. The pilot must work with the direct control of the helicopter attitude, and thus is required to anticipate the velocity response that will result, providing a significant lead in the pitch feedback to achieve stability.

Consider a short period approximation for the hover longitudinal dynamics. Since control inputs at first produce primarily pitch motion of the helicopter, the longitudinal velocity can be neglected for a short time analysis. If $\dot{x}_B$ is set equal to zero, the equation of motion for pitch becomes

$$(s - M_q)\dot{\theta}_B = M_\theta \theta_{1s} + M_u u_G$$

So initially cyclic pitch commands the helicopter pitch rate, with a first-order lag given by the pole $s = M_q$. (This pole is an approximation to the pitch root of the longitudinal dynamics, but it is not a very good approximation for articulated rotors, where the speed stability increases the magnitude of the root significantly.) The initial response is here

$$\ddot{\theta}_B = M_\theta \theta_{1s} + M_u u_G$$

which is the same as obtained with the complete equations. The steady-state limit of this approximation is the pitch rate $\dot{\theta}_B/\theta_{1s} = -M_\theta/M_q$. The pitch or roll rate response to cyclic is generally high because of the low damping. For still larger times the longitudinal velocity motion enters the dynamics and the complete equations must be considered.

15-3.4.6 Examples

As an example, consider the longitudinal dynamics of a hovering helicopter with the following parameters: Lock number $\gamma = 8$, mast height $h = 0.3$, solidity $\sigma = 0.1$, and a blade loading of $C_T/\sigma = 0.1$. Assuming a rotor radius of $9\ m$ and tip speed of $200\ m/\text{sec}$, $g/\Omega^2 R = 0.0022$ and $M^* = 127$; and a radius of gyration of $k_y^2 = 0.1$ gives $I_y^* = 12.7$. The poles and zeros will be examined for an articulated rotor ($\nu = 1$) and a hingeless rotor ($\nu = 1.15$). The vertical motion of this helicopter has a single pole $s = -0.012$

(using $C' = 0.7$ for the lift deficiency function).

The dimensionless poles, zeros, and eigenvectors of the longitudinal dynamics are given in Table 15-2. For the articulated rotor, the real root

Table 15-2 Example of the longitudinal dynamics of a hovering helicopter

	Articulated Rotor	Hingeless Rotor
	$\nu = 1$	$\nu = 1.15$
Pitch mode		
root, s	−0.023	−0.074
eigenvector, $\dot{x}_B/\theta_B$	0.11	0.03
Oscillatory mode		
root, s	$0.0076 \pm i0.018$	$0.0027 \pm i0.021$
eigenvector, $\lvert\dot{x}_B/\theta_B\rvert$	0.12	0.11
phase	64°	81°
Zeros		
$\dot{x}_B/\theta_{1s}$	$\pm i0.083$	$-0.015 \pm i0.169$
$\dot{x}_B/u_G$	$\pm i0.083$	$-0.017 \pm i0.160$
θ_B/θ_{1s}	0	−0.0001
θ_B/u_G	0	0

has a time to half amplitude of $t_{1/2} = 1.4$ sec, and the oscillatory root has a period $T = 17$ sec (frequency 0.06 Hz) and time to double amplitude $t_2 = 4.2$ sec. For the hingeless rotor the real root has a time to half amplitude $t_{1/2} = 0.4$ sec, while for the oscillatory root $T = 14$ sec (frequency 0.07 Hz) and $t_2 = 12$ sec. Fig. 15-7 shows the roots of the longitudinal dynamics for $\nu = 1$ to 1.15.

The period and time to double amplitude of the oscillatory mode decrease with the blade loading C_T/σ because of the increase of the speed stability with the rotor coefficient M_μ. The vertical mode time constant is proportional to C_T/σ and hence increases with the blade loading. For articulated rotors, both the rotor pitch damping moment and the helicopter inertia increase with C_T/σ (assuming a constant radius of gyration k_y), so there is little variation of M_q or the real root with blade loading. For hingeless rotors, however, the rotor hub moment contribution to the pitch

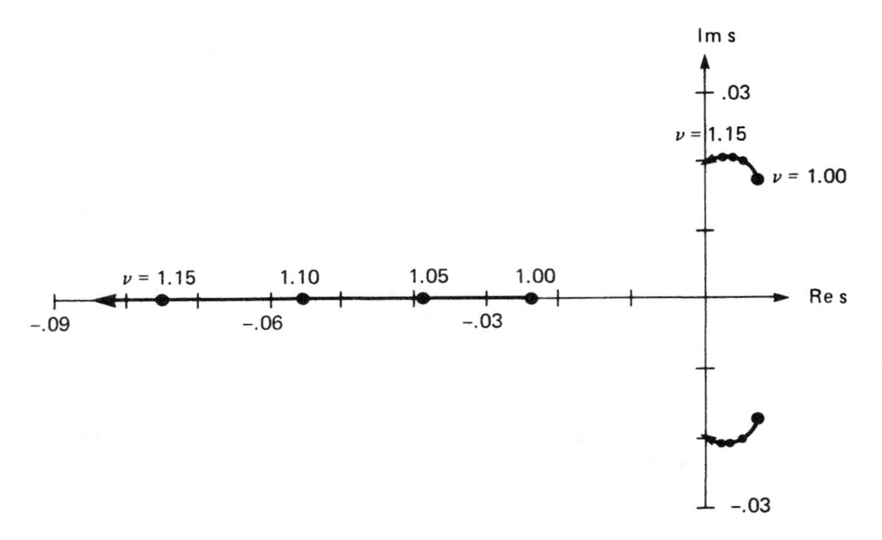

Figure 15-7 Example of the poles of the longitudinal dynamics of a hovering helicopter, shown as a function of the flap frequency ν.

damping varies little with blade loading, so the time constant of the real root increases with C_T/σ because of the effect of I_y^* on M_q.

Assuming that the tip speed and blade loading are fixed and that the geometric parameters scale with the rotor radius, the influence of the helicopter size on the roots is as follows. The dimensionless vertical and pitch roots increase with size because of the decrease of the normalized inertia M^*. The dimensionless oscillatory root of an articulated rotor helicopter increases in frequency (roughly in proportion to $R^{1/2}$) and real part. For a hingeless rotor the real part of the oscillatory root is relatively unaffected by size. The dimensional vertical root is independent of the helicopter size (see section 15-3.2), while the dimensional time constant of the pitch root increases with size. The dimensional pitch root varies relatively little with size for hingeless rotors, though. Both the period and time to double amplitude of the dinensional oscillatory root increase with size, so the instability is milder.

A number of approximations for the hover roots can be derived; see for example Hohenemser (1946a) and Bramwell (1957). These approximations are at best fair for hingeless rotors, and generally poor for articulated rotors. Because the pitch and longitudinal velocity motions are fundamentally coupled, there is no approximation that can be made to obtain simple yet consistently accurate expressions for the roots. For reliable quantitative results it is necessary to solve the complete third-order characteristic equation.

15-3.4.7 Flying Qualities Characteristics

The longitudinal dynamics of a hovering helicopter are described by a stable real root due to the pitch damping, and a mildly unstable oscillatory root due to the speed stability. The pilot has good control over the angular acceleration of the helicopter, but the direct control of translation is poor. The control sensitivity (the pitch and roll rate commanded by cyclic) is high in hover because of low damping. The combination of high sensitivity and only indirect control of translational velocity is conducive to pilot induced oscillations, and increases the difficulty of the control task. The pilot must provide attitude feedback with a fairly large lead in order to stabilize the helicopter motion. The handling qualities are improved with offset flap hinges or with a hingeless rotor because of the increased pitch damping and control power, which reduce the instability of the oscillatory mode and reduce the control sensitivity. However, the gust sensitivity is also increased with a hingeless rotor.

Miller (1948) investigated the hover handling qualities and concluded that the helicopter has low pitch and roll damping, high control sensitivity, and neutral static stability with angle of attack (see section 15-3.4.5). For articulated rotors he found low control power to deal with the unstable oscillation. Miller (1950) concluded that the unstable oscillatory mode has a period long enough to be manageable, but short enough to influence the control response. The low damping leads to an overshoot following neutralization of the controls. He also found a large lateral motion due to longitudinal cyclic stick inputs.

Reeder and Gustafson (1949) investigated helicopter flying qualities and found a high roll sensitivity in hover (the roll rate commanded by lateral cyclic), which can lead to over-controlling or even a short period pilot-induced oscillation. They found possibly unacceptable stick forces in both lateral and longitudinal maneuvers, including an unstable force gradient or zero force required to hold the roll or pitch attitude, and a lateral-longitudinal coupling of the control forces. The rotor speed stability results in a sensitivity to gusts, and therefore a drift relative to the ground in hover. The indirect nature of the control over translational velocity gives an impression of a control lag, which is undesirable. They suggested increasing the roll damping to reduce the control sensitivity. Reeder and Gustafson also noted that partial power vertical descent in the vortex ring state is accompanied by unfavorable flying qualities because of the large random variations of the flow field. Operation in the vortex ring state involves a loss of collective and cyclic control effectiveness, high vibration, large fuselage motions, and large rotor speed variations. By maintaining some forward speed in partial power descents the vortex ring state can be avoided, however.

15-3.5 Lateral Dynamics

Now let us consider the lateral dynamics of the hovering helicopter, still assuming a separation of the longitudinal and lateral motions. The degrees of freedom involved are the lateral velocity $\dot{y}_B$ and roll angle ϕ_B; the lateral cyclic control θ_{1c} and lateral gust velocity v_G are also included. When the rotor reactions are expanded in terms of stability derivatives, the differential equations of motion for the helicopter lateral dynamics become

$$\begin{aligned} M^*\ddot{y}_B &= \gamma \frac{2C_Y}{\sigma a} + M^* g\phi_B \\ &= M^*(Y_\theta \theta_{1c} + Y_v(\dot{y}_B + v_G) + Y_p\dot{\phi}_B) + M^* g\phi_B \\ I_x^*\ddot{\phi}_B &= \gamma\left(-\frac{2C_{M_x}}{\sigma a} + h\frac{2C_Y}{\sigma a}\right) \\ &= I_x^*(L_\theta\theta_{1c} + L_v(\dot{y}_B + v_G) + L_p\dot{\phi}_B) \end{aligned}$$

or

$$\begin{bmatrix} s - Y_v & -Y_p s - g \\ -L_v & s^2 - L_p s \end{bmatrix} \begin{pmatrix} \dot{y}_B \\ \phi_B \end{pmatrix} = \begin{pmatrix} Y_\theta \\ L_\theta \end{pmatrix} \theta_{1c} + \begin{pmatrix} Y_v \\ L_v \end{pmatrix} v_G$$

The rotor in hover is entirely axisymmetric. The only physical difference between the longitudinal and lateral dynamics of the hovering helicopter is that the roll moment of inertia I_x^* is generally much smaller than the pitch moment of inertia I_y^*. It follows that the lateral stability derivatives are equal to the corresponding longitudinal derivatives, except that in the moment derivatives k_y^2 must be replaced by k_x^2: $Y_v = X_u$, $Y_p = -X_q$, $Y_\theta = X_\theta$, $L_v = -M_u$, $L_p = M_q$, and $L_\theta = -M_\theta$. The sign change is due to the reversal of the orientation of the angular motion relative to the linear motion when transforming from the longitudinal dynamics to the lateral. The effect of the smaller roll inertia is to increase the magnitude of the roll stability derivatives relative to the pitch derivatives.

Thus the lateral dynamics are described by a real convergence mode due to the roll damping L_p, and an unstable oscillatory mode due to the rotor dihedral effect or speed stability L_v. For an articulated rotor the roll mode typically has a time to half amplitude of $t_{1/2} = 0.4$ to 0.8 sec, and the lateral oscillatory mode has a period $T = 7$ to 15 sec and time to double amplitude $t_2 = 4$ to 8 sec. With a hingeless rotor the roll damping is much higher, and the oscillatory mode will have a larger time to double amplitude and somewhat larger period than with an articulated rotor. Compared with the longitudinal dynamics, the roll damping is larger than the pitch damping, because of the smaller roll inertia. The lateral oscillatory mode tends to have a higher frequency than the longitudinal mode and hence is the more objectionable instability.

To stabilize the lateral motion, rate plus attitude feedback of the helicopter roll to lateral cyclic is required. Although the rotor dynamics are axisymmetric in hover, there are two factors that generally make the lateral motion more difficult to control than the longitudinal motion. First, the smaller roll inertia leads to a smaller period and less damping of the lateral mode. Secondly, the pilot has more difficulty sensing the helicopter roll motion and applying the appropriate lateral control than he has with the similar tasks for the longitudinal dynamics. Hence the lateral motion of the helicopter is particularly susceptable to pilot-induced oscillations.

A short period approximation for the lateral dynamics consists of just the roll motion:

$$(s - L_p)\dot{\phi}_B = L_\theta \theta_{1c}$$

The single pole of this equation is not too bad an approximation for the actual roll mode, even for articulated rotors, because of the smaller inertia than for the longitudinal motion. The steady-state response is $\dot{\phi}_B/\theta_{1c} = -L_\theta/L_p$, so lateral cyclic commands a roll rate with a small first order lag.

As an example, the lateral dynamics will be examined for the same helicopter considered in section 15-3.4.6, with roll inertia $I_x^* = 2.5$ ($k_x^2 = 0.02$). The dimensionless roots and eigenvectors of the lateral dynamics for hover are given in Table 15-3. For an articulated rotor, the real root has a time to half amplitude of $t_{1/2} = 0.6$ sec, while the oscillatory root has a period of $T = 11$ sec (frequency 0.09 Hz) and time to double amplitude of $t_2 = 4.1$ sec. For a hingeless rotor the real root has $t_{1/2} = 0.1$ sec, while the oscillatory root has $T = 13$ sec and $t_2 = 80$ sec (frequency 0.08 Hz).

Table 15-3 Example of the lateral dynamics of the hovering helicopter

	Articulated Rotor	Hingeless Rotor
	$\nu = 1$	$\nu = 1.15$
Roll mode		
root, s	−0.051	−0.339
eigenvector, $\dot{y}_B/\phi_B$	−0.05	−0.01
Oscillatory mode		
root, s	$0.0079 \pm i0.027$	$0.0004 \pm i0.022$
eigenvector, $\lvert\dot{y}_B/\phi_B\rvert$	0.08	0.10
phase	−73°	−87°

15-3.6 Coupled Longitudinal and Lateral Dynamics

Let us examine now the coupled longitudinal and lateral motions of the single main rotor helicopter in hover. The longitudinal and lateral dynamics are in fact strongly coupled by the rotor forces. Using the low frequency response (section 15-3.1), the rotor reactions can be expanded in terms of the stability derivatives as follows:

$$\begin{pmatrix} -\dfrac{\gamma}{M^*}\dfrac{2C_H}{\sigma a} \\ \\ \dfrac{\gamma}{M^*}\dfrac{2C_Y}{\sigma a} \end{pmatrix} = \begin{bmatrix} X_{\theta_s} & X_{\theta_c} \\ \\ Y_{\theta_s} & Y_{\theta_c} \end{bmatrix} \begin{pmatrix} \theta_{1s} \\ \\ \theta_{1c} \end{pmatrix} + \begin{bmatrix} X_q & X_p \\ \\ Y_q & Y_p \end{bmatrix} \begin{pmatrix} \dot{\theta}_B \\ \\ \dot{\phi}_B \end{pmatrix} +$$

$$+ \begin{bmatrix} X_u & X_v \\ Y_u & Y_v \end{bmatrix} \begin{pmatrix} \dot{x}_B + u_G \\ \dot{y}_B + v_G \end{pmatrix}$$

$$\begin{pmatrix} \dfrac{\gamma}{I_y^*}\left(\dfrac{2C_{M_y}}{\sigma a} + h\,\dfrac{2C_H}{\sigma a}\right) \\ \dfrac{\gamma}{I_x^*}\left(-\dfrac{2C_{M_x}}{\sigma a} + h\,\dfrac{2C_Y}{\sigma a}\right) \end{pmatrix} = \begin{bmatrix} M_{\theta_s} & M_{\theta_c} \\ L_{\theta_s} & L_{\theta_c} \end{bmatrix} \begin{pmatrix} \theta_{1s} \\ \theta_{1c} \end{pmatrix}$$

$$+ \begin{bmatrix} M_q & M_p \\ L_q & L_p \end{bmatrix} \begin{pmatrix} \dot{\theta}_B \\ \dot{\phi}_B \end{pmatrix} + \begin{bmatrix} M_u & M_v \\ L_u & L_v \end{bmatrix} \begin{pmatrix} \dot{x}_B + u_G \\ \dot{y}_B + v_G \end{pmatrix}$$

The diagonal stability derivatives were given in sections 15-3.4.2, 15-3.4.4, and 15-3.5. The off-diagonal derivatives, which couple the lateral and longitudinal motions, are:

$$X_{\theta_c} = \frac{\gamma}{M^*(1 + N_*^2)}\left[N_*\,\frac{2C_T}{\sigma a} - \frac{\nu^2 - 1}{\gamma/8}\left(H_\theta + N_* R_\theta\right)\right]$$

$$X_v = \frac{\gamma}{M^*(1 + N_*^2)}\left[N_*\,\frac{2C_T}{\sigma a} - \frac{\nu^2 - 1}{\gamma/8}\left(H_\theta + N_* R_\theta\right)\right] 8M_\mu$$

$$+ \frac{\gamma}{M^*}\left(R_\theta\, 8M_\mu - R_r\right)$$

$$X_p = \frac{\gamma}{M^*(1 + N_*^2)}\left[\left(\frac{2C_T}{\sigma a} + H_{\dot\beta}^* - \frac{\nu^2 - 1}{\gamma/8} R_\theta\right)\left(\frac{16}{\gamma} N_* - 1\right) + H_\theta K_P\left(\frac{16}{\gamma} + N_*\right)\right]$$

$$+ \frac{\gamma}{M^*}\left(\frac{C_T}{\sigma a} + H_{\dot\beta}^* + \frac{16}{\gamma} R_\theta\right) + hX_v$$

$$M_{\theta_c} = -\frac{(\nu^2 - 1)N_*}{k_y^2 M^*(1 + N_*^2)} - \frac{h}{k_y^2} X_{\theta_c}$$

$$M_v = -\frac{(\nu^2 - 1)N_* 8M_\mu}{k_y^2 M^*(1 + N_*^2)} - \frac{h}{k_y^2} X_v$$

$$M_p = -\frac{(\nu^2 - 1)\left(\frac{16}{\gamma} N_* - 1 + N_* h\, 8M_\mu\right)}{k_y^2 M^*(1 + N_*^2)} - \frac{h}{k_y^2} X_p$$

Again as a result of the rotor axisymmetry, the side force and roll moment derivatives are equal to the corresponding drag force and pitch moment derivatives, except that in the moment derivatives k_y^2 must be replaced by k_x^2: $Y_u = -X_v$, $Y_q = X_p$, $Y_{\theta_s} = -X_{\theta_c}$, $L_u = M_v$, $L_q = -M_p$, and $L_{\theta_s} = M_{\theta_c}$. For an articulated rotor with no flap hinge offset ($\nu = 1$), these force derivatives reduce to

$$X_{\theta_c} = 0$$

$$X_v = \frac{\gamma}{M^*} R_\theta\, 8M_\mu - R_r \cong 0$$

$$X_p = \frac{\gamma}{M^*}\left(-\frac{C_T}{\sigma a} + \frac{16}{\gamma} R_\theta\right) + hX_v$$

and the moment derivatives are related to the force derivatives by $M = -(h/k_y^2)X$ again. Hence the coupling consists principally of the derivatives M_p and L_q.

Miller (1946) examined the lateral-longitudinal coupling and found that one of the oscillatory modes is stabilized and the other destabilized. If the roll inertia is small enough, the former mode can even be slightly stable. Table 15-4 compares the roots obtained from the coupled and uncoupled equations of motion for the same numerical example considered in sections 15-3.4.6 and 15-3.5. The coupling tends to destabilize the longitudinal oscillation and stabilize the lateral oscillation for these cases, with some effect on the frequency as well as the damping. The pitch and roll real roots are given fairly well by the uncoupled equations, particularly for the hingeless rotor. Generally, the uncoupled equations correctly give the basic characteristics and most of the quantitative results of the coupled dynamics. The eigenvectors, however, show that even when the roots are not influenced much, the coupling still introduces considerable roll motion in the longitudinal dynamics and pitch motion in the lateral dynamics.

Table 15-4 Comparison of coupled and uncoupled solutions for the longitudinal-lateral dynamics of a hovering helicopter

	Uncoupled Roots	Coupled Roots	Eigenvector (Pitch-roll Ratio)
Articulated rotor ($\nu = 1$)			
longitudinal	−0.023	−0.027	$\phi_B/\theta_B = -0.84$
	$0.0076 \pm i0.017$	$0.0085 \pm i0.016$	−0.58
lateral	−0.051	−0.040	$\theta_B/\phi_B = -0.39$
	$0.0079 \pm i0.027$	$0.0036 \pm i0.031$	−0.38
Hingeless rotor ($\nu = 1.15$)			
longitudinal	−0.074	−0.078	$\phi_B/\theta_B = -0.27$
	$0.0027 \pm i0.021$	$0.0065 \pm i0.020$	−0.84
lateral	−0.339	−0.334	$\theta_B/\phi_B = -0.06$
	$0.0004 \pm i0.022$	$-0.0041 \pm i0.022$	−0.78

15-3.7 Tandem Helicopter

There are major differences between the flying qualities of helicopters with two main rotors, and the single main rotor and tail rotor configuration. With coaxial, contrarotating main rotors the helicopter behaves as if it had a single main rotor with truly decoupled longitudinal and lateral dynamics. The yaw control and damping are obtained from the main rotor torque, however, instead of from a tail rotor (see section 15−1). The most common twin main rotor configuration is the tandem rotor helicopter, in which the main rotors have a typical longitudinal separation of 1.5 to 1.8R between the shafts (hence 20% to 50% overlap of the rotor disks). The tandem helicopter in hover has longitudinal symmetry (about the x-z plane), if it is possible to ignore such differences as the vertical rotor separation (the rear rotor is elevated above the front rotor to avoid the wake of the latter), the inertial and aerodynamic effects of the rear rotor pylon, and offset of the helicopter center of gravity from midway between the rotors. Consequently, the tandem helicopter dynamics separate into symmetric and antisymmetric motions, at least to a better approximation than the decoupling of the lateral and longitudinal motions of a single main rotor helicopter. The symmetric motion consists of the helicopter roll and side velocity (the lateral dynamics), and the vertical velocity; in hover it is also possible to separate the lateral and vertical dynamics. The antisymmetric motion of the tandem rotor helicopter consists of the longitudinal dynamics (pitch and longitudinal velocity) and yaw, which also separate for hover. The vertical dynamics of the tandem helicopter are identical to those of the single main rotor helicopter, as analyzed in section 15-3.2. The lateral dynamics (roll and side velocity) are equivalent to the truly uncoupled lateral dynamics of a single main rotor helicopter, but with quantitative differences because the fuselage of a tandem helicopter usually has a higher roll inertia. The longitudinal and yaw dynamics of the tandem helicopter involve new phenomena. The pitch control is by differential collective, and an additional source of pitch damping is the differential thrust of the main rotors due to the rotor axial velocity during pitch motions. Yaw control of the tandem rotor helicopter is obtained by differential lateral cyclic (see section 15–1); the yaw damping is provided by the drag damping forces of the rotors.

The side-by-side helicopter configuration has true lateral symmetry, so there is a separation of the symmetric and antisymmetric motions in both hover and forward flight. In hover the dynamics are basically the same as for the tandem rotor helicopter except for the interchange of the pitch and roll axes. Hence the longitudinal and vertical dynamics (the symmetric motions) are similar to the dynamics of a single main rotor helicopter. The lateral and yaw dynamics of the side-by-side configuration are similar to the longitudinal and yaw dynamics of the tandem rotor configuration. The interchange of the pitch and roll axes has a major impact on the handling qualities, though, since different requirements are placed on lateral and longitudinal dynamics.

Consider now the longitudinal dynamics of a tandem rotor helicopter in hover. Complete longitudinal symmetry is assumed so that the symmetric and antisymmetric motions decouple, and it is assumed that the yaw and longitudinal motions can also be analyzed separately. The longitudinal degrees of freedom then consist of the pitch angle θ_B and longitudinal velocity $\dot{x}_B$; excitation is by longitudinal gust u_G. The pitch control is by differential main rotor collective $\Delta\theta_0$ ($\frac{1}{2}\Delta\theta_0$ at the front rotor and $-\frac{1}{2}\Delta\theta_0$ at the rear rotor), and a differential vertical gust velocity Δw_G is also included. The two main rotors are separated by a distance ℓ, and it is assumed that the center of gravity is midway between the rotors. The equation of motion for pitch and longitudinal velocity are then

$$M\ddot{x}_B = -H_F - H_R - Mg\theta_B$$

$$I_y \ddot{\theta}_B = M_{y_F} + M_{y_R} + h(H_F + H_R) + \frac{\ell}{2}(T_F - T_R)$$

where the subscript F means the front rotor and R means the rear rotor. Dividing by NI_b to normalize the equations and inertias gives

$$M^* \ddot{x}_B = -\frac{\gamma}{2}\left(\frac{2C_H}{\sigma a}\right)_F - \frac{\gamma}{2}\left(\frac{2C_H}{\sigma a}\right)_R - M^* g \theta_B$$

$$I_y^* \ddot{\theta}_B = \frac{\gamma}{2}\left(\frac{2C_{M_y}}{\sigma a} + h\frac{2C_H}{\sigma a}\right)_F + \frac{\gamma}{2}\left(\frac{2C_{M_y}}{\sigma a} + h\frac{2C_H}{\sigma a}\right)_R$$

$$+ \ell\frac{\gamma}{2}\left(\frac{C_T}{\sigma a}\right)_F - \ell\frac{\gamma}{2}\left(\frac{C_T}{\sigma a}\right)_R$$

The conditions for vertical force equilibrium give $Mg = 2T$, and therefore $M^*g = \gamma(2C_T/\sigma a)_{trim}$ again. The longitudinal hub velocity of the rotors is $\dot{x}_h = -\dot{x}_B + h\dot{\theta}_B$, and the vertical hub velocity is $\dot{z}_h = \frac{1}{2}\ell\dot{\theta}_B$ for the front rotor and $\dot{z}_h = -\frac{1}{2}\ell\dot{\theta}_B$ for the rear rotor. The difference between the tandem and single rotor analyses is that now it is necessary to consider the differential thrust and vertical velocity perturbations of the two rotors. Using the low frequency response, the rotor reactions are expanded in terms of the stability derivatives, so the equations of motion become

$$M^* \ddot{x}_B = M^*(X_\theta \Delta\theta_0 + X_u(\dot{x}_B + u_G) + X_q \dot{\theta}_B) - M^* g \theta_B$$

$$I_y^* \ddot{\theta}_B = I_y^*(M_\theta \Delta\theta_0 + M_u(\dot{x}_B + u_G) + M_q \dot{\theta}_B + M_{\Delta w} \Delta w_G)$$

These differential equations are identical to those for the longitudinal dynamics of the single main rotor helicopter. With tandem rotors, however, the differential rotor thrust forces contribute to the pitch moments. Also, the pitch inertia of the tandem helicopter fuselage is greater than that of a single main rotor helicopter. Thus there are significant differences in the values of the stability derivatives for the two configurations.

With articulated main rotors ($\nu = 1$), the stability derivatives of the tandem helicopter are:

$$X_\theta = 0$$

$$X_u = -g8M_\mu$$

$$X_q = g\frac{16}{\gamma}\left(1 + \frac{H_{\dot\beta}^*}{2C_T/\sigma a}\right) - hX_u$$

$$M_\theta = \frac{\gamma \ell C'/12}{k_y^2 M^*} = \frac{g\ell C'}{12k_y^2(2C_T/\sigma a)}$$

$$M_u = -(h/k_y^2)X_u$$

$$M_q = -(h/k_y^2)X_q - \frac{\gamma \ell^2 C'/16}{k_y^2 M^*} = -(h/k_y^2)X_q - \frac{g\ell^2 C'}{16k_y^2(2C_T/\sigma a)}$$

$$M_{\Delta w} = \frac{\gamma \ell C'/16}{k_y^2 M^*} = \frac{g\ell C'}{16k_y^2(2C_T/\sigma a)}$$

where C' is the lift deficiency function for rotor thrust perturbations in hover. Compared with the single main rotor case, the derivatives X_u, M_u, and X_q are unchanged, although the speed stability M_u will be numerically smaller because of the large pitch inertia. The control by differential collective produces a pure pitch moment ($X_\theta = 0$). The control derivative M_θ for the tandem helicopter typically has a value around three times that possible with the longitudinal cyclic control of a single articulated main rotor with no flap hinge offset and is comparable to the control power possible with a single hingeless rotor. The pitch damping M_q is dominated by the differential thrust term because of the large rotor thrust change produced by an axial velocity perturbation. The pitch damping derivative M_q of a tandem helicopter is typically four times that of a single main rotor helicopter with no flap-hinge offset, or about twice that possible with a single offset-hinge articulated rotor, and moreover it varies little with the rotor loading. Thus the longitudinal dynamics of a tandem helicopter are characterized by high control power and pitch damping. The longitudinal handling qualities in hover should be noticeably better than those of a single main rotor helicopter of comparable size.

Because of the dominance of the differential thrust term, it is useful to write the pitch damping as $M_q = -(h/k_y^2)X_q + \Delta M_q$, where $\Delta M_q = -\gamma\ell^2 C'/16I_y^*$. With articulated rotors, $X_\theta = 0$ and $M_u = -(h/k_y^2)X_u$ as well. Thus the characteristic equation for the longitudinal dynamics of a tandem helicopter is

$$\Delta = s^3 - (X_u + M_q)s^2 + X_u \Delta M_q s + gM_u = 0$$

and the response to control and gust is

$$\begin{pmatrix} \dot{x}_B \\ \theta_B \end{pmatrix} = \frac{1}{\Delta}\begin{pmatrix} M_\theta (X_q s - g) \\ M_\theta (s - X_u) \end{pmatrix} \Delta\theta_0 + \frac{1}{\Delta}\begin{pmatrix} X_u (s^2 - \Delta M_q s + gh/k_y^2) \\ M_u s \end{pmatrix} u_G$$

$$+ \frac{1}{\Delta}\begin{pmatrix} M_{\Delta w}(X_q s - g) \\ M_{\Delta w}(s - X_u) \end{pmatrix} \Delta w_G$$

Since $X_u + M_q \cong \Delta M_q$, the characteristic equation can be written as

$$s^2(s - \Delta M_q) + M_u\left[-(k_y^2/h)\Delta M_q s + g\right] = 0$$

which is equivalent to a feedback system with gain M_u, three "open loop poles" (two at the origin and a negative real root at $s = \Delta M_q$), and a single negative real "open loop zero" at $s = gh/(k_y^2 \Delta M_q)$. The root locus for varying speed stability can be constructed, as in Fig. 15-8, and hence the hover roots of the tandem helicopter can be located for the actual value of M_u. Alternatively, the characteristic equation can be written as a root locus in the pitch damping:

$$(s^3 + gM_u) - \Delta M_q(s - X_u)s = 0$$

again neglecting all but the ΔM_q term in the s^2 coefficient. The locus for varying pitch damping has three "open loop poles" and two "open loop zeros," as sketched in Fig. 15-9. The gain ΔM_q is perhaps high enough to stabilize the oscillatory mode. Note that the "open loop zero" $s = X_u = -g8M_\mu$ gives a lower limit on the real part of the oscillatory root. Dimensionally, the time to half amplitude must be greater than about 35 sec, so the mode is still not very stable.

The locus for varying speed stability gives some useful quantitative information about the hover roots of the tandem helicopter longitudinal dynamics. The open loop pole $s = \Delta M_q$ (which is the root of the uncoupled

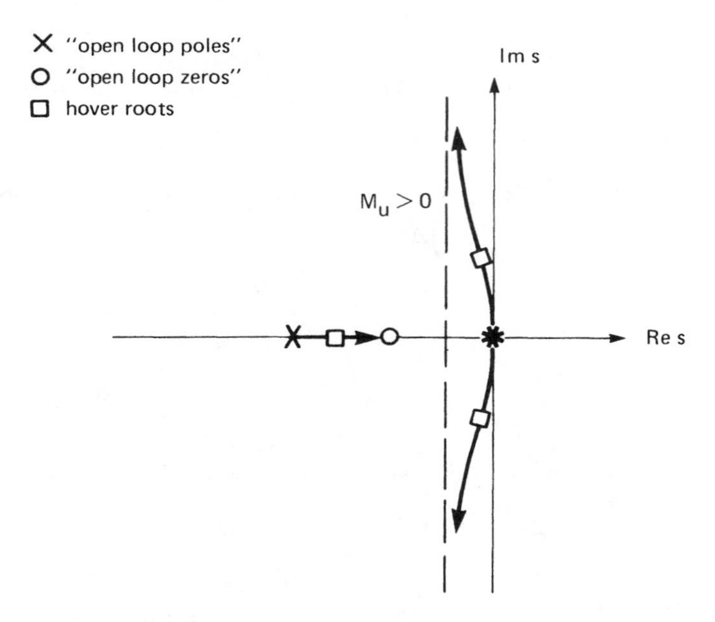

Figure 15-8 Influence of speed stability ($M_u > 0$) on the longitudinal roots of a tandem helicopter.

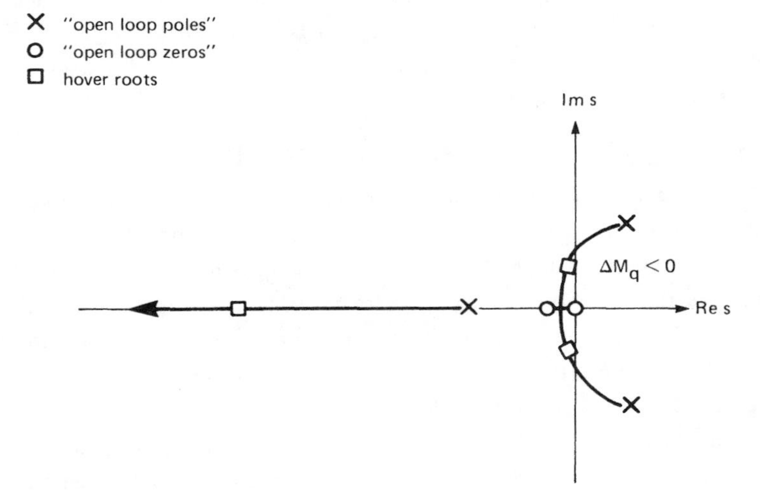

Figure 15-9 Influence of pitch damping ($\Delta M_q < 0$) on the longitudinal roots of a tandem helicopter.

pitch motion) is a good approximation for the actual pitch root, because the damping is high and the "gain" M_u is low. If the vertical asymptote of the M_u root locus is in the left half-plane, the hover oscillatory mode of the helicopter will be stable (see Fig. 15-8). The asymptote is at

$$\begin{aligned} \text{Re}\, s &= \tfrac{1}{2}(\Sigma \text{ open loop poles} - \Sigma \text{ open loop zeros}) \\ &= \tfrac{1}{2}(\Delta M_q - gh/(k_y^2 \Delta M_q)) \\ &= \frac{1}{2\Delta M_q}\left((\Delta M_q)^2 - gh/k_y^2\right). \end{aligned}$$

Alternatively, by Routh's criterion the oscillatory mode will be stable if

$$\begin{aligned} -(X_u + M_q)X_u \Delta M_q - gM_u &= -X_u[(X_u + M_q)\Delta M_q - gh/k_y^2] \\ &\cong -X_u[(\Delta M_q)^2 - gh/k_y^2] > 0 \end{aligned}$$

or $|\Delta M_q| > (gh/k_y^2)^{1/2}$, from which

$$\frac{2C_T}{\sigma a} < \frac{\ell^2 C'}{16}\sqrt{\frac{g}{hk_y^2}}$$

This limit is typically $C_T/\sigma = 0.05$ or 0.08, increasing with the helicopter size roughly as $R^{1/2}$. Therefore, the oscillatory mode of the tandem helicopter longitudinal dynamics may well be stable, if only slightly so.

For the tandem helicopter, the longitudinal velocity response to control $\dot{x}_B/\Delta\theta_0$ has a single real zero, which is so large that it has negligible influence on the flight dynamics. The pitch response to differential collective control $\theta_B/\Delta\theta_0$ has a single real, negative zero at $s = X_u$, which is fairly small although not exactly at the origin as for a single main rotor helicopter using cyclic control. The pole-zero configuration of the hover dynamics of the tandem helicopter is thus basically the same as for the single rotor helicopter (see section 15-3.4.3), from which it follows that the root loci for the various loop closures are similar. The higher pitch damping and control power of the tandem helicopter ease the control tasks somewhat.

As an example, consider the helicopter described in section 15-3.4.6 again, with longitudinal separation of the tandem rotors $\ell = 1.8R$. A larger pitch inertia $I_y^* = 38.2$ ($k_y^2 = 0.3$) typical of a tandem helicopter fuselage is assumed. The poles of the hover longitudinal dynamics are then $s = -0.035$

and $s = 0.0005 \pm i0.0082$, and the corresponding eigenvectors $\dot{x}_B/\theta_B = 0.07$ and $|\dot{x}_B/\theta_B| = 0.28$ (phase 80°). The pitch damping root has a time to half amplitude of $t_{1/2} = 0.9$ sec. The oscillatory mode has a period of $T = 35$ sec (frequency 0.03 Hz) and time to double amplitude of $t_2 = 63$ sec. The zeros are $s = 1.03$ for $\dot{x}_B/\Delta\theta_0$, and $s = -0.001$ for $\theta_B/\Delta\theta_0$. As the gross weight or rotor loading C_T/σ is increased, the damping and period of the oscillatory mode decrease. The oscillatory mode is unstable for $C_T/\sigma > 0.07$ in this case.

Assuming that the yaw motion of the tandem helicopter in hover can be analyzed separately, the conditions for equilibrium of yaw moments about the center of gravity give the differential equation for ψ_B,

$$I_z \ddot{\psi}_B = \frac{\ell}{2}(Y_F - Y_R)$$

or, dividing by NI_b,

$$I_z^* \ddot{\psi}_B = \ell \frac{\gamma}{2}\left(\frac{2C_Y}{\sigma a}\right)_F - \ell \frac{\gamma}{2}\left(\frac{2C_Y}{\sigma a}\right)_R$$

The hub lateral velocity is $\dot{y}_h = \frac{1}{2}\ell\dot{\psi}_B$ at the front rotor, and $\dot{y}_h = -\frac{1}{2}\ell\dot{\psi}_B$ at the rear rotor. The yaw damping is due to the rotor side forces produced by this lateral velocity through the rotor drag damping. Yaw control is by the rotor side forces due to differential lateral cyclic θ_{1s}; a differential side gust velocity Δv_G is also included. Using the low frequency response for the rotor forces, the yaw equation of motion is

$$I_z^* \ddot{\psi}_B = I_z^*(N_\theta \theta_{1c} + N_{\Delta v}\Delta v_G + N_r \dot{\psi}_B)$$

or

$$\ddot{\psi}_B - N_r \dot{\psi}_B = N_\theta \theta_{1c} + N_{\Delta v}\Delta v_G$$

which has the time constant $\tau_r = -1/N_r$. With articulated rotors the stability derivatives are

$$N_\theta = -\frac{\gamma}{I_z^*}\frac{2C_T}{\sigma a}\ell = -\frac{g\ell}{k_z^2}$$

$$N_{\Delta v} = \frac{\gamma}{I_z^*}\frac{2C_T}{\sigma a}\frac{\ell}{2} 8M_\mu = \frac{g\ell}{2k_z^2} 8M_\mu$$

$$N_r = -\frac{\gamma}{I_z^*}\frac{2C_T}{\sigma a}\frac{\ell^2}{2}\,8M_\mu = -\frac{g\ell^2}{2k_z^2}\,8M_\mu$$

Compared to the yaw control provided by a tail rotor, the tandem helicopter tends to have a smaller control derivative N_θ because of the large yaw inertia. Also, with articulated rotors the yaw control power of a andem helicopter is proportional to the rotor loading. The yaw damping moment of a tandem helicopter is typically half that possible with a tail rotor, and it depends on the rotor loading. The damping derivative N_r is further reduced by the larger yaw inertia. Thus the yaw time to half amplitude is typically about $t_{1/2} = 7$ sec, much larger than with a tail rotor. There is actually some coupling between the yaw and longitudinal dynamics of the tandem helicopter. For example, differential collective will produce a net torque increment on the helicopter, so the pedal control must be coordinated with the longitudinal cyclic stick to maintain the heading during pitch maneuvers.

In summary, the longitudinal dynamics of a tandem helicopter in hover are described by a stable real root due to the large pitch damping, and a mildly unstable low-frequency oscillatory mode (for the case of identical tilt of the main rotor shafts). While the longitudinal oscillation of the helicopter might even be stable, the time to half amplitude is large enough that the pilot must still provide feedback for adequate handling qualities. The large pitch damping does ease the longitudinal control task. Generally, the yaw handling qualities of a tandem helicopter are not as good as those with a tail rotor, because of the lower damping and slower response. Bramwell (1960) concluded that the longitudinal dynamics of the tandem rotor helicopter in hover depend mainly on the speed stability and pitch damping (M_u and M_q). He found little influence of the center-of-gravity position, and none of differential pitch-flap coupling. However, the speed stability depends significantly on longitudinal "dihedral" of the rotors, i.e. on tilt of the rotor shafts toward or away from each other. By tilting the shafts outward, the speed stability can be reduced, because of the in-plane component of the rotor thrust damping. Such effects can produce handling qualities substantially different from those obtained from the basic analysis of this section.

15–4 Flying Qualities in Forward Flight

15-4.1 Equations of Motion

Next let us examine the flying qualities of the helicopter in forward flight. Forward speed introduces new forces acting on the helicopter: centrifugal forces due to the rotation of the trim velocity vector by the angular velocity of the body axes; aerodynamic forces on the fuselage and tail; and major rotor forces that are proportional to the advance ratio. As a result, the handling qualities in forward flight differ significantly from those in hover. In forward flight, the vertical and lateral-longitudinal dynamics are coupled by both the rotor forces and the body accelerations. However, it will again be assumed that it is possible to analyze the longitudinal dynamics (longitudinal velocity, pitch attitude, and vertical velocity) and the lateral dynamics (lateral velocity, roll attitude, and yaw rate) separately. Such an analysis provides a reasonable description of the helicopter flight dynamics, although in fact all six degrees of freedom are coupled.

A body axis coordinate frame with origin at the center of gravity is used for the analysis of the helicopter motion in forward flight (see Fig. 15-1). To simplify the equations of motion, the coordinate axes are aligned with the rotor shaft and hub plane, and it is assumed that the helicopter center of gravity is directly below the rotor hub. The forces and moments about the helicopter center of gravity are then obtained from the hub reactions simply by a translation along the z-axis, with no rotations. For numerical work it is possible to use an arbitrary reference axis system in the body, perhaps the principal axes; and in general the center of gravity will be offset from the shaft axis. Aligning the x-axis with the trim velocity vector is not appropriate because of the difficulties with this convention at low speeds or in hover. The conditions for the equilibrium of forces and moments give the differential equations for motion perturbed from the trim flight condition. For body axes with origin at the center of gravity, the six equations of motion are

$$\vec{F} = M(\dot{\vec{u}} - \vec{u} \times \vec{\omega})$$
$$\vec{M} = I\dot{\vec{\omega}} + \vec{\omega} \times I\vec{\omega}$$

where $\vec{u}$ is the aircraft velocity in body axes, $\vec{\omega}$ is the angular velocity, M is the aircraft mass, and I is the moment-of-inertia matrix:

$$I = \begin{bmatrix} I_x & -I_{xy} & -I_{xz} \\ -I_{xy} & I_y & -I_{yz} \\ -I_{xz} & -I_{yz} & I_z \end{bmatrix}$$

in which $I_x = \int (y^2 + z^2) dm$, $I_{xz} = \int xz dm$, etc. It is assumed that the helicopter inertia has lateral symmetry, so $I_{xy} = I_{yz} = 0$. The mass and moments of inertia include the rotor mass.

The rigid body equations of motion must be linearized about the trim flight state. It is assumed that the helicopter is in steady level flight at velocity V, so that the angular velocity in the trim state is zero. The trim linear velocity of the helicopter has dimensionless components μ and $\mu \tan \alpha_{HP}$ in the hub plane coordinate frame, where α_{HP} is the tilt of the hub plane relative to the helicopter velocity V (positive for forward tilt of the disk), giving a trim velocity of $\vec{u}_0 = \mu \vec{i} - \mu \tan\alpha_{HP} \vec{k}$. The linearized equations of motion are thus

$$\vec{F} = M(\dot{\vec{u}} - \vec{u}_0 \times \vec{\omega})$$

$$\vec{M} = I \dot{\vec{\omega}}$$

The forces and moments acting on the helicopter are produced by the main rotor, tail rotor, fuselage and tail aerodynamics, and gravity. The axis system has been chosen specifically to simplify the role of the main rotor forces and moments. The only tail rotor contribution considered is the yaw moment produced by its thrust. The aerodynamic forces on the fuselage and tail will be omitted for now; their contributions to the stability derivatives can be added later as required. For level flight the hub plane incidence angle α_{HP} also determines the attitude of the axes relative to the gravitational force (vertical). The equations of motion for the six rigid body degrees of freedom are then as follows:

$$M\ddot{x}_B - M\mu \tan \alpha_{HP} \dot{\theta}_B = -H - Mg \cos\alpha_{HP}\theta_B$$

$$M\ddot{y}_B + M\mu\dot{\psi}_B + M\mu \tan\alpha_{HP}\dot{\phi}_B = Y + Mg \cos\alpha_{HP}\phi_B - Mg \sin\alpha_{HP}\psi_B$$

$$M\ddot{z}_B - M\mu\dot{\theta}_B = -T + Mg \sin\alpha_{HP}\theta_B$$

$$I_x\ddot{\phi}_B - I_{xz}\ddot{\psi}_B = -M_x + hY$$

$$I_y\ddot{\theta}_B = M_y + hH$$

$$I_z\ddot{\psi}_B - I_{xz}\ddot{\phi}_B = Q - \ell_{tr}T_{tr}$$

These equations are normalized as usual by dividing by the characteristic inertia $\frac{1}{2}NI_b$. For hover ($\mu = 0$ and $\alpha_{HP} = 0$) the equations reduce to those given in section 15-3.1. Forward flight influences the inertial forces by introducing the centrifugal acceleration required to turn the trim velocity vector when the body axes are rotated by the angular velocity. Principally there is a vertical acceleration due to the pitch rate and a lateral acceleration due to the yaw rate (note that these forces couple the vertical and lateral-longitudinal dynamics). Since the purpose of this analysis is to demonstrate the basic characteristics of the helicopter handling qualities in forward flight, a number of further approximations are acceptable. The inertial cross-coupling of roll and yaw will be neglected ($I_{xz} = 0$), as will the $\mu \tan \alpha_{HP}$ acceleration terms and the $g \sin \alpha_{HP}$ gravitational terms. The trim Euler angles have already been neglected in writing $p = \dot{\phi}_B$, $q = \dot{\theta}_B$, and $r = \dot{\psi}_B$ for the angular velocity components.

As for hover, the rotor forces and moments acting on the helicopter in forward flight will be obtained from the low frequency response, and therefore the rotor dynamics do not add degrees of freedom to the system. Usually the low frequency response is a good representation of the rotor for the flight dynamics, but there are circumstances where this approximation is not adequate. In section 12–1 the quasistatic rotor forces and moments, including the effects of the flap motion, were derived. Forward flight introduces changes of order μ^2 in the rotor stability derivatives present in hover, so there is no radical change in those derivatives (up to about $\mu = 0.5$). In forward flight there are also derivatives of order μ that couple the vertical and lateral-longitudinal dynamics of the rotor. The most important of these new stability derivatives is the pitching moment due to angle-of-attack perturbations of the helicopter. In section 12-3.1 the low frequency response of the longitudinal flapping to rotor vertical velocity in forward flight was found to be $\Delta\beta_{1c} = 2\mu\dot{z}_h = -2\mu\dot{z}_B$ (to order μ for an articulated rotor). The downward vertical velocity $\dot{z}_B$ produces an increase of the rotor blade angle of attack. Because of the larger dynamic pressure on the advancing side in forward flight, this blade angle-of-attack change produces a lateral moment on the rotor disk (toward the retreating side) that is proportional to μ. The rotor responds 90° later with rearward tip-path-plane tilt until the lateral moment due to the flapping velocity is sufficient to establish equilibrium again. The thrust vector tilts with the tip-path plane to produce a moment about the helicopter center of gravity:

$$\Delta\left(\frac{2C_{M_y}}{\sigma a}\right)_{CG} = -h\left(\frac{2C_T}{\sigma a}\right)\Delta\beta_{1c} = h\left(\frac{2C_T}{\sigma a}\right)2\mu\dot{z}_B$$

So downward vertical velocity results in a nose-up pitch moment on the helicopter. For hingeless rotors there is also the direct hub moment term, which greatly increases the pitch moment. In forward flight, the helicopter angle-of-attack perturbation is $\dot{z}_B/\mu$. An angle-of-attack increase then produces a pitch-up moment on the helicopter, which would tend to increase the angle of attack further. Hence the rotor is the source of an angle-of-attack instability of the helicopter in forward flight.

The derivation of the aerodynamic forces acting on the helicopter fuselage and tail can be obtained from an airplane stability and control text. All of the airframe contributions to the stability derivatives are proportional to the forward speed and therefore are zero in hover. The fuselage drag in forward flight produces damping forces X_u and Z_w; the fuselage trim aerodynamic pitch moment produces a contribution (often unstable) to the speed stability M_u. The helicopter fuselage also produces unstable moments due to angle-of-attack changes: a pitch moment M_w and a yaw moment N_v. The horizontal and vertical tails give the remaining contributions to the stabilityderivatives, assuming that the helicopter does not have a wing. The horizontal tail gives a pitch moment due to angle of attack M_w, which contributes to the static angle-of-attack stability, countering the destabilizing contribution of the main rotor. The horizontal tail also gives a pitch moment due to the helicopter pitch rate M_q (by the same mechanism as M_w) that adds to the rotor pitch damping; and corresponding contributions to the vertical force derivatives Z_w and Z_q, since the tail lift is involved. Finally, the horizontal tail contributes to the helicopter speed stability M_u and produces forces X_w and X_u. Since these last three derivatives are proportional to the tail lift coefficient, they depend on how the horizontal tail is used to trim the helicopter. The vertical tail contributions to the lateral derivatives are similar to the horizontal tail contributions to the longitudinal derivatives. The major effects of the vertical tail are the yaw moments due to lateral velocity N_v and yaw rate N_r, which add to the directional stability and yaw damping of the helicopter. There are also the corresponding side force derivatives, and as for the horizontal tail derivatives due to the vertical tail lift coefficient. (The vertical tail is often given a non-zero incidence angle so that it develops a yaw moment in forward flight to counter the main rotor torque.) The aerodynamic environment of the helicopter fuselage and tail is very complex, which makes estimation of the stability derivatives difficult. The fuselage is usually aerodynamically blunt and rough, and the tail surfaces must operate in the wake of the main rotor, tail rotor, and fuselage. It is therefore best, and often essential, to use experimental data for the aerodynamic characteristics of the helicopter airframe.

The present analysis will not consider in detail the influence of the helicopter aerodynamic forces on the flight dynamics, except for the angle-of-attack stability produced by the horizontal tail. It will be seen that the rotor angle-of-attack instability has an adverse effect on the forward flight handling qualities, and therefore a horizontal tail is used to reduce this instability. The pitch moment produced by the horizontal tail lift due to the vertical velocity is $M_y = -\frac{1}{2}\rho V S_t \ell_t a_t \dot{z}_B$, or in helicopter coefficient form

$$\frac{2C_{M_y}}{\sigma a} = -\mu \frac{S_t \ell_t a_t}{\sigma A R a} \dot{z}_B$$

where S_t is the horizontal tail area, ℓ_t the tail length, and a_t the tail lift curve slope (including the effects of the wing or rotor wake at the tail). Roughly, the angle-of-attack instability due to the main rotor is

$$\frac{2C_{M_y}}{\sigma a} = \left(\frac{\nu^2 - 1}{\gamma} + h \frac{2C_T}{\sigma a}\right) 2\mu \dot{z}_B$$

The ratio of the horizontal tail and main rotor contributions is thus

$$\frac{(M_w)_{\text{tail}}}{(M_w)_{\text{rotor}}} = -\frac{a_t}{2a} \frac{S_t \ell_t}{\sigma A R} \left(\frac{\nu^2 - 1}{\gamma} + h \frac{2C_T}{\sigma a}\right)^{-1}$$

Since both the tail and rotor moments are proportional to μ, their relative contributions to the angle-of-attack stability are independent of speed. (The dependence on speed is actually more complex, because of the wake interference effects on the tail and the rotor terms of higher order in μ.) This expression can be used to estimate the tail area required to completely offset the rotor instability, reducing to zero the net pitch moment due to angle-of-attack perturbations.

The linear acceleration relative to inertial space is important to the pilot and passengers. In terms of the body axis degrees of freedom, the inertial acceleration is $\vec{a} = \dot{\vec{u}} - \vec{u} \times \vec{\omega}$, or

$$a_x = - \ddot{x}_B + \mu \tan \alpha_{HP} \dot{\theta}_B$$

$$a_y = \ddot{y}_B + \mu \dot{\psi}_B + \mu \tan \alpha_{HP} \dot{\phi}_B$$

$$a_z = - \ddot{z}_B + \mu \dot{\theta}_B$$

In forward flight, therefore, a pitch rate produces a vertical acceleration a_z, which is the centripetal acceleration required to turn the velocity vector. Similarly, yaw rate produces a lateral inertial acceleration. As in hover, the short time response to longitudinal cyclic is primarily a pitch rate. Thus in forward flight longitudinal cyclic controls the helicopter vertical acceleration, giving the pilot control over the flight path trajectory as desired. The collective stick is used mainly for lift trim in forward flight. The helicopter roll axis and velocity vector coincide, so there is no inertial acceleration due to rolling of the body axes. Then lateral cyclic still commands roll rate, as desired.

15-4.2 Longitudinal Dynamics

15-4.2.1 Equations of Motion

Consider the uncoupled longitudinal dynamics of a helicopter in forward flight, consisting of three degrees of freedom: longitudinal velocity $\dot{x}_B$, pitch attitude θ_B, and vertical velocity $\dot{z}_B$. The controls are collective and longitudinal cyclic, and excitation by vertical and longitudinal gust velocities is included. Neglecting terms of order α_{HP}, the equations of motion are

$$M^* \ddot{x}_B = -\gamma \frac{2C_H}{\sigma a} - M^* g \theta_B$$

$$I_y^* \ddot{\theta}_B = \gamma\left(\frac{2C_{M_y}}{\sigma a} + h\,\frac{2C_H}{\sigma a}\right)$$

$$M^* \ddot{z}_B - M^* \mu \dot{\theta}_B = -\gamma\,\frac{2C_T}{\sigma a}$$

Using the low frequency rotor response, the reactions on the helicopter can be expanded in terms of the stability derivatives as follows:

$$-\gamma\,\frac{2C_H}{\sigma a}$$

$$= M^*\left[X_\theta \theta_{1s} + X_u(\dot{x}_B + u_G) + X_q \dot{\theta}_B + X_{\theta_0}\theta_0 + X_w(\dot{z}_B + w_G)\right]$$

$$\gamma\left(\frac{2C_{M_y}}{\sigma a} + h\,\frac{2C_H}{\sigma a}\right)$$

$$= I_y^*\left[M_\theta \theta_{1s} + M_u(\dot{x}_B + u_G) + M_q \dot{\theta}_B + M_{\theta_0}\theta_0 + M_w(\dot{z}_B + w_G)\right]$$

$$-\gamma\,\frac{2C_T}{\sigma a}$$

$$= M^*\left[Z_\theta \theta_{1s} + Z_u(\dot{x}_B + u_G) + Z_q \dot{\theta}_B + Z_{\theta_0}\theta_0 + Z_w(\dot{z}_B + w_G)\right]$$

The aerodynamic forces on the fuselage and tail can be included in these stability derivatives as well. Forward flight introduces the vertical acceleration due to pitch rate of the body axes, and rotor stability derivatives coupling the vertical and pitch motions. The most important new term is the angle-of-attack derivative M_w, which has unstable contributions ($M_w > 0$) from the rotor and fuselage, and a stable contribution from the horizontal tail. The other new derivatives are X_{θ_0}, M_{θ_0}, X_w, Z_θ, Z_u, and Z_q. For an articulated rotor $Z_q = 0$. It is also found that Z_u is very small, and that the force X_w is much less important than the pitch moment M_w.

The equations of motion for the three degrees of freedom of the helicopter longitudinal dynamics, in Laplace form, are:

$$\begin{bmatrix} s - X_u & -X_q s + g & -X_w \\ -M_u & s^2 - M_q s & -M_w \\ -Z_u & -Z_q s - \mu s & s - Z_w \end{bmatrix} \begin{pmatrix} \dot{x}_B \\ \theta_B \\ \dot{z}_B \end{pmatrix} = \begin{bmatrix} X_\theta & X_{\theta_0} & X_u & X_w \\ M_\theta & M_{\theta_0} & M_u & M_w \\ Z_\theta & Z_{\theta_0} & Z_u & Z_w \end{bmatrix} \begin{pmatrix} \theta_{1s} \\ \theta_0 \\ u_G \\ w_G \end{pmatrix}$$

The characteristic equation is

$$
\begin{aligned}
(s-Z_w)[s^3 \; - \; (X_u+M_q)s^2 \; + \; (X_uM_q-X_qM_u)s \; + \; gM_u] \\
- M_w[(Z_q+\mu)s^2 \; + \; (Z_uX_q-Z_qX_u-\mu X_u)s \; - \; gZ_u] \\
- X_w[Z_us^2 \; + \; (Z_qM_u-Z_uM_q+\mu M_u)s] \; = \; 0
\end{aligned}
$$

The first term is the characteristic equation for hover (for $M_w = X_w = 0$), which decouples into the vertical and longitudinal motions.

15-4.2.2 Poles

The influence of forward flight is most easily found in the case of an articulated rotor with no flap hinge offset, for which all the drag force and pitch moment derivatives are related by $M = -(h/k_y^2)X$. The characteristic equation then reduces to

$$
\begin{aligned}
(s-Z_w)[s^3 \; - \; (X_u+M_q)s^2 \; + \; gM_u] \\
- \; M_w[(\mu \; + \; Z_q \; - \; (k_y^2/h)Z_u)s^2 \; - \; gZ_u] \; = \; 0
\end{aligned}
$$

For the articulated rotor, $Z_q = 0$ and Z_u is small, so the last term is nearly just $-M_w\mu s^2$. For an offset-hinge articulated rotor or a hingeless rotor it is still found that the term $-M_w\mu s^2$ dominates the influence of forward flight on the characteristic equation. In general, therefore, the characteristic equation is approximately

$$
(s-Z_w)[s^3-(X_u+M_q)s^2+(X_uM_q-X_qM_u)s+gM_u] \; - \; M_w\mu s^2 \; = \; 0
$$

The influence of forward flight on the longitudinal dynamics of the helicopter thus consists primarily of the following forces: the pitch moment due to vertical velocity; the vertical acceleration due to pitch rate; and the longitudinal inertia of the helicopter. Their product gives the $-M_w\mu s^2$ term in the characteristic equation. The influence of forward flight on the roots is easily obtained by viewing this as the characteristic equation of a feedback system with "gain" M_w. The "open loop poles" are the hover roots (more correctly, the roots of the hover characteristic equation obtained using the forward flight stability derivatives), and there are two "open loop zeros" at the origin. The hover poles consist of the two real, negative roots for the vertical and pitch motions and a long period, mildly unstable oscillation. The "gain" can alternatively be considered μ^2, since M_w is proportional to μ. The root locus for varying M_w then is essentially the locus for varying speed as well. Fig. 15-10 shows the influence of forward

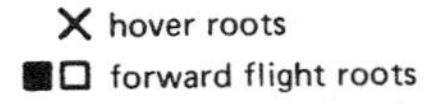

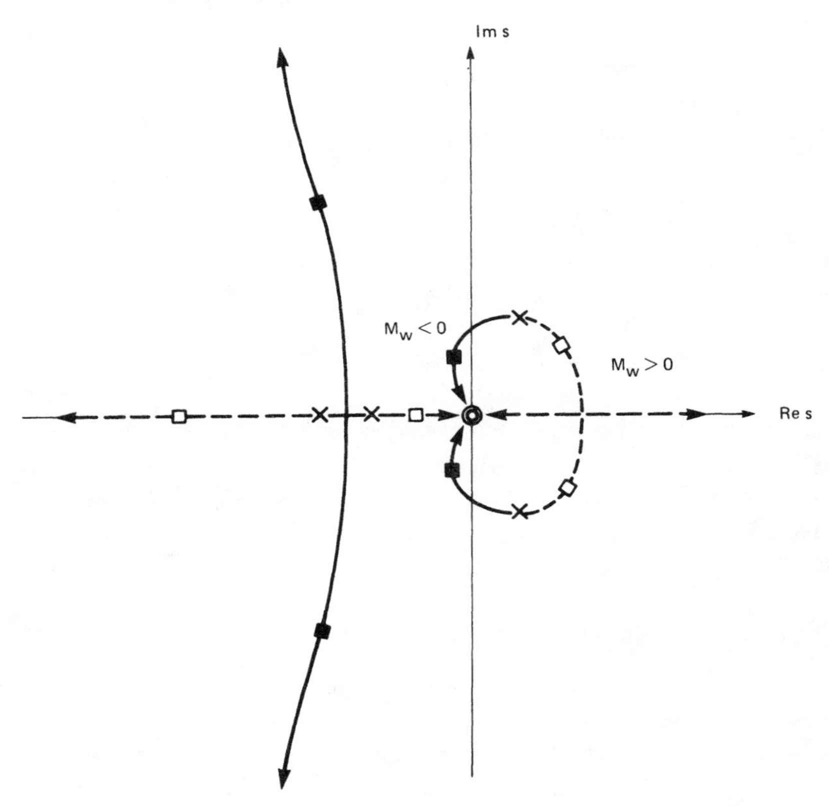

Figure 15-10 Influence of forward flight on the longitudinal roots for a helicopter without a horizontal tail ($M_w > 0$) and with a horizontal tail ($M_w < 0$).

flight on the roots of the longitudinal dynamics when there is a net angle-of-attack instability due to the rotor ($M_w > 0$), and when there is a large enough horizontal tail for angle-of-attack stability ($M_w < 0$).

With the main rotor alone, the angle-of-attack instability in forward flight reduces the damping of the smaller real root (usually the vertical mode) and increases the damping of the larger real root. The influence of forward speed on the oscillatory mode is to increase the period and decrease the time to double amplitude (reduced damping). Hence the forward flight dynamics of a helicopter without a horizontal tail are characterized by two real roots and an unstable oscillatory mode, with a degradation of the handling qualities due to the angle-of-attack instability. With a hingeless rotor M_w can be large enough at very high speed to replace the oscillatory mode with two positive real roots, one with an unacceptably small time to double amplitude. The helicopter can have net static stability with respect to angle of attack by using a large enough horizontal tail. In that case, forward speed transforms the pitch and vertical roots of hover into an oscillatory mode with a short period and high damping; the long period hover mode is usually moved into the left half-plane, with the period increased somewhat and the damping also increased. The forward flight longitudinal dynamics of a helicopter with a horizontal tail are thus characterized by a short period mode due to the damping of vertical and pitch motion, and a long period mode stabilized by the static stability with respect to angle of attack. A horizontal tail large enough to produce a high level of static stability is not always practical, particularly with hingeless rotors. Moreover, the tail effectiveness is reduced at low speeds by interference with the rotor and fuselage wakes. The improvement of the handling qualities is so significant, though, that most single main rotor helicopter designs have a horizontal tail.

Nonuniform inflow can be an important factor in the forward flight dynamics, producing significant changes in the stability derivatives. For example, the speed stability derivative is particularly sensitive to longitudinal variations of the inflow. It has been assumed in the present analysis that the rotor speed is constant. For helicopters in autorotation, in partial-power descent, or without a tight governor there will be significant rotor speed perturbations, which will have a major influence on the forward flight dynamics. It is found that the autorotating rotor has neutral speed stability ($M_u = 0$) and positive angle-of-attack stability ($M_w < 0$).

15-4.2.3 Short Period Approximation

Let us consider a short period approximation for the helicopter longitudinal dynamics. The initial response to control and gusts is primarily vertical and pitch acceleration, with little longitudinal acceleration. The control over the longitudinal motion is indirect, so it takes a while for a significant $\dot{x}_B$ response to develop. Therefore, as an approximation for short times, the longitudinal velocity degree of freedom will be neglected. The equations of motion then reduce to

$$\begin{bmatrix} s-M_q & -M_w \\ -\mu & s-Z_w \end{bmatrix} \begin{pmatrix} \dot{\theta}_B \\ \dot{z}_B \end{pmatrix} = \begin{bmatrix} M_\theta & M_{\theta_0} & M_u & M_w \\ Z_\theta & Z_{\theta_0} & Z_u & Z_w \end{bmatrix} \begin{pmatrix} \theta_{1s} \\ \theta_0 \\ u_G \\ w_G \end{pmatrix}$$

(neglecting also the small vertical force due to $\dot{\theta}_B$, produced by the Z_q derivative). These equations retain the principal coupling of forward flight: the angle-of-attack derivative M_w and the vertical acceleration due to pitch rate. Since the gravitational spring term on the pitch motion appears in the longitudinal force equation, the short period approximation gives a second-order system for the two degrees of freedom $\dot{\theta}_B$ and $\dot{z}_B$. The equations of motion invert to

$$\begin{pmatrix} \dot{\theta}_B \\ \dot{z}_B \end{pmatrix} = \frac{1}{\Delta} \begin{bmatrix} s-Z_w & M_w \\ \mu & s-M_q \end{bmatrix} \begin{bmatrix} M_\theta & M_{\theta_0} & M_u & M_w \\ Z_\theta & Z_{\theta_0} & Z_u & Z_w \end{bmatrix} \begin{pmatrix} \theta_{1s} \\ \theta_0 \\ u_G \\ w_G \end{pmatrix}$$

with the characteristic equation

$$\Delta = (s-M_q)(s-Z_w) - M_w\mu = 0$$

In hover the pitch and vertical motions decouple, and the two solutions of the characteristic equation are $s = Z_w$ and $s = M_q$. The first solution is exactly the hover vertical pole. The second solution is the pole of the short period approximation for the hover longitudinal dynamics (see section 15-3.4.5); which is an approximation for the pitch root in hover. The locus of the short period roots for varying speed or M_w is sketched in Fig. 5-11. When Fig. 15-11 is compared with the root locus for the complete dynamics (Fig. 15-10), it is observed that this approximation neglects the helicopter long period mode. Note also that if M_w has a large enough negative value, one of the roots goes through the origin into the right half-plane, indicating

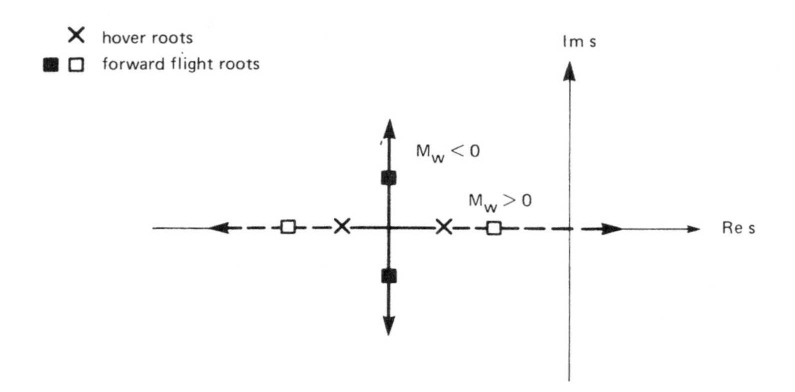

Figure 15-11 Influence of forward flight on the short period approximation for helicopter longitudinal roots.

a static divergence of the short period motions due to the angle-of-attack instability. Actually, this branch of the locus only approaches the origin (see Fig. 15-10), so it is necessary to consider the complete dynamics for M_w that large. The characteristic equation of the short period approximation can be written $\Delta = (s - s_z)(s - s_\theta)$, where s_z and s_θ are the two roots, which actually are a complex conjugate pair if $M_w < 0$ in forward flight. Note that in the limit of very small time (s approaching infinity), the short period approximation gives exactly the same vertical and pitch acceleration response as the complete model.

The principal concern in the short period longitudinal dynamics is the normal acceleration response of the helicopter. Recall that in terms of the body axis degrees of freedom, the vertical acceleration in inertial space is $a_z = -\ddot{z}_B + \mu\dot{\theta}_B$. The pitch rate is the main source of normal acceleration in forward flight. The response of $a_z = -s\dot{z}_B + \mu\dot{\theta}_B$ to longitudinal cyclic, as given by the short period approximation, is

$$\frac{a_z}{\theta_{1s}} = \frac{1}{\Delta}\left[-sZ_\theta(s - M_q) + \mu(Z_\theta M_w - M_\theta Z_w)\right]$$

$$\cong -Z_\theta + \frac{-\mu M_\theta Z_w}{(s - s_z)(s - s_\theta)}$$

The initial response is $a_z/\theta_{1s} = -Z_\theta$. This is a small vertical acceleration produced immediately after the control application by the thrust increment due to cyclic. The response of the pitch rate is zero initially, but as it builds up it contributes to the normal acceleration. The steady-state response (of the short period approximation) is

$$\frac{a_z}{\theta_{1s}} = -Z_\theta - \frac{\mu M_\theta Z_w}{s_z s_\theta}$$

The second term is the acceleration due to the steady-state pitch rate response to cyclic.

The following behavior of the helicopter normal acceleration in response to longitudinal cyclic in forward flight has been found. Consider a step input of θ_{1s} (Fig. 15-12). By using the low frequency rotor response, the lag in

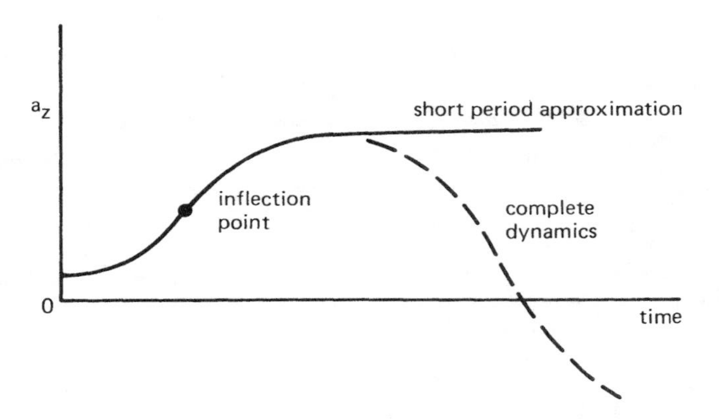

Figure 15-12 Normal acceleration response to a step input of longitudinal cyclic.

the development of the rotor forces has been neglected. Thus immediately after the application of the cyclic control there is a thrust increment that produces a small vertical acceleration. The pitch angular velocity of the helicopter is initially zero, but it builds up to the steady-state values given by the short period approximation, with a second-order transient defined by the two short period poles. This body axis pitch rate produces the major portion of the normal acceleration in forward flight. For times beyond the validity of the short period approximation, the long period mode enters the response. A low frequency, slowly decaying (or growing) oscillation of the normal acceleration develops (see Fig. 15-12). The pilot must then take the corrective action required to bring the helicopter back to the trim state.

The longitudinal cyclic command of the normal acceleration of the helicopter, and hence of its flight path trajectory, is an important aspect of the handling qualities. The helicopter pitch control sensitivity is high, and a reasonable normal acceleration is achieved eventually. However, there is a delay after the application of the cyclic control (which initially produces only the small vertical acceleration Z_θ) until the pitch rate develops enough to give that normal acceleration. Thus the helicopter is characterized by a lag before the maximum normal acceleration in response to longitudinal cyclic is achieved, which can make control difficult if the lag is too long. A basic requirement of the helicopter handling qualities in forward flight is that the maximum acceleration be achieved within a certain time after the application of the cyclic control. This requirement has been quantified by using the inflection point on the trace of a_z as a function of time (Fig. 15-12). The smaller the inflection point time t_I, the sooner the maximum acceleration will be achieved; moreover, the existence of an inflection point means that the normal acceleration response is not divergent. Note that beyond t_I the normal acceleration curve is concave downward. Thus the following specification has been developed: the time history of the normal acceleration due to a step input of longitudinal cyclic should be concave downward within a time t_I. The usual specification calls for the acceleration to be concave downward within $t_I = 2$ sec after application of the control.

To investigate the implications of the concave downward requirement, consider the time history of the response to cyclic using the short period approximation. The second derivative of the acceleration is $\ddot{a}_z = s^2 a_z$, and a step cyclic input is $\theta_{1s} = \overline{\theta}_{1s} s^{-1}$. Hence

$$\frac{\ddot{a}_z}{\overline{\theta}_{1s}} = \frac{-\mu M_\theta Z_w}{s_z - s_\theta}\left(\frac{s_z}{s - s_z} - \frac{s_\theta}{s - s_\theta}\right)$$

For zero initial conditions, the response in the time domain is then

$$\frac{\ddot{a}_z}{\overline{\theta}_{1s}} = \frac{-\mu M_\theta Z_w}{s_z - s_\theta}\left(s_z e^{s_z t} - s_\theta e^{s_\theta t}\right)$$

The inflection point is by definition where $\ddot{a}_z = 0$. The concave downward requirement is then that $\ddot{a}_z \leqslant 0$ for $t = t_I$, or

$$\frac{1}{s_z - s_\theta}\left(s_z e^{s_z t} - s_\theta e^{s_\theta t}\right) \leqslant 0$$

For a complex pair of short period roots, this criterion is

$$\mathrm{Re}\, s < \frac{-\mathrm{Im}\, s}{\tan\left(t_I\, \mathrm{Im}\, s\right)}$$

which defines a region on the s-plane that is forbidden to the short period roots if the concave downward requirement is to be satisfied (Fig. 15-13).

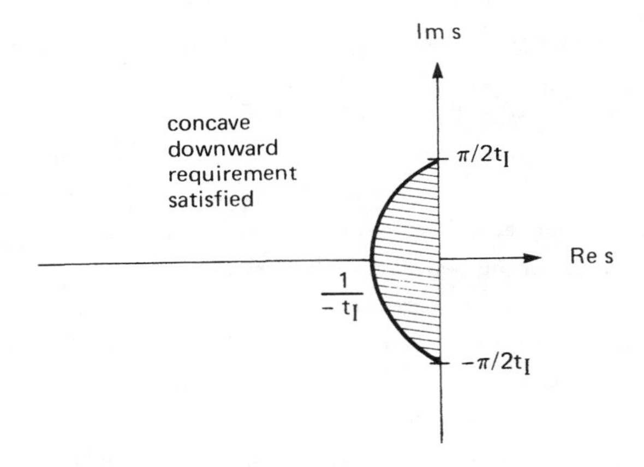

Figure 15-13. Concave downward requirement in terms of the short period roots (complex roots only).

For $t_I = 2$ sec, the imaginary axis intercept is at a frequency of 0.125 Hz and the real axis intercept corresponds to a time constant of 2 sec. If there are two real short period roots, the criterion is

$$t_I \geqslant \frac{\tau_z \tau_\theta}{\tau_z - \tau_\theta} \ln \frac{\tau_z}{\tau_\theta}$$

where τ_z and τ_θ are the time constants of the two roots. If one root has a time constant greater than t_I, the requirement can still be satisfied if the time constant of the other root is sufficiently smaller than t_I. The stability derivatives involved in the short period response are Z_w, M_q, and M_w. The vertical damping is essentially fixed by the basic rotor design parameters. Increasing the pitch damping improves the normal acceleration response, as does increasing the static stability with respect to angle of attack. As a simpler treatment of the concave downward requirement, consider the normal acceleration response for very small time: $a_z/\theta_{1s} = -\mu M_\theta Z_w/s^2$. The response to a step cyclic input thus has an initial growth proportional to the time squared: $a_z/\bar{\theta}_{1s} = -\mu M_\theta Z_w t^2/2$. Now the intersection of this quadratic curve with the steady-state response will be taken as an estimate of the inflection point. Requiring that the time at this intersection be less than the specified inflection time t_I is usually conservative. The steady-state response is $a_z/\bar{\theta}_{1s} = -\mu M_\theta Z_w/(s_z s_\theta)$. Equating this to the initial transient gives $t = (2/s_z s_\theta)^{1/2}$. The concave downward requirement applied to this intersection time is $t \leqslant t_I$, or $s_z s_\theta \geqslant 2/t_I^2$, which defines a forbidden region on the s-plane that is a circle about the origin with radius $\sqrt{2}/t_I$. The intercept of the imaginary axis is at a frequency 0.113 Hz, and the real axis intercept corresponds to a time constant of 1.4 sec for $t_I = 2$ sec. Substituting for $s_z s_\theta$ gives the criterion

$$Z_w M_q - \mu M_w \geqslant 2/t_I^2$$

which expresses the concave downward requirement directly in terms of the stability derivatives. The favorable influence of pitch damping and angle-of-attack stability is clear. Both the initial and steady-state response are proportional to M_θ, so the criterion does not involve the control power. Note also that increasing $s_z s_\theta$ does not change the initial response, but rather it reduces the inflection time by decreasing the control sensitivity. Amer (1954) and Bramwell (1957) present charts relating the concave downward criterion to the helicopter stability derivatives and roots.

The inverse of the normal acceleration response gives the longitudinal cyclic stick displacement per g of normal acceleration. The steady-state response of the short period approximation gives

$$\frac{\theta_{1s}}{a_z/g} = \frac{g s_z s_\theta}{-\mu M_\theta Z_w} = \frac{g(Z_w M_q - \mu M_w)}{-\mu M_\theta Z_w}$$

At low speed this stick gradient is a measure of the pitch control sensitivity M_q/M_θ, and at high speed it is a measure of the angle-of-attack stability M_w.

Using the rough criterion obtained above, $s_z s_\theta \geqslant 2/t_l^2$, a minimum stick gradient criterion is obtained:

$$\frac{\theta_{1s}}{a_z/g} \geqslant \frac{g}{-\mu M_\theta Z_w} \frac{2}{t_l^2}$$

Thus the longitudinal cyclic stick gradient per normal load factor is directly related to the concave downward requirement (since both are related to the steady-state response). Alternatively, the concave downward requirement can be viewed as a specification of the helicopter maneuver margin in forward flight.

Considering either the concave downward requirement or the stick gradient with load factor, it is concluded that the helicopter maneuver capability is an important factor in sizing the horizontal tail. The angle-of-attack stability due to the tail is an effective means of achieving the required normal acceleration response. Indeed, a helicopter without a horizontal tail will often not have acceptable short period handling qualities. Increasing the pitch damping by using rate feedback or a hingeless rotor will also improve the short period response by reducing the control sensitivity, particularly at low speeds (before the angle-of-attack derivative becomes large).

In summary, the concave downward requirement is a specification on the longitudinal handling qualities that is intended to ensure acceptable maneuver characteristics of the helicopter: the time history of the normal acceleration in response to a step longitudinal cyclic displacement should be concave downward within 2 sec. The problem this specification addresses is the delay in development of the normal acceleration after control application, which must not be too long for satisfactory handling qualities. The longitudinal stick gradient with normal load factor is directly related to the concave downward requirement. The stick gradient must be sufficiently positive in order to meet the maneuver requirement.

15-4.2.4 Static Stability

Static stability can be defined as a tendency for a system to return toward the equilibrium position when disturbed, which implies a force or moment opposing static perturbations from equilibrium. Static stability is related to divergence of a system. The divergence stability boundary is defined by the criterion that one pole of the system be at the origin, so divergence stability is assured if the constant term of the characteristic equation is positive. In contrast, dynamic stability means that all disturbances from equilibrium die out eventually, which requires that all poles of the system be in the left half-plane. Static stability can also be related to the steady-state response of the system to control. The presence of a force or moment opposing perturbations from equilibrium (i.e. static stability) implies that in order to change the equilibrium trim state it is necessary to apply a force or moment to the aircraft, by means of a control deflection. The amount of the control required (the control gradient) is a measure of the force or moment produced by the perturbation from trim and hence of the static stability. The sign of the control gradient indicates whether the system is statically stable or unstable. For simple systems all these definitions of static stability are equivalent and have elementary interpretations. For complex systems, however, the definition and interpretation of static stability is more difficult. For the helicopter, more than one stability derivative is involved even in most static effects, so it is difficult to relate stick gradients, static stability, and the dynamic stability characteristics.

The helicopter in hover has neutral static stability with respect to pitch or roll perturbations, since no moments to oppose the motion are generated directly by such perturbations (see section 15-3.4.5). The hovering helicopter does have positive static stability with respect to longitudinal or lateral velocity perturbations, because of the speed stability derivatives M_u and L_v. Analogous behavior is found in airplane lateral dynamics, where static stability with respect to lateral velocity perturbations (sideslip) is produced by the wing dihedral effect, but there is neutral static stability with respect to roll angle perturbations. The rotor in forward flight is statically unstable with respect to angle of attack, as discussed in the preceding sections. The fuselage and tail contribute significantly to the overall helicopter static stability with respect to angle of attack and speed in forward flight. Perturbations of the rotor rotational speed can also radically alter the helicopter static stability.

The important control gradients for the helicopter are the longitudinal cyclic displacements required to change speed and to change the normal load factor. For static stability with respect to speed, a forward displacement of the longitudinal cyclic stick must be required in order to increase speed, $\partial\theta_{1s}/\partial\mu < 0$. This stick gradient is primarily related to the speed stability derivative M_u. Generally, as the forward speed of the helicopter increases, the rotor tip-path plane flaps back, and a forward tilt of the control plane is needed to maintain the aircraft trim (see also section 15–1). At low forward speeds, however, some helicopters exhibit an unstable stick gradient with velocity. For acceptable maneuver characteristics, an aft displacement of the longitudinal cyclic stick must be required in order to increase the helicopter vertical load factor in forward flight, $\partial\theta_{1s}/\partial a_z > 0$. The analysis in the last section showed that this control gradient is related to the angle-of-attack and pitch damping derivatives M_w and M_q, and thus to the concave downward requirement. A minimum gradient or a maximum control sensitivity is required for acceptable maneuver characteristics.

The simplest means of calculating the control gradients is to plot the stick position as a function of speed or load factor, and then evaluate the derivative numerically or graphically. The control position for the required trim or maneuver condition is best calculated using a sophisticated aeroelastic analysis of the helicopter. The gradients can be obtained directly from the static response of a perturbation analysis such as that developed here, but approximations are usually introduced in the process of linearizing the equations of motion.

15-4.2.5 Example

As an example, consider the helicopter described in section 15-3.4.6, at a forward speed of $\mu = 0.35$ ($V = 135$ knots). A fuselage parasite drag of $f/A = 0.015$ is assumed, and the horizontal tail (when used) has area $S_t/\sigma a = 0.05$, moment arm $\ell_t/R = 1.15$, and lift curve slope $a_t = 4$. Table 15-5 gives the dimensionless roots and eigenvectors of the forward flight longitudinal dynamics with an articulated rotor. For an articulated rotor helicopter with no horizontal tail, the real roots have times to half amplitude of $t_{1/2} = 0.7$ and 2.7 sec, and the oscillatory mode has a period $T = 22$ sec and a time to double amplitude $t_2 = 3.2$ sec. For an articulated rotor helicopter with a horizontal tail, the short period mode has $T = 5.8$ sec and $t_{1/2} = 1.4$ sec, while the long period mode has $T = 40$ sec and $t_{1/2} = 21$ sec. With the horizontal tail, therefore, this helicopter has a well-damped short period mode and a mildly stable long period mode in forward flight.

Table 15-5 Example of helicopter longitudinal dynamics in forward flight

	Root, s	Eigenvector	
		$\dot{x}_B/\theta_B$	$\dot{z}_B/\theta_B$
Without horizontal tail			
vertical mode	−0.046	0.07	0.91
pitch mode	−0.012	0.22	−0.26
oscillatory mode	0.0099 ± i 0.013	0.13	0.14
phase		13°	147°
With horizontal tail			
short period mode	−0.023 ± i 0.050	0.05	0.38
phase		113°	148°
long period mode	−0.0015 ± i 0.0073	0.32	0.08
phase		86°	92°

Note that the short period mode is primarily coupled θ_B and $\dot{z}_B$ motion with little longitudinal velocity, as assumed in the short period analysis. The longitudinal roots of a hingeless rotor show a similar behavior in forward flight, but a larger tail volume is required to counter the larger angle-of-attack instability of the rotor.

15-4.2.6 Flying Qualities Characteristics

Gustafson and Reeder (1948) noted in one helicopter at least that the maximum normal acceleration due to longitudinal control displacement in forward flight occurred several seconds after the maximum control displacement. This represented an angle-of-attack instability, and meant that after initiating the maneuver the pilot had to move the control in the opposite direction, beyond the trim position, in order to limit the acceleration to the desired value. They also found a high vibration level and a lack of trim control forces.

Reeder and Gustafson (1949) investigated the longitudinal dynamics of a tailless helicopter and concluded that the major problems regarding the flying qualities in forward flight were the instability with respect to angle of attack, and the control forces accompanying maneuvers. The angle-of-attack instability resulted in unacceptable normal acceleration response to longitudinal cyclic in a pull-up maneuver. A divergent behavior of the normal acceleration was found, and as in hover undesirable stick forces during both lateral and longitudinal maneuvers were noted. There was an extreme deterioration of the stability of the long period oscillation in forward flight, also due to the angle-of-attack instability, which increased with speed. Reeder and Gustafson suggested the use of a horizontal tail to achieve angle-of-attack stability in forward flight.

Gustafson, Amer, Haig, and Reeder (1949) conducted flight tests of the longitudinal flying qualities of helicopters. The pilot is very sensitive to the normal acceleration, which is the primary measure and control of the flight path trajectory in forward flight. Therefore they used a pull-up maneuver as a standard test of the normal acceleration characteristics. For a single main rotor and tail rotor helicopter without a horizontal tail, they found the response to a step longitudinal cyclic was a steadily growing pitch rate. There was a pause in the development of the normal acceleration, and then it grew steadily with no tendency to achieve a constant value. Moreover, the stick-fixed dynamics showed an unstable oscillatory mode. With a horizontal tail, the helicopter quickly developed a pitch rate after a step input longitudinal cyclic, and the pitch rate tended to reach a constant value. There was a pause in the development of the normal acceleration, but at about 2 sec after the application of control it tended to reach a constant value. Less control was required to recover from the maneuver, and the stick-fixed oscillation was lightly damped. The flying qualities with the horizontal tail were much more acceptable. In an attempt to quantify the desired characteristics, they developed the concave downward requirement for the longitudinal dynamics. They concluded that the most important factor in the longitudinal dynamics is whether or not a prolonged stick-fixed divergence of the normal acceleration will occur; the characteristics are further improved if there is a continuous development of the normal acceleration, without a pause during the first second. Their analysis showed that the effect of the tail is primarily due to its influence on the angle-of-attack stability derivative M_w, increasing it from the positive value (unstable) for the rotor and fuselage alone to a negative value of about half the magnitude in their case.

Reeder and Whitten (1952) investigated the influence of the pitch and roll damping on the flying qualities of a single main rotor helicopter. They used a gyro stabilizer bar giving lagged feedback to increase the damping M_q to a maximum three times the value without the stabilizer bar. In pull-up maneuvers, the lagged rate feedback considerably improved the longitudinal flying qualities. Without the stabilizer bar the normal acceleration showed a divergent tendence for too long, the angular acceleration was constant for the first 1.5 sec, and the normal acceleration was concave upward for about 2.5 sec. When the pitch damping was increased by a factor of two or three, acceptable handling qualities were obtained. There was then a rapid reduction in angular acceleration and development of the steady-state pitch rate; there was also a rapid development of the normal acceleration, which was concave downward in less than 2 sec. The increased pitch damping also reduced the frequency and increased the time to double amplitude of the long period oscillation; when the damping was about 2.7 times the level without augmentation, the mode was even slightly stable. The forward flight roll control characteristics remained satisfactory with the lagged rate feedback. The increased roll damping decreased the steady-state response to control, which is usually too high anyway. The initial value of the roll angular acceleration was unchanged and the long time response was improved with the feedback, a more constant roll rate due to lateral cyclic being achieved.

15-4.3 Lateral Dynamics

Consider next the uncoupled lateral dynamics of the helicopter in forward flight, consisting of three degrees of freedom: side velocity, roll attitude, and yaw rate. The control of the lateral motion is by main rotor cyclic and tail rotor collective, and the lateral gust velocity is included. Neglecting the inertial and gravitational terms of order α_{HP}, the equations of motion are

$$M^* \ddot{y}_B + M^* \mu \dot{\psi}_B = \gamma \frac{2C_Y}{\sigma a} + M^* g \phi_B$$

$$I_x^* \ddot{\phi}_B = \gamma \left(-\frac{2C_{M_x}}{\sigma a} + h \frac{2C_Y}{\sigma a} \right)$$

$$I_z^* \ddot{\psi}_B = \gamma \left(\frac{2C_Q}{\sigma a} - \ell_{tr} \frac{(\sigma a A (\Omega R)^2)_{tr}}{(\sigma a A (\Omega R)^2)_{mr}} \left(\frac{2C_T}{\sigma a} \right)_{tr} \right)$$

In forward flight there is a lateral acceleration due to the yawing velocity of the body axes. Using the rotor low frequency response, the stability derivatives are defined as follows:

$$\gamma \frac{2C_Y}{\sigma a} = M^*(Y_\theta \theta_{1c} + Y_v(\dot{y}_B + v_G) + Y_p \dot{\phi}_B + Y_r \dot{\psi}_B)$$

$$\gamma \left(-\frac{2C_{M_x}}{\sigma a} + h \frac{2C_Y}{\sigma a} \right) = I_x^*(L_\theta \theta_{1c} + L_v(\dot{y}_B + v_G) + L_p \dot{\phi}_B + L_r \dot{\psi}_B)$$

$$\gamma \left(\frac{2C_Q}{\sigma a} - \ell_{tr} \frac{(\sigma a A(\Omega R)^2)_{tr}}{(\sigma a A(\Omega R)^2)_{mr}} \left(\frac{2C_T}{\sigma a} \right)_{tr} \right)$$

$$= I_z^*(N_\theta \theta_{1c} + N_v(\dot{y}_B + v_G) + N_p \dot{\phi}_B + N_r \dot{\psi}_B + N_{\theta_{tr}} \theta_{0_{tr}})$$

The main rotor side force and roll moment due to yaw rate (Y_r and L_r) are small and will be neglected. Also, since the main rotor torque perturbations are small compared to the tail rotor thrust contributions to the yaw moments, the N_θ, N_p, and N_r derivatives will be neglected. The equations of motion for the helicopter lateral dynamics, in Laplace form, are then

$$\begin{bmatrix} s - Y_v & -Y_p s - g & \mu \\ -L_v & s^2 - L_p s & 0 \\ -N_v & 0 & s - N_r \end{bmatrix} \begin{pmatrix} \dot{y}_B \\ \phi_B \\ \dot{\psi}_B \end{pmatrix} = \begin{bmatrix} Y_\theta & 0 & Y_v \\ L_\theta & 0 & L_v \\ 0 & N_{\theta_{tr}} & N_v \end{bmatrix} \begin{pmatrix} \theta_{1c} \\ \theta_{0_{tr}} \\ v_G \end{pmatrix}$$

In this approximation, the only stability derivatives retained are those present in hover as well as in forward flight. The coupling of the yaw and lateral dynamics is due to the lateral acceleration produced by yaw rate, and the directional stability N_v (yaw moment produced by lateral velocity). The directional stability is due to the change in tail rotor thrust during lateral velocity of the helicopter. It does not influence the hover roots though, but rather it is responsible for a yaw response during lateral motions of the hovering helicopter. In forward flight the rotor is not axisymmetric, so the side force and roll moment derivatives cannot be evaluated from the corresponding longitudinal stability derivatives, as is possible in hover.

The characteristic equation of the helicopter lateral dynamics in forward flight is

$$(s - N_r)[s^3 - (Y_v + L_p)s^2 + (Y_v L_p - Y_p L_v)s - g L_v] + \mu N_v (s^2 - L_p s) = 0$$

The first term is the product of the characteristic equations for uncoupled yaw and roll plus side velocity. The second term is due to the coupling in forward flight by the lateral acceleration and directional stability. If the advance ratio μ is viewed as the "gain," the root locus for varying speed can be constructed. The "open loop poles" are the hover roots: the uncoupled yaw root, the roll root, and the unstable lateral oscillation. There are two "open loop zeros," one at the origin and the other at $s = L_p$. Note that $s = L_p$ is the pole of the uncoupled roll motion, which is somewhat to the right of the hover roll root. Fig. 15-14 shows the locus of roots of the lateral dynamics in forward flight that follows from this characteristic equation. The directional stability is always positive ($N_v > 0$). Moreover, the tail rotor gives a high level of directional stability, and as a result the "gain" is high in forward flight. Thus in forward flight the two real roots are very near the "zeros," which are the poles of the uncoupled roll response. Since the other two roots approach the vertical asymptote, the inertial coupling in forward flight transforms the hover oscillatory mode into a stable, short period oscillation.

The vertical asymptote of the forward flight root locus gives a good estimate of the damping of the short period oscillation, since with the high directional stability the roots are close to the asymptote. The asymptote is given by the vertical line with rear part

$$\begin{aligned} \mathrm{Re}\, s &= \tfrac{1}{2}[\Sigma \text{ open loop poles} - \Sigma \text{ open loop zeros}] \\ &= \tfrac{1}{2}[(Y_v + L_p + N_r) - L_p] \\ &\cong \tfrac{1}{2}N_r \end{aligned}$$

which is just half the hover yaw root.

In forward flight, the roll root approaches $s = L_p$, the value given by the uncoupled damping. This roll root typically has a time to half amplitude of $t_{1/2} = 0.5$ to 1 sec with articulated rotors, and much smaller with hingeless rotors. The derivative L_p does vary somewhat with speed, so while the forward flight root has nearly the value of the uncoupled roll damping, that value varies with advance ratio. The other real root (the yaw mode in hover) approaches the origin, indicating that the roll response becomes a rate change rather than an attitude change. Finally, forward speed transforms the unstable, long period oscillation of hover into a well-damped short period mode. The time to half amplitude of this mode is about double

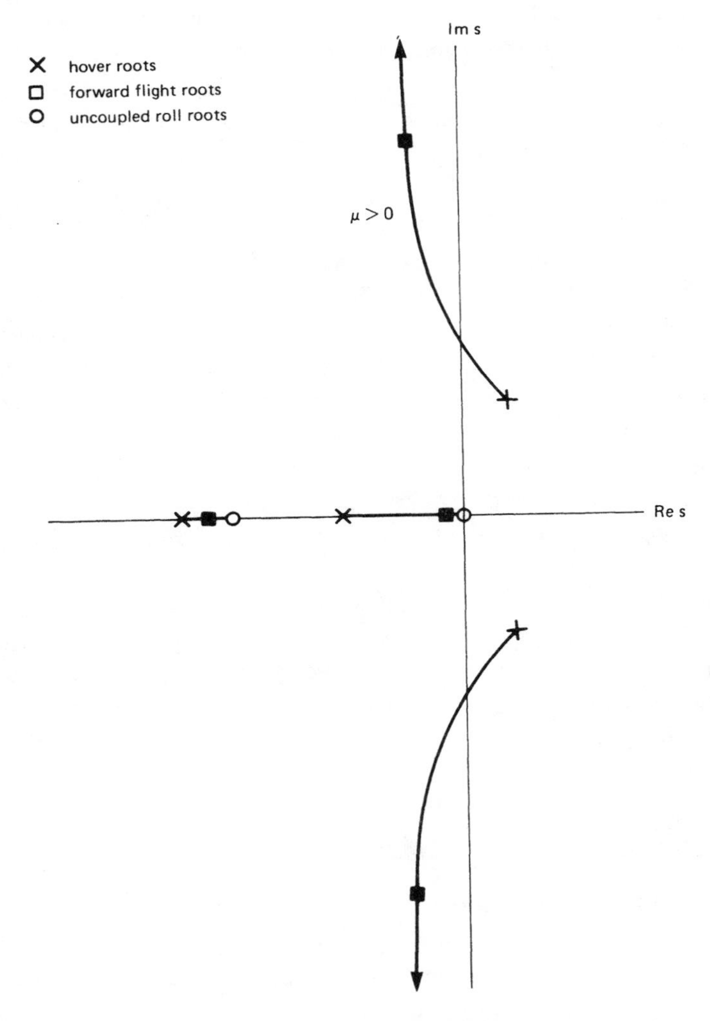

Figure 15-14 Influence of forward flight on the helicopter lateral roots.

that of the hover yaw root. Sideslip to the right produces a yaw to the right through the directional stability. In forward flight this yaw motion implies a lateral centrifugal acceleration on the helicopter, or equivalently an inertial force acting to the left. In contrast with hover, therefore, the lateral velocity of the helicopter produces a force opposing the motion, and the oscillatory mode is stable in forward flight.

As a short period approximation to the lateral dynamics in forward flight, neglect the lateral velocity, since it builds up much more slowly than the roll or yaw motion. Then, since roll and yaw are not coupled in the present model, the equations of motion for the short period approximation reduce to simply the uncoupled roll response: $(s - L_p)\dot{\phi}_B = L_\theta \theta_{1c}$. In forward flight this is indeed a good approximation for the roll dynamics, because the directional stability tends to decouple the lateral velocity from the roll motion. Hence lateral cyclic commands the roll rate, with a small first-order time lag due to the inertia. The steady-state response of the short period approximation gives $\dot{\phi}_B/\theta_{1c} = -L_\theta/L_p$, which is usually high because of low roll damping. With a hingeless rotor the control sensitivity is reduced, since the damping is increased more than the control moment.

This elementary analysis of the helicopter lateral dynamics is sufficient to show the basic influence of forward speed and the tail rotor on the motion. A better analysis must also include the details of the fuselage inertial and aerodynamic forces, the vertical tail aerodynamics, and the tail rotor position and direction of rotation. The mutual aerodynamic interference between the fuselage, the vertical tail, the tail rotor, and the main rotor can greatly influence the flight dynamics, and hence is an important consideration in the aircraft design.

As an example, consider the lateral dynamics of the helicopter described in sections 15-3.4.6 and 15-3.5, at a forward speed of $\mu = 0.35$ ($V = 135$ knots). For the tail rotor a moment arm $\ell_{tr} = 1.15$ and blade area $\sigma A_{tr}/\sigma A_{mr} = 0.05$ are assumed. The normalized roll and yaw inertias are $I_x^* = 2.5$ and $I_z^* = 10.2$ ($k_x^2 = 0.02$ and $k_z^2 = 0.08$). Table 15-6 gives the dimensionless roots and eigenvectors of the lateral dynamics in forward flight. For the articulated rotor the real roots have times to half amplitude of $t_{1/2} = 0.07$ and 15 sec. The short period oscillatory mode has a period $t = 3.2$ sec, and time to half amplitude $t_{1/2} = 2.9$ sec. For the hingeless rotor the real roots have $t_{1/2} = 0.1$ and 20 sec, while the short period oscillation has $T = 3.1$ sec and $t_{1/2} = 2.4$ sec. The roll control sensitivity is $\dot{\phi}_B/\theta_{1c} = -15$ deg/sec/deg for this helicopter with an articulated rotor, and $\dot{\phi}_B/\theta_{1c} = -10$ deg/sec/deg with a hingeless rotor.

Table 15-6 Example of helicopter lateral dynamics in forward flight

	Root, s	Eigenvector $\dot{y}_B/\phi_B$	$\dot{\psi}_B/\phi_B$
Articulated rotor ($\nu = 1$)			
roll mode	-0.046	-0.021	0.004
spiral mode	-0.0021	0.007	0.006
oscillatory mode	$-0.011 \pm i0.090$	0.66	0.17
phase		-168°	-93°
Hingeless rotor ($\nu = 1.15$)			
roll mode	-0.313	-0.015	-0.004
spiral mode	-0.0016	0.007	0.007
oscillatory mode	$-0.013 \pm i0.094$	0.52	0.14
phase		-118°	143°

15-4.4 Tandem Helicopters

In hover the longitudinal handling qualities of the tandem helicopter are somewhat better than those of the single main rotor configuration because of the higher pitch damping and control power; the lateral handling qualities are somewhat worse because of the lower yaw damping and higher yaw and roll inertias. In forward flight the tandem helicopter has a large angle-of-attack instability due to the main rotors (and the fuselage), but a large horizontal tail is not very practical. Thus there is a degradation of the longitudinal handling qualities in forward flight, the angle-of-attack instability producing an unstable oscillation or even a real divergence. The tandem helicopter does not have much directional stability even in hover, although some can be obtained with the center of gravity forward of the midpoint between the rotors. There is a large unstable contribution to N_v from the fuselage in forward flight, and the rear rotor pylon is not very effective as a vertical tail. Hence a directional instability is likely, and the lateral dynamics retain the unstable long period oscillation in forward flight.

There are in forward flight a number of unfavorable effects on the handling qualities of tandem helicopters that arise from the aerodynamic interference between the two rotors. There is often an instability with respect to speed. Each rotor has speed stability as usual, but the change of the rear rotor thrust with speed due to the wake of the front rotor produces an unstable moment. A speed increase reduces the induced downwash of the front rotor and hence reduces the downwash of the front rotor wake at the rear rotor ($v_{R/F} \cong 2v_F$). The resulting increase of the rear rotor thrust produces a nose-down pitch moment, which is a speed instability. Since this speed instability due to the rotor thrust perturbation is large, the tandem helicopter can easily have a net speed instability. The rear rotor is closer to stall because of the downwash of the front rotor, and therefore the speed instability is reduced at high loadings. The speed stability can be improved by using the longitudinal dihedral of the shafts or swashplate, so the tip-path planes are tilted toward each other. The thrust perturbations due to the axial components of the helicopter longitudinal velocity perturbation produce a nose-up moment, increasing the speed stability. The effectiveness of such dihedral is reduced somewhat by the higher collective required to trim the rear rotor in forward flight when its shaft is tilted forward. Also, the amount of allowable dihedral is limited by interference between the rotors and fuselage.

The helicopter instability with respect to angle of attack is also increased by the aerodynamic interference between the rotors. An angle-of-attack increase (hence greater downward vertical velocity of the helicopter) increases the thrust of the rotors and therefore also their induced velocities. In addition, the larger downwash of the front rotor wake at the rear rotor produces a decrease of the rear rotor thrust and hence a net nose-up pitch moment on the helicopter. Since the rear rotor is closer to stall than the front rotor, the rear rotor thrust increases are further limited, giving a larger angle-of-attack instability at high loading. The use of pitch-flap coupling on the front rotor improves the angle-of-attack stability by reducing the lift curve slope of the front rotor relative to the rear rotor. The angle-of-attack instability is also reduced with a forward center-of-gravity position, or a reduction in thrust coefficient.

Amer (1951) concluded that the primary problem regarding the longitudinal flying qualities of the tandem helicopter was the angle-of-attack instability due to the rotors. He suggested the use of pitch-flap coupling on the front rotor to increase the stability. The tandem helicopter he tested also had an instability with respect to speed. Tapscott and Amer (1956) conducted a theoretical and flight test investigation of the tandem helicopter speed instability in forward flight, produced by the variation with speed of the front rotor downwash at the rear rotor. Calculations using $2v_F \cong C_{T_F}/\mu$ for the interference-induced velocity at the rear rotor gave an approximate prediction of the speed instability. They found that longitudinal dihedral of the swashplates such that the tip-path planes are tilted toward each other improves the speed stability. The helicopter was slightly stable with respect to speed with 4.5° of dihedral. Blake, Clifford, Kaczynski, and Sheridan (1958) concluded that for tandem helicopters the problems regarding longitudinal flying qualities are principally due to the aerodynamic interference between the rotors, the angle-of-attack instability being the most severe problem. Bramwell (1960) investigated tandem helicopter dynamics and found that the wake interference effects decreased at high speed because the induced velocity was reduced. He also found that forward movement of the center of gravity was effective in stabilizing the motion in forward flight, as was differential pitch-flap coupling (positive on the front rotor and negative on the rear rotor).

Amer and Tapscott (1954) concluded that to improve the lateral flying qualities of the tandem helicopter in forward flight, the lateral speed stability (dihedral effect) should be reduced, to increase the stability of the lateral oscillation. The lateral speed stability can be reduced by using a wing on the helicopter, which also improves the roll control; or by using elastic twist of the blade, since a blade torsion moment about the aerodynamic center produces a 1/rev pitch variation proportional to the rotor lateral velocity perturbation (see section 15–6). They also concluded that the roll damping should be increased, and that a roll moment to the right due to yawing to the right ($L_r > 0$) is desirable. They found that directional stability or yaw damping increases are much less effective in improving the flying qualities of a tandem helicopter. The fuselage aerodynamic moment has an important role in the helicopter directional stability, however. Blake, Clifford, Kaczynski, and Sheridan (1958) found that the tandem helicopter has nearly neutral directional stability, and strong roll-sideslip coupling because of the dihedral effect (lateral speed stability). Bramwell (1960) concluded that the fuselage contributions to the lateral stability derivatives are very important, particularly the dihedral effect (L_v) of the rear rotor pylon. The increased speed stability due to the airframe aerodynamics, with no corresponding increase of the roll damping, resulted in an unstable long period mode in forward flight that was stable if the rotor contribution alone was considered. With a wing or large horizontal tail to increase the roll damping, the characteristics of the oscillatory mode are considerably improved, as is the roll response to lateral control.

15-4.5 Hingeless Rotor Helicopters

The capability of the hingeless rotor to transmit large hub moments to the helicopter has a major impact on its handling qualities. The articulated rotor in contrast can achieve only a limited hub moment with offset hinges, roughly comparable to the moment about the center of gravity due to the rotor thrust tilt. The hingeless rotor gives the helicopter a high control power compared to the articulated rotor, and the damping in pitch and roll are increased by an even larger factor. The high damping also means an increased gust sensitivity, however, so a high-speed hingeless rotor helicopter often requires some sort of automatic control system for gust alleviation. The lateral-longitudinal coupling of the control response is also increased substantially, but can be handled satisfactorily by proper phasing of the swashplate. The increased lateral-longitudinal coupling of the transient motion and response to external disturbances remains, however. The angle-of-attack instability of the hingeless rotor in forward flight is much larger than that of an articulated rotor and requires a larger horizontal tail volume or an automatic control system to prevent a degradation of the handling qualities. The hingeless rotor is able to maintain its control power and damping at low load factor, in contrast to the articulated rotor, which produces moments on the helicopter primarily by tilting the thrust vector.

It is often necessary to include in the analysis the blade lag and torsion degrees of freedom as well as the flap motion in order to accurately predict the flight dynamics of a hingeless rotor helicopter. The inertial and structural couplings involved in the hingeless rotor blade dynamics can have a major impact on the handling qualities. The use of the low frequency rotor response is generally an acceptable approximation even with a hingeless rotor, though.

15–5 Low Frequency Rotor Response

Our analysis of helicopter flight dynamics has been based on the low frequency quasistatic rotor response. This approximation reduces the order of the dynamics problem to just the six rigid body degrees of freedom, with the rotor influence taking the form of stability derivatives. A low-order system is desirable for analytical work, and often for numerical investigations as well. Generally, the low frequency response is quite a good model for the rotor in a flight dynamics analysis. This approximation is consistent with the very low frequencies involved in the aircraft rigid body motions, as the numerical examples for the roots given in the preceding sections illustrate. The use of the low frequency rotor response is justified by the fast decay of the rotor flap motion transients (see section 12-1.3). The small time constant is due to the large flap damping of the rotor blade. In section 12–1, the low frequency rotor response was derived directly from the differential equations of motion for the flapping relative to the the shaft and the hub reactions. In the nonrotating frame, to lowest order in the Laplace variables all the time derivatives of the flap degrees of freedom are omitted, so the equations reduce to simply the quasistatic flap response to control, shaft motion, and gusts. There are cases when the quasistatic representation of the rotor is not satisfactory, even for the helicopter flight dynamics. In particular, with high-gain feedback systems it may be necessary to retain the rotor dynamics, both to correctly calculate the flight dynamics roots and to account for feedback-induced instabilities of the rotor motion. It is best to always verify the validity of the approximation for a particular application, by comparing it with the results obtained using the complete rotor dynamics.

Often it is necessary to consider more rotor degrees of freedom than just the first mode flapping, but still the low frequency response may be acceptable for the flight dynamics. The low frequency response can be obtained by deriving the complete differential equations of motion in the nonrotating frame for the rotor degrees of freedom involved. The quasistatic approximation then drops the acceleration and velocity terms from these equations (if the rotor motion is defined relative to the shaft). Alternatively, the steady-state (periodic) rotor response can be calculated using the required degrees of freedom. Such an analysis is an extension of those discussed in section 5–25, with the control and shaft motion inputs included one at a time in order to obtain the steady-state hub reactions, which then give the rotor stability derivatives.

Hohenemser (1939) originated the use of the quasistatic rotor response in helicopter flying qualities investigations, on the basis of the very low frequency of the rigid body motions compared to the rotor rotational speed.

Miller (1948) compared the roots obtained for the helicopter longitudinal dynamics, using the complete rotor dynamics and the low frequency response. He considered the longitudinal dynamics of a hovering helicopter, consisting of four degrees of freedom: longitudinal velocity $\dot{x}_B$, pitch angle θ_B, longitudinal flapping β_{1c}, and lateral flapping β_{1s}. The quasistatic rotor approximation reduced the model to just the two body degrees of freedom, $\dot{x}_B$ and θ_B. On the basis of a comparison of the roots of the helicopter longitudinal dynamics with and without the flapping degrees of freedom for both articulated and hingeless rotors, and also a comparison of the frequency response (up to 0.14/rev), Miller concluded that the quasistatic approximation gives a good representation of the rotor for flight dynamics analyses.

Kaufman and Peress (1956) investigated the longitudinal dynamics in forward flight, considering a six degree of freedom model including the flap dynamics ($\dot{x}_B$, θ_B, $\dot{z}_B$, β_0, β_{1c}, and β_{1s}) as well as the three-degree-of-freedom model obtained using the low frequency rotor response ($\dot{x}_B$, θ_B, and $\dot{z}_B$). Below about 0.1/rev they found that the frequency response of θ_B/θ_{1s} obtained from the two models was essentially identical, and the two results were close even up to 1/rev (particularly the magnitude of the response).

Hohenemser and Yin (1974a, 1974b) examined the question of whether a quasistatic representation of the rotor (particularly a hingeless rotor) is satisfactory for flight dynamics analyses. They considered a six-degree-of-freedom model consisting of the body pitch, roll, and vertical motions, and the rotor flap degrees of freedom (θ_B, ϕ_B, $\dot{z}_B$, β_0, β_{1c}, and β_{1s}). The roll motion was included because it is coupled with the longitudinal dynamics, especially in the case of a hingeless rotor. The longitudinal and lateral velocities were neglected as important only for the long period mode. For a hingeless rotor at high speed ($\nu = 1.2$, $\gamma = 5$, and $\mu = 0.8$) they calculated the roots of the system, with and without a feedback control system; and the helicopter body response to step inputs of rotor cyclic. They examined the following cases: the complete system; the quasistatic rotor approximation; and a first-order approximation in which the flapping acceleration terms were dropped but the velocity terms (in the nonrotating frame) were retained. They also considered the complete system including the periodic coefficients due to the rotor aerodynamics in forward flight. It was concluded that for the rotor stability it is necessary to retain the periodic coefficients, but the constant coefficient approximation gives good results for the body roots and response even at this high advance ratio. From their results the quasistatic rotor model appears to be an adequate representation of the dynamics, giving nearly the same roots and response as the complete model.

15–6 Stability Augmentation

Helicopter handling qualities can be improved by using an automatic control system. For certain operations, such as IFR flight, a stability and control augmentation system is essential. The use of such systems naturally increases the cost and complexity of the helicopter. A gyroscope is frequently the basic element of a helicopter automatic control system. Since the rotor itself can be considered a gyro, a control gyro can be used to sense the same inertial forces acting on the rotor. Such a control system can be entirely mechanical, or the gyro can just be the sensor with the control inputs provided by electro-hydraulic servos.

Consider a gyro gimballed to the rotor shaft and rotating at speed Ω_G, which for mechanical systems will usually be the same as the rotor rotational speed Ω. The undisturbed gyro plane is parallel to the rotor hub plane. The gyro tilt relative to the shaft is described in the nonrotating frame by the pitch angle β_{GC} and the roll angle β_{GS} (positive for tilt forward and to the left respectively). The gyro responds to the rotor shaft pitch and roll motions (α_y and α_x). It is assumed that the gyro is axisymmetric, consisting of three or more identical, equally spaced radial elements. Let I_G be the gyro pitch and roll inertia. Damping C_R in the rotating frame and damping C_F in the nonrotating frame are included, but act only on the motion of the gyro relative to the shaft (i.e. the damping is mechanical, not aerodynamic). The equations of motion for the gyro degrees of freedom follow from the conditions for equilibrium of pitch and roll moments. Rather than deriving the equations, we can simply note the similarity between the gyro and the flapping rotor (see section 12-1.2), from which it follows that the equations of motion are

$$\begin{bmatrix} I_G s^2 + (C_R + C_F)s & 2\Omega_G I_G s + C_R \Omega_G \\ -(2\Omega_G I_G s + C_R \Omega_G) & I_G s^2 + (C_R + C_F)s \end{bmatrix} \begin{pmatrix} \beta_{GC} \\ \beta_{GS} \end{pmatrix}$$

$$= \begin{bmatrix} I_G s & 2\Omega_G I_G \\ -2\Omega_G I_G & I_G s \end{bmatrix} \begin{pmatrix} \dot{\alpha}_y \\ -\dot{\alpha}_x \end{pmatrix} + \begin{pmatrix} -M_{G_y} \\ M_{G_x} \end{pmatrix}$$

where M_{G_y} and M_{G_x} are pitch and roll moments acting on the gyro. There is no structural spring, so the gyro natural frequency is $\nu_G = \Omega_G$.

Typically the gyro is connected to the rotor cyclic pitch in such a way that the control plane tilt is proportional to the gyro tilt:

$$\begin{pmatrix} -\theta_{1s} \\ \theta_{1c} \end{pmatrix} = K_G \begin{bmatrix} \cos \Delta\psi_G & \sin \Delta\psi_G \\ -\sin \Delta\psi_G & \cos \Delta\psi_G \end{bmatrix} \begin{pmatrix} \beta_{GC} \\ \beta_{GS} \end{pmatrix}$$

where K_G is the gain and $\Delta\psi_G$ is the azimuthal phasing. There may be a direct mechanical link from the gyro to the blade pitch horns that produces this control input. (In that case the blade feathering moments are also transmitted to the gyro, which at least increases its effective inertia and damping.) Alternatively, an electro-hydraulic system can provide the swashplate inputs proportional to the gyro tilt, with the appropriate gain and compensation network.

To order s, the gyro response to helicopter angular velocity is

$$\begin{pmatrix} \beta_{GC} \\ \beta_{GS} \end{pmatrix} \cong \frac{2I_G/C_R}{4I_G s + C_R} \begin{bmatrix} 2I_G s + C_R & \frac{1}{\Omega_G}\left(\frac{C_R}{2} + C_F\right)s \\ -\frac{1}{\Omega_G}\left(\frac{C_R}{2} + C_F\right)s & 2I_G s + C_R \end{bmatrix} \begin{pmatrix} \dot{\alpha}_y \\ -\dot{\alpha}_x \end{pmatrix}$$

Neglecting the helicopter angular acceleration terms as well then gives

$$\begin{pmatrix}\beta_{GC}\\ \beta_{GS}\end{pmatrix} \cong \frac{2I_G/C_R}{4\dfrac{I_G}{C_R}s+1}\begin{pmatrix}\dot{\alpha}_y\\ -\dot{\alpha}_x\end{pmatrix}$$

Hence with control plane tilt proportional to the gyro tilt, this control system provides lagged rate feedback of the helicopter pitch and roll, which greatly improves the helicopter handling qualities. Note that this response results from the damping in the rotating frame, C_R. A mechanical system in the rotating frame must provide the same feedback for both the pitch and roll axes. If there is no damping in the rotating frame ($C_R = 0$), the low frequency response of the gyro to shaft pitch and roll is

$$\begin{pmatrix}\beta_{GC}\\ \beta_{GS}\end{pmatrix} \cong \frac{I_G}{2C_F I_G s + 4I_G^2\Omega_G^2 + C_F^2}\begin{bmatrix}C_F s + 4\Omega_G^2 I_G & 2\Omega_G C_F\\ -2\Omega_G C_F & C_F s + 4\Omega_G^2 I_G\end{bmatrix}\begin{pmatrix}\alpha_y\\ -\alpha_x\end{pmatrix}$$

If there is no damping in either the rotating or nonrotating frames, the response is exactly

$$\begin{pmatrix}\beta_{GC}\\ \beta_{GS}\end{pmatrix} = \begin{pmatrix}\alpha_y\\ -\alpha_x\end{pmatrix}$$

In this case the gyro remains fixed in space, and thus it provides pure attitude sensing. The rate feedback obtained with damping in the rotating frame is most beneficial, however (see section 15-3.4.3).

The low frequency response of the gyro to applied moments is

$$\begin{pmatrix}\beta_{GC}\\ \beta_{GS}\end{pmatrix} \cong \frac{-1}{C_R\Omega_G(4I_G s + C_R)}\begin{bmatrix}2I_G s + C_R & \dfrac{1}{\Omega_G}(C_F + C_R)s\\ -\dfrac{1}{\Omega_G}(C_F + C_R)s & 2I_G s + C_R\end{bmatrix}\begin{pmatrix}M_{G_x}\\ M_{G_y}\end{pmatrix}$$

$$\cong -\frac{1}{C_R\Omega_G}\begin{pmatrix} M_{G_x} \\ M_{G_y} \end{pmatrix}$$

So the gyro tilt senses the applied moments if there is damping in the rotating frame. With no damping, the response is

$$\begin{pmatrix} \beta_{GC} \\ \beta_{GS} \end{pmatrix}^{\bullet} = -\frac{1}{I_G(s^2+4\Omega_G^2)}\begin{bmatrix} 2\Omega_G & s \\ -s & 2\Omega_G \end{bmatrix}\begin{pmatrix} M_{G_x} \\ M_{G_y} \end{pmatrix}$$

$$\cong -\frac{1}{I_G 2\Omega_G}\begin{pmatrix} M_{G_x} \\ M_{G_y} \end{pmatrix}$$

so the gyro senses the integral of the moments and responds with an angular velocity proportional to the applied moment after a 90° azimuthal lag.

Consider now a hub moment feedback control system utilizing a gyro. Such a system can be used with a hingeless rotor to alleviate the large hub moments produced by gusts. The hub moment of a hingeless rotor is

$$\begin{pmatrix} -\gamma\dfrac{2C_{M_y}}{\sigma a} \\ \gamma\dfrac{2C_{M_x}}{\sigma a} \end{pmatrix} = \left(\nu^2 - 1\right)\begin{pmatrix} \beta_{1c} \\ \beta_{1s} \end{pmatrix}$$

where ν is the rotor blade flap frequency. Hub moment feedback then is equivalent to tip-path-plane tilt feedback, which can be used with articulated rotors as well. The hub moment or tip-path-plane tilt is sensed and transmitted to the gyro by some means, producing moments on the gyro according to

$$\begin{pmatrix} M_{G_x} \\ M_{G_y} \end{pmatrix} = K_\beta\begin{bmatrix} \cos\Delta\psi_\beta & -\sin\Delta\psi_\beta \\ \sin\Delta\psi_\beta & \cos\Delta\psi_\beta \end{bmatrix}\begin{pmatrix} \beta_{1c} \\ \beta_{1s} \end{pmatrix}$$

where K_β is the feedback gain and $\Delta\psi_\beta$ is the azimuthal phase. Then a control plane tilt is produced proportional to the gyro tilt as a response to this moment. Thus a control can be applied to the rotor to cancel the hub moment due to external disturbances. This hub moment feedback will act as well to reduce the control power and damping of the rotor, which also produce hub moments. The gyro, however, can sense the helicopter angular velocity and hence replace the pitch and roll damping, and the pilot can control the helicopter with such a system by applying moments directly to the gyro. The performance of such a tip-path-plane tilt feedback system can be analyzed using the low frequency response of the gyro and rotor. The analysis for the hingeless rotor is complicated by the azimuthal phase shifts. The fundamental behavior of such a control system can be established by considering an articulated rotor, however. The low frequency response of a hovering articulated rotor ($\nu = 1$) is

$$\begin{pmatrix} \beta_{1c} \\ \beta_{1s} \end{pmatrix} = \begin{pmatrix} -\theta_{1s} \\ \theta_{1c} \end{pmatrix} + 8M_\mu \begin{pmatrix} \dot{x}_h - u_G \\ \dot{y}_h + v_G \end{pmatrix} + \frac{16}{\gamma} \begin{pmatrix} \dot{\alpha}_y \\ -\dot{\alpha}_x \end{pmatrix} + \begin{pmatrix} \dot{\alpha}_x \\ \dot{\alpha}_y \end{pmatrix}$$

(see section 12-1.3). The low frequency gyro response with damping in the rotating frame has been found to be

$$\begin{pmatrix} \beta_{GC} \\ \beta_{GS} \end{pmatrix} = -\frac{1}{C_R \Omega_G} \begin{pmatrix} M_{G_x} \\ M_{G_y} \end{pmatrix} + \frac{2I_G}{C_R} \begin{pmatrix} \dot{\alpha}_y \\ -\dot{\alpha}_x \end{pmatrix}$$

For an articulated rotor, the control plane tilt should be proportional to the gyro tilt with no phase shift:

$$\begin{pmatrix} -\theta_{1s} \\ \theta_{1c} \end{pmatrix} = \begin{pmatrix} -\theta_{1s} \\ \theta_{1c} \end{pmatrix}_{\text{con}} + K_G \begin{pmatrix} \beta_{GC} \\ \beta_{GS} \end{pmatrix}$$

Direct cyclic input from the pilot has been included, although it will be found that such control is washed out by the feedback. Finally, the moments applied to the gyro are proportional to the rotor tip-path-plane tilt, again with no phase shift required:

$$\begin{pmatrix} M_{G_x} \\ M_{G_y} \end{pmatrix} = \begin{pmatrix} M_{G_x} \\ M_{G_y} \end{pmatrix}_{\text{con}} + K_\beta \begin{pmatrix} \beta_{1c} \\ \beta_{1s} \end{pmatrix}$$

The applied moments from the pilot, by which the rotor can be controlled, have also been included. The complete control law for this system (using the gyro low frequency response) is thus

$$\begin{pmatrix} -\theta_{1s} \\ \theta_{1c} \end{pmatrix} = \begin{pmatrix} -\theta_{1s} \\ \theta_{1c} \end{pmatrix}_{con} + K_G \left\{ \frac{2I_G}{C_R} \begin{pmatrix} \dot{\alpha}_y \\ -\dot{\alpha}_x \end{pmatrix} - \frac{1}{C_R \Omega_G} \left[K_\beta \begin{pmatrix} \beta_{1c} \\ \beta_{1s} \end{pmatrix} + \begin{pmatrix} M_{Gx} \\ M_{Gy} \end{pmatrix}_{con} \right] \right\}$$

The total gain on the tip-path-plane tilt is $K_G K_\beta / C_R \Omega_G$. The moments on the gyro should be kept low, so a high gain should be obtained not from K_β but from K_G/C_R. Hence a low but finite damping in the rotating frame and a high ratio of the swashplate tilt to gyro tilt are desired. On substituting for this control law, the rotor flap response becomes

$$\begin{pmatrix} \beta_{1c} \\ \beta_{1s} \end{pmatrix} = \frac{1}{1+K} \left\{ \begin{pmatrix} -\theta_{1s} \\ \theta_{1c} \end{pmatrix}_{con} + 8M_\mu \begin{pmatrix} \dot{x}_h - u_G \\ \dot{y}_h + v_G \end{pmatrix} + \frac{16}{\gamma} \begin{pmatrix} \dot{\alpha}_y \\ -\dot{\alpha}_x \end{pmatrix} + \begin{pmatrix} \dot{\alpha}_x \\ \dot{\alpha}_y \end{pmatrix} \right\}$$

$$- \frac{K}{1+K} \frac{1}{K_\beta} \begin{pmatrix} M_{Gx} \\ M_{Gy} \end{pmatrix}_{con} + \frac{K}{1+K} \frac{2I_G \Omega_G}{K_\beta} \begin{pmatrix} \dot{\alpha}_y \\ -\dot{\alpha}_x \end{pmatrix}$$

where $K = K_G K_\beta / C_R \Omega_G$. For large gain the response reduces to

$$\begin{pmatrix} \beta_{1c} \\ \beta_{1s} \end{pmatrix} = -\frac{1}{K_\beta} \begin{pmatrix} M_{Gx} \\ M_{Gy} \end{pmatrix}_{con} + \frac{2I_G \Omega_G}{K_\beta} \begin{pmatrix} \dot{\alpha}_y \\ -\dot{\alpha}_x \end{pmatrix}$$

A hub moment feedback control system therefore reduces the direct rotor response to control, shaft motion, and gusts. The gust alleviation and, generally, the reduction of the speed stability are beneficial. In forward flight the rotor angle-of-attack instability is also reduced, which substantially improves the longitudinal flying qualities. The response to direct cyclic pitch inputs is also reduced, but the pilot can control the rotor by applying moments to the gyro instead. The hub moment feedback reduces the rotor angular damping, but it also reduces the response to the shaft angular velocity that couples the lateral and longitudinal motions. With damping in the rotating system, the gyro supplies feedback of the pitch and roll rate to replace the rotor damping of the helicopter. The performance of the hub moment feedback system is similar with a hingeless rotor. The rotor response to external disturbances is reduced, as are the direct rotor forces due to the helicopter motion (including the speed stability and angle-of-attack instability), but angular damping is provided to replace the reduced damping from the rotor. The principal additional consideration with hingeless rotors is the selection of the azimuthal phase angles in the feedback loop, such that the helicopter longitudinal and lateral dynamics and control response are not coupled. With high gain, it may not be sufficient to consider only the rotor and gyro low frequency response (i.e. just the static performance of the system) in order to appraise the control system behavior. Moreover, such a feedback control system is generally unstable at high gain, and this is an important factor in its design.

The Bell stabilizer bar, developed for two-bladed teetering rotors, is a two-arm gyro mounted on the rotor hub at right angles to the rotor blades. Both the rotor and the gyro dynamics are actually described by periodic coefficient differential equations, but the low frequency response is identical to that described above (see section 12-1.5). The gyro arms are linked to the pitch horns, where their input is mechanically mixed with the pilot's control input from the swashplate. There is mechanical damping in the rotating frame, between the gyro and the rotor shaft. Hence this stabilizer bar provides lagged rate feedback of the helicopter pitch and roll, as shown above. Such a system is simple mechanically. The same feedback is provided for both pitch and roll, which is not really desirable since the inertia is lower in roll than in pitch. Miller (1950) showed that such a stabilizer bar is equivalent to lagged rate feedback for low frequency. Sissingh (1967) discusses this system and compares it with others, including hub moment feedback systems.

The Hiller control rotor, also developed for two-bladed rotors, is a two-arm gyro with a small airfoil on each arm. The airfoils provide aerodynamic damping in the rotating frame, so the gyro gives pitch and roll rate feedback. The gyro tilt produces cyclic pitch of the main rotor, and the pilot controls the rotor by means of cyclic pitch of the airfoils on the gyro arms, thus producing a moment on the gyro. With this configuration the cyclic stick control forces are less sensitive to the main rotor conditions, and the mechanical gyro damper is eliminated. Stuart (1948) described this control rotor and its influence on the handling qualities of the helicopter. Miller (1950) showed that such a system provides lagged rate feedback of pitch and roll, and also feedback of longitudinal velocity and main rotor flapping due to the aerodynamic forces on the gyro.

The Lockheed gyro stabilizer is a hub moment feedback system developed for three- and four-bladed hingeless rotors. The gyro arms are linked to the blade pitch horns, so that gyro tilt provides the cyclic pitch. The pilot controls the helicopter by applying moments to the gyro. One version of this gyro stabilizer used feathering moment feedback. The blades were swept forward of the pitch axis, so that a flap moment had a component about the pitch axis. The feathering moment was transmitted through the pitch links to the gyro, providing hub moment feedback. The gyro damping was due to the flap damping of the main rotor. A later design used direct hub moment feedback. Arms on the blades detected the flap motion, which was then transmitted to springs that applied a moment to the gyro. The gyro had mechanical damping. The gyro response was sensed and then fed to the swashplate by hydraulic actuators, so the blade feathering moments did not influence the gyro at all. Sissingh (1967) analyzed the quasistatic performance of the Lockheed hub moment feedback system. Johnson and Hohenemser (1970) analyzed thrust and tip-path-plane-tilt integral feedback systems, particularly their influence on the rotor flap and lag dynamics, including the high gain stability. In a review of this paper, Sissingh analyzed the high-gain stability of hub moment feedback systems.

The blade elastic pitch motion produced by inertial and aerodynamic feathering moments when the center of gravity is offset from the aerodynamic center can be used to provide stability augmentation for the helicopter flight dynamics. A forward shift of the blade center of gravity will increase the helicopter pitch damping. A pitch angular velocity $\dot{\theta}_B$ of the helicopter and rotor combined with the blade velocity Ωr gives a Coriolis force on the blade, downward on the advancing side and upward on the retreating side. This Coriolis force acts at the blade center of gravity to produce a pitch moment. The response of a torsionally flexible blade with center of gravity forward of the pitch axis will thus be a 1/rev variation equivalent to longitudinal cylic, $\theta_{1s} < 0$ for $\dot{\theta}_B > 0$, which implies increased pitch damping. An aft shift of the aerodynamic center of the blade also increases the helicopter pitch damping. With a pitch rate $\dot{\theta}_B$, the rotor flaps forward to provide a lateral moment on the disk (toward the retreating side), which precesses the rotor to follow the shaft. The lift forces producing this moment act at the aerodynamic center of the blade, producing a feathering moment also. When the aerodynamic center is aft of the pitch axis, longitudinal cyclic $\theta_{1s} < 0$ is produced, which increases the pitch damping. Miller (1948) analyzed the rigid flap and rigid pitch dynamics of an articulated rotor, obtaining the low frequency response of the blade pitch to the helicopter motions when the center of gravity and aerodynamic center are offset from the pitch axis (by x_I and x_A respectively). He found that the elastic torsion response of the blade provides pitch and roll rate feedback proportional nearly to $x_A - x_I$. For increased damping the center of gravity should be forward of the aerodynamic center ($x_A > x_I$, which is favorable for flutter and divergence stability also). If $x_A \neq 0$, the blade pitch also responds to the longitudinal and lateral velocity of the helicopter ($\dot{x}_B$ and $\dot{y}_B$), and hence the speed stability will be influenced. When $x_A = 0$, the longitudinal feedback reduces to

$$\theta_{1s} = \dot{\theta}_B \frac{3x_I/R}{K_\theta/I_b\Omega^2}$$

where K_θ is the control system stiffness. Damping in the nonrotating cyclic control system introduces a lag in the feedback; damping in the rotating system gives a lag and also couples the longitudinal and lateral feedback. Hence chordwise offset of the blade center of gravity from the aerodynamic center will provide lagged rate feedback of the helicopter pitch and roll motion. A high gain requires torsionally flexible blades, however, or a large center-of-gravity shift forward, which means a large leading edge balance weight. The influence of the blade torsional moments on the control loads and stick forces must also be considered with such a design. The use of the blade torsion dynamics to provide stability augmentation for the flight dynamics is discussed further by Miller (1950), McIntyre (1962), and Reichert and Huber (1971).

15–7 Flying Qualities Specifications

The helicopter user or purchaser, or the appropriate regulatory agency, must determine what characteristics are required for acceptable flying qualities of the aircraft, and establish quantitative measurements of the desired characteristics. The specification and evaluation of flying qualities are concerned with many properties of the helicopter, among them control displacement and force gradients, and static stability; dynamic stability, particularly the long period roots; transient response characteristics, especially the short period behavior; control power, damping, and control sensitivity; and control coupling. The specifications vary greatly with the use intended for the vehicle, such as VFR or IFR flight. Moreover, stricter specifications are continually developed as more is learned about measuring helicopter handling qualities and designing helicopters to achieve the desired characteristics.

The military specification MIL-H-8501A defines the flying and ground handling qualities required for military helicopters. This specification is somewhat dated, but is still the most complete general statement of helicopter flying qualities requirements available. For static stability, MIL-H-8501A specifies the minimum and maximum initial force gradient of the longitudinal and lateral sticks, and requires that the gradient always be positive. The longitudinal stick should have a stable force and position gradient with respect to speed; at low speed (transition) a moderate degree of instability is permitted for VFR operations, but is not desirable. A stable gradient of the pedal and lateral cyclic stick with sideslip angle is required, and positive directional stability and effective dihedral (lateral speed stability) are required in forward flight. For IFR operations, the directional and lateral controls must have stable force and displacement gradients. The transient control forces, control force coupling, control margin, and other factors are also addressed. The dynamic stability characteristics in forward flight are specified by MIL-H-8501A in terms of the period and damping of the long period modes; Fig. 15-15 summarizes the requirements for VFR and IFR operations. So that there is no excessive delay in the development of the helicopter angular velocity in response to control, the specification requires that the roll, pitch, and yaw acceleration be in the proper direction within 0.2 sec after the control displacement. To insure acceptable maneuver stability characteristics (normal acceleration in forward flight, pitch rate in hover and at low speeds) the concave downward requirement is used: the time history of the normal acceleration and angular velocity of the helicopter should be concave downward within 2 sec after a step displacement of the longitudinal stick. Preferably, the normal acceleration should be concave downward throughout the maneuver (until

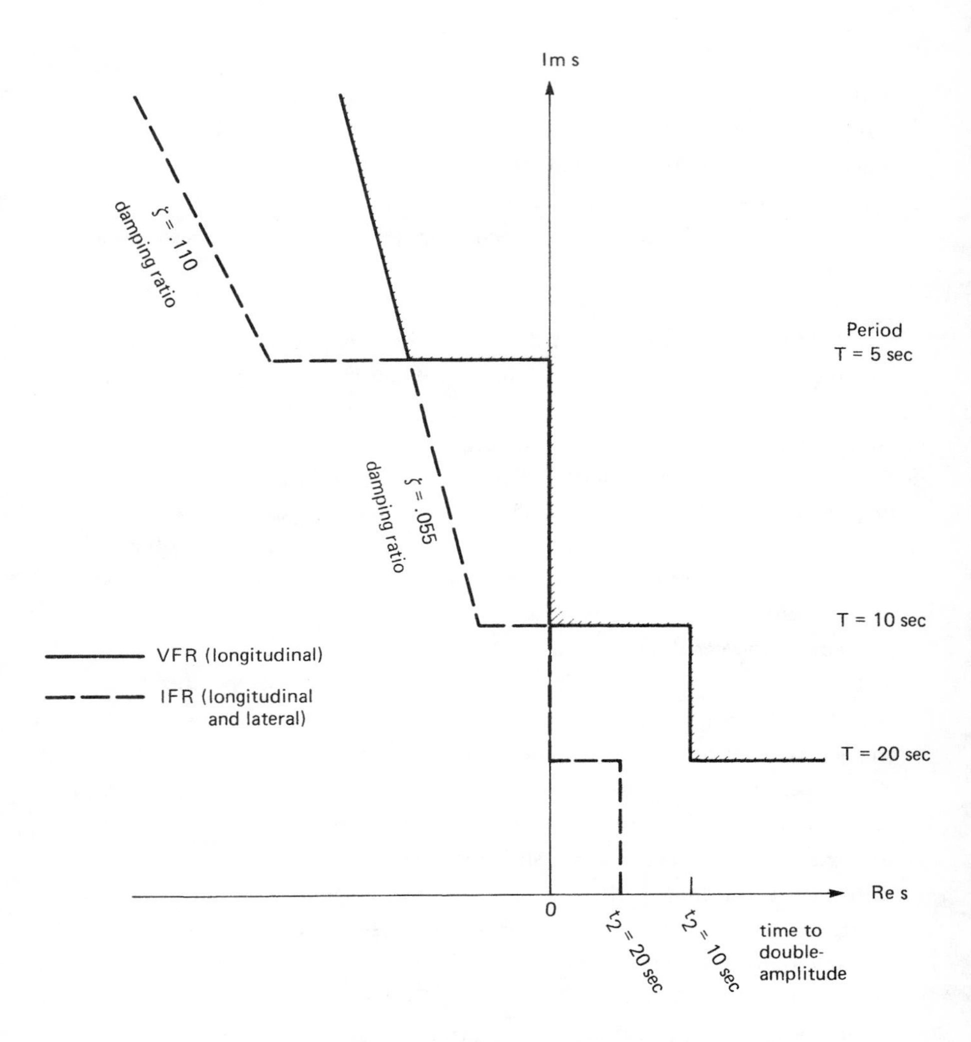

Figure 15-15 MIL-H-8501A specification of helicopter dynamic stability (long period modes) in forward flight.

the maximum acceleration), and the angular velocity should be concave downward after 0.2 sec. To insure that the pilot has a reasonable time for corrective action following perturbations from the trim attitude, the following requirement is used: within 10 sec after a longitudinal cyclic pulse lasting at least 0.5 sec (to simulate a disturbance) the normal acceleration should not increase by more than 0.25 g, and during the subsequent nose-down motion the normal acceleration should not decrease more than 0.25 g below the trim value. The minimum helicopter control power in hover is specified by requiring that the attitude change be at least α_{min} one second after a 1 inch step control displacement from trim (0.5 sec for roll), where the angle α_{min} depends on the axis (pitch, roll, or yaw) and the helicopter gross weight. A similar requirement is given for the response to maximum control displacement. The directional control sensitivity for hover, and the roll control sensitivity at all speeds, should not be so high that the pilot tends to over-control the helicopter. In any case, it is required that the roll sensitivity be less than 20 deg/sec/in, and that the yaw sensitivity be low enough that the yaw angle change after one second is less than 50 deg/in. For satisfactory initial control response characteristics, a minimum level of pitch, roll, and yaw damping that depends on the moment of inertia about the corresponding axis is required. The minimum control power and damping required in pitch and roll is increased for IFR operations. The specifications of hover control power, damping, and control sensitivity can be expressed in terms of requirements on the ratio of damping to inertia and of control power to inertia [with units of (ft-lb-sec)/(slug-ft^2) = sec^{-1} and (ft-lb/in)/(slug-ft^2) = rad/sec^2/in, respectively]; Fig. 15-16 summarizes the requirements for the pitch, roll, and yaw stability derivatives. (Note that the ratio of damping to inertia is the inverse of the time constant, the ratio of control power to inertia is the initial angular acceleration, and the ratio of control power to damping is the control sensitivity.) MIL-H-8501A also provides some specifications for the ride qualitity (maximum vibration) at the crew and passenger stations, and for the control stick vibrations. Other major specifications of helicopter flying qualities are the Federal Aviation Regulations defining the airworthiness standards for civilian certification of rotorcraft; and MIL-H-83300, which is an Air Force specification for V/STOL aircraft, including helicopters.

Gustafson, Amer, Haig, and Reeder (1949) introduced the concave downward requirement, based on flight test investigations of helicopter longitudinal handling qualities. They concluded that the most important consideration is prevention of a prolonged stick-fixed divergence of the

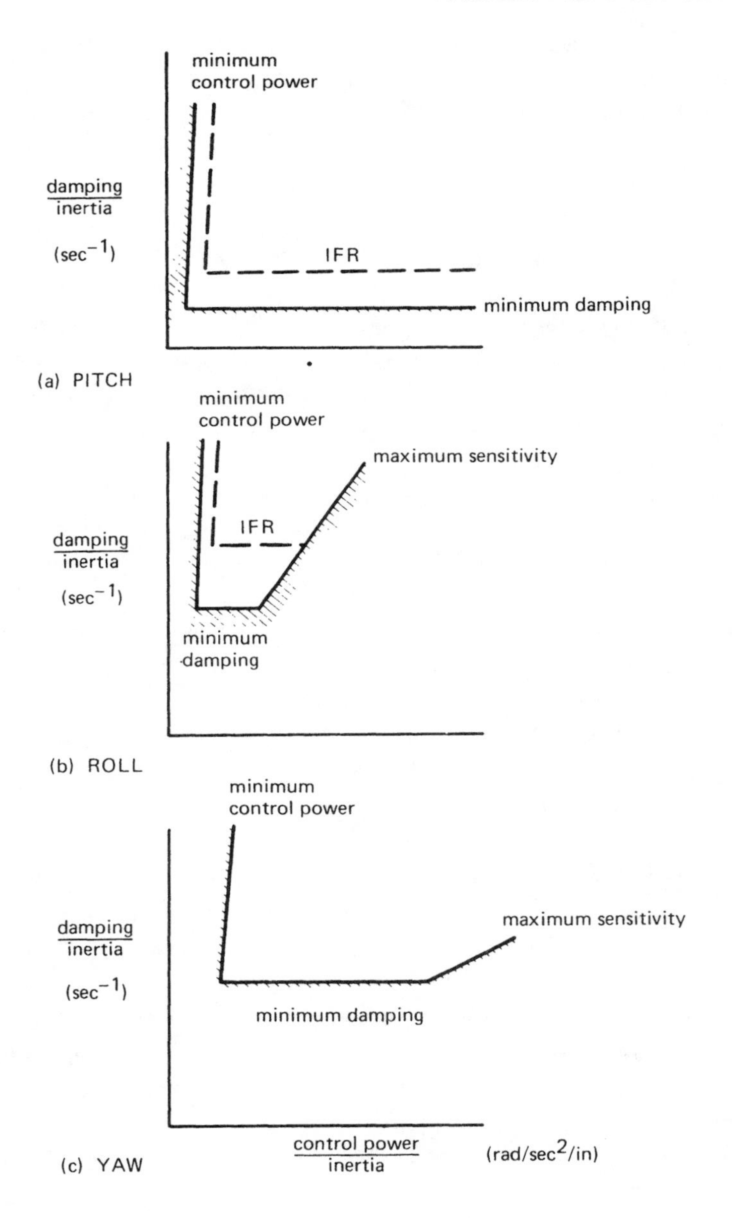

Figure 15-16 MIL-H-8501A specification of hover control power, damping, and control sensitivity.

normal acceleration in response to cyclic control. In addition, a continuous development of the normal acceleration is desirable, rather than a pause in the development during the first second of the maneuver. Therefore, they developed a requirement to preclude the divergent tendencies of the normal acceleration: the time history of normal acceleration should be concave downward within 2 sec after a sudden displacement of the longitudinal control stick. To reduce the difficulty of anticipating the results of a control deflection they developed the requirement that the time history of the normal acceleration should preferably be concave downward from the beginning of the maneuver to the maximum acceleration after a step displacement of the longitudinal stick, or at least that the slope of the normal acceleration should be positive throughout the maneuver. A supplementary requirement tending to insure that an oscillation rather than a sudden divergence occurs in response to disturbances was also developed: after a pulse displacement of the longitudinal stick lasting at least 0.5 sec the normal acceleration should not increase by more than 0.25 g within 10 sec; and during the first 10 sec of the subsequent nose-down motion after the load factor first returns to 1 g, the normal acceleration should not decrease by more than 0.25 g (i.e. the vertical load factor should remain between 1.25 g and 0.75 g). Amer (1951) investigated the flying qualities of tandem helicopters. He found that the maneuver requirement developed by Gustafson et al. is also applicable to tandem helicopters, although a stricter criterion might be needed because of the possible instability with respect to speed. Reeder and Whitten (1952) investigated the influence of the helicopter pitch damping on the longitudinal maneuver characteristics by using a stabilizer bar to provide increased damping due to lagged rate feedback. Correlating the pilot opinion of the handling qualities with the concave downward requirement, they concluded that the requirement as developed by Gustafson et al. is applicable for variations of the pitch damping as well. (The original development was concerned primarily with the influence of the helicopter angle-of-attack instability.) Marginal handling qualities correlated with an inflection point of the normal acceleration occurring 1.85 to 2.1 sec after the control application. Amer and Tapscott (1954) developed flying qualities criteria for the lateral dynamics of both single main rotor and tandem helicopters in forward flight. They found that pedal-fixed directional stability is required. For a reasonably damped stick-fixed oscillation, if the period is less than 10 sec, the time to half amplitude should be less than 2 cycles (giving a damping ratio greater than 5.5% critical) and there should be no residual oscillation. Finally, there should be no reversal of the rolling rate within 6 sec after a small step displacement of the lateral stick (pedals fixed).

Crim, Reeder, and Whitten (1953) conducted instrument flight trials of a helicopter. They concluded that the flying qualities requirements based on VFR operations were adequate for instrument flight at forward speeds above minimum power, although a close and constant attention to the instruments was necessary. At low speeds and during precision maneuvers, lateral directional problems were encountered, making instrument flight possible only for very short periods. Since even small unbalanced control forces were found to be objectionable, a means of trimming the stick forces would be required for instrument flight.

Salmirs and Tapscott (1959) conducted a flight test investigation of the influence of damping and control power on the helicopter handling qualities. A definite improvement of the handling qualities (based on pilot opinion) was found with increased damping. They presented charts summarizing the results in terms of the ratio of damping to inertia and of control power to inertia (analogous to Fig. 15-16), and defining the requirements for good IFR, acceptable IFR, acceptable VFR, and unacceptable characteristics. They also give the results in terms of the damping required as a function of the helicopter attitude change 1 sec after a 1 inch control displacement; and in terms of the roll damping required as a function of the roll control sensitivity.

Seckel, Traybar, and Miller (1961, 1962) conducted a flight test investigation of criteria for the longitudinal handling qualities of a hovering helicopter. They found an important influence of the speed stability on the handling qualities for precision hovering or low-speed flight in turbulence. A high value of speed stability is unfavorable principally because of the increased pitch response to gusts and turbulence it produces. A secondary detrimental effect is the increased longitudinal cyclic stick displacement required to trim the helicopter. They concluded that the range of satisfactory control power is rather large (not the sharp optimum found by others), and that the desirable damping and control power levels are somewhat higher than the requirements suggested by others.

Garren, Kelly, and Reeder (1964) conducted a flight test investigation of helicopter directional handling qualities in hover and low speed. They established a minimum requirement for the static directional stability, and an optimum yaw damping level.

Clark and Miller (1965), and Miller and Clark (1965), presented the results of analytical and flight simulator investigations of the helicopter pitch damping and control power requirements. They also found a significant influence of the speed stability and the level of atmospheric turbulence on the handling qualities. An increase in speed stability requires increased control power, mainly to handle the pitch response to gusts.

Kelly and Garren (1968) investigated the influence of pitch damping, control power, speed stability, and angle-of-attack stability on helicopter longitudinal handling qualities. The optimum values of these parameters were found to be largely independent of changes in the other parameters, and of the operating conditions considered. A minimum level of pitch damping was established, although the improvement in handling qualities continued as the damping was increased further. They concluded that the helicopter should have neutral or slightly positive stability with respect to angle of attack. A minimum control power and an optimum speed stability were also established.

15–8 Literature

On helicopter stability and control: Crocco (1923), Kussner (1937), Schrenk (1938), Hohenemser (1939, 1946a, 1946b, 1950, 1974), Donovan and Goland (1944), Gustafson and Reeder (1948), Miller (1948, 1950), Stewart (1948, 1950), Stuart (1948), Zbrozek (1948, 1949a), Gustafson, Amer, Haig, and Reeder (1949), Reeder and Gustafson (1949), Reeder and Haig (1949), Amer (1950, 1951, 1954), Cannon and Niehaus (1950), Gessow and Amer (1950a, 1950b), Lange and McLemore (1950), Smith (1950), Amer and Gustafson (1951), Gustafson (1951), Gessow and Myers (1952), Reeder and Whitten (1952), Crim, Reeder, and Whitten (1953), Ellis (1953), Sanders (1953), Amer and Tapscott (1954), Crim (1954, 1959), Williams (1954, 1958), Amer and Gessow (1955), Payne (1955c, 1957), Whitten, Reeder, and Crim (1955), Kaufman and Peress (1956), Leone (1956a), Tapscott and Amer (1956), Bramwell (1957, 1960, 1969), Wernicke (1957),

Blake, Clifford, Kaczynski, and Sheridan (1958), Gustafson and Tapscott (1958), Tapscott (1958, 1960, 1964), Heffron, Bristow, Gass, and Brown (1959), Salmirs and Tapscott (1959), Anderson (1960), Connor and Tapscott (1960), Liberatore (1960), Mallick and Reeder (1960), Perlmutter, Kisielowski, Miller, and George (1960), Sweet (1960a), Amer et al. (1961), Curtiss (1961, 1965, 1970), Garren (1961a, 1961b, 1962a), Klinar and Craig (1961), Reeder (1961), Seckel, Traybar, and Miller (1961, 1962), Segner (1961), Garren and Assadourian (1962), Gerdes and Weick (1962), Jenkins, Winston, and Sweet (1962), Lean et al. (1962), Lynn (1962), McIntyre (1962), Sweet and Jenkins (1962), Brown and Schmidt (1963), Price (1963), Seckel and Curtiss (1963), Biggers and McCloud (1964), Curry and Matthews (1964), Garren, Kelly, and Reeder (1964), Seckel (1964), Stutz and Price (1964), Toler et al. (1964), Ward and Huston (1964), Clark and Miller (1965), Drees and McGuigan (1965), Feistel and Drinkwater (1965), Garren and Kelly (1965), Jenkins (1965), Johnston, Culver, and Friend (1965), Kelly and Winston (1965), Miller and Clark (1965), Huston (1966), Reeder, Tapscott, and Garren (1966), Reichert (1966, 1968, 1973b), Smith (1966), Tapscott and Sommer (1966), Walton and Stapleford (1966), Ward (1966a), Azuma (1967), Emery, Sonneborn, and Elan (1967), Kisielowski, Perlmutter, and Tang (1967), Lytwyn (1967), Streiff (1967), Walton and Ashkenas (1967), Blackburn (1968), Griffin and Bellaire (1968), Kelley, Pegg, and Champine (1968), Kelly and Garren (1968), Lollar and Kriechbaum (1968), Miller and Vinje (1968), Reichert and Oelker (1968), Rempfer and Stevenson (1968), Sardanowsky and Harper (1968), Albion, Leet, and Mollenkof (1969), Harper and Sardanowsky (1969), Livingston and Murphy (1969), Blake, Albion, and Redford (1970), Harper, Sardanowski, and Scharpf (1970), Huston and Morris (1970, 1971), Jenkins and Deal (1970), La Forge and Rohtert (1970), Matheny and Wilkerson (1970), Rohtert and La Forge (1970, 1971), Shupe (1970), Sinacori (1970a, 1970b), Balke (1971), Curtiss and Shupe (1971), Dmitriyev and Yesanlov (1971), Gabel, Henderson, and Reed (1971), Garay and Kisielowski (1971), Johnson (1971), Kaplita (1971), Reichert and Huber (1971), Reschak (1971), Szustak and Jenney (1971), Tuck (1971), Welch and Warren (1971b), Wolkovitch and Hoffman (1971), Green (1972), Kefford (1972), Livingston (1972), Robbins (1972), Rudolph (1972), Gould and Hinson (1973, 1974), Molusis (1973a, 1973b, 1974), Monteleone (1973), Takasawa (1973), Wells and

Wood (1973), Albion and Larson (1974), Callan, Houck, and Di Carlo (1974), Flemming and Ruddell (1974), Foster, Kidwell, and Wells (1974), Gorenberg and Harvick (1974), Hansen, McKeown, and Gerdes (1974), Kelley and West (1974), Kessler, Murakoshi, and Sinacori (1974), Vinje (1974), Wood, Ford, Brigman (1974), Yeager, Young, and Mantay (1974), Hohenemser, Banerjee, and Yin (1975, 1976), Horst and Reschak (1975), Merkley (1975), Sorensen, Mohr, and Cline (1975), Houck (1976), Houck and Bowles (1976), Moen, Di Carlo, and Yenni (1976), Oehrli (1976), Ostroff, Downing, and Rood (1976), Pausder and Jordan (1976), Simon and Savage (1976), Simons (1976), Tomaine (1976), Blake and Alansky (1977), Freeman and Yeager (1977), Houck et al. (1977), Johnson and Chopra (1977), Mineck (1977a), 1977b), Mineck and Freeman (1977a, 1977b), Niessen et al. (1977), Parrish, Houck, and Martin (1977), Rix, Huber, and Kaletka (1977), Talbot and Corliss (1977), Amer, Prouty, Walton, and Engle (1978), Chen and Talbot (1978), Cooper (1978), Hall, Gupta, and Hansen (1978), Reaser (1978).

On the dynamics of a helicopter with a suspended load: Kaufman and Schultz (1962), Lucassen and Sterk (1965), Lytwyn (1967), Bricainski and Karas (1971), Szustak and Jenney (1971), Dukes (1972, 1973), Asseo and Whitbeck (1973), Gupta and Bryson (1973, 1976), Liu (1973), Micale and Poli (1973), Smith, Allen and Vensel (1973), Di Carlo, Kelley, and Yenni (1974), Gera and Farmer (1974), Kisielowski, Smith, and Spittle (1974), Watkins, Sinacori, and Kesler (1974), Davis, Landis, and Leet (1975), Hone (1975), Campbell et al. (1976), Garnett and Smith (1976), Garnett, Smith, and Lane (1976), Alansky, Davis, and Garnett (1977a, 1977b), Feaster, Poli, and Kirchoff (1977), Sheldon (1977), Prabhakar (1978).

On helicopter gust response: Payne (1954a), Maglieri and Reisert (1955), McCarty, Brooks, and Maglieri (1959b), Drees and McGuigan (1965), Webber (1965), Harvey, Blankenship, and Drees (1969), Drees and Harvey (1970), Bergquist (1973), Arcidiacono, Bergquist, and Alexander (1974), Ham, Bauer, Lawrence, and Yasue (1975), Whitaker and Cheng (1975), Briczinski (1976), Cheng (1976), Judd and Newman (1976), Amos and Alexander (1977), Johnson (1977b), Mantay, Holbrook, Campbell, and Tomaine (1977), Vehlow (1977), Yasue (1977), Jenkins and Yeager (1978), Yasue, Vehlow, and Ham (1978).

On helicopter stability and control augmentation, and automatic control systems: Meyers, Vanderlip, and Halpert (1951), Salmirs and Tapscott (1957), Wernicke (1957), Blake, Clifford, Kaczynski, and Sheridan (1958), Peress and Kaufman (1958), Vague and Seibel (1959), Gerstine and Blake (1961), Kaufman and Schultz (1963), Naumann and Schmier (1964), Trueblood, et al. (1964, 1967), Kaufman (1965), McBrayer and Robinson (1965), Garren, DiCarlo, and Driscoll (1966), Hill (1966), Buffum and Robertson (1967), George, Kisielowski, and Perlmutter (1967), Sissingh (1967), Curties (1968), Rempfer, Stevenson, and Collins (1968), Livingston and Murphy (1969), Murphy and Narendra (1969), Cocakyne, Rusnak, and Shub (1970), Dukes (1970), Griffith and Crosby (1970), Johnson and Hohenemser (1970), Shupe (1970), Trueblood, Bryant, and Cattel (1970), Bryson, Chasteen, Hall, and Mohr (1971), Dmitriyev and Yesanlov (1971), Heimbold and Griffith (1971, 1972), Hoffman, Zvara, Bryson, and Ham (1971), Hooper (1971), Ogren (1971, 1974), Konig and Schmitt (1971), Schmitz (1971), Szustak and Jenney (1971), Welch and Warren (1971a), Born, Carico, and Durbin (1972), Born, Durbin, and Schmitz (1972), Borodin and Ryl'skiy (1972), Crossley and Porter (1972), Graham (1972), Hall and Bryson (1972, 1973), O'Connor (1972), O'Connor and Fowler (1972), Narendra and Tripathi (1972, 1973), Deardorff, Freisner, and Albion (1973), Donham and Cardinale (1973), Kilmer and Sklaroff (1973), McElreath, Klein, and Thomas (1973), Potthast and Blaha (1973), Puri and Niemala (1973), Cotton (1974), Diamond and Davis (1974), Hoffman, Hollister, and Howell (1974), Hohenemser and Yin (1974a), Kelly, et al. (1974), Ogren, Satanski, and Genan (1974), Potthast and Kerr (1974, 1975), Bengston, Dickovich, and Helfinstine (1975), Bengtson, Hedeen, and Helfinstine (1975), Blake and Alansky (1975), Born and Kai (1975), Briczinski and Cooper (1975), Bryant, Cattel, Russell, and Trueblood (1975), Bryant and Trueblood (1975), Meier, Groth, Clark, and Verzella (1975), Briczinski (1976), McManus and Gonsalves (1976), Banaszak and Posingies (1977), Corliss and Talbot (1977), Hess (1977), Hollister and Hoffman (1977), Niven, Sanders, and McManus (1977), Stengel, Broussard, and Berry (1977), Chen, Talbot, Gerdes, and Dugan (1978), Hofmann et al. (1978), McManus (1978), Miyajima (1978).

Chapter 16

STALL

Stall of a helicopter rotor blade is characteristically manifested as high control system loads and helicopter vibration. The alternating control loads show a gradual increase with speed until stall occurs, and then a sharp and large rise in magnitude (Fig. 16-1). The rapid growth of the blade torsion and control system loads associated with stall is a major constraint on the helicopter speed, lift, and maneuver capability. Because

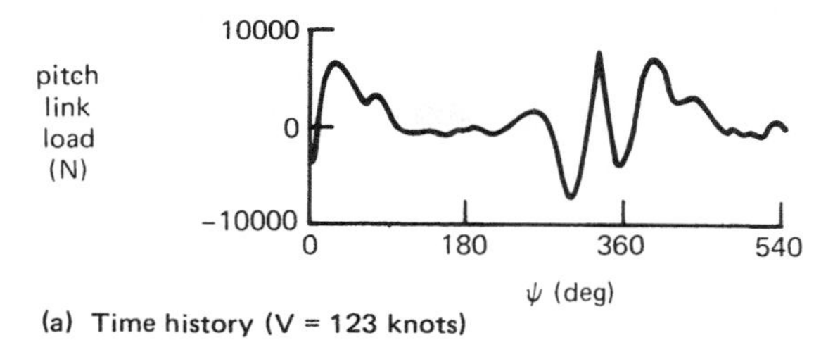

(a) Time history (V = 123 knots)

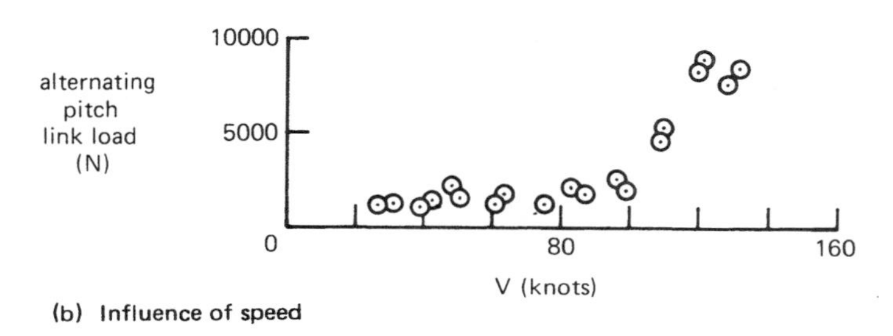

(b) Influence of speed

Figure 16-1 Influence of blade stall on rotor pitch link loads; from Tarzanin (1972).

of the flow axisymmetry, stall of a hovering rotor occurs as a limit cycle torsional oscillation of the blade, called stall flutter. In the periodically varying aerodynamic environment of the rotor in forward flight, stall occurs on the retreating blade, producing torsional loads such as those shown in Fig. 16-1. This stall phenomenon of forward flight is often called stall flutter as well. Sometimes stall has a role in the dynamic stability of the rotor, but usually the problem is the extremely high loads on the blade and control system. Because of the high rate of increase of the stall loads with speed or load factor, it is not very effective to increase the structural flight envelope by strengthening the control system. Stall also limits the aerodynamic performance of the helicopter. When the rotor blade stall in forward flight persists into the highly loaded fourth quadrant of the disk, there is an increase in power required, a loss of lift and propulsive force capability, and a loss of control power. Rotor stall produces a significant vibration of the helicopter, which serves as a signal to the pilot of the onset of stall. With a mechanical control system, the stall-induced torsional loads also produce a significant vibration in the cyclic and collective control sticks. Such vibration can be severe enough to be a limiting factor in itself. As a principal limitation of rotor performance and source of high loads and vibration in extreme operating conditions, stall is a major factor in the aerodynamic and structural design of the helicopter rotor and control system. Stall must be included in the analysis of the helicopter performance and aeroelastic behavior (see Chapters 6 and 14). It is difficult, however, to predict stall and all its effects, particularly in the complex three-dimensional and unsteady aerodynamic environment of the rotor blade. A complete understanding and theoretical description of rotary wing stall is still the subject of research.

16–1 Rotary Wing Stall Characteristics

In order to maintain low drag and high lift, the flow over an airfoil section must remain smooth and attached to the surface. This flow has a rapid acceleration around the nose of the airfoil to the point of maximum suction pressure, and then a slow deceleration along the remainder of the upper surface to the trailing edge. The deceleration must be very gradual for the flow to remain attached to the surface. At a high enough angle of attack, stall will occur; the deceleration will be too large for the boundary layer to support, and the flow will separate from the airfoil surface. Typically, the stall angle of attack is around 12° (corresponding to a lift coefficient of roughly $c_\ell = 1.2$), but it is highly dependent on Reynolds number, Mach number, and the airfoil shape. The unstalled airfoil has a low drag, and a lift coefficient linear with angle of attack. The airfoil in stall at high angles of attack has high drag, a loss of lift, and an increased nose-down pitch moment caused by a rearward shift of the center of pressure. The aerodynamic flow field of an airfoil or wing in stall is extremely complex, and for the rotary wing there are important three-dimensional and unsteady phenomena as well. The effects of stall on the aerodynamics of the entire rotor follow from the effects on the blade section. The limit on the blade lift capability together with the drag increase results in a limit on the rotor thrust C_T, a sharp increase in the rotor profile power C_{P_o} at high loading, a reduced propulsive force capability, and high flapwise and chordwise bending stresses. The section center-of-pressure shift produces the high blade torsional and control system loads. The helicopter vibration increase is due to the rapid change of the aerodynamic forces when the blade stalls. Stall over a limited portion of the disk can be tolerated, but as the stall area increases these effects eventually become objectionable.

Significant stall of the rotary wing is encountered when the helicopter is operating in a condition requiring the blades to work at high angles of attack. The two primary flight conditions requiring high blade angles of attack are high thrust and high speed. Recall that the blade mean lift coefficient in hover is $\bar{c}_\ell = 6C_T/\sigma$ (section 2-6.4). This expression is of little value as a stall criterion, but it does show that the ratio of the rotor thrust coefficient to solidity is a measure of the blade angle of attack, and hence is a primary parameter of the rotor stall characteristics. Note that $C_T/\sigma = T/\rho A_{\text{blade}}(\Omega R)^2$, so for a given thrust the blade working lift coefficient can be reduced by increasing the blade area or tip speed. Forward flight has a great influence on the angle-of-attack distribution over the disk, and therefore on the rotor stall behavior. As speed increases, the dynamic pressure increases on the advancing side and decreases on the retreating side of the disk. Balance of the aerodynamic roll moment on the rotor then requires that the blade work at low angles of attack on the advancing side and at high angles on the retreating side. This lateral asymmetry increases with the rotor speed, so that at a given thrust stall eventually occurs on the retreating side. Thus the limiting thrust of the rotor, C_T/σ, will be a decreasing function of the advance ratio. With a hingeless rotor the roll moment could be partially reacted by a structural hub moment, thus reducing the required lateral asymmetry of angle of attack, and delaying stall. The effect is significant only with very stiff blades, however, and with a single main rotor such a hub moment means a large trim roll angle of the fuselage. Note that the load on the sides of the rotor disk is limited by the maximum lift coefficient at stall and the low dynamic pressure of the retreating blade, so that as the helicopter speed increases the working area of the disk becomes concentrated at the front and back (around $\psi = 0°$ and $\psi = 180°$).

In forward flight, stall is also encountered because of the reverse flow region. Near the reverse flow boundary, the small normal velocity u_T produces a large inflow angle $\phi = \tan^{-1} u_P/u_T$, and hence a large angle of attack. Sufficiently near the reverse flow boundary the angle of attack will always be above the stall value, but the dynamic pressure is so low that the effects of this stall near the boundary are not great. Moreover, at low advance ratio the reverse flow region occupies only a small region of the disk near the blade root. At very high advance ratio, the aerodynamic forces on the blade inside the reverse flow region become important, particularly the effects of stall and the shift of the aerodynamic center to the three-quarter chord as a result of the flow reversal.

The rotor can also encounter stall in maneuvers of the helicopter or in aerodynamic turbulence. Turns, pull-ups, and similar maneuvers involve an increase in the rotor thrust and therefore will be limited by stall. A gust will change the blade angle of attack, perhaps producing stall. Both the maneuver and the gust encounter cases benefit from the transient nature of the angle-of-attack changes, which tends to delay the occurrence of stall. As a result, the transient maneuver capability of the rotor is usually greater than the static load capability.

The blade angle of attack or lift coefficient (the actual value or some representative value) is the primary criterion for stall of rotary wings. Basically, the rotor will exhibit the effects of stall when the blade section angle of attack is above the stall angle over a significant portion of the rotor disk. Translation of such a criterion into rotor operating limits is a complicated task, however. The angle of attack is not uniform over the disk and is difficult to calculate accurately, especially at extreme flight conditions. Moreover, the stall of rotor blades is a more complex aerodynamic phenomenon than the stall of two-dimensional airfoils. Therefore, the useful stall criteria that have been developed all have some empirical basis. Stall criteria have been developed using global parameters, such as C_T/σ and μ, but usually some local parameter is preferable since stall is generally restricted to only part of the rotor disk. A number of criteria have been developed for predicting the occurrence of stall on the rotor blade based on the angle of attack at some specific (and presumably critical) point on the disk. A still better procedure is to calculate the detailed aerodynamic loading of the rotor in a given flight state, including a model of the blade section forces in stall (see Chapter 14). Even such large scale aeroelastic analyses do not yet provide a consistently reliable prediction of rotor stall, however, because our knowledge of the blade aerodynamics is not complete.

In hover the flow is axisymmetric, so stall occurs in an annulus on the rotor disk (ignoring the effects of blade motion and unsteady aerodynamics). As the thrust is increased, the blade angle-of-attack change is greatest at the tip, because the induced velocity increase with thrust produces the smallest inflow angle change there. It follows that the tip region stalls first on a hovering rotor. In autorotation, where the net inflow is upward through the disk, the angle of attack is largest at the root. It is expected then that for the autorotating rotor the root sections stall first. When the blade stalls at one section first, it means that at the rotor performance limit part of the blade is still working below its maximum capability. The rotor efficiency is improved if the angle of attack is uniform over the blade span at high loading, which requires negative built-in twist to counter the higher tip angles of attack of the hovering rotor.

In forward flight, the largest angles of attack occur on the retreating blade, and increase with speed. As the rotor thrust increases, the angle-of-attack change due to the larger induced velocity is smallest at the tip, as in hover. Consequently, the blade stalls first on the tip of the retreating blade. Frequently, therefore, the rotor stall criterion is based on the angle of attack at $r/R = 1$ and $\psi = 270°$, denoted $\alpha_{1,270}$. Finally, consider an unpowered rotor in forward flight (an autogyro). As speed or thrust increases, the rotor losses $(D/L)_r = (D/L)_i + (D/L)_o$ increase, requiring a larger rearward tilt of the tip-path plane to supply the rotor power. The resulting increased upward inflow velocity produces the greatest angle-of-attack increase on the inboard sections of the blade. It may be concluded that the autogyro rotor stalls first on the inboard sections of the retreating blade. The sections near the reverse flow region and at the root are at low dynamic pressure; the stall is of concern at radial stations where the normal velocity u_T is not too small. Frequently, therefore, the rotor stall criterion is based on the angle of attack at $u_T = 0.4$ and $\psi = 270°$, denoted $\alpha_{\mu+.4,270}$.

Nonuniform inflow has a major influence on the angle-of-attack distribution over the rotor disk (see the examples of section 13–2) and hence on the rotor stall behavior. When nonuniform inflow is taken into account, somewhat different conclusions are reached about the detailed characteristics of the rotor blade environment at high loading. Generally, the induced velocity tends to be higher at the retreating tip than is indicated by the mean inflow value, which alleviates the high angles of attack at the tip. The stall region is thus moved inboard and into the third quadrant of the disk, especially for blades with low twist. Nonuniform inflow also tends to increase the maximum angle of attack on the disk and to increase the rate of change of α near stall. The conclusion that $\alpha_{1,270}$ is the maximum angle of attack of the rotor blade in forward flight is therefore not correct. Indeed, poor agreement is found between the stall areas measured experimentally and predicted using uniform inflow. The importance of nonuniform inflow to the actual rotor aerodynamics does not necessarily invalidate a stall criterion using some elementary parameter, such as $\alpha_{1,270}$ evaluated with uniform inflow theory. The use of such a criterion is based on correlation between the parameter and the measured stall effects. Good correlation can be obtained sometimes despite the limitations of the model from which the stall parameter was deduced.

The aeroelastic response of the blade to the high aerodynamic loads in stall is another important factor in the rotor stall behavior. The high blade loads, control loads, and vibration characteristic of stall are manifestations of the blade dynamic response. Moreover, the blade motion in turn influences the angle of attack and hence the aerodynamic forces. In particular, the large nose-down pitch moments due to stall produce substantial torsional motion of the blade, which directly changes the angle of attack. Since the control system is usually the softest element, the torsion response is predominantly rigid pitch motion due to control system flexibility. It follows that a calculation of the effects of stall on the helicopter rotor cannot be concerned with the aerodynamic forces alone, but instead it requires a complete aeroelastic analysis that includes the dynamic response of the blades. The local angle of attack must be calculated from the actual wake-induced nonuniform inflow and the elastic torsion motion of the blade. The use of uniform inflow or a torsionally rigid blade can lead to quite incorrect predictions of the stall behavior of rotors.

In helicopter aeroelastic analyses, the section loading is generally obtained from tabular data based on two-dimensional, static airfoil tests. The aerodynamic environment of the rotor blade is actually three-dimensional and unsteady, and for stall in particular the effects of this complex environment must be accounted for if accurate predictions of the aeroelastic behavior are to be obtained. One important aspect of the three-dimensional flow that must be included in the analysis is the radial velocity or yaw angle at the blade section (see section 5–12). Yawed flow has the effect of delaying the occurrence of stall, and also has considerable influence on the nature of the separated flow.

The rotor blade section in forward flight has a large 1/rev variation in angle of attack (and higher harmonics as well), which has a major impact on the rotor stall characteristics. The stall of an airfoil in unsteady flow, called dynamic stall, is very different from that in steady flow. A finite rate of increase in the angle of attack ($\dot{\alpha} > 0$) has the effect of delaying the occurrence of stall, so that the dynamic stall angle of attack is larger than the angle for static stall. It follows that the section is capable of higher lift in unsteady conditions than it can sustain under static conditions, since the range of linear aerodynamics is larger. After dynamic stall does occur, the transient lift and nose-down moment are much greater than the static stall loads. From the delay of stall it follows that when unsteady angle-of-attack changes are involved, as in maneuvers or the retreating blade stall of forward flight, the rotor is capable of a higher thrust with no stall effects than is implied by the static airfoil characteristics. This behavior has been verified in wind tunnel and flight rests of rotors. The large transient loads on the section when dynamic stall occurs are the source of the high vibration and loads associated with rotor stall, particularly the blade torsion and control system loads in response to the pitch moment of dynamic stall.

While stall in unsteady flow is not yet completely understood, the experimental and theoretical investigations of recent years present the following picture of dynamic stall. Consider an airfoil with a large periodic variation of angle of attack, from a value well below static stall to a value above dynamic stall. Such an angle-of-attack variation is typical of the 1/rev variation of the rotor blade in forward flight, with a large mean angle corresponding to a high rotor loading C_T/σ. As the angle of attack increases, there is a delay in the occurrence of stall due to the unsteady flow, so the linear lift and low moment are maintained to an angle of attack larger than the static stall angle. When the dynamic stall angle is reached (which itself depends on the pitch rate α), there is a loss of leading edge suction, accompanied by the shedding of a strong vortex from the vicinity of the leading edge of the airfoil. This vortex moves aft over the upper surface of the airfoil, at a velocity considerably lower than the free stream value. The vortex induces a pressure disturbance on the airfoil upper surface, an area of high suction moving aft. This pressure disturbance produces the high transient lift, moment, and drag forces on the airfoil that characterize dynamic stall. There is a large peak lift coefficient (as high as $c_\ell = 3.0$ for large $\dot{\alpha}$), followed by a large peak nose-down moment (as high as $c_m = -0.7$). After the passage of the leading edge vortex over the upper surface, the flow progresses to the fully separated state, and hence to the static stall loads. The flow at this point depends greatly on the transient blade motion, including the magnitude of the mean and oscillatory angles of attack. For example, there may be secondary vortices shed from the leading edge. The high initial loads of dynamic stall usually produce transient pitch motion of the blade, which also influences the stall phenomena at this point. If the angle of attack decreases now, as for the rotor blade in forward flight, the flow eventually reattaches to the airfoil surface. The unsteady flow delays this reattachment to an angle below the static stall angle of attack.

Dynamic stall experiments have been conducted on both two-dimensional airfoils and on rotors. A convenient experimental arrangement consists of a two-dimensional airfoil oscillating about a pitch axis in a wind tunnel. The mean angle, the oscillation amplitude, and the oscillation frequency should be chosen typical of the aerodynamic environment of a rotary wing. The mean and oscillatory angles should be large and about equal, and the oscillation frequency should correspond to the rotor speed (for a 1/rev angle-of-attack variation). The pressure, section loads, and other quantities are measured during the oscillation cycle. Fig. 16-2 gives an

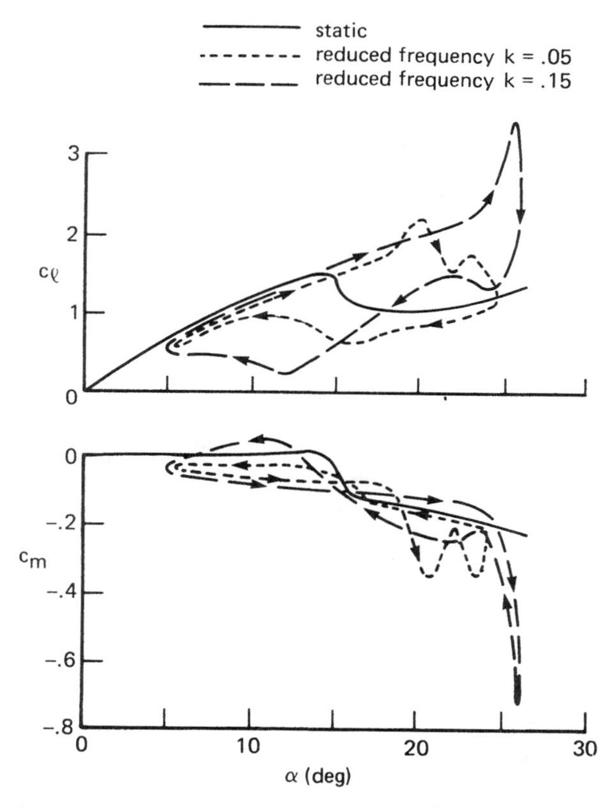

Figure 16-2 Unsteady lift and moment on an airfoil oscillating in pitch; from Carr, McAlister, and McCroskey (1977).

example of such oscillatory airfoil data (there is actually a large scatter in the loads measured for decreasing angle of attack). The delay of stall due to the airfoil pitch rate is seen, as are the higher loads than in the static case. Such a presentation also shows the hysteresis of the unsteady loads: the lift and moment depend not just on the current angle of attack but also on the past history of the motions.

Because the aerodynamic environment of the rotor in hover is axisymmetric, stall is expected to occur in an annulus on the rotor disk. The section angle of attack and loads are independent of ψ, so static stall data are applicable. When the blade motion is considered, a quite different stall phenomenon is possible. Consider a hovering rotor operating at high lift. A gust or other disturbance may trigger dynamic stall of the blade. The resulting large transient moment will then twist the blade nose-down. If the blade is sufficiently flexible in torsion, this nose-down motion will reduce the angle of attack enough for the flow to reattach. With the return of attached flow loads, the blade rebounds up in pitch, overshooting the static stall level because of the small damping of the torsional motions. The overshoot in pitch increases the angle of attack so that the blade stalls, and the cycle begins again. An oscillation of the blade in and out of stall is thus established. The energy to sustain the oscillation comes from the hysteresis of the moment coefficient as a function of angle of attack during dynamic stall (see Fig. 16-2); the loops represent a net amount of work performed on the blade during a cycle. The oscillation is a limit cycle in which the balance of the negative damping in stall and the positive damping below stall determine the oscillation amplitude. This single-degree-of-freedom limit cycle instability is called stall flutter.

Rotor stall is a major consideration in the design of a helicopter. The limit on the thrust coefficient to solidity ratio C_T/σ is determined by the requirement for an adequate stall margin in forward flight (the forward flight limit is much lower than the hover limit). For a given gross weight, then, the quantity $A_{\text{blade}}(\Omega R)^2$ is determined. The combination of an advancing-tip Mach number limit (due to compressibility effects on performance and noise) and an advance ratio limit (due to stall and other factors) constrains the tip speed ΩR. Then the minimum blade area that must be provided to meet the stall margin requirement is defined. The fact that the blade loading limit decreases with speed suggests using a fixed wing on the helicopter to reduce the rotor lift required in forward flight. Unloading the rotor also reduces its propulsive force capability and control power, however. Stall is also a major concern in selecting the blade airfoil section (see section 7–7). An airfoil with high maximum lift coefficient at low to moderate Mach number is desired for the retreating blade stall environment. Reducing the helicopter drag (hence the rotor propulsive requirement) is also effective in improving the rotor stall characteristics, raising the limits on C_T/σ and μ.

16–2 NACA Stall Research

The NACA conducted a series of investigations directed at developing a means of predicting helicopter rotor stall; these have been summarized by Gessow and Myers (1952). Following the arguments in section 16–1, it was concluded that in forward flight the largest angle of attack occurs at the retreating blade tip; $\alpha_{1,270}$ was therefore chosen as the basis of the stall criterion. For the autorotating rotor, the largest angle of attack occurs inboard, so $\alpha_{\mu+.4,270}$ was used. After solving for the rotor performance and flapping motion in forward flight, using for example the analysis of Bailey (1941), the angle of attack at the retreating blade tip can be evaluated. The angle-of-attack distribution depends on the advance ratio μ, the rotor loading C_T/σ, and the rotor power C_P. Alternatively, the rotor parasite power or the parasite drag-to-lift ratio $(D/L)_p = (D/L)_{total} - (D/L)_r = \alpha_{TPP}$ can be used as a parameter in place of the total power. Thus the angle of attack at the retreating blade tip can be calculated as a function of $(D/L)_p$ and C_T/σ for a given μ. Fig. 16-3 shows the results of such calculations for a rotor with zero twist. The contours of $\alpha_{1,270} = 12°$ and $16°$ are presented on the plane of $(D/L)_p$ vs. C_T/σ. Near autorotation the inboard angle $\alpha_{\mu+.4,270}$ becomes larger than $\alpha_{1,270}$; the boundary is based on the more critical of $\alpha_{1,270}$ and $\alpha_{\mu+.4,270}$. Fig. 16-3 shows that the angle of attack of the retreating blade tip depends primarily on C_T/σ and μ, with some sensitivity to the tip-path-plane incidence $(D/L)_p = \alpha_{TPP}$. [To include climb and descent flight states, the ordinate should be interpreted as $(D/L)_p + (D/L)_c = \alpha_{TPP}$.]

On the basis of rotor tuft behavior and the pilot's assessment of the vibration and control characteristics, Gustafson and Myers (1946) found that the angle of attack of the retreating blade tip correlated well with the helicopter stall behavior. This correlation with flight test results is the basis for using $\alpha_{1,270}$ calculated by a simple theory (uniform inflow, no stall in the airfoil lift and drag) as a criterion for the rotor stall. It was found that $\alpha_{1,270} = \alpha_{ss}$ corresponds to incipient stall, and that $\alpha_{1,270} = \alpha_{ss} + 4°$ corresponds to excessive stall (where α_{ss} is the airfoil static stall

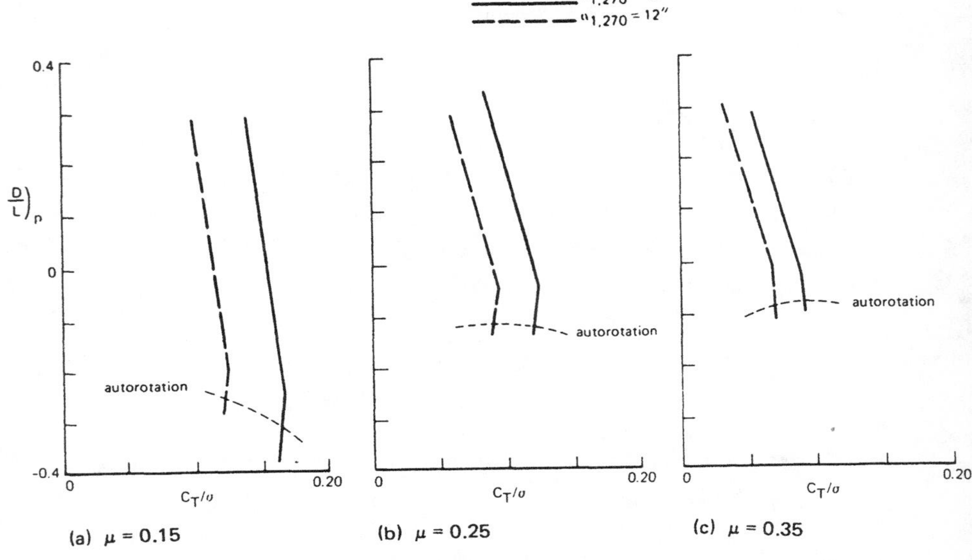

Figure 16-3 Retreating blade stall boundaries for stall inception ($\alpha_{1,270} = 12^\circ$) and excessive stall ($\alpha_{1,270} = 16^\circ$); from Gustafson and Myers (1946).

angle of attack). For the NACA 0012 airfoil of the helicopter involved, $\alpha_{ss} = 12^\circ$ and hence the boundaries of interest are $\alpha_{1,270} = 12^\circ$ and 16°. Thus the following criterion for helicopter rotor stall was developed. Below the boundary of incipient stall, $\alpha_{1,270} = 12^\circ$, there are no noticeable stall effects; above this boundary the vibration, loads, and power increase because of stall. At the boundary of excessive or objectionable stall, $\alpha_{1,270} = 16^\circ$, the helicopter is still controllable but the loads and vibration have reached the limit for practical operation. Near autorotation these limits are applicable to $\alpha_{\mu+.4,270}$. After this correlation with measured stall behavior is established, the calculated retreating blade tip angle of attack can be used to predict rotor stall. Gustafson and Gessow (1947) found by flight tests that the increase in rotor profile power also correlated well with $\alpha_{1,270}$. The ratio of the measured rotor profile power to the calculated value (obtained using a theory without stall effects) was around unity until $\alpha_{1,270}$ reached the static stall angle (12°). Above α_{ss} there was a sharp rise in the measured profile losses, to about twice the predicted (no-stall) value at $\alpha_{1,270} = \alpha_{ss} + 4^\circ = 16^\circ$. At still higher angles of attack there are control difficulties as well. The observed effects of stall on the rotor were first a large increase in the rotor profile power loss, and then vibration and loads large enough to limit the helicopter operation.

The stall boundary is presented more conveniently in terms of the maximum loading C_T/σ as a function of speed μ, for a given power C_P or $(D/L)_{\text{total}}$. Fig. 16-4 presents in this fashion the stall boundaries obtained

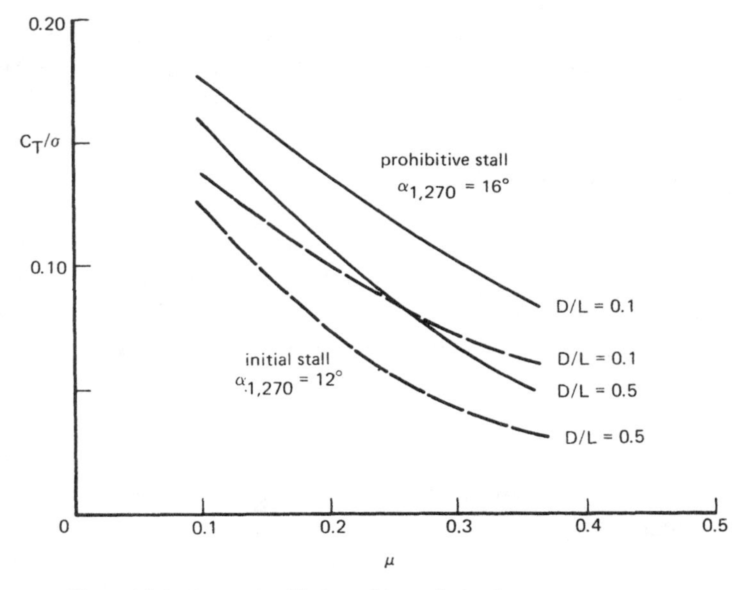

Figure 16-4 Retreating blade stall boundaries for an untwisted rotor; from Gessow and Myers (1952).

for an untwisted rotor by calculating the retreating blade tip angle of attack. The restriction on the rotor loading at high speed is quite severe according to these results.

Bailey and Gustafson (1939) determined the stall regions of an autogyro rotor in forward flight by means of photographs of tuft behavior. Both the size of the stall region and its rate of growth with speed were larger than predicted using the analysis of Wheatley (1937a), although there was general agreement with regard to the shape and location of the stalled area.

Bailey (1941) developed a rotor performance analysis using a section profile drag polar of the form $c_d = \delta_0 + \delta_1\alpha + \delta_2\alpha^2$. He presented a method for evaluating the constants δ_0, δ_1, and δ_2 for a given airfoil, and the limiting angle of attack α_{limit} for which this polar is valid (see section 7–8). The actual drag characteristics diverge from this polar at high angle of attack, primarily as a result of stall. Bailey therefore suggested as the limit of validity of the theory the criterion $\alpha_{\mu+.4,270} = \alpha_{limit}$, which was basically a stall criterion. (Bailey was concerned with autogyro performance, hence the use of $\alpha_{\mu+.4,270}$.)

Gustafson and Myers (1946) conducted a flight test and theoretical investigation of rotor stall. They calculated the blade angle-of-attack distribution using the analysis of Bailey (1941) and found that the stall area appears first on the retreating blade tip and increases in size with μ. The stall area was measured in flight using photographs of tuft behavior, and they concluded that the actual stall area and growth could be at least roughly predicted. They found that the measured rotor stall characteristics correlated well with the calculated retreating blade tip angle of attack, and so developed the stall criterion discussed above. Gustafson and Gessow (1947) measured the rotor performance in flight at conditions of high loading and high speed involving stall. They found that the rotor profile power increase correlated well with $\alpha_{1,270}$, as discussed above.

Gessow (1948d) conducted a flight test investigation of the effect of blade twist on rotor forward flight performance, specifically the stall-limited maximum speed and other stall effects. He compared the stall behavior for rotors with twist of 0° and -8°. The blade twist was effective in increasing the maximum speed and reducing the power losses due to stall at a given thrust and speed. About a 10% increase in maximum speed was found with -8° of twist, a result of the performance improvement and a delay in the vibration rise.

Gessow and Tapscott (1956) present charts of calculated rotor performance (see section 6–6). For a stall criterion they use the retreating blade tip angle of attack $\alpha_{1,270}$ (or $\alpha_{\mu+.4,270}$ near autorotation, and present lines of $\alpha_{1,270} = 12^\circ$ and 16° on the performance charts. They also give separate charts of the stall limit as a function of the rotor operating parameters. The charts are based on calculations using Bailey's drag polar, so the performance predictions are not valid when any significant stall is indicated. Thus the stall boundaries in this case also are the limits of the validity of the theory.

Ludi (1958b) measured in flight tests the effects of retreating blade stall on the rotor blade bending and torsion moments. He found that in stall the magnitude of the higher harmonics of the moments increased, so that they were almost as important for the fatigue life as the lower harmonics. The higher harmonics of the loads were also responsible for the increased control loads and vibration that restrict the maximum speed of the helicopter. Pull-up maneuvers produced essentially the same results as high speed, steady flight conditions. The maximum normal acceleration obtainable tends to be limited by stall. Ludi found that the bending and torsion moments in stall were up to three times the unstalled loads. There was an abrupt rise in the torsion and flapwise bending moments when $\alpha_{1,270}$ exceeded the static stall angle, and a smaller rise in the chordwise bending moments.

McCloud and McCullough (1958) conducted wind tunnel tests of the stall characteristics of a rotor in forward flight at $\mu = 0.3$ to 0.4. They investigated the performance gains possible when blade airfoil sections with increased maximum lift due to camber were used to delay retreating blade stall. The stall boundary was defined by a marked change in the rotor torque and in the characteristics of the blade torsional moment. These two criteria gave essentially the same boundaries of maximum C_T/σ as a function of μ. The rotor blades with cambered sections had stall characteristics superior to those of the symmetrical sections. The lifting capability at a given speed was increased by about 15%, and the maximum speed capability at a given lift was increased by 20% to 25%.

McCloud and McCullough (1959) compared the calculated and measured influence of stall on a helicopter rotor using the analysis of Gessow (1956) with static airfoil data, and the test results of McCloud and McCullough (1958). The calculated and measured stall boundaries based on the rotor torque rise were roughly parallel, but the calculated boundary was at a lower lift (by as much as 10%; typically $\Delta C_T/\sigma = 0.01$ lower over $\mu = 0.3$ to 0.4). In operating states free of stall the agreement between the calculated and measured thrust was good, but the calculated propulsive force was larger and the torque smaller than measured. They suggest possible sources of the lower predicted lift boundary of the rotor, including the effects of radial flow, nonuniform inflow, unsteady flow, and the elastic blade motion.

Gessow and Tapscott (1960) present tables and charts of calculated rotor performance, including flight conditions well into stall. The theory of Gessow and Crim (1955) and Gessow (1956) was used, with static airfoil data. The performance was calculated over the speed range $\mu = 0.1$ to 0.5 for a rectangular, articulated blade with -8° of twist.

16–3 Dynamic Stall

Halfman, Johnson, and Haley (1951) measured the loads on a two-dimensional airfoil oscillating in pitch and translation. Rainey (1956) investigated the stall flutter of thin nonrotating wings and, by correlating the occurrence of stall flutter with negative damping, demonstrated the importance of the aerodynamic torsional damping moments due to stall. Rainey (1957) measured the loads on an airfoil oscillating in pitch about the midchord, including oscillations through stall where negative torsional damping is encountered.

Brooks and Baker (1958) conducted an experimental investigation of the flutter of a model helicopter rotor in hover, including stall flutter at high collective. Classical flutter at low collective and stall flutter exhibited distinctly different behavior. The flutter speed (Ω/ω_θ) in stall was about one-third the instability speed below stall. The flutter frequency was around ω_θ in stall (compared to about $0.7\omega_\theta$ for classical flutter), indicative of a single-degree-of-freedom instability. The chordwise center-of-gravity position had no appreciable effect on stall flutter, in contrast to the large effect on classical flutter.

Ham (1962) conducted an experimental investigation of the retreating blade stall of a model rotor in forward flight, with the emphasis on the torsional oscillations in response to the stall loads. A large transient torsion motion at a frequency around the pitch natural frequency ω_θ was found to occur after the blade entered the stall region on the retreating side. The amplitude of this torsional oscillation was increased by increasing the speed or by an aft shift of the blade center of gravity; and the amplitude was decreased by increasing the pitch damping or the pitch spring rate. The mechanism for this oscillation was identified as the reduction in the aerodynamic pitch damping due to stall, leading to a large amplitude torsional motion in response to the high loads produced at stall.

Amer and La Forge (1965) developed a procedure for calculating blade bending and torsion moments under conditions of retreating blade stall, including the negative aerodynamic torsional damping and the lift hysteresis in stall. They used the results of Halfman, Johnson, and Haley (1951) to express the section lift and moment as linear functions of the pitching rate and vertical translation velocity. The calculated bending moments were only about half the measured moments when static lift data were used. When the lift hysteresis was included, predictions the same order as the measurements were obtained. See also La Forge (1965).

Ham and Young (1966) conducted an investigation of stall flutter. With a model rotor in hover, a single-degree-of-freedom, limit-cycle torsional oscillation at a frequency around ω_θ was found at high collective. The torsion amplitude increased with collective and varied with the reduced frequency $k = \omega_\theta b/(.75\Omega R)$, a maximum amplitude occurring at $k = 0.2$ to 0.5, depending on the collective. This frequency corresponded to the maximum negative damping due to stall. The chordwise pressure data showed a large suction peak over the whole section after the maximum pitch angle was reached, resulting in a nose-down moment phased with the nose-down pitch velocity (negative damping). This suction peak was apparently caused by the shedding of concentrated vorticity from the leading edge of the blade at stall, which occurred at an angle of attack considerably above the angle for static stall. The limit-cycle amplitude is determined by the balance of the positive aerodynamic damping below stall (which increases with the pitch rate and hence with the pitch amplitude at a given frequency) and the negative aerodynamic damping above stall. By an elementary analysis it was concluded that the net damping ratio ζ_θ of an airfoil oscillating in pitch should correlate as

$$\frac{\zeta_\theta}{\zeta_0} = f\left(\frac{1}{k}, \alpha_0\right)$$

where α_0 is the mean pitch angle and ζ_0 is the potential flow (unstalled) damping ratio. Such a correlation is presented for an NACA 0012 airfoil, showing that negative damping was possible for $\alpha_0 > 8^\circ$ at frequencies around $k = 0.3$.

Carta (1967) developed an analysis of helicopter stall flutter, based on the measured unsteady aerodynamic loads of an NACA 0012 airfoil oscillating in pitch about the quarter chord. The moment coefficient data show hysteresis loops, as in Fig. 16-5. For oscillations below or in stall

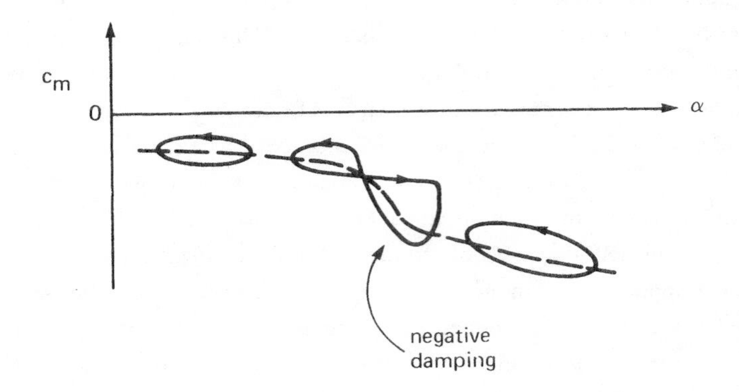

Figure 16-5 Typical unsteady moment coefficient data for an airfoil oscillating in pitch below, through, and in stall.

there is positive damping, but for oscillations about a mean angle of attack near stall the net pitch damping is negative. A two-dimensional aerodynamic damping parameter Ξ_{a_2} related to the work performed per oscillation cycle is defined:

$$\Xi_{a_2} = -\frac{1}{\pi\bar{\alpha}^2}\oint c_m \, d\alpha$$

where $\bar{\alpha}$ is the oscillation amplitude. Tables of Ξ_{a_2} are given based on the experimental data, which show negative damping is possible for mean angles of attack $\alpha_M = 15^\circ$ to 25° and reduced frequencies $k \cong 0.2$ to 0.5. Carta discusses retreating blade stall in terms of this negative damping. The motion of the blade is stable overall, but several cycles of a stall-flutter torsional oscillation can occur in the stall region under the proper conditions. Hence a three-dimensional aerodynamic damping parameter for the blade torsion motion is obtained by integrating over the span:

$$\Xi_{a_3} = \int_0^1 \Xi_{a_2}\left(\frac{U}{u_T}\right)^2 \xi^2 \, dr$$

where ξ is the torsion mode shape. Calculations of Ξ_{a_3} as a function of ψ

for a full scale rotor showed that regions of net negative damping were possible at highly loaded conditions. The most extreme case had negative damping extending over 90° or 100° of azimuth on the retreating side, which corresponds to two or three cycles of torsion motion. These results also indicate that in hover or forward flight the rotor will be most susceptible to stall flutter when the maximum angle of attack is near the stall angle, since that is where the negative damping occurs. When the maximum angle of attack is well above stall there will be positive damping again. See also Carta and Niebanck (1969).

Ham and Garelick (1968) conducted an experimental investigation of dynamic stall. They measured the loads on a two-dimensional airfoil undergoing a transient angle-of-attack change, linearly increasing with time. Extremely high transient lift and moment coefficients were found when the section stalled, at an angle well above the static stall angle of attack. At dynamic stall a loss of leading-edge suction occurred, and simultaneously a suction pressure disturbance moved aft over the upper surface of the airfoil. The character of this disturbance suggested that a vortex was shed from the leading edge at dynamic stall. A transient lift much higher than the static loading was produced as a result of the vortex-induced pressures and also the stall delay. A large, transient, nose-down moment was produced by the pressure disturbance moving aft over the blade. At low pitch rates, the vortex-induced loading is negligible, so the only effect of the unsteady flow is the maximum lift increase due to the stall delay. The measured peak lift and moment coefficients correlated well with the dimensionless pitch rate at the instant of stall, as shown in Fig. 16-6. The dynamic stall angle of attack increased with the pitch rate, and was also sensitive to the pitch axis location. The maximum dynamic stall angle found was for pitching about the leading edge, but the sensitivity to pitch axis location decreased for high pitch rates. The pitch axis had little influence on the peak loading. It was concluded that for high rates of angle-of-attack change the aerodynamic loading on an airfoil is dominated by the effects of an intense vortex shed from the leading edge following dynamic stall. For the case of oscillatory pitch motion, this vortex-induced loading results in negative pitch damping and hence a limit-cycle stall flutter. In the case of transient blade motion, the vortex produces a peak dynamic lift and moment substantially higher than the maximum static loads.

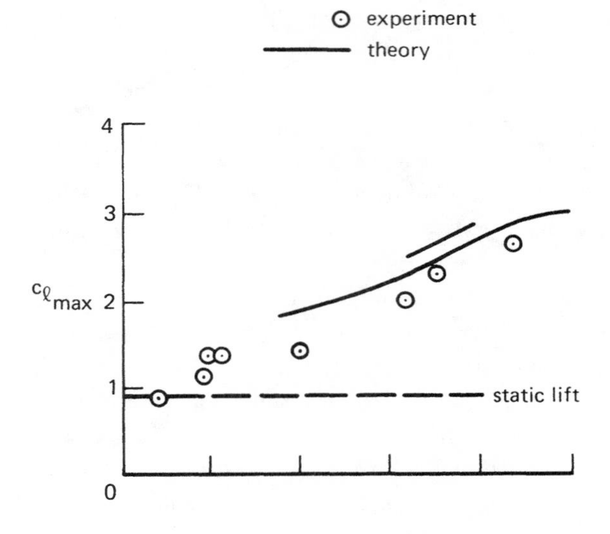

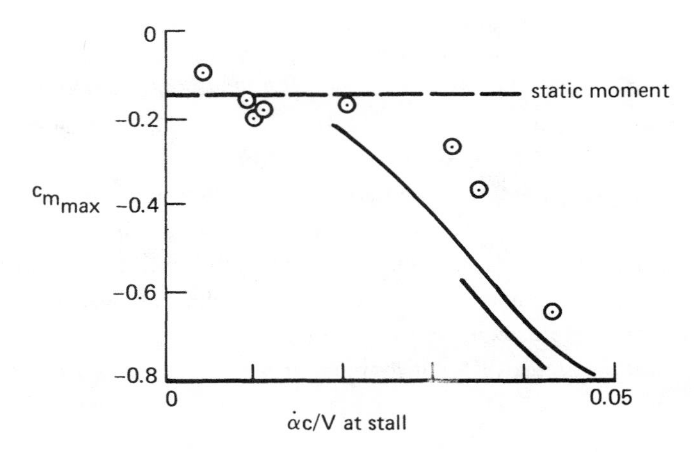

Figure 16-6 Peak lift and moment coefficients for a two-dimensional airfoil during a linear, transient increase in angle of attack through stall; from Ham and Garelick (1968).

Ham (1968) developed an analysis of the unsteady aerodynamic loading on an airfoil during dynamic stall, using a potential flow model for the vorticity shed from the leading edge. For a thin airfoil undergoing a transient angle-of-attack change, shedding of vorticity from the leading edge was begun at a specified dynamic stall angle. The criteria of a stagnation point at the leading edge, the Kutta condition at the trailing edge, and the boundary condition at the airfoil surface determined the bound circulation and the strength of the vorticity shed from the leading and trailing edges at each point in time. The self-induced velocity of the shed vorticity determined its trajectory. It was necessary to prescribe the angle of attack to start the stall process, but then the analysis gave the vortex-induced loading of dynamic stall. The results of the analysis were compared with experimental data for a step increase in angle of attack, harmonic pitch motion, and a linear angle-of-attack increase. Satisfactory correlation with the measured pressure distributions, vortex trajectories, and loads was found. Fig. 16-6 shows the peak lift and moment predicted by this analysis for a linearly increasing angle of attack.

Johnson (1969, 1970b) developed a model for incorporating dynamic stall into the calculation of the aerodynamic loading on a rotating wing, based on the experimental data of Ham and Garelick (1968) for the peak transient loads. It was assumed that the leading-edge vortex shed at dynamic stall produces a large increase in the lift and moment, with a short rise time to the peak loads and then a short decay time to the static loads. Thus when dynamic stall occurs, there is an impulsive lift and nose-down moment increase, which produces the blade motion and loads characteristic of rotor stall. Based on the experimental data (Fig. 16-6) the peak lift and moment coefficients depend on the pitch rate at stall as follows:

$$c_\ell = \begin{cases} c_{\ell_{ss}} + (3 - c_{\ell_{ss}})(20\dot{\alpha}c/V) & \dot{\alpha}c/V < 0.05 \\ 3 & \dot{\alpha}c/V > 0.05 \end{cases}$$

$$c_m = \begin{cases} c_{m_{ss}} & \dot{\alpha}c/V < 0.02 \\ c_{m_{ss}} - (0.8 + c_{m_{ss}})(33.3\dot{\alpha}c/V - 0.667) & 0.02 < \dot{\alpha}c/V < 0.05 \\ -0.8 & \dot{\alpha}c/V > 0.05 \end{cases}$$

where $c_{\ell_{ss}}$ and $c_{m_{ss}}$ are the static stall coefficients. A dynamic stall angle of attack about three degrees above the static stall angle ($\alpha_{ds} = \alpha_{ss} + 3°$) generally gives good results. Thus the following model was used. When the transient angle of attack is below α_{ds}, the flow remains attached. When the angle of attack exceeds α_{ds}, dynamic stall occurs. The lift and nose-down pitching moment coefficients then rise linearly (in an azimuth increment of $\Delta\psi = 10°$ to $15°$) to the peak values given above, which depend on the pitch rate $\dot{\alpha}c/V$ at the instant of stall. Then the lift and moment coefficients decay linearly (in an azimuth increment of $\Delta\psi = 10°$ to $15°$ again) to the static stall values. Flow reattachment takes place when the angle of attack again falls below the static stall angle.

Liiva and Davenport (1969), and Liiva (1969), measured the loads on two-dimensional airfoils oscillating in pitch and vertical translation (NACA 0012 and 0006 sections and modified NACA 23010 and 23006 sections). The oscillating airfoil data showed the dynamic stall delay, which produces maximum lift coefficients greater than the static values; and the negative pitch damping for oscillations through stall. This negative damping was found to be sensitive to the Mach number. Data on the unsteady drag are also given. The cambered airfoils had better characteristics than the symmetrical sections. The maximum lift during the oscillation was larger and the negative damping occurred at larger mean angles of attack. It was found that by mounting the airfoil with a pitch spring characteristic of rotor blades (a natural frequency typically around 4 to 6/rev) and oscillating the angle of attack at 1/rev, stall behavior characteristic of the rotor in forward flight could be reproduced in this two-dimensional test. Liiva and Davenport suggest an analytical model for the lift behavior in dynamic stall, using a second-order differential equation to incorporate the stall delay, peak lift overshoot, and reattachment lag. See also Liiva, Davenport, Gray, and Walton (1968), and Gray, Liiva, and Davenport (1969).

Gross and Harris (1969) developed a method for predicting the dynamic stall loads, based on oscillating airfoil data. An empirical expression for the dynamic stall angle was obtained:

$$\alpha_{ds} = \alpha_{ss} + C_1(\dot{\alpha}c/V)^{1/2}$$

where the parameter C_1 was a function of Mach number, determined from the oscillating airfoil data. It was found that a further increase in the dynamic stall angle by about 2.5° was required to obtain a good prediction of the stall of rotor blades; this increment in the stall angle was attributed to three-dimensional effects, such as those due to the yawed flow.

Arcidiacono, Carta, Casellini, and Elman (1970) developed an analysis of rotor dynamic stall, based on data measured for an NACA 0012 airfoil oscillating in pitch. The lift and moment coefficient data were correlated as functions of α, $A = \dot{\alpha}c/2V$, and $B = \ddot{\alpha}(c/2V)^2$. It was assumed therefore that the loads were independent of the past history of the airfoil motion, except for the recent motion as accounted for by the current velocity and acceleration. The moment coefficient hysteresis loops reconstructed from this tabulated data, and also the negative aerodynamic damping, showed good correlation with the original oscillatory data. The effects of compressibility were accounted for by scaling the incompressible data. This analytical model was used to preduct the effects of dynamic stall on a helicopter rotor in forward flight. A blade torsional oscillation characteristic of retreating blade stall at high loading, not predicted using static airfoil data, was indeed obtained. Quantitative correlation with flight test data was less successful, however. The rotor thrust had to be increased by 30% to obtain the torsional loads measured in flight. There were also significant differences in the waveform; for example, the oscillation began at about $\psi = 180°$ in flight but began at about $\psi = 270°$ in the predictions. Carta, Commerford, and Carlson (1973) obtained additional measurements of unsteady lift and moment on an NACA 0012 airfoil oscillating in pitch, and tabulated the data as a function of α, $A = \dot{\alpha}c/2V$, and $B = \ddot{\alpha}(c/2V)^2$. Good correlation was found when this tabular data was used to reconstruct the original oscillatory loads. The use of this new unsteady airfoil data significantly improved the calculation of the rotor response. Comparison with flight test results showed a reasonably accurate prediction of the blade root torsional amplitude and waveform. The improvement was perhaps due to the change in aspect ratio and Reynolds number in the oscillating airfoil tests; it was probably partly due to the use of nonuniform inflow in the analysis as well. See also Carta, Casellini, Arcidiacono, and Elman (1970), and Carta, Commerford, Carlson, and Blackwell (1972).

Harris, Tarzanin, and Fisher (1970) conducted a theoretical investigation of the influence of three-dimensional and unsteady flow on the rotor stall behavior in forward flight. They developed an analytical model based on static, two-dimensional airfoil data, with the angle of attack corrected for unsteady aerodynamic effects. The maximum lift coefficient of a section was increased by the local flow yaw angle Λ according to

$$c_{\ell_{max}} = \frac{(c_{\ell_{max}})_{\Lambda=0}}{\cos \Lambda}$$

which is a reasonable approximation to the measured effects of sweep. The drag force was assumed to act in the direction of the resultant velocity at the section. Unsteady thin airfoil theory with Theodorsen's lift deficiency function was used to obtain the unsteady aerodynamic loads below stall. The unsteady flow had little influence on the rotor performance below stall, but it did affect the flapping motion. Dynamic stall was accounted for by using an empirical modification of the angle of attack:

$$\alpha_{dyn} = \alpha - C_1 (|\dot{\alpha} c/V|)^{1/2} \operatorname{sign}(\dot{\alpha})$$

where C_1 was a function of Mach number and the airfoil section, obtained from oscillating airfoil data. The lift and drag coefficients were thus evaluated from the static airfoil data, at the effective angle of attack α_{dyn}. The lift was further corrected for yawed flow as follows:

$$c_\ell = \alpha_{dyn} * \min \left\{ \frac{c_\ell(\alpha_{dyn})}{\alpha_{dyn} \cos \Lambda}, c_{\ell a} \right\}$$

The lift predicted by this model for an oscillating airfoil showed good correlation with the measured data. The use of this model improved the prediction of the rotor performance. In the case of no stall the theory greatly overpredicted the lift capability at high loading, and with static stall underpredicted the lift. Including both the unsteady and three-dimensional flow effects gave good correlation, with about 40% of the improvement over static stall a result of the yawed flow effects and 60% due to dynamic stall. Tarzanin (1972) further developed this aerodynamic theory for retreating blade stall by incorporating the blade torsion dynamics. An effective angle of attack for the moment coefficient was defined, similar to α_{dyn} for the lift coefficient but with a different value of the parameter C_1. It was found that the calculated pitch link loads and other blade loads correlated fairly well with flight test measurements; a good prediction of the peak-to-peak amplitude was obtained, and the waveform was qualitatively correct in general. The performance calculation at high loading was also improved. It was found that while the yawed flow had an important influence on performance, it had negligible influence on the control loads due to stall. For the case considered, the dynamic stall delay at the tip led to simultaneous stall over the outer 40% of the blade. The result was a very large control load at stall, which was also amplified by the subsequent torsional response of the blade. Gormont (1973) continued the development of this theoretical model for the rotor blade aerodynamic forces.

McCroskey and Fisher (1972) conducted an experimental investigation of the retreating blade stall on a model rotor blade. The pressure, skin friction, surface streamline direction, and angle of attack were measured. The following sequence of stall events was observed at high rotor loading, over an azimuth increment of about 60°: a dynamic overshoot of the static stall angle, the beginning of a rapid rearward shift of the center of pressure and increase in the moment, then the collapse of the leading edge suction peak, achievement of the maximum lift, and finally achievement of the maximum moment. This stalling process appeared to be associated with the formation and shedding of a vortex from the leading edge of the airfoil. The blade then entered a region of deep stall, perhaps with a torsional oscillation characteristic of stall flutter. Finally the flow reattached (at about $\psi = 360^\circ$ for very high loading), with the occurrence in quick succession of re-establishment of the leading edge suction, stabilization of the center of pressure, and re-establishment of the normal boundary layer character. It was found that the behavior of the moment coefficient vs. lift coefficient curve for the stalled rotor blade section compared quite well with the two-dimensional data of Ham and Garelick (1968) for the same pitch rate $\dot{\alpha}c/V$ at stall. Hence the three-dimensional nature of the rotor flow does not significantly affect the section stall behavior.

Martin, Empey, McCroskey, and Caradonna (1974) conducted an experimental investigation of dynamic stall on a large two-dimensional NACA 0012 airfoil oscillating in pitch at high amplitudes and frequencies representative of rotor operating conditions (1/rev). The formation and shedding of the leading-edge vortex was observed in pressure, hot wire, and smoke visualization data. It was found that the dynamic stall angle of attack decreased, while the angle for maximum lift increased, with a Reynolds number increase. The stall delay increased with the frequency. It was also noted that a leading-edge vortex was always shed when the angle of attack reached its maximum value in the oscillation; the process entailed a somewhat different mechanism than the vortex shedding at stall with increasing α.

Johnson (1974b) conducted a comparison of the rotor performance, airloads, and blade bending and torsion moments calculated using three analytical models of the dynamic stall loads: the methods developed by Arcidiacono, Carta, Casellini, and Elman (1970), by Johnson (1970b), and by Gormont (1973), which have been discussed above. The only difference in the calculation was in the evaluation of the lift, drag, and moment coefficients of a blade section in stalled conditions. The rotor performance and loads were evaluated for an articulated rotor at a high speed, high thrust condition. The cases of no stall and static stall were also considered. With no stall the rotor power was underestimated by 40%, and with static stall the flight condition chosen was really beyond the predicted capability of the rotor. It was concluded that the three dynamic stall methods predicted essentially the same rotor performance and trim, but gave differences of 25% to 40% in the predicted peak-to-peak torsion and bending moments. Moreover, there were significant differences in the details of the aerodynamic loading, particularly the pitch moment, due to the major differences in the stall forces predicted by the three methods.

Bielawa (1975) developed a method for predicting the lift and moment on an airfoil during dynamic stall, using analytical expressions synthesized from experimental oscillating airfoil data. A correction of static airfoil data of the following form was adopted:

$$c_m = c_{m_{\text{static}}}(\alpha - \Delta\alpha_m) + a_{0_m}\Delta\alpha_m + \Delta c_m$$

where $a_{0_m} = \partial c_m/\partial\alpha$ is evaluated for the static data at $\alpha = 0^\circ$; and $\Delta\alpha_m$ and Δc_m are functions of α, $A = \dot{\alpha}c/2V$, and $B = \ddot{\alpha}(c/2V)^2$:

$$\Delta\alpha_m = \alpha_{s_m} f_1(\alpha/\alpha_{s_m}, A, B)$$

$$\Delta c_m = f_2(\alpha/\alpha_{s_m}, A, B)$$

(α_{s_m} is the angle for static stall). A similar expression was used for the section lift coefficient. This method was shown to be as accurate as that of Carta, Commerford, and Carlson (1973), while also more efficient.

Its form also permits an approximate calculation of the unsteady loads on an airfoil for which only static data are available.

McCroskey, Carr, and McAlister (1976) conducted an experimental investigation of stall and boundary layer separation on several airfoils, at low Mach number but relatively high Reynolds numbers. Shedding of a vortex from the leading edge during dynamic stall was observed in all cases. Provided the flow separated over most of the airfoil upper surface, a vortex-like disturbance developed and was convected aft over the airfoil surface at one-third to one-half the free stream velocity. For most airfoils, including the NACA 0012, the leading-edge vortex was produced by a breakdown of the turbulent boundary layer. Although the vortex shedding phenomenon appears to be a general feature of dynamic stall, significant differences were found in the unsteady loads of the various airfoils.

Carr, McAlister, and McCroskey (1977) conducted an experimental investigation of dynamic stall on airfoils oscillating in pitch, considering the basic NACA 0012 profile and several leading-edge modifications. They describe the stall process in detail. The major features of dynamic stall were qualitatively the same regardless of whether the separation phenomenon observed was a gradual boundary layer reversal from the trailing edge, an abrupt breakdown due to laminar leading-edge bubble bursting, or an abrupt breakdown due to turbulent separation at the leading edge. In all cases a leading-edge vortex was formed, and it moved aft to produce the large transient lift and pitch moment. Significant flow reversal always occurred before stall was noticeable in the section lift or moment. In the case of the NACA 0012 airfoil, the observed dynamic stall mechanism at typical helicopter Reynolds numbers and low Mach numbers was turbulent leading-edge separation initiated by a progressive forward movement of trailing-edge separation. See also McAlister, Carr, and McCroskey (1978), and McAlister and Carr (1978).

16–4 Literature

Tanner (1964a) constructed charts of rotor performance, which were discussed in section 6–6. Stall effects were identified by a rapid increase in the blade profile torque at any azimuth on the retreating side. Thus the maximum value of $C_{Q_o}/\sigma = \int_0^1 \frac{1}{2} c_d u_T^2 r dr$ was found over the region

$180^\circ < \psi < 360^\circ$. Then the lower stall limit $C_{Q_0}/\sigma = 0.004$ was associated with the onset of significant stall effects, and beyond the upper stall limit $C_{Q_0}/\sigma = 0.008$ operation was undesirable.

Taylor, Fries, and MacDonald (1973) present analytical and flight test results demonstrating that damping in the nonrotating control system reduces the control loads due to stall. A reduction in peak-to-peak control system loads of up to 50% was found. The fixed system damping had no effect on the peak-to-peak blade torsional moments, which were dominated by the 1/rev loading. The impulsive torsion moment at stall was significantly reduced by the damping, however.

Adams (1973) conducted an experimental investigation of the use of damping in the rotating control system to reduce the stall-induced control loads. When the pitch links were replaced by spring-damper units, the flight test results showed a reduction in the rotating control system loads by almost 50%. The increased pitch damping reduced the higher harmonic pitch motion in response to the stall loads. The oscillating loads in the nonrotating control system were reduced by about 40%. No significant influence on the helicopter performance or handling qualities was observed.

Baskin (1975) conducted flight tests on a tandem helicopter with damping of the nonrotating control system. The rotating control system loads were reduced by about 16%, and the nonrotating control system loads by 22% to 29%.

Landgrebe and Bellinger (1974b) conducted wind tunnel tests of a model rotor to determine the effects of blade twist, camber, taper, torsional frequency, and number of blades on the occurrence and extent of stall. The rotor performance, blade response, and boundary layer separation were measured. The changes in rate of growth of C_T/σ, blade flapping, and vibratory edgewise stress were found to give an indication of the stall threshold generally the same as the measurement of flow separation on the blade. Stall appeared initially on the retreating side, at an azimuth angle between 260° and 330° and a radial station around $r = .75R$. With increasing C_T/σ, the onset of stall progressed rapidly to $\psi = 180^\circ$, and the azimuth for stall recovery progressed rearward to about $\psi = 20^\circ$. At moderate lift, the flow separation propagated forward from the trailing edge of the blade to the leading edge essentially instantaneously. At high lift, the separation progressed from the leading edge, presumably because of a vortex shed at stall. At a given speed the stall boundary depended primarily on the thrust, not on the combination of collective and tip-path-plane incidence used to obtain that lift. The C_T/σ value for initiation of stall decreased with μ. Camber and taper provided the greatest improvement of the rotor stall characteristics, increasing the C_T/σ at the stall threshold and improving the performance in stall. Lowering the torsional frequency tended to delay the onset of stall flutter, but twist, torsional frequency, and number of blades had little or no effect on either the stall threshold or the performance in stall. See also Bellinger, Patrick, Greenwald, and Landgrebe (1974).

Blackwell and Mirick (1974) conducted an experimental and analytical investigation of the effect of blade design parameters on the helicopter stall boundaries, as determined by the build-up of vibratory control loads. The stall-induced torsional moments were reduced by decreasing the torsion stiffness or inertia, and by moving the aerodynamic center of the blade aft. Increasing the blade twist was found to delay the stall speed at high C_T/σ based on the control loads, but it increased the rate of build-up of control loads with speed. However, increased twist also increased the blade flapwise bending moments substantially. See also Blackwell and Commerford (1975).

McHugh and Harris (1976) re-examined the potential high-speed capability of the conventional helicopter configuration. Earlier investigations indicated that at around $\mu = 0.5$ to 0.6 the lift rotor encounters a propulsive force and loads limitation, mainly due to stall, suggesting that for speeds above 200 knots a compound helicopter or other advanced configuration would be required. However, current technology gives us an improved knowledge of the influence of stall on the rotor performance and loads, especially the unsteady stall effects; improved airfoil designs for better stall characteristics; improved control system design for reduced loads at high speed; and low parasite drag designs for the fuselage and hub, to reduce the propulsive force requirement. Hence it was concluded that efficient pure helicopter designs (aircraft utilizing a lifting and propelling rotor) with cruise speeds up to 250 or 300 knots might be possible with current technology. Experimental confirmation of this conjecture is required. See also McHugh (1978).

Further work on the stall of helicopter rotors: Gustafson and Gessow

(1948), Halfman (1951), Stewart (1952b), Sisto (1958), Amer (1955), Carpenter (1958), Payne (1958), Powell and Carpenter (1958), Shivers and Carpenter (1958), Evans and Mort (1959), Powell (1959), McCloud, Hall, and Brady (1960), Shivers (1960, 1961), Arcidiacono (1961, 1964), Shivers and Monahan (1962), Sissingh (1962), Sweet and Jenkins (1963), Wernicke and Drees (1963), Sweet, Jenkins, and Winston (1964), Davenport (1965), Moss and Murdin (1965), Harris (1966a, 1966b), Ericsson (1967), Ham (1967, 1972, 1973), Pruyn et al. (1967), Scheiman and Kelley (1967), Harris and Pruyn (1968), McCroskey and Yaggy (1968), Paglino (1968), Ericsson and Reding (1969, 1971a, 1971b, 1972, 1973, 1977), Ham and Johnson (1969), Loewy (1969), Niebanck (1969), Patay (1969), Niebanck and Bain (1970), Windsor (1970), Bellinger (1971, 1972a, 1972b), Carta (1971), Dwyer and McCroskey (1971), Fisher, Tompkins, Bobo, and Child (1971), Ward (1971), Bobo (1972), Crimi and Reeves (1972, 1976), Gabel and Tarzanin (1972), Isogai (1972), Johnson and Ham (1972), McCroskey (1972a, 1972b, 1973, 1975), Beno (1973), Benson, Dadone, Gormont, and Kohler (1973), Crimi (1973a, 1973b, 1974, 1975), Lang (1973), Carlson, Blackwell, Commerford, and Mirick (1974), Dadone and Fukushima (1974), Dugundji and Aravamudan (1974), Parker and Bicknell (1974), Scarpati, Sandford, and Powell (1974), Tarzanin and Mirick (1974), Tarzanin and Ranieri (1974), Dugundji and Chopra (1975), McCroskey and Philippe (1975), Taylor and Gabel (1975), Van Gaasbeek (1975), Beddoes (1976), Costes (1976), Parker (1976), Riley (1976), Amer and La Forge (1977), Dadone (1977), Fukushima and Dadone (1977), McHugh, Clark, and Solomon (1977), Beddoes (1978), Brotherhood and Riley (1978), Maskew and Dvorak (1978), Pierce, Kunz, and Malone (1978), Rao, Maskew, and Dvorak (1978).

Chapter 17

NOISE

17–1 Helicopter Rotor Noise

The helicopter is the quietest VTOL aircraft, but its noise level can still be high enough to compromise its utility unless specific attention is given to designing for low noise. As the restrictions on aircraft noise increase, the rotor noise becomes an increasingly important factor in helicopter design. Helicopter rotor noise tends to be concentrated at harmonics of the blade passage frequency $N\Omega$ (see Fig. 17-1), because of the periodic nature

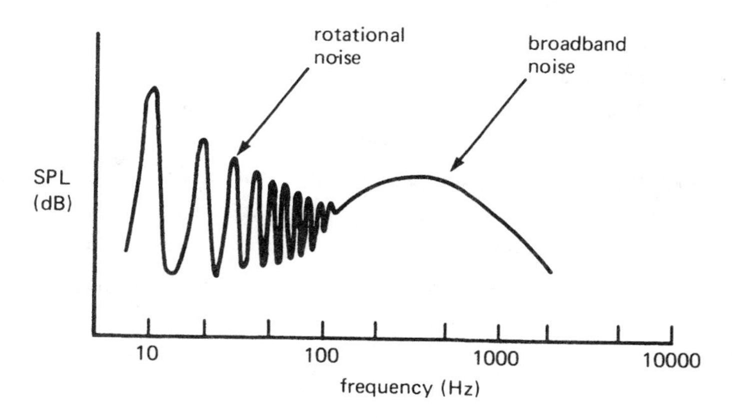

Figure 17-1 Helicopter rotor sound spectrum.

of the rotor as seen in the nonrotating frame. There is sound radiated because the mean thrust and drag forces rotate with the blades, and because of the higher harmonic loading as well. The spectral lines are broadened at the higher harmonics because of the random character of the rotor flow, particularly variations in the wake-induced loads. The acoustic pressure signal is basically periodic in time (the period is $2\pi/N\Omega$), with sharp impulses due to localized aerodynamic phenomena such as compressibility effects and vortex-induced loads. The contributions to helicopter rotor noise can be classified as vortex or broadband noise, rotational noise, and blade slap. While the distinction between these types of rotor noise is not as sharp as was once thought, the classification remains useful for purposes of exposition.

Vortex or broadband noise is a high frequency swishing sound produced by the rotor and modulated in frequency and amplitude at the blade passage frequency. Vortex noise is random sound radiated as a result of random fluctuations of the forces on the blades. The sound energy is distributed over a substantial portion of the spectrum in the audible range, typically extending for main rotors from about 150 Hz to 1000 Hz, with a peak around 300 or 400 Hz. Vortex noise of rotors is principally produced by the random lift fluctuations resulting from operation of the blade in the turbulent wake, especially the random blade loads induced by the tip vortices. Other sources of broadband noise are the forces on the blade due to vortex shedding from the trailing edge, turbulence in the free stream, and boundary layer turbulence and separation. (The random noise of rotating wings was originally associated with the shedding of vortices, as from a cylinder, hence the name "vortex noise.")

Rotational noise is a thumping sound at the blade passage frequency (or at a multiple of $N\Omega$ if the fundamental is inaudible). As the higher harmonic content increases, the thumps sharpen into bangs, and eventually into blade slap. The spectrum varies greatly with the rotor geometry and the operating condition. Rotational noise is a purely periodic sound pressure radiated as a result of the periodic forces exerted by the blade on the air. Its spectrum thus consists of discrete lines at harmonics of the blade passage frequency $N\Omega$. Rotational noise dominates the low frequency end of the spectrum, for the main rotor from below audible frequencies to about 150 Hz. (The rotational noise is found to extend even higher if a narrow band measurement is made.) The fundamental frequency $N\Omega$ is typically 10 to 20 Hz for a main rotor, so the fundamental and perhaps also the first or second harmonic will be below the threshold of hearing. (The sound identified at the blade passage frequency is the higher harmonics, and the vortex noise modulated at N/rev.) It is therefore the higher harmonics of main rotor rotational noise that are important subjectively. For propellers or a tail rotor the fundamental frequency is much higher, typically around 100 Hz, which so increases the importance of the rotational noise that it becomes the dominant component. Rotational noise is produced by the force periodically exerted on the air at any fixed point on the rotor disk because of the rotation of the lift and drag with the blade. The higher harmonic blade loading is responsible for the large high frequency content of helicopter rotor rotational noise.

Blade slap is a sharp cracking, popping, or slapping sound occurring at the blade passage frequency and produced by the main rotor in certain flight conditions. Blade slap is a periodic, impulsive sound pressure disturbance, so it may be considered an extreme case of rotational noise. The impulsive character of blade slap results in a substantial increase of the sound level over the entire spectrum (covering a range of about 20 to 1000 Hz for a main rotor). When it occurs, blade slap is the dominant noise source; it has a high overall sound pressure level, and its impulsive character is most objectionable. Blade slap tends to occur most often in such maneuvers as flare to landing, shallow descents, and decelerating steep turns; and at high forward speeds. It can also occur with some helicopters in level flight at moderate speeds. The most likely sources of blade slap are vortex-blade interactions and the effects of thickness at high Mach number. Such aerodynamic phenomena produce a large, localized, transient force on the blade, which results in impulsive sound radiation. Local stall and supersonic flow probably play a role also. The tandem helicopter has blade slap as a result of the interaction of the rear rotor blades with the tip vortices in the wake of the front rotor.

The order of importance of the sources of main rotor noise is blade slap (when it occurs), vortex noise, and rotational noise. The rotational noise is most intense at very low frequency (the first few harmonics may even be below the threshold of hearing). Thus, while rotational noise is the primary determinant of the overall sound pressure level, it is not the most important source in terms of subjective annoyance. When the sound level is corrected for frequency content, it is often the vortex noise that dominates. However, rotational noise can be important when its level increases at high frequencies, i.e. in cases approaching blade slap, in particular at high tip speed with a small number of blades. Rotational noise can also produce acoustic fatigue and vibration of the helicopter structure. Moreover, the low frequencies propagate best in the air, the high frequencies being attenuated most with distance. Consequently, at very large distances from the helicopter the blade slap and rotational noise of the main rotor are most important. The acoustic detectability of the helicopter is often determined by the rotational noise.

On many helicopters the tail rotor is the source of the most noticeable and disturbing noise (next to blade slap), unless some effort is made to quiet it. The design limitations for a quiet tail rotor (mainly low tip speed) impose little performance degradation, because the tail rotor accounts for a small part of the total power loss. Basically, the tail rotor noise mechanisms are the same as the main rotor mechanisms but have a higher fundamental frequency (40 to 120 Hz typically). Hence tail rotor rotational noise is the most important subjectively. The helicopter transmission and engine are sources of high frequency sound, which is important mainly for the internal and near field noise. Since the high frequencies attenuate most with distance, the rotors are the primary source of far field noise.

Let us examine the statistics of the rotor sound field. Because sound consists of pressure waves radiated in a fluid, the noise analysis is concerned with the perturbation of the pressure from the atmospheric level. The rotor aerodynamic loading and hence the sound pressure are random signals with periodic (hence nonstationary) statistics. The fundamental frequency is the blade passage frequency, so the period is $2\pi/N\Omega$, or in terms of the rotor azimuth $2\pi/N$. The expected value of the sound pressure perturbation generated by the rotor is $\bar{p}(\psi) = Ep(t = \psi/\Omega)$. Since the signal is nonstationary, the expected value $\bar{p}$ is not constant. It is a periodic function of ψ, however, so $\bar{p}$ can be written as a Fourier series:

$$\bar{p} = \sum_{m=-\infty}^{\infty} p_m e^{imN\Omega t}$$

where p_m are the harmonics of the discrete spectrum of $\bar{p}$. For the autocorrelation, let $\tilde{p} = p - \bar{p}$, so

$$R_p(t_1,t_2) = E(\tilde{p}(t_1)\tilde{p}(t_2))$$

Again since the rotor noise is nonstationary, R_p is not a function of $t_1 - t_2$

alone, but rather is a periodic function when $t_1 - t_2$ is a constant. Define therefore the new variables $s_1 = \frac{1}{2}(t_2 - t_1)$ and $s_2 = \frac{1}{2}(t_2 + t_1)$. The autocorrelation R_p is then periodic in s_2, and hence may be written as follows:

$$
\begin{aligned}
R_p(s_1, s_2) &= E[\tilde{p}(t_1 = s_2 - s_1)\tilde{p}(t_2 = s_2 + s_1)] \\
&= \sum_{m=-\infty}^{\infty} R_p^m(s_1)\, e^{imN\Omega s_2} \\
&= \sum_{m=-\infty}^{\infty} e^{imN\Omega s_2} \int_{-\infty}^{\infty} e^{i\omega s_1}\, S_p^m(\omega) d\omega
\end{aligned}
$$

The standard deviation of the pressure at a particular instant in the period can be obtained from the autocorrelation:

$$
\sigma_p^2(\psi) = E\tilde{p}^2 = R_p(t_1 = t_2) = R_p(s_1 = 0)
$$

Hence the spectral decomposition of σ_p^2 is

$$
\begin{aligned}
\sigma_p^2(\psi) &= \sum_{m=-\infty}^{\infty} R_p^m(s_1 = 0)\, e^{imN\psi} \\
&= \sum_{m=-\infty}^{\infty} \left(\int_{-\infty}^{\infty} S_p^m(\omega)\, d\omega \right) e^{imN\psi}
\end{aligned}
$$

It follows that $Ep^2 = \bar{p}^2 + \sigma_p^2$, and the rms sound pressure is given by

$$
\begin{aligned}
(\text{rms}\, p)^2 &= \lim_{T \to \infty} \frac{1}{T} \int_0^T p^2\, dt \\
&= \lim_{K \to \infty} \frac{1}{K} \sum_{k=1}^{K} \frac{N\Omega}{2\pi} \int_{\frac{2\pi(k-1)}{N\Omega}}^{\frac{2\pi k}{N\Omega}} p^2\, dt
\end{aligned}
$$

$$= \frac{N}{2\pi} \int_0^{\frac{2\pi}{N}} Ep^2 \, d\psi$$

$$= (\bar{p}^2 + \sigma_p^{\;2})_{\text{average over } \psi}$$

$$= \sum_{m=-\infty}^{\infty} |p_m|^2 + \int_{-\infty}^{\infty} S_p^{\;0}(\omega) d\omega$$

The expected value of the sound pressure $\bar{p}$ is the rotational noise (and blade slap if it is impulsive enough); the autocorrelation R_p is the broad-band noise. Included in R_p is the standard deviation of the periodic pressure waveform, $\sigma_p^{\;2}$. The root-mean-square pressure, which is generally used for noise assessment, is then the average of $Ep^2 = \bar{p}^2 + \sigma_p^{\;2}$ over one period. The spectral decomposition of the rotational noise gives the discrete harmonics p_m. The spectral analysis of the rms pressure involves in addition the spectrum of the average of the autocorrelation, $S_p^{\;0}(\omega)$. Because the rotor noise is not stationary, there is more information about the broad-band noise that is not considered by the rms pressure (such as the spectrum of $\sigma_p^{\;2}$). This additional information is likely to be important for the subjective perception of rotor noise.

Sound is measured in units of decibels (dB), defined as

$$10 \log_{10} \frac{\text{sound power}}{\text{reference power}}$$

A logarithmic scale is used because of the need to handle orders-of-magnitude differences in the sound levels encountered, and because the human ear has basically a logarithmic response to sound. The energy flux at a given point in the sound field is given by the acoustic intensity $I = E(pu)$. Here p is the perturbation pressure and u is the velocity due to the sound waves, so the instantaneous intensity pu is the power radiated per unit area. In the far field, the velocity and pressure disturbances of a sound wave are related by $u = p/\rho_0 c_s$, so the intensity is

$$I = E(pu) = \frac{E(p^2)}{\rho_0 c_s} = \frac{p_{\text{rms}}^{\;2}}{\rho_0 c_s}$$

where ρ_0 is the mean air density and c_s is the speed of sound. Thus the rms pressure is a measure of the acoustic intensity. Moreover, the human ear and the aircraft structure respond to the pressure deviations from the mean atmospheric value. Hence noise is measured in terms of the sound pressure level, defined as

$$\mathrm{SPL} = 10 \log \frac{p_{rms}^2}{p_{ref}^2} = 20 \log \frac{p_{rms}}{p_{ref}}$$

in units of dB (re p_{ref}). For the reference pressure, $p_{ref} = 0.0002$ dyne/cm^2 is normally used. The spectrum of the rms pressure can then be regarded as the distribution of the sound energy over frequency.

The overall sound pressure level (OSPL) is the total rms pressure. It is common practice to measure and present the noise data in terms of its spectrum, either by octave band, third-octave band, or narrow band measurements. Since the subjective perception of sound depends on the frequency content, a number of frequency-weighted measures of the sound pressure level have been developed, notably the perceived noise level (PNdB) and the A-weighted sound level (dBA). Further corrections have been developed for aircraft noise, for such effects as the sound duration and the presence of pure tones. The estimate of the subjective level of helicopter noise is difficult, however, because no standard has yet been formulated that is really applicable to the noise characteristic of rotors: a random pressure disturbance with periodic statistics, composed of low frequency rotational noise, high frequency broadband noise, and often significant impulsive noise.

17–2 Vortex Noise

Rotor vortex noise is a high frequency sound produced by random fluctuations of the forces on the blade. The principal source of vortex noise appears to be the lift fluctuations resulting from operation of the blades in the turbulent rotor wake; the random loads induced by the tip vortices in the wake are especially important. For an elementary analysis of vortex noise, consider a blade of length ℓ in a flow at speed V with a random lift force (per unit length) $F_z(t)$ induced by the turbulence and vorticity in the wake. Assuming impulsive loading over the chord, the section force is a vertical dipole, for which the sound pressure p is

$$p = -\frac{1}{4\pi}\frac{\partial}{\partial z}\frac{F_z(t - s/c_s)}{s}$$

where s is the distance from the dipole F_z to the observer, and z is the vertical coordinate. The effect of the blade motion on the retarded time has been neglected. Using the relation $\partial s/\partial z = z/s$, the sound pressure in the far field (s very large) is

$$p = -\frac{1}{4\pi c_s}\frac{z}{s^2}\left[\dot{F}_z(t - s/c_s) - \frac{c_s}{s}F_z(t-s/c_s)\right]$$

$$\cong -\frac{1}{4\pi c_s}\frac{z}{s_0^2}\dot{F}_z$$

$$= -\frac{1}{4\pi c_s}\frac{\sin\theta_0}{s_0}\dot{F}_z$$

where s_0 is the distance from the blade to the observer, and θ_0 is the angle of the observer from the disk plane ($\sin\theta_0 = z/s_0$). Integrating over the length of the rod gives the total sound pressure:

$$p = -\frac{1}{4\pi c_s}\frac{\sin\theta_0}{s_0}\int_0^{\ell}\dot{F}_z\,d\ell_1$$

The rms pressure is then

$$p_{rms}^2 = E(p^2) = \left(\frac{\sin\theta_0}{4\pi c_s s_0}\right)^2\int_0^{\ell}\int_0^{\ell}E(\dot{F}_z(\ell_1)\dot{F}_z(\ell_2))\,d\ell_1\,d\ell_2$$

Now write $\dot{F}_z = \omega F_z$ for a force at frequency ω, and define

$$R_{ff} = E(F_z(\ell_1)\,F_z(\ell_2)) = E(F_z(\ell_2 + \ell_3)F_z(\ell_2))$$

so that

$$p_{rms}^2 = \left(\frac{\sin\theta_0}{4\pi c_s s_0}\right)^2\omega^2\int_0^{\ell}\int_0^{\ell}R_{ff}\,d\ell_1\,d\ell_2$$

Assuming that the force correlation is homogeneous over the length of the blade, it depends only on $\ell_3 = \ell_1 - \ell_2$. Specifically, $R_{ff} = \bar{F}^2 R(\ell_3)$, where $\bar{F}$ is the rms level of the section lift, so $R(0) = 1$. Then

$$\int_0^\ell \int_0^\ell R_{ff}\, d\ell_1\, d\ell_2 = \tfrac{1}{2}\bar{F}^2 \int_{-\ell}^{\ell} \int_{|\ell_3|}^{2\ell - |\ell_3|} R(\ell_3)\, d\ell_4\, d\ell_3$$

$$= \bar{F}^2 \int_{-\ell}^{\ell} (\ell - |\ell_3|)\, R(\ell_3)\, d\ell_3$$

where $\ell_4 = \ell_1 + \ell_2$. A perfect correlation of the force over the blade length would give $R(\ell_3) = 1$. A small correlation length over the blade is more appropriate, however, so assume $R(\ell_3)$ is small except for distances of order ℓ_c from the origin:

$$\int_0^\ell \int_0^\ell R_{ff}\, d\ell_1\, d\ell_2 = \bar{F}^2 \ell \ell_c$$

where the correlation length $\ell_c = 2 \int_0^\ell (1 - \ell_3/\ell) R(\ell_3)\, d\ell_3 \cong 2 \int_0^\ell R(\ell_3) d\ell_3$ is here much less than ℓ. Hence

$$p_{rms}^2 = \left(\frac{\sin\theta_0}{4\pi c_s s_0}\right)^2 \omega^2 \bar{F}^2 \ell \ell_c$$

In terms of the rms lift coefficient $C_L = F/\tfrac{1}{2}\rho V^2 d$ and the Strouhal number $S = \omega d/2\pi V$ (where d is a characteristic dimension of the section), the result is

$$p_{rms}^2 = \frac{\rho^2}{16 c_s^2} \left(\frac{\sin\theta_0}{s_0}\right)^2 C_L^2\, S^2\, V^6\, \ell \ell_c$$

Yudin (1947) obtained essentially this result. He interpreted vortex noise on a rotating rod as being due to the oscillating forces induced on the section by the shed vortices, such as in a von Kármán vortex street. (By considering the solution for a vortex street, he related the oscillatory lift force produced by the vortices to the steady drag force on the rod. That step is not required for the present analysis.) Yudin quotes experimental data that indicate the sound is proportional to $(\sin\theta_0)^2 V^{5.5}\ell$, verifying the dipole character of the vortex noise and the assumption of small correlation length (a large correlation length would give $p^2 \sim \ell^2$). Yudin attributed the slightly smaller growth with speed to the dependence of $C_L{}^2$ on V (actually, to the dependence of the drag coefficient on V for his solution).

For the rotating blade, we now specify that s_0 is the distance from the hub to the observer, at an angle θ_0 above the disk plane; the section lift coefficient is proportional to the blade loading C_T/σ; the tip speed ΩR is used for the speed V; and the Strouhal number is assumed constant (so the noise frequency scales with V/d). The blade radius R is used for the length ℓ; it is assumed that the correlation length ℓ_c is proportional to the blade chord c; and the sound power must be multiplied by the number of blades N. Thus the sound is proportional to $N\ell\ell_c \sim NRc = A_b$, the total blade area. The rotor vortex noise is then

$$p_{rms}^2 = \text{constant} * \frac{\rho^2}{c_s^2}\left(\frac{\sin\theta_0}{s_0}\right)^2 A_b(\Omega R)^6 (C_T/\sigma)^2$$

A constant Strouhal number implies that the rotor vortex noise frequency scales with $\Omega R/c$, but since the blade velocity varies linearly along the span and varies in direction relative to the observer, the noise is spread over a considerable frequency range. The assumption that the sound is due to lift fluctuations has given it the directivity of a vertical dipole: maximum on the rotor axis ($\theta_0 = 90°$) and zero in the rotor disk plane ($\theta_0 = 0$). The far field sound varies with distance from the rotor as $s_0{}^{-2}$, which is required for constant total radiated power. For fixed C_T/σ and blade area, the vortex noise is proportional to the tip speed to the sixth power (because of the scaling of F_z with speed). Alternatively, in terms of the rotor thrust, $p^2 \sim T^2(\Omega R)^2/A_b$. The above expression for the vortex noise can be generalized to

$$p_{rms}^2 = \text{constant} * \frac{\rho^2}{c_s^2}\left(\frac{\sin\theta_0}{s_0}\right)^2 A_b(\Omega R)^6 f(C_T/\sigma)$$

Widnall (1969) correlated the existing data on main rotor vortex noise in the form of

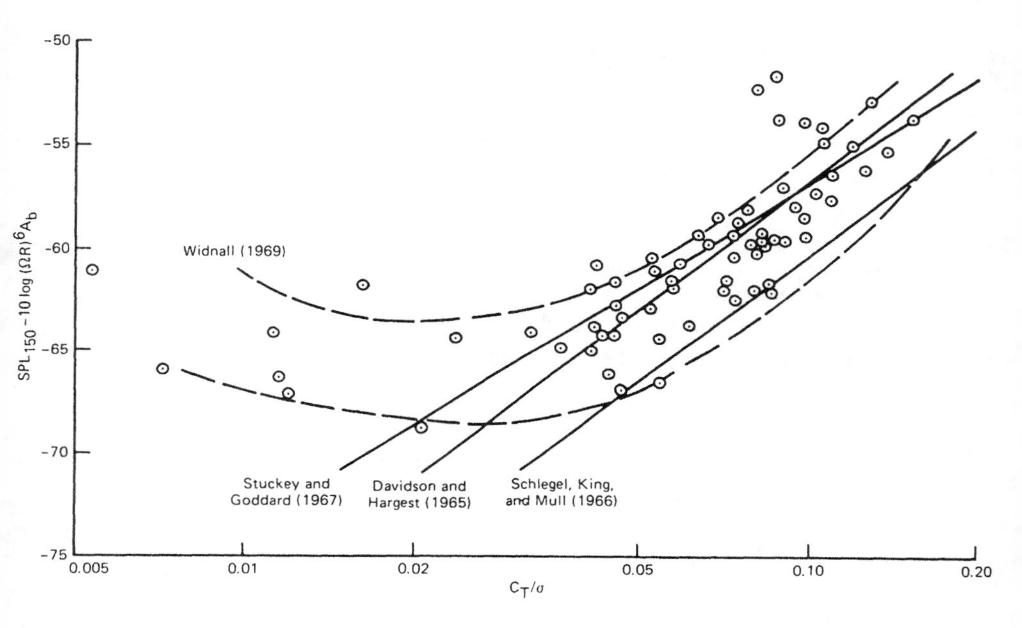

Figure 17-2 Correlation of rotor vortex noise with C_T/σ, adapted from Widnall (1969). Shown are measured noise data (O), Widnall's correlation, and several empirical vortex noise expressions.

$$\mathrm{SPL}_{150} - 10 \log (\Omega R)^6 A_b$$

as a function of C_T/σ (Fig. 17-2), where SPL_{150} is the sound pressure level on the rotor axis at 150 m from the hub (with ΩR in m/sec and A_b in m^2). The sound level at an arbitrary observer point is then

$$\mathrm{SPL} = \mathrm{SPL}_{150} + 20 \log \left(\frac{150\,\mathrm{m}}{s_0} \sin \theta_0 \right)$$

The data correlated well with C_T/σ, with a scatter typical of the original measurements. At blade loadings typical of helicopter rotor operation, the relationship $p^2 \sim (C_T/\sigma)^2$ is found. For a highly loaded rotor the noise increases faster with C_T/σ, while for low loading the noise is constant or even increases again, probably as a result of the proximity of the rotor wake. A number of empirical expressions have been developed to predict vortex noise, based on correlation of measured rotor noise and on the parameter dependence indicated by simple theories. Fig. 17-2 shows the results of three such expressions, which predict the same overall sound pressure level to within a few decibels. Davidson and Hargest (1965) obtained from measurements of rotor noise in hover:

$$SPL_{150} = 10 \log [(\Omega R)^6 A_b (C_T/\sigma)^2] - 36.7 \text{ dB}$$

(with ΩR in m/sec and A_b in m^2). They also suggest corrections for hovering in a wind, for forward flight, and for perceived noise level (PNdB). This expression gives about the best correlation over the range $C_T/\sigma = 0.035$ to 0.15.

Stuckey and Goddard (1967) found that the vortex noise measurements of a rotor on a whirl tower correlated best with the expression

$$SPL_{150} = 10 \log [(\Omega R)^6 A_b (C_T/\sigma)^{1.66}] - 39.7 \text{ dB}$$

The broadband noise spectrum peak frequency was given by $f = 0.45\Omega R/c$ Hz (where c is the blade chord), and the spectrum fell off at about a constant rate of 7.5 dB per octave on either side of the peak. On a third-octave band spectrum, the peak was then 5.3 dB below the overall sound pressure level.

Schlegel, King, and Mull (1966) developed an empirical expression equivalent to

$$SPL_{150} = 10 \log [(\Omega R)^6 A_b (C_T/\sigma)^2] - 40.2 \text{ dB}$$

for predicting vortex noise. The vortex noise spectrum was found to extend from 150 to 9600 Hz, with the greatest intensity over 300 to 600 Hz. The peak frequency was found to correlate as $f = 0.20\Omega R/h$, where h is the projected thickness of the blade. They gave generalized octave band spectra for the vortex noise below and above stall conditions, scaled to this peak frequency. When the rotor stalled there was a sharp increase in the 1200 to 2400 Hz octave band, and the above empirical expression underpredicted the vortex noise. See also Schlegel and Bausch (1969).

Lowson and Ollerhead (1969a, 1969b) found that with very narrow band measurements the rotational noise harmonics are identifiable up to 400 Hz at least, in contrast with wide band measurements, which lead to the impression that the random noise dominates the main rotor sound above about 150 Hz. Thus there is really a gradual transition from harmonic (rotational) to broadband (vortex) noise as the frequency increases. They found that the vortex noise directivity is basically that of a dipole, but not exactly zero in the disk plane. They suggested using

$$D = 10 \log \frac{\sin^2 \theta_0 + 0.1}{1.1}$$

instead of a $10 \log \sin^2 \theta_0$ directivity factor. As a result, the sound in the disk plane is 10.4 dB below that on the rotor axis. They noted that in general the rotor noise was strongly dependent on the tip speed, and that there was increased noise at high collective because of stall and at low collective because of wake interference. See also Ollerhead and Lowson (1969).

Leverton (1973) obtained noise measurements of a helicopter rotor on a whirl tower, and found using a narrow band analysis that the spectrum in the frequency range generally associated with vortex noise was actually a combination of rotational and broadband components. He found that the broadband noise spectrum did not show a peak, but was flat out to a fundamental frequency of 325 to 450 Hz (depending on the thrust), and fell off at a roughly constant decibel-per-octave rate after that.

There has been some work toward the development of a rigorous theory for the broadband noise produced by the random loading and motion of helicopter rotor blades. In particular, there have been recent investigations of isolated airfoil noise due to boundary layer turbulence and vortex shedding, and of the noise produced by rotating or nonrotating wings in a turbulent flow field. Also, a high-frequency range of main rotor broadband noise, occurring at around 2 to 5 kHz, has been identified. A complete definition of the origin of helicopter rotor broadband noise, and the development of analyses to predict it, are still subjects of research.

17–3 Rotational Noise

Rotor rotational noise is a periodic sound pressure disturbance. The rotational noise spectrum consists therefore of discrete lines at the harmonics of the blade passage frequency $N\Omega$, which dominate the low frequency portion of the rotor noise. With a narrow band analysis many harmonics of the rotational noise are identifiable. As the frequency increases, the random pressure fluctuations become more important, until the discrete harmonics can no longer be discriminated from the broadband noise spectrum.

Rotational noise is produced by the periodic lift and drag forces acting on the blade. The blade exerts equal and opposite reaction forces on the air and these rotate with the blade, so that at any fixed point in the rotor disk plane there is a periodic application of a force as the blades pass (at the fundamental frequency $N\Omega$). Such an unsteady force results in dipole radiation of a periodic pressure disturbance into the fluid, which is the rotational noise. The force acting on the air is unsteady because of the rotation of the blades and the periodic variation of the blade loading. For the rotational noise analysis, the section forces rotating with the blade (F_x, F_z, and F_r; respectively the chordwise, vertical, and radial forces) will be replaced by an equivalent distribution of periodic forces on the surface of the rotor disk in the nonrotating frame (with components G_x, G_y, and G_z). Fig. 17-3 shows the rotor coordinate axes used; as usual, the

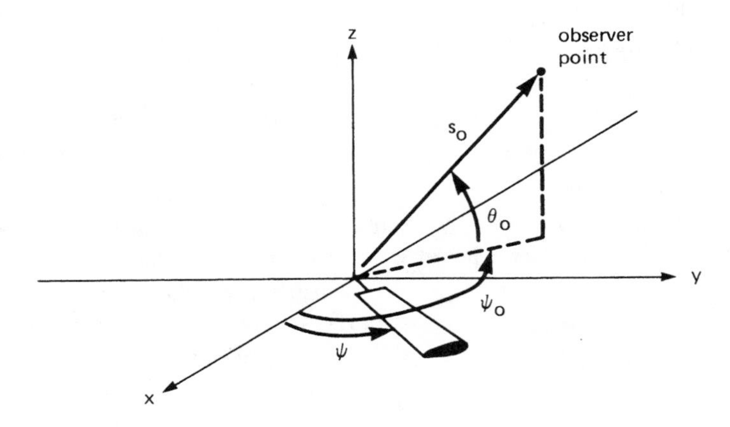

Figure 17-3 Definition of the rotor disk plane coordinate axes, and the observer position.

x-axis is aft, the y-axis is to the right, and the z-axis is upward. The rotor disk is in the x-y plane, with the blade at an azimuth angle $\psi = \Omega t$. The forces on the rotating blades can thus be represented by a distribution over the rotor disk of dipoles that do not rotate but have periodically varying strength. The rotor disk with the nonrotating dipole distribution then moves with the helicopter at a steady velocity in vertical or forward flight. Fig. 17-3 also defines the position of an observer, where the rotor noise is detected, by the range s_0, aximuth angle ψ_0, and elevation θ_0.

17-3.1 Rotor Pressure Distrubution

The differential pressure on the surface of the rotating blade, $\Delta p(r,x,\psi)$, is a periodic function of the blade azimuth angle ψ for steady-state operation of the rotor. Here x and r are rectangular coordinates rotating with the blade, chordwise (positive aft) and spanwise, respectively. Thus Δp is nonzero only within the area bounded by the blade leading and trailing edges ($x_{le} < x < x_{te}$), and the root and tip ($r_R < r < R$). This pressure will be written in terms of the section lift L and a chordwise loading function ℓ:

$$\Delta p(r, x, \psi) = L(r, \psi)\,\ell(x, \psi)$$

where by definition $\int_{x_{le}}^{x_{te}} \ell dx = 1$. The rotational noise analysis will require a knowledge of the pressure distribution over the rotor disk $\Delta p_1(r_1, \psi_1, t)$, where r_1 and ψ_1 are the polar coordinates of the point on the disk. With N blades, Δp_1 will be periodic in time t with period $2\pi/N\Omega$. It is necessary then to relate Δp_1, the pressure at a fixed point (r_1, ψ_1) on the rotor disk as a function of time t, to the pressure Δp at a point (x, r) on the rotating blade at azimuth angle ψ.

Each of the N blades produces an identical pressure pulse of duration $c/\Omega r$ as, one at a time, they sweep past a given point on the rotor disk. Thus it is only necessary to consider the pressure produced by the reference blade when its azimuth angle ψ is in the vicinity of the disk point at ψ_1. The pressure on the disk is just $\Delta p_1(r_1, \psi_1, t) = \Delta p(r, x, \psi)$, and it is necessary to determine the blade coordinate and azimuth (r, x, ψ) for a given time and point on the rotor disk (r_1, ψ_1, t). The reference blade is at azimuth angle $\psi = \Omega(t - t_0)$, where t_0 is an arbitrary initial time. Now given ψ and the disk point polar coordinates r_1 and ψ_1, the chordwise and radial coordinates of the corresponding point on the disk are

$$x = r_1 \sin(\psi - \psi_1)$$

$$r = r_1 \cos(\psi - \psi_1)$$

For a high-aspect-ratio blade x/r is small, so $x \cong r_1(\psi - \psi_1)$ and $r \cong r_1$. Then the distribution of pressure over the rotor disk is

$$\Delta p_1(r_1, \psi_1, t) = \Delta p\left\{r_1, x = r_1[\Omega(t - t_0) - \psi_1], \psi = \Omega(t - t_0)\right\}$$

This determines Δp_1 for a time interval of length $T = 2\pi/N\Omega$ about $t = t_0 + \psi_1/\Omega$, and hence for all time, since Δp_1 is periodic.

Now write the disk pressure distribution as a Fourier series in time:

$$\Delta p_1 = \sum_{m=-\infty}^{\infty} \Delta p_m \, e^{imN(\Omega t - \psi_1)}$$

where the harmonics are obtained from

$$\Delta p_m = \frac{N\Omega}{2\pi}\int \Delta p\left\{r_1, x = r_1[\Omega(t - t_0) - \psi_1], \psi = \Omega(t - t_0)\right\} e^{-imN(\Omega t - \psi_1)} dt$$

(the integral is over the period $T = 2\pi/N\Omega$). Transforming the integration variable to $x = r_1(\Omega(t - t_0) - \psi_1)$ gives

$$\Delta p_m = \frac{N}{2\pi r_1} e^{-imN\Omega t_0} \int_{x_{le}}^{x_{te}} \Delta p(r_1, x, \psi = \psi_1 + x/r_1)\, e^{-imNx/r_1} dx$$

Hence the harmonics of the pressure distribution on the rotor disk can be obtained from an integral of the blade pressure distribution over the chord. Separating the blade loading into the section lift and a chordwise distribution factor, write $\Delta p = L(r, \psi)\ell(x, \psi)$. For steady loading, L and ℓ are independent of ψ, so

$$\Delta p_m = \frac{N}{2\pi r_1} e^{-imN\Omega t_0} L(r_1)\ell_{mN}$$

where

$$\ell_{mN} = \int_{x_{le}}^{x_{te}} \ell(x)\, e^{-imNx/r_1} dx$$

Thus even a steady lift on the rotating blade produces an unsteady pressure on the air in the nonrotating frame. For the general case of periodic blade loading,

$$L = \sum_{n=-\infty}^{\infty} L_n(r)\, e^{in\psi},$$

the lift still factors out of the chordwise integration, giving

$$\Delta p_m = \frac{N}{2\pi r_1} e^{-imN\Omega t_0} \sum_{n=-\infty}^{\infty} L_n(r_1)\, e^{in\psi_1}\, \ell_{mN-n}(\psi_1)$$

where now

$$\ell_{mN} = \int_{x_{le}}^{x_{te}} \ell(x, \psi{=}\psi_1 + x/r_1)\, e^{-imNx/r_1}\, dx$$

$$\cong \int_{x_{le}}^{x_{te}} \ell(x, \psi_1)\, e^{-imNx/r_1}\, dx$$

The factors ℓ_{mN} account for the influence of the blade chordwise loading distribution on the pressure at a disk point. If the chordwise loading factor $\ell(x, \psi)$ is independent of the blade azimuth position, then so is ℓ_{mN}; ℓ_{mN} always varies with the radial station, however. Impulsive chordwise loading, $\Delta p = L(r, \psi)\delta(x)$, gives

$$\Delta p_m = \frac{N}{2\pi r_1} L(r_1, \psi_1)\, e^{-imN\Omega t_0}$$

So the disk loading at the fixed point (r_1, ψ_1) is then simply an N/rev impulse of strength $NL/2\pi r_1$, as expected.

Consider the chordwise distribution factor ℓ_n for some specific cases. Note that in general

$$\ell_0 = \int_{x_{te}}^{x_{le}} \ell(x)dx = 1$$

by the definition of $\ell(x)$; and for small nx/r, $\ell_n \cong 1 - inx_{CP}/r$, where $x_{CP} = \int \ell x dx$ is the location of the center of pressure. For impulsive loading or in the limit of small chord, $\ell(x) = \delta(x)$, so that $\ell_n = 1$ for all n. A rectangular distribution of pressure over the blade chord, $\ell(x) = c^{-1}$, gives

$$\ell_n = e^{-inx_m/r} \frac{2r}{cn} \sin \frac{cn}{2r}$$

where x_m is the coordinate of the blade midchord. A loading distribution more approporate for the lift on an airfoil is

$$\ell(x) = \frac{2}{\pi c} \sqrt{\frac{1-\xi}{1+\xi}}$$

from thin airfoil theory $[\xi = 2(x - x_m)/c]$, which gives

$$\ell_n = e^{-inx_m/r} \left[J_0\left(\frac{nc}{2r}\right) + iJ_1\left(\frac{nc}{2r}\right) \right]$$

where J_0 and J_1 are Bessel functions. Note that when the loading is distributed over a finite chord, the factors ℓ_n are reduced in magnitude relative to the impulsive loading result $\ell_n = 1$. The approximation of impulsive loading is useful because it is simple and conservative, but for best accuracy the actual chordwise distribution must be accounted for.

17-3.2 Hovering Rotor with Steady Loading

Consider the rotational noise of a hovering rotor. The rotor disk is stationary in space, and with axisymmetric flow the loading on the blades is steady. The section aerodynamic forces are $F_z(r)$ normal to the disk plane and $F_x(r)$ in the disk plane. For steady loading it is convenient to write F_z and F_x in terms of the rotor thrust and torque:

$$F_z = \frac{1}{N} \frac{dT}{dr}$$

$$F_x = \frac{1}{Nr} \frac{dQ}{dr}$$

(since the sound is due to the pressure reaction on the fluid, the rotor profile torque should not be included in Q). Following section 17-3.1, the corresponding pressure distribution on the rotor disk in the nonrotating frame is

$$g_z = \sum_{m=-\infty}^{\infty} \frac{N}{2\pi r} F_z \ell_{mN} e^{imN(\Omega t - \psi)}$$

$$g_x = \sum_{m=-\infty}^{\infty} \frac{N}{2\pi r} F_x \ell_{mN} e^{imN(\Omega t - \psi)}$$

Here g_z and g_x are the normal and in-plane components of the pressure acting at the point (r, ψ) on the disk (g_x is still in the chordwise direction along the blade). Note that the initial time $t_0 = 0$ has been used, so the reference blade azimuth is $\psi = 0$ at $t = 0$. The normal and in-plane pressures would have different chordwise distributions in general, but for now impulsive loading is assumed, which gives $\ell_{mN} = 1$. Considering just the mth harmonic, the pressure on the rotor disk is thus

$$g_z = \frac{1}{2\pi r} \frac{dT}{dr} e^{imN(\Omega t - \psi)}$$

$$g_x = \frac{1}{2\pi r^2} \frac{dQ}{dr} e^{imN(\Omega t - \psi)}$$

The components of the pressure in the rectangular coordinate system (Fig. 17-3) are then $G_x = -g_x \sin\psi$, $G_y = g_x \cos\psi$, and $G_z = -g_z$. This pressure is equivalent to a distribution of dipoles over the surface of the rotor disk.

The dipole solution for the sound pressure p detected at an observer point (x, y, z) and due to a concentrated force with components G_x, G_y, and G_z at the point (x_1, y_1, z_1) is

$$p = -\frac{1}{4\pi}\left[\frac{\partial}{\partial x}\frac{G_x(t-s/c_s)}{s} + \frac{\partial}{\partial y}\frac{G_y(t-s/c_s)}{s} + \frac{\partial}{\partial z}\frac{G_z(t-s/c_s)}{s}\right]$$

where s is the distance from the dipole to the observer,

$$s^2 = (x-x_1)^2 + (y-y_1)^2 + (z-z_1)^2,$$

and $t - s/c_s$ is the retarded time which accounts for the finite time s/c_s required for a sound wave emitted at the source to travel to the observer. In the present case, the forces acting on the rotor disk are periodic with fundamental frequency $N\Omega$. It follows that the sound pressure is also periodic:

$$p = \sum_{m=-\infty}^{\infty} p_m e^{imN\Omega t}$$

Since G_x, G_y, and G_z have the time dependence $e^{imN\Omega t}$, the mth harmonic of the sound pressure is

$$p_m = -\frac{1}{4\pi}\left[G_x \frac{\partial}{\partial x} + G_y \frac{\partial}{\partial y} + G_z \frac{\partial}{\partial z}\right]\frac{e^{-iks}}{s}$$

where $k = mN\Omega/c_s$, and the factor e^{-iks} is due to the retarded time. The force is at the disk point $x_1 = r\cos\psi$, $y_1 = r\sin\psi$, and $z_1 = 0$, so

$$-\left[G_x \frac{\partial}{\partial x} + G_y \frac{\partial}{\partial y}\right]\frac{e^{-iks}}{s} = g_x\left[\sin\psi \frac{\partial}{\partial x} - \cos\psi \frac{\partial}{\partial y}\right]\frac{e^{-iks}}{s}$$

$$= g_x\left[-\sin\psi \frac{\partial}{\partial x_1} + \cos\psi \frac{\partial}{\partial y_1}\right]\frac{e^{-iks}}{s}$$

$$= g_x \frac{1}{r}\frac{\partial}{\partial \psi}\frac{e^{-iks}}{s}$$

Then integrating over the rotor disk area $dS = rdrd\psi$ gives the total mth harmonic of the rotational noise:

$$p_m = \frac{1}{8\pi^2}\int_0^{2\pi}\int_0^R\left[\frac{dT}{dr}\frac{\partial}{\partial z}\frac{e^{-iks}}{s} + \frac{1}{r^2}\frac{dQ}{dr}\frac{\partial}{\partial \psi}\frac{e^{-iks}}{s}\right]e^{-imN\psi}\,dr\,d\psi$$

Now

$$\frac{\partial}{\partial z}\frac{e^{-iks}}{s} = \frac{e^{-iks}}{s}\left[-\frac{ikz}{s} - \frac{z}{s^2}\right]$$

and the torque term is integrated by parts with respect to ψ, giving

$$p_m = -\frac{imN\Omega}{8\pi^2 c_s}\int_0^{2\pi}\int_0^R\left[\frac{dT}{dr}\frac{z}{s}\left(1 - \frac{i}{ks}\right) - \frac{c_s}{\Omega r^2}\frac{dQ}{dr}\right]\frac{e^{-iks}}{s}\,e^{-imN\psi}\,dr\,d\psi$$

The rotational noise has a discrete spectrum, with harmonics at the frequencies $mN\Omega$. The sound pressure level is obtained by summing the contributions from all harmonics:

$$p_{rms}^2 = \frac{N\Omega}{2\pi} \int_0^{2\pi/N\Omega} p^2\, dt = \sum_{m=-\infty}^{\infty} |p_m|^2$$

The above spectrum is two sided, extending from $m = -\infty$ to $m = \infty$. It is also common to work with a one-sided spectrum, defined for $m > 0$ only. Since p_m and p_{-m} are complex conjugates, for the same sound pressure level p_{rms}^2 the one-sided spectrum is obtained by multiplying the harmonics of the two-sided spectrum by $\sqrt{2}$:

$$|p_m| = \frac{mN\Omega}{4\sqrt{2}\pi^2 c_s} \left| \int_0^{2\pi} \int_0^R \left[\frac{dT}{dr} \frac{z}{s}\left(1 - \frac{i}{ks}\right) - \frac{c_s}{\Omega r^2} \frac{dQ}{dr} \right] \frac{e^{-iks}}{s} e^{-imN\psi}\, dr\, d\psi \right|$$

This is the rotational noise spectrum of a hovering rotor with steady thrust and torque loading.

The far field approximation allows the integration over the rotor azimuth to be evaluated analytically. Assume that the observer is far from the rotor, so that $s \gg R$. Then to order R/s_0,

$$\begin{aligned} s &= \sqrt{z^2 + (x - r\cos\psi)^2 + (y - r\sin\psi)^2} \\ &\cong s_0 - \frac{xr\cos\psi}{s_0} - \frac{yr\cos\psi}{s_0} \end{aligned}$$

where $s_0 = (z^2 + x^2 + y^2)^{1/2}$ is the distance from the hub to the observer. For the rotor thrust term the following approximation is made:

$$\frac{z}{s}\left(1 - \frac{1}{ks}\right) \cong \frac{z}{s_0}$$

which requires not just that $s \gg R$, but also that s be much greater than the wavelength of the sound, $ks \gg 1$. Since $k = mN\Omega/c_s$, this criterion can be written as $s/R \gg 1/(mNM_{tip})$. Since M_{tip} is of order 1 for helicopter rotors, the criterion reduces to $s/R \gg 1$ again. Furthermore, we make the approximation

$$\frac{1}{s} e^{-iks} \cong \frac{1}{s_0} e^{-iks_0} e^{ikr(x/s_0)\cos\psi}$$

The last factor accounts for the difference in retarded time over the rotor disk; note that it is the only influence of the order R/s_0 term in s to be retained. Since the hovering rotor is axisymmetric, it has been assumed that the observer is at $y = 0$, in the x-z plane. The sound pressure in the far field is thus

$$p_m = -\frac{imN\Omega e^{-iks_0}}{8\pi^2 c_s s_0} \int_0^R \left[\frac{dT}{dr}\frac{z}{s_0} - \frac{c_s}{\Omega r^2}\frac{dQ}{dr}\right] \int_0^{2\pi} e^{ikr(x/s_0)\cos\psi - imN\psi} \, d\psi \, dr$$

The integral over ψ can be evaluated in terms of Bessel functions using the relation

$$\int_0^{2\pi} e^{iz\cos\psi - in\psi} \, d\psi = 2\pi i^n J_n(z)$$

Hence the far field sound pressure is

$$p_m = -\frac{imN\Omega e^{-imN\Omega s_0/c_s} i^{mN}}{4\pi c_s s_0} \int_0^R \left[\frac{dT}{dr}\frac{z}{s_0} - \frac{c_s}{\Omega r^2}\frac{dQ}{dr}\right] J_{mN}\left(\frac{mN\Omega r}{c_s}\frac{x}{s_0}\right) dr$$

The far field approximation eliminates the integration over ψ, and introduces the Bessel functions. In terms of the elevation angle θ_0 of the observer above the disk plane (see Fig. 17-3) $z = s_0 \sin\theta_0$, so

$$p_m = -\frac{imN\Omega e^{-imN\Omega s_0/c_s} i^{mN}}{4\pi c_s s_0} \int_0^R \left[\frac{dT}{dr}\sin\theta_0 - \frac{c_s}{\Omega r^2}\frac{dQ}{dr}\right] J_{mN}\left(\frac{mN\Omega r}{c_s}\cos\theta_0\right) dr$$

The far field approximation is usually valid beyond four or five rotor radii from the rotor hub.

As a further approximation, evaluate the integrand at an effective radius r_e, which is equivalent to assuming that the loading is all concentrated at r_e. Then the integration over the blade span can be eliminated, giving for the one-sided spectrum

$$|p_m| = \frac{mN\Omega}{2\sqrt{2}\pi c_s s_0}\left[T\sin\theta_0 - \frac{c_s}{\Omega r_e^2}Q\right]J_{mN}\left(\frac{mN\Omega r_e}{c_s}\cos\theta_0\right)$$

or in terms of the tip Mach number $M_{tip} = \Omega R/c_s$,

$$|p_m| = \frac{mNM_{tip}}{2\sqrt{2}\pi R s_0}\left[T\sin\theta_0 - \frac{R}{M_{tip}r_e^2}Q\right]J_{mN}\left(mNM_{tip}\frac{r_e}{R}\cos\theta_0\right)$$

Using $r_e = 0.8R$ for the effective radius is generally satisfactory.

The far field sound pressure level p_{rms}^2 is proportional to s_0^{-2}, as required for energy conservation. The rotational noise due to the thrust is zero on the rotor axis, where $\cos\theta_0 = 0$ and therefore $J_{mN} = 0$, and in the disk plane, where $\sin\theta_0 = 0$. The sound thus has a broad maximum at an angle between the disk plane and rotor axis, typically around $\theta_0 = \pm 30°$. The noise due to the torque is zero on the rotor axis, with the same phase as the thrust noise below the disk ($\sin\theta_0 < 0$) and the opposite phase above the disk. The sound is thus greatest below the disk plane. For the helicopter rotor the effect of the torque term is small ($Qc_s/\Omega R^2 T \ll 1$), however, except near the disk plane, where the thrust noise is small. The above results show that the rotational noise of the hovering rotor has the functional form

$$|p_m|^2 = \frac{T^2/A}{s_0^2} f(mN, M_{tip}, \theta_0)$$

Thus the noise is proportional to the rotor thrust and disk loading. Because of the Bessel function behavior, the notational noise harmonics fall off rapidly with harmonic number m (for steady loading of the blades). It also follows that increasing the number of blades N will reduce the rotational noise harmonics, besides increasing the fundamental frequency $N\Omega$. At a constant thrust coefficient C_T, the sound pressure level is proportional to $(\Omega R)^6$ (neglecting the effect of the Bessel function). Hence the rotational noise will increase with about the sixth power of the tip speed or tip Mach number.

To account for the actual chordwise distribution of the blade loading it is simply necessary to introduce in the rotational noise expressions above the factor

$$\ell_{mN} = \int_{x_{le}}^{x_{te}} \ell(x)\, e^{-imNx/r}\, dx$$

inside the integral over the blade span (since ℓ_{mN} depends on r). Garrick and Watkins (1954) compared ℓ_{mN} for several elementary distributions. In section 17-3.1, expressions are given for rectangular and thin-airfoil chordwise pressure distributions. In general it is found that the factors ℓ_{mN} decrease in magnitude as $mNc/2r$ increases. Above about the 10th harmonic the reduction in the rotational noise by the chordwise pressure distribution is significant. If a rectangular pressure distribution is assumed the rotational noise will be underpredicted, however, since the harmonics will be reduced proportionally to $2r/cmN$ at high harmonic number. With the more appropriate thin airfoil theory pressure distribution, ℓ_{mN} decreases much more slowly, proportionally to $(2r/cmN)^{1/2}$ at high harmonic number. In a numerical evaluation of the rotational noise, the factors ℓ_{mN} can be easily incorporated.

Gutin (1948) developed a theory for the rotational noise of propellers, obtaining the above result for the far field noise of a stationary rotor with steady loading. Deming (1940) extended Gutin's result by considering the distribution of the loading over the blade span, instead of using an effective radius. Hubbard and Regier (1950) extended Gutin's analysis to include the near field noise calculation as well. They found that the far field result significantly underestimates the sound pressure in the near field, which extends to about $5R$ from the rotor hub.

The Gutin result is generally accurate for the rotational noise of static propellers. Reasonable agreement with the measured noise is found for the first few harmonics, and it gives an adequate estimate of the overall sound pressure level. For the rotor of a hovering helicopter it is greatly in error. Stuckey and Goddard (1967) found that the Gutin result significantly underestimated all but the first harmonic of the rotational noise of a hovering rotor, although the trends with tip speed and thrust were given correctly. They found that eliminating the effective radius and far field approximations by integrating numerically over the rotor disk did not improve the predictions. The predicted rotational noise falls off rapidly with harmonic number, while the measured noise falls slowly or is even constant, apparently because of periodic blade loads that exist in even a nominally hovering condition. Schlegel, King, and Mull (1966) found the rotational noise predicted by the Gutin theory was about 4 dB low for the fundamental harmonic of a helicopter rotor, and deteriorated rapidly with harmonic number; see also Schlegel and Bausch (1969). Ollerhead and Lowson (1969) concluded that the Gutin theory significantly underestimates the rotational noise of a helicopter rotor because of the neglect of the higher harmonic airloads.

17-3.3 Vertical Flight and Steady Loading

Consider a helicopter rotor in vertical flight, still with steady blade loading. For a helicopter, the Mach number of the vertical velocity, $M = V_c/c_s = \lambda_c M_{\text{tip}}$, is small. The observer is assumed to be moving with the rotor. The rotor is represented by a distribution of forces over the disk, but now it is necessary to use the solution for the sound generated by a dipole moving with uniform velocity in the z-direction:

$$p = -\frac{1}{4\pi}\left[\frac{\partial}{\partial x}\frac{G_x(t-\sigma/c_s)}{S} + \frac{\partial}{\partial y}\frac{G_y(t-\sigma/c_s)}{S} + \frac{\partial}{\partial z}\frac{G_z(t-\sigma/c_s)}{S}\right]$$

where

$$S^2 = (z-z_1)^2 + \beta^2(x-x_1)^2 + \beta^2(y-y_1)^2$$

$$\sigma = \frac{1}{\beta^2}\left[S + M(z-z_1)\right]$$

and $\beta^2 = 1 - M^2$. The mth harmonic of the rotational noise is then

$$p_m = -\frac{1}{4\pi}\left[G_x\frac{\partial}{\partial x} + G_y\frac{\partial}{\partial y} + G_z\frac{\partial}{\partial z}\right]\frac{e^{-ik\sigma}}{S}$$

$$= \frac{1}{4\pi}\left[g_z\frac{\partial}{\partial z} + g_x\frac{1}{r}\frac{\partial}{\partial\psi}\right]\frac{e^{-ik\sigma}}{S}$$

where $k = mN\Omega/c_s$. The total sound is obtained by integrating over the rotor disk area. Using

$$\frac{\partial}{\partial z}\frac{e^{-ik\sigma}}{S} = \frac{e^{-ik\sigma}}{S}\left(-\frac{ikM}{\beta^2} - \frac{ikz}{\beta^2 S} - \frac{z}{S^2}\right)$$

and integrating the torque term by parts with respect to ψ gives the sound pressure for the rotor with steady loading in axial flight:

$$p_m = -\frac{imN\Omega}{8\pi^2 c_s}\int_0^{2\pi}\int_0^R\left[\frac{dT}{dr}\left(\frac{M}{\beta^2}+\frac{z}{\beta^2 S}-\frac{iz}{kS^2}\right)-\frac{c_s}{\Omega r^2}\frac{dQ}{dr}\right]\frac{e^{-ik\sigma}}{S}\,e^{-imN\psi}\,dr\,d\psi$$

When $M = 0$, this reduces to the result of the last section (multiply by $\sqrt{2}$ to obtain the one-sided spectrum).

The far field approximation now gives

$$S \cong S_0 - \frac{\beta^2 x r\cos\psi}{S_0} - \frac{\beta^2 y r\sin\psi}{S_0}$$

where $S_0^2 = z^2 + \beta^2 x^2 + \beta^2 y^2$. Assuming the observer is at $y = 0$, then

$$\frac{e^{-ik\sigma}}{S} \cong \frac{e^{-ik\sigma_0}}{S_0}\,e^{ikr(x/S_0)\cos\psi}$$

and the far field sound pressure is

$$p_m = -\frac{imN\Omega i^{mN}e^{-imN\Omega\sigma_0/c_s}}{4\pi c_s S_0}\int_0^R\left[\frac{dT}{dr}\left(M+\frac{z}{S_0}\right)\frac{1}{\beta^2} - \frac{c_s}{\Omega r^2}\frac{dQ}{dr}\right]J_{mN}\left(\frac{mN\Omega r}{c_s}\frac{x}{S_0}\right)dr$$

Now in terms of the range and elevation angle of the observer (s_0 and θ_0 in Fig. 17-3), $z = s_0 \sin\theta_0$ and $x = s_0\cos\theta_0$. Hence $S_0^2 = s_0^2(1 - M^2\cos^2\theta_0)$ and then

$$p_m = -\frac{imN\Omega i^{mN}e^{-imN\Omega\sigma_0/c_s}}{4\pi c_s s_0\sqrt{1-M^2\cos^2\theta_0}}\int_0^R\left[\frac{dT}{dr}\left(M+\frac{\sin\theta_0}{\sqrt{1-M^2\cos^2\theta_0}}\right)\frac{1}{\beta^2} - \frac{c_s}{\Omega r^2}\frac{dQ}{dr}\right]J_{mN}\left(\frac{mN\Omega r}{c_s}\frac{\cos\theta_0}{\sqrt{1-M^2\cos^2\theta_0}}\right)dr$$

The principal influence of the rotor axial velocity is an order M increase in the thrust-induced noise above the disk when the helicopter is climbing.

There is also a small (order M^2) increase in the magnitude of the noise since $S_0 < s_0$, and a shift in the directivity pattern.

Garrick and Watkins (1954) derived the rotational noise of a rotor in axial flight. They were actually considering a propeller in forward flight, for which the axial Mach number can be large. Watkins and Durling (1956) extended the analysis to include more general chordwise loading distributions. Van de Vooren and Zandbergen (1963) examined a rotor in vertical flight and analyzed the rotational noise due to the blade lift and thickness. However, they considered elementary dipoles and sources moving with the blade in helical paths, instead of a distribution of dipoles over the rotor disk as in the present derivation.

17-3.4 Stationary Rotor with Unsteady Loading

Consider next a stationary rotor with unsteady loading. The helicopter rotor in forward flight has periodic aerodynamic forces acting on the blades. If the effects of the helicopter translation on the sound radiation are neglected, the present model is obtained. In any case, it is useful to examine separately the influence of unsteady loads before the effects of the forward motion are included as well. Moreover, the existence of unsteady loads in a nominally hovering condition may be responsible for the difficulties in predicting the higher harmonics of helicopter rotor rotational noise (although such loading would not be truly periodic, as is considered here).

Assuming impulsive chordwise loading of the blades, section 17-3.1 gives the pressure distribution on the rotor disk as

$$g_z = \sum_{m=-\infty}^{\infty} \frac{N}{2\pi r} L(r, \psi) e^{imN(\Omega t - \psi)}$$

where $L(r, \psi)$ is the blade section lift, now depending on the blade azimuth as well as the radial station. The sound due to in-plane blade forces will no longer be included in the analysis, since it is small compared to the sound due to the lift. Using the stationary dipole solution, the analysis proceeds just as in section 17-3.2 to obtain the mth harmonic of the sound pressure,

$$p_m = -\frac{imN^2\Omega}{8\pi^2 c_s}\int_0^{2\pi}\int_0^R L(r,\psi)\frac{z}{s}\left(1-\frac{i}{ks}\right)\frac{e^{-iks}}{s}\,e^{-imN\psi}\,dr\,d\psi$$

The section lift is a periodic function of ψ, so

$$L = \sum_{n=-\infty}^{\infty} L_n(r)\,e^{in\psi}$$

For the far field approximation,

$$s \cong s_0 - \frac{r}{s_0}(x\cos\psi + y\sin\psi) = s_0 - r\cos\theta_0\cos(\psi-\psi_0)$$

where ψ_0 is the azimuth angle of the observer (see Fig. 17-3; the loading is not axisymmetric, so it is not possible to examine as in section 17-3.2 only observer points in the x-z plane, where $\psi_0 = 0$). The rotational noise in the far field is then

$$p_m = -\frac{imN^2\Omega\sin\theta_0\,e^{-imN\Omega s_0/c_s}}{8\pi^2 c_s s_0}\sum_{n=-\infty}^{\infty}\int_0^R L_n$$

$$\int_0^{2\pi} e^{ikr\cos\theta_0\cos(\psi-\psi_0)-imN\psi+in\psi}\,d\psi\,dr$$

$$= -\frac{imN^2\Omega\sin\theta_0\,e^{-imN\Omega s_0/c_s}}{4\pi c_s s_0}\sum_{n=-\infty}^{\infty} e^{i(n-mN)[\psi_0-(\pi/2)]}$$

$$\int_0^R L_n J_{mN-n}\left(\frac{mN\Omega r}{c_s}\cos\theta_0\right)dr$$

So every harmonic of the blade loading contributes to the mth harmonic of the sound pressure. In particular, it is found that the maximum sound produced by the loading harmonic L_n occurs at the harmonic $mN = n$. The higher harmonic loading contributes substantially more than the mean loading to the high frequency of the rotational noise. It also follows that to predict the rotational noise, which is significant up to $m = 20$ or 30, accurate data (measured or calculated) for the blade loads to very high harmonic number are required. At such high harmonics a deterministic loading does not really exist, however. A combined analysis of rotational and broadband noise is required to properly calculate the high frequency noise.

17-3.5 Forward Flight and Steady Loading

Consider next the helicopter rotor in steady forward flight at advance ratio μ. The higher harmonic loads on the blade are large, and important to the rotational noise, but for the moment only the mean loading will be considered in order to examine the influence of forward motion of the rotor on the sound radiation. The rotor disk is represented by a distribution of vertical dipoles moving in the negative x-direction with Mach number $M = \mu M_{tip}$. The observer is moving with the rotor velocity also. The sound pressure due to a moving vertical dipole is

$$p = -\frac{1}{4\pi}\frac{\partial}{\partial z}\frac{G_z(t-\sigma/c_s)}{S}$$

where here

$$S^2 = (x-x_1)^2 + \beta^2(y-y_1)^2 + \beta^2(z-z_1)^2$$

$$\sigma = \frac{1}{\beta^2}[S - M(x-x_1)]$$

and $\beta^2 = 1 - M^2$. The mth harmonic of the pressure is

$$p_m = -\frac{1}{4\pi} G_z \frac{\partial}{\partial z}\frac{e^{-ik\sigma}}{S}$$

$$= \frac{1}{4\pi} G_z \frac{e^{-ik\sigma}}{S}\left(\frac{ikz}{S} + \frac{\beta^2 z}{S^2}\right)$$

where

$$G_z = -\frac{1}{2\pi r}\frac{dT}{dr}\, e^{imN(\Omega t-\psi)}$$

Integrated over the rotor disk, the sound pressure due to the rotor thrust in forward flight is

$$p_m = -\frac{imN\Omega}{8\pi^2 c_s} \int_0^{2\pi} \int_0^R \frac{dT}{dr} \frac{z}{S}\left(1 - \frac{\beta^2 i}{kS}\right) \frac{e^{-ik\sigma}}{S} e^{-imN\psi}\, dr\, d\psi$$

The far field approximation here gives

$$S \cong S_0 - \frac{xr\cos\psi}{S_0} - \frac{\beta^2 yr\sin\psi}{S_0}$$

so in the retarded time

$$\sigma \cong \sigma_0 - \frac{1}{\beta^2}\left(\frac{x}{S_0} - M\right) r\cos\psi - \frac{y}{S_0}\, r\sin\psi$$

where $\sigma_0 = (S_0 - Mx)/\beta^2$. Note that

$$\frac{1}{\beta^2}\left(\frac{x}{S_0} - M\right) = \frac{x - M\sigma_0}{S_0}$$

and

$$\sqrt{\left(\frac{x - M\sigma_0}{\sigma_0}\right)^2 + \left(\frac{y}{\sigma_0}\right)^2} = \sqrt{1 - \left(\frac{z}{\sigma_0}\right)^2}$$

Then

$$\sigma \cong \sigma_0 - \sqrt{\left(\frac{x - M\sigma_0}{S_0}\right)^2 + \left(\frac{y}{S_0}\right)^2}\; r\cos(\psi - \psi_r)$$

$$= \sigma_0 - \frac{\sigma_0}{S_0}\, r\cos\theta_r \cos(\psi - \psi_r)$$

where $\psi_r = \tan^{-1} y/(x - M\sigma_0)$ and $\theta_r = \sin^{-1} z/\sigma_0$. Here σ_0 is the distance from the rotor hub to the observer at the time the sound was emitted, i.e. at the retarded time $t - \sigma_0/c_s$ (if the observer had been stationary while the rotor moved). Hence θ_r and ψ_r are the elevation and azimuth angle of the (fixed) observer at the retarded time. Finally, we can write

$$S_0 = \beta^2\sigma_0 + Mx = \sigma_0\left(1 - M\,\frac{-x + M\sigma_0}{\sigma_0}\right) = \sigma_0(1 - M\cos\delta_r)$$

where δ_r is the angle between the observer and the forward velocity of the rotor at the retarded time, so that $M \cos \delta_r$ is the Mach number of the forward speed component in the direction of the observer. The sound pressure in the far field is thus

$$p_m = -\frac{imN\Omega \sin \theta_r \, e^{-imN\Omega\sigma_0/c_s - imN[\psi_r - (\pi/2)]}}{4\pi c_s \sigma_0 (1 - M \cos \delta_r)^2} \int_0^R \frac{dT}{dr} J_{mN}\left(\frac{mN\Omega r \cos \theta_r}{c_s (1 - M \cos \delta_r)}\right) dr$$

For the stationary rotor ($M = 0$) the elevation angle $\theta_r = \theta_0$, and the result of section 17-3.2 is recovered.

Recall that $\sin \theta_r / \sigma_0 (1 - M \cos \delta_r)^2 = z/S_0^2$, so in forward flight there is an increase in the magnitude of the rotational noise harmonics because $S_0 < s_0$. The effect of the $(1 - M \cos \delta_r)$ factor in the argument of the Bessel function is to increase the sound radiated forward of the rotor, and decrease the sound behind the rotor. Comparing the present result for the rotor in forward flight with the hovering rotor result in section 17-3.2, we note that

$$|p_m|_{\substack{\text{forward}\\ \text{flight}}} = |p_m|_{\text{hover}}$$

if the hover expression is evaluated at the range $S_0 = \sigma_0 (1 - M \cos \delta_r)$ and the elevation θ_r, and if an effective Mach number

$$M_{\text{eff}} = \frac{M_{\text{tip}}}{1 - M \cos \delta_r}$$

is used in the hover expression. Lowson and Ollerhead (1969a) suggest that this relationship can be used to obtain with reasonable accuracy an estimate of the noise in forward flight from a calculation (or measurement) of the hover noise, which must of course be calculated using the unsteady loading of the actual forward flight condition. (The approximation lies in applying this relationship to the noise due to the unsteady loading as well as the mean loading.) Ahead of the rotor the effective tip Mach number is greater than M_{tip}, so the noise is increased, and behind the rotor $M_{\text{eff}} < M_{\text{tip}}$, so the noise is decreased. It is probably consistent with this approximation to neglect the retarded time in evaluating the range and elevation of the observer.

17-3.6 Forward Flight and Unsteady Loading

Now let us consider the case of a helicopter rotor in forward flight, with periodic loading on the rotating blades. With impulsive chordwise loading, the normal pressure distribution on the rotor disk is the same as with steady loading:

$$G_z = -\frac{N}{2\pi r} L(r, \psi)\, e^{imN(\Omega t - \psi)}$$

except that now the section lift L varies with ψ. Hence the spectrum of the rotational noise in forward flight is

$$p_m = -\frac{imN^2\Omega}{8\pi^2 c_s} \int_0^{2\pi} \int_0^R L \frac{z}{S}\left(1 - \frac{\beta^2 i}{kS}\right) \frac{e^{-ik\sigma}}{S}\, e^{-imN\psi}\, dr\, d\psi$$

as in the last section. When the section lift is written as a Fourier series,

$$L = \sum_{n=-\infty}^{\infty} L_n(r)\, e^{in\psi},$$

the far field sound pressure is

$$p_m = -\frac{imN^2\Omega}{4\pi c_s} \frac{\sin\theta_r}{\sigma_0(1 - M\cos\delta_r)^2} e^{-imN\Omega\sigma_0/c_s} \sum_{n=-\infty}^{\infty} e^{-i(mN-n)[\psi_r - (\pi/2)]} \int_0^R L_n J_{mN-n}\left(\frac{mN\Omega r\cos\theta_r}{c_s(1 - M\cos\delta_r)}\right) dr$$

Accounting for the actual chordwise pressure distribution gives instead

$$G_z = -\frac{N}{2\pi r} \sum_{m=-\infty}^{\infty} L_n \ell_{mN-n}\, e^{imN(\Omega t - \psi) + in\psi}$$

(see section 17-3.1) so the factor ℓ_{mN-n} must be included in the spanwise integration to evaluate the far field noise (assuming the chordwise distribution factor ℓ is independent of ψ). Alternatively, the result of section 17-3.1 for the mth harmonic of the disk pressure can be used directly in the form

$$G_z = -\frac{N}{2\pi r} e^{imN(\Omega t-\psi)} \int_{x_{le}}^{x_{te}} \Delta p(r,x,\psi+x/r)\, e^{-imNx/r}\, dx$$

$$= -\frac{N}{2\pi r} e^{imN(\Omega t-\psi)} \sum_{n=-\infty}^{\infty} G_{z_n} e^{in\psi}$$

where

$$G_{z_n} = \frac{1}{2\pi}\int_0^{2\pi} e^{-in\psi} \int_{x_{le}}^{x_{te}} \Delta p(r,x,\psi+x/r)\, e^{-imNx/r}\, dx\, d\psi$$

which can be evaluated numerically, given the actual variation of the pressure over the surface of the blade and around the azimuth. With this form, L_n in the present result for impulsive loading is simply replaced by G_{z_n}.

Loewy and Sutton (1966a, 1966b) developed a theory for helicopter rotor rotational noise in forward flight, including a treatment of unsteady airloads on the blades. They numerically integrated over the rotor disk to obtain the sound due to the dipole distribution at an arbitrary point in the near or far field. The blade flap motion and disk tilt were accounted for in determining the orientation of the dipoles. The rotor blade loading was assumed to be an input to the analysis. Simple chordwise distributions of the lift and drag were used, not impulsive loading. It was found that an azimuth increment of 1° or less was required in the numerical integration, and that the far field result significantly underestimates the noise in the near field. The principal effect of forward flight was to raise the level of the higher harmonics. The directivity was found to remain nearly axisymmetric. The correlation with measured noise was good for the low harmonics, but the predicted noise harmonics (based on measured loadings) fell off rapidly with harmonic number while the measured values did not.

Schlegel, King, and Mull (1966) developed a theory for calculating helicopter rotor rotational noise in forward flight. They considered a stationary rotor (i.e. the stationary dipole solution), but included the unsteady airloads, as in section 17-3.4. The measured or calculated blade loading was assumed to be given, and a rectangular chordwise distribution of the lift was used. The sound pressure at an arbitrary field point was calculated, by numerically integrating over the rotor disk. When a comparison was made with flight test measurements of the rotational noise, it was found that the prediction of the first harmonic in forward flight had been improved (compared to predictions using the Gutin theory, which is accurate for the first harmonic in hover but underestimates the noise in forward flight). However, the prediction of the third, fourth, and higher harmonics was still poor. Schlegel and Bausch (1970) modified this analysis to use the actual chordwise loading distribution. Measured data for the pressure distribution over the rotating blade were converted to a distribution of pressure on the rotor disk, which was then harmonically analyzed. With this approach, good correlation with the measured noise was found up to at least the fourth harmonic, in both forward flight and hover. (Recall that with a rectangular distribution of the load over the chord, the factor ℓ_{mN} falls off in magnitude too fast.) They presented examples of the theoretical influence the higher harmonic airloads have on the noise, and concluded that at least mN harmonics of the loads were required to obtain the mth harmonic of the rotational noise. See also Schlegel and Bausch (1969).

Lowson and Ollerhead (1969a, 1969b) developed a theory for rotational noise of a rotor in forward flight, including the effects of unsteady airloads and the rotor motion. The derivation was based on the solution for the sound radiated by a rotating and translating dipole. The total rotational noise was calculated by representing the pressure distribution on the rotating blade as a distribution of such dipoles, and then integrating over the surface of the blade. Assuming an impulsive chordwise loading reduces the calculation to an integral over the blade span of the section lift, drag, and radial forces. The time history of the rotational noise was calculated for a single period and then harmonically analyzed. Lowson and Ollerhead also developed a simplified analysis that made the far field approximation and assumed a certain behavior of the higher harmonic airloads. An analytical evaluation of the integrals was then possible, and the computation required was reduced by about a factor of 100. They examined the mth harmonic of the sound pressure due to the nth harmonic of the rotor loading, and concluded that to calculate p_m the loading harmonics in the range

$$mN(1 - 0.8M_{\text{tip}}) < n < mN(1 + 0.8M_{\text{tip}})$$

are required. (However, the loading harmonics decrease rapidly in magnitude as n increases, reducing the number of harmonics required.) For high tip Mach numbers and a large number of blades, therefore, a large number of loading harmonics are required to predict the rotational noise, many more than are usually measured or calculated in investigations of the rotor aerodynamics. They concluded that the discrepancies in earlier analyses were due to the neglect of the very high harmonics of the loading, but that a practical calculation procedure should not rely on having such high frequency aerodynamic data, both to minimize the computation required and because the data are often not available. In their simplified analysis, they therefore considered impulsive chordwise loading (which is conservative); used an equivalent radius (i.e. the loads were concentrated at a single radial point as well, since their calculations indicated the noise was not too sensitive to the radial loading distribution); and from an examination of measured rotor loads assumed that the magnitude of the higher harmonics of the airloads varied with harmonic number n according to $F_n = F_0 n^{-k}$, where F_0 is the mean loading. The best correlation (for all outboard stations of the blade and for operating conditions from hover to $\mu = 0.2$) was obtained with this expression for $k = 2$, based on 10 harmonics of the airloading. There was some indication that $k = 1$ should be used for rough operating conditions. The spanwise variation of the loading was assumed to have a correlation length proportional to n^{-1}, which had the approximate effect of increasing the exponent of the loading law to $k = 2.5$, giving $F_n = F_0 n^{-2.5}$. Since no trend was observed in the phase of the measured loading, the phase was assumed to be random. It followed from this random characteristic that the square of the mth harmonic sound-pressure level was given by the sum of the squares of the contributions from each of the loading harmonics: $|p_m|^2 = \sum_n |p_{mn}|^2$. With this approximation for the higher harmonic loads, the calculated rotational noise was found to be proportional to $(T^2/A)(\Omega R)^2$ for the lower harmonics, and $p^2 \sim (T^2/A)(\Omega R)^{2N}$ at high frequency. The dependence on harmonic number was approximately $p^2 \sim (mN)^{2-2k}$. Little influence of the blade motion was found, or of the blade drag and radial forces (except in the plane of the disk). In hover there was a broad maximum of the rotational noise about 30° below the disk plane, and a distinct minimum about 10° above. Forward flight tended to shift this directivity pattern forward. Using their simplified analysis, Lowson and Ollerhead constructed design charts for estimating rotor rotational noise in the form

$$|p_m|^2 = \frac{T^2/A}{s_0^{\,2}} f(mN, M_{tip}, \theta_0)$$

The influence of forward flight was accounted for by using an effective Mach number $M_{eff} = M_{tip}/(1 - M \cos \delta_r)$ (see section 17-3.5). Lowson and Ollerhead found that the calculated and measured rotational noise agreed fairly well for the first harmonic. In forward flight the predicted levels of the higher harmonics tended to be lower than measured, but the shape of the spectrum from about $m = 3$ to $m = 30$ was given well. See also Ollerhead and Lowson (1969).

Wright (1968, 1969b) developed a theory for rotational noise that was based on the solution for a rotating dipole and included unsteady airloads but was applicable to a stationary rotor. He derived the mth harmonic of the sound pressure due to the nth blade lift harmonic, using the far field approximation, an effective radius, and impulsive chordwise loading. He concluded that the magnitude of p_m due to L_n is significant when

$$\frac{n}{1 + \frac{r_e}{R} M_{tip} \cos\theta_0} < mN < \frac{n}{1 - \frac{r_e}{R} M_{tip} \cos\theta_0}$$

(note that this is the range of the Doppler shift of the frequency, due to the rotor rotation). The magnitude of p_m is greatest at $mN = n$ and increases with harmonic number as $n^{1/2}$ because of the asymptotic behavior of the Bessel function. Examining experimental rotor airloads data, Wright concluded that the lift harmonics decrease with n, roughly as n^{-1}, giving a net reduction of the sound pressure harmonics with increasing frequency: $p_m \sim (mN)^{-1/2}$. Tanna (1970b) investigated the influence of the chordwise and spanwise loading distribution on the predicted rotational noise. The chordwise loading and its variation with azimuth had a significant influence, particularly on the higher harmonics of the noise. The spanwise loading and its variation with azimuth affected mainly the directivity of the rotational noise. See also Wright (1969a), Wright and Tanna (1969), and Tanna (1970a).

Arndt and Borgman (1971) developed a rotational noise theory including the effects of thickness and drag divergence due to compressibility. Measured rotor noise showed a large change in the spectrum and overall sound pressure level for small rotor or helicopter speed changes at high advancing-tip Mach number. Including the blade drag rise due to compressibility, the theory indeed predicted a sharp increase in the rotational noise when $M_{1,90}$ exceeded the critical Mach number of the section. With this compressibility effect, the correlation between the predicted and measured noise was improved, particularly for the higher harmonics. They also concluded that the noise could be reduced significantly at high Mach number by using a thin-tip blade.

Sternfeld, Bobo, Carmichael, Fukushima, and Spencer (1972) evaluated the rotational noise theory developed by Lowson and Ollerhead. They found that the predicted noise was sensitive to the assumed blade loading law. The best correlation was obtained with $F_n = F_0(n+1)^{-2}$. They also extended the Lowson and Ollerhead analysis to include a general radial load distribution, although the equivalent radius approximation was found to give reasonably accurate results ($r_e = 0.8R$ was best). Leverton (1973) compared rotational noise measurements with the predictions from the theory of Lowson and Ollerhead. He found that the directivity correlated well, but that the higher harmonics of the noise were greatly underpredicted.

17-3.7 Thickness Noise

The helicopter rotor produces periodic (rotational) noise due to the thickness of the blades as well as noise due to the pressure forces. By periodically pushing the air aside, each blade produces a pressure disturbance. The sound pressure is linearly related to the blade lift and thickness, so the rotational noise due to the two sources can be evaluated separately and then added. Consider therefore a rotor blade with finite thickness but no lift. A symmetric section is assumed, so that it is not necessary to consider the pressure forces due to camber, either. The thickness of the blade produces a velocity of the air normal to the section, first upward and then downward as the blade passes (considering only the air above the disk plane). This displacement velocity is determined by the boundary condition that the flow must be tangent to the airfoil surface. Let $v(r, x)$ be the velocity normal to the surface of the rotating blade, due to thickness. Such a velocity can be produced by a distribution of sources over the blade, with strength proportional to v. For simplicity it has been assumed that v is independent of ψ, although in forward flight it will be periodic since it is proportional to the blade tangential velocity u_T. A source is a fundamentally more effective radiator of sound than a dipole, but here the net source strength over the blade chord is zero (for a closed airfoil section), so the same order noise as from a dipole may be expected.

The sources on the rotating blade will now be transformed to a distribution of fixed sources on the rotor disk, as is required to give the periodic normal displacement velocity due to thickness. Following the results of section 17-3.1, the distribution of normal velocity over the disk is

$$v_1(r, \psi, t) = \sum_{m=-\infty}^{\infty} \frac{N}{2\pi r} v_m(r)\, e^{imN(\Omega t - \psi)}$$

where

$$v_m = \int_{x_{le}}^{x_{te}} v(r, x)\, e^{-imNx/r}\, dx$$

When a hovering rotor is considered, these sources on the rotor disk are stationary. The sound pressure due to a stationary but time-varying source of strength $\rho 2v_1 dA$ (considering both upper and lower surfaces of the rotor disk) is

$$p = \frac{\rho}{2\pi s}\frac{\partial}{\partial t} v_1(t - s/c_s)dA$$

On integrating over the rotor disk ($dA = r\,dr\,d\psi$), the mth harmonic of the sound becomes

$$p_m = \rho\frac{imN^2\Omega}{4\pi^2}\int_0^{2\pi}\int_0^R v_m(r)\frac{e^{-iks}}{s}e^{-imN\psi}\,dr\,d\psi$$

In the far field, $s \cong s_0 - r\cos\theta_0\cos(\psi - \psi_0)$ as usual, so

$$p_m = \rho \frac{imN^2\Omega}{4\pi^2 s_0} e^{-imN\Omega s_0/c_s} \int_0^R v_m \int_0^{2\pi} e^{-ikr\cos\theta_0\cos(\psi-\psi_0)-imN\psi}\, d\psi\, dr$$

$$= \rho \frac{imN^2\Omega}{2\pi s_0} e^{-imN\Omega s_0/c_s} e^{-imN[\psi_0-(\pi/2)]} \int_0^R v_m J_{mN}\left(\frac{mN\Omega r}{c_s}\cos\theta_0\right) dr$$

is the spectrum of the rotational noise due to thickness. This noise is zero on the rotor axis, because of the $\cos\theta_0$ factor in the Bessel function. Fig. 17-4 summarizes the directivity patterns of the rotor noise sources analyzed.

To complete the analysis of the thickness noise, it is necessary to evaluate v_m. For thin sections, the velocity normal to the airfoil is given by the slope of the surface times the free stream velocity:

$$v(r,x) = \Omega r \frac{1}{2}\frac{dt}{dx}$$

where $t(x)$ is the thickness of the section. Then

$$v_m = \int_{x_{le}}^{x_{te}} \Omega r \frac{1}{2}\frac{dt}{dx} e^{-imNx/r}\, dx$$

$$= imN\Omega \frac{1}{2} \int_{x_{le}}^{x_{te}} t\, e^{-imNx/r}\, dx$$

$$= imN\Omega \frac{A_{xs}}{2} a_{mN}$$

where

$$a_{mN} = \frac{1}{A_{xs}} \int_{x_{le}}^{x_{te}} t\, e^{-imNx/r}\, dx$$

and A_{xs} is the area of the blade cross-section. Substituting for v_m then gives

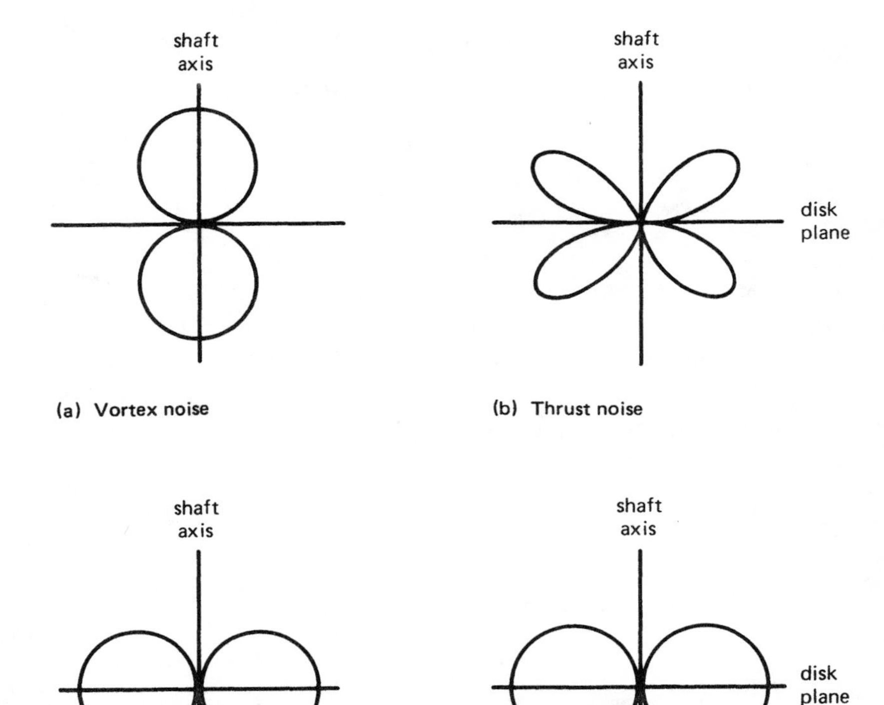

Figure 17-4. Schematic of the directivity patterns for rotor vortex noise, and for rotational noise due to thrust, torque, and thickness.

$$p_m = -\frac{\rho(mN\Omega)^2 N}{4\pi s_0} e^{-imN\Omega s_0/c_s} e^{-imN[\psi_0 - (\pi/2)]}$$

$$\int_0^R A_{xs}\, a_{mN}\, J_{mN}\left(\frac{mN\Omega r}{c_s}\cos\theta_0\right) dr$$

In general, a_{mN} can be evaluated for a given thickness distribution $t(x)$. For an impulsive thickness distribution (small chord), $t(x) = A_{xs}\delta(x)$, so $a_{mN} = 1$. Then only the section area is required. As an example, the area of the NACA 4- and 5-digit airfoils is $A_{xs} = 0.685\tau c^2$, where c is the chord and τ is the maximum thickness ratio (t_{max}/c). Generally, the thickness noise is small compared to the lift noise for the lower harmonics, but for high harmonic number and high tip Mach number the thickness noise can be large. It follows that the thickness noise consists of an impulsive type of pressure disturbance which is significant at high speeds.

Deming (1937, 1938) analyzed the rotational noise due to thickness, for a propeller at zero thrust. He found the theory predicted the propeller noise fairly well for the first five harmonics, and he found good agreement with the measured directivity. Diprose (1955) developed a theory for the lift and thickness noise of a propeller in axial flight.

17-3.8 Rotating Frame Analysis

The rotational noise analysis developed in the previous sections has been based on dipoles that were stationary or moving with constant velocity. It was therefore necessary to transform the distribution of pressure on the blade into an equivalent distribution of fixed dipoles on the rotor disk. An alternative approach is to use the pressures on the rotating blade directly, by using the solution for a rotating and translating dipole. The sound pressure due to dipoles and sources in arbitrary motion has been derived by Lowson (1965), Ffowcs Williams and Hawkings (1969), and Farassat (1974, 1975). Farassat's result takes the form

$$
\begin{aligned}
4\pi p(\vec{x}, t) = {} & \rho \frac{\partial}{\partial t} \int \left[\frac{v_n}{s(1 - M_r)} \right]_T dS(\vec{y}) \\
& + \frac{1}{c_s} \frac{\partial}{\partial t} \int \left[\frac{p\vec{n} \cdot (\vec{x} - \vec{y})}{s^2 (1 - M_r)} \right]_T dS(\vec{y}) \\
& + \int \left[\frac{p\vec{n} \cdot (\vec{x} - \vec{y})}{s^3 (1 - M_r)} \right]_T dS(\vec{y})
\end{aligned}
$$

where p is the sound pressure due to a moving body, detected at the observer location $\vec{x}$; $\vec{n}$ is the outward normal on the body; and

$$s^2 = |\vec{x} - \vec{y}|^2$$

$$\vec{M} = \frac{1}{c_s}\frac{d\vec{y}}{dt}$$

$$M_r = -\frac{1}{c_s}\frac{ds}{dt} = \vec{M}\cdot\frac{(\vec{x}-\vec{y})}{s}$$

$$\tau = t - \frac{s(\tau)}{c_s}$$

The integrals are over the surfaces of all N blades, with the integrands evaluated at the retarded time τ. The first term is the thickness noise, produced by the velocity normal to the surface, v_n. The second term is the lift noise due to the surface pressure p, and the third term is a near field lift noise. The time derivative can be applied to the integrand using

$$\frac{\partial}{\partial t}\int f(\tau)dS = \int \frac{\partial f/\partial\tau}{\partial t/\partial\tau}\,dS = \int\left[\frac{\partial f/\partial\tau}{1-M_r}\right]_\tau dS$$

since $\partial t/\partial\tau = \partial(\tau + s/c_s)/\partial\tau = 1 - M_r$, where M_r is the Mach number of the source velocity component in the direction of the observer. Then the far field noise is

$$4\pi p(\vec{x}, t) = \rho\int\left[\frac{1}{s(1-M_r)^2}\left(\frac{\partial v_n}{\partial t} + \frac{v_n}{1-M_r}\frac{\partial M}{\partial t}\right)\right]_\tau dS + \frac{1}{c_s}\int\left[\frac{\vec{n}\cdot(\vec{x}-\vec{y})}{s^2(1-M_r)^2}\left(\frac{\partial p}{\partial t} + \frac{p}{1-M_r}\frac{\partial M}{\partial t}\right)\right]_\tau dS$$

which is Lowson's result. A more convenient form of the far field noise for the present purposes is

$$4\pi p(\vec{x}, t) = \rho\int\left[\frac{1}{s(1-M_r)}\frac{\partial}{\partial\tau}\frac{v_n}{1-M_r}\right]_\tau dS + \frac{1}{c_s}\int\left[\frac{\vec{n}\cdot(\vec{x}-\vec{y})}{s^2(1-M_r)^2}\frac{\partial}{\partial\tau}\frac{p}{1-M_r}\right]_\tau dS$$

Consider a thin blade, so that the upper and lower surfaces coincide. Then only the difference in pressure and normal velocity between the upper and lower blade surfaces is required. The differential pressure on the blade surface, in rotating coordinates x and r, is written $\Delta p = L(r, \psi)\ell(x)$ (see section 17-3.1). It is assumed that the chordwise pressure distribution does not vary over the disk. The normal velocity due to the blade thickness (see section 17-3.7) has the same magnitude on the upper and lower surfaces, so

$$\Delta v_n = V_T A_{xs} \frac{da}{dx}$$

where $a(x) = t(x)/A_{xs}$ is the normalized thickness distribution, and $V_T = \Omega r + V_x \sin\psi$ is the free stream velocity seen by the blade. The velocity normal to the blade surface actually is

$$v_n = V_T \frac{t'/2}{\sqrt{1 + (t'/2)^2}}$$

(at zero angle of attack). Since the arc length along the airfoil chord is $ds = [1 + (t'/2)^2]^{1/2} dx$, the monopole strength is $\Delta v_n dS = V_T t' dx\, dr$, which gives $a(x) = t(x)/A_{xs}$ as above. The surface of the blade is defined by the root and tip, and the leading and trailing edges in the rotating coordinate system with axes r and x. The blade azimuth position is $\psi = \Omega\tau$. Instead of integrating over all the blades, it is equivalent to multiply the rotational noise due to one blade by N. The rotor is assumed to have a forward velocity V_x, and a vertical velocity V_z. The position of a source or dipole on the blade surface is then

$$\vec{y}(\tau) = \begin{bmatrix} x_1 - V_x\tau \\ y_1 \\ z_1 + V_z\tau \end{bmatrix} = \begin{bmatrix} r\cos\Omega\tau + x\sin\Omega\tau - V_x\tau \\ r\sin\Omega\tau - x\cos\Omega\tau \\ V_z\tau \end{bmatrix}$$

$$= \begin{bmatrix} \sqrt{r^2 + x^2}\cos[\Omega\tau - \tan^{-1}(x/r)] - V_x\tau \\ \sqrt{r^2 + x^2}\sin[\Omega\tau - \tan^{-1}(x/r)] \\ V_z\tau \end{bmatrix}$$

where the components x_1, y_1, and z_1 represent the position relative to a tip-path-plane axis system moving with the rotor (as shown in Fig. 17-3). The Mach number of a point on the blade is then

$$\vec{M} = \begin{bmatrix} -\dfrac{\Omega}{c_s}\sqrt{r^2+x^2}\,\sin\left[\Omega\tau - \tan^{-1}(x/r)\right] - M_x \\ \dfrac{\Omega}{c_s}\sqrt{r^2+x^2}\,\cos\left[\Omega\tau - \tan^{-1}(x/r)\right] \\ M_z \end{bmatrix}$$

The observer position will also be defined relative to the axis system moving with the rotor hub:

$$\vec{x}(t) = \begin{bmatrix} x_0 - V_x t \\ y_0 \\ z_0 + V_z t \end{bmatrix}$$

The observer should actually be fixed in space, so that the time derivative does not operate on $\vec{x}$ when the sound pressure is evaluated by the above expressions, but for the present far field analysis a moving observer can be used with no difficulty. Note however that the observer position is evaluated at the present time t, not at the retarded time τ. The observer location relative to the rotor hub can be defined in terms of the range, elevation, and azimuth (see Fig. 17-3); we will have more need for these quantities at the retarded time, however.

The radial distance at the retarded time $s(\tau)$ is

$$\begin{aligned} s^2 &= |\vec{x}(t) - \vec{y}(\tau)|^2 \\ &= [x_0 - x_1 - V_x(t-\tau)]^2 + (y_0 - y_1)^2 + [z_0 - z_1 + V_z(t-\tau)]^2 \\ &= (x_0 - x_1 - M_x s)^2 + (y_0 - y_1)^2 + (z_0 - z_1 + M_z s)^2 \end{aligned}$$

since $s(\tau) = c_s(t-\tau)$. The solution of this quadratic equation for s is

$$s = \frac{1}{\beta^2}\left[S - M_x(x_0 - x_1) + M_z(z_0 - z_1)\right]$$

where $\beta^2 = 1 - M_x{}^2 - M_z{}^2$ and

$$S^2 = \beta^2\left[(x_0 - x_1)^2 + (y_0 - y_1)^2 + (z_0 - z_1)^2\right] + \left[M_x(x_0 - x_1) - M_z(z_0 - z_1)\right]^2$$

Define $\sigma_0 = s$ and $S_0 = S$ as the values at the center of the rotor ($x_1 = y_1 = z_1 = 0$), so that

$$\sigma_0 = \sqrt{(x_0 - M_x\sigma_0)^2 + y_0{}^2 + (z_0 + M_z\sigma_0)^2} = \frac{1}{\beta^2}(S_0 - M_x x_0 + M_z z_0)$$

$$S_0{}^2 = \beta^2(x_0{}^2 + y_0{}^2 + z_0{}^2) + (M_x x_0 - M_z z_0)^2$$

Also, let $M \cos\delta_r$ be the value of $M_r = \vec{M} \cdot (\vec{x} - \vec{y})/s$ at the center of the rotor:

$$M\cos\delta_r = \frac{1}{\sigma_0}\begin{pmatrix} -M_x \\ 0 \\ M_z \end{pmatrix} \cdot \begin{pmatrix} x_0 - M_x\sigma_0 \\ y_0 \\ z_0 + M_z\sigma_0 \end{pmatrix} = M_x{}^2 + M_z{}^2 - \frac{M_x x_0 - M_z z_0}{\sigma_0}$$

Note then that $S_0 = \sigma_0(1 - M\cos\delta_r)$. As in section 17-3.5, σ_0 is the range of the observer from the rotor at the retarded time, and $M\cos\delta_r$ is the Mach number of the rotor toward the observer.

The acoustic far field of the rotor is defined by the condition $S_0/R \gg 1$ [$\sigma_0 = S_0/(1 - M\cos\delta_r) > S_0$, so the condition $\sigma_0/R \gg 1$ is less critical]. Expanding S and then s for large S_0/R gives

$$s \cong \sigma_0 - \frac{1}{S_0}\left[(x_0 - M_x\sigma_0)x_1 + y_0 y_1 + (z_0 + M_z\sigma_0)z_1\right]$$

$$= \sigma_0 - \frac{1}{S_0}\sqrt{r^2+x^2}\Big\{(x_0 - M_x\sigma_0)\cos[\Omega\tau - \tan^{-1}(x/r)]$$

$$+ y_0 \sin[\Omega\tau - \tan^{-1}(x/r)]\Big\}$$

$$= \sigma_0 - \frac{1}{S_0}\sqrt{r^2+x^2}\sqrt{(x_0 - M_x\sigma_0)^2 + y_0^2}\cos[\Omega\tau - \tan^{-1}(x/r) - \psi_r]$$

$$= \sigma_0 - \frac{\cos\theta_r}{1 - M\cos\delta_r}\sqrt{r^2+x^2}\cos[\Omega\tau - \tan^{-1}(x/r) - \psi_r]$$

where $\psi_r = \tan^{-1} y_0/(x_0 - M_x\sigma_0)$ and $\theta_r = \sin^{-1}(z_0 + M_z\sigma_0)/\sigma_0$ are the azimuth and elevation of the observer at the retarded time, as in section 17-3.5. The far field approximation also gives

$$\frac{\vec{n}\cdot(\vec{x}-\vec{y})}{s} \cong \frac{z_0 + V_z(t-\tau)}{\sigma_0} = \frac{z_0 + M_z\sigma_0}{\sigma_0} = \sin\theta_r$$

The relative Mach number M_r is

$$M_r = \frac{\vec{M}\cdot(\vec{x}-\vec{y})}{s} = \frac{\Omega}{c_s}\sqrt{r^2+x^2}\Big\{-\frac{x_0 - M_x s}{s}\sin[\Omega\tau - \tan^{-1}(x/r)]$$

$$+ \frac{y_0}{s}\cos[\Omega\tau - \tan^{-1}(x/r)]\Big\}$$

$$- \frac{M_x}{s}\sqrt{r^2+x^2}\cos[\Omega\tau - \tan^{-1}(x/r)]$$

$$- M_x\frac{x_0 - M_x s}{s} + M_z\frac{z_0 + M_z s}{s}$$

so in the far field

$$1 - M_r \cong 1 - M\cos\delta_r + \frac{\Omega}{c_s}\sqrt{r^2+x^2}\cos\theta_r\sin[\Omega\tau - \tan^{-1}(x/r) - \psi_r]$$

In summary, the far field approximation gives

$$s \cong \sigma_0$$

$$(1 - M_r) \cong (1 - M\cos\delta_r)(1 + \alpha\sin\psi)$$

$$\frac{\vec{n}\cdot(\vec{x} - \vec{y})}{s} \cong \sin\theta_r$$

$$\tau = t - \frac{s}{c_s} \cong t - \frac{\sigma_0}{c_s} + \frac{\alpha}{\Omega}\cos\psi$$

where

$$\psi = \Omega\tau - \tan^{-1}(x/r) - \psi_r$$

$$\alpha = \frac{\Omega\sqrt{r^2 + x^2}}{c_s}\,\frac{\cos\theta_r}{1 - M\cos\delta_r}$$

and so

$$\Omega t = \psi - \alpha\cos\psi + \frac{\Omega\sigma_0}{c_s} + \tan^{-1}(x/r) + \psi_r$$

The far field rotational noise due to the thickness and lift of the rotating blade is thus

$$p(\vec{x}, t) = \frac{N\Omega\rho}{4\pi\sigma_0(1 - M\cos\delta_r)^2}\int_0^R\int_{x_{le}}^{x_{te}} \frac{A_{xs}a'}{1 + \alpha\sin\psi}\,\frac{\partial}{\partial\psi}\,\frac{V_T}{1 + \alpha\sin\psi}\,dx\,dr$$

$$- \frac{N\Omega\sin\theta_r}{4\pi c_s\sigma_0(1 - M\cos\delta_r)^2}\int_0^R\int_{x_{le}}^{x_{te}} \frac{\ell}{1 + \alpha\sin\psi}\,\frac{\partial}{\partial\psi}\,\frac{L}{1 + \alpha\sin\psi}\,dx\,dr$$

The sound pressure is periodic with fundamental frequency $N\Omega$, so

$$p = \sum_{m=-\infty}^{\infty} p_m\, e^{imN\Omega t}$$

where the mth harmonic is

$$p_m = \frac{\Omega}{2\pi}\int_0^{2\pi/\Omega} e^{-imN\Omega t}\, p\,dt$$

$$= \frac{1}{2\pi} \int_0^{2\pi} e^{-imN\Omega t} \, p(1 + \alpha \sin \psi) \, d\psi$$

since $\Omega dt = (1 + \alpha \sin \psi) d\psi$. The blade section lift can be expanded as a Fourier series:

$$L = \sum_{n=-\infty}^{\infty} L_n(r) \, e^{in\Omega\tau}$$

and similarly for the velocity V_T:

$$V_T = \Omega r + V_x \sin \Omega\tau = \Omega r \sum_{n=-\infty}^{\infty} V_n(r) \, e^{in\Omega\tau}$$

[where $V_0 = 1$ and $V_{\pm 1} = V_x / (\pm 2 i \Omega r)$]. It is assumed that the chrodwise loading distributions ℓ and a are independent of time. The mth harmonic of the rotational noise is thus

$$p_m = \frac{N\Omega}{4\pi\sigma_0 (1 - M \cos \delta_r)^2} \int_0^R \int_{x_{le}}^{x_{te}} e^{-imN[\Omega\sigma_0/c_s + \psi_r + \tan^{-1}(x/r)]}$$

$$\sum_{n=-\infty}^{\infty} e^{in[\psi_r + \tan^{-1}(x/r)]} \left[\rho A_{xs} \, a' \Omega r V_n - \frac{\sin \theta_r}{c_s} \, \ell L_n \right]$$

$$\frac{1}{2\pi} \int_0^{2\pi} e^{-imN(\psi - a \cos \psi)} \, \frac{\partial}{\partial \psi} \, \frac{e^{in\psi}}{1 + \alpha \sin \psi} \, d\psi \, dx \, dr$$

Integrating by parts gives

$$\frac{1}{2\pi} \int_0^{2\pi} e^{-imN(\psi - a \cos \psi)} \, \frac{\partial}{\partial \psi} \, \frac{e^{in\psi}}{1 + \alpha \sin \psi} \, d\psi$$

$$= imN \, \frac{1}{2\pi} \int_0^{2\pi} e^{-imN(\psi - a \cos \psi) + in\psi} \, d\psi$$

$$= imN\, i^{mN-n}\, J_{mN-n}(mN\alpha)$$

using the Bessel function relation

$$\frac{1}{2\pi}\int_0^{2\pi} e^{iz\cos\psi - in\psi}\, d\psi = i^n J_n(z)$$

Now, assume $x/r \ll 1$ so that the chordwise and spanwise integrals separate. [Since $\tan^{-1}x/r \cong x/r$ and $(r^2 + x^2)^{1/2} \cong r$, the Bessel functions do not depend on x.] In addition, the thickness term must be integrated by parts with respect to x. Define the chordwise loading factors ℓ_n and a_n as before:

$$\ell_n = \int_{x_{\text{le}}}^{x_{\text{te}}} \ell(x)\, e^{-inx/r}\, dx$$

$$a_n = \int_{x_{\text{le}}}^{x_{\text{te}}} a(x)\, e^{-inx/r}\, dx$$

For impulsive loading, $\ell_n = a_n = 1$ for all n.

Then the far field rotational noise due to the rotor blade thickness and lift is

$$p_m = -\frac{(mN\Omega)^2 N\rho}{4\pi\sigma_0 (1 - M\cos\delta_r)^2}\, e^{-imN\Omega\sigma_0/c_s} \sum_{n=-\infty}^{\infty}\left[e^{-i(mN-n)[\psi_r - (\pi/2)]}\left(1 - \frac{n}{mN}\right) \int_0^R A_{xs}\, a_{mN-n}\, V_n J_{mN-n}\, dr\right]$$

$$-\frac{imN^2\Omega\sin\theta_r}{4\pi c_s \sigma_0 (1 - M\cos\delta_r)^2}\, e^{-imN\Omega\sigma_0/c_s} \sum_{n=-\infty}^{\infty}\left[e^{-i(mN-n)[\psi_r - (\pi/2)]} \int_0^R \ell_{mN-n}\, L_n J_{mN-n}\, dr\right]$$

where the argument of the Bessel function is

$$\frac{mN\Omega r}{c_s} \frac{\cos\theta_r}{1 - M\cos\delta_r}$$

This result agrees with the solutions for the lift and thickness rotational noise that were derived in sections 17-3.6 and 17-3.7. Working directly with the loading and motion of the rotating blade as here is more convenient for developing advanced analyses of the rotor noise.

17-3.9 Doppler Shift

The rotational noise of the rotor in forward or vertical flight has been derived for an observer moving with the helicopter. For a fixed observer, these solutions can still be used to evaluate the time history of the sound pressure, by using the instantaneous observer position relative to the rotor. In addition, the relative motion between the observer and the rotor produces a shift of the frequencies of the perceived sound, which is the Doppler effect. We have seen that an acoustic source at frequency ω produces a sound pressure proportional to $e^{i\omega\tau}$, where $\tau = t - s(\tau)/c_s$ is the retarded time. The frequency of this sound at the observer then is

$$\omega_{\text{obs}} = \frac{\partial}{\partial t}\,\omega\tau = \omega\frac{d\tau}{dt} = \frac{\omega}{1 - M_r}$$

since $dt/d\tau = d(\tau + s/c_s)/d\tau = 1 - M_r$ (see sections 17-3.6 and 17-3.8). Hence

$$\omega_{\text{obs}} = \frac{\omega}{1 - M\cos\delta_r}$$

where $M\cos\delta_r$ is the Mach number of the helicopter in the direction of the observer. A fixed observer hears a frequency increase as the helicopter approaches and a frequency decrease as the helicopter recedes.

17–4 Blade Slap

Rotor blade slap is an impulsive type of sound pressure disturbance occurring at the blade passage frequency $N\Omega$. The resulting slapping or banging noise is very objectionable, and when it occurs it is the dominant rotor noise source. Blade slap increases the overall sound pressure level because of an increase in the noise spectrum level over a wide range at high frequencies, and the impulsive character makes the sound more annoying. Blade slap can be considered an extreme case of rotational noise that is best investigated in the time domain because of the impulsive character of the sound pressure. The cause of blade slap can be any aerodynamic phenomenon that produces rapidly changing loading on the blade, such as compressibility and thickness effects at the tip, interaction of the blades with the vortices in the wake, and probably also blade stall. The occurrence of blade slap is sensitive to the rotor design parameters and the helicopter operating condition. At high tip speeds or high forward speed, the compressibility and thickness effects probably dominate. In flight conditions involving close encounters of the blades with their own wake or the wake of another rotor, the vortex-blade interaction is an important source of blade slap.

Cox and Lynn (1962) found that the occurrence of blade slap was associated with evidence of significant vortex-induced loads in the rotor airloads measurements. They concluded that slap at low flight speeds is probably due to vortex-blade interaction on the advancing side. The dependence of slap on the flight condition was therefore explained by the variation of the wake geometry. They also found blade slap at high speeds, possibly due to the formation of a local shock wave on the advancing tip. See also Cox (1963).

Schlegel, King, and Mull (1966) concluded that blade slap was a high amplitude rotational noise, with a highly modulated broadband noise component as well. They found that the slap was associated with the retreating blade, suggesting that it was due to stall. For tandem rotors they attributed blade slap to the interaction of the rear rotor with the vortex wake of the front rotor. They also suggested the local shock waves produced at drag divergence as a source of blade slap.

Leverton and Taylor (1966) simulated a vortex-blade encounter by directing jets at the disk plane of a hovering rotor to approximate the downwash profile of a vortex. They simulated both parallel vortex-blade encounters (as with a tandem helicopter) and perpendicular encounters (as with a single main rotor). They found that both the spectrum and the time history of the noise produced by this simulation correlated well with the actual blade slap of a helicopter in flight. Thus the interaction of the blades with the tip vortices from a preceding blade or rotor is a likely cause of blade slap. The simulated slap experiments, and a theory developed to estimate the slap noise, showed that the sound pressure level scaled with the fourth power of the tip speed and the square of the vortex strength: $p^2 \sim (\Omega R)^4 \Gamma^2$. Leverton (1968b) continued this investigation of blade slap. Of the mechanisms postulated for blade slap (primarily the unsteady loading due to vortex-blade interaction or stall, and shock wave formation due to local supersonic flow at high tip speed or perhaps vortices), he concluded that vortex-blade interaction was the most likely source. Since the tip vortex strength Γ is proportional to $T/\rho N\Omega R^2$, the sound power of the vortex interaction should scale as

$$W_B \sim (\Omega R)^4 \Gamma^2 \sim \frac{(\Omega R)^2 T^2}{N^2 A}$$

Leverton found fairly good correlation between W_B and subjective assessment of the severity of slap. See also Leverton (1972a).

Widnall (1971) developed a theoretical analysis of the impulsive sound generated by a vortex-blade interaction on a helicopter rotor. Good agreement was found with experimental results, for both the wave form and the peak magnitude (in dB) of the impulsive sound pressure.

Bausch, Munch, and Schlegel (1971) conducted an experimental investigation of rotor impulsive noise, measured in hover and forward flight. They concluded that the impulsive noise in hover was due to the high frequency airloads produced by vortex-blade interactions. The blade slap in forward flight was due to a combination of the effects of the high advancing-tip Mach number on the propagation of the acoustic pressure waves, and the effect is on the blade drag forces.

Sternfeld, Bobo, Carmichael, Fuskushima, and Spencer (1972) investigated the impulsive noise of a hovering rotor. They found that the lift divergence boundary (where $dc_\ell/dM = 0$) of the airfoil section they used agreed with the subjectively determined boundary for excessive blade slap. For this airfoil, shock waves formed on the upper surface at the lift divergence boundary. Therefore, the following mechanism was suggested for the blade slap of a single main rotor in hover. The tip vortex from a blade produces transient flow changes at the following blades, and these flow changes either produce a shock wave or move an existing shock. The pressure disturbance due to the localized phenomenon is the blade slap. They found that calculations based on this mechanism gave a reasonable estimate of the magnitude of the sound pressure pulse. It was also concluded that to reduce the blade slap noise, an airfoil with a high lift divergence Mach number should be used.

Isom (1975) examined the scaling of noise from a hovering rotor at high tip Mach number. The sound contributions from thickness, lift, drag, and shock waves were considered. He identified a narrow zone of intense noise carried by each blade tip. This narrow zone would produce an N/rev impulsive noise at a fixed observer. The sound pressure was examined as the thickness ratio τ approached zero. Simultaneously, the blade aspect ratio AR was made to approach infinity in such a way that $AR\tau$ remained order 1 (the beamlike character of the noise disappears if AR approaches infinity while τ is fixed); and the Mach number M approaches 1 in such a way that the transonic similarity parameter $(1 - M)/\tau^{2/3}$ remained order one. The sound pressure amplitude inside and outside the narrow zone scaled with the blade thickness ratio τ [or equivalently with AR^{-1} or $(1 - M)^{3/2}$] as follows:

	Inside Narrow Zone	Outside Narrow Zone
thickness noise	τ	τ^2
lift noise	τ	$\tau^{5/3}$
drag noise	$\tau^{5/3}$	$\tau^{8/3}$
shock noise	$\tau^{5/3}$	τ^2

It was concluded that the lift dominates the rotational noise, while the thickness as well is important for impulsive noise. The thickness noise energy is almost all radiated in the narrow beam, while the lift noise energy is about equally divided inside and outside the narrow zone.

Farrasat (1975) calculated the thickness noise of a rotor blade, and concluded that it exhibits the characteristics of blade slap. He found that the thickness noise consisted of an impulsive acoustic pressure, with a sharp negative peak at high Mach number. The thickness noise is strongly dependent on Mach number and has maximum directivity in the plane of the rotor disk.

Boxwell, Schmitz, and Hanks (1975) obtained in-flight measurements of helicopter impulsive noise. In these measurements they identified three distinct types of impulsive noise, which occurred sequentially in the pressure time history. There was a series of positive pressure pulses, appearing at all speeds but sensitive to the helicopter rate of descent. These pulses were associated with blade-vortex interaction. Secondly, there was a negative pressure disturbance, with a directivity concentrated in the disk plane. The amplitude of this pulse increased rapidly with the forward velocity of the helicopter until at maximum speed it dominated the pressure signature. This negative pulse was associated with compressibility effects because of its dependence on Mach number. Thirdly, at high speeds there was a narrow positive pressure spike, closely following the negative pulse. They also concluded that judging the occurrence and severity of slap from cabin-based noise measurements can be misleading. See also Schmitz and Boxwell (1976).

17–5 Rotor Noise Reduction

By far the most important parameter influencing the noise level of a rotor is the tip speed. The vortex noise, rotational noise, and blade slap can all be significantly reduced by operating the rotor at a lower tip speed. At high Mach number, decreasing the tip speed is particularly effective in reducing the rotational noise and blade slap. The vortex noise generally decreases slowest, so at low tip speed it will become the dominant noise source. The vortex noise can also be reduced by lowering the rotor thrust or the blade loading T/A_b, and the rotational noise and blade slap can be reduced by lowering the thrust or the disk loading T/A. Decreasing the rotational speed of the rotor also lowers the frequency range of the rotor noise, which can be beneficial for large rotors. Increasing the number of blades tends to decrease the magnitude of the rotational noise harmonics, although it also increases the fundamental frequency. Decreasing the higher harmonic content of the rotor airloads will reduce the rotational noise; in particular, this requires decreasing the vortex-blade interaction loads, which will also reduce some kinds of blade slap. The parameters influencing the rotor noise (especially the tip speed, disk loading, and number of blades) are also major factors in determining the rotor performance. It follows that any improvement in the aerodynamic efficiency of the rotor can be utilized in designing a quieter helicopter while maintaining the current level of performance and cost, rather than being used exclusively to improve the helicopter performance.

The blade tip planform and section can influence the rotor noise by altering the tip aerodynamic loads and the structure of the trailed tip vortex. The airfoil section shape and thickness ratio at the tip should be chosen for good characteristics at high Mach number, because of the evident importance of compressibility effects in rotational noise and blade slap. An appropriate planform shape can minimize the tip vortex roll-up, and thus reduce the vortex-blade interaction loads.

Literature on the design of quiet helicopters, including the performance penalties involved: Miller (1968), Cox (1970), Faulkner (1971, 1974a), Barlow, McCluskey, and Ferris (1972), Henderson, Pegg, and Hilton (1973), Pegg, Henderson and Hilton (1973), Bowes (1973, 1977), Faulkner and Swan (1976), Gibs, Stepniewski, and Spencer (1976). Concerning the influence of the blade tip on rotor noise: Spencer, Sternfeld, and McCormick (1966), Schlegel and Bausch (1969), Spencer and Sternfeld (1969), Cox (1970), Soderman (1973), Tangler (1975), Mantay, Shidler, and Campbell (1977), Lee and Mosher (1978).

17–6 Literature

Summaries and reviews of helicopter rotor noise: Hubbard and Lassiter (1954), Cox and Lynn (1962), Cox (1963, 1973), Davidson and Hargest (1965), Hargest (1966), Lowson (1966, 1973), Richards and Mead (1968), Lowson and Ollerhead (1969a), Ollerhead and Lowson (1969), Marte and Kurtz (1970), Cheeseman (1971), Hubbard, Lansing, and Runyan (1971), Leverton (1971), Munch and King (1974), Magliozzi, Metzger, Bausch, and King (1975), George (1978).

Additional work on helicopter rotor noise: Stowell and Deming (1935a, 1935b), Ernstausen (1937), Hicks and Hubbard (1947), Hubbard and Regier (1947), Regier and Hubbard (1947), Hubbard (1948, 1950), Jewel and Harrington (1958), Hubbard and Maglieri (1960), Jewel (1960), Maglieri, Hilton, and Hubbard (1961), Loewy (1963), Sternfeld (1967), Bragg (1968), Laskin, Orcutt, and Shipley (1968), Leverton (1968, 1969, 1972b, 1974,

1975a, 1975b), Matveev and Mel'nikov (1968), Shapiro and Healy (1968), Tanna (1968), Cox (1969), King and Schlegel (1969), Ollerhead and Taylor (1969), Sadler and Loewy (1969), Spencer and Sternfeld (1969), Sternfeld and Schairer (1969), Wright and Leverton (1969), Badgley and Laskin (1970), Malone, Schweichert, and Ketchel (1970), Spencer (1970), Spivey and Morehouse (1970), Brown (1971), Hawles (1971), Kiang and Ng (1971), Lince (1971), Lyon (1971), Ollerhead (1971), Paterson, Vogt, Fink, and Munch (1971, 1973), Scheiman, Hilton, and Shivers (1971), Sternfeld, Spencer, and Schairer (1971), Gray and Pierce (1972), Hirsh and Ferris (1972), Johnson and Katz (1972), Leverton and Amor (1972, 1973), Lowson (1972), Nagel (1972), Schmitz, Stepniewski, Gibs, and Hinterkeuser (1972), Soderman (1972), Sternfeld, Schairer, and Spencer (1972), Broll (1973), Filotas (1973b), Gibs, Stepniewski, Spencer, and Kohler (1973), Giordano and Keane (1973), Hinterkeuser (1973), H. K. Johnson (1973, 1974), Kasper (1973), Leverton and Pollard (1973), Lowson, Whatmore, and Whitfield (1973), Lyon, Mark, and Pyle (1973), Robinson (1973), Sadler, Johnson, and Evans (1973), Schmitz and Stepniewski (1973), Schwind and Allen (1973), Stave (1973), Stepniewski and Schmitz (1973), Tangler, Wohlfeld, and Miley (1973), Wagner (1973), Whatmore and Lowson (1973), Clark (1974), Faulkner (1974b), Hawkings and Lowson (1974, 1975), Hersh, Soderman, and Hayden (1974), Hinterkeuser and Sternfeld (1974), Homicz and George (1974), Hosier and Ramakrishnan (1974), Hosier, Ramakrishnan, and Pegg (1974), Kinney, Pierce, and Rickley (1974), Munch and King (1974), Scheiman (1974), Widnall, Harris, Lee, and Drees (1974), Chalupnik and Clark (1975), Charles (1975), Davis (1975), Farassat, Pegg, and Hilton (1975), Fink (1975a, 1975b), Hall (1975b), Lee (1975,1977, 1978), Mall and Farassat (1975), Munch (1975), Munch, Paterson, and Day (1975), Parks (1975), Paterson, Amiet, and Munch (1975), Pegg, Hosier, Balcerak, and Johnson (1975), White (1975, 1977), Balcerak (1976), Burton and Blevin (1976), Clark, Chalupnik, and Hodder (1976), Fink, Schlinker, and Amiet (1976), Lawton (1976), Levine (1976), MAN-Acoustics and Noise (1976), Ramakrishnan, Randall, and Hosier (1976), Vause, Schmitz, and Boxwell (1976), Virchis and Wright (1976), Amiet (1977, 1978), Farassat and Brown (1977), George and Kim (1977), Galloway (1977, 1978), Howlett, Clevenson, Rupf, and Synder (1977), Lee, Harris, and Widnall (1977), Leverton, Pollard, and Wills (1977), Paterson and

Amiet (1977), Patterson, Mozo, Schomer, and Camp (1977), Scheiman and Hoad (1977), Schmitz, Boxwell, and Vause (1977), Schmitz and Yu (1977), Shahady et al. (1977), Shockey, Cox, and Williamson (1977), Tangler (1977), True and Letty (1977), True and Rickley (1977), Whitfield (1977), Aravamudan, Lee, and Harris (1978a, 1978b), Boxwell, Yu, and Schmitz (1978), d'Amdra and Damongeot (1978), Fink (1978), Hanson and Fink (1978), Hawkings (1978), Hayden and Aravamudan (1978), Hoad and Greene (1978), Homans, Little, and Schomer (1978), Johnson and Lee (1978), Lee and Mosher (1978), Leverton, Southwood, and Pike (1978), Mantay, Campbell, and Shidler (1978), Nakamura and Azuma (1978), Powell (1978a, 1978b), Sternfeld and Doyle (1978), Wolf and Widnall (1978).

CITED LITERATURE

The following is a bibliography of helicopter literature, arranged alphabetically by author. The material is classified according to subject when it is cited in the text. The books on helicopter analysis include: Glauert (1935), Stepniewski (1950), Nikolsky (1951), Gessow and Myers (1952), Dommasch (1953), Shapiro (1955), Payne (1959), Legrand (1964), Seckel (1964), Mil′ (1966), McCormick (1967), Wood (1968), Richards and Mead (1968), and Bramwell (1976). Bibliographies of helicopter work: Gessow (1948c, 1952), Gessow and Myers (1952), and United States Defense Documentation Center (1968, 1972, 1973). The following abbreviations are used for the sources of reports:

AFFDL	U.S. Air Force Flight Dynamics Laboratory
AGARD	NATO Advisory Group for Aerospace Research and Development
AHS Forum	Annual National V/STOL Forum, American Helicopter Society, Washington, D.C.
ARC	British Aeronautical Research Council
CAL	Cornell Aeronautical Laboratory
JAHS	Journal of the Americal Helicopter Society
JAS	Journal of the Aeronautical Sciences
J. Sound Vib.	Journal of Sound and Vibration
NACA	National Advisory Committee for Aeronautics
NASA	National Aeronautics and Space Administration
TCREC	U.S. Army Transportation Research Command
TRECOM	U.S. Army Transportation Research Command
USAAMRDL	U.S. Army Air Mobility Research and Development Laboratory
USAAVLABS	U.S. Army Aviation Materiel Laboratories
USARTL	U.S. Army Research and Technology Laboratories

Adamczyk, J. J. "Analytical Investigation of Compressibility and Three-Dimensionality on the Unsteady Response of an Airfoil in a Fluctuating Flow Field." Journal of Aircraft, 11:5 (May 1974).

Adams, D. O. "The Evaluation of a Stall-Flutter Spring-Damper Pushrod in the Rotating Control System of a CH-54B Helicopter." USAAMRDL TR 73-55, August 1973.

Akeley, C. R., and Carson, G. W. "Fan-in-Fuselage Advanced Antitorque System." USAAMRDL TR 74-89, November 1974.

Alansky, I. B.; Davis, J. M.; and Garnett, T. S., Jr. "Limitations of the CH-47 Helicopter in Pertorming Terrain Flying with External Loads," USAAMRDL TR 77-21, August 1977.

Alansky, I. B.; Davis, J. M.; and Garnett, T. S., Jr. Limitations of the UTTAS Helicopter in Performing Terrain Flying with External Loads," USAAMRDL TR 77-22, September 1977.

Albion, N., and Larson, W. R. "A Frequency Response Approach to Flying Qualities Criteria and Flight Control System Design." JAHS, 19:1 January 1974).

Albion, N.; Leet, J. R.; and Mollenkof, P. A. "Ground-Based Flight Simulation of the CH-47C Helicopter." USAAVLABS TR 69-71, October 1969.

Albrecht, C. O. "Factors in the Design and Fabrication of Powered Dynamically Similar V/STOL Wind Tunnel Models." American Helicopter Society Mideast Symposium, Philadelphia, Pennsylvania, September 1972.

Alexander, H. R.; Eason, W.; Gillmore, K.; Morris, J.; and Spittle, R. "Tilt Rotor Flight Control Program Feedback Studies." NASA CR 114600, March 1973.

Alexander, H. R.; Hengen, L. M.; and Weiberg, J. A. "Aeroelastic Stability Characteristics of a V/STOL Tilt-Rotor Aircraft with Hingeless Blades: Correlation of Analysis and Test." JAHS, 20:2 (April 1974).

Alexander, H. R., et al. "V/STOL Dynamics and Aeroelastic Rotor-Airframe Technology." AFFDL TR 72-40, January 1973.

Allen, R. W. "Flapping Characteristics of Rigid Rotor Blades." JAS, 13:4 (April 1946).

Allen, R. E., and Calcaterra, P. C. "Design, Fabrication, and Testing of Two Electrohydraulic Vibration Isolation Systems for Helicopter Environments." NASA CR 112052, 1972.

Amer, K. B. "Theory of Helicopter Damping in Pitch or Roll and a Comparison with Flight Measurements." NACA TN 2136, October 1950.

Amer, K. B. "Some Flying-Qualities Studies of a Tandem Helicopter." NACA RM L51H20a, October 1951.

Amer, K. B. "Method for Studying Helicopter Longitudinal Maneuver Stability." NACA Report 1200, 1954.

Amer, K. B. "Effect of Blade Stalling and Drag Divergence on Power Required by a Helicopter Rotor at High Forward Speed." AHS Forum, 1955.

Amer, K. B., and Gessow, A. "Charts for Estimating Tail-Rotor Contribution to Helicopter Directional Stability and Control in Low-Speed Flight." NACA Report 1216, 1955.

Amer, K. B., and Gustafson, F. B. "Charts for Estimation of Longitudinal-Stability Derivatives for a Helicopter Rotor in Forward Flight." NACA TN 2309, March 1951.

Amer, K. B., and La Forge, S. "Effects of Blade Stall on Helicopter Rotor Blade Bending Loads." AHS Forum, 1965.

Amer, K. B., and La Forge, S. V. "Maximum Rotor Thrust Capabilities, Articulated and Teetering Rotors." JAHS, 22:1 (January 1977).

Amer, K. B., and Neff, J. R. "Vertical-Plane Pendulum Absorbers for Minimizing Helicopter Vibratory Loads." JAHS, 19:4 (October 1974).

Amer, K. B.; Prouty, R. W.; Walton, R. P.; and Engle, J. E. "Handling Qualities of Army/Hughes YAH-64 Advanced Attack Helicopter." AHS Forum, 1978.

Amer, K. B., and Tapscott, R. J. "Studies of the Lateral-Directional Flying Qualities of a Tandem Helicopter in Forward Flight." NACA Report 1207, 1954.

Amer, K. B., et al. "Fundamentals of Helicopter Stability and Control." Boeing Vertol Company, Report R242, February 1961.

American Helicopter Society. "Helicopter Vibration Reduction." JAHS, 2:3 (July 1957).

Amiet, R. K. "Noise Produced by Turbulent Flow into a Propeller or Helicopter Rotor." AIAA Journal, 15:3 (March 1977).

Amiet, R. K. "Noise Due to Rotor-Turbulence Interaction." NASA CP 2052, May 1978.

Amos, A. K., and Alexander, H. R. "Simulation Study of Gust Alleviation in a Tilt Rotor Aircraft." NASA CR 152050 and CR 152051, June 1977.

Anderson, S. B. "Examination of Handling Qualities Criteria for V/STOL Aircraft." NASA TN D-331, July 1960.

Anderson, W. D. "Investigation of Reactionless Mode Stability Characteristics of a Stiff Inplane Hingeless Rotor System." AHS Forum, 1973.

Anderson, W. D.; Connor, F.; and Kerr, A. W. "Application of an Interdisciplinary Rotary-Wing Aircraft Analysis to the Prediction of Helicopter Maneuver Loads." USAAMRDL TR 73-83, December 1973.

Anderson, W. D.; Connor, F.; Kretsinger, P.; and Reaser, J. S. "REXOR Rotorcraft Simulation Model." USAAMRDL TR 76-28, July 1976.

Anderson, W. D., and Gaidelis, J. A. "Vibration Inputs and Response of a Hingeless Rotor Compound Helicopter." Lockheed Report LR-26523, June 1974.

Anderson, W. D., and Watts, G. A. "Rotor Blade Wake Flutter." Lockheed Report LR-26213, December 1973.

Anderson, W. D., and Watts, G. A. "Rotor Blade Wake Flutter, A Comparison of Theory and Experiment." JAHS, 21:2 (April 1976).

Anderson, W. D., and Wood, E. R. "AH-56A (AMCS) Compound Helicopter

Vibration Reduction." AHS Forum, 1974.

Anoshchenko, N. D., ed. "History of Aviation and Cosmonautics, Vol. 5." NASA TT F-11851, August 1968.

Aravamudan, K. S.; Lee, A.; and Harris, W. L. "A Simplified Mach Number Scaling Law for Helicopter Rotor Noise." J. Sound Vib., 57:4 (1978).

Aravamudan, K. S.; Lee, A.; and Harris, W. L. "Wind Tunnel Investigation of Model Rotor Noise at Low Tip Speeds." NASA CP 2052, May 1978.

Arcidiacono, P. J. "Theoretical Performance of Helicopters Having Second and Higher Harmonic Feathering Control." JAHS, 6:2 (April (1961).

Arcidiacono, P. J. "Aerodynamic Characteristics of a Model Helicopter Rotor Operating under Nominally Stalled Conditions in Forward Flight." JAHS, 9:3 (July 1964).

Arcidiacono, P. J. "Prediction of Rotor Instability at High Forward Speeds—Steady Flight Differential Equations of Motion for a Flexible Helicopter Blade with Chordwise Mass Unbalance." USAAVLABS TR 68-18A, February 1969.

Arcidiacono, P. J.; Bergquist, R. R.; and Alexander, W. T., Jr. "Helicopter Gust Response Characteristics Including Unsteady Aerodynamic Stall Effects." JAHS, 19:4 (October 1974).

Arcidiacono, P. J., and Carlson, R. G. "Helicopter Rotor Loads Prediction." AGARD Conference Proceedings No. 122, Milan, Italy, March 1973.

Arcidiacono, P. J.; Carta, F. O.; Casellini, L. M.; and Elman, H. L. "Investigation of Helicopter Control Loads Induced by Stall Flutter." USAAVLABS TR 70-2, March 1970.

Arcidiacono, P., and Zincone, R. "Titanium UTTAS Main Rotor Blade." JAHS, 21:2 (April 1976).

Arents, D. N. "An Assessment of the Hover Performance of the XH-59A Advancing Blade Concept Demonstrator Helicopter." USAAMRDL TN 25, May 1977.

Arndt, R. E. A., and Borgman, D. C. "Noise Radiation from Helicopter Rotors Operating at High Tip Mach Number." JAHS, 16:1 (January 1971).

Asher, N. J.; Donelson, J.; and Higgins, G. F. "Changes in Helicopter Reliability/Maintainability Characteristics over Time." Institute for Defense Analysis, Report S-451, March 1975.

Ashley, H.; Brunelle, E.; and Moser, H. H. "Unsteady Flow through Helicopter Rotors." Journal of Applied Mathematics and Physics, 9b:5-6 (March 1958). (Special issue.)

Ashley, H.; Moser, H. H.; and Dugundji, J. "Investigation of Rotor Response to Vibratory Aerodynamic Inputs—Three-Dimensional Effects on Unsteady Flow through a Helicopter Rotor." Wright Air Development Center, WADC TR 58-87, Part III, October 1958.

Asseo, S. J., and Whitbeck, R. F. "Control Requirements for Sling-Load

Stabilization in Heavy Lift Helicopters." JAHS, 18:3 (July 1973).

Astill, C. J., and Niebanck, C. F. "Prediction of Rotor Instability at High Forward Speeds—Classical Flutter." USAAVLABS TR 68-18B, February 1969.

Austin, E. E., and Vann, W. D. "General Description of the Rotorcraft Flight Simulation Computer Program (C-81)." USAAMRDL TN 11, June 1973.

Austin, R. G.; Rogers, V. A. B.; and Smith, A. H. "The Concept, Design, and Development of the WG-13." The Aeronautical Journal, 78:757 (January 1974).

Autry, C. P. "Helicopter Power Requirements." Aircraft Engineering, 18:211 (October 1946).

Azuma, A. "Dynamic Analysis of the Rigid Rotor System." Journal of Aircraft, 4:3 (May-June 1967).

Azuma, A., and Nakamura, Y. "Pitch Damping of Helicopter Rotor with Nonuniform Inflow." Journal of Aircraft, 11:10 (Ooctober 1974).

Azuma, A., and Obata, A. "Induced Flow Variation of the Helicopter Rotor Operating in the Vortex Ring State." Journal of Aircraft, 5:4 (July-August 1968).

Badgley, R. H., and Laskin, I. "Program for Helicopter Gearbox Noise Prediction and Reduction." USAAVLABS TR 70-12, March 1970.

Bailey, F. J., Jr. "A Study of the Torque Equilibrium of an Autogiro Rotor." NACA Report 623, 1938.

Bailey, F. J., Jr. "Flight Investigation of Control-Stick Vibration of the YG-1B Autogyro." NACA TN 764, June 1940.

Bailey, F. J., Jr. "A Simplified Theoretical Method of Determining the Characteristics of a Lifting Rotor in Forward Flight." NACA Report 716, 1941.

Bailey, F. J., Jr., and Gustafson, F. B. "Observations in Flight of the Region of Stalled Flow over the Blades of an Autogiro Rotor." NACA TN 741, December 1939.

Bailey, F. J., Jr., and Gustafson, F. B. "Charts for Estimation of the Characteristics of a Helicopter Rotor in Forward Flight." NACA ACR L4H07, August 1944.

Bailey, R. G., and Hammer, J. M. "Helicopter Application Studies of the Variable Deflection Thruster Jet Flap." Honeywell Document 12153-FR1(R), November 1970.

Bain, L. J. "Comparison of Theoretical and Experimental Model Rotor Blade Vibratory Shear Forces." USAAVLABS TR 66-77, October 1967.

Bain, L. J., and Landgrebe, A. J. "Investigation of Compound Helicopter Aerodynamic Interference Effects." USAAVLABS TR 67-44, November 1967.

Baird, E. F.; Bauer, E. M.; and Kohn, J. S. "Model Tests and Analysis of Proprotor Dynamics for Tilt-Rotor Aircraft." American Helicopter Society Mideast Region Symposium, Philadelphia, Pennsylvania, October 1972.

Balaban, K. "Evolution of the Helicopter." NACA TN 196, March 1923.

Balcerak, J. C. "A Method for Predicting the Aerodynamic Loads and Dynamic Response of the Rotor Blades of a Tandem-Rotor Helicopter." USAAVLABS TR 67-38, June 1967.

Balcerak, J. C. "Parametric Study of the Noise Produced by the Interaction of the Main Rotor Wake with the Tail Rotor." NASA CR 145001, 1976.

Balcerak, J. C., and Erickson, J. C., Jr. "Suppression of Transmitted Harmonic Vertical and Inplane Rotor Loads by Blade Pitch Control." USAAVLABS TR 69-39, July 1969.

Balcerak, J. C.; Erickson, J. C., Jr.; and McGarvey, J. H. "Higher Harmonic Pitch Control." CAL/AVLABS Symposium, Buffalo, New York, June 1969.

Balcerak, J. C., and Feller, R. F. "Vortex Modification by Mass Injection and by Tip Geometry Variation." USAAMRDL TR 73-45, June 1973.

Balcerak, J. C., and Feller, R. F. "Effect of Sweep Angle on the Pressure Distributions and Effectiveness of the Ogee Tip in Diffusing a Line Vortex." NASA CR 132355, 1973.

Balch, D. T. "Full-Scale Wind Tunnel Tests of a Modern Helicopter Main Rotor – Correlation with Model Rotor Test Data and with Theory." AHS Forum, 1978.

Baldock, J. C. A. "Some Calculations for Air Resonance of a Helicopter with Non-Articulated Rotor Blades." ARC R&M 3743, April 1972.

Balke, R. W. "A Review of Turbine Engine Vibration Criteria for VTOL Aircraft." JAHS 16:4 (October 1971).

Balmford, D. E. H. "The Control of Vibration in Helicopters." The Aeronautical Journal, 81:794 (February 1977).

Banaszak, L. J., and Posingies, W. M. "Hydrofluidic Stability Augmentation System (HYAS) Operational Suitability Demonstration." USAAMRDL TR 77-31, October 1977.

Banerjee, D.; Crews, S. T.; Hohenemser, K. H.; and Yin, S. K. "Identification of State Variables and Dynamic Inflow from Rotor Model Dynamic Tests." JAHS, 22:2 (April 1977).

Banerjee, D., and Hohenemser, K. H. "Optimum Data Utilization for Parameter Identification with Application to Lifting Rotors." Journal of Aircraft, 13:12 (December 1976).

Banks, W. H. H., and Gadd, G. E. "Delaying Effect of Rotation on Laminar Separation." AIAA Journal, 1:4 (April 1963).

Barlow, W. H.; McCluskey, W. C.; and Ferris, H. W. "OH-6A Phase II Quiet

Helicopter Program." USAAMRDL TR 72-29, September 1972.

Barnaby, R. S.; Berkowitz, S. M.; and Colcord, W. H. "Convertible Aircraft," Aeronautical Engineering Review, 8:4 (April 1949).

Bartlett, F. D., Jr., and Flannelly, W. G. "Application of Antiresonance Theory of Helicopters." JAHS, 19:1 (January 1974).

Bartsch, E. A., and Sweers, J. E. "Inflight Measurement and Correlation with Theory of Blade Airloads and Responses on the XH-51A Compound Helicopter Rotor." USAAVLABS TR 68-22, May 1968.

Baskin, J. M. "CH-47C Fixed System Stall-Flutter Damping." USAAMRDL TR 75-29, August 1975.

Baskin, V. E.; Vil'dgrube, L. S.; Vozhdayev, Ye. S.; and Maykapar, G. I. "Theory of the Lifting Airscrew." NASA TT F-823, February 1976.

Bass, R. L., III; Johnson, J. E.; and Unruh, J. F. "Aerodynamic Damping of Vibrating Helicopter Rotors." Southwest Research Institute, Report 02-2865, January 1974.

Bausch, W. E.; Munch, C. L; and Schlegel, R. G. "An Experimental Study of Helicopter Rotor Impulsive Noise." USAAVLABS TR 70-72, June 1971.

Bazov, D. I. "Helicopter Aerodynamics." NASA TT F-676, May 1972.

Beavan, J. A., and Lock, C. N. H. "The effect of Blade Twist on the Characteristics of the C.30 Autogiro." ARC R&M 1727, April 1936.

Beddoes, T. S. "A Synthesis of Unsteady Aerodynamic Effects Including Stall Hysteresis." Vertica, 1:2 (1976).

Beddoes, T. S. "Onset of Leading Edge Separation Effects under Dynamic Conditions and Low Mach Number." AHS Forum, 1978.

Bell Helicopter Company. "Advancement of Proprotor Technology – Design Study Summary." NASA CR 114682, September 1969.

Bell Helicopter Company. "Advancement of Proprotor Technology – Wind Tunnel Test Results." NASA CR 114363, September 1971.

Bell Helicopter Company. "V/STOL Tilt Rotor Study – Conceptual Design." NASA CR 114441, 1972.

Bell Helicopter Company. "V/STOL Tilt Rotor Study – Research Aircraft Design." NASA CR 114442, 1972.

Bell Helicopter Company. "Large Scale Wind Tunnel Investigation of a Folding Tilt Rotor." NASA CR 114464, May 1972.

Bell Helicopter Company. "Rotor Systems Research Aircraft." NASA CR 112156 and CR 112157, 1972.

Bell Helicopter Company. "Full Scale Hover Test 25-foot Low Disk Loading Tilt Rotor." NASA CR 114626, May 1973.

Bellinger, E. D. "Analytical Investigation of the Effects of Blade Flexibility, Unsteady Aerodynamics, and Variable Inflow on Helicopter Rotor Stall Characteristics." NASA CR 1769, September 1971.

Bellinger, E. D. "Experimental Investigation of Effects of Blade Section

Camber and Planform Taper on Rotor Hover Performance." USAAMRDL TR 72-4, March 1972.

Bellinger, E. D. "Analytical Investigation of the Effect of Blade Flexibility, Unsteady Aerodynamics, and Variable Inflow on Helicopter Rotor Stall Characteristics." JAHS, 17:3 (July 1972).

Bellinger, E. D.; Patrick, W. P.; Greenwald, L. E.; and Landgrebe, A. J. "Experimental Investigation of the Effects of Helicopter Rotor Design Parameters on Forward Flight Stall Characteristics." USAAMRDL TR 74-1, April 1974.

Bengtson, D.; Dickovich, T.; and Helfenstine, R. "Roll-Axis Hydrofluidic Stability Augmentation System Development." USAAMRDL TR 75-43, September 1975.

Bengtson, D.; Hedeen, J.; and Helfenstine, R. "Development and Flight Test of an Advanced Hydrofluidic Stabilization System." USAAMRDL TR 76-2, February 1976.

Bennett, J. A. J. "Vertical Descent of the Autogiro." NACA TM 673, June 1932.

Bennett, J. A. J. "The Flight of an Autogiro at High Speed." NACA TM 729, December 1933.

Bennett, J. A. J. "Rotary Wing Aircraft." Aircraft Engineering, 12:131 (January 1940) through 12:138 (August 1940).

Bennett, R. L. "Rotor System Design and Evaluation Using a General Purpose Helicopter Flight Simulation Program." AGARD Conference Proceedings No. 122, Milan, Italy, March 1973.

Bennett, R. L. "Fully Coupled Natural Frequencies and Mode Shapes of a Helicopter Rotor Blade." NASA CR 132662, March 1975.

Beno, E. A. "Analysis of Helicopter Maneuver-Loads and Rotor-Loads Flight-Test Data." NASA CR 2225, March 1973.

Benson, R. G.; Bumstead, R.; and Hutto, A. J. "Use of Helicopter Flight Simulation for Height-Velocity Test Predictions and Flight Test Risk Reduction." AHS Forum, 1978.

Benson, R. G.; Dadone, L. U.; Gormont, R. E.; and Kohler, G. R. "Influence of Airfoils on Stall Flutter Boundaries of Articulated Helicopter Rotors." JAHS, 18:1 (January 1973).

Bergquist, R. R. "Helicopter Gust Response Including Unsteady Aerodynamic Stall Effects." USAAMRDL TR 72-68, May 1973.

Berman, A. "Response Matrix Method of Rotor Blade Analysis." JAS, 23:2 (February 1956).

Berman, A. "Some Applications of the Response Matrix Method to Rotor Blade Bending Response." JAHS, 1:3 (July 1956).

Berman, A. "A New Approach to Rotor Blade Dynamic Analysis." JAHS, 10:3 (July 1965).

Berrington, D. K. "Design and Development of the Westland Sea Lynx." JAHS, 19:1 (January 1974).

Betz, A. "Development of the Inflow Theory of the Propeller." NACA TN 24, November 1920.

Betz, A. "The Theory of the Screw Propeller." NACA TN 83, February 1922.

Betz, A. "Windmills in the Light of Modern Research." NACA TM 474, August 1928.

Betz, A. "Considerations of Propeller Efficiency." NACA TM 481, September 1928.

Betz, A. "The Ground Effect on Lifting Propellers." NACA TM 836, August 1937.

Bielawa, R. L. "Synthesized Unsteady Airfoil Data with Applications to Stall Flutter Calculations." AHS Forum, 1975.

Bielawa, R. L. "Blade Stress Calculations – Mode Deflection vs. Force Integration." JAHS 23:3 (July 1978).

Bielawa, R. L. "Aeroelastic Analysis for Helicopter Rotor Blades with Time-Variable, Nonlinear Structural Twist and Multiple Structural Redundancy." NASA CR 2638, 1976.

Bielawa, R. L. "Aeroelastic Characteristics of Composite Bearingless Rotor Blades." JAHS, 22:4 (October 1977).

Bielawa, R. L.; Cheney, M. C., Jr.; and Novak, R. C. "Investigation of a Bearingless Helicopter Rotor Concept Having a Composite Primary Structure." NASA CR 2637, 1976.

Bies, D. A. "Feasibility Study of a Hybrid Vibration Isolation System." USAAVLABS TR 68-54, August 1968.

Biggers, J. C. "Some Approximations to the Flapping Stability of Helicopter Rotors." JAHS, 19:4 (October 1974).

Biggers, J. C.; Lee, A.; Orloff, K. L.; and Lemmer, O. J. "Measurements of Helicopter Rotor Tip Vortices." AHS Forum, 1977.

Biggers, J. C.; Lee, A.; Orloff, K. L; and Lemmer, O.J. "Laser Velocimeter Measurements of Two-Bladed Helicopter Rotor Flow Fields." NASA TM X-73238, May 1977.

Biggers, J. C., and McCloud, J. L. III. "A Note on Multicyclic Control by Swashplate Oscillation." NASA TM 78475, April 1978.

Biggers, J. C.; McCloud, J. L. III; and Patterakis, P. "Wind Tunnel Tests of Two Full-Scale Helicopter Fuselages." NASA TN D-1548, October 1962.

Biggers, J. C., and Orloff, K. L. "Laser Velocimeter Measurements of the Helicopter Rotor-Induced Flow Field." JAHS, 20:1 (January 1975).

Bilezikjian, V., et al. "Parametric Analysis and Preliminary Design of a Shaft-Driven Rotor System for a Heavy-Lift Helicopter." USAAVLABS TR 66-48, August 1966.

Bingham, G. J. "An Analytical Evaluation of Airfoil Sections for Helicopter Rotor Applications." NASA TN D-7796, February 1975.

Blackburn, W. E. "Methods for Improving Flying Qualities of Compound Aircraft." JAHS, 13:1 (January 1968).

Blackburn, W. E., and Rita, A. D. "Flight Research Program to Evaluate Methods of Improving Compound Helicopter Maneuver Capability." USAAVLABS TR 67-59, April 1968.

Blackwell, R. H. "Investigation of the Compliant Rotor Concept." USAAMRDL TR 77-7, June 1977.

Blackwell, R. H., and Commerford, G. L. "Investigation of the Effects of Blade Structural Design Parameters on Helicopter Stall Boundaries." USAAMRDL TR 74-25, May 1975.

Blackwell, R. H., and Merkley, D. J. "The Aeroelastically Conformable Rotor Concept." AHS Forum, 1978.

Blackwell, R. H., and Mirick, P. H. "Effect of Blade Design Parameters on Helicopter Stall Boundaries." AHS Forum, 1974.

Blake, B. B., and Alansky, I. B. "Stability and Control of the YUH-61A," JAHS, 22:1 (January 1977).

Blake, B. B.; Albion, N.; and Radford, R. C. "Flight Simulation of the CH-46 Helicopter." JAHS, 15:1 (January 1970).

Blake, B. B; Burkam, J. E.; and Loewy, R. G. "Recent Studies of the Pitch-Lag Instabilities of Articulated Rotors." JAHS, 6:3 (July 1961).

Blake, B.; Clifford, J. M.; Kaczynski, R.; and Sheridan, P. F. "Recent Advances in Flying Qualities of Tandem Helicopters." AHS Forum, 1958.

Blankenship, B. L., and Harvey, K. W. "A Digital Analysis for Helicopter Performance and Rotor Blade Bending Moments." JAHS, 7:4 (October 1962).

Blaser, D. A., and Velkoff, H. R. "Pressure Distribution and Angle of Attack Variation on a Helicopter Rotor Blade." JAHS, 13:2 (April 1968).

Blaser, D. A., and Velkoff, H. R. "Boundary Layer Velocity Profiles on a Helicopter Rotor Blade in Hovering and Forward Flight." AHS Forum, 1971.

Blaser, D. A., and Velkoff, H. R. "A Preliminary Analytical and Experimental Investigation of Helicopter Rotor Boundary Layers." AIAA Journal, 11:12 (December 1973).

Boatwright, D. W. "Measurements of Velocity Components in the Wake of a Full-Scale Helicopter Rotor in Hover." USAAMRDL TR 72-33, August 1972.

Boatwright, D. W. "Three-Dimensional Measurements of the Velocity in the Near Flow Field of a Full-Scale Hovering Rotor." Mississippi State University, Report EIRS-ASE-74-4, January 1974.

Bobo, C. J. "Theory/Test Correlation of Helicopter Rotor Blade Element

Airloads in the Blade Stall Regime." NASA CR 114489, August 1972.

Boeing Vertol Company. "V/STOL Tilt Rotor Aircraft Study – Conceptual Design of Useful Military and/or Commercial Aircraft." NASA CR 114437, March 1972.

Boeing Vertol Company. "V/STOL Tilt Rotor Aircraft Study – Prliminary Design of Research Aircraft." NASA CR 114438, March 1972.

Boeing Vertol Company. "Heavy Lift Helicopter Advanced Technology Component Program – Hub and Upper Controls." USAAMRDL TR 77-37, September 1977.

Boeing Vertol Company. "Heavy Lift Helicopter Advanced Technology Component Program – Rotor Blade." USAAMRDL TR 77-41, September 1977.

Boirun, B. H. "Generalized Helicopter Flight Test Performance Data." AHS Forum, 1978.

Boirun, B. H.; Jefferis, R. P.; and Holasek, R. S. "Rotor Flow Survey Program – UH-1M Helicopter." Army Aviation Systems Test Activity, USAASTA Project 72-05, May 1974.

Bolanovich, M., and Marks, M. D. "Experimental Downwash Velocity, Static Pressure, and Temperature Distributions in Ground Effect for a 75-foot Jet Driven Rotor." JAHS, 4:2 (April 1959).

Born, G. J.; Carico, D.; and Durbin, E. J. "A Dynamic Helicopter Performance and Control Model." U.S. Army Electronics Command, ECO, 02412-11, August 1972.

Born, G. J.; Durbin, E. J.; and Schmitz, F. H. "Take Off of Heavily Loaded Helicopters." U.S. Army Electronics Command, ECOM 02412-7, January 1972.

Born, G. J., and Kai, T. "Optimal Control Theory Applied to a Helicopter in the Hover and Approach Phase." NASA CR 152135, January 1975.

Borodin, V. T., and Ryl'skiy, G. I. "Flight Control of Airplanes and Helicopters." Foreign Technology Division, Air Force Systems Command, FTD-MT-24-318-73, 1972.

Borst, H. V., et al. "Summary of Propeller Design Procedures and Data." USAAMRDL TR 73-34, November 1973.

Bossler, R. B., Jr. "A Main Rotor System for Shaft Driven Heavy Lift Helicopters." USAAVLABS TR 65-52, October 1966.

Bott, L. R. "A Statistical Method of Predicting Fatigue Lives of Helicopter Dynamic Components." JAHS, 9:4 (October 1964).

Bousman, W. G. "An Experimental Investigation of Hingeless Helicopter Rotor-Body Stability in Hover, NASA TM 78489, June 1978.

Bousman, W. G.; Sharpe, D. L.; and Ormiston, R. A. "An Experimental Study of Techniques for Increasing the Lead-Lag Damping of Soft Inplane Hingeless Rotors." AHS Forum, 1976.

Bowden, T. H., and Shockey, G. A. "A Wind-Tunnel Investigation of the Aerodynamic Environment of a Full-Scale Helicopter Rotor in Forward Flight." USAAVLABS TR 70-35, July 1970.

Bowes, M. A. "Test and Evaluation of a Quiet Helicopter Configuration HH-43B." The Journal of the Acoustical Society of America, 54:5 (November 1973).

Bowes, M. A. "Helicopter Noise Reduction Design Trade-Off Study." Federal Aviation Administration, Report FAA-AEQ-77-4, January 1977.

Bowes, M. A. "Engine/Airframe/Drive Train Dynamic Interface Documentation." USARTL TR 78-14, June 1978.

Boxwell, D. A.; Schmitz, F. H.; and Hanks, M. L. "In-Flight Far-Field Measurements of Helicopter Impulsive Noise." European Rotorcraft and Powered Lift Aircraft Forum, Southampton, England, September 1975.

Boxwell, D. A.; Yu, Y. H.; and Schmitz, F. H. "Hovering Impulsive Noise — Some Measured and Calculated Results." NASA CP 2052, May 1978.

Bradley, F. E. "An Expression for Rotor Blade Section Loading Including Reversed Flow Effects." JAHS, 1:4 (October 1956).

Bragg, T. S. "Acoustical Study of the CH-47B (Chinook) Helicopter," U.S. Army Human Engineering Laboratories, TN 4-68, March 1968.

Bramwell, A. R. S. "Longitudinal Stability and Control of the Single-Rotor Helicopter." ARC R&M 3104, January 1957.

Bramwell, A. R. S. "The Longitudinal Stability and Control of the Tandem-Rotor Helicopter. The Lateral Stability and Control of the Tandem Rotor Helicopter." ARC R&M 3223, January 1960.

Bramwell, A. R. S. "A Theory of the Aerodynamic Interference between a Helicopter Rotor Blade and a Fuselage and Wing in Hovering and Forward Flight." J. Sound Vib., 3:3 (1966).

Bramwell, A. R. S. "A Method for Calculating the Stability and Control Derivatives of Helicopters with Hingeless Rotors." London University, Research Memorandum Aero 69/4, 1969.

Bramwell, A. R. S. "Some Remarks on the Induced Velocity Field of a Lifting Rotor and on Glauert's Formula." ARC CP No. 1301, October 1971.

Bramwell, A. R. S. "An Introduction to Helicopter Air Resonance." ARC R&M 3777, September 1972.

Bramwell, A. R. S. "Helicopter Dynamics." London: Edward Arnold Publishers, 1976.

Bramwell, A. R. S. "On the Static Pressure in the Wake of a Hovering Rotor." Vertica, 1:3 (1977).

Brandt, D. E. "Vibration Control in Rotary Wing Aircraft." AGARD Conference Proceedings No. 7, Paris, January 1966.

Bratanow, T., and Ecer, A. "Sensitivity Analysis of Torsional Vibration

Characteristics of Helicopter Rotor Blades." NASA CR-2379, CR-2380, March 1974.

Bratanow, T., and Ecer, A. "Sensitivity of Rotor Blade Vibration Characteristics to Torsional Oscillations." Journal of Aircraft, 11:7 (July 1974).

Braun, J. F., and Giessler, F. J. "CH-54A Skycrane Helicopter Flight Loads Investigation Program." USAAVLABS TR 66-58, June 1966.

Breguet, L. "The Gyroplane – Its Principles and Its Possibilities." NACA TM 816, January 1937.

Briardy, F. J.; La Forge, S. V.; and Neff, J. R. "Rotor/Wing Final Technical Report." Hughes Tool Company, Report HTC-AD 69-12A, March 1970.

Briczinski, S. J. "Validation of the Rotorcraft Flight Simulation Program (C81) for Articulated Rotor Helicopters through Correlation with Flight Data." USAAMRDL TR 76-4, May 1976.

Briczinski, S. J. "Analytical Investigation of Gust Suppression Techniques for the CH-53 Helicopter." NASA CR 145013, 1976.

Briczinski, S. J., and Cooper, D. E. "Flight Investigation of Rotor/Vehicle State Feedback." NASA CR 132546, 1975.

Briczinski, S. J., and Karas, G. R. "Criteria for Externally Suspended Helicopter Loads." USAAMRDL TR 71-61, November 1971.

Brightwell, V. L.; Peters, M. D.; and Sanders, J. C. "Comparison of Hovering Performance of Helicopters Powered by Jet-Propulsion and Reciprocating Engines." NACA RM E7K21, June 1948.

Broll, C. "Noise Measurements in the Modane Large Wind Tunnel." Royal Aircraft Establishment, Library Translation 1683, March 1973.

Brooks, G. W. "Analytical Determination of the Natural Coupled Frequencies of Tandem Helicopters." JAHS, 1:3 (July 1956).

Brooks, G. W., and Baker, J. E. "An Experimental Investigation of the Effect of Various Parameters Including Tip Mach Number on the Flutter of Some Model Helicopter Rotor Blades." NACA TN 4005, September 1958.

Brooks, G. W., and Leonard, H. W. "An Analysis of the Flapwise Bending Frequencies and Mode Shapes of Rotor Blades Having Two Flapping Hinges to Reduce Vibration Levels." NASA TN D-633, December 1960.

Brooks, G. W., and Silviera, M. A. "Some Recent Studies in Structural Dynamics of Rotor Aircraft." NASA TN D-737, March 1961.

Brooks, G. W., and Sylvester, M. A. "The Effect of Control System Stiffness and Forward Speed on the Flutter of a 1/10-Scale Dynamic Model of a Two-Blade Jet-Driven Helicopter Rotor." NACA TN 3376, April 1955.

Brotherhood, P. "Flight Measurements of Helicopter Rotor Airfoil Characteristics and Some Comparisons with Two-Dimensional Wind Tunnel Results." AGARD Conference Proceedings No. CP-187, Valloire, France, 1975.

Brotherhood, P. "An Investigation in Flight of the Induced Velocity Distribution under a Helicopter Rotor When Hovering." ARC R&M 2521, June 1947.

Brotherhood, P. "Flow through a Helicopter Rotor in Vertical Descent." ARC R&M 2735, July 1949.

Brotherhood, P. "Some Aerodynamic Measurements in Helicopter Flight Research." The Aeronautical Journal, 79:778 (October 1975).

Brothrrhood, P., and Riley, M.J. "Flight Experiments on Aerodynamic Features Affecting Helicopter Blade Design." Vertica, 2:1 (1978).

Brotherhood, P., and Stewart, W. "An Experimental Investigation of the Flow through a Helicopter Rotor in Forward Flight." ARC R&M 2734, September 1949.

Brown, D. "Baseline Noise Measurements of Army Helicopters." USAAMRDL TR 71-36, July 1971.

Brown, E. L., and Fischer, J. N. "Comparative Projections of Low-Disk-Loading VTOL Aircraft for Civil Applications." Journal of Aircraft, 5:5 (September-October 1968).

Brown, E. L., and Schmidt, P. S. "The Effect of Helicopter Pitching Velocity on Rotor Lift Capability." JAHS, 8:4 (October 1963).

Brunelle, E. J., Jr. "Investigation of Rotor Response to Vibratory Aerodynamic Inputs – An Aerodynamic Strip Theory for Rotary Wings Executing Simple Harmonic Motions Near a Ground Plane." Wright Air Development Center, WADC TR 58-87, Part II, October 1958.

Brunelle, E. J., Jr., and Robertson, S. R. "The Transient Aeroelastic Response of Rotor Systems in Axial Flight." U.S. Army Watervliet Arsenal, TR WVT-7141, June 1971.

Bryant, W. B.; Cattel, J. J.; Russell, W. A.; and Trueblood, R. B. "VTOL Advanced Flight Control System Studies for All-Weather Flight." USAAMRDL TR-75-13, July 1975.

Bryant, W. B., and Trueblood, R. B. "Use of Programmable Force Feel for Handling Qualities Improvement in a Helicopter Velocity Flight Control System." AHS Forum, 1975.

Bryson, A. E.; Chasteen, L. H.; Hall, W. E.; and Mohr, R. L. "Studies of Control and Guidance for Rotary-Wing VTOL Vehicles." Stanford University, SUDAAR No. 419, March 1971.

Buffum, R. S., and Robertson, W. T. "A Hover Augmentation System for Helicopters." Journal of Aircraft, 4:4 (July-August 1967).

Burkam, J. E.; Capurso, V.; and Yntema, R. T. "A Study of the Effect of Systematic Variations of Rotor Blade Planform, Twist and Mass Distribution on Hub Loads for a Three-Bladed Helicopter Rotor." Boeing Vertol Report R175, July 1959.

Burkam, J. E., and Miao, W.–L. "Exploration of Aeroelastic Stability

Boundaries with a Soft-Inplane Hingeless-Rotor Model." JAHS, 17:4 (October 1972).

Burpo, F. B., and Lynn, R. R. "Measurement of Dynamic Airloads on a Full-Scale Semirigid Rotor." TCREC 62-42, December 1962.

Burton, T. E., and Blevins, R. D. "Vortex Shedding Noise from Oscillating Cylinders." Journal of the Acoustical Society of America, 60:3 (September 1976).

Calcaterra, P. C., and Schubert, D. W. "Isolation of Helicopter Rotor-Induced Vibrations Using Active Elements." USAAVLABS TR 69-8, June 1969.

Callan, W. M.; Houck, J. A.; and DiCarlo, D.J. "Simulation Study of Intercity Helicopter Operations under Instrument Conditions to Category I Minimums." NASA TN D-7786, December 1974.

Campbell, J. P. "Status of V/STOL Research and Development in the United States." Journal of Aircraft, 1:3 (May-June 1964).

Campbell, R., et al. "Heavy Lift Helicopter – Cargo Handling ATC Program." USAAMRDL TR 74-97, January 1976.

Cannon, J. A., and Niehaus, C. Q. "Practical Instrument Flying in the Helicopter." Aeronautical Engineering Review, 9:6 (June 1950).

Capurso, V.; Yntema, R. T.; and Gabel, R. "Helicopter Rotor Hub Vibratory Forces Systematic Variation of Flexible Rotor Blade Parameters." Boeing Vertol Company, Report R244, May 1961.

Caradonna, F. X., and Isom, M. P. "Subsonic and Transonic Potential Flow over Helicopter Rotor Blades." AIAA Journal, 10:12 (December 1972).

Caradonna, F. X., and Isom M. P. "Numerical Calculation of Unsteady Transonic Potential Flow over Helicopter Rotor Blades." AIAA Journal, 14:4 (April 1976).

Caradonna, F. X., and Philippe, J. J. "The Flow over a Helicopter Blade Tip in the Transonic Regime." Vertica, 2:1 (1978).

Cardinale, S. V. "Soft Inplane Matched-Stiffness/Flexure-Root-Blade Rotor System Summary Report." USAAVLABS TR 68-72, August 1969.

Carlson, R. G.; Blackwell, R. H.; Commerford, G. L.; and Mirick, P. H. "Dynamic Stall Modeling and Correlation with Experimental Data on Airfoils and Rotors." NASA SP-352, February 1974.

Carlson, R. G., and Cassarino, S. J. "Aeroelastic Analysis of a Telescoping Rotor Blade." USAAMRDL TR 73-48, August 1973.

Carlson, R. G., and Hilzinger, K. D. "Analysis and Correlation of Helicopter Rotor Blade Response in a Variable Inflow Environment." USAAVLABS TR 65-51, September 1965.

Carlson, R. M., and Kerr, A. W. "Integrated Rotor/Body Loads Prediction." AGARD Conference Proceedings No. 122, Milan, Italy, March 1973.

Carlson, R. M., and Wong, J. T. "An Exact Solution for the Static Bending of Uniform Rotating Beams." JAHS, 23:4 (October 1978).

Carpenter, P. J. "Effect of Wind Velocity on Performance of Helicopter Rotors As Investigated with the Langley Helicopter Apparatus." NACA TN 1698, October 1948.

Carpenter, P. J. "Effects of Compressibility on the Performance of Two Full-Scale Helicopter Rotors." NACA Report 1078, 1952.

Carpenter, P. J. "Lift and Profile-Drag Characteristics of an NACA 0012 Airfoil Section As Derived from Measured Helicopter Rotor Hovering Performance, NACA TN 4357, September 1958.

Carpenter, P. J., and Fridovich, B. "Effect of a Rapid Blade-Pitch Increase on the Thrust and Induced-Velocity Response of a Full-Scale Helicopter Rotor." NACA TN 3044, November 1953.

Carpenter, P. J., and Paulnock, R. S. "Hovering and Low-Speed Performance and Control Characteristics of an Aerodynamic-Servo Controlled Helicopter Rotor System As Determined on the Langley Helicopter Tower." NACA TN 2086, May 1950.

Carpenter, P. J., and Peitzer, H. E. "Response of a Helicopter Rotor to Oscillatory Pitch and Throttle Movements." NACA TN 1888, June 1949.

Carr, L. W.; McAlister, K. W.; and McCroskey, W. J. "Analysis of the Development of Dynamic Stall Based on Oscillating Airfoil Measurements." NASA TN D-8382, January 1977.

Carta, F. O. "An Analysis of the Stall Flutter Instability of Helicopter Rotor Blades." JAHS, 12:4 (October 1967).

Carta, F. O. "Effect of Unsteady Pressure Gradient Reduction on Dynamic Stall Delay." Journal of Aircraft, 8:10 (October 1971).

Carta, F. O.; Casellini, L. M.; Arcidiacono, P. J.; and Elman, H. L. "Analytical Study of Helicopter Rotor Stall Flutter." AHS Forum, 1970.

Carta, F. O.; Commerford, G. L.; and Carlson, R. G. "Determination of Airfoil and Rotor Blade Dynamic Stall Response." JAHS, 18:2 (April 1973).

Carta, F. O.; Commerford, G. L.; Carlson, R. G.; and Blackwell, R. H. "Investigation of Airfoil Dynamic Stall and Its Influence on Helicopter Control Loads." USAAMRDL TR 72-51, September 1972.

Carta, F. O., and Niebanck, C. F. "Prediction of Rotor Instability at High Forward Speeds – Stall Flutter." USAAVLABS TR 68-18C, February 1969.

Carter, E. S., Jr. "Technological Contributions of the CH-53A Transport Helicopter Development Program." Journal of Aircraft, 2:5 (September-October 1965).

Cassarino, S. J. "Effect of Root Cutout on Hover Performance." AFFDL TR 70-70, June 1970.

Cassarino, S. J. "Effect of Rotor Blade Root Cutout on Vertical Drag."

USAAVLABS TR 70-59, October 1970.

Castles, W., Jr. "A Direct Method of Estimating the Performance of a Helicopter in Powered Flight." JAS, 12:4 (October 1945).

Castles, W., Jr. "Approximate Solution for Streamlines about a Lifting Rotor Having Uniform Loading and Operating in Hovering or Low-Speed Vertical-Ascent Flight Conditions." NACA TN 3921, February 1957.

Castles, W., Jr. "Flow Induced by a Rotor in Power-On Vertical Descent." NACA TN 4330, July 1958.

Castles, W., Jr., and De Leeuw, J. H. "The Normal Component of the Induced Velocity in the Vicinity of a Lifting Rotor and Some Examples of its Application." NACA Report 1184, 1954.

Castles, W., Jr., and Ducoffe, A. L. "Static Thrust Analysis for Helicopter Rotors and Airplane Propellers." JAS, 15:5 (May 1948).

Castles, W., Jr., and Durham, H. L., Jr. "The Effect of Rotor Blade Planform on the Tip-Stall-Limited Top Speed of a Helicopter." JAHS, 1:4 (October 1956).

Castles, W., Jr., and Durham, H. L., Jr. "Distribution of Normal Component of Induced Velocity in Lateral Plane of a Lifting Rotor." NACA TN 3841, December 1956.

Castles, W., Jr., and Durham, H. L., Jr. "Tables for Computing the Instantaneous Velocities Induced at the Blade Axes of a Lifting Rotor in Forward Flight by the Skewed Helical Wake Vortices and a Method for Calculating the Resultant Airloads." Georgia Institute of Technology, June 1962.

Castles, W., Jr.; Durham, H. L., Jr.; and Kevorkian, J. "Normal Component of Induced Velocity for Entire Field of a Uniformly Loaded Lifting Rotor with Highly Swept Wake As Determined by Electromagnetic Analog." NASA TR R-41, 1959.

Castles, W., Jr., and Gray, R. B. "Empirical Relation between Induced Velocity, Thrust, and Rate of Descent of a Helicopter Rotor As Determined by Wind-Tunnel Tests on Four Model Rotors." NACA TN 2474, October 1951.

Castles, W., Jr., and New, N. C. "A Blade-Element Analysis for Lifting Rotors That Is Applicable for Large Inflow and Blade Angles and Any Reasonable Blade Geometry." NACA TN 2656, July, 1952.

Chalupnik, J. D., and Clark, L. T. "A Study of Sound Generation in Subsonic Rotors." NASA CR 146349, CR 146420, 1975.

Chang, T. T. "A Method For Predicting the Trim Constants and the Rotor Blade Loadings and Responses of a Single-Rotor Helicopter." USAAVLABS TR 67-71, November 1967.

Charles, B. D. "An Experimental/Theoretical Correlation of Model and Full-Scale Rotor Performance at High Advancing-Tip Mach Numbers and

Advance Ratios." USAAVLABS TR 70-69, January 1971.

Charles, B. D. "Acoustic Effects of Rotor-Wake Interaction during Low-Power Descent." American Helicopter Society Northeast Region Symposium, Hartford, Connecticut, March 1975.

Charles, B. D., and Tanner, W. H. "Wind Tunnel Investigation of Semirigid Full-Scale Rotors Operating at High Advance Ratios." USAAVLABS TR 69-2, January 1969.

Cheeseman, I. C. "Circulation-Controlled Rotor Aircraft." Aircraft Engineering, 41:7 (July 1969).

Cheeseman, I. C. "The Noise of Rotorcraft and Other VTOL Aircraft – A Review." The Aeronautical Journal, 75:726 (June 1971).

Cheeseman, I. C., and Bennett, W. E. "The Effect of the Ground on a Helicopter Rotor in Forward Flight." ARC R&M 3021, September 1955.

Cheeseman, I. C., and Seed, A. R. "The Application of Circulation Control by Blowing to Helicopter Rotors." Journal of the Royal Aeronautical Society, 71:679 (July 1967).

Chen, R. T. N., and Talbot, P. D. "An Exploratory Investigation of the Effects of Large Variations in Rotor System Dynamics Design Parameters on Helicopter Handling Characteristics in Nap-of-the-Earth Flight." JAHS, 23:3 (July 1978).

Chen, R. T. N.; Talbot, P. D.; Gerdes, R. M.; and Dugan, D. C. "A Piloted Simulator Investigation of Augmentation Systems to Improve Helicopter Nap-of-the-Earth Handling Qualities." AHS Forum, 1978.

Cheney, M. C., Jr. "The ABC Helicopter." JAHS, 14:4 (October 1969).

Cheney, M. C., Jr. "Results of Preliminary Studies of a Bearingless Helicopter Rotor Concept." JAHS, 17:4 (October 1972).

Cheng, Y. "Application of Active Control Technology to Gust Alleviation System for Tilt Rotor Aircraft." NASA CR 137958, November 1976.

Chigier, N. A., and Corsiglia, V. R. "Tip Vortices – Velocity Distributions." AHS Forum, 1971.

Chopra, I. "Nonlinear Dynamic Response of Wind Turbine Rotor." Sc.D. Thesis, Massachusetts Institute of Technology, February 1977.

Chopra, I., and Dugundji, J. "Nonlinear Dynamic Response of a Wind Turbine Rotor under Gravitational Loading." AIAA Journal, 16:8 (August 1978).

Chopra, I., and Johnson, W. "Flap-Lag-Torsion Aeroelastic Stability of Circulation-Controlled Rotors in Hover." AHS Forum, 1978.

Chou, H. F., and Fanucci, J. B. "Helicopter Lifting Surface Theory with Force Free Wakes." West Virginia University, Aerospace Engineering Report TR-44, February 1975.

Chou, H. F., and Fanucci, J. B. "Nonlinear Helicopter Rotor Lifting Surface Theory." West Virginia University, Aerospace Engineering Report TR-48, April 1976.

Chou, P. C. "Pitch-Lag Instability of Helicopter Rotors." JAHS, 3:3 (July 1958).

Chu, S., and Widnall, S. E. "Lifting-Surface Theory for a Semi-Infinite Wing in Oblique Gust." AIAA Journal, 12:12 (December 1974).

Churchill, G. B., and Harrington, R. D. "Parasite-Drag Measurements of Five Helicopter Rotor Hubs." NASA Memo 1-31-59L, February 1959.

Clark, D. R. "Can Helicopter Rotors Be Designed for Low Noise and High Performance." AHS Forum, 1974.

Clark, D. R. "Aerodynamic Design Rationale for the Fan-in-Fin on the S-67 Helicopter." AHS Forum, 1975.

Clark, D. R., and Arnoldi, D. R. "Rotor Blade Boundary Layer Calculation Programs." USAAVLABS TR 71-1, March 1971.

Clark, D. R., and Lawton, T. D. "Rotor Blade Boundary Layer Measurement Hardware Feasibility Demonstration." NASA CR 112194, 1972.

Clark, D. R., and Leiper, A. C. "The Free Wake Analysis, A Method for the Prediction of Helicopter Rotor Hovering Performance." JAHS, 15:1 (January 1970).

Clark, J. W., and Miller, D. P. "Research on Factors Influencing Handling Qualities for Precision Hovering and Gun Platform Tasks." AHS Forum, 1965.

Clark, L. T.; Chalupnik, J. D.; and Hodder, B. "Wake-Related Sound Generation from Isolated Airfoils." Journal of the Acoustical Society of America, 59:1 (January 1976).

Clarke, A. E., and Bramwell, A. R. S. "Selected Aspects of the Aerodynamics of Rotorcraft." The Aeronautical Journal, 72:686 (February 1968).

Clay, L. E.; Braun, J. F.; Chestnutt, D.; and Bartek, L. "UH-1B Helicopter Flight Loads Investigation Program." USAAVLABS TR 66-46, May 1966.

Cockayne, W.; Rusnak, W.; and Shub, L. "Digital Flight Control and Landing System for the CH-46C Helicopter." NASA CR 111024, May 1970.

Coleman, R. P. "A Preliminary Theoretical Study of Helicopter-Blade Flutter Involving Dependence upon Coning Angle and Pitch Setting." NACA MR L6G12, July 1946.

Coleman, R. P., and Feingold, A. M. "Theory of Self-Excited Mechanical Oscillations of Helicopter Rotors with Hinged Blades." NACA Report 1351, 1958.

Coleman, R. P.; Feingold, A. M.; and Stempin, C. W. "Evaluation of the Induced-Velocity Field of an Idealized Helicopter Rotor." NACA ARR L5E10, June 1945.

Coleman, R. P., and Stempin, C. W. "A Preliminary Theoretical Study of Aerodynamic Instability of a Two-Blade Helicopter Rotor." NACA RM L6H23, March 1947.

Condon, G. W. "Rotor Systems Research Aircraft." AGARD Conference

Proceedings No. 172, Hampton, Virginia, November 1974.

Connor, A. B. "A Summary of Operating Conditions Experienced by Three Military Helicopters and a Mountain-Based Commercial Helicopter." NASA TN D-432, October 1960.

Connor, A. B., and Ludi, L. H. "A Summary of Operating Conditions Experienced by Two Helicopters in a Commercial and Military Operation." NASA TN D-251, April 1960.

Connor, A. B., and O'Bryan, T. C. "A Brief Evaluation of Helicopter Wake As a Potential Operational Hazard to Aircraft." NASA TN D-1227, March 1962.

Connor, A. B., and Tapscott, R. J. "A Flying Qualities Study of a Small Ram-Jet Helicopter." NASA TN D-186, April 1960.

Cook, C. V. "The Structure of the Rotor Blade Tip Vortex." AGARD Conference Proceedings No. 111, Marseilles, France, September 1972.

Cook, W. L., and Poisson-Quinton, P. "A Summary of Wind Tunnel Research on Tilt Rotors from Hover to Cruise Flight." AGARD Conference Proceedings No. 111, Marseilles, France, September 1972.

Cooper, D. E. "YUH-60A Stability and Control." JAHS, 23:3 (July 1978).

Cooper, D.; Hansen, K. C.: and Kaplita, T. T. "Single Rotor Helicopter Dynamics Following Power Failure at High Speeds." USAAVLABS TR 66-30, June 1966.

Corliss, L. D., and Talbot, P. D. "A Failure Effects Simulation of a Low Authority Flight Control Augmentation System on a UH-1H Helicopter." NASA TM 73258, August 1977.

Costes, J. J. "Computation of Unsteady Aerodynamic Forces on Helicopter Rotor Blades." NASA TT F-15039, August 1973.

Costes, J. J. "Introduction of Unsteady Separation into Acceleration Potential Theory – Application to Helicopters." European Space Agency, ESA TT-307, July 1976.

Costes, J. J. "Rotor Response Prediction with Non-Linear Aerodynamic Loads on the Retreating Blade." Vertica, 2:2 (1978).

Cotton, L. S. "Three-Axis Fluidic/Electronic Automatic Flight Control System Flight Test Report." USAAMRDL TR 74-62, August 1974.

Cox, C. R. "Helicopter Noise and Passive Defense." AHS Forum, 1963.

Cox, C. R. "Rotor Noise Measurements in Wind Tunnels." CAL/AVLABS Symposium, Buffalo, New York, June 1969.

Cox, C. R. "Helicopter Noise Reduction and Its Effects on Operations." JAHS, 15:1 (January 1970).

Cox, C. R. "Aerodynamic Sources of Rotor Noise." JAHS, 18:1 (January 1973).

Cox, C. R., and Lynn, R. R. "A Study of the Origin and Means of Reducing Helicopter Noise." TCREC TR 62-73, November 1962.

Cox, T. L.; Johnson, R. B.; and Russell, S. W. "Dynamic Loads and Structural Criteria." USAAMRDL TR 75-9, April 1975.

Craig, R. R., Jr. "Rotating Beam with Tip Mass Analyzed by a Variational Method." Journal of the Acoustical Society of America, 35:7 (July 1963).

Cresap, W. L. "Development and Tests of Multi-Bladed Semi-Rigid Rotor Systems." JAHS, 5:2 (April 1960).

Cresap, W. L. "Rigid Rotor Development and Flight Tests." JAHS, 7:2 (April 1962).

Cresap, W. L., and Lynn R. R. "Research Flight Tests of the High Performance Iroquois." AHS Forum, 1963.

Cresap, W. L.; Myers, A. W.; and Viswanthan, S. P. "Design and Development Tests of a Four-Bladed Light Helicopter Rotor System." AHS Forum, 1978.

Crews, S. T.; Hohenemser, K. H.; and Ormiston, R. A. "An Unsteady Wake Model for a Hingeless Rotor." Journal of Aircraft, 10:12 (December 1973).

Crim, A. D. "Gust Experience of a Helicopter and an Airplane in Formation Flight." NACA TN 3354, December 1954.

Crim, A. D. "Hovering and Low Speed Controllability of VTOL Aircraft." JAHS, 4:1 (January 1959).

Crim, A. D., and Hazen, M. E. "Normal Accelerations and Operating Conditions Encountered by a Helicopter in Air-Mail Operations." NACA TN 2714, June 1952.

Crim, A. D.; Reeder, J. P.; and Whitten, J. B. "Initial Results of Instrument Flying Trials Conducted in a Single-Rotor Helicopter." NACA Report 1137, 1953.

Crimi, P. "Theoretical Preduction of the Flow in the Wake of a Helicopter Rotor." Cornell Aeronautical Laboratory, Report CAL BB-1994-S, September 1965.

Crimi, P. "Prediction of Rotor Wake Flows." CAL/AVLABS Symposium, Buffalo, New York, June 1966.

Crimi, P. "A Method for Analyzing the Aeroelastic Stability of a Helicopter Rotor in Forward Flight." NASA CR 1332, August 1969.

Crimi, P. "Stability of Dynamic Systems with Periodically Varying Parameters." AIAA Journal, 8:10 (October 1970).

Crimi, P. "Dynamic Stall." AGARD AG 172, November 1973.

Crimi, P. "Analysis of Stall Flutter of a Helicopter Rotor Blade." NASA CR-2322, November 1973.

Crimi, P. "Analysis of Helicopter Rotor Blade Stall Flutter." Journal of Aircraft, 11:7 (July 1974).

Crimi, P. "Analysis of Helicopter Rotor Blade Torsional Oscillations Due

to Stall." NASA CR 2573, September 1975.
Crimi, P., and Reeves, B. L. "A Method for Analyzing Dynamic Stall of Helicopter Rotor Blades." NASA CR 2009, May 1972.
Crimi, P., and Reeves, B. L. "Analysis of Leading Edge Separation Bubbles on Airfoils." AIAA Journal, 14:11 (November 1976).
Crimi, P., and White, R. P., Jr. "Investigation of the Aeroelastic Characteristics of a Jet-Flap Helicopter Rotor in Hovering Flight." JAHS, 7:2 (April 1962).
Critzos, C. C.; Heyson, H. H.; and Boswinkle, R. W., Jr. "Aerodynamic Characteristics of NACA 0012 Airfoil Section at Angles of Attack from 0° to 180°." NACA TN 3361, January 1955.
Crocco, G. A. "Inherent Stability of Helicopters." NACA TM 234, October 1923.
Cronkhite, J. D. "Development, Documentation, and Correlation of a NASTRAN Vibration Model of the AH-1G Helicopter Airframe." NASA TM X-3428, October 1976.
Crossley, T. R., and Porter, B. "Synthesis of Helicopter Stabilization Systems Using Modal Control Theory." Journal of Aircraft, 9:1 (January 1972).
Cruz, E. S.; Gorenberg, N. B.; and Kerr, A. W. "A Flight Envelope Expansion Study for the XH-51A Compound Helicopter." USAAVLABS TR 69-78, October 1969.
Csencsitz, T. A.; Fanucci, J. B.; and Chou, H. F. "Nonlinear Helicopter Rotor Lifting Surface Theory." West Virginia University, Aerospace Engineering Report TR-35, September 1973.
Culhane, K. V. "The Soviet Attack Helicopter." U.S. Army Institute for Advanced Russian and East European Studies, March 1977.
Curry, P. R., and Matthews, J. T., Jr. "Advanced Rotary-Wing Handling Qualities." AHS Forum, 1964.
Curties, M. C., "Helicopter All-Weather Operation – Equipment for the Transport Role." The Aeronautical Journal, 72:687 (March 1968).
Curtiss, H. C., Jr. "Some Basic Considerations Regarding the Longitudinal Dynamics of Aircraft and Helicopters." Princeton University, Report No. 562, July 1961.
Curtiss, H. C., Jr. "An Analytical Study of the Dynamics of VTOL Aircraft in Unsteady Flight." AHS Forum, 1965.
Curtiss, H. C. Jr. "Dynamic Stability of V/STOL Aircraft at Low Speeds." Journal of Aircraft, 7:1 (January-February 1970).
Curtiss, H. C., Jr. "Complex Coordinates in Near Hovering Rotor Dynamics." Journal of Aircraft, 10:5 (May 1973).
Curtiss, H. C., Jr. "Sensitivity of Hingeless Rotor Blade Flap-Lag Stability in Hover to Analytical Modeling Assumptions." NASA CR 137967, January 1975.

Curtiss, H. C., Jr., and Putman, W. F. "An Experimental Investigation of the Flap-Lag Stability of a Hingeless Rotor with Comparable Levels of Hub and Blade Stiffness in Hovering Flight." NASA CR 151924, June 1976.

Curtiss, H. C., Jr., and Shupe, N. K. "A Stability and Control Theory for Hingeless Rotors." AHS Forum, 1971.

Dadone, L. U. "U.S. Army Helicopter Design Datcom, Volume I, Airfoils." U.S. Army Air Mobility Research and Development Laboratory, May 1976.

Dadone, L. U. "Two-Dimensional Wind Tunnel Test of an Oscillating Rotor Airfoil." NASA CR 2914 and CR 2915, December 1977.

Dadone, L. "Rotor Airfoil Optimization: An Understanding of the Physical Limits." AHS Forum, 1978.

Dadone, L. U., and Fukushima, T. "Investigation of Rotor Blade Element Airloads for a Teetering Rotor in the Blade Stall Regime." NASA CR 137534, September 1974.

Dadone, L. U., and Fusushima, T. "A Review of Design Objectives for Advanced Helicopter Rotor Airfoils." American Helicopter Society Northeast Region Symposium, Hartford, Connecticut, March 1975.

Dajani, J. S.; Warner, D.; Epstein, D.; and O'Brien, J. "The Role of the Helicopter in Transportation." NASA CR 146351, January 1976.

d'Amdra, F., and Damongeot, A. "Annoyance of Helicopter Impulsive Noise." NASA CP 2052, May 1978.

Dat, R. "Representation of a Lifting Line in an Arbitrary Motion by a Line of Acceleration Doublets." NASA TT F-12952, May 1970.

Dat, R. "Lifting Surface Theory Applied to Fixed Wings and Propellers." European Space Research Organization, ESRO TT-90, September 1974.

Dat, R. "Unsteady Aerodynamics of Helicopter Blades." European Space Agency, ESA TT-327, October 1976.

Daughaday, H. "Suppression of Transmitted Harmonic Rotor Loads by Blade Pitch Control." USAAVLABS TR 67-14, November 1967.

Daughaday, H. "Suppression of Transmitted Harmonic Rotor Loads by Blade Pitch Control." JAHS, 13:2 (April 1968).

Daughaday, H., and Du Waldt, F. "The Effect of Blade Root Properties on the Natural Mode Shapes, Bending Moments, and Shears of a Model Helicopter Rotor Blade," JAHS 1:2 (April 1956).

Daughaday, H.; Du Waldt, F.; and Gates, C. "Investigation of Helicopter Blade Flutter and Load Amplification Problems." JAHS, 2:3 (July 1957).

Daughaday, H., and Piziali, R. A. "An Improved Computational Model for Predicting the Unsteady Aerodynamic Loads of Rotor Blades." JAHS, 11:4 (October 1966).

Davenport, F. J. "A Method for Computation of the Induced Velocity Field of a Rotor in Forward Flight, Suitable for Application to Tandem Rotor Configurations." JAHS, 9:3, (July 1964).

Davenport, F. J. "Rotor Performance in the Light of Recent Advances in Aerodynamics Methodology." AHS Forum, 1965.

Davenport, F. J., and Front, J. V. "Airfoil Sections for Helicopter Rotors – A Reconsideration." AHS Forum, 1966.

Davidson, I. M., and Hargest, T. J. "Helicopter Noise." Journal of the Royal Aeronautical Society, 69:653 (May 1965).

Davidson, J. K.; Havey, C. T.; and Sherrieb, H. E. "Fan-in-Fin Antitorque Concept Study." USAAMRDL TR 72-44, July 1972.

Davis, J. M.; Landis, K. H.; and Leet, J. R. "Development of Heavy Lift Helicopter Handling Qualities for Precision Cargo Operations." AHS Forum, 1975.

Davis, J. M.; Kannon, J. F.; Leone, P. F.; and McCafferty, H. A. "Study of Tandem Rotor Helicopter Dynamics Following Power Failure at High Speed." USAAVLABS TR 65-72, November 1965.

Davis, J. M.; Bennett, R. L.; and Blankenship, B. L. "Rotorcraft Flight Simulation with Aeroelastic Rotor and Improved Aerodynamic Representation." USAAMRDL TR 74-10, June 1974.

Davis, S. J., and Stepniewski, W. Z. "Documenting Helicopter Operations from an Energy Standpoint." NASA CR-132578, November 1974.

Davis, S. J., and Wisniewski, J. S. "User's Manual for HESCOMP, The Helicopter Sizing and Performance Computer Program." Boeing Vertol Company, Report D210-10699-1, September 1973.

Davis, S. S. "Theory of Discrete Vortex Noise." AIAA Journal, 13:3, (March 1975).

Deardorff, J. C.; Freisner, A. L.; and Albion, N. "Flight Test Development of the Tactical Aircraft Guidance System." AHS Forum, 1973.

de Bothezat, G. "The General Theory of Blade Screws." NACA Report 29, 1919.

Deckert, W. H., and Ferry, R. G. "Limited Flight Evaluation of the XV-3 Aircraft." Air Force Flight Test Center, AFFTC TR 60-4, May 1960.

Deckert, W. H., and Hickey, D. H. "Summary and Analysis of Feasibility-Study Designs of V/STOL Transport Aircraft." Journal of Aircraft, 7:1 (January-February 1970).

Deckert, W. H., and McCloud, J. L. III. "Considerations of the Stopped Rotor V/STOL Concept." JAHS, 13:1 (January 1968).

de Guillenchmidt, P. "Calculation of the Bending Stresses in Helicopter Rotor Blades." NACA TM 1312, March 1951.

De Larm, L. N. "Whirl Flutter and Divergence Aspects of Tilt-Wing and Tilt-Rotor Aircraft." Air Force V/STOL Technology and Planning Conference, Las Vegas, Nevada, September 1969.

Deming, A. F. "Noise from Propellers with Symmetrical Sections at Zero Blade Angle." NACA TN 605, July 1937.

Deming, A. F. "Noise from Propellers with Symmetrical Sections at Zero Blade Angle, II." NACA TN 679, December 1938.

Deming, A. F. "Propeller Rotational Noise Due to Torque and Thrust." NACA TN 747, January 1940.

Desjardins, R. A., and Hooper, W. E. "Rotor Isolation of the Hingeless Rotor B0-105 and YUH-61 Helicopters." Vertica, 2:2 (1978).

Desjardins, R. A., and Hooper, W. E. "Antiresonant Rotor Isolation for Vibration Reduction." AHS Forum, 1978.

De Tore, J. A., and Brown, E. L. "Summary of Design Studies and Results of Model Tests of The Folding-Proprotor Aircraft Concept." AFFDL TR 72-81, July 1972.

De Tore, J. A., and Gaffey, T. M. "The Stopped-Rotor Variant of the Proprotor VTOL Aircraft." JAHS, 15:3 (July 1970).

De Tore, J. A., and Sambell, K. W. "Conceptual Design Study of 1985 Commercial Tilt Rotor Transports." NASA CR 2544, CR 137602, CR 137765, May 1975.

Deutsch, M. L. "Ground Vibrations of Helicopters." JAS, 13:5 (May 1946).

Diamond, E. D., and Davis, J. M. "Heavy-Lift Helicopter Flight Control System Design." AGARD Conference Proceedings No. 148, Stuttgart, Germany, 1974.

Di Carlo, D. J. "A Summary of Operational Experiences of Three Light Observation Helicopters and Two Large Load-Lifting Military Helicopters." NASA TN D-4120, September 1967.

Di Carlo, D. J. "Operational Experiences of a Commercial Helicopter Flown in a Metropolitan Area." NASA TN D-8000, August 1975.

Di Carlo, D. J.; Kelley, H. L.; and Yenni, K. R. "An Exploratory Flight Investigation of Helicopter Sling-Load Placements Using a Closed-Circuit Television as a Pilot Aid." NASA TN D-7776, November 1974.

Dietz, C. R. "Simplified Aircraft Performance Methods: Power Required for Single and Tandem Rotor Helicopters in Hover and Forward Flight." U.S. Army Materiel Systems Analysis Agency, TR No. 78, August 1973.

Dingeldein, R. C. "Wind-Tunnel Studies of the Performance of Multirotor Configurations." NACA TN 3236, August 1954.

Dingeldein, R. C. "Considerations of Methods of Improving Helicopter Efficiency." NASA TN D-734, April 1961.

Dingeldein, R. C., and Schaefer, R. F. "Static-Thrust Tests of Six Rotor-Blade Designs on a Helicopter in the Langley Full-Scale Tunnel." NACA ARR L5F25b, September 1945.

Dingeldein, R. C., and Schaefer, R. F. "Full-Scale Investigation of the Aerodynamic Characteristics of a Typical Single-Rotor Helicopter in Forward

Flight," NACA Report 905, 1948.

Diprose, K. V. "Some Propeller Noise Calculations, Showing the Effect of Thickness and Planform." Royal Aircraft Establishment, Tech. Note No. M.S. 19, January 1955.

Dmitriyev, I. S., and Yesanlov, S. Yu. "Control Systems for Single-Rotor Helicopters." NASA TT F-636, March 1971.

Doblhoff, F. L. "Some Characteristics and Limitations of the Unloaded Rotor Compound Helicopter." JAHS, 4:1 (January 1959).

Dommasch, D. O. "A Method for the Stress Analysis of Helicopter Blades." JAS, 13:4 (April 1946).

Dommasch, D. O. Elements of Propeller and Helicopter Aerodynamics. New York: Pitman Publishing Corp., 1953.

Done, G. T. S. "A Simplified Approach to Helicopter Ground Resonance." The Aeronautical Journal, 78:761 (May 1974).

Done, G. T. S., and Hughes, A. D. "Reducing Vibration by Structural Modification." Vertica, 1:1 (1976).

Donham, R. E., and Cardinale, S. V. "Flight Test and Analytical Data for Dynamics and Loads in a Hingeless Rotor." Lockheed Report LR-26215, December 1973.

Donham, R. E.; Cardinale, S. V.; and Sachs, I. B. "Ground and Air Resonance Characteristics of a Soft Inplane Rigid Rotor System." JAHS, 14:4 (October 1969).

Donham, R. E., and Harvick, W. P. "Analysis of Stowed Rotor Aeroelastic Characteristics." JAHS, 12:1 (January 1967).

Donham, R. E.; Watts, G. A.; and Cardinale, S. V. "Dynamics of a Rigid-Rotor Controlled by a High-Speed Gyro As It Slows/Stops at High Forward Speed." AHS Forum, 1969.

Donovan, A. F., and Goland, M. "The Response of Helicopters with Articulated Rotors to Cyclic Blade Pitch Control." JAS, 11:4 (October 1944).

Donovan, R. F., and Leoni, R. D. "Flying Crane Parametric Study." JAHS, 8:1 (January 1963).

Dooley, L. W. "Rotor Blade Flapping Criteria Investigation." USAAMRDL TR 76-33, December 1976.

Dooley, L. W., and Ferguson, S. W. III. "Effect of Operational Envelope Limits on Teetering Rotor Flapping." USARTL TR 78-9, July 1978.

Dorand, R., and Boehler, G. D. "Application of the Jet-Flap Principle to Helicopters." JAHS, 4:3 (July 1959).

Douglas, L. L. "Development Problems of the Large Helicopter." Aeronautical Engineering Review, 13:4 (April 1954).

Douglas, L. L. "The Development of the Tandem Helicopter." JAHS, 3:1 (January 1958).

Dowell, E. H. "A Variational-Rayleigh-Ritz Modal Approach for Non-Uniform Twisted Rotor Blades Undergoing Large Bending and Torsional Motion." Princeton University, AMS Report No. 1193, November 1974.

Dowell, E. H., and Traybar, J. "An Experimental Study of the Nonlinear Stiffness of a Rotor Blade Undergoing Flap, Lag, and Twist Deformations." NASA CR 137968, January 1975; NASA CR 137969, December 1975.

Dowell, E. H.; Traybar, J.; and Hodges, D. H. "An Experimental-Theoretical Correlation Study of Nonlinear Bending and Torsion Deformations of a Cantilever Beam." J. Sound Vib., 50:4 (1977).

Drees, J. M. "A Theory of Airflow Through Rotors and Its Application to Some Helicopter Problems." Journal of the Helicopter Association of Great Britain, 3:2 (July-September 1949).

Drees, J. M. "High Speed Helicopter Rotor Design." JAHS, 8:3 (July 1963).

Drees, J. M., and Harvey, K. W. "Helicopter Gust Response at High Forward Speed." Journal of Aircraft, 7:3 (May-June 1970).

Drees, J. M., and Hendal, W. P. "Airflow Patterns in the Neighborhood of Helicopter Rotors." Aircraft Engineering, 23:266 (April 1951).

Drees, J. M., and McGuigan, M. J. "High Speed Helicopters and Compounds in Maneuvers and Gusts." AHS Forum, 1965.

Drees, J. M., and Wernicke, R. K. "An Experimental Investigation of a Second Harmonic Feathering Device on the UH-1A Helicopter." TCREC TR 62-109, June 1963.

Duberg, J. E., and Luecker, A. R. "Comparisons of Methods of Computing Bending Moments in Helicopter Rotor Blades in the Plane of Flapping." NACA ARR L5E23, August 1945.

Dugundji, J., and Aravamudan, K. "Stall Flutter and Nonlinear Divergence of a Two-Dimensional Flat Plate Wing." Massachusetts Institute of Technology, ASRL TR 159-6, July 1974.

Dugundji, J., and Chopra, I. "Further Studies of Stall Flutter and Nonlinear Divergence of Two-Dimensional Wings." NASA CR 144924, August 1975.

Duhon, J. M.; Harvey, K. W.; and Blankenship, B. L. "Computer Flight Testing of Rotorcraft." JAHS, 10:4 (October 1965).

Duke, F. H., and Hooper, W. E. "The Boeing Model 347 Advanced Technology Helicopter Program." AHS Forum, 1971.

Dukes, T. A. "Feedback Control of VTOL Aircraft." USAAVLABS TR 69-96, April 1970.

Dukes, T. A. "Elements of Helicopter Hovering and Near-Hover Operations with a Sling Load." U.S. Army Electronics Command, ECOM 02412-12, September 1972.

Dukes, T. A. "Maneuvering Heavy Sling Loads Near Hover." JAHS, 18:2

(April 1973), 18:3 (July 1973).

Dumond, R. C., and Simon, D. R. "Flight Investigation of Design Features of the S-67 Winged Helicopter." JAHS, 18:3 (July 1973).

Dutton, W. J. "Parametric Analysis and Preliminary Design of a Shaft-Driven Rotor System for a Heavy Lift Helicopter." USAAVLABS TR 66-56, February 1967.

Duvivier, J. F. "Study of Helicopter Rotor-Rotor Interference Effects on Hub Vibration." Air Force Systems Command, ASD TR 61-601, June 1962.

Du Waldt, F. A. "Wakes of Lifting Propellers (Rotors) in Ground Effect." Cornell Aeronautical Laboratory, Report CAL BB-1665-S-3, November 1966.

Du Waldt, F. "Flutter of Lightly Loaded Rotors in Forward Flight." Air Force Conference, Las Vegas, Nevada, September 1969.

Du Waldt, F. A.; Gates, C. A.; and Piziali, R. A. "Investigation of Helicopter Rotor Blade Flutter and Flapwise Bending Response in Hovering." Wright Air Development Center, WADC TR 59-403, August 1959.

Du Waldt, F. A., and Piziali, R. A. "Comparison of Theoretical and Experimental Flutter Results for a Full-Scale Helicopter Rotor." Wright Air Development Center, WADD TR 60-692, July 1960.

Du Waldt, F. A., and Statler, I. C. "Derivation of Rotor Blade Generalized Airloads from Measured Flapwise Bending Moment and Measured Pressure Distributions." USAAVLABS TR 66-13, March 1966.

Dwyer, H. A., and McCroskey, W. J. "Crossflow and Unsteady Boundary-Layer Effects on Rotating Blades." AIAA Journal, 9:8 (August 1971).

Eastman, S. E. "Comparative Cost and Capacity Estimates of Vertiports and Airports, 1975–1985." Journal of Aircraft, 8:8 (August 1971).

Edenborough, H. K. "Investigation of Tilt-Rotor VTOL Aircraft Rotor-Pylon Stability." Journal of Aircraft, 5:2 (March-April 1968).

Edenborough, H. K.; Gaffey, T. M.; and Weiberg, J. A. "Analyses and Tests Confirm Design of Proprotor Aircraft." AIAA Paper No. 72-803, August 1972.

Edwards, W. T., and Miao, W. "Bearingless Tail Rotor Loads and Stability." USAAMRDL TR 76-16, November 1977.

Ekquist, D. G. "Design and Wind Tunnel Test of a Model Helicopter Rotor Having an Independently Movable Inboard Panel." USAAVLABS TR 65-63, October 1965.

Ellis, C. W. "Effects of Articulated Rotor Dynamics on Helicopter Automatic Control System Requirements." Aeronautical Engineering Review, 12:7 (July 1953).

Ellis, C. W. "Alternatives to Large Single Lifting Rotors." AGARD Conference Proceedings No. 7, Paris, January 1966.

Ellis, C. W.; Acurio, J.; and Schneider, J. J. "Helicopter Propulsion Trends." AGARD Conference Proceedings No. 31, Ottawa, Canada, June 1968.

Ellis, C. W., and Jones, R. "Application of an Absorber to Reduce Helicopter Vibration Levels." JAHS, 8:3 (July 1963).

Elman, H. L.; Niebanck, C. F.; and Bain, L. J. "Prediction of Rotor Instability at High Forward Speeds – Flapping and Flap-Lag Instability." USAAVLABS TR 68-18E, February 1969.

Emery, J. H.; Sonneborn, W. G. O.; and Elam, C. B. "A Study of the Validity of Ground-Based Simulation Techniques for the UH-1B helicopter." USAAVLABS TR 67-72, December 1967.

Empey, R. W., and Ormiston, R. A. "Tail-Rotor Thrust on a 5.5-foot Helicopter Model in Ground Effect." AHS Forum, 1974.

Erickson, J. C., Jr. "A Continuous Vortex Sheet Representation of Deformed Wakes of Hovering Propellers." CAL/AVLABS Symposium, Buffalo, New York, June 1969.

Erickson, J. C., Jr., and Hough, G. R. "Fluctuating Flow Field of Propellers in Cruise and Static Operation." Journal of Aircraft, 7:1 (January-February 1970).

Ericsson, L. E. "Comment on Unsteady Airfoil Stall." Journal of Aircraft, 4:5 (September-October 1967).

Ericsson, L. E., and Reding, J. P. "Unsteady Airfoil Stall." NASA CR 66787, July 1969.

Ericsson, L. E., and Reding, J. P. "Dynamic Stall Simulation Problems." Journal of Aircraft, 8:7 (July 1971).

Ericsson, L. E., and Reding, J. P. "Unsteady Airfoil Stall, Review and Extension." Journal of Aircraft, 8:8 (August 1971).

Ericsson, L. E., and Reding, J. P. "Dynamic Stall of Helicopter Blades." JAHS, 17:1 (January 1972).

Ericsson, L. E., and Reding, J. P. "Stall-Flutter Analysis." Journal of Aircraft, 10:1 (January 1973).

Ericsson, L. E., and Reding, J. P. "Further Considerations of 'Spilled' Leading Edge Vortex Effects on Dynamic Stall." Journal of Aircraft, 14:6 (June 1977).

Ernsthausen, W. "The Source of Propeller Noise." NACA TM 825, May 1937.

Evans, W. T., and McCloud, J. L. III. "Analytical Investigation of a Helicopter Rotor Driven and Controlled by a Jet Flap." NASA TN D-3028, September 1965.

Evans, W. T., and Mort, K. W. "Analysis of Computed Flow Parameters for a Set of Sudden Stalls in Low-Speed Two-Dimensional Flow." NASA TN D-85, August 1959.

Fagan, C. H. "Elastomeric Bearing Application to Helicopter Tail Rotor

Designs." JAHS, 13:4 (October 1968).

Fagan, C. H. "Flight Evaluation of Elastomeric Bearings in an AH-1 Helicopter Main Rotor." USAAVLABS TR 71-16, March 1971.

Fail, R. A., and Eyre, R. C. W. "Loss of Static Thrust Due to a Fixed Surface under a Helicopter Rotor." Royal Aircraft Establishment, Tech. Note Aero 2008, July 1949.

Fail, R. A., and Eyre, R. C. W., "Downwash Measurements behind a 12-ft Diameter Helicopter in the 24-ft Wind Tunnel." ARC R&M 2810, September 1949.

Fail, R., and Squire, H. B. "24-ft Wind Tunnel Tests on Model Multi-Rotor Helicopters." Royal Aircraft Establishment, Report No. Aero 2207, June 1947.

Falabella, G., Jr., and Meyer, J. R., Jr. "Determination of Inflow Distributions from Experimental Aerodynamic Loading and Blade-Motion Data on a Model Helicopter Rotor in Hovering and Forward Flight." NACA TN 3492, November 1955.

Farassat, F. "The Acoustic Far-Field of Rigid Bodies in Arbitrary Motion." J. Sound Vib., 32:3 (1974).

Farassat, F. "Theory of Noise Generation from Moving Bodies with an Application to Helicopter Rotors." NASA TR R-451, December 1975.

Farassat, F., and Brown, T. J. "A New Capability for Predicting Helicopter Rotor and Propeller Noise Including the Effect of Forward Motion." NASA TM X-74037, June 1977.

Farassat, F.; Pegg, R. J.; and Hilton, D. A. "Thickness Noise of Helicopter Rotors at High Tip Speeds." AIAA Paper 75-453, March 1975.

Faulkner, H. B. "The Cost of Noise Reduction in Helicopters." Massachusetts Institute of Technology, FTL Report R71-5, November 1971.

Faulkner, H. B. "The Cost of Noise Reduction in Intercity Commercial Helicopters." Journal of Aircraft, 11:2 (February 1974).

Faulkner, H. B. "The Cost of Noise Reduction in Commercial Tilt Rotor Aircraft." NASA CR 137552, August 1974.

Faulkner, H. B. "A Computer Program for the Design and Evaluation of Tilt Rotor Aircraft." Massachusetts Institute of Technology, FTL TM 74-3, 1974.

Faulkner, H. B., and Swan, W. M. "The Cost of Noise Reduction for Departure and Arrival Operations of Commercial Tilt Rotor Aircraft." NASA CR 137803, June 1976.

Feaster, L.; Poli, C.; and Kirchhoff, R. "Dynamics of a Slung Load." Journal of Aircraft, 14:2 (February 1977).

Feistel, T. W., and Drinkwater, F. J. III. "Flight Tests of a One-Man Helicopter and a Comparison of Its Handling Qualities with Those of Larger VTOL Aircraft." NASA TN D-3060, October 1965.

Fenaughty, R. R., and Beno, E. A. "Airload, Blade Response, and Hub Force Measurements on the NH-3A Compound Helicopter." Journal of Aircraft, 6:5 (September-October 1969).

Fenaughty, R. R., and Noehren, W. L. "Composite Bearingless Tail Rotor for UTTAS." JAHS, 22:3 (July 1977).

Fertis, D. G. "Dynamic Response of Nonuniform Rotor Blades." Journal of Aircraft, 14:5 (May 1977).

Ffowcs Williams, J. E., and Hawkings, D. L., "Sound Generation by Turbulence and Surfaces in Arbitrary Motion." Philosophical Transactions of the Royal Society of London, 246A:1151 (May 1969).

Filotas, L. T. "Vortex Induced Wing Loads." AIAA Journal, 10:7 (July 1972).

Filotas, L. T. "Finite Chord Effects on Vortex Induced Wing Loads." AIAA Journal, 11:6 (June 1973).

Filotas, L. T. "Vortex Induced Helicopter Blade Loads and Noise." J. Sound Vib., 27:3 (1973).

Fink, M. R. "Prediction of Airfoil Tone Frequencies." Journal of Aircraft, 12:2 (February 1975).

Fink, M. R. "Experimental Evaluation of Theories for Trailing Edge and Incidence Fluctuation Noise." AIAA Journal, 13:11 (November 1975).

Fink, M. R. "Minimum On-Axis Noise for a Propeller or Helicopter Rotor." Journal of Aircraft, 15:10 (October 1978).

Fink, M. R.; Schlinker, R. H.; and Amiet, R. K. "Prediction of Rotating-Blade Vortex Noise from Noise of Nonrotating Blades." NASA CR 2611, March 1976.

Fisher, R. R., Jr.; Tompkins, J. E.; Bobo, C. J.; and Child, R. F. "An Experimental Investigation of the Helicopter Rotor Blade Element Airloads on a Model Rotor in the Blade Stall Regime." NASA CR 114424, September 1971.

Flannelly, W. G. "The Dynamic Antiresonant Vibration Isolator." AHS Forum, 1966.

Flannelly, W. G.; Bartlett, F. D., Jr.; and Forsberg, T. W. "Laboratory Verification of Force Determination." USAAMRDL TR 76-38, January 1977.

Flannelly, W. G.; Berman, A.; and Barnsby, R. M. "Theory of Structural Dynamic Testing Using Impedance Techniques." USAAVLABS TR 70-6, June 1970.

Flannelly, W. G.; Berman, A.; and Giansante, N. "Research on Structural Dynamic Testing by Impedance Methods." USAAMRDL TR 72-63, November 1972.

Flannelly, W. G., and Giansante, N. "Experimental Verification of System Identification." USAAMRDL TR 74-64, August 1974.

Flax, A. H. "The Bending of Rotor Blades." JAS, 14:1 (January 1947).

Flax, A. H., and Goland, L. "Dynamic Effects in Rotor Blade Bending." JAS, 18:12 (December 1951).

Flemming, R., and Ruddell, A. "RSRA Sixth-Scale Wind Tunnel Test." NASA CR 144964, November 1974.

Focke, H. "The Focke Helicopter." NACA TM 858, April 1938.

Focke, E. H. H. "German Thinking on Rotary Wing Development." Journal of the Royal Aeronautical Society, 69:653 (May 1965).

Fogarty, L. E. "The Laminar Boundary Layer on a Rotary Blade." JAS, 18:4 (April 1951).

Fogarty, L. E., and Sears, W. R. "Potential Flow around a Rotating, Advancing Cylindrical Blade." JAS, 17:9 (September 1950).

Foster, R. D. "A Rapid Performance Prediction Method for Compound Type Rotorcraft." JAHS, 2:4 (October 1957).

Foster, R. D.; Kidwell, J. C.; and Wells, C. D. "Analysis of Maneuverability Effects on Rotor/Wing Design Characteristics." USAAMRDL TR 74-26, February 1974.

Foulke, W. K. "Exploration of High-Speed Flight with the XH-51A Rigid Rotor Helicopter." U.S. Army Aviation Materiel Laboratories, USAAML TR 65-25, June 1965.

Fradenburgh, E. A. "The Helicopter and the Ground Effect Machine." JAHS, 5:4 (October 1960).

Fradenburgh, E. A. "High Performance Single Rotor Helicopter Study." TRECOM TR 61-44, April 1961.

Fradenburgh, E. A. "Aerodynamic Factors Influencing Overall Hover Performance." AGARD Conference Proceedings No. 111, Marseilles, France, September 1972.

Fradenburgh, E. A. "Application of a Variable Diameter Rotor System to Advanced VTOL Aircraft." AHS Forum, 1975.

Fradenburgh, E. A. "Aerodynamic Design of the Sikorsky S-76 Helicopter." AHS Forum, 1978.

Fradenburgh, E. A., and Chuga, G. M. "Flight Program on the NH-3A Research Helicopter." JAHS, 13:1 (January 1968).

Fradenburgh, E. A.; Hager, L. N.; and Keffort, N. F. K. "Evaluation of the TRAC Variable Diameter Rotor: Preliminary Design of a Full-Scale Rotor and Parametric Mission Analysis Comparisons." USAAMRDL TR 75-54, February 1976.

Fradenburgh, E. A., and Kiely, E. F. "Development of Dynamic Model Rotor Blades for High Speed Helicopter Research." JAHS 9:1 (January 1964).

Fradenburgh, E. A.; Murrill, R. J.; and Kiely, E. F. "Dynamic Model Wind Tunnel Tests of a Variable-Diameter, Telescoping-Blade Rotor System

(TRAC Rotor)." USAAMRDL TR 73-32, July 1973.

Fradenburgh, E. A., and Rabbott, J. P., Jr. "High Speed Helicopter Research." AHS Forum, 1962.

Fradenburgh, E. A., and Segel, R. M. "Model and Full Scale Compound Helicopter Research." AHS Forum, 1965.

Free, F. W. "Russian Helicopters." The Aeronautical Journal, 74:717 (September 1970).

Freeman, C. E., and Yeager, W. T., Jr. "Wind Tunnel Investigation of an Unpowered Helicopter Fuselage Model with V-Type Empennage." NASA TM X-3476, March 1977.

Freeman, F. D., and Bennett, R. L. "Application of Rotorcraft Flight Simulation Program (C81) to Predict Rotor Performance and Bending Moments for a Model Four-Bladed Articulated Rotor System." USAAMRDL TR 74-70, November 1974.

Frick, J. K., and Johnson, W. "Optimal Control Theory Investigation of Proprotor/Wing Response to Vertical Gust." NASA TM X-62384, September 1974.

Friedmann, P. "Investigation of Some Parameters Affecting the Stability of a Hingeless Helicopter Blade in Hover." NASA CR 114525, August 1972.

Friedmann, P. "Aeroelastic Instabilities of Hingeless Helicopter Blades." Journal of Aircraft, 10:10 (October 1973).

Friedmann, P. "Some Conclusions Regarding the Aeroelastic Stability of Hingeless Helicopter Blades in Hover and in Forward Flight." JAHS, 18:4 (October 1973).

Friedmann, P. "Recent Developments in Rotary-Wing Aeroelasticity." Journal of Aircraft, 14:11 (November 1977).

Friedmann, P. "Aeroelastic Modeling of Large Wind Turbines." JAHS, 21:4 (October 1976).

Friedmann, P. "Influence of Modeling and Blade Parameters on the Aeroelastic Stability of a Cantilevered Rotor." AIAA Journal, 15:2 (February 1977).

Friedmann, P.; Hammond, C.E.; and Woo, T.-H. "Efficient Numerical Treatment of Periodic Systems with Application to Stability Problems." International Journal for Numerical Methods in Engineering, 11:7 (1977).

Friedmann, P., and Shamie, J. "Aeroelastic Stability of Trimmed Helicopter Blades in Forward Flight." Vertica, 1:3 (1977).

Friedmann, P., and Silverthorn, L. J. "Flap-Lag Dynamics of Hingeless Helicopter Blades at Moderate and High Advance Ratios." NASA SP-352, February 1974.

Friedmann, P., and Silverthorn, L. J. "Aeroelastic Stability of Periodic Systems with Application to Rotor Blade Flutter." AIAA Journal, 12:11 (November 1974).

Friedmann, P., and Silverthorn, L. J. "Aeroelastic Stability of Coupled Flap-Lag Motion of Hingeless Helicopter Blades at Arbitrary Advance Ratios." J. Sound Vib., 39:4 (1975).

Friedmann, P., and Tong, P. "Dynamic Nonlinear Elastic Stability of Helicopter Rotor Blades in Hover and in Forward Flight." NASA CR 114485, May 1972.

Friedmann, P., and Tong, P. "Nonlinear Flap-Lag Dynamics of Hingeless Helicopter Blades in Hover and in Forward Flight." J. Sound Vib., 30:1 (1973).

Friedmann, P., and Yuan C. "Effect of Modified Aerodynamic Strip Theories on Rotor Blade Aeroelastic Stability." AIAA Journal, 15:7 (July 1977).

Fry, B. L. "Design Studies and Model Tests of the Stowed Tilt Rotor Concept." AFFDL TR 71-62, July 1971.

Fukushima, T., and Dadone, L. U. "Comparison of Dynamic Stall Phenomena for Pitching and Vertical Translation Motions." NASA CR 2793, July 1977.

Gabel, R. "In-Flight Measurement of Steady and Oscillatory Rotor Shaft Loads." CAL/TRECOM Symposium, Buffalo, New York, June 1963.

Gabel, R. "Current Loads Technology for Helicopter Rotors." AGARD Conference Proceedings No. 122, Milan, Italy, March 1973.

Gabel, R. "Pendulum Absorbers Reduce Transition Vibration." AHS Forum, 1975.

Gabel, R., and Capurso, V. "Exact Mechanical Instability Boundaries as Determined from the Coleman Equation." JAHS, 7:1 (January 1962).

Gabel, R.; Henderson, B. O.; and Reed, D. A. "Pilot and Passenger Vibration Environment Sensitivity." JAHS, 16:3 (July 1971).

Gabel, R., and Tarzanin, F., Jr. "Blade Torsional Tuning to Manage Rotor Stall Flutter." Journal of Aircraft, 11:8 (August 1974).

Gablehouse, C. Helicopters and Autogiros, New York: J. B. Lippincott Co., 1969.

Gaffey, T. M. "The Effect of Positive Pitch-Flap Coupling (Negative δ_3) on Rotor Blade Motion Stability and Flapping." JAHS, 14:2 (April 1969).

Gaffey, T. M., and Maisel, M. D. "Measurement of Tilt Rotor VTOL Rotor Wake-Airframe Ground Aerodynamic Interference for Application to Real Time Flight Simulation." AGARD Conference Proceedings No. 143, Delft, Netherlands, April 1974.

Gaffey, T. M.; Yen, J. G.; and Kvaternik, R. G. "Analysis and Model Tests of the Proprotor Dynamics of a Tilt-Proprotor VTOL Aircraft." Air Force V/STOL Technology and Planning Conference, Las Vegas, Nevada, September 1969.

Gallant, R.; Scully, M.; and Lange, W. "Analysis of V/STOL Aircraft

Configurations for Short Haul Air Transportation Systems." Massachusetts Institute of Technology, FTL FT-66-1, November 1966.
Gallot, J. "Prediction of Helicopter Rotor Loads." AGARD Conference Proceedings No. 122, March 1973. (Also NASA TT F-14845.)
Galloway, W. J. "Physical Analysis of the Impulsive Aspects of Helicopter Noise." Federal Aviation Administration, Report FAA-EQ-77-8, April 1977.
Galloway, W. J. "Subjective Evaluation of Helicopter Blade Slap Noise." NASA CP 2052, May 1978.
Gangwani, S. T. "The Effect of Helicopter Main Rotor Blade Phasing and Spacing on Performance, Blade Loads, and Acoustics." NASA CR 2737, September 1976.
Ganzer, V. M., and Rae, W. H., Jr. "An Experimental Investigation of the Effect of Wind Tunnel Walls on the Aerodynamic Performance of a Helicopter Rotor." NASA TN D-415, May 1960.
Gaonkar, G. H. "Interpolation of Aerodynamic Damping of Lifting Rotors in Forward Flight from Measured Response Variance." J. Sound Vib., 18:3 (1971).
Gaonkar, G. H. "A General Method with Shaping Filters to Study the Random Vibration Statistics of Lifting Rotors with Feedback Controls." J. Sound Vib., 21:2 (1972).
Gaonkar, G. H. "A Study of Lifting Rotor Flapping Response Peak Distribution in Atmospheric Turbulence." Journal of Aircraft, 11:2 (February 1974).
Gaonkar, G. H. "Peak Statistics and Narrow-Band Features of Coupled Torsion-Flapping Rotor Blade Vibrations to Turbulence." J. Sound Vib., 34:1 (1974).
Gaonkar, G. H. "Random Vibration Peaks in Rotorcraft and the Effects of Nonuniform Gusts." Journal of Aircraft, 14:1 (January 1977).
Gaonkar, G. H., and Hohenemser, K. H. "Flapping Response of Lifting Rotor Blades to Atmospheric Turbulence." Journal of Aircraft, 6:6 (November-December 1969).
Gaonkar, G. H., and Hohenemser, K. H. "Stochastic Properties of Turbulence Excited Rotor Blade Vibrations." AIAA Journal, 9:3 (March 1971).
Gaonkar, G. H., and Hohenemser, K. H. "An Advanced Stochastic Model for Threshold Crossing Studies of Rotor Blade Vibrations." AIAA Journal, 10:8 (August 1972).
Gaonkar, G. H.; Hohenemser, K. H.; and Yin, S. K. "Random Gust Response Statistics for Coupled Torsion-Flapping Rotor Blade Vibration." Journal of Aircraft, 9:10 (October 1972).
Gaonkar, G. H., and Subramanian, A. K. "A Study of Feedback, Blade and Hub Parameters on Flap Bending Due to Nonuniform Rotor Disk

Turbulence." J. Sound Vib., 51:4 (1977).

Geray, E. K., and Kisielowski, E. "Stability and Control Handbook for Compound Helicopters." USAAVLABS TR 70-67, February 1971.

Garnett, T. S., Jr.; Smith, J. H.; and Lane, R. "Design and Flight Test of the Active Arm External Load Stabilization System." AHS Forum, 1976.

Garnett, T. S., Jr., and Smith, J. H. "Active Arm (External Cargo) Stabilization System Flight Demonstration." USAAMRDL TR 76-23, September 1976.

Garren, J. F., Jr. "Effects of Gyroscopic Cross Coupling between Pitch and Roll on the Handling Qualities of VTOL Aircraft." NASA TN D-812, April 1961.

Garren, J. F., Jr. "Effects of Gyroscopic Cross Coupling between Pitch and Yaw on the Handling Qualities of VTOL Aircraft." NASA TN D-973, November 1961.

Garren, J. F., Jr. "Effects of Coupling between Pitch and Roll Control Inputs on the Handling Qualities of VTOL Aircraft." NASA TN D-1233, March 1962.

Garren, J. F., Jr., and Assadourian, A. "VTOL Height-Control Requirements in Hovering As Determined from Motion Simulator Study." NASA TN D-1488, October 1962.

Garren, J. F., Jr.; Di Carlo, D. J.; and Driscoll, N. R. "Flight Investigation of an On-Off Control for V/STOL Aircraft under Visual Conditions." NASA TN D-3436, June 1966.

Garren, J. F., Jr., and Kelly, J. R. "Application of the Model Technique to a Variable-Stability Helicopter for Simulation of VTOL Handling Qualities." NASA TM X-56821, 1965.

Garren, J. F., Jr.; Kelly, J. R.; and Reeder, J. P. "Effects of Gross Changes in Static Directional Stability on V/STOL Handling Characteristics Based on a Flight Investigation." NASA TN D-2477, October 1964.

Garrick, I. E., and Watkins, C. E. "A Theoretical Study of the Effect of Forward Speed on the Free-Space Sound-Pressure Field Around Propellers." NACA Report 1198, 1954.

Gates, C. A., and Du Waldt, F. A. "Experimental and Theoretical Investigation of the Flutter Characteristics of a Model Helicopter Rotor Blade in Forward Flight." Air Force Systems Command, ASD TR 61-712, February 1962.

Gates, C. A., and Du Waldt, F. A. "Aeroelastic Investigation of a Model Helicopter Rotor." Air Force Systems Command, ASD-TDR 63-299, April 1963.

Gates, C. A.; Piziali, R. A.; and Du Waldt, F. A. "Comparison of Theoretical and Experimental Flutter Characteristics for a Model Rotor in Translational Flight." JAHS, 8:2 (April 1963).

Gaukroger, D. R., and Cansdale, R. "Rotor Impedance Measurements at Model Scale." Vertica, 1:1 (1976).

Gaukroger, D. R., and Hassal, C. J. W. "Measurement of Vibratory Displacements of a Rotating Blade." Vertica, 2:2 (1978).

George, A. R., "Helicopter Noise: State of the Art." Journal of Aircraft, 15:11 (November 1978).

George, A. R., and Kim, Y. N. "High Frequency Broadband Rotor Noise." AIAA Journal, 15:4 (April 1977).

George, M.; Kisielowski, E.; and Perlmutter, A. A. "Dynagyro—A Mechanical Stability Augmentation System for Helicopter." USAAVLABS TR 67-10, March 1967.

Gera, J., and Farmer, S. W., Jr. "A Method of Automatically Stabilizing Helicopter Sling Loads." NASA TN D-7593, July 1974.

Gerdes, R. M., and Weick, R. F. "A Preliminary Piloted Simulator and Flight Study of Height Control Requirements for VTOL Aircraft." NASA TN D-1201, February 1962.

Gerstenberger, W., and Wood, E. R. "Analysis of Helicopter Aeroelastic Characteristics in High-Speed Flight." AIAA Journal, 1:10 (October 1963).

Gerstine, M. I., and Blake, B. B. "Helicopter Automatic Control through Integration of Separate Functional Units." JAHS, 6:4 (October 1961).

Gessow, A. "Effect of Rotor-Blade Twist and Planform Taper on Helicopter Hovering Performance." NACA TN 1542, February 1948.

Gessow, A. "Standard Symbols for Helicopters." NACA TN 1604, June 1948.

Gessow, A. "Bibliography of NACA Papers on Rotating-Wing Aircraft." NACA RM L7J30, July 1948.

Gessow, A. "Flight Investigation of Effects of Rotor-Blade Twist on Helicopter Performance in the High-Speed and Vertical-Autorotative-Descent Conditions." NACA TN 1666, August 1948.

Gessow, A. "An Analysis of the Autorotative Performance of a Helicopter Powered by Rotor-Tip Jet Units." NACA TN 2154, July 1950.

Gessow, A. "Bibliography of NACA Papers on Rotating-Wing Aircraft." NACA RM L52B18a, January 1952.

Gessow, A. "Review of Information on Induced Flow of a Lifting Rotor." NACA TN 3238, August 1954.

Gessow, A. "Equations and Procedures for Numerically Calculating the Aerodynamic Characteristics of Lifting Rotors." NACA TN 3747, October 1956.

Gessow, A. "A Note on the Calculation of Helicopter Performance at High Tip-Speed Ratios." NASA TN D-97, September 1959.

Gessow, A., and Amer, K. B. "An Explanation of Some Important Stability

Parameters That Influence Helicopter Flying Qualities." Aeronautical Engineering Review, 9:8 (August 1950).

Gessow, A., and Amer, K. B. "An Introduction to the Physical Aspects of Helicopter Stability." NACA Report 993, 1950.

Gessow, A., and Crim, A. D. "An Extension of Lifting Rotor Theory to Cover Operation at Large Angles of Attack and High Inflow Conditions." NACA TN 2665, April 1952.

Gessow, A., and Crim, A. D. "A Method for Studying the Transient Blade-Flapping Behavior of Lifting Rotors at Extreme Operating Conditions." NACA TN 3366, January 1955.

Gessow, A., and Crim, A. D. "A Theoretical Estimate of the Effects of Compressibility on the Performance of a Helicopter Rotor in Various Flight Conditions." NACA TN 3798, October 1956.

Gessow, A., and Gustafson, F. B. "Effect of Blade Cutout on Power Required by Helicopters Operating at High Tip-Speed Ratios." NASA TN D-382, September 1960.

Gessow, A., and Myers, G. C., Jr. "Flight Tests of a Helicopter in Autorotation, Including a Comparison with Theory." NACA TN 1267, April 1947.

Gessow, A., and Myers, G. C., Jr. Aerodynamics of the Helicopter. New York: Frederick Ungar Publishing Co., 1952.

Gessow, A., and Tapscott, R. J. "Charts for Estimating Performance of High-Performance Helicopters." NACA Report 1266, 1956.

Gessow, A., and Tapscott, R. J. "Tables and Charts for Estimating Stall Effects on Lifting-Rotor Characteristics." NASA TN D-243, May 1960.

Ghareeb, N. "Programs for Machine Computation of Rotor Blade Downwash." Massachusetts Institute of Technology, ASRL TR 107-1, August 1964.

Giansante, N. "Rapid Estimation of the Effects of Material Properties on Blade Natural Frequencies." JAHS, 16:1 (January 1971).

Gibs, J.; Stepniewski, W. Z.; and Spencer, R. "Effects of Noise Reduction on Characteristics of a Tilt-Rotor Aircraft." Journal of Aircraft, 13:11 (November 1976).

Gibs, J.; Stepniewski, W. Z.; Spencer, R.; and Kohler, G. "Noise Reduction of a Tilt-Rotor Aircraft Including Effects on Weight and Performance." NASA CR 114648, June 1973.

Gibson, I. S. "On the Velocity Induced by a Semi-Infinite Vortex Cylinder: With Extension to the Short Solenoid." The Aeronautical Journal, 78:762 (June 1974).

Giessler, F. J.; Clay, L. E.; and Nash, J. F. "Flight Loads Investigation of OH-6A Helicopters Operating in Southeast Asia." USAAMRDL TR 71-60, October 1971.

Gillespie, J., Jr. "Streamline Calculations Using the *XYZ* Potential Flow Program." USAAMRDL TN 16, May 1974.

Gillespie, J., Jr., and Windsor, R. I. "An Experimental and Analytical Investigation of the Potential Flow Field, Boundary Layer, and Drag of Various Helicopter Fuselage Configurations." USAAMRDL TN 13, January 1974.

Gillmore, K. B. "Helicopter Flight Vibration Problems." Aeronautical Engineering Review, 14:11 (November 1955).

Gillmore, K. B., and Schneider, J. J. "Design Considerations of the Heavy Lift Helicopter." JAHS, 8:1 (January 1963).

Gilmore, D. C., and Gartshore, I. S. "The Development of an Efficient Hovering Propeller/Rotor Performance Prediction Method." AGARD Conference Proceedings No. 111, Marseilles, France, September 1972.

Giordano, T. A., and Keane, G. C. "The Effect of Helicopter Noise on Communication and Hearing." U.S. Army Electronics Command Report ECOM-4140, August 1973.

Giurgiutiu, V., and Stafford, R. O. "Semi-Analytic Methods for Frequencies and Mode Shapes of Rotor Blades." Vertica, 1:4 (1977).

Gladwell, G. M. L., and Stammers, C. W. "Prediction of the Unstable Regions of a Reciprocal System Governed by a Set of Linear Equations with Periodic Coefficients." J. Sound Vib., 8:3 (1968).

Glauert, H. "An Aerodynamic Theory of the Airscrew." ARC R&M 786, January 1922.

Glauert, H. "The Analysis of Experimental Results in the Windmill Brake and Vortex Ring States of an Airscrew." ARC R&M 1026, February 1926.

Glauert, H. "A General Theory of the Autogyro." ARC R&M 1111, November 1926.

Glauert, H. "Lift and Torque of an Autogyro on the Ground." ARC R&M 1131, July 1927.

Glauert, H. "On the Vertical Ascent of a Helicopter." ARC R&M 1132, November 1927.

Glauert, H. "On the Horizontal Flight of a Helicopter." ARC R&M 1157, March 1928.

Glauert, H. "Airplane Propellers." Aerodynamic Theory. Edited by W. F. Durand. New York: Dover, 1935.

Glauert, H., and Lock, C. N. H. "A Summary of the Experimental and Theoretical Investigations of the Characteristics of an Autogiro." ARC R&M 1162, April 1928.

Gockel, M. A. "Practical Solution of Linear Equations with Periodic Coefficients." JAHS, 17:1 (January 1972).

Goland, L., and Perlmutter, A. A. "A Comparison of the Calculated and

Observed Flutter Characteristics of a Helicopter Rotor Blade." JAS, 24:4 (April 1957).

Goldstein, S. "On The Vortex Theory of Screw Propellers." Proceedings of the Royal Society of London, Series A, 123:440 (1929).

Goodier, J. N. "Elastic Torsion in the Presence of Initial Axial Stress." Journal of Applied Mechanics, 17:4 (December 1950).

Goorjian, P. M. "An Invalid Equation in the General Momentum Theory of the Actuator Disk." AIAA Journal, 10:4 (April 1972).

Gorenberg, N. B., and Harvick, W. P. "Analysis of Maneuverability Effects on Rotor/Wing Design Characteristics." USAAMRDL TR 74-24, February 1974.

Gormont, R. E. "A Mathematical Model of Unsteady Aerodynamics and Radial Flow for Application to Helicopter Rotors." USAAVLABS TR 72-67, May 1973.

Gormont, R. E., and Wolfe, R. A. "The U.S. Army UTTAS and AAH Programs." AGARD Conference Proceedings No. 233, Moffett Field, California, May 1977.

Gould, D. G., and Hindson, W. S. "Estimates of the Lateral-Directional Stability Derivatives of a Helicopter from Flight Measurements." National Research Council of Canada, Report NRC No. 13882, December 1973.

Gould, D. G., and Hindson, W. S. "Estimates of the Stability Derivative of a Helicopter and a V/STOL Aircraft from Flight Data." AGARD Conference Proceedings No. 172, Hampton, Virginia, November 1974.

Graham, D. "System Study of Landing." U.S. Army Electronics Command, ECOM 02412-8, March 1972.

Grant, B. E. "A Method for Measuring Aerodynamic Damping of Helicopter Rotors in Forward Flight." J. Sound Vib., 3:3 (1966).

Gray, L.; Liiva, J.; and Davenport, F J. "Wind Tunnel Tests of Thin Airfoils Oscillating Near Stall." USAAVLABS TR 68-89, January 1969.

Gray, R. B. "Experimental Smoke and Electromagnetic Analog Study of Induced Flow Field about a Model Rotor in Steady Flight within Ground Effect." NASA TN D-458, August 1960.

Gray, R. B., and Pierce, G. A. "Exploratory Investigation of Sound Pressure Level in the Wake of an Oscillating Airfoil in the Vicinity of Stall." NASA CR 1948, February 1972.

Gray, R. B., et al. "Helicopter Hovering Performance Studies." Georgia Institute of Technology, October 1976.

Green, D. L. "A Review of MIL-F-83300 for Helicopter Applications." AHS Forum, 1972.

Greenberg, J. M. "Airfoil in Sinusoidal Motion in a Pulsating Stream." NACA TN 1326, June 1947.

Griffin, J. M., and Bellaire, R. G. "A Graphical Summary of Military

Helicopter Flying and Ground Handling Qualities of MIL-H-8501A." Air Force Systems Command, ASNF TN 68-3, September 1968.

Griffin, M. J. "The Transmission of Triaxial Vibration to Pilots in the Scout AH MkI Helicopter." Institute of Sound and Vibration Research, ISVR TR 58, August 1972.

Griffin, M. J. "The Evaluation of Human Exposure to Helicopter Vibration." The Aeronautical Journal, 81:795 (March 1977).

Griffith, C. D., and Crosby, V. M. "A Self-Monitored Stability Augmentation System for the CX-84 V/STOL Aircraft." JAHS, 15:2 (April 1970).

Grina, K. I. "Helicopter Development at Boeing Vertol Company." The Aeronautical Journal, 79:777 (September 1975).

Gross, D. W., and Harris, F. D. "Prediction of Inflight Stalled Airloads from Oscillating Airfoil Data." AHS Forum, 1969.

Grumm, A. W., and Herrick, G. E. "Advanced Antitorque Concepts Study." USAAMRDL TR 71-23, July 1971.

Gupta, B. P., and Lessen, M. "Hydrodynamic Stability of the Far Wake of a Hovering Rotor." AIAA Journal, 13:6 (June 1975).

Gupta, B. P., and Loewy, R. G. "Analytical Investigation of the Aerodynamic Stability of Helical Vortices Shed from a Hovering Rotor." USAAMRDL TR 73-84, October 1973.

Gupta, B. P., and Loewy, R. G. "Theoretical Analysis of the Aerodynamic Stability of Multiple, Interdigitated Helical Vortices." AIAA Journal, 12:10 (October 1974).

Gupta, N. K., and Bryson, A. E., Jr. "Guidance for a Tilt-Rotor VTOL Aircraft during Takeoff and Landing." NASA CR 132043, December 1972.

Gupta, N. K., and Bryson, A. E., Jr. "Automatic Control of a Helicopter with a Hanging Load." NASA CR 136504, June 1973.

Gupta, N. K., and Bryson, A. E., Jr. "Near-Hover Control of a Helicopter with a Hanging Load." Journal of Aircraft, 13:3 (March 1976).

Gustafson, F. B. "Effect on Helicopter Performance of Modifications in Profile-Drag Characteristics of Rotor-Blade Airfoil Sections." NACA ARC L4H05, August 1944.

Gustafson, F. B. "Flight Tests of the Sikorsky HNS-1 (Army YR-4B) Helicopter, Part I." NACA MR L5C10, March 1945.

Gustafson, F. B. "A Summary of the Effects of Blade Twist on Helicopter Performance." NACA MR L5H24, 1945.

Gustafson, F. B. "Notes on the Application of Airfoil Studies to Helicopter Rotor Design." NACA RM L8C26, September 1948.

Gustafson, F. B. "The Application of Airfoil Studies to Helicopter Rotor Design." NACA TN 1812, February 1949.

Gustafson, F. B. "Desirable Longitudinal Flying Qualities for Helicopters and Means to Achieve Them." Aeronautical Engineering Review, 10:6 (June 1951).

Gustafson, F. B. "Charts for Estimation of the Profile Drag-Lift Ratio of a Helicopter Rotor Having Rectangular Blades with −8° Twist." NACA RM L53G20a, October 1953.

Gustafson, F. B. "History of NACA/NASA Rotary Wing Aircraft Research, 1915-1970." Vertiflite, 16:6 (June 1970) through 16:12 (December 1970).

Gustafson, F. B.; Amer, K. B.; Haig, C. R.; and Reeder, J. P. "Longitudinal Flying Qualities of Several Single-Rotor Helicopters in Forward Flight." NACA TN 1983, November 1949.

Gustafson, F. B., and Crim, A. D. "Flight Measurements and Analysis of Helicopter Normal Load Factors in Maneuvers." NACA TN 2990, August 1953.

Gustafson, F. B., and Gessow, A. "Flight Tests of the Sikorsky HNS-1 (Army YR-4B) Helicopter, Part II." NACA MR L5D09a, April 1945.

Gustafson, F. B., and Gessow, A. "Effect of Rotor-Tip Speed on Helicopter Hovering Performance and Maximum Forward Speed." NACA ARR L6A16, March 1946.

Gustafson, F. B., and Gessow, A. "Effect of Blade Stalling on the Efficiency of a Helicopter Rotor As Measured in Flight." NACA TN 1250, April 1947.

Gustafson, F. B., and Gessow, A. "Analysis of Flight-Performance Measurements on a Twisted, Plywood-Covered Helicopter Rotor in Various Flight Conditions." NACA TN 1595, June 1948.

Gustafson, F. B., and Myers, G. C., Jr. "Stalling of Helicopter Blades." NACA Report 840, 1946.

Gustafson, F. B., and Reeder, J. P. "Helicopter Stability." NACA RM L7K04, April 1948.

Gustafson, F. B., and Tapscott, R. J. "Methods for Obtaining Desired Helicopter Stability Characteristics and Procedures for Stability Predictions." NACA Report 1350, 1958.

Gutin, L. "On the Sound Field of a Rotating Propeller." NACA TM 1195, October 1948.

Hafner, R. "The Bristol 171 Helicopter." Journal of the Royal Aeronautical Society, 53:460 (April 1949).

Hafner, R. "The Domain of the Helicopter." Journal of the Royal Aeronautical Society, 58:526 (October 1954).

Hafner, R. "Domain of the Convertible Rotor." Journal of Aircraft, 1:6 (November-December 1964).

Hafner, R. "The Case for the Convertible Rotor." Aeronautical Journal 75:728 (August 1971).

Hale, R. W.; Tan, P.; Stowell, R. C.; Iwan, L. S.; and Ordway, D. E. "Preliminary Investigation of the Role of the Tip Vortex in Rotary Wing

Aerodynamics through Flow Visualization." Sage Action Inc., Ithaca, New York, December 1974.

Halfman, R. L. "Experimental Aerodynamic Derivatives of a Sinusoidally Oscillating Airfoil in Two-Dimensional Flow." NACA TN 2465, November 1951.

Halfman, R. L.; Johnson, H. C.; and Haley, S. M. "Evaluation of High-Angle-of-Attack Aerodynamic-Derivative Data and Stall-Flutter Prediction Techniques." NACA TN 2533, November 1951.

Hall, G. F. "Unsteady Vortex Lattice Techniques Applied to Wake Formation and Performance of the Statically Thrusting Propeller." NASA CR 132686, 1975.

Hall, G. F. "Transient Airload Computer Analysis for Simulating Wind Induced Impulsive Noise Conditions of a Hovering Helicopter Rotor." NASA CR 137772, October 1975.

Hall, W. E., Jr. "Prop-Rotor Stability at High Advance Ratios." JAHS, 11:2 (April 1966).

Hall, W. E., Jr., and Bryson, A. E., Jr. "Synthesis of Hover Autopilots for Rotary-Wing VTOL Aircraft." NASA CR 132053, June 1972.

Hall, W. E., Jr., and Bryson, A. E., Jr. "Inclusion of Rotor Dynamics in Controller Design for Helicopters." Journal of Aircraft, 10:4 (April 1973).

Hall, W. E., Jr.; Gupta, N. K.; and Hansen, R. S. "Rotorcraft System Identification Techniques for Handling Qualities and Stability and Control Evaluation." AHS Forum, 1978.

Halley, D. H. "ABC Helicopter Stability, Control, and Vibration Evaluation on the Princeton Dynamic Model Track." AHS Forum, 1973.

Halwes, D. R. "Flight Operations to Minimize Noise." Vertiflite, 17:2 (February 1971).

Ham, N. D. "An Experimental Investigation of Stall Flutter." JAHS, 7:1 (January 1962).

Ham, N. D. "An Experimental Investigation of the Effect of a Non-Rigid Wake on Rotor Blade Airloads in Transition Flight." CAL/TRECOM Symposium, Buffalo, New York, June 1963.

Ham, N. D. "Some Conclusions from an Experimental Investigation of Rotor Harmonic Airloads." JAHS, 10:2 (April 1965).

Ham, N. D. "Stall Flutter of Helicopter Rotor Blades: A Special Case of the Dynamic Stall Phenomenon." JAHS, 12:4 (October 1967).

Ham, N. D. "Aerodynamic Loading on a Two-Dimensional Airfoil during Dynamic Stall." AIAA Journal, 6:10 (October 1968).

Ham, N. D. "Some Recent MIT Research on Dynamic Stall." Journal of Aircraft, 9:5 (May 1972).

Ham. N. D. "Helicopter Blade Flutter." AGARD Report 607, 1973.

Ham, N. D. "Some Preliminary Results from an Investigation of Blade-Vortex Interaction." JAHS, 19:2 (April 1974).

Ham, N. D. "Some Conclusions from an Investigation of Blade-Vortex Interaction," JAHS, 20:4 (October 1975).

Ham, N. D.; Bauer, P. H.; Lawrence, T. H.; and Yasue, M. "A Study of Gust and Control Response of Model Rotor-Propellers in a Wind Tunnel Airstream." NASA CR 137756, August 1975.

Ham, N. D., and Garelick, M. S. "Dynamic Stall Considerations in Helicopter Rotors." JAHS, 13:2 (April 1968).

Ham, N. D., and Johnson, W. "A Comparison of Dynamically Scaled Model Rotor Test Data with Step-by-Step Calculations of Rotor Blade Motion." AHS Forum, 1969.

Ham, N. D., and Madden, P. A. "An Experimental Investigation of Rotor Harmonic Airloads Including the Effects of Rotor-Rotor Interference and Blade Flexibility." U.S. Army Aviation Materiel Laboratories, USAAML TR 65-13, May 1965.

Ham, N. D.; Moser, H. H.; and Zvara, J. "Investigation of Rotor Response to Vibratory Aerodynamic Inputs – Experimental Results and Correlation with Theory." Wright Air Development Center, WADC TR 58-87, Part I, October 1958.

Ham, N. D., and Young, M. I. "Torsional Oscillation of Helicopter Blades Due to Stall." Journal of Aircraft, 3:3 (May-June 1966).

Ham, N. D., and Zvara, J. "Experimental and Theoretical Analysis of Helicopter Rotor Hub Vibratory Forces." Wright Air Development Center, WADC TR 59-522, October 1959.

Hammond, C. E. "An Application of Floquet Theory to Prediction of Mechanical Instability." JAHS, 19:4 (October 1974).

Hammond, C. E., and Pierce, G. A. "A Compressible Unsteady Aerodynamic Theory for Helicopter Rotors." AGARD Conference Proceedings No. 111, Marseilles, France, September 1972.

Hancock, G. J. "Aerodynamic Loading Induced on a Two-Dimensional Wing by a Free Vortex in Incompressible Flow." The Aeronautical Journal, 75:726 (June 1971).

Hansen, K. C. "Single Rotor Helicopter Transient Following a Power Failure at High Speeds." AHS Forum, 1966.

Hansen, K. C.; McKeown, J. C.; and Gerdes, W. R. "Analysis of Maneuverability Effects on Rotor/Wing Design Characteristics." USAAMRDL TR 74-23, February 1974.

Hanson, D. B., and Fink, M. R. "The Importance of Quadrupole Sources in Prediction of Transonic Tip Speed Propeller Noise." NASA CP 2052, May 1978.

Hansford, R. E., and Simons, I. A. "Torsion-Flap-Lag Coupling on Helicopter Rotor Blades." JAHS, 18:4 (October 1973).

Harendra, P. B.; Joglekar, M. J.; Gaffey, T. M.; and Marr, R. L. "V/STOL Tilt Rotor Study – A Mathematical Model for Real Time Flight Simulation of the Bell Model 301 Tilt Rotor Research Aircraft." NASA CR 114614, April 1973.

Hargest, T. J. "Noise of VTOL Aircraft." J. Sound Vib., 4:3 (1966).

Harper, H. P., and Sardanowsky, W. "A Study of Task Performance and Handling Qualities Evaluation Techniques at Hover and in Low-Speed Flight." USAAVLABS TR 69-47, July 1969.

Harper, H. P.; Sardanowsky, W.; and Scharpf, R. "Development of VTOL Flying and Handling Qualities Requirements Based on Mission-Task Performance." JAHS, 15:3 (July 1970).

Harrington, R. D. "Full-Scale-Tunnel Investigation of the Static-Thrust Performance of a Coaxial Helicopter Rotor." NACA TN 2318, March 1951.

Harrington, R. D. "Reduction of Helicopter Parasite Drag." NACA TN 3234, August 1954.

Harris, F. D. "High Performance Tandem Helicopter Study." TRECOM TR 61-42 and TR 61-43, April 1961.

Harris, F. D. "Spanwise Flow Effects on Rotor Performance." CAL/AVLABS Symposium, Buffalo, New York, June 1966.

Harris, F. D. "Preliminary Study of Radial Flow Effects on Rotor Blades." JAHS, 11:3 (July 1966).

Harris, F. D. "Articulated Rotor Blade Flapping Motion at Low Advance Ratio." JAHS, 17:1 (January 1972).

Harris, F. D.; Cancro, P. A.; and Dixon, P. G. C. "The Bearingless Main Rotor." European Rotorcraft and Powered Lift Aircraft Forum, Aix-en-Provence, France, 1977.

Harris, F. D., and Pruyn, R. R. "Blade Stall – Half Fact, Half Fiction." JAHS, 13:2 (April 1968).

Harris, F. D.; Tarzanin, F. J., Jr.; and Fisher, R. K., Jr. "Rotor High Speed Performance, Theory vs. Test." JAHS, 15:3 (July 1970).

Harrison, J. M., and Ollerhead, J. B. "The Nature of Limitations Imposed on the Performance of a Helicopter Rotor." J. Sound Vib., 3:3 (1966).

Harvey, K. W.; Blankenship, B. L.; and Drees, J. M. "Analytical Study of Helicopter Gust Response at High Forward Speeds." USAAVLABS TR 69-1, September 1969.

Hawkings, D. L. "Theoretical Models of Helicopter Rotor Noise." NASA CP 2052, May 1978.

Hawkings, D. L., and Lowson, M. V. "Theory of Open Supersonic Rotor Noise." J. Sound Vib., 36:1 (1974).

Hawkings, D. L., and Lowson, M. V. "Noise of High Speed Rotors." AIAA Paper 75-450, March 1975.

Hayden, J. S. "The Effect of the Ground on Helicopter Hovering Power Required." AHS Forum, 1976.

Hayden, R. E., and Aravamudan, K. S. "Prediction and Reduction of Rotor Broadband Noise." NASA CP 2052, May 1978.

Hazen, M. E. "A Study of Normal Accelerations and Operating Conditions Experienced by Helicopters in Commercial and Military Operations." NACA TN 3434, April 1955.

Head, R. E. "A Comparison of Rotor Blade Loads Measured in Flight and on a Quarter-Scale Wind Tunnel Model." JAHS, 7:3 (July 1962).

Heffron, W. G.; Bristow, T. R.; Gass, W. C.; and Brown, J. C. "Simulation of a Helicopter Rotor." JAHS, 4:3 (July 1959).

Heimbold, R. L., and Griffith, C. D. "Synthesis of an Electromechanical Control System for a Compound Hingeless Rotor Helicopter." JAHS, 17:2 (April 1972).

Helmbold, H. B. "Goldstein's Solution of the Problem of the Aircraft Propeller with a Finite Number of Blades." NACA TM 652, December 1931.

Henderson, H. R.; Pegg, R. J.; and Hilton, D. A. "Results of the Noise Measurement Program on a Standard and Modified OH-6A Helicopter." NASA TN D-7216, September 1973.

Hersh, A. S.; Soderman, P. T.; and Hayden, R. E. "Investigation of Acoustic Effects of Leading-Edge Serrations on Airfoils." Journal of Aircraft, 11:4 (April 1974).

Herkovitz, A., and Steinmann, H. "CH-47A Design and Operational Flight Loads Study." USAAMRDL TR 73-40, November 1973.

Hess, R. A. "Application of a Model-Based Flight Director Design Technique to a Longitudinal Hover Task." Journal of Aircraft, 14:3 (March 1977).

Heyson, H. H. "Preliminary Results from Flow-Field Measurements around Single and Tandem Rotors in the Langley Full-Scale Tunnel." NACA TN 3242, November 1954.

Heyson, H. H. "Induced Velocity Near a Rotor and Its Application to Helicopter Problems." AHS Forum, 1958.

Heyson, H. H. "An Evaluation of Linearized Vortex Theory As Applied to Single and Multiple Rotors Hovering in and out of Ground Effect." NASA TN D-43, September 1959.

Heyson, H. H. "Ground Effect for Lifting Rotors in Forward Flight." NASA TN D-234, May 1960.

Heyson, H. H. "A Note on the Mean Value of Induced Velocity for a Helicopter Rotor." NASA TN D-240, May 1960.

Heyson, H. H. "Measurements of the Time-Averaged and Instantaneous Induced Velocities in the Wake of a Helicopter Rotor Hovering at High Tip Speeds." NASA TN D-393, July 1960.

Heyson, H. H. "Equations for the Induced Velocities Near a Lifting Rotor with Nonuniform Azimuthwise Vorticity Distribution." NASA TN D-394, August 1960.

Heyson, H. H. "Jet-Boundary Corrections for Lifting Rotors Centered in Rectangular Wind Tunnels." NASA TR R-71, 1960.

Heyson, H. H. "Wind-Tunnel Wall Interference and Ground Effect for VTOL-STOL Aircraft." JAHS, 6:1 (January 1961).

Heyson, H. H. "Tables and Charts of the Normal Component of Induced Velocity in the Lateral Plane of a Rotor with Harmonic Azimuthwise Vorticity Distribution." NASA TN D-809, April 1961.

Heyson, H. H. "Nomographic Solution of the Momentum Equation for VTOL-STOL Aircraft." NASA TN D-814, April 1961.

Heyson, H. H. "The Flow throughout a Wind Tunnel Containing a Rotor with a Sharply Deflected Wake." CAL/AVLABS Symposium, Buffalo, New York, June 1969.

Heyson, H. H. "Theoretical Study of Conditions Limiting V/STOL Testing in Wind Tunnels with Solid Floor." NASA TN D-5819, June 1970.

Heyson, H. H. "A Momentum Analysis of Helicopters and Autogyros in Inclined Descent, With Comments on Operational Restrictions." NASA TN D-7917, October 1975.

Heyson, H. H. "Theoretical Study of the Effect of Ground Proximity on the Induced Efficiency of Helicopter Rotors." NASA TM X-71951, May 1977.

Heyson, H. H. "A Brief Survey of Rotary Wing Induced-Velocity Theory." NASA TM 78741, June 1978.

Heyson, H. H., and Katzoff, S. "Induced Velocities Near a Lifting Rotor with Nonuniform Disk Loading." NACA Report 1319, 1957.

Hickey, D. H. "Full-Scale Wind-Tunnel Tests of the Longitudinal Stability and Control Characteristics of the XV-1 Convertiplane in the Autorotating Flight Range." NACA RM A55K21a, May 1956.

Hicks, C. W., and Hubbard, H. H. "Comparison of Sound Emission from Two-Blade, Four-Blade, and Seven-Blade Propellers." NACA TN 1354, July 1947.

Hicks, J. G., and Nash, J. F. "The Calculation of Three-Dimensional Turbulent Boundary Layers on Helicopter Rotors." NASA CR 1845, May 1971 .

Hill, T. G. "Some Fundamentals for Efficient VTOL Aircraft Hover Control." JAHS, 11:1 (January 1966).

Hiller Aircraft Company. "Heavy-Lift Tip Turbojet Rotor System." USAAVLABS TR 64-68, October 1965.

Hinterkeuser, E. "Acoustical Properties of a Model Rotor in Non-Axial Flight." NASA CR 114749, September 1973.

Hinterkeuser, E. G., and Sternfeld, H., Jr. "Civil Helicopter Noise Assessment Study, Boeing Vertol Model 347." NASA CR 132420, May 1974.

Hirsch, H. "The Contribution of Higher Mode Resonance to Helicopter Rotor-Blade Bending." JAS, 20:6 (June 1953).

Hirsch. H.; Hutton, R. E.; and Rasumoff, A. "Effect of Spanwise and Chordwise Mass Distribution on Rotor Blade Cyclic Stresses." JAHS, 1:2 (April 1956).

Hirsh, N. B., and Ferris, H. W. "Design Requirements for a Quiet Helicopter." AHS Forum, 1972.

Hoad, D. R., and Greene, G. C. "Helicopter Noise Research at the Langley V/STOL Tunnel." NASA CP 2052, May 1978.

Hodges, D. H. "Nonlinear Bending and Torsion of Rotating Beams with Application to Linear Stability of Hingeless Helicopter Rotors.:: Ph.D. dissertation, Stanford University, December 1972.

Hodges, D. H. "Nonlinear Equations of Motion for Cantilever Rotor Blades in Hover with Pitch Link Flexibility, Twist, Precone, Droop, Sweep, Torque Offset, and Blade Root Offset." NASA TM X-73,112, May 1976.

Hodges, D. H. "Aeromechanical Stability of Helicopters with a Bearingless Main Rotor." NASA TM 78459 and TM 78460, February 1978.

Hodges, D. H. "An Aeromechanical Stability Analysis for Bearingless Rotor Helicopters." AHS Forum, 1978.

Hodges, D. H., and Dowell, E. H. "Nonlinear Equations of Motion for the Elastic Bending and Torsion of Twisted Nonuniform Rotor Blades." NASA TN D-7818, December 1974.

Hodges, D. H., and Ormiston, R. A. "Stability of Elastic Bending and Torsion of Uniform Cantilevered Rotor Blades in Hover." AIAA Paper No. 73-405, March 1973.

Hodges, D. H., and Ormiston, R. A. "Nonlinear Equations for Bending of Rotating Beams with Application to Linear Flap-Lag Stability of Hingeless Rotors." NASA TM X-2770, May 1973.

Hodges, D. H., and Ormiston, R. A. "Stability of Elastic Bending and Torsion of Uniform Cantilever Rotor Blades in Hover with Variable Structural Coupling." NASA TN D-8192, April 1976.

Hodges, D. H., and Ormiston, R. A. "Stability of Hingeless Rotor Blades in Hover with Pitch-Link Flexibility." AIAA Journal, 15:4 (April 1977).

Hodges, D. H., and Peters, D. A. "On the Lateral Buckling of Uniform Slender Cantilever Beams." International Journal of Solids and Structures, 11:12A (1975).

Hoerner, S. F. Fluid-Dynamic Drag. New Jersey: Published by the Author, 1965.

Hoffman, J. D., and Velkoff, H. R. "Vortex Flow over Helicopter Rotor Tips." Journal of Aircraft, 8:9 (September 1971).

Hoffman, W. C.; Hollister, W. M.; and Howell, J. D. "Navigation and Guidance Requirements for Commercial VTOL Operations." NASA CR 132423, January 1974.

Hoffman, W. C.; Zvara, J.; Bryson, A. E., Jr.; and Ham, N. D. "Automatic Guidance Concept for VTOL Aircraft." Journal of Aircraft, 8:8 (August 1971).

Hoffstedt, D. J., and Swatton, S. "Advanced Helicopter Structural Design Investigation." USAAMRDL TR 75-56, March 1976.

Hofmann, L. G., et al. "Development of Automatic and Manual Flight Director Landing Systems for the XV-15 Tilt Rotor Aircraft in Helicopter Mode." NASA CR 152040, January 1978.

Hohenemser, K. "Performance of Rotating-Wing Aircraft." NACA TM 871, July 1938.

Hohenemser, K. "Dynamic Stability of a Helicopter with Hinged Rotor Blades." NACA TM 907, September 1939.

Hohenemser, K. "Stability in Hovering of the Helicopter with Central Rotor Location." Air Materiel Command, Translation F-TS-687-RE, August 1946.

Hohenemser, K. "Longitudinal Stability of the Helicopter in Forward Flight." Air Materiel Command, Translation F-TS-688-RE, August 1946.

Hohenemser, K. "A Type of Lifting Rotor with Inherent Stability." JAS, 17:9 (September 1950).

Hohenemser, K. H. "Aerodynamic Aspects of the Unloaded Rotor Convertible Helicopter." JAHS, 2:1 (January 1957).

Hohenemser, K. H. "On a Type of Low-Advance-Ratio Blade Flapping Instability of Three-or-More Bladed Rotors without Drag Hinges." AHS Forum, 1957.

Hohenemser, K. H. "Remarks on Dynamic Problems of Composite Aircraft." JAHS, 13:1 (January 1968).

Hohenemser, K. H. "Hingeless Rotorcraft Flight Dynamics." AGARD-AG-197, 1974.

Hohenemser, K. H., and Banerjee, D. "Application of System Identification to Analytic Rotor Modeling from Simulated and Wind Tunnel Dynamic Test Data." NASA CR 152023, June 1977.

Hohenemser, K. H.; Banerjee, D.; and Yin, S. K. "Methods Studies on System Identification from Transient Rotor Tests." NASA CR 137965, June 1975.

Hohenemser, K. H.; Banerjee, D.; and Yin, S. K. "Rotor Dynamic State and Parameter Identification from Simulated Forward Flight Transients." NASA CR 137963, June 1976.

Hohenemser, K. H., and Crews, S. T. "Concepts for a Theoretical and Experimental Study of Lifting Rotor Random Loads and Vibrations." NASA CR 114388, June 1971; NASA CR 114481, June 1972; NASA CR 114711, June 1973.

Hohenemser, K. H., and Crews, S. T. "Model Tests on Unsteady Rotor Wake

Effects." Journal of Aircraft, 10:1 (January 1973).
Hohenemser, K. H., and Crews, S. T. "Experiments with a Four-Bladed Cyclic Pitch Stirring Model Rotor." NASA CR 137572, June 1974.
Hohenemser, K. H., and Crews, S. T. "Additional Experiments with a Four-Bladed Cyclic Pitch Stirring Model Rotor." NASA CR 137966, June 1975.
Hohenemser, K. H., and Crews, S. T. "Rotor Dynamic State and Parameter Identification from Hovering Transients." NASA CR 137964, June 1976.
Hohenemser, K. H., and Crews, S. T. "Unsteady Hovering Wake Parameters Identified from Dynamic Model Tests." NASA CR 152022, June 1977.
Hohenemser, K. H., and Gaonkar, G. H. "Concepts for a Theoretical and Experimental Study of Lifting Rotor Random Loads and Vibrations." Phase I, NASA CR 114707, September 1967; Phase II, NASA CR 114708, August 1968; Phase III, June 1969; Phase IV, June 1970.
Hohenemser, K. H., and Gaonkar, G. H. "Effects of Torsional Blade Flexibility on Single Blade Random Gust Response Statistics." NASA CR 114386, June 1971.
Hohenemser, K. H., and Heaton, P. W., Jr. "Aeroelastic Instability of Torsionally Rigid Helicopter Blades." JAHS, 12:2 (April 1967).
Hohenemser, K. H., and Perisho, C. H. "Analysis of the Vertical Flight Dynamic Characteristics of the Lifting Rotor with Floating Hub and Off-Set Coning Hinges." JAHS, 3:4 (October 1958).
Hohenemser, K. H., and Prelewicz, D. A. "Identification of Lifting Rotor System Parameters from Transient Response Data." NASA CR 114710, June 1973.
Hohenemser, K. H., and Yin, S. K. "Analysis of Gust Alleviation Methods and Rotor Dynamic Stability." NASA CR 114387, June 1971.
Hohenemser, K. H., and Yin, S. K. "Effects of Blade Torsion, of Blade Flap Bending Flexibility and of Rotor Support Flexibility on Rotor Stability and Random Response." NASA CR 114480, June 1972.
Hohenemser, K. H., and Yin, S.-K. "Some Applications of the Method of Multiblade Coordinates." JAHS, 17:3 (July 1972).
Hohenemser, K. H., and Yin, S.-K. "On the Question of Adequate Hingeless Rotor Modeling in Flight Dynamics." AHS Forum, 1973.
Hohenemser, K. H., and Yin, S. K. "The Effects of Some Rotor Feedback Systems on Rotor-Body Dynamics." NASA CR 114709, June 1973.
Hohenemser, K. H., and Yin, S. K. "On the Use of First Order Rotor Dynamics in Multiblade Coordinates." AHS Forum, 1974.
Hohenemser, K. H., and Yin, S. K. "Methods Studies toward Simplified Rotor-Body Dynamics." NASA CR 137570, June 1974.
Hohenemser, K. H., and Yin, S. K. "Computer Experiments in Preparation of System Identification from Transient Rotor Model Tests." NASA

CR 137571, June 1974.

Hohenemser, K. H., and Yin, S. K. "Finite Element Stability Analysis for Coupled Rotor and Support Systems." NASA CR 152024, June 1977.

Hollister, W. M., and Hoffman, W. C. "Guidance Logic for Spiral Approaches." Journal of Aircraft, 14:10 (October 1977).

Homans, B.; Little, L.; and Schomer, P. D. "Rotary Wing Aircraft Operations Noise Data." Construction Engineering Research Laboratory, Technical Report N-38, February 1978.

Homicz, G. F., and George, A. R. "Broadband and Discrete Frequency Radiation from Subsonic Rotors." J. Sound Vib., 36:2 (1974).

Hone, H. T. "Flight Load Investigation of Helicopter External Loads." USAAMRDL TR 74-104, February 1975.

Hooper, W. E. "Helicopter Ground Resonance." Aircraft Engineering, 31:360 (February 1959).

Hooper, W. E. "UREKA – A Vibration Balancing Device for Helicopters." JAHS, 11:1 (January 1966).

Hooper, W. E. "Boeing Model 347 Flying Qualities Demonstrator." The Aeronautical Journal, 75:732 (December 1971).

Horst, T. J., and Reschak, R. J. "Designing to Survive Tail Rotor Loss." AHS Forum, 1975.

Horvay, G. "Vibrations of a Helicopter on the Ground." JAS, 13:11 (November 1946).

Horvay, G. "Stress Analysis of Rotor Blades." JAS, 14:6 (June 1947).

Horvay, G. "Rotor Blade Flapping Motion." Quarterly of Applied Mathematics, 5:2 (July 1947).

Horvay, G. "Chordwise and Beamwise Bending Frequencies of Hinged Rotor Blades." JAS, 15:8 (August 1948).

Horvay, G., and Yuan S. W. "Stability of Rotor Blade Flapping Motion When the Hinges Are Tilted." JAS, 14:10 (October 1947).

Hosier, R. N., and Ramakrishnan, R. "Helicopter Rotor Rotational Noise Predictions Based on Measured High-Frequency Blade Loads." NASA TN D-7624, December 1974.

Hosier, R. N.; Ramakrishnan, R.; and Pegg, R. J. "The Prediction of Rotor Rotational Noise Using Measured Fluctuating Blade Loads." JAHS, 20:2 (April 1974).

Houbolt, J. C., and Brooks, G. W. "Differential Equations of Motion for Combined Flapwise Bending, Chordwise Bending, and Torsion of Twisted Nonuniform Rotor Blades." NACA Report 1346, 1958.

Houbolt, J. C., and Reed, W. H. III. "Propeller-Nacelle Whirl Flutter." JAS, 29:3 (March 1962).

Houck, J. A. "Computational Aspects of Real-Time Simulation of Rotary Wing Aircraft." NASA CR 147932, May 1976.

Houck, J. A., and Bowles, R. L. Effects of Rotor Model Degradation on the Accuracy of Rotorcraft Real-Time Simulation." NASA TN D-8378, December 1976.

Houck, J. A., et al. "Rotor Systems Research Aircraft Simulation Mathematical Model." NASA TM 78629, November 1977.

Hough, G. R., and Ordway, D. E. "The Generalized Actuator Disk." Developments in Theoretical and Applied Mechanics, Vol. 2. Edited by W. A. Shaw. New York: Pergamon Press, 1965.

Howarth, R. M., and Jones, C. H. "Ground Resonance of the Helicopter." Journal of the Helicopter Society of Great Britain, 7:4 (April 1954).

Howlett, J. T.; Clevenson, S. A.; Rupf, J. A.; and Synder, W. J. "Interior Noise Reduction in a Large Civil Helicopter." NASA TN D-8477, July 1977.

Hsu, C. S. "On Approximating a General Linear Periodic System." Journal of Mathematical Analysis and Applications, 45:1 (January 1974).

Hubbard, H. H. "Sound from Dual-Rotating and Multiple Single-Rotating Propellers." NACA TN 1654, July 1948.

Hubbard, H. H. "Sound Measurements for Five Shrouded Propellers at Static Conditions." NACA TN 2024, April 1950.

Hubbard, H. H.; Lansing, D. L.; and Runyan, H. L. "A Review of Rotating Blade Noise Technology." J. Sound Vib., 19:3 (1971).

Hubbard, H. H., and Lassiter, L. W. "Some Aspects of the Helicopter Noise Problem." NACA TN 3239, August 1954.

Hubbard, H. H., and Maglieri, D. J. "Noise Characteristics of Helicopter Rotors at Tip Speeds up to 900 Feet per Second." Journal of the Acoustical Society of America, 32:9 (September 1960).

Hubbard, H. H., and Regier, A. A. "Propeller-Loudness Charts for Light Airplanes." NACA TN 1358, July 1947.

Hubbard, H. H., and Regier, A. A. "Free-Space Oscillating Pressures Near the Tips of Rotating Propellers." NACA Report 996, 1950.

Huber, H. "Parametric Trends and Optimization – Preliminary Selection of Configuration – Prototype Design and Manufacture." AGARD Lecture Series No. 63, Brussels, April 1973.

Huber, H. B. "Effect of Torsion-Flap-Lag Coupling on Hingeless Rotor Stability." AHS Forum, 1973.

Huber, H. B., and Strehlow, H. "Hingeless Rotor Dynamics in High Speed Flight." Vertica, 1:1 (1976).

Hufton, P. A.; Woodward Nutt, A. E.; Bigg, F. J.; and Beavan, J. A. "General Investigation into the Characteristics of the C.30 Autogiro." ARC R&M 1859, March 1939.

Hughes, C. F., and Wernicke, R. K. "Flight Test of a Hingeless Flexbeam Rotor System." UAAAMRDL TR 74-38, June 1974.

Hunt, G. K. "Similarity Requirements for Aeroelastic Models of Helicopter Rotors." ARC CP No. 1245, January 1972.

Huston, R. J. "Wind Tunnel Measurements of Performance, Blade Motions, and Blade Air Loads for Tandem-Rotor Configurations with and without Overlap." NASA TN D-1971, October 1963.

Huston, R. J. "An Exploratory Investigation of Factors Affecting the Handling Qualities of a Rudimentary Hingeless Rotor Helicopter." NASA TN D-3418, May 1966.

Huston, R. J.; Jenkins, J. L., Jr.; and Shipley, J. L. "The Rotor Systems Research Aircraft – A New Step in the Technology and Rotor System Verification Cycle." AGARD Conference Proceedings No. 233, Moffett Field, California, May 1977.

Huston, R. J., and Morris, C. E. K., Jr. "A Note on a Phenomenon Affecting Helicopter Directional Control in Rearward Flight." JAHS, 15:4 (October 1970).

Huston, R. J., and Morris, C. E. K., Jr. "A Wind Tunnel Investigation of Helicopter Directional Control in Rearward Flight in Ground Effect." NASA TN D-6118, March 1971.

Huston, R. J., and Shivers, J. P. "A Wind Tunnel and Analytical Study of the Conversion from Wing Lift to Rotor Lift on a Composite-Lift VTOL Aircraft." NASA TN D-5256, June 1969.

Hutchins, C. W. "Navy Vehicle Design and Construction: Measurement of Triaxial Vibration at Significant Human Interface Points on the CH-47C and SH-3A Helicopters.:: Naval Air Development Center, Report No. NADC-72226-CS, December 1972.

Ibrahim, R. A., and Barr, A. D. S. "Parametric Vibration." The Shock and Vibration Digest, 10:1 (January 1978) to 10:5 (May 1978).

Ichikawa, T. "Linear Aerodynamic Theory of Rotor Blades." Journal of Aircraft, 4:3 (May-June 1967).

Isaacs, R. "Airfoil Theory for Flows of Variable Velocity." JAS, 12:1 (January 1945).

Isaacs, R. "Airfoil Theory for Rotary Wing Aircraft." JAS, 13:4 (April 1946).

Isakson, G., and Eisley, J.G. "Natural Frequencies in Bending of Twisted Rotating and Nonrotating Blades." NASA TN D-371, March 1960.

Isakson, G., and Eisley, J. G. "Natural Frequencies in Coupled Bending and Torsion of Twisted Rotating and Nonrotating Blades." NASA CR 65, July 1964.

Isay, W. H. "A Vortex Model Dealing with the Airstream at the Rotor Blade of a Helicopter." NASA TT F-14228, April 1972.

Isay, W. H. "A Vortex Model for the Study of the Flow at the Rotor Blade of a Helicopter." NASA TT F-14637, December 1972.

Isay, W. H. "Transonic Rotor Aerodynamics – Fundamentals of the Theory." NASA TT F-17395, February 1977.

Isogai, K. "An Experimental Study on the Unsteady Behavior of a Short Bubble on an Airfoil during Dynamic Stall with Special Reference to the Mechanism of the Stall Overshoot Effect." Massachusetts Institute of Technology, ASRL TR 130-2, June 1970.

Isogai, K. "Numerical Study of Transonic Flow over Oscillating Airfoils Using the Full Potential Equation." NASA TP 1120, April 1978.

Isom, M. P. "Unsteady Subsonic and Transonic Potential Flow over Helicopter Rotor Blades." NASA CR 2463, October 1974.

Isom, M. P. "The Theory of Sound Radiated by a Hovering Transonic Helicopter Blade." Polytechnic Institute of New York, Report POLY-AE/AM 75-4, May 1975.

Izakson, A. M. "Soviet Helicopter Construction." U.S. Air Force Foreign Technology Division, FTD-MT-65-90, November 1966.

Jenkins, B. Z., and Marks. A. S. "Rotor Downwash Velocities about the UH-1M Helicopter." U.S. Army Missile Research, Development, and Engineering Laboratory, Report RD-75-27, January 1975.

Jenkins, J. L., Jr. "Wind-Tunnel Investigation of a Lifting Rotor Operating at Tip-Speed Ratios from 0.65 to 1.45." NASA TN D-2628, February 1965.

Jenkins, J. L., Jr. "Trim Requirements and Static-Stability Derivatives from a Wind-Tunnel Investigation of a Lifting Rotor in Transition." NASA TN D-2655, February 1965.

Jenkins, J. L., Jr. "A Numerical Method for Studying the Transient Blade Motions of a Rotor with Flapping and Lead-Lag Degrees of Freedom." NASA TN D-4195, 1967.

Jenkins, J. L., Jr. "An Analysis of Blade-Motion Stability for an Articulated Rotor." AHS Forum, 1968.

Jenkins, J. L., Jr., and Deal, P. L. "Investigation of Level-Flight and Maneuvering Characteristics of a Hingeless-Rotor Compound Helicopter." NASA TN D-5602, January 1970.

Jenkins, J. L., Jr.; Winston, M. M.; and Sweet, G. E. "A Wind Tunnel Investigation of the Longitudinal Aerodynamic Characteristics of Two Full-Scale Helicopter Fuselage Models with Appendages." NASA TN D-1364, July 1962.

Jenkins, J. L., Jr., and Yeager, W. T., Jr. "An Analysis of the Gust-Induced Overspeed Trends of Helicopter Rotors." NASA TP 1213, July 1978.

Jenney, D. S.; Arcidiacono, P. J.; and Smith, A. F. "A Linearized Theory for the Estimation of Helicopter Rotor Characteristics at Advance

Ratios above 1.0." AHS Forum, 1963.

Jenney, D. S.; Olson, J. R.; and Landgrebe, A. J. "A Reassessment of Rotor Hovering Performance Prediction Methods." JAHS, 13:2 (April 1968).

Jepson, W. D. "Some Consideration of the Landing and Take Off Characteristics of Twin Engine Helicopters." JAHS, 7:4 (October 1962).

Jewel, J. W., Jr. "Compressibility Effects on the Hovering Performance of a Two-Blade 10-foot-diameter Helicopter Rotor Operating at Tip Mach Numbers up to 0.98." NASA TN D-245, April 1960.

Jewel, J. W., Jr., and Carpenter, P. J. "A Preliminary Investigation of the Effects of Gusty Air on Helicopter Blade Bending Moments." NACA TN 3074, March 1954.

Jewel, J. W., Jr., and Harrington, R. D. "Effect of Compressibility on the Hovering Performance of Two 10-foot-diameter Helicopter Rotors Tested in the Langley Full-Scale Tunnel." NACA RM L58B19, April 1958.

Jewel, J. W., Jr., and Heyson, H. H. "Charts of the Induced Velocities Near a Lifting Rotor." NASA Memo 4-15-59L, May 1959.

Joglekar, M., and Loewy, R. "An Actuator-Disc Analysis of Helicopter Wake Geometry and the Corresponding Blade Response." USAAVLABS TR 69-66, December 1970.

Johansson, B. C. A. "Lifting-Line Theory for a Rotor in Vertical Climb." The Aeronautical Research Institute of Sweden, Report 118, July 1971.

Johansson, B. C. A. "Disk Approximation for a Helicopter Rotor in Forward Flight." The Aeronautical Research Institute of Sweden, Report 123, July 1972.

Johansson, B. C. A. "Evaluation of the Accuracy of Two Helicopter Rotor Theories." The Aeronautical Research Institute of Sweden, Report 124, May 1973.

Johansson, B. C. A. "Compressible Flow about Helicopter Rotors." Vertica, 2:1 (1978).

Johnson, H. B. "The Effects of Semi-Rigid Rotors on Helicopter Autostabiliser Design." AGARD Conference Proceedings No. 86, Constance, Germany, June 1971.

Johnson, H. K. "Development of a Technique for Realistic Prediction and Electronic Synthesis of Helicopter Rotor Noise." USAAMRDL TR 73-8, March 1973.

Johnson, H. K. "Development of an Improved Design Tool for Predicting and Simulating Helicopter Rotor Noise." USAAMRDL TR 74-37, June 1974.

Johnson, H. K., and Katz, W. M. "Investigation of the Vortex Noise Produced by a Helicopter Rotor." USAAMRDL TR 72-2, February 1972.

Johnson, R. L., and Hohenemser, K. H. "On the Dynamics of Lifting Rotors with Thrust or Tilting Moment Feedback Controls." JAHS, 15:1 (January 1970).

Johnson, W. "The Effect of Dynamic Stall on the Response and Airloading of Helicopter Rotor Blades." JAHS, 14:2 (April 1969).

Johnson, W. "A Lifting Surface Solution for Vortex Induced Airloads and Its Application to Rotary Wing Airloads Calculations." Massachusetts Institute of Technology, ASRL TR 153-2, April 1970.

Johnson, W. "The Response and Airloading of Helicopter Rotor Blades Due to Dynamic Stall." Massachusetts Institute of Technology, ASRL TR 130-1, May 1970.

Johnson, W. "A Comparison between Experimental Data and Helicopter Airloads Calculated Using a Lifting-Surface Theory." Massachusetts Institute of Technology, ASRL TR 157-1, July 1970.

Johnson, W. "A Comparison between Experimental Data and a Lifting-Surface Theory Calculation of Vortex Induced Loads." NASA CR 112769, August 1970.

Johnson, W. "A Lifting-Surface Solution for Vortex-Induced Airloads." AIAA Journal, 9:4 (April 1971).

Johnson, W. "Application of a Lifting-Surface Theory to the Calculation of Helicopter Airloads." AHS Forum, 1971.

Johnson, W. "A Perturbation Solution of Helicopter Rotor Flapping Stability." NASA TM X-62165, July 1972.

Johnson, W. "A Perturbation Solution of Rotor Flapping Stability." AIAA Paper No. 72-955, September 1972.

Johnson, W. "A Perturbation Solution of Helicopter Rotor Flapping Stability." Journal of Aircraft, 10:5 (May 1973).

Johnson, W. "Theory and Comparison with Tests of Two Full-Scale Proprotors." NASA SP 352, February 1974.

Johnson, W. "Dynamics of Tilting Proprotor Aircraft in Cruise Flight." NASA TN D-7677, May 1974.

Johnson, W. "Perturbation Solutions for the Influence of Forward Flight on Helicopter Rotor Flapping Stability." NASA TM X-62361, August 1974.

Johnson, W. "Analytical Model for Tilting Proprotor Aircraft Dynamics, Including Blade Torsion and Coupled Bending Modes, and Conversion Mode Operation." NASA TM X-62369, August 1974.

Johnson, W. "Comparison of Three Methods for Calculation of Helicopter Rotor Blade Loading and Stresses Due to Stall." NASA TN D-7833, November 1974.

Johnson, W. "The Influence of Engine/Transmission/Governor on Tilting Proprotor Aircraft Dynamics." NASA TM X-62455, June 1975.

Johnson, W. "Analytical Modeling Requirements for Tilting Proprotor

Aircraft Dynamics." NASA TN D-8013, July 1975.

Johnson, W. "Predicted Dynamic Characteristics of the XV-15 Tilting Proprotor Aircraft in Flight and in the 40- by 80-ft Wind Tunnel." NASA TM X-73158, June 1976.

Johnson, W. "Elementary Applications of a Rotorcraft Dynamic Stability Analysis." NASA TM X-73161, June 1976.

Johnson, W. "The Influence of Pitch-Lag Coupling on the Predicted Aeroelastic Stability of the XV-15 Tilting Proprotor Aircraft." NASA TM X-73213, February 1977.

Johnson, W. "Optimal Control Alleviation of Tilting Proprotor Gust Response." Journal of Aircraft, 14:3, (March 1977).

Johnson, W. "Helicopter Optimal Descent and Landing after Power Loss." NASA TM-73244, May 1977.

Johnson, W. "Flap/Lag/Torsion Dynamics of a Uniform, Cantilever Rotor Blade in Hover." NASA TM-73248, May 1977.

Johnson, W. "Calculated Dynamic Characteristics of a Soft-Inplane Hingeless Rotor Helicopter." NASA TM-73262, June 1977.

Johnson, W. "Aeroelastic Analysis for Rotorcraft in Flight or in a Wind Tunnel." NASA TN D-8515, July 1977.

Johnson, W. "Comprehensive Helicopter Analyses: A State-of-the Art Review." NASA TM 78539, November 1978.

Johnson, W., and Chopra, I. "Calculated Hovering Helicopter Flight Dynamics with a Circulation Controlled Rotor." NASA TM 78443, September 1977.

Johnson, W., and Ham, N. D. "On The Mechanism of Dynamic Stall." JAHS, 17:4 (October 1972).

Johnson, W., and Lee, A. "Comparison of Measured and Calculated Helicopter Rotor Impulsive Noise." NASA TM 78472, March 1978.

Johnson, W., and Scully, M. P. "Aerodynamic Problems in the Calculation of Helicopter Airloads." American Helicopter Society Mideast Region Symposium, Essington, Pennsylvania, October 1972.

Johnston, J. F., and Conner, F. "The Reactionless Inplane Mode of Stiff-Inplane Hingeless Rotors." Lockheed Report LR-26214, December 1973.

Johnston, J. F., and Cook, J. R. "AH-56A Vehicle Development." AHS Forum, 1971.

Johnston, J. F.; Culver, I. H.; and Friend, C. F. "Study of Size Effects on VTOL Handling Qualities Criteria." USAAVLABS TR 65-24, September 1965.

Johnston, R. A. "Parametric Studies of Instabilities Associated with Large, Flexible Rotor Propellers." AHS Forum, 1972.

Johnston, R. A. "Rotor Stability Prediction and Correlation with Model and Full-Scale Tests." JAHS, 21:2 (April 1976).

Johnston, R. A., and Cassarino, S. J. "Aeroelastic Rotor Stability Analysis." USAAMRDL TR 75-40, January 1976.

Johnston, R. A., and Kefford, N. F. K. "Parametric Investigations of Instabilities Associated with Propeller Mode Operation of Large Rotor/Propellers." Sikorsky Aircraft Report SER-60583, December 1970.

Jonda, W., and Frommlet, H. "IDS – An Advanced Hingeless Rotor System." Vertica, 2:1 (1978).

Jones, J. P. "Helicopter Rotor Blade Flapping and Bending." Aircraft Engineering, 29:337 (March 1957); and 29:338 (April 1957).

Jones, J. P. "The Influence of the Wake on the Flutter and Vibration of Rotor Blades." The Aeronautical Quarterly, 9:3 (August 1958).

Jones, J. P. "Helicopter Vibrations." The Journal of the Royal Aeronautical Society, 64:600 (December 1960).

Jones, J. P. "The Use of an Analogue Computer to Calculate Rotor Blade Motion." JAHS, 9:2 (April 1964).

Jones, J. P. "An Extended Lifting Line Theory for the Loads on a Rotor Blade in the Vicinity of a Vortex." Massachusetts Institute of Technology, ASRL TR 123-3, December 1965.

Jones, J. P. "An Actuator Disk Theory for the Shed Wake at Low Tip Speed Ratios." Massachusetts Institute of Technology, ASRL TR 133-1, 1966.

Jones, J. P. "Rotor Aerodynamics – Retrospect and Prospect." AGARD Advisory Report 13, Gottingen, Germany, September 1967.

Jones, J. P. "The Helicopter Rotor." The Aeronautical Journal, 74:719 (November 1970).

Jones, J. P. "The Rotor and Its Future." The Aeronautical Journal, 77:751 (July 1973).

Jones, J. P., and Bhuta, P. G. "Vibrations of a Whirling Rayleigh Beam." Journal of the Acoustical Society of America, 35:7 (July 1963).

Jones, R. "An Analytical and Model Test Research Study on the Kaman Dynamic Antiresonant Vibration Isolator (DAVI)." USAAVLABS TR 68-42, November 1968.

Jones, R. "The Exploratory Development of the Three-Dimensional Dynamic Antiresonant Vibration Isolator for Rotary-Wing Application." USAAVLABS TR 70-30, August 1970.

Jones, R. "Flight Test Results of a DAVI Isolated Platform." USAAVLABS TR 70-57, November 1970.

Jones, R. "A Full-Scale Experimental Feasibility Study of Helicopter Rotor Isolation Using the Dynamic Antiresonant Vibration Isolator." USAAVLABS TR 71-17, June 1971.

Jones, R., and Flannelly, W. G. "Application of the Dynamic Antiresonant Vibration Isolator to Helicopter Vibration Control." Shock and Vibration Bulletin 37, Part 6, January 1968.

Jones, R., and McGarvey, J. H. "Advanced Development of a Helicopter Rotor Isolation System for Improved Reliability." USAAMRDL TR 77-23, December 1977.

Jones, W. P., and Rao, B. M. "Compressibility Effects on Oscillating Rotor Blades in Hovering Flight." AIAA Journal, 8:2 (February 1970).

Jones, W. P., and Rao, B. M. "Tip Vortex Effects on Oscillating Rotor Blades in Hovering Flight." AIAA Journal, 9:1 (January 1971).

Judd, M., and Newman, S. J. "An Analysis of Helicopter Rotor Response Due to Gusts and Turbulence." Vertica, 1:3 (1977).

Kaman, C. H. "Aerodynamic Considerations of Rotors in Hovering and Vertical Climb Conditions." JAS, 10:7 (July 1943).

Kana, D. D., and Chu, W.-H. "The Response of a Model Helicopter Rotor Blade to Random Excitation during Forward Flight." Southwest Research Institute, SwRI Report No. 02-1732, August 1972.

Kanning, G., and Biggers, J. C. "Application of a Parameter Identification Technique to a Hingeless Helicopter Rotor." NASA TN D-7834, December 1974.

Kaplita, T. T. "Investigation of the Stabilizer on the S-67 Aircraft." USAAMRDL TR 71-55, October 1971.

Kasper, P. K. "Determination of Rotor Harmonic Blade Loads from Acoustic Measurements." NASA CR 2580, October 1975.

Katzenberger, E. F., and Rich, M. J. "An Investigation of Helicopter Descent and Landing Characteristics Following Power Failure." JAS, 23:4 (April 1956).

Kaufman, L. A. "A Concept for the Development of a Universal Automatic Flight Control System for VTOL Aircraft." JAHS, 10:1 (January 1965).

Kaufman, L., and Peress, K. "A Review of Methods for Predicting Helicopter Longitudinal Response." JAS, 23:3 (March 1956).

Kaufman, L., and Schultz, E. R. "The Stability and Control of Tethered Helicopters." JAHS, 7:4 (October 1962).

Kaufman, L., and Schultz, E. R. "VTOL Automatic Flight Control." AHS Forum, 1963.

Kawakami, N. "Dynamics of an Elastic Seesaw Rotor." Journal of Aircraft, 14:3 (March 1977).

Kaza, K. R. V. "Effect of Steady State Coning Angle and Damping on Whirl Flutter Stability." Journal of Aircraft, 10:11 (November 1973).

Kaza, K. R. V. "Rotation in Vibration, Optimization, and Aeroelastic Stability Problems." Ph.D. thesis, Stanford University, May 1964.

Kaza, K. R. V., and Kvaternik, R. G. "A Critical Examination of the Flap-Lag Dynamics of Helicopter Rotor Blades in Hover and in Forward Flight." AHS Forum, 1976.

Kaza, K. R. V., and Kvaternik, R. G. "Nonlinear Flap-Lag-Axial Equations of a Rotating Beam." AIAA Journal, 15:6 (June 1977).

Kaza, K. R. V., and Kvaternik, R. G. "Nonlinear Aeroelastic Equations for Combined Flapwise Bending, Chordwise Bending, Torsion, and Extension of Twisted Nonuniform Rotor Blades in Forward Flight." NASA TM 74059, August 1977.

Kee, R. M. "Main Rotor Blade Design and Development." JAHS, 4:4 (October 1959).

Kefford, N. F. K. "Investigation of the Speed Brakes on the S-67 Aircraft." USAAMRDL TR 72-22, May 1972.

Kefford, N. F. K., and Munch, C. L. "Conceptual Design Study of 1985 Commercial VTOL Transports that Utilize Rotors." NASA CR 2532, May 1975.

Kelley, B. "Response of Helicopter Rotors to Periodic Forces." NACA ARR 5A09, March 1945.

Kelley, B. "Contributions of the Bell Helicopter Company to Helicopter Development." The Aeronautical Journal, 76:735 (March 1972).

Kelley, H. L.; Pegg, R. J.; and Champine, R. A. "Flying Qualities Factors Currently Limiting Helicopter Nap-of-the-Earth Maneuverability As Identified by Flight Investigation." NASA TN D-4931, December 1968.

Kelley, H. L., and West, T. C. "Flight Investigation of Effects of a Fan-in-Fin Yaw Control Concept on Helicopter Flying-Qualities Characteristics." NASA TN D-7452, April 1974.

Kelly, J. R., and Garren, J. F., Jr. "Study of the Optimum Values of Several Parameters Affecting Longitudinal Handling Qualities of VTOL Aircraft." NASA TN D-4624, July 1968.

Kelly, J. R.; Niessen, F. R.; Thibodeaux, J. J.; Yenni, K. R.; and Garren, J. F., Jr. "Flight Investigation of Manual and Automatic VTOL Decelerating Instrument Approaches and Landings." NASA TN D-7524, July 1974.

Kelly, J. R., and Winston, M. M. "Stability Characteristics of a Tandem-Rotor Transport Helicopter As Determined by Flight Test." NASA TN D-2847, June 1965.

Kemp, L. D. "An Analytical Study for the Design of Advanced Rotor Airfoils." NASA CR 112297, March 1973.

Kenigsberg, I. J. "CH-53A Flexible Frame Vibration Analysis/Test Correlation." Sikorsky Aircraft SER-651195, March 1973.

Kerr, A. W. "Effect of Helicopter Drag Reduction on Rotor Dynamic Loads and Blade Life." AHS Forum, 1975. (Rotorcraft Parasite Drag Supplement.)

Kesler, D. F.; Murakoshi, A. Y.; and Sinacori, J. B. "Flight Simulation of the Model 347 Advanced Tandem-Rotor Helicopter." USAAMRDL TR

74-21, November 1974.

Ketchel, J. M.; Danaher, J. W.; and Morrissey, C. J. "Effects of Vibration on Navy and Marine Corps Helicopter Flight Crews." Matrix Research Company, August 1969.

Keys, C. N., and Rosenstein, H. J. "Summary of Rotor Hub Drag Data." NASA CR 152080, March 1978.

Keys, C. N., and Wiesner, R. "Guidelines for Reducing Helicopter Parasite Drag." JAHS, 20:1 (January 1975).

Kfoury, D. J. "A Routine Method for the Calculation of Aerodynamic Loads on a Wing in the Vicinity of Infinite Vortices." Massachusetts Institute of Technology, ASRL TR 133-2, May 1966.

Kiang, D., and Ng, C. Y. "Rotational Noise Characteristics of Helicopters." Massachusetts Institute of Technology, FTL TM 71-8, December 1971.

Kidd, D. L.; Spivey, R. F.; and Lawrence, K. L. "Control Loads and Their Effects on Fuselage Vibrations." JAHS, 12:4 (October 1967).

Kiessling, F. "Some Problems in Research on Whirl Flutter in V/STOL Aircraft." European Space Research Organization, ESRO TT-160, May 1975.

Kiessling, F. "Ground Vibration Test – A Tool for Rotorcraft Dynamic and Aeroelastic Investigations." Vertica, 1:2 (1976).

Kilmer, F. G., and Sklaroff, J. R. "Redundant System Design and Flight Test Evaluation for the TAGS Digital Control System." AHS Forum, 1973.

King, R. J., and Schlegel, R. G. "Prediction Methods and Trends for Helicopter Rotor Noise." CAL/AVLABS Symposium, Buffalo, New York, June 1969.

Kinney, W. A.; Pierce, A. D.; and Rickley, E. J. "Helicopter Noise Experiments in an Urban Environment." Journal of the Acoustical Society of America, 56:2 (August 1974).

Kisielowski, E.; Bumstead, R.; Fissel, P.; and Chinsky, I. "Generalized Rotor Performance." USAAVLABS TR 66-83, February 1967.

Kisielowski, E.; Perlmutter, A. A.; and Tang, J. "Stability and Control Handbook for Helicopters." USAAVLABS TR 67-63, August 1967.

Kisielowski, E.; Smith, J. H.; and Spittle, R. W. "Design and Optimization Study of the Active Arm External Load Stabilization System for Helicopters." USAAMRDL TR 74-55, August 1974.

Klemin, A. "An Introduction to the Helicopter." NACA TM 340, 1925.

Klemin, A.; Walling, W. C.; and Wiesner, W. "Rotor Bending Moments in Plane of Flapping." JAS, 9:11 (September 1942).

Klinar, W. J., and Craig, S. J. "Study of VTOL Control Requirements during Hovering and Low-Speed Flight under IFR Conditions." JAHS, 6:4 (October 1961).

Klingloff, R. F.; Sardanowsky, V.; and Baker, R. C. "The Effect of VTOL Design Configuration on Power Required for Hover and Low Speed Flight." JAHS, 10:3 (July 1965).

Knight, M. "Analytical Comparison of Helicopter and Airplane in Level Flight." JAS, 5:11 (September 1938).

Knight, M., and Hefner, R. A. "Static Thrust Analysis of the Lifting Airscrew." NACA TN 626, December 1937.

Knight, M., and Hefner, R. A. "Analysis of Ground Effect on the Lifting Airscrew." NACA TN 835, December 1941.

Knight, V. H., Jr.; Haywood, W. S., Jr.; and Williams, M. L. "A Rotor-Mounted Digital Instrumentation System for Helicopter Blade Flight Research Measurements." NASA TP 1146, April 1978.

Kocurek, J. D., and Tangler, J. L. "A Prescribed Wake Lifting Surface Hover Performance Analysis." JAHS, 22:1 (January 1977).

Koenig, D. G.; Greif, R. K.; and Kelly, M. W. "Full-Scale Wind-Tunnel Investigation of the Longitudinal Characteristics of a Tilting-Rotor Convertiplane." NASA TN D-35, December 1959.

Konig, H., and Schmitt, H. "Optimization of Automatic Flight Control Concepts for Light Helicopters with All-Weather Capability." AGARD Conference Proceedings, No. CP-86, Constance, Germany, June 1971.

Koo, J., and Oka, T. "Experimental Study on the Ground Effect of a Model Helicopter Rotor in Hovering." NASA TT F-13938, December 1971.

Kramer, K. N. "The Induced Efficiency of Optimum Propellers Having a Finite Number of Blades." NACA TM 884, January 1939.

Kretz, M. "Research in Multicyclic and Active Control of Rotary Wings." Vertica, 1:2 (1976).

Kretz, M.; Aubrun, J.-N.; and Larche, M. "Wind Tunnel Tests of the Dorand DH 2011 Jet Flap Rotor." NASA CR 114693 and CR 114694, June 1973.

Kuczynski, W. A. "Experimental Hingeless Rotor Characteristics at Full Scale First Flap Mode Frequencies." NASA CR 114519, October 1972.

Kuczynski, W. A.; Sharpe, D. L.; and Sissingh, G. J. "Hingeless Rotor – Experimental Frequency Response and Dynamic Characteristics with Hub Moment Feedback Controls." AHS Forum, 1972.

Kuczynski, W. A., and Sissingh, G. J. "Research Program to Determine Rotor Response Characteristics at High Advance Ratios." NASA CR 114290, February 1971.

Kuczynski, W. A., and Sissingh, G. J. "Characteristics of Hingeless Rotors with Hub Moment Feedback Controls Including Experimental Rotor Frequency Response." NASA CR 114427 and CR 114428, January 1972.

Kuhn, R. E. "Review of Basic Principles of V/STOL Aerodynamics." NASA

TN D-733, March 1961.

Kunz, D. L. "Effects of Unsteady Aerodynamics on Rotor Aeroelastic Stability." NASA TM-78434, September 1977.

Kussner, H. G. "Helicopter Problems." NACA TM 827, May 1937.

Kvaternik, R. G. "Studies in Tilt Rotor VTOL Aircraft Aeroelasticity." NASA TM X-69497 and TM X-69496, June 1973.

Kvaternik, R. G. "Experimental and Analytical Studies in Tilt-Rotor Aeroelasticity." NASA SP-352, February 1974.

Kvaternik. R. G., and Kaza, K. R. V. "Nonlinear Curvature Expressions for Combined Flapwise Bending, Chordwise Bending, Torsion, and Extension of Twisted Rotor Blades." NASA TM X-73997, December 1976.

Kvaternik, R. G., and Kohn, J. S. "An Experimental and Analytical Investigation of Proprotor Whirl Flutter." NASA TP 1047, December 1977.

La Forge, S. V. "Effects of Blade Stall on Helicopter Rotor Blade Bending and Torsional Loads." Hughes Tool Company, Report HTC-AD 64-8, May 1965.

La Forge, S. V., and Rohtert, R. E. "Aerodynamic Tests of an Operational OH-6A Helicopter in the Ames 40- by 80-foot Wind Tunnel." Hughes Tool Company, Report 369-A-8020, May 1970.

Laing, E. J. "Army Helicopter Vibration Survey Methods and Results." JAHS, 19:3 (July 1974).

Laitone, E. V., and Talbot, L. "Subsonic Compressibility Corrections for Propellers and Helicopter Rotors." JAS, 20:10 (October 1953).

Lakshmikantham, C., and Rao, C. V. J. "Response of Rotor Blades to Random Inputs – Combined Bending and Torsion." U.S. Army Materials and Mechanics Research Center, AMMRC TR 72-15, May 1972.

Lakshmikantham, C., and Rao, C. V. J. "Response of Helicopter Rotor Blades to Random Loads Near Hover." The Aeronautical Quarterly, 23:4 (November 1972).

Lambermont, P., and Pirie, A. Helicopters and Autogyros of the World. 2nd ed. New York: A. S. Barnes and Co., 1970.

Landgrebe, A. J. "Simplified Procedures for Estimating Flapwise Bending Moments on Helicopter Rotor Blades." NASA CR 1440 and CR 1441, October 1969.

Landgrebe, A. J. "An Analytical Method for Predicting Rotor Wake Geometry." JAHS, 14:4 (October 1969).

Landgrebe, A. J. "An Analytical and Experimental Investigation of Helicopter Rotor Hover Performance and Wake Geometry Characteristics." USAAMRDL TR 71-24, June 1971.

Landgrebe, A. J. "Simplified Procedures for Estimating Flapwise Bending Moments on Helicopter Rotor Blades." JAHS, 16:3 (July 1971).

Landgrebe, A. J. "The Wake Geometry of a Hovering Helicopter Rotor and Its Influence on Rotor Performance." JAHS, 17:4 (October 1972).

Landgrebe, A. J., and Bellinger, E. D. "An Investigation of the Quantitative Applicability of Model Helicopter Rotor Wake Patterns Obtained from a Water Tunnel." USAAMRDL TR 71-69, December 1971.

Landgrebe, A. J., and Bellinger, E. D. "Experimental Investigation of Model Variable-Geometry and Ogee Tip Rotors." AHS Forum, 1973.

Landgrebe, A. J., and Bellinger, E. D. "Experimental Investigation of Model Variable-Geometry and Ogee Tip Rotors." NASA CR-2275, February 1974.

Landgrebe, A. J., and Bellinger, E. D. "A Systematic Study of Helicopter Rotor Stall Using Model Rotors." AHS Forum, 1974.

Landgrebe, A. J., and Bennett, J. C., Jr. "Investigation of the Airflow of a Hovering Model Helicopter at Rocket Trajectory and Wind Sensor Locations." United Technologies Research Center, Report No. R77-912573-15, July 1977.

Landgrebe, A. J., and Cheney, M. C., Jr. "Rotor Wakes – Key to Performance Prediction." AGARD Conference Proceedings No. 111, Marseilles, France, September, 1972.

Landgrebe, A. J., and Egolf, T. A. "Prediction of Rotor Wake Induced Flow along Rocket Trajectories of an Army AH-1G Helicopter." Picatinny Arsenal Report TR 4797, March 1975.

Landgrebe, A. J., and Egolf, T. A. "Rotorcraft Wake Analysis for the Prediction of Induced Velocities." USAAMRDL TR 75-45, January 1976.

Landgrebe, A. J., and Egolf, T. A. "Prediction of Helicopter Induced Flow Velocities Using the Rotorcraft Wake Analysis." AHS Forum, 1976.

Landgrebe, A. J., and Johnson, B. V. "Measurement of Model Helicopter Rotor Flow Velocities with a Laser Doppler Velocimeter." JAHS, 19:3 (July 1974).

Landgrebe, A. J.; Moffitt, R. C.; and Clark, D. R. "Aerodynamic Technology for Advanced Rotorcraft." JAHS, 22:2 (April 1977); 22:3 (July 1977).

Lang, J. D. "On Predicting Leading-Edge Bubble Bursting on an Airfoil in Unsteady Incompressible Flow." Cranfield (England) Institute of Technology, Memo 109, April 1973.

Lange, R. H., and McLemore, H. C. "Static Longitudinal Stability and Control of a Convertible-type Airplane As Affected by Articulated and Rigid Propeller Operation." NACA TN 2014, February 1950.

Laskin, I.; Orcutt, F. K.; and Shipley, E. E. "Analysis of Noise Generated by UH-1 Helicopter Transmission." USAAVLABS TR 68-41, June 1968.

Laufer, T. "A Vibration Absorber for Two-Bladed Helicopters." NASA TT F-43, November 1960.

Laurenson, R. M. "Modal Analysis of Rotating Flexible Structures." AIAA

Journal, 14:10 (October 1976).

Law, H. Y. H. "Two Methods of Prediction of Hovering Performance." U.S. Army Aviation Systems Command, USAAVSCOM TR 72-4, February 1972.

Lawton, B. W. "Subjective Assessment of Simulated Helicopter Blade-Slap Noise." NASA TN D-8359, December 1976.

Lean, D., et al. "Recommendations for V/STOL Handling Qualities." AGARD Report 408, 1962.

Lee, C.; Charles, B.; and Kidd, D. "Wind-Tunnel Investigation of a Quarter-Scale Two-Bladed High-Performance Rotor in a Freon Atmosphere." USAAVLABS TR 70-58, February 1971.

Lee, C. D., and White, J. A. "Investigation of the Effect of Hub Support Parameters on Two-Bladed Rotor Oscillatory Loads." NASA CR 132435, May 1974.

Lee, A. "An Experimental and Theoretical Study of Helicopter Rotor Noise." Ph.D. thesis, Massachusetts Institute of Technology, September 1975.

Lee, A. "High Speed Helicopter Noise Sources." NASA CR 151997 and CR 151996, January 1977.

Lee, A. "Acoustical Effects of Blade Tip Shape Changes on a Full-Scale Helicopter Rotor in a Wind Tunnel." NASA CR 152082, April 1978.

Lee, A.; Aravamudan, K.; Bauer, P.; and Harris, W. R. "An Experimental Investigation of Helicopter Rotor High Frequency Broadband Noise." AIAA Paper 77-1339, October 1977.

Lee, A.; Harris, W. L.; and Widnall, S. E. "A Study of Helicopter Rotor Rotational Noise." Journal of Aircraft 14:11 (November 1977).

Lee, A., and Mosher, M. "A Study of the Noise Radiation from Four Helicopter Rotor Blades." NASA CP 2052, May 1978.

Legrand, F. "Rotorcraft." NASA TT F-11530, April 1968.

Lehman, A. F. "Model Studies of Helicopter Rotor Flow Patterns." USAAVLABS TR 68-17, April 1968.

Lehman, A. F. "Model Studies of Helicopter Rotor Flow Patterns in a Water Tunnel." AHS Forum, 1968.

Lehman, A. F. "Model Studies of Helicopter Tail Rotor Flow Patterns in in and out of Ground Effect." USAAVLABS TR 71-12, April 1971.

Lehman, A. F., and Besold, J. A. "Test Section Size Influence on Model Helicopter Rotor Performance." USAAVLABS TR 71-6, March 1971.

Lemnios, A. Z. "Rotary Wing Design Methodology." AGARD Conference Proceedings No. 122, Milan, Italy, March 1973.

Lemnios. A. Z., and Dunn, F. K. "Theoretical Study of Multicyclic Control of a Controllable Twist Rotor." NASA CR 151959, April 1976.

Lemnios, A. Z., and Howes, H. E. "Wind Tunnel Investigation of the

Controllable Twist Rotor Performance and Dynamic Behavior." USAAMRDL TR 77-10, June 1977.

Lemnios, A. Z., and Smith, A. F. "An Analytical Evaluation of the Controllable Twist Rotor Performance and Dynamic Behavior." USAAMRDL TR 72-16, May 1972.

Lemnios, A. Z.; Smith, A. F.; and Nettles, W. E. "The Controllable Twist Rotor, Performance and Blade Dynamics," AHS Forum, 1972.

Le Nard, J. M. "A Theoretical Analysis of the Tip Relief Effect on Helicopter Rotor Performance." USAAMRDL TR 72-7, August 1972.

Le Nard, J. M., and Boehler, G. D. "Inclusion of Tip Relief in the Prediction of Compressibility Effects on Helicopter Rotor Performance." USAAMARDL TR 73-71, December 1973.

Lentine, F. P.; Groth, W. P.; and Oglesby, T. H. "Research in Maneuverability of the XH-51A Compound Helicopter." USAAVLABS TR 68-23, June 1968.

Leone, P. F. "Theory of Rotor Blade Uncoupled Lag Bending Aero-Elastic Vibrations." AHS Forum, 1955.

Leone, P. F. "Experimental Study of a Tandem Helicopter Employing Cocked Vertical Blade Hinges." JAHS, 1:3 (July 1956).

Leone, P. F. "Mechanical Stability of a Two-Bladed Cantilever Helicopter Rotor." JAS, 23:7 (July 1956).

Leone, P. F. "Theoretical and Experimental Study of the Coupled Flap Bending and Torsion Aeroelastic Vibrations of a Helicopter Rotor Blade." AHS Forum, 1957.

Leverton, J. W. "Helicopter Noise." AGARD Conference Proceedings No. 31, Ottawa, Canada, June 1968.

Leverton, J. W. "Helicopter Noise – Blade Slap. Part 1: Review and Theoretical Study." NASA CR 1221, October 1968.

Leverton, J. W. "Helicopter Rotor Noise Final Report – Experimental Study of Rotor Noise." NASA CR 66868, June 1969.

Leverton, J. W. "The Sound of Rotorcraft." The Aeronautical Journal, 75:726 (June 1971).

Leverton, J. W. "Helicopter Noise – Blade Slap. Part 2: Experimental Results." NASA CR 1983, March 1972.

Leverton, J. W. "The Noise Characteristics of a Large 'Clean' Rotor." AGARD Conference Proceedings No. 111, Marseilles, France, September 1972.

Leverton, J. W. "The Noise Characteristics of a Large 'Clean' Rotor." J. Sound Vib., 27:3 (1973).

Leverton, J. W. "Helicopter Noise – Are Existing Methods Adequate for Rating Annoyance or Loudness." JAHS, 19:2 (April 1974).

Leverton, J. W. "Helicopter Noise: Can It Be Adequately Rated?" J. Sound Vib., 43:2 (1975).

Leverton, J. W. "Discrete Frequency Rotor Noise." AIAA Paper 75-451, March 1975.

Leverton, J. W., and Amor, C. B. "An Investigation of the Impulsive Rotor Noise Using a Model Rotor." American Helicopter Society, Mideast Region Symposium, Philadelphia, Pennsylvania, October 1972.

Leverton, J. W., and Amor, C. B. "An Investigation of Impulsive Rotor Noise of a Model Rotor." J. Sound Vib., 28:1 (1973).

Leverton, J. W., and Pollard, J. S. "A Comparison of the Overall and Broadband Noise Characteristics of Full-Scale and Model Helicopter Rotors." J. Sound Vib., 30:2 (1973).

Leverton, J. W.; Pollard, J. S.; and Wills, C. R. "Main Rotor Wake/Tail Rotor Interaction." Vertica, 1:3 (1977).

Leverton, J. W.; Southwood, B. J.; and Pike, A. C. "Rating Helicopter Noise." NASA CP 2052, May 1978.

Leverton, J. W., and Taylor, F. W. "Helicopter Blade Slap." J. Sound Vib., 4:3 (1966).

Levine, L. S. "An Analytic Investigation of Techniques to Reduce Tail Rotor Noise." NASA CR 145014, July 1976.

Levinsky, E. S., and Strand, T. "A Method for Calculating Helicopter Vortex Paths and Wake Velocities." AFFDL TR 69-113, July 1970.

Levinsky, E. S.; Thommen, H. U.; Yager, P. M.; and Holland, C. H. "Lifting Surface Theory for V/STOL Aircraft in Transition and Cruise." Journal of Aircraft, 6:6 (November-December 1969); and 7:1 (January-February 1970).

Levy, H., and Forsdyke, A. G. "The Steady Motion and Stability of a Helical Vortex." Proceedings of the Royal Society of London, Series A, 120:428 (1928).

Lewis, R. B. II. "Army Helicopter Performance Trends." JAHS, 17:2 (April 1972).

Liberatore, E. K. "ISR – Inherently Stable Rotor Systems." JAHS, 5:4 (October 1960).

Lichten, R. L. "Helicopter Performance." JAS, 13:7 (July 1946).

Lichten, R. L. "Some Aspects of Convertible Aircraft Design." JAS, 16:10 (October 1949).

Lichten, R. L. "The Future VTOL Transport." JAHS, 4:1 (January 1959).

Lightfoot, R. B. "Single Rotor Merits." JAHS, 3:2 (April 1958).

Lightfoot, R. B. "Large Rotors with Mechanical Drive." AGARD Conference Proceedings No. 7, Paris, France, January 1966.

Lightfoot, R. B., Immenschuh, W. T., et al. "VTOL – 1968." Journal of Aircraft, 6:4 (July-August 1969).

Liiva, J. "Unsteady Aerodynamic and Stall Effects on Helicopter Rotor Blade Airfoil Sections." Journal of Aircraft, 6:1 (January-February 1969).

Liiva, J., and Davenport, F. J. "Dynamic Stall of Airfoil Sections for High-Speed Rotors." JAHS, 14:2 (April 1969).

Liiva, J.; Davenport, F. J.; Gray, L.; and Walton, I. C. "Two-Dimensional Tests of Airfoils Oscillating Near Stall." USAAVLABS TR 68-13, April 1968.

Lince, D. L. "Baseline Noise Measurements of the OH-58A Helicopter." U.S. Army Human Engineering Laboratories, TN 3-71, April 1971.

Linden, A. W., et al. "Variable Diameter Rotor Study." AFFDL TR 71-170, January 1972.

Linden, A. W., et al. "Rotor System Research Aircraft." NASA CR 112152, CR 112153, CR 112154, CR 112155, October, 1972.

Linville, J. C. "An Experimental Investigation of the Effects of Rotor Head Configuration and Fuselage Yaw on the Wake Characteristics and Rotor Performance of a 1/8th Scale Helicopter." USAAVLABS TR 69-94, February 1970.

Linville, J. C. "An Experimental Investigation of High-Speed Rotorcraft Drag." USAAMRDL TR 71-46, February 1972.

Lipeles, J. L. "Vibrations of a Rotating Beam." JAHS, 11:4 (October 1966).

Lipson, S. "Static-Thrust Investigation of Full-Scale PV-2 Helicopter Rotors Having NACA 0012.6 and 23012.6 Airfoil Sections." NACA MR L6D24, May 1946.

Liu, D. T. "Rotor Stability Derivatives at High Inflow Ratio Conditions for Tilt Rotor VTOL Aircraft." JAHS, 7:2 (April 1962).

Liu, D. T. "In Flight Stabilization of Externally Slung Helicopter Loads." USAAMRDL TR 73-5, May 1973.

Livingston, C. L. "Wind Tunnel Tests of a Full-Scale Rotor at High Speeds." USAAVLABS TR 65-42, July 1965.

Livingston, C. L. "Prediction of Stability and Control Characteristics of Rotorcraft." American Helicopter Society, Mideast Region Symposium, Philadelphia, Pennsylvania, October 1972.

Livingston, C. L., and Murphy, M. R. "Flying Qualities Considerations in the Design and Development of the Hueycobra." JAHS, 14:1 (January 1969).

Livingston, C. L., et al. "A Stability and Control Prediction Method for Helicopters and Stoppable Rotor Aircraft." AFFDL TR 69-123, February 1970.

Lo, C.-F. "Wind-Tunnel Boundary Interference on a V/STOL Model." Journal of Aircraft, 8:3 (March 1971).

Lo, C.-F., and Binion, T. W., Jr. "A V/STOL Wind-Tunnel Wall Interference Study." Journal of Aircraft, 7:1 (January-February 1970).

Lock, C. N. H. "Further Development of Autogyro Theory." ARC R&M 1127, March 1927.

Lock, C. N. H. "Photographs of Streamers Illustrating the Flow Around an Airscrew in the 'Vortex Ring State.'" ARC R&M 1167, April 1928.

Lock, C. N. H. "The Application of Goldstein's Theory to the Practical

Design of Airscrews." ARC R&M 1377, November 1930.

Lock, C. N. H. "Note on the Characteristic Curve for an Airscrew or Helicopter." ARC R&M 2673, June 1947.

Lock, C. N. H., and Bateman, H. "Experiments with a Family of Airscrews, Part III." ARC R&M 892, December 1923.

Lock, C. N. H.; Bateman, H.; and Townend, H. C. H. "An Extension of the Vortex Theory of Airscrews with Applications to Airscrews of Small Pitch, Including Experimental Results." ARC R&M 1014, September 1925.

Lock, C. N. H., and Yeatman, D. "Tables for Use in an Improved Method of Airscrew Strip Theory Calculation." ARC R&M 1674, October 1934.

Lockheed-California Company. "Investigation of Elastic Coupling Phenomena of High Speed Rigid Rotor Systems." TRECOM TR 63-75, June 1964.

Lockheed-California Company. "Wind Tunnel Tests of an Optimized, Matched Stiffness Rigid Rotor." TRECOM TR 64-56, November 1964.

Lockheed-California Company. "Stopped-Stowed Rotor Composite Research Aircraft." USAAVLABS TR 68-40, January 1969.

Loewy, R. G. "A Two-Dimensional Approximation to the Unsteady Aerodynamics of Rotary Wings." JAS, 24:2 (February 1957).

Loewy, R. G. "Aural Detection of Helicopters in Tactical Situations." JAHS, 8:4 (October 1963).

Loewy, R. G. "A Review of Rotary-Wing V/STOL Dynamic and Aeroelastic Problems." JAHS, 14:3 (July 1969).

Loewy, R. G., and Sutton, L. R. "A Theory for Predicting the Rotational Noise of Lifting Rotors in Forward Flight, Including a Comparison with Experiment." USAAVLABS TR 65-82, January 1966.

Loewy, R. G., and Sutton, L. R. "A Theory for Predicting the Rotational Noise of Lifting Rotors in Forward Flight, Including a Comparison with Experiment." J. Sound Vib., 4:3 (1966).

Logan, A. H. "Evaluation of a Circulation Control Tail Boom for Yaw Control." USARTL TR 78-10, April 1978.

Loiselle, J. W. "Generalized Helicopter Rotor Performance Predictions." Naval Postgraduate School, Monterey, California, September 1977.

Lollar, T. W., and Kriechbaum, G. K. L. "VTOL Handling Qualities Criteria and Control Requirements – Analysis and Experiment." JAHS, 13:3 (July 1968).

London, R. J.; Watts, G. A.; and Sissingh, G. J. "Experimental Hingeless Rotor Characteristics at Low Advance Ratio with Thrust." NASA CR 114684, December 1973.

Losch, F. "Calculation of the Induced Efficiency of Heavily Loaded Propellers Having Infinite Number of Blades." NACA TM 884, January 1939.

Lowis, O. J. "The Stability of Rotor Blade Flapping Motion at High Tip Speed Ratios." ARC R&M 3544, January 1963.

Lowson, M. V. "The Sound Field for Singularities in Motion." Proceedings of the Royal Society of London Series A, 286:1407 (August 1965).

Lowson, M. V. "Basic Mechanisms of Noise Generation by Helicopters, V/STOL Aircraft and Ground Effect Machines." J. Sound Vib., 3:3 (1966).

Lowson, M. V. "Fundamental Considerations of Noise Radiation by Rotary Wings." AGARD Conference Proceedings No. 111, Marseilles, France, September 1972.

Lowson, M. V. "Helicopter Noise: Analysis – Prediction and Methods of Reduction." AGARD Lecture Series No. 63, Brussels, April 1973.

Lowson, M. V., and Ollerhead, J. B. "Studies of Helicopter Rotor Noise." USAAVLABS TR 68-60, January 1969.

Lowson, M. V., and Ollerhead, J. B. "A Theoretical Study of Helicopter Rotor Noise." J. Sound Vib., 9:2 (1969).

Lowson, M. V.; Whatmore, A. R.; and Whitfield, C. E. "Source Mechanisms for Rotor Noise Radiation." NASA CR 2077, August 1973.

LTV Aerospace Corporation, "Research Report: Jet-Flap Rotor Preliminary Application Study." NASA CR 73319 and CR 73320, February 1969.

Lucassen, L. R., and Sterk, F. J. "Dynamic Stability Analysis of a Hovering Helicopter with a Sling Load." JAHS, 10:2 (April 1965).

Lucassen, L. R., and Vodegel, H. J. C. C. "Lift Distribution and Induced Drag of an Infinite Wing, Perpendicularly Crossed by a Free Line Vortex." National Aerospace Laboratory (Netherlands), Memo VH-70-016, 1970.

Ludi, L. H. "Flight Investigation of Effects of Atmospheric Turbulence and Moderate Maneuvers on Bending and Torsional Moments Encountered by a Helicopter Rotor Blade." NACA TN 4203, February 1958.

Ludi, L. H. "Flight Investigation of Effects of Retreating-Blade Stall on Bending and Torsional Moments Encountered by a Helicopter Rotor Blade." NACA TN 4254, May 1958.

Ludi, L. H. "Flight Investigation of Effects of Transition, Landing Approaches, Partial-Power Vertical Descents, and Droop-Stop Pounding of the Bending and Torsional Moments Encountered by a Helicopter Rotor Blade." NASA Memo 5-7-59L, May 1959.

Ludi, L. H. "Flight Investigation of Effects of Selected Operating Conditions on the Bending and Torsional Moments Encountered by a Helicopter Rotor Blade." NASA TN D-759, April 1961.

Ludi, L. H. "Composite Aircraft Design." JAHS, 13:1 (January 1968).

Ludi, L. H., and Yeates, J. E., Jr. "Flight Measurements of the Effects of Blade out of Track on the Vibration Levels on a Tandem Rotor Helicopter." NASA TN D-364, May 1960.

Lynn, R. R. "Dynamic Absorbers in the Rotating System of Helicopters." JAHS, 5:4 (October 1960).

Lynn, R. R. "New Control Criteria for VTOL Aircraft." Aerospace Engineering, 21:8 (August 1962).

Lynn, R. R. "Wing-Rotor Interactions." Journal of Aircraft, 3:4 (July-August 1966).

Lynn, R. R., and Drees, J. M. "Promise of Compounding." JAHS, 12:1 (January 1967).

Lynn, R. R., et al. "Tail Rotor Design." JAHS, 15:4 (October 1970).

Lyon, R. H. "Radiation of Sound by Airfoils That Accelerate Near the Speed of Sound." Journal of the Acoustical Society of America, 49:3, part 2 (1971).

Lyon, R. H.; Mark, W. D.; and Pyle, R. W., Jr. "Synthesis of Helicopter Rotor Tips for Less Noise." Journal of the Acoustical Society of America, 53:2 (1973).

Lytwyn, R. T. "An Analysis of the Divergent Vertical Helicopter Oscillations Resulting from the Physical Presence of the Pilot in the Collective Control Loop." JAHS, 12:1 (January 1967).

Lytwyn, R. T.; Miao, W.; and Woitsch, W. "Airborne and Ground Resonance of Hingeless Rotors." JAHS, 16:2 (April 1971).

McAlister, K. W., and Carr, L. W. "Water-Tunnel Experiments on an Oscillating Airfoil at Re = 21000." NASA TM 78446, March 1978.

McAlister, K. W.; Carr, L. W.; and McCroskey, W. J. "Dynamic Stall Experiments on the NACA 0012 Airfoil." NASA TP 1100, January 1978.

McBrayer, J. S., and Robinson, R. W. "A Fail-Functional Stability Augmentation System." JAHS, 10:4 (October 1965).

McCall, C. D.; Field, D. M.; and Reddick. H. "Advanced Technology As Applied to the Design of the HLH Rotor Hub." JAHS, 18:4 (October 1973).

McCarty, J. L., and Brooks, G. W. "A Dynamic-Model Study of the Effect of Added Weights and Other Structural Variations on the Blade Bending Strains of an Experimental Two-Blade Jet-Driven Helicopter in Hovering and Forward Flight." NACA TN 3367, May 1955.

McCarty, J. L.; Brooks, G. W.; and Maglieri, D. J. "Wind Tunnel Investigation of the Effect of Angle of Attack and Flapping-Hinge Offset on Periodic Bending Moments and Flapping of a Small Rotor." NASA Memo 3-3-59L, March 1959.

McCarty, J. L.; Brooks, G. W.; and Maglieri, D. J. "A Dynamic Model Investigation of the Effect of a Sharp-Edge Vertical Gust on Blade Periodic Flapping Angles and Bending Moments of a Two-Bladed Rotor." NASA TN D-31, September 1959.

McClements, A., and Armitage, A. "Helicopter Developments during the Post-War Years." Aircraft Engineering, 28:331 (September 1956).

McCloud, J. L. III. "Studies of a Large-Scale Jet-Flap Rotor in the 40- by 80-foot Wind Tunnel." American Helicopter Society Mideast Region Symposium, Philadelphia, Pennsylvnia, October 1972.

McCloud, J. L. III. "An Analytical Study of a Multicyclic Controllable Twist Rotor." AHS Forum, 1975.

McCloud, J. L. III, and Biggers, J. C. "Full-Scale Wind-Tunnel Tests of a Nonarticulated Helicopter Rotor." NASA TN D-2392, July 1964.

McCloud, J. L. III; Biggers, J. C.; and Maki, R. L. "Full-Scale Wind Tunnel Tests of a Medium-Weight Utility Helicopter at Forward Speeds." NASA TN D-1887, May 1963.

McCloud, J. L. III; Biggers, J. C.; and Stroub, R. H. "An Investigation of Full-Scale Helicopter Rotors at High Advance Ratios and Advancing Tip Mach Numbers." NASA TN D-4632, July 1968.

McCloud, J. L. III; Hall, L. P.; and Brady, J. A. "Full-Scale Wind-Tunnel Tests of Blowing Boundary-Layer Control Applied to a Helicopter Rotor." NASA TN D-335, September 1960.

McCloud, J. L. III, and Kretz, M. "Multicyclic Jet-Flap Control for Alleviation of Helicopter Blade Stresses and Fuselage Vibration." NASA SP 352, February 1974.

McCloud, J. L. III, and McCullough, G. B. "Wind Tunnel Tests of a Full-Scale Helicopter Rotor with Symmetrical and with Cambered Blade Sections at Advance Ratios from 0.3 to 0.4." NACA TN 4367, September 1958.

McCloud, J. L. III, and McCullough, G. B. "Comparison of Calculated and Measured Stall Boundaries of a Helicopter Rotor at Advance Ratios from 0.3 to 0.4." NASA TN D-73, September 1959.

McCloud, J. L. III, and Weisbrich, A. L. "Wind-Tunnel Test Results of a Full-Scale Multicyclic Controllable Twist Rotor." AHS Forum, 1978.

McCormick, B. W., Jr. "The Effect of a Finite Hub on the Optimum Propeller." JAS, 22:9 (September 1955).

McCormick, B. W., Jr. "On the Initial Vertical Descent of a Helicopter Following Power Failure." JAS, 23:12 (December 1956).

McCormick, B. W., Jr. Aerodynamics of V/STOL Flight. New York: Academic Press, 1967.

McCormick, B. W., Jr., and Surendraiah, M. "A Study of Rotor Blade-Vortex Interaction." AHS Forum, 1970.

McCroskey, W. J. "Measurements of Boundary Layer Transition, Separation and Streamline Direction on Rotating Blades." NASA TN D-6321, April 1971.

McCroskey, W. J. "Dynamic Stall of Airfoils and Helicopter Rotors." AGARD

Report No. 595, April 1972.

McCroskey, W. J. "Recent Developments in Rotor Blade Stall." AGARD Conference Proceedings No. 111, Marseilles, France, September 1972.

McCroskey, W. J. "Inviscid Flowfield on an Unsteady Airfoil." AIAA Journal, 11:8 (August 1973).

McCroskey, W. J. "Recent Developments in Dynamic Stall." Symposium on Unsteady Aerodynamics, Tucson, Arizona, March 1975.

McCroskey, W. J.; Carr, L. W.; and McAlister, K. W. "Dynamic Stall Experiments on Oscillating Airfoils." AIAA Journal, 14:1 (January 1976).

McCroskey, W. J., and Fisher, R. K., Jr. "Detailed Aerodynamic Measurements on a Model Rotor in the Blade Stall Regime." JAHS, 17:1 (January 1972).

McCroskey, W. J.; Nash, J. F.; and Hicks, J. G. "Turbulent Boundary-Layer Flow over a Rotating Flap-Plate Blade." AIAA Journal, 9:1 (January 1971).

McCroskey, W. J., and Philippe, J. J. "Unsteady Viscous Flow on Oscillating Airfoils." AIAA Journal, 13:1 (January 1975).

McCroskey, W. J., and Yaggy, P. F. "Laminar Boundary Layers on Helicopter Rotors in Forward Flight." AIAA Journal, 6:10 (October 1968).

McCutcheon, R. "S-67 Flight Test Program." AHS Forum, 1972.

McDaniel, T. J., and Murthy, V. R. "Bounds on the Dynamic Characteristics of Rotating Beams." AIAA Journal, 15:3 (March 1977).

McElreath, K. W.; Klein, J. A.; and Thomas, R. C. "Pilot-in-the-Loop Control Systems." AHS Forum, 1973.

McHugh, F. J. "What Are the Lift and Propulsive Force Limits at High Speed for the Conventional Rotor?" AHS Forum, 1978.

McHugh, F.; Clark, R.; and Solomon, M. "Wind Tunnel Investigation of Rotor Lift and Propulsive Force at High Speed." NASA CR 145217, October, 1977.

McHugh, F. J.; Eason, W.; Alexander, H. R.; and Mutter, H. "Performance and Stability Test of a 1/4.622 Froude Scaled Model Tilt Rotor Aircraft." NASA CR 114603, October 1973.

McHugh, F. J., and Harris, F. D. "Have We overlooked the Full Potential of the Conventional Rotor?"JAHS, 21:3 (July 1976).

McHugh, F. J., and Shaw, J., Jr. "Benefits of Higher-Harmonic Blade Pitch: Vibration Reduction, Blade-Load Reduction, and Performance Improvement." American Helicopter Society Mideast Region Symposium, Essington, Pennsylvania, August 1976.

McHugh, F. J., and Shaw, J., Jr. "Helicopter Vibration Reduction with Higher Harmonic Blade Pitch." JAHS, 23:4 (October 1978).

McIntyre, H. H. "Longitudinal Dynamic Stability of a Helicopter with Torsionally Flexible Blades and Servo-Flap Control." AHS Forum, 1962.

McIntyre, H. H. "Note on the Reduction of Harmonic Vertical Hub Forces." JAHS, 13:2 (April 1968).

McIntyre, H. H. "A Simplified Study of High Speed Autorotation Entry Characteristics." AHS Forum, 1970.

McLarty, T. T.; Van Gaasbeek, J. R.; and Hsieh, P. Y. "Rotorcraft Flight Simulation with Coupled Rotor Aeroelastic Stability Analysis." USAAMRDL TR 76-41, May 1977.

McKee, J. W. "Experimental Investigation of the Pressure Fluctuations on a Flat Plate and a Cylinder in the Slipstream of a Hovering Rotor." NASA TN D-112, September 1959.

McKee, J. W., and Naeseth R. L. "Experimental Investigation of the Drag of Flat Plates and Cylinders in the Slipstream of a Hovering Rotor." NACA TN 4239, April 1958.

McKee, J. W., and Naeseth, R. L. "A Wind Tunnel Investigation of Rotor Behavior under Extreme Operating Conditions with a Description of Blade Oscillations Attributed to Pitch-Lag Coupling." NASA Memo 1-7-59L, January 1959.

McKenzie, K. T., and Howell, D. A. S. "The Prediction of Loading Actions on High Speed Semirigid Rotor Helicopters." AGARD Conference Proceedings No. 122, Milan, Italy, March 1973.

McManus, B. L. "Fly-by-Wire for Vertical Lift." AHS Forum, 1978.

McManus, B., and Gonsalves, J. "CH-47C Vulnerability Reduction Modification Program – Fly-by-Wire Backup Demonstration." USAAMRDL TR 76-22, August 1976.

MacNeal, R. H. "Direct Analog Method of Analysis of the Vertical Flight Dynamic Characteristics of the Lifting Rotor with Floating Hub." JAHS, 3:4 (October 1958).

MacNeal, R. H., and Hedgepeth, J. M. "Helicopters for Interplanetary Space Flight." AHS Forum, 1978.

McNeill, L. H.; Plaks, A.; and Blackburn, W. E. "Analysis of Unmanned, Tethered, Rotary-Wing Platforms." USAAMRDL TR 74-56, July 1974.

McVeigh, M. A. "Preliminary Simulation of an Advanced, Hingeless Rotor XV-15 Tilt Rotor Aircraft." NASA CR 151950, December 1976.

McVeigh, M. A. "Pilot Evaluation of an Advanced Hingeless Rotor XV-15 Simulation." NASA CR 152034, June 1977.

McVeigh, M. A., and Widdison, C. A. "A Mathematical Simulation Model of a 1985-Era Tilt-Rotor Passenger Aircraft." NASA CR 151949, August 1976.

Mack, J. C. "Heavy Lift Helicopter Drive System." USAAMRDL TR 77-38, September 1977.

Madden, P. A. "Angle-of-Attack Distributions of a High Speed Helicopter Rotor." JAHS, 12:2 (April 1967).

Magee, J. P., and Alexander, H. R. "V/STOL Tilt Rotor Aircraft Study – Wind Tunnel Tests of a Full Scale Hingeless Prop/Rotor." NASA CR 114664, October 1973.

Magee, J. P., and Alexander, H. R. "Wind Tunnel Test on a 1/4.662 Froude Scale, Hingeless Rotor, Tilt Rotor Model" NASA CR 151936, CR 151937, CR 151938, and CR 151939, September 1976.

Magee, J. P.; Clark, R.; and Alexander, H. R. "Conceptual Design Studies of 1985 Commercial VTOL Transports That Utilize Rotors." NASA CR 137599, CR 137600, November 1974.

Magee, J. P.; Clark, R. D.; and Widdison, C. A. "Conceptual Engineering Design Studies of 1985-Era Commercial VTOL and STOL Transports That Utilize Rotors," NASA CR 2545, May 1975.

Magee, J. P.; Maisel, M. D.; and Davenport, F. J. "The Design and Performance Prediction of Propeller/Rotors for VTOL Applications." AHS Forum, 1969.

Magee, J. P., and Pruyn, R. R. "Prediction of the Stability Derivatives of Large Flexible Prop/Rotors by a Simplified Analysis." AHS Forum, 1970.

Mager, A. "Generalization of Boundary-Layer Momentum-Integral Equations to Three-Dimensional Flows Including Those of Rotating Systems." NACA Report 1067, 1952.

Maglieri, D. J.; Hilton, D. A.; and Hubbard, H. H. "Noise Considerations in the Design and Operation of V/STOL Aircraft." NASA TN D-736, April 1961.

Maglieri, D. J., and Reisert, T. D. "Gust-Tunnel Investigation of the Effect of a Sharp-Edge Gust on the Flapwise Blade Bending Moments of a Model Helicopter Rotor." NACA TN 3470, August 1955.

Magliozzi, B.; Metzger, F. B.; Bausch, W.; and King, R. J. "A Comprehensive Review of Helicopter Noise Literature." Federal Aviation Administration, Report No. FAA-RD-75-79, June 1975.

Makofski, R. A. "Charts for Estimating the Hovering Endurance of a Helicopter." NACA TN 3810, October 1956.

Makofski, R. A., and Menkick, G. F. "Investigation of Vertical Drag and Periodic Airloads Acting on Flap Panels in a Rotor Slipstream." NACA TN 3900, December 1956.

Mall, G. H., and Farassat, F. "A Computer Program for the Determination of the Acoustic Pressure Signature of Helicopter Rotors Due to Blade Thickness." NASA TM X-3323, January 1976.

Mallick, D. L., and Reeder, J. P. "Flight Evaluation of Several Spring Force Gradients and a Bobweight in the Cyclic-Power-Control System of a Light Helicopter." NASA TN D-537, October 1960.

Malone, T. B.; Schweichert, G. A., Jr.; and Ketchel, J. M. "Effects of Noise and Vibration on Commercial Helicopter Pilots." NASA CR 117181, April 1970.

Maloney, P. F., and Porterfield, J. D. "Elastic Pitch Beam Tail Rotor." USAAMRDL TR 76-35, December 1976.

Maloy, R. B. "The Development of CAA Helicopter Flight Testing." Aeronautical Engineering Review, 8:11 (November 1949).

MAN-Acoustics and Noise, Inc. "Noise Certification Considerations for Helicopters Based on Laboratory Conditions." Federal Aviation Administration, Report No. FAA-RD-76-116, July 1976.

Mangler, K. W. "Calculation of the Induced Velocity Field of a Rotor." Royal Aircraft Establishment, Report Aero 2247, February 1948.

Mangler, K. W., and Squire, H. B. "The Induced Velocity Field of a Rotor." ARC R&M 2642, May 1950.

Mantay, W. R.; Campbell, R. L.; and Shidler, P. A. "Full-Scale Testing of an Ogee Tip Rotor." NASA CP 2052, May 1978.

Mantay, W. R.; Holbrook, G. T.; Campbell, R. L.; and Tomaine, R. L. "Helicopter Response to an Airplane's Trailing Vortex." Journal of Aircraft, 14:4 (April 1977).

Mantay, W. R.; Shidler, P. A.; and Campbell, R. L. "Some Results of the Testing of a Full Scale Ogee Tip Helicopter Rotor: Acoustics, Loads, and Performance." AIAA Paper 77-1340, October 1977.

Maresca, C.; Favier, D.; and Rebont, J. "Instantaneous Velocity Measurements in the Near Wake of a Helicopter Rotor." AIAA Journal, 12:8 (August 1974).

Margoulis, W. "Propeller Theory of Professor Joukowski and His Pupils." NACA TM 79, April 1922.

Marinescu, A. "Control Effects on the Elastic Blade of a Helicopter Rotor." NASA TT F-9371, May 1965.

Marks, M. D. "Flight Test Development of XV-1 Convertiplane." JAHS, 2:1 (January 1957).

Marks, M. D. "Comparison of Current Operational Operational Rotor Systems and a Rotor Having Floating Hub and Offset Coning Hinges." JAHS, 5:4 (October 1960).

Marr, R. L. "Wind Tunnel Test Results of 25-ft Tilt Rotor during Autorotation." NASA CR 137824, February 1976.

Marr, R. L.; Ford, D. G.; and Ferguson, S. W. "Analysis of the Wind Tunnel Test of a Tilt Rotor Powered Force Model." NASA CR 137529, June 1974.

Marr, R. L., and Neal, G. T. "Assessment of Model Testing of a Tilt-Proprotor VTOL Aircraft." American Helicopter Society Mideast Region Symposium Essington, Pennsylvania, October 1972.

Marr, R. L., and Roderick, W. E. B. "Handling Qualities Evaluation of the

XV-15 Tilt Rotor Aircraft." JAHS, 20:2 (April 1974).

Marr, R. L.; Sambell, K. W.; and Neal, G. T. "V/STOL Tilt Rotor Study – Hover, Low Speed, and Conversion Tests of a Tilt Rotor Aeroelastic Model." NASA CR 114615, May 1973.

Marr, R. L.; Willis, J. M.; and Churchill, G. B. "Flight Control System Development for the XV-15 Tilt Rotor Aircraft." AHS Forum, 1976.

Marte, J. E., and Kurtz, D. W. "A Review of Aerodynamic Noise from Propellers, Rotors, and Lift Fans." NASA CR 107568, January 1970.

Martin, J. M.; Empey, R. W.; McCroskey, W. J.; and Caradonna, F. X. "A Detailed Experimental Analysis of Dynamic Stall on an Unsteady Two-Dimensional Airfoil." JAHS, 19:1 (January 1974).

Maskew, B., and Dvorak, F. A. "The Prediction of $C_{l_{max}}$ Using a Separated Flow Model." JAHS, 23:2 (April 1978).

Matheny, W. G., and Wilderson, L. E. "Functional Requirements for Ground Based Trainers: Helicopter Response Characteristics." U.S. Army Human Resources Research Organization, TR 70-17, October 1970.

Matthys, C. G.; Joglekar, M. M.; and Ksieh, P. Y. "An Analysis of Fixed Wing-Proprotor Interference for Folding Proprotor Aircraft." AFFDL TR 72-115, March 1973.

Matveev, Yu. G., and Mel'nikov, B. N. "Noise Characteristics of the Mi-8 and Mi-4 Passenger Helicopters." Soviet Physics–Acoustics, 14:2 (October-December 1968).

Mayo, A. P. "Matrix Method for Obtaining Spanwise Moments and Deflections of Torsionally Rigid Rotor Blades with Arbitrary Loadings." NACA TN 4304, August 1958.

Mayo, A. P. "Comparison of Measured Flapwise Structural Bending Moments on a Teetering Rotor Blade with Results Calculated from the Measured Pressure Distribution." NASA Memo 2-28-59L, March 1959.

Meier, W.; Groth, W. P.; Clark, D. R.; and Verzella, D. "Flight Testing of a Fan-in-Fin Antitorque and Directional Control System and a Collective Force Augmentation System." USAAMRDL TR 75-19, June 1975.

Meirovitch, L., and Hale, A. L. "Dynamic Synthesis and Modal Characteristics of Helicopters." NASA CR 158909, July 1978.

Merkley, D. J. "An Analytical Investigation of the Effects of Increased Installed Horsepower on Helicopter Agility in the Nap-of-the-Earth Environment." USAAMRDL TN-21, December 1975.

Metzger, R. F. "Mechanical Instability Ground Dynamics Program." Kaman Aerospace Corporation, Report No. R-1249, March 1974.

Metzger, R. F.; Plaks, A.; Meier, R. C.; and Berman, A. "Development of a Method for the Analysis of Improved Helicopter Design Criteria." USAAMRDL TR 74-30, July 1974.

Meyer, J. R., Jr. "An Investigation of Bending-Moment Distribution on a

Model Helicopter Rotor Blade and a Comparison with Theory." NACA TN 2626, February 1952.

Meyer, J. R., Jr., and Falabella, G., Jr. "An Investigation of the Experimental Aerodynamic Loading on a Model Helicopter Rotor Blade." NACA TN 2953, May 1953.

Meyers, D. N.; Tompkins, L. V.; and Goldberg, J. M. "16H-1A Flight Test Research Program." USAAVLABS TR 67-58, August 1968.

Meyers, D. N.; Vanderlip, E. G.; and Halpert, P. "Helicopter Stability and Automatic Control." Aeronautical Engineering Review, 10:7 (July 1951).

Miao, W.-L., and Huber, H. B. "Rotor Aeroelastic Stability Coupled with Helicopter Body Motion." NASA SP-352, February 1974.

Micale, E. C., and Poli, C. "Dynamics of Slung Bodies Using a Rotating Wheel for Stability." Journal of Aircraft, 10:12 (December 1973).

Michel, P. L. "Research and Design Progress toward High Performance Rotary Wing Aircraft." JAHS, 5:2 (April 1960).

Migotsky, E. "Full-Scale-Tunnel Performance Tests of the PV-2 Helicopter Rotor." NACA MR L5C29a, April 1945.

Migotsky, E. "Full-Scale Investigation of the Blade Motion of the PV-2 Helicopter Rotor." NACA TN 1521, March 1948.

Mil', M. L., et al. "Helicopters – Calculation and Design." NASA TT F-494, 1966; and NASA TT F-519, 1967.

Miller, D. P., and Clark, J. W. "Research on VTOL Aircraft Handling Qualities Criteria." Journal of Aircraft, 2:3 (May-June 1965).

Miller, D. P., and Vinje, E. W. "Fixed-Base Flight Simulator Studies of VTOL Aircraft Handling Qualities in Hovering and Low-Speed Flight." AFFDL TR 67-152, January 1968.

Miller, N.; Tang, J. C.; and Perlmutter, A. A. "Theoretical and Experimental Investigation of the Instantaneous Induced Velocity Field in the Wake of a Lifting Rotor." USAAVLABS TR 67-68, January 1968.

Miller, R. H. "Jet Propulsion Applied to Helicopter Rotors." JAS, 13:12 (December 1946).

Miller, R. H. "Helicopter Control and Stability in Hovering Flight." JAS, 15:8 (August 1948).

Miller, R. H. "A Method for Improving the Inherent Stability and Control Characteristics of Helicopters." JAS, 17:6 (June 1950).

Miller, R. H. "Rotor Blade Harmonic Air Loading." Institute of the Aerospace Sciences, IAS Paper No. 62-82, January 1962.

Miller, R. H. "On the Computation of Airloads Acting on Rotor Blades in Forward Flight." JAHS, 7:2 (April 1962).

Miller, R. H. "Unsteady Air Loads on Helicopter Rotor Blades." Journal of the Royal Aeronautical Society, 68:640 (April 1964).

Miller, R. H. "Rotor Blade Harmonic Air Loading." AIAA Journal, 2:7 (July 1964).

Miller, R. H. "Theoretical Determination of Rotor Blade Harmonic Air-Loads." Massachusetts Institute of Technology, ASRL TR 107-2, August 1964.

Miller, R. H. "Notes on Cost of Noise Reduction in Rotor/Prop Aircraft." Massachusetts Institute of Technology, FTL Memo M68-9, August 1968.

Miller, R. H., and Ellis, C. W. "Helicopter Blade Vibration and Flutter." JAHS, 1:3 (July 1956).

Mineck, R. E. "Tail Contribution to the Directional Aerodynamic Characteristics of a 1/6-Scale Model of the Rotor Systems Research Aircraft with a Tail Rotor." NASA TM X-3501, May 1977.

Mineck, R. E. "Effect of Rotor Wake on Aerodynamic Characteristics of a 1/6-Scale Model of the Rotor Systems Research Aircraft." NASA TM X-3548, September 1977.

Mineck, R. E., and Freeman, C. E. "Airframe, Wing, and Tail Aerodynamic Characteristics of a 1/6-Scale Model of the Rotor Systems Research Aircraft with the Rotors Removed." NASA TN D-8456, May 1977.

Mineck, R. E., and Freeman, C. E. "Aerodynamic Characteristics of a 1/6-Scale Powered Model of the Rotor Systems Research Aircraft." NASA TM X-3489, June 1977.

Mirandy, L. "A Dynamic Loads Scaling Methodology for Helicopter Rotors." Journal of Aircraft, 15:2 (February 1978).

Miyajima, K. "Analytical Design of a High Performance Stability and Control Augmentation System for a Hingeless Rotor Helicopter." AHS Forum, 1978.

Moen, G. C.; Di Carlo, D J.; and Yanni, K. R. "A Parametric Analysis of Visual Approaches for Helicopters." NASA TN D-8275, December 1976.

Moffitt, R. C., and Sheehy, T. W. "Prediction of Helicopter Rotor Performance in Vertical Climb and Sideward Flight." AHS Forum, 1977.

Molusis, J. A. "Helicopter Stability Derivative Extraction from Flight Data Using the Bayesian Approach to Estimation." JAHS, 18:2 (April 1973).

Molusis, J. A. "Analytical Study to Define a Helicopter Stability Derivative Extraction Method." NASA CR 132371 and CR 132372, May 1973.

Molusis, J. A. "Rotorcraft Derivative Identification from Analytical Models and Flight Test Data." AGARD Conference Proceedings No. 172, Hampton, Virginia, November 1974.

Monnerie, B., and Toguet, A. "Effect of the Vortex Springing from a Helicopter Blade Tip on the Flow around the Next Blade." NASA TT F-14462, February 1972.

Montana, P. S. "Experimental Investigation of Three Rotor Hub Fairing Shapes." Naval Ship Research and Development Center, Report ASED 333, May 1975.

Montana, P. S. "Experimental Evaluation of Analytically Shaped Helicopter

Rotor Hub-Pylon Configurations." David W. Taylor Naval Ship Research and Development Center, Report ASED 355, July 1976.
Montana, P. S. "Experimental Evaluation of the Effect of Rotation on the Aerodynamic Characteristics of Two Helicopter Rotor Hub Fairing Shapes." David W. Taylor Naval Ship Research and Development Center, Report ASED 364, September 1976.
Monteleone, R. A. "Investigation of the Maneuverability of the S-67 Winged Helicopter." USAAMRDL TR 73-51, June 1973.
Morduchow, M. "On Internal Damping of Rotating Beams." NACA TN 1996, December 1949.
Morduchow, M. "A Theoretical Analysis of Elastic Vibrations of Fixed-Ended and Hinged Helicopter Blades in Hovering and Vertical Flight." NACA TN 1999, January 1950.
Morduchow, M., and Hinchey, F. G. "Theoretical Analysis of Oscillations in Hovering of Helicopter Blades with Inclined and Offset Flapping and Lagging Hinge Axes." NACA TN 2226, December 1950.
Morduchow, M., and Muzyka, A. "Analysis of Harmonic Forces Produced at Hub by Imbalances in Helicopter Rotor Blades." NACA TN 4226, April 1958.
Moreno-Caracciolo, M. "The Autogiro." NACA TM 218, July 1923.
Morris, C. E. K., Jr. "Rotor-Airfoil Flight Investigation: Preliminary Results." AHS Forum, 1978.
Morris, C. E. K., Jr., and Yeager, W. T., Jr. "Theoretical Analysis of Aerodynamic Characteristics of Two Helicopter Rotor Airfoils." NASA TM 78680, March 1978.
Moss, G. F., and Murdin, P. M. "Two Dimensional Low-Speed Tunnel Tests on the NACA 0012 Section Including Measurements Made during Pitching Oscillations at the Stall." ARC CP 1145, May 1965.
Mouille, R. "The Fenestron Shrouded Tail Rotor of the SA.341 Gazelle." JAHS, 15:4 (October 1970).
Mouille, R. "New Concepts for Helicopter Main Rotors." AHS Forum, 1975.
Munch, C. L. "A Study of Noise Guidelines for Community Acceptance of Civil Helicopter Operations." JAHS, 20:1 (January 1975).
Munch, C. L., and King, R. J. "Community Acceptance of Helicopter Noise: Criteria and Application." NASA CR 132430, June 1974.
Munch, C. L.; Paterson, R. W.; and Day, H. "Rotor Broadband Noise Resulting from Tip Vortex/Blade Interaction." Sikorsky Aircraft, Report SER 50909, February 1975.
Munk, M. M. "Notes on Propeller Deisgn." NACA TN 91 and TN 94, April 1922; NACA TN 95 and TN 96, May 1922.
Munk, M. M. "Wind-Driven Propellers." NACA TM 201, April 1923.
Munk, M. M. "General Theory of Windmills." NACA TN 164, October 1923.

Munk, M. M. "Model Tests on the Economy and Effectiveness of Helicopter Propellers." NACA TN 221, July 1925.

Murphy, R. D., and Narendra, K. S. "Design of Helicopter Stabilization Systems Using Optimal Control Theory." Journal of Aircraft, 6:2 (March-April 1969).

Murthy, V. K., and Swartz, G. B. "Annotated Bibliography on Cumulative Fatigue Damage and Structural Reliability Models." Aerospace Research Laboratories, ARL 72-0161, December 1972.

Murthy, V. R. "Dynamic Characteristics of Rotor Blades." J. Sound Vib., 49:4 (1976).

Murthy, V. R. "Dynamic Characteristics of Rotor Blades: Integrating Matrix Method." AIAA Journal, 15:4 (April 1977).

Murthy, V. R., and Pierce, G. A. "Effect of Phase Angle on Multi-Bladed Rotor Flutter." J. Sound Vib., 48:2 (1976).

Myers, G. C., Jr. "Flight Measurements of Helicopter Blade Motion with a Comparison between Theoretical and Experimental Results." NACA TN 1266, April 1947.

Nagel, R. T. "The Influence of Leading Edge Serrations on the Noise Radiation from a Statically Thrusting Rotor." Pennsylvania State University, ORL TR 72-188, August 1972.

Nakamura, Y., and Azuma, A. "Improved Methods for Calculating the Thickness Noise." NASA CP 2052, May, 1978.

Narendra, K. S., and Tripathi, S. S. "Identification and Optimization of Aircraft Dynamics." Yale University, Becton Center Technical Report CT-50, July 1972.

Narendra, K. S., and Tripathi, S. S. "Identification and Optimization of Aircraft Dynamics." Journal of Aircraft, 10:4 (April 1973).

NACA. "The Oehmichen-Peugeot Helicopter." NACA TM 13, March 1921.

NACA. "Recent European Developments in Helicopters." NACA TN 47, April 1921.

Naumann, E. A., and Schmier, H. "Design and Development of the Stability Augmentation System for the Lockheed/Army XV-4A." AHS Forum, 1964.

Nazarov, V. A. "Helicopters on the Baykal-Amur Line." NASA TT F-16869, February 1976.

Needham, J. F., and Banerjee, D. "Engine/Airframe/Drive Train Dynamic Interface Documentation." USARTL TR 78-12, May 1978.

Newman, E. M. "A New Approach to the Calculation of the Effect of the Ground on the Performance of Rotary Wing Aircraft." Naval Air Systems Command, PDC Special Project 71-02, November 1971.

Nichols, J. B. "The Pressure-Jet Helicopter Propulsion System." The

Aeronautical Journal, 76:741 (September 1972).

Niebanck, C. F. "A Comparison of Dynamically Scaled Model Rotor Test Data with Discrete Azimuth Aeroelastic Stability Theory." AHS Forum, 1969.

Niebanck, C. F. "Model Rotor Test Data for Verification of Blade Response and Rotor Performance Calculations." USAAMRDL TR 74-29, May 1974.

Niebanck, C. F., and Bain, L. J. "Rotor Aeroelastic Instability and Transient Characteristics." USAAVLABS TR 69-88, February 1970.

Niebanck, C. F., and Elman, H. L. "Prediction of Rotor Instability at High Forward Speeds – Torsional Divergence." USAAVLABS TR 68-18D, February 1969.

Niebanck, C., and Girvan, W. "Sikorsky S-76 Analysis, Design, and Development for Successful Dynamic Characteristics." AHS Forum, 1978.

Niessen, F. R., et al. "The Effect of Variations in Controls and Displays on Helicopter Instrument Approach Capability." NASA TN D-8385, February 1977.

Nikolsky, A. A. Helicopter Analysis. New York: John Wiley and Sons, 1951.

Nikorsky, A. A., and Seckel, E. "An Analytical Study of the Steady Vertical Descent in Autorotation of Single-Rotor Helicopters." NACA TN 1906, June 1949.

Nikolsky, A. A., and Seckel, E. "An Analysis of the Transition of a Helicopter from Hovering to Steady Autorotative Vertical Descent." NACA TN 1907, June 1949.

Niven, A. J.; Sanders, T. H.; and McManus, B. L. "Heavy Lift Helicopter Flight Control System." USAAMRDL TR 77-40, September 1977.

Noonan, K. W., and Bingham, G. J. "Two Dimensional Aerodynamic Characteristics of Several Rotorcraft Airfoils at Mach Numbers from 0.35 to 0.90." NASA TM X-73990, January 1977.

Norman, D. C., and Somsel, J. R. "Determination of Helicopter Rotor Blade Compressibility Effects – Prediction vs. Flight Test." AHS Forum, 1967.

Norman, D. C., and Sultany, D. J. "An Empirical Method for Calculating the Power Required Due to Compressibility on a Single Rotor Helicopter." JAHS, 10:3 (July 1965).

O'Bryan, T. C. "An Investigation of the Effect of Downwash from a VTOL Aircraft and a Helicopter in the Ground Environment." NASA TN D-977, October 1961.

O'Connor, S. J. "Feel Augmentation and Sensitivity Control in High Speed Helicopters." JAHS, 17:3 (July 1972).

O'Connor, S. J., and Fowler, D. W. "S-67 Aircraft Feel Augmentation System Flight Evaluation." USAAMRDL TR 72-41, August 1972.

Oehrli, R. R. "Computer Programs for Helicopter Aerodynamic Stability Evaluation." U.S. Army Materiel Systems Analysis Activity, AMSAA TR 130, August 1976.

Oemichen, E. "My Experiments with Helicopters." NACA TM 199, April 1923.

Ogren, H. "A Three-Axis Fluidic Stability Augmentation System." USAAMRDL TR 71-30, October 1971.

Ogren, H. D. "Hydrofluidic Yaw SAS Analysis Design and Development." USAAMRDL TR 74-7, March 1974.

Ogren, H.; Sotanski, R.; and Genaw, L. "Yaw Axis Stability Augmentation System Flight Test Report." USAAMRDL TR 74-39, June 1974.

Ohtsuka, M. "Untwist of Rotating Blades." Transactions of the ASME – Journal of Engineering for Power, 97A:2 (April 1975).

Ollerhead, J. B. "Helicopter Aural Detectability." USAAMRDL TR 71-33, July 1971.

Ollerhead, J. B., and Lowson, M. V. "Problems of Helicopter Noise Estimation and Reduction." AIAA Paper No. 69-195, February 1969.

Ollerhead, J. B., and Taylor, R. B. "Description of a Helicopter Rotor Noise Computer Program." USAAVLABS TR 68-61, January 1969.

Ormiston, R. A. "Comparison of Several Methods for Predicting Loads on a Hypothetical Helicopter Rotor." JAHS, 19:4 (October 1974).

Ormiston, R. A. "Techniques for Improving the Stability of Soft-Inplane Hingeless Rotors." NASA TM X-62390, October 1974.

Ormiston, R. A. "Application of Simplified Inflow Models to Rotorcraft Dynamic Analysis." JAHS, 21:3 (July 1976).

Ormiston, R. A. "Aeromechanical Stability of Soft Inplane Hingeless Rotor Helicopters." European Rotorcraft and Powered Lift Aircraft Forum, Aix-en-Provence, France, September 1977.

Ormiston, R. A., and Bousman, W. G. "A Theoretical and Experimental Investigation of Flap-Lag Stability of Hingeless Helicopter Rotor Blades." NASA TM X-62179, August 1972.

Ormiston, R. A., and Bousman, W. G. "A Study of Stall-Induced Flap-Lag Instability of Hingeless Rotors." JAHS, 20:1 (January 1975).

Ormiston, R. A., and Hodges, D. H. "Linear Flap-Lag Dynamics of Hingeless Helicopter Rotor Blades in Hover." JAHS, 17:2 (April 1972).

Ormiston, R. A., and Peters, D. A. "Hingeless Helicopter Rotor Response with Nonuniform Inflow and Elastic Blade Bending." Journal of Aircraft, 9:10 (October 1972).

Ostroff, A. J.; Downing, D. R.; and Rood, W. J. "A Technique Using a Nonlinear Helicopter Model for Determining Trims and Derivatives."

NASA TN D-8159, May 1976.

Padakannaya, R. "Experimental Study of Rotor Unsteady Airloads Due to Blade-Vortex Interaction." NASA CR 1909, November 1971.

Padakannaya, R. "The Vortex Lattice Method for the Rotor-Vortex Interaction Problem." NASA CR-2421, July 1974.

Paglino, V. M. "High-Speed Rotor Performance Correlation." AHS Forum, 1968.

Paglino, V. M. "Yawed Blade Element Rotor Performance Method." Sikorsky Aircraft, Report SER-50620, September 1969.

Paglino, V. M. "Forward Flight Performance of a Coaxial Rigid Rotor." AHS Forum, 1971.

Paglino, V. M., and Beno, E. A. "Full-Scale Wind-Tunnel Investigation of the Advancing Blade Concept Rotor System." USAAMRDL TR 71-25, August 1971.

Paglino, V. M., and Clark, D. R. "A Study of the Potential Benefits of Advance Airfoils for Helicopter Applications." American Helicopter Society Northeast Region Symposium, Hartford, Connecticut, March 1975.

Paglino, V. M., and Logan, A. H. "An Experimental Study of the Performance and Structural Loads of a Full Scale Rotor at Extreme Operating Conditions." USAAVLABS TR 68-3, July 1968.

Parker, A. G. "Force and Pressure Measurements on an Airfoil Oscillating through Stall." Journal of Aircraft, 13:10 (October 1976).

Parker, A. G., and Bicknell, J. "Some Measurements on Dynamic Stall." Journal of Aircraft, 11:7 (July 1974).

Parks, C. L. "A Computer Program for Helicopter Rotor Noise Using Lowson's Formula in the Time Domain." NASA TM X-72759, July 1975.

Parkus, H. "The Disturbed Flapping Motion of Helicopter Rotor Blades." JAS, 15:2 (February 1948).

Parrish, R. V.; Houck, J. A.; and Martin, D. J., Jr. "Empirical Comparison of a Fixed-Base and a Moving-Base Simulation of a Helicopter Engaged in Visually Conducted Slalom Runs." NASA TN D-8424, May 1977.

Patay, S. A. "Leading Edge Separation on an Airfoil During Dynamic Stall." Massachusetts Institute of Technology, ASRL TR 156-1, October 1969.

Patel, M. H., and Hancock, G. J. "Some Experimental Results of the Effect of a Streamwise Vortex on a Two-Dimensional Wing." The Aeronautical Journal, 78:760 (April 1974).

Paterson, R. W., and Amiet, R. K. "Noise and Surface Pressure Response of an Airfoil to Incident Turbulence." Journal of Aircraft, 14:8 (August 1977).

Paterson, R. W.; Amiet, R. K.; and Munch, C. L. "Isolated Airfoil-Tip

Vortex Interaction Noise." Journal of Aircraft, 12:1 (January 1975).

Paterson, R. W.; Vogt, P. G.; Fink, M. R.; and Munch, C. L. "Vortex Shedding Noise of an Isolated Airfoil." United Aircraft Corporation Research Laboratories, Report No. K910867-6, December 1971.

Paterson, R. W.; Vogt, P. G.; Fink, M. R.; and Munch, C. L. "Vortex Noise of Isolated Airfoils." Journal of Aircraft, 10:5 (May 1973).

Patterson, J. H., Jr.; Mozo, B. T.; Schomer, P. D.; and Camp, R. T., Jr. "Subjective Ratings of Annoyance Produced by Rotary-Wing Aircraft Noise." U.S. Army Aeromedical Research Laboratory, USAARL Report No. 77-12, May 1977.

Paul, W. F. "A Self-Excited Rotor Blade Oscillation at High Subsonic Mach Numbers." JAHS, 14:1 (January 1969).

Paul, W. F. "Development and Evaluation of the Main Rotor Bifilar Absorber." AHS Forum, 1969.

Paul, W. F. "Developments in Helicopter Vibration and Aeroelastic Loads Technology." Air Force Conference, Las Vegas, Nevada, September 1969.

Pausder, H.-J., and Jordan, D. "Handling Qualities Evaluation of Helicopters with Different Stability and Control Characteristics." Vertica, 1:2 (1976).

Payne, P. R. "A Method of Estimating Helicopter Performance." Aircraft Engineering, 25:297 (November 1953).

Payne, P. R. "Rotor Blade Motion in a Vertical Sharp-Edged Gust." Aircraft Engineering, 26:299 (January 1954); 26:300 (February 1954).

Payne, P. R. "A General Theory of Helicopter Rotor Dynamics." Aircraft Engineering, 26:306 (August 1954).

Payne, P. R. "High Offset Flapping Pin Rotor Analysis." Aircraft Engineering, 26:309 (November 1954).

Payne, P. R. "Hub Moments and Forces of a High Offset Rotor." Aircraft Engineering, 27:311 (January 1955).

Payne, P. R. "Helicopter Rotor Vibration in the Tip Path Plane." Aircraft Engineering, 27:316 (June 1955).

Payne, P. R. "Helicopter Stability in Hovering Flight." Journal of the Royal Aeronautical Society, 59:537 (September 1955).

Payne, P. R. "The Stiff-Hinged Helicopter Rotor." Aircraft Engineering, 27:321 (November 1955).

Payne, P. R. "Induced Aerodynamics of Helicopters." Aircraft Engineering, 28:324 (February 1956) to 28:327 (May 1956).

Payne, P. R. "Helicopter Longitudinal Stability." Aircraft Engineering, 29:339 (May 1957); 29:340 (June 1957).

Payne, P. R. "Higher Harmonic Rotor Control." Aircraft Engineering, 30:354 (August 1958).

Payne, P. R. Helicopter Dynamics and Aerodynamics. London: Sir Isaac Pitman and Sons, 1959.

Payne, P. R. "Dynamics of a Rotor Controlled by Aerodynamic Servo Flaps." Aircraft Engineering, 31:369 (November 1959).

Pearcey, H. H.; Wilby, P. G.; Riley, M. J.; and Brotherhood, P. "The Derivation and Verification of a New Rotor Profile on the Basis of Flow Phenomena; Airfoil Research and Flight Tests." AGARD Conference Proceedings No. 111, Marseilles, France, September 1972.

Peck, W. C. "Landing Characteristics of an Autogiro." NACA TN 508, November 1934.

Pegg, R. J. "An Investigation of the Helicopter Height-Velocity Diagram Showing Effects of Density Altitude and Gross weight." NASA TN D-4536, May 1968.

Pegg, R. J. "A Flight Investigation of a Lightweight Helicopter to Study the Feasibility of Fixed-Collective-Pitch Autorotations." NASA TN D-5270, June 1969.

Pegg, R. J.; Henderson, H. R.; and Hilton, D. A. "Results of the Flight Noise Measurement Program Using a Standard and Modified SH-3A Helicopter." NASA TN D-7330, December 1973.

Pegg, R. J.; Hosier, R. N.; Balcerak, J. C.; and Johnson, H. K. "Design and Preliminary Tests of a Blade Tip Air Mass Injection System for Vortex Modification and Possible Noise Reduction on a Full-Scale Helicopter Rotor." NASA TM X-3314, December 1975.

Peress, K. E., and Kaufman, L. "A Simplified Simulation of the Helicopter in Automatic Stabilization Analyses." JAHS, 3:2 (April 1958).

Perisho, C. H. "Analysis of the Stability of a Flexible Rotor Blade at High Advance Ratio." JAHS, 4:2 (April 1959).

Perisho, C. H., and de Garcia, H. J., Jr. "A Comparison of Detailed and Simplified Methods of Analysis of Rotor Stability in Forward Flight with Model Test Results." AHS Forum, 1962.

Perlmutter, A. A.; Kisielowski, E.; Miller, N.; and George, M. "Stability and Control Handbook for Helicoopers." TCREC TR 60-43, August 1960.

Perlmutter, A. A.; Yackle, A. R.; Chou, P. C.; and Miller, N. "Helicopter Structural Design Criteria Analytical Solutions of Flight and Landing Maneuvers." Air Force Systems Command, WADD TR 60-734, June 1962.

Peters, D. A. "An Approximate Solution for the Free Vibrations of Rotating Uniform Cantilever Beams." NASA TM X-62299, September 1973.

Peters, D. A. "Hingeless Rotor Frequency Response with Unsteady Inflow." NASA SP-352, February 1974.

Peters, D. A. "An Approximate Closed-Form Solution for Lead-Lag Damping of Rotor Blades in Hover." NASA TM X-62425, April 1975.

Peters, D. A. "Flap-Lag Stability of Helicopter Rotor Blades in Forward Flight." JAHS, 20:4 (October 1975).

Peters, D. A., and Hohenemser, K. H. "Application of the Floquet Transition Matrix to Problems of Lifting Rotor Stability." JAHS, 16:2 (April 1971).

Peters, D. A., and Ormiston, R. A. "The Effects of Second Order Blade Bending on the Angle of Attack of Hingeless Rotor Blades." JAHS, 18:4 (October 1973).

Peters, D. A., and Ormiston, R. A. "Flapping Response Characteristics of Hingeless Rotor Blades by a Generalized Harmonic Balance Method." NASA TN D-7856, February 1975.

Pfluger, A. "Aerodynamics of Rotating-Wing Aircraft with Blade-pitch Control." NACA TM 929, February 1940.

Phelps, A. E., and Mineck, R. E. "Aerodynamic Characteristics of a Counter-Rotating, Coaxial, Hingeless Rotor Helicopter Model with Auxiliary Propulsion." NASA TM 78705, May 1978.

Piarulli, V. J. "The Effects of Nonuniform Swashplate Stiffness on Coupled Blade-Control System Dynamics and Stability." NASA CR 1817 and CR 1818, August 1971.

Piarulli, V. J., and White, R. P., Jr. "A Method for Determining the Characteristic Functions Associated with the Aeroelastic Instabilities of Helicopter Rotors in Forward Flight." NASA CR 1577, June 1970.

Pierce, G. A.; Kunz, D. L.; and Malone, J. B. "The Effect of Varying Freestream Velocity on Airfoil Dynamic Stall Characteristics." JAHS, 23:2 (April 1978).

Pierce, G. A., and White, W. F., Jr. "Unsteady Rotor Aerodynamics at Low Inflow and Its Effect on Flutter." AIAA Paper No. 72-959, September 1972.

Piper, R. R. "A Note on Estimating Acceleration and Deceleration Capability of Rotary Wing Aircraft." JAHS, 10:3 (July 1965).

Piziali, R. A. "Tables of Two-Dimensional Oscillating Airfoil Coefficients for Rotary Wings." TRECOM TR 64-53, October 1964.

Piziali, R. A. "A Method for Predicting the Aerodynamic Loads and Dynamic Response of Rotor Blades." USAAVLABS TR 65-74, January 1966.

Piziali, R. A. "Method for the Solution of the Aeroelastic Response Problem for Rotary Wings." J. Sound Vib., 4:3 (1966).

Piziali, R. A. "An Investigation of the Structural Dynamics of Helicopter Rotors." USAAVLABS TR 70-24, April 1970.

Piziali, R.; Daughaday, H.; and DuWaldt, F. "Rotor Airloads." CAL/TRECOM Symposium, Buffalo, New York, June 1963.

Piziali, R. A., and Du Waldt, F. A. "A Method for Computing Rotary Wing

Airload Distribution in Forward Flight." TCREC TR 62-44, November 1962.

Piziali, R. A., and Trenka, A. R. "A Theoretical Study of the Application of Jet Flap Circulation Control for Reduction of Rotor Vibratory Forces." NASA CR 137515, May 1974.

Platt, H. H. "The Helicopter: Propulsion and Torque." JAS, 3:11 (September 1936).

Porterfield, J. D., and Alexander, W. T. "Measurement and Evaluation of Helicopter Flight Loads Spectra Data." JAHS, 15:3 (July 1970).

Porterfield, J. D., and Clark, F. B. "Elastic Pitch Beam Tail Rotor Study for LOH Class Helicopters." USAAMRDL TR 75-41, July 1976.

Porterfield, J. D., and Maloney, P. F. "Evaluation of Helicopter Flight Spectrum Data." USAAVLABS TR 68-68, October 1968.

Potash, M. L. "Fan-in-Tailcone Vehicle Definition Resulting from Engine/Transmission/Airframe Integration Analysis." USAAMRDL TR 75-28, July 1975.

Potthast, A. J., and Blaha, J. T. "Handling Qualities Comparison of Two Hingeless Rotor Control System Designs." AHS Forum, 1973.

Potthast, A. J., and Kerr, A. W. "Flying Qualities of a Gyro-Controlled Hingeless-Rotor Compound Helicopter." Lockheed Report LR-26216, December 1973.

Potthast, A. J., and Kerr, A. W. "Rotor Moment Control with Flap-Moment Feedback." JAHS, 20:2 (April 1974).

Powell, C. A. "Annoyance Due to Simulated Blade-Slap Noise." NASA CP 2052, May 1978.

Powell, C. A. "A Subjective Field Study of Helicopter Blade-Slap Noise." NASA TM 78758, July 1978.

Powell, R. D., Jr. "Hovering Performance of a Helicopter Rotor Using NACA 8-H-12 Airfoil Sections." NACA TN 3237, August 1954.

Powell, R. D., Jr. "Compressibility Effects on a Hovering Helicopter Rotor Having an NACA 0018 Root Airfoil Tapering to an NACA 0012 Airfoil." NACA RM L57F26, September 1957.

Powell, R. D., Jr. "Maximum Mean Lift Coefficient Characteristics at Low Tip Mach Numbers of a Hovering Helicopter Rotor Having an NACA 64_1A012 Airfoil Section." NASA Memo 1-23-59L, February 1959.

Powell, R. D., Jr., and Carpenter, P. J. "Low Tip Mach Number Stall Characteristics and High Tip Mach Number Compressibility Effects on a Helicopter Rotor Having an NACA 0009 Tip Airfoil Section." NACA TN 4355, July 1958.

Prabhakar, A. "Stability of a Helicopter Carrying an Underslung Load." Vertica, 2:2 (1978).

Prewitt, R. H. "Possibilities of the Jump-Off Autogiro." JAS, 6:1 (November 1938).

Prewitt, R. H. "The Design of Rotor Blades." JAS, 9:7 (May 1942).

Prewitt, R. H., and Wagner, R. A. 'Frequency and Vibration Problems of Rotors." JAS, 7:10 (August 1940).

Price, H. L. "The Avoidance of Ground Resonance." Aircraft Engineering, 32:376 (June 1960) to 32:377 (July 1960).

Price, H. L. "Simplified Helicopter Ground Resonance Stability Boundaries." Aircraft Engineering, 34:404 (October 1962) to 34:405 (November 1962).

Price, H. L. "Rotor Dynamics and Helicopter Stability." Aircraft Engineering, 35:3 (March 1963) to 35:12 (December 1963); 36:3 (March 1964); 36:4 (April 1964); 37:11 (November 1965).

Prinz, R. "Determination of the Life of Helicopter Structural Members." NASA TT F-14280, May 1972.

Prouty, R. W. "A State-of-the-Art Survey of Two-Dimensional Airfoil Data." JAHS, 20:4 (October 1975).

Pruyn, R. R., and Alexander, W. T., Jr. "USAAVLABS Tandem Rotor Airloads Measurement Program." Journal of Aircraft, 4:3 (May-June 1967).

Pruyn, R. R., and Swales, T. G. "Development of Rotor Blades with Extreme Chordwise and Spanwise Flexibility." AHS Forum, 1964.

Pruyn, R. R., and Taylor, R. B. "Design Considerations for Tilt-Rotor VTOL Aircraft to Minimize the Effects of the Recirculating Downwash Environment." JAHS, 16:4 (October 1971).

Pruyn, R. R., et al. "Inflight Measurement of Rotor Blade Airloads, Bending Moments, and Motions, Together with Rotor Shaft Loads and Fuselage Vibration, on a Tandem Rotor Helicopter." USAAVLABS TR 67-9, May 1967.

Puri, N. N., and Niemela, R. J. "Optimal Design of Helicopter Precision Hover Control Systems." U.S. Army Electronics Command, ECOM 4109, April 1973.

Putman, W. F., and Curtiss, H. C., Jr. "An Experimental and Analytical Investigation of the Hovering and Forward Flight Characteristics of the Aerocrane Hybrid Heavy Lift Vehicle." Naval Air Development Center, Report NADC 76201-30, September 1977.

Putman, W. F., and Traybar, J. J. "An Experimental Investigation of Compound Helicopter Aerodynamics in Level and Descending Forward Flight and in Ground Proximity." USAAMRDL TR 71-19, July 1971.

Quigley, H. C., and Keonig, D. G. "A Flight Study of the Dynamic Stability of a Tilting-Rotor Convertiplane." NASA TN D-778, April 1961.

Rabbott, J. P., Jr. "Static-Thrust Measurements of the Aerodynamic Loading on a Helicopter Rotor Blade." NACA TN 3688, July 1956.

Rabbott, J. P., Jr. "Comparison of Theoretical and Experimental Model Helicopter Rotor Performance in Forward Flight." TCREC TR 61-103, July 1961.

Rabbott, J. P., Jr. "A Study of the Optimum Rotor Geometry for a High Speed Helicopter." TCREC TR 62-53, May 1962.

Rabbott, J. P., Jr., and Churchill, G. B. "Experimental Investigation of the Aerodynamic Loading on a Helicopter Rotor Blade in Forward Flight." NACA RM L56I07, October 1956.

Rabbott, J. P., Jr.; Lizak, A. A.; and Paglino, V. M. "Tabulated Sikorsky CH-34 Blade Surface Pressures Measured at the NASA/Ames Full-Scale Wind Tunnel." Sikorsky Aircraft, Report SER-58399, January 1966.

Rabbott, J. P., Jr.; Lizak, A. A.; and Paglino, V. M. "A Presentation of Measured and Calculated Full-Scale Rotor Blade Aerodynamic and Structural Loads." USAAVLABS TR 66-31, July 1966.

Rabbott, J. P., Jr., and Niebanck, C. "Experimental Effects of Tip Shape on Rotor Control Loads." AHS Forum, 1978.

Rabbott, J. P., Jr., and Paglino, V. M. "Aerodynamic Loading of High Speed Rotors." CAL/AVLABS Symposium, Buffalo, New York, June 1966.

Radford, R. C., et al. "Evaluation of XV-15 Tilt Rotor Aircraft for Flying Qualities Research Application." NASA CR 137828, April 1976.

Rae, W. H., Jr. "Limits on Minimum-Speed V/STOL Wind-Tunnel Tests." Journal of Aircraft, 4:3 (May-June 1967).

Rae, W. H., Jr., and Shindo, S. "Comments on V/STOL Wind Tunnel Data at Low Forward Speeds." CAL/AVLABS Symposium, Buffalo, New York, June 1969.

Rae, W. H., Jr., and Shindo, S. "Limits on Low Speed Wind Tunnel Tests of Rotors." U.S. Army Research Office – Durham, September 1977.

Rainey, A. G. "Preliminary Study of Some Factors Which Affect the Stall-Flutter Characteristics of Thin Wings." NACA TN 3622, March 1956.

Rainey, A. G. "Measurement of Aerodynamic Forces for Various Mean Angles of Attack on an Airfoil Oscillating in Pitch and on Two Finite-Span Wings Oscillating in Bending with Emphasis on Damping in the Stall." NACA Report 1305, 1957.

Raitch, F. "Summary of Antitorque Devices Other Than Tail Rotors." American Helicopter Society Northeast Region Symposium, Hartford, Connecticut, March 1975.

Ramakrishnan, R.; Randall, D.; and Hosier, R. N. "A Computer Program to Predict Rotor Rotational Noise of a Stationary Rotor from Blade Loading Coefficients." NASA TM X-3281, February 1976.

Rao, B. M.; Maskew, B.; and Dvorak, F. A. "Theoretical Prediction of Dynamic Stall on Oscillating Airfoils." AHS Forum, 1978.

Rao, B. M., and Schatzle, P. R. "Analysis of Unsteady Airloads of Helicopter

Rotors in Hover." Journal of Aircraft, 15:4 (April 1978).
Rao, C. V. J. "The Helicopter Rotor." U.S. Army Materials and Mechanics Research Center, AMMRC MS-71-3, December 1971.
Rao, J. S., and Carnegie, W. "Non-Linear Vibration of Rotating Cantilever Blades Treated by the Ritz Averaging Process." The Aeronautical Journal, 76:741 (September 1972).
Razak, K. "Blade Section Variation on Small-Scale Rotors." JAS, 11:1 (January 1944).
Reaser, J. S. "Rotorcraft Linear Simulation Model.' NASA CR 152079, January 1978.
Reaser, J. S., and Kretsinger, P. H. "REXOR II Rotorcraft Simulation Model." NASA CR 145331, CR 145332, CR 145333, June 1978.
Rebont, J.; Soulez-Lariviere, J.; and Valensi, J. "Response of Rotor Lift to an Increase in Collective Pitch in the Case of Descending Flight, the Regime of the Rotor Being Near Autorotative." NASA TT F-18, April 1960.
Rebont, J.; Velensi, J.; and Soulez-Lariviere, J. "Wind Tunnel Study of the Response in Lift of a Rotor to an Increase in Collective Pitch in the Case of Vertical Flight Near the Autorotative Regime." NASA TT F-17, April 1960.
Rebont, J.; Valensi, J.; and Soulez-Lariviere, J. "Response of a Helicopter Rotor to an Increase in Collective Pitch for the Case of Vertical Flight." NASA TT F-55, January 1961.
Reed, W. H. III "Propeller-Rotor Whirl Flutter: A State-of-the-Art Review." J. Sound Vib., 4:3 (1966).
Reed, W. H. III. "Review of Propeller-Rotor Whirl Flutter." NASA TR R-264, July 1967.
Reeder, J. P. "Handling Qualities Experience with Several VTOL Research Aircraft." NASA TN D-735, March 1961.
Reeder, J. P., and Gustafson, F. B. "On the Flying Qualities of Helicopters." NACA TN 1799, January 1949.
Reeder, J. P., Jr., and Haig, C. R., Jr. "Some Tests of the Longitudinal Stability and Control of an H-13B Helicopter in Forward Flight." NACA RM L9E25a, August 1949.
Reeder, J. P., Jr.; Tapscott, R. J.; and Garren, J. F., Jr. "The Case for Inherent Stability of Helicopters." AGARD Conference Proceedings No. 7, Paris, January 1966.
Reeder, J. P., Jr., and Whitten, J. B. "Some Effects of Varying the Damping in Pitch and Roll on the Flying Qualities of a Small Single-Rotor Helicopter." NACA TN 2459, January 1952.
Regier, A. A., and Hubbard, H. H. "Factors Affecting the Design of Quiet Propellers." NACA RM L7H05, September 1947.

Reichert, G. "The Flying Qualities of Helicopters with Hingeless Rotors." Royal Aircraft Establishment, Library Translation 1147, January 1966.

Reichert, G. "Handling Qualities of Helicopters with Elastically Attached Rotor Blades." NASA TT F-11374, July 1968.

Reichert, G. "Loads Prediction Methods for Hingeless Rotor Helicopters." AGARD Conference Proceedings No. 122, Milan, Italy, March 1973.

Reichert, G. "Basic Dynamics of Rotors; Control and Stability of Rotary Wing Aircraft; Aerodynamics and Dynamics of Advanced Rotary-Wing Configurations." AGARD Lecture Series No. 63, Brussels, April 1973.

Reichert, G., and Huber, H. "Influence of Elastic Coupling Effects on the Handling Qualities of a Hingeless Rotor Helicopter." AGARD Conference Proceedings No. 121, Hampton, Virginia, September 1971.

Reichert, G., and Oelker, P. "Handling Qualities with the Bolkow Rigid Rotor System." AHS Forum, 1968.

Reichert, G., and Wagner, S. N. "Some Aspects of the Design of Rotor-Airfoil Shapes." AGARD Conference Proceedings No. 111, Marseilles, France, September 1972.

Reissner, H. "On the Vortex Theory of the Screw Propeller." JAS, 5:1 (November 1937).

Reissner, H. "A Generalized Vortex Theory of the Screw Propeller and Its Application." NACA TN 750, February 1940.

Reissner, H., and Morduchow, M. "A Theoretical Study of the Dynamic Properties of Helicopter-Blade Systems." NACA TN 1430, November 1948.

Rempfer, P. S., and Stevenson, L. "Two Sets of Linearized Aircraft Equations of Motion for Control System Analysis." NASA TM X-66493, August 1968.

Rempfer, P. S.; Stevenson, L.; and Collins, D. "The Use of Automatic Control Theory and a Pilot Math Model in Helicopter Manual Control System Synthesis." NASA TM X-66492, October 1968.

Reschak, R. J. "Suggested VTOL Handling Qualities Criteria for Civil IFR Qualifications." JAHS, 16:4 (October 1971).

Rhodes, J. E., and Gaidelis, J. A. "Loads Prediction for Structural Rigid Rotor Systems." AHS Forum, 1963.

Ribner, H. S. "Formulas for Propellers in Yaw and Charts of the Side-Force Derivative." NACA Report 819, 1945.

Ribner, H. S. "Propellers in Yaw." NACA Report 820, 1945.

Rich, M. J. "Investigation of Advanced Helicopter Structural Designs." USAAMRDL TR 75-59, May 1976.

Rich, M. J.; Kenigsberg, I. J.; Israel, M. H.; and Cook, P. P. "Power Spectral Density Analysis of V/STOL Aircraft Structures." JAHS, 13:4 (October 1968).

Richards, E. J., and Mead, D. J., eds. Noise and Acoustic Fatigue in Aeronautics. London: John Wiley and Sons, 1968.

Richardson, D. A. "The Application of Hingeless Rotors to Tilting Prop/Rotor Aircraft." JAHS, 16:3 (July 1971).

Richardson, D. A., and Alwang, J. R. "Engine/Airframe/Drive Train Dynamic Interface Documentation." USARTL TR 78-11, April 1978.

Richardson, D. A.; Liiva, J.; et al. "Configuration Design Analysis of a Proprotor Aircraft." AFFDL TR 70-44, April 1970.

Ricks, R. G. "A Study of Tandem Helicopter Fuselage Vibration." Air Force Systems Command, ASD-TDR 62-284, September 1962.

Riley, M. J. "A Flight Investigation of the Spanwise Lift Requirements of a Helicopter Rotor Blade." ARC R&M 3812, September 1976.

Riley, M. J., and Brotherhood, P. "Comparative Performance Measurements of Two Helicopter Blade Profiles in Hovering Flight." ARC R&M 3792, February 1974.

Rinehart, S. A. "Effects of Modifying a Rotor Tip Vortex by Injection on Downwash Velocities, Noise, and Airloads." JAHS, 16:4 (October 1971).

Rinehart, S. A.; Balcerak, J. C.; and White, R. P., Jr. "An Experimental Study of Tip Vortex Modification by Mass Flow Injection." Rochester Applied Science Associates, Report 71-01, January 1971.

Rita, A. D.; McGarvey, J. H.; and Jones, R. "Helicopter Rotor Isolation Evaluation Utilizing the Dynamic Antiresonant Vibration Isolator." JAHS, 23:1 (January 1978).

Rix, O.; Huber, H.; and Kaletka, J. "Parameter Identification of a Hingeless Rotor Helicopter." AHS Forum, 1977.

Robbins, T. "An Analog Computer Simulation of a Generalized Helicopter Rotor System." U.S. Army Electronic Command, ECOM-4043, November 1972.

Robinson, D. W., Jr. "Use of Thrust and Lift Augmentation for Increasing Helicopter Operating Speed." JAHS, 8:2 (April 1963).

Robinson, D. W., Jr.; Nettles, W. E.; and Howes, H. E. "Design Variables for a Controllable Twist Rotor." AHS Forum, 1975.

Robinson, F. "Increasing Tail Rotor Thrust and Comments on Other Yaw Control Devices." JAHS, 15:4 (October 1970).

Robinson, F. "Component Noise Variables of a Light Observation Helicopter." NASA CR 114761, 1973.

Rohtert, R. E., and La Forge, S. V. "Aerodynamic Tests of an Operational OH-6A Helicopter in the Ames 40- by 80-ft Wind Tunnel." Naval Air Systems Command, May 1970.

Rohtert, R. E., and La Forge, S. V. "Rotor Stability Derivatives Determined from Wind Tunnel Tests of an Instrumented Helicopter." AHS Forum, 1971.

Rorke, J. B. "Hover Performance Tests of Full Scale Variable Geometry Rotors," NASA CR 2713, August 1976.

Rorke, J. B., and Moffitt, R. C. "Measurement of Vortex Velocities over a Wide Range of Vortex Age, Downstream Distance, and Free Stream Velocity." NASA CR 145213, July 1977.

Rorke, J. B.; Moffitt, R. C.; and Ward, J. F. "Wind Tunnel Simulation of Full-Scale Vortices." AHS Forum, 1972.

Rorke, J. B., and Wells, C. D. "The Prescribed Wake-Momentum Analysis." CAL/AVLABS Symposium, Buffalo, New York, June 1969.

Rose, R. E.; Hammer, J. M.; and Kizilos, A. P. "Feasibility Study of a Bi-directional Jet Flap Device for Application to Helicopter Rotor Blades." NASA CR 114359, July 1971.

Rosen, A., and Friedmann, P. "Nonlinear Equations of Equilibrium for Elastic Helicopter or Wind Turbine Blades Undergoing Moderate Deformation." University of California, Los Angeles, Report ENG-7718, January 1977.

Rosenberg, R. "Aeroelastic Instability in Unbalanced Lifting Rotor Blades." JAS, 11:4 (October 1944).

Rosenstein, H.; McVeigh, M. A.; and Molienkof, P. A. "V/STOL Tilt Rotor Aircraft Study – Piloted Simulation Evaluation of the Boeing Vertol Model 222 Tilt Rotor Aircraft." NASA CR 114602 and CR 114601, 1973.

Ross, R. S. "An Investigation of the Airflow Underneath Helicopter Rotors." JAS, 13:12 (December 1946).

Rott, N., and Smith, W. E. "Some Examples of Laminar Boundary-Layer Flow on Rotating Blades." JAS, 23:11 (November 1956).

Ruddell, A. J. "Advancing Blade Concept Development." JAHS, 22:1 (January 1977).

Rudhman, W. E. "A Numerical Solution of the Unsteady Airfoil with Application to the Vortex Interaction Problem" NASA CR 111843, December 1970.

Rudolph, J. F. "V/STOL Certification." Journal of Aircraft, 9:3 (March 1972).

Rudy, M. D. "Structural Dynamics of a Helicopter Rotor Blade." M.S. thesis, Pennsylvania State University, June 1973.

Ryan, J. P.; Berens, A. P.; Coy, R. G.; and Roth, G. J. "Helicopter Fatigue Load and Life Determination Methods." USAAMRDL TR 75-27, August 1975.

Sadler, S. G. "A Method for Predicting Helicopter Wake Geometry, Wake Induced Flow, and Wake Effects on Blade Airloads." AHS Forum, 1971.

Sadler, S. G. "Development and Application of a Method for Predicting

Rotor Free Wake Positions and Resulting Rotor Blade Airloads." NASA CR 1911 and CR 1912, December 1971.

Sadler, S. G. "Blade Frequency Program for Nonuniform Helicopter Rotors, with Automated Frequency Search." NASA CR 112071, 1972.

Sadler, S. G. "Main Rotor Free Wake Geometry Effects on Blade Air Loads and Response for Helicopters in Steady Maneuvers." NASA CR 2110 and CR 2111, September 1972.

Sadler, S. G.; Johnson, H. K.; and Evans, T. D. "Determination of the Aerodynamic Characteristics of Vortex Shedding from Lifting Airfoils for Application to the Analysis of Helicopter Noise." Rochester Applied Science Associates, Report 73-02, February 1973.

Sadler, S. G., and Loewy, R. G. "A Theory for Predicting the Rotational and Vortex Noise of Lifting Rotors in Hover and Forward Flight." NASA CR 1333, May 1969.

Salmirs, S., and Tapscott, R. J. "Instrument Flight Trials with a Helicopter Stabilized in Attitude about Each Axis Individually." NACA TN 3947, January 1957.

Salmirs, S., and Tapscott, R. J. "The Effects of Various Combinations of Damping and Control Power on Helicopter Handling Qualities during Both Instrument and Visual Flight." NASA TN D-58, October 1959.

Sambell, K. W. "Proprotor Short-Haul Aircraft—STOL and VTOL." Journal of Aircraft, 9:10 (October 1972).

Sanders, J. C. "Influence of Rotor-Engine Torsional Oscillation on Control of Gas-Turbine Engine Geared to Helicopter Rotors." NACA TN 3027, October 1953.

Sardanowsky, W., and Harper, H. P. "A Study of Handling Qualities Requirements of Winged Helicopters." USAAVLABS TR 68-39, July 1968.

Scarpati, T.; Sandford, R.; and Powell, R. "The Heavy Lift Helicopter Rotor Blade." JAHS, 19:2 (April 1974).

Schad, J. L. "Small Autogyro Performance." JAHS, 10:3 (July 1965).

Schaefer, R. F.; Loftin, L. K., Jr.; and Horton, E. A. "Two-Dimensional Investigation of Five Related NACA Airfoil Sections Designed for Rotating-Wing Aircraft." NACA TN 1922, July 1949.

Scheiman, J. "A Tabulation of Helicopter Rotor-Blade Differential Pressures, Stresses, and Motions As Measured in Flight." NASA TM X-952, 1964.

Scheiman, J. "Further Analysis of Broadband Noise Measurements for a Rotating Blade Operating with and without Its Shed Wake Blown Downstream." NASA TN D-7623, September 1974.

Scheiman, J.; Hilton, D. A.; and Shivers, J. P. "Acoustical Measurements of the Vortex Noise for a Rotating Blade Operating with and without Its Shed Wake Blown Downstream." NASA TN D-6364, August 1971.

Scheiman, J., and Hoad, D. R. "Investigation of Blade Impulsive Noise on a Scaled Fully Articulated Rotor System." NASA TM X-3528, June 1977.

Scheiman, J., and Kelley, H. "Comparison of Flight Measured Helicopter Rotor Blade Chordwise Pressure Distributions and Two-Dimensional Airfoil Characteristics." CAL/TRECOM Symposium, Buffalo, New York, June 1963.

Sheiman, J., and Kelley, H. L. "Comparison of Flight-Measured Helicopter Rotor-Blade Chordwise Pressure Distributions with Static Two--Dimensional Airfoil Characteristics." NASA TN D-3936, May 1967.

Scheiman, J., and Ludi, L. H. "Qualitative Evaluation of Effect of Helicopter Rotor-Blade Tip Vortex on Blade Airloads." NASA TN D-1637, May 1963.

Schlegel, R. G., and Bausch, W. E. "Helicopter Rotor Noise Prediction and Control." JAHS, 14:3 (July 1969).

Schlegel, R. G., and Bausch, W. E. "Helicopter Rotor Rotational Noise Prediction and Correlation." USAAVLABS TR 70-1, November 1970.

Schlegel, R.; King, R.; and Mull, H. "Helicopter Rotor Noise Correlation and Propagation." USAAVLABS TR 66-4, October 1966.

Schmitz, F. H. "Optimal Takeoff Trajectories of a Heavily Loaded Helicopter." Journal of Aircraft, 8:9 (September 1971).

Schmitz, F. F., and Boxwell, D. A. "In-Flight Far-Field Measurement of Helicopter Impulsive Noise." JAHS, 21:4 (October 1976).

Schmitz, F. H.; Boxwell, D. A.; and Vause, C. R. "High-Speed Helicopter Impulsive Noise." JAHS, 22:4 (October 1977).

Schmitz, F. H., and Stepniewski, W. Z. "Reduction of VTOL Operational Noise through Flight Trajectory Management." Journal of Aircraft, 10:7 (July 1973).

Schmitz, F. H.; Stepniewski, W. Z.; Gibs, J.; and Hinterkeuser, E. "A Comparison of Optimal and Noise-Abatement Trajectories of a Tilt-Rotor Aircraft." NASA CR 2034, May 1972.

Schmitz, F. H., and Vause, C. R. "Near-Optimal Takeoff Policy for Heavily Loaded Helicopters Exiting from Confined Areas." Journal of Aircraft, 13:5 (May 1976).

Schmitz, F. H., and Yu, Y. H. "Theoretical Modeling of High Speed Helicopter Impulsive Noise." European Rotorcraft and Powered Lift Aircraft Forum, Aix-en-Provence, France, September 1977.

Schneider, J. J. "The Influence of Propulsion Systems on Extremely Large Helicopter Design." JAHS, 15:1 (January 1970).

Schrage, D. P., and Peskar, R. E. "Helicopter Vibration Requirements." AHS Forum, 1977.

Schrenk, M. "Aerodynamic Principles of the Direct Lifting Propeller." NACA TM 733, January 1934.

Schrenk, M. "Static Longitudinal Stability and Longitudinal Control of

Autogiro Rotors." NACA TM 879, 1938.

Schroder, R. "A Spacial Theory for the Ground Resonance of Helicopters." European Space Research Organization, ESRO TT-108, November 1974.

Schuett, E. P. "Passive Helicopter Rotor Isolation Using the Kaman Dynamic Antiresonant Vibration Isolator (DAVI)." USAAVLABS TR 68-46, December 1968.

Schuett, E. P. "Application of Passive Helicopter Rotor Isolation for Alleviation of Rotor Induced Vibration." JAHS, 14:2 (April 1969).

Schulz, G. "Concepts for the Design of a Completely Active Helicopter Isolation System Using Output Vector Feedback." NASA TM 75161, October 1977.

Schumacher, H. "Dynamic Testing of Helicopter Components." NASA TT F-14282, May 1972.

Schumacher, W. J. "An Investigation of the Trailing Vortex System Generated by a Jet-Flapped Wing Operating at High Wing Lift Coefficients.' JAHS, 16:4 (October 1971).

Schwartzberg, M. A. "Rotor Induced Power." U.S. Army Aviation Systems Command, USAAVSCOM TR 75-10, May 1975.

Schwartzberg, M. A., et al. "Single-Rotor Helicopter Design and Performance Estimation Programs." U.S. Army Air Mobility Research and Development Laboratory, Report No. SRIO 77-1, June 1977.

Schwind, R. G., and Allen, H. J. "The Effects of Leading-Edge Serrations on Reducing Flow Unsteadiness about Airfoils – An Experimental and Analytical Investigation." NASA CR 2344, November 1973.

Scully, M. P. "Approximate Solutions for Computing Helicopter Harmonic Airloads." Massachusetts Institute of Technology, ASRL TR 123-2, December 1965.

Scully, M. P. "A Method of Computing Helicopter Vortex Wake Distortion." Massachusetts Institute of Technology, ASRL TR 138-1, June 1967.

Scully, M. P. "Computation of Helicopter Rotor Wake Geometry and Its Influence on Rotor Harmonic Airloads." Massachusetts Institute of Technology, ASRL TR 178-1, March 1975.

Scully, M. P., and Faulkner, H. "Helicopter Direct Operating Cost Comparison." Massachusetts Institute of Technology, FTL Memo 71-2, February 1971.

Scully, M. P., and Faulkner, H. "Helicopter Design Program Description." Massachusetts Institute of Technology, FTL Memo 71-3, March 1971.

Scully, M. P., and Sullivan, J. P. "Helicopter Rotor Wake Geometry and Airloads and Development of Laser Doppler Velocimeter for Use in Helicopter Rotor Wakes." Massachusetts Institute of Technology, AL TR 179, August 1972.

Sears, W. R. "The Boundary Layer of Yawed Cylinders." JAS, 15:1 (January 1948).

Sears, W. R. "Potential Flow around a Rotating Cylindrical Blade." JAS, 17:3 (March 1950).

Seckel, E. Stability and Control of Airplanes and Helicopters. New York: Academic Press, 1964.

Seckel, E., and Curtiss, H. C. Jr. "Aerodynamic Characteristics of Helicopter Rotors – Rotor Contributions to Helicopter Stability Parameters." Princeton University, Report No. 659, December 1963.

Seckel, E.; Traybar, J. J.; and Miller, G. E. "Longitudinal Handling Qualities for Hovering." Princeton University, Report No. 594, December 1961.

Seckel, E.; Traybar, J. J.; and Miller, G. E. "Longitudinal Handling Qualities for Hovering." AHS Forum, 1962.

Segel, L. "Air Loadings on a Rotor Blade as Caused by Transient Inputs of Collective Pitch." USAAVLABS TR 65-65, October 1965.

Segel, L. "A Method for Predicting Nonperiodic Airloads on a Rotary Wing." Journal of Aircraft, 3:6 (November-December 1966).

Segel, L. "Rotor Airloads, Blade Motion, and Stress Caused by Transient Inputs of Shaft Torque As Related to Stoppable Rotor Operation." USAAVLABS TR 67-18, May 1967.

Segel, R. M., and Bain, L. J. "Experimental Investigation of Compound Helicopter Aerodynamic Interference Effects." CAL/AVLABS Symposium, Buffalo, New York, June 1966.

Segner, D. R. "Thoughts and Basic Techniques Concerning Helicopter Flying Qualities Testing." JAHS, 6:1 (January 1961).

Seibel, C. "Periodic Aerodynamic Forces on Rotors in Forward Flight." JAS, 11:4 (October 1944).

Seibel, C. M. "The U.S. Army HueyCobra Configuration and Design Considerations." JAHS, 13:1 (January 1968).

Seiferth, R. "Testing a Windmill Airplane (Autogiro)." NACA TM 394, January 1927.

Selic, Z. R. "Measurement of Vortex Properties during Wing-Vortex Interaction." Massachusetts Institute of Technology, ASRL TR 178-3, March 1975.

Shahady, P. A., et el. "Quiet Propeller/Rotor Concept Evaluation." Air Force Aero-Propulsion Laboratory, AFAPL TR 77-56, October 1977.

Shamie, J., and Friedmann, P. "Aeroelastic Stability of Complete Rotors with Application to a Teetering Rotor in Forward Flight." J. Sound Vib., 53:4 (1977).

Shamie, J., and Friedmann, P. "Effect of Moderate Deflections on the Aeroelastic Stability of a Rotor Blade in Forward Flight." European Rotorcraft and Powered Lift Symposium, Aix en-Provence, France, September 1977.

Shapiro, J. Principles of Helicopter Engineering. New York: McGraw-Hill Book Co., 1955.

Shapiro, N., and Healy, G. J. "A Realistic Assessment of the Vertiport/Community Noise Problem." Journal of Aircraft, 5:4 (July-August 1968).

Shapley, J. J., Jr.; Kyker, R. A.; and Ferrell, K. R. "The Development of an Improved Method of Conducting Height-Velocity Testing on Rotary Wing Aircraft." JAHS, 15:2 (April 1970).

Shaw, J., Jr. "Higher Harmonic Blade Pitch Control for Helicopter Vibration Reduction." Massachusetts Institute of Technology, ASRL TR 150-1, December 1968.

Shaw, J., Jr., and Edwards, W. T. "The YUH-61A Tail Rotor: Development of a Stiff-Inplane Bearingless Flexstrap Design." JAHS, 23:2 (April 1978).

Shaydakov, V. I. "Aerodynamic Calculation for Helicopter Lifting Rotors in Steep Descent (Vortex Ring Method)." U.S. Army Foreign Science and Technology Center, FSTC-HT-23-707-71, October 1971.

Shaydakov, V. I. "Aerodynamic Calculation for Helicopter Lifting Rotors in Vertical Descent (Vortex Ring Method)." U.S. Army Foreign Science and Technology Center, FSTC-HT-23-708-71, October 1971.

Shaydakov, V. I. "Application of the Ring Vortex Method to Aerodynamic Design of Lifting Rotor Systems." U.S. Army Foreign Science and Technology Center." FSTC-HT-23-709-71, November 1971.

Sheehy, T. W. "A Simplified Approach to Generalized Helicopter Configuration Modeling and the Prediction of Fuselage Surface Pressures." JAHS, 21:1 (January 1976).

Sheehy, T. W. "A General Review of Helicopter Rotor Hub Drag Data." JAHS, 22:2 (April 1977).

Sheehy, T. W., and Clark, D. R. "A General Review of Helicopter Rotor Hub Drag Data." AHS Forum, 1975. (Rotorcraft Parasite Drag Supplement.)

Sheehy, T. W., and Clark, D. R. "A Method for Predicting Helicopter Hub Drag." USAAMRDL TR 75-48, January 1976.

Sheldon, D. F. "An Appreciation of the Dynamic Problems Associated with the External Transportation of Loads from a Helicopter – State of the Art." Vertica, 1:4 (1977).

Sheridan, P. F. "Interactional Aerodynamics of the Single Rotor Helicopter Configuration." USARTL TR 78-23, September 1978.

Sheridan, P. F., and Wiesner, W. "Aerodynamics of Helicopter Flight Near the Ground." AHS Forum, 1977.

Shipman, D. P.; White, J. A.; and Cronkhite, J. D. "Fuselage Nodalization." AHS Forum, 1972.

Shipman, K. W. "Effect of Wake on the Performance and Stability

Characteristics of Advance Rotor Systems." USAAMRDL TR 74-45, September 1974.

Shipman, K. W., and Wood, E. R. "A Two-Dimensional Theory for Rotor Blade Flutter in Forward Flight." Journal of Aircraft, 8:12 (December 1971).

Shivananda, T. P.; McMahon, H. M.; and Gray, R. B. "Surface Pressure Measurements at the Tip of a Model Helicopter Rotor in Hover." Journal of Aircraft, 15:8 (August 1978).

Shivers, J. P. "High-Tip-Speed Static-Thrust Tests of a Rotor Having NACA $63_{(215)}$A018 Airfoil Sections with and without Vortex Generators Installed." NASA TN D-376, May 1960.

Shivers, J. P. "Hovering Characteristics of a Rotor Having an Airfoil Section Designed for Flying-Crane Type of Helicopter." NASA TN D-742, April 1961.

Shivers, J. P. "Hovering Performance Characteristics of a Rotor with and without Blade-Pitch Servo-Control Flaps." NASA TN D-3786, January 1967.

Shivers, J. P. "Wind-Tunnel Investigation of Various Small-Scale Rotor/Wing Configurations for VTOL Composite Aircraft in the Cruise Mode." NASA TN D-5945, October 1970.

Shivers, J. P., and Carpenter, P. J. "Experimental Investigation on the Langley Helicopter Test Tower of Compressibility Effects on a Rotor Having NACA 63_2-015 Airfoil Sections." NACA TN 3850, December 1956.

Shivers, J. P., and Carpenter, P. J. "Effects of Compressibility on Rotor Hovering Performance and Synthesized Blade-Section Characteristics Derived from Measured Rotor Performance of Blades Having NACA 0015 Airfoil Tip Sections." NACA TN 4356, September 1958.

Shivers, J. P., and Monahan, W. J. "Hovering Characteristics of a Rotor Having an Airfoil Section Designed for a Utility Type of Helicopter." NASA TN D-1517, December 1962.

Shockey, G. A.; Cox, C. R.; and Williamson, J. W. "AH-1G Helicopter Aerodynamic and Structural Loads Survey." USAAMRDL TR 76-39, February 1977.

Shockey, G. A.; Williamson, J. W.; and Cox, C. R. "Helicopter Aerodynamics and Structural Loads Survey." AHS Forum, 1976.

Shovlin, M. D., and Gambucci, B. J. "Effect of High Lift Flap Systems on the Conceptual Design of a 1985 Short-Haul Commercial STOL Tilt-Rotor

Transport." NASA TM 78474, April 1978.

Shrager, J. J. "A Summary of Helicopter Vorticity and Wake Turbulence Publications with an Annotated Bibliography." Federal Aviation Administration Report No. FAA-RD-74-48, May 1974.

Shulman, Y. "Stability of a Flexible Helicopter Rotor Blade in Forward Flight." JAS, 23:7 (July 1956).

Shupe, N. K. "A Study of the Dynamic Motions of Hingeless Rotored Helicopters." U.S. Army Electronics Command, ECOM 3323, August 1970.

Shupe, N. K. "Simulation of the Induced Flow through a Rotor in Descending Flight." U.S. Army Electronics Command, ECOM 3579, June 1972.

Shutler, A. G., and Jones, J. P. "The Stability of Rotor Blade Flapping Motion." ARC R&M 3178, May 1958.

Sibley, J. D., and Jones, G. H. "Some Design Aspects of Tandem Rotor Helicopters." Journal of the Helicopter Association of Great Britain, 13:10 (October 1959).

Sikorsky, I. A. "Aerodynamic Parameters Selection in Helicopter Design." JAHS 5:1 (January 1960).

Sikorsky, I. I. "Progress of the Vought-Sikorsky Helicopter Program in 1942." Aeronautical Engineering Review, 2:4 (April 1943).

Sikorsky, I. I. The Story of the Winged-S. New York: Dodd, Mead, and Company, 1967.

Sikorsky, I. I. "Sixty Years in Flying." The Aeronautical Journal, 75:731 (November 1971).

Silviera, M. A. "An Investigation of Periodic Forces and Moments Transmitted to the Hub of Four Lifting Rotor Configurations." NASA TN D-1011, March 1962.

Silviera, M. A., and Brooks, G. W. "Analytical and Experimental Determination of the Coupled Natural Frequencies and Mode Shapes of a Dynamic Model of a Single-Rotor Helicopter." NASA Memo 11-5-58L, December 1958.

Silviera, M. A., and Brooks, G. W. "Dynamic-Model Investigation of the Damping of Flapwise Bending Modes of Two-Blade Rotors in Hovering and a Comparison with Quasistatic and Unsteady Aerodynamic Theories." NASA TN D-175, December 1959.

Simon, D. R., and Savage, J. C. "Flight Test of the Aerospatiale SA-22 Helicopter." USAAMRDL TR 75-44, August 1976.

Simons, I. A. "Some Aspects of Blade/Vortex Interaction on Helicopter Rotors in Forward Flight." J. Sound Vib., 4:3 (1966).

Simons, I. A. "Advanced Control Systems for Helicopters." Vertica, 1:1 (1976).

Simons, I. A.; Pacifico, R. E.; and Jones, J. P. "The Movement, Structure, and Breakdown of Trailing Vortices from a Rotor Blade." CAL/AVLABS Symposium, Buffalo, New York, June 1966.

Simpkinson, S. H.; Eatherton, L. J.; and Millenson, M. B. "Effect of Centrifugal Force on the Elastic Curve of a Vibrating Cantilever Beam." NACA Report 914, 1948.

Simpson, J. R. "Preliminary Design of a Rotor System for a Hot Cycle Heavy-Lift Helicopter." USAAVLABS TR 67-1, March 1967.

Sinacori, J. B. "A Study of V/STOL Ground-Based Simulator Techniques." USAAVLABS TR 70-16, April 1970.

Sinacori, J. B. "Validation of Ground Based Simulation." JAHS, 15:3 (July 1970).

Sissingh, G. "Contribution to the Aerodynamics of Rotating-Wing Aircraft." NACA TM 921, 1939.

Sissingh, G. "Contribution to the Aerodynamics of Rotating-Wing Aircraft, Part II." NACA TM 990, October 1941.

Sissingh, G. J. "Contributions to the Dynamic Stability of Rotary-Wing Aircraft with Articulated Blades." Air Materiel Command, Translation F-TS-690-RE, July 1946.

Sissingh, G. J. "The Frequency Response of the Ordinary Rotor Blade, the Hiller Servo Blade, and the Young-Bell Stabiliser." ARC R&M 2860, May 1950.

Sissingh, G. J. "Variation of Rotor Dynamic Response by Self-Contained Mechanical Feedback." Institute of the Aerospace Sciences, IAS Paper No. 61-25, January 1961.

Sissingh, G. J. "Delay of Rotor Blade Stall by Dynamic Tuning." AHS Forum, 1962.

Sissingh, G. J. "Dynamics of a Hinged Rotor Submitted to Steady Precession." JAHS, 9:1 (January 1964).

Sissingh, G. J. "Effect of Tip Speed Ratio on the Longitudinal and Lateral Response Characteristics of a Lifting Rotor." JAHS, 9:2 (April 1964).

Sissingh, G. J. "Comparative Response of Rotor Tip Path Plane to Longitudinal Cyclic Pitch and Changes in Rotor Angle of Attack." JAHS, 10:2 (April 1965).

Sissingh, G. J. "Response Characteristics of the Gyro-Controlled Lockheed Rotor System." JAHS, 12:4 (October 1967).

Sissingh, G. J. "Dynamics of Rotors Operating at High Advance Ratios." JAHS, 13:3 (July 1968).

Sissingh, G. J., and Kuczynski, W. A. "Investigations on the Effect of Blade Torsion on the Dynamics of the Flapping Motion." JAHS, 15:2 (April 1970).

Sisto, F. "Stall-Flutter in Cascades." JAS, 20:9 (September 1953).

Skingle, C. W.; Gaukroger, D. R.; and Taylor, G. A. "A Preliminary Experiment in Resonance Testing a Rotating Blade." ARC CP 1070, April 1970.

Skujins, O., and Walters, R. E. "The Stability of Helical Vortex Filaments in the Wake of a Hovering Rotor." West Virginia University, Aerospace Engineering TR-36, May 1973.

Slaymaker, S. E., and Gray, R. B. "Power-Off Flare-Up Tests of a Model Helicopter Rotor in Vertical Autorotation." NACA TN 2870, January 1953.

Slaymaker, S. E.; Lynn, R. R.; and Gray, R. B. "Experimental Investigation of Transition of a Model Helicopter Rotor from Hovering to Vertical Autorotation." NACA TN 2648, March 1952.

Sloof, J. W.; Wortmann, F. X.; and Duhon, J. M. "The Development of Transonic Airfoils for Helicopters." AHS Forum, 1975.

Smith, A. F. "Effects of Parasite Drag on Rotor Performance and Dynamic Response." AHS Forum, 1975. (Rotorcraft Parasite Drag Supplement.)

Smith, C. C., Jr. "Static Directional Stability of a Tandem-Helicopter Fuselage." NACA RM L50F29, August 1950.

Smith, C. R. "Hot Cycle Rotor/Wing Composite Research Aircraft." USAAVLABS TR 68-31, August 1968.

Smith, J. H.; Allen, E. M.; and Vensel, D. "Design, Fabrication, and Flight Test of the Active Arm External Load Stabilization System for Cargo Handling Helicopters." USAAMRDL TR 73-73, September 1973.

Smith, R. H. "VTOL Control Power Requirements Reappraised." Journal of Aircraft, 3:1 (January-February 1966).

Smith, R. L. "Closed-Form Equations for the Lift, Drag, and Pitching-Moment Coefficients of Airfoil Sections in Subsonic Flow." NASA TM 78492, August 1978.

Smollen, L. E.; Marshall, P.; and Gabel, R. "Active Vibration Isolation of Helicopter Rotors." JAHS 7:2 (April 1962).

Snyder, W. J. "A Summary of Rotor-Hub Bending Moments Encountered by a High-Performance Hingeless-Rotor Helicopter during Nap-of-the-Earth Maneuvers." NASA TN D-4574, May 1968.

Soderman, P. T. "Aerodynamic Effects of Leading-Edge Serrations on a Two-Dimensional Airfoil." NASA TM X-2643, September 1972.

Soderman, P. T. "Leading-Edge Serrations Which Reduce the Noise of Low-Speed Rotors." NASA TN D-7371, August 1973.

Sonneborn, W. G. O. "High Mach Number/High Advance Ratio Flight Test Program with the High-Performance UH-1 Compound Helicopter." USAAVLABS TR 71-2, February 1971.

Sonneborn, W. G. O., and Drees, J. M. "The Scissors Rotor." JAHS, 20:3 (July 1975).

Sonneborn, W. G. O., and Hartwig, L. W. "Results of High Speed Flight Research with the High Performance UH-1 Compound Helicopter." AHS Forum, 1971.

Soohoo, P.; Morino, L.; Noll, R. B.; and Ham, N. D. "Aerodynamic Interference Effects on Tilting Proprotor Aircraft." NASA CR 152053, January 1977.

Soohoo, P.; Noll, R. B.; Morino, L.; and Ham, N. D. "Rotor Wake Effects on Hub/Pylon Flow." USARTL TR 78-1, May 1978.

Sopher, R. "Three-Dimensional Potential Flow past the Surface of a Rotor Blade." AHS Forum, 1969.

Sopher, R. "Derivation of Equations of Motion for Multi-Blade Rotors Employing Coupled Modes and Including High Twist Capability." NASA CR 137810, February 1975.

Sorensen, J. A.; Mohr, R. L.; and Cline, T. B. "Instrumentation Requirements for Aircraft Parameter Identification with Application to the Helicopter." NASA CR 132675, June 1975.

Soule, V. "V/STOL Tilt Rotor Aircraft Study—Definition of Stowed Rotor Research Aircraft." NASA CR 114598, March 1973.

Soule, V. "V/STOL Tilt Rotor Aircraft Study – Preliminary Design of a Composite Wing for Tilt Rotor Research Aircraft." NASA CR 114599, March 1973.

Soule, V. H., nnd Clark, R. D. "A Comparison of Static and Cruise Test Results from a Series of Tests on 13-foot Diameter Low Disk Loading Rotors." NASA CR 114625, March 1973.

Soulez-Lariviere, J. "Method of Calculation to Determine Helicopter Blade Life." JAHS, 6:2 (April 1961).

Southwell, R. V., and Gough, B. S. "On the Free Transverse Vibration of

Air-Screw Blades." ARC R&M 766, October 1921.

Sowyrda, A. "On the Feasibility of Replacing a Helicopter Tail Rotor." Cornell Aeronautical Laboratory, Report CAL BB-2584-S-1, July 1968.

Spangler, S. B., and Smith, C. A. "Theoretical Study of Hull-Rotor Aerodynamic Interference on Semibuoyant Vehicles." NASA CR 152127, April 1978.

Spencer, R. H. "Application of Vortex Visualization Test Techniques to Rotor Noise Research." AHS Forum, 1970.

Spencer, R. H., and Sternfeld, H., Jr. "Measurement of Rotor Noise Levels and Evaluation of Porous Blade Tips on a CH-47A Helicopter." USAAVLABS TR 69-18, September 1969.

Spencer, R. H.; Sternfeld, H., Jr.; and McCormick, B. W. "Tip Vortex Core Thickening for Application to Helicopter Rotor Noise Reduction." USAAVLABS TR 66-1, September 1966.

Spivey, R. F. "Blade Tip Aerodynamics – Profile and Planform Effects." AHS Forum, 1968.

Spivey, W. A., and Morehouse, C. G. "New Insights into the Design of Swept-Tip Rotor Blades." AHS Forum, 1970.

Spreuer, K. R.; Snackenberg, S. J.; and Roeck, G. D. "Dynamic Loads and Structural Criteria Study." USAAMRDL TR 74-74, September 1974.

Spreuer, W. E. "Experimental Flight Tests of the XH-51A Compound Helicopter." JAHS, 13:3 (July 1968).

Squire, H. B. "The Flight of a Helicopter." ARC R&M No. 1730, November 1935.

Squire, H. B.; Fail, R. A.; and Eyre, R. C. W. "Wind-Tunnel Tests on a 12-foot-Diameter Helicopter Rotor." ARC R&M 2695, April 1949.

Staley, J. A. "Validation of Rotorcraft Flight Simulation Program through Correlation with Flight Data for Soft-in-Plane Hingeless Rotors." USAAMRDL TR 75-50, January 1976.

Stammers, C. W. "The Flapping Torsion Flutter of a Helicopter Blade in Forward Flight." Institute of Sound and Vibration Research, ISVR TR 3, February 1968.

Stammers, C. W. "The Bending Torsion Flutter of a Hingeless Rotor Blade." Institute of Sound and Vibration Research, ISVR TR 9, October 1968.

Stammers, C. W. "The Flutter of a Helicopter Rotor Blade in Forward Flight." The Aeronautical Quarterly, 21:1 (February 1970).

Statler, W. H.; Heppe, R. R.; and Cruz, E. S. "Results of the XH-51A Rigid Rotor Research Helicopter Program." AHS Forum, 1963.

Stave, A. M. "Effects of Helicopter Noise and Vibration on Pilot Performance." NASA CR 132347, December 1973.

Stengel, R. F.; Broussard, J. R.; and Berry, P. W. "Digital Flight Control Design for a Tandem-Rotor Helicopter." AHS Forum, 1977.

Stepniewski, W. Z. Introduction to Helicopter Aerodynamics. Revised edition. Morton, Pennsylvania: Rotorcraft Publishing Co., 1956.

Stepniewski, W. Z. "Some Aerodynamic Problems in Helicopter Design." Aircraft Engineering, 24:283 (September 1952).

Stepniewski, W. Z. "A Simplified Approach to the Aerodynamic Rotor Interference of Tandem Helicopters." American Helicopter Society West Coast Region Meeting, September 1955.

Stepniewski, W. Z. "Energy Aspects of Helicopters in Comparison with Other Air and Ground Vehicles." JAHS, 23:1 (January 1978).

Stepniewski, W. Z., and Schmitz, F. H. "Possibilities and Problems of Achieving Community Noise Acceptance of VTOL." The Aeronautical Journal, 77:750 (June 1973).

Stepniewski, W. Z., and Schneider, J. J. "Design Philosophy and Operational Requirements of Subsonic VTOL Aircraft." Journal of Aircraft, 1:3 (May-June 1964).

Sternfeld, H., Jr. "Influence of the Tip Vortex on Helicopter Rotor Noise." AGARD Conference Proceedings No. 22, Gottingen, Germany, September 1967.

Sternfeld, H.; Bobo, C.; Carmichael, D.; Fukushima, T.; and Spencer, R. "An Investigation of Noise Generation on a Hovering Rotor, Part II." Boeing Vertol Company, Report D210-10550-1, November 1972.

Sternfeld, H., Jr., and Doyle, L. B. "Evaluation of the Annoyance Due to Helicopter Rotor Noise." NASA CR 3001, June 1978.

Sternfeld, H., Jr., and Schairer, J. O. "Study of Rotor Blade Tip Vortex Geometry for Noise and Airfoil Applications." Boeing Vertol Company, Report D8-2464-1A, December 1969.

Sternfeld, H.; Schairer, J.; and Spencer, R. "Investigation of Helicopter Transmission Noise Reduction by Vibration Absorbers and Damping." USAAMRDL TR 72-34, August 1972.

Sternfeld, H.; Spencer, R. H.; and Schairer, J. O. "An Investigation of Noise Generation on a Hovering Rotor, Part I." Boeing Vertol Company, Report D210-10229-1, January 1971.

Stewart, W. "Flight Testing of Helicopters." The Journal of the Royal Aeronautical Society, 52:449 (May 1948).

Stewart, W. "Helicopter Control to Trim in Forward Flight." ARC R&M 2733, March 1950.

Stewart, W. "Higher Harmonics of Flapping on the Helicopter Rotor." ARC CP 121, March 1952.

Stewart, W. "Second Harmonic Control on the Helicopter Rotor." ARC R&M 2997, August 1952.

Stewart, W. "Recent Technical Progress in Aeronautics – Helicopter Developments." Aircraft Engineering, 34:403 (September 1962).

Stewart, W. "Research and Development of Rotating Wing Aircraft." Journal of the Royal Aeronautical Society, 66:623 (November 1962).

Stowell, E. Z., and Deming, A. F. "Vortex Noise from Rotating Cylindrical Rods." NACA TN 519, February 1935.

Stowell, E. Z., and Deming, A. F. "Noise from Two-Bladed Propellers." NACA Report 526, 1935.

Strand, T.; Levinsky, E. S.; and Wei, M. H. Y. "Unified Performance Theory for V/STOL Aircraft in Equilibrium Flight." Journal of Aircraft, 4:2 (March-April 1967); 4:3 (May-June 1967).

Stratton, W. K.; Scarpati, T. S.; and Feenan, R. J. "UTTAS/HLH Rotor Blade Fabrication and Testing." JAHS, 22:4 (October 1977).

Streiff, H. G. "Study, Survey of Helicopter and V/STOL Aircraft Simulator Trainer Dynamic Response." Naval Training Device Center, NAVTRADEVCEN Report 1753, August 1966 through June 1967.

Stroub, R. H. "Full-Scale Wind Tunnel Test of a Modern Helicopter Main Rotor – Investigation of Tip Mach Number Effects and Comparison of Four Tip Shapes." AHS Forum, 1978.

Stroub, R. H.; Falarski, M. D.; McCloud, J. L. III; and Soderman, P. T. "An Investigation of a Full-Scale Advancing Blade Concept Rotor System at High Advance Ratio." NASA TM X-62081, August 1971.

Stuart, Joseph III. "The Helicopter Control Rotor." Aeronautical Engineering Review, 7:8 (August 1948).

Stuckey, T. J., and Goddard, J. O. "Investigation and Prediction of Helicopter Rotor Noise." J. Sound Vib., 5:1 (1967).

Stutz, R. G., and Price, G. "Agility Performance of Several V/STOL and STOL Aircraft." Journal of Aircraft, 1:5 (September-October 1964).

Sullivan, J. P. "An Experimental Investigation of Vortex Rings and Helicopter Rotor Wakes Using a Laser Doppler Velocimeter." Massachusetts Institute of Technology, AL TR 183, June 1973.

Sullivan, R. J.; La Forge, S.; and Holchin, B. W. "A Performance Application

Study of a Jet-Flap Helicopter Rotor." NASA CR 112030, May 1972.

Summa, J. M. "Potential Flow about Impulsively Started Rotors." Journal of Aircraft, 13:4 (April 1976).

Sutton, L. R. "Development of an Analysis for the Determination of Coupled Helicopter Rotor/Control System Dynamic Response." NACA CR 2452, CR 2453, January 1975.

Sutton, L. R. "Documentation of Helicopter Aeroelastic Stability Analysis Computer Program (HASTA)." USARTL TR 77-52, December 1977.

Sutton, L. R., and Gangwani, S. T. "The Development and Application of an Analysis for the Determination of Coupled Tail Rotor/Helicopter Air Resonance Behavior." USAAMRDL TR 75-35, August 1975.

Sutton, L. R.; Gangwani, S. T.; and Nettles, W. E. "Rotor/Helicopter System Dynamic Aeroelastic Stability Analysis and Its Application to a Flex-strap Tail Rotor." AHS Forum, 1976.

Swain, R. L. "Minimum Control Power for VTOL Aircraft Stability Augmentation." Journal of Aircraft, 7:3 (May-June 1970).

Sweet, G. E. "Static-Stability Measurements of a Stand-On Type Helicopter with Rigid Blades, Including a Comparison with Theory." NASA TN D-189, February 1960.

Sweep, G. E. "Hovering Measurements for Twin-Rotor Configurations with and without Overlap." NASA TN D-534, November 1960.

Sweep, G. E., and Jenkins, J. L., Jr. "Wind-Tunnel Investigation of the Drag and Static Stability Characteristics of Four Helicopter Fuselage Models." NASA TN D-1363, July 1962.

Sweet, G. E., and Jenkins, J. L., Jr. "Results of Wind Tunnel Measurements on a Helicopter Rotor Operating at Extreme Thrust Coefficients and High Tip-Speed Ratios." JAHS, 8:3 (July 1963).

Sweet, G. E.; Jenkins, J. L., Jr.; and Winston, M. M. "Wind-Tunnel Measurements on a Lifting Rotor at High Thrust Coefficients and High Tip-Speed Ratios." NASA TN D-2462, September 1964.

Szustak, L. S., and Jenney, D. S. "Control of Large Crane Helicopters." JAHS, 16:3 (July 1971).

Takasawa, K. "On the Pitch Damping Moment in Hovering of a Rigid Helicopter Rotor." NASA TT F-15010, September 1973.

Talbot, P. D., and Corliss, L. D. "A Mathematical Force and Moment Model of a UH-1H Helicopter for Flight Dynamics Simulations." NASA TM 73254, June 1977.

Talbot, P. D., and Schroers, L. G. A Simple Method for Estimating Minimum Autorotative Descent Rate of Single Rotor Helicopters." NASA TM 78452,, March 1978.

Talkin, H. W. "Charts for Helicopter Performance Estimation." NACA ACR L5E04, August 1945.

Talkin, H. W. "Charts Showing Relations among Primary Aerodynamic Variables for Helicopter-Performance Estimation." NACA TN 1192, February 1947.

Tan, H. S. "On Laminar Boundary Layer over a Rotating Blade." JAS, 20:11 (November 1953).

Tangler, J. L. "The Design and Testing of a Tip to Reduce Blade Slap." AHS Forum, 1975.

Tangler, J. L. "Schlieren and Noise Studies of Rotors in Forward Flight." AHS Forum, 1977.

Tangler, J. L., Wohlfeld, R. W. and Miley, S. J. "An Experimental Investigation of Vortex Stability, Tip Shapes, and Noise for Hovering Model Rotors." NASA CR 2305, September 1973.

Tanna, H. K. "Computer Program for the Prediction of Rotational Noise Due to Fluctuating Loading on Rotor Blades." Institute of Sound and Vibration Research, ISVR TR 13, December 1968.

Tanna, H. K. "Helicopter Rotor Noise Final Report, Part I: Theoretical Investigation of Rotational Noise." NASA CR 66870, March 1970.

Tanna, H. K. "Theoretical Study of High-Frequency Helicopter Rotor Rotational Noise." Institute of Sound and Vibration Research, ISVR CR 70/3, September 1970.

Tanner, W. H. "Generalized Rotor Performance Method." Sikorsky Aircraft, Report SER-50309, May 1964.

Tanner, W. H. "Charts for Estimating Rotary Wing Performance in Hover and at High Forward Speeds." NASA CR 114, November 1964.

Tanner, W. H. "Tables for Estimating Rotary Wing Performance at High Forward Speeds." NASA CR 115, November 1964.

Tanner, W. H. "Rotary Wing Boundary Layer and Related Researches." AGARD Conference Proceedings No. 22, Gottingen, Germany, September 1967.

Tanner, W. H., and Bergquist, R. R. "Some Problems of Design and Operation of a 250-knot Compound Helicopter Rotor." Journal of Aircraft 1:5 (September-October 1964).

Tanner, W. H., and Buettiker, P. "The Boundary Layer of the Hovering Rotor." CAL/AVLABS Symposium, Buffalo, New York, June 1966.

Tanner, W. H., and Van Wyckhouse, J. F. "Wind Tunnel Tests of Full-Scale Rotors Operating at High Advancing Tip Mach Numbers and Advance Ratios." USAAVLABS TR 68-44, July 1968.

Tanner, W. H.; Van Wyckhouse, J. F.; Cancro, P.; and McCloud, J. L. III. "The Helicopter Rotor at High Mach Numbers." JAHS, 13:2 (April 1968).

Tanner, W. H., and Wohlfeld, R. M. "New Experimental Techniques in Rotorcraft Aerodynamics and Their Application." JAHS, 15:2 (April 1970).

Tanner, W. H., and Yaggy, P. F. "Experimental Boundary Layer Study on Hovering Rotors." JAHS, 11:3 (July 1966).

Tapscott, R. J. "Some Static Longitudinal Stability Characteristics of an Overlapped-Type Tandem-Rotor Helicopter at Low Airspeeds." NACA TN 4393, September 1958.

Tapscott, R. J. "Helicopters and VTOL Aircraft – Criteria for Control and Response Characteristics in Hovering and Low-Speed Flight." Aerospace Engineering, 19:6 (June 1960).

Tapscott, R. J. "Review of Helicopter Handling-Qualities Criteria and Summary of Recent Flight Handling-Qualities Studies." AHS Forum, 1964.

Tapscott, R. J., and Amer, K. B. "Studies of the Speed Stability of a Tandem Helicopter in Forward Flight." NACA Report 1260, 1956.

Tapscott, R. J., and Gessow, A. "Charts for Estimating Rotor-Blade Flapping Motion of High-Performance Helicopters." NACA TN 3616, March 1956.

Tapscott, R. J., and Gustafson, F. B. "Helicopter Blade Flapping." Aeronautical Engineering Review, 14:9 (September 1955).

Tapscott, R. J., and Sommer, R. W. "A Flight Study with a Large Helicopter Showing Trends of Lateral and Longitudinal Control Response with Size." NASA TN D-3600, September 1966.

Tarzanin, F. J., Jr. "Prediction of Control Loads Due to Blade Stall." JAHS, 17:2 (April 1972).

Tarzanin, F. J., Jr., and Mirick, P. H. "Control Envelope Shaping by Live Twist." NASA SP-352, February 1974.

Tarzanin, F. J., Jr., and Ranieri, J. "Investigation of the Effect of Torsional Natural Frequency on Stall-Induced Dynamic Loads." USAAMRDL TR 73-94, February 1974.

Taylor, J. L. "Helicopter Design." Aircraft Engineering, 14:155 (January 1942).

Taylor, J. L. "Natural Vibration Frequencies of Flexible Rotor Blades." Aircraft Engineering, 30:357 (November 1958).

Taylor, M. K. "A Balsa-Dust Technique for Air-Flow Visualization and Its Application to Flow through Model Helicopter Rotors in Static Thrust." NACA TN 2220, November 1950.

Taylor, R.; Fries, J.; and MacDonald, H. I., Jr. "Reduction of Helicopter Control System Loads with Fixed System Damping." AHS Forum, 1973.

Taylor, R. B., and Gabel, R. "Prediction of Helicopter Control Load Structural Limits." AHS Forum, 1975.

Taylor, R. B., and Teare, P. A. "Helicopter Vibration Reduction with Pendulum Absorbers." JAHS, 20:3 (July 1975).

Teleki, A. "High Speed Flight Tests with the Bo.105." The Aeronautical Journal, 79:778 (October 1975).

Tencer, B., and Cosgrove, J. P. "Heavy Lift Helicopter Program: An Advanced Technology Solution to Transportation Problems." Journal of Aircraft, 9:11 (November 1972).

Theodorsen, T. "The Theory of Propellers." NACA Report 775, Report 776, Report 777, and Report 778, 1944.

Theodorsen, T. "Theoery of Static Propellers and Helicopter Rotors." AHS Forum, 1969.

Thibert, J. J., and Gallot, J. "A New Airfoil Family for Rotor Blades." European Rotorcraft and Powered Lift Aircraft Forum, Aix-en-Provence, France, September 1977.

Thompson, A. R.; Kulkarni, S. B.; and Lee, D. H. "Design Criteria for Elastomeric Bearings." USAAMRDL TR 75-39, March 1976.

Tiller, F. E., Jr., and Nicholson, R. "Stability and Control Considerations for a Tilt-Fold-Proprotor Aircraft." JAHS, 16:3 (July 1971).

Timm, C. K. "Obstacle-Induced Flow Recirculation." JAHS, 10:4 (October 1965).

Timman, R., and van de Vorren, A. I. "Flutter of a Helicopter Rotor Rotating in Its Own Wake." JAS, 24:9 (September 1957).

Toler, J. R., et al. "Simulation of Helicopter and V/STOL Aircraft." U.S. Naval Training Device Center, NAVTRADEVCEN 1205, December 1964.

Tomaine, R. L. "Flight Data Identification of Six Degree-of-Freedom Stability and Control Derivatives of a Large Crane Type Helicopter." NASA TM X-73958, September 1976.

Toms, C. F. "The Performance of Rotating-Wing Aircraft Rotors." Aircraft Engineering, 19:219 (May 1947).

Tong, P. "Nonlinear Instability of Rotor Blades in Flap-Lag Using Hohenemser's Equations." Massachusetts Institute of Technology, ASRL TR 166-1, May 1971.

Tong, P. "The Nonlinear Instability in Flap-Lag of Rotor Blades in Forward Flight." NASA CR 114524, October 1971.

Tong, P. "Nonlinear Instability of a Helicopter Blade." AIAA Paper No. 72-956, September 1972.

Tong, P. "Nonlinear Instability of a Helicopter Blade in Hovering." AIAA Journal, 12:3 (March 1974).

Torres, M. "A Wing on the SA.341 Gazelle Helicopter and Its Effects." Vertica, 1:1 (1976).

Toussaint, A. "Drag or Negative Traction of Geared-Down Supporting Propellers in the Downward Vertical Glide of a Helicopter." NACA TN 21, September 1920.

Tran, C. T., and Renaud, J. "Theoretical Predictions of Aerodynamic and Dynamic Phenomena on Helicopter Rotors in Forward Flight." Vertica, 2:2 (1978).

Tran, C. T.; Twomey, W.; and Dat, R. "On the Use of Branch Modes for the Calculation of Helicopter Structural Dynamic Characteristics." NASA TT F-15713, July 1974.

Trenka, A. R. "The Aerodynamic and Aeroelastic Characteristics of a Full-Scale Rotor Operating at Very High Advance Ratios and during Start/Stop Operation." USAAVLABS TR 70-60, May 1971.

Trenka, A. R. "A Theoretical Study of the Application of Jet Flap Circulation Control for Reduction of Rotor Vibratory Forces – Addendum." NASA CR 137779, October 1975.

Trenka, A. R., and White, R. P., Jr. "Theoretical Investigation of the Flutter Characteristics of a Jet-Flap Helicopter Rotor in Hovering and Forward Flight." TRECOM TR 63-9, April 1963.

True, H. C., and Rickley, E. J. "Noise Characteristics of Eight Helicopters." Federal Aviation Administration, Report No. FAA-RD-77-94, July 1977.

True, H. C., and Letty, R. M. "Helicopter Noise Measurements Data Report." Federal Aviation Administration, Report No. FAA-RD-77-57, April 1977.

Trueblood, R. B., et al. "Advanced Flight Control System Concepts for VTOL Aircraft—Phase I." TRECOM TR 64-50, October 1964.

Trueblood, R. B., et al. "Advanced Flight Control System Concepts for VTOL Aircraft—Phase II." USAAVLABS TR 66-74, May 1967.

Trueblood, R. B.; Bryant, W. B.; and Cattel, J. J. "Advanced Flight Control System Concepts for VTOL Aircraft – Phase III." USAAVLABS TR 69-95, July 1970.

Tuck, D. A. "IFR Airworthiness Standards for VTOL Aircraft." JAHS, 16:4 (October 1971).

Tung, C., and Du Waldt, F. A. "Analysis of Measured Helicopter Rotor Pressure Distributions." USAAVLABS TR 70-47, September 1970.

Tung, C.; Erickson, J. C., Jr.; and Du Waldt, F. A. "The Feasibility and Use of Anti-Torque Surfaces Immersed in Helicopter Rotor Downwash." Cornell Aeronautical Laboratory, Report CAL BB-2584-S-2, February 1970.

Turner, M. J., and Duke, J. B. "Propeller Flutter." JAS, 16:6 (June 1949).

Twelvetrees, W. N. "The Evolution of the Rotor Blade." Aircraft Engineering, 41:7 (July 1969).

Twomey, W. J., and Ham, E. H. "Review of Engine/Airframe/Drive Train

Dynamic Interface Development Problems." USARTL TR 78-13, June 1978.

Unger, G. "Nary/Marine 1980 Rotary Wing Candidates." AHS Forum, 1975.

United States Defense Documentation Center. "A DDC Bibliography on Helicopter Engines and Rotors." DDC-TAS-68-57, November 1968.

United States Defense Documentation Center. "Heavy Lift Helicopters – A DDC Bibliography." DDC-TAS-72-16-1, April 1972.

United States Defense Documentation Center. "DDC Bibliography – Helicopter Engines and Rotors." DDC-TAS-73-48, July 1973.

United States Department of Defense. "Military Specification MIL-H-8501A: Helicopter Flying and Ground Handling Qualities Requirements." September 1961.

United States Department of Defense, U.S. Air Force. "Military Specification MIL-F-83300: Flying Qualities of Piloted V/STOL Aircraft." December 1970.

United States Federal Aviation Administration. "Federal Aviation Regulations: FAR Part 27 – Airworthiness Standards, Normal Category Rotorcraft. FAR Part 29 – Airworthiness Standards, Transport Category Rotorcraft."

Vague, M., and Seibel, C. M. "Helicopter Stabilization and Handling Characteristics Improvement by Mechanical Means." JAHS, 4:4 (October 1959).

van der Harten, R. J. "Minimizing N-per-rev Vibrations." Vertiflite, 22:6 (November-December 1976).

van de Vooren, A. I., and Zandbergen, J. J. "Noise Field of a Rotating Propeller in Forward Flight." AIAA Journal, 1:7 (July 1963).

Van Gaasbeek, J. R. "An Investigation of High-G Maneuvers of the AH-1G Helicopter." USAAMRDL TR 75-18, April 1975.

Van Gaasbeek, J. R., and Austin, E. E. "Digital Simulation of the Operational Loads Survey Flight Tests." AHS Forum, 1978.

van Holten, T. "The Computation of Aerodynamic Loads on Helicopter Blades in Forward Flight, Using the Method of the Acceleration Potential." Technishe Hogeschool Delft (Netherlands), Report VTH-189, March 1975; also International Council of the Aeronautical Sciences, ICAS Paper No. 74-54, August 1974.

van Holten, T. "On the Validity of Lifting Line Concepts in Rotor Analysis." Vertica, 1:3 (1977).

Vann, W. D.; Mirick, P. H.; and Austin, E. E. "Use of Computer Math Models for Aircraft Evaluation." USAAMRDL TN 12, August 1973.

Van Wyckhouse, J. F. "High-Performance UH-1 Compound Helicopter

Maneuver Flight Test Program." USAAVLABS TR 66-17, February 1966.

Van Wyckhouse, J. F., and Cresap, W. L. "High Performance Helicopter Program." TRECOM TR 63-42, September 1963; TRECOM TR 64-61, October 1964.

Vause, C. R.; Schmitz, F. H.; and Boxwell, D. A. "High-Speed Helicopter Impulsive Noise." AHS Forum, 1976.

Vehlow, C. A. "Experimental Investigation of Gust Response of Hingeless Helicopter Rotors." M.S. thesis, Massachusetts Institute of Technology, June 1977.

Velazquez, J. L. "Advanced Anti-Torque Concepts Study." USAAMRDL TR 71-44, August 1971.

Velkoff, H. R. "A Preliminary Study of the Effect of a Radial Pressure Gradient on the Boundary Layer of a Rotor Blade." CAL/AVLABS Symposium, Buffalo, New York, June 1966.

Velkoff, H. R.; Blaser, D. A.; and Jones, K. M. "Boundary-Layer Discontinuity on a Helicopter Rotor Blade in Hovering." Journal of Aircraft, 8:2 (February 1971).

Velkoff, H. R.; Blaser, D. A.; Shumaker, G. C.; and Jones, K. M. "Exploratory Study of the Nature of Helicopter Rotor Blade Boundary Layers." USAAVLABS TR 69-50, July 1969.

Velkoff, H. R.; Hoffman, J. D.; and Blaser, D. A. "Investigation of Boundary Layers and Tip Flows of Helicopter Rotor Blades." USAAMRDL TR 71-73, May 1972.

Virchis, V. J., and Wright, S. E. "Radiation Characteristics of Acoustic Sources in Circular Motion." J. Sound Vib., 49:1 (1976).

Vinje, E. W. "Flight Simulator Evaluation of Control Moment Usage and Requirements for V/STOL Aircraft." JAHS, 19:2 (April 1974).

von Hardenberg, P. W., and Saltanis, P. B. "Ground Test Evaluation of the Sikorsky Active Transmission Isolation System." USAAMRDL TR 71-38, September 1971.

von Kármán, T. Aerodynamics. New York: McGraw-Hill Book Co., Inc., (1954).

von Kerczek, C., and David, S. H. "Calculation of Transition Matrices." AIAA Journal, 13:10 (October 1975).

Wachs, M. A., and Rabbott, J. P., Jr. "Rotary Wing Aircraft Design Trends." JAHS, 9:2 (April 1964).

Wadsworth, M., and Wilde, E. "Differential Eigenvalue Problems with Particular Reference to Rotor Blade Bending." The Aeronautical Quarterly, 19:2 (May 1968).

Wagner, R. A. "Noise Levels of Operational Helicopters of the OH-6 Type Designed to Meet the LOH Mission." NASA CR 114760, 1973.

Wald, Q. "A Method for Rapid Estimation of Helicopter Performance." JAS, 10:4 (April 1943).

Walters, R. E., and Skujins, O. "A Schlieren Technique Applied to Rotor Wake Studies." American Helicopter Society Mideast Region Symposium, Philadelphia, Pennsylvania, October 1972.

Walton, R. P., and Ashkenas, I. L. "Analytical Review of Military Helicopter Flying Qualities." Systems Technology Inc., August 1967.

Walton, R. P., and Stapleford, R. L. "An Analytical Study of Helicopter Handling Qualities in Hover." AHS Forum, 1966.

Wan, F. Y. M., and Lakshmikantham, C. "Rotor Blade Response to Random Loads: A Direct Time-Domain Approach." AIAA Journal, 11:1 (January 1973).

Wan, F. Y. M., and Lakshmikanthan, C. "Spacial Correlation Method and a Time-Varying Flexible Structure." AIAA Journal, 12:5 (May 1974).

Ward, J. F. "A Summary of Hingeless-Rotor Structural Loads and Dynamics Research." J. Sound Vib., 4:3 (1966).

Ward, J. F. "Exploratory Flight Investigation and Analysis of Structural Loads Encountered by a Helicopter Hingeless Rotor System." NASA TN D-3676, November 1966.

Ward, J. F. "The Dynamic Response of a Flexible Rotor Blade to a Concentrated Force Moving from Tip to Root." NASA TN D-5410, September 1969.

Ward, J. F. "Helicopter Rotor Periodic Differential Pressures and Structural Response Measured in Transient and Steady-State Maneuvers." JAHS, 16:1 (January 1971).

Ward, J. F., and Huston, R. J. "A Summary of Hingeless-Rotor Research at NASA-Langley." AHS Forum, 1964.

Ward, J. F., and Snyder, W. J. "The Dynamic Response of a Flexible Rotor Blade to a Tip-Vortex Induced Moving Force." AIAA Paper No. 69-203, February 1969.

Warmbrodt, W. G. "Aeroelastic Response and Stability of a Coupled Rotor/Support System with Application to Large Horizontal Axis Wind Turbines." Ph.D. thesis, University of California, Los Angeles, 1978.

Warming, T. "Some New Conclusions about Helicopter Mechanical Instability." JAHS, 1:3 (July 1956).

Warner, E. P. "The Problem of the Helicopter." NACA TN 4, May 1920.

Warsi, Z. U. A. "One Parameter Solution of the Spanwise Rotating Blade Boundary-Layer Equation." AIAA Journal, 12:10 (October 1974).

Washizu, K.; Azuma, A.; Koo, J.; and Oka, T. "Experimental Study on the Unsteady Aerodynamics of a Tandem Rotor Operating in the Vortex Ring State." AHS Forum, 1966.

Washizu, K.; Azuma, A.; Koo, J.; and Oka, T. "Experiments on a Model

Helicopter Rotor Operating in the Vortex Ring State." Journal of Aircraft, 3:3 (May-June 1966).

Watkins, C. E., and Durling, B. J. "A Method for Calculation of Free-Space Sound Pressures Near a Propeller in Flight Including Considerations of the Chordwise Blade Loading." NACA TN 3809, November 1956.

Watkins, T. C.; Sinacori, J. B.; and Kesler, D. F. "Stabilization of Externally Slung Helicopter Loads." USAAMRDL TR 74-42, August 1974.

Watts, G. A., and Biggers, J. C. "Hingeless Rotor Vibration and Loads at High Advance Ratio." AHS Forum, 1972.

Watts, G. A., and London, R. J. "Vibration and Loads in Hingeless Rotors." NASA CR 114562 and CR 114568, September 1972.

Watts, G. A.; London, R. J.; and Snoddy, R. J. "Trim, Control, and Stability of a Gyro-Stabilized Hingesless Rotor at High Advance Ratio and Low Rotor Speed." NASA CR 114362, May 1971.

Wax, C. M., and Tocci, R. C. "Study of the Heavy-Lift Helicopter Rotor Configuration." USAAVLABS TR 66-61, November 1966.

Webber, D. A. "Comparison of Helicopter and Airplane Vertical Accelerations in Turbulence." ARC CP 878, August 1965.

Wei, F.-S., and Peters, D. A. "Lag Damping in Autorotation by a Perturbation Method." AHS Forum, 1978.

Weick, F. E. "Propeller Design." NACA TN 235, TN 236, and TN 237, May 1926.

Weiland, E. F. "Development and Test of the Bo. 105 Rigid Rotor Helicopter." JAHS, 14:1 (January 1969).

Weisbrich, A.; Perley, R.; and Howes, H. "Design Study of a Feedback Control System for the Multicyclic Flap System Rotor." NASA CR 151960, January 1977.

Weiss, H. "Vibration Treatment of Bo.105 Helicopter." European Rotorcraft and Powered Lift Aircraft Forum, Southampton, England, September 1975.

Welch, A. J., and Warren, E. L. "Analysis and Design Study of a Pilot Assist System for Helicopters." USAAVLABS TR 71-11, April 1971.

Welch, A. J., and Warren, E. L. "Comparison of UH-1C Flight Test Data with MOSTAB-C Small Perturbation Math Model." USAAMRDL TR 71-66, December 1971.

Weller, W. H. "Load and Stability Measurements on a Soft-Inplane Rotor System Incorporating Elastomeric Lead-Lag Dampers." NASA TN D-8437, July 1977.

Weller, W. H., and Lee, B. L. "Wind-Tunnel Tests of Wide-Chord Teetering Rotors with and without Outboard Flapping Hinges." NASA TP 1046, November 1977.

Weller, W. H., and Mineck, R. E. "An Improved Computational Procedure for Determining Helicopter Rotor Blade Natural Modes." NASA TM 78670, August 1978.

Wells, C. D., and Wood, T. L. "Maneuverability – Theory and Application." JAHS, 18:1 (January 1973).

Wernicke, K. G. "Helicopter Longitudinal Stability in Forward Flight with Control Feedback and Fuselage Aerodynamics." JAHS, 2:1 (January 1957).

Wernicke, K. G. "Tilt Proprotor Composite Research Aircraft." USAAVLABS TR 68-32, November 1968.

Wernicke, K. G. "Tilt Proprotor Composite Aircraft, Design State of the Art." JAHS, 14:2 (April 1969).

Wernicke, K. G., and Edenborough, H. K. "Full Scale Proprotor Development." JAHS, 17:1 (January 1972).

Wernicke, R. K., and Drees, J. M. "Second Harmonic Control." AHS Forum, 1963.

Whatmore, A. R., and Lowson, M. V. "Some Effects of Ground and Side Planes on the Acoustic Output of a Rotor." NASA CR 132306, July 1973.

Wheatley, J. B. "Lift and Drag Characteristics and Gliding Performance of an Autogiro As Determined in Flight." NACA Report 434, 1932.

Wheatley, J. B. "Simplified Aerodynamic Analysis of the Cyclogiro Rotating Wing System." NACA TN 467, August, 1933.

Wheatley, J. B. "The Aerodynamic Analysis of the Gyroplane Rotating-Wing System." NACA TN 492, March 1934.

Wheatley, J. B. "Choice of Airfoils for Rotating-Wing Aircraft." JAS, 1:2 (April 1934).

Wheatley, J. B. "Wing Pressure Distribution and Rotor-Blade Motion of an Autogiro As Determined in Flight." NACA Report 475, 1934.

Wheatley, J. B. "An Aerodynamic Analysis of the Autogiro Rotor with a Comparison between Calculated and Experimental Results." NACA Report 487, 1934.

Wheatley, J. B. "The Influence of Wing Setting on the Wing Load and Rotor Speed of a PCA-2 Autogiro As Determined in Flight." NACA Report 523, 1935.

Wheatley, J. B. "A Study of Autogiro Rotor Blade Oscillations in the Plane of the Rotor Disk." NACA TN 581, September 1936.

Wheatley, J. B. "An Analytical and Experimental Study of the Effect of Periodic Blade Twist on the Thrust, Torque, and Flapping Motion of an Autogiro Rotor." NACA Report 591, 1937.

Wheatley, J. B. "An Analysis of the Factors That Determine the Periodic Twist of an Autogiro Rotor Blade, with a Comparison of Predicted and Measured Results." NACA Report 600, 1937.

Wheatley, J. B., and Bioletti, C. "Wind-Tunnel Tests of a 10-foot-Diameter Gyroplane Rotor." NACA Report 536, 1935.

Wheatley, J. B., and Bioletti, C. "Analysis and Model Tests of Autogiro Jump Take-Off." NACA TN 582, October 1936.

Wheatley, J. B., and Bioletti, C. "Wind-Tunnel Tests of 10-foot-Diameter Autogiro Rotors." NACA Report 552, 1936.

Wheatley, J. B., and Hood, M. J. "Full-Scale Wind-Tunnel Tests of a PCA-2 Autogiro Rotor." NACA Report 515, 1935.

Wheatley, J. B., and Windler, R. "Wind-Tunnel Tests of a Cyclogiro Rotor." NACA TN 528, May 1935.

Whitaker, H. P., and Cheng, Y. "Use of Active Control Systems to Improve Bending and Rotor Flapping Responses of a Tilt Rotor VTOL Airplane." NASA CR 137815, October 1975.

White, R. P., Jr. "VTOL Periodic Aerodynamic Loadings: The Problems, What Is Being Done, and What Needs To Be Done." J. Sound Vib., 4:3 (1966).

White, R. P., Jr. "The Aerodynamic and Aeroelastic Characteristics of a Full-Scale Rotor Blade Stopped in Flight." USAAVLABS TR 69-7, April 1969.

White, R. P., Jr. "Instabilities Associated with a Rotor Blade Stopped in Flight." JAHS, 14:2 (April 1969).

White, R. P., Jr. "An Investigation of the Vibratory and Acoustic Benefits Obtainable by the Elimination of the Blade Tip Vortex." JAHS, 18:4 (October 1973).

White, R. P., Jr. "V/STOL Rotor and Propeller Noise – Its prediction and Analysis of Its Aural Characteristics." AIAA Paper 75-452, March 1975.

White, R. P., Jr. "Wind Tunnel Tests of a Two Bladed Model Rotor to Evaluate the TAMI System in Descending Forward Flight." NASA CR 145195, May 1977.

White, R. P., Jr., and Balcerak, J. C. "The Nemesis of the Trailed Tip Vortex – Is It Now Conquered?" AHS Forum, 1972.

White, R. P., Jr., and Balcerak, J. C. "Investigation of the Dissipation of the Tip Vortex of a Rotor Blade by Mass Injection." USAAMRDL TR 72-43, August 1972.

White, R. P., Jr., and Crimi, P. "Theoretical Investigation of the Flutter Characteristics of a Jet-Flap Rotor System in Hovering Flight." TCREC TR 61-142, November 1961.

White, R. P., Jr., and Du Waldt, F. A. "Helicopter Vibration – A Major Source, Its Prediction, and an Approach to Its Control." Shock and Vibration Bulletin 37, Part 6, January 1968.

White, R. P., Jr.; Sutton, L. R.; and Nettles, W. E., "Examination of the Air Resonance Stability Characteristics of a Bearingless Main Rotor." AHS Forum, 1978.

White, R. W. "Investigation of Helicopter Airframe Normal Modes." Vertica, 1:2 (1976).

White, W. F., Jr. "Importance of Helicopter Dynamics to the Mathematical Model of the Helicopter." AGARD Conference Proceedings No. 172, Hampton, Virginia, November 1974.

White, W. F., Jr., and Malatino, R. E. "A Numerical Method for Determining the Natural Vibration Characteristics of Rotating Nonuniform Cantilever Blades." NASA TM X-72751, October 1975.

Whitfield, C. E. "An Investigation of Rotor Noise Generation by Aerodynamic Disturbance." NASA CR 157571, September 1977.

Whitten, J. B.; Reeder, J. P.; and Crim, A. D. "Helicopter Instrument Flight and Precision Maneuvers as Affected by Changes in Damping in Roll, Pitch, and Yaw." NACA TN 3537, November 1955.

Widdison, C. A.; Magee, J. P.; and Alexander, H. R. "Conceptual Design Study of a 1985 Commercial STOL Tilt Rotor Transport." NASA CR 137601, November 1974.

Widnall, S. E. "A Correlation of Vortex Noise Data from Helicopter Main Rotors." Journal of Aircraft, 6:3 (May-June 1969).

Widnall, S. E. "Helicopter Noise due to Blade-Vortex Interaction." Journal of the Acoustical Society of America, 50:1 (July 1971).

Widnall, S. E. "The Stability of a Helical Vortex Filament." Journal of Fluid Mechanics, 54:4 (August 1972).

Widnall, S. E.; Harris, W. L.; Lee, A.; and Drees, H. M. "The Development of Experimental Techniques for the Study of Helicopter Rotor Noise." NASA CR 137684, 1974.

Wiesner, W., and Kohler, G. "Tail Rotor Design Guide." USAAMRDL TR 73-99, January 1974.

Wiesner, W., and Kohler, G. "Tail Rotor Performance in Presence of Main Rotor, Ground, and Winds." JAHS, 19:3 (July 1974).

Wilby, P. G. "Effect of Production Modifications to Rear of Westland Lynx Rotor Blade on Sectional Aerodynamic Characteristics." ARC CP 1362, August 1973.

Wilby, P. G.; Gregory, N.; and Quincey, V. G. "Aerodynamic Characteristics of NPL 9626 and NPL 9627, Further Aerofoils Designed for Helicopter Rotor Use." ARC CP No. 1262, November 1969.

Wilde, E.; Bramwell, A. R. S.; and Summerscales, R. "The Flapping Behavior of a Helicopter Rotor at High Tip-Speed Ratios." ARC CP 877, April 1965.

Wilde, E., and Price, H. L. "A Theoretical Method for Calculating Flexure and Stresses in Helicopter Rotor Blades." Aircraft Engineering, 37:2 (February 1965).

Wilford, E. B. "Performance and Control of Rotary-Wing Aircraft." JAS,

5:7 (May 1938).

Wilkerson, J. B. "NATFREQ – A Computer Program for Calculating the Natural Frequency of Rotating Cantilevered Beams." David W. Taylor Naval Ship Research and Development Center, Report ASED 370, January 1977.

Wilkerson, J. B. "An Assessment of Circulation Control Airfoil Development." David Taylor Naval Ship Research and Development Center, Report 77-0084, August 1977.

Wilkerson, J. B. "Static Stability Derivatives in Pitch and Roll of a Model Circulation Control Rotor." David W. Taylor Naval Ship Research and Development Center, Report 77-0066, October 1977.

Wilkerson, J. B., and Linck, D. W. "A Model Rotor Performance Validation for the CCR Technology Demonstrator." AHS Forum, 1975.

Williams, J. C. III, and Young, W. H., Jr. "The Laminar Boundary Layer on a Rotating Blade of Symmetrical Airfoil Shape." USAAVLABS TR 70-64, December 1970.

Williams, J. L. "Directional Stability Characteristics of Two Types of Tandem Helicopter Fuselage Models." NACA TN 3201, May 1954.

Williams, J. L. "Wind-Tunnel Investigation of Effects of Spoiler Location, Spoiler Size, and Fuselage Nose Shape on Directional Characteristics of a Model of a Tandem-Rotor Helicopter Fuselage." NACA TN 4305, July 1958.

Williams, R. M. "Application of Circulation Control Rotor Technology to a Stopped Rotor Aircraft Design." Vertica, 1:1 (1976).

Williams, R. M., and Montana, P. S. "A Comprehensive Plan for Helicopter Drag Reduction." AHS Forum, 1975. (Rotorcraft Parasite Drag Supplement.)

Willmer, M. A. P. "The Loading of Helicopter Rotor Blades in Forward Flight." ARC R&M 3318, April 1959.

Wilson, J. C., and Mineck, R. E. "Wind-Tunnel Investigation of Helicopter-Rotor Wake Effects on Three Helicopter Fuselage Models." NASA TM X-3185 (and supplement), March 1975.

Wilson, J. C.; Mineck, R. E.; and Freeman, C. E. "Aerodynamic Characteristics of a Powered Tilt-Proprotor Wind-Tunnel Model." NASA TM X-72818 (and supplement), March 1976.

Wimperis, H. E. "The Rotating Wing in Aircraft." ARC R&M 1108, August 1926.

Windsor, R. I. "Measurement of Aerodynamic Forces on an Oscillating Airfoil, USAAVLABS TR 69-98, March 1970.

Winson, J. "Motion of an Unarticulated Helicopter Blade." JAS, 14:9 (September 1947).

Winson, J. "The Testing of Rotors for Fatigue Life." JAS, 15:7 (July 1948).

Winston, M. M. "An Investigation of Extremely Flexible Lifting Rotors." NASA TN D-4465, April 1968.

Wolf, T. L., and Widnall, S. E. "The Effect of Tip Vortex Structure on Helicopter Noise Due to Blade/Vortex Interaction." NASA CR 152150, March 1978.

Wolkovitch, J. "Analytical Prediction of Vortex-Ring Boundaries for Helicopters in Steep Descents." JAHS, 17:3 (July 1972).

Wolkovitch, J., and Hoffman, J. A. "Stability and Control of Helicopters in Steep Approaches." USAAVLABS TR 70-74, May 1971.

Wong, P. K. "Far-Field Wake Structure Due to Wing-Vortex Interaction." Massachusetts Institute of Technology, ASRL TR 178-2, March 1975.

Wood, E. R., and Buffalano, A. C. "Parametric Investigation of the Aerodynamic and Aeroelastic Characteristics of Articulated and Rigid (Hingeless) Helicopter Rotor Systems." TRECOM TR 64-15, April 1964.

Wood, E. R., and Hilzinger, K. D. "A Method for Determining the Fully Coupled Aeroelastic Response of Helicopter Rotor Blades." AHS Forum, 1963.

Wood, E. R.; Hilzinger, K. D.; and Buffalano, A. C. "An Aeroelastic Study of Helicopter Rotor Systems in High-Speed Flight." CAL/TRECOM Symposium, Buffalo, New York, June 1963.

Wood, K. D. Aerospace Vehicle Design, Vol. 2: Aircraft Design. 3rd ed. Boulder, Cororado: Johnson Publishing Co., 1968.

Wood, T. L. "High Energy Rotor System." AHS Forum, 1976.

Wood, T. L.; Ford, D. G.; and Brigman, G. H. "Maneuver Criteria Evaluation Program." USAAMRDL TR 74-32, May 1974.

Woodley, J. G. "The Influence of Near Wake Assumptions on the Lifting Characteristics of a Rotor Blade." Royal Aircraft Establishment, Tech. Report 71046, March 1971.

Wortmann, F. X., and Drees, J. M. "Design of Airfoils for Rotors." CAL/AVLABS Symposium, Buffalo, New York, June 1969.

Wright, S. E. "Sound Radiation from a Lifting Rotor Generated by Asymmetric Disc Loading." Institute of Sound and Vibration Research, ISVR TR 5, April 1968.

Wright, S. E. "Theoretical Study of Rotational Noise." Institute of Sound and Vibration Research, ISVR TR 14, May 1969.

Wright, S. E. "Sound Radiation from a Lifting Rotor Generated by Asymmetric Disk Loading." J. Sound Vib., 9:2 (1969).

Wright, S. E., and Leverton, J. W. "Helicopter Rotor Noise Generation." CAL/AVLABS Symposium, Buffalo, New York, June 1969.

Wright, S. E., and Tanna, H. K. "A Computational Study of Rotational Noise." Institute of Sound and Vibration Research, ISVR TR 15, May 1969.

Wu, J. C., and Sigman, R. K. "Optimum Performance and Potential Flow Field of Hovering Rotors." NASA CR 137705, May 1975.

Wu, J. C.; Sigman, R. K.; and Goorjian, P. M. "Optimum Performance of Hovering Rotors." NASA TM X-62138, March 1972.

Wyrick, D. R. "Extension of the High-Speed Flight Envelope of the XH-51A Compound Helicopter." USAAVLABS TR 65-71, November 1965.

Yasue, M. "A Study of Gust Response for a Rotor-Propeller in Cruising Flight." NASA CR 137537, August 1974.

Yasue, M. "Gust Response and Its Alleviation for a Hingeless Helicopter Rotor in Cruising Flight." Massachusetts Institute of Technology, ASRL TR 189-1, January 1977.

Yasue, M.; Vehlow, C. A.; and Ham, N. D. "Gust Response and Its Alleviation for a Hingeless Helicopter Rotor in Cruising Flight." European Rotorcraft and Powered Lift Aircraft Forum, Stresa, Italy, September 1978.

Yatsunovich, M. S. "Aerodynamics of the Mi-4 Helicopter." Air Force Foreign Technology Division, FTD-MT-24-449-68, March 1969.

Yatsunovich, M. S. "Practical Aerodynamics of the Mi-6 Helicopter." U.S. Army Foreign Science and Technology Center, FSTC-HT-23-1172-70, November 1970.

Yeager, W. T., Jr., and Mantay, W. R. "Correlation of Full-Scale Helicopter Rotor Performance in Air with Model-Scale Freon Data." NASA TN D-8323, November 1976.

Yeager, W. T., Jr.; Young, W. H., Jr.; and Mantay, W. R. "A Wind Tunnel Investigation of Parameters Affecting Helicopter Directional Control at Low Speeds in Ground Effect." NASA TN D-7694, November 1974.

Yeates, J. E., Jr. "A Discussion of Helicopter Vibration Studies Including Flight Test and Analysis Methods Used to Determine the Coupled Response of a Tandem Type." JAHS, 1:3 (July 1956).

Yeates, J. E., Jr. "Flight Measurements of the Vibration Experienced by a Tandem Helicopter in Transition, Vortex Ring State, Landing Approach, and Yawed Flight." NACA TN 4409, September 1958.

Yeates, J. E., Jr.; Brooks, G. W.; and Houbolt, J. C. "Flight and Analytical Methods for Determining the Coupled Vibration Response of Tandem Helicopters." NACA Report 1326, 1957.

Yen, J. G.; Weber, G. E.; and Gaffey, T. M. "A Study of Folding Proprotor VTOL Aircraft Dynamics." AFFDL TR 71-7, September 1971.

Yntema, R. T. "Simplified Procedures and Charts for the Rapid Estimation of Bending Frequencies of Rotating Beams." NACA TN 3459, June 1955.

Young, C. "The Prediction of Helicopter Rotor Hover Performance Using a Prescribed Wake Analysis." ARC CP No. 1341, June 1974.

Young, C. "A Theoretical Study of the Effect of Blade Ice Accretion on the Power-Off Landing Capability of a Wessex Helicopter." Vertica, 2:1 (1978).

Young, H. R., and Simon, D. R. "The Advancing Blade Concept (ABC) Rotor Program." AGARD Conference Proceedings No. 233, Moffett Field, California, May 1977.

Young, M. I. "A Simplified Theory of Hingeless Rotors with Application to Tandem Helicopters." AHS Forum, 1962.

Young, M. I. "On the Kinematics and Dynamics of Large Amplitude Forced Lead-Lag Rotor Systems." JAHS, 7:3 (July 1962).

Young, M. I. "A Theory of Rotor Blade Motion Stability in Powered Flight." JAHS, 9:3 (July 1964).

Young, M. I. "An Engineering Approximation for the Aerodynamic Loading of Helicopter Rotor Blades." J. Sound Vib., 4:3 (1966).

Young, M. I. "Coriolis Coupled Bending Vibrations of Hingeless Rotor Blades." Journal of Aircraft, 8:8 (August 1971).

Young, M. I. "Scale Effects in the Bending Vibrations of Helicopter Rotor Blades." J. Sound Vib., 21:1 (1972).

Young, M. I., and Bailey, D. J. "Stability and Control of Hingeless Rotor Helicopter Ground Resonance." Journal of Aircraft, 11:6 (June 1974).

Young, M. I., and Lytwyn, R. T. "The Influence of Blade Flapping Restraint of the Dynamic Stability of Low Disk Loading Propeller-Rotors." JAHS, 12:4 (October 1967).

Young, W. H., Jr., and Williams, J. C. "Boundary Layer Separation on Rotating Blades in Forward Flight." AIAA Journal, 10:12 (December 1972).

Yuan, S. W. "Bending of Rotor Blade in the Plane of Rotation." JAS, 14:5 (May 1947).

Yuan, S. W. "Jet Circulation Control Airfoil for VTOL Rotors." Journal of Aircraft, 7:5 (September-October 1970).

Yudin, E. Y. "On the Vortex Sound from Rotating Rods." NACA TM 1136, March 1947.

Zbrozek, J. "Ground Effect on the Lifting Rotor." ARC R&M 2347, July 1947.

Zbrozek, J. "Investigation of Lateral and Directional Behavior of Single-Rotor Helicopter (Hoverfly Mk I)." ARC R&M 2509, June 1948.

Zbrozek J. K. "Stability and Control of Single Rotor Helicopter with Hinged Blades." Aircraft Engineering, 21:240 (February 1949).

Zbrozek, J. K. "The Simple Harmonic Motion of a Helicopter Rotor with Hinged Blades." ARC R&M 2813, April 1949.

Zimmer, H. "The Rotor in Axial Flow." AGARD Conference Proceedings

No. 111, Marseilles, France, September 1972.

Zvara, J., and Ham, N. D. "Helicopter Rotor Model Research at Massachusetts Institute of Technology." JAHS, 5:1 (January 1960).

INDEX

actuator disk, 29–30
air resonance, 668, 693
airfoil, rotor, 332–37; drag, 54, 337–39
airframe structural dynamics, 708
airloads data, 276–77
articulated rotor, 8, 313
autogiro, 10–11, 15–16, 301–302
autorotation, 10, 325–30; forward flight, 132–33, 286, 295–96; vertical, 101, 105–14, 282. *See also* power-off landing

bending: in-plane, 397–99; out-of-plane, 384–91, 558–89; out-of-plane and in-plane, 399–403, 444–56
bending moments, 392–93, 450–51, 699–706, 708–709
blade element theory, 45–51, 167–71, 205, 469–71, 549–56; definition, 45–46; forward flight, 167–71, 255; history, 46–49; vertical flight, 49. *See also* lifting-line theory
body axes, 442, 589, 822–24
boundary layers, 755
broadband noise, 904, 909–15

ceiling, 298
coaxial rotor helicopter, 119, 143, 318, 770–71
Coleman diagram, 677–80, 688–89
collective pitch, 159, 193, 195, 769, 773
combined blade element and momentum theories, 56–57, 69–71, 86, 121
compound helicopter, 324, 341
comprehensive helicopter analyses, 757–60, 767
compressibility, 262–64, 322, 332–34, 697
constant coefficient approximation, 568–74, 583, 609–11, 626–27, 635
control, 769–74
cyclic pitch, 159, 193, 769, 773

delta-three, 238
dimensionless quantities, 20–21
divergence instability, 640–42
drag, blade profile, 54, 337–39; helicopter, 331; rotor, 288–89; vertical, 114–18
dynamic stall, 879–82, 888–99

eigenvalues and eigenvectors of rotor modes, 361–64, 615–17
endurance, 299–300
equivalence of flap and feather, 161–63
equivalent solidity, 63–64

fatigue life, 708
figure of merit, 34–36, 55, 290, 292, 332
flap dynamics, 602–37; constant coefficient approximation, 609–11, 626–27, 635; forward flight, 363–64, 605–11; hover, 603–605, 612–13, 615–22; low frequency response, 622–35; nonrotating frame, 615–22; roots, 363–64, 605–11, 615–17; stability, 609, 636–37; transfer function, 612–13, 617–22; two-bladed rotor, 632–36
flap-lag motion, 393–97, 653-68; aerodynamics, 560–64, 585–86; articulated rotor stability, 657–58; coupled bending, 444–56; equation, 394–95, 653–55; hingeless rotor stability, 658–63
flap motion, 6, 8, 23–24, 149–50, 156–57, 381–82; aerodynamics, 187, 256–59, 556–58, 585, 596–600; and lag, 393–97, 653–67; and pitch, 403–408, 421–22, 596–600, 637–52; dynamics, 602–37; equation, 187, 223, 230–31, 256–59, 382–83, 390–91, 602, 614; forward flight, 189–93, 196–97, 209, 224–26, 232, 240–42,

flap motion *(cont.)*
261, 270, 773–74; frequency, 186, 223, 236, 238, 240, 383; hover, 190, 212–13, 225, 241–42; higher harmonics, 210–13, 271; low frequency response, 622–28, 780–82; nonrotating frame, 390–91, 437–38, 564–74, 586, 589, shaft motion, 436–38, 441, 585–86; weight moment, 206; with hinge offset, 227–33, 382–84, 558; with hinge spring, 222–27; with nonuniform inflow, 209, 268, 273–74, 728; with pitch-flap coupling, 238–42; with reverse flow, 260–61
flight dynamics, 774–75; concave downward requirement, 834–38, 842–43, 863; coupled lateral-longitudinal dynamics, 810–13; equations, 775–79, 787–88, 808, 810–11, 815, 822–24, 828, 844; forward flight, 822–52; hingeless rotor, 800–802, 851–52; horizontal tail, 825–26, 831, 838; hover, 775–821; lateral dynamics, 808–10, 843–47; longitudinal dynamics, 787–808, 814–20, 827–43; loop closure, 794–800, 802; normal acceleration, 833–35; poles, 788–93, 802, 805–806, 809–10, 812–13, 817–20, 829–32, 840–41, 845–48; response to control, 803–804; short period, 804, 809–10, 831–38, 842; specifications, 862–69; stability augmentation, 854–62; static stability, 838–40; tandem helicopter, 813–21, 848–51; vertical dynamics, 782–84; yaw dynamics, 784–86, 820–21; zeros, 793, 802, 805
Floquet theory, 369–71, 606–607, 636–37
flow states in axial flight, 98–102, 107
flutter, 637–53; divergence, 640–42; equations, 638–39; instability, 642–46; shed wake influence, 646–48. *See also* pitch-flap motion
force equilibrium, lateral, 245–46; longitudinal, 132, 182, 243–45
Fourier coordinate transformation, 349–61, 571–74; eigenvalues and eigenvectors, 361–64; flap, 390–91, 437–38, 564–74, 586, 614–17; hub reactions, 431–34, 439–40, 580–82; lag, 437–38, 586; pitch, 421–22; shaft motion, 437–40, 586
Fourier series, 153–54, 354–56
frequency: and loads, 706; and vibration, 697–98; bending, 456–60; coupled out-of-plane and in-plane bending, 456; flap,186, 223, 230, 236, 240, 383; in-plane bending, 398–99; lag, 253, 395–96; out-of-plane bending, 388–90; pitch, 407; torsion, 417

gimballed rotor, 8, 235–36, 314
ground effect, 122–24, 146–48, 298
ground resonance, 668–93; Coleman diagram, 677–80, 688–89; damping for stability, 681–84, 691–92; equations, 669–71, 685–87; two-bladed rotor, 685–93

harmonic analysis, 348–49
harmonics, sum, 347–48
height-velocity diagram, 328
higher harmonic control, 708
hinge offset: flap, 155, 227–28, 382; lag, 155, 251, 393
hingeless rotor, 8, 234–35, 275, 314, 800–802, 851–52
history: of autogiro development, 15–16; of blade element theory, 46–49; of helicopter development, 11–20; of momentum theory, 28; of vortex theory,73–74
hub forces, 166, 423, 432–34, 440, 581–83, 588–90, 695; forward flight, 172–73, 177–78, 208, 217–20; low frequency response, 629–34, 781–82
hub moment feedback, 857–61
hub moments, 166, 173–74, 423, 431–32, 434, 439–40, 696; with hinge offset, 232–33; with hinge spring, 226–27
hub reactions, 422–23, 429–34, 579–83, 694–96; aerodynamics, 579–83, 588–90; low frequency response, 628–34, 779–82; shaft motion, 439–40, 442, 588–90. *See also* hub forces; hub moments; thrust; torque
hub plane, 164

ideal precone, 224
ideal rotor, 64-65, 70
ideal twist, 52–53, 55, 57, 64
induced power: climb, 33, 132–33, 286, 295–96; forward flight, 126, 130–33, 141, 183; hover, 31–32, 34, 54; twin rotor, 118–22, 142–46; vertical flight, 93–94, 102–105, 107. *See also* power
induced velocity: climb, 33, 56; forward flight, 128–30, 195, 721–22; hover, 31, 43–44, 52–53, 57, 78–79; linear variation, 139–41, 207, 521–23; vertical flight, 95–96, 107. *See also* nonuniform inflow
inertial acceleration, 442, 827
inflow dynamics, 520–26, 595, 637. *See also* unsteady aerodynamics
in-plane bending, 397–99; and out-of-plane bending, 399–403, 444–56; equation, 398, 402, 453–54; frequency, 398–99, 456–59; modal equation, 398, 455
in-plane shear, root, 423, 427–29, 439, 576–79, 586–87

lag motion, 8, 23–24, 154–55, 157–58, 250–55, 363, 393–97; aerodynamics, 560–64, 585; and flap, 393–97, 560–64, 585, 653–67; equation, 252, 394–95; forward flight, 254–55; frequency, 253, 395–96; ground resonance, 668–93; shaft motion, 437–38, 441, 585–86
lift coefficient, mean, 62–63
lift deficiency function, inflow dynamics, 525–26; lifting-line approximation, 487–88, 513; time-varying free stream, 495–97; two-dimensional rotary wing theory, 502, 505–10, 513, 515; unsteady airfoil theory, 481, 487–88; unsteady vortex theory, 519–20. *See also* unsteady aerodynamics
lift, rotor, 288–89
lifting-line theory, 205, 469–71. *See also* blade element theory
lifting-surface theory, 471, 739–44, 751–55
linear system theory, 365–66; constant coefficients, 366–69; periodic coefficients, 369–76
literature: air resonance, 693; airfoil design and selection, 336–37; airfoil theory with returning shed wake, 512; airframe structural dynamics, 708; antitorque devices, 341; automatic control and stability augmentation, 872; blade bending modes and frequencies, 459–60; boundary layers, 755; compound helicopter, 341; comprehensive helicopter analyses, 767; dynamics with suspended load, 871; flap-lag dynamics, 667–68; flapping response, 636; flapping stability in forward flight, 636–37; flow states in axial flight, 107; forward flight, 275–76; ground effect, 124; ground resonance, 693; gust response, 871; helicopter design, development, and construction, 340; helicopter operating conditions, 341; higher harmonic control, 708; hingeless rotor modelling, 235; history, 20; hover and vertical flight, 91–92; induced velocity, 107, 191; inflow dynamics, 637; lifting-surface theory, 755; noise, 957–59; nonuniform inflow, 735; other rotary wing configurations, 342; performance, 311–12; pitch-flap flutter, 652–53; power-off landing, 330; propellers, 91; rotor airloads data, 276–77; rotor fatigue life, 708; rotor loads, 708–709; rotor response to random turbulence, 637; stability and control, 869–71; stall, 901–902; tilting prop-rotor aircraft, 341–42; twin rotor performance, 122, 146; vertical autorotation, 114; vertical drag, 118; vibration and vibration reduction, 707–708; vortices and wakes, 753–54; vortex-induced loads, 753; vortex theory, 91, 139; wake flow field, 107, 141; wake geometry, 749; wind tunnel and rotor interference, 276
load factor, 341
loads: blade, 699–706; analysis methods, 708; data 709
loads: blade root, 422–29, 574–79; aerodynamics, 574–79, 586–87; shaft motion, 438–39, 586–87. *See also* in-plane shear; vertical shear

loads hub, *see* hub reactions
Lock number, 22, 186, 267, 383
low frequency response, 622–35, 779–82, 824–25, 852–54; flap, 622–28, 780–81; hub reactions, 628–34, 779, 781–82

modal equation: coupled out-of-plane and in-plane bending, 455; in-plane bending, 398; out-of-plane bending, 396; torsion, 416
modes: coupled out-of-plane and in-plane bending, 455; flap, 228, 234, 275, 382, 458–59; in-plane bending, 398, 458; lag, 251, 393; out-of-plane bending, 385–86, 390, 458; torsion, 416–18, 421
moment equilibrium, pitch, 247–48; roll, 248–49
momentum theory, 28–33, 36–45, 126–130; climb, 32–33, 37–40, 94–96; descent, 96; differential form, 40, 80, 520, 522–23; forward flight, 126–30; history, 28; hover, 30–32, 37–40; vertical flight, 94–98; with swirl in wake, 41–45

no-feathering plane, 160–61, 164, 167
noise, 903–59; blade slap, 906, 952–56; broadband or vortex, 904, 909–15; forward flight, 931–39, 943–52; hover, 920–27, 929–31; reduction, 956–57; steady loading, 920–29, 931–34; tail rotor, 906; thickness, 939–52, 955; unsteady loading, 929–31, 934–39, 943–52; vertical flight, 927–29; vortex-blade interaction, 904–905
nonuniform inflow, 88–91, 273–75, 697, 713–35, 878
normal working state, 97–99
numerical integration, 760–67

optimum hovering rotor, 65–67, 70
out-of-plane bending, 384–91; aerodynamics, 558–59; and in-plane bending, 399–403, 444–56; and torsion, 412–21; equation, 387, 402, 419–20, 558–59; frequency, 388–90, 456–59; modal equation, 386, 455

parasite power, 183–84, 250, 284–85, *See also* power
performance charts, 279, 304, 306–11
pilot's controls, 769–70
pitch-flap coupling, 238–39, 408, 603–605; structural, 408–12
pitch-flap motion, 403–408, 421–22; aerodynamics, 596–600; elastic, 412–21; equation, 404, 407, 638–39. *See also* flutter
pitch-lag coupling, 657–58, 662–63; structural, 408–12
pitch motion, 8, 23–24, 158–59, 315, 769; aerodynamics, 596–600; and flap, 403–408, 421–22, 596–600; equation, 407; frequency, 407; nonrotating frame, 421–22
power, climb, 114–16, 282, 132–33, 286, 295–96; D/L formulation, 286–89; forward flight, 179–84, 209–10, 249–50, 284–90, 293–95; hover, 34, 51, 53–55, 68–69, 105, 280–81, 290–91; induced, 31–34, 54, 93–94, 102–105, 107; 126, 130–33, 141, 183; minimum for hover, 3, 32, 39–40, 82–83, 291–93; P/T formulation, 289–90; parasite, 183–84, 250, 284–85; profile, 54–55, 183–84, 217–20, 875; speed limitation, 296–98, 321; twin rotor, 118–22, 142–46
power-off landing, 325–30. *See also* autorotation
precone, 223
predictive capability, 706, 759–60
profile power, forward flight, 183–84, 217–20, 875; hover, 34, 54–55, 67. *See also* power
propeller moment, 405–406

radial drag, 170, 214–16; and profile power, 213–21
range, 299–300
reverse flow, 152, 218–19, 255–62, 275, 534–35, 876
root cutout, 62, 206
root loads, *see* loads
rotational noise, 904–905, 915–52

shaft motion, 435, 555; aerodynamics, 583–90; blade root loads, 438–39, 586–87; flap equation, 436–38, 441, 585–86; hub reactions, 439–40, 442,

588–90; lag equation, 437–38, 441, 585–86; two-bladed rotor, 441–42
side-by-side rotor helicopter, 143, 146, 317, 770–71
single main rotor and tail rotor helicopter, 9, 316, 770–71
slap, blade, 906, 952–56
soft in-plane rotor, 253
speed, limitations, 321–24; maximum, 296–98
stability augmentation, 794–800, 802, 854–62, 872; and gyro, 854–57, 860; hub moment feedback, 857–61
stability derivatives, 778–79, 783, 785, 789, 800–801, 809, 811–12, 815–16, 820–21
stall, 322, 332–34, 873–902; and non-uniform inflow, 878; and vibration, 697; criteria, 875–78, 883–84, 900; dynamic, 879–82, 888–99; flutter, 877, 889; loads due to, 874–75
stiff in-plane rotor, 253
Sturm-Liouville theory, 378–81
swashplate, 159–60

tail rotor, 9, 264–65, 302–303, 316, 341, 784–86, 906
tandem main rotor helicopter, 9, 119–21, 143–46, 317–18, 770–71, 813–21, 848–51
teetering rotor, 8, 237–38, 314–15
thrust, 166, 423, 429–31, 439, 580, 583, 588, 590, 694; forward flight, 172–73, 176, 206, 208, 229, 262; hover, 51–52, 60; low frequency response, 628–29, 779
tilting proprotor aircraft, 324, 341–42
tip loss factor, 59–60, 86, 133–34
tip losses, 34, 58–59, 61, 81, 86, 133–34, 206
tip-path plane, 157, 164, 167, 178
tip vortices, *see* vortex-induced loads; vortex-induced velocity; wake
torque, 166, 172–73, 179, 423, 434, 440, 580, 688, 695
torsion, 412-21; equation, 419–20; frequency, 417; modal equation, 416
trim, 132, 182, 243–49, 758–59, 772–74
turbulent wake state, 97–98, 101
twin rotor interference, 118–22, 142–46
twist, blade, 52, 55, 68, 111, 173, 195–99, 723–24. *See also* ideal twist
two-bladed rotor: aerodynamics, 582, 586; flap dynamics, 632–36; flap equation, 441; ground resonance, 685–93; hub reactions, 434, 442, 582; lag equation, 441; shaft motion, 441–42, 586

unsteady aerodynamics, for rotary wing, 526–35; inflow dynamics, 520–26, 595; lifting-line approximation, 484–92, 513; two-dimensional airfoil theory, 471–84; vortex theory, 515–20; with rotor returning shed wake, 498–512; with time-varying free stream, 492–98, 534–35

vertical drag, 114–18
vertical shear, root, 423–26, 438, 575, 586
vibration, 694–98, 707–708; and hub reactions, 694–96; reduction, 697–98, 707–708
vortex-blade interaction, 713, 739–44, 749–53; and noise, 904–905; and vibration, 697
vortex core, 536, 537–40, 753–54
vortex-induced loads, 713–14, 727, 739–44, 749–53
vortex-induced velocity, 535–47, 753–54
vortex noise, *see* broadband noise
vortex ring state, 97–101, 106
vortex theory, 72–74, 134–41; finite number of blades, 81–88; forward flight, 134–41; history, 73–74; hover, 76–81, 91; unsteady, 515–20
vortex wake, *see* wake

wake, 710–13, 735–36; forward flight, 134–36, 141; hover, 74–76, 107
wake geometry, 75, 89, 134–35, 275, 712, 735–49; free, 739–49
windmill brake state, 97–98, 101–102

A CATALOG OF SELECTED

DOVER BOOKS

IN SCIENCE AND MATHEMATICS

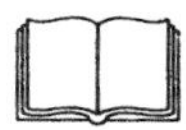

Engineering

DE RE METALLICA, Georgius Agricola. The famous Hoover translation of greatest treatise on technological chemistry, engineering, geology, mining of early modern times (1556). All 289 original woodcuts. 638pp. 6¾ x 11. 0-486-60006-8

FUNDAMENTALS OF ASTRODYNAMICS, Roger Bate et al. Modern approach developed by U.S. Air Force Academy. Designed as a first course. Problems, exercises. Numerous illustrations. 455pp. 5⅜ x 8½. 0-486-60061-0

DYNAMICS OF FLUIDS IN POROUS MEDIA, Jacob Bear. For advanced students of ground water hydrology, soil mechanics and physics, drainage and irrigation engineering and more. 335 illustrations. Exercises, with answers. 784pp. 6⅛ x 9¼. 0-486-65675-6

THEORY OF VISCOELASTICITY (SECOND EDITION), Richard M. Christensen. Complete consistent description of the linear theory of the viscoelastic behavior of materials. Problem-solving techniques discussed. 1982 edition. 29 figures. xiv+364pp. 6⅛ x 9¼. 0-486-42880-X

MECHANICS, J. P. Den Hartog. A classic introductory text or refresher. Hundreds of applications and design problems illuminate fundamentals of trusses, loaded beams and cables, etc. 334 answered problems. 462pp. 5⅜ x 8½. 0-486-60754-2

MECHANICAL VIBRATIONS, J. P. Den Hartog. Classic textbook offers lucid explanations and illustrative models, applying theories of vibrations to a variety of practical industrial engineering problems. Numerous figures. 233 problems, solutions. Appendix. Index. Preface. 436pp. 5⅜ x 8½. 0-486-64785-4

STRENGTH OF MATERIALS, J. P. Den Hartog. Full, clear treatment of basic material (tension, torsion, bending, etc.) plus advanced material on engineering methods, applications. 350 answered problems. 323pp. 5⅜ x 8½. 0-486-60755-0

A HISTORY OF MECHANICS, René Dugas. Monumental study of mechanical principles from antiquity to quantum mechanics. Contributions of ancient Greeks, Galileo, Leonardo, Kepler, Lagrange, many others. 671pp. 5⅜ x 8½. 0-486-65632-2

STABILITY THEORY AND ITS APPLICATIONS TO STRUCTURAL MECHANICS, Clive L. Dym. Self-contained text focuses on Koiter postbuckling analyses, with mathematical notions of stability of motion. Basing minimum energy principles for static stability upon dynamic concepts of stability of motion, it develops asymptotic buckling and postbuckling analyses from potential energy considerations, with applications to columns, plates, and arches. 1974 ed. 208pp. 5⅜ x 8½. 0-486-42541-X

BASIC ELECTRICITY, U.S. Bureau of Naval Personnel. Originally a training course; best nontechnical coverage. Topics include batteries, circuits, conductors, AC and DC, inductance and capacitance, generators, motors, transformers, amplifiers, etc. Many questions with answers. 349 illustrations. 1969 edition. 448pp. 6½ x 9¼. 0-486-20973-3

ROCKETS, Robert Goddard. Two of the most significant publications in the history of rocketry and jet propulsion: "A Method of Reaching Extreme Altitudes" (1919) and "Liquid Propellant Rocket Development" (1936). 128pp. 5³/₈ x 8¹/₂. 0-486-42537-1

STATISTICAL MECHANICS: PRINCIPLES AND APPLICATIONS, Terrell L. Hill. Standard text covers fundamentals of statistical mechanics, applications to fluctuation theory, imperfect gases, distribution functions, more. 448pp. 5³/₈ x 8¹/₂. 0-486-65390-0

ENGINEERING AND TECHNOLOGY 1650–1750: ILLUSTRATIONS AND TEXTS FROM ORIGINAL SOURCES, Martin Jensen. Highly readable text with more than 200 contemporary drawings and detailed engravings of engineering projects dealing with surveying, leveling, materials, hand tools, lifting equipment, transport and erection, piling, bailing, water supply, hydraulic engineering, and more. Among the specific projects outlined-transporting a 50-ton stone to the Louvre, erecting an obelisk, building timber locks, and dredging canals. 207pp. 8³/₈ x 11¹/₄. 0-486-42232-1

THE VARIATIONAL PRINCIPLES OF MECHANICS, Cornelius Lanczos. Graduate level coverage of calculus of variations, equations of motion, relativistic mechanics, more. First inexpensive paperbound edition of classic treatise. Index. Bibliography. 418pp. 5³/₈ x 8¹/₂. 0-486-65067-7

PROTECTION OF ELECTRONIC CIRCUITS FROM OVERVOLTAGES, Ronald B. Standler. Five-part treatment presents practical rules and strategies for circuits designed to protect electronic systems from damage by transient overvoltages. 1989 ed. xxiv+434pp. 6¹/₈ x 9¹/₄. 0-486-42552-5

ROTARY WING AERODYNAMICS, W. Z. Stepniewski. Clear, concise text covers aerodynamic phenomena of the rotor and offers guidelines for helicopter performance evaluation. Originally prepared for NASA. 537 figures. 640pp. 6¹/₈ x 9¹/₄. 0-486-64647-5

INTRODUCTION TO SPACE DYNAMICS, William Tyrrell Thomson. Comprehensive, classic introduction to space-flight engineering for advanced undergraduate and graduate students. Includes vector algebra, kinematics, transformation of coordinates. Bibliography. Index. 352pp. 5³/₈ x 8¹/₂. 0-486-65113-4

HISTORY OF STRENGTH OF MATERIALS, Stephen P. Timoshenko. Excellent historical survey of the strength of materials with many references to the theories of elasticity and structure. 245 figures. 452pp. 5³/₈ x 8¹/₂. 0-486-61187-6

ANALYTICAL FRACTURE MECHANICS, David J. Unger. Self-contained text supplements standard fracture mechanics texts by focusing on analytical methods for determining crack-tip stress and strain fields. 336pp. 6¹/₈ x 9¹/₄. 0-486-41737-9

STATISTICAL MECHANICS OF ELASTICITY, J. H. Weiner. Advanced, self-contained treatment illustrates general principles and elastic behavior of solids. Part 1, based on classical mechanics, studies thermoelastic behavior of crystalline and polymeric solids. Part 2, based on quantum mechanics, focuses on interatomic force laws, behavior of solids, and thermally activated processes. For students of physics and chemistry and for polymer physicists. 1983 ed. 96 figures. 496pp. 5³/₈ x 8¹/₂. 0-486-42260-7

Mathematics

FUNCTIONAL ANALYSIS (Second Corrected Edition), George Bachman and Lawrence Narici. Excellent treatment of subject geared toward students with background in linear algebra, advanced calculus, physics and engineering. Text covers introduction to inner-product spaces, normed, metric spaces, and topological spaces; complete orthonormal sets, the Hahn-Banach Theorem and its consequences, and many other related subjects. 1966 ed. 544pp. 6⅛ x 9¼. 0-486-40251-7

DIFFERENTIAL MANIFOLDS, Antoni A. Kosinski. Introductory text for advanced undergraduates and graduate students presents systematic study of the topological structure of smooth manifolds, starting with elements of theory and concluding with method of surgery. 1993 edition. 288pp. 5⅜ x 8½. 0-486-46244-7

VECTOR AND TENSOR ANALYSIS WITH APPLICATIONS, A. I. Borisenko and I. E. Tarapov. Concise introduction. Worked-out problems, solutions, exercises. 257pp. 5⅝ x 8¼. 0-486-63833-2

AN INTRODUCTION TO ORDINARY DIFFERENTIAL EQUATIONS, Earl A. Coddington. A thorough and systematic first course in elementary differential equations for undergraduates in mathematics and science, with many exercises and problems (with answers). Index. 304pp. 5⅜ x 8½. 0-486-65942-9

FOURIER SERIES AND ORTHOGONAL FUNCTIONS, Harry F. Davis. An incisive text combining theory and practical example to introduce Fourier series, orthogonal functions and applications of the Fourier method to boundary-value problems. 570 exercises. Answers and notes. 416pp. 5⅜ x 8½. 0-486-65973-9

COMPUTABILITY AND UNSOLVABILITY, Martin Davis. Classic graduate-level introduction to theory of computability, usually referred to as theory of recurrent functions. New preface and appendix. 288pp. 5⅜ x 8½. 0-486-61471-9

AN INTRODUCTION TO MATHEMATICAL ANALYSIS, Robert A. Rankin. Dealing chiefly with functions of a single real variable, this text by a distinguished educator introduces limits, continuity, differentiability, integration, convergence of infinite series, double series, and infinite products. 1963 edition. 624pp. 5⅜ x 8½. 0-486-46251-X

METHODS OF NUMERICAL INTEGRATION (SECOND EDITION), Philip J. Davis and Philip Rabinowitz. Requiring only a background in calculus, this text covers approximate integration over finite and infinite intervals, error analysis, approximate integration in two or more dimensions, and automatic integration. 1984 edition. 624pp. 5⅜ x 8½. 0-486-45339-1

INTRODUCTION TO LINEAR ALGEBRA AND DIFFERENTIAL EQUATIONS, John W. Dettman. Excellent text covers complex numbers, determinants, orthonormal bases, Laplace transforms, much more. Exercises with solutions. Undergraduate level. 416pp. 5⅜ x 8½. 0-486-65191-6

RIEMANN'S ZETA FUNCTION, H. M. Edwards. Superb, high-level study of landmark 1859 publication entitled "On the Number of Primes Less Than a Given Magnitude" traces developments in mathematical theory that it inspired. xiv+315pp. 5⅜ x 8½. 0-486-41740-9

CALCULUS OF VARIATIONS WITH APPLICATIONS, George M. Ewing. Applications-oriented introduction to variational theory develops insight and promotes understanding of specialized books, research papers. Suitable for advanced undergraduate/graduate students as primary, supplementary text. 352pp. 5⅜ x 8½.
0-486-64856-7

MATHEMATICIAN'S DELIGHT, W. W. Sawyer. "Recommended with confidence" by *The Times Literary Supplement*, this lively survey was written by a renowned teacher. It starts with arithmetic and algebra, gradually proceeding to trigonometry and calculus. 1943 edition. 240pp. 5⅜ x 8½. 0-486-46240-4

ADVANCED EUCLIDEAN GEOMETRY, Roger A. Johnson. This classic text explores the geometry of the triangle and the circle, concentrating on extensions of Euclidean theory, and examining in detail many relatively recent theorems. 1929 edition. 336pp. 5⅜ x 8½. 0-486-46237-4

COUNTEREXAMPLES IN ANALYSIS, Bernard R. Gelbaum and John M. H. Olmsted. These counterexamples deal mostly with the part of analysis known as "real variables." The first half covers the real number system, and the second half encompasses higher dimensions. 1962 edition. xxiv+198pp. 5⅜ x 8½. 0-486-42875-3

CATASTROPHE THEORY FOR SCIENTISTS AND ENGINEERS, Robert Gilmore. Advanced-level treatment describes mathematics of theory grounded in the work of Poincaré, R. Thom, other mathematicians. Also important applications to problems in mathematics, physics, chemistry and engineering. 1981 edition. References. 28 tables. 397 black-and-white illustrations. xvii + 666pp. 6⅛ x 9¼.
0-486-67539-4

COMPLEX VARIABLES: Second Edition, Robert B. Ash and W. P. Novinger. Suitable for advanced undergraduates and graduate students, this newly revised treatment covers Cauchy theorem and its applications, analytic functions, and the prime number theorem. Numerous problems and solutions. 2004 edition. 224pp. 6½ x 9¼. 0-486-46250-1

NUMERICAL METHODS FOR SCIENTISTS AND ENGINEERS, Richard Hamming. Classic text stresses frequency approach in coverage of algorithms, polynomial approximation, Fourier approximation, exponential approximation, other topics. Revised and enlarged 2nd edition. 721pp. 5⅜ x 8½. 0-486-65241-6

INTRODUCTION TO NUMERICAL ANALYSIS (2nd Edition), F. B. Hildebrand. Classic, fundamental treatment covers computation, approximation, interpolation, numerical differentiation and integration, other topics. 150 new problems. 669pp. 5⅜ x 8½. 0-486-65363-3

MARKOV PROCESSES AND POTENTIAL THEORY, Robert M. Blumental and Ronald K. Getoor. This graduate-level text explores the relationship between Markov processes and potential theory in terms of excessive functions, multiplicative functionals and subprocesses, additive functionals and their potentials, and dual processes. 1968 edition. 320pp. 5⅜ x 8½. 0-486-46263-3

ABSTRACT SETS AND FINITE ORDINALS: An Introduction to the Study of Set Theory, G. B. Keene. This text unites logical and philosophical aspects of set theory in a manner intelligible to mathematicians without training in formal logic and to logicians without a mathematical background. 1961 edition. 112pp. 5⅜ x 8½.
0-486-46249-8

Physics

OPTICAL RESONANCE AND TWO-LEVEL ATOMS, L. Allen and J. H. Eberly. Clear, comprehensive introduction to basic principles behind all quantum optical resonance phenomena. 53 illustrations. Preface. Index. 256pp. 5⅜ x 8½.
0-486-65533-4

QUANTUM THEORY, David Bohm. This advanced undergraduate-level text presents the quantum theory in terms of qualitative and imaginative concepts, followed by specific applications worked out in mathematical detail. Preface. Index. 655pp. 5⅜ x 8½.
0-486-65969-0

ATOMIC PHYSICS (8th EDITION), Max Born. Nobel laureate's lucid treatment of kinetic theory of gases, elementary particles, nuclear atom, wave-corpuscles, atomic structure and spectral lines, much more. Over 40 appendices, bibliography. 495pp. 5⅜ x 8½.
0-486-65984-4

A SOPHISTICATE'S PRIMER OF RELATIVITY, P. W. Bridgman. Geared toward readers already acquainted with special relativity, this book transcends the view of theory as a working tool to answer natural questions: What is a frame of reference? What is a "law of nature"? What is the role of the "observer"? Extensive treatment, written in terms accessible to those without a scientific background. 1983 ed. xlviii+172pp. 5⅜ x 8½.
0-486-42549-5

AN INTRODUCTION TO HAMILTONIAN OPTICS, H. A. Buchdahl. Detailed account of the Hamiltonian treatment of aberration theory in geometrical optics. Many classes of optical systems defined in terms of the symmetries they possess. Problems with detailed solutions. 1970 edition. xv + 360pp. 5⅜ x 8½. 0-486-67597-1

PRIMER OF QUANTUM MECHANICS, Marvin Chester. Introductory text examines the classical quantum bead on a track: its state and representations; operator eigenvalues; harmonic oscillator and bound bead in a symmetric force field; and bead in a spherical shell. Other topics include spin, matrices, and the structure of quantum mechanics; the simplest atom; indistinguishable particles; and stationary-state perturbation theory. 1992 ed. xiv+314pp. 6⅛ x 9¼.
0-486-42878-8

LECTURES ON QUANTUM MECHANICS, Paul A. M. Dirac. Four concise, brilliant lectures on mathematical methods in quantum mechanics from Nobel Prize-winning quantum pioneer build on idea of visualizing quantum theory through the use of classical mechanics. 96pp. 5⅜ x 8½.
0-486-41713-1

THIRTY YEARS THAT SHOOK PHYSICS: THE STORY OF QUANTUM THEORY, George Gamow. Lucid, accessible introduction to influential theory of energy and matter. Careful explanations of Dirac's anti-particles, Bohr's model of the atom, much more. 12 plates. Numerous drawings. 240pp. 5⅜ x 8½. 0-486-24895-X

ELECTRONIC STRUCTURE AND THE PROPERTIES OF SOLIDS: THE PHYSICS OF THE CHEMICAL BOND, Walter A. Harrison. Innovative text offers basic understanding of the electronic structure of covalent and ionic solids, simple metals, transition metals and their compounds. Problems. 1980 edition. 582pp. 6⅛ x 9¼.
0-486-66021-4